advanced
CHEMISTRY

*an enquiry-based
approach*

advanced

CHEMISTRY

*an enquiry-based
approach*

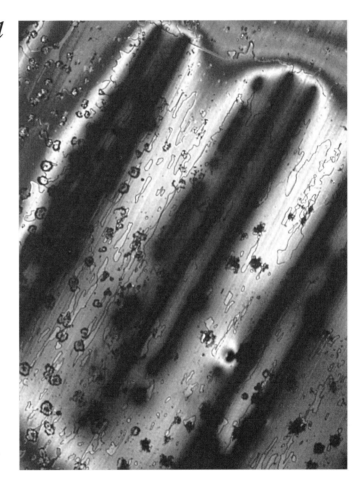

JAMES
MAPLE

JOHN MURRAY

© James Maple 1996

First published in 1996
by John Murray (Publishers) Ltd
50 Albemarle Street
London W1X 4BD

Boxed illustrations by Mike Humphries.
Text illustrations by Wearset, Boldon, Tyne and Wear.

Design and layout by Can Do Design.

Typeset in 9.5/11 pt Galliard by Wearset, Boldon, Tyne and Wear.

Printed and bound in Great Britain by Butler & Tanner Ltd, Frome and London.

A catalogue entry for this title may be obtained from the British Library.

ISBN 0-7195-5359-8

Contents

4 CHEMICAL BONDING 1: IONIC AND COVALENT BONDS

5 CHEMICAL BONDING 2: 'IMPURE' AND METALLIC BONDS

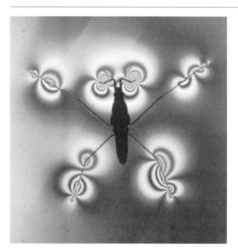

6 INTERMOLECULAR FORCES

7 OXIDATION AND REDUCTION REACTIONS

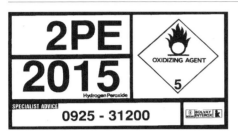

8 THERMODYNAMICS 1: ENERGY EXCHANGE AND CHEMICAL SYSTEMS

9 THERMODYNAMICS 2: THE DIRECTION OF CHEMICAL CHANGE

12 ACIDS, BASES AND ALKALIS – COMPETITION FOR PROTONS

13 ORGANIC CHEMISTRY 3: ALDEHYDES, KETONES, CARBOXYLIC ACIDS AND THEIR DERIVATIVES

14 CHEMICAL KINETICS: THE STUDY OF RATES OF REACTION

15 ORGANIC MOLECULES CONTAINING NITROGEN

16 ELECTROCHEMISTRY

17 THE TYPICAL ELEMENTS 1: THE S-BLOCK

18 THE TYPICAL ELEMENTS 2: THE P-BLOCK

19 THE ELEMENTS OF THE D-BLOCK

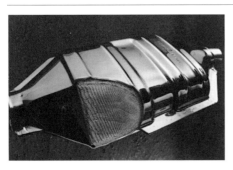

From the author

The genesis of this book lies in a number of sources. The first seeds were sown by David Pickup, a brilliant student at my school in Dunstable in the mid-1980s, and currently a PhD student at Reading University. David came back from one of our ancient universities after a year of study and said that chemistry had lost a lot of its vividness for him, because, unlike in school, the university tutors made no use of metaphor in their teaching. From that moment on, I was confirmed in my belief that chemistry should be a story, which like any other story needed telling with a strong narrative drive and with literary devices like, of course, metaphor. That belief has been embodied in this book.

A short time after this conviction had taken root, I took a year out from teaching to do an MSc at the Oxford University Department of Education. It was during that year that I met the other two people who were to lead me to embark on this writing project. Firstly, there was Joan Solomon, who was my tutor for the year. Joan is a major figure in the field of science education, but what impressed me most was the triple life she led. Monday might be an interview on TV with a government minister, Tuesday might be a tutorial with the likes of me, and Wednesday she was teaching Year 10 at Wheatley Park Comprehensive. Anyway, Joan's contribution to the story of this book was to encourage me in the belief that I could write. If you've ever been told by someone you respect that you really can write, dance, play football etc., you'll know what a great boost it is. I just hope that after using the book, the reader won't consider Joan's judgement to have been made on one of her rare off-days.

The other Oxford Department of Education figure who helped this book into existence was Terry Allsop. It was Terry, with his great knowledge of the world of science education in general, and science education publishing in particular, who first put me on to John Murray Publishers, and thus at a stroke cancelled out the need to write lots of begging letters, and at the same time allowed me to establish my bona-fides by name-dropping a name of the highest value. Terry showed faith in the future of the book when it was no more than a sketchy prototype. It is my fondest hope that the finished product doesn't fall too far short of justifying his, and Joan's support.

If those three people were midwives at the birth of the project, lots more people gave invaluable help as the writing proceeded. Some supplied specific ideas – for instance my students Andrew Mayer and Angelina Selvaratnam came up with two that I seized on gratefully. Andrew is the inventor of the 'bit' concept in the shapes-of-molecules section, while from Angelina I borrowed the neat $x = a - b$ method of absorbing all the detail of a Hess's Law calculation into one compact relationship. Two industrial chemists supplied their own internal articles which have been merged with the text – they are John Goddard of Kodak and William Wood of ICI, whom I thank

for their excellent pieces on photography and dyes. In addition, Bob Larbey of Associated Octel, Rob Potter of Johnson Matthey and Peter Taylor of Tungstone Batteries were very helpful with information on electrochemistry. Thanks also to Pete Biggs of Oxford University, Department of Physical Chemistry, who helped with photographs, and Sandra Purdey of the University of Luton who generated many of the spectra.

I'm very grateful to all my 'handlers' at John Murray, – that's Katie Mackenzie Stuart, Camilla Cochrane, Jo Ray and Nikki Taylor, who always struck the right balance between nagging and *laissez-faire*. And thanks for the free lunch.

I came to have a lot of respect for my editors, Hendrina Ellis and Ruth Holmes. I'd never been edited before, and expected it to be rather a brutal experience, but I soon came to value their judgement and knowledge of chemistry.

Thanks also to my proofreaders, Mahesh Uttamlal of University College Oxford, and Mike Bullivant and Charlotte Sweeney of the Open University, for their diligent tracking-down of my many errors in the first draught. Those that remain are my responsibility alone.

Next I'd like to say thank you to Brenda Boyle, Joan Spall and Deirdre Hawkins, the secretaries at Manshead School, for dealing patiently with a lot of the communication between myself and the publishers. Deirdre, what's more, had a brilliant idea for the cover of the book, which was to use that three-dimensional 'magic-eye' technique, and have a 3-D methane molecule on the front. Considering how hard it is to represent the third dimension on paper in chemistry books, this would have been superbly appropriate, and a good selling gimmick too. Unfortunately the man who creates those things charges US$5000 even if you only want the rights to use a picture that already exists, so we had to let the matter drop – but it was a nice idea while it lasted.

Finally I offer my thanks to people who have had to put up with me over the last few years while I've been writing. To my colleagues and friends in the science department of Manshead School, Brian Circuit, Roger Seldon, Hilary Saunders, John Batten, Cynthia Hayhurst, Al Bowley, Alan Heyes, Lesley Charters, Nasreen Mecci, Gerry Braby and Hilary Swain, for politely not mentioning the fact that they weren't getting quite the usual quality of dynamic 'Gus Hedges'-style leadership that they were used to. To my friends Richard, Denis, Nick and Norm, who may have noticed I was always just too late to buy that first, expensive round. To my grown-up sons Tim, Matt and Tom, who get phoned once a year. To my wife Di, whose exercise regime has been based around stair-climbing with cups of tea. To my small daughter Rosie, who has had to put up with sub-standard parenting, featuring a complete lack of the stuff other kids expect as birthright, like trips to Alton Towers (so it's not all bad news).

An introductory note to students and teachers

This book is aimed at the student of A-level chemistry, which means it is designed to be compatible with a number of different syllabuses. However, it isn't practicable to cover every optional bit of every syllabus, because the book would be too big and expensive, and there would be too much irrelevant matter in it for any one single reader. So the finished product is a compromise. The core material that is common to all syllabuses is here, and the specialised areas featured in the book are those that seemed to be most popular with a range of examination boards.

Among the non-core topics that receive coverage are advanced thermodynamics, advanced kinetics, spectroscopic analytical methods, dyes, paints, photography, a lot of biochemistry, a good deal of materials science, fats, foods, corrosion, and the chlor-alkali, rubber, PVC, battery and automobile catalyst industries. Teachers will obviously advise their students about the applicability of each section to their course.

When you come to select what to read, it'll be quite easy to skip a whole section on, say, enzymes, but it'll be a little harder for you when you get into advanced thermodynamics in Chapter 9. There is a big fundamental idea in thermodynamics, involving quantities called *entropy* and *free energy*. I've opted to recognise the importance of this idea and woven it centrally into the text, thereby matching the approach of at least three major boards (AEB, NEAB and London Nuffield). However, other boards have omitted entropy completely. So in Chapter 9 there are small signpost paragraphs, to pilot you through the text in either an 'entropy-free' or an 'entropy-full' way. A similar ploy is used in Chapter 14 with some of the maths of chemical kinetics. What can be said about all this business of selecting or omitting text is that most students will be working with a teacher, who will be their final guide and counsellor.

The book has been written with one simple psychological model in mind – that students will become more engaged with the text if it interacts with them. In other words you'll find that the text is littered with embedded questions, which form part of the narrative and which require the student to respond before he or she proceeds to the next passage. This strategy has a risk factor attached, in that students may be irked by the interruptions, and worse, may derail their own progress by getting something wrong. Not much can be done about the first problem, but the second one is addressed by having answers to the in-text questions at the back of the book, rather in the style of Open University 'self-assessment questions'. Teachers might like to consider the ploy of setting homework that uses the in-text questions, and asking students to self-mark, having used that old line about 'you're only cheating yourself'. The students then merely have to draw the teacher's attention to any bits that they still don't understand even after reading the answers. For teachers who feel uneasy with an honesty-based scheme; there's still another resort – a pool of old exam questions (without answers) drawn from a range of boards, placed at the end of Chapter 19.

Allow me finally to express the hope that you might actually enjoy reading at least some bits of this book. When I was at university the image of chemistry was of a dowdy activity carried out by prosaic literal-minded people, and yes, if truth be told, anoraks and open-toed sandals may have been involved. The implicit idea, and one which to some extent the chemists themselves helped to propagate, was that chemistry, as a way of seeing how the world works, was somehow less transcendent than the arts, or politics or philosophy. One of the purposes of this book is to redress this self-deprecating tendency of chemists, and to try to confirm in the minds of chemistry students a sense of the worthwhileness of their subject choice. I have tried to represent chemistry as a strong story, and a vivid and mind-expanding path to a view of the world, a view which is just as important, amusing, exciting, profound, noble and satisfying as the view from any other field of academic knowledge.

JAMES MAPLE

1

The behaviour of gaseous matter

Many authors devote several pages, a mini-chapter even, at the front of their book to explain to the student their overall aims, methods and philosophy. Many students know exactly how to react to this ploy; they skip the introduction and jump to the first 'real' chapter. This introduction, however, sets out certain ideas about the nature of scientific activity which illuminate every other aspect of the course. It is therefore quite short and embedded in this first chapter. Those ideas are then immediately deployed on the first section of course content, concerning the behaviour of gases.

1.1 Introduction – chemistry and the mind

Chemistry should be a feat of imagination. If you want to get a proper feel for the subject you really *must* use your imagination.

One of these imaginative skills is the power to make your mind do the sort of journey which used to be performed, in a different context, by the opening graphic on ITV's *News at Ten*. At first the camera took in the whole world, and then swooped down through levels of scale – continent, country, city – until it ended up staring at the fine detail on the face of Big Ben (Figure 1.1). As a chemist, you might begin to mimic this process by doing exactly what you are doing now – looking at this book. Your 'scale swoop'

would resolve the page into fibres of paper, then fibres into molecules, then molecules into atoms, and even further into sub-atomic particles.

Figure 1.1 Swooping down through levels of scale

There are two important points here – one is that some people reading this book will already know some of the chemical facts about, in this case, paper, but they will be in a much more powerful position if they have the mental picture too: it is one of the marks of a good scientist. The second point is that chemistry must inevitably stay rooted in the imagination. You can in principle see all the stages of the *News at Ten* swoop with your own eyes; but the journey to the heart of chemistry must be done without sight. There is evidence, which we are able to see, touch and smell, and there are some very clever instruments which can go where our senses cannot reach, but in the end we are like a blind person building a mental picture of the Sun, from evidence of what the Sun can do.

The reason for the existence of this barrier to sight is that chemistry is a science that concerns itself at heart with the very, very small.

1.2 A look at the evidence for the existence of molecules

Historically, curious people have always been concerned with the question 'What are things made of?' These days we are brought up with the atomic answer to that question, so that it is quite hard to imagine back into a time before such ideas were common currency. However thinking back into states of historical ignorance is time well spent – we can learn from the way that scientific knowledge grew and is still growing, rather than just accepting the set of facts and theories that are in place at the time.

When as chemists we look at water we are not instantly reminded of tiny lumpy nuggets of matter. A person unfamiliar with the concept of atoms might, understandably, consider water as a continuous fluid – in the equivalent of a *News at Ten* scale swoop, that person would expect never to arrive at a *unit* of water.

Question

1 Can you think of any evidence (without using very expensive instruments) which supports the idea that water actually does consist of tiny discrete particles, rather than continuous homogeneous matter?

Note: This text contains many questions like question 1, as distinct from and in addition to past examination questions. The best use for them is to try to answer them before turning to the answers at the end of the book, and to keep a file of your answers that can run as a parallel record with the text. The hope for this policy is that it will force you to participate actively in the learning process, rather than be a passive reader.

1.3   A comment about models

Science teachers find that there are two complaints often made by students. One is 'Yes, but how do we *know* that is true?' In this case, how do we actually *know* that water consists of molecules? To answer that question evasively, do not worry if it is literally true. The more you get involved with the world of the very small, the more it becomes apparent that to try to picture what things are 'really like' is to try to track down a mirage. The best a scientist can hope for is to have a mental model to work with. (This is the mind's-eye picture mentioned earlier.) The best that the model can do is to account for the existing evidence and to make predictions which may subsequently turn out to be correct.

So far everything we know about water can be explained using the idea of

molecules. If something were to be found out about water which conflicted with that idea, the model would have to be modified. A famous philosopher of science, Karl Popper, has said that one of the characteristics of a scientific theory is that it can never be proved *true*, however many times it is found to agree with the facts; the only possibility is that it may be proved *false*.

Question

2 One hundred years ago the Law of Conservation of Energy and the Law of Conservation of Mass would have been given 'absolute truth' status in physics and chemistry respectively. What has happened to these two 'laws' in the meantime, especially at the hands of Einstein? Why are they still to be found in current textbooks?

The second student complaint, and one which is linked to the first, is about the multiplicity of models. Figure 1.2 shows a number of brands of molecular-model kits, with a methane molecule built from each one. 'Which one of those is true?' asks the student. 'Why do teachers use

different models according to whichever one suits the point they are making at the time?' The answer to that is that moving from model to model is just what scientists do. None of these models is exactly right. (For example, as we shall see in later chapters (pp. 39–42), it is all but impossible to picture electrons.) However, one model is useful for discussing bond angles, because it emphasises bond directions and because it fits the evidence from machines which measure bond angles. Another style of model conforms to the atomic radius evidence, and yet another one is cheap.

This may sound anarchic, but even 'overthrown' versions of a theory, the sort that have fallen foul of new conflicting bits of evidence, can still be very useful. Physicists have not discarded Newton's laws just because they do not apply very well to electrons. After all, Newtonian physics still describes to an adequate degree of accuracy the behaviour of bricks and snooker balls.

I want to encourage you to have as many and as vivid pictures of chemical objects as you can muster, but I want you to see the true status of these pictures – that they are thinking tools rather than absolute realities.

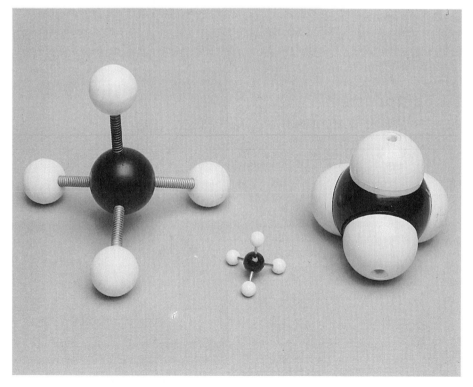

Figure 1.2 Different sorts of molecular model

1.4 The behaviour of gases

In your answer to question 1 (p. 2) you may have said, justifiably, that one of the factors which makes it easier to accept a molecular view of water is the evidence that it comes from the combination of oxygen and hydrogen. The idea that something is made up of ingredients suggests that a *News at Ten* scale swoop would eventually reach a level where you could see bits of those ingredients. In other words, it is evidence for matter being 'lumpy', rather than smooth and continuous.

Even more significant is the evidence about how those two ingredients combine to form water. Hydrogen and oxygen, when mixed together and sparked, explode to form water; but the really surprising thing is that they do it in a strict and simple **ratio of volumes** (assuming that the temperature and pressure of the gases are the same). Two volumes of hydrogen always combine with one volume of oxygen, and any mixtures which deviate from this ratio result in left-over unreacted gas. When the unreacted gas is discounted, the volumes that actually react are again seen to reveal the 2 : 1 ratio.

Joseph Louis Gay-Lussac (Figure 1.3) first published the observation about gases reacting in simple whole-number volume ratios. He had studied several reactions between gases, and although they did not all follow the 2 : 1 pattern of water (Figure 1.4), they did all display a whole-number ratio of some kind.

What would you, had you been a scientist of the time, have made of this? If these two gases were just clouds of invisible continuous matter, why were they

Joseph Louis Gay-Lussac

Gay-Lussac (1778–1850), a French chemist and physicist, was born in Haute Vienne. In 1801 he began to research gases, temperature, hygrometry and the behaviour of vapours. He made balloon ascents to collect samples of air for analysis, which resulted in his famous memoir to the Academy of Sciences on the 2 : 1 combining ratio of hydrogen and oxygen. In 1808 he published his law of volumes.

Figure 1.3 Joseph Gay-Lussac

bound by such strict laws governing their interaction? And in any case, why should those laws have anything to say about the *space* the gases took up? What possible model can account for this evidence, other than the one involving millions of particles of each ingredient performing simple operations with each other?

Gay-Lussac accepted a particle model of the universe. Here is a possible thought sequence he might have had:

1 If gases are made of 'atoms', then reactions between gases must mean *breakings* and *joinings* between 'atoms'.

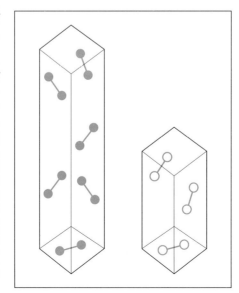

Figure 1.4 Two volumes of hydrogen molecules and one volume of oxygen molecules

2 If gas reactions follow reliable patterns, it is probably because there are simple rules controlling these atomic joinings; for example, an 'atom' of oxygen, O, might naturally combine with two hydrogen, H, 'atoms'.

3 So, in the water example, the most probable explanation for the volume ratios is that **two volumes of hydrogen 'atoms' contain exactly twice as many 'atoms' as there are in one volume of oxygen 'atoms'**, so that a 2 : 1 volume ratio provides exactly the right relative numbers of 'atoms' to fit the natural joining habits of hydrogen and oxygen (Figure 1.5).

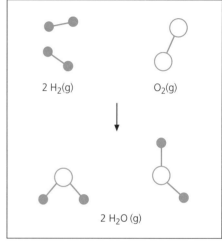

Figure 1.5 The reaction of hydrogen molecules and oxygen molecules to form water

4 This in turn suggests that **equal volumes of hydrogen and oxygen contain the same number of 'atoms'**.

5 Since this same argument could be applied to any of the gases which Gay-Lussac studied, it seems safe to generalise that **equal volumes of all gases contain the same number of 'atoms'**.

1.5 Gay-Lussac's Law and Avogadro's modification

The word 'atom' has been put carefully in quotation marks while treading in Gay-Lussac's footsteps. The reason is that scientists in the early 1800s were hazy about the distinction between atoms and molecules, and used the word 'atom' for both. This mistake cost Gay-Lussac the chance of an even bigger place in history. Even so, the experimental finding mentioned above is called Gay-Lussac's Law, namely that:

Gases react together in simple whole-number ratios of volumes, the volumes being measured at the same temperature and pressure.

Question

3 Why is it necessary to add that last bit about measuring gas volumes at the same temperature and pressure?

Gay-Lussac's experimental finding is more closely associated, with one small but important word change, with his contemporary, the Italian chemist Amedeo Avogadro (Figure 1.6).

Avogadro was aware that one of the gas reactions which had been studied at the time did not sit comfortably with the

Amedeo Avogadro (1776–1856)
Avogadro was born in Turin, where he was Professor of Physics from 1834 to 1850. In 1811 he formulated the famous hypothesis which carries his name.

Figure 1.6 Amedeo Avogadro

prevailing version of chemical 'truth'. At first sight this particular reaction fitted Gay-Lussac's pattern very well – the gases hydrogen and chlorine reacted in a 1 : 1 volume ratio to give two volumes of something which at the time was known as muriatic gas. (We now call it hydrogen chloride.) 1 : 1 : 2 was certainly a classic example of a simple whole-number volume ratio.

However, another theory in place at the time was the idea that atoms were *indivisible*. This was one of the hypotheses of the very influential British chemist John Dalton, whose classic set of laws (of constant composition of gases, multiple proportions, etc.) has been somewhat eroded over time, but like Newton's laws still provides good approximations for most normal chemical situations.

John Dalton (1766–1844)
Unquestionably one of the great chemists, Dalton (Figure 1.7) was 'grave and reserved but kindly, never finding the time to marry'. Born in a small Cumbrian village, he was the son of a Quaker weaver. After being a proprietor of a school in Kendal, where he developed an interest in mathematics, he moved to New College, Manchester.

Avogadro may have thought like this:

1 The 1 : 1 : 2 ratio of the hydrogen–chlorine reaction must mean that one 'atom' of hydrogen reacted with one 'atom' of chlorine to form two 'atoms' of muriatic gas.

Figure 1.8 The molecules involved in the hydrogen–chlorine reaction

Figure 1.7 John Dalton

2 So each of the two 'atoms' of muriatic gas must contain half an atom each of hydrogen and chlorine.
3 Yet how could that be, if atoms are indivisible?

Avogadro's answer to this puzzle was to realise that some 'atoms' might in fact be 'double atoms', and therefore be divisible without offending Dalton's unsplittable atom theory. He proposed correctly that the hydrogen and chlorine 'atoms' were both diatomic molecules, and thus made possible an era in which the concepts of atom and molecule were clearly distinguished. Figure 1.8 shows the equation for the reaction that had caused all the trouble.

Avogadro's contribution to Gay-Lussac's Law was only a small change of words – he changed the word 'atom' to 'molecule'. This change, however, is so

important that far from it being called 'Avogadro's minor modification' it is one of the central ideas of chemistry known everywhere as Avogadro's hypothesis:

Equal volumes of gases contain equal numbers of molecules, all gases being measured at the same temperature and pressure.

Surprisingly (and unfortunately) his fellow chemists did not comprehend his major contribution to chemical thought, and stumbled along in confusion for several more decades. When finally Avogadro's idea was accepted and its power recognised, it was too late for his self-esteem, as he was already dead.

(He might have felt comforted if he could have foreseen that every scientific calculator has a button permanently programmed with something called the Avogadro number – of which more on p. 10.)

1.6 Why gases behave so simply

When Michael Faraday set up the tradition of Christmas lectures at the Royal Institution, he dedicated them to the cause of overcoming the 'anaesthetic of familiarity'. He was concerned that the growth of knowledge led people to take remarkable things for granted. A similar complacency might cloud your vision of this truth about gases. It really is quite remarkable, that a million gaseous mercury atoms should occupy the same space as a million helium atoms, in complete disregard of the fact that one kind of atom is some fifty times heavier than the other (Figure 1.9).

Question

4 Try to explain why, in the matter of gas volumes, 'fat' atoms and molecules take up as much space as 'thin' ones. Bear in mind that a gas occupies a *volume* because it exerts a *pressure*. Then recall that pressure is caused by the force of molecular collisions with the walls of the container. You are looking therefore for compensatory mechanisms whereby 'fatties' and 'thinnies' can exert the same force per molecular collision (both at the same temperature).

1.7 Making use of Avogadro's hypothesis

Avogadro and Gay-Lussac, between them, have opened our eyes to the fact that equal volumes of gases contain equal numbers of molecules. We can add to the hypothesis by saying that two volumes of a gas will contain twice as many molecules as one volume of another gas, or indeed generally that:

The ratio of any two gas volumes is numerically the same as the ratio of their numbers of molecules.

So gases, by the simplicity of the way they fill space, allow an easy insight into their chemical structure. Because of the truth embodied in Avogadro's hypothesis, we have a way of 'counting' particles. It is an indirect and relative way, but the truth about a chemical reaction can be

very adequately told in terms of the *relative* number of particles taking part. This means that if we measure the **volume ratio** of two reacting gases, we have their **molecular ratio** as well, and can therefore deduce the chemical equation for the reaction.

For instance, two volumes of hydrogen react with one volume of oxygen. Therefore two molecules of hydrogen react with one molecule of oxygen. We are entitled to visualise the reaction happening as in Figure 1.5, or to write the more formal version, the chemical equation, as:

$$2H_2(g) + O_2(g) \rightarrow 2H_2O(l)$$

The numbers in a chemical equation that express the ratios of reactant particles destroyed, and of product particles formed, are called the **stoichiometric numbers**.

Question

5 Is the following statement true or false? 'If the above reaction were carried out above 100 °C the water product would be a gas too. Its volume would be the sum of the volumes of the two reacting gases, because of the principle that material is never destroyed in a chemical change.' As well as saying 'true' or 'false', explain your answer fully.

Here are two more questions. In these, the starting point is the volume data for the reaction, and your target is the chemical equation.

Questions

6 10 cm³ of the gas butadiene react with 20 cm³ of hydrogen, H_2, to produce 10 cm³ of butane, C_4H_{10}. Which of the following would be a correct chemical equation for this change? (They are all correctly balanced – that is not the sort of error you are searching for.)

   A $C_4H_4(g) + 3H_2(g) \rightarrow C_4H_{10}(g)$
   B $C_4H_6(g) + 2H_2(g) \rightarrow C_4H_{10}(g)$
   C $2C_4H_7(g) + 3H_2(g) \rightarrow 2C_4H_{10}(g)$
   D $C_4H_8(g) + H_2(g) \rightarrow C_4H_{10}(g)$
   E $2C_4H_9(g) + H_2(g) \rightarrow 2C_4H_{10}(g)$

Figure 1.9 Equal volumes of mercury and helium gases (hot enough for mercury to be gaseous)

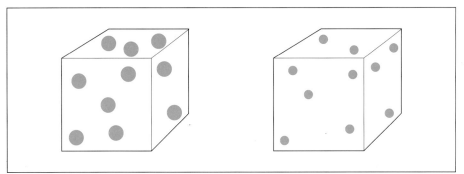

7 Not many gases react in the cold on contact, but two that do are nitrogen monoxide, NO, and oxygen, O_2, to form the brown oxide of nitrogen (also a gas). In the apparatus in Figure 1.10, there are 50 cm³ of nitrogen monoxide in the left-hand syringe and 50 cm³ of oxygen in the right. The oxygen is added, 5 cm³ at a time, to the nitrogen monoxide, by pushing in the oxygen syringe and letting the other syringe do what it will. The data are given in Table 1.1.

Table 1.1 Data for question 7

Volume of oxygen added(cm³)	Volume of gas in left syringe after addition(cm³)	Total volume in both syringes(cm³)
0	50	100
5	50	95
10	50	90
15	50	85
20	50	80
25	50	75
30	55	75
35	60	75
40	65	75
45	70	75
50	75	75

How many cm³ of oxygen have reacted with the 50 cm³ of nitrogen monoxide? (It may help to plot the first column (x) against the third column (y), and spot the point at which the reaction ceased.) From that information, deduce:

a the molecular ratio of the two reactants

b how much oxygen remained unreacted

Figure 1.10 Apparatus for question 7

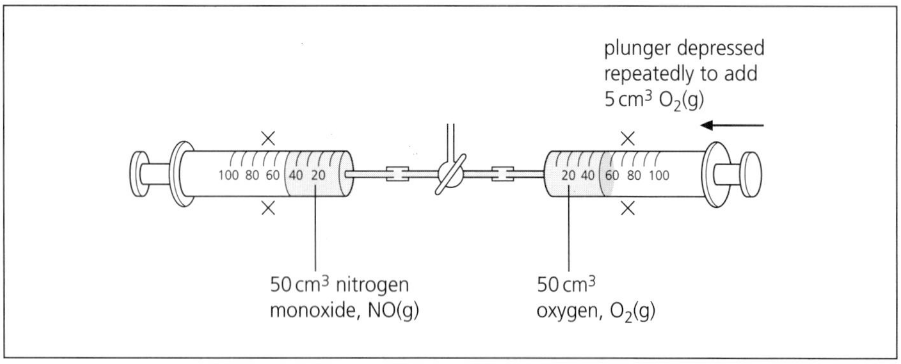

plunger depressed repeatedly to add 5 cm³ O_2(g)

100 80 60 40 20

20 40 60 80 100

50 cm³ nitrogen monoxide, NO(g)

50 cm³ oxygen, O_2(g)

c how much of the gas at the end was the new product gas

d the molecular ratio of the product to the reactants

e finally, the complete chemical equation for the reaction, including the formula of the 'brown gas'.

f On a practical point, why does it help the accuracy of the experiment to keep the connecting tubes as narrow as possible?

The next question also begins with volume data, and ends by asking for a molecular ratio. This time it is in the context of a *mixture* of gases (air) rather than a reaction between gases.

Question

8 Carbon dioxide is present in ordinary air at a level of about 0.03 % *by volume*. When discussing gases, people who are monitoring pollution more often use the unit parts per million (ppm), meaning how many out of a million molecules are of the named sort. Apply Avogadro's hypothesis and convert the volume unit so as to work out the level of carbon dioxide in air in ppm.

The remaining questions probe the link between volumes and molecules from the other direction – in these questions the chemical equation is either a known factor, or else you have to derive it from a knowledge of reactants and products. Then you are asked to make predictions about reacting *volumes* (and in some cases continue through to masses).

Questions

9 Work out the perfect volume ratio for these explosive reactions (assuming complete oxidation to carbon dioxide and water).

a Methane, CH_4 (ordinary 'gas-tap' gas), and oxygen, O_2.

b Ethyne, C_2H_2, and oxygen, O_2.

c Practically, what would be the best way of filling a vessel with an accurately measured volume of gas?

10 The following gas reactions took place. Which one would have products which took up *more* space than the reactants (assuming all volumes are measured at the same temperature and pressure)? Show how you reached your answer.

A $N_2(g) + 3H_2(g) \rightarrow 2NH_3(g)$

B $C_6H_6(g) + 3H_2(g) \rightarrow C_6H_{12}(g)$

C $4NO_2(g) + O_2(g) \rightarrow 2N_2O_5(g)$

D $4NH_3(g) + 3O_2(g) \rightarrow 2N_2(g) + 6H_2O(g)$

E $2SO_2(g) + O_2(g) \rightarrow 2SO_3(g)$

11 Assume for the moment that petrol is octane, C_8H_{18}.

a Write a balanced equation for the complete combustion (to carbon dioxide and water) of octane in oxygen, assuming all the fuel is in the gas phase.

b What *volume* ratio does this equation predict for the two reactants, assuming all the fuel is in the gas phase?

c Look up in a data book the percentage of oxygen by volume in air. Express this value as a volume ratio [oxygen] : [total air].

d Put answers (b) and (c) together to get the volume ratio fuel : air.

Designers of carburettors and other 'breathing' devices in car engines need to know these ratios so as to get the mixture right. (The matter of what constitutes 'right' is considered in greater detail in the section on the catalytic converter in Chapter 19, p. 498.) However, designers prefer to work in masses rather than volumes, mainly because the fuel is delivered to the carburettor in liquid form rather than as a gas.

e The density of air at 298 K and 1 atmosphere pressure is 1.18 g dm^{-3}. Octane is a liquid under these conditions, but if it were a gas its density would be 4.66 g dm^{-3}. Use these data to convert answer (d) into a mass ratio. Have this answer available – you will need it in (h). Parts (f) and (g) are a slight diversion.

f You might think it odd that we can include in a calculation a value for the density of a non-existent gas, as we did with that 4.66 g dm^{-3} for octane in (e). The reason is that this question is all about *ratios*. The volume ratios of fuel : oxygen and oxygen : air will hold good at all temperatures, even those at which octane is only a hypothetical gas (as long as the gases are all at the same temperature). Can you use Avogadro's hypothesis to defend this claim?

g Even the density ratio will stay constant, as I hope you will now convince yourself: (i) Density is mass per volume. Which of these two quantities is independent of temperature? (ii) Show, by use of Avogadro's hypothesis, that if you start with 1 dm^3 of air and take it to a temperature at which it occupies 2 dm^3, that the same volume change would have happened to 1 dm^3 of octane. (iii) Therefore show that the ratio of the two densities will be unaffected by the change in temperature.

h You will need your ratio from (e). The answer to (e) is slightly larger than the one used by engine designers, who usually accept a ratio of 14.7 : 1 air : fuel by mass. Say which of the following suggestions could be an explanation of the difference (any or all may be true), and explain your reason: (i) Real petrol is a mixture of molecules whose average carbon and/or hydrogen content is slightly *less* per mole than that of octane. (For example, more C_7 than C_8. Imagine what that would do to the equation in (a) and how that would have affected the rest of the calculations.) (ii) Real petrol is a mixture of molecules whose average carbon and/or hydrogen content is slightly *greater* per mole than that of octane (for example, more C_9 than C_8). (iii) The liquid petrol enters the combustion chamber as a fine mist of liquid droplets, so not all the molecules are in the gas phase by the time of the burn.

Conclusion

We have now had our first experience of seeing into chemistry. In this case the key to sight is the knowledge that gas volumes run parallel to numbers of molecules. Results from relatively simple volume-measuring apparatus can give an insight into the 'unseeable'.

However, it turns out that the gas phase is the only one that is so ready to surrender its secrets. When people tried to extend the same methods into the study of liquids and solids, they were unsuccessful. They looked for simple whole-number ratios of volumes, and found none. Reacting masses likewise failed to show any simple patterns. The way these problems were unravelled is the subject matter of Chapter 2.

Summary

This chapter is the simplest in the book, since it is based on just one big idea, that *equal volumes of gases*, measured at the same temperature and pressure, *contain equal numbers of molecules*. But you need to realise that there are a number of parallel ways of expressing this idea, each with a different viewpoint, and each useful in different problem-solving contexts.

• The **stoichiometric numbers** in gas-reaction equations (those are the numbers in front of the individual molecular formulae) **are in the same ratio as the reacting volumes**.

• The volume of, say, one billion molecules of *any* gas, will be the same.

• If molecules of gas A are twice as heavy as molecules of gas B, it follows that gas A is twice as *dense* as gas B (since they both take up the same volume per molecule).

2

Chemical reactions and the mole

- ▷ **Problems with solids and liquids**
- ▷ **The need for a mole unit**
- ▷ **More on the mole concept**
- ▷ **Using the mole concept – exercises in translation**
- ▷ **The molar volumes of gases**
- ▷ **Reactions occurring in solution – molar concentration units**
- ▷ **Problems featuring the concentration variable**
- ▷ **Problems featuring reactions in solution**
- ▷ **Summary**

2.1 Problems with solids and liquids

In Chapter 1 we saw how the behaviour of gases offers vital clues to the student of chemistry. The fact that gas *volumes* revealed information about particle *numbers* enabled the (relative) numbers of particles taking part in a reaction to be counted. Thus the single most significant fact about a gaseous chemical reaction – its chemical equation – could be found using apparatus for containing, measuring and bringing about reaction between volumes of gases.

Unfortunately the great majority of chemical substances are solids or liquids at convenient temperatures. There is certainly not any simple relationship in these cases between volume and number of particles. It is never true (except by rarest coincidence) that equal volumes of two different solids or liquids contain the same number of particles.

So what has the chemist to go on? An easily measurable quantity associated with a portion of a solid or liquid is its **mass**. Do relative masses of reacting chemicals reveal any simple ratios? Here are some data relevant to that question.

Questions

1 When magnesium is burnt in oxygen, it gains mass and forms

Table 2.1 Data for question 1

Mass of magnesium (g)	Mass of magnesium oxide (g)	Mass of oxygen (g)
0.50	0.85	
1.01	1.68	
1.52	2.51	
1.99	3.31	
2.44	4.10	
2.86	4.76	

magnesium oxide. Here are some 'before-and-after' measurements of mass for this reaction.

Copy the table above and complete the third column.

a What is the ratio by mass of magnesium to oxygen (Mg : O) reacting together?

b Does this ratio tell us anything directly about the relative *numbers* of atoms reacting? Or is there another explanation?

There is every possibility that the imbalance in the mass ratio could be due to one atom being intrinsically heavier than the other. It was no secret that different atoms had different masses – the study of gases would have revealed that.

2 From Chapter 1 you know that equal volumes of gases contain equal

numbers of molecules. Here are some mass data for two gases: 1 dm³ of hydrogen gas has a mass of 0.083 g; 1 dm³ of helium under the same conditions has a mass of 0.166 g.

a What is the mass ratio of the particles in these two gases?

b Chemists know that helium atoms have a mass *four* times as great as hydrogen atoms. How could you explain the discrepancy?

In the early 1800s, atomic masses were the subject of heated academic debate, because of discrepancies like that in question 2b. Avogadro's hypothesis helped to settle the matter, but it was the invention of the mass spectrometer in 1919 which led to definitive determinations of atomic masses (p. 29).

But let us go back to the situation in the 1800s, when there was no access to data books containing the atomic masses of all the elements to three decimal places. There were simply experimental facts like the one we discovered in question 1; namely that 3 g of an element called magnesium reacted with 2 g of oxygen to form a white powder (which we now call magnesium oxide).

Question

3 The outstanding worker in the field of atomic masses was the Englishman John Dalton (p. 4). His table of elements, published along with his atomic theory in 1803, is shown in Figure 2.1. The errors in this table illustrate the fact that even a brilliant

scientist is constrained by circumstances prevailing at the time of his or her work. At the time there was no technique for decomposing some of the most resolutely bonded compounds into their elements. There was also a lack of apparatus for accurate weighing of gases. In addition there was one false assumption.

a Which 'elements' listed in the table are really compounds?

b Which technique for the extraction of reactive elements from their compounds had yet to be invented?

c The false assumption was this: in the absence of other evidence, it

was assumed that a compound had a 1:1 ratio of *numbers* of atoms. Dalton knew that hydrogen was the lightest element, and would therefore make a good standard by which to express the atomic masses of all the other elements. He studied water, and announced that on a scale of hydrogen = 1, the atomic mass of oxygen was 8. Can you deduce how his assumption might have led him to propose an atomic mass for oxygen of 8, instead of 16?

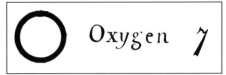

Figure 2.2 Limitations of accuracy led Dalton to a value of 7 for the atomic mass of oxygen, but question 3 uses 8

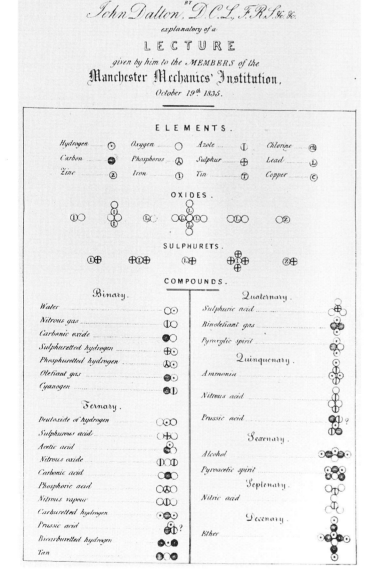

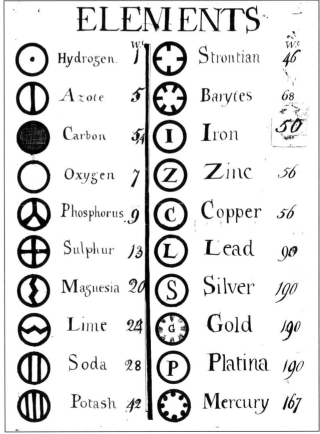

Figure 2.1 Dalton's table of the elements (1806–1807)

d Dalton worked without the important gas-volume results of Gay-Lussac and Avogadro. How might he have avoided his error if he had focused on volume when studying the reaction to make water?

e To return to the magnesium oxide data from question 2, p. 8, and accepting a correct relative atomic mass of 16 for oxygen, show how you would deduce the relative atomic mass of magnesium. State any assumptions you use (and thereby acquire some sympathy for Dalton's position).

f (Harder) It is also possible to interpret the results as meaning that magnesium oxide is really Mg_2O, and the relative atomic mass of magnesium is 12. Show how a belief that magnesium = 12 would affect your interpretation of the fact that 12 g of magnesium react with 35.5 g of chlorine. (In other words, what would be the formula of magnesium chloride?)

The mass spectrometer (Chapter 3, p. 29) is an apparatus whose ability to take atoms singly and make them perform according to their mass puts an end to all argument over atomic masses. In the case of magnesium there is a preponderance of signals from atoms with a mass/charge ratio of 24, confirming a relative atomic mass of 24, not 12.

2.2   The need for a mole unit

The mass spectrometer revolutionised 'seeing' atoms and their relationships. There is no longer any doubt that magnesium atoms have relative atomic masses of 24 (or 25 or 26 in some cases), nor that magnesium oxide is a 1:1 compound, formula MgO. Atomic masses are quoted to four decimal places in data books, and mass spectrometers can distinguish between carbon monoxide, CO, and nitrogen, N_2, (whose atomic masses are identical to two significant figures) on the strength of those decimal places.

Knowing relative atomic masses so accurately, it is now possible in effect to 'count' particles. This has two practical uses. On the one hand we can assemble the correct relative numbers of particles exactly to react with each other in a known reaction to give a known compound. Alternatively we can 'count' the number of particles reacting in an unknown reaction, and deduce what has happened. The next example is of the first type.

Example: The tin iodide reaction

It is known that tin reacts directly with iodine to make a compound with formula SnI_4. If we wanted to put together the correct amounts of these two elements to produce complete reaction and no waste, what masses would we use?

The relative atomic masses of the two elements are tin, Sn, = 118.69 and iodine, I, = 126.9. Iodine is slightly heavier than tin, so a pile of iodine containing the same number of atoms as a pile of tin would be slightly the heavier of the two.

Questions

4 What is the mass ratio of these two piles of equal numbers of atoms of tin and iodine?

5 To make the compound we need a pile of iodine atoms containing *four times as many* atoms as are in the pile of tin atoms. What mass ratio should the two piles now be in?

6 What actual masses might you put in a 'recipe' for making tin iodide, suitable for apparatus holding *tens* of grams?

In question 6 you carried out an 'information translation', by going from ratios between atoms (1:4), via mass ratios, and finally into actual masses in grams. The easiest way to achieve the final step is to add the word 'grams' to the mass-ratio numbers. The mass-ratio is 118.69:507.60. So the masses 118.69

grams and 507.60 *grams* contain the correct numbers of atoms.

This simple step is also an important one. By taking it we have answered the challenge set at the beginning of the chapter. We have found a way of taking two solids, and translating between their masses in grams and their relative numbers of particles. In other words, we have a particle-*counting* method for solids equivalent to the comparisons of volumes which were used for gases.

The tin iodide example demonstrates the usefulness of the quantity the **atomic mass in grams**. We can see that 118.69 g of tin contains the *same number of atoms* as 126.9 g of iodine.

Question

7 Revisiting an earlier case, what could you say about the relative number of atoms in 24 g of magnesium and 16 g of oxygen?

This rule will work for *any* example:

The relative atomic masses in grams of all the elements contain the same number of atoms.

The Avogadro number

If a mole of any element contains the same number of atoms, what is that number? The answer to that question has been determined to many decimal places by numerous ingenious methods, and the actual number has been given a name in recognition of the work of Avogadro (p. 4).

Avogadro number
$L = 6.0222045 \times 10^{23} \text{ mol}^{-1}$

This suggests a new way of describing amounts of chemical elements. It is quite laborious to say 'Take the atomic mass in grams of tin and react it with four times the atomic mass in grams of iodine ...'. There is a word for 'the atomic mass in grams': it is called a **mole**. The 'recipe' for tin iodide becomes 'React one mole of tin atoms with four moles of iodine atoms' (or 0.1 mole with 0.4 mole; 10 moles with 40 moles). This form of language allows us to focus on the real heart of chemistry – that is, the simple whole-number ratios in which chemical couplings usually take place.

The mole unit acts as a *translator*. In the tin iodide case, you can use it to translate the simple chemical truth contained in the 1:4 ratio into the practical truth which answers the question 'How much of each shall I actually weigh out?'. The reason it works is because **the mole ratio is the same as the atomic ratio**, and yet a mole of any element is **expressible in grams**.

Example: What is the formula of the compound formed between phosphorus and oxygen?

White phosphorus spontaneously reacts with oxygen, and this reaction forms the basis of incendiary devices (Figure 2.3). Mass measurements reveal that 5 g of phosphorus form 11.45 g of an oxide of phosphorus.

Mass of oxygen reacted with the phosphorus

$$= 11.45 - 5.00$$

$$= 6.45 \text{ g}$$

So the mass ratio of the reactants is 5:6.45, but this is no help in finding the formula because the two atoms have different atomic masses. Clearly we need to get out of the language of grams, and into the language of moles.

Here are two 'translation formulae':

$$\text{Mass in grams} = \text{number of moles} \times \text{relative atomic mass} \quad (2.1)$$

$$\text{Number of moles} = \frac{\text{mass in grams}}{\text{relative atomic mass}} \quad (2.1, \textit{rearranged})$$

These two versions of formula (2.1) are useful in two distinct types of problem. The first is used in 'tin iodide'-type situations, where the chemical equation is already known and the only question concerns how much of each reactant to weigh out. It also applies when the question is the closely related one of 'What mass of product will be obtainable from a given amount of reactant?'. In both cases we are translating **moles into grams**.

The rearranged version on the other hand suits situations like the present example, in which the formula of the product, far from being known, is the actual target of the enquiry. In such cases the reacting masses would be known and the translation must work in the direction **grams into moles**.

In the phosphorus/oxygen case, what we want is the mole ratio, so we need to translate from grams into moles (that is, by using (2.1, *rearranged*)).

For phosphorus, number of moles used

$$= \frac{\text{mass in grams used}}{\text{relative atomic mass}}$$

$$= \frac{5}{31} = 0.161$$

For oxygen, number of moles used

$$= \frac{\text{mass in grams used}}{\text{relative atomic mass}}$$

$$= \frac{6.45}{16}$$

$$= 0.403$$

So the mole ratio is 0.161:0.403 or 1:2.5 or 2:5.

And since a mole of phosphorus contains as many atoms as a mole of oxygen, the atom ratio is 2:5 also.

So the formula of this particular oxide of phosphorus is P_2O_5.

Figure 2.3 Phosphorus was included in the bomb-loads dropped during World War II

2.3 More on the mole concept

The examples so far have referred to moles of *atoms*. Can the idea be extended to cover other types of chemical entity?

A mole of atoms of an element has been defined so far as a mass of the element equal to the relative atomic mass in grams.

So it seems reasonable that a mole of a molecular compound should be a mass of the compound equal to its **relative molecular mass in grams**. We will test this proposition in the case of tin iodide.

Example: The tin iodide reaction (continued)

The tin iodide reaction (p. 10) can be analysed in three parallel ways. The chemical equation for the change is:

$$Sn(s) + 2I_2(s) \rightarrow SnI_4(s)$$

The three ways are:

Masses:
$$118.69 \text{ g} + 507.6 \text{ g} \rightarrow 626.3 \text{ g}$$

Numbers:
$$6 \times 10^{23} \text{ Sn atoms } +$$
$$24 \times 10^{23} \text{ I atoms} \rightarrow$$
$$6 \times 10^{23} \text{ SnI}_4 \text{ molecules}$$

Moles: $1 + 4 \rightarrow ?$

Can we talk about a mole of tin iodide, SnI_4, *molecules*? There are two reasons why the answer to that question is yes. First, there is exactly the same *number* of SnI_4 molecules at the end of reaction as there were tin atoms at the start of the reaction. Second, the mass of tin iodide is also its relative *molecular* mass in grams. So the expression 'a mole of tin iodide' means 6×10^{23} SnI_4 molecules and is also a way of counting relative numbers of reactant and product particles.

Finally, there is the same facility for translating between grams and moles as exists for single atoms, using the formulae (*2.1*) and (*2.1, rearranged*) (p. 11). The only change is that the word 'atom' is replaced by the word 'molecule'.

Generalising from the last example, the word 'mole' can be attached to any collection of 6×10^{23} chemical entities; the pile in question will have a mass equal to the relative formula mass in grams.

We need to look closely at the word 'formula' in the last sentence, which is related to the choice of 'chemical entity'. In an ionic solid, such as sodium chloride, there is no such thing as a sodium chloride molecule: the structures of ionic solids are continuous three-dimensional arrays (p. 65). 'NaCl' is used to express the *relative* numbers of the two individual ions in the structure. Nevertheless, if you choose to call NaCl the formula, then you can have a mole of NaCl. What is meant by that is 6×10^{23} lots of formula as you have written it (that is, 6×10^{23} sodium atoms paired with 6×10^{23} chlorine atoms). So a mole of NaCl has a mass equal to the formula mass as written, that is, a mass of $(23 + 35.5)$ g or 58.5 g.

It is very important to be clear about the 'chemical entity'. To return yet again to the tin iodide system, one mole of tin reacts with four moles of iodine:

$$Sn(s) + 2I_2(s) \rightarrow SnI_4(s)$$

| 1 mole of atoms | 4 moles of atoms | 1 mole of molecules |

That is, one mole of tin *atoms* reacts with four moles of iodine *atoms*. But it would be equally true to substitute the words 'two moles of iodine molecules' for 'four moles of iodine atoms'. These are both versions of the same truth, because of course one mole of iodine molecules, I_2, contains two moles of iodine, I, atoms. But one mole of iodine *molecules* has a mass of 253.8 g while one mole of iodine *atoms* has a mass of 126.9 g. We must never talk loosely about 'a mole of iodine', because that could mean I or I_2, either 126.9 g or 253.8 g. It is best to always write out a 'mole of iodine molecules, I_2', or 'a mole of sodium chloride, NaCl', etc.

Question

8 This is a question about the mole concept applied to those elements which exist as giant structures of atoms: a theoretically perfect diamond would be *one* continuous molecule, since the diamond structure is a continuous three-dimensional array of covalently bonded carbon atoms (p. 88).

a How many carbon atoms would there be in a perfect diamond weighing 12 g?

b If a mole of tin iodide molecules, SnI_4, comprised 6×10^{23} actual tin iodide molecules, would a mole of C(s) (diamond) be 6×10^{23} actual diamonds? Explain your answer.

Moles and chemical equations

The statement:

'One mole of tin atoms reacts with two moles of iodine molecules'

is very similar to the equally true:

'One atom of tin reacts with two molecules of iodine'.

This is of course because the truth about chemical reactions resides in *ratios*, and those ratios are unaltered whether it is one, or 6×10^{23} atoms which are taking part. An important result of this is that chemical equations can be read either in moles or in individual particles. However, only the mole interpretation allows any sense to be made out of:

$$Mg(s) + \tfrac{1}{2}O_2(g) \rightarrow MgO(s)$$

because there is no such thing as half an oxygen, O_2, molecule. To make sense in terms of particles it has to be written:

$$2Mg(s) + O_2(g) \rightarrow 2MgO(s)$$

The 'stoichiometric numbers' in chemical equations (that is, the numbers written ahead of each chemical species in the equation) are also the relative numbers of moles taking part.

2.4   Using the mole concept – exercises in translation

As a chemist, you have the gram as the language of your measuring instrument, and the mole as the language of the chemical equation. You have to be fluent in hopping between these two language systems, in either direction, depending on the nature of the unknown factor in any given problem. Like any linguist, you will get better with practice, so there follows a number of problems involving the interconversion of moles and grams. Throughout these exercises you will need the following translation formula:

$$\text{Number of moles} = \frac{\text{mass in grams}}{\text{RFM}} \quad (2.2)$$

where RFM is a useful abbreviation for 'relative formula mass'. This formula supersedes (*2.1*), because it is the most general expression of the key translation idea, free of any particular references to atoms, molecules or ions.

Questions

9 How many grams would you need to weigh out to get:

a 3 moles of iodine atoms?

b 0.5 moles of iodine, I_2, molecules?

To do (b), you need to know that the formula mass of a molecule is equal to the sum of the relative atomic masses of its constituent atoms. That is equally true of those formulae (whether ionic or molecular) in which the constituent atoms are not the same. Now carry on:

c 0.1 moles of sodium chloride, NaCl?

d 5 moles of ethanol, C_2H_5OH?

Some ionic crystals have water molecules built into the structure, in definite stoichiometric numbers. These therefore become part of the formula, and contribute to the formula mass. Carry on.

e 2.5 moles of sodium carbonate-10-water, $Na_2CO_3 \cdot 10H_2O$?

f 0.2 moles of Plaster of Paris, $CaSO_4 \cdot \frac{1}{2}H_2O$?

g 0.4 moles of calcium nitrate-4-water, $Ca(NO_3)_2 \cdot 4H_2O$?

10 How many moles would be contained in:

a 50 g of silver?

b 1 kg of lead?

c 100 g of malachite, $Cu_2CO_3(OH)_2$?

d 50 g of 1-bromobutane, C_4H_9Br?

Parents and daughters

The formula of malachite in (c) above reveals that one mole of malachite contains *two* moles of copper atoms. There will be situations in which you will be weighing out the 'parent' substance (in this case malachite), but are more interested in the 'daughter' or constituent atom or ion (in this case copper atoms).

Question

11a How many grams of malachite would have to be taken to provide 1 mole of copper atoms?

b What percentage *by mass* of malachite is copper?

Situations in which the chemical equation is known

Often the chemist needs to work out the right mass of starting material to get a desired mass of product, or to predict how much product to expect. Here is an example showing how it is done.

Example: Limestone

At the heart of a cement works is a furnace in which some form of limestone is heated (Figure 2.4). The most significant ingredient in the limestone is calcium carbonate. One of the chemical reactions taking place is:

$$CaCO_3(s) \rightarrow CaO(s) + CO_2(g)$$

What mass of calcium oxide could be expected from the decomposition of 1000 kg of calcium carbonate?

Step 1: translate from grams into moles, using (2.2):

Moles in 1000 kg $CaCO_3$

$$= \frac{\text{mass in grams}}{\text{RFM}} \qquad (2.2)$$

$$= \frac{1000 \times 10^3}{40 + 12 + 3 \times 16}$$

$$= \frac{1\,000\,000}{100}$$

$$= 10\,000 \text{ moles}$$

Step 2: look at the chemical equation and extract the crucial mole ratio,

Figure 2.4 A cement works (top). Building a road (bottom)

thereby finding moles of calcium oxide produced. The equation reveals a −1:+1:+1 ratio (the signs discriminate between moles of product *created* (+), and moles of reactant *destroyed* (−)), meaning that for every mole of calcium carbonate destroyed, one mole of each of calcium oxide and carbon dioxide were created. (There is *no* rule saying that because two moles of product were created therefore two moles of starting material must have been destroyed.) We can now forget about the carbon dioxide, and simply say:

If 10 000 moles of calcium carbonate were destroyed, 10 000 moles of calcium oxide would have been created.

Step 3: now translate back from moles into grams to fit the required format of the answer, using (2.2) again:

Mass (g) of 10 000 moles of CaO
= number of moles × RFM
(2.2, rearranged)

(Notice that this time we want the relative formula mass of CaO.)

= 10 000 × (40 + 16)

= 560 000 g

= 560 kg

A note on style of presentation of calculations

There are two reasons for maximum explanation when you are doing a calculation. One is for the reader's benefit, but the more important one is for you – in talking to yourself you help to keep the progress of your calculation clear in your own mind. So your calculation should appear as a series of 'sentences' in words and numbers, with each new line carrying a new meaning. When using a formula, it is a good idea to quote the formula in words and/or symbols on one line, and 'plug in' the numbers in the next line, preferably exactly underneath the equivalent symbols.

Finally, a warning against a specifically *bad* practice. It is easy to make bits of the

calculation 'grow' on the same line. So for instance you might write, for the conversion of a mass in grams to moles:

$$\text{Mass of } CaCO_3 = \frac{1\ 000\ 000\ g}{100}$$

This line began life without the '/100' bit, and at that stage was correct. What happened next was that the student built the conversion to moles on to the existing line. What is so bad about this? First, the final version of the line is *not true* (the mass of $CaCO_3$ is *not* equal to (1 000 000 g/100)), and second it is the mathematical equivalent of bad grammar. But the most important and practical objection is that the student can easily lose grip on where the calculation is going. 'Careful talking to oneself' is what is needed and one of the hallmarks of this method is the use of a new line of working for each new idea or meaning.

The three-step format for solving mole problems

The contexts of mole problems vary widely, but basically most mole problems (at least those in which the equation is part of the prior knowledge) are similar. The three-step format of the calcium carbonate example can be used in most mole problems. Here are generalised versions of those three steps:

Step 1: identify as 'first reagent' the chemical (either reactant or product) whose amount is already specified. Find the moles of first reagent created or destroyed by use of (2.2). (In the limestone example, $CaCO_3$ was the first reagent, and 10 000 moles were destroyed.)

Step 2: identify as 'second' or 'target' reagent' the chemical whose amount is required. (It, too, may be either a reactant or a product.) Go to the chemical equation, and extract the relevant *mole ratio* which links the first and the target reagent. Use it, along with the answer to *step 1*, to deduce the number of moles of target reagent created or destroyed. (In the limestone example, the mole ratio was −1:+1, and the mole numbers were −10 000:+10 000.)

Step 3: finally, convert the moles of target reagent into grams using (2.2).

In the examples which follow, think of this structure, and identify the chemicals playing the roles of first and target reagent.

Questions

12a The element bromine is prepared from bromide ions in sea-water. The bromide ions are oxidised by chlorine gas. The equation is:

$$Cl_2(g) + 2Br^-(aq) \rightarrow Br_2(aq) + 2Cl^-(aq)$$

Calculate the mass of bromine obtainable from 1 tonne (10^6 g) of chlorine.

b In contrast iodine is obtained from deposits of potassium iodate(v) found in the rocks of Chile. The iodate is 'reduced' (Chapter 7, p. 121) by sodium hydrogensulphite:

$$2KIO_3(aq) + 5NaHSO_3(aq) \rightarrow$$
$$2SO_4{}^{2-}(aq) + 3HSO_4{}^-(aq) + 5Na^+(aq) +$$
$$2K^+(aq) + H_2O(l) + I_2(s)$$

Calculate the mass of iodine obtainable from 50 kg of potassium iodate. (Try to tease out the relevant simple stoichiometric ratio from this complicated equation – it is the first time you have met a ratio that is not 1:1.)

13 One version of the polymer **nylon** (Figure 2.5) is made by mixing equimolar (that is, 1:1 by mole) amounts of two monomers: 1,6-diaminohexane, $H_2N(CH_2)_6NH_2$, and decanedioyl chloride, $ClOC(CH_2)_8COCl$. What mass of the chloride would be needed to react with 1 kg of the diamine? (This problem can be done by a closely parallel method to the limestone example (p. 13), the only difference being that the formula masses are more laborious to work out, and that here it is a second *reactant* which provides the unknown.)

14 In the nylon reaction, for every mole of each of the two reactants that joins up, two moles of hydrogen chloride gas are expelled. (Those who

Figure 2.5 Manufacture of nylon (top), a light micrograph of a piece of nylon stocking (centre) and one use of nylon (bottom)

have already studied polymerisation may spot that this is not exactly true, as hydrogen and chlorine atoms survive on the very ends of the chains, but since the chains are hundreds of monomers long it is a good approximation.) The equation for the polymerisation could therefore be written:

n(1,6-diaminohexane) + n(decanedioyl chloride) → nylon(s) + $2n$HCl(g)

where n is an unknown but consistent number of moles, dependent on how long the chains are.

By using the idea that mass is always conserved in a chemical reaction, and therefore that the mass of nylon produced will equal the mass of the reactants used *minus* the mass of

hydrogen chloride liberated, calculate the mass of nylon produced in the previous equation. (This is a harder question, but its core is the same as the limestone example, with hydrogen chloride playing the part of the 'target' reagent, parallel to calcium oxide. The little subtraction sum on the end gives this question its added complication.)

Why could we not do this question more directly, say by treating the nylon in the same way as calcium oxide?

15 Now for a problem in which the *known* factor is a desired amount of product, and the unknown is a reactant: what mass of iron oxide, Fe_2O_3 would be needed to provide 1000 kg of iron metal, assuming 100 % conversion? Notice here that only the crucial part of the equation is specified, namely:

$$Fe_2O_3(s) + \text{some reducing agent} \rightarrow 2Fe(s) + \text{'something oxide'}$$

We have already seen, in question 14, that problems can have the same basic structure at heart, but appear different because of peripheral 'twiddly bits'. (In that case it was the subtraction sum by which the mass of hydrogen chloride was used to find the mass of nylon.)

A classic example of a 'twiddly bit' happens in questions about percentage composition. In these, the basic idea is that you take a portion of a mixture of ingredients, put it through a process which will cause *one* ingredient to undergo a reaction that the rest cannot do, and deduce from the outcome of the reaction how much of the reacting stuff was in the original mixture.

The next question exemplifies this technique.

Question

16 A company buys some land with a natural spring on it, and decides to exploit the vogue for natural mineral waters. The label information needs to state which ions are present and at what concentrations. One of the ions present in the water is the sulphate

Figure 2.6 Mineral composition of bottled water is usually found on the label

ion, SO_4^{2-} (Figure 2.6). Barium ions (added in the form of $BaCl_2$(aq)) will join with sulphate ions to form an insoluble (and therefore weighable) white precipitate. When 1 kg of the water is treated with an excess of barium ions, 29 mg of barium sulphate are produced. The reaction for the precipitation (excluding 'spectator' Cl^- ions) is:

$$Ba^{2+}(aq) + SO_4^{2-}(aq) \rightarrow BaSO_4(s)$$

a What factor(s) influence the choice of barium ions for this technique?

b What does the word 'excess' mean in this context, what is it that the barium ions are in excess *of*, and why does it not matter about the exact amount used?

c Calculate the sulphate content of the water, expressing the answer in mg of sulphate per kg of water, and also as a percentage by mass of sulphate. In working out the relative formula mass of the sulphate ion you can ignore the '2–'.

Situations in which the equation is unknown

Example: Antimony oxide

Antimony, Sb, forms more than one oxide. What is the chemical equation for the formation of the oxide produced when 20.0 g of antimony react with 3.95 g of oxygen?

Moles in 20.0 g of antimony

$$= \frac{\text{mass in g}}{\text{RFM}} \qquad (2.2)$$

$$= \frac{20.0}{121.8}$$

$$= 0.164$$

Moles of *atoms* in 3.94 g of oxygen

$$= \frac{3.95}{16}$$

$$= 0.247$$

Moles ratio $Sb : O$

$$= 0.164 : 0.247$$

$$= 1 : 1.5$$

$$= 2 : 3$$

From this we deduce that the formula for antimony oxide is Sb_2O_3, and that the only balanced equation which fits the formation of an oxide of this formula is:

$$4Sb(s) + 3O_2(g) \rightarrow 2Sb_2O_3(s)$$

Note 1: This time moles of oxygen *atoms* were specified, to avoid ambiguity between O and O_2. This was not necessary for antimony, because it exists, like all metals, as a giant structure of atoms, and it is assumed to be a mole of atoms.

Note 2: It is worth noting that the answer could be *wrong*. At least one data book gives the formula as Sb_4O_6, indicating the existence of an actual molecule, rather than a continuous ionic lattice. That debate need not concern you now, but it is a salutary reminder that this method, on its own, will only tell you a mole *ratio*.

Before solving some of these unknown equation/formula problems for yourself, there is one more new idea for you to absorb.

2.5   The molar volumes of gases

There are already strong links between the subject matter of Chapters 1 and 2. In both chapters we are seeking the truth about *ratios* of reacting particles. In Chapter 1 the evidence was obtained from the volumes of reacting gases, while in Chapter 2, so far concentrating on solids and liquids, we have had to convert from grams into moles before looking for those all-important ratios. It is now time to go a step further, and forge a more quantitative and direct link between the chapters.

One of the rules in Chapter 1 (p. 5) was that equal volumes of gases, measured at the same temperature and pressure, contained equal numbers of particles (be they atoms or molecules). It must also be true therefore that equal numbers of particles of two gases would occupy the same amount of space. In turn therefore it follows that 6×10^{23} particles of any gas would occupy the same space, or that:

A mole of any gas occupies the same volume.

This is true to a reasonable degree of approximation for all real gases, and the further above their boiling temperatures they are the truer it becomes. The actual volume in question is 24.45 dm^3 at 298 K and 1 atmosphere pressure (but of course this figure varies with temperature and pressure). This is the basis for another conversion formula, this time translating between volumes and moles of gases:

Number of moles of a gas

$$= \frac{\text{volume of gas in dm}^3 \text{ at 298 K, 1 atm}}{24.45 \text{ dm}^3}$$

$$(2.3)$$

Gas volume conversion

Equation (2.3) would seem to suffer from the problem that it only works if the gas volume in question has been measured under one specific set of conditions. In fact there is no problem, since the regularity of gas behaviour allows us to calculate the volume it *would have had*

at 298 K and 1 atmosphere, from any other combination of temperature and pressure. These gas formulae rely on the fact that volume V is **proportional** to temperature T (in kelvins) and **inversely proportional** to pressure P (in atmospheres):

$$\frac{V_1}{V_2} = \frac{T_1}{T_2} \qquad (2.4)$$

$$\frac{V_1}{V_2} = \frac{P_2}{P_1} \qquad (2.5)$$

There is a combination equation for situations in which both T and P are measured at non-standard values:

$$\frac{V_1}{V_2} = \frac{P_2}{P_1} \times \frac{T_1}{T_2} \qquad (2.6)$$

Here is an example of a gas volume conversion using this last formula.

Example: Octane

A quantity of octane, C_8H_{18}, occupies 330 cm^3 at $150\ ^\circ C$ and 10 atmospheres. What volume would it have occupied at 298 K, 1 atm, and what fraction of a mole is this?

Using (2.6) and denoting the standard conditions by suffix '1',

$$\frac{V_1}{V_2} = \frac{P_2}{P_1} \times \frac{T_1}{T_2} \qquad (2.6)$$

Figure 2.7 Octane is one of the products from this crude oil refining plant

Therefore

$$\frac{V_1}{330} = \frac{10}{1} \times \frac{298}{150 + 273}$$

$$V_1 = 330 \times \frac{10}{1} \times \frac{298}{150 + 273}$$

$$= 2325 \text{ cm}^3$$

$$= 2.325 \text{ dm}^3$$

To calculate the fraction of a mole we use (2.3):

Moles of gas

$$= \frac{V (\text{dm}^3 \text{ at } 298 \text{ K, 1 atm})}{24.45} \qquad (2.3)$$

Therefore the fraction of a mole

$$= \frac{2.325}{24.45} = 0.095.$$

Notice that this calculation works even though octane is not a gas at 298 K and 1 atm. This fact does not affect the right-hand side of (2.3), and does not matter since we are interested in ratios of numbers to work out the number of moles present. Thus the *ratio* of the two numbers would be unaltered.

The reason these gas volumes have been included at this particular stage is that they can be used in conjunction with mass evidence in the unravelling of unknown-equation and/or unknown-formula problems. At the end of Section 2.4 (opposite), we faced the limit to our knowledge of the formula of antimony oxide, in that we could not decide between Sb_2O_3 and Sb_4O_6 (or indeed any other 2 : 3 ratio). We needed corroborative evidence which found out where the molecular unit ended (if such a thing did actually exist). Now, at least in the case of gases, we have the extra evidence we need. On top of knowing the 'ratio formula' of a gas, we can say what fraction of a mole we are dealing with. If the mass of that fraction of a mole is known, then the mass of a whole mole, and therefore the true molecular formula, can be found.

There is an opportunity to use this concept in questions 17 and 18.

Question

17 From the mixture which makes up crude oil, a particular hydrocarbon ingredient (that is, one containing hydrogen and carbon atoms only) has been isolated. 10 g of this liquid are burned in excess oxygen, and the products are 31.4 g of carbon dioxide and 12.9 g of water. We want to find the ratio of C : H in the substance. This is tricky, but it can be broken up into sub-tasks.

a Find out how many moles of carbon dioxide and of water there are in those two masses.

b Express this as a simple whole-number ratio.

c Note that this is not itself the answer, since while there is only *one* mole of carbon atoms in a mole of carbon dioxide there are *two* moles of hydrogen atoms in a mole of water. What is the ratio of C : H in *atoms*?

The answer you should have at the end of question 17c is still *not* the actual formula of the hydrocarbon molecule. What you should have found is the simplest whole-number ratio of C : H atoms in the molecule (the so-called **empirical formula**). The real numbers of carbon and hydrogen atoms could be any multiple of this ratio, as in the Sb_2O_3 case. This is the point where a knowledge of the molar volumes of gases comes in useful. If we can find the mass of 24.45 dm³ of this hydrocarbon (at 298 K, 1 atm), we will have the actual molecular mass, and will then know what multiple of the empirical formula constitutes the molecular formula. Here are some extra data to enable you to do this.

Question

18 0.20 g of the same hydrocarbon as in question 17, when turned into gas at 373 K, 1 atm, occupied 72.8 cm³.

a Find the volume the gas would have occupied at 298 K, 1 atm.

b Work out, using (2.3), the number of moles of gas in this mass, and

then calculate the mass of one mole.

c Give a molecular formula for the hydrocarbon which fits both the relative molecular mass answer from (b), and the empirical formula from question 17.

d Then write a full chemical equation for the combustion of the hydrocarbon to carbon dioxide and water, as described in question 17.

e People in professional laboratories would probably not have to go through the process described in the last two questions, in order to find the molecular mass of the unknown hydrocarbon. What method would have provided the answer much more directly (and expensively)?

2.6 Reactions occurring in solution – molar concentration units

For reactions to happen, particles must bump into each other. Of the various circumstances in which this event can happen, the gas phase suffers from the drawback that particles are widely spread apart, while the solid phase, even when the solids in question are finely powdered, has the drawback that particles can only meet at the outside of grains. So a lot of chemistry is conducted with the reactants dispersed as **solutes** in a **liquid solvent**, in other words in **solution**. In this section we will look at reactions taking place in solution, using the mole concept.

Question

19 Just how widely spread apart the molecules in a gas are can be appreciated by a little calculation. Liquids, like solids, have molecules which more or less touch each other. So the volume of a liquid is a reasonable indication of the volume of the actual molecules in it. Thus 18 cm³ are approximately the volume of 18 g (a mole) of water molecules (Figure 2.8, overleaf).

Figure 2.8 Contrasting 1 mole of liquid water with 1 mole of gaseous water

a What volume would the same set of molecules have if they existed as a gas at 298 K and 1 atm?

b How many times more space do water molecules occupy as a gas than as a liquid at the same temperature and pressure?

The liquid phase clearly offers the best medium for the free and frequent random meeting of particles. In many cases the solvent is of course *water*.

The great advantages of solution chemistry are offset by a few drawbacks, one of which is that the user is restricted to the temperature range in which the solvent is actually a liquid. In the context of the present chapter, however, there is another problem. By introducing a second inert 'carrier' substance (the solvent), we have made it more difficult to 'count' the particles of the reacting substance (the solute). Think back to our other techniques for counting particles: if we were dealing with a gas, we had only to measure a volume (at known temperature and pressure) and we knew (via (2.3)) how many moles we had; if we were dealing with a solid, we had only to measure a mass, look up a relative formula mass, and we knew (via (2.2)) how many moles we had. In contrast, imagine a bottle of salt solution – you could weigh it or measure its volume quite easily, but those efforts would be pointless if you could not answer the question 'How much solvent is there for a given amount of solute?'.

The need for a concentration unit

Dealing with solutions therefore requires extra information. We need to know not just how many grams/moles there are in the solution, but into what volume of solvent they were put. However, even this knowledge is *not* the best piece of information about a solution. This is because solvents 'swell' when they take in solutes, and the 1 dm^3 of solvent, to which (let us say) one mole of solute is added, will occupy rather more than 1 dm^3 by the time it has dissolved the solute.

This has practical implications for people making up solutions, and it also guides us to the best choice for a concentration unit. On the practical side, you would weigh out your mole (or whatever) of solute, dissolve it in a quantity of solvent at least *10 % less* than the desired total, and *once the solute has fully dissolved* make up the volume to the desired mark. (There are flasks specially made for this procedure.) Suppose that in this example the chosen volume was 1 dm^3. You would now know that 1 dm^3 (and not '1 dm^3-and-a-bit') contained exactly 1 mole of solute, and therefore that each cm^3 contained 1/1000 of a mole of solute, and so on for every sub-multiple of the original cubic decimetre.

The concentration unit that clearly fits the described procedure is the 'mole per cubic decimetre *of solution*' (as opposed to 'of solvent'). This distinction is assumed when it comes to the abbreviation for the unit, which is just mol dm^{-3}.

We are also ready for a formula to relate the concentration of a solution to its component variables. The units of concentration virtually write the formula for you:

$$\text{Concentration (mol dm}^{-3})$$
$$= \frac{\text{number of moles}}{\text{volume of solution (dm}^3)} \qquad (2.7)$$

Notice that when making up a solution of a prescribed concentration you need to use (2.2) in conjunction with (2.7) to calculate an actual mass to weigh out.

2.7 Problems featuring the concentration variable

In the problems which follow, you will need either or both of (2.2) and (2.7). You may need to rearrange them.

Making up solutions of specified concentration

Questions

20 Calculate the mass of substance needed to make up each of the solutions in Table 2.2. (Use (2.7) first – remember it only works when the volume is in dm^3 – and then use (2.2) to go from moles to masses.)

Table 2.2 Data for question 20

Substance	Volume	Concentration (mol dm^{-3})
NaCl	1 dm^3	1
NaOH	1 dm^3	0.5
$BaCl_2$	250 cm^3	0.1
$K_2Cr_2O_7$	500 cm^3	0.2
$CuSO_4 \cdot 5H_2O$	2 dm^3	0.1

21 Work out the concentrations in mol dm^{-3} of the solutions in Table 2.3. (Use (2.2) first, to translate the masses, then (2.7).)

Table 2.3 Data for question 21

Substance	Mass	Volume of solution
KI	1 kg	10 dm^3
$FeSO_4$	100 g	250 cm^3
*Ethanol (C_2H_5OH)		50:50 by volume in water

* This last entry is a slightly unusual example, but can be done if we assume the solution has a total volume equal to the sum of the two liquids' individual volumes, and if we take the densities of ethanol and water to be 0.789 and 1.00 g cm^{-3} respectively

22 Work out the volumes needed to deliver the desired number of grams of each substance. (Use (2.2) first, then (2.7).)

a 100 g of sulphuric acid, H_2SO_4, from a solution which is 1 mol dm^{-3} in sulphuric acid.

b 100 g of bromine, Br^-, from sea-water in which the concentration of bromine is 8×10^{-4} mol dm^{-3}.

Independent concentrations in the same solution

That last sea-water question demonstrates an interesting point about concentrations in mixed solutions. Each solute in the solution, and even each *ion* from those solutes, can have an independent concentration in the same solution. For instance, the chloride and sodium ion concentrations in sea-water are about 0.54 and 0.47 mol dm^{-3} respectively. Notice that these data give the lie to the idea that sea-water is an aqueous solution of sodium chloride. It is in fact a 'soup' of many ions, of which sodium and chlorine are the commonest. The origins of the ions are mineral deposits in rocks, and the only controlling factor governing the mixture is that **overall cation charge = overall anion charge**.

Even if there is a single ionic solute in a solution, it is known from electrolytic experiments that the ions lead semi-independent lives. This gives justification for the idea of quoting independent concentrations for each ion. In the case of a single solute, however, unlike sea-water, there are strict mole relationships between parent ion concentrations and the 'daughters', as the following problem will show.

Question

23a Give the chloride, Cl^- ion concentration of a solution which is 0.1 mol dm^{-3} with respect to calcium chloride, $CaCl_2$.

b If you want a solution which is 0.5 mol dm^{-3} with respect to sodium ions, Na^+, and your source of the ions is sodium sulphate, Na_2SO_4, what parent concentration would you need?

2.8 Problems featuring reactions in solution

The reagents in solutions are bound by the fundamental laws of chemistry, among which is the fact that reactions take place between whole-number ratios of molecules (and moles). But there is *no* law which governs the ratios of *volumes* of solutions reacting with each other, because that depends on how concentrated the operator made them. The more dilute a solution, the more of it you need to deliver any required number of moles.

Quite often reactions between solutions are conducted for the purpose of discovering the concentration of one of them from a knowledge of the concentration of the other. This is called **titration**, and is conducted in standard items of calibrated glassware (Figure 2.9). A titration example follows.

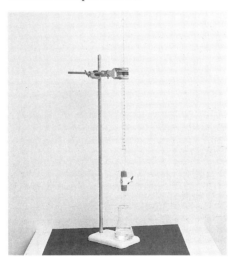

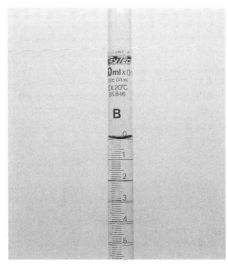

Figure 2.9 Calibrated glassware for titration

Example: Titration with dilute hydrochloric acid

Dilute hydrochloric acid is a familiar aqueous-solution reagent, and in most laboratories the concentration of the bench solution is approximately 2 mol dm^{-3}. However, when technicians are making it up its concentration is difficult to gauge precisely, because the solute (hydrogen chloride gas) cannot be conveniently weighed out like a solid salt. Normally the solution is prepared from commercial concentrated aqueous hydrogen chloride, which is about 10 mol dm^{-3}, by a ×5 dilution.

Question

24 How would you bring about this ×5 dilution in order to prepare 1 dm^3 of end product? In other words, what volume of concentrated aqueous hydrogen chloride would you mix with what volume of water?

One way to find the concentration of a solution of hydrochloric acid is to make up a second solution with which it will react, and which *can* be prepared by weighing to be a precise concentration. An ideal substance is borax, $Na_2B_4O_7 \cdot 10H_2O$. This is a free-running powder which does not go sticky by absorbing atmospheric moisture while being weighed out, and which dissolves freely in water. (RFM of borax = 381.4.) It reacts with aqueous hydrogen chloride thus:

$$Na_2B_4O_7(aq) + 2HCl(aq) + 5H_2O(l) \rightarrow 4H_3BO_3(aq) + 2NaCl(aq)$$

(*Note:* This reaction can be compared with those between acids and carbonates, except that carbonic acid decomposes further to water and carbon dioxide, whereas boric acid does not. It is therefore an example of the broad group of reactions in which a stronger acid forces hydrogen ions (H^+) on to the conjugate base of a weaker acid. These reactions are discussed in detail in Chapter 12, p. 256.)

Suppose a solution of borax has been accurately made up to a concentration of 0.100 mol dm^{-3}. A sample of 'dilute hydrochloric acid' is diluted by a (precise) factor of $\times 10$ (by taking $20\ cm^3$ and making it up with water to $200\ cm^3$).

(*Note:* You have to do this dilution, because the best titrations are achieved by having reagents whose concentrations are at least within the same order of magnitude. Otherwise you have 'burette deliveries' which are either very small (losing accuracy on the burette) or very big (making end-points vague). It so happens that the solubility of borax limits its concentration to the $0.1\ mol\ dm^{-3}$ order of magnitude, and that is why we have to bring the hydrochloric acid concentration down to match.)

The dilution produces an acid solution which is *approximately* $0.2\ mol\ dm^{-3}$, the approximation arising from the vagueness of the concentration of the original acid. It is found by titration (Figure 2.9) that $25.0\ cm^3$ ($= 25/1000\ dm^3$) of borax solution needs $23.6\ cm^3$ of hydrochloric acid (in the burette) to react exactly with no excess of either reagent. (This is called the end-point, and is shown by using a coloured indicator.) The unknown hydrochloric acid concentration is found in three steps as follows:

Example: Titration

Step 1: find how many moles of borax are used (note the units).

Moles of borax used

$=$ concentration $\times$ volume
$\quad (mol\ dm^{-3}) \qquad (dm^3)$
$\qquad\qquad\qquad$ (*2.7, rearranged*)

$= 0.100 \times \dfrac{25}{1000}$

$= 2.5 \times 10^{-3}$

Step 2: look at the stoichiometry and extract the relevant mole ratio.

From the equation on p.19, we can see that 2 moles HCl react with 1 mole borax. So moles of hydrogen chloride reacting exactly with 2.5×10^{-3} moles borax

$= 2 \times 2.5 \times 10^{-3}$

$= 5 \times 10^{-3}$

Step 3: scale up from the batch quantity used to find the concentration.

Question

25 So $1\ dm^3$ of this solution contains 2.12 moles of hydrochloric acid. How much water would you need to add to this $1\ dm^3$ to obtain exactly $2.00\ mol\ dm^{-3}$ hydrochloric acid?

General points about titration problems

A further note on style

Notice again the use of carefully constructed whole sentences of words, symbols and numbers in this example. Remember how important this is.

Another three-step structure

The example has a three-step structure which is common to all titration problems in which the equation is known. (It is very similar to the structure already highlighted in the discussion of straight mole calculations in Section 2.4, p.12). That three-step structure is something you must try to to employ in your own problem-solving. To repeat the steps:

This number of moles of hydrogen chloride was present in $23.6\ cm^3$ of solution ($= 23.6/1000 = 0.0236\ dm^3$ of solution).

So concentration ($mol\ dm^{-3}$)

$= \dfrac{\text{number of moles}}{\text{volume }(dm^3)}$ $\quad$ (*2.7, rearranged*)

$= \dfrac{5 \times 10^{-3}}{0.0236}$

$= 0.212\ mol\ dm^{-3}$

This particular problem has an extra bit on the end – in this case the unusual circumstance that the aqueous hydrogen chloride titrated was a diluted version of the original solution. It remains therefore to add another step which is not typical of all titration problems:

Concentration of original 'dilute hydrochloric acid'

$= 0.212 \times 10$

$= 2.12\ mol\ dm^{-3}$

Step 1: this involves the solution you know *everything* about (both its volume and its concentration). From these two bits of data you derive the **number of moles of first reagent**.

Step 2: this leads from knowledge of the moles of the first reagent to knowledge of the moles of the second 'target' reagent – the bridge being the mole ratio in which they react, found from the *chemical equation*.

Step 3: still concerned with the second reagent, you scale from moles-in-the-batch-used to some other reference point, normally a concentration in moles per cubic decimetre.

Significant figures

The (seven-figure) calculator gives 2.118644 as the numerical answer to the borax example. Calculators are programmed to go on dividing until they run out of screen, but that does not mean all those extra figures are important. The factor restricting the level of accuracy of our knowledge is the precision of the glassware. A person with a burette and normal eyesight can reasonably judge 23.6 (three significant figures) but beyond that, the 23.6 could have been anything between 23.55 and 23.64. If you use these extreme possibilities in the calculation, the answers would be 2.123142 and 2.115059, from which you can see that 2.12 is about the limit of certainty. This example offers a general rule – **give an answer with the same number of significant figures as the worst item of data.**

Units

It is easy to make mistakes with formulae, especially when some rearrangement is necessary. However, there is an excellent trick for checking formulae, involving units. Take *step 1* of the recent example – had you been uncertain about the rearranged version of (*2.7*) which reads 'moles = concentration $\times$ volume', you could have visualised the calculation taking place with no numbers, just units. Like this:

$$\text{moles} = (\text{moles}/dm^3) \times dm^3$$

The 'dm^3's cancel, so moles = moles and the rearrangement must be correct.

Logically an equation has to have the

same units on either side of the equals sign. The only thing a certain number of moles can be equal to is the same number *of moles*. Suppose you had picked a false version of the equation, like 'moles = concentration/volume', you would have had right-hand side units of mol dm^{-6}, and left-hand side units of moles, which tells you it is wrong. This is a very quick and useful technique for insuring against lapses of memory of formulae.

A final warning

Do not look for simple whole-number ratios between the *volumes*. They are very unlikely. The place for whole-number ratios is between reacting *moles*.

Questions

26 The ingredient which gives vinegar its acidity is ethanoic acid, an organic acid with formula CH_3CO_2H. Only the last hydrogen is acidic (p. 293), so the mole ratio is 1 : 1, thus:

$$CH_3CO_2H(aq) + NaOH(aq) \rightarrow$$
$$CH_3CO_2Na(aq) + H_2O(l)$$

A sample of white wine vinegar is titrated against a solution of 1.00 mol dm^{-3} sodium hydroxide, NaOH, and 25.0 cm^3 of the sodium hydroxide required 35.6 cm^3 of the vinegar to neutralise it. Calculate the concentration of ethanoic acid in the vinegar, in mol dm^{-3}.

27 Ion-exchange resins are both interesting and useful. Their domestic and industrial value is in their ability to remove ions from water supplies (water 'softening') (Figure 2.10) to save on detergent use or minimise lime scaling in pipes and boilers. However, certain types of resin have an analytical value too, in that they can turn a neutral salt into an equivalent amount of acid, by exchanging metal cations for hydrogen ions (Figure 2.11). This has the effect of rendering salt solutions accessible to acid–alkali titrations. The technique here is being used to estimate the solubility of gypsum (calcium sulphate, the substance that plasterboard is made from) in water.

25 cm^3 of a saturated solution of gypsum ($CaSO_4 \cdot 2H_2O$) in water are run through an ion exchange resin, and a subsequent batch of water is also collected and added to the washings. The washings are titrated against 1.00 mol dm^{-3} aqueous sodium hydroxide, NaOH(aq), and 35.0 cm^3 of the alkali are required for exact reaction. The relevant equations are:

In the resin:
$$CaSO_4(aq) + H_2\text{-Resin} \rightarrow$$
$$\text{Ca-Resin} + H_2SO_4(aq)$$

In the titration:
$$2NaOH(aq) + H_2SO_4(aq) \rightarrow$$
$$Na_2SO_4(aq) + 2H_2O$$

Figure 2.10 An ion-exchange resin for use in a water softener

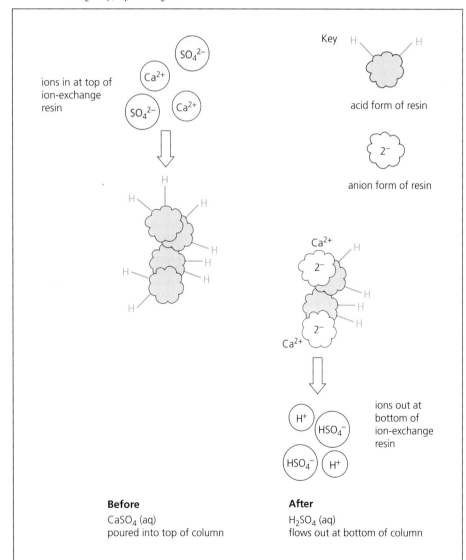

Before
$CaSO_4$ (aq)
poured into top of column

After
H_2SO_4 (aq)
flows out at bottom of column

Figure 2.11 A gypsum ion exchange

a Calculate the number of moles of sodium hydroxide used.

b Deduce the number of moles of sulphuric acid present in all the washings.

c Link the answer from (b) to the number of moles of calcium sulphate present in the original 25 cm^3.

d Translate your last answer to a solubility in mol dm^{-3}.

e Why is there a water washing in addition to the 25 cm^3 of original solution?

f Why was it never necessary to calculate the concentration of the sulphuric acid solution?

28 We first met the idea of percentage purity calculations on p. 15. The idea was to carry out a reaction with just one ingredient in a mixture. Here is another example of the type, but this time in solution.

In the cement industry it is necessary to know the percentage of actual calcium carbonate in the various limestone raw materials. The impurities in limestones are generally sands and clays, which unlike the carbonate are inert to acids. This is the basis for a purity test. But more is required than carrying out a liquid-on-solid titration and recognising the end-point when the fizzing stops.

a Why would such an end-point be unusable in practice?

An alternative is to carry out the reaction in stages.

- Soak the lump of limestone in *more than enough* acid, and leave it until *all* reaction has ceased.
- Then find (by titration) how much acid there is left.
- If you compare that with the amount of acid present at the start, you will find how much acid was removed by the limestone.
- In turn this tells you how much limestone there was in the original lump. A moles-to-mass conversion

and a percentage-of-original-lump calculation brings you to the answer. Here are the data:

A 5.13 g lump of limestone is left in contact with 100 cm^3 of 2.00 mol dm^{-3} aqueous hydrochloric acid, until all reaction has ceased. Then a 10.0 cm^3 portion of the acid is removed, and titrated against 1.00 mol dm^{-3} NaOH(aq). 11.1 cm^3 of the alkali are needed for neutralisation. The relevant equations are as follows.

Acid on carbonate:
$$CaCO_3(s) + 2HCl(aq) \rightarrow$$
$$CaCl_2(aq) + H_2O(l) + CO_2(g)$$

Acid on alkali:
$$HCl(aq) + NaOH(aq) \rightarrow$$
$$NaCl(aq) + H_2O(l)$$

To find the percentage of calcium carbonate in the stone, answer parts (b) to (g) which break the calculation down into steps.

b The first thing many students would do with this question is to use (*2.3*) and thus write 5.13/100. (100 is the RFM of $CaCO_3$.) Why would this be a mistake?

c The true starting point is with the sodium hydroxide, which is once again playing 'first reagent', so use (*2.7*) to calculate the number of moles of alkali used.

d Hence find the number of moles of acid in the 10 cm^3 batch (using the chemical equation) and then (an easily forgotten stage) scale up to find the numbers of moles of acid remaining in the original 100 cm^3.

e Now compare your answer from (d) with the number of moles of acid which would have been in 100 cm^3 of acid *before* it was poured on to the lump. The difference between these two figures is the acid destroyed by reaction with the lump.

f The number of moles of acid destroyed by reaction with the lump should now be used (via the mole ratio 'bridge' of the chemical

equation) to reveal the number of moles of actual calcium carbonate present in the lump.

g Finally turn this number into its equivalent in grams of calcium carbonate, and express this as a percentage of 5.13.

If you can do this question you can cope with almost any mole-based problem!

This method of using excess of a reagent and titrating to find what is left is known as **back titration**. It is used wherever the end-point of a direct reaction would be hard to see or slow to develop. For example, it is used to estimate the oxidisable material in samples of sewage.

A question in which the equation is unknown

In questions of this sort, there are no 'first' and 'second' or 'target' reagents. The task is to find the number of moles of both reagents used in the reaction with each other, and extract a whole-number mole ratio with which to construct the equation.

Question

29 An organic chemical is known to be acidic, and mass spectrometer data give its RFM as 146. Combustion analysis (the technique which involves catching and weighing the carbon dioxide and water from the burning of the sample, p. 17) gives a molecular formula of $C_6H_{10}O_4$. How many of those hydrogens are acidic ones? Here are some more data:

2.00 g of the acid react with 13.7 cm^3 of 2.00 mol dm^{-3} aqueous sodium hydroxide.

Find a mole ratio for the reaction between the alkali and the acid, and thereby deduce how many acidic hydrogens the acid has. Try to write a chemical equation which expresses the limit of your knowledge about the structure of the organic molecule – this will vary according to your prior knowledge of organic acids and your chemical intuition.

Summary

- Chemical reactions do *not* take place between simple whole-number ratios of masses of reactants.

- So there is a need for a quantity which translates between the mass of a substance and the number of particles in it.

- That quantity is the **mole**. A mole of any named chemical formula is a quantity of substance containing approximately 6×10^{23} repeats of that formula.

- The formula in question can be anything which it is sensible to express – often it will be an actual molecule (methane, CH_4) but in ionic compounds it will only represent a ratio between the entities which make up the giant molecular structure (magnesium chloride, $MgCl_2$).

- The mass of a mole of a substance is called the **relative formula mass** (**RFM**) and is equal to **the sum of the constituent relative atomic masses expressed in grams**.

- A useful standard for approximate work with relative formula masses is achieved by defining H = 1. Thus the relative formula mass is the number of times by which the named formula is heavier than an atom of hydrogen.

- The stoichiometric numbers in a chemical equation express the reacting ratios in terms of both numbers of molecules and numbers of moles. Thus two molecules or moles of dihydrogen react with one molecule or mole of dioxygen to create two molecules or moles of water.

$$2H_2 + O_2 \rightarrow 2H_2O$$

- One mole of a compound M_xN_y contains x moles of M atoms and y moles of N atoms.

- Measurements of mass can be converted into numbers of moles by means of the formula:

Number of moles

$$= \frac{\text{mass in grams}}{\text{RFM}} \qquad (2.2)$$

- A mole of any ideal gas (p. 16) occupies the same volume under the same conditions of temperature and pressure. Two common standard sets of conditions are 'stp' (0 °C, 1 atm), and 298 K, 1 atm. The volumes of a mole of gas under these two sets of conditions are $22.4 \, dm^3$ and $24.45 \, dm^3$ respectively.

Conversion between gas volumes and moles is given by:

Number of moles

$$= \frac{\text{volume in } dm^3 \text{ at 298 K, 1 atm}}{24.45 \, dm^3}$$

$$(2.3)$$

- The concentrations of solutions may be expressed in 'moles per cubic decimetre' (sometimes just expressed as capital M and called 'molarity'). This quantity is related to moles of solute and volume of *solution* (not solvent) by the relationship:

Concentration (mol dm^{-3})

$$= \frac{\text{number of moles}}{\text{volume } (dm^3)} \qquad (2.7)$$

- There is a three-step structure that lies at the heart of most mole and molarity problems:

Step 1: translate mass of given reagent into moles.

Step 2: use the stoichiometric ratio from the equation to deduce the number of moles of target reagent.

Step 3: reconvert to the required format (mass, concentration, percentage, purity, etc.).

- Purity may be assessed by making the 'target' ingredient in a mixture perform a reaction to which other ingredients are inert.

3

Structure within the atom

3.1   The few

William Prout (1785–1850)

Prout, a chemist and physiologist born near Chipping Sodbury (Figure 3.1), was a man ahead of his time. In 1815 he put forward an idea which followed from his observation that many atomic weights were nearly whole-number multiples of that of hydrogen. Therefore, he wrote; 'We may almost consider the *prima materia* of the ancients (philosophers) to be realised in hydrogen'. Prout's hypothesis was buried for years under layers of conflicting evidence, but in the end something very like it turned out to be the truth – there are indeed very few primary building blocks of matter. The truth took more than a hundred years to emerge, and this chapter picks out some of the key events in that intellectually turbulent period.

Figure 3.1 William Prout, whose ideas on the constituents of matter were so well vindicated by the gradually unfolding story of atomic structure

Chemists are fortunate – ours is a subject whose *roots* are understood. It was very different in the nineteenth century, however – by the middle of that century chemical data were accumulating at a vast rate, due mainly to the energetic practical organic chemists of Germany, who recorded the melting and boiling temperatures and chemical properties of hundreds of compounds. What chemistry had in abundance were items of data – important in their own way but essentially 'single sentences' in the story of the subject; what chemistry lacked was the *big idea* which overarched and guided *all* the sentences.

Now those big ideas are more or less in place. They were generated in a thrilling 40-year period in which the principal protagonists – Thomson, Rutherford, the Curies, Bohr, de Broglie, Schrödinger, Heisenberg, Planck, Einstein, Fermi – acquired heroic historical status. The truly remarkable thing about the 'plot' that these workers uncovered was its *simplicity*. The enormous diversity of chemical science, the science which embraces cement, chlorine, ion-exchange resins, saliva, asbestos, air, PTFE, PVC, DDT and TNT, could be seen as deriving directly from the behaviour of just *three* particles.

Of the three particles, **neutrons** occupy a position in the margin of chemistry, and even **protons** can be seen as supporting actors in the background. The behaviour of atoms is largely due to just one sort of particle – the **electron**.

To understand the importance of these concepts more clearly we need to return to the period in which 92 different sorts of atom were thought to represent the ultimate root from which chemistry grew.

At the start of this chapter chemists were said to be lucky because the *roots* of chemistry are established. You may think, conversely, that the lucky ones were the scientists who were 'midwives' at the birth of the solution. If so, then you are the sort of person who ought to direct their career either to a technology (where the application of science will always generate problems), or to one of the 'younger' sciences like molecular biology. At the present time the field of cancer medicine, for example, resembles nineteenth-century chemistry, with a large body of knowledge about individual cancers, and some awareness of the role of viruses, carcinogens (cancer inducers) like nicotine, and genes. However, the big idea which would enable us to bring the whole problem of uncontrolled cell reproduction under one roof remains elusive.

Figure 3.2 J. J. Thomson

3.2 Thomson discovers the electron

Discovering the electron is not like discovering, say, the source of the Nile. The source of the Nile had genuinely never been seen by Europeans before, but its existence had of course been inferred. With the electron it was the other way round – it had been 'seen' but people did not know what they were looking at. For centuries people had been looking at the effects of build-ups of electrons – for instance, when charged rods were seen to attract pieces of paper. It was J. J. Thomson in Cambridge (Figure 3.2), who identified the particle responsible for many already familiar phenomena.

His research was into the conduction of electricity by gases. Gases are not of course conventional conductors, but under extreme conditions of low pressure and high voltage, they will allow currents to pass. Thomson enclosed his gases in glass tubes (Figure 3.3). He varied the identity of the gas, and the material of the electrodes. He noticed exotic patterns of colour developing in the gases as their pressures became low enough to allow conduction, with each gas producing a different colour. Then as he went

to lower pressures still, he noticed a decline in the colour of the gas itself, but a luminosity at the end of the glass tube remote from the cathode.

He suspected this luminosity was due to something coming *from* the cathode. He constructed special tubes with anodes with holes in them. He was then able to gather and study these **cathode rays** in a large space beyond the anode (Figure 3.4). He varied the materials of the tube and the nature of the gas, and he investigated the effects on the cathode rays of external magnetic and electrostatic fields. He also studied 'positive rays' which were attracted *by* the cathode.

Figure 3.3 Diagram of one of Thomson's tubes, to investigate the nature of cathode rays. In this tube the pressure would be very low, and the majority of the space in the tube is dedicated to the region beyond the anode, where the beams of cathode rays could be studied and manipulated

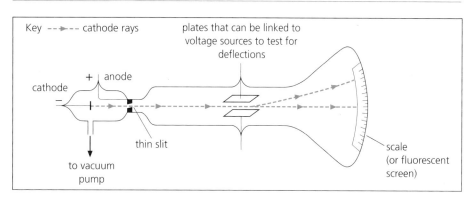

Figure 3.4 Thomson's original apparatus

Experiments on cathode rays	Result	Details of apparatus
A Pass them through an electric field	Rays deflect to *positive* plate	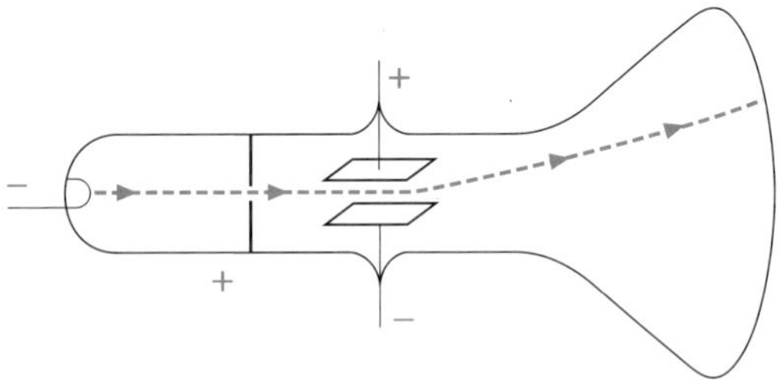
B Pass them through a magnetic field oriented at 90° to line of rays	Rays deflect up or down depending on polarity of magnets	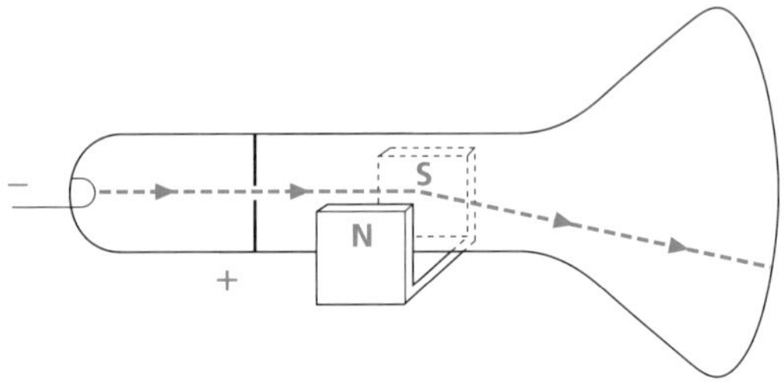
C Change materials of apparatus, e.g. cathode metal, trace of gas in tube, anode metal	Nothing alters	
Experiments on positive rays	**Result**	**Details of apparatus**
D Attempt to deflect them with electric and magnetic fields	Deflections much *less* than that of cathode rays and in *opposite directions*	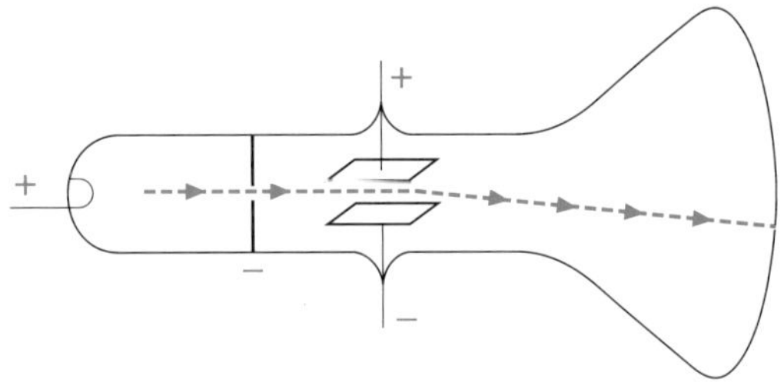
E Change gas in tube (at very low pressure)	Degree of deflection is different for each gas	

Figure 3.5 Summary of Thomson's results

Thomson's results are summarised in Figure 3.5. The conclusion he drew was that cathode rays were a sort of disembodied version of an electric current, consisting of a stream of negatively charged particles. He concluded that they were not peculiar to any one type of material, but were a universal constituent of matter, driven off from the cathode material by the potential difference of the supply. He was even able to calculate that a cathode-ray particle had a mass of about one two-thousandth that of a hydrogen atom.

Question

1 a Which results offer evidence for the cathode-ray particles being negatively charged?

b Which results offer evidence for the idea that cathode-ray particles are a universal constituent of matter?

c Thomson's experiments had the side-effect of shaking up people's assumptions about the nature of electric currents. In particular, what assumption about currents had to be *reversed*?

d Assuming positive rays are made of particles created between the electrodes, and remembering that they vary with the choice of trace gas in the tube, suggest what they *are*.

Plum puddings

Thomson built a famous false hypothesis on his revolutionary results. Correctly realising he was looking into a sub-atomic world, he proposed a model for the internal structure of an atom. He

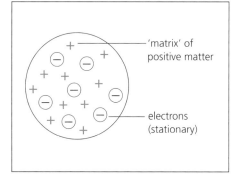

Figure 3.6 Thomson's 'plum-pudding' model of the internal structure of an atom

Figure 3.7 Cavendish Laboratory, 1898

knew that atoms contained electrons, and that atoms were electrically neutral, so he proposed that there was an **equal and opposite positive charge** enveloping the electrons in a matrix. It has come to be known as the 'plum pudding' model, after a type of Victorian dessert in which bits of plums were surrounded by a matrix of pudding (Figure 3.6). Thomson proposed an atom that was both *static* and relatively *homogeneous*. In each respect he was soon to be proved wrong.

3.3 Rutherford and the nucleus

J. J. Thomson's false model does not lessen the importance of his experimental discovery. And Professor Thomson compensated for his own false move by creating the teams of people who did eventually obtain the extra evidence that he himself was lacking.

Ernest Rutherford (1871–1937)

Rutherford was a New Zealander who worked with Thomson in the Cavendish laboratory at Cambridge. The 'team photo' (Figure 3.7) of the research group shows both men together – not close together, however, for Thomson sits in the 'captain's' position (front row, arms crossed) while Rutherford is a fresh-faced junior at the back (middle row, fourth from left).

A decade later Rutherford was running his own research group in Manchester, and curiously had grown to look rather like his former boss (Figure 3.8a). One of his research projects was the launching of 'alpha (α) particles' (from a

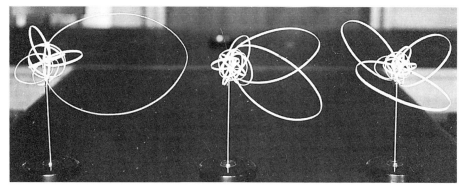

Figure 3.8 (**a**, top) Ernest Rutherford, seated in the centre and holding a pipe. (**b**, bottom) Models of atomic structure according to Rutherford's theory

radioactive source) at exceptionally thin sheets of gold. He was in effect using these particles to rip through the insides of the gold atoms, in the hope that it would help him to understand internal atomic structure (see Figure 3.8b). You might say he was feeling around inside Thomson's plum pudding to see how lumpy it was.

To appreciate the meaning of this line of research, we need to understand how the α particle was seen then. Now we see it as a helium nucleus, just two protons and two neutrons; then, before concepts of neutron, proton and nucleus existed, it must have looked like a totally different sort of matter, notable for its incredibly high *density*.

Question

2 Modern knowledge can show how much greater is the density of 'nucleus-alone' matter (such as an α particle) than ordinary atomic matter. Here are some (very approximate) data:

Volume of helium nucleus
$\approx 5 \times 10^{-39}$ cm^3

Mass of helium nucleus
$\approx 10^{-23}$ g

Volume of gold atom
$\approx 1.5 \times 10^{-23}$ cm^3

Mass of gold atom
$\approx 3 \times 10^{-22}$ g

Calculate the two densities in g cm^{-3} and express them as a ratio.

This calculation shows why people expected α particles to rip through gold atoms. There is far less difference between air and a bullet than there is between gold atoms and α particles – assuming of course that the density of a gold atom is *evenly* distributed. It had not occurred to anyone that the α particle itself could be, as it were, a spare part from a plum pudding. For one thing, it was known that α particles were positively charged, and according to plum-pudding theory the positive component of atomic matter was amorphous pudding.

Those who imagine that science is a totally rational human activity do insufficient justice to the role of intuition. For no logical reason, Rutherford suggested to his co-worker Geiger (of '-counter' fame) that he might look for a mass concentration within the gold atom, by searching for α particles *rebounding* from the gold film. His diaries reveal that he had no reason to expect a positive result.

One day Geiger came to me and said, 'Don't you think that young Marsden, whom I am training in radioactive methods, ought to begin a small research?' Now I had thought that too, so I said, 'Why not let him see if any alpha particles can be scattered through a large angle?' I may tell you in confidence that I did not believe they would be, since we knew that the alpha particle was a very fast massive particle, with a great deal of energy, and you could show that if the scattering was due to the accumulated effect of a number of small scatterings, the chance of an alpha particle being scattered backwards was very small. Then I remember two or three days later Geiger coming to me in great excitement and saying, 'We have been able to get some of the alpha particles coming backwards.' It was quite the most incredible event that has ever happened to me in my life. It was almost as incredible as if you fired a 15-inch shell at a piece of tissue paper and it came back and hit you.

E. Rutherford 'The Development of the Theory of Atomic Structure'

Rutherford was quick to appreciate the significance of Geiger and Marsden's result. Having realised that the rebounding α particles had met something even more massive than themselves inside the gold atom, he set about some calculations as to the nature of the thing in the middle (soon to be renamed the **nucleus**). He assumed that the rebounds were the result of single collisions between nucleus and α particle, and that the nucleus itself must have been of **like charge** to the α particle (Figure 3.9). The plum-pudding model was thus shown to be doubly wrong. In the first place atomic matter was not homogeneous. In fact it was so unhomogeneous that gold

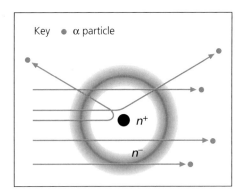

Figure 3.9 Geiger and Marsden's result. The α-particles mostly pass through the cloud of electrons undeflected. But a few come close enough to the nucleus to bounce back. The size of the nucleus relative to the whole atom has been greatly exaggerated in this drawing

atoms contained something which could 'out-α-particle' an α particle. In the second place, it appeared that the electrons were not embedded in anything at all, and therefore must be performing some sort of planet-like **orbit** of the nucleus, which would be the only way to stop the electrostatic forces drawing them into the nucleus.

Rutherford now focused his attention on the make-up of the nucleus, both in terms of charge and mass. He calculated values for the charges on various nuclei, as multiples of the charge on a single electron, and found that in every case the number was the same as the **atomic number**. This caused a rapid re-evaluation of the significance of the atomic number, which had previously been thought no more than the 'catalogue number' in a list of atoms of increasing atomic mass.

Just when a sceptics-versus-believers debate was about to erupt over the atomic number issue, an extra bit of evidence was found which proved that the atomic number was more than just a catalogue number. The experiment that provided the evidence (performed by Henry Moseley, a young student at the time, in Oxford (Figure 3.10a)) was concerned with the X-ray emission spectra of various elements. The importance of the result is simply this – that Moseley found a relationship which linked the frequencies of X-rays produced by each element, to its **atomic number** (Figure 3.10b). Here was more support for the idea that atomic number was an expression of something in the natural world, and not just a human artefact.

Figure 3.10 a Henry Moseley and **b** Moseley's apparatus and results

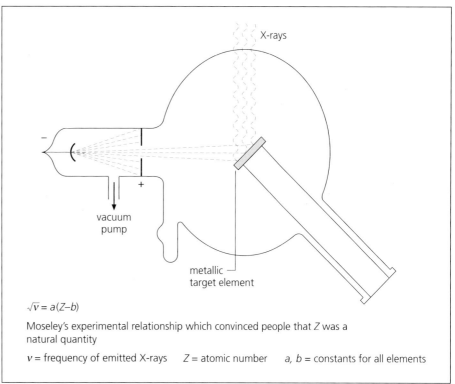

$\sqrt{v} = a(Z-b)$

Moseley's experimental relationship which convinced people that Z was a natural quantity

v = frequency of emitted X-rays Z = atomic number a, b = constants for all elements

It was thus much easier to grasp the idea that atomic number might actually be the number of positive charge-carriers on the nucleus. Another piece of the story now followed logically – the electrons must be equal in number to the protons (the name coined for the positive charge-carriers in the nucleus). New discoveries began to mount up, some concerned more with the nucleus, and others with the electrons. The next section looks at a breakthrough in the former area. What emerged was a picture of the nucleus which is still serving chemistry well.

However, there were a number of non-integer atomic masses which needed explaining. Did an atom like chlorine (atomic number 17, relative atomic mass 35.5) have 17 protons and 18.5 mass-units-worth of some other stuff? Was the other stuff as 'particulate' as protons, or was it more like some sort of splurgy glue holding the protons together? (This latter suggestion gains credibility when you recall that the only atom that does without extra stuff in its nucleus, namely hydrogen itself, contains only a single proton and so would not need any 'glue'.) The answer came quickly and unequivocally as a result of the invention of the **mass spectrometer** by F. W. Aston in 1919.

3.4 Mass spectrometry, isotopes and the neutron

Atomic masses were well established by the 1920s. Many were quite close to simple multiples of the atomic mass of hydrogen, which had given rise to the use of H = 1 as the standard for relative atomic mass. Atomic (that is, proton) number was found to increase along with atomic mass, but with a value of approximately *half* the atomic mass. It was possible therefore to visualise a universe of atoms, all of which were made up from multiples of the proton–electron combination, and with the chemical personalities of each atom deriving from that combination. Prout's hypothesis of 1815 (p. 24) was close to vindication.

The mass spectrometer – principles and components

In the early part of the twentieth century, near the top of any chemist's shopping list would have been a machine for separating atoms and molecules by virtue of their mass, and then weighing them (albeit relatively). Years which would have been spent trying to deduce the relative atomic masses by techniques such as the meticulous weighing of precipitates could be freed for other work.

When the discovery came, it had at its heart a principle that is familiar to anyone who has had any experience of bodily-contact sports. That is, if a person is running past you at any given speed, then the heavier they are, the harder it is to deflect them from their path (see Figure 3.11). So the idea was to make particles 'run', and then try to deflect them.

The difficulties of turning this basic idea into reality are obvious – first, atomic and molecular motions are random and multidirectional so it is hard to get them running as a deflectable 'beam', and second, how can they be deflected?

Both these challenges can be met by means of the same ploy, namely to put the particles into an **ionized** state (by removal or addition of an electron or two). Ionization is the stone which

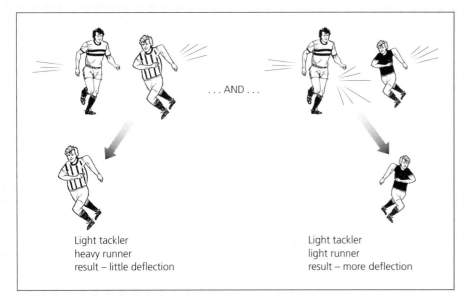

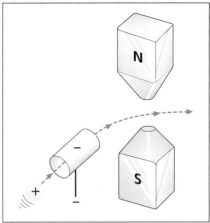

Figure 3.12 Showing how a positive ion can be both accelerated and deflected

Figure 3.11 A rugby tackle, showing the principle behind the mass spectrometer

kills the two birds: a collection of ionized particles can be turned into a beam by accelerating it with charged plates, and once moving, the beam of ionized matter can be deflected by a **magnetic field** (Figure 3.12).

The force which causes the deflection derives from the same source as the forces felt by current-carrying wires in magnetic fields. So we are talking about the **electric motor** principle, which should be familiar from your GCSE electromagnetism studies. There are two differences: the charge-carriers are the flying positive ions instead of lots of identical electrons, and the ion-current runs in space rather than within a solid conductor. In an electric motor every individual moving electron is acted upon separately by its own deflecting force, but since they are all 'trapped' within a wire it is the whole wire which moves, thereby spinning the armature. In the mass spectrometer the individual ions are all free to respond to the deflecting forces as separate particles. The crucial aspect of this freedom to respond as separate entities is that each particle will respond according to its individual **mass**.

We will now look at the anatomy of a mass spectrometer section by section, and identify the processes at the heart of its operation. Figure 3.13 shows the layout of the components in the simplest design of machine. The whole interior of the machine is maintained at a pressure of about 10^{-10} atmospheres, so that the ions can fly without risk of collision with other molecules.

Stage 1: introducing the sample to the machine

Mass spectrometry is a very sensitive technique, and sample sizes need be no more than micrograms. But the sample must be in the vapour phase by the time it reaches the ionization chamber. The introduction technique will vary slightly from solid to liquid to gas.

Figure 3.13 A simple mass spectrometer

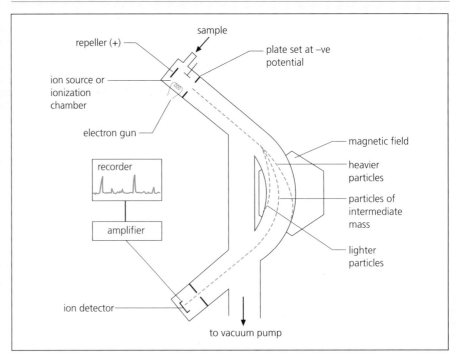

In all cases there is the problem of introducing any new matter into a system at high vacuum. The solution is to use an arrangement rather like the airlock on a spaceship, with a double vacuum-tight seal neither half of which is open at the same time.

Gases, once introduced beyond the first seal, can diffuse into the main body of the machine. Volatile liquids can be treated similarly, as they will boil under reduced pressure. Non-volatile liquids and solids are injected on a lance with a heated tip. This heat, together with external heating of the whole sample inlet/ionization region of the spectrometer, and the very low pressures, can vaporise most substances with boiling points of 500 °C or less at normal pressures.

Stage 2: ionization

The injected sample molecules, in the vapour phase, have 'wandered' into the ionization chamber. At this stage their motions are those typical of gas molecules – random collisions with the walls and each other, with kinetic energies set by the temperature in the chamber. But either side of the ionization chamber are the electrodes that are going to change the character of the particles. A heated filament (cathode) acts as a source of fast electrons, and a potential difference of 70 volts pulls them towards an anode. The sample molecules are caught in a 'crossfire' of electrons (Figure 3.14). Each electron has a collision energy equivalent, on a mole scale, to several thousand

kJ mol^{-1}, well in excess of the energy needed to knock out electrons from the sample molecules (see Section 3.10 on ionization energies), and so if a particle X is hit, it becomes X$^+$; the original molecule or single atom minus one electron:

$$X(g) + e^-(fast) \rightarrow X^+(g) + 2e^-$$

where '2e$^-$' is the ejected electron and the 'collider' electron, now slower.

The collision energies are quite capable of breaking chemical bonds, which are typically around hundreds of kJ mol^{-1}. So if the sample is originally molecular rather than atomic (which the vast majority are), the result is a heavily damaged jumble of whole molecules and broken-off fragments of molecules. This is nothing like the disaster it sounds, because the purpose of mass spectrometry has expanded far beyond its origins as a probe into the nature of sub-atomic matter.

The common use of mass spectrometry now is simply to find out what an unknown molecule is, and the mixture of broken molecular bits is uniquely characteristic of the particular original molecule, thus providing a 'fingerprint' method of substance identification (Figure 3.15).

Stages 3 and 4: acceleration and deflection

The newly ionized particles are pushed into their flight down the tube of the

mass spectrometer by a positive repellerplate (Figure 3.14), and pulled and focused by one or more negatively charged plates with holes in (the 'slit system'). Now they approach the bend in the tube where they are to be deflected. At the bend is an electromagnet, and the ions experience the deflecting force as they go past.

The heavier the particle the less it will be deflected. The *angles* of deflection (at any given fixed value of the magnetic field) will be a spectrum of angles dependent on mass. However, there is only one angle of deflection which will match the actual bend angle of the tube. So with fixed values of magnetic field and bend angle, only one particle-mass will be just right for landing on the detector.

All that is necessary to make the machine a spectrometer is to introduce a variable, and the most obvious one is the magnetic field strength. When the machine is in use, the electromagnetic field on the bend is gradually increased, and so heavier and heavier particles arrive at the detector. The *x*-axis of the graph of the final spectrum is linked to the changing magnetic field. The manufacturers of the machine know which value of magnetic field causes a particular mass of ion to be detected, so the graph can be read directly in mass units rather than as field strengths.

Figure 3.14 What happens in the ionization chamber of a mass spectrometer

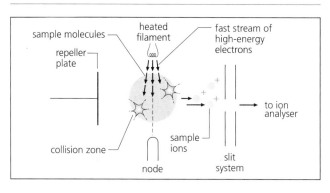

Figure 3.15 Mass spectrum of a (relatively) complicated molecule, RFM 122, showing all the broken-off fragments which will help to identify it

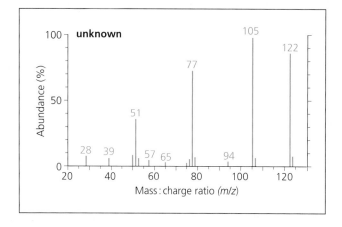

Questions

3 There is one other variable that could have been used to create the mass spectrum, while keeping magnetic field strength constant (indeed this method is used on some instruments). What is it?

4 The actual horizontal (*x*) axis is labelled not just as mass but as *m/z*, or mass : charge ratio. This recognises the fact that some collisions in the ionization chamber result in double electron removal. How would a doubly charged particle fare on the deflection stage compared to a singly charged one of identical mass? Why is it necessary to identify a peak with a mass : charge *ratio* rather than an individual mass?

5 You will have noticed that we are measuring the masses of particles after they have had a bit of themselves knocked out. You might think that would make the measured values obtained for the masses untrustworthy. Why can we nevertheless ignore the effect of the events which happen in the ionization chamber?

magnetic field strength. The final spectrum emerges as a series of lines, each of the bigger lines labelled with its *m/z* number, and the biggest line being set automatically to an abundance value of 100 % (Figure 3.17).

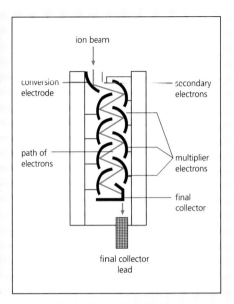

Figure 3.16 The ion detector and amplifier system

Figure 3.17 Typical mass spectrum of an organic molecule, showing labelled fragment peaks and '100%' setting for largest one

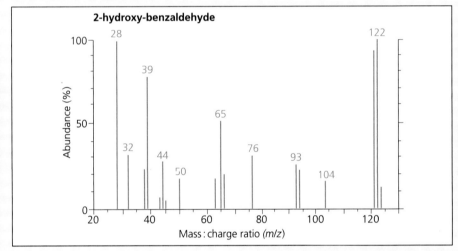

2-hydroxy-benzaldehyde

Stage 5: detection

When the flying ion finally reaches the detector it collides with a surface. In modern machines the surface is made of a copper–beryllium alloy which is negatively charged, and the ion's arrival triggers the release of a little shower of electrons. This little shower is then magnified into a downpour by a cascade on to a series of similar electrodes (Figure 3.16) so that there are about a million electrons at the collector lead for every original ion strike.

The final output depends on computer technology, with the signal's abundance (the value on the vertical axis, or *y*-axis) and its *m/z* ratio (*x*-axis value) being calculated respectively from the size of the collector current and the simultaneous size of the

Isotopes

Early mass spectra clearly showed the existence of different versions of the same elemental atoms. Such a state of affairs had already been anticipated by workers in the field of radioactivity, when they found that the element lead, Pb, from two different radioactive decay paths had showed two different atomic masses.

The next question is about different versions of the element chlorine.

Question

6 Figure 3.18 shows the mass spectrum of chlorine molecules, Cl_2. One cluster of peaks derives from chlorine molecules themselves, and the other cluster relates to broken-off chlorine atoms.

a Which cluster refers to chlorine atoms?

b How many different versions of chlorine atoms appear to exist?

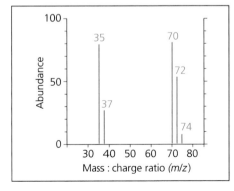

Figure 3.18 Mass spectrum of Cl_2(g)

c Why is the overall atomic mass of chlorine atoms 35.5?

d Why are there *more* different versions of chlorine molecules than of chlorine atoms?

e *(Difficult.)* Can you account for the relative heights of the chlorine molecule peaks?

So the mass spectrometer shows that you can have chemically identical atoms with different atomic masses, and gives a clue as to the underlying cause. The word **isotope** was created to describe different versions of an element, because all these versions had to occupy the same place in the Periodic Table (*'isos topos'* means equal place in Greek). The atomic mass of each isotope was found to be very close to an exact whole-number multiple of the mass of a hydrogen atom, and this number became referred to as the isotope's **mass number** (or its **nucleon number**). This implied that the missing nuclear ingredient which made up the mass of an atom must show two features:

• It must be **particulate** and have very nearly the same mass as a proton. This is because each and every isotope has a mass very close to a whole number on the relative mass scale, on which scale a proton itself has a mass of 1 (by present-day standards of atomic mass, the mass of a proton is 1.007276 and the mass of a neutron – the name eventually given to the new particle – is 1.008665).

• It must be **neutral**, since an extra charged particle on the nucleus would require a change in the number of electrons, and that would run counter to the fact that isotopes of a given element are chemically identical.

Questions

7 Justify the reasoning in the last paragraph.

a Why would an extra charged particle on the nucleus require a change in the number of electrons?

b Why would a change in number of electrons prevent an isotope from being chemically identical to another isotope of the same element?

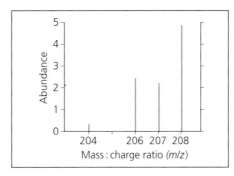

Figure 3.19 The mass spectrum of a sample of a single element – lead

8 The non-integer relative atomic masses of naturally occurring elements can now be seen as weighted averages of their isotope mass numbers. From Figure 3.19, calculate the relative atomic mass of naturally occurring lead.

9 Lead from a mineral deposit in Sri Lanka, a mineral which also contains the radioactive element thorium, has a relative atomic mass of 207.77. Lead from a Norwegian source, in which it is accompanied by the radioactive element uranium, has a relative atomic mass of 206.08. Can you explain this discrepancy? What has it to do with the two named radioactive elements?

The discovery of the neutron

Although the existence of isotopes was well recognised by 1920, the particle responsible for them, the neutron, was not isolated until 1932. Neutrons were produced by the impact of α particles on certain light elements such as beryllium or boron. The resulting radiation made no tracks in a cloud chamber, but showed great ability to penetrate other sorts of matter. It caused great upheaval in molecules which were rich in hydrogen, producing a shower of emitted protons, and this was how Chadwick demonstrated the neutron's existence (Figure 3.20).

Question

10 At the height of the Cold War, there was much talk about the use of neutron bombs – bombs with low blast but high levels of neutron radiation. Here was a kind of radiation which had radical effects upon hydrogen-containing matter, but tended to pass straight through other materials such as ceramics: a bomb which destroyed life but which did not leave brick, concrete or stone with much residual radiation.

a Why do *neutral* particles have a better chance of passing straight through other atoms than do *charged* particles? (Neutrons are far more penetrating than α particles, for instance.)

b Why do neutrons leave no tracks in a cloud chamber?

c Why would people be so at risk from neutron showers? (Remember the kind of chemicals which make up human bodies, and link that to the information in the introduction to the question.)

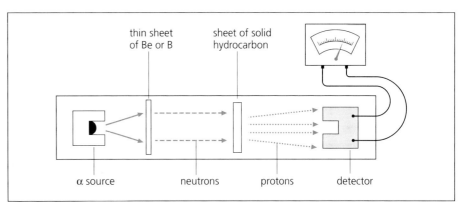

Figure 3.20 Chadwick's way of detecting neutrons – by proxy. He made them disrupt molecules of a hydrocarbon and then detected the resulting protons

d What were the perceived political and military advantages of a neutron bomb?

High-resolution mass spectrometery

To return to the workings of the mass spectrometer (p. 29), ions of similar m/z value will arrive at the detector roughly simultaneously, when the conjunction of magnetic field strength and bend-angle of the tube is just right for them. This 'roughly' arises because not all ions of the same mass will have quite the same **kinetic energy** (KE) when they get to the bend.

Question

11 Looking back to Figure 3.14 you may be puzzled by the fact there is a spread of kinetic energies. After all, these kinetic energies will be the result of 'force × distance' work done on the ions by the plates. Can you see where that variation might have come from?

This small uncertainty in kinetic energy leads to a corresponding uncertainty in the mass of the arriving particles. (Is it a slightly lighter ion with a slightly higher kinetic energy, or a slightly heavier ion with a slightly lower kinetic energy? You cannot tell because they both arrive together.) So simple mass spectrometers cannot distinguish between (or 'resolve') ions of closely similar but slightly different mass. The best they can do is to discriminate between different whole-number mass numbers.

So a generation of refined machines was developed, with a *double* curve in the tube. The inventors introduced a kinetic-energy-sorting bend, using charged plates, which caused particles with higher or lower kinetic energies to fail to get through a set of slits (Figure 3.21). The kinetic energies of all particles getting through the slits were now identical, so that the x-axis variations were solely due to mass variations. With this high-resolution technique, it is possible to measure atomic masses to many decimal places.

The ^{12}C = 12 scale of relative atomic mass

Before high-resolution mass spectrometry was available, the masses of isotopes seemed all to be whole-number multiples of the mass of ^1H, giving support to the view that protons and neutrons both had a mass of '1', and the electron masses could be considered negligible. *After* the advent of high-resolution mass spectrometry scientists realised that every isotope

Table 3.1 Accurate relative formula masses.

Atom	Number of protons	Number of neutrons	Number of electrons	Old RFM	New RFM
C	6	6	6	12	12.000 00
H	1	0	1	1	1.007 82
N	7	7	7	14	14.003 07
O	8	8	8	16	15.994 91

of every atom had a characteristic mass slightly different, if only in the third or fourth decimal place, from that predicted on the old simple model. It was at about this time that a new standard of atomic mass was set. The new system was based on calling the relative atomic mass of the isotope ^{12}C 12.0000000. Some results from high-resolution mass spectrometry are shown in Table 3.1. It is possible to turn these small variations to advantage in identifying unknown molecules, as in the following question.

Question

12 a Calculate the relative atomic masses of carbon monoxide, CO, and dinitrogen, N_2, to 5 dp.

b Explain why, on a low-resolution mass spectrometer without the kinetic energy screening stage, it is possible that a carbon monoxide molecule could reach the detector at a slightly *higher* field strength than a dinitrogen molecule. Hence explain why such a machine could not resolve the peaks from carbon monoxide and nitrogen molecules.

c What reading would you get for the relative atomic masses of CO and N_2 from a low-resolution machine?

d Even on a low-resolution machine, carbon monoxide might still have given a mass spectrum which was distinguishable from that of dinitrogen. Can you suggest why?

e An old definition of a mole of substance might have been 'a collection of particles containing the same number of particles as there are hydrogen atoms in 1 gram of hydrogen atoms'. Reword this definition so as to conform to the new standard of atomic mass.

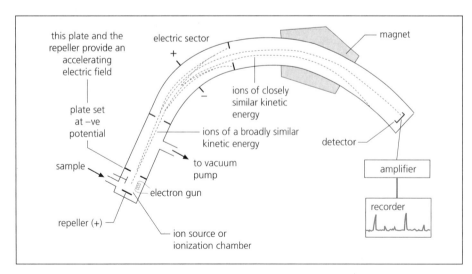

Figure 3.21 The more accurate 'two-bend' mass spectrometer

At this point it is time to take stock:

• Our picture of an atom is as shown in Figure 3.22.

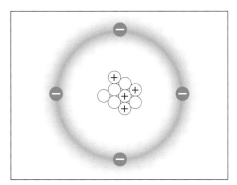

Figure 3.22 The story so far – a historical staging-post on the road to knowledge of atomic structure

• We can write some relationships which hold good for all neutral atoms:

Atomic number = number of protons
= number of electrons

[But for ions (number of protons – number of electrons) = ion charge]

Mass number   = number of + number of
of an isotope   protons   neutrons

• We can tabulate the relative masses and charges of the three sub-atomic particles (Table 3.2).

Table 3.2 Masses and charges of three sub-atomic particles

Particle	Relative mass	Relative charge
Proton	1	+1
Neutron	1	0
Electron	$\dfrac{1}{1850}$	−1

Question

13 Copy and complete the following table:

Table 3.3 For question 13

Species	At. no.	At. mass	No. of protons	No. of electrons	No. of neutrons
^{75}As	33	75			
^{79}Br$^-$	35	79			
^{208}Pb	82	208			
^{208}Pb^{2+}	82	208			

Now we turn our attention to the study of the electrons. The story of the nucleus was dominated by English-speaking physicists with inclinations towards experiment-driven modes of working ('fire the particles and see what happens'). The story of the electrons, however, is dominated by German speakers, and rooted in the theoretical tradition in which experiments play a subsequent confirming role. The new way of looking at matter that emerged, called the **quantum theory**, ranks as one of the great revolutions in scientific thought, along with the ideas of Copernicus, Darwin and Newton.

3.5 The evidence of spectral lines

Advanced chemistry students regularly peer through spectroscopes at the emission spectra of hot or electrically excited gases (Figure 3.23), or at flame tests of metal chlorides, and are then told that they are looking at evidence for the arrangements of electrons in atoms. The students can be forgiven for not immediately seeing the link. They would certainly be following in the footsteps of the first scientists to observe spectral lines, because they also did not realise the source of the phenomenon. The one thing that was obvious to everyone was that these emission spectra, with their sharp restricted lines of colour, stood in stark contrast to the continuous **band-style** spectra of hot bodies like the Sun, a flame and, in our own day, the filament of a light bulb.

Figure 3.23 (top) Looking at a discharge tube through a spectroscope and (bottom) neon lights

Fraunhofer lines

The first precise spectroscopy was conducted by Fraunhofer in the early nineteenth century. He was working for a glass-making firm, and wanted to obtain a quantitative way to measure how much light was bent (refracted) by various glasses. In his subsequent experiments on sunlight he noticed that the spectrum of light from the Sun was continually, if minutely, interrupted with a series of dark lines (later called Fraunhofer lines). Fraunhofer was delighted because he had found that sunlight miraculously came with its own internal graduations, ideal for testing refractive power of grades of glass. This was therefore an early case of a pure-science spin-off from applied research.

The rest of the science world took notice. Flame tests were already current (Table 3.4), and the line spectra of flame colours were known. But now it was found that if a bright source of 'white' light was placed *behind* a sodium flame, the flame acted as an absorber of exactly those lines which it normally emitted (Figure 3.24). It was realised that each element had a characteristic set of places to itself within the spectral range of visible light, and that it would either be a net absorber *or* emitter at these places, depending on the relative amounts of incoming and outgoing radiation. The Fraunhofer lines, then, were indications of the elements present in the gases of the Sun, acting in this case as absorbers.

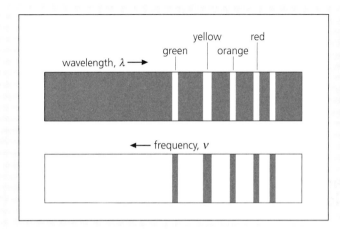

Figure 3.24
Emission (upper) and absorption (lower) spectra of Na(g). The former is obtained from an electrical discharge tube against a dark background. The latter is obtained by observing a bright 'white' source through a sodium flame

Table 3.4 Flame colours of cations in Groups I and II

Group I	Colour	Group II	Colour
Li^+	red	Be^{2+}	none
Na^+	yellow	Mg^{2+}	none
K^+	lilac	Ca^{2+}	vermilion
Rb^+	red	Sr^{2+}	red
Cs^+	blue	Ba^{2+}	green

are better understood with a wave model (as in the case of interference), while at other times a particle model is better. Expressing this wave/particle duality, the German physicist Max Planck proposed that a photon of light (the particle) possessed energy E dependent upon its frequency v (a wave property), according to:

$$E = h \times v \qquad (3.1, \text{ Planck's equation})$$

where h is now called the Planck constant. Thus a spectral line in an emission spectrum was to be visualised as a stream of photons all with a **single frequency** (giving them their single colour) but also all with the **same energy** (Figure 3.26).

This is contrasted with the spectrum of light emitted by hot bodies like suns and filament light bulbs, in which photons of a continuous band of frequencies are emitted. (A dark spectral line in an **absorption** spectrum would be seen as the complete *absence* of photons of a particular frequency/energy.)

Question

14a If the source of photons in Figure 3.26 were observed through a spectroscope, how would it look?

b How could you adapt the picture in Figure 3.26 to symbolise 'white light'?

Having realised that the line spectra of elements were in effect their 'fingerprints', chemists looked for the reason for the complex yet organised patterns that these lines made. Figure 3.25 shows the **Balmer series** of lines in the (absorption) spectrum of atomic hydrogen, as seen in the form of dark lines in the light from a distant star. Clearly such a pattern derived from a distinctly non-random process.

To follow the unravelling of this story we need to revise a little of the nature of light itself. Light is one of those phenomena in science which forces the student into realisation that a single mental picture cannot always explain a situation. The behaviour of light needs two parallel mental pictures, which do not sit comfortably with each other, or with our need to visualise. Sometimes the properties of light

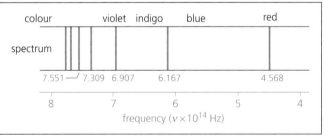

Figure 3.25 Hydrogen in the star Zeta Tauri causes the dark lines in the band spectrum of visible light – the Balmer series (v = frequency)

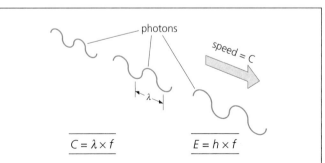

Figure 3.26 Possible mental picture of the wave/particle dual personality of light. These are all photons of a single frequency and wavelength and thus a single energy. They would be perceived as a single colour

How were photons of these precise energies produced (or absorbed) within the atom? The answer to this question was provided by one of the greatest scientists of the twentieth century, a football-playing Dane whose research career included some time spent as a student on Rutherford's team in Manchester.

3.6 Niels Bohr and the theory of electron orbits

Figure 3.27 Niels Bohr

Niels Bohr (1885–1962)

Bohr (Figure 3.27) became Professor in Copenhagen in 1916 after working with J. J. Thomson (p. 25) at Cambridge and Lord Rutherford (p. 27) at Manchester. He explained the spectrum of hydrogen by using the atomic model and quantum theory in 1913. During World War II he escaped from German-occupied Denmark and assisted atom bomb research in America. He was awarded the Nobel prize in 1922.

There were two major strands (one might almost say planks) to Bohr's picture of the origin of spectral lines:

• He accepted Planck's ideas that, for some types of particle, energy could only be absorbed or emitted in discrete packets called **quanta**. Applied to electrons, this meant that electrons could only exist at certain allowed energy levels. This was a revolutionary idea, because people had to start thinking about electrons not as Newtonian bodies (which can have an infinite number of different values of, say, kinetic energy), but as a branch of matter ruled over by a new set of physical laws – laws which seemed to say that bodies possessed energy like a person on a ladder possesses height: you can exist stably at any rung, but you cannot stay in between.

• Bohr proposed that a spectral absorption line is produced when an electron absorbs a photon of the correct energy (and therefore frequency) to send it from one of its allowed energy states to another (Figure 3.28). (In the ladder analogy, this is moving from any rung to any higher one.) An emission line is produced by the reverse process, that is, by an electron 'relaxing' back down to a lower energy level, and emitting a photon of the correct energy. The resulting

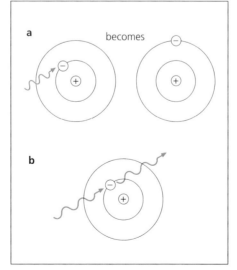

Figure 3.28 a A photon is absorbed, and results in an electron being 'promoted' to a higher energy level. b A photon of the wrong frequency fails to match the energy needs of the promotion, and passes through unabsorbed

spectra appear as lines (rather than the continuous rainbow band of colours of the spectrum of white light) because there is only a finite number of possible jumps between the finite number of allowed energy levels for the electrons. The exchange of energy between electron and photon can be expressed by:

$$E_2 - E_1 = h \times \nu \qquad (3.2)$$

where E_1 and E_2 are electron energies before and after absorption of a photon, and ν is the frequency of the absorbed photon. h is the Planck constant, which expresses the relationship for a photon between our human-set frequency unit (the s^{-1} or Hz) and our human-set energy unit (the joule). The value of the Planck constant is 6.626×10^{-34} J Hz^{-1}. The implication of (3.2) is that the bigger the jump between electron energy levels, the nearer the blue (high-frequency) end of the spectrum the absorbed photon will be.

Bohr set about vindicating his theory by using it to make predictions about real hydrogen atoms. He mixed classical mechanics and electrostatics with the new quantum mechanics, to work out where the electron in a hydrogen atom might be if it were to set up a stable orbit around the nucleus. His value of 0.05 nm for the radius of a hydrogen atom was of the right order of magnitude.

Bohr's technique was to use classical equations concerned with the energies of bodies in circular motion, and then introduce integer (whole) numbers, starting at 1, 2 . . ., which he called quantum numbers. Thus the energy of the electron was constrained to have only certain set values. At the time, he did not understand why such quantum numbers existed – only that their existence could be inferred from the pattern of spectral lines. The answer to the 'why' question had to await the concept of electron waves.

Question

15 Name two physical constants which Bohr had to put into his calculations on the hydrogen electron.

Bohr then calculated the energy separations between the various allowed states for a hydrogen electron, and worked out a series of energy levels or orbits in which a hydrogen electron might be found. The innermost orbit was (obviously) the most stable orbit, where the electron would be found under normal circumstances, in the absence of any absorption of energy from outside. Less obviously, the higher-energy orbits, more remote from the nucleus, turned out to have energies which got closer and closer together, that is, the gap between them became less. He gave each orbit a **quantum number**, beginning with 1 for the innermost one (Figures 3.29 and 3.30).

The decrease in energy gap as the quantum number increases can be understood by analogy. If you hold a nail 1 cm from a magnet, and then pull it an extra centimetre away, you have to input a certain amount of energy. However, if you pull the nail from 10 cm to 11 cm, far away from the magnet's influence, there is much less work to do. (Nucleus = magnet, electron = nail, Figure 3.31.)

Question

16 The 'nail' analogy is not perfect, because it contains no equivalent of the quantisation of electron energy levels. Can you adapt the analogy, or indeed invent a better one, to include this missing factor?

Figure 3.32 shows again the Bohr energy levels for an electron in a hydrogen atom. The arrows on the diagram represent the jumps between levels which the electron might perform when energy was input from outside. The bottom part of the diagram shows how each jump could have given rise to a spectral line – it works on the principle that the bigger the jump, the higher the frequency (see equation (3.2)), so the further towards the high-frequency end of the spectrum the line will appear on paper. Bohr was able to predict a hypothetical spectrum

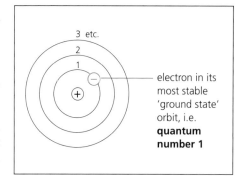

Figure 3.29 Quantum numbers attached to allowed orbitals in a hydrogen atom

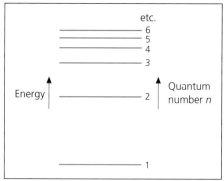

Figure 3.30 The energy levels of the orbits in atomic hydrogen, shown on a vertical axis. Note how the energy differences get less as n increases

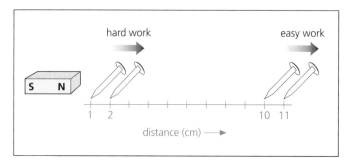

Figure 3.31 The 'nail' analogy, to explain differences in energy separations

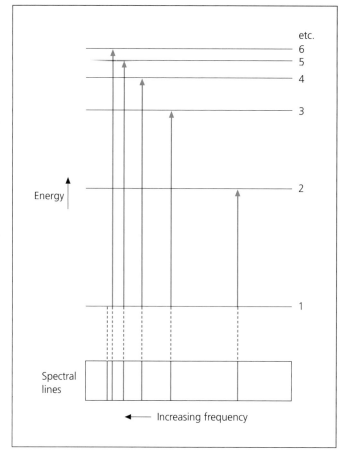

Figure 3.32 Energy levels and the spectrum of atomic hydrogen (Lyman series)

which, it turned out, matched the real **Lyman series** of lines for atomic hydrogen, both in the general pattern of convergence, and in the quantitative aspects of frequency. He had thus obtained persuasive and triumphant evidence that his revolutionary new physics gave the best description of the way electrons behave.

Question

17a Would the spectral lines in Figure 3.32 be dark lines taken from a spectrum of white light, or bright lines?

b In other words, is Figure 3.32 a diagram of an absorption or an emission spectrum?

c How would I have had to redraw Figure 3.32 to represent the other type of spectrum?

d Why does the decreasing spacing of the energy levels lead to the convergence of the spectral lines?

It was soon realised that the frequency at which the lines converged carried a special significance. There is very little difference between promoting an electron from quantum level 1 to, say, 7, and *removing it altogether*. Quantum level 7 is interesting because it is around that level that the blurring of the lines starts. This is the same as saying that:

• Beyond about quantum level 7, the electron of a hydrogen atom can barely feel the nucleus.

• Quantum level 7 or greater might just as well be quantum level *infinity*.

• Therefore the frequency or energy represented by the convergence 'blur' is the same as the frequency or energy necessary completely to remove an electron from its ground state, or in other words to **ionize** a hydrogen atom.

Question

18 The Lyman series of lines in the spectrum of atomic hydrogen converges at a frequency of 3.27×10^{15} Hz. In other words, this is the frequency of a photon whose energy is capable of lifting a ground-

state hydrogen electron all the way to 'quantum level infinity'. Calculate the ionization energy of a mole of hydrogen atoms, bearing in mind equation (*3.1*), and the fact that the value of *h*, the Planck constant, quoted (6.63×10^{-34} J Hz^{-1}), only relates frequency to the energy of a single photon, and therefore to the energy of a single electron ionization.

The Lyman series, in which the quantum level 1 (ground state) is involved in every electron jump, is not the only series of lines in the spectrum of atomic hydrogen. Where did the Balmer series come from (Figure 3.25, p. 36)? To answer this question you have to be aware that Figure 3.25 shows the Balmer series from the light of a *star*. At the high temperatures in the 'atmosphere' of a star, atoms of hydrogen might well have their electron already partially 'excited' (that is, above ground state), so that instead of promotion happening from level 1, it can begin at other levels. Any conceivable jump is allowed. Perhaps now you can answer the question yourself.

Question

19 Suggest a series of jumps which would give a general pattern of lines similar to those in Figure 3.32, but in which all the lines were at lower frequencies.

3.7 The second revolutionary new physics – wave mechanics

Bohr's model of sub-atomic events survived less than ten years before it was the subject of important modifications. Like Thomson, Bohr was in no way diminished by this turn of events, because the scientists concerned owed a great deal to the school of thought that Bohr himself had pioneered.

The Nobel Physics Laureate in 1923 was the Frenchman Louis de Broglie, whose contribution was to recognise *why* there were such things as quantum numbers and allowed energy levels for elec-

trons. The idea that photons could only be adequately described by allowing for wave/particle 'dual personality' had been in currency ever since Planck's work; de Broglie now proposed that the same duality might be true of electrons. In other words, he was asking people to believe in **matter as waves**.

It then followed that restrictions would have to be placed on electron orbits, argued de Broglie, because otherwise according to the wave side of their natures, they would interfere themselves out of existence. Only electrons whose orbits allowed a whole number of wavelengths would be able to exist (Figure 3.33).

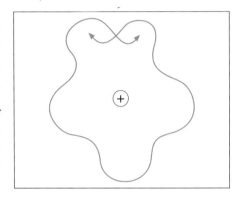

Figure 3.33a Possible mental picture which might describe an electron interfering itself out of existence

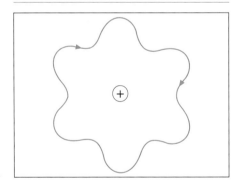

Figure 3.33b Allowed standing wave for an electron

Fortunately, as in the case of Moseley's contribution to the atomic-number debate, a piece of evidence was soon found which helped people to accept this revolutionary idea. If electrons had wave-like properties as de Broglie proposed, then they ought to show wave-like behaviour – diffraction and interference, for instance. Dead on cue, electron diffraction pictures were indeed produced.

It turned out that electrons had wave-lengths similar to those of X-rays, and therefore could be diffracted by crystal lattices just like X-rays (Figure 3.34). (One of the first people to study electron diffraction was G. P. Thomson, the son of J. J. Thomson, which suggests that early twentieth-century science was quite a closed social activity.)

Even with the diffraction evidence, the idea of matter waves does not leap easily into mind or picture, and has left chemists with some difficult concepts, compared to the comfortable planets-around-the-Sun atom of the Bohr model. One subsequent principle of this new physics was the **principle of uncertainty** (due to Werner Heisenberg) – that you could not simultaneously know *where* a wave/particle was and *what* its momentum was; indeed the more exact your knowledge of the one, the greater your uncertainty about the other.

(In fact these wave mechanical ideas are not the exclusive preserve of electrons. All matter is thought to be subject to wave mechanical rules, but only the very small particles show any detectable wave properties. People have amused themselves by calculating the wavelength of a football, for instance, which is around 10^{-25} nm. Clearly it is not crucial for goalkeepers to have a working knowledge of the uncertainty principle.)

What we are left with, as far as electrons are concerned, is the idea of an **orbital**. An orbital only tells you where you are *likely to find* an electron, and it is less certain in time and space than an orbit. Indeed, the shapes of these orbitals (as shown in Figure 3.37, p. 41) should be seen as **probability densities**. In order to give a clearer idea of the meaning of this term, we will consider some actual orbitals calculated for the ground state and the excited states of the hydrogen electron.

The pioneer mathematician in this field, and the third of the younger inheritors of Bohr's tradition, was Erwin Schrödinger (Figure 3.35), an Austrian physicist who was born and died in Vienna, but spent his working life all over Europe from Dublin to Berlin to Oxford. What he did in effect was to calculate the distributions in space of various **standing waves of electrons**. The reason electron states are quantised is because they must not interfere them-

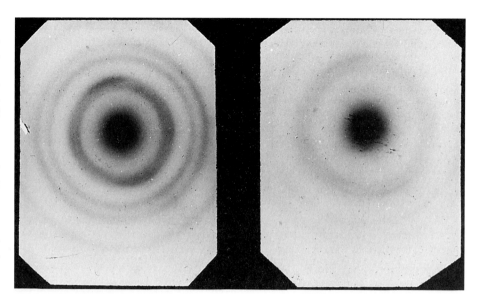

Figure 3.34 Early electron diffraction photograph obtained by G. P. Thomson

Figure 3.35 Erwin Schrödinger

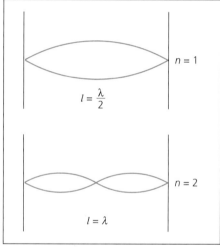

Figure 3.36 First two 'quantum states' of a guitar string of length *l*

selves out of existence (p. 39), and that condition is only met by standing waves. We have therefore an accessible, if distant, analogy for the behaviour of electron waves, in the standing waves of a guitar string.

The sound a guitar string emits depends on the standing wave set up in it. For a given string length), there is only a very exclusive set of wavelengths which will fit into the string. In this case the limiting condition is that the ends of the string must be static, which translates into a mathematical relationship between string length *l* and wavelength λ:

$$\frac{n\lambda}{2} = l \quad \text{or} \quad \lambda = \frac{2l}{n}$$

where *n* is an integer

It may be helpful to restate these equations in a word-equivalent form: they express the idea that the length of the string must be equal to a whole number of 'single loops', or **half-wavelengths**.

The first two standing waves allowed in a string are shown in Figure 3.36. The first of these waves, in which *n* = 1, and

$l = \lambda/2$, is called by musicians the **fundamental**, but for the purposes of our analogy we can see it as the **ground state** wave. n is playing the part of a **quantum number**, equivalent to the numbers on the energy levels of the Bohr atom. It only remains to imagine that when the string is plucked harder, energy is absorbed by the string and some of the waves with higher quantum numbers are set up, only to relax back down to the ground state again by the emission of energy.

This is a useful mental picture, but an imperfect match with the hydrogen electron case, for the following reasons:

• The electron wave does not have ends. Its restrictive condition comes from the need to avoid destructive interference with its own 'second lap of the circuit', whereas the guitar wave is interfering with its own *reflection*.

• Whereas a string can have several waves existing in it at once (**overtones**), the electron must be found in only one of the quantum states at a time.

• A real guitar string absorbs the energy of a harder pluck mainly by increasing its **amplitude** (wave height from mid-line to crest), rather than going into overtone waveforms.

Even with these drawbacks, the guitar string analogy is a useful comparison and learning tool, not least because it shows how wave character leads naturally to quantum numbers and quantum restrictions. Keep the analogy in mind as we go on to unveil Schrödinger's solutions to the wave equation for the hydrogen atom.

3.8   s, p and d orbitals

Schrödinger fed into his equation the known facts about the charges on a proton and an electron, and the mass of the electron, and added the new equations

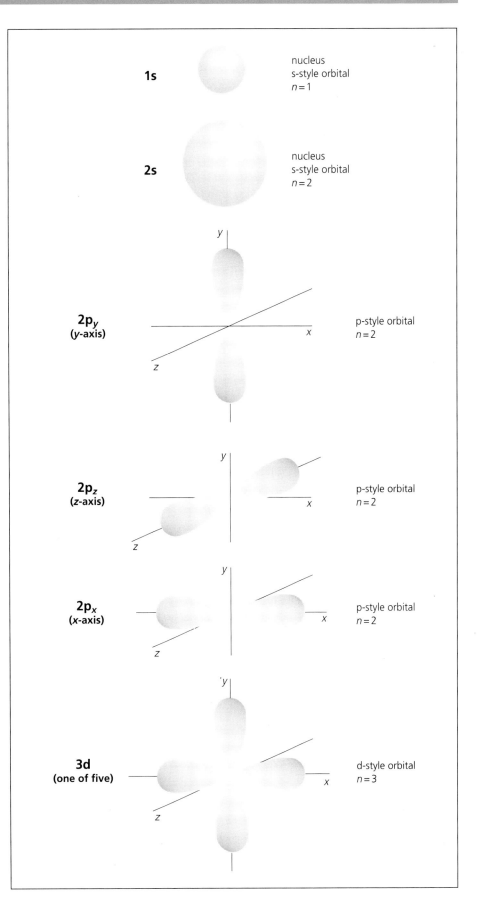

Figure 3.37 Electron orbital probability densities, as calculated by the Schrödinger wave equation. All these orbitals exist as a potential place for an electron on all atoms. Whether there is an electron 'at home' depends on the atom and on its energy state. If you are not comfortable with the 'probability density' idea, think of them as electron 'clouds'

demanded by the wave side of the electron's behaviour. His calculations suggested a number of possible waveforms for the hydrogen electron, each with different quantum numbers and different energies. These are the things which can be seen as analogous to the fundamental and overtones of the plucked guitar string. You might think that each waveform was description enough for the path taken by the electron, just as we can use the guitar string waveforms to describe the path taken by bits of the string. Unfortunately, in the strange world of wave/particle duality and wave mechanics, things are not that simple.

The uncertainty principle means that the electron's position can never be pinpointed. What we can draw, following Schrödinger's calculations, are envelopes of probability. In the drawings in Figure 3.37 (previous page), the position of the electron is represented as a cloud. This does not mean however that the electron has become 'smeared'. The best way of visualising Figure 3.37 is to see the cloud as being a region in space, outside of which there is only a 5 % (say) chance of finding the electron. Another way of visualising it is the slightly less correct 'pointilliste' model – imagine you could place a dot at the electron's position at a moment in time. If you marked 1000 such moments you would have a pattern of dots like the clouds in Figure 3.37. From now on we will refer to the clouds as **electron orbitals.**

The electron orbitals in Figure 3.37 have special names and significances. The ground state for the hydrogen electron, analogous to the fundamental wave in the guitar string, turns out to have a spherical orbital. It is called a **1s orbital**, where the '1' is a quantum number expressing the fact that this is the lowest, most stable energy level allowed, and the 's' indicates the sphere.

The second calculated orbital, quantum number 2, also turns out to be spherical (2s), but with a larger radius – its higher energy deriving from the fact that it is further from the nucleus. However, there are other solutions to the wave equation which share quantum number (and energy level) 2. These solutions have orbitals which have two lobes, and a very directional character. There are three of them, pointing at 90 ° to each other, and they are called 2p

orbitals. In all, therefore, there are four allowed orbitals at quantum level 2, one spherical and three lobed.

In fact, we refer from now on to **principal quantum number 2**, since the new shapes have themselves been assigned **secondary** quantum numbers. (The s shape is given the secondary quantum number 1, the p shape is given 2, the d shape (still to come) is 3, and so on.) By and large these are matters beyond Advanced level, but are relevant to an understanding of the naming system.

Within principal quantum level 3 there is an even bigger range of available orbitals. There is a single 3s orbital, three more p orbitals (called 3p orbitals), and a new set of five equivalent related shapes called 3d orbitals (Figure 3.38). The pattern thus established continues with each new principal quantum level: a pattern of more shapes becoming available, with more equivalent versions. (The next shape is the f shape, with seven equivalent, but different, orientations in space. This is where we call a halt, for reasons which will become apparent in later sections on the Periodic Table. Our main concern will be confined to electrons in s, p and d orbitals, and with principal quantum levels up to about 7.)

If you feel this bizarre world of wave mechanics is getting out of your mental control, keep hold of two ideas: first that these shapes are like (if only a bit like) the fundamental and overtones of a plucked guitar string, and second that the electron in any given hydrogen atom inhabits one or other of these shapes. If there is no significant energy input to the atom (from, say, flames, photons or high-voltage discharge tubes), then the electron will be in its 1s orbital.

The orbital model explains the line spectrum of hydrogen as follows: in Figure 3.32 (p. 38) it is now the **principal** quantum number which defines the energy level of the orbital, so that photons of the same energy could bring about a 1s → 3s transition, or a 1s → 3d transition.

So far we have restricted ourselves to talking about an atom with only *one* electron. When we move on to consider the multi-electron atom, the rules change slightly, as we shall see in the next section.

3.9 The multi-electron atom

When more than one electron is present in any atom, the force of repulsion between electrons is introduced as a new factor, modifying the forces of attraction on each electron from the nucleus. But the most significant fact about an atom with many electrons is this:

An orbital can be shared by no more than two electrons.

This is equivalent to saying that although the 1s orbital is the most stable one, only two electrons are privileged to occupy it, and after that they have to start filling up the 2s, 2p and higher-energy orbitals.

In fact the electrons were found to have another quantised property, namely **spin**, so that each electron behaves like a tiny magnet. The electron has only two possible values of the spin quantum number, and if two electrons are sharing an orbital they must be **spin opposed**. Each electron is thus characterised by four quantum numbers, namely:

- principal – based on distance from the nucleus.
- describing the original shape – s, p, d, etc
- describing the orientation in space of the orbital – *x*-axis, *y*-axis, etc
- describing the spin.

The last of the great ideas of quantum mechanics in this chapter, after those of Bohr, de Broglie, Schrödinger and Heisenberg, belongs to W. Pauli, another Viennese-born physicist who also worked in Munich and Copenhagen. His important **exclusion principle** states that:

No two electrons in any one atom can have the same value of all four quantum numbers.

Although no Advanced level question assumes a knowledge of the Pauli exclusion principle, it is an interesting background to the earlier idea that every orbital is full up once it contains two electrons.

We are now approaching a very big prize indeed in our search for knowledge, namely a thorough understanding of the periodic behaviour of the elements. The periodic behaviour of the elements is entirely dependent on the behaviour of the electrons, and they in turn are governed by the rules of quantum mechanics. It is certainly possible to discuss the properties of atoms without a knowledge of wave mechanics, simply by stating the rules which govern electron arrangements. For that reason you will probably be able to follow the ideas of the next section even if, understandably, you were bemused by the last one. However, the quantum mechanical basis of particle physics is one of the authentic roots of chemistry, and it is important to recognise that it is there, even if it is imperfectly understood.

3.10 Filling the orbitals

Let us consider the Periodic Table from the point of view of some god-like being who could create the elements from a stock of protons, neutrons and electrons. We would create our nuclei first, each one with an additional proton compared to the last (and variable numbers of neutrons), and then feed in the electrons one by one to keep the overall charge on the atom at zero. We know that the electrons will 'seek out' the innermost orbitals first – the ones with the lowest principal quantum numbers. We know that each element differs from the last by (among other things) the possession of one extra electron. We know that each orbital can house a maximum of two electrons. Hence we can predict the number of elements which will occur while filling up the orbitals of a particular principal quantum number.

Questions

20 Why do electrons 'seek out' the innermost orbital first?

21 The existence of, and numbers of, the various types of orbitals are shown in Table 3.5. Predict how many elements will be formed while we are filling up:

a principal quantum number 1

b principal quantum number 2

c principal quantum number 3.

Table 3.5 Numbers of the respective orbitals. Other orbitals exist (for example, 5g) but are not occupied in the ground state of any known element

Orbital	How many of them?
1s	1
2s	1
2p	3
3s	1
3p	3
3d	5
4s	1
4p	3
4d	5
4f	7
5s	1
5p	3
5d	5
5f	7
6s	1
6p	3
6d	5
7s	1

d Does the pattern of your answers in (a), (b) and (c) seem to be linked to anything in the arrangement of the real Periodic Table?

e Is there any clash between your pattern and the arrangement of the Periodic Table?

So the number of spaces in the orbitals of principal quantum levels 1 and 2 (namely 2 and 8), exactly matches the lengths of the rows of the Periodic Table. (This pattern breaks down when row 3 of the Periodic Table turns out to be another 8 and not an 18, for reasons set out below.) In other words, the elements from hydrogen to neon are made in the cosmic workshop by filling up the first two principal quantum levels. However, what we do not know at this stage is this:

• whether electrons have any preference, within principal quantum level 2, for either a p orbital or an s orbital?

• whether electrons voluntarily pair up in an orbital, or do they stay in different orbitals if there is space available?

There are both theoretical and practical clues to help us answer these questions.

Evidence from ionization energies

The ionization energy of an atom is the energy necessary to remove an electron completely from the influence of its nucleus.

It can be calculated from measurements of the frequencies of certain of the spectral lines of that atom. It is usually assumed when talking about ionization energies that the atom is in the monatomic gaseous state, and that the electron in question is the one which would naturally be the first and easiest to remove. The units are **kilojoules per mole**.

The ionization energies (for a single electron removal) of the elements from hydrogen to neon give an insight into how strong the attraction is between a succession of nuclei and their outer electrons – the **nuclear pull**. Do these data contain some clues to the relative 'comfort' of electrons in s and p orbitals? Making some predictions, prior to looking at the ionization energy data, will help us to prepare our thoughts.

Question

22 How would the ionization energy vary in going from hydrogen to neon, if each of the following factors was the *only* one being considered?

a the change in nuclear pull

b the distance between the nucleus and the removed electron

c the repulsive forces between electrons.

These three factors play crucial roles in deciding the sizes of ionization energies, but they do not operate *alone*. Sometimes the factors pull in different directions. For instance a particular element may have more nuclear pull than another, but because of quantum restrictions on the arranging of electrons, its outer electrons may be further away – they cannot *all* be in the 1s orbital. Clearly under these circumstances the ionization energy will be set by the relative influences of these conflicting trends. Table 3.6 shows the first ionization energies for the ten elements in question. The patterns in the data will be examined in the next question.

Table 3.6 The pattern of ionization energies across a row of the Periodic Table

Element	First ionization energy (kJ mol⁻¹)
H	1312
He	2372
Li	520
Be	900
B	801
C	1086
N	1402
O	1314
F	1681
Ne	2081

Questions

23a Plot a graph of these data.

b After which element is the 1s orbital full up, and what does your graph show is happening to the pattern of ionization energies after it is full?

c The element lithium has a greater nuclear charge than helium (one more proton), yet its first ionization energy is much lower. (That is, its outer electron is much easier to remove.) Name two factors which must have 'overruled' the influence of the increasing nuclear pull.

24 The ionization energies of the elements from lithium to neon show a generally upward trend. However, X-ray data have shown that the atoms from lithium to neon actually get *smaller*. Which one or more of the following factors must have been the dominant influence(s) on the change of the ionization energies as we cross this row?

a increasing nuclear pull (caused by extra protons)

b decreasing distance from the nucleus

c increasing electron–electron repulsion.

The factors behind the Periodic Table are becoming clearer. Increasing nuclear charge holds sway as you cross a row, holding the electrons tighter against the counter influence of electron–electron repulsion. However, when quantum rules force an electron into a new principal quantum level, the extra distance involved brings about a sharp reduction in the attraction of the nucleus, despite the extra proton.

Question

25 How would you expect the graph of atomic radius against atomic number of the first ten elements to look? Look up values of r_{cov} (where 'cov' means covalent) in a data book and draw the graph.

3.11 The influence of shielding on ionization energies

It has been found that increasing distance alone cannot account for the sharpness of the decrease in ionization energy in passing from helium to lithium, and indeed all the elements of the second principal quantum level have ionization energies less than would be predicted on distance grounds. It seems therefore that the intervening 1s electrons act as a **shield**, reducing the amount of nuclear pull felt by the 2s and 2p electrons. This leads to the principle that inner electrons in general have a shielding effect on those outside them.

(It is possible to see shielding as just another word for electron–electron repulsion pushing outward, as Figure 3.38 shows. The shielding effect of electrons in the *same* principal quantum level on each other appears to be smaller than that of inner electrons on outer electrons. That conclusion is forced on us by the steady increase in ionization energy across a row, since increasing ionization energies suggest that shielding effects are being overruled by the nuclear pull factor (Figure 3.39).)

We still have no answers to our more subtle questions on p. 43 – which is more stable out of 2s and 2p, and do electrons like to pack in 'two to a bed'? To answer these questions we must study the same data (in Table 3.5) in more detail.

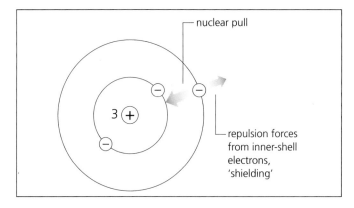

Figure 3.38 Forces operating on the outer electron in lithium. The shielding force is what makes the electron fairly easy to remove

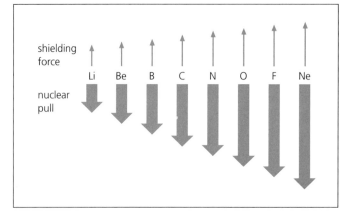

Figure 3.39 Showing how the balance between shielding forces and nuclear pull changes across a row of the Periodic Table. Note how shielding stays fairly constant, depending as it does on the constant factor of the 1s² electrons

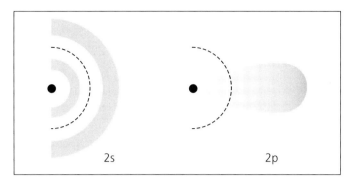

Figure 3.40 Probability densities (electron clouds if you like) of the 2s and 2p electron orbitals. The dotted line shows the extent of the 1s orbital, so we see that the 2s electron can partially penetrate the 1s cloud, increasing its stability

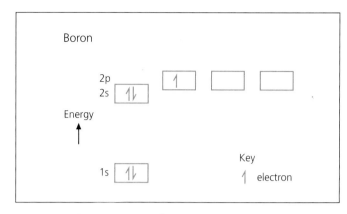

Figure 3.41 The 'energy-boxes' format, with the example of boron. (The direction of the arrowheads represent electron spins)

The 'hiccup' between beryllium and boron

Look at your graph from question 24 – the general trend of ionization energy across a row is up. Why then are there two small reversals in this pattern, dividing the row of 8 into sub-groups of 2, 3 and 3? A clue came from theoretical quantum mechanics, when people working out the probability density of the 2s electron orbit found that their model predicted a strange behaviour. As Figure 3.40 shows, the 2s electron has a finite if small probability of 'burrowing' into the 1s cloud, whereas the 2p electron lacks this facility, which is known as **penetration**. This suggests that the 2s orbital is a more stable home for an electron than the 2p, and that therefore the electron structures of the elements lithium and beryllium are probably:

Li: $1s^2 2s^1$ (i.e. the 1s orbital contains two electrons and the 2s orbital has one electron)
Be: $1s^2 2s^2$

In the case of the hydrogen atom with its one electron, all orbitals of the same principal quantum number are as good as each other. The advantage of the 2s orbital over the 2p disappears when there is no 1s cloud for a 2s electron to burrow into.

Question

26 The spectrum of atomic hydrogen would be a good deal more complex if every sub-shell were really of slightly different energy (and indeed the spectra of other elements are very complex). Can you explain why there would be extra lines?

Going back to the beryllium–boron 'hiccup', we can now see why it occurs. The electron structures of beryllium and boron are:

Be: $1s^2 2s^2$
B: $1s^2 2s^2 2p^1$

So the first ionization energy of boron is the energy needed to remove a 2p electron, which because of its lack of penetration ability is not held quite so tightly as a 2s electron (as in beryllium). All the remaining elements in the row have 2p electrons as their 'first removal', so the rest of the graph is displaced downwards by a fairly constant amount, relative to the two 2s-only elements lithium and beryllium.

Another useful way of representing orbitals is to show them as boxes, each box with a vacancy for a maximum of two electrons, on a vertical energy axis (Figure 3.41). This 'energy box' format is particularly useful for emphasising the energy levels of electrons.

The 'hiccup' between nitrogen and oxygen

What is it that causes the break in increasing ionization energy between nitrogen and oxygen? Oxygen, fluorine and neon form one sub-group; boron, carbon and nitrogen another. Their electron structures are:

B $1s^2 2s^2 2p^1$
C $1s^2 2s^2 2p^2$

N $1s^2 2s^2 2p^3$
O $1s^2 2s^2 2p^4$
F $1s^2 2s^2 2p^5$
Ne $1s^2 2s^2 2p^6$

It appears that three of the 2p electrons are easier to remove than the other three. The reason is that in boron, carbon and nitrogen the electrons can stay apart and reside 'one in a bed' in the three separate 2p orbitals. In oxygen, fluorine and neon the electrons have to share an orbit, which brings them closer to another electron. The extra repulsion thus incurred is enough to displace the ionization energy line downwards. Figure 3.42 shows an 'energy box' version of this story.

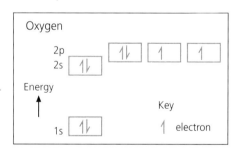

Figure 3.42 'Two-in-a-bed' repulsion effect, lowering the first ionization energy of oxygen

Studying these two 'hiccups' as case studies has given us some additional principles with which to extend our knowledge of the way electrons fill up the orbitals.

• We know that within a principal quantum level, the relative stability of orbitals is: s > p > d > f (... if they exist), so that the s orbitals will fill up first, followed by p, d etc., and electrons will be pulled out in the reverse order.

• We know that electrons will not pair up in an orbital if an equivalent orbital is vacant.

• We already knew that normally the lowest principal quantum levels fill up first.

Questions

27 Use some of these principles to continue the pattern of electron arrangements begun in Table 3.7:

Table 3.7 For question 27

Element	Electron arrangement
H	$1s^1$
He	$1s^2$
Li	$1s^2 2s^1$
Be	$1s^2 2s^2$
B	$1s^2 2s^2 2p^1$
C	$1s^2 2s^2 2p^2$
N	
O	
F	
Ne	
Na	
Mg	
Al	
Si	
P	
S	
Cl	
Ar	

28 Predict how the ionization energy graph from question 24 would look, if you were to extend it as far as argon.

Successive ionizations

So far in our discussions of ionization energies we have dealt with removal of a *single* electron from a series of elements. There are also data for the energies needed to remove one electron after another from a single element. These are called **successive** ionization energies, and there is one overriding pattern – each successive ionization energy is bigger than the last. In other words, it gets more and more difficult to remove successive electrons as the ion left behind becomes increasingly positive.

However, it does not get *steadily* harder. The pattern of big jumps, little jumps and mere upward drifts is as

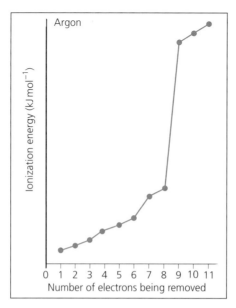

Figure 3.43 Eleven successive ionization energies of argon

revealing as the jagged shape of the first ionization energy graph, and reveals some of the same underlying reasons. (See Figure 3.43, which shows the successive ionization energies of argon.) Some questions now follow.

Question

29a What is the electron arrangement of argon?

b The first three electrons in argon are all removed from equivalent positions. From which orbitals are they being removed?

c If all these first three electrons are coming from equivalent orbitals, why is it that removal gets harder?

d Account for the big jump between the ionization energies of electrons 8 and 9.

e Account for the modest jump between the ionization energies of electrons 6 and 7.

f Account for the slight extra difficulty of removing electrons 4, 5 and 6, relative to 1, 2 and 3.

The transition elements

You will remember from the work on the hydrogen spectrum that the separations of the energies of the principal quantum levels get successively *less* (p. 38). When, in the multi-electron atoms, you have to add on the effect of relative stabilities of s, p and d orbitals within a principal quantum level, the pattern of energy levels can become confused. In question 22, you may have noticed that although principal quantum level 3 contains space for 18 electrons, the third row of the Periodic Table contains only 8 elements. The reason is that the 4s orbital actually offers a *more* stable home than the 3d, so a new row of elements begins with a whole sub-shell left unfilled (Figure 3.44). Thus it happens that the next two elements in our sequence, potassium and calcium, have the electronic structures:

K $1s^2 2s^2 2p^6 3s^2 3p^6 4s^1$
Ca $1s^2 2s^2 2p^6 3s^2 3p^6 4s^2$

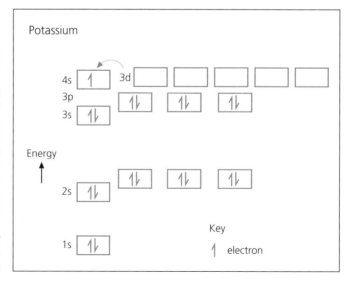

Figure 3.44 The moment when the 3d orbitals get 'left out'. Potassium takes an electron into its 4s orbital, this putting itself in Group I and beginning a new row

even though from Figure 3.37, the last sub-shell should be 3d, not 4s.

Only now, at scandium, when the choice is between 3d and 4p, do the 3d orbitals begin to fill up, and as there are five 3d orbitals the result is ten new elements, the **first transition series**. The chemical implications of this deferred filling of the 3d orbitals will be discussed along with all the other periodic phenomena in Chapter 4, p. 53. The rest of the elements have electronic structures that follow this delayed-filling pattern. With a few minor reversals, the order of filling up, after the end of the first transition series, is:

4p, 5s, 4d, 5p, 6s, 4f (long delayed), 5d, 6p, 7s, 5f, 6d.

3.12 Electron-filling and the Periodic Table

The modern Periodic Table is constructed so as to recognise all of the current ideas on orbitals. The various **blocks** of the Periodic Table are called s, p, d and f, highlighting the type of orbital which is being filled at that point in the development of the table. This arrangement makes it easy to deduce the electron arrangement of any element. You simply trace the filling path, beginning at hydrogen, moving in atomic number order, and ending at the target element. As you traverse a particular block, the name of the block and its electron contents are noted. For instance, to get to iron from hydrogen, you have to go two squares across the 1s box, followed by two 2s squares, six 2p, two 3s, six 3p, two 4s, and six 3d. Hence the electron structure of iron is:

Fe $1s^2 2s^2 2p^6 3s^2 3p^6 4s^2 3d^6$

There is a strange result of the finely poised balance of preference between the 3d orbitals and 4s orbitals. It is that although the 3d are the last to fill, they seem to acquire stability in the filling. So when it comes to ionization, the 4s orbitals are the first to empty. The iron, Fe^{2+} ion, for example, has the electron structure.

Fe^{2+} $1s^2 2s^2 2p^6 3s^2 3p^6 3d^6$

This switch in relative stability after the filling leads some books to place the 3d

electrons before the 4s electrons in writing the electron structure of transition metal elements. You need to be aware of the two versions you might meet, and why there is a problem. The underlying physical reason for the switch need not concern you.

Question

30 Use the Periodic Table to deduce the electron structures of the following elements:

Ge Sr Ag Sn

and the following ions (which have lost or gained electrons relative to the parent atoms):

Mg^{2+} Na^+ Cl^- Cu^{2+}

We have now reached the stage at which we know the 'anatomy' of atoms. That is, we know what is inside them, and how those structures show patterns of repetition which are closely allied to the modern Periodic Table. We can see that the elements within the s, p, d and f blocks are united in having a common sub-shell in the process of filling. Every time we find ourselves in the same vertical column, or **group** of the Periodic Table, we

know we will return to a similar pattern of outer electrons. This strong repeating rhythm in the Periodic Table is called **periodicity**. Each row of the table, which constitutes one complete cycle of outer electron filling, is called a **period**.

If that was anatomy, what we now progress to is the 'physiology' of atoms, or in other words what happens when atoms use their working parts to 'do' some chemistry. We will see that this aspect is totally dependent on the outer electron structure, and so shows the same periodicity. This is the subject of Chapter 4.

Question

32 For each of the groups of the Periodic Table, give the common feature of their outer electron structures which unites all members of the group:

Group I:	(H) Li Na K Rb Cs
Group II:	Be Mg Ca Sr Ba
Group III:	B Al Ga In Tl
Group IV:	C Si Ge Sn Pb
Group V:	N P As Sb Bi
Group VI:	O S Se Te Po
Group VII:	F Cl Br I At
Group VIII:	He Ne Ar Kr Xe Rn

Summary

• **Electrons** were first discovered and characterised by J. J. Thomson using cathode-ray tubes. They were recognised as negative charge-carriers with a mass 1/1850 of a hydrogen atom, and as universal constituents of all atoms.

• Early assumptions about atomic structure visualised lots of electrons embedded in a matrix of positive charge – the plum-pudding model.

• Rutherford's experiments with alpha particles revealed the existence of the **nucleus**, where resided most of the mass, and all of the positive charge, of the atom (made up of **protons**). The electrons were seen as orbiting particles of very small mass, but each electron had a negative charge equal in mag-

nitude to the positive charge on the proton.

• The **mass spectrometer** ionizes the sample under test, and then deflects a beam of its ions. A variable magnetic field is used to deflect the ions, and bring them to a detector. Scanning through the magnetic field strength brings ions of successively higher mass to the detector, creating a mass/charge spectrum.

• Mass spectrometers provide a quick and reliable way of measuring relative atomic mass. They also reveal the existence of different masses within collections of the same sort of atom. These differing species are known as **isotopes** of the atom.

• **High-resolution mass spectrometers**, with a second deflection process, can measure atomic masses to seven decimal places.

- The **neutron** is identified as the second major mass carrier in the nucleus. Its mass is nearly identical to that of a proton but it carries no charge and has virtually no effect on chemical character (as opposed to radioactive character on which it can have a major influence). Atoms of isotopes of only in neutron number.

The **mass number** of an atom is the sum of its proton (p) and neutron (n) numbers. The **atomic number** of an atom is the number of its protons (and of its electrons (e) if no ionization has occurred).

- So

$p + n$ = mass number

$\quad$ = relative atomic mass
$\quad\quad$ (to nearest whole number)

and

$p = e$ = atomic number, Z

Therefore

relative atomic mass $- Z = n$

- The new standard of relative atomic mass is achieved by setting the mass of the ^{12}C isotope of carbon at 12.0000000. The differences between relative atomic masses on this scale and on the old H = 1 scale are very slight.

- The spectrum of light emitted from energetically excited matter is often caused by electrons jumping between allowed energy levels.

- The energy of the emitted or absorbed photon is related to its frequency by **Planck's equation**:

$$E = h \times \nu \qquad (3.1)$$

where E is energy (J), ν is frequency (Hz), and h is the Planck constant (6.626×10^{-34} J Hz^{-1}).

- This behaviour was recognised by Niels Bohr as evidence for the need for a **quantum theory** to describe the way electrons could possess energy. The overriding theme of the new theory was one of *restriction* on allowed energies.

- de Broglie, Schrödinger, Pauli and Heisenberg showed that the roots of quantum restrictions lay in the **wave properties** of all matter, a feature specially significant in small particles like electrons. Confirmation came with the discovery of electron diffraction. (Diffraction and interference are wave phenomena.)

- So the idea of an electron **orbital** came into being. An orbital was an allowed space in which an electron could set up a sort of standing wave, not unlike the allowed standing waves in a guitar string.

- Just as allowed standing waves in guitar strings can be characterised by numbers – fundamental, first overtone, etc., so electron orbitals could be characterised by **quantum numbers**.

- Every electron in every atom occupies an allowed orbital describable in terms of its quantum numbers.

- The idea that no two electrons in an atom could have the same value for all four quantum numbers (**principal**, **shape**, **directional orientation** and **spin**), led to the idea of s orbitals, p orbitals and d orbitals, each capable of occupation by a spin-opposed pair of electrons.

- For example the symbol '$2p_x^1$', refers to one electron in an orbital in principal quantum level 2, with a shape called 'p' (propeller shape – although I do not think that is where the p came from), pointing in direction 'x' (there are two other directions, y and z). Another electron could possibly occupy the same orbital, but it would have to have opposite spin, hence the 'two opposing arrows in a box' symbolism for an electron pair.

- 'Energy box' diagrams express the numbers of allowed orbitals of each type, their energies, and their state of occupation. Potential energies are negative relative to a zero set by an electron so far out as not to be able to feel the nucleus.

- The Periodic Table is now seen as being shaped by quantum rules, with s, p, d and f blocks.

- Each element in the Periodic Table has an electron configuration which can be expressed as $1s^2 2s^2 2p^6$, etc. To deduce the electron configuration of any element it is only necessary to trace from hydrogen to the target element on a Periodic Table, noting the electron blocks passed.

- Ionization energies represent the work done to remove an electron from an atom. They are equal and opposite to the potential energies of electrons.

- Ionization energies can be calculated from (among other things) the frequencies of spectral lines.

- Ionization energies were seen to confirm ideas about quantum levels, and to be at the heart of chemical character.

- The 'periodicity' of chemical properties is seen to have strong links with the periodicity of ionization energies, with (for example) Group I metals having the lowest ionization energies in any row, and Group VIII 'noble' gases having the highest. These extremes correlate with their extremes of reactivity.

- More generally, we can see why metallic (electron loss) behaviour occurs on the left-hand side of each row.

- The size of ionization energies can be seen as being under the control of three parameters:

1 the size of nuclear charge
2 the distance from nucleus to electron
3 the repulsive effects of other electrons (shielding).

For example, the dominance of the last two factors over the first one explains the increasing ease of electron loss from the lower members of groups in the Periodic Table.

4

Chemical bonding 1: Ionic and covalent bonds

4.1 Introduction – the origins of chemical bonding

The Universe is full of atoms, and as long as they stay cool (less than about 1000 K), these atoms are nearly always found combined with each other, rather than as single atoms. It is one of the fundamental ideas of chemistry that atoms like to combine, and they do so in a bewildering array of formats; as short clumps, as long chains, as three-dimensional crystalline arrays, and with their own kind or with sharply contrasting other kinds.

Once they are combined in a chemical **bond**, a pair of atoms can only be parted by a force, in much the same way as a pair of attractive magnets are parted. What is the origin of this force – in other words, what 'glues' atoms together? The answer across a wide range of different bonding processes, is **electrostatic force**, and indeed all chemical bonds can be traced back to the attraction between nuclei and electrons, that is, between positive and negative charges.

However, the part played by electrons in the bonding process is far more active than that played by nuclei. The nucleus of an atom is of course what maintains the electrons in their orbitals and what creates the electrostatic environment in which different orbitals have different energy levels (see Chapter 3). But beyond that the story of bonding lies mainly with the electrons – in particular how various pairs of atoms manipulate their electrons to make it better for them to be together than to be apart.

4.2 The hydrogen molecule – a simple covalent bond

It takes 436 kJ of work to separate one mole of dihydrogen, H_2, molecules into separate atoms. This is not a trivial amount of work – it is equivalent to lifting a 1 kg mass from the laboratory floor on to the bench approximately 43 600 times! (Or lifting a large articulated lorry just the once.) It is no wonder that dihydrogen molecules do not break up until molecular collisions get very violent indeed (around 1700 K). What is it that makes the hydrogen atoms in the molecules so stable? The simple answer is that the hydrogen atoms *share* electrons. In fact, X-ray diffraction experiments show the electrons inhabiting the space between the two atoms. A bond constructed from a shared pair of electrons is called a **covalent bond**.

Why is this arrangement so much more stable than two separate hydrogen atoms? We can begin by considering the arrangement from the electron's point of view. There is a strong attraction between the electron and the nucleus in hydrogen atoms; the ionization energy of hydrogen (the energy to remove an electron from its ground state 1s orbital) is a very considerable 1312 kJ mol⁻¹ (p. 39). So if the electron could spend its time close to two such nuclei, it would presumably be even more firmly held. Since every hydrogen atom has a single vacancy in its 1s orbital, the optimum benefit can be achieved by just two atoms coming together, each one offering the other the use of its 1s orbital (Figure 4.1).

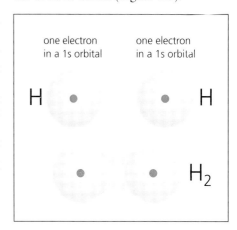

one electron in a 1s orbital   one electron in a 1s orbital

H   H

H_2

Figure 4.1 Two 1s orbitals overlapping, to allow two electrons to make the bond in the hydrogen, H_2, molecule

From the point of view of the nuclei, there are now two electrons occupying positions which, for a substantial portion of their time, lie between the two nuclei, serving to bind them together. There is an optimum (ideal) distance at which these two nuclei will want to reside. Too far apart and an electron finds it hard to feel the effects of both nuclei, and starts to relocalise on one – too close together and the positive charges on the nuclei begin to repel each other.

We can represent these effects in the form of a graph (Figure 4.2). Distance between nuclei is plotted on the *x*-axis. Potential energy is plotted on the *y*-axis. When the hydrogen atoms are so far apart that the two electrons cannot feel each other's nuclei, the potential energy is zero. As the distance between the nuclei grows less than this, the potential energy reduces – rather like a north and a south magnet being inched towards each other until they begin to draw each other in.

The line on the graph describing the change of potential energy of the bond with changing distance now plunges into a trough, reaching a minimum of potential energy at the point X, Y. This point represents the stable state of the bond.

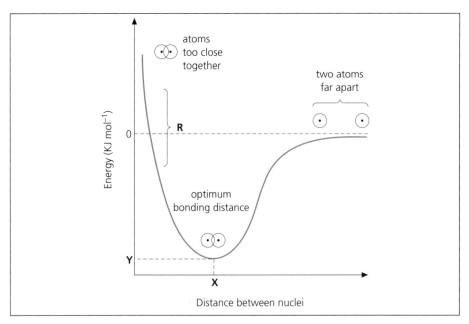

Figure 4.2 Potential energy/distance graph for the H_2 molecule

Question

1 By finding the appropriate data in a data book:

a Give the value on the *y*-axis at *Y* in Figure 4.2.

b Give the value on the *x*-axis at *X* in Figure 4.2.

c Explain the steep rise in potential energy of the bond in the region R, as the inter-nuclear distance gets less and less.

Bonding diagrams

There are several possible diagrammatic representations of covalent bonds, each of which explains a particular aspect of bonding. These diagrams will be used throughout this chapter (and later in the book). Here they are used to explain the bonding in a hydrogen molecule.

Dot-and-cross diagram

In a covalent bond there are two electrons, one from each atom, so we represent them as a dot and a cross. They are effective as a bond insofar as they spend time between the nuclei, so we draw them between two 'H's, symbolising the two nuclei. We have thereby created a dot-and-cross diagram of the covalent bond in H_2 (Figure 4.3).

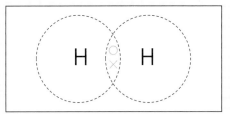

Figure 4.3 Dot-and-cross diagram of H_2. The dotted line circles are optional

Figure 4.4 Letters-and-sticks diagram of H_2

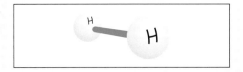

Figure 4.5 Balls-and-sticks diagram of H_2

Letters-and-sticks or displayed formula

This is a simplified version of the dot-and-cross format, in which a single line symbolises the shared electron pair (Figure 4.4). Another name for this style is 'displayed formula'.

Three-dimensional model

There are many styles of model available, and many of them are 'balls-and-sticks', in which each stick is a covalent bond (Figure 4.5). The third dimension is a great advantage in discussing true molecular shapes, as a two-dimensional flat paper representation is often rather misleading.

The value of three dimensions will be more evident when we have to discuss inherently three-dimensional molecules like methane (Figure 10.5, p. 193)

Box diagram

These emphasise the 'vacancies' for electrons on each atom, but they suffer from a drawback – they tend to view the situation existing on one atom at a time, rather than look at the bond as a whole. Figure 4.6 shows the box diagram for one of the hydrogens in H_2, showing the 'home' electron and the 'visitor' sharing the 1s box.

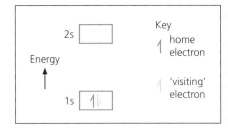

Figure 4.6 Box diagram of one H atom within an H_2 bond

Cloud pictures

These are pictures of the electron probability density in the covalent bond, and as such are the equivalent of the orbital shapes shown in Chapter 3, p. 41. Figure 4.1 is a cloud diagram, which illustrates the two orbitals coming close enough together to overlap.

Electron density maps

These are more 'real' than the other representations, in that they come directly from measurements, without a 'modelling' stage. When X-rays fall on a regularly arranged solid lattice of particles, they produce an interference pattern (Figure 4.7). This pattern becomes the raw data for a complex computation which is thankfully none of our business. The end product is a contour map like those of physical geography, except that the contours are electron densities. The map shows important information such as bond lengths and angles, as well as showing which atoms have the greatest density of electrons around them. This is useful evidence for bond polarity (Chapter 5, p. 73). The main drawback is that hydrogen atoms have a low electron density and therefore can only just be seen (Figure 4.8).

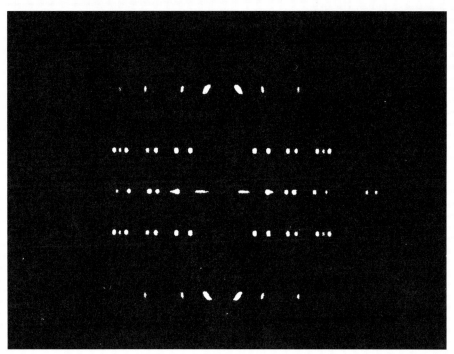

Figure 4.7 X-ray diffraction pattern produced when a beam of X-rays falls on a crystal of urea

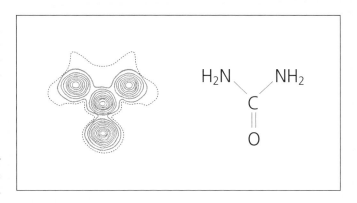

Figure 4.8 Electron density map of urea

Molecular orbital theory

This theory, although not part of the Advanced course, is worth a brief mention because it is the most complete description of how quantum mechanical principles and wave mechanics can be applied to molecules. It proposes whole new solutions to the problem of setting up a 'standing-wave' electron orbital, this time for the two-nucleus, molecular, environment. It has only been quantitatively developed for one molecule, the near non-existent H_2^+. There is a qualitative version which says that every time two atomic orbitals overlap, two new molecular orbitals are created, each capable of holding two electrons (Figure 4.9). The two electrons in a dihydrogen molecule therefore occupy the bonding molecular orbital, and we have a picture of a shared electron pair, not dissimilar to the output of all the other bonding models. When you delve deeper into the theory, and encounter difficult concepts like the 'anti-bonding orbital', you will probably be glad you need go no further.

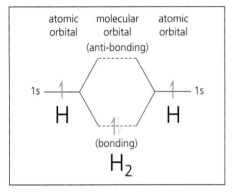

Figure 4.9 A taste of what molecular orbital theory is all about

4.3 Rules governing covalent bonds

You might assume that covalent bonding was a very simple matter – any two atoms come together and allow their outer electrons to 'mix'. This impression is misleading, since there are many rules and restrictions operating to govern covalent bonding. We cannot escape the quantum-mechanical restrictions which we met in Chapter 3. In the hydrogen case, the bond worked because there was a one-electron **vacancy** on each hydrogen

in principal quantum level 1. So each electron had an allotted place on its 'visited' atom.

But now consider the problems facing two helium atoms contemplating a covalent bond. They have twice as many electrons as hydrogen, but one big problem – *each 1s orbital is already full.* The only way of creating a vacancy would be to promote one electron on each atom to a 2s orbital, and have one molecular orbital made of overlapping 1s orbitals, and the other of 2s orbitals. You could even make a dot-and-cross diagram for such an arrangement (Figure 4.10).

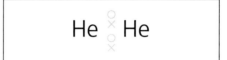

Figure 4.10 The non-existent He_2

However, look at the box diagram for helium (Figure 4.11). This reminds us that there is a large energy penalty to be paid for the $1s \rightarrow 2s$ promotion, which is much greater than most bond energies.

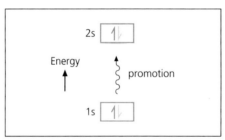

Figure 4.11 Box diagram, showing what a helium atom would have to do if it wanted to form He_2

As Figure 4.12 shows, it is more energetically profitable for the electrons to abandon bonding and drop back down to their ground states in the 1s orbital.

This leads to a generalisation. We must formulate a theory which recognises how hard it is to make use of already-paired electrons in bonding. The helium case suggests that promotion of electrons into higher vacant orbitals will not be useful, yet without that promotion already-paired electrons are out of the bonding game. We can arrive therefore at a model which says the limit of covalent bonding activity is reached when an element has

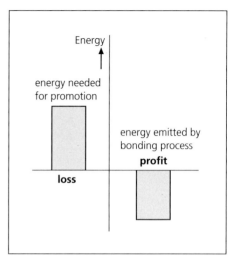

Figure 4.12 Another view of why He_2 does not exist

filled all the vacancies in its existing outermost principal quantum level. (Here 'filled' means filled by sharing, even though in reality this means offering a half-share in one of the atom's own electrons to get a half-share in one of the partner's.)

Let us apply this model to the water molecule, and use it to account for the water, H_2O formula. Oxygen has an electron arrangement of $1s^2 2s^2 2p^4$, so it has *two* vacancies in the $n - 2$ level (actually, in two 2p orbitals). This means it can accept (a share in) two incoming electrons, which means partnership with two separate hydrogen atoms. From the hydrogens' points of view, they fill their own 1s vacancy, just as they did when making hydrogen, H_2. Figures 4.13 and 4.14 show dot-and-cross diagrams and box diagrams of the bonding in water. We have not only accounted for why the molecular formula of water is H_2O, but we have arrived at the point at which we can predict how oxygen is going to behave in all its covalent bonding activities.

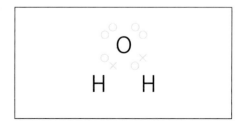

Figure 4.13 Dot-and-cross diagram of H_2O

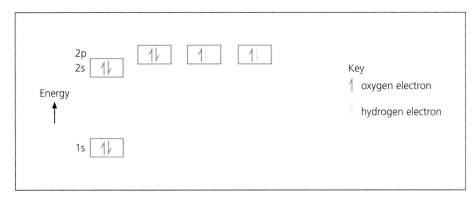

Figure 4.14 Box diagram of the state of occupation of the orbitals on the oxygen atom in H_2O

Question

2 a How many electrons does oxygen have (either by total possession or by sharing) in its outer shell (that is, the whole of principal quantum level 2) in the water molecule?

b How many electrons does a bare oxygen atom have in its outer shell?

c How many single electron-pair bonds will oxygen need to make with other elements in any bonding situation? (This number is called the **valency** of oxygen, from a word meaning 'going off in directions'.)

d By similar arguments, what would you predict as the natural valency of hydrogen?

The rule of eight

Familiarity with the use of this method of predicting bonding arrangements leads to some even broader generalisations. Elements all seem to be seeking filled outer shells, which in the earlier part of the Periodic Table means (at least a share in) eight electrons (apart from hydrogen, which needs two). So the **rule of eight** provides a guide to bonding and valency. In the following examples, you will have to arrange the bonding partners so that each fulfils the other's need for filled sets of eight (or two) electrons. You will need to know which Periodic Table group each atom comes from, in order to know how many electrons each starts out with.

Questions

3 Draw dot-and-cross diagrams for the following molecules. (The main skill needed here is the ability to count. Each atom must start with the correct number of outer electrons – nitrogen, N, 5; oxygen, O, 6; chlorine, Cl, 7; etc.)

Cl_2, NH_3, CH_4, CCl_4, HF, C_2H_6, CH_3OH, H_2O_2

(The last compound is called hydrogen peroxide, and its atoms are arranged HOOH.)

4 a What is the normal valency of the following atoms:

H, C, N, O, F, Ne?

b What pattern links the valency of each of these elements with its group number in the Periodic Table?

It is tempting to think from question 4 that we have resolved the question of valency for every element in the Periodic Table, each one inexorably linked to its group number. This is not so – my narrow choice of elements was because of the rule of eight, while a useful tool with considerable applicability throughout the Periodic Table, is only a hard-and-fast rule for elements as far as neon. (Notice too that I have yet to consider elements from the low number groups like I and II, since they employ a different method of bonding.)

The deviation from rule-of-eight behaviour occurs after neon, because the elements after neon have more available

vacant orbitals to 'play with'. The contrast between second- and third-row elements (between oxygen and sulphur, say) is due to the fact that there is no such thing as a 2d orbital, but there is such a thing as a 3d orbital. Examples from the chemistry of sulphur illustrate this. Sulphur has a number of compounds which obey the rule of eight, and yet also some which violate it.

Question

5 a To judge by its group number, what would be sulphur's rule of eight valency?

b Which of the following compounds violate the rule?

SF_6, H_2S, C_2H_5SH, S_2Cl_2 (this is arranged Cl—S—S—Cl)

How did sulphur manage to generate a valency of six in sulphur hexafluoride, SF_6? Box diagrams will help here. You will see that sulphur, which has the electron structure $1s^2 2s^2 2p^6 3s^2 3p^4$, has plenty of vacant orbitals available at energies not so very far above the filled ones (Figure 4.15). So the energy needed for unpairing electrons and promoting single electrons into orbitals like 4s and 3d is small. The crucial point is that this energy is *less* than the benefits in energy to be made by extra bonding. Figures 4.16 and 4.17 (overleaf) show a box and a dot-and-cross diagram for sulphur hexafluoride.

Questions

6 Can you now put into words why sulphur, from Group VI, can break the rule of eight and show valencies up to six, while oxygen, also in Group VI, cannot?

7 Would six represent a maximum valency for sulphur?

8 What general relationship might there be between the maximum valency of a rule-of-eight-breaking element and its group number, and why is there this relationship?

9 Now for a rather more subtle point – when sulphur breaks the rules it does it

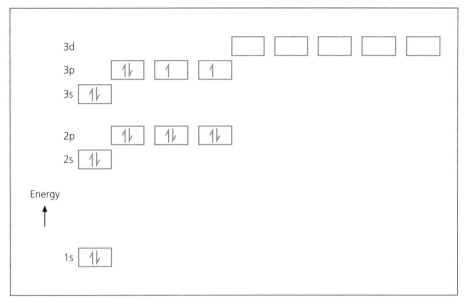

Figure 4.15 Box diagram of a lone S atom

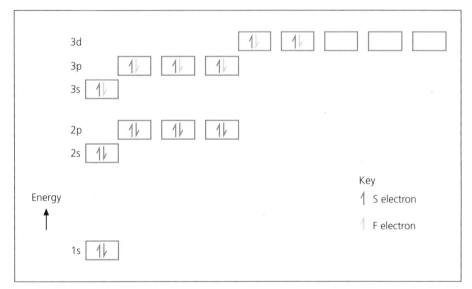

Figure 4.16 Box diagram of S atom in a sulphur hexafluoride, SF_6, molecule

Figure 4.17 Dot-and-cross diagram of sulphur hexafluoride, SF_6

with fluorine. There is no compound 'sulphur hexahydride', SH_6, for example. In fact most rule-breaking elements do their rule-breaking in compounds in which either fluorine or oxygen is the other element. Do you have any idea why this is?

Unpairing electrons within a principal quantum level

So far we have accepted that carbon is a reliable obeyer of the rule of eight. It has four outer electrons of its own, and therefore needs to make four electron-pair bonds to fill its outer shell. This happens in the vast majority of carbon's compounds with other elements. Yet this conformity to the rules conceals a little internal arrangement which in some ways resembles the tricks which sulphur gets up to. The fact is that the outer electron arrangement in carbon is $2s^2 2p^2$. The box diagram is shown in Figure 4.18. From this point of view carbon has only two unpaired electrons, and seems incapable both of getting up to eight and of displaying a valency of four.

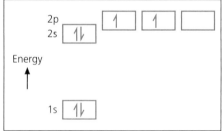

Figure 4.18 Box diagram of a lone carbon atom

Question

10 How does carbon gain access to its full valency of four, and why is it energetically worthwhile?

Multiple bonds

Sometimes a pair of atoms can share more than one pair of electrons. Consider, for example, the case of the dioxygen molecule, O_2. Each oxygen atom needs to acquire a share in two more electrons – that is, two single bonds – so they use the neat ploy of making two bonds with each other. See Figure 4.19 for a dot-and-cross version of the oxygen molecule. The $O{=}O$ bond is more than twice as strong, and a good deal shorter, than the $O{-}O$ bond, as befits a construction made from four electrons rather than two. Even though the style of bonding is new, the valency of oxygen is still its predictable two, as shown in the letters-and-sticks diagram (Figure 4.20). Table 4.1 shows the strengths and lengths of some multiple bonds.

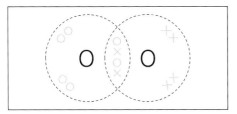

Figure 4.19 Dot-and-cross diagram of an oxygen molecule

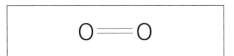

Figure 4.20 Letters-and-sticks diagram of oxygen

Table 4.1 Some bond lengths and energies

Bond	Energy needed to break it (kJmol^{-1})	Length (nm)
O—O	144	0.148
O=O	498	0.121
N—N	158	0.145
N=N	410	0.120
N≡N	945	0.110
C—C	347	0.154
C=C	612	0.134

Multiple bonds can be either double or triple, and can happen equally probably to rule-of-eight-obeyers and rule-of-eight-breakers. A multiple bond can even occur *between* a rule-breaker and a rule-obeyer. Bear in mind in answering the following question that every single bond or single part of a multiple bond will be constructed from a shared pair of electrons, one from each partner. Remember also that elements up to neon have completely reliable rule of eight valencies, irrespective of whether those valencies are part of a multiple-bond or single-bond arrangement:

H 1, C 4, N 3, O 2, F 1

Questions

11 Draw dot-and-cross diagrams for the following multiple-bonded structures (both of which obey the rule of eight): nitrogen, N_2 (this is the ordinary nitrogen gas in air); carbon dioxide, CO_2.

12 The following structures have mixtures of single and multiple bonds, but all obey the rule of eight. The atoms are arranged in the order shown. Draw dot-and-cross diagrams (the first one is presented as an example in Figure 4.21): H_2CCH_2 (ethene), HONO (nitrous acid), HCN (hydrogen cyanide), H_2CO (methanal), $(HO)_2CO$ (better known as H_2CO_3, carbonic acid).

Figure 4.21 Dot-and-cross diagram of ethene

13 The following structures have multiple bonds *and* break the rule of eight. Draw dot-and-cross diagrams (again the first one is done for you, in Figure 4.22): SO_2 (sulphur dioxide), SO_3 (sulphur trioxide), $(HO)_2SO_2$ (better known as H_2SO_4, sulphuric acid).

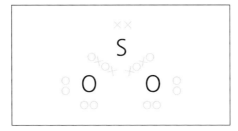

Figure 4.22 Dot-and-cross diagram of sulphur dioxide

4.4   Dative covalency

So far, we have investigated covalent bonds which consist of a shared pair of electrons, one electron coming from each bond member. However, there are some bonds in which *both* the electrons come from one donor. The word for this type of bond is **dative** (from a Latin root to do with giving).

To find out more about dative covalency, we will look at the molecule H_3NBF_3. To appreciate the true oddness of this molecule we must take a brief detour into the chemistry of the element boron, and also look at the problems of covalent bonding in the whole of Group III.

Question

14 The outer electron structure of boron is $2s^2 2p^1$.

a   Draw a box diagram for these electrons.

b   Is unpairing of the $2s^2$ pair energetically feasible?

c   What is the maximum number of electrons which boron can have access to by conventional electron sharing (that is, its own plus other elements)?

d   Is this a number less than, or equal to, eight?

e   Are there any vacant orbitals in the outer (valence) shell of boron, even when it is fully covalently bonded?

f   Draw a dot-and-cross diagram of boron trifluoride, BF_3.

The previous question should have convinced you the boron atom, having exhausted the possibilities for conventional covalent bonding, still has one totally vacant orbital, and so could still be involved in dative covalent bonding. Boron accepts a *pair* of electrons in these cases. So any molecule in which there is an atom with a **lone pair** of electrons – that is, a pair that were 'original residents' on their atom and have so far played no part in bonding – will be a candidate for the job of filling the vacant site on boron.

15 Give two examples from previous questions of molecules which include at least one atom with a lone pair.

The lone pair on the nitrogen atom in nitrogen–hydrogen compounds is one of the most active in this donating role. Ammonia, for instance, reacts readily, offering its lone pair as occupants of the double vacancy on the boron trifluoride, BF_3, molecule. So the formation of the H_3NBF_3 compound is due to a dative covalent bond between the N and the B. The symbol for a dative covalent bond is →, and the direction of the arrow indicates the direction of the 'gift' of electrons (Figure 4.23). However, once given, there is nothing to distinguish those electrons from the others, so a dative covalent bond looks just the same as a conventional covalent bond on an electron density map.

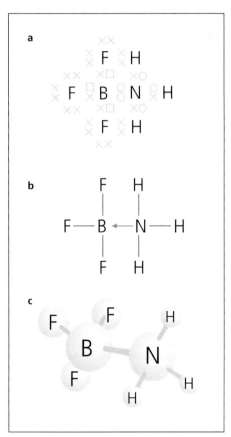

Figure 4.23 Three representations of the H_3NBF_3 addition compound: **a** dot-and-cross, **b** displayed formula and **c** balls-and-sticks

16 Nitric acid is a common enough molecule, yet its bonding is quite difficult to analyse. It can only be explained by proposing a structure which includes a dative covalent bond.

a Redraw, from question 12, the dot-and-cross diagram of the rule-of-eight-obeying nit*rous* acid, HONO.

b Nitric acid, $HONO_2$ (more conventionally written HNO_3), contains an extra oxygen, and yet nitrogen undoubtedly belongs in the group which unswervingly obeys the rule of eight. Try to identify therefore what is wrong with attaching the oxygen on a normal double bond, as follows:

c Can you form the nitric acid molecule without breaking the rule of eight, by employing instead a dative covalent bond? Draw a dot-and-cross diagram to prove it.

d Draw the letters-and-sticks equivalent of (c).

e Bearing in mind that bond lengths show up on electron density maps, why would an electron density map of the nitric acid molecule distinguish between (c) and Figure 4.24?

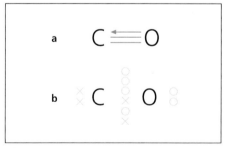

Figure 4.24 Two versions of bonding in a molecule of carbon monoxide: **a** displayed formula and **b** dot-and-cross diagram

There are other important examples of dative covalent bonds, especially in the chemistry of acids and bases, but they occur in ionic structures, and they will be discussed after ionic bonding. To end this section, we will look at the peculiar bonding in the carbon monoxide molecule (Figure 4.24). As you can see, it manages to obey the rule of eight by the unlikely route of mixing a conventional double bond with a dative covalent bond. Yet remember that in terms of bond length and electron density we are looking at a carbon–oxygen triple bond.

17 Which other molecule has exactly the same number of electrons in the same places, but with a conventional triple bond and no dative covalent bond? (These two molecules are said to be **isoelectronic**.)

4.5 The shapes of molecules

Nature has dealt a piece of good fortune to the student who embarks on a study of molecular shape. Despite the complexity of shapes of atomic orbitals, especially the later ones like 3d or 4f, the shapes of the actual molecules held together by the overlapping of these orbitals turn out to be governed by rules of great simplicity. Let us illustrate the point via a study of methane.

Figure 4.25 is a box diagram of the outer electrons of carbon, after they have gone through the unpairing process which was uncovered in question 10.

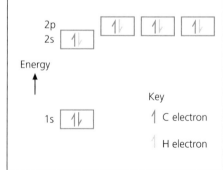

Figure 4.25 Box diagram of the carbon atom in a methane molecule, CH_4

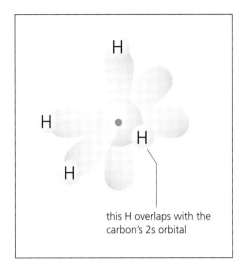

Figure 4.26 A non-existent structure for methane, if carbon used its 'pure' 2s and 2p orbitals, which it does not

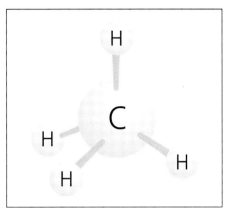

Figure 4.27 The real structure for methane – perfect regular tetrahedron

The hydrogen electrons have been inserted, and these will be offering up their 1s orbitals as their part of the sharing deal. You can see that the hydrogen atoms would be sharing with one s-shape carbon orbital, and three p-shape ones. This suggests that the shape of such a molecule might look like Figure 4.26. Three C—H bonds would be at 90 ° to each other, since three of the hydrogens would be sharing the three 2p orbitals, which are themselves at 90 °. The fourth hydrogen atom could go almost anywhere, since there is no directional constraint on the overlap of the hydrogen 1s and carbon 2s spherical orbits. It has been drawn trying to 'sit' between the other-end lobes of the 2p orbitals.

However, in reality methane is not like this. In fact, as shown by electron density maps, methane is a perfect regular tetrahedron (Figure 4.27). The theory chemists use to explain how Nature turns three p orbitals and an s orbital into the four points of a tetrahedron is called **hybridisation** – but it is not part of the Advanced course.

In fact, molecular shapes can be predicted and explained by a much simpler theory, based on the idea that electrons repel each other. The student has to decide:

• how many electron clusters are repelling each other at the atom in question

• which geometric figure is represented by the number of clusters above.

In the case of methane, the answer to the first question above is *four*, and geometry tells us that four things trying to stay apart will make up a tetrahedron, in which bond angles will all be 109.5 °.

Questions

18 Imagine each electron cluster to be like a spoke. The spokes radiate out from the atom at the centre of the molecule, and carry other atoms on their ends like the hydrogen atoms in methane. What geometric shapes would be obtained by the following numbers of spokes all repelling each other equally? (In each case give the bond angle between the spokes.)

2, 3, 5, 6.

19 Suppose the molecule has more than one centre? The answer is that it would then have a composite shape, with each centre following the above repulsion rules. This sounds complicated to explain, but is easy by example. What shape would you predict for the molecule that is a sort of double-headed methane, namely ethane, H_3CCH_3? It may help to do a dot-and-cross diagram first, so as to clarify the number and positions of the electron pairs.

However, not every case is quite as clear-cut as methane. There are two complicating features, and the second one depends on the first.

Repulsive 'bits'

The first complicating feature is expressed in the question 'What constitutes an electron cluster?' The answer is that an electron cluster can mean any one of the following, which (to borrow from one of my own students) we will call 'bits'. Any of these electron clusters is one bit:

• one single bond pair

• one lone pair

• one double bond 'foursome'

• one triple bond 'sixsome'.

It seems odd at first that a triple bond only counts as much as a single bond, but the crucial thing is that both single and triple bonds are independent movers in the repulsion calculation. The three single bonds within the triple bond are not independent movers, because they are not free to flee each other.

Example: Molecular shapes

Predict the shapes of the following molecules:

a water

b carbon dioxide.

Answer: **a** The dot-and-cross diagram of water (Figure 4.28) shows two single bond pairs and two lone pairs of electrons. So we can see that water is truly a four-bit molecule. So it belongs to the same set of shapes as methane, namely tetrahedral, bond angle 109.5 °.

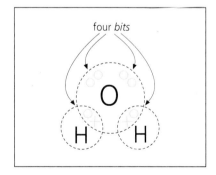

Figure 4.28 Counting the *bits* on water

However, there is one difference between water and methane, and that is that two of the spokes (the lone pairs) carry no atom. Hence water is a tetrahedron with two 'invisible' points (invisible, that is, unless you count a cloud of electron density). If you take away two points from a tetrahedron you are left with the final answer to the question of the shape of water, namely:

Bent, HOH, angle 109.5 ° (Figure 4.29)

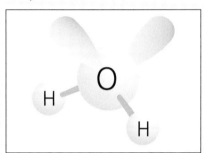

Figure 4.29 The four *bits* again, spread tetrahedrally

(This angle is a few degrees out, but we will come to that shortly.)

Answer: **b** The dot-and-cross diagram of carbon dioxide (Figure 4.30) shows two double bonds and no lone pairs. So carbon dioxide is a two-bit molecule. So, as you showed in your answer to question 18, the two bits will be at 180 °, and carbon dioxide will be **linear** (Figure 4.31).

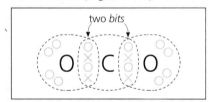

two *bits*

Figure 4.30 Dot-and-cross diagram of CO_2

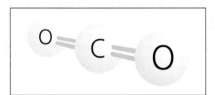

Figure 4.31 Balls-and-sticks model of CO_2

As you can see from the examples, it is best to begin with a dot-and-cross diagram of the molecule in question, then count up the 'bits', then select the appropriate shape, and finally mentally remove any 'invisible points' produced by lone pairs. See if you can carry through all

four of those stages in the next question. (Remember that to do a dot-and-cross diagram you need to know what Periodic Table group the atoms are in.)

Questions

20 Work out the shapes and central bond angles of the following molecules:

H_2C══O, BF_3, PCl_5*, NH_3, SO_2*, SO_3*, HONO (nitrous acid), $HONO_2$ (nitric acid).

(* rule-of-eight-breakers. In the last two find the shape at the nitrogen atom – just treat the hydroxide, OH, as a single group.)

21 Here are two harder examples. In recognition of their hardness the question has been switched so that you have to explain shape rather than predict it.

a Bromine trifluoride, BrF_3, is T-shaped (Figure 4.32). Why?

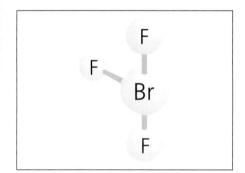

Figure 4.32 See question 21a

b Xenon tetrafluoride, XeF_4, is square-planar (Figure 4.33). Why? (It *is* the noble gas xenon, and it *does* form a few compounds, confined to fluorides and oxides.)

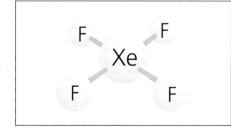

Figure 4.33 See question 21b

Small distortions of shape

This is the second complicating feature referred to earlier. So far we have got to the stage where we can allocate any molecule to one of the shape sets, like 'triangular planar, 120 °' (three bit), or 'tetrahedral 109.5 °' (four bit), before if necessary removing points to take account of the lone pairs. However, as mentioned in the water example (p. 57) there are sometimes deviations of a few degrees from the bond angles predicted by the repulsive-bit method.

Fortunately, these deviations are not completely random. The simple fact is that some of our repulsive 'bits' are more repulsive than others. For example, a lone pair is slightly more laterally spread out than a bond pair (perhaps because it has no other atom to orbit), and therefore is more repulsive. In water, there are two lone pairs and two bond pairs, so the two bond pairs will be squeezed together a trifle by the extra repulsions of the lone pairs. The actual HOH bond angle in water is 104.5 °.

Question

22 Use this theory of small distortions to account for the gradually changing central bond angles in the following set of molecules (all of which are members of the four-bit set).

methane, CH_4	109.5 °
ammonia, NH_3	107 °
water, H_2O	104.5 °

So far we have looked at small distortions caused by the unequal repulsive pushes of lone pairs and single bond pairs. What about when one of the bits is a multiple bond? It seems reasonable to assume that a double bond, containing as it does more electron density, will demand more space to itself than a single bond pair. This prediction is supported by the bond angles in ethene, in which the bond angle between the two single bonds is forced to shrink a little from its calculated value of 120 °.

H   H
  \   /
118° C══C 121°
  /   121° \
H   H

Question

23 Predict approximate values for the bond angles in the methanal molecule, whose letters-and-sticks formula is displayed below.

$$
\begin{array}{c}
\text{H} \\
\diagdown \\
\quad \text{C} = \text{O} \\
\diagup \\
\text{H}
\end{array}
$$

The final repulsion interaction which we need to consider is that between double bonds and lone pairs. They are both more repulsive than a single bond pair, but which 'out-repels' the other? The next question is about this situation in the case of sulphur dioxide.

Questions

24 Sulphur dioxide, SO_2 is a bent molecule, O==S==O bond angle exactly 120 °. Draw a dot-and-cross diagram. Then decide what the 120 ° angle tells you about the relative sizes of the repulsive effects of double bonds and lone pairs.

25 Predict the bond angles in the sulphurous acid molecule. The displayed formula (letters and sticks) is shown as follows.

$$
\begin{array}{c}
\text{H} - \text{O} \\
\diagdown \\
\quad\quad \text{S} = \text{O} \\
\diagup \\
\text{H} - \text{O}
\end{array}
$$

(Displayed formulae are not very helpful in questions like this, because they only show which atom is joined to which. They are locked into two dimensions on the page, and so give deceptively flat indications of shape, and what is more they do not show lone pairs. This is a long-winded way of saying this is quite a hard question.)

4.6 Delocalisation

The underlying theme of this chapter could be seen in retrospect as 'rules and their exceptions'. Each time we have met a usable pattern of chemical behaviour, it has been necessary to issue a list of 'get-out' clauses. First there was the rule of eight, which had plenty of violators, and then there was the idea of the one-electron-per-partner-per-bond, which was undermined by the recognition of the dative covalent bond. Now we look at two more principles both so far untouched by exceptions:

• every single bond features a pair of electrons from somewhere (and doubles use two pairs, and triples three)

• electrons stay around and between the atoms which donate and share them.

As we shall see these rules break down in certain cases.

It turns out that not all electrons are located permanently between the same two atoms, and the result is that the average number of bonding electrons need *not* be a multiple of two. The idea of delocalised bonding electrons came late in the study of bonding, and arose from the study of benzene.

Benzene is an organic molecule whose molecular formula is C_6H_6. It presented intellectual problems to chemists right from its initial identification, and a major breakthrough was achieved in the nineteenth century when Kekulé realised the molecule was a ring. (A fuller version of this story appears in Chapter 10, p. 220.) The structure which Kekulé proposed nevertheless followed normal bonding rules, with alternating double and single carbon–carbon bonds around the ring (Figure 4.34).

However, electron density maps of molecules containing the benzene ring show that Kekulé's structure cannot be exactly correct (Figure 4.35).

Question

26 Which feature of Figure 4.35 is incompatible with alternating double and single bonds?

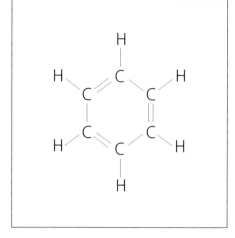

Figure 4.34 Kekulé's formula for benzene

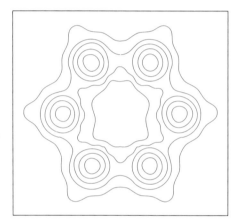

Figure 4.35 Electron density map of benzene

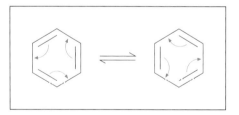

Figure 4.36 Resonance model for benzene

The first model which was used to explain the real structure of benzene was the **resonance** model. Benzene was considered to be in a state of rapid oscillation between two equally likely structures (sometimes called **hybrids**). To put it another way, each carbon–carbon bond was rapidly alternating between double and single-bond status. The curly arrows in Figure 4.36 show how the second-bond electrons would have to move to bring about the change. According to the resonance model, what the X-ray pictures were showing was an average over time of the electron density as the bonds repeatedly switch status.

This oscillating molecule model was not entirely convincing. Evidence that some of the electrons in benzene were free to roam, creating what amounted to tiny electric currents in the ring, conflicted with the model. The description of benzene we favour now is that the six electrons from the three supposedly double bonds are actually **delocalised** around the ring. That is to say, each carbon–carbon bond has three electrons in it at any one time, and the third electron in each bond is one of six electrons which are *free to visit all the atoms on the ring*. Figure 4.37a shows a dot-and-cross picture of the localisation, in contrast to the delocalised version in Figure 4.37b.

Calculations in Chapter 10 (p. 220, question 57) show that benzene is more than 200 kJ mol^{-1} lower in energy (more stable) than if it did not have delocalised bonds. Let us consider where this stabilisation energy might have come from. One possibility is that, although the same number of bonding electrons is involved in the ring, they are more spread out when delocalised. This therefore *reduces the electron–electron repulsion factor*.

The structure below shows the well-known symbol for benzene which expresses delocalisation in the ring.

(To be consistent with other symbolisms, it would be better to draw it as below, where the dotted line indicates something greater than single, but less than double bonding. However, convention for the present uses the ring-in-a-hexagon symbol.)

The older resonance theory still has its uses. For example it provides a handy diagnostic test for when to expect to find delocalisation in a molecule. A set of steps to test for potential delocalisation, using nitric acid as an example follows.

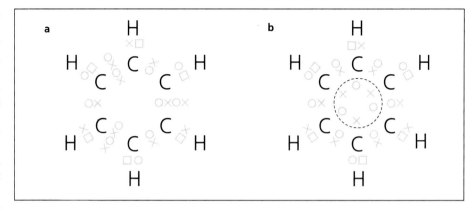

Figure 4.37 Dot-and-cross diagrams of **a** localised Kekulé benzene and **b** delocalised real benzene

Example: Testing for delocalisation

1 Draw out a displayed diagram or dot-and-cross diagram of the molecule in question, with all the bonds localised between atoms.

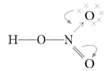

2 See if you can move (a minimum of) two pairs of electrons, so that on the one hand a lone pair comes in to a bonding position to create the second part of a double bond; while on the other hand proper valency rules are preserved by the second part of a double bond simultaneously becoming a lone pair. (This is a complex description, but the curly arrows in the above structure should clarify it.)

3 Alternatively, see if you can move second parts of double bonds around a ring (as in benzene) while preserving normal valencies. (Look back to Figure 4.36.)

4 Check the new structure to see if it is likely to be of similar stability and viability to the one you began with.

5 If it is, then you can assume that the real molecule will be a delocalised version of the localised structures.

Question

27a Using a data book, make estimates of the bond lengths and bond energies of the two N——O bonds involved in the delocalised system in nitric acid.

 b Why can the third N——O bond not get involved in delocalisation?

Hint: If you are ever faced with explaining why a molecule is more symmetrical than can be deduced from its standard displayed formula – suspect delocalisation. We will revisit the subject of delocalisation after the ionic bond has been introduced.

A note on 'curly arrows'

Figure 4.36 (p. 59) introduced the 'curly arrow' symbol. You will meet it again, as it features prominently in the subject of reaction mechanisms. The curly arrow nearly always represents *the movement of a pair of electrons*. The movement in question can be *between* molecules or *within* them, and it can be a lone pair moving to become a bond pair (as in the second formula in the example opposite), or a bond pair becoming a lone pair (as in the second formula in that example), or even a double bond changing places, as in benzene.

4.7 Electron-cloud pictures

The cloud way of representing bonds has been neglected since our early discussions on the hydrogen molecule. The problem is that a full understanding of an electron-cloud picture (which is really a picture of electron probability distribution) demands some acquaintance with **hybridisation theory**, a set of concepts which is not in the Advanced syllabus.

There is a mystery lying at the heart of the subject of molecular shape (at least, shapes more complex than hydrogen, H_2). How can carbon, for instance, offer vacancies to prospective partners in one s orbital and three p orbitals, and yet end up with the four bonds sticking out tetrahedrally? Equally oddly, when carbon includes a double bond in its bonding arrangements, why do all the bonds end up at 120 ° to each other?

In Section 4.5, p. 56, you were advised to forget the shapes of the atomic orbitals; they apply only to single atoms in the gaseous state. Once into molecules, the overlapped orbitals follow the principle of mutual repulsion, and stay as far apart as possible (give or take the odd small distortion).

And yet it is still worthwhile for students to have a more pictorial view of bonds – more pictorial, that is, than the skeletal scaffolding of the three-dimensional pictures, and the coded symbolism of the dot and cross and displayed formulae. The electron-cloud model fits here, and provides an explanation for a number of other facts. If we omit the underlying references to hybrid orbitals, the best way for you to learn about this model is as (another) set of rules.

Tetrahedral arrangements

As you have seen, single-bonded molecules obeying the rule of eight belong to the tetrahedral-shape group, for the reason that eight electrons make four 'bits'. If we begin with methane, as perhaps the simplest member of the group, then the molecular orbitals in methane are seen as four tetrahedrally arranged directional clouds (Figure 4.38). The hydrogen 1s clouds and the four carbon clouds (actually called sp^3 hybrids) are physically *overlapped*, and it therefore becomes easy to imagine the shared electrons drifting in and out of each other's nuclear influence. If anything more complex than hydrogen is on the other end, then another set of tetrahedra is needed. See Figure 4.39 for

ethanol, where three tetrahedral arrangements are overlapping. Even in molecules without the familiar pattern of 'four bits, arranged tetrahedrally', the electron-cloud picture still applies, with adjustments of bond angle; for example, see boron trifluoride, BF_3, in Figure 4.40.

All these bonds have one unifying stylistic feature – the directional lobes on participating atoms overlap in a fashion

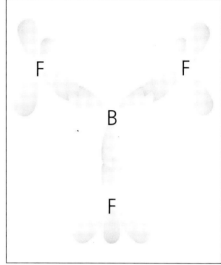

Figure 4.40 Electron-cloud picture of boron trifluoride

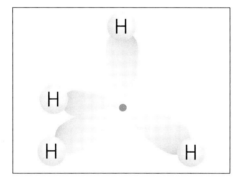

Figure 4.38 Electron-cloud picture of methane

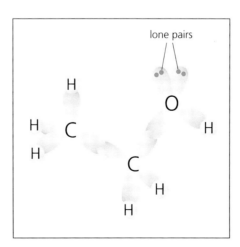

Figure 4.39 Electron-cloud picture of ethanol, CH₃CH₂OH

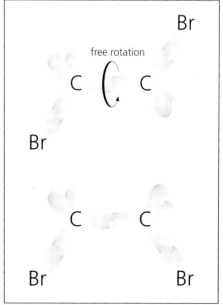

Figure 4.41 Free rotation about the axis of a σ-bond (sigma bond)

that might be called nose-to-nose. The important thing about this style of overlap is that it allows bits of molecule on either end of the bond to *rotate freely*, without breaking or even weakening the bond. For example, the two drawings in Figure 4.41 represent the same molecule, because the molecule in one drawing can be converted into the other by rotation alone. Displayed formulae on flat paper are only too effective at disguising the equivalence of different orientations, as Figure 4.42 shows. Any bond which allows this freedom of rotation with no bond-rupturing is called a σ ('**sigma**') bond. All single bonds (apart from a few 'freaks' from boron hydride chemistry) are of this σ type.

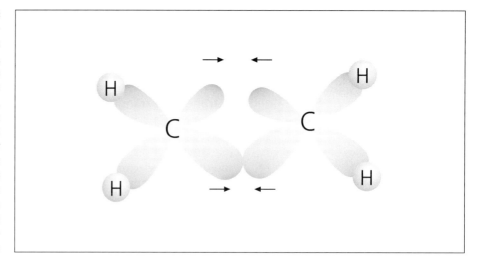

Figure 4.43 Two 'bent' style overlaps getting ready to make a double bond. A credible idea but *not true*

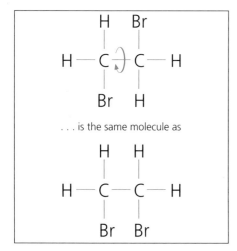

Figure 4.42 Don't be deceived by apparently different molecules. Rotation about the C—C bond renders them identical

Double bonds

Eth*ane*, C_2H_6, is a typical σ-bonded double tetrahedron. Eth*ene*, C_2H_4, has two less hydrogen atoms and a double bond between the carbons. It is tempting to assume that the removal of the hydrogen atoms has prompted the two tetrahedra to perform a mild reorientation, so that the two newly de-partnered carbon lobes can overlap each other, and thereby restore themselves to rule of eight correctness. A double bond, by this picture, is seen simply as two somewhat bent single bonds (Figure 4.43).

This model, like any model seeking to describe what double bonds are like, has to fit the evidence. One of those pieces of evidence is the fact that doubly substituted ethene derivatives (such as

$C_2H_2Br_2$) really do show *two* different versions. For example, the two molecules below have different properties and are genuinely different substances. Such pairs of molecules are called **isomers** of each other; their interconversion requires actual bond breaking.

The occurrence of isomers in the derivatives of ethene implies that there is a fundamental difference about the extent to which the central C═C bond tolerates rotation, compared to its C—C equivalent.

Question

28a What is this fundamental difference?

b Does the model for the C═C bond in Figure 4.43 (the two-bent-single-bonds model) fit in with this experimental evidence about isomers, and why?

c The bond lengths and strengths of the C—C and C═C bonds are given in Table 4.1 (p. 55). Can the two-bent-single-bonds model account for why the C═C bond is shorter, and less than double the strength?

You will have seen from your answers to (b) and (c) that the two-bent-single-bonds model does well at explaining several features of C═C bonds. The problem with accepting the two-bent-single-bonds picture comes from the measurement of bond angles, made possible by X-ray and electron diffraction. The HCH bond angle in ethene is only a small distortion away from 120°, at 118°.

Question

29a What would the HCH bond angle in ethene be on the two-bent-single-bonds model?

b Is it possible to retain the two-bent-single-bonds model, and explain away the difference in bond angle between your answer in (a) and 118° by holding a distortion responsible? Justify your answer.

The theory that delivers all the agreements with real bond lengths and angles uses the idea that the second bond is very different from the first. The framework of the molecule, in this version, is made from three directional lobes on each carbon, using σ bonding. So far we have used only three of the four electrons on carbon. The bonds would be at 120°, by repulsion theory. But where is the fourth electron on each carbon?

The answer places them not in directional-lobe type orbitals, but in ordinary

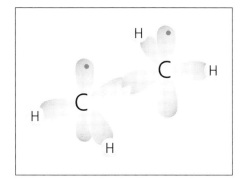

Figure 4.44 Electrons (shown as •) waiting in pure p orbitals to complete the cheek-to-cheek π-bond in ethene

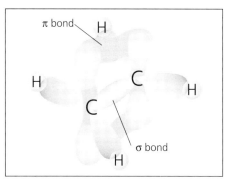

Figure 4.45 The double-lobed π-bond contains only *two* electrons and only counts as one bond

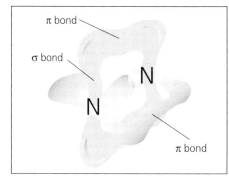

Figure 4.46 A triple bond (N₂). One σ-bond and two π-bonds (total six electrons)

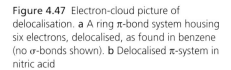

Figure 4.47 Electron-cloud picture of delocalisation. **a** A ring π-bond system housing six electrons, delocalised, as found in benzene (no σ-bonds shown). **b** Delocalised π-system in nitric acid

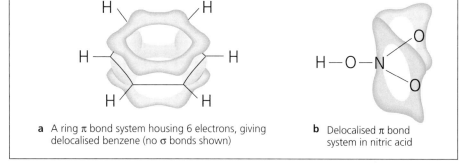

a A ring π bond system housing 6 electrons, giving delocalised benzene (no σ bonds shown)

b Delocalised π bond system in nitric acid

2p atomic orbitals (Figure 4.44). These two orbitals then overlap in a way that can only be described, in contrast to the nose-to-nose style of the σ-bond, as cheek-to-cheek (Figure 4.45). We use the Greek letter π (pi) for this bond. All the double parts of the double bonds are constructed in this style, and triple bonds have two sets (Figure 4.46).

Delocalised systems

Delocalised systems, as we have seen, involve movements of electrons in what otherwise would have been lone pairs and what otherwise would have been double bonds, or between alternate double bonds (Figures 4.36 and the nitric acid structures on pp. 59–60). The electron-cloud model does rather well in picturing these delocalised bonds, for it sees the delocalised electrons migrating freely within a π-shape molecular orbital which runs the entire length of the zone of delocalisation. It thus captures the spirit of what delocalisation is like rather better than do dot-and-cross, curly-arrow, or dotted-line models. Figure 4.47 shows the π electron cloud in the two molecules in which we have so far found delocalisation, namely benzene and nitric acid.

4.8   The ionic bond

Consider a possible covalent bond between lithium and fluorine. Their electron configurations are as follows:

Li: $1s^2 2s^1$

F: $1s^2 2s^2 2p^5$

The element fluorine offers a very tempting vacancy for a prospective incoming electron – 'prime site, featuring p-style orbital, and (most crucially) featuring excellent views of well-supplied oppositely charged nucleus'. There are no vacancies in any neutral atom anywhere in the Periodic Table to compare with that single vacancy on fluorine.

The reasons are not too hard to see, and they apply to all the members of fluorine's group – every Group VII element's vacancy is offering the maximum possible nuclear pull within that row of the Periodic Table. One place to the left (to Group VI) and there is one less proton on the nucleus, hence less pull.

Question

30 (See if you can carry on this sequential argument: complete the sentences which follow, from your memory of electron arrangements in Chapter 3.)

a One place to the right (in Group O), and the nuclear pull has increased but there are no . . .

b Two places on from Group VII (into the next row and Group I), the number of protons on the nucleus has increased still further and there are vacancies, but they are . . .

If the above sentences make the case for each Group VII element having the most welcoming vacancy in each row, then within Group VII the fluorine vacancy is the ultimate in warm welcomes, because it has the least distance to the nucleus and minimal inner-shell shielding.

Figure 4.48 shows a possible dot-and-cross diagram of lithium fluoride. The lithium is offering its grateful electron an experience of spending some time around the fluorine nucleus. However, at this point things go wrong. What can the lithium offer the fluorine electron? The answer is, it can offer the fluorine electron a vacancy in a shell equivalent to

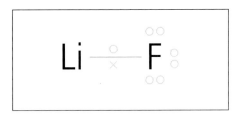

Figure 4.48 Another diagram of a non-existent arrangement. A proposed covalent bond in lithium fluoride

that from which it has just come (principal quantum level 2), but with *only 3 protons on the nucleus instead of 9.*

As far as covalent bonding is concerned, it simply does not make energetic sense for a fluorine electron to be wrenched from its stable home position to go and orbit the lithium nucleus in the $n = 2$ shell, where the nuclear pull is so enfeebled. However, there is another plan which makes a lot of energetic sense. The very feebleness of the nuclear pull on the outer lithium electron means that it is no great sacrifice for it to be completely lost, and to spend all its time on the fluorine atom (Figure 4.49).

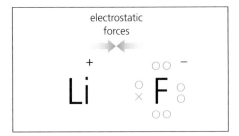

Figure 4.49 This arrangement *does* happen. Ionic bond between Li⁺ and F⁻ ions

There is an added advantage to this ploy – the lithium and fluorine atoms have now acquired fully fledged single charges, due to the complete loss or gain of an electron, while the nuclei (of course) remained unchanged in number of protons. The two particles are now no longer lithium and fluorine atoms, but have become Li⁺ and F⁻, a cation and an anion respectively. Because they are ions of opposite charge, they are attracted to each other by electrostatic force, and so we have the creation of a new type of bond. This is a bond which relies not on electron sharing, but on complete loss and gain of electrons. This type of bond is called an **ionic bond**.

Note on nomenclature
Any charged particle is called an ion, but the 'cat-' and 'an-' prefixes indicate the direction of migration of an ion if electrolysis were to occur. Since opposite charges attract, it follows that cations are positively charged ions, and anions are negative.

4.9   The influence of bonding style on physical properties

Ionic bonding has one very marked effect on the overall structure of a material. Since electron donation and reception are complete, any *directional nature* has disappeared from the ionic bond. So there is no discrete molecule with a beginning and an end. An ion deploys its electric field evenly in the space all around it, so other ions of opposite charge can be attracted on all sides. This means that when a mass of ions get together, they arrange themselves in continuous three-dimensional lattices of alternating ions, all held together by electrostatic force.

Covalently bonded substances, on the other hand, may contain strong bonds, but there is nothing substantial holding one molecule to the next because covalent bonds lie in particular directions. A mass of covalent molecules therefore is a mass of alternating strong attractions and weak ones.

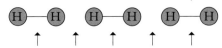

strong weak strong weak strong

This has a big impact on physical properties. A collection of hydrogen molecules (covalent) ceases to be a solid at 14 K, but does not break into single atoms until about 1500 K. In contrast, lithium fluoride LiF (ionic) resists melting up to 1118 K.

Also, the two major groups of bonded structures show marked differences in

their behaviour towards attempted electrolysis. This will be discussed more fully in Chapter 16, but basically ionic substances will undergo electrolysis (providing the ions are mobilised by melting or dissolving), while covalent ones will not.

4.10   Formulae of ionic compounds

Ionic bonds are possible whenever the two elements in the prospective bonding relationship have very unequal abilities to entice in and hold on to electrons. Thus ionic bonding is not the exclusive preserve of combinations of elements from Groups I and VII.

Question

31 Generalise about when you would expect a given pair of elements to favour ionic bonding. You will need to consider the sorts of pairings which would fit the description in the first sentence of this section. You might want to use data on ionization energies to guide your judgement. Give your answer in terms of Periodic Table groups.

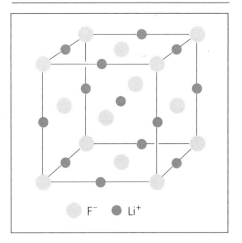

Figure 4.50 A typical ionic structure – the lithium fluoride crystal lattice (identical to the sodium chloride structure, see Figure 4.52)

Lithium fluoride turns out to be a 1:1 compound, and the reasons are entirely determined by which periodic groups lithium and fluorine come from. There is the perfect match between the number of loosely held electrons on the lithium atom (1), and the number of desirable vacancies in the outer shell of the fluorine atom (1) (Figure 4.50). As you can see, this combination of circumstances would be the same for any ionic bond between a Group I element and a Group VII one. But as your answer to question 31 showed, other groups can do it too, and there is no necessity for 1:1 stoichiometry. It is quite feasible for an element with two vacancies – oxygen, say – to accommodate lithium electrons, only it will need *two* lithium atoms to supply them. The formula of lithium oxide is therefore Li_2O (Figure 4.51).

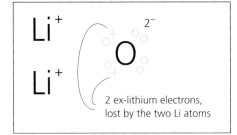

Figure 4.51 A 2:1 ionic compound

This line of thinking needs only a small adjustment to take in situations in which the cation is the species which causes and introduces the 'two'. Members of Group II, for instance, have two outer electrons which are not too difficult to remove, and therefore whose removal is energetically worthwhile if ionic bonding is the reward. Again, in parallel with the lithium oxide case, this does not mean they are banned from combining with, say, Group VII anions. The only restriction is this: that the cation:anion ratio is such that the total number of electrons being given is equal to the total number being received into vacancies. For lithium oxide this means a cation:anion ratio of 2:1. In the next question you are asked to apply the rule to other possible ionic compounds.

Question

32 Predict the formulae of the following named ionic compounds. (You need to know the group numbers of participating elements. Note that all the compounds are '-ides'. This indicates they are all *binary* compounds made from two elements

only, as opposed to '-ates' which have extra oxygen.)

> magnesium chloride
>
> barium bromide
>
> sodium oxide (in fact, there are two sodium oxides – see Chapter 17, p. 426; I want the predictable one)
>
> caesium sulphide
>
> calcium oxide
>
> aluminium fluoride
>
> rubidium nitride
>
> aluminium oxide
>
> magnesium nitride.

4.11 Structures of ionic compounds

Formulae of ionic compounds, like those you worked out in your answers to question 32, do not carry the same quality of information as formulae of covalent ones. To take a covalent example, when we write CO_2 for an oxide of carbon, we are really proposing the existence of a discrete particle with that composition. In other words there really is a carbon dioxide, CO_2 molecule. When we turn to ionic bonding and write CaO for calcium oxide, we mean rather less. There is no Ca—O molecule. Nor does any Ca^{2+} cation have a single O^{2-} anion as its sole 'partner'. In fact the symbolism 'CaO' means nothing more than the fact that there are as many Ca^{2+} anions as there are O^{2-} ions in the overall solid mass that we call the crystal lattice. Ionic formulae are thus no more than stoichiometric ratios.

This difference between the meanings of ionic and covalent formulae stem from a fact mentioned previously – that ionic bonds are **non-directional**, and each ion tries to surround itself with as many of the other kind as it can. If I tell you that each calcium ion is surrounded by six oxide ions, you might think I was violating the electrical neutrality of the compound. If I go on to say 'However, each oxide ion is surrounded by six calcium ions', you might think I had answered one problem by setting up a harder one. We have entered the strange three-dimensional geometric world of ionic

crystal lattices, a world in which understanding is impossible without pictures, and in which pictures are a poor second to three-dimensional models.

The sodium chloride structure

The naming system for the structures of ionic compounds relies upon selecting a well-known example of a particular structure and using it as an archetype. Thus 'the sodium chloride structure' is a name used for an ionic formation found in many ionic crystals, the most familiar example of which is sodium chloride, NaCl, common salt itself. Figure 4.52 shows how the ions are arranged in space. The positions occupied by sodium ions and chloride ions in the archetypal structure are of course occupied by other ions in the rest of the set of compounds which share the same structure.

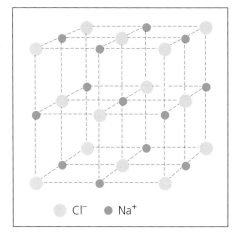

Figure 4.52 The sodium chloride structure

Questions

33 Name a feature which must be common to all the formulae of compounds sharing the sodium chloride structure.

34 Is it possible to imagine a compound with a 2+ cation having the sodium chloride structure?

There are certain key terms and concepts which people use to describe ionic structures, and they are listed overleaf. The sodium chloride structure will be used to illustrate each concept.

The unit cell, single-ion lattices, sites and co-ordination number

The unit cell

The bonding in ionic structures is, as we have seen, non-directional. Therefore, with no obvious beginning and end points, ionic structures can extend over millions of particles. The least part of an ionic structure which can convey the complete pattern of its formation is called a **unit cell**. Figure 4.52 is the unit cell of the sodium chloride structure.

Single-ion lattices

It sometimes helps in the analysis of ionic structures to focus on one ion at a time. For example, if you ignore the sodium ions, Na^+, you will see that there is a chloride ion, Cl^-, at every corner of the cube which makes up the unit cell. There is another one at the centre of every face of the cube. This formation of chloride ions is known as **face-centred cubic**, and is a recurrent motif in all sorts of solid structures.

Now let us focus on the sodium ions, Na^+. As Figure 4.53 shows, they also make up a face-centred cubic lattice, albeit with its unit cell displaced

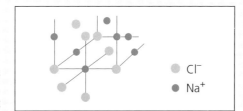

Figure 4.53 A fragment of the sodium chloride structure, focusing on the site of an Na^+ ion

relative to the chloride one. Thus we can think of sodium chloride as beingmade of **two interlocking face-centred cubic lattices**.

Sites

There is one other point to make about the 'two interlocking lattices' view of sodium chloride, and that concerns the *way* they interlock. We can describe the style of 'hole' occupied, say, by each Na^+ ion in the Cl^- lattice. Figure 4.53 shows the 'nearest neighbours' arranged around a single Na^+ ion, in a drawing which is really a selected extract from Figure 4.52, except that some ions have been added from adjacent unit cells. The shape made by the six Cl^- ions in Figure 4.53 is an octahedron, and so the hole in which the Na^+ rests is called an **octahedral site**.

Question

35 What can you say about the sites occupied by each of the Cl^- ions within the lattice of Na^+ ions?

At this stage in our analysis, we can already describe the sodium chloride structure with some accuracy. We would call it 'a pair of interlocking face-centred cubic ionic lattices, each lattice's ions occupying octahedral sites in the other lattice'. (You can see why it is called the sodium chloride structure for short.)

Co-ordination number

This is a concept very closely allied to site.

The co-ordination number of an ion in a lattice is the number of nearest neighbours of the opposite charge.

This is something we have already considered in naming the sites. The co-ordination in sodium chloride is described as 6:6, meaning that each ion has six nearest neighbours of the other kind.

Stoichiometry

At the start of this section I predicted you would have some trouble with the idea that one ion could have six of the other sort around it and yet still be a 1:1 compound. Now with the help of the diagrams and models you can see how it happens. Similarly, the ratio of co-ordination numbers is the same as the stoichiometric ratio – all 1:1 ionic compounds have 1:1 co-ordination ratios.

There is another feature of a unit cell which must match the stoichiometric ratio, and that, fairly obviously, is the number of ions in the unit cell. After all, if the unit cell is meant to stand for the whole structure then it ought to reproduce one of the whole structure's key features, namely its stoichiometry. However, if you inspect the unit cell of sodium chloride (Figure 4.52), you will see that the number of chloride symbols

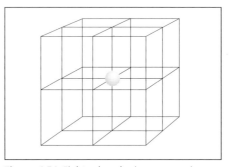

Figure 4.54 Eight cubes sharing a corner ion

in the picture is 14, while the sodium symbols number 13. Is there a mistake?

The solution to the anomaly lies in the fact that *not all the ions in the unit cell model are completely and uniquely within that unit cell*. In fact there is only *one* ion in the whole of Figure 4.52 which is totally within the unit cell. Consider the chloride ions at the corners: a corner ion is playing that corner role in a total of

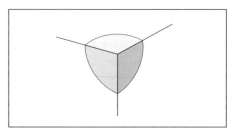

Figure 4.55 'Slicing' the corners off a corner ion, to show one-eighth of a sphere inside the cell

eight unit cells – eight is the number of cubes that can share a corner (Figure 4.54). By that logic you can say that only one-eighth of the ion is actually in any single unit cell. Another style of diagram helps to make this clearer – Figure 4.55 shows the unit cell with a corner ion sliced flush with the cube. As you can see, only one-eighth of the corner ion has 'survived'.

The same operation may be performed on all the positions occupied by ions in the sodium chloride structure. In each case the same question must be asked: 'How many unit cells are sharing this ball?' To complete the overall audit of numbers of ions in the unit cell it only remains for us to multiply the number of ions in each type of position by the fraction of each one which is truly inside the cell.

So for example:

Chloride ions:

8 at corners = $8 \times \frac{1}{8} = 1$

6 at centres of faces = $6 \times \frac{1}{2} = 3$

Total $Cl^- = 4$

Question

36 Complete the following audit of the Na^+ ions:

Sodium ions:

12 at centres of edges $12 \times \frac{1}{4} = ?$

1 wholly within cell = ?

Total Na^+ in cell = ?

So this accounting process shows that the overall numbers of ions in the unit cell does indeed match the stoichiometry of sodium chloride. We will be considering two other examples of lattices with $1:1$ stoichiometry – the caesium chloride structure and the zinc blende structure.

Problems with lattice building

Ionic bonding has something in common with financial investment. The initial removal of the electrons to form the cation is like the initial outlay, because (as you will recall from Chapter 3) ionization energies are always positive, which is the same as saying that even the most loosely held electron must be worked on to remove it from the influence of its nucleus. You begin to get some return on your investment when the transferred electrons arrive in their new orbitals on the anion, but the major part of the return on the initial investment of energy happens when the new bonds between ions are formed, creating the crystal lattice (Figure 4.56).

From the above argument, the most 'profitable' ionic bonding processes, and therefore the most stable ionic compounds, should be those in which as many oppositely charged ions are as intimate with each other as possible – that is, a high co-ordination number. Why has sodium chloride, and indeed the many other compounds which share this layout, an apparently modest co-ordination number of six?

The problem with maximising co-ordination number is this – if you pack too many ions of one charge-type around one of the other kind, then you begin to incur energy penalties in the form of like-charge-ion **repulsions**. But, strangely enough, this tends only to be a problem in one direction, because on the whole, cations are smaller than anions. It is quite surprising what the loss of an outer shell can do to the size of a cation, whereas the addition of an electron into an existing shell has a much less dramatic effect on the size of an anion.

Question

37a Look up the relevant ionic radii in a data book, and quote the cation:anion radius ratios of the following compounds:

NaF, NaCl, NaBr, NaI, CsF, CsCl, CsBr, CsI

b On the whole, which is harder – getting a lot of anions around a cation, or getting a lot of cations around an anion?

The caesium chloride structure

From your researches in question 37, you will have seen that there are a handful of ionic compounds, no more, which have cations bigger than their anions. These are the ones which might be able to accommodate more than six anions around a cation. The archetype of this group, and the one whose name has become its label, is caesium chloride. The structure is shown in Figure 4.57. Question 38 takes you through the analysis of the unit cell of caesium chloride, in a similar way to the analysis of sodium chloride.

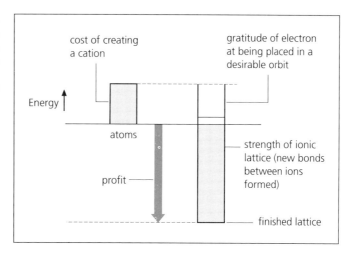

Figure 4.56 'Profit and loss' account of the making of an ionic lattice (this is actually an extract from something called a Born–Haber cycle – see Chapter 12)

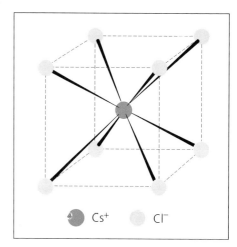

Figure 4.57 The caesium chloride structure

Question

38a Figure 4.57 shows Cl^- ions make up what is known as a simple cubic lattice (as opposed to the face-centred cubic lattice of sodium chloride). The caesium ions, Cs^+, occupy cubic 'sites' in the Cl^- lattice. Take a corner Cl^-

and work out where its nearest Cs^+ neighbours are. Is it also true that each Cl^- ion also occupies a cubic site in an entirely equivalent lattice of Cs^+ ions?

b What is the co-ordination number ratio of caesium chloride?

c Show, by doing an audit of total ions in the unit cell, that the stoichiometry of the structure is 1:1.

d To revise our discussion during and after question 37, why do you think caesium chloride has this structure while sodium chloride has its own archetypal structure?

e Suggest one other ionic compound which might be expected to crystallise in the caesium chloride structure, and justify your choice.

The zinc blende structure

Zinc blende (a form of zinc sulphide, see Figure 4.58) is the last of the three 1:1 ionic structures commonly studied at Advanced level. It shares more than its stoichiometry with sodium chloride, as a close inspection of Figure 4.58 will reveal. As in sodium chloride, we are again dealing with two interlocking face-centred cubic single-ion lattices. There must be something different about it, and that is what the next question is about.

Question

39a In the zinc blende structure, how would you describe the holes occupied by each type of ion in the other's lattice? (Use a word for the geometric shape made by joining up the nearest neighbours around a chosen ion – look at the dotted lines, centring on the Zn^{2+} ions.)

b Hence work out the co-ordination ratio of the structure.

c Make a 'total number of ions' audit, and show that it conforms to the requirements of a 1:1 structure.

d Look up the ionic radii of Zn^{2+} and S^{2-}, and discuss whether zinc blende fits the pattern established by caesium chloride and sodium chloride, namely that the smaller the cation:anion size ratio, the lower the co-ordination number.

Stoichiometries of 1:2

The classic 1:2 structure is that of the mineral **fluorite**, a form of calcium fluoride, CaF_2. (It is quite common for ionic compounds to have two crystal structures with different unit cells. For instance, there is another form of zinc sulphide distinct from zinc blende, called **wurtzite**.) Figure 4.59 shows the unit cell of fluorite. Structures with 1:2 stoichiometries need twice as many of one ion as of the other. This is achieved

in a way which bears a close resemblance with zinc blende. As in zinc blende, one ion (the Ca^{2+}) makes up a face-centred cubic lattice; again as in zinc blende, the other ion (F^-) occupies tetrahedral sites in the Ca^{2+} lattice. The difference is that *every* tetrahedral hole is filled, as opposed to just half. Certain other aspects of the fluorite structure are the subject of the next question.

Question

40a Figure 4.59 includes some extra fluoride, F^-, ions which really belong in the next unit cell. The purpose is to make it easier to see what sort of hole the Ca^{2+} ions occupy in the F^- lattice. What sort is it?

b One of the big differences in 1:2 structures is that the ion in the minority has the *bigger* co-ordination number. In this case you have to get twice as many F^- ions around a Ca^{2+} as Ca^{2+} ions around an F^-, so it is the co-ordination number of the Ca^{2+} which is the higher. Find both the co-ordination numbers from the unit cell of fluorite, and say what you notice about the co-ordination ratio compared to the stoichiometric ratio.

c Make a 'total number of ions' audit – say what you notice in contrast to all the previous structures we have studied.

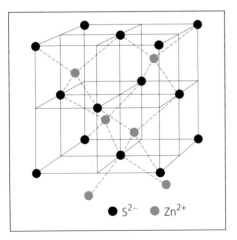

Figure 4.58 The zinc blende structure

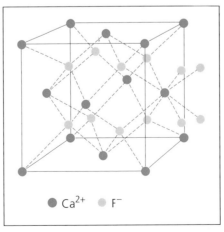

Figure 4.59 The fluorite structure

Summary

• Electrons bind atoms together because in the bonded atoms they can achieve greater energetic stability than in the separate atoms, and at the same time they can pull nuclei together. There are three main styles of electron arrangement which result in bonding, called **covalent**, **ionic** and **metallic** (the last dealt with in Chapter 5).

• **Covalent bonds** depend upon electrons from both bonded atoms occupying molecular orbitals which bring them within the influence of *both* nuclei.

• By occupying space between the nuclei, the bonding electrons create electrostatic forces which bind the nuclei together.

• Covalent bonds have optimum lengths. Too far apart and the electrons cannot interact with each other's nuclei; too close together and the bond is destabilised by inter-nuclear repulsion (Figure 4.60).

• Molecular orbitals are subject to restrictive quantum mechanical rules, just like atomic orbitals. Full molecular orbital theory lies outside Advanced chemistry, but we can say that **molecular orbitals are made when atomic orbitals overlap**.

• The quantum mechanical restrictions are as follows: each atomic orbital entering into a covalent bond must contain a single electron, so the overlapping orbital formed contains two electrons. This pair of electrons is visualised as spending some time in both atomic orbitals, orbiting both nuclei. Hence if an atom cannot offer a half-filled orbital, it cannot bond covalently with another atom. This model for a covalent bond can be represented by a dot-and-cross diagram.

$$\text{etc.} \ \overset{\circ\circ}{\underset{\times}{A}} \ \overset{\circ}{} \ \overset{\times\times}{B} \ \text{etc.}$$

Example of dot-and-cross diagram

• Covalent bonding therefore depends upon how many unpaired electrons (and what amounts to the same thing, how many vacancies) an atom brings to the bonding situation. This in turn is linked to its position in the Periodic Table. For example, an element in Group V can offer s^2p^3, which amounts to three unpaired electrons and three vacancies. Three covalent bonds are therefore possible with other atoms. For example, a nitrogen atom bonds with three different atoms.

$$A \ \overset{\circ\circ}{\underset{\times}{N}} \ \overset{\circ}{\underset{\circ\times}{}} \ C$$
$$B$$

• The **rule of eight** states that elements will try to achieve partial or total ownership of a full set of eight electrons in their outer shell, with an s^2p^6 arrangement.

• The rule of eight can be expressed in equation form. If the number of bonds possible for any given atom is called its **valency**, then:

valency + group number = 8
(or 2 in the case of the first row).

• This rule-of-eight model predicts the valencies of the first ten elements from hydrogen to neon with a good degree of accuracy (Table 4.2).

Table 4.2 Valencies of the first ten elements

Element	Valency
H	1
He	0
Li	*
Be	*
B	*
C	4
N	3
O	2
F	1
Ne	0

* Not applicable, either because all bonds are ionic, or because covalent arrangements break the rule of eight

• Rearrangements of electrons to maximise the number of vacancies are permitted, as long as the energy cost is not too high. For instance, carbon enters into bonding as s^1p^3 rather than s^2p^2.

• The rule of eight is broken because of electron rearrangements. Elements beyond neon can utilise d orbitals as vacancies to relocate newly unpaired electrons. When this happens the maximum valency becomes identical with the group number, because every outer electron can be used in bonding.

• These 'orbital-utilisation' aspects of bonding are best represented by box diagrams (Figure 4.61).

Figure 4.61 Box diagram method of showing how phosphorus (P) can achieve a valency of 5, despite a ground state electron configuration of $1s^2 2s^2 2p^6 3s^2 3p^3$

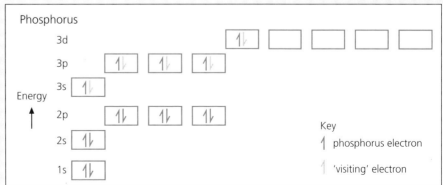

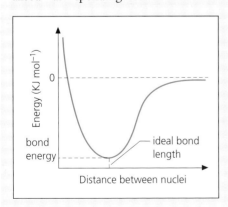

Figure 4.60 Figure 4.2 revisited

• The bonding in certain molecules can only be made to fit simple dot-and-cross or box diagram models if it is assumed that it is possible for *both* electrons of a bonding pair to come from the *same* member of the bond. Such bonds are called **dative covalent**. Well-known examples are H_3NBF_3 (Figure 4.62) and nitric acid, HNO_3.

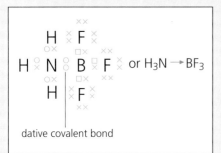

dative covalent bond

Figure 4.62 The dative covalent bond in H_3NBF_3

• Instead of having shapes derived from putting the atoms into positions where the atomic orbitals can best overlap, molecules seem instead to obey a simple but incompatible rule that says **bonds (and lone pairs) like to stay as far apart as possible**. We introduced the terminology of 'bits' to predict the shapes of molecules (Table 4.3).

Table 4.3 Molecular shapes as predicted by mutual repulsion

No. of 'bits'	*Shape	Angle (°)
2	Linear	180
3	Trigonal plane	120
4	Tetrahedron	109.5
5	Trigonal bipyramid	90/120
6	Octahedron	90

* Assuming no lone pairs, so no 'invisible' points.

• X-ray crystallography provides us with **electron density maps** which confirm that shapes of molecules follow the 'mutual repulsion rule'.

• To call a bond 'σ' implies that the two atomic orbitals whose overlap created it have overlapped with their direction axes in line (or 'nose-to-nose'). (s orbitals do not have direction axes, of course, but their overlaps are counted as σ anyway.) This type of overlap is unaffected by rotation about the axis of the bond, so σ bonds can rotate freely.

• **Double bonds** (such as the familiar $C=C$ bond in alkenes) are seen as consisting of two distinctly different types of orbital overlap. The first type features two hybrid style (sp^2) orbitals overlapping 'nose-to-nose', and is called a σ bond. The second is made up of unhybridised p orbitals with a 'cheek-to-cheek' overlap, called a π bond.

• This σ/π model for a double bond explains why double bonds *do not allow rotation*, and therefore offers an explanation for the existence of *cis* and *trans* isomers.

• Small distortions of the predicted shapes of molecules are due to a 'hierarchy' of repulsion. Thus triple bonds repel more strongly than double bonds, which repel more strongly than single bonds. In fact, a double bond will have the effect of forcing single bonds closer to each other. The league table of 'interbit' repulsions is shown in Table 4.4.

Table 4.4 Hierarchy of repulsive effects, most repulsive first

Triple bond
Double bond ≈ Lone pair
Single bond

• Some molecules, for which predictions have foreseen single and double bonds adjacent to each other, in fact turn out to have bonds of identical length and electron density intermediate between the predicted values. This is due to **delocalisation** in the π-bonding system of the molecule; delocalised π-electrons are free to 'visit' more than just two atomic centres. The situations in which to expect delocalisation are summarised in Table 4.5.

• Some pairs of atoms have such unequal abilities to attract electrons, that covalent-style sharing of a pair of electrons becomes impossible. Instead the pair of electrons spends *all* the time on the more electronegative member of the bond. The result is the complete donation of a single electron from one atom to another, and the resulting bond is called an **ionic bond**.

Table 4.5 Summary of delocalisation

Mode of delocalisation	Example
Single and double bonds alternate around a ring	benzene
Lone pair can come in to be a π-bond, and π-bond pair can go out to become a lone pair	nitric(v) acid

• The bonding force in an ionic bond is electrostatic attraction between anion (the receiver of the transferred electron) and cation (the donor).

• Ionic bonds occur between left-hand and right-hand elements in the Periodic Table, between metallic cations and non-metallic anions. (In contrast, covalent bonds occur between fellow right-hand elements.)

• It is possible for multiple electron transfers to take place between atoms – the only requirement is that the overall cation and anion charges should add up to zero.

• For anions, group number + (negative) charge on ion = 8. For cations, group number = (positive) charge on ion.

• Like the rule of eight, these rules lose their universality in parts of the Periodic Table. (They have little meaning for d-block cations, or for oxyanions like sulphate, nitrate and carbonate.) And yet, along with the 'total charge = zero' rule, they enable us to predict the formulae of simple binary compounds such as $(Al^{3+})_2(O^{2-})_3$.

• The influence of bonding style on properties is profound, and is summarised in Table 4.6.

• Ionic structures in the solid state are (mostly) organised interlocking arrays of the two sets of ions.

Table 4.6 Influence of bonding on properties of substances

Bond type	Likely melting point and boiling point	Electrical properties	Likely pattern of solubility
Covalent	low – often liquid or gas at room temperature (r.t.)	insulators	soluble in non-polar solvents
Ionic	high – solid at r.t.	insulators when solid; conductors when molten or in solution	insoluble or otherwise soluble only in polar solvents
Metallic	high – solid at r.t. (except Hg)	conductors both when solid and when liquid	insoluble in all solvents (unless there is a reaction)

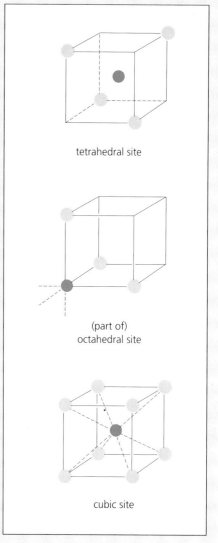

tetrahedral site

(part of) octahedral site

cubic site

Figure 4.63 Different sorts of site

• They are described using the three following terminologies – **co-ordination number**, **stoichiometric ratio**, and **occupied sites** (or holes).

• There are three simple structures adopted by ionic crystals of 1:1 stoichiometry, as shown in Table 4.7. The particular structure adopted depends on the cation:anion radius ratio.

fluorite structures exist in which the cation is the '2'.

• The **unit cell** is a diagram of a section of an ionic crystal lattice of sufficient extent to capture the pattern of the whole lattice.

• An audit process can be used on unit cell diagrams to find how many ions of each type are *within* the unit cell. This audit process uses the findings of Table 4.8.

Table 4.7 Some 1:1 ionic structures

Structure	Co-ordination no.	Cation:anion radius ratio
CsCl	8:8	0.94
NaCl	6:6	0.57
ZnS (zinc blende)	4:4	0.41

• The most important (at Advanced level) of the structures adopted by ionic crystals with 2:1 stoichiometry is the **fluorite** (CaF_2) structure. Inverse

Table 4.8 A guide to the audit process for unit cells of the cubic type

Position	Fraction within unit cell
Corner	$\frac{1}{8}$
Centre of edge	$\frac{1}{4}$
Centre of face	$\frac{1}{2}$
Middle	1

• Site terminology is illustrated by Figure 4.63.

5

Chemical bonding 2: 'Impure' and metallic bonds

5.1 Intermediate styles of bonding

Chapter 4 presented ionic and covalent bonding as if they were two very distinct methods of holding atoms together. There is plenty of justification for this, some to do with the tactics of teaching, and some to do with reality. On the tactical front, it is easier to teach black-and-white differences to the student before moving on to shades of grey.

On the reality front, there is certainly a genuine division between a large group of substances with high melting points and electrical conductivities when molten (the ionics), and another large group which are liquids, gases or low-melting point solids and which have very low conductivities (the covalents).

Furthermore, there is certainly such a thing as a 'pure' covalent bond. A bond between two identical non-metallic atoms, Cl—Cl say, or H—H, has to have equal electron sharing by sheer logic, let alone physical evidence. Equally, a bond between two atoms such as sodium and chlorine, with such different arrangements of their outer electrons, would seem logically to be an ionic give-and-take.

However, there is plenty of physical evidence to suggest that many instances of bonding lie somewhere *in between* these two 'pure' archetypes. It is more useful to visualise the range of chemical bonding as a continuous spectrum, rather than as two separate camps. Figure 5.1 shows how a gradually increasing bias in electron-pair possession by one atom of the partnership could bring about a smooth transition from a pure covalent bond to a pure ionic one.

It needs to be said before this chapter goes any further that the words 'pure' and 'impure' are here being used in a specific context. 'Impure', for example, is not meant to imply any mixing or contamination of one chemical by another. The words are merely saying something about how closely the bonding in compounds fits the two ideal pictures.

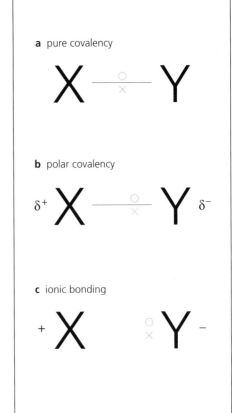

a pure covalency

b polar covalency

c ionic bonding

Figure 5.1 Symbolising the continuous spectrum of bond types. Key: δ = 'a little bit' or 'partial'

The causes of intermediate styles of bonding

So bond types are a spectrum ranging from pure ionic to pure covalent. In any spectrum there is a quantity whose variation defines the *x*-axis, as it were. In the case of the electromagnetic spectrum, that quantity is frequency (or wavelength). What is it that plays this *x*-axis role in the bonding spectrum?

The quantity that places the hydrogen molecule, H_2, at one end of the spectrum and sodium chloride at the other is this: *the difference between the electron-attracting power of the two participants in the bond.* (Sodium chloride has a big difference, while the hydrogen molecule has a difference equal to zero.)

The quantity in italics is not precisely set out in data books. However, we can connect it with two other quantities which are. An atom with a high electron-attracting power *within* a bond will be the same as one with a high positive first ionization energy (p. 150), and with a large negative **electron affinity** (p. 150) (fluorine, for example). In fact, both these quantities refer to single atoms in the gaseous state, which is rather different from being in a bond. However, both these quantities relate back to the pull of the nucleus on the electrons.

So the rule for estimating where a particular bond might be on the spectrum between the two extremes of ionic and covalent is this:

If the atom on one end of a bond has a very different ionization energy from the atom on the other end (and if, as follows almost inevitably, they also have very different electron affinities), then the bond between them will tend towards the ionic.

The greater these differences, the more ionic will be the bond. As these differences approach zero, the bond will approach perfect covalence.

Questions

1 Draw on your knowledge of the way atoms are made up to explain why, if two atoms have very unequal ionization energies, they will also have very unequal electron affinities.

(Remember that ionization energy measures how hard it is to remove an electron, while electron affinity measures how strongly a visiting electron is held.)

2 Which pair of elements in the Periodic Table provide the 'purest' example of ionic bonding? Justify your answer.

Electronegativity

There have been at least three attempts to set up a scale of numbers for the elements which would express each element's power to bias a bonding electron-pair in its own direction. The most famous version is that of Linus Pauling (Figure 5.2). Linus Pauling

Figure 5.2 Linus Pauling

(1901–1994) was born in Portland, Oregon, and began his career as a bio-chemist. He became a hyperactive, high-profile chemist, whose interests ranged from electronegativity to the hunt for DNA structure to the benefits of vitamin C consumption. He was the first person to win two full Nobel prizes: one for chemistry and the other for peace. Pauling's electronegativity scale is based on experimental data such as ionization energies and electron affinities, mathematically interwoven in an intuitive way in order to produce the final number. Some Pauling electronegativities are shown in Table 5.1 (overleaf).

Pauling also proposed a way in which a *difference* in electronegativities of the two atoms in a bond could be translated into a **percentage ionic character** for the bond (Figure 5.3, overleaf).

Table 5.1 Electronegativities of selected elements

Element	Group in Periodic Table	N_p Pauling electronegativity activity index
Li	I	1.0
Na		0.9
K		0.8
Rb		0.8
Cs		0.7
(H)		2.1
Mg	II	1.2
Ca		1.0
Al	III	1.5
C	IV	2.5
Si		1.8
N	V	3.0
P		2.1
O	VI	3.5
S		2.5
F	VII	4.0
Cl		3.0
Br		2.8
I		2.5
Fe	d-block	1.8
Co		1.8
Ni		1.8

For a graph of how to use these data to estimate 'percentage ionic character' in a bond see Figure 5.3.

Question

3 Work out the percentage ionic character in the bonds in the following molecules.

Cl_2, CH_4, NH_3, H_2O, $AlCl_3$, $FeBr_3$, $NaCl$, $CsCl$, CsF

Do these values sit comfortably with what you know of the character of these substances? If not, comment on any apparent anomalies.

The symptoms of 'impure' bonding

It is convenient to deal with the symptoms of impure bonding by dividing up compounds into two groups: 'mainly covalent but with ionic tendencies',

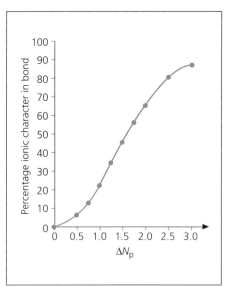

Figure 5.3 How to use N_p values to estimate how polar a bond is. Key: ΔN_p = difference in electronegativity between atoms on either end of a bond

and 'mainly ionic but with covalent tendencies'.

Of the 'mainly covalent' group, water is a prime example. We will look at its deviant behaviour under three headings.

Attraction by a charged rod

A stream of water can be deflected by a charged rod (Figure 5.4). This suggests that there are electrostatic **poles** on the molecule. If electron sharing is *even*, then each atom gains and loses a half-share in two electrons, giving a net charge of zero. However, according to Pauling's theory, the sharing is far from even. In each bond, the oxygen gives away *less*

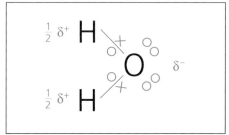

Figure 5.4 Attraction of a stream of water by a charged rod

Figure 5.5 Electrostatic 'poles' in a water molecule

than a half share in its electron, and gains *more* than a half-share in the hydrogen electron. The situation is shown in Figure 5.5, with a net negativity associated with the oxygen atom. So the water deflection trick is explained in Figure 5.6, with each molecule behaving like the electrostatic equivalent of a tiny magnet.

(Note the use of the lower case Greek letter 'δ' (delta), which is the conventional symbol for denoting charge build-up in a bond. It roughly translates as 'a little bit of'.)

Figure 5.6 What happens in the water deflection trick

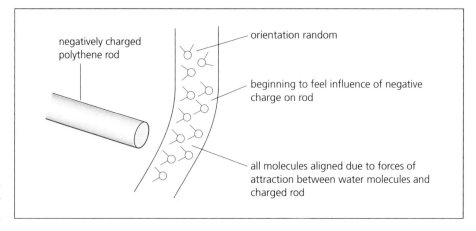

Dipole moment

A moment is the product of a turning force, F, and its distance, d, from the pivot point (Figure 5.7). To transfer the

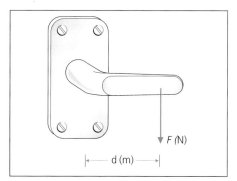

Figure 5.7 Definition of the moment of a force. Moment (N m) = force (N) × distance (m)

idea to a water molecule, you need to imagine the molecule at right angles to an imposed electrostatic field. The molecule will experience a force attempting to align it in the lowest energy direction, according to the 'unlike poles attract' rule (Figure 5.8). The **moment** of that turning force will be proportional to:

1 the size of the charges which have built up in the impure bond (measured in coulombs)
2 the separation between the poles (measured in metres) – see Figure 5.9.

This explains why dipole moment is measured in coulomb metres (C m). (However, many data books quote dipole moments in debyes (D), where $1\,D = 3.34 \times 10^{-30}\,C\,m$.) The dipole moment is as an index of the 'impurity', that is, the polarity – in an apparently covalent bond.

Questions

4 Comment on the trends in dipole moment in the following series:

Series 1	Dipole moment(D)
Chloromethane, CH_3Cl	1.86
Bromomethane, CH_3Br	1.79
Iodomethane, CH_3I	1.64

Series 2	
Chloromethane, CH_3Cl	1.86
Chloroethane, CH_3CH_2Cl	1.98
1-Chloropropane, $CH_3CH_2CH_2Cl$	2.10

5 Dipoles can often occur in several bonds within the same molecule, in which case they add like vectors in mathematics. For example, in water:

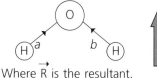

Where $\vec{R}$ is the resultant, or overall dipole $\qquad \vec{R} = \vec{a} + \vec{b}$
Explain the following sequence of overall dipole moments. (Remember that the dipoles can be in *three* dimensions.):

	Dipole moment(D)
Chloromethane, CH_3Cl	1.86
Dichloromethane, CH_2Cl_2	1.54
Trichloromethane, $CHCl_3$	1.02
Tetrachloromethane, CCl_4	0.00

6 Some of the biggest dipole moments are associated with multiple bonds. For example:

	Dipole moment(D)
Propane, $CH_3\overset{\displaystyle O}{\overset{\displaystyle \|}{C}}CH_3$	2.95
Ethanenitrile, $CH_3C{\equiv}N$	3.92
Hydrogen cyanide, $HC{\equiv}N$	2.80

Can you suggest *why* they are so big?

Partial ionic cleavage

The third symptom of impure covalent bonding is in some ways the most striking. Some polar covalent compounds allow some of their polarised bonds to break ionically, so that the negative end of the dipole gains total custody of the pair of electrons.

There are two features of this phenomenon – first it may only happen to a very low percentage of any given batch of molecules, in which case the bonds are constantly being broken and remade in a 'dynamic equilibrium'. This way there are always a few fractured ionized molecules, but not the *same* fractured ionized molecules. Second, the bonds would not break without help from adjacent molecules, either of their own kind, or of a solvent. Again, water is a classic case. Even the purest water has a low finite measurable electrical conductivity, and the reason is this reaction, which it does with itself:

$$H_2O + H_2O \rightarrow H_3O^+ + OH^-$$

This is illustrated below. Only a few molecules do this at any one time (about 1 in 10^9), and *not* the same molecules all the time.

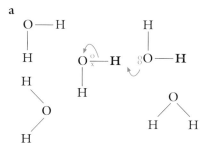

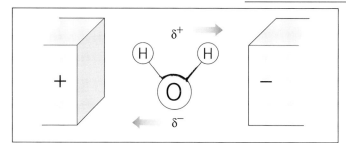

Figure 5.8 Twisting forces attempting to align a water molecule with an imposed electric field

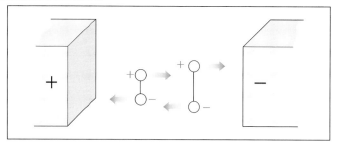

Figure 5.9 Showing how a bigger charge separation gives a molecule a bigger dipole moment, even though the size of the charges themselves may be the same (See example in question 4.)

b

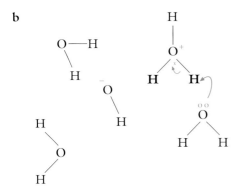

c

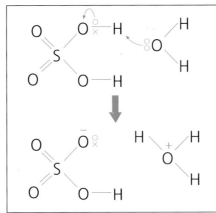

Figure 5.10 Ionization of sulphuric acid by contact with water (Note that the second H atom is much less easy to pull off once the first one has gone, so HSO_4^- ions are commoner than SO_4^{2-} ions.)

Question

7 Draw on your knowledge from Chapter 4 to describe the bond between one water molecule and the donated H^+ ion from the other.

Acids and bases

There are a number of 'covalent' substances which undergo more ionic breakup than water. The sulphuric acid molecule, for example, will *totally* shed one of its hydrogens as an H^+ ion at the first contact with a water molecule (Figure 5.10). This behaviour is characteristic of the compounds we call **acids**. (Acids and bases are treated in much more detail in Chapter 12.)

The bond which breaks ionically in sulphuric acid (H_2SO_4) is an O—H bond, just as in water, and yet the O—H bond in the acid is millions of times more willing to break. This is caused by the way the combination of dipoles in the molecule interact. Thus if an atom (the sulphur in this case) is the victim of several electron 'pulls', its willingness to give up electron density to any one puller is reduced.

Question

8 Consider the sulphuric acid molecule:

a What sort of dipoles will be in the S═O bonds? Give your answers by showing the positive and negative ends of the dipoles, marked δ⁺ and δ⁻, and by suggesting whether the dipoles would be big or little.

b Even without the S═O bonds, the sulphur atoms would still be surrendering a more-than-half share of the electron pairs in the S—O single bonds. What extra effect will the S═O dipoles have on the willingness of the sulphur atom to give up its usual share in the S—O single bonds?

c How will this in turn alter the way the two single-bonded oxygens affect the two hydrogens?

d Does this explain why sulphuric acid is so much more complete in its ionization than water?

The other two well-known mineral acids of inorganic chemistry, hydrochloric acid, HCl(aq), and nitric acid, HNO_3(aq), both show 100 % ionic breakage in water, as shown in Figure 5.11. In fact the best known of all chemical reactions – 'acid plus alkali produces salt plus water' – can now be seen as the induced

breakage of an H—X bond under the influence of an hydroxide ion, OH^-. Even quite reluctant breakers like ethanoic acid will react under these circumstances (Figure 5.12).

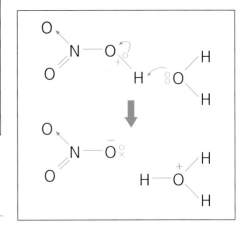

Figure 5.11 Ionization of nitric acid in water

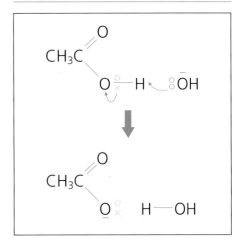

Figure 5.12 Neutralisation of a weak acid by aqueous OH^- (that is, by an alkali)

Question

9 Why is an hydroxide ion, OH^-, successful in attracting all the H^+s in situations (like ethanoic acid) where water only succeeds in receiving a few? In other words, why do you think OH^- is a better H^+ 'grabber' than water?

This study of acids has the effect of introducing us to the anions left behind when the H^+ ion has split off. 'Molecular ions' like nitrate and sulphate were not mentioned in the ionic section of Chapter 4, because a knowledge of impure covalent bonding was needed. However, the nitrate, sulphate and carbonate ions will provide useful revision of some earlier theories such as those on molecular shapes and delocalisation.

Questions

10 The sulphate ion is an exactly regular tetrahedron with four bonds the same length. Explain this finding using delocalisation and 'bit theory' (pp. 59 and 57 respectively).

11 Predict the shape of the carbonate and nitrate ions, using the same two models.

12 The ammonium ion, NH_4^+, is formed at low concentration when ammonia is dissolved in water. Show how the ammonium ion is formed from ammonia plus water, and draw a dot-and-cross diagram of the ion. Does your answer also show why ammonia solutions are weakly alkaline?

13 Hydrogen chloride exists as a gas, and is obviously therefore molecular, showing no sign of the ionization which happens so readily in water (except that it does have a quite healthy dipole moment of 1.050 D). Why does its bonding state depend on its circumstances in this way?

Now that we have established a clear picture of polarised or 'impure' covalent bonding, let us look at a phenomenon which owes its existence to that polarity.

5.2 Infrared spectroscopy and bond vibration

This section is about how the bonds in molecules interact with photons of light; but first we must prepare the ground by looking more closely at bond behaviour.

In Chapter 4, we studied a graph which showed the potential energy of the covalent bond in hydrogen, H_2, plotted against inter-nuclear distance (p. 50). Figure 5.13 shows the equivalent graph for the bond in hydrogen chloride, HCl. Graphs like this encourage a static picture of a covalent bond – we can imagine the

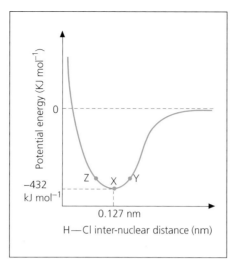

Figure 5.13 Potential energy curve for the H—Cl bond

bond existing in its most stable state, which for hydrogen chloride is at the point X, with an inter-nuclear distance of 0.127 nm, and a potential energy of −432 kJ mol⁻¹.

Real bonds are not like that. Instead, even at very low temperatures, they are in a state of *vibration*. In this picture of the hydrogen chloride bond, the true significance of the point X (0.127 nm, −432 kJ mol⁻¹) is that it is a *mean position* about which vibration takes place. However, we do not need to abandon the graph in order to accommodate this new way of thinking: in fact the graph offers an ideal way of showing the energetics of the vibration process. The following question is about this aspect.

Question

14 It will help you to think about bond vibrations if you can picture the HCl bond as behaving in some ways like a *spring*.

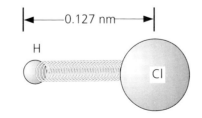

a What would be the 'natural length' (to borrow an expression from the physics of real springs) of the HCl 'spring'?

b Consider the point Y on the graph in Figure 5.13. Describe in words what has happened to the HCl 'spring' in moving from point X to point Y, explaining how the change in length causes the change in potential energy.

c At point Y there will be a force in the 'spring', seeking to change its shape. In which direction will the force operate?

d Describe in words why the force will not succeed in restoring the natural resting condition of the spring at point X, but instead will cause the spring to go through to point Z.

e What happens next?

The last question should have helped you appreciate the fact that there is no rest for bonds. As long as we accept bond vibration as a fact of life (albeit a fact coming from outside Advanced chemistry) then we can see that bonds like those in hydrogen chloride molecules will be for ever oscillating between their own individual versions of points Y and Z. (Of course a real spring will eventually come to rest due to effects like air-damping, of which there is no equivalent in our story of bond vibrations.)

The absorption of energy by vibrating bonds

Now for a second vibration metaphor. There is a danger in hitching too many different metaphors to the task of describing one phenomenon: and yet although the spring metaphor was best as a direct picture of a bond, in your actual experience of systems which vibrate, springs may not be the most familiar. For that reason, and because they fit the energy absorption situation better, we will now switch back to thinking about an analogue system we used once before (in a different context) – *guitar strings*.

You have probably seen the phenomenon of **resonance** occurring to a guitar string. The string appears to start vibrating all on its own (detectable by the unseating of a tiny paper 'rider'), but what has really happened is that it has absorbed energy from another source of sound vibration nearby. One of the key aspects of resonance is that not just any sound will make it happen – the source must be at the *same* frequency as the **natural frequency** of the guitar string. When that condition is satisfied, energy can be transferred from the source to the string, which ends up vibrating at larger amplitude, although still at the *same* frequency.

So it is with bonds. The bond in a molecule like hydrogen chloride has its natural vibrational frequency, and it can absorb energy from a source of the same frequency, and move into an excited state in which it vibrates harder.

Question

15 Place two more points on the graph in Figure 5.13 to represent the inner and outer limits of a 'harder' vibration – harder, that is, than the Y–Z vibration in question 14.

What *sort* of source will provide the energy? In the case of the guitar string, it was fairly obvious that the source would be a *sonic* one, since the receiver was itself a sonic device. Like absorbs like, as it were. But what source is 'like' a vibrating bond? The answer to this question will explain why this section on infrared spectroscopy is in the present chapter.

In fact, the sources of energy are photons in the *infrared* region of the electromagnetic spectrum (Figure 5.14). Spectroscopy in which 'light' interacts with matter is not a new idea to us – we have already seen, in Chapter 3, that photons in the *visible* region of the elec-

tromagnetic spectrum can provide energy for the excitation of *electron energy levels* in atoms (p. 36). We saw the reverse process too, in which electron 'relaxations' *create* photons in the visible region, as in flame colours. We met Planck's equation, in which the energy which the photon had to offer was linked to its frequency:

$$E = h \times \nu \qquad (3.1)$$

But let us return to the 'like absorbs like' idea. *Electrons* can interact with light because electrons have electrical, magnetic and wave-like character, just like photons. Exactly *what*, therefore, allows a *vibrating bond* to absorb photons? A clue to the answer lies in the fact that *some* vibrating bonds do *not* absorb photons, despite being perfectly good oscillators. We will consider them now.

Question

16 Among the molecules with *no* infrared absorptions are H_2, Cl_2 and O_2. They surely cannot fail to vibrate, since their bonds are no different in principle from, say, the bond in HCl, which *does* absorb. Or are they? See if you can pick an item which other molecules have and which these lack, and which would assist the HCl-like ones to

interact with electromagnetic radiation.

So the necessary condition for a molecular vibration to absorb a photon is that the vibration must cause oscillation in those partially developed opposite-charge-centres which we have called *dipoles*. For instance, the vibration of the bond in HCl will cause the dipole in the bond to oscillate as in Figure 5.15. In this case we see that there is only one movement which can create an oscillating dipole, so there is only one peak in the infrared spectrum of HCl.

Now that we have in place the fundamental reason why infrared absorptions happen, let us look at some of the variations in the infrared spectra of different molecules, and how chemists have made use of the information. Two points of interest to chemists are the *number* of the peaks, and their *positions* in the spectrum (which are determined by the frequencies of the photons absorbed). We will consider each in turn.

Number of peaks

Figure 5.16 shows the infrared absorption spectrum of carbon dioxide. It has two main peaks: at 2360 cm^{-1}, and around 670 cm^{-1}. Let us consider the dipoles in the bonds of the carbon

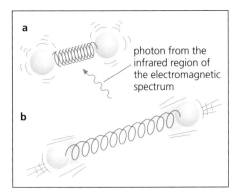

Figure 5.14 Vibrational excitation by absorption of a photon. The frequency of the photon is equal to the frequency of vibration of the bond. Excitation has the effect of increasing the *amplitude*, not the frequency

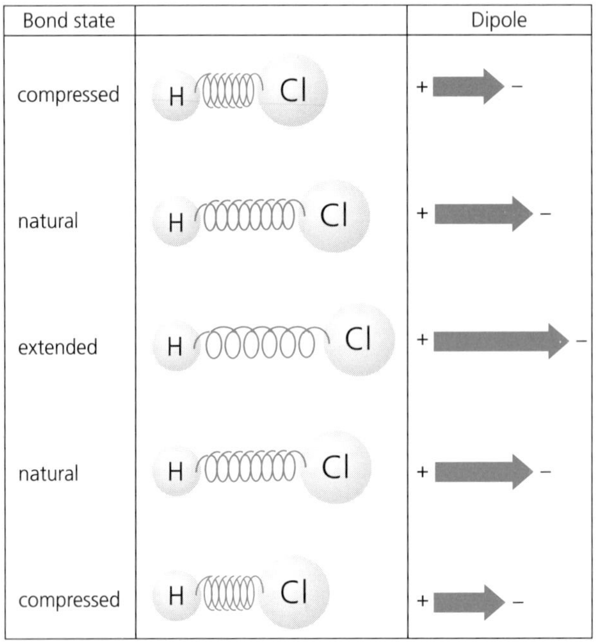

Bond state		Dipole
compressed	H ((((Cl	+ ➡ –
natural	H (((((((Cl	+ ➡ –
extended	H (((((Cl	+ ➡➡ –
natural	H (((((((Cl	+ ➡ –
compressed	H ((((Cl	+ ➡ –

Figure 5.15 Showing how an oscillating bond leads to an oscillating dipole

dioxide molecule, and how many ways the molecule can make them oscillate. The most significant fact is that the molecule has two symmetrically opposed dipoles, one in each bond, and therefore the overall dipole moment of the molecule is *zero*.

There is one type of vibration which leaves this symmetry untouched. This is the **symmetrical stretch** (Table 5.2).

Table 5.2 How the overall dipole of carbon dioxide remains at zero throughout the symmetrical stretch vibration

Bond condition	Dipoles in individual bonds	Overall dipole
O=C=O		0
O=C=O		0
O=C=O		0

Both the individual dipoles extend as the bonds stretch, but they remain at all times equal and opposite, and because this vibration produces no overall oscillating dipole, it *produces no absorption*. On the other hand, the **asymmetrical stretch** (Table 5.3), and the **bending vibration** do create overall dipole moments when they are in the deformed positions, and therefore do give rise to absorption peaks.

Table 5.3 How the asymmetrical stretch vibration *does* lead to an overall oscillating dipole

Bond condition	Dipoles in individual bonds	Overall dipole
O=C=O		$- \Leftarrow +$
O=C=O		0
O=C=O		$+ \Rightarrow -$

Questions

17 Draw tables similar to Table 5.3, to show the way the bending vibration causes an oscillating dipole moment.

18 As we found in Chapter 4 (p. 59, question 24), sulphur dioxide SO_2, unlike carbon dioxide, is a *bent* molecule. Its infrared spectrum is shown in Figure 5.17. Two of the absorptions are equivalent to peaks in the carbon dioxide spectrum: 1360 cm^{-1} is the asymmetrical stretch and 540 cm^{-1} is the bend. Can you explain the existence of the third peak, showing clearly the contrast with carbon dioxide?

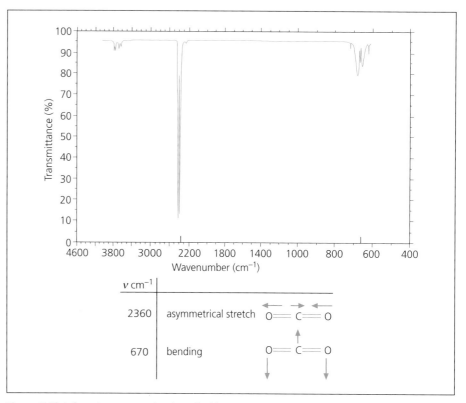

Figure 5.16 Infrared spectrum of carbon dioxide

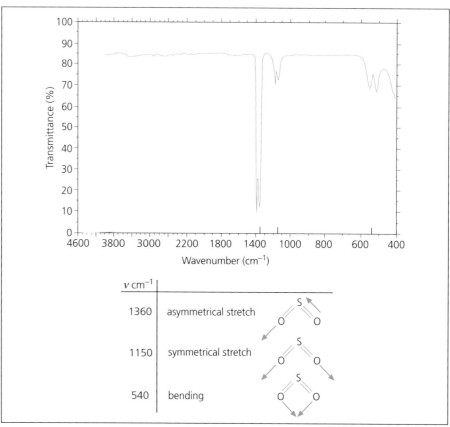

Figure 5.17 Infrared spectrum of sulphur dioxide

As molecules get more complex, the number of ways which a molecule has of 'wobbling' its many individual dipoles increases, so the spectrum increases greatly in complexity. However this is partially offset by two beneficial aspects.

First, the very complex bits of the spectrum have a 'fingerprint' value, since each is unique to its source molecule.

Second, when each individual bond is a smaller part of a larger heavier whole, it becomes possible to treat the vibrations of each bond in isolation, rather than, as in the carbon dioxide and sulphur dioxide cases, treating the molecule as a unit. To take a slightly eccentric parallel, look at Figure 5.18. The vibration of the antenna would be pretty much independent of the person and would leave her unaffected, because the two things on either end of the spring (that is, the girl and the ball) are of such unequal mass. In other words, the spring would have the same natural frequency no matter which human being was wearing it. But if the same device were strapped to, say, a chihuahua, then every time the antenna wobbled it would wobble the dog's backside as well, affecting the frequency of the overall process.

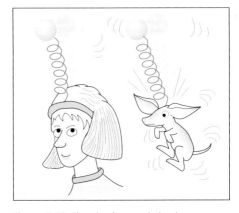

Figure 5.18 Showing how variation in masses on either end of a spring/bond can affect the degree of independence of the vibration

The moral of the tale is this: if each bond in a big molecule vibrates more or less independently, it means that every, say, C——H bond in every (big) molecule will always absorb photons of roughly the same frequency. We can therefore build up a table equating particular positions of absorption peaks with particular bonds. Such a table is very useful in analysing unknown compounds, and we

shall see it applied in the chapters on organic chemistry later in the book.

Positions of peaks

Why do particular bonds absorb at different positions on the x-axis? We can again use the spring analogy for an explanation. You can probably visualise that a floppy spring would have a natural frequency of oscillation which was lower than a stiff spring – that is, a slower, more leisurely vibration. 'Floppy' and 'stiff' translate into bond language as 'weak' and 'strong'. So, for example, a $C{=}O$ bond needs higher frequency photons with which to interact than a $C{-}O$ bond. Another factor affecting bond oscillation frequency is *mass*: if the atoms on the end of the bond are heavier, they will vibrate slower than lighter ones. For instance, bonds to hydrogen atoms all vibrate (and so absorb) at particularly high frequencies.

So what exactly does the x-axis measure? You might expect it to be frequency, but a closer look at the units shows 'cm^{-1}'. This quantity is called the **wavenumber**, and it is the conventional x-axis for all infrared spectra. (Another convention is that high wavenumbers are on the *left*.) The wavenumber is the **reciprocal of the wavelength** of the photons (that is, 1/wavelength) and a little mathematics will show that it is proportional to, and so serves as a good substitute for, frequency:

$$\text{Speed of light} = \lambda \times \nu$$

where λ = wavelength and

ν = frequency

So   wavenumber = 1/wavelength
$$= 1/\lambda$$
$$= f/\text{Speed of light}$$

Because the speed of light is a constant, wavenumber $\propto f$

The advantage of using wavenumber is the 'handiness' of the magnitudes – in wavenumber units the infrared region runs from about 400 to 4000 cm^{-1}, which is easier to remember than the equivalent frequencies (1.2–12×10^{13} Hz).

Questions

19 Table 5.4 shows the regions of the infrared spectrum in which various types of vibration absorb energy. Explain why:

a C——H *bends* are found at much lower wavenumbers than C——H *stretches*

b *double bonds* are found at lower wavenumbers than equivalent *triple bonds*.

Table 5.4 Identifying bond vibrations

Type of bond vibration	Region in which absorption occurs
Stretches of single bonds to hydrogen (e.g. C——H)	$2500 \rightarrow 4000$ cm^{-1}
Stretches of triple bonds (e.g. C≡N)	$2000 \rightarrow 2500$ cm^{-1}
Stretches of double bonds (e.g. C=O)	$1500 \rightarrow 2000$ cm^{-1}
Bending of C——H bonds	$800 \rightarrow 1500$ cm^{-1}

20 The three hydrogen halide gases have only single absorption peaks. The wavenumbers of the absorptions are shown in Table 5.5. Try to explain the variation.

Table 5.5 Absorption wavenumbers

Molecule	Wavenumber of peak (cm^{-1})
HCl(g)	2886
HBr(g)	2000*
HI(g)	1650*

* guesses

21 Figure 5.19 shows the infrared spectrum of trichloromethane, CHCl$_3$, with each of the major peaks accounted for. Predict which peaks might be in different places if the spectrum had been of *trichloro-deuteromethane*, CDCl$_3$, and suggest in which general direction they might have moved. (Deuterium is the isotope of hydrogen with one neutron, so a D atom is twice as heavy as an H.)

22 You will have perhaps realised that infrared spectroscopy is a lower-energy phenomenon than atomic absorption/

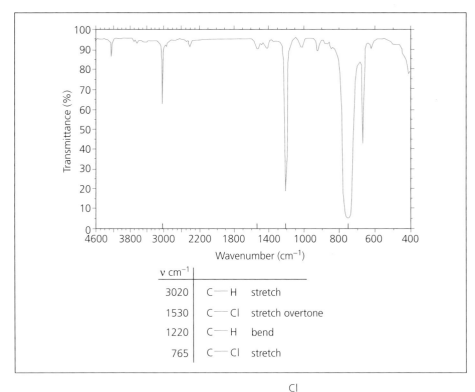

Figure 5.19 Infrared spectrum of trichloromethane

v cm^{-1}		
3020	C——H	stretch
1530	C——Cl	stretch overtone
1220	C——H	bend
765	C——Cl	stretch

5.3 'Impure' ionic bonding

The sure sign that the ionic model is breaking down, for any given compound, is when quantitative predictions based on electrostatic principles begin to give inaccurate answers. As we shall see in Chapter 8, the quantity in which this 'wobble' usually shows up is the **lattice enthalpy**, a thermodynamic quantity which gives a value for the strength of the crystal lattice. If lattice enthalpies calculated on the ionic model start to deviate seriously from the real ones, then you know you have 'model breakdown' (Table 5.6). As you can see, the deviations happen more markedly when either:

1 the cation is small and/or highly charged, or
2 the anion is large and/or not very electronegative.

Different descriptions of impure ionic bonding

The results in Table 5.6, and illustrations such as Figure 5.21, have prompted scientists to offer explanations as to why small, highly charged cations and large, not-very-electronegative anions conspire to provide impure ionic bonding. Accounts tend to belong to one or other of two schools of thought, which might be seen as the 'ions were never really there' school, and the 'they were there to start with, but they affected each other' school. As we shall see, the differences between the two models boil down to no more than arguments about the easiest mental pictures, and both

emission spectroscopy – in which electrons change energy levels (p. 38). After all, photons have energies given by Planck's equation *(3.1)*, and the photons in the infrared region are of lower frequency than those in the visible and ultraviolet which are necessary to excite electrons. Explain, in terms of what the electrons have to go through in the two types of excitation, why it needs more energy to excite an electron than to make a molecule vibrate harder.

Molecular vibrational energy, like electron energy, is **quantised**. A vibrating bond absorbs (and emits) photons one at a time and it cannot absorb a fraction of a photon. So the vibrational energy can only increase (and decrease) by whole-number multiples of the energy of the relevant photons. So possible vibrational energies are on an energy 'ladder' whose 'rungs' are $h \times v$ apart where v is the frequency of the photon absorbed (see

Figure 5.20). The higher vibrational quantum numbers signify more violent vibrations.

Infrared spectroscopy is very important in organic chemistry, since it is so useful for identifying which bonds are present in molecules. So we will be meeting it again in the organic chapters, together with some information about how the spectra are obtained. Now it is time to get back to the main narrative of Chapter 5, and shift the focus from molecules with dipoles (that is, molecules with 'impure' covalent bonding), to compounds with 'impure' ionic bonding.

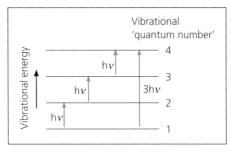

Figure 5.20 An 'energy ladder'

Table 5.6 A comparison between real lattice enthalpies (LE) and theoretical ones, as a test of whether the ionic-bonding model suits a particular compound

Substance	Real LE (kJ mol^{-1})	Theoretical LE (kJ mol^{-1})	Percentage deviation
NaCl	780	770	1.28
CsF	747	739	1.07
MgI$_2$	2327	1944	16.5
MgCl$_2$	2526	2326	7.9

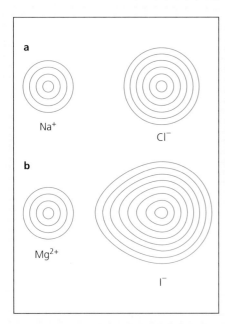

Figure 5.21 Electron density maps: **a** a 'pure' ionic bond in sodium chloride and **b** an 'impure' ionic bond in magnesium iodide, showing how the high charge density of the Mg^{2+} ion polarises an adjacent I^- ion

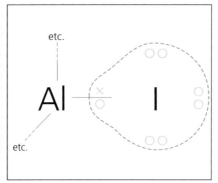

Figure 5.22 Is it a polarised covalent bond? Is it a distorted ionic bond? Does it really matter?

models come up with the same story in the end.

Let us allow both versions to describe the same situation, the bonding in the very impurely bonded aluminium iodide.

They were never really ions in the first place

In this version we would note the great difficulty in removing three sucessive electrons from a Group III element, especially one from quite high in the group, like aluminium. Secondly we would recognise that the degree of nuclear pull being offered by the electron-vacancy in the iodide 5p orbital is far weaker than the sort of reception an electron would get from the 2p orbital in fluorine. From these facts we would deduce that electron exchange never really happens, and that what we are really dealing with is just a highly polarised covalent bond – the electrons are polarised towards the iodine, admittedly, but nowhere near to the extent of a true ionic donation (Figure 5.22). In fact if we wanted, we

could calculate its 'percentage ionic character' using Pauling's ΔN_P method, exactly as we did for bonds which were labelled as 'impure covalent'.

They affected each other

In this version the story begins with true Al^{3+} and I^- ions actually existing, and sitting in a lattice. It then recognises what a high charge-density the Al^{3+} ion has, what with its being so small and highly charged. This intense positive electrostatic field will of course influence nearby electrons, especially if they're not under too strong a 'nuclear discipline', as is certainly the case with I^-. The effect will be to attract the I^- ion's electron density towards the Al^{3+} ion, thus creating an uneven distribution of electrons around the I^- (Figure 5.22). Anions such as I^-, whose outer electrons are easily influenced by cations of high charge-density, are said to be **polarisable**.

Of course the argument between the two schools of thought is hollow, since they end up by supporting the same picture of electron density (Figure 5.22). Perhaps it might be said in favour of the first model that it does a good job of seeing the whole of bonding as a continuous spectrum of bond types. Meanwhile, in the second model's favour, the ideas of charge-density and polarisability do seem to create strong visual images in people's minds.

Questions

23 Put the following compounds on one continuous spectrum from 'most covalent' to 'most ionic'.

NaCl, CH_4, $AlCl_3$, CsF, S_8, CsF, H_2O, $SiCl_4$.

24 From the discussion above it seems that an Al^{3+} ion would be hard to obtain. Which binary compound of Al (where 'binary' indicates a compound featuring Al bonded to *one* other element) would be the one most likely to contain a genuine Al^{3+} ion?

25 Magnesium chloride, which clearly belongs in the 'impure ionic' camp, nevertheless can be electrolysed when molten, as if it were purely ionic, with magnesium and chlorine appearing at cathode and anode respectively. This looks like one up for the 'charge-density/polarisability' school of thought, which does see the ions as really existing. But can you offer an alternative explanation of this electrolysis, from the point of view of the 'they were not proper ions to start with' school?

26 Considerable mention was made of size, both atomic and ionic, in the course of the 'charge-density' discussion. What would you predict was the trend in size of these series (justify your answer)?

a The atoms Na to Ar.

b The ions Na^+, Mg^{2+}, Al^{3+}.

c The ions N^{3-}, O^{2-}, F^-.

27 What do all the ions in series (b) and (c) in question 26 have in common?

28 Why is the chemistry of carbon totally free of ions of either sort? (This is a question which can be approached in either of the two parallel vocabularies. You can either consider the electronegativity differences of various C——X covalent bonds, or alternatively consider what would happen to hypothetical ionic versions of say methane, $C^{4-}(H^+)_4$, or tetrachloromethane, $C^{4+}(Cl^-)_4$.)

As a footnote to this section on impure ionic bonding, here is a warning not to be too rigid in equating likelihood of bond impurity with the size of an anion. Admittedly, in the series made up by the fluoride, chloride, bromide and iodide of the same cation, the degree of partial covalent character does indeed increase in that order, and therefore also in order of anion size. But the biggest anions of all are the molecular ions like sulphate, and they are just exactly as electronegative (or as polarisable) as the *site where the negative charge resides*. In sulphates, this is the O^- part, which is nearly as uncompromisingly strong in its grip on electrons as F^- itself. Thus, harking back to question 24, a genuine Al^{3+} ion is likely to be found in aluminium sulphate, or one of the mixed sulphates of aluminium and a 1+ cation (M), which have the general formula $MAl(SO_4)_2 \cdot 12H_2O$, go by the common name of **alums**, and are beloved of crystal growers everywhere (Figure 5.23).

Figure 5.23 A potash alum crystal

Question

29a As a piece of revision, suggest the *formulae* of the following compounds, assuming ionic character, and remembering that for all its internal complexity, the sulphate ion is still just a 2– anion.
Sodium sulphate, magnesium sulphate, aluminium sulphate.

b By similar reasoning, suggest the formulae of sodium nitrate, magnesium nitrate, aluminium nitrate.

c Suggest the formulae of sodium carbonate, magnesium carbonate, aluminium carbonate.

5.4   The metallic bond

We have named and characterised two types of bond which hold atoms in clusters – ionic and covalent bonds. As we have seen, one type (ionic) works for sets of atoms in which one sub-group is willing to release electrons and the other anxious to gain them. The other (covalent) works for groups of atoms all of which are more or less committed to electron gain, and so end up more or less sharing. It might have occurred to you that there is a third situation – in which all the atoms are more or less committed to electron *loss*.

This is the situation with *metals*, where all the atoms are identical (in pure single metals anyway), and in which every atom has a number of loosely held outer electrons, and quite a few not very attractive vacancies.

This is a good time to think about why metals are where they are in the Periodic Table. *Where* they are is on the left – that is, Groups I and II, with parts of III and lower-IV, plus the whole of the d- and f-blocks (Figure 5.24). *What* they are is

elements with at the most three electrons (more normally one or two) in the outermost principal quantum level. The link here is fairly obvious. Since a new row of the Periodic Table begins with the opening of residence in a new principal quantum level, it naturally means that the early (left-hand) elements will have small numbers of outer electrons.

Yet the word 'metal' means more than just a style of electron arrangement. It also conveys a package of properties:

• high electrical conductivity

• high thermal conductivity

• malleability (literally, beatable into thin sheets) and ductility (literally, drawable into wires), both of which mean the opposite of brittle

• (generally) high melting and boiling points – all metals bar one (Hg) are solids at room temperature, and only five more (Na, K, Rb, Cs and Ga) have melted by 100 °C

• high latent heats of melting and (especially) boiling (also known as enthalpies of fusion and vaporisation).

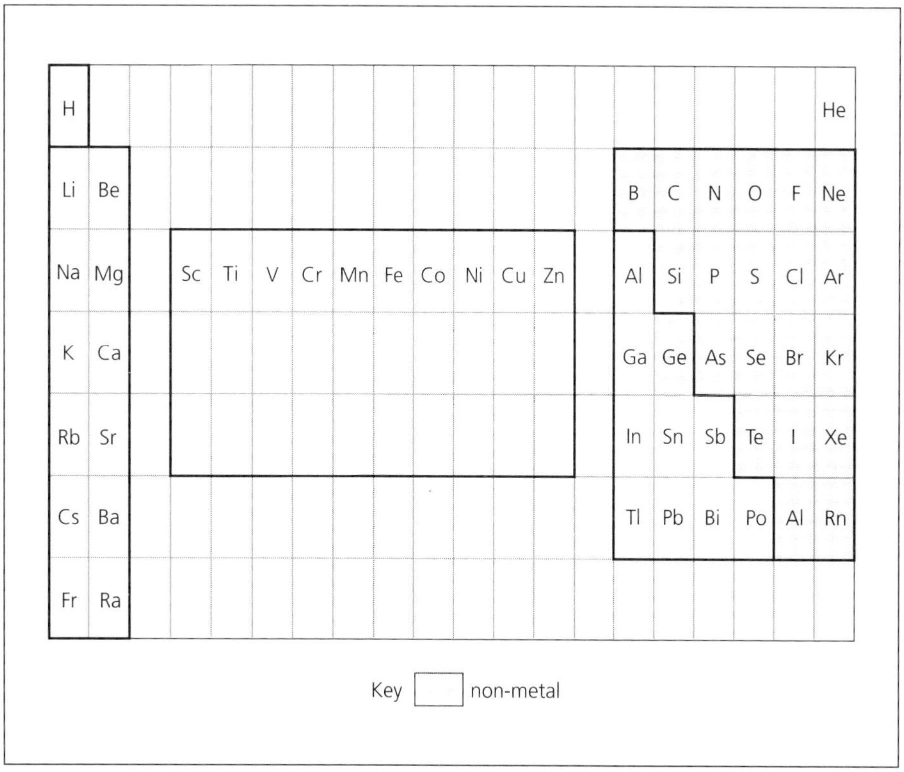

H																	He
Li	Be											B	C	N	O	F	Ne
Na	Mg	Sc	Ti	V	Cr	Mn	Fe	Co	Ni	Cu	Zn	Al	Si	P	S	Cl	Ar
K	Ca											Ga	Ge	As	Se	Br	Kr
Rb	Sr											In	Sn	Sb	Te	I	Xe
Cs	Ba											Tl	Pb	Bi	Po	At	Rn
Fr	Ra																

Key ☐ non-metal

Figure 5.24 Distribution of metallic and non-metallic elements. (Germanium is an element with intermediate qualities.)

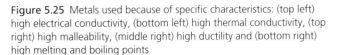

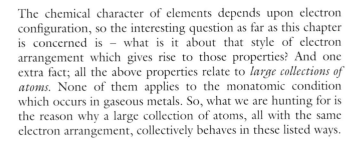

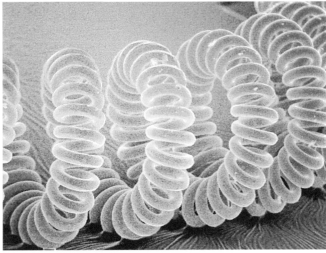

Figure 5.25 Metals used because of specific characteristics: (top left) high electrical conductivity, (bottom left) high thermal conductivity, (top right) high malleability, (middle right) high ductility and (bottom right) high melting and boiling points

The chemical character of elements depends upon electron configuration, so the interesting question as far as this chapter is concerned is – what is it about that style of electron arrangement which gives rise to those properties? And one extra fact; all the above properties relate to *large collections of atoms*. None of them applies to the monatomic condition which occurs in gaseous metals. So, what we are hunting for is the reason why a large collection of atoms, all with the same electron arrangement, collectively behaves in these listed ways.

Question

30 You may be able to sense one of the connections already. Why might elements with only a few outer electrons be good conductors of electricity?

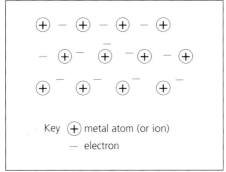

Figure 5.26 A simple model of metallic bonding

The full story of metallic bonding is complicated, and can be found in undergraduate texts under the heading 'Band theory'. For our purposes a much simpler picture will do, that of 'ions in a sea of electrons' (see Figure 5.26). The 'sea' comprises the outer electrons of each atom. They are seen to be delocalised and free to jump from atom to atom, especially under the influence of an applied potential difference (hence the electrical conductivity). The incentive for visiting another atom, from the electron's point of view, is not very great (weak nuclear pull), but then neither is the penalty for leaving its 'home base' (for the same reason). The great benefit of the arrangement, as with covalent bonding, is that it puts electrons between atoms, thus pulling the nuclei together.

Going back to the defining phrase, the 'ions' are the atoms which have donated their outer electrons to the 'sea'. If they are considered as separate entities to the electrons, then of course they would be 'cations' equal in charge to the normal cation which would represent the metal in its ionic compounds.

How well does the 'sea of electrons' model account for the set of properties common to metals and listed above? The key is to remember that metallic bonding has a crucial difference from both the other two styles.

Metallic bonding contrasted to covalent bonding

Both styles depend on there being electron density between atoms, but in the covalent case the electrons are *dedicated to the space between particular pairs*. Conversely, in metallic bonding (to stretch the sea analogy) the electrons

'slop around' indiscriminately, that is, they are *not* dedicated to particular atoms. For metals, this means that the sliding of one 'plate' of atoms over another is *not* a major structural upheaval. It also means that there is not a huge difference in the binding energy of the ordered *solid* state, and the disordered *liquid* one, since the atoms still stay more or less just as close to each other. In contrast the change from liquid to gas requires profound bond breaking.

Question

31 Which of the metallic properties (p. 83) have just been accounted for? Justify your answer.

Metallic bonding contrasted to ionic bonding

Both styles share one feature – the non-directional nature of the bonding. In neither of these styles is there anything like the *pairing* of covalent bonding. In both cases, the nearest thing to a 'partner' in bonding is any and all of an atom's nearest neighbours. But there *are* some important physical contrasts between metals and ionic solids. Metals are much less *brittle* than ceramics (ionic solids), and Figure 5.27 may help to explain this. The sliding of one plate of atoms over another requires that the system goes through rather a tricky moment when the structure is at its least stable. In this respect the moment is an 'energy hump' as well as a real one. Figure 5.27 shows the tricky moment for both an ionic solid and a metallic solid.

Question

32 Why is the metal's tricky moment a good deal less tricky than the ceramic's? How does this account for the fact that a ceramic would rather break than bend?

I must now confess that this account of the deformability of metals is about seventy years out-of-date (but it is still useful). In the 1920s it was realised that the real properties of metals, especially

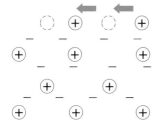

a pushing the top 'plate' of atoms in a metal structure. The original positions are shown by dotted lines. The atoms will come to rest in the next dotted line

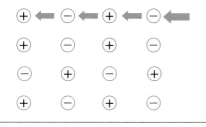

b pushing the top 'plate' of ions in an ionic structure. The ions will not be 'at rest' again until they have moved one more position

Figure 5.27 Trying to deform solids: **a** metallic and **b** ceramic.

the way they behave when compressed and stretched, depend crucially on *imperfections* in the array of atoms in the lattice. It is interesting to note that imperfections, such as a missing atom, occur very commonly when metal atoms are settling into their resting places during solidification. By comparison an imperfection in an ionic crystal is rare. Figure 5.28 shows a 'vacancy' as it might look in a metallic and in an ionic crystal. The metallic structure has sagged inwards, partially to fill the vacancy.

Question

33 The moment when the vacancy is created is the moment when a settling atom or ion 'puts the lid' on the hole. Why is the 'lid' far less likely to settle on the hole in the ionic case than in the metallic one?

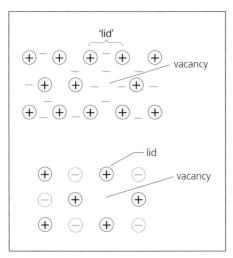

Figure 5.28 Various types of crystal imperfections. **a** Metallic lattice with vacancy – a common occurrence. **b** Ionic lattice with vacancy – a rare occurrence

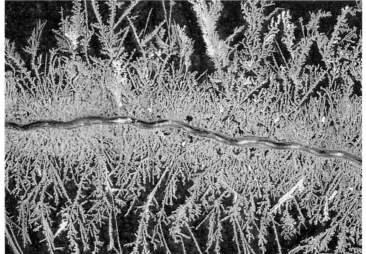

Figure 5.29 Metal crystals can be visible to the naked eye, as in the galvanised coating on steel (left) and dendritic growths of silver on copper wire (bottom)

5.5 Different metal crystal lattices

The juxtaposition of the words 'metal' and 'crystal' may seem strange. Yet metal crystals are visible to the naked eye in the zinc coatings (galvanising) on steel railings and dustbins (lying flat on the surface of course) (Figure 5.29). You may have thought that the drawings such as Figure 5.12 of atomic arrangements in metal crystals told the whole story. In fact, there are *three* naturally occurring ways of packing identical metal atoms into solid lattices. These ways have been found by X-ray diffraction methods, and their description shares some of the language and concepts which were developed for ionic crystals. Thus we shall meet again the unit cell and the atom audit, the co-ordination number and the site.

The truth, as any snooker player knows, is that there is only one way of close-packing a single plane of identical spheres. It is the same as the packing of reds into the triangular frame at the beginning of the game (Figure 5.30).

The three different structures come from two sources:

1 there are variations in the way *successive* layers can lie on top of each other
2 the structure can simply ignore the close-packed layer format altogether.

A close-packed structure – hexagonal close-packed (HCP)

Of all the ideas dealt with in this book, this is about the least appropriate for transmission by words and pictures on paper. Real models are almost essential. As we have seen, there is only one way of close-packing a single plane, and that is to make balls fit into the recesses between balls in adjacent rows. A similar principle attends the placing of the next layer of balls: they do not sit on top of the balls in the first layer, but rather nestle down into the recesses. The possibilities for variation only occur at the moment of constructing the *third* layer.

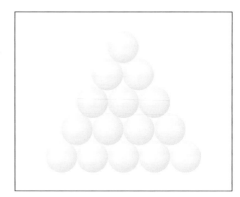

Figure 5.30 The close-packing of identical spheres in a plane

As Figure 5.31 shows, there are two sets of recesses in layer 3. The new balls can either rest above balls in layer 1, or offset from them. And they cannot do *both at once*.

The hexagonal close-packed arrangement is the one in which the layer sequence is ABABAB. The 'hexagonal' part of the name comes from the unit cell. There is a real art to picking out a unit cell, that is mentally extracting from the endless featureless planes of atoms the minimum number necessary to describe the structure. Figure 5.32 shows how the hexagon-shaped unit cell captures the idea of 'ABA'.

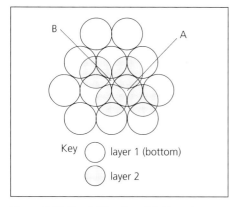

Key ◯ layer 1 (bottom)
     ◯ layer 2

Figure 5.31 Possibilities for a third layer in a close-packed array of identical spheres.

A represents a site in layer 3, where a ball would be directly above a ball in layer 1.

B represents a site in layer 3, where a ball would be above a gap in layer 1. If you start building layer 3 at A, **you cannot use B-type sites**

Figure 5.32 (below left) The unit cell of the hexagonal close-packed (HCP) structure, in which balls in layer 3 lie directly above balls in layer 1, hence the reference to ABABAB
Figure 5.33 (below right) The unit cell of the face-centred cubic (FCC) structure, with an attempt to show how it delivers the ABCABC layer sequence

The other close-packed structure – face-centred cubic (FCC)

There is no problem in accepting that this is the *other* layer sequence. The order is ABCABC, meaning that a ball does not sit directly above a ball in a previous layer until the fourth layer up. The really mind-stretching thing about this structure is trying to see the link between the layer picture and the unit cell. It is much more difficult than the HCP case. There, the close-packed layers ran horizontally across the unit cell and are therefore easy to see. In the FCC case, the close-packed layers slope diagonally through the cell as shown in Figure 5.33. Try it out with a set of models if you can. If you have access to models you will find the following questions easier to answer.

Questions

34 The co-ordination number (p. 66) of both close-packed structures is the same, so I would recommend answering this question by considering Figure 5.32.

a How many nearest neighbours does an atom have in its own plane?

b How many nearest neighbours does an atom have in the layer above, and below, itself?

c Hence what is the total co-ordination number of an atom in a close-packed structure?

35 How many atoms are there in a unit cell of the FCC structure, if you use the 'audit' method of counting (p. 67)?

36 Figure 5.33 shows one set of close-packed planes slanting through the unit cell of the FCC structure. Indicate another set of planes running at a different angle through the same unit cell.

FCC is the structure adopted by three of the most malleable metals of all, copper, silver and gold. In contrast the HCP metals tend to be rather more brittle. (Imagine hitting a toy (zinc alloy) car with a hammer, compared with hitting a piece of gold jewellery.) It is said that the extra malleability of metals with FCC structures is due to having three directions of 'plane-slip' as possibilities, rather than one. This could well be a helpful factor; remember that a single lump of gold, say, consists of many different crystals with their planes all stacked at random angles, whereas the hammer is pushing at just one angle.

A non-close-packed structure – body-centred cubic (BCC)

In this structure (Figure 5.34) the arrangement of the atoms resembles the caesium chloride structure from the ionic family (p. 67, Figure 4.57), except that all the atoms are identical. (It is *not* true to describe the caesium chloride structure as BCC

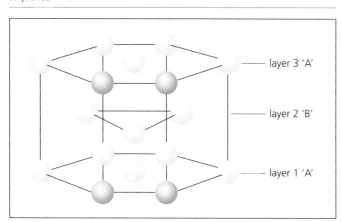

layer 3 'A'
layer 2 'B'
layer 1 'A'

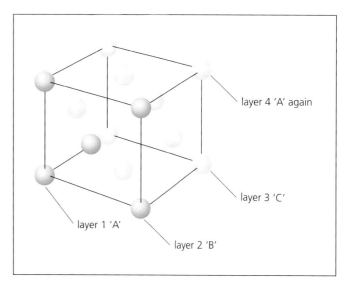

layer 4 'A' again
layer 3 'C'
layer 1 'A'
layer 2 'B'

however; it is best seen as two interlocking simple cubic arrays of contrasting ions.) Body-centred cubic is the structure adopted by the alkali (Group I) metals, and it may be no coincidence that this *non*-close-packed array occurs in some of the *least dense* metals in the Periodic Table. Of the three common metallic structures, BCC is the easiest to analyse, as you should find in the following question.

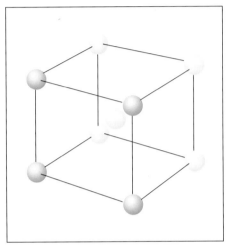

Figure 5.34 The unit cell of the body-centred cubic (BCC) structure, which has no close-packed layers

Question

37 a How many atoms are in the unit cell of a body-centred cubic metal (by the audit method)?

b What is the co-ordination number of atoms in a BCC array?

5.6 Periodicity of properties 1: Trends in the properties of elements

Some chemistry books construct a whole chapter around this heading. We have put in so much groundwork on the underlying principles, that it will be reduced to an applications-and-revision session.

To summarise this chapter so far, chemical character in an atom stems from what it does with its own outer electrons and with other atoms' outer electrons. These two factors are determined by the sizes of ionization energies and electron affinities. These in turn are closely linked with the atom's position in the Periodic Table.

Questions

38 (*Revision, from Chapter 3*) What is the general trend of ionization energy across the Periodic Table?

39 (*Revision*) What is the general trend in electronegativity across the Periodic Table?

40 In answering question 38 you probably thought about the trend in *first* ionization energy. But a fairer comparison might be to consider the *total* ionization energy needed to bring a cation to its normal bonding condition. How would this affect the general trend, as applied to, say, the series Na, Mg, Al?

Position in the Periodic Table

1   Structures of the elements

The left-hand elements in any row are metals, because that is where the willing electron-losers are found. As we have seen, their structures are 'giants' – in other words, continuous bonding systems showing long-range order. The metallic bond, and the arrangements of atoms in metallic structures, have been dealt with extensively in Sections 5.4 and 5.5.

On the right-hand side of the Periodic Table are the fierce electron-grabbers, so all they can do with each other is make covalent bonds, producing small molecules like F_2.

In the middle, the elements of Group IV, upper Group III, and lower Group V, show an interesting compromise. They use covalent bonds, yet make them into giant structures, like diamond, graphite (Figure 5.35) and silicon. Figure 5.36 shows the zones of the Periodic Table in which these styles of structure occur.

Questions

41 It is possible, or at least it does not break the rule of eight, to imagine a quadruple bond between two silicon atoms. Why do you think such a molecule would be unstable?

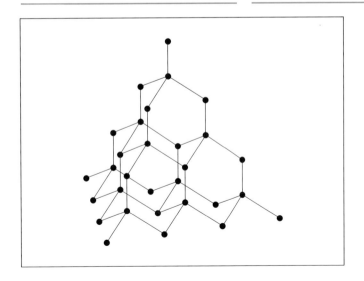

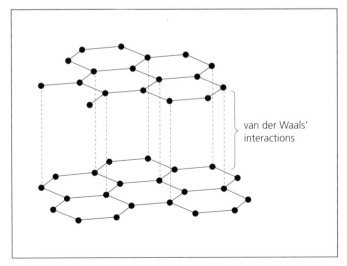

Figure 5.35 The structures of the allotropes of carbon: diamond (left) and graphite (right)

van der Waals' interactions

Figure 5.35
(continued) Uses of
Group IV elements are
shown a diamond-
studded dentist's drill
(top left), a silicon
microchip (top right)
and graphite drawing
materials (bottom left)

B	C	N	O	F
Al	Si	P	S	Cl
Ga	Ge	As	Se	Br
In	Sn	Sb	Te	I
Tl	Pb	Bi	Po	At

Key | Al | metallic

| C | giant molecules

| I | small molecules

Figure 5.36 Where the various types of
structures occur amongst elements of the p-
block of the Periodic Table (All the s-block and
d-block are metallic, and the noble gases are
monatomic.)

Questions

43 This anomaly gives hydrogen what you could call 'honorary membership' of *two* of the groups of the Periodic Table. Which two?

44 Nevertheless, when it comes to structure, hydrogen goes very definitely into the 'small molecules' category. Why do you think there is no metallic form of hydrogen? (This is quite a hard question with no clear-cut answer, and plenty of room for speculation.)

2   Trends within a group

Group I elements are all metals, and Group VII elements are all non-metals, but most of the other groups, while retaining a degree of likeness, show marked variations in character. For instance, formulae of compounds may stay the same in all the members of a group, while the elements themselves appear as 'distant cousins'. All the elements of Group IV, for example, form a covalent chloride whose formula is XCl_4, where X is the Group IV element, yet the elements carbon and lead are quite remote from each other in chemical character. You will have noticed from the structure chart in Figure 5.36 that metallic behaviour is found lower down in groups, and non-metallic behaviour is at the top of groups.

Questions

45 Name a group in which the top element is the only non-metal. Name the element.

46 Name a group that has more than one of both metals and non-metals. Name them.

After all, metallic character is all to do with lowish ionization energies – the fundamental source of metallic character is the ability to release electrons to allow electrical conduction. So the conclusion we must draw is that ionization energies must get less down every group, allowing the lower elements to be freer with their electrons, and in some cases enter the metals club.

42 From the above reasoning, where in the Periodic Table would you expect to find gaseous elements?

Hydrogen does not fit your answer to question 42 – but then hydrogen is in a unique position anyway. It is one electron away from a full-up shell, yet one electron away from an empty shell.

Questions

47 Look up the first ionization energies for three elements in each of Groups III, IV and V, and comment on whether this explanation has agreed with the evidence.

48 Some people talk about a *diagonal relationship*, whereby the top member of each group resembles the second member of the next group along. For example, boron seems closer to silicon in character than it does to its true 'sibling' aluminium. (For example, boron has a giant molecular structure and a covalent chloride, like silicon but not like aluminium.) Was there anything in your data search from question 47 which might offer an explanation for this effect?

3 Chemical trends across a row

At a level of broad generalisation, the chemical properties of the elements can be summed up very succinctly. The elements of the left are strong reducing agents, and the elements of the right are strong oxidisers. More detail will be added to this picture as the rest of the chapters unfold. The meanings of the terms oxidation and reduction are addressed in Chapter 7, where many examples will be drawn from the reactions of elements. Further aspects of the chemistry of individual elements will appear in Chapters 17, 18 and 19 on the s, p and d blocks.

5.7 Periodicity of properties 2: Trends in the character of compounds

This section is a broad overview of the landscape of compounds. A good strategy for obtaining this overview is to consider a single element, and then look at all its compounds with a series of other elements from a single row. The single element chosen is chlorine, and we will consider the chloride of an element X. Figure 5.37 shows an X—Cl bond. It could be the complete molecule, or just one bond within a molecule; it could

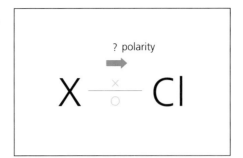

Figure 5.37 The degree of even-ness of the sharing will decide the character of the compound

even be non-existent. One thing is certain, the nature of that bond will determine the character of 'X-chloride'.

As a typical member of Group VII chlorine is an uncompromising 'electron-grabber'. So the properties of X—Cl will depend on the willingness of X to let chlorine have its way, or conversely to fight the pull of chlorine and retain its electron.

Questions

49 a If X is a very willing electron-giver, what will be the nature of the bond from X to Cl?

b So where in any given row of the Periodic Table would you get chlorides of this type?

c If X were just as irresistible a 'grabber' as chlorine itself, what would be the nature of the X—Cl bond?

d So where in any given row of the Periodic Table would you expect to find chlorides of this type?

50 The answers to question 49 are summarised in Table 5.7 for a complete Periodic Table row of X-chlorides. Complete the table by putting in rows describing the trends in:

a melting point (qualitative answer – high/low)

b electrical conductivity when molten (yes/no)

c dipole moment in an X—Cl bond, if the bond is at least partially covalent. (Qualitative only – words like big, small, none.)

Table 5.7 Periodic variations in properties of chlorides

Element	Na	Mg	Al	Si	P	S	Cl	Ar
Formula of rule-of-eight chloride	NaCl	$MgCl_2$	$AlCl_3$	$SiCl_4$	PCl_3	SCl_2	Cl_2	–
Any other chlorides?	–	–	–	–	PCl_5	S_2Cl_2	–	–
Bonding in chlorides	I	I	(v)PC	PC	PC	PC	C	–
Melting point of chloride								
Does molten chloride conduct?								
Dipole moment of X—Cl bond								
Reaction with H_2O	D	D	R/H	R/H	R/H	R/H	Redox	–

I = ionic; PC = polar covalent; v = very; C = covalent; D = dissolves without reacting; R/H = reacts with hydrolysis (i.e. 'breaking by water')

Questions 49 and 50 have focused on the physical character of the chlorides (ionic or covalent, high or low melting point, etc.). We will now consider a couple of chemical trends in compounds, namely the reactions of chlorides and of oxides with water (these are broad trends, fine detail is given in chapters on specific substances).

1   Reactions of X——Cl with water

If the chlorine atom has total possession of the electron pair, we have an ionic compound, and the Cl^- ion will either do nothing with water or dissolve. Most ionic chlorides *are* soluble in water.

On the other hand, polar covalent chlorides are a tempting target for particles which are lovers of poles. There is a particular group of attackers called **nucleophiles**, whose name means 'lovers of nuclei', but which perhaps are better (if less snappily) described as lovers of the positively charged ends of dipoles. Water is one of these nucleophiles, or at least its negatively charged (oxygen) end is. The attack by water on aluminium chloride, $AlCl_3$, is typical of this sort of attack, which ends up with the aluminium–chlorine bonds 'breaking ionically' to let chlorine have what it wanted all along, namely the electron pair. Such reactions are accurately described either as **nucleophilic substitutions**, or as **hydrolyses** (singular *hydrolysis*) meaning breakage by water. The water *takes the place* of the ejected chlorine atom (see Figure 5.38).

2   Reactions of the X——O bond with water

Left-wing and transition metal oxides (like chlorides) are nearly purely ionic. If they dissolve at all in water, and many of them are very insoluble, the oxide ion immediately forms an alkaline solution.

$$MgO(s) \rightleftharpoons Mg^{2+}(aq) + O^{2-}(aq)$$

Only a few MgO molecules dissolve in this way, but as soon as they do:

$$O^{2-} + H_2O \rightarrow 2OH^-(aq)$$

Figure 5.38 Behaviour typical of the 'polar covalent' style of chloride

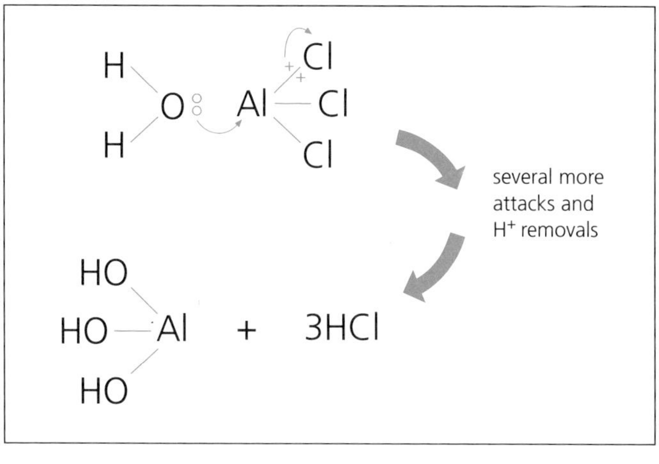

several more attacks and H^+ removals

On the other hand, right-wing oxides are polar covalent, and the oxygen does not come away as a free O^{2-} ion. But its bonds to X are sufficiently polar to bring out the nucleophile side of water's behaviour, resulting in some compounds which also look like hydroxides at first sight. The following shows sulphur trioxide reacting with water.

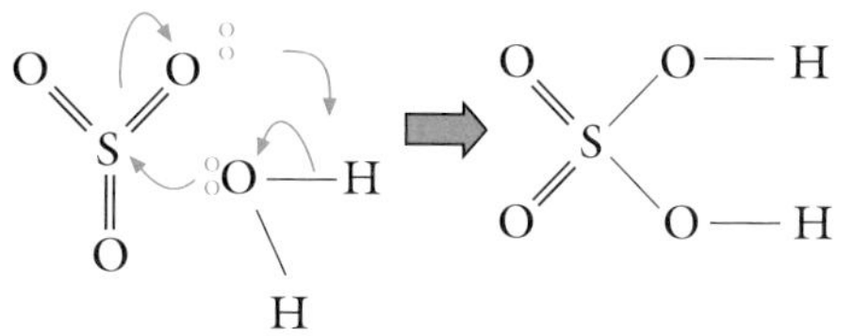

But a closer look reveals that the sulphuric acid molecule has been made, and that if there is any ionic breakage, it will not be in the S——OH bond, but in the bond with the biggest dipole.

Question

51 Which bond is this, and why does its breakage result in acidic behaviour?

The trend in oxide character is shown in Table 5.8 (overleaf). Aluminum oxide, because of its ability to dissolve in both acids and alkalis, is described as **amphoteric**. In acidic media the oxide dissolves conventionally to give $[Al(H_2O)_6]^{3+}$ or $Al^{3+}(aq)$. But as the medium gets more alkaline, the hydration sheath of water molecules either drops off or surrenders H^+s. At a certain pH close to 7, the preferred species is $Al(OH)_3(s)$, which precipitates out. But in fully alkaline media, the precipitate redissolves as $[Al(OH)_4]^-$.

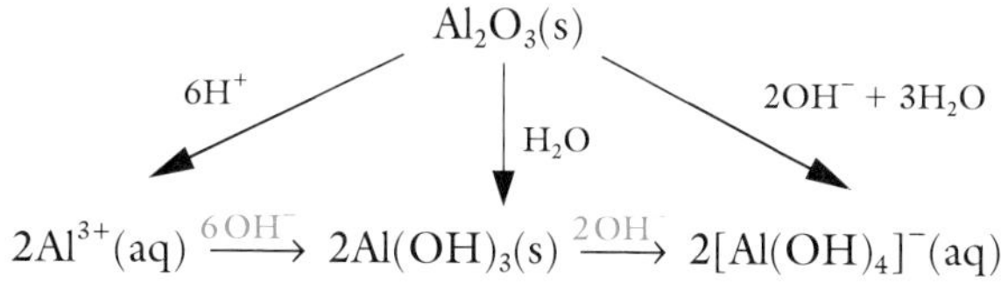

$$Al_2O_3(s)$$

$$6H^+ \qquad\qquad H_2O \qquad\qquad 2OH^- + 3H_2O$$

$$2Al^{3+}(aq) \xrightarrow{6OH^-} 2Al(OH)_3(s) \xrightarrow{2OH^-} 2[Al(OH)_4]^-(aq)$$

Amphoteric behaviour is shown by several atoms whose electronegativity gives the X——O bond a position right in the middle of the 'bond-purity spectrum', for example, zinc and gallium as well as aluminium.

Conclusion

These last two chapters on bonding have really been two subsections of a big and important story. They have been responsible for establishing many conceptual foundation stones. The story of chemical bonds is preparation for much of the rest of the book. When we classify reactions as redox (Chapter 7), we will be considering the transfer of electrons. When we talk about acids and alkalis (Chapter 12), we are concerned with ionic breakage to release hydrogen ions. Complexation (Chapter 19) is the story of dative covalent bonds. The whole saga of organic chemistry (Chapters 10, 11, 13 and 15) has bond polarity as one of its main themes. Finally the big topic of thermodynamics (Chapters 8 and 9) focuses on the energy changes associated with bond making and breaking.

Mastery of the last two chapters, then, is a necessary passport to the rest of the book.

Table 5.8 Trends in oxide behaviour

Element	Formula of oxide	Bonding in oxide	Solubility of oxide in water	Acidity of resulting solution
Na	Na_2O	I	high	alkaline
Mg	MgO	I	low	slightly alkaline
Al	Al_2O_3	–	very low but dissolves both acids and alkalis	
		–		
Si	SiO_2	GM(PC)	none, but dissolves in concentrated alkalis	
P	P_2O_5	M(PC)	high	acidic
S	SO_2, SO_3	M(PC)	high	acidic
Cl	Cl_2O etc.	M(PC)	?	acidic
Ar	–	–	–	–

I = ionic; GM = giant molecular; M = molecular; PC = polar covalent

Summary

- There is a continuous spectrum of bond types ranging from completely even-handed electron sharing between atoms ('pure covalent') to complete donation of one or more electrons from atom to atom ('pure ionic').

- Each pair of atoms will enter into a style of bonding somewhere on that spectrum, the position depending on factors deriving from the nuclear hold over the electrons. These factors are summed up in a single parameter, characteristic of each type of atom, called the **electronegativity**.

- The greater the **electronegativity difference** between two prospective members of a bond, the more ionic will be the style of bonding.

- For convenience, we can consider each individual compound, other than those whose bonding is perfectly covalent or near-perfectly ionic, either as 'impure covalent' or 'impure ionic'. The word 'impure' in this context is not meant to suggest contamination.

'Impure covalent' bonding

- 'Impure covalent' substances tend, like 'pure covalent' ones, to be gases, liquids, or low-melting solids (for example, HCl(g), H_2O(l), glucose(s)).

- The 'impure covalent' bond is characterised by the development of an **electrostatic dipole**, caused by the asymmetrical distribution of electrons. The charge values of the poles are less than whole units of charge (where a

unit is one electron's worth) because the 'electron-grab' at one end of the bond stops short of complete ownership.

- The existence of these dipoles can be shown in a number of ways:

Streams of liquids whose molecules have dipoles may be attracted by charged rods.

Molecules with dipoles experience a **turning force** in an electrostatic field. The moment of the force depends on the product (charge × separation distance) for the particular dipole. This product is called the **dipole moment** (measured in debyes, D).

Bonds which are already dipolar may go further and allow a fully ionic exchange of electrons. This happens when certain 'impure covalent' molecules meet and dissolve in water – the resulting solutions are found to be electrolytes.

- All the **acids** belong to this group, acidity being caused when a dipolar covalent bond to hydrogen, H—X, breaks ionically to leave the X with both the 'dot' and the 'cross' electrons. The solution now contains H_3O^+(aq) and X^-(aq). (X represents whatever is the rest of the molecule.)

- Harking back to Chapter 4, we see new possibilities for **delocalisation** on the X^- or X^{2-} anions after H^+ ions have left in this way (notable examples of such ions being NO_3^-, SO_4^{2-} and CO_3^{2-}).

- Looking forward to Chapter 6, we should note that the existence of dipoles

in a molecule has important effects upon factors such as solubility in water, and on melting and boiling temperatures.

- The overall dipole moment of a molecule is the sum of the dipole moments of its individual bonds. The sum can be done by the mathematical technique of **vector addition**.

Infrared spectroscopy

- All molecules' bonds are in a perpetual state of **vibration**. The vibration of a bond naturally affects its length, so if the bond is dipolar the charge separation distance will oscillate along with the bond. This will in turn set up an oscillating electromagnetic field. It is thus possible for 'resonance' effects to occur between the oscillating electromagnetic fields of the bond and of a photon. This is the mechanism which enables molecules to absorb photons in the **infrared** region of the electromagnetic spectrum and pass into excited vibrational energy states.

- Because different types of bond differ in their vibrational frequencies, they absorb photons at **characteristic wavenumbers**, and thus give rise to the analytical technique of **infrared spectroscopy**. (Wavenumber is the reciprocal of wavelength; it is proportional to frequency, and is measured by convention in cm^{-1}.)

- These differences of wavenumber of absorption between one type of bond and another are due to variations in **bond strength**, and variations in the **mass of the vibrating atoms**.

'Impure ionic' bonding

• 'Impure ionic' bonding can be imagined as the **incomplete donation** of electrons from cation to anion.

• Factors which increase the possibility of 'impure ionic' bonding are:

– a high charge density on the part of the cation (as in Al^{3+}, Fe^{3+})

– a poorly retained set of outer electrons on the part of the anion (as in I^-, S^{2-}), a condition often described as **easy polarisability**.

• If one or both of these two conditions apply, the cation is said to have 'polarised' the electron cloud on the anion.

• Another way of saying this is to say that the kind of atoms in question (Al, I) were exactly the ones which were least willing to give and to receive electrons in the first place. 'Impure ionic' bonding is to be expected when electronegativity differences go below about 1.5.

• Symptoms of 'impure ionic' bonding include a discrepancy between the real value of the **lattice energy** (see Chapter 8) and the theoretical one calculated on the basis of the ionic model. Also, 'impure ionic' compounds can sometimes show melting temperatures lower than is typical of 'proper' ionic compounds (AlI_3, for instance, melts at 191 °C).

• 'Impure ionic' compounds can still undergo electrolysis, since the final cleavage of the dubious bond into genuine ions is prompted by the electrical energy of the source. (This is even true to a small extent in water.)

Metallic bonding

• Metallic bonding has a sophisticated post-Advanced-level theoretical explanation called 'band theory', but here we will limit our model of metallic bonding to the phrase 'ions in a sea of electrons'. The electrons in question are the loosely held outer-shell ones, of which all metals have one, two or three.

• This model explains the **electrical conductivity** of metals, and to some degree their **malleability**.

• There are only three metal structures to be studied, and they can all be modelled as different ways of arranging identical spheres in space.

• Two structures feature identical planes of atoms in 'close-packed' formation. They only differ in the way these planes are laid on top of each other.

• If the planes alternate, so that the atoms in plane 3 lie directly above those in plane 1, then the unit cell is a hexagon and the structure is called **hexagonal close-packed** (**HCP**) (p. 87).

• If the superimposition of plane 1 does not happen until you reach plane 4, then the unit cell can be seen (with difficulty) to be face-centred cube, giving this structure the name **face-centred cubic** (**FCC**) or occasionally **cubic close-packed**.

• This is not to be confused with the non-close-packed member of the metals set, whose unit cell gives it the self-evident name of **body-centred cubic (BCC)**.

• There is not much pattern about the way these structures occur in the Periodic Table, except that the alkali metals (Group I) are all BCC, while all the very malleable coinage metals (copper, silver, gold) are FCC. HCP metals (for example, zinc) tend to be more brittle than the others.

Periodicity of properties in elements and compounds

• There are many properties which can be predicted for elements and their compounds purely on the basis of their positions in the Periodic Table.

• All of the major trends can be seen as relating in some way to the variation in **electronegativity** as a row is crossed.

• Bonding in the elements varies from **metallic** on the left (for example, Na, Mg), through **giant molecular** in the middle (often with semiconductor properties, for example, Si, Ge), to **small molecules with pure covalent bonds** on the right (for example, O_2, Cl_2). Melting and boiling temperatures begin lowish on the left, with the Group I elements, reach a maximum with the giant-lattice semiconductors in Group IV, and then dive to extreme lows with the gaseous small-molecule elements on the right.

• d-block elements are all metals, and show much less variation in properties (see Chapter 19).

• The compounds of the elements, on crossing a row of s- and p-block elements (the so-called 'typical elements'), show regular patterns of variation of properties.

• For example, the chlorides of these elements vary from **ionic crystalline solids** (at room temperature) like NaCl, through the dubious chlorides of Group III ('impure ionic' such as $AlCl_3$) to the **small-molecule** chlorides of Groups V, VI and VII (for example, PCl_3, S_2Cl_2, ICl). The degree of dipolarity decreases across the last three groups.

• These trends are highlighted by considering behaviour towards water. Left-hand elements' chlorides dissolve to form **neutral aqueous electrolytes**, middle elements' chlorides tend to **undergo vigorous attack** at the positive end of the X—Cl dipole, while right-hand elements' chlorides may do some quiet **disproportionations** (see Chapter 7).

• A comparison of oxides across a row of typical elements (s- and p-block) shows a similarly marked variation. Left-hand oxides (such as CaO) are **ionic and basic** (in the acid–base sense, Chapter 12), middle oxides (such as Al_2O_3) tend to be **amphoteric** (having both acidic and basic tendencies depending on environment), while right-hand oxides are small molecules with **acidic** (SO_2) or **neutral** (N_2O) reactions in water.

6

Intermolecular forces

6.1   Introduction

Let us begin this chapter by reviewing one of the fundamental differences between ionic and covalent styles of bonding. This difference relates to the direction of the forces in the two styles of bond.

First, we consider an ionic structure. As Figure 6.1 shows, once ions have been created, there is a force field around them in all directions. Any oppositely

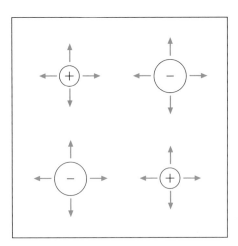

Figure 6.1 The force field around an ion is multi-directional

charged ion within that region will experience that force, and indeed will be the source of a force field of its own, so that mutual forces of attraction are set up.

The crucial point is this: that having attracted one ion, our central ion has not exhausted its capacity to attract others. As we have seen (p. 67), Nature takes this as an invitation to cluster as many anions around a cation (and vice versa) as possible. The only directional restriction on an ionic bond is that set by the problems of packing identically-charged spheres close to each other.

The picture is very different when we consider a covalent structure. For there to be a covalent bond between, for example, the oxygen and hydrogen atoms in water, there must be an overlap between the two orbitals containing the bonding electrons. There must also be vacancies in the overlapping orbitals for the visiting electrons to be accommodated on each atom. For these reasons, as we saw in Chapter 4, the bonding in water has a highly specific directional character, producing a bond angle of 104 ° (Figure 6.2).

Furthermore, once the two O—H bonds in a water molecule have been made, the oxygen atom has exhausted its attractiveness for further bonding to

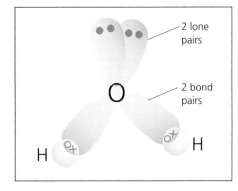

Figure 6.2 Covalent bonds are localised between a pair of atoms, and operate in one direction only

hydrogens. It has not become a fully fledged oxide ion, attracting any hydrogen which falls within its electrostatic field; if it wanted to attract more than two hydrogen atoms it would have to offer another electron in another orbital with another vacancy, the impossibility of which you will understand from your study of Chapter 4. Thus a water molecule is a self-contained self-sufficient bonding unit, with a beginning and an end (Figure 6.3a). The next water molecule is the same, and is neither dependent on nor structurally linked to the first. Ions, in contrast, are caught up in

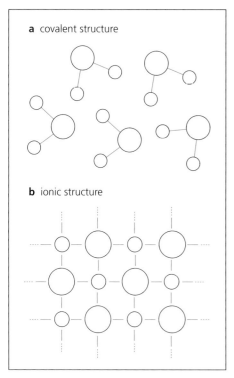

a covalent structure

b ionic structure

Figure 6.3 Covalent structures are small and local, while ionic structures go on 'for ever' with no weak spots

has made itself too useful already in the course of the last two chapters to be rejected. It worked well on shapes of molecules and formulae of compounds, for example, and explained the high melting and boiling points of ionic compounds. Yet clearly the theory needs a major tweak in relation to covalent structures. The statement in the introduction about how the oxygen in water has 'exhausted' its capacity to attract further hydrogen atoms once it has taken two on board, must be carefully inspected again. Although the classical covalent routes to bonding *are* exhausted at H_2O, perhaps there remains some bonding force beyond the horizon of our previous theories, by which a bit of one water molecule can hang on to a bit of the next one.

Let us look at some data to support the need for new thinking. Why is 41 kJ of energy required to boil a mole of water, *after* it has reached the temperature of 100 °C (373 K)? Certainly it is nothing to do with breaking the bonds *within* the molecule, since steam is as much H_2O as either of the other forms

of water. Similarly, why is there not enough thermal energy in the molecules of water to break free of the ice structure long before reaching the comparatively high temperature of 273 K?

The answer takes the form of a logical necessity – there *must* be something holding water molecules together, one to another. One set of forces must give way to thermal agitations at 273 K, to allow melting, while the rest give way at 373 K, allowing water molecules to roam free in the gas phase. Yet both sets of forces must be distinct from the classical covalent bonds in the water molecule, since these remain unaffected throughout the changes.

This chapter is about the nature of these intermolecular forces, not just in water but in every covalent or monatomic substance that exists. Even helium is a liquid below 4 K, so the Universe's least attractive atoms must find something to interest each other, at least until the extremely modest thermal jostlings at 4 K (–269 °C) prise and wobble them apart.

the vast continuous force fields which turn an ionic lattice into one big single bonding system (Figure 6.3b). And yet, can there really be *no* interaction between molecules in water? Let us now see why that idea cannot be allowed to stand unmodified.

6.2 There must be a force between molecules

If everything you have just read in the introduction were true, all covalent substances would be *gases*: and what is more, they would be gases right down to just above the absolute zero of temperature. If there is really *nothing* by way of a force between one water molecule and the next, then the slightest thermal agitations, meeting no resistance, would manifest themselves as translational kinetic energy (that is, free-moving particles in the gas phase). When a theoretical model projects a picture of the Universe that is clearly nonsense, it is time to re-inspect the model, and either reject or refine it.

Of those two options, refinement seems the better. Our bonding model

Boiling

Before we pass on to the next section, there is something to be said on the subject of boiling temperatures. The value of 373 K for the

boiling temperature of water is *not* a constant (Figure 6.4). A liquid boils when its vapour pressure (which depends upon the *escaping tendency*

Figure 6.4a Water boils at a temperature lower than 100 °C at high altitude

Figure 6.4b Water boils at a temperature higher than 100 °C in a pressure cooker

of the molecules) reaches a pressure equal to the external atmospheric pressure. Boiling is distinct from evaporation, in that boiling can even take place in a closed syringe, as long as the vapour can 'push back the atmosphere' (Figure 6.5). But this in turn depends upon the external atmospheric pressure, so boiling temperatures are lower when there is less atmosphere to push back. 373 K refers to the boiling point of water at 'standard' pressure, which is 101.325 kPa.

(At 2 atmospheres water boils at 120 °C; at 0.5 atmosphere, or 400 mmHg, it boils at 83 °C.)

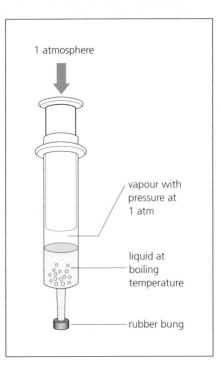

1 atmosphere

vapour with pressure at 1 atm

liquid at boiling temperature

rubber bung

Figure 6.5 Showing that a liquid can boil even in a closed syringe, and that what we call boiling temperature is the temperature at which the liquid's vapour pressure reaches 1 atmosphere

6.3   Sources of intermolecular forces 1: Dipoles

It is a good idea on any search to have some idea of the magnitude of your quarry. The 41 kJ figure quoted in the last section gives us a clue. It is the work needed to separate a mole of water molecules one from another. From one point of view it is a big amount – to use the mechanical-work yardstick, it would lift about 4000 bags of sugar from the floor to the table. Yet when set beside the amounts of energy associated with classical bonding, it is very modest. It is about *one tenth* the size of the work needed to break one of the two hydrogens away from the oxygen – in other words, to break a classical covalent bond. This point is emphasised when you consider that water boils at 100 °C (373 K), but does not start to decompose into atoms until the temperature is well over 1000 °C (1300 K). These considerations indicate that we are looking for a force which is much smaller than the forces produced by the exchange of electrons, and which can work at slightly greater distances.

And yet we cannot be on the track of anything too exotic. After all, how many different types of force are there? Physicists have listed the types of force at large in the Universe, and are busy trying to show that they are all versions of a single unifying force. As things stand at present for the ordinary chemistry student, there is **gravity**, and there is **electromagnetic force**. Of the two, the second must be the prime candidate, because gravitational forces are too *small* to account for the phenomena we are explaining. The inter-atomic bonding theories we have met are based on the fact that matter at a sub-atomic level is made up of particles with electrostatic charges. Therefore it seems probable that *intermolecular* bonding must tap the same source, but in a way we have not yet encountered.

What new variation on the theme of electrostatic force might there be? We will begin by looking at **dipoles**. As we saw in Chapter 5, many covalent molecules are known to have dipoles, that is, sites where electrostatic charge has built up. What more reasonable source for an intermolecular force than the attractions

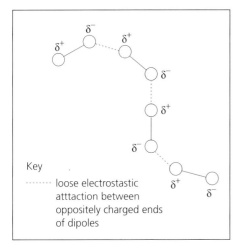

Key
.... loose electrostastic
atttaction between
oppositely charged ends
of dipoles

Figure 6.6 Dipoles as a source of intermolecular attraction

between dipoles on adjacent molecules (Figure 6.6)? We can put this theory to the test by comparing data from molecules with and without serious dipoles, but with similar molecular masses.

Questions

1 Which of the molecules below would you expect to have dipoles? How does this agree with the data in Table 6.1?

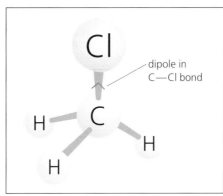

propanone   propanal   butane

Table 6.1 Evidence for dipoles as a source of intermolecular force (see Figure 6.6)

Molecule	Relative molecular mass	Boiling point (K)
Propanone	58.1	329
Propanal	58.1	322
Butane	58.1	272

2 Make a drawing to show the dipole on a propanone molecule. Then show how two molecules might orient themselves relative to each other so as to maximise the attractions between their dipoles.

dipole in C—Cl bond

Figure 6.7 The rather compact and 'blob-shaped' chloromethane molecule

So a clear picture emerges from this data search, suggesting that dipoles are indeed a source of intermolecular force. However, it must be admitted that the molecules in Table 6.1 were rather carefully chosen. First, any molecule with an electropositive hydrogen atom (that is, a hydrogen atom made '$\delta+$' by polarisation of the covalent bond to its partner) has been avoided because it has an important system of intermolecular bonding all to itself. Second, the message would be much less clear-cut if substances were included such as chloromethane (Figure 6.7), relative atomic mass 50.5, a healthy dipole moment (1.86 D), and yet a boiling point even lower (249 K) than that of butane.

Why does chloromethane not fit the pattern which links dipole, as cause, to intermolecular force, as effect? It appears that we have to control not only the variable of relative molecular mass, but *physical size and shape* as well. (Chloromethane is a much less *linear* molecule than the others.) So what other intermolecular force depends on size and shape as well as molecular mass, and works alongside dipole–dipole interaction?

6.4 Sources of intermolecular forces 2: Van der Waals' forces

There obviously has to be another intermolecular bonding force, since we have to explain attractions between molecules with no dipoles at all. At the end of Section 6.3 we suggested it had to do with the size, shape and molecular mass of molecules. First, we will discuss the molecular mass.

You might think that it is obvious that heavy molecules should have higher boiling points than light ones, because it would take more energy to get them moving – 'thermally agitated'. This argument is not quite as strong as it may seem. Temperature is related to average kinetic energy per molecule, and so molecules at the same temperature will have the same kinetic energy. The heavier molecules may be slower, which would hamper escape, but they are compensated by their greater mass, which makes them harder to stop once they're moving. So when it comes to using kinetic energy to overcome intermolecular forces, the heavier molecules have just as much chance as the lighter ones. On the other hand, boiling points do correlate with molecular mass (consider the series of halogen molecules, or alkanes). It appears that something which goes up alongside molecular mass is the real cause of increasing intermolecular forces.

Figure 6.8 Johannes Didente van der Waals (1837–1923)

This is where shape and size come in. The so-called **van der Waals' force** is indeed dependent on the size and shape of molecules, and yet the actual origin of the force is still electrostatic. Van der Waals (Figure 6.8) suggested that even in molecules with no *permanent* dipole, there will still be small *transient* dipoles produced by uneven distribution of electron density around each atom. What is more, if a dipole sets itself up by chance in one atom, it will influence the electron density in the next atom in a way that favours attraction (see Figure 6.9). The

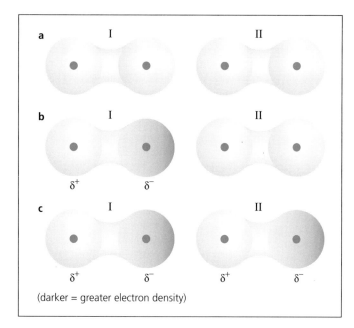

(darker = greater electron density)

Figure 6.9 Steps in the sequence of events leading to a van der Waals' force. The dipoles are temporary, and the whole process is just as liable to go in reverse.
a Two non-polar molecules with even distributions of electrons.
b By chance, an uneven electron distribution occurs in molecule I, creating a slight dipole.
c Electrons in II are repelled by the δ– in I, so a matching dipole is induced in II. The two dipoles are now a source of attraction

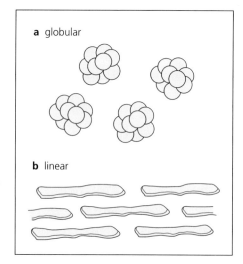

Figure 6.10 Implying how linear molecules can interact with each other more intimately than globular ones

situation is short-lived, but constantly self-renewing, because just as one van der Waals' attraction collapses another begins, somewhere else within the mass of molecules. Overall, the effect produces bonding.

If van der Waals' forces depend upon the uneven distribution of electrons, it follows that the molecules with highest van der Waals' forces will be those in which it is easiest to set up such a distribution. At first sight a prime candidate might be hydrogen, since a one-electron atom has a perpetually uneven distribution of electrons. Unfortunately there is no data on the behaviour of collections of single hydrogen atoms, since they do not exist under normal conditions. However, similar arguments can be put forward concerning helium.

The chances of two electrons 'both being on the same side' of a helium atom are quite high, much higher for example than the chance of having a similar arrangement in neon. (Also in neon some of the electrons are p-type, which are inevitably spaced at 90° to each other.) Yet as we know there is no substance in the Universe which has *less* inter-particle bonding than helium.

We have allowed ourselves to be drawn into this false hypothesis by a model which sees the electrons behaving as passively as, say, tiddlywinks being moved on the throw of a die. First, remember that electrons repel each other and so resist getting 'heavily clustered', espe-

cially when they are in the same orbital. Second, the electrons in helium are the most 'tightly disciplined' (by their nucleus) in the entire Periodic Table (being even more firmly bound than those of fluorine).

So this leads towards a more realistic idea – that the atoms and molecules which are better at setting up van der Waals' attractions are those whose outer electrons are *less* 'well disciplined', or in other words are more polarisable. So for instance molecules containing iodine will be more polarisable (as distinct from permanently polarised) than those containing fluorine, and a similar pattern should hold in other groups. Now at last we see the true reason for the connection between van der Waals' forces and molecular size. The big multi-electron molecules are on the whole going to be the ones with the most polarisable electron clouds, and so if they depend on van der Waals' forces for their intermolecular attractions, they will have the highest melting and boiling points.

There is still the shape variable to consider. If a dipole which is small and short-lived is to have the effect of inducing a similar dipole in its neighbours before it itself dies away, then the degree of intimacy between molecules is clearly an important factor. Molecules which can lie close to each other, and have, as it were, a high area of contact, will be better at setting up van der Waals' attractions than globular near-spherical molecules (Figure 6.10).

Questions

3 Using a data book, select a series of molecules in which the shape and size of dipole are held reasonably constant, while the number of electrons varies. (The elements might be a fruitful place to look.) For each one of this series, record an item of data which will act as a measure of intermolecular forces. Is there evidence that number of electrons is a significant factor in determining van der Waals' forces?

4 Now consider a series of molecules in which number of electrons and degree of dipolarity are constant, and shape is the only variable. Look in the organic section of the data book, where you will find a trio of isomers with molecular formulae C_4H_9Br, differing only in the way those atoms are arranged. They are shown below.

$$CH_3—CH_2—CH_2—CH_2—Br$$

1-bromobutane

$$CH_3—CH_2—CH—CH_3$$
$$|$$
$$Br$$

2-bromobutane

$$CH_3$$
$$|$$
$$CH_3—C—CH_3$$
$$|$$
$$Br$$

2-bromo-2-methylpropane

a Which variables are controlled by selecting a trio of isomers?

b Record and comment on any evidence which supports the idea that shape is a significant factor in determining van der Waals' forces.

c Which of the three molecules is playing the part of the linear molecule in Figure 6.10, and which is the globular one?

5 The most spectacular demonstration of van der Waals' forces is in polymers, many of which have no other intermolecular force and yet have melting points over 200 °C. Poly(ethene) (whose brandname is polythene, and which is used to make washing-up bowls etc.), for example, belongs to the same alkane family of hydrocarbons as does methane, and yet because of its length has a softening point in excess of 150 °C.

$$\cdots CH_2—CH_2—CH_2—CH_2—CH_2 \cdots$$

a Explain why the molecular shapes of polymers equip them to take advantage of the van der Waals' style of bonding.

b Figure 6.11 shows, in a rather schematic way, the different molecular arrangements in two forms of poly(ethene). Can you suggest reasons for the differences in physical properties between two substances which are after all chemically identical?

6 Although van der Waals' forces make a more significant contribution to intermolecular forces in some molecules than in others, they do contribute something in *all* situations where any atoms or molecules gather together. Can you explain why any particle has the necessary equipment for van der Waals' bonding, whereas dipole–dipole interactions are 'selective'?

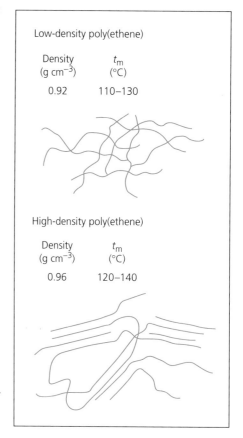

Low-density poly(ethene)

Density (g cm^{-3})	t_m (°C)
0.92	110–130

High-density poly(ethene)

Density (g cm^{-3})	t_m (°C)
0.96	120–140

Figure 6.11 Different types of packing of poly(ethene) molecules

A footnote on van der Waals' radii
An electron orbital does not have an exact boundary. From Chapter 3, p. 41, an orbital is no more than an envelope of probability. Hence if you want to talk about the 'size' of an atom, you have a logical difficulty, because there may be a low but finite chance of finding the outer electrons some way from the nucleus. One of the ways round this is to say that the radius of an atom is half the distance to the next identical atom, in a solid lattice of the atoms.

However, even this approach has its problems, as Figure 6.12 demonstrates, using the example of a portion of the lattice of solid iodine, I_2. Which distance shall we divide by two – the distance between covalently bonded

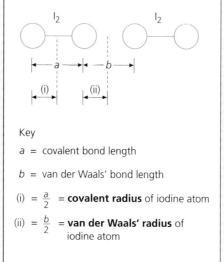

Key

a = covalent bond length

b = van der Waals' bond length

(i) = $\frac{a}{2}$ = **covalent radius** of iodine atom

(ii) = $\frac{b}{2}$ = **van der Waals' radius** of iodine atom

atoms (which is somewhat artificial because the two atoms will have been pulled together slightly by the bonding process) or the distance between two non-bonded atoms? In fact, both distances are used. Distance (i) in the diagram is called the **covalent radius** of iodine, and distance (ii) is called (and this is what this footnote is about) the **van der Waals' radius** of iodine, on the basis that the two atoms in question are sitting at a separation distance dictated by van der Waals' attractions.

Figure 6.12 Using various concepts of atomic radius. The covalent radius is smaller than the van der Waals' radius

6.5   Sources of intermolecular forces 3: Hydrogen bonds

We now move from the intermolecular force which is universal – the van der Waals' force – to the one which needs the most narrowly specific conditions. Some molecules have such peculiar melting and boiling points that no explanations in terms of dipoles or van der Waals' forces will do. Water is a case in point – a tiny molecule with an unpromising V-shape, it is still a liquid at temperatures which would have sent every other molecule of its size into the gas phase – methane and ammonia to name but two.

The most famous demonstration of the odd behaviour of certain molecules, water included, is given by the graph of boiling point against row in the Periodic Table for the hydrides of Groups IV, V, VI and VII, and for the noble gas elements (which of course do not form hydrides). The graph is shown in Figure 6.13. The noble gases provide a sort of 'baseline', showing the effect of van der Waals' forces on a typical group's boiling points.

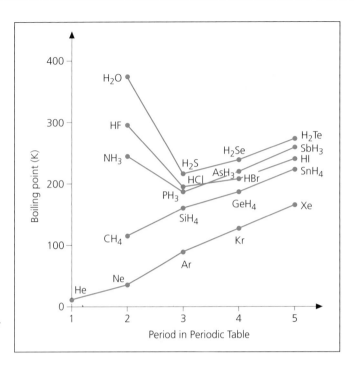

Figure 6.13
Spectacular pattern-breaking behaviour by water, ammonia and hydrogen fluoride, in a survey of boiling points

Question

7 First we will justify using the noble gases as a van der Waals' baseline. Remember that van der Waals' forces depend upon number and polarisability of electrons and on molecular shape.

a Using the model for the origins of van der Waals' forces, account for the fact that the noble gases' boiling points increase as you descend the group. The answer is more than just 'because the atoms get bigger' because (see Section 6.4, p. 97) it is not sheer size which delays boiling.

b Consider a vertical column of points on the graph in Figure 6.13. (Notice that this graph works in the opposite way to the Periodic Table, in that the vertical columns are periods and the horizontal lines are groups.) What happens to *polarisability* as you go through the series CH_4, NH_3, H_2O, HF, Ne?

c So if van der Waals' forces were the only source of intermolecular bonding in the series CH_4, NH_3, H_2O and HF, place them in order of boiling temperature, lowest first. Point out ways in which this sequence differs from the real one.

d If van der Waals' forces were the only contributor to the intermolecular forces in those hydrides, what gradient would the 'across' line for each group show?

I hope that your own answers to question 7 have convinced you that any deviation from the slope set by the inert gases represents a contribution from a style of intermolecular bonding *other than* van der Waals' forces. You can see from Figure 6.13 that deviations seem to fall into two groups, the moderates (such as HCl) and the monsters (H_2O, HF, NH_3). The moderate deviations can perhaps be put down to a contribution from dipole–dipole interactions, as we know that HCl has a well-developed dipole. A whole new type of explanation is needed for the behaviour of H_2O, HF and NH_3.

The model that accounts for these anomalous properties proposes an 'élite' form of intermolecular force called a **hydrogen bond**. It is the only one of the three intermolecular forces dignified by the word 'bond', and indeed it is suggested that lone-pair electrons may play a part. Its range of bond strengths is quite respectable too, of the order of tens of kilojoules per mole, with the very strongest hydrogen bonds (that is, those involving fluorine) breaking three figures. (Compare this with typical classical covalent bonds, whose bond energies are always in three figures, and typically around 350 kJ mol^{-1}.) However, it is the *relatively* high bond energies of hydrogen bonds, compared to other types of intermolecular force, which give substances like water, ammonia and hydrogen fluoride their anomalously high boiling points.

There are two criteria which must be met before a hydrogen bond can form between two molecules, a state of affairs which emphasises the 'élite' nature of the bond:

1 One of the molecules must contain a covalently bonded, yet heavily positively polarised, hydrogen atom. This condition is met most frequently in hydrogen atoms attached directly to fluorine, oxygen or nitrogen atoms, although some unusual groupings like —CCl_3 can do it too (Figure 6.16).

2 The molecule on the other end of the bond must contain a very electronegative atom featuring a lone pair. This

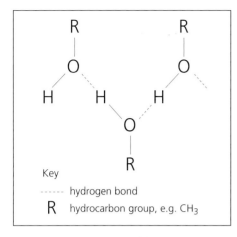

Figure 6.14 Typical chain of hydrogen-bonded alcohol molecules. In the liquid state these bonds could be broken and remade quite often. (It is assumed here that there is only one hydrogen bond per oxygen atom, although a maximum of two may be possible, as in ice.)

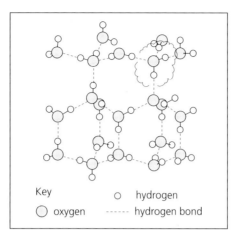

Figure 6.15 The ice structure, obtained from X-ray evidence. Each oxygen atom is in the same formation as the carbon atoms in diamond. One typical oxygen atom in the structure has been highlighted with a 'cloud' symbol. Notice how the oxygen atom has four bonds, arranged tetrahedrally, two of which are covalent and two hydrogen bonds

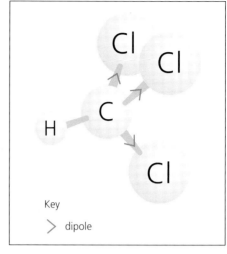

Figure 6.16 Trichloromethane. If there are strong dipoles in each C——Cl bond, what effect will that have on the H——C bond? See question 11

again narrows the field to fluorine, oxygen or nitrogen (not even chlorine gets through this test), and suggests that the nature of a hydrogen bond is a cross between some sort of unofficial dative covalency and a big dipole attraction.

The presence of a group like OH in a molecule *guarantees* the existence of hydrogen bonding, because the very same atom, oxygen, which creates the polarised hydrogen atom on one molecule is waiting on the next molecule to be the other end of the bond (Figure 6.14). Thus the entire group of organic chemicals known as **alcohols** display hydrogen bonding in their solid and liquid states, and have significantly higher boiling points than other similar-sized molecules.

Questions

8 Make a data search to show, by comparison with suitable non-alcohols, the contribution to boiling point made by hydrogen bonding in alcohols.

9 Give another example of a group of substances all of which would, like the alcohols, show hydrogen bonding.

10 Redraw Figure 6.14, replacing the 'R' groups with H, thus creating water. Then consider why there is an extra *dimension* to the hydrogen bonding in

water, compared with the alcohols. Extend your diagram to show this extra dimension.

The suspicion that lone pairs are involved in hydrogen bonding is founded on evidence like the structure of ice (Figure 6.15). It appears that each oxygen atom can make *two* hydrogen bonds, which, by coincidence or otherwise, is the same number as the lone pairs on the oxygen. This means that each oxygen atom has a total of four 'bonds' coming out of itself, two classical ones to the hydrogen atoms in its own molecule, and two hydrogen bonds to hydrogen atoms on adjacent molecules. The fact that these four bonds are arranged tetrahedrally means that the two hydrogen bonds are pointing in exactly the directions which, in a single isolated water molecule, would be the directions of the two orbitals containing the lone pairs.

A curious result of this tetrahedral arrangement of bonds around the oxygen atom is that ice has a structure which echoes that of diamond, except that each 'bond' is really a composite of a classic covalent and a hydrogen bond, which we might symbolise as O——H···O. The O——H···O structure, which appears in all molecules containing O——H, seems

always to have an O——H···O bond angle of 180°. (At least, it does in *solids*. In the disordered liquid state you cannot be sure of the strict linearity of such a bond.)

6.6 More on the nature of hydrogen bonds – liquid mixtures

The hydrogen bond has been defined using the 'two-criteria' model (positively polarised hydrogen meets oxygen, nitrogen or fluorine). The historical path which led to that definition included some interesting experiments with two-component liquid mixtures.

Among the cleverest of the liquid-mixture experiments were those which sought to isolate the two ends of the hydrogen bond by putting them on two different molecules which could not hydrogen bond to their own kinds. This was no easy task, because most molecules with strongly positively polarised hydrogen atoms only possess them by virtue of their being joined to an oxygen, nitrogen or fluorine atom. As we have already seen with alcohols, this is a recipe for hydrogen bonds between molecules of the same type.

However, trichloromethane (Figure 6.16) can be used as the provider of the polarised hydrogen atom. Chlorine is not as high up the electronegativity table as the 'big three', but three chlorine atoms working together on a single carbon atom have a noticeable effect on the polarity of the C—H bond.

Question

11 Explain how the polarising effects in the C—Cl bonds end up producing a positively polarised hydrogen atom.

Under the proposed model for a hydrogen bond, trichloromethane molecules are incapable of hydrogen bonding with their own kind, simply because they do not have one of the exclusive trio of electronegative atoms (fluorine, oxygen, nitrogen) in the molecule. There would be a certain amount of dipole–dipole interaction, plus the inevitable van der Waals' forces contribution.

The plan for testing the hydrogen-bond theory now depended on the selection of the other component of the test mixture. The required molecule should possess an oxygen, nitrogen or fluorine atom, but *no polarised hydrogen*. This means making sure that all the covalent bonds coming from the electronegative atom go to something *other* than a hydrogen, thus insulating the hydrogen atoms from the fierce polarising force. Organic chemistry has a number of oxygen-containing groups of compounds which fit this description, either because the oxygen atom is in the middle of a carbon chain (the ethers and esters), or because the oxygen atom is double bonded to carbon (the esters again, and the ketones and aldehydes). Examples of these molecules are shown below.

$$CH_3—CH_2—O—CH_2—CH_3$$

ethoxyethane – an ether

$$CH_3—C \overset{\displaystyle O}{\underset{\displaystyle O—CH_2—CH_3}{\big\langle}}$$

ethyl ethanoate – an ester

$$CH_3—\overset{}{\underset{\displaystyle \underset{O}{\|}}{C}}—CH_3$$

propanone – a ketone

$$CH_3—\overset{}{\underset{\displaystyle \underset{O}{\|}}{C}}—H$$

ethanal – an aldehyde

Let us now consider the mixing of a two-component system which fits all our needs – trichloromethane and ethoxyethane. When it comes to intermolecular bonding to its *own* kind, neither molecule can muster more than dipole–dipole attractions. Yet if our model of a hydrogen bond is correct, they will interact strongly with each other, as shown below by the formation of a hydrogen bond between pairs of molecules.

$$Cl—\overset{\displaystyle Cl}{\underset{\displaystyle Cl}{C}}—H\text{----}O\overset{\displaystyle CH_2CH_3}{\underset{\displaystyle CH_2CH_3}{\big\langle}}$$

trichloromethane   ethoxyethane

How can we tell whether a hydrogen bond has formed or not? We have already used elevated boiling points as a detector of greater-than-expected intermolecular bonding (Figure 6.13). Let us now use the same detector in our consideration of this mixture of liquids. We might expect the boiling point of the mixture to vary with composition as the straight line on Figure 6.17, that is, to be somewhere between the boiling point of ethoxyethane (308 K) and the boiling

point of trichloromethane (335 K). However, as you can see, the boiling point of the mixture rises as high as 350 K.

Question

12 Offer your own interpretation of the hump in the boiling point–composition graph, saying whether you think that it supports the notion of hydrogen bonds having formed between the two sets of molecules.

This experimental system can act as a baseline for further variation. For instance, the experiment can be repeated with, say, chloromethane, CH_3Cl, in the trichloromethane role, keeping ethoxyethane the same. The single chlorine in CH_3Cl should not polarise the hydrogen atoms to the same extent. Disappearance of the tell-tale maximum in the boiling point/composition graph would support the idea that hydrogen bonds need strongly polarised H atoms, and this disappearance is indeed what happens.

Questions

13 Suggest a molecule which could operate in place of ethoxyethane in a mixture with trichloromethane, and offer proof that only oxygen, nitrogen

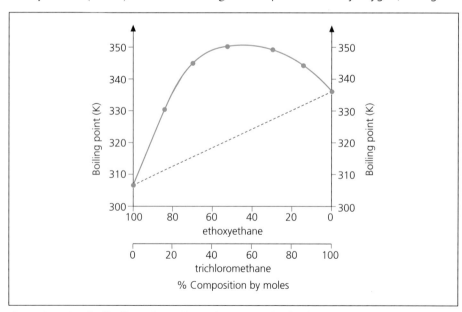

Figure 6.17 Graph of boiling point against molar composition for the ethoxyethane–trichloromethane system, showing a significant rise in the middle

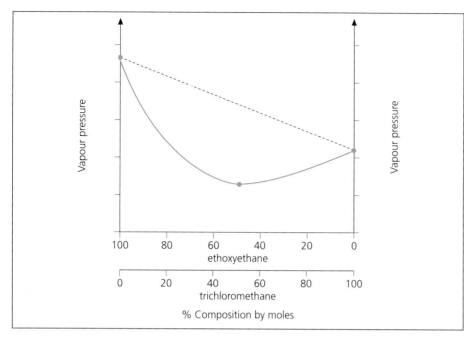

Figure 6.18 Sketch graph to show how the vapour pressure of mixtures of ethoxyethane and trichloromethane would vary with composition. (Temperature and actual values of vapour pressure are unspecified.)

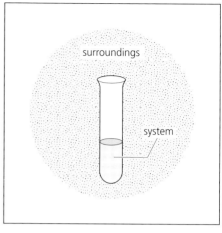

Figure 6.19 System and surroundings

or fluorine can function when it comes to hydrogen bonding.

14 Figure 6.18 shows the same system as in Figure 6.17, except that the *y*-axis has been changed from boiling temperature to *vapour pressure*. Can you explain why this change causes all the 'highs' to become 'lows', and the maximum to become a trough?

An alternative source of data from mixtures – calorimetry

You should already have met the idea that chemical systems can exchange energy with their surroundings – as in combustion reactions. The heat energy has been created at the expense of chemical energy in the system itself (where the word 'system' means the molecules taking part in the reaction, Figure 6.19). Such reactions, in which the surroundings end up as net receivers of heat, are called **exothermic reactions**.

The study of the heat changes of reaction gives us insight into chemical bonding. The reason is that the size and sign of the heat flow between the system and its surroundings indicate the size of the potential energy of the bonds in the molecules of the system.

Potential energy is usually something you associate with, say, height in a gravity field – things want to be *down*, so *up* becomes the high energy state. There is quite a strong parallel with chemical bonding – things want to be *together*, so *apart* is the high energy state. Normally a chemical reaction features bond breaking followed by bond making, so the system goes from *together* to *apart* to *together again*, which in potential energy terms means *low/high/low*. But if the bonds in the product molecules are stronger than those in the reactant molecules (and that is indeed the case in all exothermic reactions) then it would be more exact to say the change was *low/high/lower*. It is this loss of potential energy (shown in Figure 6.20) which appears as heat. (A fuller account of the field of chemical potential energy, whose official name is **enthalpy**, is given in Chapters 8 and 9.)

An act of mixing will involve the breaking of intermolecular bonds in the

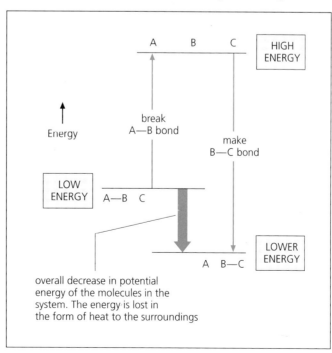

overall decrease in potential energy of the molecules in the system. The energy is lost in the form of heat to the surroundings

Figure 6.20 A symbolic representation of the way chemical systems exchange heat with the surroundings. This example shows an exothermic change because bond B—C is stronger than bond A—B

pure liquids, and a making of new ones between the liquids. In a system comprising molecules A and B, if the A–B interactions are stronger than the A–A and B–B ones, then heat will be created. The only drawback is that these bonds are an order of magnitude weaker than 'proper' bonds, so the heat exchanges with the surroundings will be small, necessitating some fairly accurate calorimetry, but still well within the scope of students.

Figure 6.21 shows some calorimetric results from the same system whose boiling points we studied earlier. The y-axis is a series of temperature rises recorded when mixtures of various compositions shown on the x-axis) were mixed.

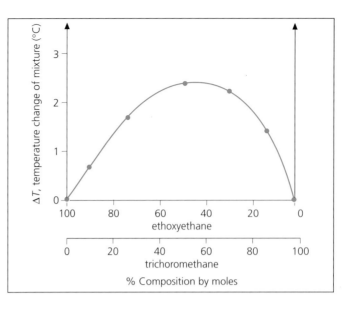

Figure 6.21 Graph of temperature change on mixing against molar composition for the ethoxyethane–trichloromethane system

Question

15a What is the predominant type of intermolecular force in pure trichloromethane? (We will call this the A–A interaction.)

b What is the predominant type of intermolecular force in pure ethoxyethane (the B–B interaction)?

c What have we previously suggested is the predominant type of A–B interaction?

d So why is the mixing 'reaction' exothermic?

e Why also does the maximum of the boiling-point graph occur at the same x-axis position as the maximum of the temperature-change graph?

Chemical potential energy has one feature which is not shared with its gravitational equivalent. Things on planets never spontaneously move uphill, but chemical systems *can* have a natural inclination to increase their chemical potential energy. In other words, the products have *weaker* bonds than the reactants (Figure 6.22). Such a system is called **endothermic**.

The detailed causes of this phenomenon are fully investigated in Chapters 8 and 9. For now, let us just say that Nature appears to have a second major drive to set against the drift towards low potential energy, and that is the drift towards high **chaos**. If a reaction can produce a more chaotic arrangement of

particles in space, it may proceed even though the potential energy shift is upwards. Many mixing processes fit this description. The process of mixing always increases the chaos factor in a chemical system, and thus many mixtures still happen even though the A–B interactions are somewhat weaker than the A–A or B–B ones (Figure 6.22). (Never lose sight of the fact that in this chapter we are dealing with *inter*molecular forces rather than *intra*molecular bonds, so the uphill energy changes are not big ones.)

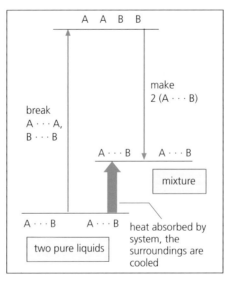

Figure 6.22 Energy profile for an endothermic mixing 'reaction' between two pure liquids, molecules A and B. In this case the A——B intermolecular bonds are somewhat weaker than the A · · · A and B · · · B intermolecular bonds

Question

16 Sketch the general shapes of the graphs of boiling point against composition, and of temperature-of-mixture against composition, which you would expect from an endothermic mixing process. Explain their shapes.

We will now look at specific pairs of liquids to pinpoint the structural features which might give rise to endothermic mixing behaviour. We should anticipate systems in which the molecules are giving up a stronger form of intermolecular bonding in the pure liquids and replacing it with a weaker one in the mixture.

Question

17 One such system is cyclohexane–ethanol, shown below.

$$H_2C—CH_2$$
$$H_2C \qquad CH_2 \qquad CH_3CH_2OH$$
$$H_2C—CH_2$$

cyclohexane   ethanol

a What type of intermolecular bonding predominates in pure ethanol?

b What type of intermolecular bonding predominates in pure cyclohexane?

c Then why are the A–B intermolecular forces weaker than those in the pure liquids?

Stoichiometry in the intermolecular bonding

We have not yet extracted *all* the information which is carried by the plots of boiling point and temperature against composition. There is also a message in the position of the *maximum of the curve*. This is determined by the relative numbers of each type of molecule taking part in intermolecular bonding.

We can assume that the trichloromethane–ethoxyethane intermolecular complex has a $1:1$ composition. (The word 'complex' is being used rather loosely here, since the association between partners is probably subject to constant breaking and remaking.) This proposed ratio is supported by the fact that the maximum in both the composition graphs (Figures 6.18 and 6.21) is in the middle. But our theoretical model for hydrogen bonding allows for the possibility that any oxygen might accommodate *two* hydrogen bonds to itself, since there is the implication that hydrogen bonds involve lone pairs, of which oxygen has two.

Questions

18 How would the boiling point/composition graph be altered if the intermolecular complex is as shown below?

19 The system **di**chloromethane–ethoxyethane (dichloromethane is CH_2Cl_2) also mixes exothermically, but the stoichiometric possibilities are more diverse and less clear-cut than in the 'tri' case. Draw diagrams in the style of

the one above of the 'complexes' which might be obtained in the following situations:

a A single hydrogen bond forms between the two molecules.

b Each molecule of either sort forms *two* hydrogen bonds, producing a chain-like complex.

c Both hydrogen atoms on the dichloromethane form hydrogen bonds, but only one bond forms on the oxygen atom of the ether.

d Only one of the above situations would give an off-centre maximum on a graph of boiling point or temperature against composition. Which one, and why would the other two be indistinguishable?

Advanced chemistry examinations often contain questions about how a particular pair of liquids will behave on mixing. The key is to ask yourself three questions:

1 What is the predominant A–A intermolecular interaction?

2 What is the corresponding B–B interaction?

3 What can they offer each other, and how does the strength of an A–B intermolecular bond compare with the other two?

Once this 'trial of strength' is complete, you will know whether to expect endothermic or exothermic mixing, and raised or lowered boiling points. Use this technique on the following question.

Question

20 Predict and explain the boiling-point shift and thermal behaviour of the following pairs of substances when mixed. The structural formulae are shown.

a Propanone–dichloromethane.

b Propanone–trichloromethane.

c Propanone–tetrachloromethane.

d Propanone–water.

e Ethanenitrile–trichloromethane.

f Propan-1-ol–butan-1-ol. (This last system offers a situation which has yet to be discussed, but you should notice that A and B are closely related, so their interactions with each other will be similar in type and strength to their interactions with their own kind.)

As a footnote to question 20(f), it is worth noting that there is a law formulated for this kind of situation, in which the A–B interactions are on a par with those in A–A and B–B. Such mixtures are referred to as **ideal solutions**.

Raoult's Law states the vapour pressures of ideal solutions should vary linearly with composition (expressed in mole fraction).

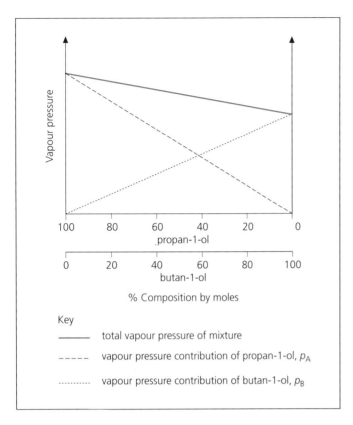

Key

—————— total vapour pressure of mixture

‐ ‐ ‐ ‐ ‐ vapour pressure contribution of propan-1-ol, p_A

············· vapour pressure contribution of butan-1-ol, p_B

Figure 6.23 Vapour pressure–composition graph for a mixture of propan-1-ol and butan-1-ol at an unspecified temperature

Figure 6.23 shows the vapour pressure against composition graph for the mixture of propan-1-ol and butan-1-ol, in which the dotted lines indicate the contributions of the individual vapour pressures of each component to the total vapour pressure.

Question

21a Sketch a version of Figure 6.23 with *boiling point* as the *y*-axis instead of vapour pressure.

b The dotted lines express proportionalities, one of which is $p_A \propto x_A$ (where p and x are vapour pressure and mole fraction of component A). To turn a proportionality into an equality, you must have a constant, as in $p_A = kx_A$. Deduce the value of k, and write the finished version of the equation.

(Try looking at what happens to the equation when $x_A = 1$. In this special case, the right-hand side has become just 'k'. So whatever you judge the left-hand side to have become in this special case, will be the value of the proportionality constant k. In other

words, k is equal to the value of p_A in the special case when $x_A = 1$. Those are enough clues for one little part-question. Check your result with the answers, because the finished equation is one way of expressing Raoult's Law.)

Immiscible liquids

The phrase 'oil and water don't mix' is used to enrich the stock of metaphor in the English language, along the same lines as 'chalk and cheese'. One might ask if it is possible to accommodate systems made of liquids which do *not* mix within our theoretical model of liquids which *do* mix? Luckily it is. We do not need to build any major extensions on to our model, since non-mixing is merely a situation in which the A–B interactions are so weak by comparison with the A–A and/or B–B ones that neither set of molecules is tempted to disengage with its own kind.

Oil is an umbrella word covering a large range of substances, many of which are hydrocarbons, and those that are not still have a large hydrocarbon component. Hydrocarbons are virtually non-polar, as we saw in Chapter 5, and so groups of hydrocarbon molecules cling

together by the only intermolecular force available to such molecules, namely the van der Waals' force. However, in long-chain hydrocarbons the van der Waals' force can be quite big, as is the case in hydrocarbon polymers like poly(ethene) and poly(phenylethene).

Van der Waals' forces may be respectably big between hydrocarbon chains, but they are very weak between a hydrocarbon chain and a water molecule. So for oil and water to mix, water would have to give up a well-organised system of hydrogen bonding, and the oil would have to sacrifice its van der Waals' arrangements, and all for not very much in return. This proposal is so energetically 'uphill' that not even the prospect of increasing the randomness of molecular distributions can compensate. The whole detergent industry is dedicated to getting round this natural antipathy between oil and water, and is discussed in Chapter 13.

Question

22 The shorter-chain alcohols, like ethanol and methanol, are miscible (mixable) with water in all mutual proportions. As you move to longer-chain alcohols, this solubility becomes finite and by the time you reach decan-1-ol it has become quite small, so that the two liquids have an oil-and-water relationship.

$$CH_3(CH_2)_8CH_2OH$$

Explain this variation in terms of the changing nature of the intermolecular forces in the alcohol series.

6.7 Solubility of solids in liquids – solvation

There is no real difference between the principles underlying the dissolving of a molecular *solid* (like sugar) in water, and the ideas already used to describe liquid–liquid mixing. The sugar may look different because it is a solid, and you cannot go all the way along the composition axis. (90 % sugar 10 % water at 25 °C is soggy sugar, rather than a true **single phase** solution. There is an upper limit on the amount of sugar which will dissolve in a given amount of water.)

Nevertheless, to predict the behaviour of the system, you have to ask the same questions, about A–A, B–B and A–B interactions, as you would for a liquid–liquid system, using the same palette of intermolecular forces (dipole–dipole, van der Waals', hydrogen bonding).

Questions

23 Everyone knows that sugars are very freely soluble in water. A molecule of the domestic sugar sucrose ($C_{12}H_{22}O_{11}$) is shown below. (Each junction of four lines implies the presence of a C atom.)

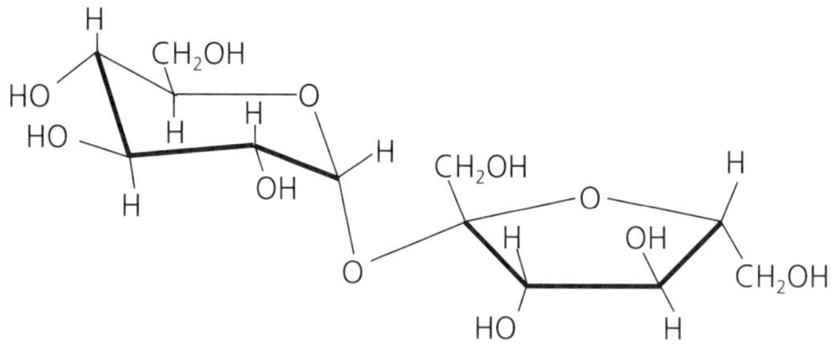

What type of interaction will be predominant between molecules of water and of sucrose? Can you suggest why the sucrose–water interaction is so favourable?

24 Anyone with experience of washing clothes will have been struck by the difference in drying times of, on the one hand, garments made of natural fibres like wool and cotton, and on the other hand garments made of synthetic fibres like acrylic and polyester. Look at the structures of these polymers below, and suggest a reason for the difference.

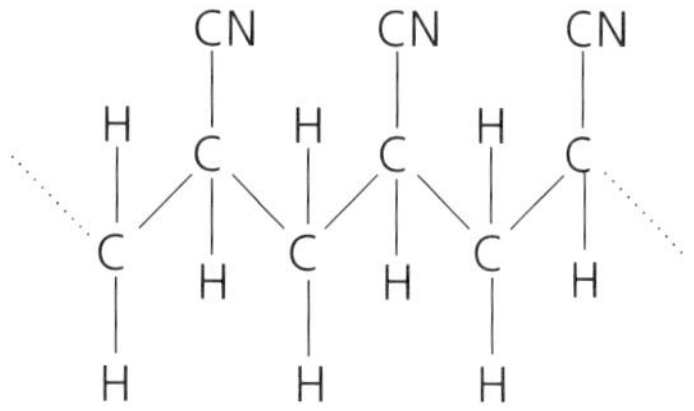

a part of the molecule of an acrylic fibre

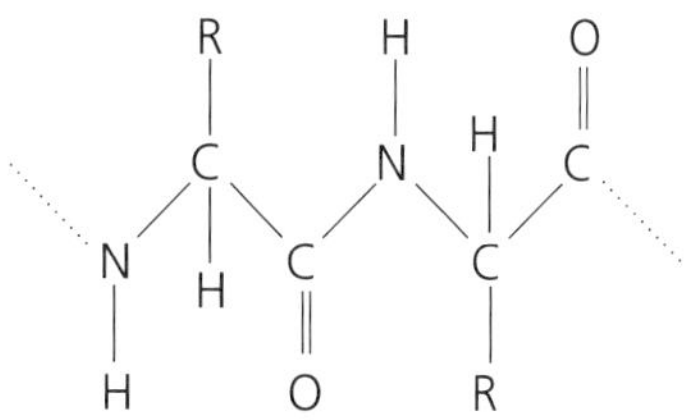

b small section of a protein chain of the sort found in wool – the R and R' indicate side-chains made up of various organic groups

c part of cellulose – the polymer found in cotton

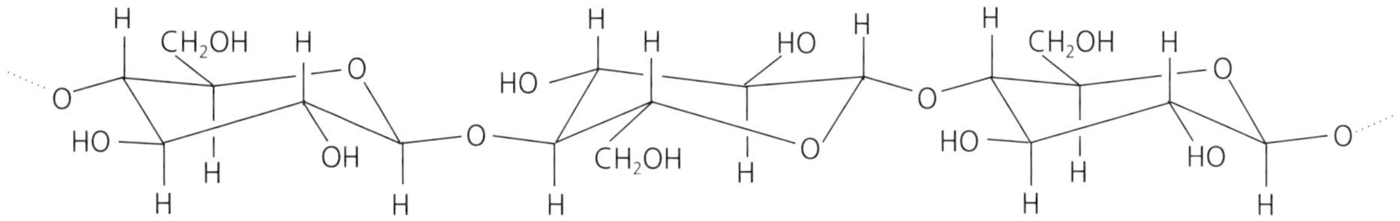

The dissolving of ionic solids in water

So far the entire chapter has been about meetings between *molecules*, and about inter*molecular* bonding. But salt, which is a classic ionic compound, dissolves just as well as sugar. What is the mechanism in this case?

Most ionic solids which *do* dissolve do so with a mildly endothermic heat exchange with the surroundings. When you consider how strong ionic bonds are in the solid crystal lattice (for which there is the evidence of the high melting points of ionic solids), it is amazing that the process is not *more* endothermic.

Question

25 If salt (NaCl) dissolves in water with an almost zero heat exchange with the surroundings (which it does), which two sets of forces must be of similar size?

What then is the nature of these impressively large forces between water molecules and ions? The answer is neither exotic nor obscure, but simply the force of electrostatic attraction.

The force of electrostatic attraction between the ion and the dipole in the solvent is called an **ion–dipole interaction**.

We believe that water molecules cluster round ions in solution, using whichever end of the water dipole is attractive to the ion in question (Figure 6.24, overleaf). Hydrogen bonds may be involved as well in the case of anions containing oxygen, like sulphate or carbonate.

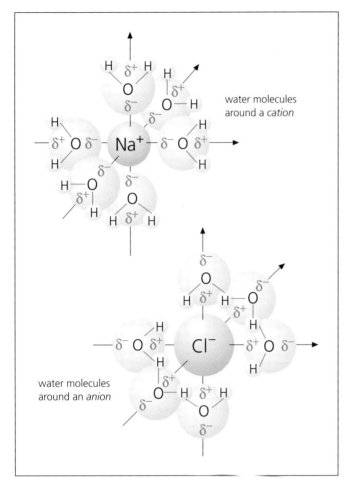

Figure 6.24 Showing the ion–dipole interactions between water molecules and ions in solution. The ions are said to be solvated

The process of bond formation between a solute and its solvent is called **solvation**.

The bonds that form may be of formidable strength.

The truth of this story is supported by the fact that sometimes the ions take some water molecules with them when they crystallise. These 'frozen-in' water molecules (the so-called **water of crystallisation**) can be studied in X-ray pictures of crystals, where their mode of clustering can be seen. In copper sulphate, for instance, it is clear that four of the five molecules of water are associated with the Cu^{2+} cation, and are indeed pointing the oxygen (negative) end of their dipoles in towards the cation (Figure 6.25).

Charge density

It seems that those ions which are most 'popular' with water molecules are those with the most intense electrostatic fields around them. In other words they have a high **charge density**, which means a high ratio of charge to size. This may explain why cations seem to get the lion's share of solvating water molecules, since anions are usually bigger than their cation partners.

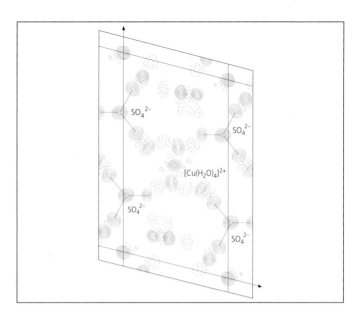

Figure 6.25a Electron density map of solid crystalline $CuSO_4 \cdot 5H_2O$ (ordinary blue copper sulphate). This portion of the picture shows four water molecules around the copper atom, in a square plane. The three-pointed stars are SO_4^{2-} ions (viewed from above). Remember that hydrogen atoms do not show up on electron density maps. Nevertheless it is easy to imagine hydrogen bonds between water molecules and sulphate ion oxygen atoms (The fifth water molecule is associated only with SO_4^{2-} ions.)

Figure 6.25b A polarised light micrograph of a copper(II) sulphate crystal

Questions

26 Consider the sulphates of the elements of Group II (Mg, Ca, Sr, Ba). Record their formulae, including the maximum number of water molecules of crystallisation. Bearing in mind that all the cations have the same charge (2+), what evidence is there to support the idea that water 'likes' cations of high charge density?

27 Sometimes in chemistry you need to be prepared to argue in opposite directions, like a barrister. For instance, the compounds of Group II are on the whole much less soluble than their equivalents in Group I, despite the obviously higher charge density of Group II cations (which would have ensured good solvation). What other factor might be pulling the other way, to ensure that more Group II compounds resist the process of dissolving in water?

28 There is a well-known maxim in chemistry that 'like dissolves like'. It has the merit of simplicity, important in a good maxim, but the question is – is it true? Discuss the merits of this maxim, with reference to the type of bonding involved, as applied to:

a petrol dissolving grease

b water dissolving sugar

c water failing to dissolve grease (in the absence of detergents)

d water dissolving salt.

6.8 How intermolecular bonding affects properties of liquids

Viscosity

Lyle's Golden Syrup is **viscous**. It can take a matter of seconds to drip off a spoon. It so happens that syrup is a very strong solution of sucrose in water, and we have already seen what extensive intermolecular bonding there is in that system. Can it be that viscosity and intermolecular bonding are linked?

Figure 6.26 is an imaginative attempt to supply the answer. If molecules are to flow over each other, then clearly the intermolecular forces must be broken and re-made, as the upper layer of molecules progresses over the lower layer. If those intermolecular forces are quite large, there may be a problem getting the flow started. Admittedly once molecule A has moved more than halfway from its initial position, it will be pulled by molecule C, but it will never get to halfway if there is not enough thermal energy in the system to do the initial breaking. (This is another topic which gets a fuller treatment later in the book, in Chapter 14.) So strong intermolecular bonds can make for sticky liquids.

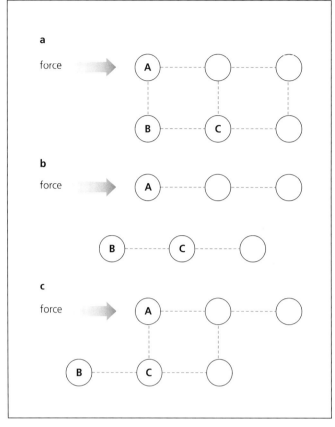

Figure 6.26a Schematic model of how intermolecular bonding might affect viscosity. Making layer A slide over layer BC involves breaking and making intermolecular bonds

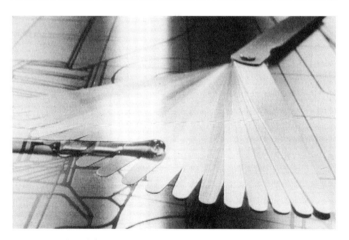

Figure 6.26b Oil flowing over an object

Questions

29 How and why would you expect viscosity to change with increasing temperature?

30 Another liquid famous for its high viscosity is propane-1,2,3-triol, or glycerol.

Explain why a structure like that of glycerol (below) would give rise to a liquid of high viscosity.

$$CH_2—CH—CH_2$$
$$\;|\qquad|\qquad|$$
$$OH\quad OH\quad OH$$

'Viscostatic' oils

At first sight it is slightly puzzling that the liquids which we use to help things *slip* are themselves quite viscous. But what moving engine parts need is a sort of compromise:

- If the lubricating oil is too thick, energy will be wasted doing work against the intermolecular bonds, in forcing the molecules to flow over each other.

- If the lubricating oil is too runny, it cannot prevent metal touching metal, which would be a huge waste of energy and lead to the rapid erosion of the metal surfaces themselves.

So one of the crucial variables that oil technologists seek to control is viscosity, and for them the big challenge is to *reduce the temperature dependence of viscosity*. Question 29 was about viscosity decreasing with temperature. Cold oils will be 'treacly' and drag on the moving parts, but when heated they become less viscous. If you select an oil for its free flow as a cold lubricant, the chances are it will become too thin at the engine's working temperature. For an oil scientist, finding a way of keeping an oil at a *constant* viscosity as an engine goes from cold to working, would be a major achievement.

In fact, the goal has already been partially achieved. It was discovered that the decrease in the viscosity of engine oil as it grew hotter could be reduced by mixing certain additives (called viscosity index improvers) into the oil. These additives consisted of long-chain molecules, one of which was poly(2-methylpropene).

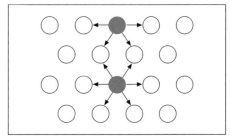

Chains of this substance assume tangled, approximately spherical, masses when cold, but tend to untangle into more linear conformations when hot. Question 31 asks you to consider how this behaviour of the additive helps to regulate the viscosity.

Question

31 You should be able to work through the answer yourself, with the following hints and some imagination (and some help from Figure 6.27):

a If oils are mainly hydrocarbons, what sort of intermolecular force will prevail in them?

b What type of intermolecular bonding would occur between the molecules of oil and of additive, bearing in mind that it too is a hydrocarbon?

c How will the degree of 'tangledness' of the additive molecule affect its ability to bond to molecules of the oil?

d So why does its gradual untangling as it gets hotter actually oppose the natural tendency of the oil to get runnier?

Note: I can remember the pleasure I felt when I first read about this piece of chemical technology. Here was a discovery of first-rank importance in the real world, achieved by people at the cutting edge of the subject, yet which could be understood with homespun mental models drawn from anything from spaghetti to balls of wool. The pleasure came from realising that not all success in under-

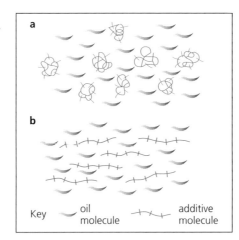

Figure 6.27 Imaginative idea of how a viscosity index improver might work.
a Oil cold – additive molecules in globular tangled shape.
b Oil hot – additive molecules untangle and extend into linear shapes

standing chemistry depends on mastery of thermodynamics or quantum mechanics, or on a compendious memory. Sometimes all it takes is a little imagination, and the mental agility to jump between the behaviour of very small objects and their big-world analogues.

Surface tension

There is a natural drive pushing systems to adopt arrangements of minimum potential energy. In systems of molecules this amounts to a drive to adopt situa-

Figure 6.28 Showing how a molecule in the middle of a liquid is more strongly bonded than a molecule at the surface

tions with the maximum amount of strong bonding (providing there is no strong opposite urging from the chaos factor). In a liquid, the most secure position with the strongest intermolecular bonding is within the mass of molecules, surrounded by other molecules and bonded to them all. Molecules on the surface of a liquid are held by fewer bonds (Figure 6.28). We can draw three related conclusions from this model:

- Liquids will show a natural tendency to *minimise surface area* (that is, to move from thin planes to 'blobs').

- Molecules sticking up out of an irregular liquid surface will tend to get pulled into the mass of the liquid.

- The stronger the intermolecular forces in a liquid, the more pronounced will be the tendency to be pulled in.

Figure 6.29a Isolated soap bubbles

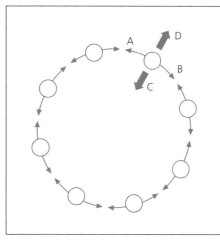

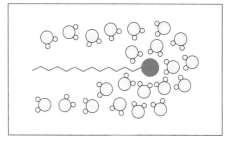

Figure 6.29b Showing the balancing of forces in a stable bubble. C represents the resultant inward force of intermolecular bonds A and B. D represents the outward force of the gas inside. In vector arithmetic:

$$\vec{A} + \vec{B} = \vec{C}$$

and

$$\vec{C} = -\vec{D}$$

Question

32 Can you explain the *spherical* shapes taken naturally by drops of water and by bubbles (Figure 6.29a)? (Someone on the Paul Daniels show produced a cubic bubble a few years ago, but it was somehow trapped inside lots of other bubbles, and the laws of Nature still applied.)

Why is a bubble a spherical shape? Of course a bubble is *not* the minimum surface area for the liquid which makes it up – that would be achieved by the bubble's collapse into a spherical droplet. But the 'blower' has supplied the extra work to push the system into this artificially high energy state. We can picture the forces in the wall of the bubble as in Figure 6.29b, with intermolecular bonds pulling the molecules inwards, opposed by and balanced against the pressure of the air inside. This balance of forces is only achieved within a rather narrow choice of liquids. In fact successful bubble-blowing solutions are liquids in which the intermolecular forces are neither too strong, nor too weak.

Question

33a Pure water is a bad liquid for bubble blowing, because its intermolecular forces are too strong. Referring back to Figure 6.29b, why might this threaten the structure of the bubble?

b Petrol is also a poor bubble-blowing liquid, because its intermolecular forces are too weak. Returning again to Figure 6.29b, what would be the reason for the collapse of the bubble this time?

Solutions of soap in water are of course the normal source of bubbles for children's games. The main account of the chemistry of soaps is in Chapter 13, p. 305 where they are viewed as organic chemicals. For now, while our focus is on intermolecular bonding, we will just say that a soap molecule is designed to do quite a simple job. It is a job which can be understood by another analogy like that used to picture what happened to viscostatic oils. A soap molecule is designed to have 'a foot in both camps'. One end is designed to go amongst van der Waals'-bonded liquids (oils and grease) and the other end is ionic, enabling it to be solvated by water (Figure 6.30). In this way the natural antipathy towards mixing of these two classes of liquid can be partially overcome. (See Figure 13.36, p. 305 for a diagram of this process happening.)

Question

34 Why does soap allow the intermolecular forces in water to be fine-tuned to the correct level for successful bubbles?

When soap molecules dissolve in water, what effect will their arrival between the water molecules have on the overall intermolecular bonding of the liquid (Figure 6.31)?

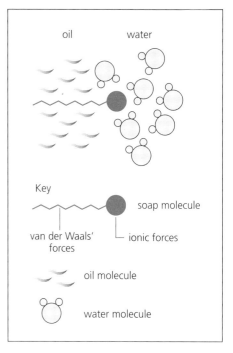

Figure 6.30 Highly schematic picture of the action of soap molecules, acting as a 'go-between' at an oil–water interface

Key

van der Waals' forces | soap molecule | ionic forces

oil molecule

water molecule

oil | water

Figure 6.31 A soap molecule in solution in water. How will it affect intermolecular bonding and surface tension?

The word *tension* in 'surface tension' has yet to be justified. The appropriateness of the word is best appreciated when you consider the behaviour of pond-surface-dwelling insects like pond skaters. They use the surface of the water as if it were an elastic skin, and it is this (entirely false) impression of a skin which best fits the description 'surface tension'. By denting the top of the water they are doing work creating extra surface area, and the water's efforts to oppose that creation of extra surface area provides the upward force which supports the insects (Figure 6.32).

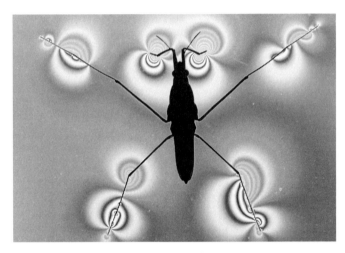

Figure 6.32 Pond skater supported by surface tension

Question

35 How would a pond skater's life be affected if its pond were polluted by:

a oil?

b detergent?

The wetting of surfaces – the meniscus

Watching the behaviour of water on glass is something anyone can do without optical aids, and yet it provides insight into a number of intermolecular phenomena. If the glass is not particularly clean, or more especially if it is greasy, the water tends to gather in blobs, as near spherical as gravity will allow. On glass which has been scrupulously de-greased, however, the water wets the whole of the glass surface smoothly (Figure 6.33).

Question

36 Which of these two behaviour patterns contradicts the 'minimum surface area' tendency?

A related phenomenon is the way water behaves in tubes and pipes. You will have seen the way aqueous solutions sit in burettes, adopting a U-shape called a **meniscus**, which is clearly not a minimum surface area (Figure 6.34). It all

Figure 6.33 If water molecules cannot bond to the surface (as is the case when the surface is a layer of hydrocarbon material such as grease or wax), they 'hug themselves' into a sphere (top). But when they can bond directly to the oxygen atom in glass, they spread themselves out (bottom)

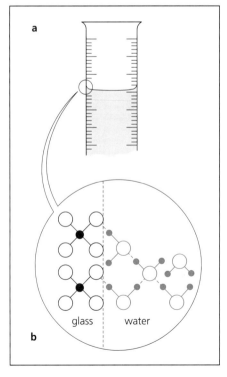

Figure 6.34
a A meniscus in a burette. **b** Hydrogen bonding creates the meniscus

depends upon intermolecular forces between the liquid and the glass surface. A clean glass surface offers oxygen atoms from the material of the glass itself, so water can *hydrogen bond* to the glass. This tendency can be sufficiently strong to cause the water molecules to prefer contact with the glass to contact with their own kind, and thus the minimum surface area criterion is overridden.

Questions

37 Can you think of a reason why water climbs higher the narrower the bore of the tubing (as in Figure 6.35)?

38 Mercury is so strongly endowed with inter-particle bonds that it makes even water look moderate. Drops of spilt mercury do almost adopt spherical shapes on the bench, heedless of gravity (Figure 6.36).

a Note that the word molecule has not been used – what are the 'particles' in liquid mercury?

b The inter-particle bonds in mercury seem a whole order of magnitude stronger even than those in water. What is the nature of the bonds in a liquid metal?

c Mercury cannot form any sort of linkage with glass molecules, and it also has these very strong inter-particle forces. Predict the shape of the meniscus of mercury in glass (or if you know it already, explain it).

Now for a series of short questions about the reasons behind some familiar liquid phenomena.

Questions

39 Like mercury, hexane does not form bonds with glass, but unlike mercury it has fairly weak intermolecular bonds. Predict the shape of its in-glass meniscus.

40 We all know the difference in appearance between rain on a new or newly cleaned car, and on a neglected one (Figure 6.37). Use your knowledge of intermolecular bonding to explain this difference.

Figure 6.35 'Capillary rise'. See question 37

41 Why does soapy water have more success in 'wetting' surfaces like the new car, compared to rain or ordinary water (Figure 6.38)?

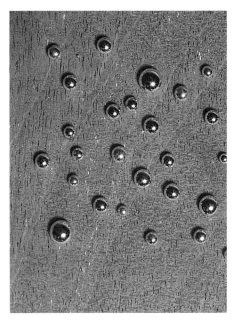

Figure 6.36 Spilt mercury

Figure 6.37 Rain on a newly waxed car (left) and on an old banger

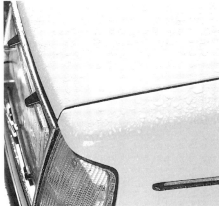

Figure 6.38 Soapy water on a new car (left) compared with rain water

6.9 Hydrogen bonds in biochemistry

We have seen already, with soaps, how certain relationships between objects have parallel examples in the big world and in the molecular world. In that case it was a story of two objects which did not have a relationship, being brought together by a go-between which could relate to both. We were able to transfer our mental imagery from concrete experience to stimulate our molecular imagination.

Now for a similar act of transference. In the ordinary world of human-sized technology there is a frequent need for joins which are strong but not that strong. Think of the way a fizzy drink can comes open, or of the perforations in sheets of postage stamps, or of Velcro or zippers. The relationship of the two objects which are in a state of not-too-firm attachment, an attachment which will adhere up to a point but is also breakable and/or reversible, is as relevant in a molecular domain as in the big world. There are two outstanding natural examples.

The DNA helix

To cut a very long story short (and to miss out most of the plot as well), **deoxyribonucleic acid** (DNA) is a molecule which needs to be able to 'unzip' itself. It exists as immensely long chains, and as a double helix of twisted strands (Figure 6.39a and b). It carries encoded information to guide the development of cells. The information is encoded in varying combinations of four nitrogen-containing molecular fragments called **bases**, and the bases are arranged in pairs like ladder-rungs between the strands of the double helix. Unlike real ladder rungs, these molecular rungs have an intentional weak point in the middle.

When cell division takes place, it is vital that the DNA code is passed on to the new daughter cells. As part of this process the double helix unzips, with each ladder-rung breaking at the weak point in its middle (Figure 6.39a). The rest of the story includes a process by which the unzipped half-ladders are able to re-synthesise the missing other half.

The purpose of having this heavily edited piece of biochemistry in this chapter is to ponder the question – how does Nature arrange for the acts of unzipping and re-zipping of molecular structures? What is needed for the rungs in DNA could be written as: 'Bond – strong enough to hold firm in normal use but must be easily breakable at short notice'. This sounds like a job for one of our intermolecular style bonding systems, since they are indeed about one power of ten weaker than classical bonds.

Figure 6.39 (**a**, left) The double helix molecule of DNA 'unzipping'. What manner of bonding holds the two sides together? (**b**, below) Tim Piggott-Smith and Jeff Goldblum as Watson and Crick in 'Life Story', the film about the discovery of the structure of DNA

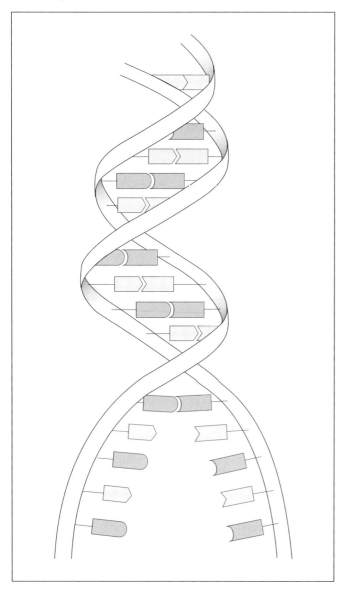

Question

42 Figure 6.40 shows the two pairings of half-rungs, using the four 'bases' of the DNA code. They have been drawn like this to help you visualise the problem. Can you suggest exactly what sorts of intermolecular bonds go where, in order to bond these two half-rungs together?

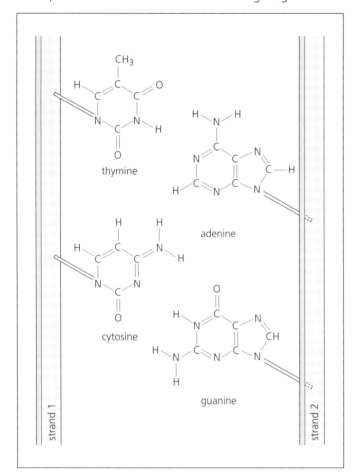

Figure 6.40 How are the two strands of the double helix connected? See question 42

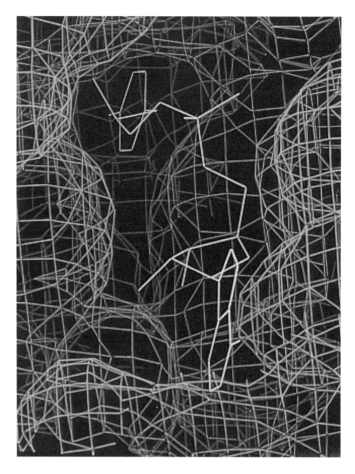

Enzymes

Enzymes, the great catalysts of Nature, have a need for reversible bonding too. An enzyme has an **active site** in which it retains its substrate. **Substrate** is the word used in biochemistry for the molecule on which the enzyme operates, usually in some destructive way. The substrate could be thought of as the enzyme's 'victim'. Bonding between an enzyme and its substrate must necessarily be of the 'not too strong' variety, or catalysis would be ineffective. Figure 6.41a shows a computer-generated image of a catalyst binding to its substrate.

Figure 6.41 (a, top right) A computer-generated image of an enzyme and its substrate. (**b**, right) How is the substrate (black) held into the cleft of the enzyme (green/shading)? See question 44

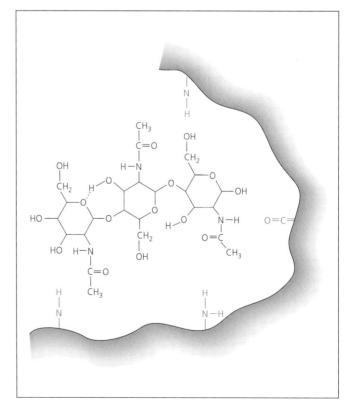

Figure 6.42a Watson and Crick's work on the structure of DNA was performed with a knowledge of Chargaff's ratios of the bases in DNA and some access to the X-ray crystallography of Maurice Wilkins and Rosalind Franklin. Combining all of this work led to the deduction that DNA exists as a double helix. Crick, Watson and Wilkins shared the 1962 Nobel Prize for medicine and physiology, Franklin having died of cancer in 1958

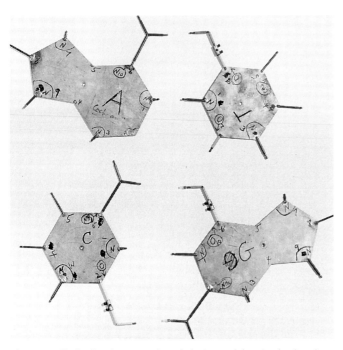

Figure 6.42b the four base templates (adenine and thymine (top) and cytosine and guanine (bottom)). Watson and Crick fitted cardboard versions together jigsaw-fashion to make the rungs of the ladder

Question

43 a If the enzyme bonded too tightly to its substrate, and to the broken fragments of the product, why would catalysis malfunction?

b Why would it be equally harmful if the enzyme and substrate had too weak a relationship?

So that is why the 'holding-on' bonds between enzymes and their substrates are of the intermolecular type. The next question asks you to pinpoint the site and type of bonding.

Question

44 Using Figure 6.41b to help you, suggest the positions and types of intermolecular bonds which might be holding the substrate to the enzyme.

Intermolecular bonds will be cropping up throughout the course. You can appreciate their significance when you think that, although the classical bonds in a molecule may be responsible for its chemistry, all those properties which relate to molecules en masse (hardness, strength, boiling point and so on) will depend upon the relationships of molecules one with another, and this of course is the province of the *inter*molecular force. In biochemistry we have seen the crucial importance of intermolecular bonds (especially hydrogen bonds) in DNA replication and enzyme–substrate binding.

Throughout the chapter we have seen the way that quite simple mental images can shed light on these intermolecular events. Figure 6.42 shows Nobel Laureates Watson (on the left) and Crick with their model of part of a DNA molecule in 1953. You might have assumed that the big breakthrough was made by working on the results of some form of sophisticated spectroscopy, and indeed X-ray crystallography did play a vital role in seeing into the double helix.

However, the real moment of revelation came while they were working with bits of cardboard and scissors. This was how they realised the bases fitted together jigsaw-fashion to make the rungs of the ladder. One of the more interesting things about chemistry is that it occasionally throws up these moments of quirky co-operation between simple and complex ideas, between advanced mathematical models and strong visual metaphors, between X-ray diffraction on the one hand, and cardboard and scissors on the other.

Question

45 Now for a question to commemorate and celebrate the lasting value of the old Watson and Crick technology. (You will need rulers and protractors.)

The sensation of **taste** depends upon receptor molecules on various regions of the surface of the tongue (Figure 6.43). There is a region for each part of the taste 'spectrum'. The relationship between the receptor molecules and the molecules being tasted is exactly analogous to the relationship between the cleft in an

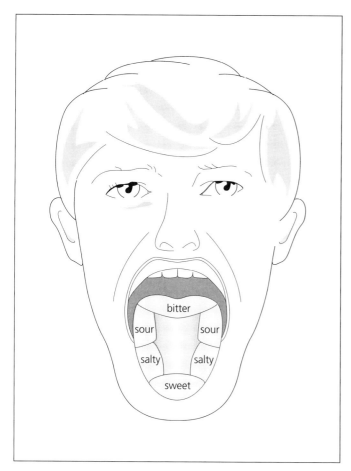

Figure 6.43 Regions of the tongue where taste receptors operate

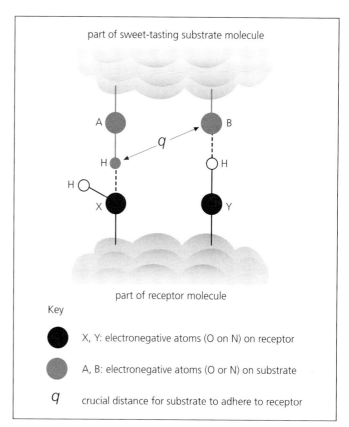

Key

● X, Y: electronegative atoms (O on N) on receptor

● A, B: electronegative atoms (O or N) on substrate

q crucial distance for substrate to adhere to receptor

Figure 6.44 Sweet-tasting substrate molecule binding to a sweet receptor site on the tongue

enzyme and the enzyme's substrate. In other words we experience the sensation of taste when a substrate molecule of the right *shape* 'lands' on the receptor molecule and bonds to it, thereby triggering a nerve impulse.

The 'bonding' must not be too permanent, or our first taste would be our last – fine for Cornish dairy ice-cream but not so good for cod liver oil. As in the enzyme and DNA cases, the need for 'disposable' bonds leads to the use of intermolecular forces. Figure 6.44 shows the receptor site for *sweet*, with a substrate molecule in temporary residence. You will see that the substrate molecule does not only have to be the right shape – it has to have the right tendencies for intermolecular bonding too.

a   Read the key to Figure 6.44 and decide what the dashed lines mean.

b   The 'size' part of the requirement for a molecule to be sweet-tasting is that the distance q must be within a few per cent of 0.3 nm. The simplest sweet-tasting molecule is ethane-1,2-diol, shown on the right.

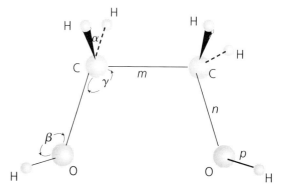

Fill in the missing angles α, β and γ by comparison with similar molecules in Chapter 4, pp. 58–63, and the distances m, n and p from a data book.

c   Draw a precise, scale version of the ethane-1,2-diol molecule above, and measure the distance q, as defined in Figure 6.44, expressing the answer in nanometres.

d   You will find from part c that ethane-1,2-diol apparently fails the sweetness test. Since it *is* sweet, the problem must be with our model. In what way is it inadequate? Which molecular movements would enable the molecule to get its q distance exactly right?

e Propane-1,3-diol is *not* sweet, but glycerol (propane-1,2,3-triol) *is*. Explain, in terms of ability to meet the q requirement and the bonding needs, this pattern of sweetness.

$$CH_2—CH_2—CH_2$$
$$|\qquad\qquad|$$
$$OH\qquad\quad OH$$

propane-1,3-diol

$$CH_2—CH_2—CH_2$$
$$|\qquad\quad|\qquad\quad|$$
$$OH\qquad OH\qquad OH$$

propane-1,2,3-triol

f *cis*-Cyclopentane-1,2-diol, despite apparently having a very similar bonding arrangement to that in ethane-1,2-diol, is *not* sweet. This is because in one important way it is not as 'free' as the simpler molecule. Can you explain further?

$$H_2C\quad\cdots\quad CH_2\quad\cdots\quad CH_2$$

(structure of cis-cyclopentane-1,2-diol with two OH groups)

g The seriously sweet molecules, like the sugars and saccharin, go further in their relationship with the receptor molecule. As well as bonding with their A and B groups to the XY site, they are held at a third point which on our diagrams would be behind the plane of the page. This third group is shown as Z in Figure 6.45, and has to be *non-polar*. What kind of group on the substrate (C in Figure 6.45) would be needed to bond at Z, and what type of bonding would be involved?

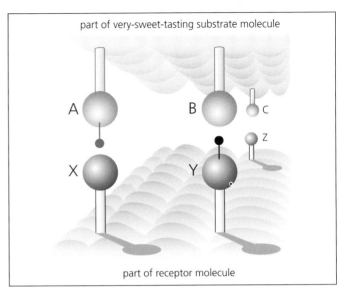

Figure 6.45 Showing the third interaction C–Z when a grossly sweet substrate is held on the sweet receptor site

h Aspartame is one of the artificial sweeteners now on sale widely and is a sugar-substitute favoured by 'calorie watchers'. Can you suggest where its A, B and C regions are?

(structure of aspartame molecule)

Summary

• Molecules are finite independent entities. That is to say, they start and end in definite places.

• However, their independence of each other cannot be total. Forces must exist between molecules or else every molecular substance would be gaseous at all temperatures above 0 K.

• Therefore the evidence of the extent of intermolecular bonding can be found in *melting* and *boiling* temperatures, and also in energy changes on *fusion* (that is, melting) and *vaporisation* (latent heats).

• This evidence reveals a strong correlation between the degree of intermolecular attraction and the existence of **dipolar bonds** within molecules. The molecules in such substances orient themselves so that the oppositely charged ends of their dipoles attract each other. This interaction is called a **dipole–dipole attraction**.

• However, plenty of molecules *without* dipoles still manage to display quite significant degrees of intermolecular force. To explain this, we propose the existence of **van der Waals'** attractions, of which all atoms and molecules are capable.

• Van der Waals' attractions depend on the short-lived existence of random asymmetric distributions of electron density within molecules. This phenomenon sets up within any given molecule a temporary dipole, which can *induce* another similar dipole in its neighbour, and via these neighbours, in other molecules in the vicinity. (A magnetic analogy is the way that iron nails can be held 'in series' on a permanent magnet. The main difference is that in van der Waals' attractions nothing is permanent.)

• The strength of van der Waals' attractions depends on molecular size and shape. Successful molecules must:

a be polarisable – which often correlates with molecular *size*

b have *shapes* which allow close contact between molecules – all the better if they can get tangled. *Linear* shapes fulfil both these requirements.

• The existence of a third intermolecular bonding force is implied by the spectacular and anomalous properties of the molecules H_2O, NH_3 and HF. These molecules have melting and boiling temperatures startlingly *above* their equivalents in the next row of the Periodic Table (H_2S, PH_3 and HCl), to an extent that cannot be explained by dipoles alone.

• The third force is known as a **hydrogen bond**, and can exist between any two molecules A and B which display these features:

a (on molecule A) one of the three most electronegative atoms in the Periodic Table – *nitrogen, oxygen* or *fluorine*

b (on molecule B) a hydrogen atom which is strongly positively polarised by its molecular environment.

The bond involves a lone pair on the nitrogen, oxygen or fluorine atom on molecule A interacting with the hydrogen atom on molecule B.

• The relative strengths of the three styles of intermolecular attraction are usually said to be:

hydrogen bonding >100–100 kJ mol^{-1}
dipole–dipole   5–10 kJ mol^{-1}
van der Waals'   variable

While this is broadly in line with thermodynamic evidence, it tends to conceal the fact that some van der Waals' structures are very durable. The hydrocarbon polymers, for example, use their extreme linearity to achieve quite high melting (or rather softening) temperatures based on nothing but van der Waals' forces.

• Mixtures of different liquids can reveal the nature of intermolecular bonding in each individual liquid. The chemist must look at the evidence of **enthalpy of mixing** and boiling temperature of the mixture.

• An **exothermic** enthalpy of mixing reveals that the intermolecular forces being made are stronger than those being broken. (If the components of the mixture are A and B, then the A–B interactions are stronger than the average of the A–A and B–B interactions.) The molecules of such a mixture will display a **reduced escaping tendency**, and therefore the mixture will have a lower vapour pressure and a higher boiling temperature than the one predicted from a weighted average of the two components (Figure 6.46).

• Mixtures which fit this pattern of behaviour are those in which hydrogen bonding is possible *between* molecules of A and molecules of B, but not *within* the pure components. This situation is uncommon because one of the components of the mixture must have a hydrogen atom which has been made electropositive by means other than being covalently bonded to an oxygen, nitrogen or fluorine atom. (Otherwise the component could hydrogen-bond to its own kind.) The molecule $CHCl_3$, trichloromethane, fits this requirement, and mixtures of this compound with esters, ketones, etc., have been used to work out the bond energy of a hydrogen bond.

• **Endothermic** enthalpies of mixing indicate that the A–B interactions are *weaker* than the average of the A–A and B–B ones. In this case the escaping tendency of the molecules in the mixture goes *up*, and the mixture shows a higher vapour pressure and a lower boiling temperature than would have been predicted (Figure 6.47).

• Mixtures fitting this latter pattern are commoner than the exothermic ones, since many liquids sacrifice a degree of hydrogen bonding when

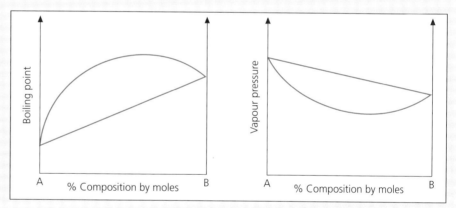

Figure 6.46 Exothermic mixing

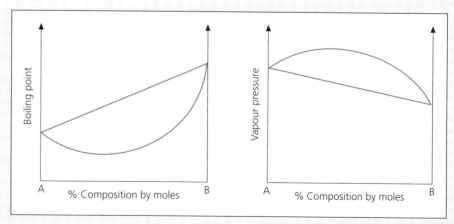

Figure 6.47 Endothermic mixing

they enter into mixtures with second components which can offer only dipoles or van der Waals' forces in return. Examples include mixtures of alcohols with hydrocarbons.

• **'Ideal' mixtures** are those in which the A–A and B–B interactions are of the *same* order of magnitude as the A–B ones. In these circumstances the mixture obeys **Raoult's Law**. This law was originally phrased to focus on the contribution of each component to the total vapour pressure of the mixture. The equations for the two vapour pressures are:

$$p_A = x_A \times P_A$$

where p_A is the vapour pressure of
        component A
        x_A is the mole fraction of A
        P_A is the vapour pressure of
        pure A
and $p_B = x_B \times P_B$

In practice it means that ideal mixtures have boiling points and vapour pressures which vary *linearly* with composition, as long as the composition is expressed in mole fractions. A 50:50 mixture, for instance, will have a boiling temperature *half-way* between those of the pure liquids. To get a system showing ideal behaviour it is necessary to choose components which are close chemical relatives, for example, propan-1-ol and butan-1-ol (Figure 6.48).

• **Solubility** (that is, how much solute dissolves in a given solvent) is heavily

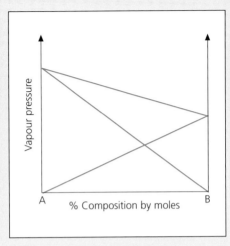

Figure 6.48 An ideal solution

dependent on intermolecular forces. In *liquid–liquid* systems, if A–B bonds are too weak relative to the bonds in the pure components, the components simply do not mix, or at best there is a solubility limit of A in B or B in A, after which two layers form. A familiar example is provided by hydrocarbon oils and water.

• The solubility of *molecular solids* in liquids follows the same principles as liquid–liquid systems, except that there is a clearer distinction between solute (the solid) and solvent (the liquid). Thus the solubility of sugars in water can be seen as being due to hydrogen bonds forming between solvent and solute. A rough-and-ready guide to predicting solubility is the maxim **'like dissolves like'** (where 'like' refers to general chemical character and the style of intermolecular force in solvent and solute).

• The solubility of *ionic solids* is dependent upon an interaction between the charge on the ion and a dipole in the solvent called an **ion–dipole interaction**. The degree of solubility will also be influenced by the strength of the ionic lattice. Water is a good solvent for ionic solids; non-polar solvents are very poor solvents for ionic solids.

• **Soaps** and **detergents** are molecules which can get round the rule about like dissolving like. They have dual character, that is, one end of the soap molecule can bond with aqueous media (usually by an ion–dipole interaction), and the other can make van der Waals bonds to non-polar oils and greases. Thus oil droplets are rendered acceptable for dispersion by water.

• Intermolecular forces affect all of the physical properties of liquids, notably **surface tension**, **viscosity** and **thermal capacity**.

• Intermolecular forces have an effect on biological systems whose importance is hard to exaggerate. No enzyme can act without them, and DNA replication depends upon them, so their influence reaches into every part of cell chemistry. They also play a major part in specialised functions such as the sense of taste.

• In some cases the feature of intermolecular bonds which makes them so useful to the working of biological systems is their *weakness*. They are used in situations where links between molecular parts need to be *reversible* (for example, between an enzyme and its substrate).

7

Oxidation and reduction reactions

7.1 Introduction

This book has already let you *meet* and *count* the building blocks of chemistry (Chapters 1, 2 and 3), and inspect ways of *sticking them together* (Chapters 4, 5 and 6). Now we turn our attention to the business of how people have organised the vast landscape of chemical reactions into useful groups. Oxidation and reduction reactions form one of the most significant of these groups, but ironically the names 'oxidation' and 'reduction' are rather unhelpful, as we shall see.

When cars were first invented, they were called 'voitures sans cheval', or horseless carriages. At the time the sense of this name was clear to all, since the most remarkable thing about cars was their lack of a horse. A century later this preoccupation with horselessness seems curious, but nevertheless we have been left the legacy of the unit 'horsepower', which has so far survived the onward drive of SI units.

So it is with 'oxidation' and 'reduction' in chemistry. The names have outgrown their historical roots, and may leave modern students wondering at the breadth of their use. Nowadays a **redox reaction** is any one in which **electron transfer** takes place, and redox reactions form one of the largest groups into which we classify chemical changes (along with acid–base, ion exchange, ligand exchange, and all the mechanistic groupings of organic chemistry – nucleophilic substitution, etc.). But it will be useful to see how the meanings of the words gradually outgrew their original boundaries, and indeed merged to form the single hybrid word 'redox'.

One thing has remained constant – the words were created for the task of classification. The early chemists realised how broad was the subject they were uncovering, and recognised the need for classification. Reactions in which oxygen was added *to*, or in which it was taken away *from*, other elements, seemed to occupy a large part of the chemical scene, and the words **oxidation** and **reduction** were coined as labels for the two classes. Here are some straightforward examples:

- Oxidation of an element (one is currently a major pollution source, Figure 7.1):

$$S(s) + O_2(g) \rightarrow SO_2(g)$$

Figure 7.1 Sulphur dioxide contributes to the formation of acid rain, which causes tree damage

• Oxidation of a compound (you do this one yourself with a Bunsen burner, Figure 7.2):

$$CH_4(g) + 2O_2(g) \rightarrow CO_2(g) + 2H_2O(l)$$

• Oxidation using an indirect source of oxygen (safety warning: this reaction is used as a primitive but effective explosive):

$$8KClO_3(s) + C_{12}H_{22}O_{11}(s) \rightarrow$$
$$12CO_2(g) + 11H_2O(l) + 8KCl(s)$$

Figure 7.2 The oxidation of hydrocarbons is a major source of air pollution. The oil wells of Kuwait destroyed by retreating Iraqi troops during the Gulf War of 1991 are a dramatic example of this pollution

Figure 7.3 Antoine Lavoisier (1743–1794) with his wife Marie. Lavoisier was instrumental in the use of accurate measurement in chemistry for the study of the composition of materials, oxidation and other combustion reactions. It was Lavoisier who correctly deduced the contribution of oxygen in the air to combustion

Question

1 Although both the compounds in the third item of the list *contain* oxygen atoms, only one of them is really *giving* them. Which one?

To continue with the list of examples:

• Reduction *to* an element (one of Lavoisier's (Figure 7.3) early experiments):

$$2HgO(s) \rightarrow 2Hg(l) + O_2(g)$$

• Reduction of one element by another (one of the reactions whose discovery was to change the course of history):

$$2CuO(s) + C(s,charcoal) \rightarrow$$
$$2Cu(s) + CO_2(g)$$

(The copper oxide would have been an intermediate from the decomposition of a carbonate ore like malachite, and the reaction itself would have helped usher in the *Bronze Age* (Figure 7.4).)

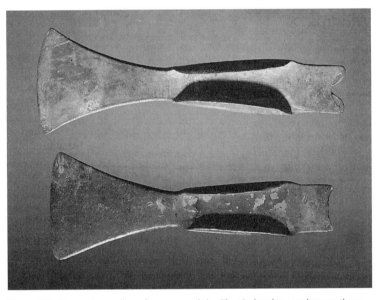

Figure 7.4 Bronze is an alloy of copper and tin. The tin hardens and strengthens the copper. The photograph shows two bronze axeheads from the Bronze Age

Questions

2 Can you see a way in which the fifth example stands apart from the other four?

3 Think of other reactions which fit *three* of the five categories, and give their historical, industrial or domestic contexts (as in the bracketed remarks).

7.2 Extending the meaning of the terms

Chemists began to realise that the role being played by oxygen in some of the above reactions was not unique to oxygen. Oxygen was historically familiar and well studied, but then was its behaviour towards other elements so fundamentally different from, for example, that of chlorine?

$$4Al(s) + 3O_2(g) \rightarrow 2Al_2O_3(s)$$
$$2Al(s) + 3Cl_2(g) \rightarrow Al_2Cl_6(s)$$

This behaviour is typical of any of the more electronegative non-metallic elements, of which oxygen is merely one. All it meant, chemists argued, was that the more metallic element was surrendering electrons to the less metallic one. The aluminium in the above example was being oxidised in *both* cases, as long as you defined oxidation simply as being *loss of electrons*. This new idea appealed to people, since it seemed to make the classification process (which after all was the main point) more embracing. It now included, for example, not just oxidations and chlorinations of *elements*, but reactions such as:

$$2FeCl_2(s) + Cl_2(g) \rightarrow 2FeCl_3(s)$$

Here the iron atom has gone from being in the form of a 2+ ion to being a 3+ ion, and has therefore lost an electron.

Perhaps more significantly, the idea had interesting things to say about reactions like:

$$Fe(s) + CuSO_4(aq) \rightarrow FeSO_4(aq) + Cu(s)$$

or to repeat the same equation, showing only the atoms and ions really taking part in the changes:

$$Fe(s) + Cu^{2+}(aq) \rightarrow Fe^{2+}(aq) + Cu(s)$$

In the above example the iron has clearly been oxidised, because it has lost electrons, and equally clearly the copper has gained electrons (been reduced). The very symmetry of such a reaction forces you to recognise the 'you-cannot-have-one-without-the-other' aspect of oxidation and reduction. We are reminded therefore that:

- If one chemical species loses electrons there must be another one gaining.
- If oxidation is redefined as electron

loss, then reduction must be seen as electron gain.

- Since electrons cannot get 'lost', oxidation and reduction must always occur simultaneously, hence the word **redox** for all reactions involving electron transfer.

This last idea, insistent though its logic is, can come as a surprise. The story began with reactions you would normally think of, on the old definition, as being purely oxidations. Now we have to find an oxidation *and* a reduction in them.

Questions

4 What is being reduced in the following reactions?

a $Mg(s) + O_2(g) \rightarrow MgO(s)$

b $2Al(s) + 3Cl_2(g) \rightarrow Al_2Cl_6(s)$

5 What is being oxidised in the following reaction?

$$2HgO(s) \rightarrow 2Hg(l) + O_2(g)$$

7.3 To the limits of credibility

In this section we shall see how chemists have stretched the meanings of oxidation and reduction still further, in order to make them apply to an even wider group of compounds.

The redefining of the words oxidation and reduction has certainly strengthened their applicability to reactions featuring ions and ionic compounds. As long as

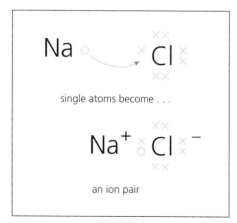

Figure 7.5 A genuine electron-transfer reaction. The making of NaCl

there is a clear-cut case to be made for electron transfer, as, for example, in the making of sodium chloride from its elements, then the new redox ideas apply unambiguously (Figure 7.5).

Question

6 What is oxidised and what is reduced in the case of the formation of sodium chloride?

However, what happens when the new-style definition, focused as it is on electron transfer, is applied to reactions of covalent or polar covalent compounds? In the sodium chloride example, there is no doubt that the sodium started out as Na and became Na⁺, while the chlorine began as covalent Cl_2 and ended up as Cl^-. Our certainty comes from our confidence that sodium chloride is well described by the ionic model. On the other hand, what would we make of the formation of a compound like tin iodide (SnI_4), where the bonding lies somewhere between ionic and covalent, and electron transfer is at best partial? Can the tin really be said to have been 'oxidised'?

$$Sn(s) + 2I_2(s) \rightarrow SnI_4(s)$$

And, in the case of a simple oxidation such as:

$$S(s) + O_2(g) \rightarrow SO_2(g)$$

(the first and most obvious example of an oxidation we met), can you recognise an electron transfer?

Chemists responded to this challenge by stretching the definition of electron transfer, some would say to the limits of credibility, and in a way which feels sometimes like a sleight of hand. Whatever the objections, the current electron-transfer definition of oxidation stretches all the way from purest ionic to almost pure covalent situations. Its justification is that it organises one large slab of chemistry – and it works. We will now see how it works.

Making covalent compounds fit electron-transfer definitions of oxidation and reduction – the polarised-bond method

The solution to the problem presented by covalent compounds, or those of dubious status between 'polarised covalent' and 'impure ionic', is to use what *polarity* there is in each bond to 'declare' one atom oxidised and the other reduced. The partial loser of electrons is defined as the oxidised member of the bonded pair of atoms, and the partial gainer is the reduced member. How, you may ask, are such losses and gains judged?

We can look at experimental evidence such as dipole moments, or indices of electronegativity like that of Pauling's (p. 73) – or even rely on intuition or arbitrary decision. (It would be hard, for example, to identify the oxidised and reduced ends of a C—H bond on either of the above criteria, and yet the carbon is said to have been reduced and the hydrogen oxidised in such a bond, relative to the condition which prevails when both are in their elemental state.)

There is a strong component of pretending in the analysis of the oxidation status of covalent compounds. You lay out the displayed formula in dot-and-cross or letters-and-stick format, look at the electron pairs sitting between atoms, and pretend that the atom that gets the greater share in each bond pair actually gets full possession of them. You can thus talk about electron gain and loss as if it had really happened. In effect you are pretending that covalent bonds have become ionic in the direction that they are polarised (Figure 7.6).

The defence of all this pretence is that it works as a mental model and classification system. In the next question you can practise this part of the redox 'game'.

Question

7 Decide which atom or atoms have been oxidised and reduced, when the following compounds are formed from their elements:

SO_2, H_2O, H_2S, NH_3, HF, BF_3.

7.4   Oxidation number

In order to extend the redox theory, a system has been evolved which describes the oxidation status of all atoms in all compounds, irrespective of the type of bonding involved. This system assigns a number, called (not unnaturally) the **oxidation number**, to every atom in

---

a hydrogen chloride molecule (normal dot-and-cross representation)

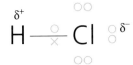

b hydrogen chloride molecule – mixture of dot-and-cross style and letter-and-stick style, showing partial dipolar charge build-up and uneven distribution of the bonding electrons

c hydrogen chloride molecule – imaginary, extending the 'unevenness' of distribution of the bond pair until the bond becomes ionic . . . (which actually happens in water, to give hydrochloric acid)

$$H^+ \quad Cl \quad ^-$$

Key $\times$ H electron $\bigcirc$ Cl electron

Figure 7.6 'Electron gain and loss' in a covalent bond

---

a a 'draw' – no winners or losers of electrons in the bond pair. Hence oxidation numbers of each Cl = 0

b clear-cut electron transfer
Na: loss of 1 e   Oxidation number = +1
Cl: gain of 1 e   Oxidation number = −1

c partial electron transfer. Electron distribution polarised towards Cl
H: partial loss of 1 e   Oxidation number = +1
Cl: partial gain of 1 e   Oxidation number = −1

Figure 7.7 Finding oxidation numbers

---

every compound. The zero, or baseline, of this number system is the *elemental* form of the atom in question, and the oxidation number is decided according to *how many electrons have been lost or gained* relative to that baseline. In other words, if an atom has started in elemental form, and enters into a compound in which it has lost one electron, it is said to have moved oxidation states from oxidation number 0 to oxidation number +1.

For ionic compounds of the sodium chloride type, electron loss and gain is clear-cut, and the oxidation numbers are each identical to the charges on the respective ions. (Thus sodium is in the +1 oxidation state, and chlorine is in the −1 state (Figure 7.7a and b).)

If electron loss or gain is not clear-cut, as in covalent bonds, then we use the 'polarised-bond' method discussed on p. 73. The rule is to pretend that both the electrons in the bond have been 'won' by the more electronegative mem-

ber. A good example of this is the covalent molecule HCl, in which the bond really is polarised to give chlorine a majority share in the two bonding electrons (as is shown by dipole moment evidence). Imagination extends this polarisation process to give H^+Cl^-, and the oxidation numbers are identified as in an ionic case – that is, hydrogen is at +1, and chlorine is at −1 (Figure 7.7c).

We now look at a more complex case, featuring several covalent bonds, namely the alcohol methanol (Figure 7.8). The diagrams show dot-and-cross and letters-and-stick versions of the molecule, and the arrows on the latter show the direction in which each of those covalent bonds is held to be polarised. Since each atom has invested a single electron per bond, the loss of a bond in the 'tug-of-war' will count as a loss of that electron (by the rules of this game), while the winning of a bond will count as an electron gain. Now if we inspect the

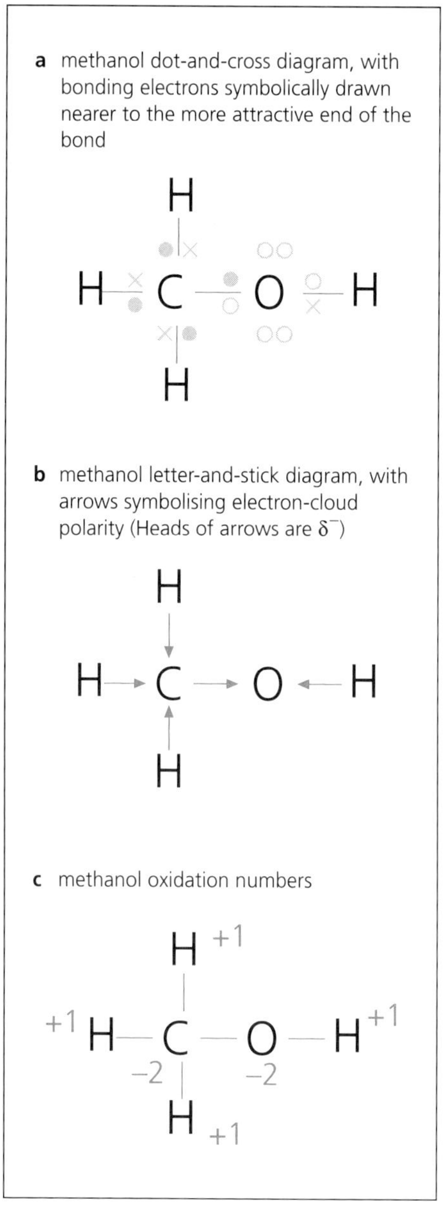

a methanol dot-and-cross diagram, with bonding electrons symbolically drawn nearer to the more attractive end of the bond

$$H$$
$$H \quad C \quad O \quad H$$
$$H$$

b methanol letter-and-stick diagram, with arrows symbolising electron-cloud polarity (Heads of arrows are δ^-)

$$H$$
$$H \rightarrow C \rightarrow O \leftarrow H$$
$$H$$

c methanol oxidation numbers

$$H \; +1$$
$$+1 \; H \text{—} C \text{—} O \text{—} H \; +1$$
$$-2 \quad -2$$
$$H$$
$$+1$$

Figure 7.8 Polarised-bond method of finding oxidation numbers

situation at the carbon atom, we will see that it has won three bonds at the expense of the hydrogens, but lost a bond to the oxygen, giving it a net gain of 2. That carbon therefore is in an oxidation state of −2, and so its oxidation number is −2. By similar reasoning, all the hydrogens have lost their single electron and are therefore in oxidation states of +1, so their oxidation number is +1.

Questions

8 What is the algebraic sum of all the oxidation numbers in the methanol molecule? Put into words an argument for *why* you get this answer, and go on to say whether you think you would get the same answer for all molecules, or only for certain cases.

9 This question is about one of the methanol's close relatives, the alcohol ethanol.

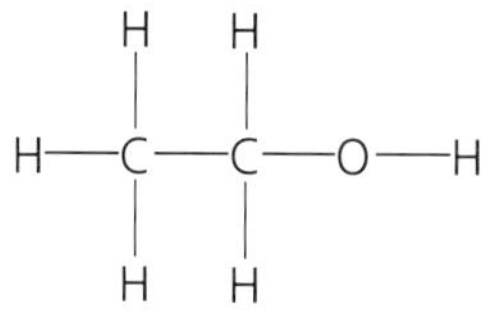

$$H \quad H$$
$$H \text{—} C \text{—} C \text{—} O \text{—} H$$
$$H \quad H$$

In many ways it is like methanol, but there is one novel feature. It has a bond between one atom of carbon and another of its own kind, which is what happens in the element carbon itself, where *all* the carbon atoms are bonded to their own kind, and all the bonds are seen as 'draws', giving oxidation state zero. With that in mind, try to decide 'winners and losers (and draw-ers)' on all the bonds in ethanol, and thus assign oxidation numbers to all the atoms in the ethanol molecule.

Multiple bonds

If an atom is involved in a double bond, then it has *two* electrons on offer, and therefore its oxidation number will go up or down by two (unless it is a bond to its own kind). This is just a consistent application of the idea that every *single* bond will have a 'plus or minus 1' effect on the atom in question. By the same criterion, triple bonds can affect oxidation number by plus or minus 3.

Question

10 Assign oxidation numbers to all three atoms in the carbon dioxide molecule.

$$O \text{==} C \text{==} O$$

'Reliable' oxidation numbers

Already from the previous three examples you can see that it is usual to have hydrogen atoms in the +1 oxidation state, and oxygen in the −2 state. These repeating patterns have their origins in two places – the position of the atom in the Periodic Table (which has a say in the *magnitude* of the oxidation number), and how electronegative that atom is relative to others it might bond with (which decides the *sign*).

Question

11 Show in the cases of hydrogen and oxygen how these two factors produce usual oxidation states of +1 and −2 respectively.

'Reliable' oxidation numbers are a help, because they eliminate the need for laying out the full displayed formulae of compounds. In many compounds there is only one atom whose oxidation number is in any doubt, and that can be simply calculated from the 'reliable' oxidation numbers, using the fact that the oxidation numbers of any neutral molecule add up to *zero*.

For instance, in the methanol case the molecular formula is CH_4O. If the oxidation number of carbon is x, then:

$$x + 4(+1) + (-2) = 0$$
$$\text{So} \qquad x = -2$$

(as we knew, but this is quicker).

Table 7.1 lists atoms which have 'reliable' oxidation numbers in compounds. These are useful in oxidation-number calculations. (Remember that the elemental forms of these atoms have oxidation numbers of zero.)

Table 7.1 Atoms with 'reliable' oxidation numbers in compounds

Atom	Usual oxidation number
H	+1
O	−2
F	−1*
Group I (Na, etc.)	+1*
Group II (Mg, etc.)	+2*
Group III (Al, etc.)	+3

* totally reliable.

Question

12 Using this method, work out the oxidation number of the named element in the following compounds:

- C in CH_4
- S in SO_3
- S in Na_2SO_4
- either of the chromium atoms in $Na_2Cr_2O_7$ (assume here that both chromium atoms are in the same oxidation state)
- N in HNO_3.

Unreliability in oxidation numbers

To repeat, reliability has one of its two sources in the fact that hydrogen (say) is nearly always partnered with elements more electronegative than itself, whereas oxygen is nearly always itself the more electronegative partner. So the system breaks down if hydrogen finds a partner *less* electronegative than itself or oxygen finds a partner *more* electronegative than itself. The arbiter in these cases is generally Pauling's electronegativity index (p. 74). The electronegativities of common elements are given in Table 7.2.

Table 7.2 Electronegativities

Element	Pauling electronegative index N_P
F	4.0
O	3.5
N	3.0
Cl	3.0
Br	2.8
S	2.5
C	2.5
H	2.1
Fe	1.8
Al	1.5
Mg	1.2
Na	0.9
Cs	0.7

From this table you can see that there are rare occasions when oxygen might bond with an even stronger electron-grabber than itself, and when hydrogen is actually the grabber.

Question

13 Work out the oxidation numbers of the atoms in the compounds F_2O and NaH. If the latter were electrolysed in the molten state, at which electrode would the hydrogen be liberated?

There is a second source of unreliability in oxidation numbers, not involving combination with extreme elements like fluorine. Consider the compound hydrogen peroxide, molecular formula H_2O_2. There are no fluorine or sodium atoms in this molecule, and yet consider this puzzle: if the two hydrogens are behaving normally (+1 each), then the two oxygens must be at −1 each (which is odd), yet if the two oxygens are behaving normally (−2 each), then the two hydrogens must be at +2 each (which is odd too).

Question

14 Sherlock Holmes said '. . . when you have eliminated the impossible, whatever remains, *however improbable*, must be the truth . . .' In this case one of the two suggestions is downright impossible. Which one, and why?

The answer to the hydrogen peroxide puzzle lies in the O—O bond, which produces an unexpected 'draw' where you expected oxygen to win. When an atom is bonded to its own kind, there will always be a deviation from the oxidation-number norms.

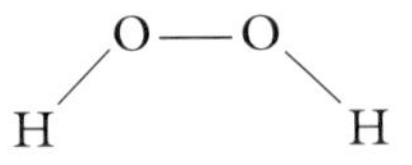

Question

15 Work out the oxidation number of sulphur in thiosulphuric acid, $H_2S_2O_3$, using the 'reliable oxidation number' method. Then draw a displayed formula, bearing in mind that the thiosulphuric acid molecule is the same as a sulphuric acid molecule (Chapter 5, Figure 5.10, p. 76), except that there is a double-bonded sulphur where one of the double-bonded oxygens would have been. Now work

out the oxidation numbers of the two sulphurs. Comment on the differences between the two results.

Oxidation numbers in compound ions

In simple ions, the oxidation numbers of atoms are identical to the ionic charge (for example, Na^+, Cl^-). In covalent molecules, the sum of the oxidation numbers of the atoms is zero. What combination of these two rules would apply, therefore, to chemical species which are both covalent molecules *and* ions – such as SO_4^{2-}, CO_3^{2-} and PO_4^{3-}?

To answer this question, imagine the formation of a sulphate ion from its 'parent' acid, sulphuric acid, H_2SO_4, such as happens during an acid–alkali titration. Figure 7.9a shows a displayed formula of

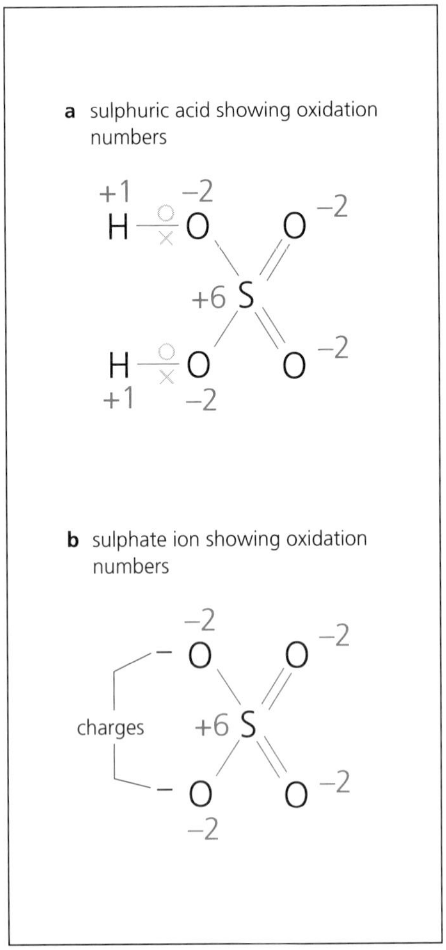

a sulphuric acid showing oxidation numbers

b sulphate ion showing oxidation numbers

Figure 7.9 The formation of the sulphate ion from its 'parent' acid, sulphuric acid

the sulphuric acid molecule, with the bonds to the hydrogen atoms highlighted by being shown in dot-and-cross form. The oxidation numbers of the various atoms (in circles) have been worked out by a bond-polarity model.

As the titration proceeds the hydrogen atoms are removed as H^+ ions, leaving behind their electrons. We are now left with a sulphate ion (Figure 7.9b). As you can see, there have been no changes in oxidation number resulting from the removal of the H^+ ions. Bonds which were polar have broken ionically (as anticipated in the oxidation-number game). However, the loss of the two hydrogen atoms means that the sum of the oxidation numbers in the whole ion is now −2. There are metal cations floating around in solution, and if you wish you can consider that (say) sodium sulphate ($Na_2SO_4(aq)$) has been formed (although really the Na^+ and SO_4^{2-} ions float around independently).

$$H_2SO_4(aq) + 2NaOH(aq) \rightarrow$$
$$Na_2SO_4(aq) + 2H_2O(l)$$

The point here is that when considering an anion in isolation, we leave out of the oxidation-number sum the oxidation numbers of cations which might be or have been its partners. All the other oxidation numbers stay unchanged because those cations have 'left behind' their electrons. Thus the sum of the oxidation numbers of the atoms in a compound ion will not be equal to zero, but will *be equal to the charge on the ion*. Armed with this formula, we can re-use the 'reliable oxidation numbers' method, except replacing 0 with the charge on the ion, as the right-hand side of the equality. So, in the case of the sulphate ion, if the oxidation number of sulphur is x, then:

$$x + 4(-2) = -2$$
$$\text{So} \quad x = -2 + 8 = +6$$

which is of course the same answer as we would have found by analysing the complete compounds H_2SO_4 or Na_2SO_4 using the zero-charge method. In fact, this is another small rule, that the central atom in a compound ion always has the same oxidation number as it does in the 'parent' acid.

Question

16 Use the 'oxidation numbers add up to the charge on the ion' method to work out the oxidation numbers of the named atom in the following cases (assuming oxygen behaves 'reliably'):

- C in CO_3^{2-}
- P in PO_4^{3-}
- V in VO_2^+ (tricky, because a cation, but the rules are the same).

Oxidation number and dative covalency

Dative covalent bonds are another challenge to the rules of assigning oxidation number, because unlike other types of single covalent bond, one atom has invested *two* electrons in the bond (p. 56). As you might expect therefore, it is possible for an atom engaged in dative covalency with a partner of greater electronegativity (greater electron-grabbing power), to 'lose' both electrons, and go up in oxidation state by 2. An example of this is provided by the nitric acid molecule. In HNO_2, nitric acid's cousin, nitrogen is obeying the rule of eight by forming the usual three bonds, and by the rules of the oxidation-number game is said to lose out in all three bonds to oxygen atoms. Hence the nitrogen in HNO_2 is said to be in oxidation state +3.

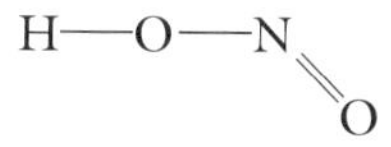

However, HNO_3 is like an HNO_2 molecule with an extra oxygen atom bound datively by the lone pair of the nitrogen atom. Those two extra electrons are held to be lost, along with the other three, so the nitrogen goes up to an oxidation state of +5.

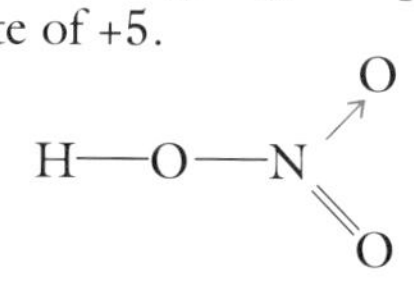

Nitric acid is a rather unusual example, in that most dative covalent donors are *more* electronegative than their receivers. (This is because most owners of donatable lone pairs come from the area to the right of Group IV in the Periodic Table, where all the big 'grabbers' are.) However, this does not mean that the oxidation numbers automatically go down by two, because the donor *owned both electrons already*. There is thus no change in the number of electrons owned by the donor, and correspondingly no change in its oxidation number.

Question

17 Assign oxidation numbers to the named atoms in the following structures (Figure 7.10):

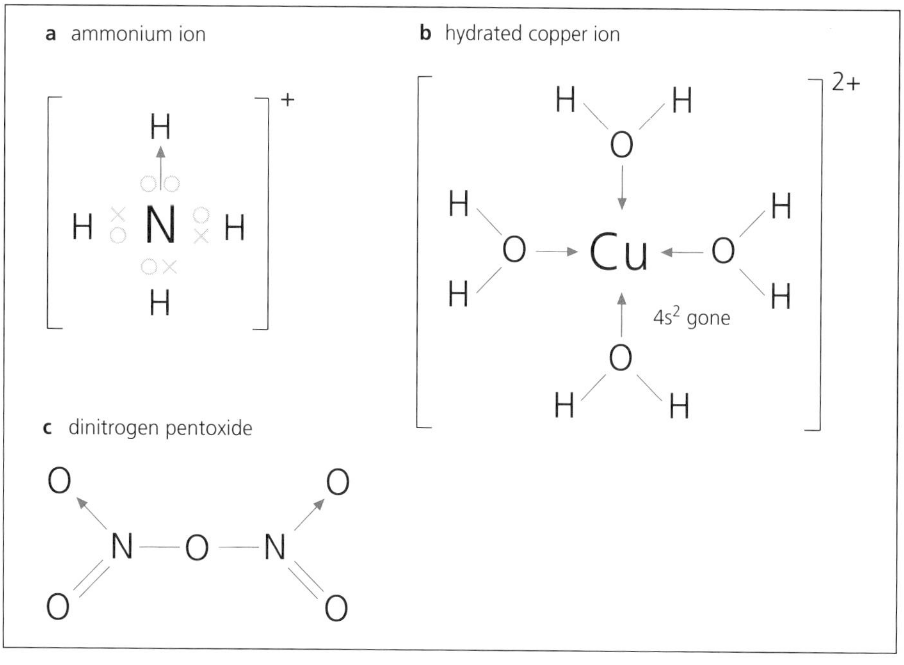

a ammonium ion

b hydrated copper ion

4s² gone

c dinitrogen pentoxide

Figure 7.10 Structures for question 17

- N in NH_4^+
- O in $[Cu(H_2O)_4]^{2+}$ (this is a 'complex' – see Chapter 19)
- N in N_2O_5.

(Note that all these answers are the same as those found by the 'reliable oxidation number' method.)

Nomenclature and oxidation number

Chemists have abandoned the earlier 'suffix' system for naming compounds, and now use oxidation numbers. If an element has more than one cationic form, the old system had the lower-charge cation ending in '-ous' and the higher one ending in '-ic', hence 'ferrous' for the Fe^{2+} ion and 'ferric' for Fe^{3+}. Now these two cations are simply called 'iron(II)' and 'iron(III)', and a compound such as $FeCl_2$ is fully addressed as iron(II) chloride.

In the case of anions and their parent acids, the old 'ite'/'ate'/'per-ate' system for identifying different amounts of oxygen in a compound ion is again simply replaced by the oxidation number of the central atom. Acids all end in '-ic acid' and anions all end in '-ate'. Thus the two acids we met earlier are now called 'nitric(III) acid' (HNO_2) and 'nitric(V) acid' (HNO_3). The compound $NaNO_3$ is officially called 'sodium nitrate(V)', while the sodium salt of the other acid, $NaNO_2$, previously called sodium nitrite, is now 'sodium nitrate(III)'. In cases where there is no variability in oxidation state, there is no need to stipulate oxidation numbers. You would not need, for instance, to write sodium(I) and carbonate(IV); just 'sodium carbonate' will do.

Question

18 Give full official names for the following compounds:

- Na_2SO_4
- $Fe(NO_3)_3$
- Na_2SO_3 (this contains the SO_3^{2-} ion, called 'sulphite').

Now here are some more problems spanning the range of ideas covered in Section 7.4.

Questions

19 Look at the displayed formulae below (Figure 7.11), and work out the oxidation number of each atom by the polarised-bonds method.

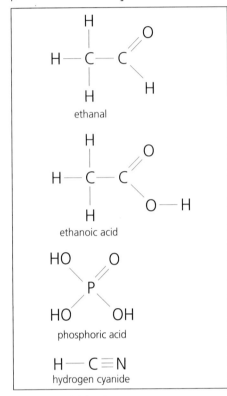

ethanal

ethanoic acid

phosphoric acid

hydrogen cyanide

Figure 7.11 Molecular structures

20 By considering the relationship between an element's place in the Periodic Table, and its possible range of oxidation numbers, work out the *highest* possible positive oxidation number of the following atoms:

C, P, S, Cl

21 By using the 'reliable oxidation number' method, allied to the 'sum of oxidation numbers equals zero' method, work out the oxidation numbers of the named atoms in the following compounds:

- P in HPO_3 (so-called metaphosphoric acid)
- Mn in $KMnO_4$ (potassium manganate)
- S in $Na_2S_2O_3$ (sodium thiosulphate)
- Cl in $HOCl$ (chloric acid).

(Some of those names in brackets are not correct, because they are missing their roman numeral oxidation numbers, but obviously that would have given the game away.)

22 Use a 'reliable oxidation number' method, allied to the 'sum of oxidation numbers equals charge on ion' rule, to work out the oxidation number of the named atom in the following compound ions, and give official names for those asterisked:

- Cl in ClO_4^-*
- Mn in MnO_4^-*
- S in $S_2O_5^{2-}$
- S in SO_3^{2-}*
- V in VO^{2+}
- N in $N_2H_5^+$ (the hydrazinium ion – assume both N atoms are at the same oxidation number)
- C in CN^- (the cyanide ion – assume N is at -3).

23a There is an ion called 'peroxydisulphate' or just 'persulphate', whose formula is $S_2O_8^{2-}$. Use a 'reliable oxidation number' method, allied to the 'sum of oxidation numbers equals charge on ion' rule, to work out the oxidation number of the sulphur atom.

b Say what makes this answer impossible.

c Try some displayed formulae for this ion, to solve the puzzle of this apparent impossibility.

24 See the displayed formulae (Figure 7.12), all of which include dative covalent bonds. Work out the oxidation numbers (by the polarised-bond method) of:

- B and N in H_3NBF_3
- H and O in H_3O^+
- Pt in $[PtCl_4]^{2-}$ (this is a 'complex' formed from 4 Cl^- ions and a Pt^{2+} ion – see Chapter 19).

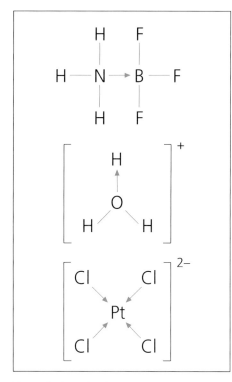

Figure 7.12 Molecular structures

25 Give official names, with oxidation numbers expressed as roman numerals, of:

CuO, Cu_2O, Fe_2O_3,   FeO,

NaClO, $NaClO_3$, $NaClO_4$,

H_2SO_3, H_3PO_4

Back to those earliest examples

The development of the oxidation number concept has taken us a long way from those early self-evident oxidations with which we began the chapter. This small section will show you that the new system does include those cases.

Question

26 Look at the reactions for the burning of methane and of sulphur in oxygen, and work out what has happened to the oxidation numbers of carbon, hydrogen, oxygen and sulphur before and after the burn, in these two cases. To repeat the equations:

$$CH_4(g) + 2O_2(g) \rightarrow CO_2(g) + 2H_2O(l)$$

$$S(s) + O_2(g) \rightarrow SO_2(g)$$

The answer to question 26 is an important one – you should have discovered that when the C, H and S atoms were oxidised, their oxidation numbers *went up*. Hardly less important, you should have noticed that as the oxygen did the oxidations, its own oxidation number *went down*. This 'you-cannot-have-one-without-the-other' feature of redox reactions is, as has been said before, fundamental, and we now can add a new idea, that:

Oxidation is the increase in an atom's oxidation number, while reduction is a decrease.

7.5 The balancing of redox changes

The last section may well have seemed like some enormously long and complicated game. Anyone who has mastered the topic of oxidation number is entitled to ask 'Is it worth it?' The answer is yes; the study of chemistry is enriched by a knowledge of oxidation number.

Assigning oxidation numbers may seem an academic exercise, but it is not an end in itself. Knowledge of oxidation numbers leads to the ability to predict what will happen in a huge range of possible reactions of the redox type.

A key feature of this predictive process is the collection of **half-equations.** Half-equations display a possible oxidation or reduction happening to a single element, but as has been said twice already, you cannot have just an oxidation or reduction on its own. So for a complete reaction, the contents of one half-equation featuring an oxidation must be 'mated' to the contents of another in which reduction is happening. Thus, armed with a library of half-equations, the chemist can predict and comprehend a massive repertoire of possible pairings. The part which oxidation numbers play is in the fitting together of the two members of the pair.

The pairing of half-equations

A half-equation represents a possibility; something that might happen to a particular atom. For instance, it is possible that copper metal might get oxidised thus:

$$Cu \rightarrow Cu^{2+}(aq) + 2e^-$$
Change in oxidation number = +2

(Notice that the potentially discarded electrons are shown as a product of the half-equation.)

But this cannot happen in isolation, because there has to be something to accept the electrons – the removal of two moles of electrons without anything else happening would be massively costly in energy terms. Some other element must go in the opposite direction, the direction of reduction, in order to accept those electrons. Here is another half-equation, this time expressing something that might happen to silver ions:

$$Ag^+(aq) + e^- \rightarrow Ag(s)$$
Change in oxidation number = −1

(Notice that in both cases – and in all others too – the change in oxidation number is identical to the number of electrons transferred.)

How can these two half-equations complete each other and work as a whole reaction? Inspection of the half-equations reveals that while the copper atom has *two* electrons 'on offer', the silver ion is only capable of accepting *one*. So we will need two moles of silver ions for every one mole of copper atoms.

$$Cu(s) + 2Ag^+(aq) \rightarrow Cu^{2+}(aq) + 2Ag(s)$$

Now you might argue that we could have balanced this equation without recourse to oxidation numbers – and indeed without ever having heard of the whole redox concept – simply by knowing the charges on the ions. Perhaps this is true, but when we move on to harder cases you will see the value of oxidation numbers. Also, you must remember that our overall aim is to bring whole families of reactions, both easy and hard ones, under a single conceptual umbrella.

The example that follows demonstrates where a knowledge of oxidation numbers really *is* a help in balancing the question.

The manganate ion, $MnO_4^-(aq)$, is always ready to take part in oxidations, being reduced to $Mn^{2+}(aq)$ in the process. It is capable of oxidising $Fe^{2+}(aq)$ ions to $Fe^{3+}(aq)$. Let us use the method of half-equations to piece together the combination of this pair of events, in the correct proportions and with the correct 'add-ons'.

The reaction of iron with manganate – a reaction balanced by the method of half-equations

Step 1: the iron half-equation is a one-electron removal, with a corresponding jump of one oxidation number.

$$Fe^{2+}(aq) \rightarrow Fe^{3+}(aq) + e^- \qquad (1)$$

Step 2: (as a preliminary to finding the manganese half-equation) display the manganese species, and work out their oxidation numbers.

In MnO_4^-, the 'reliable oxidation number' method gives (if x is the oxidation number of manganese):

$$x + 4(-2) = -1$$

So

$$x = +7$$

The Mn^{2+} ion has an oxidation number, +2, so the entire half-equation involves a reduction of $(+7) - (+2) = 5$ oxidation numbers. From this we know that five electrons will have to be given to the MnO_4^- ion, and will show up in the half-equation.

Step 3: set up the core part of the half-equation:

$$MnO_4^-(aq) + 5e^- \rightarrow Mn^{2+}(aq)$$

The immediate objection to this is that it fails the first test of anything being called an equation – both sides are not equal. (But notice that we already know one of our objectives, that one mole of MnO_4^- ions will be able to oxidise five moles of Fe^{2+} ions, because we can already see that MnO_4^- needs five electrons, while Fe^{2+} is only offering one electron.)

Step 4: make the half-equation balance. (This will seem very much like a series of tricks at first.)

The four oxygen atoms on the left are balanced by four water molecules on the right:

$$MnO_4^-(aq) + 5e^- \rightarrow Mn^{2+}(aq) + 4H_2O(l)$$

Now the oxygen atoms balance, but eight new hydrogen atoms have been introduced. So balance the hydrogen atoms on the right by adding eight H^+ ions on the left:

$$MnO_4^-(aq) + 8H^+(aq) + 5e^- \rightarrow Mn^{2+}(aq) + 4H_2O(l) \ (2)$$

In this 'add-on' part of the balancing process, you will see that no new redox changes have been introduced. The four oxygens had to show up on the other side still in the −2 state. Similarly the new hydrogens had to appear at +1 throughout, hence the use of H^+ on the left.

Step 5: now bring together the manganese half-equation and the iron half-equation, and match the number of electrons being transferred. This involves having five moles of the iron half-equation. Some students have observed that this pairing of the two half-equations is like doing simultaneous equations in maths, where you carry out addition or subtraction between a pair of equations in order to eliminate a variable. Here, it is the electrons which are playing the part of the variable. For those with experience of simultaneous equations, we can eliminate the electrons by multiplying equation *(1)* by 5 and adding it to equation *(2)*.

$$5 \times (1) + (2)$$

This results in:

$$5Fe^{2+}(aq) + MnO_4^-(aq) + 8H^+(aq) \rightarrow$$
$$5Fe^{3+}(aq) + Mn^{2+}(aq) + 4H_2O(l) \qquad (3)$$

You can now run a final check that everything balances. It is not just the atoms that must balance; the charges must equate to each other too.

Left-hand side charges:

$$5(+2) + (-1) + 8(+1) = +17$$

Right-hand side charges:

$$5(+3) + (+2) = +17$$

This shows that the final equation is correct.

There is another possible refinement: you can show what happened to the 'spectator' ions (the ones that did not take part in the reaction). Suppose that the manganate(VII) was in the form of its potassium salt, and the acid and iron compounds were sulphates:

$$5FeSO_4(aq) + KMnO_4(aq) + 4H_2SO_4(aq) \rightarrow$$
$$2.5Fe_2(SO_4)_3(aq) + MnSO_4(aq) + 0.5K_2SO_4 + 4H_2O(l)$$

There are two important points to make at the end of this example.

First, however artificial all that juggling with H^+/H_2O may have looked, it does match up to chemical reality. If you try to carry out oxidations with manganate(VII), they really do not work until there are some H^+ ions around to 'mop up' those oxygens.

Second, the balancing act would have been much more difficult without the oxidation numbers to help construct the half-equations. Here at last is the first tangible reward for all that work mastering the assignment of oxidation numbers.

Question

27 The fact that manganate(VII) ions can oxidise iron(II) ions, coupled with the intense purple colour of the manganate(VII) ion, has given rise to a titration method for analysing solutions of Fe^{2+} ions. In this equation the technique is being used to analyse some iron tablets from the pharmacist (Figure 7.13). The label on the box says '200 mg', but the question is, does that mean 200 mg of Fe ions, or 200 mg of the whole compound $FeSO_4$ (which is the active ingredient in the tablets)? These data should enable you to find an answer.

Two tablets were crushed, dissolved in 2 mol dm^{-3} sulphuric acid, and made up to 100 cm^3. (This amount of sulphuric acid was more than enough,

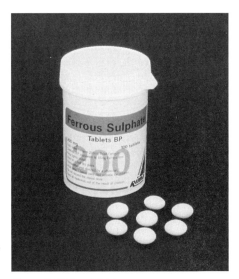

Figure 7.13 Iron tablets

not only to dissolve the tablets, but also to provide the H^+ ions needed in the redox equation.) 10 cm^3 of this solution were pipetted out, and a solution of 0.005 mol dm^{-3} KMnO$_4$ was run into it, until an end-point was observed. 9.92 cm^3 of the KMnO$_4$ solution was needed.

a   Calculate the number of moles of FeSO$_4$ present in the two tablets. (Remember that there is a factor of

ten to consider, when scaling up from a titration's worth to the whole solution.)

b   Say whether you think the '200 mg' refers to FeSO$_4$ or to Fe, and also how exactly the tablets matched their claimed content level.

c   How would the operator have seen the end-point?

(This question is really a rather hard Chapter 2 question. The only Chapter 7 knowledge it needs is the balanced redox equation, and that is the one in the example earlier.)

7.6   A second method: moles up/moles down

The method of half-equations is quite a formal and complete way of balancing redox equations. Its completeness lies in the way it formally itemises the transfer of electrons. There is another 'sketchier' way which never gets down to the electron level, but instead uses the *effects* of electron transfer.

In the formal method the number of electrons offered by one half-equation must equal the number required by the

other (as arranged by using multiples of one or both half-equation(s) – this is why we multiplied by five in the last example). But a result of this is that the total oxidation number change of one half-equation (or multiple thereof) will equal the total oxidation number change in the other half-equation.

In the iron(II)–manganate(VII) example, we had one half-equation in which *five moles of iron atoms went up one oxidation number*, while in the other half-equation *one mole of manganese atoms went down five oxidation numbers*. This kind of balance must occur in every redox equation, because it is really a re-wording of the concept that electrons lost by one species must be gained by another. This is the principle of the 'sketchier' method.

Now let us see that principle used to balance a redox equation.

If you drop a copper coin into concentrated nitric(V) acid there are several visual signals that a reaction is occurring. Most dramatic is the evolution of copious brown fumes, recognisable as nitrogen dioxide gas, $NO_2(g)$, while the solution goes a virulent limey green colour and on dilution reveals the blue of $Cu^{2+}(aq)$. If that is all that happens, what must the equation be? (See panel below for the answer.)

The reaction of copper with nitric acid – an equation balanced by the moles up/moles down method

Step 1: set out an unbalanced equation showing the known participants as reactants and products:

$$Cu(s) + HNO_3(aq) \rightarrow Cu^{2+}(aq) + NO_2(g)$$

Step 2: work out the oxidation numbers of the participants:

$$Cu = 0$$
$$Cu^{2+} = +2$$
$$N \text{ in } HNO_3 = +5$$
$$N \text{ in } NO_2 = +4$$

So copper atoms are going up two, while nitrogen atoms are coming down one.

Step 3: put the redox changes on to the equation.

Cu atom goes up 2

$$Cu(s) + HNO_3(aq) \rightarrow Cu^{2+}(aq) + NO_2(g)$$

N atom goes down 1

Step 4: put 'mole-multipliers' on the redox changes, so that the total changes in oxidation number add up to zero:

1 mole of Cu atoms goes up 2

$$Cu(s) + HNO_3(aq) \rightarrow Cu^{2+}(aq) + NO_2(g)$$

2 moles of N atoms goes down 1

Step 5: build the 'mole-multipliers' into the equation as stoichiometric numbers:

$$Cu(s) + 2HNO_3(aq) \rightarrow Cu^{2+}(aq) + 2NO_2(g)$$

Step 6: achieve complete balance using 'add-ons' of H^+, water, etc. Balance the oxygens first by adding two waters on the right-hand side, then balance the hydrogens by adding two H^+s on the left-hand side. The final equation is:

$$Cu(s) + 2HNO_3(aq) + 2H^+(aq) \rightarrow Cu^{2+}(aq) + 2NO_2(g) + 2H_2O(l)$$

(Now, rejoin the main text)

By the use of either of the two methods demonstrated, that is, the method of half-equations or the method of 'moles up/moles down', you are now in a position to construct a balanced equation for every conceivable redox event (whether or not it actually takes place).

Questions

28 By the method of half-equations, balance the following equations (H^+s and/or water may need adding on for complete balance):

a $Cu(s) + Fe^{3+}(aq) \rightarrow Cu^{2+}(aq) + Fe^{2+}(aq)$

(This is the reaction by which copper is etched off from boards to make the paths for printed circuits.)

b $Sn(s) + HNO_3(aq) \rightarrow SnO_2 + NO_2(g)$

c $SO_2(aq) + Br_2(aq) \rightarrow SO_4^-(aq) + Br^-(aq)$

29 By the 'moles up/moles down' method, balance the following equations (again, 'add-ons' may be needed):

a $KClO_3(s) + Mg(s) \rightarrow KCl(s) + MgO(s)$

b $MnO_4^-(aq) + I^-(aq) \rightarrow Mn^{2+}(aq) + I_2(aq)$

c $H_2SO_4(aq, conc.) + HI(aq) \rightarrow H_2S(g) + I_2(aq)$

30 Construct chemical equations from the following visual evidence. You may have to deduce what some of the invisible participants are. Use either of the two methods to balance the equations.

a When potassium iodide solution is added to hydrogen peroxide solution, the brown colour of iodine (the element) is seen. (Think about the oxidation numbers of the oxygen atoms in hydrogen peroxide, p. 126. Which oxidation number are they likely to go to?)

b When concentrated hydrochloric acid, HCl, is added to solid manganese oxide, MnO_2, the smell of chlorine is detected. (Assume manganese atoms go down to oxidation state +2, just as in the manganate(VII) reactions already studied.)

c When iron(III) ions are added to iodide ions, the solution goes a deeper brown colour.

Disproportionations

There are some redox reactions which, while adhering to all the rules about oxidation-number changes, still look a little odd. These are events in which atoms of the same element are simultaneously oxidised and reduced *by each other*. A famous case is the behaviour of halogen elements in cold dilute solutions of alkalis:

1 mole of Cl atoms goes down 1

$$Cl_2(aq) + 2OH^-(aq) \rightarrow Cl^-(aq) + ClO^-(aq) + H_2O(l)$$

1 mole of Cl atoms goes up 1

One of the chlorine atoms has gone from 0 to +1, while the other has gone from 0 to −1. The reaction is already balanced, since *one mole of chlorine atoms goes up one oxidation number, one mole of chlorine atoms goes down one*. This is rather trickier when the up and down journeys are unequal, as strangely enough happens between the same two substances when the alkali is hot and concentrated. The product mixture contains both chloride and chlorate(V) ions. Here is an unbalanced version of the equation:

Cl atom goes down 1

$$Cl_2(aq) + OH^-(aq) \rightarrow Cl^-(aq) + ClO_3^-(aq)$$

Cl atom goes up 5

This time chlorine atoms have gone down one and up five, so the balance is achieved by *one mole of chlorine atoms going up five oxidation numbers, five moles of chlorine atoms going down one*. Hence the ratio of products on the right-hand side is 5 : 1, and the total number of chlorine atoms involved is six. The balanced equation, with add-ons, looks like this:

5 moles of Cl atoms go down 1

$$Cl_2(aq) + 6OH^-(aq) \rightarrow 5Cl^-(aq) + ClO_3^-(aq) + 3H_2O(l)$$

1 mole of Cl atoms goes up 5

Question

31 When nitric(III) acid (HNO_2) solutions are warmed, bubbles of nitrogen appear, and the other product is nitric(V) acid. Write a balanced redox equation for this disproportionation, including 'add-ons'.

7.7 Using redox reactions as the basis for titrations

In Chapter 2 we looked at a method of making one solution react against another, using quantitative glassware and indicators to establish the exact end-point (p. 20). The object of these 'titrations' was to find out something previously unknown about one of the solutions, or to discover the equation for the reaction.

The same principles and situations apply in redox chemistry, although there are one or two differences in emphasis. For one thing, of course, the indicators of acid–alkali chemistry, geared as they are to detecting changes in pH, are useless for redox titrations. One way round this problem occurs when one of the reagents is itself fiercely coloured, as was the case with MnO_4^- in question 27, where the reaction is self-indicating. The visible end-point in that case, immediately after the last of the Fe^{2+} was oxidised, was the first permanent pink tinge, which occurs at MnO_4^- concentrations as low as 10^{-6} mol dm^{-3}. This ensures a minimal 'overshoot' beyond the real end-point.

If there is no dramatic auto-colour-change, there are indicators which are oxidised immediately after the last bit of reactant has gone, and in doing so go through a dramatic colour change.

There is a third method which is interesting and quite common. Its use is restricted to those reagents which are able to oxidise I^- ions, and relies on two facts for its success:

- Many reactants of interest *can* oxidise iodide ions, the product being iodine.

- Iodine itself has a very specific reaction with thiosulphate ions ($S_2O_3^{2-}$), which can be monitored by a crisp indicator system.

An example of this method should make these ideas clearer. Suppose we wanted to estimate the amount of chlorine in swimming-pool water. We could make the chlorine oxidise some iodide ions to iodine:

$$Cl_2(aq) + 2I^-(aq) \rightarrow 2Cl^- + I_2(aq) \qquad (4)$$

(*Two moles of iodine atoms going up one oxidation number; two moles of chlorine atoms going down one.*)

The problem here is one of end-point; the iodine is indeed coloured while none of the other three members of the reaction are, but that is of scant use when the end-point is the point at which the solution stops getting browner. That would be impossible to judge accurately by eye alone. The answer is to add an excess of iodide ions and just let the chlorine react. Afterwards you have a solution which contained Cl^- ions, unreacted left-over I^- ions, and as many I_2 molecules as there had been Cl_2 molecules.

This is where the thiosulphate ions come in; a solution of thiosulphate ions of known concentration is run in on top of the mixture, and it 'mops up' the iodine molecules:

$$I_2(aq) + 2S_2O_3^{2-}(aq) \rightarrow 2I^-(aq) + S_4O_6^{2-}(aq) \qquad (5)$$

Fortunately the end-point is now much more visible – it is the point at which the last vestige of brown colour disappears. The end-point can be improved by adding starch. The intensely coloured complex formed between starch and iodine makes iodine visible at concentrations hundreds of times lower than when it is on its own. Now the moment of the last surviving iodine can be identified very sharply. The method is complete. Here is a calculation based upon it.

Question

32 250 cm³ of swimming-pool water is shaken with 25 cm³ of 0.01 mol dm⁻³ KI(aq) (Figure 7.14). This solution provides an excess of I⁻ ions for the chlorine to oxidise – see equation (4). The iodine thus liberated is turned back to colourless iodide ions by reaction (5) above. 6.5 cm³ of 0.001 mol dm⁻³ sodium thiosulphate (Na₂S₂O₃) solution is needed for an end-point (as shown by adding starch near the end).

a How many moles of thiosulphate ions ($S_2O_3^{2-}$) are needed for the end-point?

b From this deduce (via equation (5) – look out for the 2 : 1 ratio) the number of moles of I_2 which were liberated by the chlorine in equation (4).

Figure 7.14 Testing swimming-pool water for the concentration of chlorine

c Then deduce the number of moles of chlorine present in the original sample.

d Finally express your answer as a concentration of chlorine in swimming-pool water.

There is one other rather curious feature in this system of reactions:

e Calculate the average oxidation number of the two sulphur atoms in the $S_4O_6^{2-}$ ion – see the right-hand side of equation (5). What explanation can you offer for this peculiar result?

(*Note*: The use of Cl₂(g) itself in swimming pools is now obsolete. The modern sterilisation regime is described in Chapter 12.)

Reviewing the iodine/thiosulphate method, you will see that it uses an interacting set of two reactions, with iodine as a sort of 'go-between'. First the iodine is oxidised, and then it is reduced again back to where it started. This is another new feature, compared to the titrations in Chapter 2. From that earlier phraseology (p. 20), the thiosulphate solution is playing the part of 'first reagent' (the one you know everything about), while the iodine plays a dual role: it is the 'target reagent' in its reaction with the thiosulphate, but then becomes the first reagent in the reaction by which it was created. In this second stage of the calculation, the unknown oxidiser is the 'target', and is indeed the target of the whole double operation.

7.8 Yes, but does it happen?

The story so far has brought us to the point where we can fit any two redox half-equations together, in a balanced whole equation, and use it as the basis for calculations. However, being able to write a balanced equation for a reaction does not guarantee that it will actually happen. For instance, we could write the perfectly balanced equation:

$$Cu^{2+}(aq) + 2Ag(s) \rightarrow Cu(s) + 2Ag^+(aq)$$

But we know from chemical experience that it is actually the *backward* direction of this equation that works.

Question

33a Which species is acting as electron-*giver* in the forward reaction above?

b Which species is acting as electron-*giver* in the backward reaction?

c Which species is the better electron-giver?

You can see from the answer to question 33 that there is something in the chemical personalities of those two metals which means that of two equally correct balanced redox equations, only one happens. So we need extra information in our system for organising redox knowledge – not just the ability to fit half-equations together, but which of two possible unions will be fruitful.

The block of data you need for the second bit of this task is the **electrochemical series**, which is a quantitative league table of willingness to give electrons. It empowers you to predict which way any redox event will go, but it is quite a formal system with some very strict rules, which you need to understand to use the series with confidence:

Rule 1 Every redox half-equation is written in the same way, with the oxidised form on the left (receiving electrons), and the reduced form on the right.

Rule 2 All the half-equations are arranged in order. The order is such that the really 'fierce' electron-takers, like oxygen and fluorine, are at the bottom.

Here is an extract from the series, featuring some of the more famous aqueous-solution redox couples drawn from the halogens and their halide ions, the metals and their cations, and the MnO_4^-/Mn^{2+} and Fe^{3+}/Fe^{2+} systems.

$$Mg^{2+}(aq) + 2e^- \rightarrow Mg(s) \qquad (6)$$

$$Zn^{2+}(aq) + 2e^- \rightarrow Zn(s) \qquad (7)$$

$$Cu^{2+}(aq) + 2e^- \rightarrow Cu(s) \qquad (8)$$

$$I_2(aq) + 2e^- \rightarrow 2I^-(aq) \qquad (9)$$

$$Fe^{3+}(aq) + e^- \rightarrow Fe^{2+}(aq) \qquad (10)$$

$$Br_2(aq) + 2e^- \rightarrow 2Br^-(aq) \qquad (11)$$

$$Cl_2(aq) + 2e^- \rightarrow 2Cl^-(aq) \qquad (12)$$

$$MnO_4^-(aq) + 8H^+(aq) + 5e^- \rightarrow Mn^{2+}(aq) + 4H_2O(l) \quad (13)$$

$$F_2(aq) + 2e^-(aq) \rightarrow 2F^-(aq) \qquad (14)$$

Clearly this series has a strong correlation with Pauling's electronegativity index (p. 74), but that applies only to elements. A full account of the quantitative experimental basis for the series must wait until Chapter 16 on electrochemistry. For now we will just learn to use the results. Notice that none of these reactions can happen alone (that is standard for any half-equation), yet neither can any pair of them operate in union as written. This is because you cannot have two equations both demanding electrons. Something, literally, has got to give.

A fruitful mating of two half-equations from that list requires one half-equation to run in the forward direction as written (thus accepting electrons), while another half-equation runs in the backward direction (thus offering the electrons that the forward half-equation wants). This is expressed as:

Rule 3 Any half-equation lower in the list is capable of driving any half-equation above it backwards.

For example, reaction *(13)* going forward will drive reaction *(12)* backwards. In order to construct the whole-equation from this fact, we should rewrite reaction *(12)* the other way around:

$$2Cl^-(aq) \rightarrow Cl_2(aq) + 2e^- \qquad (15)$$

Then, 'simultaneous equation style', get the electrons to cancel with a '$(2 \times (13) + 5 \times (15))$' arrangement, giving:

$$2MnO_4^-(aq) + 16H^+(aq) + 10Cl^-(aq) \rightarrow$$
$$2Mn^{2+}(aq) + 8H_2O(l) + 5Cl_2(aq) \qquad (16)$$

(Returning to an earlier point about why we learn oxidation numbers, this is the sort of case where they are invaluable.)

To retell the story of reaction *(15)* in words, it is a case of the extreme electron-'thirst' of the manganate(VII) ion forcing even such a determined electron-keeper as the chloride ion to part with its hard-won extra electron, and sending it back to elemental chlorine. It is not often that chlorine gets treated in this cavalier fashion, it being more accustomed to grabbing other species' electrons – a case of the grabber grabbed, as it were. As a half-equation from the lower part of the table, the Cl_2/Cl^- system is used to going forwards and it is only when it meets a system even lower in the table that it is forced into reverse.

If you simply want to know whether a particular reaction will go in a particular direction, the list can be used in a quicker and more casual way, using this rule:

Rule 4 If any half-equation is beneath another, the left-hand side of the lower one will react with the right-hand side of the upper one.

So the reactants are on a bottom left–top right diagonal, the products are on the other diagonal, and the whole direction of change is on the perimeter of an anti-clockwise circle (Figure 7.15). This rule is sometimes called (not un-expectedly) 'the rule of anti-clockwise circles'.

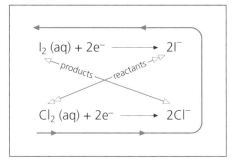

Figure 7.15 Showing how the rule of anti-clockwise circles operates on the half-equations in the electrochemical series to predict the direction of redox change

Figure 7.16 Chemical hazard information for an oxidising agent

Questions

34 Use the rule of anti-clockwise circles to answer the following challenges:

a Name three reagents in the list (p. 134) capable of oxidising bromide ions to bromine.

b Name two metals in the list which will be oxidised by Fe^{3+} ions.

c Name the only oxidised form of a redox couple in the list that cannot be reduced by zinc metal.

35 a Which of the following reactions will actually happen? (No stoichiometric ratios are implied – consider them mixed in any proportions.)

$Fe^{3+} + I_2$

$Zn^{2+} + Mg$

$Br_2 + Cl^-$

b Which one of the above trio of mixtures could you have dismissed as impossible without even looking at the electrochemical series?

'Agent' words

An oxidising agent is a substance which brings about oxidation in another species. This agent must therefore be a withdrawer of electrons, and so oxidising agents must be *reduced* when they bring about an oxidation. We also know that every reaction must contain an oxidising agent and a reducing agent. But in chemistry the words oxidising agent and reducing agent are usually reserved for extremes. Thus the MnO_4^- ion is called an oxidising agent, but not the Zn^{2+} ion, even though, in our league table, it can do one oxidation.

Because extreme oxidising agents can oxidise (among other things) most carbon- and hydrogen-rich organic molecules, there are rules for separate storage of these two classes of chemicals, and the former group have to carry hazard warning labels (Figure 7.16). Similarly, you would not necessarily think of all organic chemicals as reducing agents, even though, up against $KClO_3$, say, they would be playing that role.

Question

36 From the compounds named on either side of the half-equations in the electrochemical series list (p. 134), name:

a the two strongest oxidising agents

b the two strongest reducing agents.

One final word of warning about the electrochemical series – it is essentially a **thermodynamic** series, so its information is of the 'will it happen?' type, rather than the 'how fast will it happen?' type. The existence of an anti-clockwise circle is an absolute condition which any redox reaction must meet. However, the existence of the circle does not guarantee that the reaction will go at a measurable rate, at least at room temperature. In legalistic language, the anti-clockwise circle is a *necessary, but not sufficient* condition for reaction.

7.9   Redox reactions – just an intellectual game?

You may respond to this chapter by concluding that the whole business of awarding oxidation numbers, balancing equations, and using the rule of anti-clockwise circles is just an intellectual exercise. Above all, you may feel unconvinced that there is any benefit to be gained from trying to shovel all these rather disparate chemical changes into one conceptual compartment. This chapter ends by trying to respond to this challenge.

Consider a reaction like the burning of dihydrogen, $H_2(g)$. It sits at that covalent end of the redox spectrum where talk of electron transfer looks rather pointless. It is, if you like, a straight oxidation of the 'old school', from the days when oxidation meant what it said – namely reaction with oxygen:

$$2H_2(g) + O_2(g) \rightarrow 2H_2O(l) \quad (17)$$

(*Four moles of hydrogen atoms going up one oxidation number, two moles of oxygen atoms coming down two.*)

Although the oxidation number of hydrogen atoms changes from 0 to +1, it is one of those oxidation number changes based on bond polarity rather than real electron loss. In dihydrogen the

hydrogen atoms each possess the equivalent of one electron (as a half-share in two), whereas in water they have (by the rules of the oxidation number game) possession of no electrons at all. But whatever the rules say, the hydrogen atoms still have got some sort of grip on two electrons in both reactant and product. No electrons have gone missing, and every atom is obeying the rule of two, or the rule of eight.

This seems remote from the clear-cut electron transfers of the 'genuine electrochemical' redox reactions – ones that can be made to operate real electrochemical cells – such as:

$$Zn(s) + Cu^{2+}(aq) \rightarrow Zn^{2+}(aq) + Cu(s)$$

where there is no question that zinc atoms have lost electrons.

And yet, suppose it could be shown that even the oxidation of dihydrogen could run an electrochemical cell. Would that not be persuasive evidence that there *is* a genuine link between all redox reactions – something which makes it worth seeing them as a single class, whose linking feature is the ability (given the right apparatus) to move electrons in a wire?

There is such a cell, but at risk of being anticlimactic, a full understanding of it has to await Chapter 16, in which cells are discussed. For now, here is a foretaste. All cells work by physically separating the two half-equations which are interacting, and making them transfer their electrons by wire. The ability to write reaction *(17)* as two half-equations, therefore, is a necessary first step to envisaging it as a cell reaction.

As it stands, reaction *(17)* cannot realistically be broken into half-equations; but it can if the reaction is transferred to an aqueous solution medium. We can write:

$$H_2(g) \rightarrow 2H^+(aq) + 2e^- \quad (18)$$

$$O_2(g) + 4H^+(aq) + 4e^- \rightarrow 2H_2O(l) \quad (19)$$

Using the simultaneous equations method (p. 130):

$$2 \times (18) + (19)$$

This gives:

$$2H_2(g) + O_2(g) \rightarrow 2H_2O(l) \quad (20)$$

So it *is* possible to see oxidation of dihydrogen (along with the reduction of oxygen) as two half-equations (in aqueous solution), and therefore it is realistic to talk about hydrogen giving electrons to oxygen.

Figure 7.17 shows the outline of how a hydrogen–oxygen cell works. Because

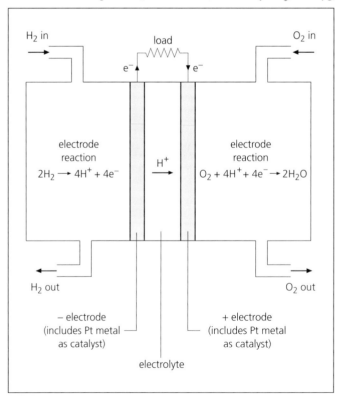

Figure 7.17 Simplified diagram of a hydrogen–oxygen fuel cell

the overall cell reaction *(20)* is the same as that for the burning of hydrogen *(17)*, this cell (like others working along similar lines) is called a **fuel cell**. The role of the platinum catalysts in the two electrode systems is vital to the operation of the two electrode half-equations. The purpose of the electrolyte is to provide a medium by which the H^+ ions may be transferred from the hydrogen electrode region to the oxygen side.

Question

37a Which species originally 'owned' the electrons that are sent around the external circuit in this fuel cell?

b There are hopes that fuel cells can be scaled up from their present modest size and electrical capacity, to the point where they can be alternative power stations. List and discuss some advantages and disadvantages of this style of energy generation.

The subject of electrochemical cells in general, including fuel cells, is revisited in greater depth in Chapter 16. The whole field of electrochemistry is inextricably linked to redox reactions, and this fact alone is probably justification enough for the effort we have invested in mastering the complicated set of rules which underpin the redox concept.

Summary

• **Oxidation**, which originally meant reaction with oxygen, has come to mean the *loss of electrons* by an atom during a chemical reaction.

• **Reduction**, which originally referred to removal of oxygen, has come to mean a *gain of electrons* by an atom during a chemical reaction.

• The losses or gains of electrons referred to in the last two sentences may be total (as in ionic bonding) or partial (as in polar covalent bonding).

• The **oxidation number** of an atom in a chemical species expresses the number of electrons possessed by that atom, relative to the number possessed in the elemental state. It therefore expresses a *difference*. A positive oxidation number indicates a loss of that number of electrons (and conversely a negative number indicates gain). Non-zero oxidation numbers are always written with plus or minus signs. Atoms in the elemental form have oxidation numbers of zero by definition.

• In ionic bonding between two atoms, the changes in oxidation number are clear-cut and unambiguous, because the losses and gains of electrons are themselves unambiguous. The oxidation numbers of simple, single-atom ions are numerically equal to the charges on the ions. (So sodium in Na^+ is showing an oxidation number of +1 (one electron lost), while oxygen in O^{2-} is showing an oxidation number of −2 (two gained).)

• In polar covalent bonding the loss or gain of electrons is at best partial. Nevertheless integer (whole) oxidation numbers are still assigned, on the basis that in a polar covalent bond the more electronegative member of the bond has kept its own electron and gained the electron of the less electronegative member. Thus in a polar covalent single bond between two atoms one atom will be held to have increased its oxidation number by 1 (the loser) and the other to have decreased its oxidation number by 1 (the gainer).

• The system described in the last sentence is equivalent to imagining that the polar covalent bond has 'gone ionic'.

• In covalent bonds between identical atoms, there is (usually) negligible polarity, and therefore no change in oxidation number.

• Assigning oxidation numbers to atoms in molecules can be done with the help of displayed formulae. Every bond (except those between identical atoms) can be declared a 'win' (electron gain) for one member and a 'loss' for the other, decided by which member has the higher electronegativity. (By these rules fluorine atoms win every bond in which they take part.) The eventual oxidation number of an atom can be found by summing the wins and losses from all its bonds. For example, a net loss of one bond means an oxidation number of +1.

• *The sum of the oxidation numbers of all the atoms in a chemical species (whether molecule or ion) adds up to the overall charge on the species.*

• Quick assignments of oxidation numbers can be made on the basis that the oxidation numbers of certain atoms are unchanging ('reliable'). An unknown oxidation number can then be deduced by using the rule in the last paragraph. Use of this rule occasionally leads to confusion, as in the case of hydrogen peroxide. The displayed formula/bond-polarity method is a safe last resort when the 'reliable oxidation numbers' method proves unreliable.

• The electrons in dative covalent bonds are held to have been either lost as a pair, or retained as a pair, by the donor atom. Thus the oxidation number of the donor changes either by +2 or not at all.

• The naming of compounds draws heavily on oxidation numbers. Polarised electropositives or fully-fledged cations of uncertain oxidation number must be named along with the oxidation state as a roman numeral (for example, copper(II) oxide, lead(IV) bromide). Oxo anions all carry the suffix '-ate' plus the roman version of the oxidation state of the central atom (for example, sodium sulphate(IV)). The extension of these rules to complexes is dealt with in Chapter 19.

• Once oxidation numbers have been assigned, oxidation and reduction can then be redefined simply as increase or decrease (respectively) in oxidation number.

• All redox reactions must logically conform to the overriding principle that *total oxidation number change in a chemical reaction is zero* (logical because electrons cannot be destroyed and so for every giver there is a taker).

• This principle gives rise to two methods of balancing redox equations.

a In the *method of half-equations* the reaction is seen as comprising two interlocking redox changes – the loss of electrons by the species being oxidised and the gain by the one being reduced. These two half-equations are individually balanced, and then combined according to the principle that total electrons 'on offer' must equal total electrons accepted.

b In *the moles up/moles down method*, the equation is never formally split into redox halves. Instead it is inspected for oxidation number changes, and then 'mole-multipliers' are applied (as stoichiometric numbers) to ensure that (in effect) the moles of redox change are the same in each direction. For example, if one mole of manganese atoms comes down by five oxidation num-

bers, it follows that five moles of iron atoms must go up one oxidation number.

• In both the above methods, the actual progress of the reaction may in some cases depend on the participation of species derived from the aqueous medium ($H^+(aq)$, $OH^-(aq)$ or H_2O itself). (This is true, for instance, in all oxidations carried out by oxyanions like manganate, chromate, etc.) To balance the equation, therefore, it means that aqueous 'add-ons' may have to be manipulated into the equation, until all oxygen atoms and hydrogen atoms are balanced.

• **Disproportionations** are redox reactions in which atoms of the same kind, and at the same oxidation number, proceed to act as electron donor and acceptor *to each other*. The classic case is that in which the atoms in dihalogen molecules oxidise *and* reduce each other in alkaline aqueous media.

• The word **agent** is applied to species with well-known or powerful redox properties. Thus manganate(VII) or difluorine could well be described as oxidising agents (or 'oxidants'), while carbon or zinc would be called reducing agents (or reductants). In reality, however, there is an oxidant and a reductant in every redox change.

• Strong oxidising agents are capable of oxidising iodide ions to iodine. This is the basis of a titration method for

estimating the concentration of solutions of these agents. A pipette-full of the oxidant liberates I_2 from an excess of I^-, and then the iodine is titrated against a solution of sodium thiosulphate of known concentration. The number of moles of thiosulphate ions used can be related back to the original moles of oxidant. Starch is used to sharpen the end-point, which is the moment of last removal of I_2.

• Every redox equation can be written reversibly, but in the majority of cases the reaction can only occur in one direction. The allowed direction may be predicted by use of the **electrochemical series** and the **rule of anti-clockwise circles**.

• Every redox reaction can in theory be made the 'engine' of an electrochemical cell, although in nearly all cases the system must be in an aqueous environment.

• Even those redox changes, like the direct literal oxidation of hydrogen, which do not seem to involve real electron transfer, can still be made to run cells. Cells in which oxygen itself is the oxidising agent, and into which the cell reagents flow continuously, are called **fuel cells**. They are heavily dependent on catalysts, usually from the platinum group, for their effectiveness, but they achieve a much more efficient conversion of chemical energy to electrical energy than conventional steam turbine/generator methods.

Thermodynamics 1: Energy exchange and chemical systems

8.1 The Law of Conservation of Energy

You are a student of chemistry. You have your own place on a bench in a laboratory, near a gas tap. This gas tap gives you the power to have instant heat, at temperatures up to about 700 °C, whenever you want it. When you go home after school or college, your house may be kept warm by the same chemical reaction as the Bunsen burner uses, or by the burning of some other hydrocarbon- or carbon-based fuel. If you go out in the evening, you may travel in a vehicle whose energy source is yet another reaction featuring the oxidation of a hydrocarbon. And on top of that we are all constantly dependent on an electricity supply industry whose main energy source is this same class of chemical substances and their reactions with oxygen. (All of these reactions are oxidations, as shown in Chapter 7.)

These are examples of reactions which we value not for their chemical out-comes, the product molecules, but for their energy outcomes. (In fact the product molecules are more likely to be sources of problem than benefit, as in the case of carbon dioxide and its 'green-house' contribution.) However, while *all* reactions have some sort of energy outcome, that outcome is not always the *production* of heat. There are also reactions in which the surroundings actually get colder, and while neither as numerous or as useful as the first kind, both kinds together make up our area of study.

This chapter is about what causes chemical systems to be capable of an energy exchange with the surroundings, and it will demonstrate how chemical systems have a place in the larger field of energy physics. This branch of science, which links up mechanics, heat and chemistry, is known as chemical **thermodynamics**.

The first thing to recognise is that chemistry does not stand outside the laws of 'greater science'. The Law of Conservation of Energy is as applicable in chemistry as it is in physics (or biology). The law forces us to ask this question: 'If large amounts of heat energy are suddenly called into being as a result of a chemical reaction, and if energy itself can neither be created nor destroyed, *where was the energy stored before the reaction?*' (Equally, if heat disappears from the surroundings, *where did it go?*) An easy answer to both these questions would be 'in the chemicals as chemical energy', but this is more of a renaming process than an act of understanding. Our answers must be at a level appropriate to the nature of our science, which means at the atomic scale.

Before answering these questions, we need to establish one of the foundations of thermodynamics – to understand the meaning of the Law of Conservation of Energy.

The meaning of the law

The Law of Conservation of Energy is an idea based on experimental observation. One place where we can see it at work is the snooker table. When two balls collide in a straight-line collision the second ball

Figure 8.1 Sir Isaac Newton

Figure 8.2 Foucault's pendulum experiment

never departs faster than the first ball arrived. In other words, there seems to be a fixed amount of energy in the system, and the two balls must share it between them. Newton (Figure 8.1) was observing similar systems when he formulated his laws of Conservation of Momentum and of the Equality of Action and Reaction.

The Law of Conservation of Energy grew, as it was realised that the concept of energy was broader than just kinetic energy. The first addition to the family

was **potential energy**. The study of pendulums made it clear that kinetic energy could be converted into the energy of a raised weight (Figure 8.2). Raising weights is normally done by work measured as 'force × distance', but in the pendulum case it seemed that a certain amount of kinetic energy could do a fixed amount of work in creating 'raised weight' potential energy. The potential energy could either be used as a way of storing that work to be used later, or it could go straight back to kinetic energy (as it does every time the pendulum swings back through its vertical position).

Let us look more closely at the nature of potential energy. For a system to have potential energy, two conditions must be fulfilled – there must be a force field in operation, and work must be done against the *force field* to put some part of the system in an *unstable position*. 'Raised weight' is just a special case of potential energy, in which the force field is *gravity* and the unstable position is *up*.

Question

1 Here are some more systems which have potential energy. In each case identify the origin of the force field, and the unstable position:

- a stretched longbow
- a trampoline at the bottom of someone's 'bounce cycle' (Figure 8.3)
- two magnets, opposite poles facing, held apart.

Figure 8.3 A trampolinist at the bottom of her 'bounce cycle'

The next section shows that what we call chemical energy is yet another form of potential energy, although disguised by the small size of the particles involved.

8.2 Chemical bond model of chemical energy

Consider a number of pairs of objects, all of which have a force relationship to each other, and one of which is chemical. The pairs are:

1 the Earth and an apple
2 two magnets, unlike poles facing, on a horizontal frictionless surface
3 two magnets, like poles facing, on a horizontal frictionless surface
4 two hydrogen atoms.

For pairs (1) and (2), Figure 8.4 shows sketch graphs of the way their potential

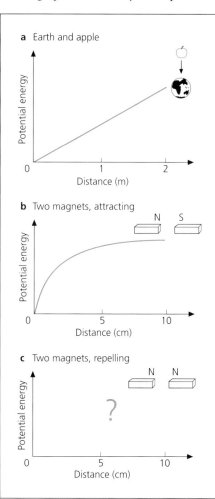

Figure 8.4 Potential energy against distance graphs for three pairs of objects

energy varies as a function of the distance between members of the pair.

In (1), as the distance between the Earth and the apple approaches zero, so does the 'raised weight' potential energy. In (2), the two attractive magnets are rather like the apple and the Earth, in that they 'want' to be together, and so their potential energy of position is at zero when they are together. (In both cases we mean that at the origin they have surrendered all their potential for doing work.)

Question

2 a At first sight it looks as if the Earth–apple system is different from the attractive magnets one in that the former has a linear graph and the latter has a curve. In fact the difference has been artificially created by the choice of distance axes in the two cases. Can you carry this explanation further?

  b In both the cases (1) and (2) the systems, if let go, end up with no potential energy. Trace the energy conversions and show where this energy will go.

  c Draw the sketch graph for case (3). Assuming you do predict some change in the potential energy of the system, state the energy conversions which would account for this change.

The fourth and most important graph in this series is that for the chemical system

(Figure 8.5). The objects are two hydrogen atoms. You will see that the graph has similarities to *both* the magnet graphs, because this time there is attraction followed at short range by repulsion. At the extreme right of the graph, there are two isolated hydrogen atoms, at a distance at which they barely 'know' the other is there. At medium distances the main effect is the attraction between the hydrogen atoms as the electrons begin to share orbits with each other and draw the nuclei closer. But at very short distances the repulsion between the two nuclei produces a sudden rise in potential energy. In between there is an optimum distance which represents the **natural bond length** of the H_2 molecule – the distance of maximum stability, at which the molecule has exhausted its capacity to change. (This distance for the H—H bond is 0.074 nm.) In some ways the bond is like a spring that can be put into both tension and compression, either side of its natural length.

Now let us consider the accompanying changes in energy. Specifically, let us consider where the potential energy goes as the bond begins to form. The system of two hydrogen atoms most closely resembles the attractive magnets. The answer to question 2b shows the two magnets accelerating towards each other and colliding. The energy conversion path in that case was: potential energy → kinetic energy → heat. So, if two hydrogen atoms should happen, in the course of their random movements, to approach close enough to 'interest' each other, they will fall towards each other with increasing kinetic energy due to the conversion of the potential energy source.

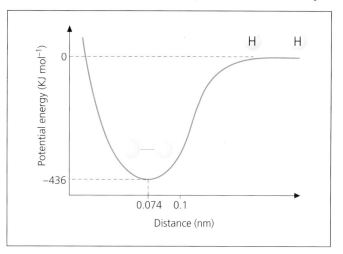

Figure 8.5 Potential energy against distance graph for two hydrogen atoms

(*Note:* Many students have trouble with the idea that the 'no interaction' condition is one of high potential energy. The two concepts do not sit together comfortably. But it is like saying that floating in space just beyond a planet's gravity is a higher potential energy situation than being on the ground.)

Most models which seek to explain the behaviour of very small particles by likening it to the behaviour of big ones have their drawbacks, and this one is no exception. It is easy enough to imagine the potential energy that was originally in the 'attractive magnets' system ending up as heat, because at the moment of collision there is a change between a kinetic energy of *organised* multi-particle movement to a kinetic energy of *chaotic* multi-particle movement (heat) of all the iron atoms in the magnets. However, in the case of the formation of the hydrogen molecule, there is only one molecule at the end to contain the former potential energy. A literal application of Newtonian principles would have the two atoms approach, accelerate, go past the optimum bond length into the repulsion part of the graph, and bounce straight out again like a pendulum returning whence it came.

So this is where the model breaks down. In the mysterious ways of small particles governed by quantum laws, the original potential energy of the two separate hydrogen atoms ends up as kinetic energy of the hydrogen molecule (Figure 8.6). Now we can extend our picture from two hydrogen atoms to two moles of hydrogen atoms, at high potential energy, joining together to form a mole of hydrogen molecules, at high kinetic energy. What an observer will record is the evolution of heat.

We now have an answer to the question of where the heat of a chemical reaction was residing before the reaction. It was in the form of molecular potential energy of position, caused by particles being in one arrangement when they would have 'preferred' to be in another. This 'preference' was a response to a force field; not a gravitational one like that experienced by a raised weight, but a force field whose origins lie in the simple fact that electrons are attracted to nuclei.

This is the logic behind seeing chemical energy as a branch of potential energy. We see that the force field giving rise to chemical potential energy is *electrostatic*, and for any two bondable atoms the high energy position is *apart*. These ideas are summed up in Table 8.1.

Stephen Hawking and others have pursued the idea that all forces are branches of a single force – that gravity and electrostatics are linked, for example. Here we have a case where a link has already been established – chemical energy is a branch of electrostatics.

Enthalpy introduced

To go back to the hydrogen atoms of case (4), p. 141:

$$2H(g) \rightarrow H_2(g)$$

If the reaction occurs on a mole scale, chemical potential energy will be destroyed, and data books tell us that 436 kJ of heat energy will be created, either to result in the faster movement of the product molecules, or to leak away to the surroundings. Chemists have a word for chemical potential energy. It is called **enthalpy**. When the reaction above has finished, the molecules have 436 kJ less

enthalpy than they had when they started. *Note:* A minor modification of this definition is given later (Section 8.9).

At this point we need to establish four important principles which will underpin the rest of the chapter:

1 Enthalpy changes in chemical systems are measured by the equal and opposite heat change in the surroundings. This sets enthalpy apart from, say, gravitational potential energy. In the latter case there is no need to wait for a mass to hit the ground and then to measure how much the ground has heated up, in order to estimate a loss of gravitational potential energy – it is readily measurable in terms of the size of the mass falling, the gravitational constant, and the distance fallen (as expressed in the formula $m \times g \times h$). But the 'fall' which occurs in the formation of a chemical bond is happening in a microworld closed off from our observation. We are forced to use a different means of energy measurement, focused on the knock-on events in the surroundings, despite the close similarities between bonding and falling.

2 The surroundings have therefore become the focus of the nomenclature for talking about energy changes in chemistry. A reaction which *gives* heat to the surroundings is called an **exothermic** reaction. A reaction which *absorbs* heat from the surroundings is called **endothermic**.

3 Thus when a reaction produces an outflow of heat (that is, is exothermic) the enthalpy change in the chemical system itself is negative, yet it is equal in magnitude to the positive heat change of

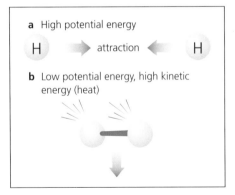

Figure 8.6 The conversion of potential energy to heat

a High potential energy

H → attraction ← H

b Low potential energy, high kinetic energy (heat)

Table 8.1 Forms of potential energy

System	Force field	Origin of force field	High potential energy position	Low potential energy position
Raised mass	gravity	?	up	down
Magnets	magnetic spin	electron identity of facing poles	depends on	
Elastic material	bond orientation	electron repulsions	deformed	not deformed
Bondable atoms	electrons and nuclear attraction		apart	together

the surroundings, and is measured in the same units (usually kilojoules per mole). (This is really just another way of expressing the Law of Conservation of Energy, which says that the energy lost by the system is equal to the energy gained by the surroundings.) In an endothermic reaction the enthalpy change in the system is positive and is equal in magnitude to a negative heat change in the surroundings. Putting this sign reversal idea into the concise language of symbols,

$$\Delta H_{\text{system}} = -(\text{heat})_{\text{surroundings}} \quad (8.1)$$

The symbol ΔH is used for enthalpy *changes* in chemical systems, and thus we might write, for the hydrogen reaction:

$$2H(g) \rightarrow H_2(g) \ \Delta H = -436 \text{ kJ mol}(H_2)^{-1}$$

4 The definition of 'surroundings' is simple in its wording, and yet quite difficult to understand. The surroundings are simply defined as the rest of the universe apart from the reaction system. In thermodynamics, the 'system' is the word used for the reactants and products of a reaction, plus any solvent medium. The difficulty arises from the obvious fact that the first thing which happens to the heat flowing to or from a reaction system, is that it lowers or raises the temperature of the molecules of the system itself, rather than the surroundings. The problem is resolved by assuming that the product molecules are allowed to return to the starting temperature, and the true enthalpy change is then the heat flow which occurs between system and surroundings in the course of this return.

Bond breaking

Question

3 What would be the chemical potential energy (enthalpy) change (ΔH), if one mole of H_2 molecules were to split up into two moles of H atoms? What heat changes would there be in the surroundings?

Here is an extra mental picture for question 3. Assume that cleavage of the molecule takes place as a result of two molecules colliding. Work has to be done, equal to the force necessary to pull the atoms apart from the natural bond length up to the 'no interaction distance', multiplied by that very distance. (This is rather like the work needed to pull two magnets apart.) This work could have been drawn from the kinetic energy of one or both of the colliding molecules, so when the collision is over, one or both will be travelling more slowly (Figure 8.7). That event, repeated in mole-sized multiples, adds up to a loss of heat energy, and therefore a fall in temperature.

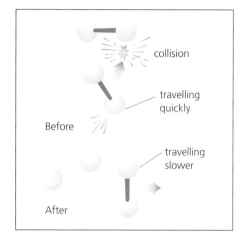

Figure 8.7 The conversion of kinetic energy (heat) to chemical potential energy, in a bond-breaking collision

The important message of this section is:

1 Bond making decreases the enthalpy (chemical potential energy) of chemical systems, and results in heat being created.
2 Bond breaking increases the enthalpy (chemical potential energy) of chemical systems, and results in heat being destroyed.

8.3 Bond making and breaking

The example used so far is an extreme one – the creation of a hydrogen molecule from two atoms is just about the simplest chemical change possible to imagine. Incidentally, imagining it is all you *can* do at any normal temperatures, since hydrogen molecules are stable below about 1000 K. However, it is by no means necessary that a reaction should actually happen before you discuss its enthalpy change. This chapter contains many reactions which do not happen.

Nearly all normal reactions which do happen are more complex than the hydrogen example, in that they feature bond breaking followed by bond making. So what sort of enthalpy changes take place when a reaction involves this combination of events?

Here is another simple example. Consider the reaction:

$$H—H + F\cdot \rightarrow H\cdot + H—F$$
(all in the gas phase)

Here the 'dots' stand for unpaired electrons (indicating single-atom free radicals, in other words). As you can see (Figure 8.8a), this reaction involves the breaking of an H—H bond, and the making of an H—F one. We have already studied the H—H bond, so now we consider similar data for the H—F bond. Figure 8.8b shows the potential energy curve for this bond – the bonded atoms are 568 kJ mol^{-1} more stable than the separate H· and F· free radicals, which means that 568 kJ mol^{-1} of heat is released when H—F forms, and the same amount of heat is absorbed when H—F

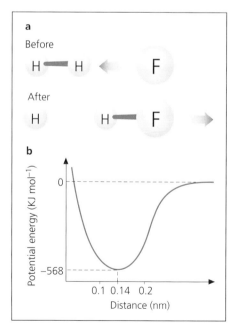

Figure 8.8 a H_2 + F· reaction. **b** Potential energy graph for an H—F bond

is broken. The reaction can be visualised as a two-step process, as follows:

1 Break one mole of H—H bonds:

$$\Delta H = +436 \text{ kJ mol}^{-1}$$

(Now we have an intermediate state in which all three atoms are free of each other, on the 'no interaction' potential energy plateau.)

2 Make one mole of H—F bonds:

$$\Delta H = -568 \text{ kJ mol}^{-1}$$

The overall 'reaction profile' (a graph with a rather ill-defined *x*-axis which is firstly concerned with enlarging the H—H distance until it becomes irrelevant, and then starts to record the shrinking H—F distance) is shown in Figure 8.9. The *y*-axis matters most, and shows that the potential energy of the chemicals ends up *lower* than it started. The change in enthalpy is equal to:

$$\Delta H_{\text{reaction}} = +436 + (-568)$$
$$= -132 \text{ kJ mol}^{-1}$$

According to the 'pretend chemicals have feelings' school of explanation, the atoms 'prefer' to be in the H—F, H· arrangement rather than in the H—H, F· one. If 132 kJ mol^{-1} is the amount of potential energy destroyed in going from H—H + F· to H—F + H·, then we would predict that 132 kJ of heat will be created in the surroundings of the reaction.

We have here a prediction that looks simple and safe, but more importantly lends itself to easy and useful generalisation. It appears that, as long as we know the potential energy curve for every bond there is, we can predict the enthalpy change for any reaction which features the breaking and making of any of those bonds – which means all reactions.

Of course, it is not quite as simple as that, because for one thing bonds do not behave identically in every molecule in which they occur. However, for fairly crude predictions, the situation described in the last paragraph does hold good.

You will have noticed that the prediction is based on the idea that the H—H bond breaks completely before the H—F bond begins to form. And yet it is highly probable that the real reaction goes by a different path. Looking at collisions setting off reactions, surely it is likely that reaction will occur when an incoming F· meets an H—H bond, in which case the H—F bond will be forming at the same time as the H—H bond is breaking? Would this invalidate the two-step mathematical model?

In fact, the mathematical model survives and stands, as shown by the following equivalent explanations.

Explanation 1: 'Same valleys/lower hill'

Imagine the F· atom coming in along the line of the H—H bond. The forming of the H—F bond will certainly bring about a lowering of potential energy for the whole molecular complex, but the simultaneous stretching of the H—H bond will raise it. We will have for a short time a species which might be symbolised thus:

$$\text{H----H----F}$$

There is never a point at which the three free atoms exist, so the top plateau of potential energy is never reached. It is as if the 'lowering' effects from the H—F side of the bond subtract from the 'raising' effects on the H—H side. (Another diagrammatic way of representing it would be to say that the two potential energy curves overlap their *x*-axes, so that the H—F distance has begun being 'interactional' before the H—H distance has ceased to be interactional.)

Another qualitative picture is of two possible routes between two valleys – one goes over the top of the mountain and needs maximum climbing effort, but by way of compensation there is a long run down; the other goes over the shoulder of the hill, with less climbing but less running down. But the way you go does not alter the relative depths of the two valleys. The difference in depth of the two valley bottoms is 132 kJ (Figure 8.10) whichever route is followed.

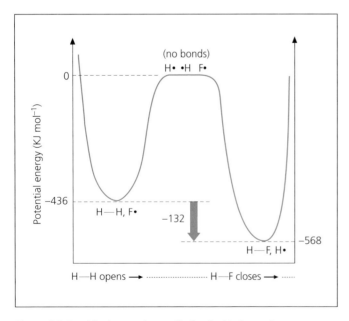

Figure 8.9 Break/make reaction profile for the H$_2$, F· reaction

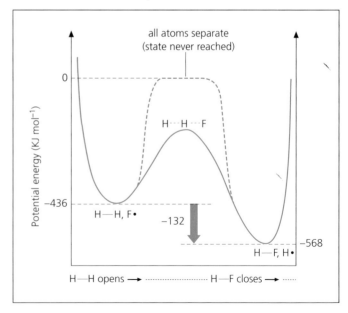

Figure 8.10 Reaction profile (realistic version) for the H$_2$, F· reaction

Explanation 2: Hess's Law

This explanation is based on the principle of Hess, which has become known as Hess's Law. It is one of the major statements of chemical thermodynamics. Hess argued that:

Any route between two arrangements of atoms must necessarily involve the same overall energy change.

He considered the alternative: suppose you could go from arrangement A to arrangement B, via two intermediate states C and D (Figure 8.11) – just as we have done in Figure 8.10 with the H—H, F· system. Suppose the route via C resulted in a lowering of 100 kJ, whereas the route via D featured a lowering of 90 kJ. We could then contemplate carrying out the forward reaction via C followed by the reverse reaction via D, and be back to the starting point A. But the overall potential energy change would be:

$$-100 + 90 = -10 \text{ kJ}$$

which would mean the calling into existence of 10 kJ of heat without any change from our starting conditions. This, said Hess, offended against the First Law of Thermodynamics (that of Conservation of Energy), and all human experience about not being able to get energy ·for nothing from anywhere. So he had to conclude that:

The enthalpy change occurring between any two chemical states is independent of the route taken.

Applied to our reaction, this means that there will be an enthalpy change of -132 kJ mol^{-1}, or that 132 kJ of heat will be given to the surroundings, irrespective of the route of the reaction.

Refining some definitions

The enthalpy change associated with a chemical reaction is defined as follows:

The enthalpy change is the heat exchanged with the surroundings in the course of the reaction when the reaction occurs at *constant pressure*.

It is normally further qualified by conditions – it is usually expressed per mole of something (either reactant destroyed or product created), and often given a standard sign, $^{\ominus}$, meaning the reactants destroyed and the products created are all in their standard states, that is, as they would be at 298 K and 1 atmosphere pressure. In other words, the standard enthalpy change of the reaction in the previous example would be the heat exchanged with the surroundings when one mole of H_2 gas at 1 atmosphere and one mole of F· gas at 1 atmosphere were destroyed, to create the respective gaseous products at 1 atmosphere, with the reaction starting and finishing at 298 K.

There is a further small refinement to the meaning of enthalpy at the end of the chapter (Section 8.9), which is to

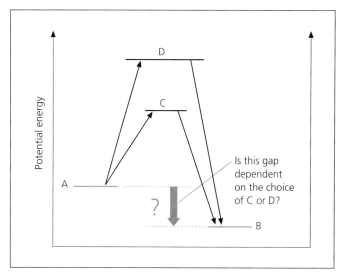

Figure 8.11 The idea that B has a single unique energy level relative to A, and independent of C and D, is an alternative way of stating Hess's Law

do with that stipulation about reactions being at 'constant pressure'. The above definition in terms of 'heat exchanged with the surroundings' is completely true, however, and can be committed to memory.

If you have done some practical thermochemistry, you probably paid little attention to these refined and precise definitions. For instance, when measuring the enthalpy change associated with acid/alkali reactions in solution, it is very rare to go to the trouble of letting the products cool back down to 298 K while releasing their liberated energy to a 'heat sink' like a water jacket. You might be forced to use a heat sink for a combustion, but if a reaction stays in aqueous solution then usually the reaction is allowed simply to heat its own solution. Admittedly the system has not stayed rigorously at 298 K, but the associated errors are very small, and are far outweighed by the convenience of the 'reaction system heats itself' technique.

Note: Students frequently find it difficult to remember the meanings of the paired concepts endothermic/exothermic, bond breaking/making, and heat from/to the surroundings. Table 8.2 restates those meanings, and summarises the connections between the correct members of the pairs.

Table 8.2 Changes associated with exothermic and endothermic reactions

Change in system	Sign of ΔH_{system}	Heat change in surroundings	Word used
weaker bonds broken, stronger bonds made	$\ominus$ (decrease in enthalpy)	$\oplus$ (surroundings get hotter)	exothermic
stronger bonds broken, weaker bonds made	$\oplus$ (increase in enthalpy)	$\ominus$ (surroundings get colder)	endothermic

8.4   Using Hess's Law and bond enthalpy tables

It is now time to put some of these ideas to the test. The principles derived from the H_2 and F· system apply to all reactions. Even if a molecule has many bonds in it, the potential energy of the system is only raised or lowered by those bonds that are actually being broken or made.

The data you will be using are the bond enthalpies associated with various bonds. They represent the depth of the potential energy 'well' for each bond, as in Figure 8.5 for H—H but also signal a decline in our concern with the x-axis of the curve. What we are interested in now is knowing that when, say, a mole of H—H bonds breaks, there is an enthalpy change in the system of +436 kJ mol^{-1}.

Bond enthalpies are always expressed by convention as signless 'scalar' quantities; so it is up to the user to realise that (using H—H again as the example) there will be an enthalpy change of +436 kJ mol^{-1} for bond breaking and –436 kJ mol^{-1} for bond making.

Another central idea throughout these examples is that we are free to imagine stepwise reaction pathways (that is, breaking first, followed by making) even though the real path may be different (a single step push–pull collision, for example). In other words, we will be applying Hess's Law.

Notes on the example

First, there is no suggestion that the reaction actually happens like that – breaking the whole double bond first and then remaking a single one – but by Hess's Law it does not matter what path you imagine, as long as you begin and end in the right places.

Second, note that Figure 8.12 reduces the significance of the x-axis compared to previous energy diagrams. All that remains is the placing of reactants on the left, intermediates in the middle, and products on the right – otherwise the graph's meaning lies solely in the enthalpy levels of the various species as shown on the y-axis.

Third, you may have felt that too much attention was given to the signs of enthalpy changes. After all, it sums up to 'plus the breaks minus the makes'. However, a key skill in enthalpy problems is *sign control*, and in more complex examples steady reproducible habits are an insurance against errors.

Example: The hydrogenation of ethene

Work out the enthalpy change when one mole of hydrogen adds on to one mole of ethene:

$$H_2C{=}CH_2(g) + H_2(g) \rightarrow H_3C{-}CH_3(g)$$
ethene   ethane

The bond enthalpy data are as follows:

Bonds broken	Bond enthalpy (kJ mol^{-1})	Bonds made	Bond enthalpy (kJ mol^{-1})
C=C	612	C—C	347
H—H	436	2 × C—H	2 × 413

Enthalpy change in system due to bond breaking
= 612 + 436 = +1048 kJ mol^{-1}

Enthalpy change in system due to bond making
= –347 + (–2 × 413) = –1173 kJ mol^{-1}

Total enthalpy change = +1048 + (–1173)
= –125 kJ mol^{-1}

Figure 8.12 summarises this situation.

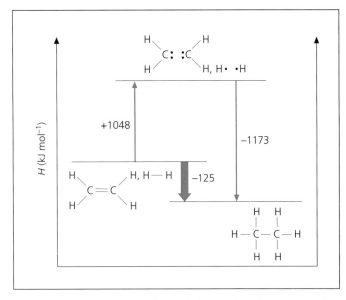

Figure 8.12 Enthalpy diagram for the hydrogenation of ethene – semi-graphical version

Hess's Law cycles

This section describes a second style of diagram for using Hess's Law in the solution of enthalpy problems. These diagrams will crop up throughout the rest of the chapter in a number of contexts, and will be referred to as **Hess's Law cycles**, a name which is meant to draw attention both to their dependence on the work of Hess, and to their essentially closed cyclic structures. To use a Hess's Law cycle, follow these stages:

1 Set up your Hess's Law cycle (Figure 8.13). In this style there is no *x*-axis, and even the *y*-axis is optional as long as the enthalpy changes on the arrows are consistent with the chemical change that occurs along that arrow *and with the direction of the arrow*.

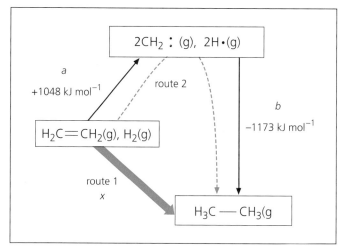

Figure 8.13 Enthalpy diagram for the hydrogenation of ethene – called a Hess's Law cycle

2 Add the enthalpy changes to the arrows, *being particularly careful about the signs*. Attach an algebraic label to each of the arrows (the *x*, *a* and *b* of Figure 8.13).
3 Call the target arrow 'route 1'.
4 All the other arrows which start at the start of route 1 and end at the end of route 1 are 'route 2'.
5 Use Hess's Law for the enthalpy changes on different routes to set up a relationship between *x*, *a* and *b*. In this case it is simply:

$$x = a + b$$

6 Insert the real values of *a* and *b* and solve for *x*.

$$x = +1048 + (-1173) = -125 \text{ kJ mol}^{-1}$$

This style of layout and algebraic method is useful for all Hess's Law problems and will be used again and again throughout the rest of the chapter. The next set of problems draws on bond enthalpy data, and will require you to try out the Hess's Law cycle style of solution.

Question

4 Work out the enthalpy changes of the following reactions, using data for bond enthalpies from a data book. In each case say what heat changes will have happened to the surroundings, both in numbers and words. Do at least two of the examples using the Hess's Law cycle layout.

a $H_2C = CH_2(g) + HBr(g) \rightarrow H_3C—CH_2Br(g)$
  ethene   bromoethane

b $C_2H_5OH(g) + HI(g) \rightarrow C_2H_5I(g) + H_2O(g)$
  ethanol   iodoethane

c $2H_2(g) + O_2(g) \rightarrow 2H_2O(g)$

d $H_2(g) + Cl_2(g) \rightarrow 2HCl(g)$

e $C \equiv O(g) + H_2(g) \rightarrow H_2C = O(g)$
                                      methanal

f $CH_4(g) + 2O_2(g) \rightarrow CO_2(g) + 2H_2O(g)$

g $2HC \equiv CH(g) + 5O_2(g) \rightarrow 4CO_2(g) + 2H_2O(g)$
  ethyne

You may have already spotted one of the flaws in the bond energy method for predicting the enthalpy changes of reactions. All the above reagents are in the *gas* phase, because bond enthalpies only tell us about *intra* molecular bonds, and nothing about *inter* molecular bonds. Gases have been used for question 4, because they have negligible intermolecular bonding. If a reaction involves a change of state, the bond enthalpy data is inadequate, for reasons that the next question will probe.

Question

5 Look at the Hess's Law cycle for the combustion of methane in Figure 8.14.

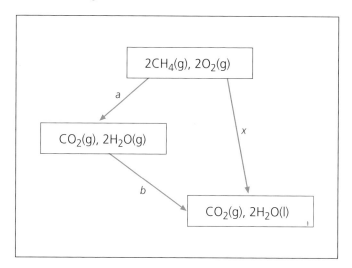

Figure 8.14 Hess's Law cycle for question 5

a Which of the arrows corresponds to the bond energy answer in question 4f?

b Which arrow corresponds to the *standard* version of the reaction (that is, the reaction in which the chemicals are in their normal states at 1 atmosphere and 298 K)?

c Which extra bit of data would be needed to translate the bond enthalpy answer into the standard one? Find it in the data book, and then calculate the standard enthalpy change for the combustion of methane.

Bond enthalpies have their shortcomings, therefore, despite their ease of use. As question 5 shows, they often need to be augmented by data about state changes. With that proviso, they are certainly a valuable if rough-and-ready resource.

However, there is a second limitation to the usefulness of bond enthalpies, already mentioned in passing, and that is the assumption that a C—H bond, say, always has the same value whatever the molecule it is in. This assumption is not strictly true, and indeed investigating the small differences in bond enthalpies of different C—H bonds in various molecular environments leads to a greater understanding of bonding. As it is, the bond enthalpies in data books should be taken as no more than averages and approximations. In the next section we turn to an alternative set of data which, when used with Hess's Law, can give much more 'molecule specific' information about enthalpy changes and bond enthalpies.

8.5 Hess's Law and specific reaction enthalpy data

Data books are full of enthalpies for specific reactions, for example:

$$C(s,graphite) + \tfrac{1}{2}O_2(g) \rightarrow CO(g)$$
$$\Delta H = -110.5 \text{ kJ mol}^{-1}$$
$$H_2(g) + C(s,graphite) + \tfrac{1}{2}O_2(g) \rightarrow HCHO(g)$$
$$\Delta H = -108.7 \text{ kJ mol}^{-1}$$

These two are both examples of a common form of available data, the **standard enthalpies of formation**. An enthalpy of formation of a compound is one in which:

1 a *mole* of the named compound is formed
2 in its *standard* state (298 K, 1 atm)
3 from its constituent *elements*
4 in *their* standard states (298 K, 1 atm)

Let us see how these two enthalpies of formation enable us to revisit question 4e. Figure 8.15 shows a Hess's Law cycle in which the 4e reaction is one of the arrows, and is marked 'route 1'. Route 2 is a combined route, featuring the two enthalpies of formation above. Here is the calculation:

Enthalpy change on route 1 = Enthalpy change on route 2

So, $x = -a + b$

$$= -(-110.5) + (-108.7)$$
$$= 1.8 \text{ kJ mol}^{-1}$$

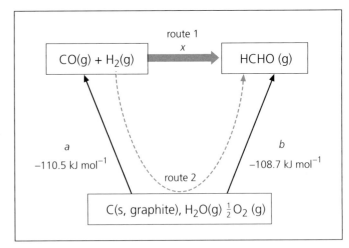

Figure 8.15 Hess's Law cycle to find the enthalpy change for the reaction of carbon monoxide and hydrogen to form methanal

(Interestingly, this is a reaction with almost no exchange of heat with the surroundings.)

Note the very important new twist in this example: one of the pieces of data related to a change which *opposed* the overall direction of route 2, so we had to employ the sign-reversed version of the enthalpy change ($-a$) to describe the reversed version of the chemical change. The general rule is – *if you have to 'drive an arrow backwards', reverse the sign of its enthalpy change.*

Question

6 Compare the answer obtained above with the one from question 4e.

a Which one would you trust more?

b How would you account for the errors in the less trustworthy one?

c What is the percentage error? Why does the percentage error look so large in this case?

d Why does the left-hand arrow on route 2 in Figure 8.15 have no enthalpy of formation of dihydrogen gas?

Let us now compare the two methods of calculating enthalpies. The second method is more accurate, because it uses data that are specific to carbon monoxide and methanal. The mathematics is no harder than before. The only trouble with the second method is that it requires a much larger data-bank.

This databank is subdivided into several sections, governed by the criteria of availability and usefulness. One category, as we have already seen, is enthalpy of formation, but there are several more, and each has its defining statement.

Databank of specific reaction enthalpies

Standard enthalpy of formation

Symbol: $\Delta H_f^{\ominus}$

Defining statement: the enthalpy change when one mole of a named compound is formed from its constituent elements, under standard conditions.

Applicability: all chemical species except the elements in their standard states – there are even enthalpies of formation for ions in solution. Notice that the enthalpy of formation of hydrogen in question 6 was zero, because it was already an element in its standard state.

Question

7 Write the equation of the reaction whose enthalpy would be the $\Delta H_f^{\ominus}$ of $Na_2CO_3 \cdot 10H_2O$ (sodium carbonate-10-water or washing soda).

Standard enthalpy of combustion

Symbol: $\Delta H_c^{\ominus}$

Defining statement: the enthalpy change when one mole of a named species is reacted with the maximum amount of oxygen, under standard conditions.

Applicability: most commonly used for organic chemicals – notice that the reaction is to $CO_2(g)$ and $H_2O(l)$ in these circumstances. Any nitrogen in the compound will appear as $N_2(g)$.

Question

8 Write the equation for the reaction whose enthalpy is the $\Delta H_c^{\ominus}$ of ethene ($H_2C{=}CH_2$).

Give two reasons why you cannot get an accurate value for this from bond enthalpy data.

Standard enthalpy of atomisation (1)

Symbol: $\Delta H_{at}^{\ominus}$

Defining statement: the enthalpy change when one mole of gaseous single atoms is formed from the element in its standard state.

Applicability: any element, but notice that the 'per mole' part of the definition here refers to single atoms on the product side. Thus $\Delta H_{at}^{\ominus}$ refers to this:

$$\tfrac{1}{2}H_2(g) \rightarrow H{\cdot}(g)$$
$$\Delta H = +218 \text{ kJ mol}^{-1}$$

and *not* this:

$$H_2(g) \rightarrow 2H{\cdot}(g)$$

Question

9 a What would be the enthalpy change in this second case?

b What is the relationship between $\Delta H_{at}[\tfrac{1}{2}H_2(g)]$ and the bond enthalpy of H—H?

Standard enthalpy of atomisation (2)

Symbol: $\Delta H_{at}^{\ominus}$

Defining statement: the enthalpy change when one mole of a named compound in its standard state is split completely into single gaseous atoms.

Applicability: any compound. Easily confused with 'reversed-formation'. For instance, this is *not* the enthalpy of atomisation of methanal:

$$HCHO(g) \rightarrow$$
$$H_2(g) + C(s,\text{graphite}) + \tfrac{1}{2}O_2(g)$$

Question

10a Write the correct equation for the atomisation of methanal.

b What would be the enthalpy change for the reaction written above? (You will need to revisit the example just before question 6.)

c Enthalpies of atomisation (both kinds) are *always* positive. Why is this?

Enthalpy of solvation (special case hydration)

Symbol: $\Delta H_{solvation}$ (or $\Delta H_{hydration}$)

Defining statement: the enthalpy change when one mole of gaseous ions becomes one mole of solvated (hydrated) ions.

Applicability: any ion. Easily confused with enthalpy of formation of aqueous ions, but distinguishable because true formation is from the standard-state elements.

Question

11 Is this equation a hydration or a formation?

$$Na^+(g) \rightarrow Na^+(aq)$$

Lattice enthalpy

Symbol: $\Delta H_{lattice}$

Defining statement: the enthalpy change when one mole of a compound is formed from its gaseous *ions*.

Note: Some data books use an inverted definition referring to the change *from* the crystal lattice *to* the gaseous ions, so check carefully which definition your particular book or exam question is using.

Applicability: any compound with any pretensions to being ionic. Sometimes confused with formation, but again the true formation is from standard-state elements, not gaseous ions.

For example:

$$Ca^{2+}(g) + 2Cl^-(g) \rightarrow CaCl_2(s)$$
$$\Delta H_{lattice} = -2237 \text{ kJ mol}^{-1}$$

Question

12 Lattice enthalpies are *always* negative (at least they are when expressed as in the first defining statement above) – why is this?

The following are not always thought about as enthalpies. However, since they do involve heat exchange with the surroundings associated with changes in systems of particles, they are in fact enthalpies.

Ionization energy

Symbol: $\Delta H_{\text{ionization}}$

Defining statement (in the context of this chapter): the enthalpy change when one mole of gaseous single atoms or ions loses one mole of electrons. We met this 'enthalpy' earlier (Chapter 3, p. 43), when it was given the symbol I.E.

Applicability: all single atoms and single-atom ions.

For example:

$$Mg^{+}(g) \rightarrow Mg^{2+}(g) + e^{-}(g)$$
$$\Delta H_{\text{2nd ionization}} = +1451 \text{ kJ mol}^{-1}$$

(This is the *second* ionization energy of magnesium.)

Question

13 Ionization energies are always positive. Why is this?

Electron affinity

Symbol: $\Delta H_{\text{electron affinity}}$

Defining statement: the enthalpy change when one mole of gaseous single atoms or ions gains one mole of

electrons. We met this quantity earlier, when it was given the symbol E.A.

Applicability: same as that of ionization energy in theory, but normally only used for those species likely to accept electrons, namely non-metal atoms or ions.

For example:

$$O(g) + e^{-}(g) \rightarrow O^{-}(g)$$
$$\Delta H_{\text{electron affinity}} = -142 \text{ kJ mol}^{-1}$$

(This is the first electron affinity of oxygen.)

Question

14 Write the equation for the change whose enthalpy change is the second electron affinity of oxygen, and look up its value.

Enthalpies of fusion and vaporisation

Symbols: $\Delta H_{\text{fus}}^{\ominus}$, $\Delta H_{\text{vap}}^{\ominus}$

Defining statement: the enthalpy change when one mole (or 1 gram or 1 kg in physics contexts) of a named substance

changes state (that is, melts or boils), at the normal temperature of that state change under standard conditions. (These are also known as latent heats of fusion and vaporisation. The enthalpy of fusion is also sometimes called the enthalpy of melting.)

Applicability: any pure chemical substance. These are the quantities you have to use in conjunction with bond enthalpies if you want to estimate enthalpy changes for reactions like:

$$C_2H_5OH(l) + 3O_2(g) \rightarrow$$
$$2CO_2(g) + 3H_2O(l)$$

Question

15 Which two enthalpies of fusion and/or vaporisation would you need to look up in order to use bond enthalpy terms to estimate the enthalpy change of the reaction above? (It would be unnecessary, however, since this reaction is covered by even a modest data book. Under what name would you look it up, that is, the enthalpy 'of what of what'?)

Using these data to find unknown enthalpy changes

Armed with this mass of data under these diverse headings, you are in a powerful position. You can use it in any combination to find unknown enthalpies. In fact we have already solved one problem of this type (Section 8.5, p. 148) – the enthalpy of the reaction by which methanal is formed from carbon monoxide and hydrogen. The trick there, and in any other related problem, is quite a simple one.

1 Call your unknown reaction 'route 1'.
2 Select from the databank whatever data exists to construct a 'route 2'. Draw both routes into a Hess's Law cycle.
3 Give algebraic symbols to the enthalpy changes on each arrow.
4 Use Hess's Law to construct a relationship between x, a and b, recognising the arrow reversal rule.

In the methanal case route 2 was constructed from enthalpies of formation. This is nearly always an option since that particular class of data exists for a very wide range of compounds. For organic reactions, enthalpies of combustion may be more easily found. In other cases enthalpies of atomisation may be best, especially since they allow you to mix specific reaction data with bond enthalpy terms (as in question 25, still to come). Generalised examples of these three classes of triangular Hess's Law cycle are

shown in Figure 8.16, along with the relationships between x, a and b.

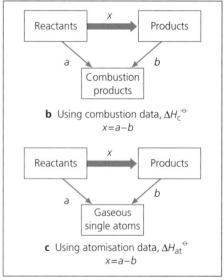

b Using combustion data, $\Delta H_c^{\ominus}$
$$x = a - b$$

c Using atomisation data, $\Delta H_{at}^{\ominus}$
$$x = a - b$$

Figure 8.16 Generalised Hess's Law cycle showing the three most commonly used types of enthalpy data

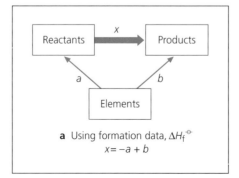

a Using formation data, $\Delta H_f^{\ominus}$
$$x = -a + b$$

The usefulness of these data is that they can give you access to enthalpy results without you actually having to do the experiment. Intriguingly, they can also give you information about reactions that would not normally happen.

Question

16 Oil industry chemists are interested in 'cracking' reactions – that is, reactions in which longer chain hydrocarbons are broken down, at high temperatures with the help of catalysts, to shorter ones, one of which is always an alkene.

a Find the enthalpy change for this reaction, using either combustion or formation data to construct route 2. Draw a Hess's Law cycle including *x*, *a* and *b*, and routes 1 and 2.

$$C_{11}H_{24}(l) \ \rightarrow \ C_9H_{20}(l) + C_2H_4(g)$$
undecane   nonane

b Why would this answer, constructed from 'standard' data, be at best an approximation to the situation existing in the industrial reaction? (There are two reasons for any discrepancy, one of which has to do with the state symbols shown above.)

Fortunately enthalpy changes are rather insensitive to changes in temperature of the system, so one of the two sources of error you may have cited in question 16 is not too serious. What it does mean, however, is that if a reaction is happening at dramatically non-standard conditions, then a bond enthalpy method may be just as accurate (or inaccurate) as a $\Delta H_f^{\ominus}$ method.

Questions

17a Graphite is the official standard state of the element carbon, because it is slightly more stable than diamond under standard conditions. Using any data you can find for route 2, construct a triangular Hess's Law cycle to find the enthalpy change of this reaction, which sadly is not easy to make happen:

$$C(s, graphite) \rightarrow C(s, diamond)$$

b If it is so difficult to make happen, why has it occurred spontaneously in the history of this planet?

18 Ozone may be a friendly shield in the upper atmosphere, but at ground level it is a serious pollutant. It is formed as a by-product of the production of oxides of nitrogen in car engines, with help (ironically, in view of ozone's role higher up) from ultra-violet radiation:

$$NO_2(g) \overset{h\nu}{\rightarrow} NO(g) + O(g)$$

$$O(g) + O_2(g) \rightarrow O_3(g)$$

a Use a variety of sources of data to construct triangular Hess's Law cycles for each of these reactions, and find their enthalpy changes. (*Hint*: use the constituent standard-state elements as the third side of the triangle in each of the two cycles.)

b Calculate the combined enthalpy change of the two reactions running in sequence, one after the other.

c It is quite unusual to have such a seriously endothermic reaction happening spontaneously. Can you see what special factor is at work in this case?

8.6 Solving enthalpy problems without diagrams

Quick formulae

From your experience of doing the problems in Section 8.5 you may have discovered some short cuts. For a start you will have discovered that the Hess's Law cycles still work whether or not you formally write out the chemical species in the 'third boxes'. The data used are still the same, even if the box carries nothing more specific than the words 'elements', 'combustion products' or 'gaseous atoms'. (There is, however, a strong case for staying with the full information in 'gaseous atoms' boxes, as a reminder of how many enthalpies of atomisation are needed.)

There is also a way to cut out the Hess's Law cycle altogether. In every Hess's Law cycle that uses elements in the third box and formation data on the arrows, the same relationship between *x*, *a* and *b* is thrown up, namely:

$$x = +b - a \qquad (8.1)$$

This is because when you decide to use elements as the third box, you have preset the directions of the arrows, and therefore the signs of *a* and *b*. So we can derive a formula which bypasses the need for the actual diagram (and which is exactly equivalent to *x* = +*b* – *a*, in more formal symbols):

$$\Delta H_{reaction}^{\ominus} = \Delta H_f^{\ominus}(\text{products}) - \Delta H_f^{\ominus}(\text{reactants}) \qquad (8.2)$$

(Remember of course that any multiples of moles will require multiples of their respective enthalpy changes.)

Questions

19 Apply similar logic to Hess's Law cycles employing enthalpies of combustion and of atomisation, and derive formulae for them in the style of the one just derived.

20 Revisit question 4b, and find its enthalpy change using a 'formula' method.

21 High-octane petrol contains a higher proportion of 'aromatic' (benzene ring) hydrocarbons than is present in the petrol fraction of naturally occurring crude oil. If we want petrol which has ignition characteristics that need no improvement by lead additives, we need to make more aromatic ingredients. The reaction known as **catalytic reforming** has been used to convert the glut of 'aliphatic' hydrocarbons (non-benzene types) in crude oil to aromatics. For example, straight chain heptane can be converted to methylbenzene ('toluene'):

$$C_7H_{16}(g) \xrightarrow{\text{Pt, heat, pressure}} \langle\!\!\bigcirc\!\!\rangle\!\!-\!\!CH_3(g) + 4H_2(g)$$

Work out the enthalpy change of this reaction using a formula method.

The additive nature of enthalpies – 'matched trios'

Sometimes there is no obvious candidate for the 'third corner' of a Hess's Law cycle, and yet the situation can be rescued by the existence of sets of 'matching trios' of reactions, two of which are known. For example, suppose we know by experiment these two enthalpies:

$$Mg(s) + Cu^{2+}(aq) \rightarrow Cu(s) + Mg^{2+}(aq)$$
$$\Delta H = -526 \text{ kJ mol}^{-1} \quad (1)$$

$$Cu(s) + 2Ag^+(aq) \rightarrow 2Ag(s) + Cu^{2+}(aq)$$
$$\Delta H = -147.2 \text{ kJ mol}^{-1} \quad (2)$$

and we are required, *from these data alone*, to derive the enthalpy change for the third one of the set:

$$Mg(s) + 2Ag^+(aq) \rightarrow 2Ag(s) + Mg^{2+}(aq)$$
$$\Delta H = x \quad (3)$$

It is possible to make up a triangular Hess's Law cycle for this reaction which uses the other two reactions, but it involves some rather imagination-stretching ideas like 'add Cu(s) to each side'. There is an easier way.

You may have noticed in the course of previous examples that when two reactions occur one after the other their enthalpies *add*, and also that the reverse of a reaction involves the reversal of the sign of the enthalpy change. Accordingly, we can imagine reaction *(1)* happening to create the copper, followed by reaction *(2)* to destroy it again. The resulting reaction will be that of reaction *(3)*, so the corresponding enthalpy change will be:

$$-526 + (-147.2) = -673.2 \text{ kJ mol}^{-1}$$

The process is similar to adding two simultaneous equations in maths, with the effect here of 'eliminating' the Cu(s).

We can generalise from this example – Table 8.3 shows a series of competition/transfer reactions, where the item being transferred might be a proton, a nucleophile, or electrons as in the last example. Possession of the transferred species is symbolised by *. We can see that any two of the equations can be added or subtracted to make the third, handling the chemical species and their signs exactly as if they were mathematical species. When the equations are added and subtracted, the associated enthalpies are added and subtracted too.

Question

22 What is the third reaction whose enthalpy change may be deduced from the enthalpy changes of reactions *(4)* and *(5)*? What is the value of that enthalpy change?

$$\text{haemoglobin} + O_2(aq) \rightarrow \text{oxyhaemoglobin}$$
$$\Delta H = -23.1 \text{ kJ mol}^{-1} \quad (4)$$

$$\text{haemoglobin} + CO(aq) \rightarrow \text{carboxyhaemoglobin}$$
$$\Delta H = -32.4 \text{ kJ mol}^{-1} \quad (5)$$

The clue here is to try to 'eliminate' the haemoglobin. There are two possible reactions which might serve as an answer, each one the reverse of the other. Whichever reaction you got, explain how your answer helps to account for the extreme toxicity of carbon monoxide.

8.7   Enthalpy changes and insight into bonding

Enthalpy changes can be used as a sort of window on to chemical bonds. From the very start of the chapter we have emphasised that enthalpy changes come from the making and breaking of bonds, so knowledge of enthalpy changes can give information about particular bond enthalpies. This experimental information can then be compared with theoretical predictions for bond enthalpies based on lattice data or average bond enthalpy data. The results have led people to reassess their ideas about bonding, and about the models on which those ideas were based.

The Born–Haber cycle

We can use a rather specialised Hess's Law cycle – called a Born–Haber cycle – to investigate how well the ionic model applies to particular cases of suspected ionic bonding. If you assume 'pure' ionic bonding in a particular compound, and if you have information about the separation in space of the particles in that compound, it is possible with the use of computers to predict how strongly the lattice ought to be held together.

Question

23 Which type of instrument might have been used to find the ionic separation distances?

The mathematicians assume that the ions can be represented electrically by points of charge located at the ionic separation distance. They have access to the charge in coulombs on those

Table 8.3 How to deal with 'matching trios' of reactions when one of the enthalpies is unknown

A + B* → A* + B	(1)
B + C* → B* + C	(2)
A + C* → A* + C	(3)
(3)	= (1) + (2)
so $\Delta H(3)$	= $\Delta H(1) + \Delta H(2)$
and also, $\Delta H(1)$	= $\Delta H(3) - \Delta H(2)$
and $\Delta H(2)$	= $\Delta H(3) - \Delta H(1)$

ions, via Millikan's results. Their calculations use equations of the form:

$$\text{force} = \frac{q_+ \, q_-}{\mu r^2}$$

where q stands for the charge on an ion, and μ is the permittivity of a vacuum.

The end-product of all this computing is a *theoretical* value for the lattice enthalpy. Let us consider the lattice enthalpy of the supposedly ionic compound calcium iodide. The lattice enthalpy equation is:

$$Ca^{2+}(g) + 2I^-(g) \rightarrow CaI_2(s)$$

The theoretical value for this lattice enthalpy is $-1905 \text{ kJ mol}^{-1}$, based on the assumption that the bonding is ionic. Chemists are interested, therefore, to compare this value with the lattice enthalpy found by experiment and thereby assess how good the ionic model is for describing calcium iodide.

But how can we find an experimental lattice enthalpy? It certainly is not the sort of reaction that can be done in a test-tube. Fortunately there is enough data in data books to construct a route 2 to link the gaseous ions with the solid lattice (Figure 8.17). The enthalpy change on route 2 is:

$$-d - c - b - a + e$$

or, in actual enthalpies:

$$-\{2 \times \Delta H_{\text{electron affinity}}\,[I(g)]\}$$
$$-\{2 \times \Delta H_{\text{at}}^{\ominus}[\tfrac{1}{2}I_2(s)]\}$$
$$-\{\text{1st and 2nd } \Delta H_{\text{ionisation}}[Ca(g)]\}$$
$$-\Delta H_{\text{at}}^{\ominus}[Ca(s)]$$
$$+\Delta H_{\text{f}}^{\ominus}[CaI_2(s)]$$

(where all the above minus signs represent arrows driven backwards)

$$\begin{aligned}
= &-2 \times (-295.4)\\
&-2 \times (+106.8)\\
&-(+590 + 1145)\\
&-(+178.2)\\
&+(-533.5)\\
= &-2069.5 \text{ kJ mol}^{-1}
\end{aligned}$$

Since by Hess's Law the two routes will have the same enthalpy changes:

$$\text{Lattice enthalpy } [CaI_2(s)] = -2069.5 \text{ kJ mol}^{-1}$$

The theoretical value deviates by 8% from this experimental figure, which looks quite large when compared with a 1.3% deviation in the case of sodium chloride. It is accepted that sodium chloride is a fairly 'pure' ionic compound, so it appears that calcium iodide is *not* so well described by the ionic model, and therefore that a degree of covalent character may be present.

Note on the calculation: The difficulty in the example above comes from the fact that CaI_2 is not a 1:1 salt. Thus you have to remember to have two electron affinities and two enthalpies of atomisation of I, and two successive ionisations of Ca. None of those complications would have arisen with, say, NaCl.

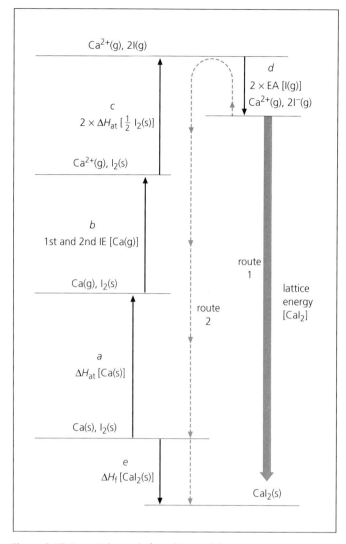

Figure 8.17 Born–Haber cycle for calcium iodide

Question

24 Work out the experimental lattice enthalpies of NaCl (easy) and Al_2O_3 (very hard – full of pitfalls). How well would you expect the Al_2O_3 value to agree with the theoretical value based on the ionic model?

Bonds with surprising strength

The fact that a study of enthalpies can uncover unexpected or anomalous bond enthalpies makes it useful in hunting down molecules with 'aromatic' character. This is the property, most famously associated with benzene, in which alternate double and single C—C bonds undergo delocalisation. The resulting molecule is considerably more stable than it would have been if the bonds had remained local. In benzene the molecule is some 160 kJ mol^{-1} harder to atomise than predicted on an alternate single/double bond model.

Question

25 The heterocyclic molecule pyridine is a liquid at room temperature. A possible structure is shown here assuming 'local' bonds.

Pyridine has an enthalpy of formation of +101.2 kJ mol⁻¹.

a Construct a Hess's Law cycle to find the enthalpy of *atomisation* of pyridine, using as one part of 'route 2' the above item of data, and as the second part of 'route 2' the enthalpies of atomisation of its constituent elements.

b Find another value for its enthalpy of atomisation by the direct use of bond enthalpies and assuming the bonds are as shown in the diagram.

c What does this evidence tell you about the closeness of real pyridine to the structure shown?

Putting bond enthalpies to the test

A third way in which we can use our 'specific reaction' enthalpy databank is to test the validity of the 'rival' databank of bond enthalpies. These are, you may recall, only useful if there is some truth in the idea that one, say, C—H bond is as strong as another, irrespective of the molecular environment in which they find themselves.

We can see how approximate that claim really is by considering the enthalpies of combustion of a series of organic compounds such as the alkanes. Every time you add an extra link to the molecular chain, you introduce an extra bit of molecule:

That extra bit of molecule will burn along with the rest, and therefore should be responsible for its own little bit of enthalpy change.

$$-CH_2 + 1\tfrac{1}{2}O_2 \rightarrow CO_2(g) + H_2O(l)$$

Bond enthalpy terms predict that this 'extra enthalpy per link of chain' factor should be equal to the enthalpy change when the following bond changes occur:

Break: $1 \times C—C, 2 \times C—H, 1\tfrac{1}{2} \times O{=}O$

Make: $2 \times C{=}O, 2 \times O—H$

Question

26a Calculate the extra enthalpy change factor, from the bond enthalpy databank, using the 'breaks' and 'makes' identified above.

b Now turn to the specific reaction enthalpy databank, and find values of this factor for some alkanes, by comparing successive members of the series.

c This extra factor should be the same for any series of organic molecules differing by successive CH_2 units (if the bond enthalpy assumption holds up). Do the same analysis on the alcohols as you did in part (c) for alkanes. Is there any difference between the two cases? If so, comment on it.

8.8   Practical methods of studying enthalpy

Enthalpies are measured using calorimeters, which in school calorimetry are of two types:

1 For reactions between pure liquids, or between solutes in solution, you measure the temperature of the reaction solution itself. The apparatus is an insulated container, with access for a thermometer, a stirrer and a heating element (Figure 8.18). (The heater is to provide the option of 'electrical compensation', p. 156.)

Figure 8.18 'Thermos flask' style of calorimeter with stirrer, thermometer and heating coil

2 For a combustion reaction, you make the reaction heat a water bath. The apparatus for this is called a combustion calorimeter (Figure 8.19), and has evolved a long way from the 'put a flame under a beaker' design. The refinements are dedicated to transferring as much heat as possible to the water, so the flame is enclosed and maintained either by suction or by an oxygen supply. ('Electrical compensation' is also an option for this apparatus.)

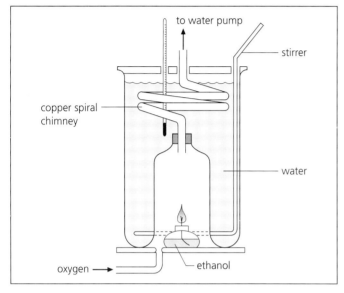

Figure 8.19 Combustion calorimeter

Question

27 Apart from the enclosure of the flame, what features of the combustion calorimeter maximise the degree of heat transfer from the flame to the water?

Note that in the first, 'thermos flask', method, the role of 'surroundings' is being played by the molecules of the system itself, whereas the second, 'combustion calorimeter', method involves a more literal application of the 'surroundings' idea. (Bear in mind that enthalpy changes are not very sensitive to temperature changes, so that there is no great problem about the reaction 'heating itself'.)

From 'raw' data to final molar enthalpies

Whichever apparatus you use, calorimetry always involves *temperature* measurements. The overall objective is to end up with *enthalpy* changes in kJ mol^{-1}, so the task is one of translation from temperature change to enthalpy change. There are two routes which the translation process may take.

1 By a knowledge of specific heat capacities

In order to translate a temperature rise into a figure for heat transferred you need to know the **specific heat capacity** of a substance – that is, the number of joules needed to raise the temperature of one gram of the named substance through 1 °C. The symbol for specific heat capacity is c_p. You can then find the enthalpy change (the heat transfer to the absorbing medium) by use of the formula:

$$-\Delta H = \begin{matrix} \text{mass of} \\ \text{absorbing} \\ \text{medium} \\ \text{(g)} \end{matrix} \times \begin{matrix} c_p \\ \\ \text{(J g}^{-1}\,°\text{C}^{-1}) \end{matrix} \times \begin{matrix} \Delta T \\ \\ (°\text{C}) \end{matrix}$$

Question

28a Explain the minus sign in the equation just given.

b Values of ΔH generated by the equation are in joules. How would you take the final step to reach kJ mol^{-1}?

Students are often confused over which specific heat capacity to use in each of the two types of calorimeter. In the 'Thermos flask' style of experiment, you need the specific heat capacity of the *system itself*; but in the combustion calorimeter you need the specific heat capacity (mainly) of the water in the heat transfer vessel.

There are three categories of pitfall in practical calorimetry. They are:

1 not quite getting the reaction you wanted to happen
2 not quite managing to keep the heat in what you wanted to call the surroundings
3 not quite knowing the correct specific heat capacity.

The most common example of the first drawback is incomplete combustion in burning experiments. Keeping the flame alight is tricky, and avoiding a bit of yellow in the flame is even trickier.

Question

29a Why is a yellow flame a sign of incomplete combustion?

b What measure might you take to minimise the problem?

The second drawback can be solved by perfect lagging. This puts a total barrier around the bit of the surroundings which you want to act as the heat receiver. However, 'perfect lagging' is easier said than done. There is a slow leakage of heat even from a Thermos flask.

The third drawback, like the second, presents profound problems. In the combustion calorimeter, for instance, there is water, copper, glass, brass (the stirrer), mercury and rubber. All absorb heat and all have different specific heat capacities. The way round this, laborious but not impossible, is to extend the equation given above with a term for every material:

Heat transfer to absorbing medium
$$= \{[\text{mass} \times c_p]_{\text{water}} + [\text{mass} \times c_p]_{\text{copper}} + \dots \text{etc} \dots\} \times \Delta T$$

However, this still leaves awkward details such as the fact that not all the stirrer was immersed.

Question

30 The main components of the combustion calorimeter are the water and the copper. Assess the percentage error introduced by ignoring all terms in the extended equation just given except the water, compared to including water *and* copper. Data are as follows:

mass of water in vessel = 600 g

c_p of water = 4.2 J g^{-1} °C^{-1}

mass of copper coil = 100 g

c_p of copper = 0.39 J g^{-1} °C^{-1}

There is another source of uncertainty about specific heat capacities which comes from the Thermos flask style of apparatus. If you are asking the system to be its own absorber, there is the problem that the absorber is often something which is hard to find in a data book. For instance, in the reaction:

$$Zn(s) + CuSO_4(aq) \rightarrow$$
$$Cu(s) + ZnSO_4(aq)$$

the absorber is a solution of zinc sulphate with bits of copper floating around in it (not to mention unreacted $CuSO_4$ if that was in excess). How can you find its specific heat capacity? The way round this is the rather crude expedient of assuming that anything vaguely 'watery' has a specific heat capacity the same as that of water.

2   Methods which do not use specific heat capacities

The translation of raw data to final enthalpy changes via a knowledge of specific heat capacities is shot through with approximations due to the three drawbacks listed above. Many of the problems can be avoided by two alternative but equally ingenious practical measures.

1 *Electrical compensation*: This technique is based on a simple idea – reproduce the same heating effect as was achieved by chemical means, in the same absorbing medium, by an electrical heating element attached to a joulemeter (Figure 8.20). Then the electrical energy recorded on the joulemeter is identical to the chemical energy released by the reaction. At one stroke you have eliminated the

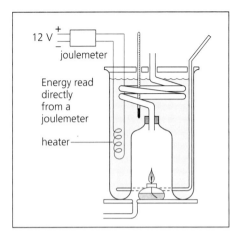

Figure 8.20 Schematic diagram of the electrical compensation method in use

problems over lagging and specific heat capacities. (Although this technique appears to solve some problems it does not always seem to give better results. However, students are obliged to be familiar with it.)

Question

31a Explain how electrical compensation gets round drawbacks 2 and 3 (p. 155) – in other words, why it no longer matters about imperfect lagging or uncertain heat capacities.

b Suppose chemical heating had raised the temperature of the

Figure 8.21 Bomb calorimeter

absorbing medium from 293 K to 298 K. Which of the two options below would give the more perfect result, and why?

- Electrically heat the apparatus from 298 K to 303 K.
- Wait for it to cool down and then electrically heat the apparatus from 293 K to 298 K.

c How would you adapt the electrical compensation method to measure the enthalpy changes of endothermic reactions (in which the heat transfer medium would have got cooler)?

2 *The bomb calorimeter*: Not quite as dramatic as its name suggests, the bomb calorimeter (Figure 8.21) nevertheless has some clever solutions to the problems of calorimetry. It is, however, normally limited to reactions which can be triggered remotely by a hot wire, and so its most common use is for combustion in oxygen. By using 20 atm pressure of oxygen it eliminates drawback 1 (p. 155), the problem of incomplete combustion, and by surrounding its water bath with *another* water bath which is electrically heated to the same temperature as the inner one, it effectively avoids drawback 2, heat loss.

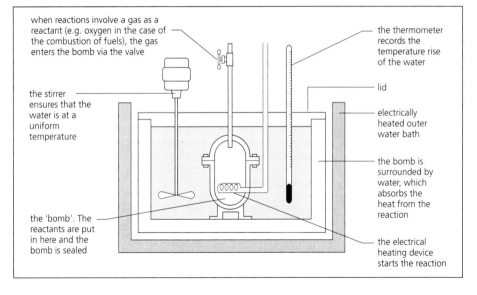

when reactions involve a gas as a reactant (e.g. oxygen in the case of the combustion of fuels), the gas enters the bomb via the valve

the thermometer records the temperature rise of the water

lid

the stirrer ensures that the water is at a uniform temperature

electrically heated outer water bath

the bomb is surrounded by water, which absorbs the heat from the reaction

the 'bomb'. The reactants are put in here and the bomb is sealed

the electrical heating device starts the reaction

Question

32 Why does this second ploy succeed?

Most bomb calorimeters are calibrated not by electrical heating but by burning a known mass of benzoic acid. The known enthalpy of combustion of benzoic acid is then used to calculate how many joules were released, and this figure is divided by the temperature rise to determine the specific heat capacity in $J\,°C^{-1}$ of the whole apparatus (that is, the heat needed to make the temperature of the whole apparatus rise by 1 °C). This enables all subsequent temperature rises achieved with the apparatus to be converted to joules.

One small drawback is that the heat released chemically does not quite correspond to the true enthalpy, because of the totally enclosed 'bomb' format. True enthalpies are measured by (the negatives of) heats exchanged with the surroundings *at constant pressure*. For instance, if a reaction like:

$$C_6H_5COOH(s) + 7\tfrac{1}{2}O_2(g) \rightarrow 7CO_2(g) + 3H_2O(l)$$

took place inside the bomb there would be a small decrease in pressure even after everything had cooled back down again.

Question

33 How can you see this from the equation?

However, if the oxygen is in excess (probable at 20 atm), the overall pressure change may not be too great.

8.9   A small confession

It must be said at the outset of this section that it contains a hard and potentially confusing concept. You will not put your Advanced qualification at any great risk if you miss the section out altogether.

Talk of 'at constant pressure' recalls a sort of white lie, which at the time was said for the sake of simplicity (Section 8.2, p. 141). If a reaction really does take place at constant pressure, and there is an increase in the number of moles of gas, then the atmosphere will be *pushed back*.

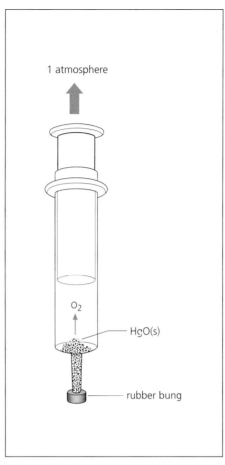

Figure 8.22 A reaction which creates gas at constant pressure does work pushing back the atmosphere

A decrease in the number of moles of gas has the opposite effect. If it helps your imagination, think of a reaction taking place in a syringe (Figure 8.22) – a syringe is thermodynamically equivalent to an open vessel insofar as it allows its contents to stay at constant pressure.

Imagine Lavoisier's famous experiment on the discovery of oxygen taking place in a syringe. The enthalpy databank tells us that the enthalpy change is:

$$HgO(s) \rightarrow Hg(l) + \tfrac{1}{2}O_2(g)$$
$$\Delta H = +90.8 \text{ kJ mol}^{-1}$$

It is completely true that 90.8 kJ of heat would be drawn from the surroundings, because that is exactly how enthalpy change is defined – as the heat exchanged with the surroundings at constant pressure. However, Section 8.2 gave the impression that the *chemical potential energy* of the molecules would increase by 90.8 kJ, and that is not quite true. In

fact, it offends against the Law of Conservation of Energy, because if 90.8 kJ of heat were destroyed and 90.8 kJ of chemical potential energy were created, *where did the work come from to push back the surroundings?*

In the 'Lavoisier' example, half a mole of oxygen would push back about 12 dm³ of air, doing work given by:

$$P \times \Delta V = 10^5 \times 12 \times 10^{-3} \text{ joules}$$
$$= 1200 \text{ J} = 1.2 \text{ kJ}$$

(Here we have converted 1 atmosphere and 12 dm³ to newtons and metres.)

Thus the 90.8 kJ of heat which were destroyed went to create not only chemical potential energy (previously claimed to be the truth and the whole truth), but also 1.2 kJ of work. Thus the actual change in chemical potential energy is only 89.6 kJ mol⁻¹.

An interesting point arises from this if you conduct the same experiment in a vessel of constant *volume*, instead of constant pressure. That way you prevent any work being done on the surroundings, so heat absorption is all channelled into the increase in potential energy of the molecules, which as we have seen is equivalent to 89.6 kJ mol⁻¹. But it is not called enthalpy any more. The heat exchanged with the surroundings at constant *volume* is known as **internal energy**, and given the symbol ΔU. Because most reactions we deal with are at constant pressure (in the end), it is enthalpy which is being measured when we measure the heat exchange, and so enthalpy remains firmly centre stage.

As you can see from the mercury example, the differences between ΔH and ΔU are not great, and the comfort of that knowledge must be your reward for reading this far.

Question

34 For anyone ready to try a question on the ideas of this section, here it is. Picture that famous novelty the 'volcano reaction', in which ammonium dichromate self-destructs exothermically via an intramolecular redox reaction, and at the same time releases lots of moles of gas per mole of reactant:

$$(NH_4)_2Cr_2O_7(s) \rightarrow Cr_2O_3(s) + N_2(g) + 4H_2O(g)$$

Under which conditions would the surroundings receive more heat, constant volume or constant pressure? Justify your answer.

8.10 The chemistry of explosives and propellants

After the academic tone of Section 8.9, this section takes a look at an intensely practical field, in which a set of exothermic reactions has played a part in the shaping of nation states on Earth, and in so doing has given such powerful expression to a facet of human nature as to threaten our existence as a social species.

A reflection on human nature

One of the fascinating threads in human history has been the interplay of technology, society, and human nature. Human technology may rule the world, but human nature is not so very different from, say, chimpanzee nature. Our big brains may enable us to extract goods of enormous value from the Earth's raw resources, but we guard our own share of those goods with a sense of territory which stays firmly rooted in the animal.

The basic mechanism for coping with this problem has been the 'tribe'. People band together on some basis – perhaps religious or geographic or racial – and agree to share resources to some degree, in return for the peace and fellowship of a place in a society. But these sharers' rights are not extended to humans outside the tribe.

When it comes to defending the tribe against foes, mastery of chemistry has always been a major source of strength. Strong materials (bronze, iron) were the original arm which gave power to confound the enemy in war and control the soils and forests in peacetime. Then for the last few hundred years (much longer in China), people have measured their military might, at least in part, by their possession of a set of chemicals which can propel bits of these strong materials through the air at high speed (Figure 8.23). Perhaps the most pithy summary of the link between politics and pyro-chemistry is the one attributed to Mao Tse Tung who said, 'Political power grows out of the barrel of a gun'.

Explosives, propellants and fuels

The reason that this discussion lies within Chapter 8 is that the hurling of metal and stone through the air requires energy, and the source has nearly always been chemical. But the use of a reaction in a particular role depends on more than just the size of its enthalpy change. Cost of reagents is clearly important, but beyond that is the issue of reaction *speed*. Very different speeds of release of energy are needed for different jobs. Table 8.4 compares straight fuels (as burnt, for instance, in power stations), with propellants (as in cartridge cases of rifle bullets and artillery shells) and explosives (in bombs and warheads).

Figure 8.23 Gunpowder was apparently discovered in the course of alchemical experiments in the Song Dynasty (960–1279). This drawing, showing arrows being fired by gunpowder, is from a 17th century treatise on the art of war

Table 8.4 Relative energy and power outputs from fuels, propellants and explosives

	Fuel (coal/air)	Propellant (black powder)	Explosive (TNT)
Energy output ($J\,g^{-1}$)	10^4	10^3	10^3
Power output ($W\,cm^{-2}$)	10	10^3	10^9

Question

35a Where in the table is there information about the rate of reaction?

b Why do propellants need to react slower than explosives?

All the reactions that are undergone by the three classes of energy-rich chemicals belong to the redox group discussed in Chapter 7, but not all of them use dioxygen as a reagent. The *fuels* are indeed all burnt with air, but the *propellants* are normally two-component mixtures in which the oxygen comes from within one member, while some of the common *explosives* work by intramolecular redox reactions.

Questions

36a In the reaction systems below, calculate the enthalpy change using direct data, bond enthalpies or enthalpies of formation and Hess's Law cycles. Convert these enthalpies into the more industrially significant units of kJ kg^{-1}. Comment on the relative sizes of the answers.

- *Fuel:* methane

$$CH_4(g) + 2O_2(g) \rightarrow CO_2(g) + 2H_2O(l)$$

- *Propellant:* black powder (one possible reaction)

$$4KNO_3(s) + 5C(s) \rightarrow 5CO_2(g) + 2N_2(g) + 2K_2O(s)$$

- *Explosive:* trinitrotoluene, or 'TNT' (structural formula and one possible reaction)

$$2C_7H_5N_3O_6(l)) \rightarrow 3N_2(g) + 5H_2O(g) + 7CO(g) + 7C(s)$$

b Identify the changes in oxidation number in the first two cases.

c Why is it no coincidence that the very fastest (the explosive) reaction is the one which occurs *intra*molecularly?

37 There is a fourth set of energy-rich redox systems which are known as pyrotechnics – close military relatives of fireworks. For example, a mixture of barium nitrate, Ba(NO$_3$)$_2$, aluminium and potassium chlorate(VII), KClO$_4$, is used as a photoflash. Write the equation for the possible reaction involving KClO$_4$ and Al in this mixture, and calculate its enthalpy change. Why would the reaction be of very limited value as an explosive?

We are reminded by the answer to the last question that a crucial aspect of an explosive reaction is the evolution of gases. But equally crucial is that they be released under very high pressure. This brings us on to two other aspects of explosive reactions: the influence of confinement, and the speed of the reaction front through the mixture.

Detonation

You might not think that a precise definition of detonation of an explosive would be necessary – after all, it is not the most subtle phenomenon in the world. Yet there *is* a formal definition. When the first bit of the explosive charge starts to react, it sends a pressure wave outwards through the material (like a sound wave – in fact it is just a single-pulse large-amplitude sound wave). In a detonation the pressure front of the sound wave triggers reaction as it passes, so that (and this is the definition) the reaction *spreads at the speed of sound* through the explosive, 'gathering' pressure as it goes (Figure 8.24). In fact the speed of the reaction front is supersonic since the compressed material will conduct sound faster than the normal, uncompressed material. In the example of TNT, the reaction front travels at about 5 miles (8 km) per second, so if you can

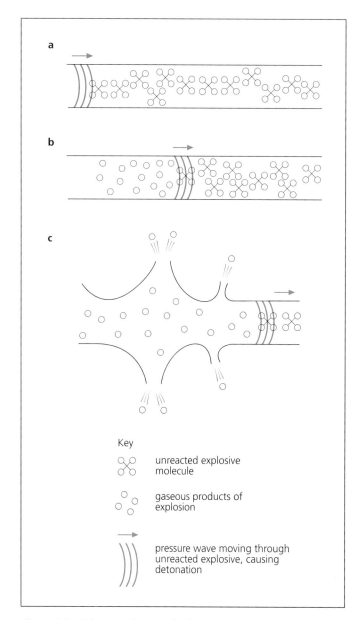

Figure 8.24 Schematic diagram of a detonation

imagine the explosive confined in a long hollow cable, you can picture the reaction front crossing the English Channel in about 4 seconds.

Yet TNT burns quietly when spread in thin layers on the ground, so clearly the degree of confinement is crucial in deciding whether detonation happens. If the pressure wave is to initiate reaction as it passes, it is important that the pressure itself cannot be dissipated by reaching the edge of the lump of explosive and escaping (rather like critical mass in nuclear fission). For that reason it is important that the explosive be confined in a container. Then the events of explosion happen in two distinct stages – a few microseconds in which the pressure wave detonates the whole lump, leaving behind a trail of gas molecules in grossly abnormal closeness to each other (see Figure 8.24) and then a subsequent stage in which all those gas molecules erupt from their confinement.

It has been estimated that the 'gases' in the immediate microsecond aftermath of an explosive detonation are in such an unnatural state that the value of the gas law function PV/nRT (normally 1) approaches 14. In other words, the gases possess much more energy due to pressure than due to temperature. Pressures of the order of 250 000 atmospheres are obtained momentarily in detonations of TNT.

Question

38a Calculate the enthalpy change of the reaction for the detonation of nitroglycerine (Figure 8.25). The Hess's Law cycle for this task is shown in Figure 8.25; you'll need bond enthalpy data for arrow 'a',

Figure 8.26 The 'hill man' and his one-legged stool

and specific enthalpies of atomisation for 'b'. For the sake of simplicity you may ignore the ΔH_{vap} of nitroglycerine.

b Under normal reaction conditions, the enthalpy loss of the system would be delivered to the surroundings as heat. But detonations, as we have seen, generate rather abnormal reaction conditions. In which forms will the energy be delivered to the surroundings? Will this mean that more or less heat will be generated by the reaction, compared to the

'standard' situation in which $\Delta H_{surr} = -\Delta H_{sys}$? Justify your answer.

c In the explosives factory, nitroglycerine was made in a room on a hill, in a batch process overseen by a man (the hill man) on a one-legged stool (Figure 8.26). Can you imagine why the stool only had one leg?

d The point of the hill was to enable liquid nitroglycerine to move from the reaction vessel to the next stage by gravity alone. Can you guess why it was not pumped?

e These days, although nitroglycerine is still made by the action of mixed nitric-sulphuric acid on glycerol, the process is computer-controlled so that only about 1 kg is in process at any one time, and once made it is rendered into the form of an emulsion in water. Why do these measures improve the safety of the process?

Blasting operations in the quarrying business do not use nitroglycerine or TNT. Instead they use ANFO, which is a mixture of ammonium nitrate and fuel oil. This mixture is very powerful (Figure 8.27). The light diesel fraction of crude oil is used (hydrocarbons with chain length C_{13}–C_{16}). The chemicals are mixed on site in a 'slurry truck' (Figure 8.28), with the ammonium nitrate in aqueous solution, and the oil in emulsion. The mixture contains a gelling agent so that it is semi-solid for insertion into the boreholes. Between 5 and 200 holes are used in blasting, and the explosions are staggered in time by a few milliseconds, using delay fuses.

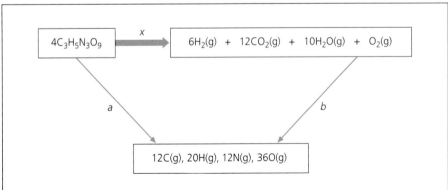

Figure 8.25 Hess's Law cycle for the detonation of nitroglycerine

Figure 8.27 The tragic bombing in Oklahoma City in the US in April 1995. The car bomb responsible contained 544 kg of ANFO

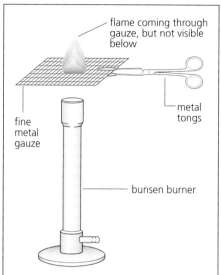

Figure 8.29 The gauze trick with a Bunsen burner

Figure 8.28 A slurry truck

Question

39a Explain how the use of the 'slurry truck' improves safety in the blasting process.

b What are the advantages of having a few milliseconds between explosions?

The final question in this section is about the piece of apparatus which is identified more closely than any other with our subject of chemistry. The question takes in fuels and explosions, and the definition of a detonation, so all in all should provide a highly suitable conclusion to the section.

Question

40 The common Bunsen burner is a machine with several subtle features, a fact which is disguised by its familiarity. Its main achievement is to establish a stable flame when the air hole is open.

When the air hole is open, the gases in the barrel have exactly the same reactant composition as the gases which are actually reacting at the flame. And yet the reaction does not spread to the barrel. This is because the reaction front spreads downwards into the unburnt region *only as fast* as the gases come up, and so stays stably in one place.

a The very fact that the barrel is made of metal helps to hold the flame's reaction front just at the nozzle, by a sort of 'negative feedback' mechanism. If there were a move by the reaction front to penetrate into the barrel, it would immediately begin to move more slowly than it had moved outside, and therefore more slowly than the gas coming up, thus moving the reaction front back upwards to the nozzle. Why would the *metal* barrel have this effect? Why is there an upper limit to the width of a single-bore Bunsen barrel?

b The famous trick of keeping a flame alight above a gauze but not below (Figure 8.29), owes its existence to the same principles. Can you explain how it works?

c The information in part (a) helps to explain why Bunsen flames rarely burn back down their own barrels, but there is no such automatic reason why flames cannot move up and be lost off the top. If the flame pulls in cold air at its base (by convection), or if you blow cold air into it, the flame begins to show signs of instability, and readiness to detach itself from the nozzle and 'lift off', thus extinguishing the burner. Clearly the balance between the speed of reactant flow upwards and the speed of reaction front downwards, mentioned earlier, has been disturbed. Which speed has changed, in which direction, and why?

d The little upside-down 'skirt' at the mouth of a Bunsen burner (Figure 8.30) is meant to help anchor the flame to the nozzle and prevent lift-off. Can you suggest how it works?

e The same gas–air mixture that burns in a controlled way on a Bunsen burner can be explosive when ignited in a plastic bottle (the effect is even more dramatic with gas and oxygen). From what you have

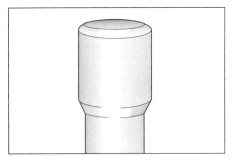

Figure 8.30 'Skirt' at the top of a Bunsen burner tube

learned in this chapter about the definition of detonation, try to explain why this is so.

8.11 Thermodynamics and prediction

All through this chapter we have seen the information which can be gleaned from a study of the enthalpy changes of reactions. However one of the big prizes

remains elusive – we have yet to reveal a way of predicting whether any given reaction will actually happen, or even more valuably, knowing how to persuade it to happen if it is reluctant.

There are two obstacles at the moment in the way of our obtaining this powerful position:

1 It is tempting to read the message of the enthalpy data as saying 'if it's exothermic it will happen', because the majority of 'happening' reactions are indeed exothermic. But endothermic spontaneous changes do exist, thereby undermining this criterion as a reliable predictor. We need a better quantity than ΔH on which to pin our trust. The search for this quantity is the subject of Chapter 9.

2 Even Chapter 9 will not be able to see off the other obstacle, and that is that the absolute limit of thermodynamic prediction is *whether* a reaction will go, not *when*. The whole matter of speed of reaction is the focus of Chapter 14.

Summary

- In thermodynamics, 'system' is the word used for the reactants and products of a reaction, plus any solvent medium. The word 'surroundings' describes the rest of the universe.

- All chemical reaction systems exchange energy with their surroundings, predominantly in the form of heat.

- The creation or destruction of heat energy occurs because of the destruction or creation of corresponding amounts of chemical potential energy.

- Chemical potential energy owes its existence to the fact that if atoms *can* be bonded together, then they have a lower potential energy when bonded than when not bonded.

- Thus when atoms enter into bonds the 'lost' potential energy appears in the surroundings as heat (well, most of it anyway – Section 8.9). Conversely when bonds are broken the surroundings *supply* heat which is converted into chemical potential energy.

- When both bond breaking *and* bond making occur (as in the vast majority of normal chemical events), then the direction of heat flow depends upon the relative strengths of the bonds being broken and made. If weaker bonds are broken and stronger bonds are made, then the system supplies heat to the surroundings. Such reactions are called **exothermic**. Conversely if stronger bonds are broken and weaker bonds are made, the surroundings supply heat to the system. Such reactions are called **endothermic**.

- Chemists are interested in changes in the potential energy of chemical systems, but they are forced to measure them via the quantities of heat going in or out. It is as if you could only measure the potential energy of a raised mass by letting it drop and seeing how much the floor warmed up. It is not necessary to do it this way because a raised mass is a conveniently big object and we can easily measure its height above the floor and calculate its potential energy from that. With molecules it is not so easy.

- **Enthalpy** (symbol H) is a quantity which is a measure of the chemical potential energy of a system (although it is not quite identical with it, as the complex argument in Section 8.9 tried to explain). Thus all reactions are accompanied by enthalpy changes (ΔH). However, as explained in the previous paragraph, the enthalpy changes are *measured* by heat exchanged with the surroundings.

- The official definition of enthalpy change is **the heat exchanged with the surroundings when the reaction occurs at constant pressure**. Enthalpy changes are not very sensitive to changes in the temperature and pressure of the surroundings, but nevertheless most enthalpy changes quoted in data books relate to *standard* conditions, which means 298 K and 1 atmosphere pressure (101 325 Pa). The symbol $^{\ominus}$ is attached as a superscript to indicate when a ΔH has been measured in standard conditions.

- Exothermic reactions have negative enthalpy changes, and endothermic ones have positive enthalpy changes (reminding us that the enthalpy change in the system is opposite to the heat change in the surroundings).

- Hess's Law states that the enthalpy change in going from a particular set of reactants to a particular set of products depends only on the potential energies of the reactant and product species, and is therefore independent of the route taken.

• Hess's Law enables us to construct reaction routes which have no practical reality, but which use readily available data for enthalpy changes, safe in the knowledge that the sum of the enthalpy changes on the unreal route is the same as on the real one. This enables us to find enthalpy changes even if real routes present practical problems or they do not happen at all.

• Hess's Law cycles look like route maps with a main road and a bypass. The bypass is constructed from available data (usually from databanks in books). Each individual stage's enthalpy change is represented by an arrow, and by an actual signed quantity in kJ mol^{-1}. If passage round the bypass requires an enthalpy change which is the reverse of one of the arrows, then the sign of the ΔH on that arrow must be reversed. This is the basis of the $x/a/b$ technique used in the text. The sum of the enthalpy changes on the bypass can then be equated to the enthalpy change on the main road.

• The data for constructing these bypasses resides in two types of databank. The first is the bank of bond enthalpies. To use bond enthalpies, a bypass must be constructed in which relevant reactant bonds are broken and product bonds made. The ΔH on the 'break' leg of the bypass will use positive bond enthalpies, while the 'make' leg will use negatives.

• The other major databank is that of specific reaction enthalpy data. These are arranged under subheadings, notably enthalpies of formation, combustion

and atomisation. Their use in Hess's Law cycles is explained diagrammatically in Figure 8.17, and their formal definitions are set out in Section 8.5, p. 148.

• There are consistent patterns in the way bypasses are constructed in Hess's Law cycles, using enthalpies of formation, combustion and atomisation, which give rise to formulae and remove the need for drawing:

$$\Delta H_{\text{reaction}}^{\ominus} = \Delta H_f^{\ominus}[\text{products}] - \Delta H_f^{\ominus}[\text{reactants}]$$

$$\Delta H_{\text{reaction}}^{\ominus} = \Delta H_c^{\ominus}[\text{reactants}] - \Delta H_c^{\ominus}[\text{products}]$$

$$\Delta H_{\text{reaction}}^{\ominus} = \Delta H_{at}^{\ominus}[\text{reactants}] - \Delta H_{at}^{\ominus}[\text{products}]$$

• Chemical equations can be treated like algebraic equations, and the associated enthalpy changes follow suit. Thus in 'matched trios' of equations, two of the equations can be added or subtracted to eliminate certain species and to generate the third equation. The enthalpy change of the third equation will be generated by a parallel addition or subtraction of the other two enthalpy changes.

• **Born–Haber cycles** are multi-stage Hess's Law cycles whose aim is to find the value of the **lattice enthalpy** of ionic crystalline solids. This value will be the *real* lattice enthalpy, which is to say the enthalpy change when the real lattice is formed from gaseous ions. It can be compared with a theoretical lattice enthalpy in which the gaseous ions are 'made' into a lattice (inside a computer) assuming pure ionic bonding. Thus the comparison of these two values for lattice enthalpy allows an estimate of how good the assumption of pure ionic bonding is.

• Practical methods of measuring enthalpies involve either (for liquids or solutions) letting the system heat itself (rather blurring the definition of 'surroundings'), or (for combustions) letting

the system heat a water bath.

• The actual enthalpy changes are derived either by use of the formula

$-\Delta H = +$ {heat transfer to absorbing medium}
$= $ mass of absorbing medium $\times C_p \times \Delta T$

or by electrical compensation, or as in the bomb calorimeter by comparing with the reaction of a standard substance. Conversion to a mole scale is the normal final step.

• Specific heat capacity (c_p) methods are riddled with problems which introduce errors and approximations – for example, aqueous solutions are normally assumed to have a specific heat capacity the same as water's. They also depend upon trustworthy lagging.

• Electrical compensation methods dispose of these problems, since the electrical heating stage is done under exactly the same conditions as the chemical heating stage. Hence any given temperature rise is due to the same number of joules either way. A joulemeter is used to 'count' the joules in the electrical heating stage.

• The bomb calorimeter eliminates the problems of incomplete combustion due to the use of an atmosphere of high-pressure oxygen.

• Fuels, explosives, pyrotechnics and propellants are all grossly exothermic reactions of a redox type. They differ in the speed of reaction and the mode of ignition.

• Explosives and propellants are triggered by pressure, while fuels and pyrotechnics are triggered by heat.

• The reaction front in an explosive moves at a speed greater than that of sound through the material.

9

Thermodynamics 2: The direction of chemical change

▷ **An important difference between mechanical and chemical systems**

▷ **The drive of probability**

▷ **Compromise between Nature's two drives**

▷ **Entropy**

▷ **The entropy of the universe**

▷ **Using entropy changes to calculate the direction of chemical change**

▷ **Manipulating the direction of chemical change for our own ends**

▷ **Free energy**

▷ **Reactions in the balance – chemical equilibrium**

▷ **The equilibrium constant**

▷ **A dynamic model of chemical equilibrium**

▷ **Manipulating equilibrium systems for our own ends**

▷ **The Contact process**

▷ **Calculations involving equilibrium constants**

▷ **The equilibrium constant in terms of partial pressures**

▷ **Summary**

9.1 An important difference between mechanical and chemical systems

(Author's note: The depth to which you need to take your studies of thermodynamics will vary depending on your syllabus, and on options selected. In particular there are differences in the emphasis that different exam boards place on the **entropy** and **free energy** concepts. My own view is that the best understanding of the factors affecting direction of chemical change comes via this entropy concept, and so that has been the foundation stone placed at the start of the chapter. Students of Nuffield, AEB and NEAB chemistry will find this approach suits their courses. But for students of courses in which entropy is marginalised, there is the option of jumping straight in at Section 9.10, and glossing over the back-references to free energy and entropy. What *is* certain is that, if you go on to do chemistry at university,

you'll have a better start if you have already come to terms with entropy.

Figure 9.1 shows a simple mechanical arrangement such as we might find on a building site – a pulley for raising buckets and bricks up to higher levels on the scaffolding. In Figure 9.1a, the bucket is in the raised position, unopposed by any mass except that of the rope, and held by a brake. In Figure 9.1b, the bucket is in the same raised position (and so has the same potential energy as in (a)), but is opposed by a wooden pallet, lighter than itself. In each case, after the appropriate warning to the people underneath (Health and Safety at Work Act 1974), the brake on the pulley is released.

Question

1 a Describe the events which follow, comparing the behaviour of the two arrangements.

  b Explain what happens to the original potential energy in each case, when the pulleys have come to rest.

In order to introduce some commonly used thermodynamic terms, we will call the bucket 'the system', and the rest of the universe (including the pallet) 'the surroundings'. On the basis of these results we can say:

1 In mechanical systems we can predict the direction of change in the system – it will be to a position of **minimum potential energy**.

2 The original potential energy in the system will have been given to the surroundings.

3 The original potential energy may appear in the surroundings solely as heat (as in the case of Figure 9.1a) or as a mixture of heat + work (as in the case of Figure 9.1b). In this latter example the work is done in raising the pallet, which counts as part of the surroundings.

Figure 9.2 is a diagram which shows how the energy from a potential energy store like our bucket might end up. Its *x*-axis is a continuum which runs from 'surroundings get nothing but heat' to 'surroundings get nothing but work'.

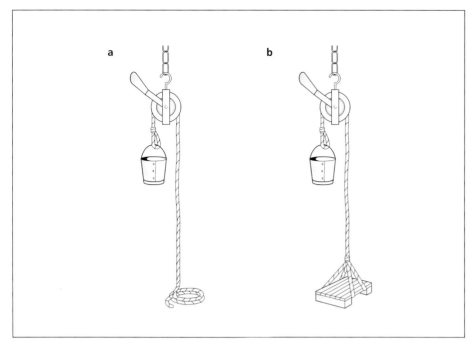

Figure 9.1 Pulley arrangements

Question

2 a Show, as closely as possible in the light of the descriptions of the two bucket situations, where Figures 9.1a and 9.1b would be placed on the x-axis in Figure 9.2.

b How could you arrange for a situation in which the work:heat ratio was greater than in Figure 9.1b? Is it possible to imagine a situation in which *all* the original potential energy in 'the system' is delivered as work to 'the surroundings'?

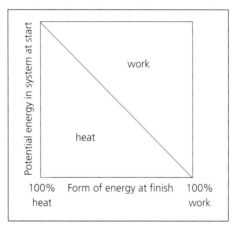

Figure 9.2 Forms of energy from a potential energy store

To review the story so far: we can see that the potential energy of a mechanical system can be translated into (almost) any combination of 'heat and work' but that no matter what combination of apparatus we attach to our falling bucket, it always obeys the law:

Potential energy destroyed
= (heat + work) created

Just to repeat a previous point, as long as the pallet has a mass less (however marginally) than the bucket, the bucket does always fall.

That's the end of our short detour into physics. It was made for the purpose of *contrast*. Let's draw up some comparisons of mechanical and chemical systems, and see where the contrasts occur. As we saw in chapter 8, there *is* in chemistry something like potential energy – the **enthalpy change** in moving from reactants to products in a chemical reaction is like a change in potential energy. And, just as with the falling bucket, in many chemical situations we do indeed find that the direction of change has been a *decrease* in the sys-

tem's potential energy, and the appearance of heat and/or work in the surroundings.

Question

3 Think of examples from everyday science in which:

a The loss of enthalpy of a chemical reaction system is converted to heat only.

b The loss of enthalpy of a chemical reaction system is converted to heat plus work. (Work in this context need not necessarily be 'raised weight' – it could equally be the work done in giving a body kinetic energy.)

So what about the contrast between physical and chemical systems referred to earlier?

Question

4 a Have you ever seen a mass move spontaneously *uphill*?

b Have you ever seen a chemical reaction move spontaneously in a direction of increased enthalpy (that is, be endothermic)?

c Have you ever seen an engine turn chemical energy into movement without also giving heat to the surroundings (Figure 9.3)?

Figure 9.3 All fuel-burning engines waste some of their chemical energy as heat to the surroundings

This last question should have brought out the contrast between chemical and mechanical systems to which I've been leading up. It is in fact a *double* contrast. On the one hand chemical systems seem freer than mechanical ones – free in the endothermic cases to move from lower to higher positions of potential energy instead of always being constrained to move lower. For instance, there are temperatures at which even a hydrogen molecule will spontaneously separate to two hydrogen atoms. (This is not to deny, however, that the majority of normal chemical reactions at normal temperatures are exothermic.) And yet in another way, chemical systems are more constrained than mechanical ones – it seems they are *not* free to convert all their potential energy to work. Even the most efficient enthalpy-run engines give up some of their energy as heat.

The central point of this chapter is to enable us to predict the direction of *chemical* change. To predict the direction of *mechanical* change it only seems necessary to recognise the drive to minimum potential energy – the descent to the bottom of the hill. But in the chemical situation, although this rule works for most reactions it does not work for them all; you cannot say that if the enthalpy change of the system is negative the reaction *will* happen. It seems another set of rules, and another drive, must be sought – a drive which explains why some reactions prefer to move 'uphill'. In the course of this search we will also find out why those reactions that *are* healthily exothermic still aren't allowed to convert all their loss of enthalpy into useful work.

9.2 The drive of probability

The most significant cause of the difference between mechanical and chemical systems is the enormous number of 'free-range' particles which make up the latter. When large numbers are involved, we have to bring in a whole new mode of thinking to guide our predictions. This new thinking can be easily illustrated by considering the situation in Figure 9.4. We have some molecules of a gas on one side of the partition in a partitioned box.

The other side is empty (a vacuum, in other words).

Question

5 Intuitively, what do you think will happen if the partition is removed?

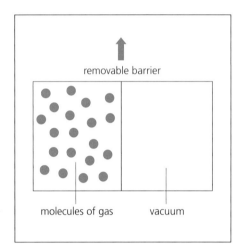

Figure 9.4 Molecules in a partitioned box

This simple example carries no less than three messages worth highlighting.

1 This was clearly a spontaneous change, and yet the expansion of a gas into a vacuum (an 'ideal' gas, at least) involves no bond breaking, and therefore has no associated enthalpy change. So, *it cannot be responding to any 'minimum potential energy' drive.*

2 While it is easy to think of gases being 'sucked' into a vacuum, it is important to realise that there is no force on any given gas molecule, and that the spreading is purely as a result of natural (thermal) molecular movement. The molecules are acting as blindly as a load of coins being tossed. The reason they end up evenly on both sides is that 'evenly' is the most *probable* distribution predictable from random chance, far more probable than finding all the molecules in one half. To extend the coin-tossing analogy, it is like predicting that the proportions of heads and tails produced by a lot of tosses would be close to 50:50.

3 Had there been, say, about five molecules in the gas in Figure 9.4, then the 50:50 prediction would have been nowhere near as safe. Nor indeed would there really have been a steady state with which to describe the 'after' condition of the system, because there is a small but perfectly reasonable chance that the five molecules might actually find themselves all back in the original half of the box at some later time (just as five coin tosses might conceivably produce five tails without divine intervention). But, were there anything remotely approaching molar numbers of gas molecules, the 50:50 prediction would be an extremely safe bet for the steady state of the system (Figure 9.5).

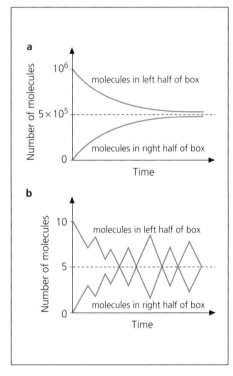

Figure 9.5 a System comprising 10^6 molecules. **b** System comprising 10 molecules

The conclusions to be drawn from this gas-in-a-box case are as follows. When the number of particles in a system is large (and energy is not an issue), the system will come to rest in a condition predictable by *the laws of probability*. Any fluctuations in the system either side of the most probable condition will be negligibly small (Figure 9.6), and the system will appear to be in a steady state.

9.3 Conflict between Nature's two drives

So we have two situations for which we can predict outcomes. Mechanical systems such as the bucket in Figure 9.1 will tend towards positions of minimum potential energy, while multi-million-particle systems, all of whose possible states are of equal potential energy, will tend towards positions of maximum probability.

The next question is – what happens if the drive to minimise potential energy and that to maximise probability are *both* relevant to a system? What if the changes required by one drive are the opposite of the changes required by the other? There is an urgent need for answers to these questions, because this conflict of drives is the rule rather than the exception in chemical systems.

Figure 9.6 In all of human history so far, no-one has ever reported a solution spontaneously concentrating itself in one corner of the beaker. It isn't actually impossible – just astronomically improbable

As an example of drives in conflict, consider the flying apart of a dihydrogen molecule into hydrogen atoms – the hydrogen dissociation reaction:

$$H_2(g) \longrightarrow 2H(g) \qquad \Delta H = +436 \text{ kJ mol}^{-1}$$

Ignoring the chemical bond for a moment, we could say that the flying apart was indeed going to happen, in response to the maximum probability drive, because particles clumping together in pairs are less likely than particles being singly distributed over space. But if we shift our focus and look at enthalpies, then the flying apart requires an uphill push of

436 kJ mol^{-1}, and therefore will not happen on minimum potential energy grounds (Figure 9.7). So, will it actually happen or won't it?

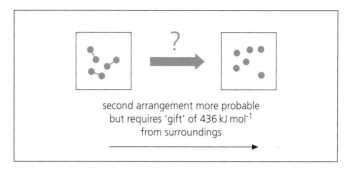

Figure 9.7 A conflict of natural tendencies in terms of distribution of atoms in space

Question

6 There are only very few reactions which manage simultaneously to go downhill in energy and yet up on the probability-of-distribution front (and they are mostly reactions involving explosives). Much more normally reactions which are exothermic ('downhill') at the same time bring about more ordered, less random and chaotic, arrangements of particles. Can you explain why this *is* the norm, and why the opposite is also true – that reactions in which there is a clear *increase* in the chaos factor are normally endothermic?

Back to our main objective, the matter of the battle between the two opposing drives. How can we 'weigh' one drive against the other? In enthalpy terms the difference between $H_2(g)$ and $2H(g)$ is a precise enough $+436 \text{ kJ mol}^{-1}$, but in probability terms we only have the vague idea that the change (from $H_2(g)$ to $2H(g)$ is *up*, due to the greater randomness associated with the distribution of the single atoms. Fortunately there is a way of expressing these two factors in the same, quantitative, unit system, and thereby seeing which one wins the battle to determine the direction of change in the system. They must both be expressed as changes in a quantity called **entropy**.

9.4 Entropy

The quantity whose measurement enables the quantitative comparison of the two drives is **entropy**. We will begin with a definition of entropy in terms of probability (9.1). This will enable us to assign an entropy value to the change from $H_2(g)$ to $2H(g)$. But we'll also eventually have an equation which relates entropy to heat changes in the surroundings (9.5), and that will enable us to convert the $+436 \text{ kJ mol}^{-1}$ into another entropy value. It will then be a simple matter to compare the two values and deduce the direction of the chemical change.

The entropy (symbol S) of a chemical system can be defined in the following terms.

1 It is a quantity which measures the degree of disorder, randomness, distribution-probability, chaos (call it what you will) of the way the particles of a system are distributed in space, and of the way energy is distributed between the particles.

2 Thus a mole of hot monatomic gas has an extremely high entropy, while all crystalline solids at 0 K (absolute zero of temperature) have zero entropy.

3 To move to a more quantitative definition, the entropy, S, of a system can be expressed as:

$$S = k \ln W \qquad (9.1)$$

and any changes in entropy, ΔS, as

$$\Delta S = k \Delta (\ln W) \qquad (9.1a)$$

where W is a number associated with the number of ways of arranging particles in space and/or the number of ways of arranging energy between particles. W is thus a big number for a hot gas and a small number for a cold crystalline solid. k is the **Boltzmann constant**, which is a constant connecting the units of work/energy and of temperature. Since logarithms (ln) have no units, the units of entropy are the same as those of the Boltzmann constant, namely $J K^{-1}$.

4 To give you a feel for entropy as a quantity, here are the entropies of a mole of various elements and compounds, at 298 K and 1 atmosphere.

Substance	Entropy $(J K^{-1} mol^{-1})$
H(g)	115
H_2(g)*	130.6
He(g)	126.0
C(s, graphite)	5.7
C(s, diamond)	2.4
Fe(s)	27.3
H_2O(l)	69.9
H_2O(g)	188.7[†]

* Although these figures make the helium and 'monohydrogen' appear less 'random' than the dihydrogen (H_2(g)), you must recognise that, per mole of *atoms*, the monatomics 'win' easily.
[†] This is the entropy a mole of H_2O *would* have, if it existed at 298 K, 1 atm.

9.5 The entropy of the universe

Why has the creation of this new quantity helped to achieve the aim of this chapter – that is, the prediction of the direction of chemical change? The connection is this:

• Entropy is a measure of probability, both of the distribution of particles in space, and of the distribution of energy between particles.

• The systems we study are multimillion-particle systems, and so are the surroundings with which they exchange energy.

• Therefore they both follow closely the laws of chance, and the most probable distributions will be a good approximation for the steady states in which they come to rest.

• So the direction of chemical change will be the direction of increasing entropy – but only if that test is applied to the whole 'universe', that is, the system *and* its surroundings. (Hence it is not enough just to look at how chaotic the molecules of the system become.) From now on, the criterion which any reaction must meet to be thermodynamically viable is:

$$\Delta S_{universe} > 0 \qquad (9.2)$$

You may have noticed that there has been a fairly dramatic development in the story. Hitherto, the direction of chemical change has been presented as being dependent on the outcome of a battle between *two* opposing drives – the drive to minimum potential energy and the drive to maximum entropy among the particles of the system. Now we are saying there is really only *one* drive and that is towards an increasing entropy of the whole universe – in other words, the supreme criterion for a change to be passed as 'viable' is that it should increase the probability of the universe. The other two drives are now seen as subordinate to the main drive and are relevant only insofar as they each contribute to the overall entropy sum.

To know whether a particular reaction will happen, therefore, all we have to do is calculate the value of the entropy change of the 'universe' – and if it is positive, the reaction will take place. But as we have already seen when considering the hydrogen dissociation reaction (Section 9.3, p. 167), there are two sides to most changes – first, what the particles do in space, and second, the energy exchange with the surroundings. We need to calculate the entropy changes associated with *both* these facets of the reaction, and then see whether the overall entropy change is positive. We can express the relationship between the overall entropy change and its two component parts, by the equation:

$$\Delta S_{universe} = \Delta S_{system} + \Delta S_{surroundings} \qquad (9.3)$$

Let's look in turn at the methods of calculating the two terms on the right-hand side of (9.3).

1 Calculating the entropy change in the system when a chemical reaction happens

For this task we are heavily dependent on data books. They carry values of the entropies of a mole of many pure chemical substances. In the course of a chemical reaction, when reactants are destroyed, the entropies of those reactants will 'die with them', to be replaced by the entropies of the products. In short:

$$\Delta S_{system} = \Sigma S_{products} - \Sigma S_{reactants} \qquad (9.4)$$

Question

7 Use (9.4) and a data book to calculate $\Delta S_{sys}^{\ominus}$ for these reactions:

a Burning of hydrogen in oxygen to form water:

$$2H_2(g) + O_2(g) \rightarrow 2H_2O(l)$$

b Burning of propane gas (as in 'Calor gas'), with water appearing as vapour:

$$C_3H_8(g) + 5O_2(g) \rightarrow 3CO_2(g) + 4H_2O(g)$$

c Displacement of copper from a solution of copper sulphate by zinc metal:

$$Cu^{2+}(aq) + Zn(s) \rightarrow Zn^{2+}(aq) + Cu(s)$$
$$S^{\ominus}Cu^{2+}(aq) = -99.6 \text{ J K}^{-1}\text{mol}^{-1}$$
$$S^{\ominus}Zn^{2+}(aq) = -112.1 \text{ J K}^{-1}\text{mol}^{-1}$$

Comment on the *signs* and *sizes* of the entropy changes you have calculated, and in particular relate them to the 'change-of-chaos' factor in each reaction.

2 Calculating the entropy change in the surroundings (ΔS_{surr}) when a chemical system exchanges heat with it

The crucial equations which define ΔS_{surr} (9.5 and 9.6) will not be derived from first principles – rather you will be asked just to accept them. But later on there will be an attempt at least to account for their general structure. For now, here they are. Of the two, (9.6) is the really useful one.

$$\Delta S_{surr} = \frac{\text{heat transferred to the surroundings}}{T} \quad (9.5)$$

But heat transferred to the surroundings is numerically equal and of opposite sign to the enthalpy change in the system (always assuming that no work is done). Therefore:

$$\Delta S_{surr} = -\frac{\Delta H_{sys}}{T} \quad (9.6)$$

There are five points to make about equation (9.6):

1 It is so very simple. It seems astonishing that entropy, which was defined quite simply in terms of statistical probability in (9.1), should have an equally simple translation into thermal quantities. (Probability and the heat/temperature relationship are not as far apart as you might think.)

2 It is made even simpler because the measured enthalpy change, ΔH_{sys}, of most reactions is a *constant* over a wide range of temperatures. (This is because bond enthalpies themselves are constants over wide temperature ranges.)

3 If equation (9.6) were expressed in words instead of symbols, its message would be this: the **entropy** change in the surroundings is bigger when the **enthalpy** change in the system is bigger. But the donation of heat *to* the surroundings by exothermic reactions (ΔH_{sys} negative) causes the entropy of the surroundings to go *up* (ΔS_{surr} positive), whereas the absorption of heat *from* the surroundings by endothermic reactions (ΔH_{sys} positive), results in a negative ΔS_{surr}. And furthermore, whatever the sign and size of the ΔH_{sys}, the entropy change in the surroundings is *less when the surroundings are already at a high temperature* (as shown by having 'T' on the bottom line of equation (9.6)).

4 Many students fall victim to a particular misconception in this field. The derivation of (9.6) uses the true fact that $\Delta(\text{heat})_{surr}$ is equal and opposite to ΔH_{sys}. It is easy to be drawn into a false comparison of enthalpy and entropy. It is specifically *not* true that ΔS_{sys} is equal and opposite to ΔS_{surr}. For one thing they are dependent on very different influences. On the one hand ΔS_{sys} relies upon the change in the 'chaos factor' of the distribution of the particles, while on the other hand ΔS_{surr} is calculated from (9.6).

5 By convention the units of ΔH in data books are kJ mol^{-1}, whereas those of entropies are J K^{-1} mol^{-1}. It is therefore necessary to convert one or other before they meet in (9.6). Normally the enthalpy is converted into joules.

For all its structural simplicity, it is still hard for most students to decide exactly what equation (9.6) *means*. Point 3 above goes part of the way, but the following analogy may possibly take you towards a deeper understanding. The analogy draws on a well-known Bible story about a poor widow who gives a mite (a low value coin) to a charitable cause (Figure 9.8).

Figure 9.8 The widow's mite parable

Various rich people express their scorn at the worthlessness of her contribution, but Jesus intervenes with what is essentially a 'proportionist' argument, reminding the Pharisees (the rich ones) that the mite was a far bigger percentage of the widow's income than it would have been of their incomes. The widow has therefore suffered a far greater loss in the number of ways she has of spending her remaining cash, than the rich men have.

Back on the chemical side of the analogy, this means that a transfer of a given number of heat quanta will mean much more to cold surroundings than it will to hot surroundings. Hence if an endothermic reaction, say, is conducted at room temperature (with the coolness representing poverty), the surroundings will suffer a much bigger loss of entropy than if the same reaction were to be conducted at 500 K (which represents riches) even though the ΔH_{sys} is the same in both cases.

Question

8 In a typical chemical system, which bit is playing the part of:

a The charitable cause?

b The widow?

c The Pharisees?

d The mite?

Note that although the analogy and the example relate to 'gifts' *from* the surroundings to the system, the ideas are reversible. If the transfer of heat is *to* the surroundings, then again the hotter the surroundings, the less will be the impact of the gift on the 'number of arrangements' (i.e. the entropy) factor.

9.6 Using entropy changes to calculate the direction of chemical change

Now we know how to calculate entropy changes in both system *and* surroundings, via equations (*9.4*) and (*9.6*) respectively, we are ready to answer the central question about the direction of chemical change. As an example, let's revisit the problem which has been hanging unresolved since the end of Section 9.3, namely, the one about whether dihydrogen molecules would decompose into hydrogen atoms at 298 K. The relevant data are:

Entropy of $H(g) = 115$ J K^{-1} mol^{-1}

Entropy of $H_2(g) = 130.6$ J K^{-1} mol^{-1}

$\Delta H_{sys} = +436$ kJ per mole of $H_2(g)$

The calculation is:

$$\Delta S_{sys} = \Sigma S_{products} - \Sigma S_{reactants} \quad (9.4)$$

$$= 2 \times 115 - 130.6$$

$$= +99.4 \text{ J K}^{-1} \text{ mol}^{-1}$$

$$\Delta S_{surr} = -\frac{\Delta H_{sys}}{T} \quad (9.6)$$

$$= -\frac{436\,000}{298}$$

$$= -1463 \text{ J K}^{-1} \text{ mol}^{-1}$$

so $$\Delta S_{universe} = \Delta S_{sys} + \Delta S_{surr} \quad (9.3)$$

$$= +99.4 - 1463$$

$$= -1363.6 \text{ J K}^{-1} \text{ mol}^{-1}$$

If you recall that the test for thermodynamic viability is that $\Delta S_{uni} > 0$, you'll see that this reaction, at this particular temperature, scores a resounding no.

The bar chart in Figure 9.9 shows the above calculation in graphic form. In probability terms the flying apart of molecular pairs is a more likely event than two atoms being close to each other (as shown by the $+99.4$ J K^{-1} mol^{-1}). However, the chances that the surroundings will part with the 436 kJ to make it happen, from a background temperature of 298 K, is overwhelmingly unlikely, as shown by the -1463 J K^{-1} mol^{-1}.

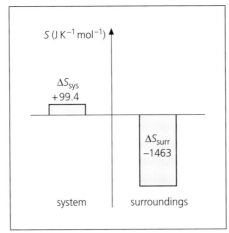

Figure 9.9 ΔS_{sys} and ΔS_{surr} for $H_2(g) \rightarrow 2H(g)$ at 298 K (not a scale drawing)

The calculation above was taken in two separate stages, to emphasise the system/surroundings structure. But it is possible, and convenient for the next question, to condense the two-stage operation into one equation. By combining (*9.3*) and (*9.6*) we get:

$$\Delta S_{universe} = \Delta S_{sys} + \Delta S_{surr} \quad (9.3)$$

$$\Delta S_{surr} = -\frac{\Delta H_{sys}}{T} \quad (9.6)$$

Therefore:

$$\Delta S_{universe} = \Delta S_{sys} - \frac{\Delta H_{sys}}{T} \quad (9.7)$$

Question

9 Test the feasibility of a few more reactions using data in the data book and equation (*9.7*). There is one extra step to take, since it will be necessary to calculate ΔH_{sys} by normal Hess's Law cycle methods, in order to use it in (*9.6*). Work out the entropy changes in the universe if molar quantities of the following reactions were to occur at 298 K and 1 atmosphere. Say whether the reaction is capable of happening or not.

a $CaCO_3(s) \rightarrow CaO(s) + CO_2(g)$

(This is the 'cement works' reaction, opposite.)

b $2NO(g) + O_2(g) \rightarrow 2NO_2(g)$

c $Fe_2O_3(s) + 2Al(s) \rightarrow Al_2O_3(s) + 2Fe(s)$

(This is the 'thermite' reaction, still used for the welding of locomotive rails on railway lines.)

9.7 Manipulating the direction of chemical change for our own ends

The answer to question 9a is surprising – the laws of probability appear to deny the viability of the cement-making process. But cement exists, so how was it that people altered the balance of probability to persuade calcium carbonate to 'co-operate'?

Of course cement has existed for much longer in human culture than thermodynamics (Figure 9.10). Trial and error is sometimes a more potent and nearly always a more handy weapon than extensive theorising. So it was that people tried the obvious ploy for most moments of chemical doubt – heat it and see what happens. Full understanding came much later. (This is not a general criticism of theories: there are plenty of useful reactions over which humans have gained importantly increased control by the application of theoretical thermodynamics.)

Let us look at *why* the decomposition of calcium carbonate occurs at higher

Figure 9.10 Cement works – the core reaction in the process is only thermodynamically feasible at high temperatures

temperatures. This seems a good moment to make an important distinction, as you may be thinking: 'All reactions go faster at higher temperatures'. This is true (with the exception of enzyme-catalysed reactions) but speed is not what this chapter is about. This chapter is about whether the 'after-reaction' condition of the universe is more *probable* than the 'before-reaction' condition, irrespective of how long it takes to effect the change.

To put it another way, this chapter and the kinetics one (Chapter 14) are like two tests that a reaction must pass to be viable.

• Thermodynamic viability – test 1 (this chapter): is the reaction possible on thermodynamic grounds (that is, is ΔS_{uni} positive)?

• Kinetic viability – test 2 (Chapter 14): is it going to go at a reasonable speed that will give a yield sometime this week?

The thermite reaction from question 9c is a classic example of a reaction for which passing one viability test is not enough. Although the reaction is definitely viable by test 1, you may know from your own classroom experience that it needs some elaborate ignition chemistry to get it to go. (Your teacher will have used a magnesium fuse and something called 'ignition powder' made from barium peroxide and magnesium powder.) In other words, it totally fails test 2 unless subjected to temperatures greater than 1000 °C.

All usable reactions must pass both these tests, and although test 2 is almost always easier to pass when hot, the hotter-the-better answer is *not* always true of test 1.

Returning at last to the calcium carbonate reaction, we'll make two assumptions. One is that ΔS_{sys} (the 'change-of-chaos' factor) will stay fairly constant over a range of temperatures. (If either of the solid reagents were to melt within the temperature range of interest, this assumption would founder, but they don't, so it doesn't.) The second assumption is that, as mentioned already, ΔH_{sys} will also remain effectively constant over the temperature range of interest.

So what is left to vary with temperature? Well, the 'system' side of the entropy bar chart in Figure 9.11 stays

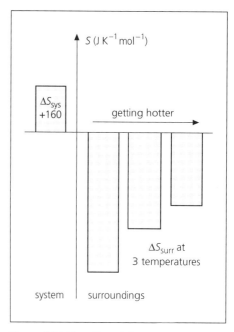

Figure 9.11 The effect of changing temperature on the feasibility of the 'cement works' reaction

steady at $+160\ \mathrm{J\ K^{-1}\ mol^{-1}}$ (assumption 1), so the only changeable factor is ΔS_{surr}, which varies because temperature is on the bottom line in equation (9.6), underneath ΔH_{sys} which does not vary (assumption 2).

Question

10 You might not have been able to predict the *size*, but the *sign* of ΔS_{sys} for this reaction is predictable just by looking at the equation (especially the state symbols). How could you tell it would be positive?

Putting the message of the bar chart into words, we get: 'The total entropy ΔS_{uni} is the sum of the entropies for the system and surroundings. The entropy change in the *system* is positive (and constant). As the reaction is endothermic, the entropy change in the *surroundings* is bound to be negative, but at least by putting up the temperature of the surroundings we can make it less negative. (In analogy-speak, this is like saying that if you are forced to give a mite, it is better – on economic grounds, if not on moral ones – to be a Pharisee than a widow.) Sooner or later we will reach a temperature at which ΔS_{surr} is of smaller magnitude than ΔS_{sys}, so ΔS_{uni} will go positive.'

Questions

11 Refer back to the data you assembled for the calcium carbonate decomposition in question 9a (that is, its ΔH_{sys} and its ΔS_{sys}). Use these numbers, plus equation (9.7), to work out whether the reaction could proceed at:

a 500 K

b 1000 K

c 1500 K.

12 Some teachers cheat when teaching younger students. If they teach about cement they will perhaps produce a bottle marked 'calcium carbonate' whose real contents are magnesium carbonate. Given that school Bunsen burners can struggle up to about 700 °C (approximately 1000 K), show why on thermodynamic grounds (if not on moral ones) this lie is quite a good idea.

9.8 Free energy

The following algebraic rearrangement creates and defines a new thermodynamic quantity. At first this may seem pointless, but bear with it, because the new quantity has quite a concrete and useful meaning.

We start with equation (9.7):

$$\Delta S_{uni} = \Delta S_{sys} - \frac{\Delta H_{sys}}{T} \qquad (9.7)$$

Multiplying through by $-T$ gives:

$$-T\Delta S_{uni} = \Delta H_{sys} - T\Delta S_{sys}$$

The new thermodynamic quantity is the subject of that final equation, $-T\Delta S_{uni}$, which is renamed ΔG_{sys} where G is a function called the **Gibbs free energy** (Figure 9.12). Thus:

$$\Delta G_{sys} = -T\Delta S_{uni} \qquad (9.8)$$

and

$$\Delta G_{sys} = \Delta H_{sys} - T\Delta S_{sys} \qquad (9.9)$$

There are two important points about the Gibbs free energy.

1 The quantity ΔG_{sys} is a very close relative of ΔS_{uni} and so gives the same quality of information about the direc-

Figure 9.12 Willard Gibbs, elucidator of the free energy concept

tion of chemical change. There *is* a shift in sign, meaning that reactions which are to be thermodynamically viable must show a *negative* ΔG_{sys} (whereas previously we focused on the equivalent necessity of a positive ΔS_{uni}).

2 By virtue of the '$-T$' multiplier, the new quantity has energy units, rather than 'energy per temperature'. As we shall soon see, it has an important energy-related *meaning* as well.

Why bother to have a Gibbs free energy?

One answer to this is trivial yet unavoidable. It is what data books deal in. They all choose to list, say, Gibbs free energies of formation rather than values of ΔS_{uni} of formation. Having a recognisable 'energy-flavoured' item of data available does help to prompt the use of Hess's Law cycle methods to find the Gibbs free energies of target reactions, in an exact parallel to previous work with enthalpies (Chapter 8). To find out whether a proposed reaction will go in the desired direction, the quickest way is to set up one of these free energy cycles. It is quicker than the method of question 11, say, which involves separate calculations of entropy and enthalpy changes. However, since ΔG data is often only available for standard conditions, we could not in fact have used it for question 11.

Question

13a Why not?

b Now for an example to show how the existence of ΔG_f data in data books makes the business of testing

a reaction for thermodynamic viability quick and easy, if only at 298 K. Using a free-energy version of equation (8.2), p. 151, find out whether the following reaction is viable:

$$CaO(s) + H_2O(l) \rightarrow Ca(OH)_2(s)$$
$$\text{'quicklime'} \qquad \text{'slaked lime'}$$

Why is it called 'free' energy? Interestingly, the explanation harks all the way back to the debate at the beginning of the chapter about the differences between chemical systems and falling buckets.

We will look at the burning of methane as an example. The equation is:

$$CH_4(g) + 2O_2(g) \rightarrow CO_2(g) + 2H_2O(l)$$

Relevant data are:

$$\Delta H_c[CH_4(g)] = -890 \text{ kJ mol}^{-1}$$
$$\Delta S_{sys} = -40 \text{ J K}^{-1} \text{ mol}^{-1}$$

From these data, ΔS_{surr} for this reaction can be calculated using equation (9.6):

$$\Delta S_{surr} = -\frac{\Delta H}{T} \qquad (9.6)$$

$$= \frac{-(-890 \times 1000)}{298}$$

$$= 2990 \text{ J K}^{-1} \text{ mol}^{-1}$$

From equation (9.3):

$$\Delta S_{uni} = \Delta S_{sys} + \Delta S_{surr} \qquad (9.3)$$
$$= -40 + 2990$$
$$\Delta S_{uni} = +2950 \text{ J K}^{-1} \text{ mol}^{-1}$$

So the reaction is thermodynamically viable, mainly because of the generous helping of heat given to the surroundings. But consider this – the only criterion a reaction has to satisfy in order to be passed fit thermodynamically is that ΔS_{uni} should be positive. It does not have to be massively positive – just a little bit will do. This means that even if the entropy of the surroundings had only increased by, say, +41 J K^{-1} mol^{-1}, the reaction would still have been viable, because ΔS_{uni} would have been (+41 – 40), which is undeniably positive.

Let us examine that +41 figure more closely, and translate it into an amount of heat. An entropy change of +41 J K^{-1} mol^{-1} would require heat being given to the surroundings to the tune of:

(heat change)$_\text{surr}$ = $T \times$ (entropy change)$_\text{surr}$ (*9.5 rearranged*)
= 298 × +41 = 12 000 J mol^{-1} = 12 kJ mol^{-1}

If the surroundings would have been satisfied with 12 kJ mol^{-1} of heat energy, why give them 890 kJ mol^{-1}? We could have kept back 878 kJ mol^{-1} for our own purposes. Here comes the important bit – that 878 kJ mol^{-1} is therefore 'free' to do work – push pistons, spin shafts, raise weights, etc. (Figure 9.13). Hence the term *free energy*, because it is the energy left over after the heat demands of the universe, so to speak, have been met.

My use of the number '41' in the previous paragraph was to make a point forcibly. 41 is very obviously just a little bit more than 40, so the entropy of the universe will go up by the obviously small amount of 1 J K^{-1} mol^{-1}. But the official calculation of free energy requires a more exact piece of maths. We must consider the reaction standing on the very brink of viability, or as the value of ΔS_uni standing on the very brink of 0. In other words, **the official free energy of a reaction is the *limiting maximum* of the available useful work which can be got from the system, as ΔS_uni approaches 0**. So we must change our figures a little. Before, we said:

(free energy change)$_\text{sys}$ = (total enthalpy change)$_\text{sys}$ −
(that part of the enthalpy which has to go as heat to the surroundings to ensure that ΔS_uni = +1)

which in figures was (in J rather than kJ):

$(-890\ 000) - (298 \times 241) = -878\ 000$ J mol^{-1} (to 3 s. f.)

Now, we must change that 41 to its limiting value of 40.

So true value for (free energy change)$_\text{sys}$
= $(-890\ 000) - (298 \times -40)$

Actually the answer to this second calculation is also 878 000 J, to three significant figures, but that's not the point. The point is that the second set of figures is the true method of calculating the free energy change, defined as the absolute limiting maximum of available work. Now let's take that last numerical equation and turn it back into symbols. If we check on the symbolic meanings of the numbers 890 000, 298 and 40, we'll see that the symbolic version of the equation is:

(free energy change)$_\text{sys}$
= $(-890\ 000) - (298 \times -40)$
= $\Delta H_\text{sys} - T\Delta S_\text{sys}$

Go back and compare this equation with (*9.9*). You will see that the right-hand sides of the above equation and of (*9.9*) are identical. What this means is that ΔG has real meaning as the upper limit of the useful work obtainable from a chemical system. Figure 9.14 summarises all this 'free energy = maximum work' argument in diagram form.

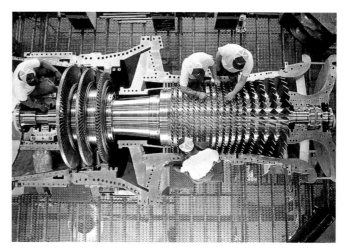

Figure 9.13 A gas turbine generator

This seems a suitable moment to pause and review the free energy landscape. Here is a collection of equations involving G.

$$\Delta G_\text{sys} = -T\Delta S_\text{uni} \tag{9.8}$$
$$\Delta G_\text{sys} = \Delta H_\text{sys} - T\Delta S_\text{sys} \tag{9.9}$$
$$\Delta G_\text{sys} = -w_\text{max} \tag{9.10}$$

The minus sign in (*9.10*) derives from the fact that G (a measure of the ability to do work) is being lost by the system, and so ΔG has an opposite sign to the positive work done (on the surroundings by the system).

Note that in our original system consisting of one bucket in a gravity field, there is no such thing as a ΔS_sys so *all* the original potential energy in the raised bucket is available for work.

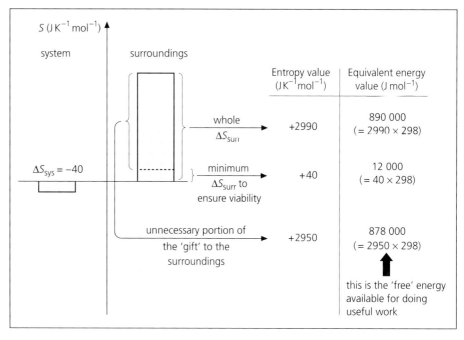

Figure 9.14 'Entropy balance sheet' bar graph for the burning of methane, showing how 878 kJ mol^{-1} is available for use in a 'non-heat' way

This is another contrast between the multi-particle world of chemistry, and the few-particles world of mechanics, to set beside the one earlier in the chapter where we noted that chemical systems can sometimes go 'uphill'.

Equation (9.9) is in many ways the keynote equation of the whole chapter. Sometimes equations can be equivalent to each other, and yet the format of one makes it a more accessible source of mental pictures. Indeed that is one of the advantages of having Gibbs free energy in the first place. In the case of (9.9) the equation says one thing clearly: as the temperature approaches zero, the second term in the equation ($-T\Delta S$) fades away. So the chance of getting a reaction with a negative ΔG (one that goes) depends increasingly on ΔH being negative. Therefore at low temperatures the only viable reactions are exothermic ones. The reverse is true too – at high temperatures it is the first term which fades into relative insignificance compared to the second, so the viable reactions are endothermic, and (9.9) shows how their positive entropies give rise to large negative free energies. Figure 9.15 is a diagrammatic summary of the last paragraph.

$$\Delta G_{sys} = \Delta H_{sys} - T\Delta S_{sys}$$

Situation 1

T is small

$T\Delta S_{sys} \longrightarrow 0$

$\Delta G_{sys} \simeq \Delta H_{sys}$

So to get a negative ΔG_{sys} we need a negative ΔH_{sys}

Situation 2

T is big

$T\Delta S_{sys} \gg \Delta H_{sys}$

$\Delta G_{sys} \simeq -T\Delta S_{sys}$

So to get a negative ΔG_{sys} we need a positive ΔS_{sys}

Figure 9.15 Summarising the messages of equation (9.9)

This last paragraph is only telling the widow's mite parable in alternative language (it costs the cold surroundings too much to 'finance' an endothermic change etc.) but it is helpful to have as many ways of picturing a thermodynamic concept as possible.

Ellingham diagrams

Equation (9.9) is a powerful predictive tool concerning whether a reaction goes or not, and it has been used in this role by the metal smelting industry. The equation is used to provide a graph of ΔG_{sys} against T for a whole series of oxidations, all of which show a straightforward reaction of a metal (or carbon, or hydrogen) with

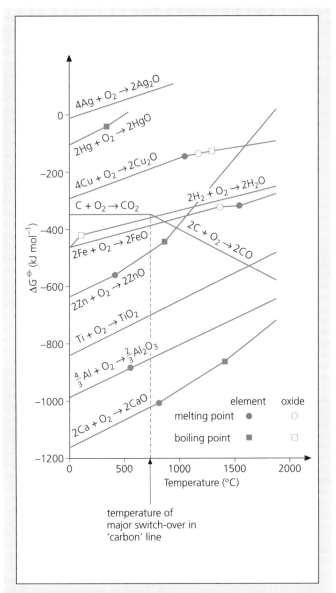

Figure 9.16 An Ellingham diagram

oxygen. A 'nest' of these lines all on the same axes is called an **Ellingham diagram**. A typical Ellingham diagram is shown in Figure 9.16.

Question

14 As mathematicians, you will know that the equation for a straight line is of the form $y = mx + c$. If a line on an Ellingham diagram is the graphical version of:

$$\Delta G_{sys} = \Delta H_{sys} - T\Delta S_{sys} \qquad (9.9)$$

can you see which items in (9.9) are playing the parts of y, m and x?

Having sorted out that the gradient of an Ellingham diagram line is the $-\Delta S_{sys}$ of the reaction, we can see why the lines have 'kinks'. A sudden shift of gradient must indicate a sudden shift of 'chaos factor' associated with a reaction.

Question

15 Can you think of a reason why a reaction might suddenly change its chaos factor at a particular temperature? Why does the line for the oxidation of zinc, for example, kink at around 420°C? (A look at the properties of zinc in a data book should provide a clue.)

You may be wondering why Ellingham diagrams are so valuable as to merit a section all to themselves. After all, they merely tell us that the vast majority of reactions of metals with oxygen have negative free energies, that the oxides are therefore stable at the temperatures on the x-axis in Figure 9.15, and that (excepting silver and mercury) you will fail to extract the metal by the use of heat on the oxide.

The really clever thing about Ellingham diagrams only emerges when two lines are considered *together*. Consider the 'carbon' line and the 'zinc' line, at 1200°C. The reactions and their free energies are as follows:

(*1*) $2Zn + O_2 \rightarrow 2ZnO$
$$\Delta G = -280 \text{ kJ mol}^{-1}$$

(*2*) $2C + O_2 \rightarrow 2CO$
$$\Delta G = -500 \text{ kJ mol}^{-1}$$

So taken individually both these reactions will take place, but CO is the stabler of the two oxides (has the more negative free energy). We can treat them like a pair of mathematical equations, using a technique developed earlier (Section 8.6). We will subtract (*1*) from (*2*):

$$(3) = (2) - (1)$$

This gives:

$2C + 2ZnO \rightarrow 2CO + 2Zn$
$$\Delta G = -220 \text{ kJ mol}^{-1}$$

So it emerges that this reaction passes the 'ΔG negative' test for the thermodynamic viability, and therefore that carbon *can* reduce zinc oxide to zinc at 1200°C. And this negative free energy can be seen on the diagram from the fact that the carbon line is *beneath* the zinc one.

So this is the true power of Ellingham diagrams. They allow quick predictions of the outcomes when any two of the reactions compete for oxygen. In all cases the reaction on the lower line can drive the upper reaction backwards, with the upper oxide giving its oxygen to the lower 'metal'. The carbon line has particular significance because, in the form of coke, it is the most readily available reducing agent for smelting. (The reason it switches product from carbon dioxide to carbon monoxide is because above 710°C carbon monoxide becomes the more stable oxide.) Ellingham diagrams can be 'read' for the following information.

- The temperature necessary for thermodynamic feasibility of a smelting reaction.

- Which reducing agent can perform the task.

- What free energy will apply to any given reaction.

Question

16 Use Figure 9.16 to answer these questions.

a Above which temperature does the smelting of iron with coke become thermodynamically feasible?

b What is the free energy of this reaction at 1200°C?

$$2C + 2FeO \rightarrow 2Fe + 2CO$$

c Why is aluminium oxide not smelted by coke, despite the fact that the carbon and aluminium lines cross? (Aluminium is actually extracted by electrolysis.)

9.9 Reactions in the balance – chemical equilibrium

So far the chapter has had as its central theme the *direction* of chemical change, with the main question couched in black and white terms – will the reaction go or not? The answer was equally black and white – if ΔS_{uni} was positive, or ΔG_{sys} was negative, then the answer was yes. However, you may have become aware from our work on the calcium carbonate decomposition reaction that it is possible to pick conditions so that a given reaction 'goes into reverse'. It is but a short step to realising that there must be a set or sets of conditions in which the function ΔS_{uni} (or its shadow ΔG_{sys}), has the value *zero*. What happens then?

Question

17 Revisit the data for ΔS and ΔH for the calcium carbonate/calcium oxide/carbon dioxide system, and find the value of T for which there would be a *zero* free energy change in passing from a mole of reactants to a mole of products.

Imagine the following simple chemical system whose ΔG_{sys} is indeed zero. The reactants and products are in their standard states, which for reagents in solution is a concentration of 1 mol dm^{-3}.

$$A(aq) \rightleftharpoons B(aq)$$
(solution volume = 1 dm^3)

This is like saying that either of the two eventualities, a mole of A, at a concentration of 1 mol dm^{-3}, or a mole of B, at a concentration of 1 mol dm^{-3}, is equally likely. Does this mean therefore that if we began with a mole of A there would be no change, since there is nothing to be gained by changing to B? It does not – and this is why. If there is nothing to be gained or lost in the transition from A to B, then there will still be one source of stimulus for change, and that is the law of random chance. Hence we should expect change to occur until the concentrations of A and B have reached 0.5 mol dm^{-3} each, for much the same reasons as those used to predict the behaviour of gas molecules in a room or coins in a multiple toss.

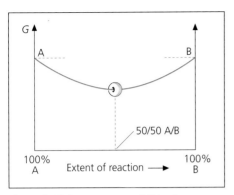

Figure 9.17 Free energy possibilities for a reaction in which $\Delta G = 0$. The 'bead' falls to the position of lowest free energy, corresponding to the equilibrium composition of the system

The situation is symbolised in Figure 9.17. What it amounts to is the statement that although there is no difference in the free energies of the standard state $(1.0 \, mol \, dm^{-3})$ versions of reactants and products, there is nonetheless a non-standard state (the mixture with each reagent at $0.5 \, mol \, dm^{-3}$) which is more likely (lower in free energy) than either of the two standard ones, and it is to this condition that the system will tend. Figure 9.17 uses the visual metaphor of the graph line being like a string on which is strung a bead. The equilibrium position of the bead is also the equilibrium mixture on the x-axis.

This example has the benefit of simplicity, but it is highly selective. Figures 9.18 and 9.19 show other chemical systems. The one in Figure 9.18 is at a temperature such that ΔG_{sys} is negative but small. In other words 'standard' B – one mole of B at $[B] = 1 \, mol \, dm^{-3}$ – is of

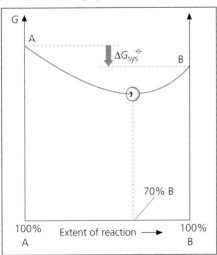

Figure 9.18 Equilibrium composition of a system in which ΔG_{sys} is small and negative

lower free energy than 'standard' A but not by much. The destruction of one mole of standard A, to be replaced by one mole of standard B, would therefore liberate this small amount of free energy.

However, as before, the system can do better than simply passing from one standard state to the other. There is still a chance-driven position of minimum free energy at which there is a mixture of reactants and products at non-standard concentrations. But this time there is an inbuilt bias, due to the inherent greater likelihood of products, so the 'bead rest-position' lies towards B.

In Figure 9.19, the bias is reversed, so that the mixture of minimum free energy contains more A than B. You will have realised from the general nature of these graphs that by putting a greater distance between the free energies of reactants and products, the 'bead' whose rest position represents the composition of the equilibrium mixture gets closer and closer to either reactants or products.

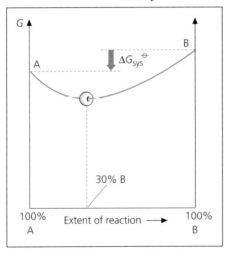

Figure 9.19 Equilibrium composition of a system in which ΔG_{sys} is small and positive

A small negative ΔG_{sys} can from now on be taken to indicate a reaction which 'goes a bit', but produces a mixture of reactants and products, in which there would be a preponderance of products. A small positive ΔG_{sys} can no longer be taken to mean the reaction is a 'non-goer', but instead that the reaction will go a little way, and come to rest with a mixture rich in reactants. When these small values of ΔG_{sys} apply, the reactions are described as 'reversible' – indicating that some degree of reaction would occur whether you began with

all-products or all-reactants, or else as 'equilibria', indicating that the rest position of the system is one of balance between reactants and products.

Despite the continuous range of possible values of ΔG_{sys} there are obviously reactions where it is far-fetched to talk about balance. Water, for example, is not in any meaningful state of equilibrium with dihydrogen and dioxygen, not at 298 K anyway. So most people use an arbitrary yardstick of plus or minus $60 \, kJ \, mol^{-1}$ to guide their use of language. Reactions whose ΔG_{sys} lie more negative than $-60 \, kJ \, mol^{-1}$ are said to 'go to completion', while those whose ΔG_{sys} values are more positive than $+60 \, kJ \, mol^{-1}$ are said 'not to go'. Those in between are the ones which get called reversible, or equilibria. Figures 9.20 and 9.21 represent the two extreme cases. But all intermediate situations are possible.

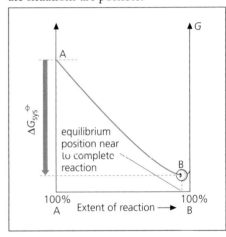

Figure 9.20 Equilibrium composition of a system in which ΔG_{sys} is big and negative

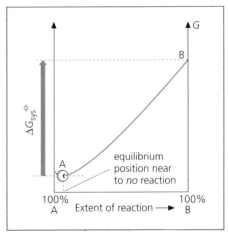

Figure 9.21 Equilibrium composition of a system in which ΔG_{sys} is big and positive

9.10 The equilibrium constant

All the graphs from 9.18 to 9.21 imply that there is a relationship between ΔG_{sys} and the composition of the equilibrium mixture. It appears that the more negative the ΔG_{sys}, the more product-heavy will be the equilibrium condition. The actual formal equation which represents this relationship is set out below:

$$\Delta G_{sys} = -RT \ln K \qquad (9.11)$$

The new function, K, is called the **equilibrium constant**. Its existence, its status as a constant, and its breakdown into concentration variables can all be derived, but this derivation is not included here as very few Advanced courses require it. So various facts about equilibrium constants are listed below, and you are asked to accept and learn them.

1 The equilibrium constant, K, expresses the composition of the equilibrium mixture. Specifically it is made up of a mathematical combination of the concentrations of the reactants and products as they exist at equilibrium. In general, for a reaction whose stoichiometric equation is:

$$a\text{A} + b\text{B} \rightleftharpoons c\text{C} + d\text{D}$$

the expression for K will be:

$$K_c = \frac{[\text{C}]_{eqm}^c \times [\text{D}]_{eqm}^d}{[\text{A}]_{eqm}^a \times [\text{B}]_{eqm}^b} \qquad (9.12)$$

Question

18 In the style of (9.12), write out the K_c expression for the following equilibrium:

$$2\text{SO}_2(g) + \text{O}_2(g) \rightleftharpoons 2\text{SO}_3(g)$$

2 The designation K_c is used whenever the concentrations are expressed in moles per dm^3, but any function that is equivalent to concentration will do. For gases, for instance, there is the option of expressing concentrations as 'partial pressures', in which case the K function is called K_p.

3 The equilibrium constant seems to be applicable to *all* situations in which the named chemicals interact. In other words, you do not have to start with reactants in their stoichiometric ratios, and neither do you have to start at 1 atmosphere or 1 mol dm^{-3}. For instance, suppose there is a reaction:

$$\text{A}(g) + 3\text{B}(g) \rightleftharpoons 2\text{C}(g)$$

You can start with an A : B ratio of 1 : 3 or 1 : 30, and you can conduct the reaction in an open vessel at 1 atmosphere or a sealed steel vessel at 20 atmospheres. In every case you end up with some combination of equilibrium concentrations, and despite the individual concentration differences, the function

$$\frac{[\text{C}]_{eqm}^2}{[\text{A}]_{eqm} \times [\text{B}]_{eqm}^3}$$

would be a constant – in fact, it would be *the* constant, K_c.

4 One meaning that resides in (9.11) is that if ΔG is big and negative, then K is big. But for K to be big, the top line in (9.12) must far exceed the bottom line. This in turn means that the equilibrium mixture is dominated by products. Linking all these factors together gives us the fact that a large negative ΔG indicates a big K and a reaction that has gone near to completion (which is exactly the prediction we'd have made from the bead-on-a-string diagrams). Conversely, a large positive ΔG means a small K and a reaction whose equilibrium mixture is dominated by reactants.

5 Equilibrium constants do not quite live up to their name in all situations. To see why not, remember that equilibrium constants are tied to free energies by (9.11), and that free energies are tied to temperatures by (9.9). So equilibrium constants are variable functions of *temperature*. This point is revisited when we look in detail at how to manipulate equilibria for our own ends.

6 Equilibrium constants can only feature reagents whose presence in a reaction system can be expressed in concentration-type units. Consider the familiar calcium carbonate decomposition equilibrium:

$$\text{CaCO}_3(s) \rightarrow \text{CaO}(s) + \text{CO}_2(g)$$

The molar entropies of the two solids cannot vary very much, since their 'chaos factors' would only be seriously affected by a change of state. However, the entropy of the carbon dioxide is a function of its pressure. (Look back to (9.3) to remind yourself of the relationship.) This means that ΔG, and therefore K, is only sensitive to the partial pressure of CO_2. The equilibrium constant for this reaction is rather peculiar looking:

$$K_c = [\text{CO}_2(g)]_{eqm}$$

or

$$K_p = p_{\text{CO}_2, eqm}$$

In general, any separate phase in an equilibrium mixture – any ingredient that cannot mix with the other ingredients and therefore can only exist 'in itself' – will be missing from the equilibrium constant expression. (For example, the concentration of calcium carbonate in lumps of calcium carbonate is more or less invariable.) Equilibria featuring such ingredients are known as **heterogeneous**.

Those equilibria in which all ingredients can have variable concentrations within the same phase, such as in the case of a group of reagents all in the same solution, or all in the gas phase, are called **homogeneous**.

9.11 A dynamic model of chemical equilibrium

The reason for the existence of an equilibrium constant is best shown using a dynamic model of equilibrium. This juxtaposition of words seems a little odd, since 'equilibrium' implies stability and stasis, while 'dynamism' is all about movement. Equilibrium is rather like a swan holding position in fast water – all looks serene on the surface, but under the water its legs are paddling frantically to push itself forward exactly as fast as the water runs backward.

The following experiment is of the type which first revealed what was going on 'under the surface' of a system in chemical equilibrium. Propanone, CH_3COCH_3, as solute has an equilibrium distribution between two immiscible solvents, trichloroethane (abbreviated to TCE) and water. The left-hand side of Figure 9.22 shows how the propanone is distributed over the two layers. For systems such as this, the equilibrium constant is an experimental fact, and is of the form:

$$K_c = \frac{[CH_3COCH_3(aq)]_{eqm}}{[CH_3COCH_3(TCE)]_{eqm}}$$

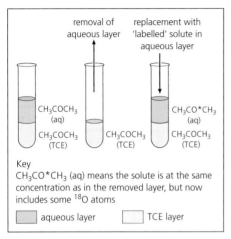

Key
$CH_3CO{*}CH_3$ (aq) means the solute is at the same concentration as in the removed layer, but now includes some ^{18}O atoms

Figure 9.22 Proving that equilibrium is a dynamic condition

In other words, there is a *fixed ratio* of the two concentrations. The top layer is then removed and replaced with another layer of identical concentration, but whose propanone molecules have been enriched in ^{18}O atoms. After some time, the propanone in the bottom layer is analysed by mass spectrometer, to test for the presence of any of the ^{18}O-containing molecules.

Question

19 Given the above situation, what results would you expect if:

a Equilibrium was a truly static condition, even on a molecular level?

b Equilibrium was a static state with regard to bulk concentrations ('on the surface'), but was constantly changing on a molecular level? (This is actually what turns out to be true.)

c Explain how molecules could be constantly passing from one layer to the other, without affecting the bulk concentrations.

A mathematical model of a dynamic equilibrium and a new meaning for 'K'

We will now analyse a system like the one in question 19 in which we can represent the reaction as:

$$A \rightleftharpoons B$$

The use of the double-headed arrow symbol is in keeping with our belief in the model of a dynamic equilibrium. It represents the fact that As are turning into Bs and Bs are turning back to As. But it carries the extra message that, at equilibrium, the rates of the two reactions are the same.

Now the number of As turning into Bs is proportional to the number of As at any given time, because the more As there are the greater are the chances of a transition. And similarly the number of Bs turning into As should be proportional to the number of Bs at any time (Figure 9.23). These ideas, turned into mathematical equations, look like this:

1 Rate of disappearance of A = $k_f[A]$
2 Rate of disappearance of B = $k_b[B]$

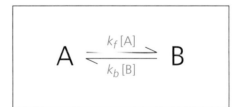

Figure 9.23 'Opposing reactions' picture of equilibrium

The f and b suffixes indicate the forward and backward reactions, and k is called a **rate constant** (capital K is reserved for the equilibrium constant). The equations show that if you start off with nothing but A, the rate of disappearance of A is at its fastest at the very beginning, whereas the rate of disappearance of B at that time is 0. As time passes, the rate of disappearance of A slows down, while that of B picks up. But the crucial thing about these two equations and the reactions they describe, is that they are *not independent of each other*. 'The rate of disappearance of B', for example, is the same as 'the rate of appearance of A' (since every lost B becomes an A). So we can restate 1 and 2 as:

3 Rate of disappearance of A = $k_f[A]$
4 Rate of appearance of A = $k_b[B]$

Up to now, [A] and [B] have stood for concentrations at any time within the course of the reaction, but now we look at the special circumstances at equilibrium, reached at time t_{eqm}. This is the time when [A] and [B] stop changing, because they are being made as fast as they disappear. They take up their specific t_{eqm} values, $[A]_{eqm}$ and $[B]_{eqm}$. But it is also the time when the expressions in 3 and 4 become equal, since appearance and disappearance of A are happening at the same rate.

So,

$$k_f[A]_{eqm} = k_b[B]_{eqm}$$

Rearranging, we get

$$\frac{[B]_{eqm}}{[A]_{eqm}} = \frac{k_f}{k_b}$$

We can see that the ratio of [A] : [B] at t_{eqm} will be equal to the ratio of rate constants of forward and back reactions. A diagrammatic retelling of this story is shown in Figures 9.24a and 9.24b, in which an actual bias in favour of B has been assumed. Each diagram tells the

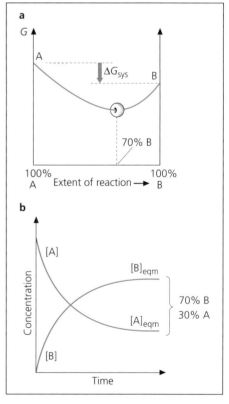

Figure 9.24 Two views of the approach to equilibrium: **a** Shows the 'minimum G' picture and **b** Shows a 'kinetic' approach, from 'time 0' to 'time (eqm)'

story with a different slant: Figure 9.24a shows how B is the more likely state for the system, as reflected by the negative ΔG_{sys}, so the equilibrium mixture is dominated by it. In Figure 9.24b we see that [B] has gone higher than [A] before they reach their equilibrium values. This is because in order to have the rate of destruction of B equal to the rate of destruction of A it has been necessary to have more of B, the 'reluctant' decomposer.

But there is another strong link between the approaches of Figures 9.24a and 9.24b. The dynamic approach has predicted that the expression $[B]_{eqm}/[A]_{eqm}$ will be equal to the ratio of two constants, which will itself be a constant. Meanwhile remember that in our previous work on ΔG and K (p. 177), we found equation (9.12), which gives another meaning to $[B]_{eqm}/[A]_{eqm}$, namely:

$$\frac{[B]_{eqm}}{[A]_{eqm}} = K_c,\ \text{the equilibrium constant}$$

So

$$K_c = \frac{k_f}{k_b} \qquad \text{(from above).}$$

So K_c is just the ratio of the forward and backward rate constants – and the use of a 'dynamic, two-opposing-reactions' model for equilibrium systems has predicted the existence of the same constant that shows up in experimental results and which can be derived from minimum free energies.

This mathematical model is so far based on a very simple system – how can it be extended to more complex systems? In a reaction of the type:

$$A + B \rightleftharpoons C + D$$

Rate of forward reaction = $k_f[A] \times [B]$

(This means that if either [A] *or* [B] alone is doubled, the chances of AB collision doubles, so doubling them both quadruples the chances of collision.)

Similarly,

Rate of reverse reaction = $k_b[C] \times [D]$

At equilibrium, the two rates are equal.

Question

20 Finish off this calculation in the style of the earlier one, ending with an expression for the equilibrium constant in terms of the four concentrations.

Does it agree with the 'formula' version of the equilibrium constant for this system as given by equation (9.12)?

The two-opposing-reactions (dynamic) model shadows and parallels the minimum-free-energy model of equilibrium. Sometimes they are alternative ways of telling the same story, and at other times one gives a clearer picture than the other. For example, we have seen that both models predict the existence, and indeed the mathematical structure, of the equilibrium constant.

One excellent aspect of the dynamic model of equilibrium is that it enables us to predict how such systems would behave if interfered with, as we shall see in the next section.

9.12 Manipulating equilibrium systems for our own ends

You might feel you have been here before, since Section 9.7 had a very similar title, also concerned with 'manipulation' and 'ends'. The difference is this: at that stage in the chapter we had a simple black-and-white view that reactions either happened or did not happen (depending on the sign of the ΔS_{uni}). Now we are not just concerned as to whether a reaction happens, but how far it goes in which direction, and how big its equilibrium constant K_c is. We need a correspondingly more complex set of explanations. However, not everything will be new to you – in particular the discussion of how temperature affects the position of equilibrium is essentially a repetition of familiar principles.

We will now propose a number of possible events that might overtake a system quietly sitting in equilibrium, and use one or other of our intellectual models to predict outcomes.

Note: Some students may find what follows rather gruelling, based as it is on quite difficult, mostly quantitative, mental modelling. You may therefore like to know that at the end of all the maths there is waiting for you something called **the principle of Le Chatelier**. It summarises in one non-mathematical sentence all the situations in the next few

pages. It is a weaker account than the following analyses, since it is based purely on experience and observation rather than the fundamentals of thermodynamics, but you may well end up thinking that it is none the worse for that. If, for now, you want to skip the maths and get to this one simple idea, it is on p. 183.

Event 1: extra reactant A is added

Model: Two opposing reactions (dynamic model)

System: $A(aq) + B(aq) \rightleftharpoons$
                  $C(aq) + D(aq)$

The addition of extra A will have the effect of diluting all the other ingredients. So while [A] might go up, [B], [C] and [D] will certainly go down.

Question

21 Here are some easy 'dummy' figures: suppose all the ingredients at t_{eqm} just happen to be at $1\ mol\ dm^{-3}$, all in the same $1\ dm^3$ volume. (This is equivalent to assuming a K_c of 1.) Now suppose an extra $1\ dm^3$ of $1\ mol\ dm^{-3}$ A(aq) is added.

a What is the total volume of the new mixture?

b What are the new concentrations of the four ingredients just after this addition? (Concentration = moles/volume (in dm^3).)

c If the rates of forward and back reactions are given by $k_f[A][B]$ and $k_b[C][D]$ respectively, which rate will be the more profoundly slowed?

d Why is the system no longer in equilibrium, and how can it get back there?

Event 1 again: extra reactant A is added

Model: 'Constant K_c'

System: $A(aq) + B(aq) \rightleftharpoons$
                  $C(aq) + D(aq)$

The approach throughout this book is to help you use a range of mental models, even if they address the same problem. However, you may think that two models (the two-opposing-reactions (dynamic) model and the minimum-free-energy model) to describe equilibrium systems is enough. Here is a third approach, with the justification that it works very well, and does not require any new ideas. The model simply assumes that the K_c value for a reaction is a constant for all situations, and that the system, if disturbed, will act to try to maintain that value. (So it only applies at constant temperature, because otherwise K_c itself would vary.)

Use the 'dummy' figures from question 21, in which the ingredients are at equilibrium at concentrations of 1 mol dm^{-3} in a system of total volume 1 dm^3.

Question

22a State the expression for K_c for this reaction, along with its units.

b Work out a value for K_c from these data.

c Now add that extra 1 dm^3 of A(aq), of concentration 1 mol dm^{-3}. You have already calculated the new temporary concentrations of the four ingredients in question 21b, so put them into the expression for K_c. Work out the result.

d The answer to (c) cannot really be called a K_c, because that name only refers to expressions made up from *equilibrium* concentrations, and as you can see the system is now out of equilibrium. What could the chemical species in the system do to restore the true value of K_c?

(Look at the parallels between this approach and that of question 21. Both models agreed that the system was out of equilibrium. One model (in question 21) saw the problem as being that the forward reaction was going too fast, while in this question the problem is seen as the bottom line of the K_c function being too big. But on either view, what must happen next is a partial destruction of A and B. The merit of the constant-K_c approach is that it is quantitative, and so we can see exactly how much destruction of A and B must occur.)

e Let the *change* in concentration of A necessary to restore the true value of K_c be $-x \text{ mol dm}^{-3}$. What would the *changes* of [B], [C] and [D] be during the same process, bearing in mind that the stoichiometry of the reaction is $-1:-1:+1:+1$? (So for every 1 mole of A destroyed, 1 mole of B would go too, and 1 mole of each of C and D would be created.)

f What would the *new* equilibrium values of [A], [B], [C] and [D] be, in terms of numbers and values of x?

g If these concentrations are values which obey the true K_c from answer (b), put them all in the expression for K_c along with answer (b), and solve for x. Finally quote the new equilibrium values of [A], [B], [C] and [D].

Having reached the end of the question, you should have an answer for the final concentration of B of 0.4 mol dm^{-3}, as opposed to the 1 mol dm^{-3} before the extra addition of A. Let us put this problem into an industrial/financial context. Suppose that B is an expensive ingredient and A is a cheap one. Suppose also that the real target of the process is C. Clearly the financial objective must be to achieve the highest percentage conversion of B to C, with A and D less important. We will now analyse the two equilibrium positions before and after adding the extra A.

Question

23a In the first equilibrium situation, pre-extra-A, there was 1 mole of each species in 1 dm^3 total volume. Bearing in mind the $-1:-1:+1:+1$ stoichiometry, how many moles of A and B must there have been at the start, when there was *only* A and B?

b What percentage conversion of moles of B had been achieved at this point?

c What if the final post-extra-A molarity of B was 0.4 mol dm^{-3}, and the total volume was 2 dm^3 – how many moles of B would there be at the end?

d What percentage conversion does this represent of the total B at the very start?

e So was the extra-A addition a good idea financially, and should it have been done right at the start?

Event 2: product removed as soon as it is produced

Model: Two opposing reactions

System: $A(g) + B(g) \rightleftharpoons C(g) + D(g)$

Conditions: All the reactants are in the gas phase, and product C has a higher boiling point than the others

The fact that product C has a higher boiling point means that it can be liquefied while the others are still gases. When (or even before) equilibrium has been reached, the mixture is passed through a condenser, and C alone is removed. Then the remaining reagents are passed back to the reaction chamber, where they are heated back up to optimum reaction temperature, maybe in the presence of a catalyst. The condense/react cycle is repeated.

Question

24a Which reaction, forward or back, will be affected by the removal of C?

b If the system has now been knocked out of equilibrium, how will it respond, when back in the reaction chamber?

c What will be the effect of repeating the react/remove-C cycle over and over again?

d Most industrial reactions of this type are conducted differently, as a constant flow process, in which fresh A and B are pumped in on each repetition. (Figure 9.25.) Why is this an improvement in practical terms on the method outlined originally?

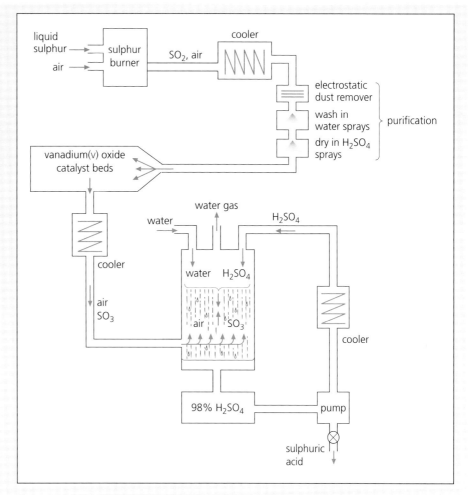

Figure 9.25 Flow diagram of the Contact process

Event 3: pressure increased

Model: Constant K

System: $A(g) + B(g) \rightleftharpoons C(g)$

Conditions: All the reactants arc gases

Equilibrium is established in a syringe-like vessel of 1 dm^3 capacity. The equilibrium concentrations of the three gases are all 1 mol dm^{-3}. (In a sample calculation it is useful to keep the figures easy.) The syringe is now pushed in and the volume of the vessel is halved.

Remember that equilibrium constants are in terms of concentrations in mol dm^{-3}, so we could re-express the K_c for this system as:

$$K_c = \frac{(c/V)}{(a/V) \times (b/V)}$$

where the lower case letters represent the number of moles of A, B and C, V stands for volume, and all values refer to t_{eqm}.

Question

25a Give the above expression in its simplest algebraic form, by cancelling Vs.

b Notice that there is still a V surviving as part of the equilibrium constant. What is the value of K_c, from the data given above?

c What will the false value of K_c be immediately after the increase of pressure has caused a halving of the volume (and before the molecules have had time to change)?

d In what general direction will changes have to occur in the number of moles of each reagent, to restore the equilibrium constant to its true value?

It is worth considering a few of the lessons from the last example. First, the system had to be chosen carefully. With $A + B \rightleftharpoons C + D$, there would have been no effect on the equilibrium mixture, irrespective of what happened to the volume/pressure.

Question

26 Why not?

Second, with an aqueous solution medium for the system, even if it had been A + B $\rightleftharpoons$ C, there would have been no pressure-related shift of moles.

Question

27 Why not? From these last two answers, can you generalise to define the types of system for which a pressure change matters? Which direction of the two possible movements of a system in equilibrium would benefit from a pressure increase?

Third, we see that there is no unique *unit* for K_c – that it depends on the reaction being studied. This system is the first one which has actually had units.

Figure 9.26 The manufacture of ammonia at high pressure in the Haber process

Question

28 What are they?

Fourth, it is very confusing for students to read about equilibria 'shifting' as a result of pressure changes, and yet there being no change in K_c. How can equilibria shift to the right or left or whatever, and moles of this or that be destroyed, and yet all the time the K_c stays *constant*? The 'shift' arises because volume (which is linked up with pressure in gas systems) is actually hidden inside the expression for K_c (p. 177) – at least, it is when there is an unequal number of moles of gas on each side. The shift is needed for the very purpose of restoring the true value of K_c.

Fifth, in an industrial context, the use of pressure changes can be just as important as removal of the product, as a way of maximising yields of desired products from equilibrium reactions (Figure 9.26).

Questions

29 Why would increased pressure have been of financial benefit to the firm running the A(g) + B(g) $\rightleftharpoons$ C(g) system? What counterbalancing drawbacks might there be about operating chemical reactions at high pressure?

30 Now try to tell the story of increasing the pressure on the A(g) + B(g) $\rightleftharpoons$ C(g) system in the alternative language of two opposing reactions – the dynamic model (p. 178).

Event 4: temperature changed in an exothermic system

Model: Entropy analysis, including

$$\Delta G_{sys} = \Delta H_{sys} - T\Delta S_{sys} \qquad (9.9)$$
$$\text{and } \Delta G_{sys} = -T\Delta S_{uni} \qquad (9.8)$$

System:
$$A(g) + B(g) \rightleftharpoons C(g)$$
$$\Delta H_{sys} = -100 \text{ kJ mol}^{-1}$$
$$\Delta S_{sys} = -200 \text{ J K}^{-1} \text{ mol}^{-1}$$

Conditions: All reactants are gases

At 500 K the entropy change in the surroundings is given by $-(-100\,000)/500 = +200$ J K^{-1} mol^{-1} (equation (9.6)). So at this temperature, ΔS_{uni} is just zero. The bar diagram in Figure 9.27 shows these entropy changes.

Question

31 Give the corresponding values for ΔG_{sys} and K_c at 500 K for this reaction.

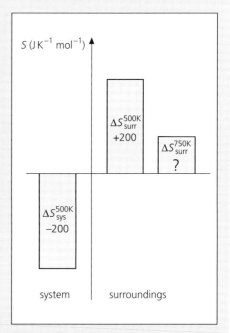

Figure 9.27 Entropy analysis method for predicting the effect of changes of temperature on a 'perfectly poised' equilibrium. Compare and contrast this situation with that described in Figure 9.11

Clearly the reaction at this temperature is poised like the bead on the necklace in Figure 9.17, hanging symmetrically.

Questions

32 Assuming ΔH_{sys} and ΔS_{sys} are not greatly affected by temperature changes, work out the effect of *increasing* the temperature on the composition of the equilibrium mixture. Use equations (*9.9*) and (*9.11*) to give a value for *K* at 750 K. Superimpose the new ΔS_{surr} on a copy of Figure 9.27. Has the yield of product at 750 K gone up or down, compared with 500 K? Why might an industrial operator, faced with such a system, be reluctant to shift to temperatures at which yields were at a maximum? (*Note*: it might help to re-read Section 9.7, p. 17, concerning the matter of kinetic viability.)

33 How will the picture change if the reaction is endothermic, and ΔS_{sys} is positive? Answer your question by sketching a version of Figure 9.27 which applies to such a system, and draw in values for the entropy of surroundings at at least two temperatures. Can you generalise to say something about the direction in which all equilibria shift when their temperatures are increased?

Notice how close the similarity is between this last piece of work and the cement works item in Section 9.7 and Figure 9.11. However, now we are not just talking about whether or not the reaction will go at certain temperatures, but about the *size* of the equilibrium constant.

The formula showing *K* as a function of temperature is:

$$\ln K = -\frac{\Delta H_{sys}}{RT} + \text{constant} \qquad (9.13)$$

This can be rearranged in a version which is convenient for finding *K* at one temperature from *K* at another:

$$\ln\left(\frac{K_1}{K_2}\right) = \frac{\Delta H_{sys}}{R}\left(\frac{1}{T_2} - \frac{1}{T_1}\right) \qquad (9.13a)$$

Question

34 An optional exercise for keen mathematicians is to derive (*9.13*) from (*9.9*) and (*9.11*), and derive (*9.13a*) from (*9.13*).

The principle of Le Chatelier

You were promised at the start of this section that there would eventually be a one-sentence model which covered all the above situations and arrived at all the same conclusions, at a tiny fraction of the cost in mental stress. The sentence came from a French scientist called Henri Le Chatelier, who stated in 1888 that:

Any change in one of the variables that determines the state of a system in equilibrium causes a shift in the position of equilibrium in a direction that tends to counteract the change in the variable under consideration.

The language of the original statement is a little opaque, so let's paraphrase; Le Chatelier was recognising a quality of 'stubbornness' in equilibrium systems – a tendency to oppose everything you do to them. So, if you pressurise them, they shift to reduce the pressure; if you make them hotter, they shift to absorb the heat; if you take an ingredient away, they make more of it; if you add an ingredient, they destroy some.

Let us see how the principle works on the 'temperature-change' system (p. 182), and confirm that it reaches the same conclusions.

Question

35 Consider the equilibrium:

$$A(g) + B(g) \rightleftharpoons C(g)$$

which we will assume is exothermic in the forward direction. Use Le Chatelier's principle to predict how the equilibrium would act under the following circumstances, and show your reasoning:

a The external pressure is increased.

b The temperature is increased.

c Ingredient C is removed as soon as it is made.

In the 'entropy conscious' Advanced-level syllabuses, you can use Le Chatelier's principle as a 'backstop' to the more mathematical models for giving a quick qualitative preview of the quantitative answer. In the other syllabuses, it is probably the only tool you will need for the analysis of equilibria.

9.13 The Contact process

Le Chatelier's principle may not be able to go deeply into why certain equilibria behave as they do, but it is good for quick insights into the operation of industrial equilibria, as we shall see. The production of sulphuric acid, in terms of sheer tonnage, is the human race's most significant chemical reaction (Figure 9.28). (The following data are taken from ICI's *Steam* magazine, edition 1.) At the heart of the process (look back to Figure 9.25 for the flow chart) lies this gaseous equilibrium:

$$2SO_2(g) + O_2(g) \rightleftharpoons 2SO_3(g)$$
$$\Delta H = -98 \text{ kJ mol}^{-1}$$

(It is this stage, catalysed as mentioned below, which bears the name 'Contact process'.)

Figure 9.29 shows the variation with temperature of the percentage conversion of SO_2 to SO_3 (the curved, green line). The degree of conversion falls off sharply at temperatures above 900 K, even though both forward and back reactions are faster.

Question

36 This fact was predictable from Le Chatelier's principle, once it was seen that the ΔH_{sys} for the forward reaction was *minus* 98 kJ mol^{-1}. Can you explain?

Figure 9.28 A sulphuric acid plant

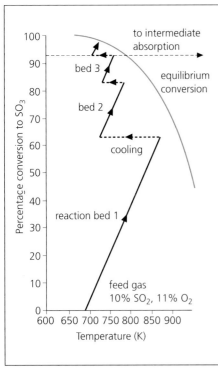

Figure 9.29 A typical % conversion against temperature equilibrium curve. The thick lines represent non-equilibrium states which exist in the actual reaction beds

The thick black lines represent the path taken by the reactants over time, from their first introduction into the catalyst vessel. Each upward thrust represents the course of the reaction within a particular bed of the $V_2O_5/SiO_2/Na_2SO_4$ catalyst, while the horizontal lines are cooling stages between beds. As time passes the mixture gets nearer and nearer to its equilibrium composition (that is, the thick lines push up nearer to the curve).

Question

37 The 'feed gas' is a mixture of SO_2 and air, in which the ratio of SO_2 to oxygen is 10 : 11 by mole.

a Show that this mixture contains an excess of oxygen compared to the stoichiometric mixture.

b Use Le Chatelier's principle, or any other model, to show what the extra oxygen does.

c Why does the reaction mixture get hotter during its time in each catalyst bed?

d Why is it a good idea to interrupt the reaction with cooling stages?

ICI operate a 'double absorption' process in which, when the conversion of SO_2 to SO_3 reaches 93%, the SO_3 is removed into aqueous solution. The remaining unreacted SO_2 and O_2 are put through one last catalyst bed, which has the effect of 'squeezing out' one last bit of conversion. These events show up on Figure 9.29 as the thin horizontal line marked 'to intermediate absorption', which is not meant to indicate a temperature rise, but merely to show the time and composition at which the removal takes place. Figure 9.30 focuses on what happens after this intermediate absorption, and its two y-axes indicate simultaneously the progress of conversion in the 'start-again' mixture, and the overall conversion from both stages. The second absorption occurs when the over-

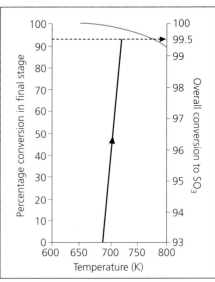

Figure 9.30 A conversion–temperature curve for the double absorption process – final stage

all conversion of SO_2 to SO_3 has reached 99.5%.

Question

38a Explain, using Le Chatelier's principle, why it is a good idea to remove the product SO_3 before the final attainment of equilibrium.

b The gases are maintained at 5–7 atmospheres during their time in the catalyst beds. Use Le Chatelier's principle to explain why this improves percentage conversion.

c Rather astonishingly the 'aqueous liquid' which is used to absorb the SO_3 gas is 98% sulphuric acid, which is allowed to concentrate up to 98.5%. It is then diluted back to 98% and so a little bit of extra sulphuric acid comes into the world. Why is ordinary water not used as the absorbing medium?

(*Note:* The book contains one other detailed study of 'equilibrium management' in an important industrial process. It is in Section 18.9, and is about the Haber process for making ammonia. It is placed in chapter 18 as a late revision of themes in this chapter, but could equally be referred to now. Most exam boards use one or both of these processes as named examples of industrial equilibria, on which questions may be set.)

9.14 Calculations involving equilibrium constants

Le Chatelier's principle cannot put a number to a specific equilibrium constant, nor can it tell you the expected composition of an equilibrium mixture. For these things we must return to rigorous mathematical analysis.

Finding experimental equilibrium constants

So far the examples, if they have been numerical at all, have had the simplest sample values. In a real situation the equilibrium concentrations would have to be found by experiments. Fortunately life is made a little easier by logical deduction. You might think that to get at an equilibrium constant like:

$$K_c = \frac{[C]_{eqm}[D]_{eqm}}{[A]_{eqm}[B]_{eqm}}$$

it would require four separate attempts to find the four equilibrium concentrations. However, it requires just *one* determination, plus a knowledge of what you began with. An example will make things clearer.

Consider the system:

$$3A(aq) + B(aq) \rightleftharpoons 2C(aq)$$

Suppose $1 \, dm^3$ of $1 \, mol \, dm^{-3}$ A was mixed with $1 \, dm^3$ of $1 \, mol \, dm^{-3}$ B. At t_{eqm} [C] was found to be $0.2 \, mol \, dm^{-3}$. What is K_c?

A suggested methodology for solving this problem is a table featuring a row each expressing the composition of the system at the two times t_0 and t_{eqm}. The numbers in it are all moles rather than molarities:

	3A	+	B	⇌	2C
Moles at t_0	1		1		0
Moles at t_{eqm}	−0.6 ↓ ?		−0.2 ↓ ?		+0.4 ↓ 0.4*

* because the total volume is 2 dm³, and [C]$_{eqm}$ = 0.2 mol dm⁻³

We need to fill in the question marks. The 'logic' referred to earlier centres around this fact: even in equilibria, and even when the starting mixture does not match the stoichiometry, the stoichiometry still has a part to play. Specifically in this case it is true that the mole ratio (A destroyed : B destroyed : C created) is

still −3 : −1 : +2. So if 0.4 mol of C is created, then:

A destroyed = $\frac{3}{2} \times 0.4 = 0.6$ mol

B destroyed = $\frac{1}{2} \times 0.4 = 0.2$ mol
(see arrows in the previous table)

So, moles of A at $t_{eqm} = (1 − 0.6) = 0.4$ and moles of B at $t_{eqm} = (1 − 0.2) = 0.8$

Therefore $[A]_{eqm} = \dfrac{0.4}{2} = 0.2 \, mol \, dm^{-3}$

and $[B]_{eqm} = \dfrac{0.8}{2} = 0.4 \, mol \, dm^{-3}$

Finally, putting all the equilibrium concentrations into the K_c expression gives us:

$$K_c = \frac{[C]_{eqm}^2}{[A]_{eqm}^3 \times [B]_{eqm}}$$

$$= \frac{0.2^2}{0.2^3 \times 0.4} = 12.5 \, mol^{-2} \, dm^6$$

Here is a similar problem for you to try.

Question

39 Esters are compounds, often sweet-smelling, whose general formula is:

R—C(=O)—O—R' , where R and R' are normally hydrocarbon chains. A certain ester can be obtained from a plant extract. It can be hydrolysed, assisted by an inorganic acid to supply H^+ ions, to produce a valuable alcohol, used in the manufacture of a drug. The reaction is −1 : −1 : +1 : +1, as shown:

$$\text{ester} + \text{water} \underset{\text{catalyst}}{\overset{H^+}{\rightleftharpoons}} \text{alcohol} + \text{acid}$$

0.15 moles of the ester are mixed with 15 g of water at t_0 and 2 cm³ of concentrated hydrochloric acid are added to catalyse the reaction. At t_{eqm} it takes 64 cm³ of 2 mol dm⁻³ sodium hydroxide to titrate the two acids in the equilibrium mixture (the created one and the catalyst acid that had been there unchanged throughout). *Assumptions:* Concentrated HCl is 10 mol dm⁻³, and 2 cm³ contains 1.5 cm³ water, which will add to the water used as reagent. The organic acid contains 1 acidic hydrogen atom.

a Calculate the volume of the titration if *only* the catalyst acid had been titrated – that is, how much 2 mol dm⁻³ alkali would 2 cm³ of 10 mol dm⁻³ HCl need?

b Hence deduce the volume of alkali which was neutralising the genuinely new organic acid, and thereby calculate the number of moles of that acid present at equilibrium.

c Set out a table, as before, and present all the information that you know so far, as moles of reagents either at t_0 or at t_{eqm}.

d Deduce the remaining numbers of moles of participants at t_{eqm}. Call the volume of the mixture V and express the amounts of the four as concentrations. (The volumes are only in there so you can see why the volume never needed to be known.)

e Put all the concentrations into the expression for K_c, and work it out, with relevant units.

f Calculate the percentage conversion of the ester by comparing the 'moles converted' figure with the 'moles at t_0'. (You will need this answer for question 41.)

Using equilibrium constants to predict outcomes

Once an equilibrium constant is known, it is valid for all pressures and starting mixtures, at that particular temperature. For example, knowledge of the ester/water equilibrium constant from question 39 would enable us to predict the effects of varying the ratio of ester : water used at the start.

Questions

40 Using Le Chatelier's principle, rather than a mathematical analysis, and giving your reasoning, say what would be the effect of increasing the water : ester ratio, and therefore what would be the most economical way of ensuring maximum conversion of the expensive starting material.

41 Now for the calculation. It will involve a quadratic equation, and your teacher may well be able to tell you if your syllabus requires such extremes of maths skill. If not, go straight to the answers. If you're still reading, let's carry on. Use the equilibrium constant you calculated in question 39. Assume the volume of water at the start had been doubled (with no other change). In other words, at t_0 there is to be 0.15 moles of ester and 30 g of water. Follow this format:

a Let the number of moles of acid and of alcohol formed by t_{eqm} be x. Set up a 'moles at t_0/t_{eqm}' table as before, and thereby work out the number of moles of all participants at t_{eqm} in terms of numbers and x.

b Go through the motions, for correctness' sake, of converting these to concentrations, using a volume variable V (which should cancel), and then set up the K_c expression, using the actual value of K_c which you found in question 39.

c Rearrange this expression to the classic quadratic format, and solve for x.

d Work out how many moles of ester remain at equilibrium, and thereby give a figure for its percentage conversion. Compare this with the corresponding figure for question 39. Would you recommend starting with the increased amount of water?

e What practical limit might stop you from recommending going to extreme ratios of water : ester?

As a postscript to these two big ester questions, consider how it was we were able to titrate the equilibrium mixture at all. After all, titration is the equivalent of removing a member of the equilibrium system (in this case a product). Our earlier models would predict that the system would frustrate our efforts by making more acid.

42 What lucky factor in this case enables us to finish the titration before nature has a chance to frustrate us?

43 Can you think of any non-intrusive methods for assessing the concentration of one or more ingredients in an equilibrium mixture – that is, methods which do not involve any removal or interference?

Solubility product

Data books do not normally list equilibrium constants, because of the enormous number of possible reactions which would have to be covered. The student is expected to extract ΔG data for the reaction by Hess's Law cycles, and derive K_c from that. But one set of equilibrium constants which is readily available is the set of **solubility products.** These are equilibrium constants of equilibria set up between a crystalline solid and its aqueous ions. Naturally these equilibria belong to the heterogeneous group, and conform to the pattern that heterogeneous 'non-mixable' phases (in this case the solids) do not appear in the K_c expression. In fact the K_c is referred to symbolically as K_{sp} in this field.

As an example let us consider iron(II) hydroxide. The full chemical equation is:

$$Fe(OH)_2(s) \rightleftharpoons Fe^{2+}(aq) + 2OH^-(aq)$$

Yet the expression for K_{sp} is only:

$$K_{sp} = [Fe^{2+}(aq)]_{eqm} \times [OH^-(aq)]_{eqm}^2$$

The value of the solubility product is given as 6×10^{-15} mol^3 dm^{-9}. One of the manipulations which can be done on a solubility product is interconversion with the normally defined solubility, in mol dm^{-3}.

Question

44 Let the solubility in mol dm^{-3} of $Fe(OH)_2$ be x.

a State the concentrations of $Fe^{2+}(aq)$ and $OH^-(aq)$ in terms of x, remembering that the stoichiometric ratio for the dissolution process is $-1 : +1 : +2$.

b Put these two expressions into the formula for the solubility product

(above) and using the data book value for K_{sp} solve for x.

That was a rather academic exercise, especially since you can often find solubilities in the same data book as solubility products. But we can turn our knowledge to more applied areas. The principle we must bear in mind is that as soon as the value of K_{sp} is exceeded, the solid will begin to precipitate, until the ion concentrations again obey the true value. This idea covers all situations in which the two ions might meet, not just those in which a portion of $Fe(OH)_2$ has been put in water. One of the places in which Fe^{2+} ions are likely to meet OH^- ions is in soil water.

Gardeners find that certain types of plant are unable to acquire enough $Fe^{2+}(aq)$ (Figure 9.31). The problem is particularly acute on alkaline soils, and prevents the growing of some plants, notably heathers, rhododendrons and azaleas, on such soils. An understanding of solubility products helps us to see why.

Question

45a Assume that in a chalky soil the pH is about 8, which means that $[OH^-(aq)]$ is about 10^{-6} mol dm^{-3}. Apply this figure to the solubility product formula for iron(II)hydroxide, and calculate the maximum allowed value of $[Fe^{2+}(aq)]$ which can coexist with the hydroxide ions. (This calculation depends on the assumption that there is plenty of available Fe^{2+} in

Figure 9.31 'Iron tonic' for plants

the soil mineral mix, and that there is no other anion with which the Fe^{2+} is more likely to precipitate, both of which are reasonable possibilities.)

b Now consider an acid soil rich in peaty materials, and assume a pH of 6. This translates to a value for $[OH^-(aq)]$ of 10^{-8} mol dm^{-3}. Calculate the new value for allowed $[Fe^{2+}(aq)]$.

So rhododendrons and other plants with urgent needs for iron are only found growing on acid or neutral soils. For example, in southern England their ideal home is the sandy soil of Surrey and Kent.

This problem has a lot in common with those earlier problems in which a reagent was added or removed. By adding OH^- we made the back reaction (the precipitation) faster than the forward, so the equilibrium could only be restored by removal of some of the ions from solution.

There will be other specialised equilibrium constants in later chapters, which will pay specific attention to acid–base equilibria and ligand–complex equilibria.

9.15 The equilibrium constant in terms of partial pressures

So far all the equilibrium constants have been expressed in terms of concentrations in mol dm^{-3}. There is an alternative unit, and it has its advantages over mol dm^{-3}, although the drawback is that it applies only to gaseous equilibria. It is a unit which has its historical roots in the work of the nineteenth-century chemist John Dalton.

What is a partial pressure?

You will remember from the very start of the book that gases are peculiarly uniform in their behaviour, obeying laws which pay no heed to their individual chemical identity. What is more, the behaviour of molecules in an 'ideal' gas (and most gases are fairly good approximations to this condition) can be explained by assuming that each molecule 'ignores' all the others completely

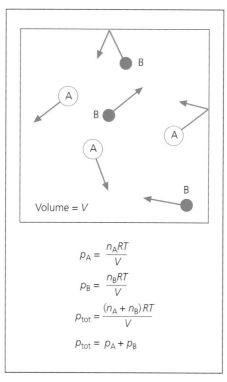

$$p_A = \frac{n_A RT}{V}$$

$$p_B = \frac{n_B RT}{V}$$

$$p_{tot} = \frac{(n_A + n_B)RT}{V}$$

$$p_{tot} = p_A + p_B$$

Figure 9.32 Partial pressures, caused by two gases in a mixture 'ignoring each other'

save for perfectly elastic kinetic-energy-conserving collisions (Figure 9.32).

Hence in a mixture of gases we can treat any of its components as if they were alone in the volume, and obeying their own ideal gas equation, $p = \dfrac{nRT}{V}$.

Equally we can apply the gas equation to the whole assembly of molecules. Here are some equations using this idea. Assume a mixture of three gases, A, B, and C. Applying the gas law to molecules of A only, we get:

$$p_A = \frac{n_A RT}{V} \qquad (9.14)$$

(n_A = number of moles of A)

The quantity p_A is called the **partial pressure** of component A, but it is actually the same as the pressure A would have exerted if it had been there alone. It is also of course the contribution that A makes to the total pressure.

Here is the expression for the total pressure of the mixture:

$$p_{tot} = \frac{(n_A + n_B + n_C)RT}{V} \qquad (9.15)$$

R, T and V are common features in these equations, so if we divide (9.14) by

(9.15), we eliminate them:

$$\frac{p_A}{p_{tot}} = \frac{n_A}{n_A + n_B + n_C} \qquad (9.16)$$

The right-hand side of (9.16) can be paraphrased as the fraction of all moles which are moles of A. This **mole fraction** is given the symbol x_A. Using this symbol and rearranging again, we get:

$$p_A = x_A p_{tot} \qquad (9.17)$$

Question

46 To give this equation some meaning, use it to find the partial pressure of A in the following situations:

a A volume of gas A is at 1 atmosphere, in a syringe. An equal volume of gas B is added, the whole mixture remaining at 1 atmosphere.

b A volume of A is at 1 atmosphere in a vessel of constant volume. An equal volume of gas B is pumped in to the same volume.

c A is carbon dioxide, which is present in air at a level of 300 ppm. Assume the whole atmosphere is at a pressure of 1 atmosphere.

Partial pressures in place of concentrations

If you look back at equation (9.14), you should be able to see the variables that make up concentration 'hiding' within the right-hand side expression. Therefore we can replace n_A/V with $[A]$, and the expression becomes:

$$p_A = [A]RT$$

So partial pressure and concentration are actually proportional to each other. Hence equilibrium constants can be expressed in both formats. For instance, in the case of the ammonia equilibrium:

$$N_2(g) + 3H_2(g) \rightleftharpoons 2NH_3(g)$$

$$\frac{[NH_3(g)]_{eqm}^2}{[N_2(g)]_{eqm} \times [H_2(g)]_{eqm}^3} = K_c$$

$$\frac{p_{NH_3,eqm}^2}{p_{N_2,eqm} \times p_{H_2,eqm}^3} = K_p$$

Question

47 What units would each of these equilibrium constants have? (Assume pressures are in atmospheres.)

Why are partial pressures sometimes preferable?

Concentrations of gases in mixtures are quite hard to measure. However, there are certain measurements which a gaseous mixture will allow you to 'get at' quite easily; for example 'apparent molecular mass' and total pressure. And it's also quite easy to calculate partial pressures from a knowledge of these two. We will take each of those types of measurement in turn and show how they lead on to equilibrium constants in terms of partial pressures. Here is an example for you to do with guidance.

Questions

48 This example concerns the equilibrium gas mixture:

$$N_2O_4(g) \rightleftharpoons 2NO_2(g)$$

a 100 cm^3 of this mixture is found to have a mass of 0.29 g, under conditions in which a mole of gas occupies 24 dm^3 (1 atm and 298 K). Work out the apparent molecular mass of the mixture, M_r' (i.e. the mass of 24 000 cm^3).

b Your answer should lie between the molecular masses of the individual gases. Can you explain why this is so?

c Let the mole fraction of N_2O_4 in the mixture be x. What would this leave as the mole fraction of the other gas in the mixture, in terms of x?

d Use the weighted average formula to find the value of x which would give rise to the apparent molecular mass found in (a).

e Now you have found the mole fractions of both components, turn them into partial pressures, using (9.17).

f Write a K_p expression for this equilibrium, and calculate its value.

g What would you predict to be the effect of increasing pressure on the mole fractions of the two gases in this mixture?

h The reaction as written is endothermic. What would happen to the value of K_p if the temperature was increased?

i This method only works if there is a difference between the *average* molecular masses of the molecules on the left and right sides of the equation. For which of the following two equations is this *not* true, and therefore for which one could you not get K_p by this method?

$$2SO_2(g) + O_2(g) \rightleftharpoons 2SO_3(g)$$

$$H_2(g) + I_2(g) \rightleftharpoons 2HI(g)$$

49 Now for another example, this time one in which the *total pressure* of the system is used as a measure of how far the reaction has progressed towards equilibrium. (The question is quite hard and is therefore 'skippable'. Advice from a teacher might help you decide.) It will show how the monitoring of total pressure can give access to partial pressures, and thence to K_p. Assume the system:

$$A + B \rightleftharpoons C \text{ (all gases)}$$

Mix 1 mole of A and 1 mole of B, the resulting mixture being at 10 atm pressure. The container is sealed and of constant volume. The system comes to rest in equilibrium, with a total pressure of 7 atm.

a What pressure would have resulted from no reaction, and from complete reaction?

b Let the number of moles of C at t_{eqm} be x. Set up a 'moles at t_0/t_{eqm}' table, and insert all the values you know or can deduce.

c Work out the total number of moles of all gases at t_{eqm} in terms of x.

d Recognise that, since the volume is constant throughout, the ratio of 10 atm : 7 atm is equal to the ratio of moles of all gases at t_0 to moles

of all gases at t_{eqm}. Thereby, using answer (c), solve for x.

e Having found x, work out the mole fractions of A, B and C at t_{eqm} and then calculate the partial pressures, using (9.14).

f Finally, calculate K_p.

g Qualitatively, what would be the effect on the yield of C of operating the reaction at 20 atm?

h In question 48i we found that the $H_2/I_2/HI$ (all g) system could not be studied by the 'apparent molecular mass' technique. Would this technique do any better?

Summary

• The study of the direction of chemical change is a branch of thermodynamics. Chemical thermodynamics can be used to say whether a reaction is viable, but it has nothing to say about how fast it may happen. This disclaimer must be applied to all the comments below about whether reactions go or not. For instance a thermodynamically viable reaction may be infinitely slow at a given temperature, and so not happen. Assessment of speed of reaction is the subject of Chapter 14.

• The direction of viable chemical change is the direction which causes the particles and energy of the universe to end up in a more probable arrangement.

• This probability factor is called the **entropy** of the universe. Chemical changes will occur if they cause a positive change in the entropy of the universe, that is, if:

$$\Delta S_{uni} > 0$$

• In obeying this rule, most chemical changes go completely from reactants to products. However, some reactions reach positions of maximum ΔS_{uni} when there is still a mixture of reactants and products. Such reactions are called **equilibria**.

• In studying the entropy changes of chemical reactions, we give separate consideration to the entropy changes in the system (the chemicals) and in the surroundings. Between them these two regions make up what we are calling the universe, and their separate entropy changes make up the total entropy change:

$$\Delta S_{uni} = \Delta S_{sys} + \Delta S_{surr} \qquad (9.3)$$

• So to run a thermodynamic viability test on a reaction, we must calculate ΔS_{sys} and ΔS_{surr} and add them up.

• The quantity ΔS_{sys} is normally calculated by reference to the standard entropies, $S^{\ominus}$, of reactants and products listed in data books. It is calculated from:

$$\Delta S_{sys}^{\ominus} = \Sigma S_{products}^{\ominus} - \Sigma S_{reactants}^{\ominus} \qquad (9.4)$$

This gives a value for the entropy change when molar quantities of reactants in their standard states at 298 K, 1 atm, are completely converted to products in their standard states. (But ΔS_{sys} is usually close to being a constant, even in non-standard conditions.)

• The quantity ΔS_{surr} is calculated from a knowledge of the heat exchanged with the surroundings, which in turn depends on the enthalpy change in the system. The formula linking these quantities is:

$$\Delta S_{surr} = -\frac{\Delta H_{sys}}{T} \qquad (9.6)$$

ΔH_{sys} is normally a constant over a wide range of temperatures to a good degree of approximation.

• The **free energy change**, ΔG_{sys}, of a reaction is a quantity closely related to ΔS_{uni} and may be used for many of the same predictive viability tests. It is defined in a number of ways, which can be expressed as the following algebraic equations:

$$\Delta G_{sys} = -T\Delta S_{uni} \qquad (9.8)$$

$$\Delta G_{sys} = \Delta H_{sys} - T\Delta S_{sys} \qquad (9.9)$$

$$\Delta G_{sys} = -w_{max} \qquad (9.10)$$

• The message of these equations in words is:

(9.8): The test for the thermodynamic viability of a chemical reaction is that the free energy change should be negative. This is equivalent to the previous test for a positive ΔS_{uni}.

(9.9): The struggle to make ΔG_{sys} negative can be seen as a conflict between the two terms on the right of the equation. If the viability of a reaction is threatened by a positive ΔH_{sys} and yet if the ΔS_{sys} is also positive (as it nearly always is in endothermic reactions), then *high temperatures* should encourage a viable reaction by stressing the $-T\Delta S_{sys}$ term. So endothermic reactions get more viable at high temperatures. Conversely, exothermic reactions are less thermodynamically viable at high temperatures.

(9.10): The free energy change of an exothermic reaction can also be seen as the amount of energy which need never appear as heat, but instead is available for doing work. It represents the theoretical maximum work which can be done by a system on its surroundings, w_{max}.

• Data books have values of $\Delta G_{sys}^{\ominus}$ for the same range of situations as those for $\Delta H_{sys}^{\ominus}$. So a quick test for thermodynamic viability of a reaction is to calculate its $\Delta G_{sys}^{\ominus}$ by Chapter 8 methods and check its sign. The problem is that viability can only be assessed at 298 K, 1 atm.

• ΔG_{sys} at other temperatures must be calculated by equation (9.9), assuming that ΔS_{sys} and ΔH_{sys} are not temperature dependent. Equation (9.9) can be expressed in a graphical format, using ΔG_{sys} as y-axis and T as x-axis. The resulting line (gradient = $-\Delta S_{sys}$) shows how viability of a reaction varies with temperature.

• Ellingham diagrams superimpose a whole 'nest' of these lines on the same axes, for the purpose of assessing the viability of competition for oxygen.

• If a reaction has a ΔG_{sys} more negative than –60 kJ mol^{-1}, then it is said to go to completion. That means the position of minimum free energy (and therefore maximum ΔS_{uni}) is achieved by going more or less all the way to products. Conversely reactions whose ΔG_{sys} are more positive than +60 kJ mol^{-1} remain almost totally as reactants.

• For a reaction whose ΔG_{sys} lies between these benchmarks, it is found

that the position of minimum free energy (and maximum ΔS_{uni}) is an equilibrium mixture of reactants and products. (The bead-on-a-string model illustrated this idea.) This means we have to modify some slight oversimplifications in the points made above. It is not quite true to say that a reaction with a positive ΔG_{sys} is non-viable. What we are now saying is that as long as it is not *too* positive, then there will be a small movement away from pure reactants, to an equilibrium mixture in which the products are in the minority (as in Figures 9.19 and 9.21). It is still true to say that all systems move towards a position of minimum free energy, but we must remember that ΔG_{sys} only tells us about the free energy change for molar quantities of complete reaction.

• There is a relationship between the sign and size of ΔG_{sys} and the composition of the equilibrium mixture (which is like saying that the positions of the ends of the 'string' affect the resting position of the bead in Figures 9.17 to 9.21).

• The composition of the equilibrium mixture can be expressed in the form of a constant, the **equilibrium constant**, K. K is expressed in units of concentration, and is then called K_c (or its analogue partial pressure, when it is called K_p), and is defined below:

$$aA + bB \rightleftharpoons cC + dD$$

$$K_c = \frac{[C]_{eqm}^{c} \times [D]_{eqm}^{d}}{[A]_{eqm}^{a} \times [B]_{eqm}^{b}} \qquad (9.12)$$

• Equilibrium constants turn out to be genuine constants, unaffected by manipulations of pressure and of starting composition. The only variable which brings about a change in the equilibrium constant is *temperature* as expressed in the equation:

$$\ln K = -\frac{\Delta H_{sys}}{RT} + constant \qquad (9.13)$$

• The relationship between K and ΔG_{sys} is:

$$\Delta G_{sys} = -RT\ln K \qquad (9.11)$$

• In *heterogeneous* equilibria, the equilibrium constant contains no mention of the phases which are incapable of mixing or of having a variable concentration.

• Solubility products (K_{sp}) are well-known examples of heterogeneous equilibrium constants. They describe the equilibrium between a solid and its aqueous ions. A solid will precipitate when the ion concentrations exceed the value of the true K_{sp}.

• Manipulations done on systems at equilibrium will upset the value of the equilibrium constant – the system then moves so as to restore the true value. This model can be used to make quantitative predictions of the way equilibria respond when interfered with.

• Another model which enables the prediction of the way equilibria will respond is based on the idea (for which there is good evidence) that the forward and reverse reactions are still going on, even when the system appears to be resting at constant composition. This concept is called **dynamic equilibrium**.

• In the dynamic equilibrium model, the equilibrium condition is when forward and reverse reactions are going at the same speed. Anything which affects the speed of one reaction more than the other (removal or addition of a reagent, or change in pressure) will throw the system out of equilibrium. The reactions will carry on in such a way as to restore the state of equilibrium (and the true value of the equilibrium constant).

• Changes in pressure only affect gaseous equilibria, and then only if there is an unequal number of reactant and product molecules.

• Two manipulations stand outside this analysis – changes in temperature, because they do actually change the value of the equilibrium constant, and the addition of a catalyst which merely speeds up reactions in both directions and gets the system to the same equilibrium position, only faster.

• All interferences with a system at equilibrium, even temperature changes, are covered by the statement known as *Le Chatelier's principle*. Paraphrased, it says that if you interfere with a system at equilibrium, it will change to nullify the effect of the interference.

• Many industrial reactions are equilibria, and much effort and energy is expended on manipulating them to give maximum yields. Examples of manipulations are:

1 Having high ratios of (cheap reactant) : (expensive reactant).
2 Removing products as they form.
3 Running exothermic reactions at low temperatures and endothermic ones at high temperatures.
4 Running gas reactions whose reactants take up more space than their products, at high pressures.

• The use of low temperatures to increase yields (point 3 above) requires a degree of compromise. If it is too cold the reactions (both forward and back) grind to a halt on kinetic grounds. Other practical limitations include the cost of maintaining high pressures and (for endothermic reactions) high temperatures.

• Equilibrium constants can be calculated experimentally from measurements of the concentrations of reagents in equilibrium systems. Independent measurement of all reagents is not necessary, since they are linked by stoichiometric relationships. A careful layout of starting and equilibrium amounts is necessary if calculations are to be successful.

• Partial pressures are versions of concentration well suited for describing gaseous systems. They can be calculated from measurements of the apparent molecular masses of gas mixtures, and then used to calculate equilibrium constants.

$$p_A = x_A \times p_{tot} \qquad (9.17)$$

10

Organic chemistry 1: Hydrocarbons

10.1 Introduction

There are 92 elements if we restrict ourselves to those which occur naturally on Earth. Why is it then that the study of molecules based on just *one* of those elements accounts for approximately one third of many chemistry courses? The answer to that question is quite simple – in chemical data books, there are three to four times as many carbon compounds as there are compounds of all the other elements put together. But that answer merely uncovers another question: *why* are there so many carbon compounds?

Let us consider some responses to this question:

1 Carbon forms chain-like molecules very easily, based on the CH_2 group as the standard 'link' in the chain, and these links can be combined in many different ways. Not only are there variations in chain length but also the links can form **branches** and **rings**. This accounts for the variety of hydrocarbon **frameworks** for molecule building (Figure 10.1). But when we

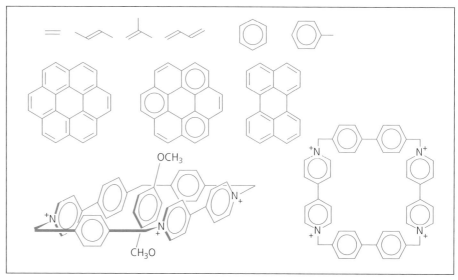

Figure 10.1 A taste of the great diversity of carbon-based compounds

realise that these frameworks can also support bits of more specialised molecular arrangement – the so-called **functional groups** like —OH or —NH_2 – and that these groups can be stuck on to *any* of the links, then we start to realise why there are huge numbers of compounds involved.

This leads to the next question: *why* is carbon so well qualified to form chains? Answers to this divide up into two groups – reasons why other elements are bad at it and reasons why carbon is good.

2 Some elements are prevented from forming chains by the sheer arithmetic

of their outer electron arrangements. For instance, once one chlorine atom has bonded to another, there is no possibility of any further bonding leading to chain formation.

3 Elements with valencies greater than 1 (that is, Groups I to VI) *are* in a position to take part in chain formation, but in most cases the structures are held together by weak bonds. Examples of this are —O—O—O—, —NH—NH—NH— and —SiH$_2$—SiH$_2$—SiH$_2$—.

4 Another problem that Groups V and VI have when forming chains, is that other events are more desirable. For instance, the —O—O—O— and —NH—NH—NH— structures in point 3 have large numbers of lone pairs of electrons which enable them to react with other molecules, rather than forming chains with themselves.

5 Carbon is faced with neither of the problems mentioned above. First, it forms strong bonds with itself and with hydrogen. Second, its basic chain-forming unit —CH$_2$— is free of lone pairs and has little bond polarity. These last two factors make hydrocarbon chains both uninterested *in* and uninteresting *for* other reactants, and they are crucial to the unique position of the element carbon on Earth.

Question

1 Use a data book to put together a short report entitled 'Bond enthalpies provide much of the evidence to explain the stability of hydrocarbon chains, compared with those of similar structures based on silicon, nitrogen or oxygen'. First, decide which bond enthalpies are relevant to the story, then look them up and present them in a table. Discuss their message, letting the discussion range not only over the bonds that would be needed to *form* chains but also over the bonds in possible alternative structures such as O$_2$ molecules.

The final reason for carbon's unique position in the chemistry of our planet lies in what has been called the 'primeval soup' which means the seas and

Figure 10.2 The 'primeval soup'

atmosphere of the young Earth (Figure 10.2). We have seen from point 5 above why the Group IV elements became the basis for a huge body of chemistry. It is not so easy to see why carbon is the 'chosen one' from Group IV. Silicon, for instance, is far more abundant than carbon in the Earth's crust. Part of the answer has already been uncovered in question 1, in the discussion on bond strengths, but that is not the whole story.

The chemistry of carbon is still called 'organic chemistry', even though it is 150 years since it was proved that organic molecules could be synthesised from totally non-organic ingredients. The name is still used, since it is recognised that the chemistry of biological systems is largely organic, and that our two main industrial sources of organic chemicals (coal and crude oil) are themselves derived from living things.

So what we are really asking is: why did life choose carbon and not silicon? Primeval soup theorists think that one of the precursors of life might have been created by the action of lightning on the early atmosphere which contained CH$_4$ and NH$_3$ (but hardly any oxygen). They then suggest that the fortunate admixture of ultra-violet photons and a generous helping of solvent (the sea) might have provided the conditions for the first proteins and then nucleic acids to self-synthesise.

Much later, the path led to the advent of the green plants, whose ability to photosynthesise led to an atmosphere

much richer in oxygen. That single factor transformed the planet's capacity for supporting respiration and ushered in the type of world ecology which we would recognise as 'modern', with the oxygen/carbon dioxide balance dependent on a balance between 'photosynthesisers' and 'respirers'. The specific role of carbon dioxide in this system is as the building-block-in-waiting at the start of the food web.

Where was silicon during this primeval creation period when the first compounds were forming? The bond enthalpy of Si—H in SiH$_4$ is 318 kJ mol^{-1}, whereas the bond enthalpy of C—H in CH$_4$ is 435 kJ mol^{-1}. So the 'stronger' methane is better suited for existence in an atmosphere under constant assault from a rain of bond-rupturing ultra-violet photons.

No less crucially, silicon found an 'alternative role', in (literally) a different sphere from its Group IV cousin. Whereas carbon became the central unit of the **biosphere**, silicon became, to a similar degree, the cornerstone of the **lithosphere**. The difference in roles is founded on the profound difference between the oxides of the two elements. The strength of silicon's single bond with oxygen (466 kJ mol^{-1}), compared with the relative weakness of its double bond with the same element (638 kJ mol^{-1}), ensures that SiO$_2$ exists in a range of extended lattices made from —O—Si—O— units. (Two Si—O bonds are worth more than one Si=O bond.) The opposite pattern in carbon/oxygen bond strengths (358 kJ mol^{-1} for C—O and 805 kJ mol^{-1} for C=O) tilts the balance in favour of a discrete small double-bonded O=C=O structure. SiO$_2$, silica, has very little in common with CO$_2$ other than an empirical formula. Silica is quartz and, together with the metal oxides of the lithosphere, it can make a bewildering range of silicates hardly less impressive, in inorganic chemistry, than the carbon compounds of the biosphere.

Question

2 a Look up the structure of SiO$_2$ and compare it with that of CO$_2$. Why is one a gas and the other a solid at normal terrestrial temperatures?

b The bond enthalpies of Si—O and Si=O are 466 and 638 kJ mol^{-1}

respectively. Yet the text refers to Si=O as being 'relatively weak'. In what sense, and relative to what, is the Si=O bond weak?

c Using average bond enthalpy terms from the data book, compare the enthalpy needed to break Si_2H_6 into gaseous atoms with the corresponding figure for C_2H_6, and comment on the result.

d Summarise briefly why life on Earth is carbon-based and not silicon-based.

10.2 Molecular arrangements and their isomeric variations

Molecules are three-dimensional, so the flat page of a book is not the ideal medium on which to display the structures of organic chemistry. A PC with a molecular graphics package or a set of plastic molecular models would give a much clearer picture. The geometries of single-bonded carbon structures are based on the tetrahedron, and yet the display of formulae on paper normally uses a square-planar layout. This can lead to confusion and can cause people to think they see different molecules when in fact they are looking at two different representations of the same one. As an example, let us consider the propane molecule, whose formula is C_3H_8.

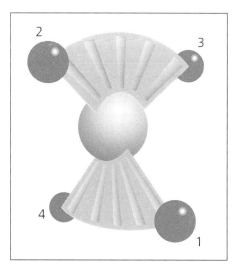

Two 'flat' representations of the molecule are shown, designed to make them look as unlike as possible. Indeed, if the geometries of the four bonds to each carbon atom were truly square-planar they *would* be two different molecules. In fact, the bonds at each carbon atom are tetrahedral and they are the same molecule.

Clearly, we need some way of representing tetrahedra. Tetrahedra are often shown as triangular-based pyramids (Figure 10.3), as indeed they are, but this mental picture has the slight drawback of giving the impression that the three base apices are somehow different from the top one.

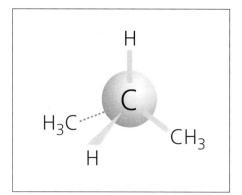

Figure 10.3 The tetrahedron – triangular-based pyramid view. Dotted lines and widening lines are codes for 'away from the viewer' and 'towards the viewer'

Sometimes it is helpful to imagine the tetrahedron as a scoop of ice-cream with two fan-shaped wafers stuck in it, one top and one bottom, with their planes at 90° to each other (Figure 10.4). Some

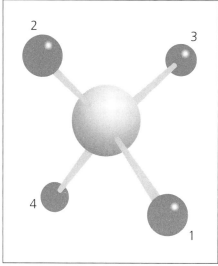

Figure 10.4 The tetrahedron – an alternative view – the ice-cream model

glacé cherries are wedged on the corners of the wafers and thus we achieve the four points of the tetrahedron. (As a further fine detail the wafer-maker would have to ensure the angles at the wafer-points were 109.5°.)

This model is a small step towards the goal of seeing the four points of the tetrahedron as equivalent, each capable of playing the 'top one' role according to how you turn it in your mind. The significance of the 109.5° is that it ensures complete regularity in the structure – that is, the angles between *all* the cherries, whether linked by wafers or not, are 109.5°. So you can equally imagine the wafers to be between cherries 2 and 4, and 1 and 3, or between 1 and 4, and 2 and 3. This being the case, the wafers have now served their purpose, and we can dissolve them out of our mental picture, leaving only sticks to represent the bonds (Figure 10.5).

Figure 10.5 Away with the 'wafers'

A further important feature must be mentioned – and that is that in chemical structures, the molecular fragments on either end of a single bond can *rotate* freely relative to each other. Figure 10.6 (overleaf) shows, as an example, atoms 1, 3 and 4 rotating around the bond between the carbon atom and substituent 2.

Now let us return to the translation of flat-page representations of molecules into 3-D tetrahedra. In the version of C_3H_8 in structure (a), the three carbons are in what might be called

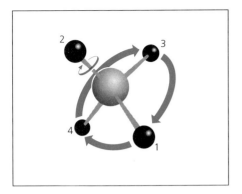

Figure 10.6 Free rotation about single bonds – all four points are equivalent

'opposite' positions, but in (b) they are 'adjacent'.

a

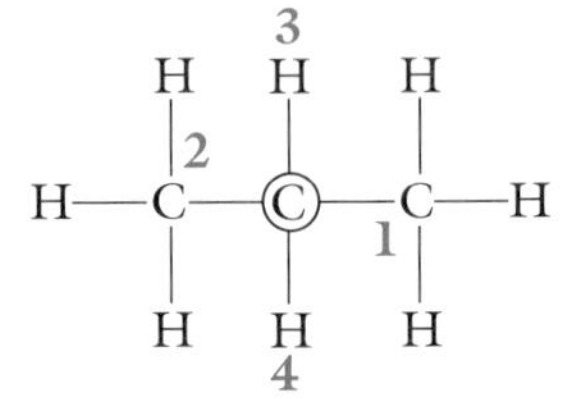

b

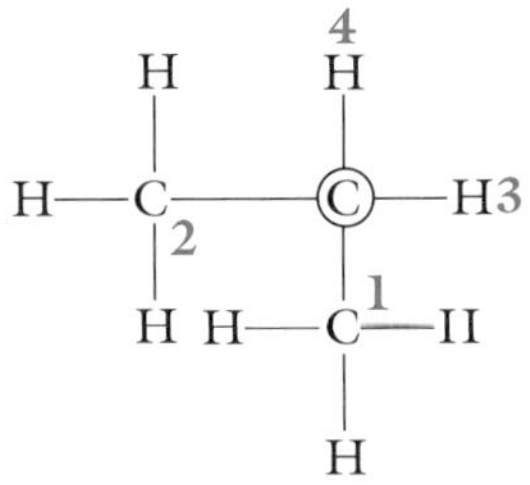

But if we move to the ice-cream model, we see that any one of substituents 1, 3 or 4 can be twisted into any of the other's positions, thus showing that the opposite and adjacent positions are only a twist away, and that therefore the two molecules are indistinguishable. What it amounts to is that, on paper, *any 'bends' you see in carbon chains are illusory* and can be twisted away. As long as each carbon atom is merely connected to one other carbon atom on each side, then it is a 'straight' chain. The only real difference comes when a branch occurs, that is, when a carbon atom is attached to three (or even four) others. For instance, (a) and (b) below are the same molecule but (c) is different.

a

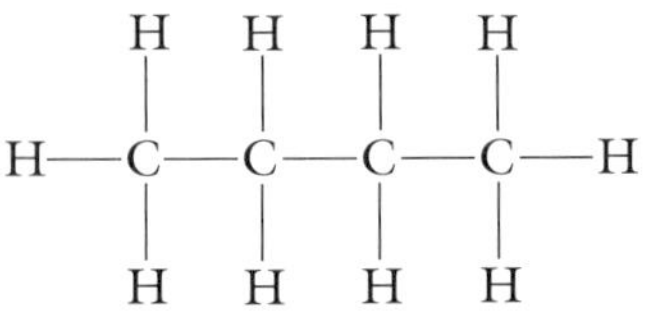

b

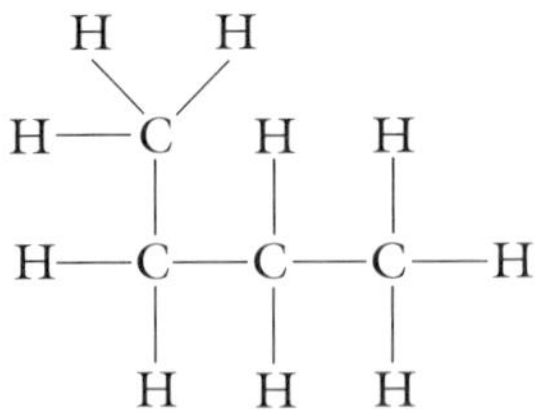

c

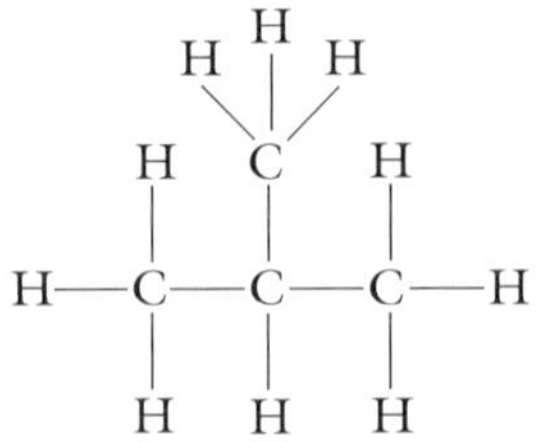

Structures whose overall number of atoms (their molecular formula) is the same but which are actually different molecules (like (a) and (c) above for instance), are called **isomers** of each other. One isomer can be converted into another only by the *breaking and re-making of bonds*, not by bond twisting.

Question

3 A student has tried to draw all of the isomers whose molecular formula is C_5H_{12}:

a $CH_3CH_2CH_2CH_2CH_3$

b $CH_3CHCH_2CH_3$
        $|$
       CH_3

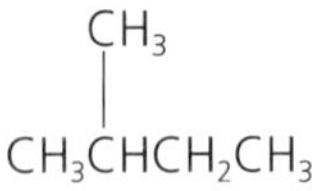

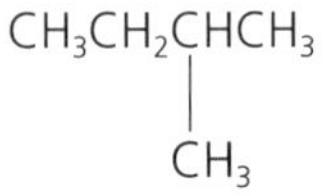

c $CH_3CH_2CHCH_3$
              $|$
             CH_3

d $CH_3CH_2CH_2CH_2$
                      $|$
                     CH_3

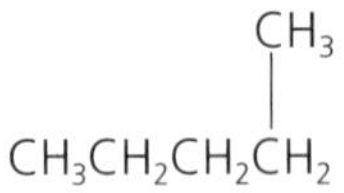

e $CH_3CH—CH_2$
        $|$ $|$
       CH_3 CH_3

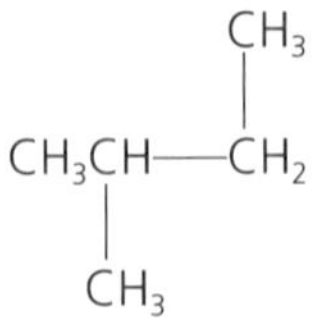

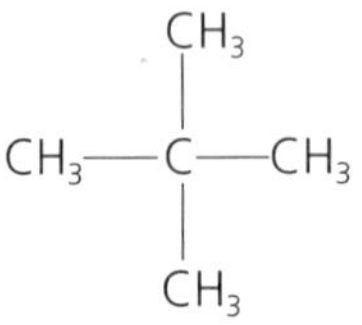

f $CH_3—C—CH_3$ g $CHCH_2CH_3$
        $|$ $|$
       CH_3 CH_3
       with top CH_3 and bottom CH_3

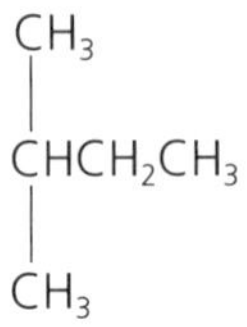

Which are actually duplicates of each other and how many truly different isomers are there?

Notice that the student had quite a good system, by beginning with the straight chain of five carbon atoms then looking at all the 4-C 'backbones' then the 3-Cs. The failures were of two types – first, failure to recognise that a carbon atom cannot be hung on the end of a 4-C chain and then be called a branch, because it has really just returned to the 5-C chain; second, failure to rotate whole structures in the mind's eye.

It actually helps to *talk* yourself through these isomer problems, because words are useful in defining the possibilities. For instance, in the case of the 4-C chain in question 3, the student could have said that there is really only one site for a branch, and that is a carbon atom one place from a chain-end. By talking, twisting or any other mental process, try the next isomer-related question.

Question

4 Draw the structures of all the isomers of C_6H_{14}.

10.3   Naming systems in organic chemistry

You may already know that structures whose molecular formula is C_6H_{14} are called 'hexane'. But with so many possible versions of hexane, we need a more refined naming system. The one at present recognised by the International Union of Pure and Applied Chemistry (IUPAC) works like this:

1 Find the longest continuous 'straight' carbon chain in the molecule, and name the molecule from that number of carbon atoms, according to:

1 = methane	8 = octane
2 = ethane	9 = nonane
3 = propane	10 = decane
4 = butane	11 = undecane
5 = pentane	12 = dodecane
6 = hexane	20 = eicosane
7 = heptane	

(Notice that only one of the C_6H_{14} isomers in question 4 is actually called 'hexane'.)

2 Number the carbon atoms along this longest chain, and use these numbers to identify the site of any substituents, using names like 'methyl' and 'ethyl' to identify the substituent. Number the carbon chain beginning at whichever end gives you the lowest number for the substituents. For example, the molecule below is 2-methylpentane (rather than 4-methylpentane).

$$CH_3$$
$$|$$
$$CH_3CH_2CH_2CHCH_3$$
$$\ \ 5\ \ \ \ 4\ \ \ \ 3\ \ \ \ 2\ \ 1$$

3 When more than one identical substituent is present, the 'site-numbers' are separated by commas, and the message is reinforced by the use of prefixes 'di', 'tri' etc., for example, 2,2-dimethylbutane.

$$CH_3$$
$$4\ \ \ \ 3\ \ \ 2|\ \ 1$$
$$CH_3CH_2CCH_3$$
$$|$$
$$CH_3$$

When the substituents are *not* identical, they are written in alphabetical order (ethyl before methyl, say) together with their site-numbers, for example, 4-ethyl-2-methylhexane.

$$1\ \ \ \ 2\ \ \ \ 3\ \ \ \ 4\ \ \ \ 5\ \ \ \ 6$$
$$CH_3CHCH_2CHCH_2CH_3$$
$$\ \ \ \ |\ \ \ \ \ \ \ \ \ \ \ |$$
$$\ \ \ \ CH_3\ \ \ \ CH_2$$
$$\ \ \ \ \ \ \ \ \ \ \ \ \ \ \ \ |$$
$$\ \ \ \ \ \ \ \ \ \ \ \ \ \ \ CH_3$$

4 The prefix 'cyclo' is used when the chain forms a ring as in cyclohexane.

$$H_2$$
$$C$$
$$H_2C\ \ \ \ \ \ \ \ CH_2$$
$$|\ \ \ \ \ \ \ \ \ \ \ \ \ \ \ \ |$$
$$H_2C\ \ \ \ \ \ \ \ CH_2$$
$$C$$
$$H_2$$

Questions

5 Draw structural formulae for the following compounds:

a 2-methylbutane

b 2,3-dimethylbutane

c 2,2,3-trimethylpentane

d 3-ethyl-3,4-dimethylheptane.

6 Name the following compounds:

a $CH_3CH_2CH_2CH_3$

b $CH_3CH_2CHCH_2CH_2CH_3$
$$|$$
$$CH_2$$
$$|$$
$$CH_3$$

c $CH_3CH_2CHCH_2CHCH_3$
$$\ \ \ \ \ \ \ \ \ |\ \ \ \ \ \ \ \ \ \ \ |$$
$$\ \ \ \ \ \ CH_3\ \ \ \ CH_2$$
$$\ \ \ \ \ \ \ \ \ \ \ \ \ \ \ \ \ \ \ |$$
$$\ \ \ \ \ \ \ \ \ \ \ \ \ \ \ \ \ CH_3$$

Other aspects of the naming system will be introduced as new groups of compounds are discussed.

10.4   Various types of formula

You will already have seen references to the adjectives 'molecular' and 'structural' modifying the noun 'formula'. There are in fact *four* different types of formula, carrying information of varying degrees of detail. Each of these four formats represents a staging-post on the path by which a new organic structure is revealed. Each stage is of course dependent on particular acts of practical chemistry.

Empirical formulae

This is the lowest level of knowledge about the formula of an organic compound, just one step above knowing which atoms are in it. It shows the *relative numbers* of atoms of each kind in the compound. It is called the **empirical formula** because 'empirical' means experimental or practical and it is derived from experiments. In the case of hydrocarbons, the practical method is combustion analysis, the technique that involves collecting and weighing the water and carbon dioxide created by the complete combustion of a known mass of the compound (Figure 10.7, and p. 17, question 17).

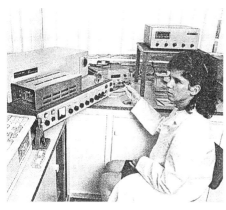

Figure 10.7 Apparatus for collecting and weighing combustion products

Question

7 How would you solve the practical problem of collecting carbon dioxide and water in a quantitative way from a combustion reaction?

Once the masses of carbon dioxide and water have been measured, it is a two-step process to derive the empirical formula.

1 Calculate the number of moles of carbon and of hydrogen present in the masses of carbon dioxide and water respectively.

2 Express these as a ratio.

You might think that we have thrown away some useful information here, in not going for the actual numbers of moles of carbon and hydrogen in the mass of the hydrocarbon. But since we do not know the relative molecular mass of the hydrocarbon we cannot find the number of moles of carbon, say, in one mole of 'parent'. The best we can do is to express the ratio of carbon to hydrogen. Two questions follow: question 8, the simpler one, presents data already processed to the stage of masses of carbon and hydrogen in the unknown compound, while question 9 presents the data at the 'rawer' stage of masses of carbon dioxide and water from the combustion analysis apparatus.

First, you need to recall the earliest work on translating grams into moles (Chapter 2, p. 12, formula (2.2)). But there is a second skill needed here, involving pure number awareness; the C:H mole ratio will be in decimal form and may not reveal on simple inspection its underlying whole-number ratio. If the ratio is not immediately obvious, try this calculator method:

1 Divide both the numbers by the smaller one (thus ensuring that at least one side of the ratio is now a whole number – namely 1). For instance, if you were working with a compound containing 0.58 moles of carbon and 1.35 moles of hydrogen, the ratio 0.58 : 1.35 is also the ratio 1 : (1.35/0.58), or 1 : 2.33.

2 If the answer is still not obvious, put 2.33 into your calculator as an *adding constant*, and then keep adding it to itself. In effect you are producing 1×, 2×, 3×, ... the number and are therefore gradually expanding both sides of the ratio. Sooner or later you should find the sought-for whole number:

$$1 : 2.33$$
$$2 : 4.66$$
$$3 : 6.99$$

And there, allowing for experimental error, is the answer – an empirical formula of C_3H_7. Armed with these techniques you're ready for the actual questions.

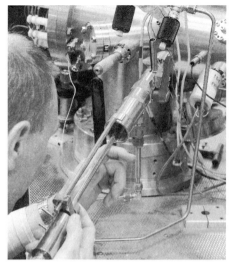

Figure 10.8 A mass spectrometer

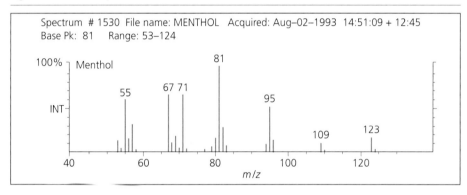

Spectrum # 1530 File name: MENTHOL   Acquired: Aug–02–1993 14:51:09 + 12:45
Base Pk: 81   Range: 53–124

Figure 10.9 A mass spectrum, (Note how the machine automatically gives the biggest peak the value of 100% on the abundance (*y*) axis)

Questions

8 A hydrocarbon is found to contain 80% by mass of carbon (and therefore 20% hydrogen). Calculate its empirical formula.

9 A hydrocarbon, having undergone combustion analysis, gave 1.285 g of carbon dioxide and 0.631 g of water. Find its empirical formula. (Remember that one mole of water contains *two* moles of hydrogen atoms.)

Molecular formulae

Neither the answer to question 8 (CH_3) nor that to the prior example (C_3H_7) are themselves feasible molecules, because they do not obey normal valency rules. However, remember the limitations of the empirical formula – it is no more than a ratio. Remember also that we need a molecular mass if we are to make real sense of an empirical formula. For instance, if we knew, in addition to the data of question 8, that the relative molecular mass of the gas was 30 then we would be confident that the molecule we were dealing with was ethane, C_2H_6.

When a formula expresses the actual numbers of atoms which we believe are in the molecule, it is called a **molecular formula**. The molecular formula is equal to a whole-number multiple of the empirical formula (as shown by CH_3/C_2H_6 in question 8). Of course the true molecular-formula mass will be the same whole-number multiple of the empirical-formula mass. In fact, dividing one of these masses into the other is the easiest way of deducing how many 'empiricals' make a 'molecular'.

The molecular formula shares two features in common with its empirical cousin – first, despite the loss of 'empirical' in the name, it still depends on experiment for its determination and, second, it, too, falls short of telling the whole story. (For instance, all the different isomers of hexane you drew in answer to question 4 have the same molecular formula.)

As for the experimental determination of molecular formulae, the two main methods vary greatly in expense, convenience and accuracy. At the convenient-and-accurate-but-expensive end of the market is the mass spectrometer (Figure 10.8), some of which are sufficiently accurate not only to yield the molecular mass but also to compare fragment patterns and isotope peaks with known spectra in a databank, and thus to identify the actual molecule, even down to a particular isomer (Figure 10.9). In the cheaper method, the molecular formula is deduced from a combination of empirical formula and relative molecular mass, the latter having been deduced from a simple physical method such as finding the molar volume of a gas.

As long as the sample is capable of

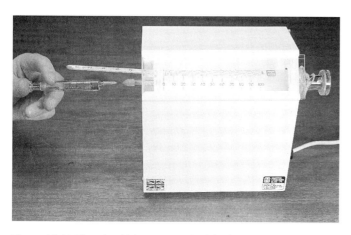
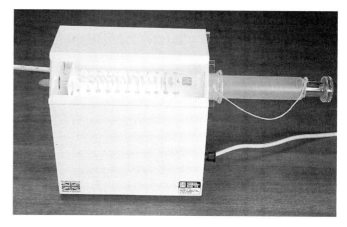

Figure 10.10 The school laboratory method for finding the molecular mass of a volatile liquid

being rendered gaseous and of having its gaseous volume measured (Figure 10.10), its molecular mass may be calculated. The method relies upon the great uniformity of behaviour of gases, which includes the fact that a gas volume at any given conditions of temperature and pressure is only dependent upon the *number* of particles present, and not their mass. At standard temperature and pressure (stp) the molar volume of any gas is 22.4 dm³, and even if the mystery substance is not gaseous at stp, we can compute the volume it would have if it were.

A worked example may make this clearer.

Example: the molecular formula of a hydrocarbon

A hydrocarbon has an empirical formula of CH_2, and when 0.27 g of it are vaporised at 100 °C and 1 atmosphere pressure, the volume is measured as 98.6 cm³. Find its molecular formula.

Conversion of volume to stp: 98.6 cm³ at 100 °C (373 K) is equivalent to an stp volume of

$$98.6 \times \frac{273}{373} = 72.2 \text{ cm}^3$$

So if 72.2 cm³ has a mass of 0.27 g, 22 400 cm³ would have a mass of

$$0.27 \times \frac{22\ 400}{72.2} = 83.8 \text{ g}$$

This value represents an experimental approximation to the molecular mass. To find how many units of CH_2 would make a number near to 83.8, divide by

14 (formula mass of CH_2), giving 83.3/14 = 5.98, that is, near enough to 6. Thus the molecular formula is C_6H_{12}, and the molecule may be cyclo-hexane or one of the isomeric hexenes.

Try this similar example, which will pull together skills from the last two sections.

Questions

10 When subjected to combustion analysis, a hydrocarbon gave 9.25 g of CO_2 and 4.25 g of H_2O from 3.00 g of original sample. Find its empirical formula.

In a separate experiment, 0.28 g of the same compound was injected into a steam-jacketed syringe at 373 K, 1 atmosphere, producing a gas of volume 75 cm³. Find the relative molecular mass and use this information, together with the empirical formula, to suggest a molecular formula.

11 In another experiment students injected an unknown hydrocarbon (not the one in question 10) into a mass spectrometer. After about a minute the machine had drawn an accurate mass spectrum, and they saw a clear 'molecular ion' peak at 98 (Figure 10.11). However, there was a small peak at 99, which a student suggested was due to the presence of naturally occurring carbon-13. The peak at 98 had a height of 29 units on the chart paper, while the one at 99 was about 2 units high.

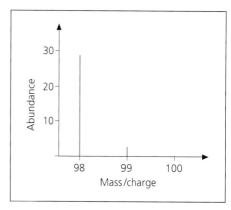

Figure 10.11 See question 11 – using isotope peaks to 'count' carbons

Look up in a data book the percentage abundance of carbon-13 in the mix of naturally occurring carbon atoms on Earth. From this information and from the height of the isotope peak as a percentage of the molecular-ion peak, work out how many carbon atoms there must be in the molecule under study, and thereby suggest a molecular formula. Why can we ignore the possibility that the peak at 99 might be due to deuterium atoms replacing hydrogen atoms?

Structural and skeletal displayed formulae

We have already mentioned the limitation of the molecular formula, in that it fails to discriminate between isomers. For instance, such dissimilar molecules as methoxymethane and ethanol have the same molecular formula, C_2H_6O. The next level in the formula sequence, the **structural formula**, does indeed show up these differences, but what extra experimental steps will be needed to pass from molecular to structural?

In the example above (C_2H_6O), simple test-tube chemistry would be enough. An ether and an alcohol have such blatant differences in their chemical behaviour (for example, towards sodium), that in practice there would be little confusion between them.

Another traditional chemical method of deciding which isomer you are dealing with (and thereby getting at the structural formula), is to make, as it were, isomers of the isomers. This method draws more upon general intellectual and logical skills than on chemical knowledge.

Question

12 A student isolated a compound whose molecular formula was C_5H_{12}. The student carried out a synthesis which substituted a single bromine atom for a hydrogen atom, and then found by chromatography that she had produced *more than one* compound of formula $C_5H_{11}Br$. Which of the following two molecules could *not* have been the original hydrocarbon, and why?

$CH_3CH_2CH_2CH_2CH_3$ $(CH_3)_4C$

Structural formulae, like the ones in the question above, discriminate between isomers. As with molecular formulae, modern instruments can make a big contribution to their determination, especially in terms of speed. The mass spectrometer gives quick answers on molecular masses, but it also gives much more information than that.

Specifically, an analysis of the fragment peaks from a mass spectrum can often eliminate certain structural formulae from among the possibilities under consideration. For instance, to return to the compounds in question 12, although they would both give a molecular ion peak at 72, only the straight-chain one could possibly give a peak due to $CH_3CH_2^+$, at 29, since 2,2-dimethylpropane cannot break up in that way (Figure 10.12).

Question

13 Two isomers have the molecular formula C_8H_{10}. The researcher suspects he has either ethylbenzene or one of several possible isomers of dimethylbenzene.

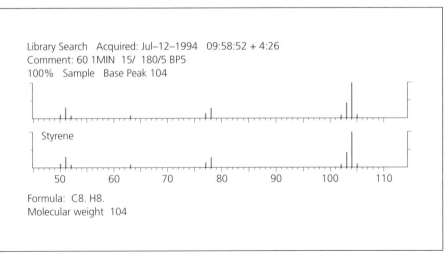

What fragment peaks might each molecule give in its mass spectrum that would help distinguish it from the other?

Every different isomeric substance has a unique mass spectrum, except, that is, for **optical isomers** (Chapter 13). So if the mass spectrometer is one of those with an electronic database of known spectra, the identification of a compound can be done by automatic comparison, using the same logical basis as police fingerprinting (Figure 10.13).

This 'fingerprint' method is even more appropriate when applied to another spectroscopic technique, that of **infrared (IR) spectroscopy** (introduced in Chapter 5). Apart from the ability conferred by infrared spectra to tell whether a molecule contains, say, a C=O bond in a ketone, they also provide information in a complex region below about $1500\ cm^{-1}$ called the 'fingerprint region' (Figure 10.14). In this region of densely packed peaks, what you lose in ability to link peaks with bonds you gain in uniqueness-to-a-single-compound. It would even be possible, harking back to the various dimethylbenzenes in question 13, to distinguish between 1,2-, 1,3- and 1,4-dimethylbenzene using infrared 'fingerprint'

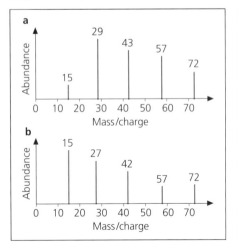

Figure 10.12 Mass spectra of two isomers of C_5H_{12}

Library Search   Acquired: Jul–12–1994   09:58:52 + 4:26
Comment: 60 1MIN 15/ 180/5 BP5
100%   Sample   Base Peak 104

Styrene

Formula: C8. H8.
Molecular weight 104

Figure 10.13 Mass spectrum of a sample (top) and the computer's best match (bottom)

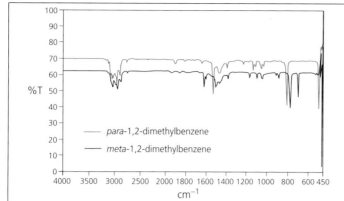

Figure 10.14 Showing how the fingerprint region of the IR spectrum of even very similar compounds can reveal differences

techniques, as long as someone had done the hard work first – that of isolating known examples of the three isomers and running their spectra.

One other spectroscopic technique that should be mentioned is **nuclear magnetic resonance (NMR) spectroscopy**, which offers a combination of 'logical' and 'fingerprint' styles of structure determination. The logical side of elucidating NMR spectra is, however, a complex business and will be visited in Chapter 13.

Finally, if the substance can be obtained in a solid crystalline form, **X-ray crystallography** can be used to determine the actual spatial layout of a molecule, and therefore reveal its structural formula.

Displayed and skeletal formulae

These are representations of molecules in which more graphic art comes into play, although they rarely convey much more information than the structural version. **Displayed formulae** show every bond in

cis

trans

a molecule, which can be useful – for example, it is the best way of showing the *cis* and *trans* isomers of certain alkenes (see Section 10.10, p. 208).

Skeletal formulae show almost nothing *but* the bonds in a molecule, leaving the rest of the molecule to be deduced by the reader using certain conventions. Chief of these is the idea that each bend in the structure represents a carbon atom, with as many hydrogens attached to it as are dictated by valency rules. The only atoms which *are* shown are those which are neither C nor H. Skeletal formulae come into their own when depicting ring structures, which would be over-full if done by 'displayed' methods.

cyclohexene

ethylbenzene

2-methylbutane

However, they are totally inadequate for simple alkanes. This dash —— is the skeletal formula of ethane, and the sentence will end with a skeletal representation of methane. (That was it – the full stop.)

Question

14 Draw displayed formulae for the *cis* and *trans* isomers of pent-2-ene (that's the 5-carbon-atom chain with the double bond between carbon atoms 2 and 3), and a skeletal formula for cyclopent*ene*.

10.5   Styles of reaction in organic chemistry

The curly arrow symbolism
There will be many examples throughout the text in which curly arrows are used to symbolise the movement of electrons as in the examples below.

a $A\!:\!B \longrightarrow A^\circ + {}_\circ B$

b ${}^{\delta-}A\!:\!B^{\delta+} \longrightarrow A{:}^- + B^+$

c ${}^{\delta+}A\!:\!B^{\delta-} \longrightarrow A^+ + {:}B^-$

If the arrow symbol is **double-headed** it means that *two* electrons are on the move. Any of the following two-electron moves can be described.

1 Two electrons move from one bonding position to another bonding position in the same molecule. An example can be found in Chapter 4, Figure 4.47, which shows the delocalisation of the π-bond system in benzene. In fact this type of movement can only occur in the π-electron systems.

2 Two electrons move from a lone-pair position to a bonding position on the same molecule (as in the delocalisation of nitric acid, Chapter 4, p. 60, upper arrow). Again only π-bonding systems are involved.

3 Two electrons move from a bonding position to a lone-pair position on the same molecule (another π-bonds-only-manoeuvre typified by the lower arrow, p. 60).

4 Two electrons move from a bonding position to a lone-pair position, thereby cleaving the molecule into fragments. This is the situation in general examples (b) and (c) at the start of the box. It is also happening in the two diagrams below, which second-time readers will recognise as being drawn from the S_N1 and S_N2 mechanisms.

5 Two electrons move from a lone-pair position to a bond position, thereby linking two molecules in a new union. This is the situation in the attack on a C—O bond by $^-$Nu (second equation, left-hand arrow).

6 Two electrons move from a bonding position on one molecule to a bonding position which links two molecules in a new union. This is the situation in the first stage of the mechanism (left-hand arrow).

If a curly arrow is **single**-headed then a single-electron move is being described. A similar set of situations can exist as in the case of two-electron moves, and an example has been seen already in this panel (in (a) at the beginning of the panel).

Classifying reactions

One of the problems encountered by students in the study of organic chemistry is the very large body of knowledge that needs to be 'filed'. What is needed therefore is a filing system, some way of putting the data into accessible packages. Traditionally, the method has been to divide up the reactions according to **functional group**, and indeed this idea has an obvious logic, in much the same way as we choose to study the Periodic Table group by group.

However, it is useful to have other methods of classification, not for the purpose of undermining the traditional one but to deepen the understanding of what makes each set of organic compounds behave in the way it does. (An inorganic analogy would be the way in which the oxidation-number concept pertains to all the periodic groups.) There are two main ways of classifying organic reactions (other than by functional group), using **reaction mechanism** and **reaction outcome**.

Reaction mechanism

Every reaction *must* involve bond making or breaking or both, and this provides us with the starting point for our classification. Imagine a covalent bond between atoms A and B. The bond might break in one of three ways, set out below.

1 One electron may be retained by each of the separating atoms (so-called **homolytic** bond-breakage).

$$A:B \rightarrow A\cdot + \cdot B$$

2 Both electrons may be retained by atom A (**heterolytic** bond-breakage).

$$^{\delta-}A:B^{\delta+} \rightarrow A:^- + B^+$$

3 Both electrons may stay with atom B (also **heterolytic** bond-breakage).

$$^{\delta+}A:B^{\delta-} \rightarrow A^+ + :B^-$$

Which of these options actually happens depends on one dominant factor, and that is the *polarity* of the bond. For instance, if a C—O bond breaks it is overwhelmingly likely that the oxygen will end up with both electrons (heterolytic fission), because the electrons are halfway there already due to the polarity of the C—O bond.

In addition, the central fact of the polarity of the bond will have two subsidiary effects. First, the polar ends of the bond will become attractive to reagents of the opposite charge and, second, the incoming reagent will have to offer the right *number* of electrons – either 0, 1 or 2 – to restore the normal bonding from the imbalance left by the breakage.

A further look at the tendencies of the C—O bond may serve to make these ideas clearer. The species labelled 'Nu' in the example in the above panel is the kind of attacker that would be interested in a C—O bond – it is attracted to the *positive* end of the C—O dipole, and it has on offer a *pair* of electrons, which is going to be needed to replace the pair destined to leave with the oxygen when the bond breaks. Thus a major theme in the chemistry of molecules containing C—O bonds is **attack by nucleophiles**. This new word 'nucleophile' is the name

given to attacking species with an interest in positive sites, and also, and more centrally to the definition, with a *lone pair* of electrons on offer.

Nucleophiles do not have to be anions, but they *must* have the pair of electrons. Common nucleophiles which we will meet in the course of this chapter will be H_2O, OH^-, NH_3, halide ions and the halogen end of hydrogen halide molecules. It is very useful to be able to look at a functional group like the alcohols, and see at once that reactions with nucleophiles will play a major part in their chemistry.

So far, we have looked at the attack by nucleophiles on the positive ends of dipoles, rather than attack by **electrophiles** on the negative ends, because that is what happens more often. (However, it is not uncommon for H^+ to attach itself to the O end of a C—O bond and act in a catalytic capacity, as we shall see on several occasions.) There *are* such things as electrophiles, but unfortunately for symmetry's sake, they do not perform mirror-image inversions of the nucleophile's tricks. An electrophile is a species with a tendency to *accept* (rather than offer) electron-pairs, but there the symmetry ends. Instead of attacking the negative ends of dipolar bonds, their main function in normal organic chemistry is to attack C=C π-bonds, as in alkenes or arenes.

The π-bond component of a C=C double bond was discussed in Chapter 4 (p. 62). Although the C=C bond is non-polar, a feature which does not interest electrophiles, it *is* a region of high electron density, a feature which most certainly *does*. The electrophile is at least being true to its name, which translates from the Greek as 'electron lover'. The first step in an electrophilic attack on a π-bond is shown opposite and you can see how the electrophile accepts both the π-electrons, destroying the doubleness of the bond and leaving a positively charged site for a second nucleophilic stage to finish the reaction off.

Once again, as in the case of nucleophiles, an electrophile does not have to be ionic. Typical electrophiles include the hydrogen end of hydrogen halide molecules, the nitronium ion NO_2^+, and, with apparent total absence of logic, the positive end of a halogen molecule. (You may argue that it has not got a positive end, but all will be revealed later in this chapter.)

Question

15 If both the reaction mechanisms mentioned so far feature heterolytic bond fission, which molecules might undergo reactions featuring homolytic bond fission? Justify your answer.

Your answer to question 15 should have suggested a class of compounds whose bonds are non-polar and single. That way they would have neither the pre-disposition to break unevenly nor the polarity to pull in passing nucleophiles, nor yet the 'electron-richness' to attract passing electrophiles. A very 'plain' group of molecules such as the alkanes would have made an ideal answer to question 15.

The main problem with this reaction mechanism is that it requires initial attack by a species with not two or zero electrons, but a *single* electron. Such species – the so-called **free radicals** – are very rare among normal stable chemicals, but this unpromising situation is rescued by the involvement of light. The absorption of a photon of the right frequency is often capable of cleaving molecules homolytically and, as we shall see below, this may be enough to start a whole reaction sequence.

An attacking free radical offers its 'victim' a single electron, and a single bond in the victim must break homolytically so as to free another single electron to make a bond pair with the newcomer. This breakage has the knock-on effect of creating another unpaired electron elsewhere in the molecule, and therefore a new free radical is born. This radical is capable of acting in just the same 'aggressive' way as the one which generated it, so you can see the possibilities of repeated cycles of events. For example, see below, where the words for the three stages are rather formal words meaning 'starting' (Initiation), 'carrying on' (Propagation) and 'stopping' (Termination). Notice the role played by a photon in creating the initial Br• radical through the homolytic cleavage of the Br—Br bond.

Stage 1: Initiation

$$Br{-}Br \xrightarrow{h\nu} 2Br\bullet$$

Stage 2: Propagation

a $Br\bullet \; H{:}CH_3 \longrightarrow Br{:}H + \bullet CH_3$

b $H_3C\bullet \; Br{:}Br \longrightarrow H_3C{:}Br + \bullet Br$

Stage 3: Termination

a $H_3C\bullet \quad \bullet Br \longrightarrow H_3C{:}Br$

or b $H_3C\bullet \quad \bullet CH_3 \longrightarrow C_2H_6$

Overall

$$CH_4 + Br_2 \xrightarrow{h\nu} CH_3Br + HBr$$

Now we've met three types of reaction mechanism.

The next questions concern classes of chemicals which have not been formally introduced. They should demonstrate the power and usefulness of classifying reactions by mechanism, enabling you to make sense of unfamiliar territory.

Questions

16 Classify the following reagents as either nucleophile, electrophile or free radical (the secret is to assess how many electrons the reagent seems to be offering to its 'victim' for new bond formation):

Cl•, H—Cl (H end), H—Cl (Cl end), CN^- (cyanide ion), OH^-, NO_2^+ (nitronium ion), NH_2NH_2 (hydrazine), and

17 What kind of reagent (nucleophile, electrophile or free radical) would be most likely to react with each of the following molecules? (This time the decisions will rest on factors such as bond polarity and accumulations of high electron density.)

$CH_3CH{=}CH_2$ (propene), CH_4, CH_3CH_2Cl (chloroethane), CH_3CH_2OH (ethanol), $(CH_2)_6$ (cyclohexane), $C_6H_5CH_2CH_3$ (ethylbenzene) – the ethyl side-chain, ethylbenzene – the ring

18 Predict the product of the reaction between bromine and methylbenzene, in the presence of *light*.

(In other words, decide which part of the molecule would be susceptible to attack by bromine radicals under these conditions.)

19 It is not organic chemistry, but it *is* a free radical reaction, when hydrogen and chlorine are exploded together:

$$H_2(g) + Cl_2(g) \xrightarrow{h\nu} 2HCl(g)$$

The reaction may be initiated by the light from burning magnesium ribbon. By comparison with the free radical mechanism (see left), try to construct the initiation, propagation and termination stages in this reaction.

10.6   Classifying organic reactions by outcome

The introduction to Section 10.5 explained that the classical method of organic classification was by functional group, but that a knowledge of reaction *mechanism* – nucleophilic, electrophilic and free-radical styles of reaction – would improve understanding, specifically by giving strong concepts on which to attach the new knowledge. We can now take this process one stage further by considering a third classification.

This third method is again designed to be a reinforcement rather than an alternative to the others, and we shall see later in this chapter the three methods working together. It uses a set of concepts that describe the *outcome* of chemical reactions – concepts such as substitution, addition and elimination, plus imported ideas such as redox and acid–base. You will see that the definitions of these words concern the start and finish of a chemical reaction, with very little attention paid to the 'journey' (unlike the previous work on reaction mechanisms).

Substitution – when one atom or molecular fragment is replaced in a molecule by another. For example:

$$HCl + C_2H_5OH \rightarrow C_2H_5Cl + H_2O$$

Addition – when a double bond opens and new molecular fragments are added to the molecule at both ends of the broken bond, with nothing being removed. For example:

$$H_2C{=}CH_2 + HCl \rightarrow CH_3CH_2Cl$$

Elimination – when atoms or molecular fragments are removed from adjacent atoms on a molecule, leaving a double bond, with nothing being added: the exact opposite of addition. For example:

$$OH^- + CH_3CH_2Cl \rightarrow H_2C{=}CH_2 + H_2O + Cl^-$$

Acid–base and redox have the same meanings as in wider chemistry. For example:

a

b $CH_3CH_2OH \xrightarrow[H^+]{Cr_2O_4{}^{2-}} CH_3C{\overset{H}{\underset{O}{}}}$

Nearly all simple organic reactions, including those you met in earlier sections, fit into one or other of these categories.

Table 10.1 Summary of the classification systems of organic chemistry. The entries describe the type of substrate likely to undergo each style of reaction

Outcome	Mechanism – attacking species is a(n) . . .		
	Nucleophile – offering 2 electrons	*Electrophile* – ready to accept 2 electrons	*Free radical* – ready to accept and to offer 1 electron
Substitution	Molecules with polar single bonds, e.g. alcohols, halogeno-alkanes. (See Chapter 11)	Molecules containing a benzene ring. (See Section 10.16)	Non-polar molecules, such as alkanes, or molecules containing alkyl groups. (See Section 10.7)
Addition	Molecules containing C=O bonds. (See Chapter 13)	Molecules containing C=C (alkenes). (See Section 10.11)	Molecules containing C=C (alkenes). (See Section 10.12 on polymerisation)
Elimination	Molecules with polar single bonds, e.g. alcohols, halogenoalkanes. (See Chapter 11)	—	—

Question

20 Into which of the 'outcome' categories would you put:

a The nucleophilic attack on the alcohol at the bottom left of the panel on p. 200?

b The free-radical attack on the alkane in the scheme on p. 201?

c The electrophilic attack on the alkene in the last column on p. 200?

The sections up to 10.6 have provided a set of concepts for the greater understanding of the chemistry of the functional groups. Table 10.1 shows how the 'mechanism' and 'outcome' concepts come together.

Section 10.7 is the first section of the chapter actually to use these concepts with a single class of organic compounds.

10.7   The alkanes

The alkanes are the 'plainest' of all the organic groups. Their structures are entirely made up of $-CH_2-$ units (except at branches or at ends-of-chains).

Question

21 Given the information above (and remembering how many ends there are to a chain), find a *general* formula for an alkane containing *n* carbon atoms. How would you adjust that formula to work for *cyclo*alkanes?

The alkanes are our first example of what is known as a **homologous series**, which means a series of molecules in which each member differs from the previous one by a single unit, in this case a $-CH_2-$ unit.

This is where we start using the new big concepts from earlier sections. The two most important features of the bonds in alkane molecules are their quite high *strength* and their almost complete *absence of polarity*.

Question

22 From these two facts, predict, with justification:

a Whether C—H bonds and C—C bonds in alkanes are likely to break *homolytically* or *heterolytically*. (Look back to Section 10.5 if you need a prompt.)

b Whether the likely attacking reagents of alkanes would be nucleophiles, electrophiles or free radicals.

c Whether the chemistry of alkanes is going to be rather limited in scope, or large and diverse.

From the answer to part (c), you will not be surprised to find that the important reactions of alkanes number just *three*. At least two of these reactions go via a free-radical mechanism, although one (combustion) is not well understood.

Reactions of alkanes

1 Halogenation (see Section 10.8). The general reaction is:

$$RCH_3 + X_2 \rightarrow RCH_2X \text{ (mixture of isomers)} + HX$$

where R = H or an alkyl group, and X = F, Cl, Br or I.

2 Combustion (see Section 10.9).
For example:

$$C_5H_{12} + 8O_2 \rightarrow 5CO_2 + 6H_2O$$

3 Cracking (see Section 10.9). The general reaction is:

$$\text{alkane} \rightarrow \text{smaller alkanes} + \text{alkenes} + H_2$$

We shall now go on to look at each of these in turn.

10.8 Halogenation of alkanes

This is a set of reactions which conforms closely to the modes of behaviour predicted by our set of concepts from earlier in the chapter. We would expect a free-radical mechanism, and this view would be reinforced if such a mechanism could explain the following phenomena.

1 The reactions between halogens and alkanes take place at room temperature in sunlight, but only at 250 °C or above in the dark.

2 Several thousand product halogenoalkane molecules are produced from each photon of light absorbed.

Looking at this evidence we might hypothesise that the first step in the mechanism is the creation of halogen free radicals (single atoms) as shown below for chlorine. (See also p. 204.) Is this feasible?

$$Cl—Cl \xrightarrow{h\nu} Cl\bullet + Cl\bullet$$

Question

23 Find the frequency of a photon whose energy would be sufficient to cleave a Cl_2 molecule using the Planck equation (below). Locate that frequency within the electromagnetic spectrum. Does your result support observation 1 above?

The Planck equation is

$$\nu = \frac{E}{h}$$

(You will need the following data: Planck constant h, Avogadro constant, Cl—Cl bond strength and frequencies in the electromagnetic spectrum.)

So, using chlorine and methane as an example, the first step is:

$$Cl_2 \rightarrow 2Cl\bullet$$

The chlorine free radical, $Cl\bullet$, is a reactive species and will not live long before it meets and reacts with some other species. The two molecules most likely to be encountered are CH_4 and Cl_2, but the latter encounter would fail to take the reaction anywhere towards a meaningful outcome.

Question

24 Why would it be meaningless?
$$Cl\bullet + Cl_2 \rightarrow ?$$

The meetings of $Cl\bullet$ with CH_4 are at the heart of the mechanism, and as we have seen we can expect homolytic cleavage, and a radical created for every radical destroyed. But *which* radical is created? There are two possibilities:

1 $Cl\bullet + CH_4 \rightarrow HCl + CH_3\bullet$

2 $Cl\bullet + CH_4 \rightarrow CH_3Cl + H\bullet$

The versions below show the electron movements involved.

$$Cl\bullet \quad H\text{:}CH_3 \rightarrow Cl\text{:}H + \bullet CH_3$$

$$Cl\bullet \quad H\text{:}CH_3 \rightarrow Cl\text{:}CH_3 + \bullet H$$

Question

25 Identify the bonds to be broken and made on these two routes. Use bond enthalpies to work out the enthalpy change for each route, and decide on the basis of your calculations which route looks the more likely.

This step turns out to be crucial to the overall rate of reaction, since it is the most difficult step with the highest activation energy – these concepts will be studied in Chapter 14. You will see from your answer to question 25 that a methyl free radical has taken the place of the chlorine free radical, and will therefore carry the mechanism to its next step.

$$H_3C\bullet \quad Cl \cdot Cl \rightarrow H_3C \cdot Cl + \bullet Cl$$

We have now regenerated the same free radical, Cl•, as began the process, so we have the conditions in place for a repeated cycle. Diagrams like the one shown below can be used to symbolise the cyclic nature of this two-step pattern, with the central rotation of alternating free radicals acting as an 'engine' to 'drive' Cl_2 and methane to the two product molecules.

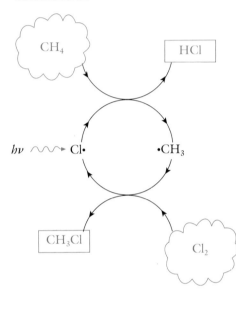

 Reactants   ☐ Products

Question

26 How does this proposed mechanism explain observation 2 at the beginning of Section 10.8?

Where will it all end? If the methane is in excess, the supply of fresh chlorine molecules for the cycle shown above will run down. The last few remaining radicals in circulation will wander around until eventually two of them will meet, and that will terminate the chain:

$$CH_3\bullet + Cl\bullet \rightarrow CH_3Cl$$

or even possibly

$$Cl\bullet + Cl\bullet \rightarrow Cl_2$$

and $$CH_3\bullet + CH_3\bullet \rightarrow CH_3CH_3$$

A summary of the whole mechanism is shown below.

Initiation

$$Cl_2 \xrightarrow{h\nu} 2Cl\bullet$$

Propagation

$$Cl\bullet + CH_4 \rightarrow CH_3\bullet + HCl$$
$$CH_3\bullet + Cl_2 \rightarrow CH_3Cl + Cl\bullet$$

Termination

$$Cl\bullet + \bullet CH_3 \rightarrow CH_3Cl$$
$$2CH_3\bullet \rightarrow C_2H_6$$
$$2Cl\bullet \rightarrow Cl_2$$

Overall (mainly)

$$CH_4 + Cl_2 \xrightarrow{h\nu} CH_3Cl + HCl$$

Questions

27 If Cl_2 is in excess, we could imagine a situation in which Cl• radicals were still being created but had run out of methanes to attack. Can you see what the Cl• radicals might turn to instead, and what extra product might result?

28a Why would the turn of events in question 28 reduce the usefulness of this reaction from the point of view of the preparative organic chemist who wanted control over which product was achieved?

b Why would the reaction still be popular with the industrial chemist who wanted to prepare chlorinated alkanes for the solvent market?

One further piece of evidence provides persuasive reinforcement of the mechanism proposed here. When a little tetraethyl-lead is added to a mixture of methane and chlorine, it brings the reaction initiation temperature *in the dark* down to 140 °C. This compound (the anti-knock agent which supplies the notorious lead in petrol) is known to break up into ethyl radicals, $C_2H_5\bullet$, at 140 °C.

$$Pb(C_2H_5)_4 \rightarrow Pb + 4C_2H_5\bullet$$

Questions

29 Show how a little of this compound could intrude into the mechanism, and thereby initiate the chlorination of methane even in the dark. Explain why the reaction still does not happen *below* 140 °C.

30 Having studied the *mechanism*, let's look at the chlorination of methane from the *outcome* viewpoint. Look at the overall starting and finishing species, and decide whether it represents a substitution, an elimination or an addition reaction. Then check your answer against Table 10.1.

10.9   Other reactions of alkanes

Combustion

'Other reactions' has a humble ring, but in fact it includes the most sociologically important chemical reaction of the last two centuries, namely, the combustion of fossil fuels. If we were still naming ages after significant technologies (bronze, iron), our own time could be dubbed the 'fossil-fuel age'. Aspects of life that people in developed countries now take as inalienable rights – such as the ability to travel, and freedom from dawn-to-dusk physical drudgery (Figure 10.15) – would still be the preserve of a small elite, were it not for the release from the limitations of muscle power granted by fossil fuels. And on the international political front, the whole agenda of the second half of this century has been interlinked with the control of access to hydrocarbons via the petroleum industry.

The reason that this reaction is not given huge amounts of space in chemistry textbooks is because, surprisingly, the actual mechanism of burning is not well understood. Once the equations are written, there is little more theoretical story to tell.

$$C_4H_{10}(g) + \tfrac{13}{2}O_2(g) \rightarrow 4CO_2(g) + 5H_2O(l)$$

There is of course a great wealth of practical knowledge about the way in

Figure 10.15 The beneficial legacy of fossil fuels – freedom from dependence on muscle-power

which hydrocarbons burn in internal combustion engines, and it is from this applied-science direction that we get one clue concerning the nature of the reaction. It comes from the use of the anti-knock agent tetraethyl-lead $(C_2H_5)_4Pb$.

Tetraethyl-lead is a compound which as we have seen readily breaks up into free radicals, and, as automobile engineers saw, by so doing reduces the tendency of petrol/air mixtures to 'go off' too early in the compression stroke. It thus seems reasonable to assume that the burning of hydrocarbons is itself a free-radical chain reaction, and that possibly the $C_2H_5\bullet$ radicals slow the reaction down by providing radical-meets-radical termination steps.

Octane number

Whatever the exact mechanism of combustion, it is known that internal combustion engines are more efficient when the fuel/air mixture is highly compressed before ignition (Figure 10.16a). Some hydrocarbons can tolerate this high compression treatment and still burn smoothly from the spark plug, whereas others, as was mentioned in the previous paragraph,

Figure 10.16a Internal combustion engines are more efficient if the fuel is mixed with air and highly compressed before ignition

'pre-ignite' before the spark. This causes the engine to fire unevenly in a manner known as 'knocking'.

The tendency of any given hydrocarbon towards pre-ignition is expressed on a scale set up by the petroleum industry and called the **octane number**. The isomer 2,2,4-trimethylpentane was given the number 100, to indicate its position as 'smoothest burner', while the bad pre-igniter heptane was given the number 0.

$$CH_3{-}CH{-}CH_2{-}C{-}CH_3$$

octane number 100

$$CH_3CH_2CH_2CH_2CH_2CH_2CH_3$$
octane number 0

All other compounds and mixtures from the petroleum fraction of crude oil were placed on this spectrum, by practical testing.

Nature is no respecter of human need for a smooth ride, and not surprisingly the petroleum fraction distilled from natural crude oil turns out *not* to have a particularly high octane number, containing as it does too many of the wrong isomers. From the early part of the century to the late 1970s the answer to the problem was to improve the octane number by adding tetraethyl-lead.

In the 1980s the pressure to produce unleaded petrol became intense, and attention shifted to methods of octane-number improvement that involved changing the hydrocarbon mix itself. **Arenes** (that is, hydrocarbons containing a benzene ring, see Section 10.15) were found to be better burners than alkanes and conversion of one to the other was possible.

The secret lay in **catalytic reformation**. This name covers a whole family of reactions in which the action of heat and pressure in the presence of a platinum catalyst is used to convert alkanes and cycloalkanes to arenes. Typical reactions are shown overleaf.

methylcyclohexane   toluene

heptane   toluene

1,2-dimethylcyclopentane   toluene

Unleaded petrol is now a major seller on garage forecourts, and governments have offered tax reductions as an incentive for motorists to convert to unleaded (Figure 10.16b). The story may still have a twist, however, since arenes are rather more carcinogenic than alkanes. Some scientists have lately been suggesting that we may have exchanged lead poisoning for cancer risks. At least the presence of arenes should discourage the old technique practised by petrol thieves of sucking up petrol by mouth prior to siphoning it.

Figure 10.16b Unleaded petrol is available on all forecourts

Cracking

Returning to the reactions of alkane hydrocarbons, we come to a class of industrially important reactions that is rather less visible than combustion, in public perception at least. The reactions, collectively called **cracking**, provide a vital step in the provision of many materials, especially plastics, that make up the physical backdrop to our modern lives. The name refers to a group of processes that usually result in the production of

smaller hydrocarbon molecules from bigger ones, in an increase in the number of molecules possessing double bonds and in the creation of some hydrogen. From a commercial viewpoint the single most important point about cracking is that *it turns less valuable compounds into more valuable ones.*

Typical cracking reactions are:

$$C_{12}H_{26}(g) \rightarrow C_8H_{16}(g) + C_4H_8(g) + H_2(g)$$

$$C_2H_6(g) \rightarrow C_2H_4(g) + H_2(g)$$

We can recognise two distinct groups of industrial operations, distinguished by temperature and the presence or absence of catalysts.

The lower-temperature/with-catalyst operation is known as **catalytic cracking** and employs silica/alumina catalysts and the relatively modest temperature of 450 °C. It is mainly used to convert the gas–oil fraction of crude oil (C_{12} hydrocarbons and above) into a slightly shorter-chain 'cocktail' suitable as a precursor of petrol.

At higher temperatures (800–850 °C), the process is known as **thermal cracking** or **steam cracking** (in which steam is used to dilute the feedstock). This version of the process is versatile and can accept feedstocks as diverse as a whole mixed fraction from crude oil, right down to a specific feed of ethane for conversion to ethene.

Figure 10.17 is a schematic diagram of the steam cracker jointly operated by ICI/BP Chemicals on Teesside. The feedstock in the diagram is the crude oil fraction called **naphtha**, a mixture of alkanes, cycloalkanes and arenes. The stages of the process, as symbolised by the boxes in Figure 10.17, are as follows.

• *Furnace*: steam and naphtha, in variable proportions, spend between 0.1 and 0.5 seconds at 800–850 °C, which is sufficient to carry out the cracking. The mixture now contains many smaller molecules in the C_1–C_4 range, as well as longer-chain compounds suitable for petrol and central heating oil. The hot gases from the furnace are quench-cooled to 400 °C, producing high-pressure steam as a by-product. This steam is used to drive the compressors for the next stage but one (Figure 10.18).

• *Primary fractionation*: here the gases are cooled to 40 °C and petrol ('gasoline') and fuel-oil fractions condense out.

• *Compression, drying and chilling*: the gases are cooled and compressed to such an extent that only hydrogen remains in the gas phase, enabling its removal.

• *Subsequent separation stages*: the remaining compounds are in the C_1–C_4 range and are separated in order of size, as gases, by gradual warming and decompression in a succession of chambers.

Almost every one of the products from the cracker is of crucial commercial significance. Taking them in order of separation, gasoline and fuel oil are fuels, hydrogen and methane are raw materials in ammonia manufacture (see the Haber process in Chapter 18), ethene and propene are monomers of vital plastics (Section 10.12), and butadiene is one of the key materials in rubber manufacture (Section 10.13) (Figure 10.19). Table 10.2 shows the yield of each product available from typical cracking runs, both on naphtha and on ethane.

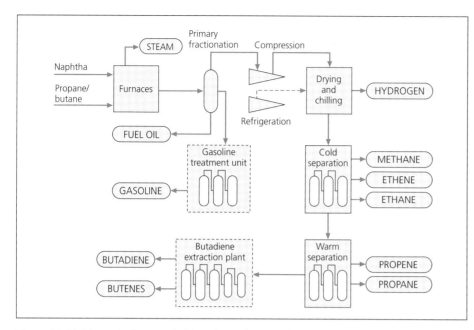

Figure 10.17 Schematic diagram of ICI/BP Chemicals' Teesside cracking plant

Table 10.2 Products of steam cracking of two 'feedstocks', naphtha and ethane (percentage by mass)

Product	Feedstock	
	Naphtha	Ethane
Hydrogen	1	6
Methane	15	8
Ethene	30	78
Propene	16	3
Butene	5	1
Butadiene	5	2
Petrol	23	2
Fuel oil	4	—

Question

31 Here are some questions to test your understanding of the cracking process:

a What variables can the plant operator manipulate to obtain control over the product mix?

b In the furnace stage there is a typical example of the 'never throw anything valuable away' principle which pervades industrial chemistry. Can you say what it is?

c Most of the stages of the cracker are separations of different sorts of molecule from each other. On the basis of which property of the various molecules is this separation done?

Figure 10.18 ICI's Wilton Teesside steam cracker

Figure 10.19 Both the garden chairs and the sportswear are made from polypropene whose monomer, propene, is available from the cracking plant

10.10   The alkenes – bonding, isomerism and nomenclature

The defining feature of an alkene is its possession of a $C{=}C$ double bond. The rest of the molecule is made up of ordinary alkyl groups.

Question

32 Because of the double bond, alkenes carry two fewer hydrogen atoms than the alk*ane* containing the same number of carbons. What then is the molecular formula of an alkene containing *n* carbon atoms?

Bonding in alkenes

This discussion of $C{=}C$ double bonds should act as a review of work done in Chapter 4.

The double bond is made up of two rather disparate parts. The 'spine' of the bond is created from a 'nose-to-nose' overlap between two orbitals, each containing a single electron – it is therefore very like the single $C{-}C$ bond in an ordinary alkane. (This style of overlap gives rise to what is referred to as a 'σ-bond'.) However, the 'double' part of the $C{=}C$ bond is produced by a very different 'cheek-to-cheek'-style overlap between two p orbitals, again containing one electron each (Figure 10.20).

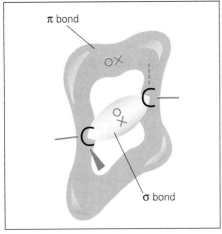

Figure 10.20
Electron-cloud picture of alkene $C{=}C$ bond

The properties of alkenes that make them distinct from alkanes are all attributable to this extra π-bond, but for now we will focus on just one aspect. The π-bond is different from the σ-bond in that it can only exist in *one plane*. Hence any attempt to twist one end of the molecule relative to the other can only succeed at the cost of breaking the double part of the bond. This is known as *hindered rotation* and it gives rise to a whole new class of isomerism.

Types of isomerism

Between some pairs of isomers, the difference is very large. For example, methoxymethane and ethanol possess different bonds, and belong to very different classes of compounds (ethers and alcohols).

$$CH_3OCH_3 \qquad CH_3CH_2OH$$

Even a rather more similar pair such as butane and methylpropane have different numbers of primary and secondary hydrogen and carbon atoms.

$$CH_3CH_2CH_2CH_3 \qquad CH_3\overset{\displaystyle CH_3}{\underset{\displaystyle |}{C}}HCH_3$$

The kind of isomerism which occurs in these two cases is referred to as **structural isomerism**.

In contrast the pair of but-2-enes shown below have none of these differences.

$$\underset{H_3C}{\overset{H}{\diagdown}}C{=}C\underset{CH_3}{\overset{H}{\diagup}} \qquad \underset{H}{\overset{H_3C}{\diagdown}}C{=}C\underset{CH_3}{\overset{H}{\diagup}}$$

The barrier preventing the one becoming the other is hindered rotation about the $C{=}C$ bond, so the only difference between the two molecules is *the arrangement of their atoms in space*. Such pairs of isomers belong to a group known as **stereoisomers**. Stereoisomers that derive their differences from hindered rotation about $C{=}C$ bonds are called **geometric isomers**.

Clearly we are going to have to add to the rules on nomenclature in order to describe these isomeric differences, as well as to carry out the more basic task of conveying the position of the double bond, and the particulars of any alkyl groups arranged around it.

Nomenclature in alkenes

The existence of a double bond somewhere in the molecule is conveyed, of course, by the suffix '-ene'. As with alkanes, the core name derives from the longest carbon chain, using eth-, prop-, but-, etc. (p. 194–5). The position of the double bond, if subject to any variability, is indicated by interrupting the molecule's name with a number, to indicate the carbon on one end of the double bond (using the 'lowest number' rule). Hence the molecules shown below are named but-1-ene and but-2-ene (rather than -4-ene and -3-ene respectively).

$$CH_3CH_2CH{=}CH_2$$

but-1-ene

$$CH_3CH{=}CHCH_3$$

but-2-ene

Question

33 Name two alkenes which have no need of a number in their names.

We now turn to the important issue of discriminating between the pairs of geometric isomers. You can see from the skeletal structures opposite that 'butene' has several isomers, some structural, but that but-2-ene has a geometric pair. The 'chair' form of the pair is referred to as the *trans* isomer, while the 'boat' form is called the *cis* isomer. Their full names are *trans*-but-2-ene and *cis*-but-2-ene, respectively.

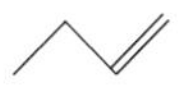

but-1-ene

methylpropene

but-2-ene

trans   *cis*

Question

34 Draw and name all of the isomeric versions of 'pentene'. (*Hint*: begin with those isomers which have all five carbons in a row, and then consider those with four-carbons-plus-a-branch. Remember, it is not a true branch when you attach the floating carbon to the end of a chain.)

10.11   Major themes in alkene chemistry 1: Heterolytic cleavage

Apart from hindered rotation, what other characteristics derive from the nature of the $C=C$ bond? The double part of the bond (the π-bond) is quite weak, so we can deduce that the alkenes are quite reactive, compared to the alkanes.

Question

35 Look up the bond enthalpies for a single and double carbon–carbon bond, and estimate the strength of the π-bond. Then look up and record the bond lengths of the two carbon–carbon bonds.

Next we should note that alkenes offer a number of attractive features to potential attackers. Their bonds are nearly as non-polar as those of alkanes, and that should interest those attacking reagents that depend upon homolytic cleavage in their 'victims' (in other words, free radicals, p. 203). And yet there is a higher concentration of electron density in the double bond that might create vulnerability to attack by a second group of intruders – the electrophiles. In fact both of these reaction types are very important in the reactions of alkenes.

Electrophilic attack on alkenes

The electron density in the $C=C$ double bond will attract the positive ends of molecules like the hydrogen halides. It is not a great energy sacrifice for the π-bond to break heterolytically, and the resulting offer to the incomer is of *both* the electrons formerly in the bond. This in turn induces extra polarity in the attacker, culminating in the heterolytic breakage of a second bond – the one in the halide itself:

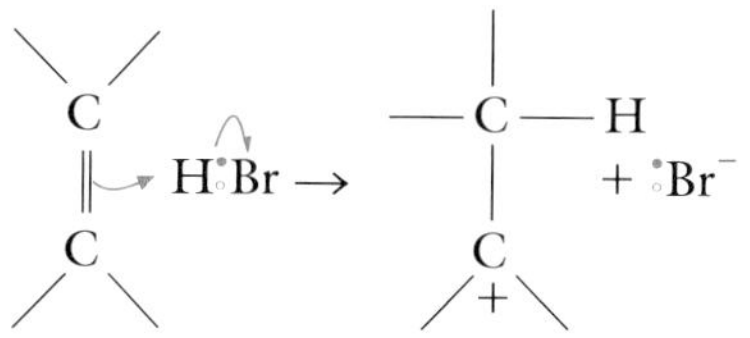

By now we have reached an intermediate state, featuring the halide anion and an unstable reactive species featuring some rather non-classical bonding (a 3-valent carbon atom) and called by the generic name **carbocation** (or 'carbonium ion'). This name is given to any molecule featuring a positive charge sited on a carbon atom. All the members of this group are inevitably short-lived reaction intermediates. The decay of the carbocation is rapid and is triggered by the approach of the anion. The fusing of anion and cation brings us to the product molecule, which is a halogenoalkane, and which has no double bonds:

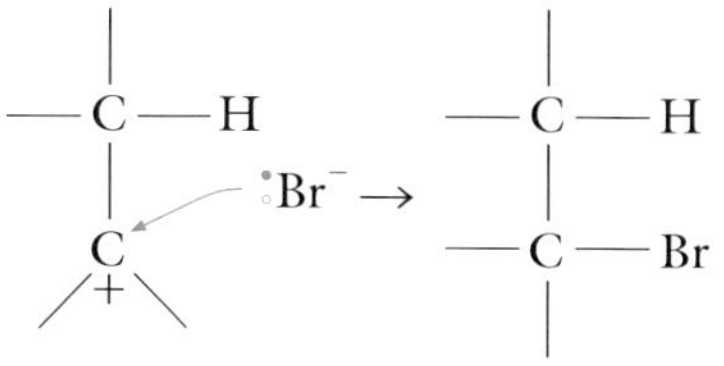

Question

36 Looking at the overall outcome of this mechanism, would you classify it as a substitution, addition or elimination reaction?

Two rather subtle points can be added to this story. If the substrate (the alkene being attacked) is asymmetric, then the reaction can take two alternative routes.

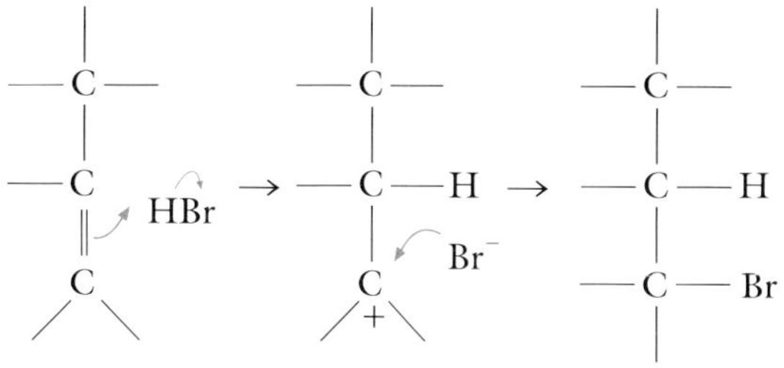

primary carbocation

or

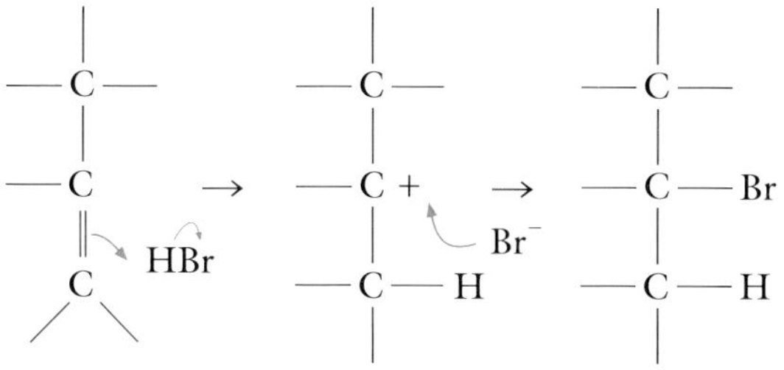

secondary carbocation

In practice it turns out that, in the example chosen, the product is almost invariably the 2-bromopropane. This is thought to be because the carbocation that lies on the path to the 2-isomer is more stable than that *en route* to the 1-isomer. The extra stability of the secondary carbocation derives from the small positive inductive effect of alkyl groups – the secondary carbocation has *two* alkyl groups pushing electron density towards it, whereas the primary has only one such alkyl group. Since these inductive pushes will have the effect of spreading the charge out round the molecule, it follows that two pushes are better than one.

So this idea predicts the way in which hydrogen halides add to asymmetric alkenes, which is that the hydrogen goes on to the carbon atom which already bears more hydrogens. This rule is **Markovnikov's rule**.

Question

37 Give the products you would predict from the addition of hydrogen bromide to the following alkenes:

a methylpropene

b but-1-ene

c 2-methylbut-1-ene.

You were promised two post-scripts to add to the basic story of electrophilic addition to alkenes. Markovnikov was the first; the second concerns the paradoxical idea of non-polar electrophiles.

Electrophilic attack on alkenes by non-electrophiles

How can a non-electrophile behave like an electrophile? The answer is 'when it comes close to the C=C bond in an alkene'. We have already seen how the proximity of the C=C bond caused the already polar H—X bond to become more polar still, to the point of heterolytic breakage. So perhaps it is not so surprising that proximity to a C=C bond can cause an initially non-polar bond to *become* polar. We are really only talking about one pair of electrons, rather concentrated in space, repelling another pair of electrons.

Thus it is that non-polar halogen molecules (except iodine) can add readily to alkenes. The rest of the mechanism (below) is identical to that which applied to hydrogen halides.

The addition of hydrogen does not appear to follow the same mechanism. Instead it requires rather more violent conditions and the crucial intervention of catalysts. The most effective catalysts come from a family of transition metals known for their ability to adsorb hydrogen and alkenes on to their surface (nickel, palladium and platinum), so we must assume that the closeness of contact of the two reactants while held on the metal surface must greatly reduce the activation energy of the reaction (Figure 10.21).

Figure 10.21 An imaginary model of how a surface (of Ni, Pd or Pt) might catalyse a reaction (addition of hydrogen), by holding the reactants in the correct orientation

10.12 Major themes in alkene chemistry 2: Homolytic cleavage

One of the most important industrial uses of the smaller alkene molecules is as the monomers of a series of familiar plastics – poly(ethene) 'polythene', poly(propene) 'polypropylene', poly(phenylethene) 'polystyrene' and poly(chloroethene) 'PVC' are famous examples. (The names in quote marks are the familiar names and those preceding them are the official IUPAC names.) The polymerisation mechanism involves free radicals and features initiation by molecules specially designed to cleave in a radical-generating way.

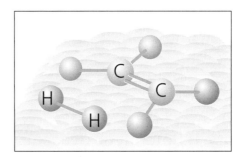

These peroxide radicals then induce the π-bond to cleave homolytically, with the now-familiar result that a new radical is formed and the propagation of the chain reaction begins. The only difference between this case and that of the alkanes is that this time the σ-bond remains after the π-bond has broken, so nothing is 'dropped off' and the new radical becomes longer and longer. In this way, molecules of relative molecular masses of the order of 20 000 can be produced before a termination step interrupts the chain.

Look carefully at what happens to the 'R' group in the generalised mechanism below (I represents an initiator radical).

Notice how it crops up on *alternate* carbons in the eventual polymer. Thus, in order to predict which polymer might result from a given monomer, it is only necessary to recognise what is playing the part of R in the monomer and substitute it into the generalised mechanism.

Questions

38 Suggest a termination step for this mechanism.

39 Give the structures of the polymers which would result from the following monomers:

a ethene

b propene

c chloroethene.

40 Give the structures of the monomers that would give rise to the following polymers:

a Orlon (fibres, fabrics)

b Teflon (non-stick cooking utensils)

c polystyrene (barrels of pens, etc.).

The properties of polymers are a massive subject in applied chemistry, because of the great diversity of outcomes arising from techniques such as co-polymerisa-

Ziegler–Natta catalysts

In the 1950s, some new methods of polymerisation were discovered which allowed for much more controlled chain growth, and therefore for much more crystalline products. Ordinary free-radical polymerisation of a monomer like ethene occasionally leads to branches in the chain, caused when a growing radical attacks an existing chain instead of a fresh alkene molecule. The 'midships' unpaired electron can now act as the growth site for a branch.

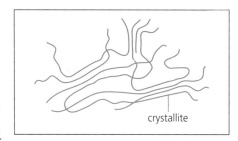

tion (alternating monomers). Some of this flavour will come across in Section 10.13 on the rubber industry. However, one simple generalisation is that the fine tuning of polymer properties derives in the main from two sources – **intermolecular forces** and **cross-linking**. The former are mainly van der Waals' forces, and are thus dependent on the degree of crystallinity (that is, molecular orderliness) in the polymer (Figure 10.22).

Figure 10.22 Schematic picture of a polymer, showing crystalline and random regions

Question

41 Do you think this branching would increase or decrease the amount of crystallinity in a polymer?

The method of Ziegler and Natta, for which they won the 1963 Nobel Prize for Chemistry, involves a mechanism in which chain growth is guaranteed to take place at only *one* site, producing linear chains as shown (overleaf).

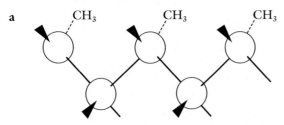

The catalysts are complexes of triethylaluminium/titanium trichloride, but behind the complexity of the name lies the simplicity of what they *do* – and that is symbolised in the schematic diagram above. The growth point is always where the polymer is 'growing out of the catalyst'. Further work showed that they could also control the stereospecificity of the polymer, so that for example it has become possible to make poly(propene) with every methyl group on the same side of the chain (structure (a) below), and a synthetic substance identical to natural rubber, in which all the double bonds are *cis* (structure (b) below). The subject of Ziegler–Natta catalysis is revisited in Chapter 19, Section 19.6.

Question

42 Given that the poly(propene) chains produced by the Ziegler–Natta method are more even and straight in structure, and remembering from Chapter 6 the factors that influence van der Waals' forces, predict how the properties of poly(propene) made by this method might differ from that made by free-radical polymerisation.

10.13   The rubber industry

Periodically throughout this book the point has been made that the chemistry student needs to have a creative 'mind's eye'. This principle is never more true than in the study of **elastomers** – that is, the class of compounds which can be deformed by modest levels of force and which then return to their original shape when the force is removed.

The word elastomer is very nearly synonymous with the word rubber. The former, newer, word carries more information about the materials which it defines (that they are elastic and polymeric). The latter, older, word was created in 1770 when the British chemist Joseph Priestley found that he could rub out pencil marks with the coagulated sap of a Brazilian tree called *Hevea brasiliensis* (Figure 10.23). The people who nowadays make the rubber and rubber products hardly bother to discriminate between the words. Although the range of modern elastomers contains materials very dissimilar in composition to the sap from those original Brazilian trees, to the rubber industry they are all just types of rubber.

Figure 10.23 Tapping the tree *Hevea brasiliensis* for its sap

Since the key aspect of rubbers is their response to deforming forces, we need to be clear about what happens to a material, on a molecular level, when it is put under stress.

Four modes of deformation of solids

A solid material is deformed when a force causes its particles to occupy different positions in space from those occupied when the material was 'at rest'. The 'at rest' positions – which of course define the natural shape of the lump of material – are set by **bonds**, which may be of both intra- and intermolecular types.

An ionic crystal represents a very simple situation – its ions are all held in position by strong ionic bonds, and to deform it you have to stretch, compress or bend those bonds. The result is that such crystals are hard, and, as explained in Section 5.3, brittle.

Question

43 There is a short distance over which ionic crystals can deform (under high force) before they surrender to brittle fracture. If the force were removed before the fracture, would the crystal behave plastically or elastically? In other words, would it stay deformed or return to its original shape? Justify your answer.

In a polymer like poly(ethene) (which is the official name for the common plastic polythene), there is much less 'position discipline' operating on the particles. Deformation can derive from not one but several sources, listed below as 'modes'.

• *Mode (a)*: the classical covalent C—C bonds can stretch, in an analogous operation to the ionic-crystal case.

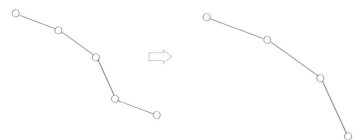

• *Mode (b)*: whole molecules can move relative to each other, in a sort of sliding process.

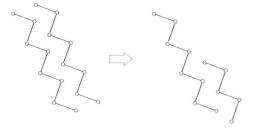

• *Mode (c)*: within molecules, the —CH₂— units can be pulled into a more linear formation from a more globular one. You could compare this to what happens to the coiled cable connecting the body of a telephone to its handset – when the telephone is at rest the cable often takes up a coil-of-coils conformation, but when you pick up the handset to speak, you pull it into a single-coil strand (Figure 10.24).

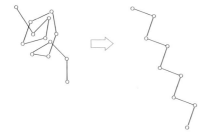

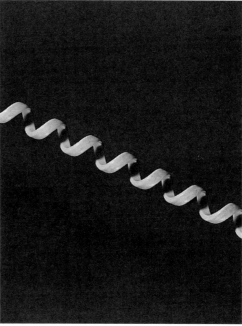

Figure 10.24 Telephone cords are good models of polymer molecules. First the coil-of-coils unravels . . . then the coils themselves open up

• *Mode (d)*: still within molecules, the C—C—C bond *angles*, rather than the bonds themselves, can stretch. Each C—C—C unit could open out from its rest angle of 109.5°. To continue with the telephone-cord analogy, this would be like a further stretch opening up the turns of the coil, after the unravelling stage described in mode (c).

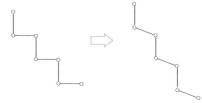

These four modes of deformation vary greatly from one another. First, they require very different sizes of force and, second, some lead to plastic and others to elastic deformation. It is your task in the next question to consider which mode fits which description.

Question

44a Which *one* of the four modes would result in permanent (plastic) deformation? Justify your answer.

b Which *one* depends for its action on the *freedom of rotation* of C—C bonds?

c Which *one* of the four modes requires work to be done mainly against *inter*molecular forces?

Figure 10.25 Polythene is stressed and heated to make plastic sheet tubing

d When a lump of polythene is stressed by a force (Figure 10.25), the four modes of response will operate in order of weakness of resistance (weakest first). The switchover between modes will happen when the scope for deformation within each mode has been exhausted. (In the analogy, the telephone cord uncoils its coil-of-coils before it begins to open the turns of its main coil.) What do you think would be the order in which the four modes of deformation of a piece of polythene would operate, as the lump is taken all the way from rest to breakage?

What makes rubber rubbery?

A rubber is a material that allows elastic deformation to happen readily at low force, and yet which at the same time resists plastic deformation. In other words, elastomers are substances in which mode (c) and (d) deformations, as defined above, occur much more readily than mode (b). Polythene fails to be an effective elastomer because the forces necessary to deform it elastically and to deform it plastically are *too close* in magnitude. (Returning yet again to the telephone cord, it is as if you had no sooner got the cord straight than you found it stretching to destruction.) It is possible to trace this failure back to two features of the polythene molecule.

1 Considering each polythene molecule in isolation, we would predict that the unravelling of the coils (mode (c)), would meet with little resistance. All it needs is rotation in the C—C bonds. But the molecules do not exist in isolation – in fact, as we saw in Figure 10.22 there are regions of considerable order ('crystallinity') in polythene. In these regions the molecules will, on the one hand, not be so tangled, by virtue of their orderly mutual alignment, and on the other hand each molecule can 'hang on' to its fellows in a show of organised resistance to the deforming force. This intermolecular mutual support also means that it is harder to achieve elastic deformation via mode (d) deformations (angles opening beyond 109.5°) than you might think from looking at a single polythene molecule. The upshot is that polythene is *harder* than most true elastomers.

2 The next problem with polythene is that shortly after it has 'grudgingly' given way to mode (c) and (d) deformations it begins to reveal its *lack* of resistance to mode (b) (intermolecular sliding) deformation. So after a brief region of elastic deformation it undergoes permanent plastic stretch (as shown in the sequence below).

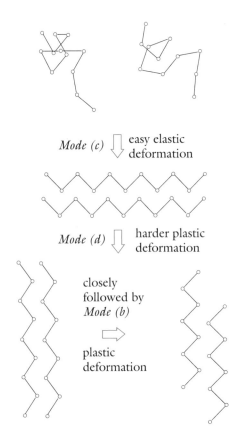

In fact there are plenty of elastomers which are *harder* than polythene, but crucially they are *much more resistant* to molecular sliding.

A pair of requirements for an elastomer can be drawn up against the backdrop of polythene's failure. The molecules in an elastomer should meet the following requirements.

1   Be not too ordered in their relationship with each other, so as to allow for easy mode (c) and (d) (elastic) deformations.
2   Strongly resist intermolecular sliding movement (plastic deformations, mode (b)).

Question

45   You are asked to tailor some molecules to show ideal elastomeric properties. Which of the following pairs of features would you consider desirable? (Justify your answers.)

• Cross-linking between molecules present/absent.

• Molecules find it easy/hard to take up orderly, crystalline arrangements.

The molecules of the rubber industry

As your answer to question 45 will have revealed, the need is for molecules which are not readily capable of being lined up and yet are heavily cross-linked (Figure 10.26). Nature can manage the first

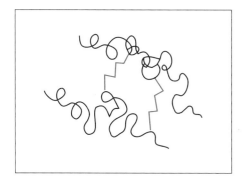

Figure 10.26 Schematic diagram of an ideal molecular arrangement for elastomeric properties. (The green bits are cross-links, which in most rubbers are provided by S_8 strands.)

specification, but it required a person called Charles Goodyear to engineer the second.

Natural rubber

The molecule of natural rubber has already been shown (p. 212). Its non-systematic name is *cis*-polyisoprene (and *all* the C=C bonds are in the *cis* conformation). It is a soft, easily deformable material. Its isomer *trans*-polyisoprene also exists, under the name of **gutta percha**, and in sharp contrast it is a hard, only slightly rubbery material formerly used to cover golf balls.

Question

46   On p. 212 the *cis* isomer is drawn as linearly as it can be. Draw the *trans* isomer, also at maximum linearity. Mention any ways in which these two drawings can be used to explain the differences between the two materials.

Pure natural rubber is too soft for any everyday use except for erasing pencil marks. In nearly every other application the chains have been artificially cross-linked. It was the discovery of this step which gained immortality for the name of Goodyear, since it was Charles Goodyear, in the US in 1839, who found that the heating of natural rubber with sulphur gave a material that was harder, stronger, less tacky when hot and less brittle when cold, than natural rubber itself (Figure 10.27). The process is now known as **vulcanisation** – a tenuous word-link via sulphur's association with volcanoes, to Vulcan the god of fire.

Sulphur atoms form bridges between molecules of the original rubber. The S_8 linkages are thought to connect the poly-isoprene chains, as shown below, and thereby achieve the resistance to plastic deformation which was the second requirement above.

Figure 10.27 Charles Goodyear

This vulcanised natural rubber is the rubber of balloons, surgeons' gloves and condoms, and it possesses one property in which it excels over all its synthetic counterparts. Perhaps because its chains are among the longest of all the elastomeric molecules, it shows a property when under stress known as 'stress crystallisation', which means that highly stressed samples of natural rubber acquire more 'polythene-like' attributes (Figure 10.28). This has the useful effect that the material opposes tearing forces by a sort

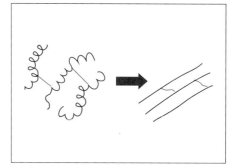

Figure 10.28 Schematic diagram of stress crystallisation

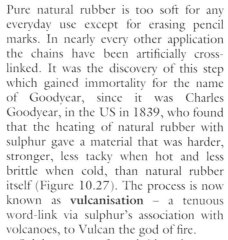

of in-built negative feedback, getting stronger as it is 'threatened'. (It is partly the effects of stress crystallisation which cause a balloon to 'fight back' as you inflate it.) The change caused by stress crystallisation is irreversible (Figure 10.29).

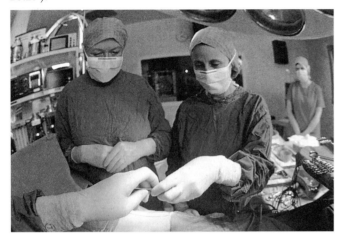

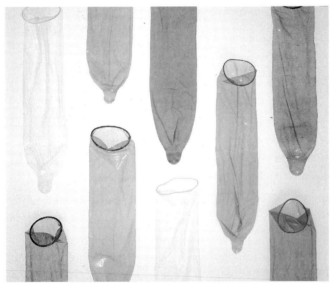

Figure 10.29 All of the above are made from natural rubber

Question

47 How would you explain the phenomenon of stress crystallisation in molecular pictures of the type shown in Figure 10.22?

Aside from erasers, balloons, condoms and gloves, many rubber objects are black. This is due to the presence of the second most important rubber additive after the cross-linking agent, namely the **filler**. The most commonly used filler is carbon, which in the form of soot or lampblack is available in such small particle sizes that it can achieve quite intimate contact with molecular chains. The inclusion of lampblack enables fine-tuning of the abrasion resistance of the rubber (important in tyres), and opposes chemical ageing by protecting the polymer molecules from ultra-violet photons.

Synthetic rubbers

Natural rubber is now only the second most commonly used rubber in industry. There are over 20 synthetic alternatives in use today, each offering special advantages. Each one is also capable of being fine-tuned to particular specifications by use of additives and by variable degrees of cross-linking. One of the simpler synthetic compounds is **butadiene rubber (BR)**, a polymer of but-1,3-diene. The monomer is polymerised by normal free-radical methods as in other polyalkenes, as shown below. It can be vulcanised just like natural rubber.

$$Initiation \quad R\cdot \quad H_2C = CH - CH = CH_2$$

$$\downarrow$$

$$RCH_2CH = CH - CH_2\cdot$$

Propagation

$$RCH_2CH = CHCH_2\cdot \quad H_2C = CH - CH = CH_2$$

$$\downarrow$$

$$[RCH_2CH = CHCH_2]CH_2 - CH = CH - CH_2\cdot$$

first monomer

. . . and so on.

Termination
Any two radicals meet

Question

48a Describe one important difference between BR and a normal polyalkene like those on p. 211.

b Use that difference to explain why BR is a rubber and not a plastic.

The rubber which is in biggest use today (and the main compound in car tyres) is **styrene butadiene rubber (SBR)**, accounting for over 45% of all production. It is a 'copolymer' of the monomers styrene and buta-1,3-diene.

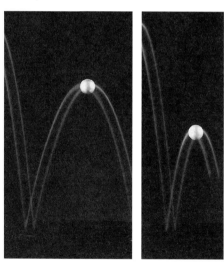

'pure' butadiene rubber

styrene butadiene copolymer

The most significant effect of having the flag-like benzene rings protruding from the chain is that SBR turns out to be a less bouncy rubber than BR. The difference is highlighted if balls of the two rubbers are dropped from the same height. Although both recover their shapes after the distortion of the bounce, the BR ball is the one that bounces back nearer to its dropping height (Figure 10.30).

Figure 10.30 BR balls are bouncier than SBR balls

Question

49a If the SBR ball fails to recover the potential energy it had before the drop, where and in what form would that energy have ended up?

b Relate this difference in action between the two rubbers to the presence of the benzene rings on the chain.

The balance between the roles of 'bouncer' and 'shock absorber' is one of the main practical defining features of a rubber. Many applications of rubber products can be placed on a spectrum from 'bounce important' to 'shock absorption important', and the rubber in Figure 10.31 fits clearly into the latter category.

Somewhere towards the shock absorber end of the spectrum is the rubber used in car tyres. Car tyres are made from SBR and of course get hot, as your answer to question 50a on the bouncing ball might have led you to predict. In every revolution of a tyre the molecules undergo a squash/release cycle, just like a bouncing ball.

A curious fact is that very big tyres for heavy vehicles like lorries are made from a blend of SBR and BR, rather than just the former, for reasons that should emerge in the next question.

Question

50 Surely we do not particularly want our lorries to bounce around, so what benefit must there be in the controlled introduction of BR into the tyre 'compound'?

Chemical resistance

One of the problems of rubbers in continual use is their vulnerability to aerial oxidation, especially when the air contains small but significant amounts of ozone (which itself can be a by-product of the use of cars). Ozone attacks C=C double bonds, in a way that will be described in item 5 of the 'reactions of alkenes' summary in Section 10.14 (p. 219).

Question

51 Explain what would happen to long-chain molecules containing C=C bonds, if they underwent the ozonolysis reaction in Section 10.14.

One of the responses from the industry to this problem was to develop a rubber *without* C=C bonds by copolymerising

Figure 10.31 A large shock absorber in a bridge

Figure 10.32 Car hose made of EPM for corrosion resistance

ethene and propene. The product is known as EPM (Figure 10.32). EPM rubbers need to be cross-linked just like the other kinds, but this cannot be done by sulphur because there are no C=C bonds in the chains. (Although sulphur atoms do not actually join the chain *at* the C=C bond, they appear to need to abstract a hydrogen *adjacent* to a C=C bond.) Instead EPM is 'cured' (cross-linked) by means of peroxides, in a reaction exactly analogous to the one shown on p. 211, except this time it is an intended outcome rather than an unwelcome side reaction.

Question

52 It seems odd that while the individual materials poly(ethene) and poly(propene) are very definitely plastics rather than elastomers, the copolymer belongs to the latter category. This is achieved by having the sequencing and stereochemistry of the copolymer as random as possible – or to put it another way, as 'un-Ziegler–Natta-ish' as possible.

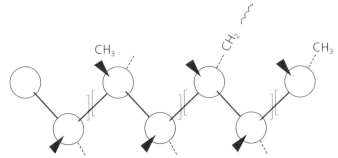

Explain why this ploy succeeds in making EPM a rubber.

In conclusion

In a section of this size it is not possible to give a full account of a major area of chemical activity like the rubber industry. Table 10.3 provides a brief reference to some of the compounds that have not been mentioned in the text above, as well as a summary of the properties of those that have.

Figure 10.33 Rollerblade wheels are made from polyurethane

Table 10.3 Some members of the rubber family

Compound	Monomers		Special assets	Applications
Natural rubber (NR)		'isoprene'	Resists tearing	Rubber gloves, balloons and condoms
Butadiene rubber (BR)		butadiene	Very bouncy	Bouncy balls
Styrene/butadiene rubber (SBR)		butadiene styrene	Versatile and hardwearing	Tyres
Ethene/propene-copolymer (EPM)		ethene propene	Resists oxidation and ageing	Car hoses
Butyl rubber		'isobutene'	Low porosity to gases	Inner tubes
Neoprene	Cl	'chloroisoprene'	Combines resistance to oxidation with good oil resistance	Car wiper blades, rubber dinghies
Nitrile rubber	CN	'acrylonitrile' butadiene	Very high oil resistance	Oil seals in engines
Polyurethanes	O=C=N—∿∿∿—N=C=O HO—∿∿∿—OH		Very versatile . . . can be poured cold as monomers and cured in moulds	Very wide . . . furniture foams, skateboard wheels (see Figure 10.33), elastic threads in 'Lycra' swimwear

10.14 Reactions of alkenes – a summary with examples

1 Catalytic addition of hydrogen

$$H_2C=CH_2 + H_2 \xrightarrow[\text{Ni, Pd or Pt}]{\text{catalyst}} H_3C-CH_3$$

2 Addition of halogens

$$CH_3CH=CHCH_3 + Br_2 \rightarrow CH_3CHBrCHBrCH_3$$

3 Addition of hydrogen halides (notice Markovnikov addition)

$$CH_3CH=CH_2 + HBr \rightarrow CH_3CHBrCH_3$$

4 Addition of water (also Markovnikov-style)

$$CH_3CH=CH_2 + HOH \xrightarrow{H^+} CH_3CH(OH)CH_3$$

5 Ozonolysis

This is a process which breaks a molecule at its carbon–carbon double bond and therefore allows researchers to deduce the location of the double bond. The fragments from the broken double bond emerge in the form of aldehydes or ketones.

$$CH_3CH=CHCH_3 + O_3 \rightarrow CH_3CH=O + O=CHCH_3$$

6 Free radical polymerisation

$$nCH_2=CH \xrightarrow[\text{heat}]{\text{peroxides}}$$
with R below the CH.

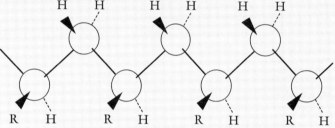

(not stereospecific, possibility of branches)

7 Ziegler–Natta polymerisation

$$nCH_2=CH \xrightarrow[\substack{\text{lower temperature} \\ \text{than 6 above}}]{\text{Al}(C_2H_5)_3, \text{TiCl}_3}$$
with R below the CH.

(all Rs on same side, no possibility of branches)

Question

53 A compound was found to have the molecular formula C_4H_8, and on ozonolysis gave two fragments of molecular mass 58 and 30. Suggest possible alternative structural formulae and names for the original compound.

10.15 The arenes – bonding and nomenclature

It is useful sometimes to relive the historical development of modern knowledge. In organic chemistry, this process is nowhere more interesting than in the story of the discovery of the structure of benzene.

Benzene had been isolated and studied experimentally since the first half of the nineteenth century. Its molecular formula was known to be C_6H_6. The problem was in deciding how these atoms were arranged – finding the structural formula, in fact. It is not too hard to suggest a few structures that agree with the molecular formula, for example:

But it's another matter to make them fit benzene's behaviour.

Alkene chemistry stressed the reactivity of compounds with double bonds (and by extension triple bonds should be the same only more so). Yet benzene reacts with bromine, for instance, at a rate thousands of times *slower* than its alleged 'sister' alkenes and alkynes.

There was also the difficult problem of the evidence of **isomer numbers**. For example, after long and painstaking synthetic chemistry, people were convinced that there was only *one* substituted chlorobenzene C_6H_5Cl. This suggested therefore that all six hydrogens were in equivalent positions.

Question

54 Which of the two structures (a) and (b) would have to be rejected on the basis of this evidence?

If the number of monosubstituted products was surprisingly low, the number of ways of putting *two* substituents on the benzene molecule was the opposite. There are *three* isomers of formula $C_6H_4Cl_2$.

Question

55 How many dichlorobenzenes could be made from the 'survivor' of the two structures on the previous page?

The breakthrough in human thought about the benzene structure finally happened as the result of a dream. The dreamer was August Kekulé (Figure 10.34), at the time the Professor of Chemistry at Ghent in Belgium. The dream was dreamed in 1865, and Kekulé recounted it later in his diaries (here in a translation by Alexander Finlay).

I turned my chair to the fire and dozed. Again the atoms were gambolling before my eyes. This time the smaller groups kept modestly in the background. My mental eye, rendered more acute by repeated visions of this kind, could now distinguish larger structures, of manifold conformation; long rows, sometimes more closely fitted together; all twining and twisting in snake-like motion. But look! What was that? One of the snakes had seized hold of its own tail, and the form whirled mockingly before my eyes. As if by a flash of lightning I awoke.

The structure that Kekulé proposed after waking up was still not quite the final answer, despite the intensity of his dream. However, his 'cyclohexatriene' molecule (see below) was a major step forward and succeeded in explaining why, for example, there is only one monochlorobenzene.

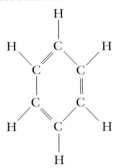

However, it still left some unanswered questions. For example, if you look care-

Figure 10.34 August Kekulé (chemist and dreamer)

fully at the structure, you will see that in fact there should be *four* disubstituted isomers instead of the three shown below.

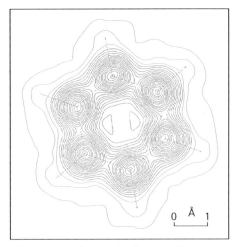

Question

56 Focus on the isomer in which the two substituents are on *adjacent* carbons – there really ought to be *two* of these, by Kekulé's ideas (and yet there are not in reality). Can you draw a fourth isomer, from these hints?

The other problem with Kekulé benzene was that it was an alkene, so there was still no answer to the surprising stability of real benzene. People who measured enthalpies of reaction produced yet more objections to Kekulé's structure, based on calculations like the one in the next question.

Question

57a Construct a Hess's Law cycle (see Chapter 9) featuring real benzene,

its constituent elements and the gaseous atoms.

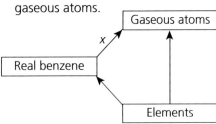

(You will need data for the enthalpies of atomisation of the elements plus the enthalpy of formation of real benzene.)
   Thereby calculate the *enthalpy of atomisation* (shown as *x*) of real benzene.

b Now switch to the Kekulé structure, and to the table of bond enthalpies in the book of data. Break the whole molecule into atoms and work out the enthalpy change, which is the enthalpy of atomisation of 'cyclohexatriene' or Kekulé benzene.

c Which molecule, the real or the false, is the more stable and by how much?

To bring the story more up-to-date, we can add the evidence of X-ray diffraction studies on solid benzene. Figure 10.35 shows how the atoms are arranged. All of the carbon-carbon bonds are the *same length*, and the length is somewhere between that of typical single and double carbon-carbon bonds. The modern

Figure 10.35 Electron density map of benzene

explanation for this is to propose *delocalisation* (Chapter 4, p. 59). A slightly older but quite helpful mental picture is to imagine a 'hybrid' of two Kekulé structures, in rapid alternation.

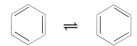

Whatever picture you favour, it is accepted that the average electron density in a benzene molecule is as shown in Figure 10.36. It has been discussed more fully in Chapter 4.

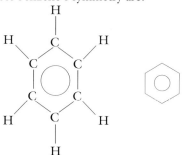

Figure 10.36 Computer-generated image of the structure of benzene

Two symbolic representations that reflect benzene's symmetry are:

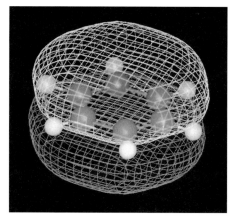

Nomenclature

This symmetry means that the naming of substituted benzene derivatives is simple. The carbons are numbered round the ring, beginning at a substituent. Examples include:

chloro-benzene   1,2-dichloro-benzene   1,4-dichloro-benzene

10.16   Major themes in benzene chemistry

Benzene displays its ring of π-electron density, exciting the attention of the same kind of reagent that attacked the alkenes with such relish.

Question

58 What kind of reagents were they?

Indeed, it even goes so far as to carry out the first step of the mechanism, in which the double bond opens and takes on the electrophile.

But just when the 'other end' of the attacker (for example, the other end of the halogen molecule) is closing in to complete the addition, the benzene molecule suddenly demonstrates its determination to maintain its π-ring. The result is that an unsuspecting hydrogen is given notice to quit and leave its electrons behind.

The whole operation is best seen as an **electrophilic substitution** reaction (rather than the anticipated addition reaction), with a single bromine atom replacing one hydrogen atom. The twist in the plot can be traced back to the central feature of the benzene molecule – namely the desirability of preserving the stable delocalised ring system. The equations above show this mechanistic drama being played out, using bromination as an example.

The above account is misleading on one point. Benzene is *not* as attractive as alkenes to electrophiles. Reaction rates for benzene, compared to alkenes, are many times slower for the same attacker, and conditions often have to be made much more violent to achieve reaction – or else some clever catalysts have to be used.

Catalysis in electrophilic substitution reactions of benzene

You will recall from the section on alkenes that they had the ability to *make* electrophiles out of neutral molecules (p. 209). Just coming near to the electron density of the double bond was enough to turn a bromine molecule, for example, into a polarised molecule with an electrophilic end. But this kind of 'auto-generation' do-it-yourself approach is completely ineffective when it comes to attacks on benzene. The reasons for this are probably two-fold: one is that the ring electrons are not so concentrated in space and so are less able to polarise the incomer; and, second, the π-bond itself is strengthened by delocalisation and more unwilling to offer an electron pair.

For reaction to occur, either there must be an order-of-magnitude increase in concentration (neat bromine, for example, instead of solutions) or a catalyst must be employed to *pre-polarise* the attacker. This second approach has been used with success in two classes of benzene substitution, in both cases making clever use of the *electron deficiency* of certain polar-covalent compounds with oxidation number +3.

Catalysis in bromination

Iron is added to bromine Br_2 to facilitate bromination, but that is not to say that iron *is* the catalyst. What happens is that the two reagents between them make some $FeBr_3$. This is a molecule that suffers from an inability to complete a rule-of-eight electron arrangement. It is thus very susceptible to dative covalent bonding with almost anything with a lone pair. The resulting $Br_2/FeBr_3$ complex is held together by a dative bond from one bromine atom: but this gift will render that bromine atom positive and cause it in turn to try to redress the imbalance by pulling on the electrons of the other 'exposed' bromine atom. *This* is the atom, now suitably polarised by the electron-drag of the catalyst, which carries out the attack on the ring, as shown below.

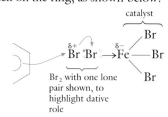

Br_2 with one lone pair shown, to highlight dative role

The Friedel–Crafts reaction

A very similar ploy, using a closely related catalyst, is used to persuade benzene to undergo 'alkylation' – that is, substitution of a hydrogen atom by an alkyl group. The equation for the change is:

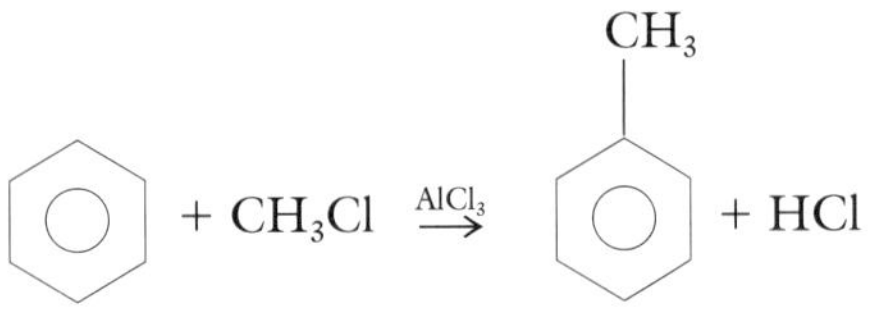

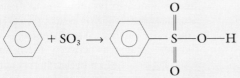

Question

59 By comparison with the bromination case, and with the aid of displayed formulae, explain how the AlCl$_3$ catalyst facilitates the Friedel–Crafts reaction.

Several other aspects of benzene ring chemistry apply to substituted rings. These issues will be dealt with in the individual sections relating to phenol, phenylamine, etc., in Chapters 11, 12 and 13. We shall see then how ring and substituent influence each other.

10.17 Reactions of benzene – a summary

1 Nitration

Generation of electrophile:
$$H_2SO_4 + HNO_3 \rightleftharpoons NO_2^+ + H_2O + HSO_4^-$$

Attack:

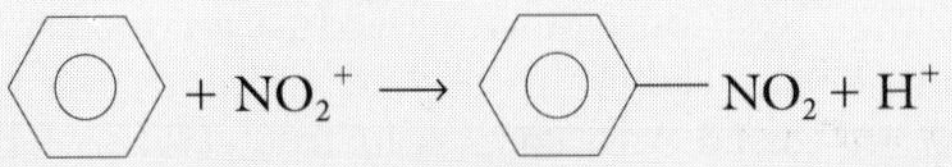

nitrobenzene

2 Sulphonation

Generation of electrophile:
$$H_2SO_4 \text{ (fuming)} \rightleftharpoons H_2O + SO_3$$

Attack:

benzenesulphonic acid

3 Halogenation

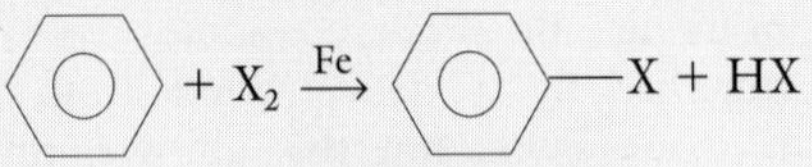

4 Friedel–Crafts alkylation

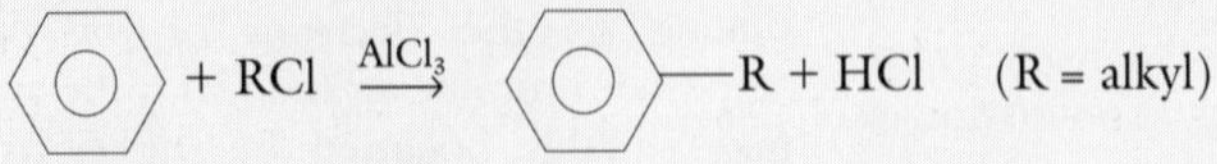

(R = alkyl)

5 Hydrogenation

This is the only member of the list which is analogous to a reaction of alkenes, but in the case of benzene the conditions have to be more severe:

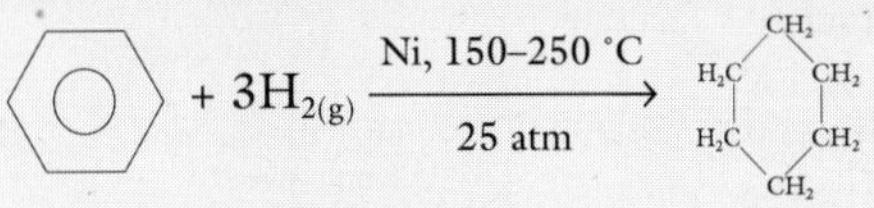

cyclohexane

Questions

60 There is also a reaction known as the Friedel–Crafts **acylation**, in which an RC=O (acyl) group is substituted into the ring. By comparison with the other version of the Friedel–Crafts reaction, what reagent(s) (other than benzene) might be needed to bring this about?

61 Some researchers had an interesting idea which they thought would shed light on our proposed mechanism of electrophilic addition. They decided to carry out a halogenation of benzene, just as in paragraph 3 above, but instead of using a normal halogen element they used an 'interhalogen' compound, iodine chloride, ICl. They were delighted when they found that, far from making a mixture of chlorobenzene and iodobenzene, they had made nearly 100% of *a single compound*.

a Which of the two possible products do you think they had made?

b Why did the result please them?

Summary

Lists of reactions of the various groups of compounds have already appeared in Sections 10.7 (alkanes), 10.14 (alkenes) and 10.17 (benzene).

Carbon's role on Earth

• The chemistry of carbon contains a very large number of compounds, far greater than the number of compounds of any other element except hydrogen – and the only reason for hydrogen's number is because nearly every carbon compound includes hydrogen atoms.

• Some of the reasons for the great number of carbon compounds are as follows.

1 Carbon's **chain-forming** ability is unrivalled. Its single bond to itself is strong, especially compared with those between other potential chain-forming elements like nitrogen and oxygen. Carbon therefore offers a 'construction kit' of great versatility.

2 Carbon's position in row *two* of the Periodic Table means it is bound by the rule of eight, and since a carbon atom has four outer electrons, it follows that carbon's normal bonding behaviour must always feature four covalent bonds. So it does not have, on the one hand, electron deficiency in the manner of boron. Neither does it have, on the other hand, any lone pairs in the manner of nitrogen and oxygen. Thus two sources of 'restlessness' that might otherwise have led to chemical activity are missing from the character of carbon compounds. The stability of carbon compounds is a contributing factor to their great diversity and number.

3 Carbon's electronegativity is not extreme and is close to that of hydrogen. This means that the C—H bond, which is one of the main building linkages of organic chemistry, is almost non-polar. Even bonds from carbon to the halogens and to oxygen and nitrogen do not have huge dipoles. This general absence of highly polar bonds is another factor which gives carbon compounds quite a 'quiet' chemistry.

4 In the evolution of life on Earth, the presence of methane and carbon dioxide gases in the primeval atmosphere enabled the original lightning- and ultra-violet-induced events in the evolution of biochemistry to be centred on carbon.

Structures, names, formulae and isomerism

• It is important to remember the **three-dimensional nature** of the bonds to a carbon atom. The fact that they adopt a tetrahedral shape means that flat-page representations are deceptive:

• There is **free rotation** about single C—C bonds. This means that any '90° bends' in carbon frameworks as shown on paper can be eliminated with a notional twist of a bond:

• A molecule may be drawn as a straight chain on the page, but models reveal that the nearest approach to straightness, consistent with tetrahedral bond angles, is a zig-zag of carbon-carbon bonds. The same molecule may be converted, by free bond rotations, from a 'straight' zig-zag into a globular clump.

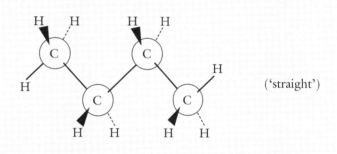

(In fact, to call bond rotations 'free' is to be slightly inaccurate, as most molecules have a preferred orientation.)

• **Branches** in a flat drawing are real molecular features. This molecule is branched, whichever way you draw it:

Branching as an alternative to straight-chain formation is one of the sources of the phenomenon of **isomerism**. Isomers (in this context) are different molecules with the same molecular formula.

- The naming system of organic chemistry has grown out of the need to convey the siting of branches. It uses the original names for the alkane series (methane, ethane, propane, etc.) to identify the longest straight chain in the molecule, and then uses numbers to identify the site on the main chain at which a branch occurs. The syntax of the name is ⟨SITE NUMBER⟩ ⟨IDENTITY OF BRANCH⟩⟨NAME OF MAIN CHAIN⟩, as in ⟨2-⟩ ⟨methyl⟩⟨propane⟩. The system can also accommodate the names of functional groups, by naming them as branches, as in 2-chloropropane. Certain subsidiary rules are needed to cope with more complicated naming problems:

1 If there is a choice of number which would tell the same story (as in 2-methylbutane and 3-methylbutane), then the lower number is used.

2 If there is more than one of the same substituent, commas are used; if there are different substituents, then alphabetical order decides the naming sequence. Both these rules can be seen operating in 2-bromo-1,1-dichloropropane.

- In a variation on the above method, certain substituents are recorded with a suffix rather than a prefix, for example, propan-2-ol.

- Formulae express the structure of chemical substances at various levels of complexity. The hierarchy of complexity is as follows, simplest first.

1 **Empirical formulae** express the simplest whole-number ratio of the atoms in a compound. For example, the empirical formula of the compound ethane is CH_3, even though the actual molecule contains two of these units.

2 **Molecular formulae** express the actual numbers of each type of atom in a compound, but no more than that. Ethane, for example, has the molecular formula C_2H_6.

3 **Structural formulae** convey extra information by grouping the atoms in the sequence which corresponds to their actual arrangement. Ethane becomes CH_3CH_3, and 2-methylbutane is $(CH_3)_2CHCH_2CH_3$.

4 **Displayed formulae** show every bond in the molecule in letters-and-sticks format, while **skeletal formulae** do the same job but without the letters and without the C—H bonds. In skeletal formulae, each apex of the skeleton (either 'bend' or 'end') is understood to represent a carbon atom, with the appropriate number of hydrogen atoms attached. The displayed formula of 2-methylbutane is

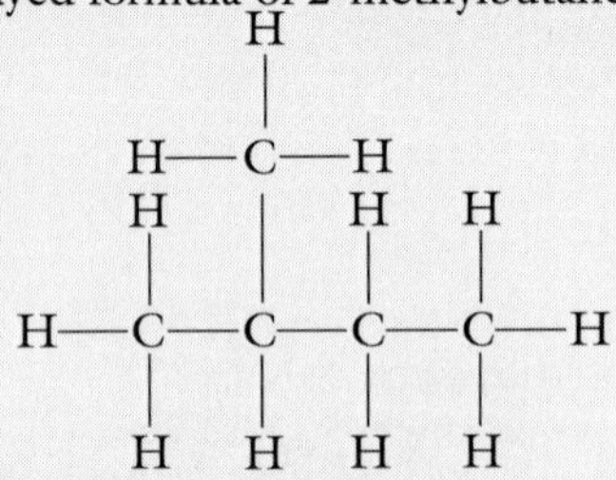

and the skeletal equivalent is

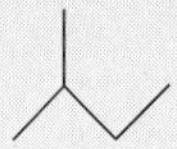

- There is a level of isomerism that corresponds to each level of formula complexity – so for example there is even a level of isomerism in which the isomers are two different molecules with the same *displayed formulae*. The only level of formula that carries so much information that no two molecules could share the same formula is the 3-D picture. Examples will help clarify these points:

1 Below are two isomers with the same empirical formula, CH_2, cyclohexane and cyclopentane (with the true differences revealed by use of the structural formulae).

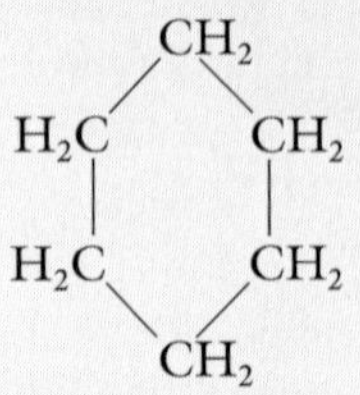

cyclohexane 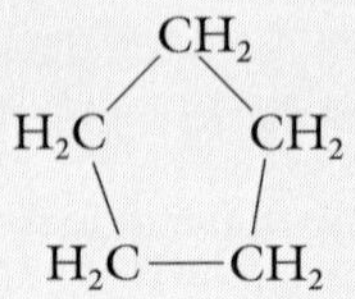   cyclopentane

2 Two isomers with the same molecular formula, C_2H_6O, are ethanol and methoxymethane (with the true differences revealed by use of the structural formulae).

CH_3CH_2OH CH_3OCH_3
ethanol   methoxymethane

3 The two isomers below have the same structural formula, $CH_3HC=CHCH_3$: *cis*- and *trans*-but-2-ene (with the true differences revealed by use of the displayed formulae).

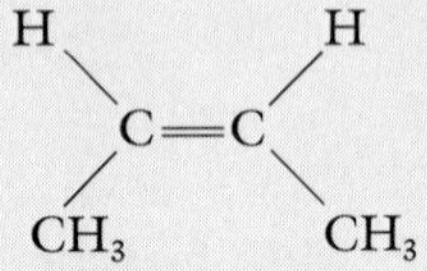

cis-but-2-ene

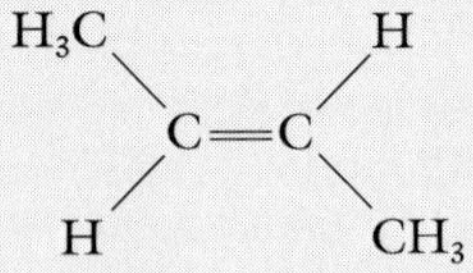

trans-but-2-ene

4 Below are two isomers with the same displayed formula – both called 2-chlorobutane (with the true differences revealed by use of 3-D formulae).

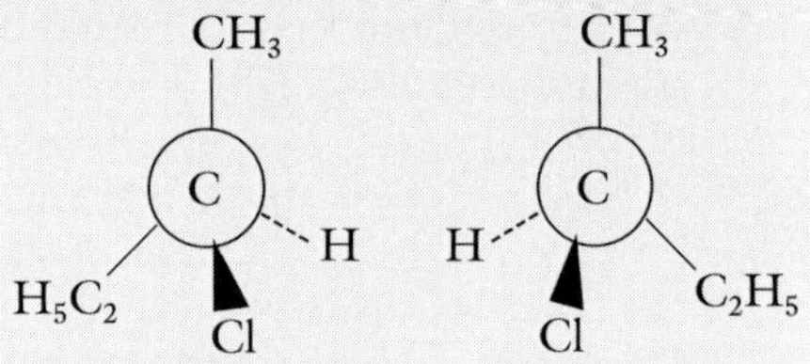

- There are different types of isomerism. Isomers which have *different* structural formulae, like the pairs in paragraphs 1 and 2 above, are called **structural isomers**. Isomers that have the *same* structural formulae – that is the same groups in the same order – but differ in the way those groups are *arranged in space*, are called **stereoisomers**. Examples 3 and 4 above represent the two sub-groups of stereoisomerism: in 3 the cause is lack of rotation about the C=C bond (and this is called **geometrical isomerism**), while 4 relates to the idea that certain structures are different from their own mirror images (and this is called **optical isomerism**). It is studied in Chapter 13.

Classifying organic reactions: 1 – by mechanism

• All chemical reactions require either bond breaking or bond making, and in the overwhelming majority of cases, both.

• We can classify bond breaking according to where the pair of electrons go after breakage. **Homolytic cleavage** sends the electrons one to each partner, while **heterolytic cleavage** involves one member of the former bond keeping both electrons.

• Most of the time we can predict what sort of bond breakage will befall each specific bond by considering two factors: its **polarity** and its **electron density**. These factors will influence aspects of possible chemical reactions.

1 **How the bond breaks**: if the bond is **polar**, the negative end of the dipole will probably keep both the electrons in a heterolytic cleavage. If the bond is **non-polar** a homolytic cleavage is more likely.

2 **Which species are attracted to the bond**: if the bond is **polar** then species called **nucleophiles** (that is, species with lone pairs of electrons, sometimes actually anionic) are likely to offer *pairs* of electrons to the positive ends of dipoles, and heterolytic breaking will ensue. If the bond is **non-polar** then species called **free radicals** (that is, species with a single unpaired electron) are likely to offer *single* electrons and initiate a chain of bond making and homolytic breaking. If the bond is **electron rich**, then species called **electrophiles** are likely to attract a pair of electrons *from* the bond (which is normally of the π type).

Classifying organic reactions: 2 – by outcome

• Apart from classifying reactions by mechanism, there is another option, which is to focus on the overall *result* or *outcome* of the chemical reaction.

• The terms used in this system are as follows.

1 **Substitution**, in which one atom or group of atoms is replaced by another.

$$
\begin{array}{c} B \\ | \\ A-C-E + X \end{array} \rightarrow \begin{array}{c} B \\ | \\ A-C-X + E \\ | \\ D \end{array}
$$

2 **Addition**, in which the π-bond of a C=C double bond is cleaved, and two new atoms or groups are added to the molecules.

$$
\begin{array}{cc} A & B \\ \diagdown & \diagup \\ C=C & + X-Y \end{array} \rightarrow \begin{array}{cc} A & B \\ | & | \\ D-C-C-E \\ | & | \\ X & Y \end{array}
$$

3 **Elimination**, in which two atoms or groups are removed from a molecule, to be replaced by the π-bond of a double bond.

$$
\begin{array}{cc} A & B \\ | & | \\ D-C-C-E \\ | & | \\ X & Y \end{array} \rightarrow \begin{array}{cc} A & B \\ \diagdown & \diagup \\ C=C & + X-Y \\ \diagup & \diagdown \\ D & E \end{array}
$$

• The outcome words can be used in collaboration with the mechanistic ones, as in 'nucleophilic substitution' or 'free-radical addition'.

• All the above concepts are now ready to organise the data associated with particular groups of compounds. In this chapter all the groups are **hydrocarbons**.

Alkanes

• The alkanes are the compounds made from single-bonded combinations of carbon and hydrogen atoms only.

• The C—H bond is almost **non-polar**, so alkanes show little reactivity towards nucleophiles. Neither do alkanes possess any outstanding concentrations of electron density, so there are no reactions with electrophiles.

• The main reactions of alkanes are, mechanistically, of the **free-radical** type. The most well understood is **free-radical substitution**, and the most important is **combustion**.

• The reactions of *halogens* with alkanes are free-radical substitutions. The mechanisms proceed via three steps which are characteristic of free-radical reaction mechanisms, namely **initiation**, **propagation** and **termination**. The products are mixtures of halogenoalkanes, with both single and multiple substitutions.

• All alkanes burn in oxygen, very exothermically. The mechanisms are not fully understood.

• The major source of alkanes is *crude oil*. The mixture of molecules in crude oil does not correspond to the profile of human demand for hydrocarbons. As a result many longer-chain alkanes are **cracked** or **catalytically reformed**, with the object of producing shorter-chain **alkenes** and **arenes**. These molecules are valuable feedstocks for the polymers and petrochemicals industries.

Alkenes

• Alkenes show no polarity, but in the C=C double bond they have a region of **high electron density**.

• Predictably, the alkenes have reactions that belong to two mechanistic groups, the **free-radical** and **electrophilic** ones. However, the outcomes of nearly all alkene reactions of whatever mechanism are **additions**.

• Many molecules, including water, hydrogen halides, and even dihalogens, can perform **electrophilic additions** on alkenes. Even if a molecule is not naturally electrophilic (as is the case with dihalogens), the very closeness of the double bond, as the attacking species approaches, induces a temporary polarity that enables one end of the attacking species to play the electrophile.

• **Free-radical additions** are the route to **polymerisation** of alkenes. The mechanisms show the classic three-step feature, but the result is the chaining together of large numbers of alkene molecules, the whole polymer ending up with the structure of an extended alk*ane*.

• A C=C double bond *will not allow free rotation*, because of the nature of the overlap in the π-bond. This reaction gives rise to **geometrical isomerism**, in which molecules can differ from each other purely by the arrangement of groups on either end of the double bond.

• The class of compounds called **rubbers** is mostly comprised of polyalkenes. Rubbers show in a very graphic way how sets of properties can arise from particular molecular layouts. The central underlying cause of soft-elastic behaviour is seen as a combination of low intermolecular forces between adjacent chains, high degrees of freedom for bond rotation, and limited cross-linking to maintain overall relative molecular-chain positions.

Arenes

• The benzene ring looks as if it should behave like an alkene. However, the apparent double-bond feature in fact turns out to involve a delocalised π-bonded system.

• The result is that benzene compounds – the arenes – are less reactive than expected. They do undergo **electrophilic attack**, but with two differences.

1 They react much more *slowly* than alkenes, under corresponding conditions.
2 They often resist the final mechanistic step that leads to addition, instead opting to undergo **substitution** of a ring hydrogen, preserving the delocalised π-bond.

• Table 10.4 shows the correlations between types of reactants, types of mechanisms, styles of bond breakage and outcomes.

Table 10.4 Types of reactant, types of mechanism, styles of bond breakage and outcomes

Group of compounds	Polarity of bonding in crucial bonds	Electron density in crucial bonds	Mechanism		Outcome
			Type of 'attacking' species attracted	Type of bond cleavage	
Alkanes	Non-polar	'Normal'	Free radicals	Homolytic	Substitution
Alkenes	Non-polar	High	Electrophiles	Heterolytic	Addition
			Free radicals	Homolytic	Addition polymerisation
Arenes	Non-polar	High-ish	Electrophiles	Heterolytic	Substitution of ring H atom

11

Organic chemistry 2: Molecules with pronounced dipoles

11.1 Introduction

There are ways of looking at organic chemistry that make it all seem very complex and daunting. One is to flick through a 1000-page undergraduate-level book on the subject. Yet equally there are ways of looking which do the exact opposite. One viewpoint presented in Chapter 10 saw the whole subject as being shaped by a few simple electrostatic concepts (Figure 11.1). For instance:

• some covalent bonds are electrostatically polarised (Figure 11.1a)

• polarity in a bond determines which of the two participating atoms gets to keep the electrons when it breaks (Figure 11.1b)

• polarity in a bond also determines which reagents will be attracted to and interact with the atoms on each end

• this in turn determines how chemical changes are likely to happen (Figure 11.1c).

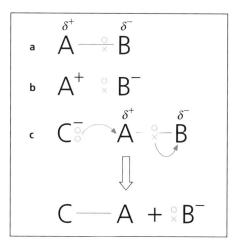

Figure 11.1 How polarity shapes mechanism

Seen from this simple set of rules, all organic reactions are just events in which two, one or zero electrons on an attacking species interact with a substrate offering zero, one or two electrons respectively, all operating under the basic 'opposites-attract' ground rule of electrostatics. There are three main types of attacking species: nucleophiles, electrophiles and free radicals.

Chapter 10 was concerned with hydrocarbons – alkanes, alkenes, arenes – that do not have polar molecules and are therefore only suitable for attack by free radicals, or, if they possess π-bonds in addition to single bonds, by electrophiles.

Question

1 a As a reminder to yourself, redefine the words 'electrophile' and 'free radical', and give examples of molecules likely to be attractive to each one.

b Which of the 'big three' attacking species had no part to play in hydrocarbon chemistry?

c Which of the 'big three' attacking species have respectively zero, one and two electrons to offer their 'victims'?

In contrast, this chapter deals with organic molecules containing single bonds between carbon and halogen atoms, and between carbon and oxygen atoms. So they all have a strongly dipolar bond as their main defining theme. Nucleophiles, which failed to play a part in hydrocarbon chemistry, now move to centre stage, and their attacks will bring about outcomes of **substitution** and **elimination**. We will even, in the alcohols section, see organic molecules themselves acting as nucleophiles towards other organic molecules.

11.2 Halogenoalkanes 1: Nomenclature and a major theme

The simplest molecules that possess pronounced dipoles are those in which a hydrocarbon framework carries one or more halogen atoms. They are named by a similar system to that used for alkanes, with prefixes for the halogens (fluoro, chloro, bromo, iodo) being used like the alkyl prefixes, methyl, ethyl, etc. as shown below.

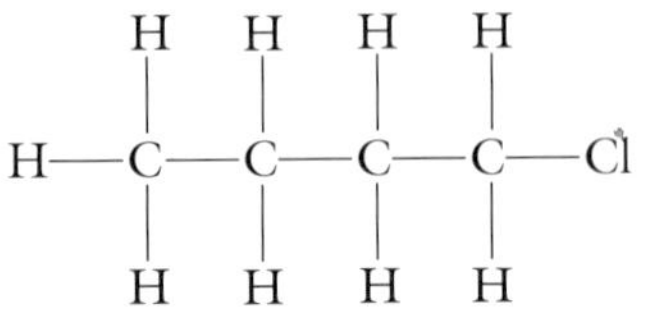

1-chlorobutane

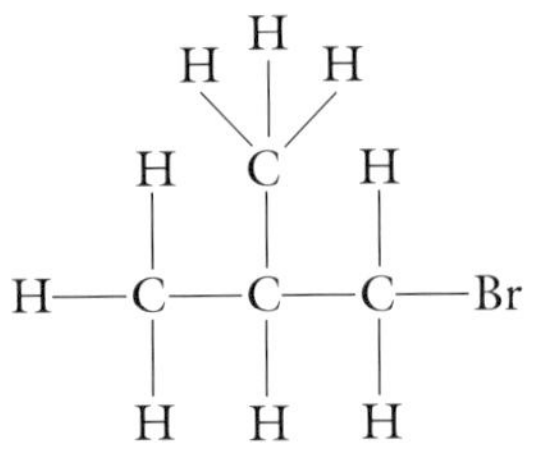

1-bromo-2-methylpropane

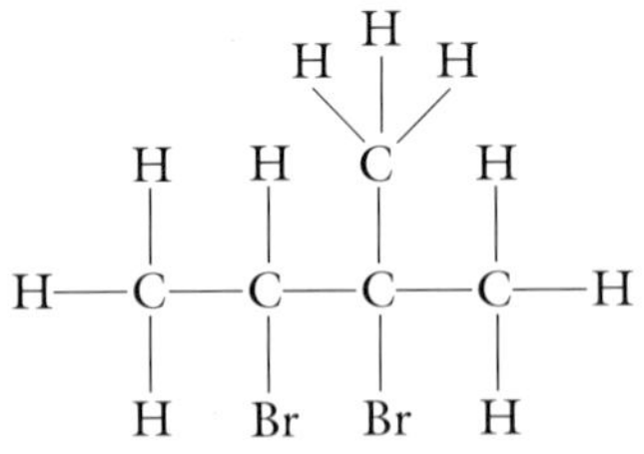

2,3-dibromo-2-methylbutane

Question

2 By comparison with the above examples, name the compounds that follow.

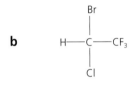

a

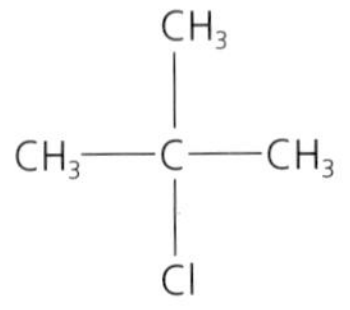

b

c $CH_3CH_2C(C_2H_5)_2$

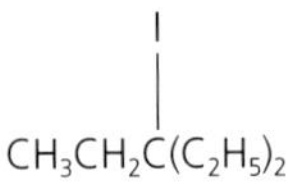

Bond polarity

All the halogens, even iodine, are more electronegative than carbon, so the C—X bond (where X is the halogen atom) will always be polarised. (See Chapter 5 to remind yourself of the definition of electronegativity).

Question

3 a Show in a diagram the polarity of the C—X bond, using the 'delta' (δ) symbolism.

b Bearing in mind this polarity, suggest how the C—X bond is likely to cleave, saying where the electrons would end up after reaction.

c How many electrons would an attacking species have to have, in order to restore the carbon atom to the rule of eight?

d What name do we give to such an attacker?

e Draw a 'curly arrows' diagram (p. 199) to show the attack of the reagent on the substrate, followed by the proposed products.

f Is the leaving group better described as a halogen atom or a halide ion?

Question 3 is designed to outline the main stages in halogenoalkane chemistry (so check your answer against the official version if there is any uncertainty). As you will have seen, nucleophilic attack at the carbon end of the C—X bond is the commonest event that can happen to a halogenoalkane. The outcome is substitution. There will be variations on that theme, which will be explored in later sections, but the attack by the pair of electrons is a constant thread running through all those variations. We begin our exploration by looking in detail at nucleophilic substitution.

11.3 Halogenoalkanes 2: Nucleophilic substitution

The archetypal nucleophilic substitution reactions of halogenoalkanes are those in which the nucleophile is either water or OH⁻. A curly arrow picture of one of these reactions is given below.

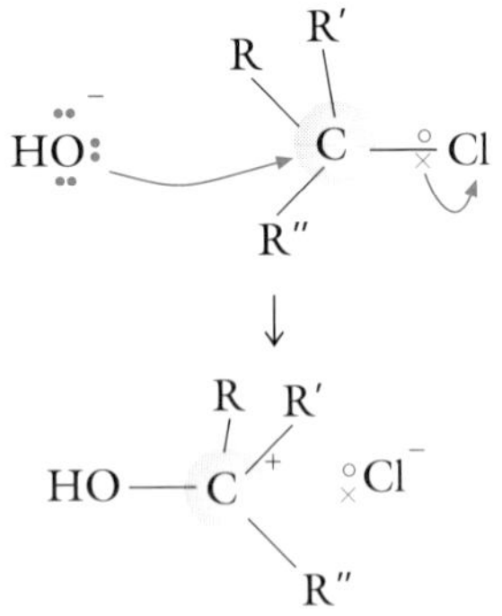

But not every nucleophilic attack conforms exactly to this pattern. For instance, there are cases in which the two stages shown above take place in sequence rather than together; and other cases in which the point of attack by the nucleophile is a hydrogen atom attached to a carbon situated one place away from the C—X bond. And even for those substitutions that do follow the simple pattern, there is a large range of reaction rates. We will look at each of these three types of variation in turn.

Variations in reaction rate – the identity-of-halogen factor

Which is the more rapidly substituted, a chloroalkane or an iodoalkane? There are two sensible lines of speculation: one is that at any given temperature, the weaker bond (C—I) will give way in a greater percentage of collisions than the stronger bond (which is, for those who have already read Chapter 14, equivalent to proposing a lower activation enthalpy for the substitution of iodine than for chlorine). Alternatively, one could imagine that the greater **dipole** in the C—Cl bond will be more attractive to the electron pair of the nucleophile and that therefore there will be *more collisions* in the chlorine case.

	Average bond enthalpy (kJ mol^{-1})	Dipole moment (D)
C—I	238	1.87 (CH$_3$I)
C—Br	276	1.81 (CH$_3$Br)
C—Cl	338	1.62 (CH$_3$Cl)

It is easy to test these rival hypotheses – aqueous silver nitrate provides an excellent nucleophilic environment in which to place the substrates to be compared. It not only supplies the nucleophile, but it also provides a handy indicator system for telling when reaction has taken place.

Question

4 a What is the most likely candidate for the role of nucleophile in AgNO$_3$(aq)?

b What is the visible sign that a halogen atom has been ejected by substitution?

c Write chemical equations for the reaction of AgNO$_3$(aq) with 1-iodo-butane, showing separately the substitution stage and then the 'indicator' reaction.

As for the answer to the question of the rival theories, the data in Table 11.1 should make things clear.

Table 11.1 Some practical results

Substrate	Approximate time for visible precipitate (s)
1-chlorobutane	75
1-bromobutane	30
1-iodobutane	12

Questions

5 In the contest to be the dominant factor in deciding the rate of substitution of halogenoalkanes, which factor, 'weaker bond' or 'bigger dipole', wins?

6 Using OH$^-$ as the nucleophile would have given a faster reaction, but OH$^-$ is impossible to use in the presence of Ag$^+$. Can you suggest explanations for *both* these observations?

Variations in mechanism – the branching factor

In the mechanism depicted in the curly-arrow diagram on p. 228, the incoming bond is forming as the outgoing bond is breaking. The whole operation takes place in a single collision. Looking back to Chapter 8, p. 144, Figure 8.10, you will be reminded that the potential energy of such a system goes through a maximum, at which point bond breaking is well advanced but bond making has not got far. At that moment we have a transition state in which the carbon is at least partially '5-valent' (Figure 11.2).

When you reach Chapter 14 you will find that not every collision is capable of

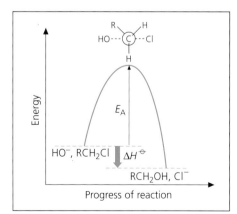

Figure 11.2 Energy profile for a one-step nucleophilic substitution

providing the energy to reach that maximum (the so-called **activation enthalpy**, E_A on Figure 11.2). That is the reason that the size of the activation enthalpy has a crucial say in the rate of reaction.

Let us consider a *different reaction pathway* between the same reactants and products (Figure 11.3). Two proposed rival paths for the reaction between iodobutane and OH$^-$(aq) are as follows. (Note that the one-step mechanism, Figure 11.3a, is a reworking of the one

One-step mechanism

$$HO^- \quad \overset{H \quad C_3H_7}{\underset{H}{\overset{|}{C}}} - I \rightarrow HO - \overset{H \quad C_3H_7}{\underset{H}{\overset{|}{C}}} + I^-$$

Two-step mechanism

$$\overset{H \quad C_3H_7}{\underset{H}{\overset{|}{C}}} - I \rightarrow \overset{H \quad C_3H_7}{\underset{H}{\overset{+|}{C}}} + I^-$$

$$HO^- \quad \overset{H \quad C_3H_7}{\underset{H}{\overset{+|}{C}}} \rightarrow HO - \overset{H \quad C_3H_7}{\underset{H}{\overset{|}{C}}}$$

given on p. 228.) The two-step mechanism features the total cleavage of the C—I bond *before* the OH$^-$ bond attacks. (This cleavage could be brought about by collisions with solvent molecules, for instance.) The resulting potential energy profile (Figure 11.3b) has a double hump, corresponding to the two steps of the mechanism. The step with the bigger hump creates a 3-valent carbon species called a 'carbocation'. (This is a general name for any species with a full positive charge centred on a carbon atom. An equivalent name for the same style of ion, as used by many texts, is 'carbonium ion'.) Because it has the bigger hump, the first step of the mechanism will, alone, be responsible for determining the rate of the reaction (see Chapter 14 on 'rate-determining steps'). The nucleophile only arrives after most of the hard work has been done, guided surely to its target by ionic attraction, and

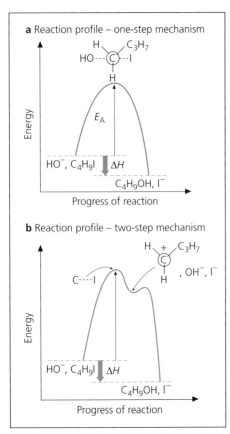

a Reaction profile – one-step mechanism

b Reaction profile – two-step mechanism

Figure 11.3 Alternative pathways to nucleophilic substitutions

Figure 11.4 A molecule of 2-bromo-2-methylpropane with a hydroxyl group

has the undemanding task of offering the carbocation the two electrons it is lacking.

At first sight this second pathway looks like a non-starter, on the grounds that it would have to have a bigger activation enthalpy than the 'pusher' mechanism.

Question

7 There is a logical reason for supposing that it would take less energy to get to the 'pusher' transition state than to the 'breaks-first' one. What is that reason?

After answering the last question, you might conclude that no halogenoalkane molecule could ever 'prefer' to undergo substitution via the two-step route. However, there are some halogeno-alkanes which do that very thing. They tend to be the ones in which the carbon carrying the halogen atom is also carrying *other alkyl groups* rather than just hydrogen atoms. In other words, the molecule is *highly branched*.

A high degree of branching of the carbon framework gives rise to two factors, both of which conspire to make the 3-valent intermediate (the one on the two-step route) preferable to its 5-valent equivalent. The first factor involves making the two-step transition state *more* energetically favourable in these highly branched cases, while the second factor involves making the one-step transition state *less* energetically favourable. We will consider each factor in turn, and the discussion points will be illustrated, in Figure 11.4, by reference to a typically highly branched substrate, 2-bromo-2-methylpropane.

1 The first factor is bound up with bond polarisation effects, which serve to stabilise the intermediate species on the two-step route – in other words, which lower the height of the 'saddle' of the reaction profile in Figure 11.3b. But the effect only works when the substrate is branched. Alkyl groups seem to have a small but definite tendency to *push* electron density away from themselves. The effect on the 3-valent intermediate derived from 2-bromo-2-methylpropane is wholly beneficial, because the original branching in the molecule means that the charge centre of the carbocation now lies at a junction of three C—C bonds. The three little electron-pushes from the alkyl groups serve to neutralise, distribute and delocalise the positive charge, by pushing electron density in to fill the electron-deficient site. Thus the carbocation is stabilised.

On the other hand, a 5-valent transition state based on 2-bromo-2-methylpropane would suffer an embarrassment of riches. The central carbon atom already has to cope with five pairs of electrons as it is and would only be destabilised by three of those pairs trying to 'crowd in' with electron 'pushes'.

Question

8 Part of the above paragraph should have reminded you of Chapter 10. Similar theories on stability of carbocations were invoked to explain Markovnikov's rule (p. 210). In what way are the two explanations similar?

2 The second factor which tilts highly branched substrates like 2-bromo-2-methylpropane towards the two-step mechanism concerns the way the branching alkyl groups occupy space. Alkyl groups are bulky, 'lumpy' in shape, and hard to move, compared with hydrogens. They therefore present the incoming nucleophile with a problem, which chemists call **steric hindrance**. From the nucleophile's point of view it must be rather like arriving at a party and trying to shake hands with the hostess while three large guards are demanding to see your invitation. With this image in mind you can perhaps see that the collision energy needed for an effective approach to the central carbon atom is increased and that therefore the percentage of successful collisions will drop. Under these circumstances a mechanism which frees up the space in front of the central carbon becomes a viable alternative. Going back to your answer to question 7, you can perhaps now see that if the bond making of the one-step transition state can only be achieved at the expense of higher collision energy, it is not energetically worthwhile.

This combination of effects results in a situation in which the one-step route actually becomes less favourable than the two-step one, as is shown in the two reaction profiles in Figure 11.5.

These two routes have symbolic names which remove the need for saying 'one-

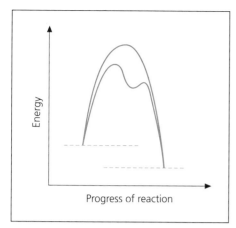

Figure 11.5 Reaction profiles for the two alternative mechanisms for nucleophilic substitution, when the substrate is a tertiary halogenoalkane

step/two-step' all the time. Oddly, the one-step mechanism, the one favoured by primary (unbranched) halogeno-alkanes, is called 'S$_N$2', while the two-step path favoured by tertiary (highly branched) substrates is called 'S$_N$1'. The S and N have the more obvious meanings of 'substitution' and 'nucleophilic', but the 1 and 2 refer to the **order of reaction** as revealed in the rate law, and a fuller discussion of that topic will be deferred till Chapter 14.

At the beginning of this section, *three* types of variation of the basic nucleophilic attack on halogenoalkanes were promised. Yet only two variations have been described (reaction rate and mechanism). The reason that only two variations fit into this section is that it is dedicated to nucleophilic *substitution*, whereas the third variation, while still concerned with nucleophilic attack, actually brings about a change in *outcome*.

11.4 Halogenoalkanes 3: Elimination

You might expect that any specific halogenoalkane, under attack from a specific nucleophile, will react along one single pathway with one single speed. In fact it is rather more complicated. In moderately branched halogenoalkanes it sometimes happens that S$_N$1 and S$_N$2 pathways can be adopted in parallel.

Both those variations normally lead to the same product, but even that is not

certain. Sometimes substitution has to compete side-by-side with **elimination**, which results in the creation of a double bond (and therefore of an alkene). This second outcome happens when the nucleophile, instead of attacking the C—X dipole, abstracts a *hydrogen atom from an adjacent carbon*. The knock-on effect of this is to liberate a pair of electrons which become the double bond's π-bond, and in so doing cause the ejection of the halogen atom (as a halide ion). The whole sequence, with curly arrows, is shown below.

$$H_2O + \quad \overset{H}{\underset{H}{C}} = \overset{H}{\underset{H}{C}} \quad + \; X^-$$

The balance between elimination and substitution – influence of the substrate

It is a fact that the more branched substrates – the same ones that are likely, when undergoing substitution, to go via S$_N$1 mechanisms – are the ones that give the highest percentages of elimination product. One possible reason for this is not too hard to deduce and is purely statistical.

Question

9 The elimination mechanism requires attack at a hydrogen atom attached to a carbon atom *adjacent to the halogen-carrying carbon*. Draw displayed formulae of 1-bromobutane and 2-bromo-2-methylpropane and try to work out why, *on probability grounds alone*, the latter molecule is more likely to undergo elimination than the former one.

As well as probability, we can re-use the steric hindrance concept to explain the bias towards elimination. We have

already seen that the more branched substrates obstruct the nucleophile's path to the crucial carbon atom, so a peripheral hydrogen is a good alternative target.

The balance between elimination and substitution – influence of the solvent

Finding a medium to act as solvent for OH$^-$ (or any other ionic or polar nucleophile) *and* a halogenoalkane at the same time, is a matter of compromise. Water alone is not a good solvent, because halogenoalkanes are not soluble in it. Reaction can only occur at the interface of two immiscible liquids (the aqueous layer and the halogenoalkane). Ethanol is a good compromise solvent, offering some solubility to both reactants – that is, the halogenoalkane and the ionic or polar nucleophile – and it must have appealed to the early organic chemists for that reason. They must have been surprised to discover that in the apparently minor matter of switching from water to ethanol they had inadvertently tipped the ratio of elimination to substitution strongly in favour of elimination.

Again, the search for reasons need not require any exotic new concepts. In ethanol the OH$^-$ ion would be solvated (surrounded) by ethanol molecules, just as in water it would be surrounded by water molecules. So in ethanol it is going to be rather a cumbersome and clumsy figure, not ideally suited to searching out the end of the C—X dipole, hidden as it is in the middle of a tetrahedron of bonds (Figure 11.6). Instead it turns its attention to the more approachable, peripheral target offered by the hydrogen atoms – hence the switch to elimination (a third use of the steric hindrance concept).

The experimenter who needs a good yield of either a substitution or an elimination product from the attack of a nucleophile on a halogenoalkane has to proceed by trial and error. He or she will have to manipulate the variable of solvent composition, using ethanol/water mixtures that combine adequate solubility with an acceptable yield of the required product. In most cases there will have to be a final separation stage anyway, to sort out the mixture of alcohols and alkenes that are liable to be formed.

Figure 11.6 An OH⁻ ion solvated by ethanol molecules is an awkward, clumsy attacker

Question

10 Imagine you had just reacted OH⁻ (as KOH) with 2-bromobutane in a 50/50 ethanol/water solvent.

a Give three possible organic products of this reaction.

b From your knowledge of intermolecular bonding, or else using a data book, suggest a method of separating these products.

c Why don't you have to worry too much about the inorganic product of the reaction, as far as separation and purification are concerned?

Table 11.2 summarises all the various combinations of mechanism and outcome, placed in an array of the two main variables – the variables of solvent and of degree of branching in the substrate.

Table 11.2 Nucleophilic attack by OH⁻ on halogenoalkanes. Summing up the variation in outcome dependent upon changes in solvent and substrates

Degree of branching substrate	Solvent	
	Minimum ethanol	Maximum ethanol
Primary	*Product mix*: Favours alcohol (i.e. the substitution product) *Mechanism*: 1-step	*Product mix*: Fairly even alcohol/ alkene *Mechanism*: 1-step
Secondary	Intermediate	Intermediate
Tertiary	*Product mix*: Fairly even alcohol/alkene *Mechanism*: 2-step	*Product mix*: Favours alkene (i.e. the elimination product) *Mechanism*: 2-step

Now let's move on to consider a certain group of halogenoalkanes which show a marked resistance towards *all* kinds of nucleophilic attack, by whatever mechanism and in whatever circumstances. They also show such an attractive range of physical properties that they became important chemicals in the late twentieth century. In 1974 a pair of American scientists asked a question about what happened to these 'very stable' halogenoalkanes when they were vented into the atmosphere after use. The result was to set in train one of the great ecological issues of the age.

11.5 CFCs and the ozone hole

When halogen atoms are attached to small carbon frameworks, the products are materials whose boiling temperatures are a little below room temperature, which means they are gases, but the sort of gases that can be liquefied by a small increase in pressure. This makes them ideal refrigerants, since only a small compressor is needed to liquefy them, and their subsequent evaporation causes refrigeration (Figure 11.7).

The same set of properties means that they make good aerosol propellants, since the pressure within the can traps them as liquids, while the operation of the button on top allows them to expand and emerge from the nozzle, projecting the product material (Figure 11.8).

Aside from their useful state-change properties, the small halogenoalkanes are also good thermal insulators. This combination of advantages means that not only are they good blowing agents for polymer foams, but that the bubbles once formed are excellent insulators (Figure 11.9).

So by the early 1970s you could find the second molecule in Table 11.3, CFC 12, in refrigerators, aerosols and foams all over the world. The initials CFC stand for **chlorofluorocarbon**(s), and Table 11.3 shows a selection of the ones that came into use in the 1960s and 1970s. A molecule like CFC 12 seemed like a perfect substance – it was stable, non-toxic,

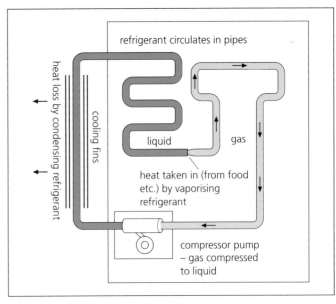

Figure 11.7 How a refrigerator works

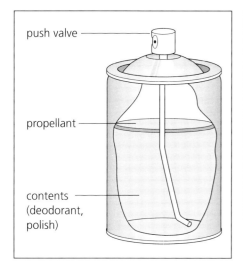

Figure 11.8 How an aerosol can works

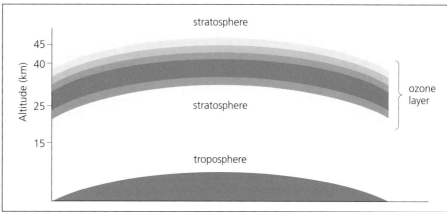

Figure 11.10 The ozone 'layer'. If all the ozone in the atmosphere were collected at the Earth's surface, it would form a layer only 3 mm thick. The boundary between troposphere and stratosphere ranges from about 8 km over the poles to 19 km over the equator

non-flammable and dispersed quietly in the atmosphere without appearing to have any harmful interaction. This represented a great improvement over, for example, its predecessor as the world's leading refrigerant – ammonia!

Now we know better. CFCs are known to be on a par with global warming and acid rain as major environmental problems. Scientists still think that CFCs are stable in the lower layers of the atmosphere (and even here they contribute to global warming), but it is what they do up in the ultra-violet-bathed stratosphere which has caused all the worry. It is thought that in those uppermost regions of the atmosphere they interfere with chemical cycles which hitherto had operated to maintain a layer of the gas *ozone* (O_3) throughout the stratosphere (Figure 11.10). The effect of the CFCs is the destruction of ozone, and with it the destruction of an important shield that filters out much of the ultra-violet radiation arriving at the solid surface of our planet. One of the fears, now that more and shorter wavelengths of ultra-violet are reaching the surface, is that humans will suffer increasingly from diseases associated with ultra-violet overdose, most notably cancers of the skin.

A simplified picture of the chemistry involved is described below:

Natural ozone balance

1 Creation

$$O_2 \xrightarrow{h\nu} 2 \cdot O \cdot$$

$$\cdot O \cdot + O_2 \rightarrow O_3$$

(*Note*: ·O· indicates an oxygen atom (free radical). The two lone pairs are not shown.)

Figure 11.9 (below) A CFC-blown foam used as a thermal insulator in roofing material. (below right) Expanded polystyrene balls, often used to protect goods from damage in transit

Table 11.3 Some CFCs

	Boiling point (°C)	Formula	Name
CFC 11	+24	$CFCl_3$	trichlorofluoromethane
CFC 12	−29	CF_2Cl_2	dichlorodifluoromethane
CFC 13	+47	CF_2ClCCl_2F	1,1,2-trichloro-1,2,2-trifluoroethane
CFC 114	+3	CF_2ClCF_2Cl	1,2-dichloro-1,1,2,2-tetrafluoroethane
CFC 115	−38	CF_3CF_2Cl	chloropentafluoroethane

2 Destruction

$$O_3 + \cdot O\cdot \rightarrow 2O_2$$

The CFCs are assumed to be involved like this (using the example of CFC 11):

$$CFCl_3 \overset{hv}{\rightarrow} CFCl_2 + Cl\cdot$$

$$Cl\cdot + O_3 \rightarrow ClO\cdot + O_2$$

So we see that the CFC molecules, via their decomposition into Cl• free radicals, are providing an extra route for the destruction of ozone, superimposed on a situation where the previous rate of destruction was just in balance with the rate of creation. Clearly the world community must act to maintain the planet's ozone layer, which means finding some substitutes for CFCs in the industries to which they have become vital (or even perhaps abandoning some technologies altogether – aerosols for instance). The question that follows is about the alternatives to CFCs.

Question

11a It seems that the problems with CFCs are concerned with their tendency to produce Cl• free radicals. Show, with reference to a data book, why the Cl• free radical is the most likely free radical to come from the decomposition of CFC 12.

b An international agreement, called the Montreal Protocol, has bound major industrial countries at least not to increase the production of CFCs above 1986 levels. The immediate result has been some stop-gap substitutions of well-known chemicals for CFCs in aerosols. One of the substitutes is butane, which is supposedly 'ozone friendly', as the current generation of aerosol cans proclaims. Suggest two reasons why butane was chosen, and two reasons why it is not all that 'friendly'.

c As for newly synthesised CFC substitutes, the leader seems to be 1,1,1,2-tetrafluoroethane, shown as HFA 134a in Table 11.4. (The initials HFA, which like CFC have been attached to a 'family' of related compounds, stand for **hydrofluoroalkane**(s).) HFA 134a is unfortunately many times the price of CFC 12, but appears not to interfere with the ozone cycle at all. How can you see by looking at its structure that it is unlikely to damage the ozone layer?

Table 11.4 Hydrofluoroalkanes

	Boiling point (°C)	Formula	Name
HFA 134a	−26	CF_3CH_2F	1,1,1,2-tetrafluoroethane
HFA 22	−42	CF_2ClH	chlorodifluoromethane
HFA 123	+28	CF_3CHCl_2	2,2-dichloro-1,1,1-trifluoroethane
HFA 141b	+32	Cl_2CFCH_3	1,1-dichloro-1-fluoroethane

■ International efforts to save the ozone layer are being hindered by a black market in ozone-destroying CFCs, according to Dr Michael Harris, a senior manager at ICI Klea, which makes CFC substitutes. Illegal imports of CFCs to Europe – thousands of tons a year – were being sold to small companies unwilling to pay the environmental premium price for substitutes, Dr Harris warned. Substitute chemicals cost about twice as much.

Figure 11.11 A black market in CFCs. *Independent*, 8 September 1994

Checking on manufacturers' claims

How can we find out if a blowing agent in a foam is a CFC? The expanded polystyrene carton in which you buy your chips might be made on the cheap by a manufacturer for whom the low cost of CFC 12 is a crucial factor – and who is going to know? (Figure 11.11.) The ideal forensic technique for detecting this eco-fraud is the gas chromatography machine connected to a mass spectrometer (GC/MS machine).

Chromatography

In chromatography mixed solutes in a moving solvent (called the **mobile phase**) will partition between the **solvent** and the **stationary phase** over which the solvent is being made to move. ('To partition' means to be divided between phases, in this case the mobile and the stationary phases.) Because of different degrees of intermolecular attraction, some components of the mixture will be more attracted to the stationary phase, while others will feel more attraction for the mobile phase. After a while there will be a physical separation in space, with the 'solventophiles' moving almost up with the mobile solvent front, while the 'stick-in-the-muds' will only just have crept off the start line. (These are figures of speech, although mud might make quite a good stationary phase, since other siliceous materials do really perform such work.)

Question

12 From your early days as a chemist, recall the chromatography of Smarties or of inks. Identify, in either one of those cases, the **solute mixture**, the **solvent** (that is, the mobile phase) and the **stationary phase**.

The principles of chromatography remain the same across the whole field, from primary school to industrial research laboratory, but the equipment becomes increasingly sophisticated and expensive.

In gas chromatography the whole separation process takes place out of sight. The *stationary phase* is housed in a coiled tube several metres in length (called, incongruously, a column) contained within a variable temperature oven (Figure 11.12). One example of a stationary phase, quite commonly used, is based on tiny silica particles which are the exoskele-

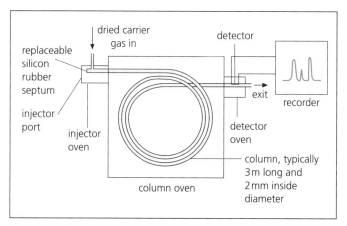

Figure 11.12 The gas chromatograph. Analytical columns tend to be narrow, and preparation columns tend to be wide to allow for the greater volumes passing through them

tons of 'diatoms', sea creatures which are members of that vaguely defined group called 'krill' (Figure 11.13). These are coated with a non-volatile liquid polymer, selected from a range of such compounds. The polymer is chosen as having the sort of projecting functional groups thought likely to separate whatever mixtures the operator is analysing. It is the polymer which counts as the stationary phase.

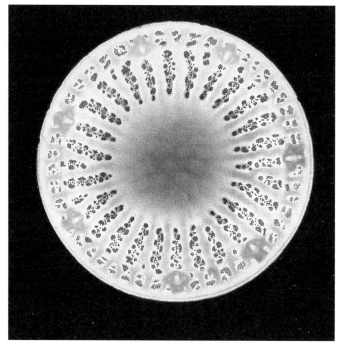

Figure 11.13 Diatoms from ocean krill. These siliceous bodies, when ground to a powder, make a stationary-phase support for GC and GLC

The *solute mixture* would be any mixture of gases (such as the CFCs we have been discussing), or any mixture of liquids which can be vaporised by the oven, the whole mix being propelled along the column by a carrier gas (the *solvent*), normally helium. The components of the mixture do not of course show up as coloured rings (as with the Smarties or inks) but come off the end of the column at different *times*.

Linking the chromatograph to a mass spectrometer

All column-style chromatography machines need detectors, because there is no visible sign, as you would get with inks on paper or with thin-layer chromatography. When gas chromatography is used alone, various forms of detector pick up the components as they emerge, and generate a signal which is sent to a chart recorder or its computer-graphic equivalent. A problem with the technique is that it is not easy to tell what the components *are* (Figure 11.14). However, in an ingenious link-up of two techniques, it is now possible to send the components from the end of a chromatographic column into a *mass spectrometer*, by having the end of the column situated in the ionisation chamber of the mass spectrometer.

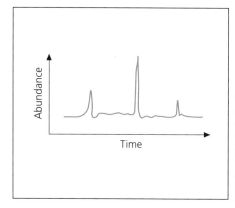

Figure 11.14 An output trace from a gas chromatography (GC) machine. You can see that there are three components in the sample mixture, but not what they are

One of the cleverer aspects of this arrangement is that two desirable objectives are achieved by one operation – the molecules 'announce their arrival' *and* simultaneously identify themselves. When the separated component molecules arrive off the column, they are ushered into the ionisation chamber of the mass spectrometer, and are ionised and beamed down the tube to have their mass spectrum recorded. The machine scans through and records a mass spectrum every second, and records the data in its memory. But (and this is the clever bit) the *total* number of ions detected each second is also logged. This acts as a measure of how much stuff is coming off the column, so the machine can put each second's worth of data together to produce a trace which serves as the conventional chromatogram (that is, a graph like Figure 11.14).

When the whole sample of test material has come off the column, the operator can display the chromatogram. He or she can then move a cursor to any point on the time axis, and request the mass spectrum which was recorded in that second. So the mass spectrum of a single peak (a single component from the mixture) can be displayed. As if all this were not enough, the computer then identifies the component by comparing its mass spectrum with those in its library.

Figure 11.15a shows a gas chromatography/mass spectrometry machine. Figure 11.15b (overleaf) shows the printout from an experiment carried out at Luton University, giving the three types of graph mentioned above (chromatogram, mass spectrum of selected peak, and library comparison), showing that this particular hamburger container was blown with butane.

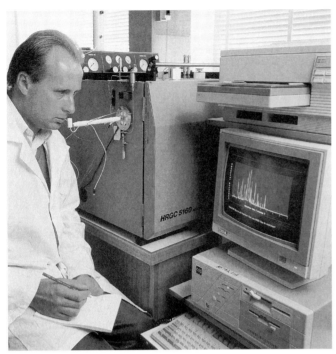

Figure 11.15a A gas chromatography/mass spectrometry machine, capable of recording both the chromatogram *and* the mass spectrum of each peak

11.6 A list of reactions of halogenoalkanes

Now that we have discussed the important concepts in halogenoalkane chemistry, you can use those concepts to make sense of the range of specific reactions of halogenoalkanes.

Nucleophilic substitution reactions

1 $R—X + OH^- \rightarrow R—OH + X^-$

2 $R—X + H_2O \rightarrow R—OH + HX$

3 $R—X + OR^- \rightarrow R—OR + X^-$
   alkoxide   an
   ion   ether

4 $R—X + NH_3 \rightarrow R—NH_2 + HX$
                       an
                       amine

5 $R—X + \text{benzene} \xrightarrow[\text{catalyst}]{AlCl_3} R—\hexagon + HX$

Questions

13 How would you expect the rates of reactions 1 and 2 to compare (assuming the same substrate in each case)?

14 How might you generate the OR$^-$ ion used in reaction 3? (This involves work later in the chapter; a clue is that it involves sodium metal and a reaction similar to that between sodium and water.)

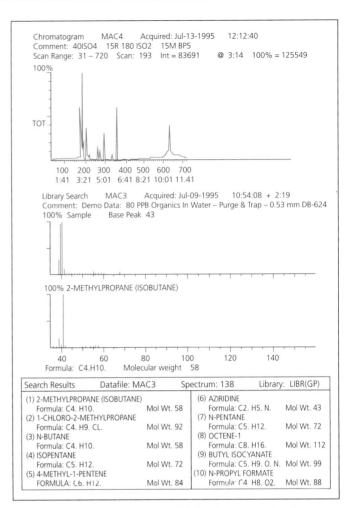

Figure 11.15b Analysis of a McDonald's hamburger container (top) The gas in the bubbles of McDonald's expanded plastic food containers consists of 10 ingredients, as shown by the printout from gas chromatography. (middle) The mass spectrum of peak no.1, the main blowing agent, shows it to be 2-methylpropane. (bottom) All 10 ingredients have been identified from their mass spectra. One (9) is a derivative of the plastic of the container, but none is a CFC.

15 You may recognise reaction 5 as the *Friedel–Crafts* reaction (p. 222), which perhaps looks rather out of place in a list of nucleophilic substitution reactions.

a Friedel–Crafts reactions are more often found in lists of *electrophilic* substitutions. From which molecule's point of view is it an electrophilic substitution? (Bear in mind that '-philic' describes the attacking species from the 'victim's' point of view.)

b From which molecule's point of view is it a nucleophilic substitution?

c You might say that it differs from other reactions in the list, because none of the other 'leaving groups' need *help* to leave. How is help offered to the halide ion? (Clue: look back to Chapter 10 and check on the role of the catalyst.)

Nucleophilic elimination reactions

1 For example:

$$CH_3CH_2CHCH_3 \ + \ KOH$$
$$\underset{Cl}{|}$$

2-chlorobutane $\downarrow$ (100% ethanol)

$$CH_3CH = CHCH_3 + CH_3CH_2CH = CH_2 + KCl + H_2O$$

       80%   20%

2 For example:

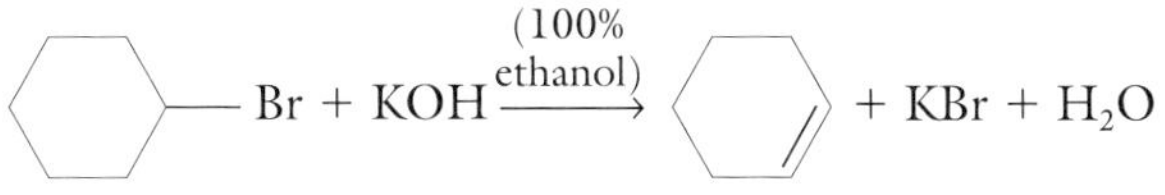

bromocyclohexane

11.7 Other halogen-substituted hydrocarbons – a comparison

Halogenoalk*enes* and halogeno*arenes* are both much less reactive than halogenoalkanes towards nucleophilic substitution. The reaction *can* be done, but conditions have to be harsh. The underlying cause is the reluctance of nucleophiles to approach a region of high electron density, which is something they have to do if a halogen atom is attached to a carbon atom that is also part of a π-bonded system. 'Harsh' conditions usually involve extremes of temperature, well in excess of the normal boiling temperatures of both the substrates and the aqueous nucleophiles, and also extremes of pressure, so that the reagents nevertheless remain liquid. One example of a reaction in such conditions is shown below:

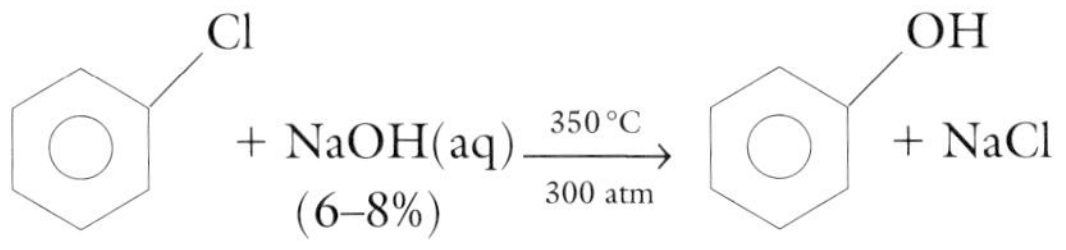

Questions

16 Less severe conditions are needed to eject the chlorine from 4-chloronitrobenzene by a nucleophilic attack, as compared with chlorobenzene. Using simple electrostatic ideas, try to account for this difference (see the equation below). In other words, explain why the nucleophile needs a little less encouragement to approach the carbon in *this* aromatic ring, compared to that in simple chlorobenzene.

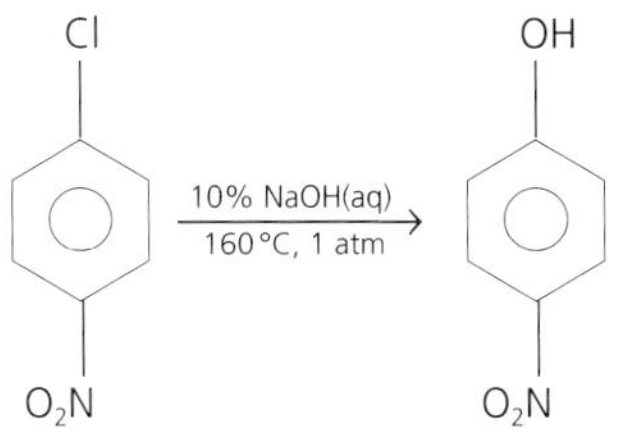

17 Evidence of a different kind to explain the relative unreactivity of the C—X bond in halogenoarenes can be found in the table of bond *lengths* in the data book. Try to search it out, and quote it. (Remember that short bonds are generally strong bonds.)

11.8 PVC – a materials case study

The compounds such as chlorobutane, which provided our subject matter for Sections 11.2 to 11.6, and which have been studied for the sake of the intricacies of their reaction mechanisms, are nevertheless only of second-division significance in the context of the chemical industry. In sharp contrast, the one organic halogen compound whose manufacture consumes 30% of all chlorine produced worldwide, comes from our 'footnote', Section 11.7. That chemical is **chloroethene**.

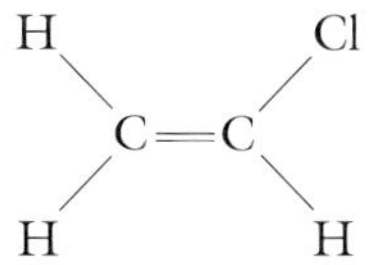

Not that you will find a bottle of chloroethene under some trade name in the cupboard under the sink. What you will find is that, unless you are reading this book in the middle of a field, you cannot turn round without seeing something made *from* chloroethene. The reason is that chloroethene is a monomer, and the polymer derived from that monomer is a substance that school and university chemistry books call poly(chloroethene), but which nearly everybody else calls **polyvinyl chloride** or **PVC** (shown below).

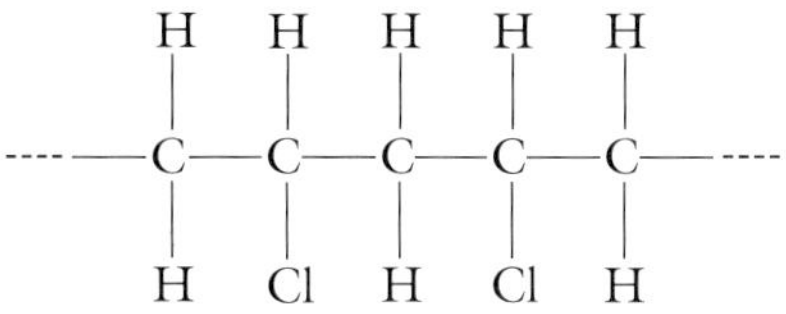

The 'cannot turn round without seeing it' test is well worth trying. Your desk is lit by a lamp whose wires are coated in PVC. On the desk is your trusty calculator whose leather-look wallet is PVC. As you spin in your executive-style anti-drowse office-type chair (with PVC seat), your gaze will take in the washable wallpaper and the silk-finish paint on your bedroom door (both PVC), finally alighting queasily on yesterday's sandwiches still uneaten in their PVC clingfilm. You pull at the curtains which swish smoothly open on their PVC rail, and throw open the window to air the room after the long intense homework session (how much less worry the new PVC window frames are than the old rotten-but-tasteful wooden ones!) to see the dawn break. A dim grey light diffuses up from the eastern horizon as the rain gurgles in the PVC downpipes of your gutter system. The mood of damp gloom demands one of those old (PVC) Leonard Cohen albums which your parents passed on to you after they got the CDs. You push two paracetamols from their (PVC) blister pack . . .

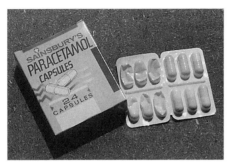

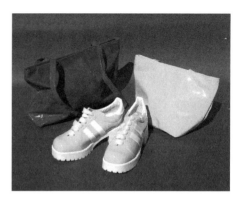

Figure 11.16 The widespread uses of PVC

Enough; you should get the idea that PVC is quite widespread (Figure 11.16). PVC is also tremendously versatile – contrast it with, for example, poly(ethene), whose derived objects are more or less similar in physical properties. The range of PVC objects that encompasses cling-film at one extreme and window frames at the other is comparatively huge (and there was no room in the story for bank cash-machine cards, imitation leather shoes and car seats, cooking oil bottles and those transparent boxes some fruits come in). So what gives PVC this breadth of physical properties which equips it for such a spectrum of uses? The short answer is its ability to accommodate character-modifying *additives*. For a longer answer we need to look at some chemistry.

From feedstock to polymer

The actual preparation of PVC uses fairly conventional chemistry, most of it within the scope of our existing concepts. The feedstock of the overall process is ethene, which is only one atom different from the desired chloroethene monomer, but a direct substitution of a chlorine atom in place of a hydrogen atom proves impossible.

Question

18a From which original source, and via which piece of industrial chemistry mentioned in Chapter 10, would the ethene be made?

b Why is it not possible to insert a chlorine atom into the ethene molecule by attack with a nucleophile like the Cl end of the HCl molecule?

Instead of a one-step substitution, the chlorine atom is introduced to the molecule in two stages. In the first stage ethene reacts with chlorine in the gas phase to produce dichloroeth*ane*, and then the dichloroethane is thermally 'cracked', which causes an HCl molecule to drop out and leaves the target molecule (below).

$$\begin{array}{ccc} H & & Cl \\ & C=C & \\ H & & H \end{array}$$

Polymer names

The chemical industry does not always use the systematic names promoted by the IUPAC. If industrial chemists are used to calling a substance by a name that has been understood by their business colleagues and customers for years, then it tends to keep that name. Chloroethene is a case in point – in ICI and other such places it is universally known as VCM (for Vinyl Chloride Monomer). Table 11.5 shows some other non-systematic names that are still in use.

Table 11.5 Official and unofficial names for monomers and polymers

Unofficial but widely used	Official IUPAC
Ethylene	Ethene
Polythene	Poly(ethene)
Propylene	Propene
Polypropylene	Poly(propene)
Vinyl chloride or VCM	Chloroethene
PVC	Poly(chloroethene)
Styrene	Phenylethene
Polystyrene	Poly(phenylethene)
Perspex, Acrylic	Poly(methyl-2-methylpropenoate)

Question

19a Which of the 'outcome' words best describes the reactions in:

i the first stage of the sequence below?

ii the second stage of the sequence below?

$$\begin{array}{ccc} H & & H \\ & C=C & + Cl_2 \\ H & & H \end{array}$$

$$\downarrow$$

$$\begin{array}{cc} H & H \\ H-C-C-H \\ Cl & Cl \end{array}$$

$$\downarrow \text{heat}$$

$$\begin{array}{ccc} H & & H \\ & C=C & + HCl \\ H & & Cl \end{array}$$

b Which of the 'mechanism' words best describes the reaction in the first stage?

When it comes to avoiding throwing anything away, the ingenuity of the chemical industry is very impressive. The problem with the overall reaction shown above was that it did produce a valuable and yet apparently unusable by-product, a by-product furthermore which could not be vented into the atmosphere. So it is now standard practice in the PVC industry to run a *second* preparation of chloroethene in parallel with the first, with the second reaction using up all the by-product from the first (as shown below).

$$H_2C=CH_2 + HCl + \tfrac{1}{2}O_2$$

$$\xrightarrow{CuCl_2 \text{ catalyst}}$$

$$H_2C=CHCl + H_2O$$

Question

20a What is the by-product from the reaction in question 19?

b The second reaction, called **oxychlorination**, has the equation shown above. Match up the equations from the two reactions so that all the by-product is exactly consumed. Give an equation for the total changes happening in the plant.

c Why is the combined process better, both in environmental and in financial terms?

d The oxychlorination reaction alone looks quite promising, yet chloroethene is always made by both reactions in tandem. Can you suggest a reason for this?

Now we have accounted for the creation of the monomer, let us turn to the *polymerising* process. The chemistry is typical of the addition polymerisation of a polyalkene, and follows the pattern laid down in Section 10.12 – that is, it is a free-radical addition polymerisation, initiated by an organic peroxide.

Question

21 As a piece of revision of the previous chapter, show the initiation, propagation and termination steps of this mechanism, using the specific case of chloroethene as monomer.

One noteworthy feature of the process is that it takes place in a *water suspension*. Chloroethene is not naturally soluble in water, but detergent-style molecules can keep it in suspension as droplets without it coagulating back into a separate layer. The initiator molecules dissolve preferentially in the droplets rather than the water, and get to work. Each droplet gradually turns into a bead of PVC. Normal bead diameters are about 10^{-4} m. The beads are then dried (any unreacted VCM is recycled), melted, mixed with any necessary additives, resolidified and ground into granules (Figure 11.17) for sale to the end-user (the firm making the window frames or clingfilm).

Figure 11.17 PVC granules, with additives already in, ready to go to the end-user

Additives and improvers

If ever a material needed improving it was prototype PVC, in the years immediately following its discovery in 1912. Several now-famous materials seem to

have gone through the 'unpromising-lump' stage (polythene, for instance) and PVC certainly was a case in point. It had many drawbacks: around the softening temperature (that is, at the temperature when it was expected to go into moulds), it decomposed giving off hydrogen chloride, which then acted as an auto-catalyst to speed up further decomposition. Also it tended, when molten, to stick to the moulds and to the giant screws which drove it into those moulds. Apart from stickiness it had a viscosity which made it behave like Plasticine that has not been worked long enough in your hands – underworked Plasticine has a tendency to fracture rather than shape smoothly. The final problem was that end-product objects turned out to be unstable on prolonged exposure to light.

Happily, by 1930 most of these problems were on their way to being resolved. Various processing aids had been discovered which made the melt flow more elastically inside the big shaping machines (Figure 11.18) and come away from the moulds or dies leaving smooth-as-liquid finishes. Other additives had been introduced to scavenge any hydrogen chloride molecules, thus greatly aiding thermal stability, and ultra-violet screening compounds protected the polymer against photolysis during its years of use, rather like sunscreens protecting human skin.

The discovery of all these improving additives carried an interesting double bonus. It did more than just solve the original batch of problems – it also alerted people to the very accommodating nature of the PVC structure as a solid. It was discovered that additives could be introduced which could stay within the structure of the finished object, and do more than just passively stop it 'going off'. In fact there were

Figure 11.18 a Extrusion: the material is heated and forced through a shaped hole, or die, to form a rod, tube, sheet or other continuous length, for example, for window frame sections and guttering, or insulation

b Moulding: the most widely used moulding method is injection moulding. The plastic material is heated and, under pressure, a measured amount is made to flow into a metal mould. When the material cools it retains the shape of the mould. This is used, for example, for footwear

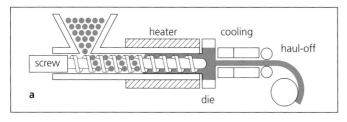

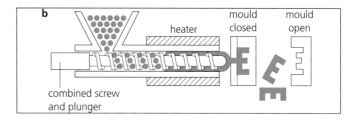

additives which could make PVC behave like a whole new material, to the extent that nowadays PVC has acquired an extra initial, either 'u' or 'p', standing for 'unplasticised' or 'plasticised'.

Fine-tuning the properties of PVC

The use of plasticisers

The idea of two plastics for the price of one is perhaps most familiar from the case of polystyrene. The unblown polymer is the material of plastics models and picnic cutlery, while the blown one is the featherlight packaging material used to encase new TV sets, etc. The effect of blowing bubbles into the polymer is obvious and easy to picture. In contrast to the polystyrene case, the 'plasticisation' of PVC is caused by events down on the molecular level, and requires more imagination to understand.

Part of that understanding might come from considering Figures 11.19 and 11.20. Figure 11.19 shows the molecular arrangement in a typical **thermoplastic** polymer – that is, a polymer in which there is no cross-linking of the chains and in which there are ordered regions (called **crystallites**) and disordered regions. Unplasticised PVC falls within this category. Such a polymer shows a profile of temperature behaviour as in Figure 11.20, with two transition temperatures T_g (the 'glass transition temperature') and T_m, the melting temperature. In between these two temperatures, the material will behave either elastically or plastically, depending on the strength of the force trying to restore the original molecular positions.

Question

22a The two transition temperatures in Figure 11.20 can be seen as the temperatures at which thermal agitations of molecules overcome sets of intermolecular forces. But simple ionic and molecular substances (for example, water and salt) only have *one* temperature at which this happens – the melting point. Can you see what aspect of the structure in Figure 11.19 gives rise to *two* transition temperatures?

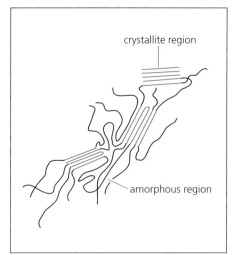

Figure 11.19 Diagram of 'crystallites' in a polymer

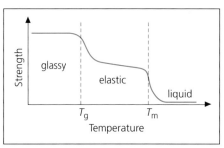

Figure 11.20 A graph of temperature against strength for a typical thermoplastic polymer. Today's polymers are designed to have T_g and T_m values which are related to the particular properties needed by a manufacturer

b Another contrast between water and salt on the one hand and thermoplastic polymers on the other is that the former melt at a sharply defined exact temperature, while the latter, even just focusing on T_m, show an imprecise 'plus-or-minus-10 K' type of melting temperature. Can you suggest why?

There is a connection between a diagram like Figure 11.20 and the 'four modes of deformation' passage from Chapter 10, in Section 10.13 on rubber. In the glassy region, the molecules are held strongly to one another, and deformation can only be by the hard routes – that is, stretching actual bonds or opening bond angles. Beyond T_g we have some scope for movement of molecules relative to each other, due to loss of intermolecular

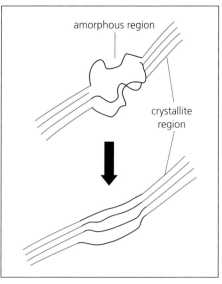

Figure 11.21 Above T_g, the amorphous regions of a polymer begin to allow elastic movement

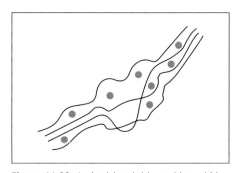

Figure 11.22 A plasticiser (•) in position within the molecular structure of a polymer

bonding in the amorphous regions. This frees molecules to perform either elastic or plastic shape changes, with the unravelling of coiled molecules giving elasticity (Figure 11.21), and any actual sliding of molecules against each other causing permanent plastic deformation.

Normal uPVC is used in situations where people want rigidity. Flexibility is an undesirable feature in a window frame. The polymer must clearly have a T_g well above its working temperature, so that no molecular movement occurs. Yet leathercloth, clingfilm, and the flex on electrical wires must *not* be rigid, so the addition of the plasticiser in pPVC must shift the material's T_g to a position below room temperature. Fairly obviously, the plasticiser molecules have to get between the PVC molecules, so as to reduce intermolecular forces. With the plasticiser molecules in place, the unravellings and bond angle stretches can now take place more freely, less impeded by intermolecular forces (Figure 11.22). We now have a material which bends into lots of different positions.

Question

23a Sheets of pPVC act quite like sheets of rubber in terms of their flexibility. Yet rubbers seem to have rather more memory of their original shapes. What difference in the molecular structures can be seen as the cause of rubber's greater 'restorability'?

b What combination of intermolecular forces holds one PVC molecule to the next?

c If flexibility involves interrupting these forces, then the plasticiser molecules cannot be too chemically similar to PVC. What would happen if they were?

d On the other hand, if the plasticiser molecules were too different from PVC (a wholly van der Waals molecule, for example), they would again fail in their task of interrupting intermolecular forces. Can you suggest why?

e An actual example of a plasticiser is dioctyl phthalate (below), which in terms of its intermolecular bonding possibilities is a bit van der Waals, a bit dipolar in the middle, rather globular in shape (when allowed to coil), and many times shorter than a PVC molecule.

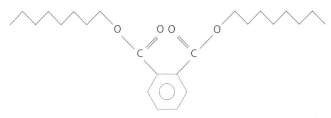

Can you see why any or all of these features make it good at its job?

f Do you think that dioctyl phthalate would plasticise polythene to the same extent? Justify your answer.

Nomenclature for discussing materials

Do plasticisers really make plastics plastic? (Or should it be elastic?) These terms may be causing some confusion. The words elastic and plastic have fairly clear-cut meanings – the first refers to materials that are restored to their original shape after the deforming force is removed (rubber being the commonest example), while the second refers to materials which stay permanently deformed (like Plasticine). But what about pairs of definitions which run across the elastic/plastic dichotomy with dichotomies of their own, like the pair flexible/stiff? And can a single material be both plastic and elastic, at different temperatures, say? Which words best describe the materials in our current section, that is, pPVC and uPVC?

Let us use the concepts on a single example, and try to generalise from there. pPVC, as used in domestic electric cable insulation, is clearly *flexible*, which is due to plasticiser molecules interrupting the intermolecular forces between PVC chains. But when you stop flexing it, what happens next? An equivalent-sized piece of rubber would return to its original linear shape, since the cross-links between chains would have helped to preserve a molecular 'memory' of the original, stablest, molecular arrangement. An equivalent-sized piece of Plasticine would stay bent, apparently because molecules have slid over each other during the flexing motion, into new positions that are energetically just as good as the original positions. Now turning to the pPVC, if you watch the behaviour of the cable, it appears to have a foot in both camps – it mainly behaves like rubber, but has a discernible tendency to not quite regain its original straightness, which is similar to Plasticine too. So the 'plasticiser' has had the effect of turning a stiff material, uPVC, in which all molecular movement is strongly opposed, into a mainly *elastic* material which under increasing stress shows a tendency to behave a bit *plastically* too (Figure 11.23).

Figure 11.23 Bent bits of pPVC behave elastically, but often don't quite regain their original shapes when unstressed

A broader sweep over the properties of materials and their descriptor words is given in Table 11.6. It has been partially filled in, by drawing on some of the ideas from the previous paragraph.

Table 11.6 Some properties of materials

	Stiff – *big* forces oppose molecular movement	Flexible – *small* forces oppose molecular movement
Elastic – molecules have *one* lowest-energy arrangement	Ceramics Metals under moderate stress Semi-crystalline plastics	Rubber
Plastic – molecules have *many* lowest-energy arrangements		Plasticine

Question

24a If the cable-insulating sheath under test in Figure 11.23 had a thick single strand of copper inside it, then once you bent it, it would certainly stay bent. Where does this place thick copper wire in Table 11.6?

b Glasses (non-crystalline amorphous solids) and ceramics (that is, ionic crystalline materials) have been placed in the 'elastic-but-stiff' box, since they would stretch a tiny bit, under severe stress, before giving way to brittle fracture. However, window glass that has aged over several centuries shows a thickening at the bottom, suggesting that over these very long times it has 'sagged' under gravity. So into which box in Table 11.6 would you put glass, from the point of view of its behaviour over very long times?

c Into which box in Table 11.6 would you place the following:

i uPVC (used for window frames and gutters) at room temperature?

ii pPVC (clingfilm and cheap car seats) at liquid nitrogen temperatures?

iii Any plastics material just below its melting temperature?

iv A metal under stress, but not so severely stressed as to cause permanent deformation (as in a spring)?

Finally in this section on nomenclature, here are some more straight definitions.

Polymer: a long molecular chain, which may or may not be cross-linked to other chains using classical bonds. A polymer can be of bio-origin (for example, starch, protein) or synthetic.

Plastics material: a synthetic polymeric material created by the linking of (usually) carbon-based monomers. What most people just call 'plastic'.

Thermoplastic: plastics material that shows a transition temperature profile like Figure 11.20, usually softening at a melting temperature of between 450 and 600 K. The polymer chains will show a low degree of cross-linking.

Thermosetting polymer: plastics material in which extensive cross-linking, usually during processing of the final object, produces a non-melting polymer.

Plastic: an adjective describing a mode of behaviour, typified by Plasticine, in which a deformed material stays deformed once the deforming force has been removed.

Elastic: another adjective describing a mode of behaviour which contrasts to plastic (see above). Elasticity is easily recognised in flexible elastics like rubber, but the word can also apply to some barely perceptible stiff/elastic materials, like metals, ceramics and glasses. The criterion for elasticity is that the material, if deformed, resumes its original shape after the deforming force is removed.

Elastomer: polymeric material with flexible/elastic properties.

Plasticiser: an additive, used to give plastics materials extra flexible-elasticity. Perhaps should more helpfully have been called 'elasticiser'.

Question

25 Say whether these sentences are true or false, justifying your answers.

a All plastics materials show plastic behaviour at room temperature.

b It is possible for a plastics material to show plastic behaviour at one temperature, and elastic behaviour at another.

c All plastics materials are polymers, but not all polymers are plastics materials.

d All materials showing plastic behaviour are polymers.

e All elastomers are polymers, but not all elastic materials are polymers.

The economics of PVC

Plastics materials, especially as used in packaging, are the visible reminders of what many people call a throw-away society. It is argued that a reduction in the use of plastics in packaging would be a significantly 'green' and desirable turn of events. Since PVC is the main material for the supermarket presentation of fruit and vegetables and all except the fizziest liquids, it is in the front line of the argument.

The PVC producers put forward the following case for carrying on the way we are:

1 Only 43% by mass of PVC is oil based (because the chlorine comes from seawater). It consumes the least energy in its production of any plastic and is beaten in Figure 11.24 only by copper and steel. (Figure 11.24 includes in its MJ kg^{-1} data the energy content of the oil which had to be 'locked up' in the polymer, as well as any energy used to heat the reactors, etc., which means that the low 43% figure works in PVC's favour.)

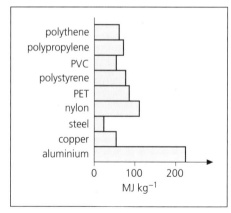

Figure 11.24 Energy consumed during manufacture (values for plastics include the burnable oil that is 'locked up' in their structures)

2 80% of crude oil is used for turning into fuels, whereas only 4% is turned into plastics. Of that 4% only 25% is PVC, and of that 25% only 12% is used in short-lived items like packaging, the assumption being that other PVC objects like window frames and wiring insulation should last for 15–100 years (Figure 11.25). So banning the use of PVC in packaging would be merely scratching the surface of the oil consumption problem.

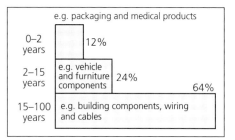

Figure 11.25 Intended service life of PVC products

3 Work is going on to develop machines which can segregate the three main bottle-making plastics – polythene, PVC and PET (polythene terephthalate, which has great impermeability to CO_2 and so gets used for the fizziest drinks). X-ray machines can 'see' the chlorine atoms in PVC, and flotation can separate polythene from PET, so there is a prospect of a useful recycling operation.

Question

26 In the light of this information, and other data you might have, try to decide whether you are happy with the packaging industry going on the way it is.

After that long detour into PVC, we now return to the main theme of the chapter, namely molecules with dipoles. The next group with pronounced dipoles that we are to study is the alcohols.

11.9 Alcohols 1: Introduction

Alcohols are molecules in which a hydrocarbon framework carries an OH group. The systematic naming of alcohols follows that of alkenes. In other words

there is a suffix system, with this time the syllable '-ol' appearing at the end of the name for the hydrocarbon framework. A number of named alcohols are shown below.

$$CH_3CH_2CH_2CH_2CH_2OH$$
pentan-1-ol

$$CH_3CH_2CHCH_3 \quad \text{butan-2-ol}$$
$$| \\ OH$$

cyclohexanol

$$(CH_3)_2CCH_2CH_3 \quad \text{2-methylbutan-2-ol}$$
$$| \\ OH$$

$$(CH_3)_2CHCH_2OH \quad \text{3-methylbutan-1-ol}$$

Alongside the systematic '-ol' naming system, there is a second, complementary, nomenclature which focuses on the degree of branching at the specific point of attachment of the functional group. An alcohol with its OH sited on the end of a chain (whether or not the chain is branched further along) is known as a **primary alcohol**. Thus all primary alcohols can be generalised as RCH_2OH. Alcohols conforming to the general formula R^1R^2CHOH are referred to as **secondary alcohols**. Those with general formulae of the type $R^1R^2R^3COH$ are called **tertiary alcohols**.

The primary/secondary/tertiary terminology is by no means exclusive to alcohols; it also finds applications in the chemistry of, for example, halogenoalkanes.

Question

27 Identify the primary, secondary and tertiary members of the collection of alcohols shown above.

We shall see plenty of parallels between the chemical tendencies of alcohols and halogenoalkanes, by virtue of the similar polarities of the bonds that retain the two functional groups. Thus, nucleophilic attack followed by heterolytic cleavage of the OH group should be expected and will indeed take place.

However, this time the creator of the polarity, the oxygen atom, is 'pulling' on *two* pairs of electrons in *two* bonds (one to carbon and one to hydrogen). As a result, we shall see a whole second group of reactions in which the O—H bond cleaves.

Just to enrich the picture still further, the oxygen atom in alcohols is much more nucleophilic, in its own right, than the halogen atom in halogenoalkanes. (For a parallel from a different but related context, compare the low basicity of Cl⁻ ions with the much more basic OH⁻ ions. This difference, too, is due to the extra 'pushiness' of the lone pair on the oxygen.) So we will be seeing reactions in which the alcohol is, as it were, the 'attacking reagent' rather than the 'substrate'. These reactions will also result in cleavage of the O—H bond, but they will be treated as a distinct sub-group of that class. The three tendencies of alcohols are summarised in the diagram below.

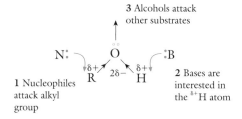

11.10 Alcohols 2: Undergoing nucleophilic attack

Like halogenoalkanes, alcohols undergo substitution and elimination reactions when attacked by nucleophiles.

Substitution

The typical nucleophilic substitution reactions of alcohols take place in acidic, or at least non-basic, conditions. In other words, species in the reaction mixture with lone pairs of electrons will tend to pick up H⁺ ions (that is, be 'protonated'). This certainly does not help the nucleophile do its job (since the best nucleophiles are strong bases, and having a proton stuck on your active site is a bit like sticking a rubber bung on your foil-tip), but it does help the leaving group leave. Protonating the OH group gives a further tug to the electrons and facilitates the cleavage of the C—O bond.

$$RCH_2OH + H^+ \rightleftharpoons RCH_2\overset{+}{O}H_2$$

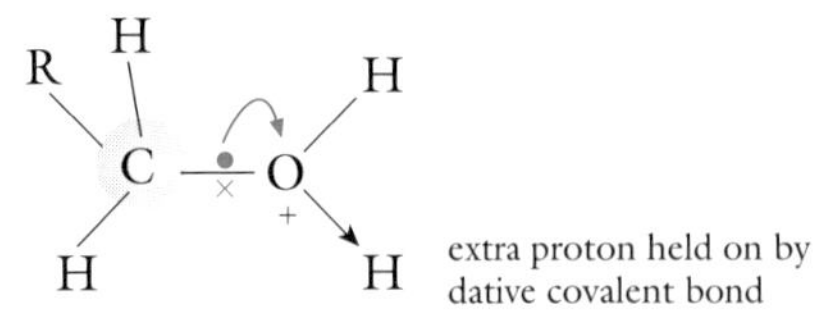

extra proton held on by dative covalent bond

Question

28 Explain why the protonation of the OH group helps the leaving group leave (give a fuller explanation than that in the last paragraph).

The following methods of substituting a halogen for the OH group illustrate this 'acid medium' idea in action.

1 Bubbling dried hydrogen halide gases through a solution of the substrate:

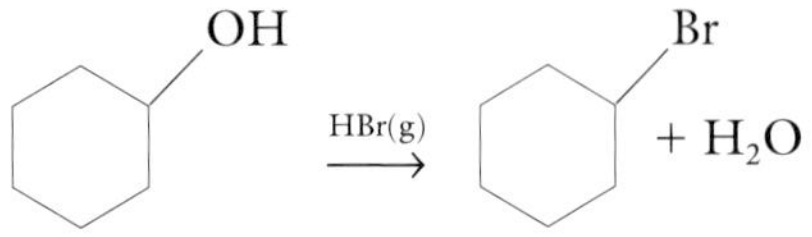

cyclohexanol   bromocyclohexane

2 Producing hydrogen halide gases *in situ* from other reagents:

$$NaBr + H_2SO_4 \rightarrow NaHSO_4 + HBr$$

$$HBr + C_4H_9OH \xrightarrow{reflux} C_4H_9Br + H_2O$$

3 Using phosphorus halides, or 'thionyl chloride':

$$PCl_3 + 3C_4H_9OH \rightarrow 3C_4H_9Cl + P(OH)_3$$

$$SOCl_2 + 2C_4H_9OH \rightarrow 2C_4H_9Cl + H_2SO_3$$

4 Making phosphorus halides *in situ* from phosphorus and the halogen:

$$2P + 3I_2 \rightarrow 2PI_3$$

$$PI_3 + 3C_3H_7OH \rightarrow 3C_3H_7I + P(OH)_3$$

One of the key experiments focused on a rather obscure alcohol, which used a mechanism different from the norm. Yet oddly this alcohol's *abnormal* behaviour shed light on what all the normal ones were doing. The alcohol in question was 2-phenylpropan-1-ol, and its special trick, when reacted with HCl, was to emerge after reaction with halogen in a *tertiary* position, despite the fact that the departing OH group had left from a primary position on the end of the chain.

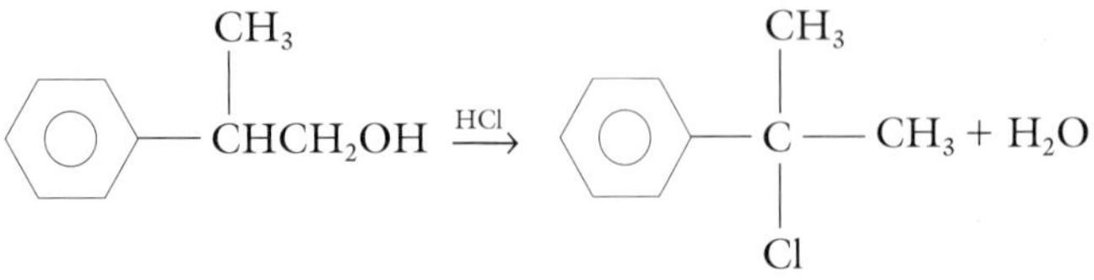

This suggested a carbocation intermediate, which had a sufficiently long lifetime to allow a hydrogen to move across and convert the less stable primary carbocation into the more stable tertiary one. The whole proposed mechanism is shown below.

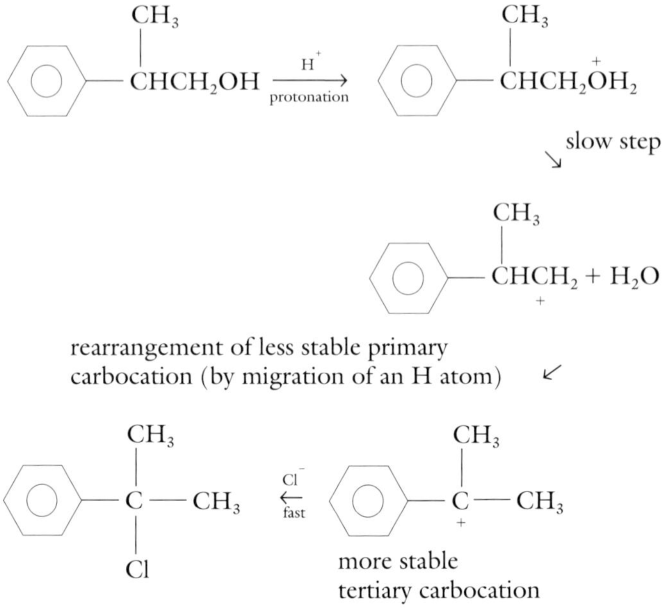

slow step

rearrangement of less stable primary carbocation (by migration of an H atom)

more stable tertiary carbocation

Questions

29a What theories (involving electron pushes) have we put forward to explain the extra stability of tertiary carbocations as compared to primary ones?

b The particular tertiary carbocation shown above would be exceptionally stable (by carbocation standards, on which scale one-thousandth of a second would be long-lived), and the reason is that it has a phenyl group (benzene ring, C_6H_5—) as one of its substituents. Can you suggest how the adjacent benzene ring can help to stabilise the ion?

30 By comparison with Section 11.3, does the mechanism above more closely resemble an S_N1 or S_N2 type? Justify your answer.

Experiments like the one described above, plus rate–law investigations (Chapter 14), established that the S_N1 mechanism was favoured in substitution reactions of alcohols. In fact it is found that only primary alcohols and methanol undergo substitution via S_N2 routes, while all the rest do what should be your answer to question 30, namely 'go S_N1'. This bias towards two-step 'leaving-group-led' mechanisms is what we could expect from the 'down-grading' of nucleophiles in acidic media. If the push of the nucleophiles is blunted, then greater responsibility falls on to the leaving group to initiate the reaction.

Question

31 Having done an in-depth study of a 'freak' mechanism, we need to make sure we don't ignore the normal ones, as used by primary and secondary alcohols. In the light of the information in the last paragraph, write out the mechanisms of the reactions between HCl gas and:

a ethanol

b propan-2-ol

  using curly arrows where appropriate. Assume the nucleophile is Cl⁻ and that the —OH groups of the alcohols become protonated.

Elimination

In the case of alcohols, elimination is also called 'dehydration', since it involves the loss of a hydrogen atom from one carbon and an OH from the next, which adds up to the removal of water. Elimination reactions in alcohols can be both compared and contrasted with substitution reactions in alcohols, and also with elimination reactions in halogenoalkanes. Here are four points, either of likeness or of contrast, linking these sets of reactions.

1 Elimination reactions of alcohols, like substitutions, tend to be first-order, two-step, leaving-group-led (call it what you will). The mechanism is as shown below.

2 As in substitution reactions of alcohols, the tertiary alcohols eliminate the most readily.
3 Unlike in halogenoalkanes, the two reactions – elimination and substitution – do not often occur in competition, 'in parallel'. Each set of reagents seems to be dedicated to doing just one job. For instance, we have seen the range of reagents which bring about substitution reactions (PI_3, etc.). Dehydrating agents also tend to be specialised, in this case as 'water grabbers' – for example, concentrated H_2SO_4 or Al_2O_3.
4 As with halogenoalkanes, the elimination products are alkenes.

11.11   Alcohols 3: As acids

If you have unwanted surplus sodium metal (and wisely ignore the advice of sensation-hungry students to 'chuck it all down the sink'), the accepted way to deal with it is to react the sodium with ethanol. The reaction mimics that between sodium and water, but with ethanol it is much slower and more controllable.

$$HOH(l) + Na(s) \rightarrow NaOH(aq) + \tfrac{1}{2}H_2(g)$$
$$C_2H_5OH(l) + Na(s) \rightarrow NaOC_2H_5(aq) + \tfrac{1}{2}H_2(g)$$

All alcohols behave in this way towards sodium, the only variation being the difference in the rate of reaction. The rate of reaction depends upon the acidity of the hydrogen atom being replaced, and this in turn is dependent on the bond-polarising effects within the molecule. Let us pursue this further: it has been a theme in these organic chapters that alkyl groups release or push electron density away from themselves (and so, in the case of alcohols, towards the oxygen). On the other hand, the ease of release of a hydrogen ion is increased by electron pull away from itself (again towards the oxygen). Put these two concepts together in answering the following question.

Questions

32a Why would water be more acidic than ethanol?

b Put the following alcohols in order of their probable rate of reaction with sodium (fastest first):

  butan-1-ol, butan-2-ol, 2-methylpropan-2-ol.

33 The RO⁻ ion that results from the reaction of an alcohol with sodium is a good nucleophile in its own right.

a Would it be a better or worse nucleophile than OH⁻? Explain your answer in terms of the inductive effect of the alkyl group.

b Being a nucleophile is more or less the same as being a base – both require a species to be 'pushy' with a lone pair of electrons. So base strength should parallel nucleophile strength. That being so, and looking back to your answer in part (a), decide in which direction the following reaction should proceed:

  RO⁻ + HOH ⇌ ROH + OH⁻

c What would be the result of reacting $NaOC_2H_5$ (in ethanol solution) with C_2H_5Br?

11.12   Alcohols 4: As nucleophiles in their own right

We saw in the last question that it is possible to make a very effective nucleophile from alcohols by converting them to OR⁻. In fact many things in alcohol chemistry suggest that an alcohol is a sort of 'alkyl-substituted water molecule', so if both OH⁻ and water can be nucleophiles, why not both OR⁻ and ROH?

  This indeed proves to be the case. Although not as effective as OR⁻, alcohols perform well as nucleophiles in certain situations; notably one famous one in which the alcohol's job is made easier by protonation of the leaving group.

The esterification reaction

Esterification, the formation of an ester, is an acid-catalysed equilibrium reaction between an alcohol and a carboxylic acid (see Chapter 12). The reaction scheme is shown below:

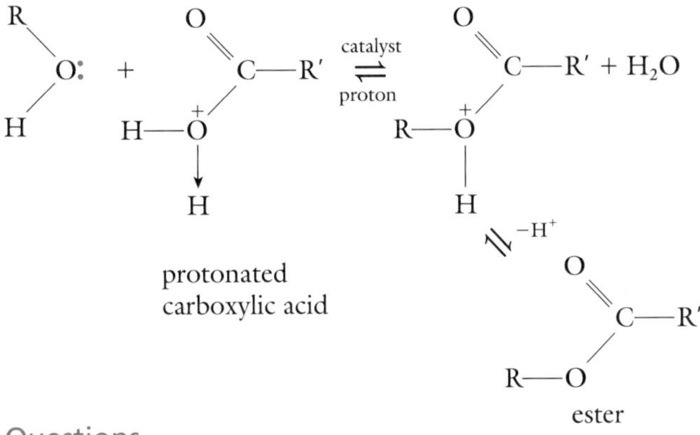

protonated carboxylic acid

ester

Questions

34 The equation above is complete, apart from the curly arrows. Add them, bearing in mind that the prime mover in this reaction is a lone pair on the alcohol's oxygen atom, and you can see what the leaving group is going to be.

35 Which is the site of attack by the oxygen lone pair, and what factors make this site such an attractive prospect for nucleophilic attack?

It is tempting to see a parallel between 'acid plus alcohol goes to ester plus water' and the familiar inorganic reaction 'acid plus alkali goes to salt plus water'. However, almost every apparent similarity is an illusion. First, the inorganic reaction is very fast and needs no catalyst. Second, the inorganic reaction goes to completion, without any equilibrium mixture. But most fundamentally of all, the organic acid in an esterification reaction is not even behaving as an acid. Look back to the answer to question 34, and try the next question.

Question

36a Which bond would you expect to break on the acid if it behaved acidically?

b Which bond *does* break on the acid according to the mechanism in question 34?

c Which bond actually breaks on the alcohol, according to the mechanism in question 34?

d Support for the question 34 mechanism came historically from some clever experiments in 'isotopic labelling'. Reaction was set up between ethanoic acid and 'labelled' ethanol containing ^{18}O atoms in the OH position. The resulting ester was isolated and subjected to mass spectrometry, to see if it contained the ^{18}O label. Redraw the question 34 mechanism with the labelled oxygen marked as O*, and show where you think it would end up.

An alternative route to esters

This second esterification process comes from a simple idea – if you want a complete non-reversible ester-formation reaction, use a better leaving group than OH^-. So what is it that makes for a good leaving group?

A good leaving group, at least in the field of nucleophilic substitution, is one which is very 'happy' to have full control of the pair of electrons in what used to be its bond to the substrate and which has no tendency, once having left, to act as the nucleophile in its own right. This description coincides with a concept from another field, that of acid–base behaviour. Our sought-for leaving group would also be the sort of species which would be poor at retaining hydrogen atoms and correspondingly good at releasing them as H^+ ions. So we are looking for the anion of a strong acid, something like, say, a halide ion. (These concepts are discussed more fully in Chapter 12, on acidity).

It was through reasoning like this that people arrived at the idea of reacting alcohols with **acyl chlorides** instead of acids. A generalised acyl chloride is shown below.

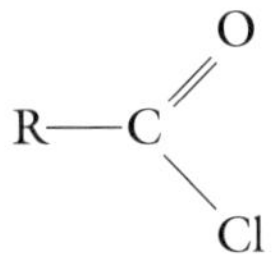

Question

37a What is the leaving group when esterification is carried out on an acyl chloride, and how do we know that it conforms to the description above of the 'good leaving group'?

b Draw a mechanism for this reaction in the style of question 34.

11.13 Alcohols 5: Undergoing oxidation

The mechanisms of oxidation reactions are incompletely understood. As a result, this section confines itself to the reagents used and the outcomes which result.

The commonest reagent used in the laboratory for oxidising alcohols is acidified potassium dichromate, containing the $Cr_2O_7^{2-}$ ion. It is reduced to the Cr^{3+} ion in the process.

Primary alcohol – to aldehyde and then carboxylic acid

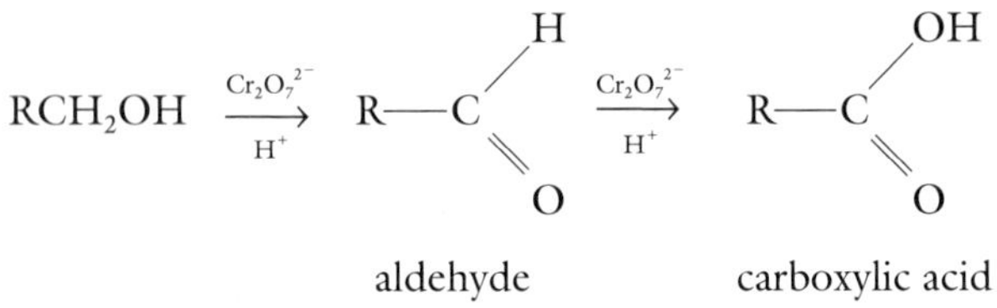

aldehyde   carboxylic acid

Secondary alcohol – to ketone

$$\overset{\displaystyle R}{\underset{\displaystyle R'}{\diagdown}}\text{CHOH} \xrightarrow[\text{H}^+]{\text{Cr}_2\text{O}_7{}^{2-}} \overset{\displaystyle R}{\underset{\displaystyle R'}{\diagup}}\text{C}=\text{O}$$

ketone

Question

38 Why is it that secondary alcohols can only undergo a single oxidation step, in contrast to primary alcohols?

Tertiary alcohols

Tertiary alcohols cannot be oxidised (at least not without rupturing the carbon framework).

Question

39 Let us revise some skills from Chapter 7 (on redox), in what is admittedly a rather hard example:

a What is the oxidation number of chromium in the dichromate ion, $Cr_2O_7{}^{2-}$?

b Write a balanced half-equation for the change in the chromium species. Remember all the 'mopping up' needed, using hydrogen ions and water molecules to make these half-equations balance. Remember also the message in Chapter 7 – these reactions will only take place if there are H⁺ ions.

c What is the oxidation number of the carbon atom bonded to the OH group in ethanol? (Use the 'who wins what bond' method, p. 124).

d What is the oxidation number of the corresponding carbon atom in ethanoic acid?

e Construct a half-equation linking ethanol and ethanoic acid, with electrons and hydrogen ions/water molecules to achieve balance.

f Now combine the two half-equations to reveal the full stoichiometric equation for the oxidation of ethanol to ethanoic acid.

If you want to stop your reaction at the aldehyde stage, in practice there are three ways to do it.

1 Distil the aldehyde off as it forms (the aldehydes nearly all have lower boiling points than the corresponding alcohols or acids).

2 Use a lower ratio of dichromate : alcohol in the reaction mixture.

Question

40 Why does that make sense?

3 Use more dilute solutions and lower temperatures.

Finally, under the heading of oxidation, it is worth noting that all alcohols can be burned in air or oxygen, but that their use as fuels is not economically sensible in the presence of the greater availability of hydrocarbons.

Energy from biologically produced ethanol

If this book were being written 100 years into the future, it is possible that the economic balance might have shifted from the position stated in the last paragraph. At the moment it does not make economic sense to hydrate ethene to make ethanol for use as a fuel, because cracked fractions of crude oil already offer an unbeatable range of fuels.

However, people are already considering a future in which crude oil has become scarce. One of the energy sources that should last for a while yet is the Sun, but its energy arrives in a very diffuse form, and concentrating solar energy into a usable package has proved difficult, especially in northern Europe.

Question

41 What methods have been tried with reasonable success in sunny countries?

Although we humans have so far had limited success in fixing solar energy, green plants have been doing it well for 1 000 000 000 years. This is because of their ability to manufacture glucose from carbon dioxide and water:

$$\overset{\text{solar energy}}{6CO_2 + 6H_2O \rightarrow \underset{\text{glucose}}{C_6H_{12}O_6} + 6O_2}$$

Question

42a From a data book and a Hess's Law cycle, work out the free energy change ΔG for this reaction. (Take the ΔG_f of glucose to be -750 kJ mol^{-1}.)

b How can the reaction still happen, despite apparently causing a decrease in the entropy of the universe? (Remember that positive values of ΔG_{sys} indicate negative values of ΔS_{uni}.)

The oldest human chemical activity is the conversion of sugars like glucose into alcohol, under the control of organisms of the yeast type – the process known as **fermentation** (Figure 11.26). It is this reaction that would be the final stage of a

Figure 11.26 The production of both wine and beer involves fermentation

future fossil-fuel-free route to energy for transport and power generation.

Naturally there will be a modern aspect to the fermentation routes to alcohol. The focus will be on finding the most efficient Sun-trapping green plant and the best enzyme for the sugar-to-alcohol stage. The chances are that the enzyme will be mass produced by genetically engineered bacteria.

11.14   Synthetic routes

We have seen that alcohols are versatile chemicals, in that they can lead on to a wide range of other organic products, from halogenoalkanes through carboxylic acids to esters and ethers. The business of interconverting chemicals belongs to a sub-section of organic chemistry known as 'synthetic routes' or just synthesis. It involves skills of two sorts – picking out the reagents and techniques to get from start to finish, and knowing which routes make economic sense.

Choosing reagents and techniques in order to build up maps of synthetic pathways is an exercise in data collection. Ethanol, versatile as it is, makes an excellent centre-point for such a map.

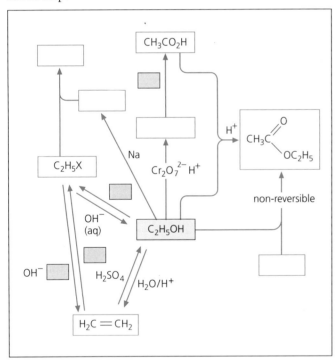

Figure 11.27 Map of synthetic routes centred on ethanol – see question 43

Question

43 Fill in the gaps in Figure 11.27, indicating either the missing reagents, substrates or products.

Multistep syntheses

When you are presented with a synthesis problem which does not have a single-step solution, you need to develop certain skills to work out the stages. One idea is to look at the target chemical (the final product), and think about the various molecules which could have been *its* precursor. Then identify one of these precursors which is capable of being made from the starting material. The method has been symbolised in Figure 11.28. Try to solve the following problems by the suggested tactic.

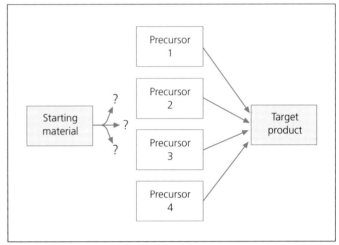

Figure 11.28 A tactical approach to synthetic problems

Question

44 Outline the steps in the following syntheses, giving any necessary inorganic reagents and conditions:

a The conversion of butan-1-ol to 2-bromobutane.

b The conversion of cyclohexanol to 1,2-dibromocyclohexane.

c The conversion of ethanol to ethyl ethanoate (an ester – see below).

$$CH_3C \overset{\displaystyle O}{\underset{\displaystyle OC_2H_5}{<}}$$

This question can be easy or hard, depending on whether you allow yourself access to ethanoic acid as a separate reagent (the easy way), or whether you have ethanol as the only organic starting material. Try it the hard way.

It would be wrong to leave a chapter on alcohols without mentioning the application which has most relevance to many adult humans. This of course is the application in which one particular alcohol, ethanol, has come to be called by the name of the whole class. Both pleasure and pain have come from human use of 'alcohol' – we will look at one aspect of its *mis*use.

11.15 Alcohol, infrared spectroscopy and the breath test

Infrared spectroscopy was discussed in Chapter 5 on polar covalent bonds (p. 77). This revealed that molecules absorb only those infrared photons whose frequency matches the vibrational frequencies of bonds within the molecules. (The proviso is that the absorbing vibration has to set up an oscillating dipole.) An infrared spectrum thus gives information about the bonds in the molecules under test.

Infrared spectroscopy of alcohols illustrates some fresh concepts, and because it plays a big part in the forensic measurement of blood alcohol levels, this section picks up the topic of infrared spectroscopy again and considers the techniques involved.

The spectrometer

Figure 11.29 is a schematic picture of an infrared spectrophotometer. A hot wire on the left (shown as 'glowing source') is radiating photons from the whole of the infrared region of the electromagnetic spectrum. Mirrors are used to direct the radiation through two cells, labelled 'sample' and 'reference'. Sample and reference beams are then compared, by means of a spinning segmented mirror (Figure 11.30). This mirror sends alternate bursts of radiation from the sample and reference paths into a common receiver, called the **monochromator grating**. It is here, and only now, that the signals are 'filtered' for wavelength. The grating is a movable device which,

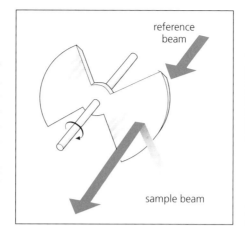

Figure 11.30 The rotating segmented mirror in an infrared spectrophotometer

for every given position, permits only photons of a single wavelength to reach the detector.

So as the grating moves, it allows a succession of wavelengths to come to the detector and, because of the spinning mirror, it deals alternately with the sample and reference beams. The detector is connected to the chart recorder pen, and a 'peak' (always upside down) is drawn whenever a *difference* is detected between sample and reference beams. Meanwhile the grating's movements are synchronised with a roll of chart paper marked out in wavenumbers (reciprocals of wavelengths, used by convention in infrared work for the benefit of their 'handier' numbers. The 'pen' and 'chart paper' are quite likely nowadays to have been replaced by a computer and a TV screen (Figure 11.31).

Let us assume that the sample contains molecules which have C—H bonds – which means that the sample will

Figure 11.31 A modern infrared machine, showing TV monitor display of the spectrum

absorb photons of wavenumber about 3000 cm^{-1}. As the grating reaches the position in which it should be transmitting photons of 3000 cm^{-1}, it gets a much reduced supply from the sample beam. The reference beam will be unaffected, so the detector will sense a difference in the two beams. This difference signal will be converted to a peak, which will occur at 3000 cm^{-1} on the chart paper. When the sample does not absorb photons, the sample and reference beams give the same signal to the detector, which responds by *not* producing a peak.

Questions

45 a The implication in the above account is that the reference beam path contains no object at all. This is sometimes the case. However, when the sample is a solute in solution, the reference beam has a new and important role. What would you suppose is put in the reference beam under these circumstances, and why?

b The cells that contain the samples for infrared spectroscopy are made of a salt – sodium chloride or sodium bromide – compressed into clear discs. The samples are smeared on one disc and trapped by another,

Figure 11.29 The infrared spectrophotometer

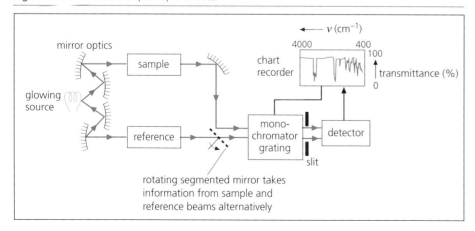

Figure 11.32 Sodium bromide 'windows' for mounting a liquid sample in infrared spectrophotometry

almost like mounting a biological slide (Figure 11.32). Why do you suppose glass cells are not used? Can you see any limitations in the use of cells made of sodium chloride or bromide?

46 Figure 11.33 shows the infrared spectrum of ethanol, with bonds attributed to relevant peaks in the table below it.

a The breadth of the O—H stretch peak suggests that not every O—H bond vibrates at quite the same frequency as every other O—H bond. Can you explain what variable factor might affect and broaden the range of their vibrations? (Answers may be found in Chapter 6 on intermolecular bonding.)

b When it comes to estimating alcohol levels in breath, the C—H peak is the one that is used, rather than the more characteristically 'alcoholic' O—H peak. Can you suggest why?

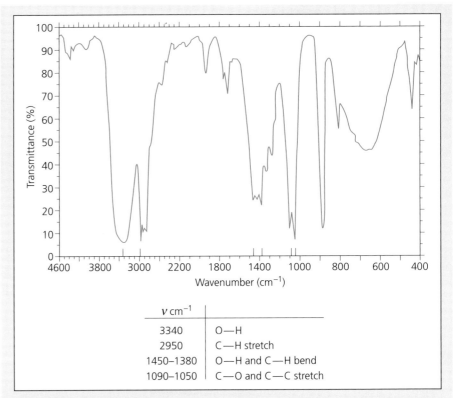

ν cm^{-1}	
3340	O—H
2950	C—H stretch
1450–1380	O—H and C—H bend
1090–1050	C—O and C—C stretch

Figure 11.33 Infrared absorption spectrum of ethanol

The intoximeter

The roadside breathalyser machine (Figure 11.34) has no connection with infrared. It works as a fuel cell (Chapter 6), in which the ethanol on the person's breath is oxidised in a way which gives a voltage. However, roadside breathalysers are not considered quite accurate enough to provide the basis for evidence in a court case. If the roadside machine suggests the suspect may be over the limit (which at present in Britain is set at 35 micrograms

Figure 11.34 A roadside breath-test machine – a type of fuel cell

of ethanol per 100 cm^3 of breath), then he or she is taken to the police station and asked to breathe into a specialised infrared spectrophotometer, called an **intoximeter**. Since it is dedicated to the analysis of a single substance, it can do away with some of the systems of the full laboratory machine. On the other hand it has the extra challenge of detecting molecules at low concentrations in gaseous samples.

Figure 11.35 Components of the infrared analyser unit of the Lion Intoximeter 3000

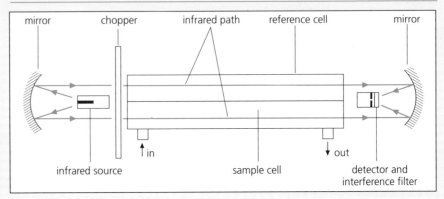

Consequently, we have the machine shown in Figure 11.35, in which the sodium chloride plates are replaced by cells which are 70 cm in length, while the grating is replaced by a simple single filter which transmits only photons of wavelength 2950 cm^{-1}.

Question

47a Explain these two design variations, comparing the intoximeter to a conventional infrared spectrophotometer.

b People suffering from diabetes occasionally have propanone (H$_3$C—CO—CH$_3$) in their breath. This is a problem for users of the intoximeter – can you see why? (They have actually found a way round this problem.)

It is of course hoped that no reader of this book ever has to face one of these machines in earnest. At present you will be prosecuted if your breath alcohol level is above 40 micrograms per 100 cm^3 of breath. If the intoximeter records you at over 50 micrograms, then prosecution proceeds on that basis alone, with no further tests. If you come out between 40 and 50 micrograms, then you are given the option of a blood or urine test. The sample of blood or urine is divided in half, and you have the right to keep one half to have it privately analysed. The police sample is further halved, and both new halves are analysed, with the higher result being discarded.

Question

48a What would you suppose are the merits of a blood or urine test relative to a breath one?

b Why under some circumstances is a breath test alone sufficient basis for conviction?

11.16 Phenols – the OH group modified by the benzene ring

Phenols are compounds in which an OH group (or groups) is attached to a benzene ring, in place of one (or more) of the hydrogen atoms. Phenols are not alcohols, even though they look like them on paper.

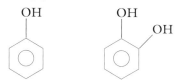

phenol   2-hydroxyphenol

They do not behave like alcohols because the neighbouring benzene ring system has a profound effect on the behaviour of the OH group. Indeed, as we shall see, the influence is mutual, since the OH group has a profound effect on the behaviour of the ring.

We have already built up a picture of delocalised bonding, and used it to explain the singular properties of benzene – its unusual stability, for instance (Chapter 10, p. 221). In fact, we have built up two pictures, one of which shows the electrons as a cloud, and the other which imagines the molecule as being a *hybrid* of more than one classical electron arrangement (p. 221). It is this second picture which we will use in this last section of the chapter. (Chapter 1 introduced the idea that scientists are happy to switch models when it suits them.)

You may remember from Section 4.6 that the 'rules' of the hybrid game are as follows:

1 Draw as many feasible structures as possible.

2 The structures with lowest energy (that is, the most stable ones) will contribute most to the hybrid mixture.

3 The more feasible hybrids of reasonably low stability there are, the more the 'delocalisation stabilisation' of the molecule.

Benzene, viewed in this way, is a hybrid of *two* structures, and the stabilisation energy is estimated (in question 57 of Chapter 10) at about 270 kJ mol^{-1}.

We can draw several more structures for phenol.

The ionic ones may be less energetically stable than the covalent ones, but they do suggest how the phenoxide ion could delocalise the negative charge around the ring. The way phenol behaves certainly supports the idea that these ionic structures contribute significantly to the way the electrons are distributed.

Question

49a If the ionic structures contribute to phenol, what effect will this have on phenol's willingness to part with an H$^+$ ion, and therefore its acidity, compared with ethanol?

b If, as the ionic structures imply, the electron density from the ruptured O—H bond is delocalised round the ring, how would this affect the reactivity of phenol towards approaching electrophiles such as NO$_2^+$?

Your answers to the last question should have prepared you to accept that, if phenol turns out to be a stronger acid than ethanol, and if it turns out to undergo electrophilic substitution more rapidly than benzene, our 'resonance hybrids' model for the bonding in phenol is vindicated. In fact, phenol turns out to be a stronger acid even than water (let alone the alcohols), although not in the same league as the carboxylic acids (Table 11.7 shows the relevant values of pK_a).

Table 11.7 Acidity of —OH compounds

Substance	pK_a
Ethanol	>14.0
Water	14.0
Phenol	9.9

Phenol shows its acidic character by a reaction with sodium (as an alcohol would), but also with sodium hydroxide (as a 'proper' acid would). Phenol is insoluble in water (or at least it does a strange thing where two immiscible layers form, one a saturated solution of a little water in phenol, and the other a saturated solution of a little phenol in water). However, it dissolves in sodium hydroxide solution and reappears (as a milky precipitate) when the alkali is neutralised.

Questions

50 From the information in the above paragraph, complete the following equations (if you think anything happens).

$C_2H_5OH + Na \longrightarrow$? $C_2H_5OH + NaOH \longrightarrow$?

51 Put the following in order of acidity:

water, ethanol, phenol

and the following in order of basicity:

hydroxide ion, ethoxide ion, phenoxide ion.

Phenol as a nucleophile

To return to the hybrid molecules on p. 251, if a lone pair on the oxygen atom is delocalised round the ring, it has less incentive to be pushy. Being pushy with lone pairs is of course a prime qualification for a successful nucleophile, so we would conclude that, just as phenoxide ion is not a very good base, and for exactly the same reasons, so phenol and phenoxide ion are not very good nucleophiles. This leads to a second major contrast between phenols and alcohols.

Question

52 Which group of alcohol reactions, dealt with in Section 11.12, required the alcohol to act as a nucleophile?

It is not totally impossible to make the esters of phenols, but you have to adapt the recipe. The weak phenol nucleophile must be given an easier target. It can just about manage to carry through an attack on the more vulnerable and reactive group of molecules known as acyl chlorides:

phenyl ethanoate

but a straight attack of phenol on a carboxylic acid would end in failure.

Questions

53 Put the curly arrows on the phenol/acyl chloride equation above, to show the movement of electrons.

54a The carbon end of the C—OH bond in phenols is also a member of the ring, and thus is 'protected' by the π-electron cloud. Which reagents will therefore not want to come near to such a carbon atom?

b Which reaction-type(s) of alcohols is/are therefore likely to be missing from the repertoire of phenols?

Phenols – reactivity of the ring

The relationship between the benzene ring and the OH group is a genuine *inter*action. In other words, not only is the alcohol behaviour of the OH group modified by the nearness of the benzene ring, but the behaviour of the ring is modified by the OH group. Look back to the answer to question 49b – you will see that it predicted that the ring in phenols should hold more attraction for electrophiles than the ring of benzene itself. This indeed proves to be the case, reaction with any given electrophile being thousands of times faster on phenol, even though the outcome, substitution of a ring hydrogen, remains the same. The benzene ring of phenol can also be attacked using much milder conditions than for benzene. For example, nitration of phenol is possible with dilute nitric acid alone, compared with the rather vicious mixture of concentrated acids needed for benzene.

A final interesting observation on the electrophilic substitution reactions of phenol concerns the mixture of isomers produced. Clearly, an incoming electrophile has five carbon atoms to choose from, as possible positions at which to replace a hydrogen.

Question

55 Look at the various hybrids on p. 251 Which carbon atoms look the most potentially attractive to an electrophile (which, remember, means 'electron-lover'), given the way the negative charge is delocalised round the ring?

Summary of reactions of phenol

1 As an acid

sodium phenoxide

2 As a nucleophile (ester formation)

3 Ring substitution

+ di- and tri-nitrated products

Summary

- If an organic molecule includes a bond with a pronounced dipole, then a dominant theme in its chemical reactions will be *attacks by nucleophiles on the positive end of the dipole*.

- Nucleophiles are species which offer a pair of electrons to the substrate.

- Halogenoalkanes contain one or more bonds between a carbon atom and a halogen atom. In all cases the C—X bond (X = halogen) will be polarised so as to make the *carbon partially positive*.

- Many halogenoalkanes undergo nucleophilic substitution, notably with the nucleophiles OH^- and H_2O. In these cases the oxygen atom offers one of its lone pairs to the carbon end of the C—X bond, while an X^- ion is ejected along with the pair of electrons which made up the original C—X bond.

- The rate of this reaction depends on the identity of X. The order of reaction rate, fastest first, is I > Br > Cl > F.

- The mechanism of this nucleophilic substitution reaction can vary. *Highly branched* halogenoalkanes follow a **two-step reaction mechanism** (called the S_N1 mechanism) in which the halide ion leaves *before* the incoming nucleophile arrives. *Primary* halogenoalkanes follow a **one-step mechanism** (called the S_N2 mechanism) in which the pair of electrons from the nucleophile begin forming the new bond to the carbon atom at the same time as the halide ion is getting ready to leave on the other side of the molecule. (The use of the words 'halide *ion*' might be confusing. The halogen atom is covalently bonded in the original substrate, but becomes an ion in the course of the ejection.)

- The reasons why one halogenoalkane follows one mechanism, while another halogenoalkane follows a different one, are to do with differences in the energetic difficulty of getting to the respective **transition states**. This in turn is dependent on factors such as degree of **steric hindrance** and **polarisation effects**. And, to trace the chain of cause and effect back to its source, these factors are determined by whether or not the molecule is highly branched.

- Not every meeting between a nucleophile like H_2O or OH^- and a halogenoalkane results in a substitution reaction. Sometimes you get the removal of a hydrogen atom *and* the halogen atom, to create a π-bond, and therefore an alkene. This type of outcome is called an **elimination**.

- Elimination reactions can actually go on alongside substitution reactions, to give mixed products. The chances of elimination are affected by two factors.

1 The choice of **solvent** – the greater the proportion of ethanol in the solvent mix, the greater the proportion of elimination product.
2 The degree of **branching** – the more highly branched the substrate, the greater the proportion of elimination product.

- Among the halogenoalkanes most resistant to substitution, or indeed to reactions of any kind, are the **chlorofluorocarbons**, or **CFCs**. Several of them also have boiling points not far below room temperature, and thus lend themselves to applications in refrigeration, aerosol propulsion and foam blowing.

- The stability of CFCs means they have long lifetimes in the atmosphere, and find their way up to the stratosphere. Here, in an environment rich in high-energy ultra-violet photons, they take part in free-radical reactions which upset the balance between the creation and destruction of the gas **ozone**, O_3. The balance is shifted towards destruction, and the result is a hole in the previously continuous ozone layer around the Earth. Since ozone acts as an ultra-violet filter, the Earth is being dangerously over-exposed to ultra-violet light.

- Scientists are looking for replacements for CFCs in the various parts of their working range. At the moment butane, though hardly ideal, is a stop-gap aerosol propellant. A compound code-named HFA 134a, whose formula is CF_3CH_2F, is a promising candidate as a CFC substitute, since its total lack of C—Cl bonds renders it incapable of getting involved in ozone destruction. At present, its use is inhibited by its cost.

- The combined technique of GC/MS (gas chromatography/ mass spectrometry) is ideal for identifying which gaseous or volatile compounds are present in a test sample. The gas chromatography section of the machine separates the sample into single compounds, while the mass spectrometer acts both as detector and identifier.

• Halogenoalk*enes* and halogeno*arenes* are less reactive than corresponding halogenoalkanes. In both cases the reason is the same – the carbon atom of the C—X bond is also one end of a region of high electron density (due to double bonding), so nucleophiles are deterred.

• One halogenoalkene in particular, **chloroethene**, has enormous significance as the monomer of the widely used polymer **poly(chloroethene)** or **PVC**.

• Chloroethene is manufactured from ethene via a two-step synthetic route, which gets round the impossibility of direct substitution of a hydrogen atom by a chlorine atom. The first step is addition of a dichlorine molecule across the double bond, and the second step is the elimination of an HCl molecule by thermal cracking.

• PVC is polymerised by a standard free-radical addition-polymerisation method, initiated by an organic peroxide. The reaction takes place in an aqueous suspension.

• PVC-based polymers can be given a wide range of physical properties by the use of **additives**. The PVC molecule seems to be particularly good at accommodating 'foreign' molecules within its structure, perhaps because its own intermolecular forces are a mixture of van der Waals' and dipole attractions.

• The most important group of additives for extending the range of uses of PVC are the **plasticisers**. So dramatic is the modification of the character of the base polymer by the use of plasticisers that PVC is often referred to according to whether or not it has been plasticised. Hence we have the names uPVC (u for unplasticised) and pPVC (p for plasticised).

• uPVC is a fairly standard thermoplastic polymer. At room temperature it is below T_g, its glass transition temperature, and is quite hard. Any response to deformation is elastic or, in the extreme, brittle. It finds uses in, among other things, packaging, window framing and guttering.

• pPVC is a soft flexible material with a degree of plastic, as well as elastic, response to deformation. The difference is brought about by plasticisers *interrupting the intermolecular forces* between individual polymer chains, and facilitating their movements. pPVC is used among other things as a waterproof clothing material, leather substitute and insulating material for electric wiring.

• Of all plastics materials, PVC is the one which uses *least crude oil and energy* in its manufacture (that is, when comparisons are made by mass). 57% of its mass comes from the sea.

• **Alcohols**, like halogenoalkanes, have a polar —C bond, and undergo some of the same reactions, notably nucleophilic substitutions. The differences between alcohols and halogenoalkanes can be summarised under three headings.

1 *Mechanistic differences*: substitution reactions on alcohols normally take place in acidic media, and S_N1-style mechanisms predominate, with the first step being the departure of the protonated OH group, leaving as water.

2 *Reagent differences*: the reagents needed to bring about substitutions in alcohols (phosphorus halides, for instance) are completely different from those which bring about elimination (which are dehydrating agents).

3 *Versatility differences*: alcohols can do things which halogenoalkanes cannot. They can use lone pairs on their own oxygen atom to become *nucleophiles* in their own right (as in esterification reactions). Then again they can release their hydrogen atoms to behave like *acids* (as in the reaction with sodium metal), and they can be *oxidised* (as opposed to burnt), by reagents like acidified potassium or sodium dichromate(VI) – that is, $K_2Cr_2O_7$ or $Na_2Cr_2O_7$.

(Now for some detail on the points from the last paragraph.)

• Alcohols react with carboxylic acids to make **esters**. The reaction is slow, even when catalysed by inorganic acids, and the outcome is an equilibrium mixture. A faster reaction, with more complete yields, is achieved by reacting alcohols with **acyl halides**.

• When oxidised by dichromate(VI) ions, alcohols show a range of outcomes, depending on the degree of branching. *Primary alcohols* are oxidised first to **aldehydes** and then on to **carboxylic acids**. *Secondary alcohols* are oxidised to **ketones**, but no further oxidation is possible. *Tertiary alcohols* cannot be oxidised at all.

• Ethanol is the alcohol of intoxicating drinks. It has an infrared spectrum which, like those of all alcohols, shows a characteristically broad absorption peak at around 3200 cm^{-1}. The peak is due to the stretching vibration of the O—H bond, and its breadth is caused by variable degrees of hydrogen bonding, which have a correspondingly variable effect on the O—H bond vibration frequency.

• A machine which detects alcohol in human breath (the **intoximeter**) looks only at the peak at 2950 cm^{-1}, where the C—H stretching absorption is situated. The O—H peak is no good for detection, because of the impossibility of discriminating between alcohol O—H bonds and water ones. Detection is made harder if other organic chemicals are present in the person's breath.

• **Phenols** are molecules with an OH group attached to a benzene ring. They resemble alcohols in some ways, but the juxtaposition of OH group and benzene ring has a modifying effect on the behaviour of both.

• Phenols *resemble* alcohols in these respects:

1 They react with **acyl halides** to make **esters**.
2 They react with metallic **sodium** to liberate **hydrogen**.

• Phenols *differ* from alcohols in these respects:

1 They show no tendency to undergo nucleophilic substitution, except under very severe conditions.

2 They are less nucleophilic in their own right than alcohols, as witness their inability to perform esterifications with ordinary carboxylic acids.

3 They are more acidic than alcohols, and will dissolve in and exchange protons with aqueous sodium hydroxide.

• These contrasts between alcohols and phenols can be explained by reference to a model which sees a lone pair on the oxygen atom being to some extent delocalised on the benzene ring's π-bonded system.

• The benzene ring itself is modified by the presence of the OH group. Again the change is attributable to the delocalisation of extra electron density on to the ring π-bond system – this transfer of charge makes the ring more attractive to electrophiles, and the substitution of ring hydrogen atoms is many times faster in phenol, for a given electrophile, than in benzene itself.

12

Acids, bases and alkalis – competition for protons

12.1   Introduction

Acidity was a concept long before there was any theory to explain it. People have known about the acidity of vinegar for as long as they have made alcohol by fermentation – and that was one of the earliest of all chemistry practicals. The acidity created by the action of water on the oxides of sulphur and phosphorus was part of the knowledge of medieval alchemists (Figure 12.1). Acids from biochemical sources, such as formic acid in the sting of a red ant, have been an unwelcome and familiar part of life all over the world.

Acids were named from a Latin word meaning 'sour' (as of taste), but their power to corrode rocks and metals, and their ability to inflict pain, means that acids were among the most high-profile members of the worlds of alchemy, chemistry and pharmacy. Yet for many centuries there was no understanding of the source of their power. Perhaps because ignorance and awe often go

Figure 12.1 Distillation of oil of vitriol (sulphuric acid), taken from *A description of new philosophical furnaces*, 1651

Figure 12.2 Antoine Lavoisier (1743–1794) (top) Lavoisier giving a demonstration at a lecture on the power of air

hand in hand, the acids acquired colourful names like 'oil of vitriol' (sulphuric acid) and '*aqua regia*' (a mixture of sulphuric and nitric acids).

One of the first chemists to attempt to classify the family of the acids was that great originator of so much early chemistry, Antoine Lavoisier (Figure 12.2). He did so by suggesting that an essential ingredient of all acids was *oxygen*.

Question

1 Suggest what methods of making acids might have given rise to this idea.

The discovery of the 'hydro-halic' acids (HCl, HBr, etc.) put an end to that line of thought and led to an improved definition – the common ingredient of acids was *hydrogen*. And yet there are many hydrogen-containing compounds which show not the least tendency towards behaving like an acid.

Question

2 Give some examples of these hydrogen-containing non-acids.

The next step was to see what made an acidic hydrogen different from a non-acidic one. One important clue lay in the fact that all the famous 'mineral' acids (that is, those not derived from organic

Figure 12.3 Electrolysing H_2SO_4 solution in a Hofmann voltameter

sources), such as hydrochloric acid, nitric acid and sulphuric acid, were **electrolytes** in aqueous solution. In other words, their solutions were all full of ions. This led to the idea that true acidity lay in the possession of *hydrogen ions*.

Question

3 There is some extra evidence for this idea.

a What event which takes place during the electrolysis of acids (Figure 12.3) points to the idea that they contain the hydrogen ion specifically, rather than just ions in general?

b The non-acidic hydrogen compounds supported the H^+ theory of acidity by what they did *not* do. If they dissolved in water at all, they did so without the creation of ionic solutions – in other words, there was no increase in electrical conductivity of the water. Which of your molecules in question 2 fit this pattern of behaviour? (If none of your question 2 molecules dissolve, think of a new one which does.)

This state of knowledge, that acids are acidic by virtue of their containing hydrogen ions, was the one with which chemistry entered the twentieth century. It is a reasonable and workable definition which covers the common acids whose bottles are on the shelves or in the cupboards of your school laboratory.

There was a companion theory which covered the substances that were able to neutralise acids. The **alkalis** were substances which could match acids ion-for-ion by providing hydroxide, OH^-, ions, which joined with the H^+ ions from the acid to form water. The Group I and lower Group II hydroxides were the archetypes of this group (for example, sodium hydroxide and barium hydroxide). Then there was a group of water-insoluble compounds, all oxides or hydroxides of metals, which behaved in a similar way to the alkalis, except that they deployed their anions from the undissolved solid lattice, rather than using the pre-mobilised $OH^-(aq)$ ion, and these acquired the name **bases**. Copper oxide

Figure 12.4 CuO solid – which doesn't dissolve in water (top) but does dissolve in acids (bottom)

is a typical example (Figure 12.4). Alkalis were held to be a special case of the larger family of bases, the water-soluble members of the family as it were, so that sodium hydroxide and copper oxide were both bases, but only the former was an alkali. The rule which governed this field was:

acid + base → salt + water only

This left carbonates and metals outside the family of bases, since although they both neutralised acids, and both produced salts, they could not pass the 'water only' test (Figure 12.5).

The fact that these definitions are still taught indicates that they are rugged and workable definitions for students of a

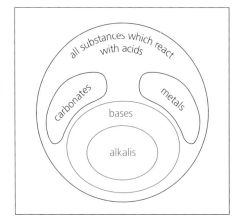

Figure 12.5 Venn diagram showing the pre-Brønsted–Lowry classification of chemicals which reacted with acids

certain age. However, the model presented so far is not quite refined enough to deal with some of the situations which occur at the margins of the field of acidity (see Section 12.2). Yet as well as being not refined enough it is in some ways not simple enough, because it has too many categoric 'pigeon-holes'. We will see, for example, that it is perfectly reasonable to include carbonates as bases (Figure 12.6).

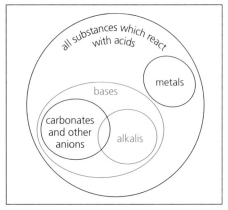

Figure 12.6 Venn diagram showing the Brønsted–Lowry classification of chemicals which reacted with acids

Finally, the cruder model misses a chance to place acidity into a larger chemical context – the general group of **competition processes**. For instance, this chapter is in some ways the other half of a matched pair with the *redox* (Chapter 7) – one chapter is about *electron* transfer, the other is about *proton* transfer, and in both areas there are league tables setting out the order of chemical dominance, and 'circle laws' to describe the direction of transfer. These extra refinements were achieved by the first major twentieth-century statement in this field, which redefined both acids and bases.

12.2 The Brønsted–Lowry definition of acidity

There are quite a few acid properties which the simple theory about acids – the one that sees acids as solutions containing hydrogen ions – does not handle too well. First, there is the matter of the *weak acids*. There are a number of substances, including the carboxylic acids from organic chemistry, that give rise to

poorly conducting solutions, but which nevertheless seem to have the same power to neutralise alkalis as the *strong* (highly conducting) *acids*. Do they or do they not contain hydrogen ions? It seems that they carry hydrogen ions in a fashion that is 'latent'.

Then chemists realised that if solvents other than water were considered, the divisions between weak and strong acids sometimes shifted. Liquid ammonia, for

Figure 12.7 J. N. Brønsted (top) and M. Lowry – the originators of the modern definition of acids and bases

instance, can be used as a solvent, as long as it is kept very cold. When dissolved in liquid ammonia, even carboxylic acids produced conducting solutions and behaved as strong hydrogen-ion providers. By 1923, the chemists J.N. Brønsted and M. Lowry (Figure 12.7) had realised that acidity was more like a *tendency*, and that there was a spectrum of this tendency which ran across a large range of hydrogen-containing molecules. They also realised that the *solvent* had an active role in deciding acidity.

At this point you should remind yourself of the work in Chapter 5 on 'impure' covalent bonds. You will recall that covalent bonds may be in a state of unequal charge distribution – they may be **polarised**. This tendency derives from the fact that atoms on either end of the bond have unequal electronegativities. We can now recognise the following features.

• The family of acids is that family of molecules which have *positively polarised hydrogen atoms*. These are the hydrogen atoms which are liable to break off acidically – in other words, to break off as hydrogen ions leaving their electrons behind.

• The spectrum of acidity is the spectrum of degree of polarisation, so that at one end of the spectrum the non-acids have —H bonds which are nearly nonpolar (as in the C—H bonds in methane below)

$$—\overset{|}{\underset{|}{C}}—H$$

whereas at the other end the highly polar H—X (where X stands for a halogen) bonds in the hydrogen halides are ripe for breaking in a way which creates acidity.

$$\overset{\delta-}{:\!Cl\!:}—\overset{\delta+}{H}$$

Question

4 Place the molecules opposite in a sequence, with those showing the maximum tendency towards providing hydrogen ions *last*. If necessary put an asterisk by the hydrogen atom which is the one likely to break off.

ethane

ethanol

ethoxyethane

sulphuric(IV) acid

sulphuric(VI) acid

However, as Brønsted and Lowry realised, even highly polar bonds do not just break by themselves, at least not at normal room temperatures. So even such a likely event as

$$H—Cl \rightarrow H^+ + Cl^-$$

would not happen in isolation. Acidic behaviour depends not only on the character of the would-be acid, but also on the character of one other crucial chemical species – *the receiver of the hydrogen ion*. By taking this step Brønsted and Lowry recognised that acidic behaviour was a **transfer reaction**, a transfer whose 'currency' was the second simplest entity in chemistry (after the electron), namely the **proton** (that is, the H^+ ion). The Brønsted–Lowry definition of an acid is therefore this:

An acid is a proton donor.

The next question is, what acts as the proton *receiver*? The immediate answer is the *solvent*, because hydrogen chloride, for example, would not be able to display its acidity were it not for the role of water.

Question

5 Fill in the missing curly arrows in the scheme below, which shows the water accepting the proton (hydrogen ion) and the chloride ion being created. What sort of covalent bond is it that holds the hydrogen ion on to the water molecule?

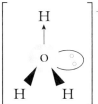

The Brønsted–Lowry model introduced a shift in emphasis. In the earlier view of acidity the focus was on hydrogen chloride, HCl, with its apparently automatic right to be an acid. It was simply seen as *containing* H^+ ions:

$$HCl(aq) \rightarrow H^+(aq) + Cl^-(aq) \qquad (1)$$

The water had no role other than to be the place where it all happened. Now in contrast we see that it all depends on water having a lone pair and being able to use it. We even have a new species, the **hydroxonium ion** H_3O^+, whose existence has been established spectroscopically.

(*Note:* the formula H_3O^+ is still often abbreviated to H^+. Even if a writer is working with the Brønsted–Lowry model, $H^+(aq)$ can be a useful abbreviation. The full formula is used when the writer wishes to lay particular stress on the basic (proton-accepting) role of water, or on the mechanism of proton transfer.)

It is useful to pinpoint the difference between the Brønsted–Lowry and the pre-Brønsted–Lowry views of acidity. In the old view, the *solution* of HCl(aq) was

the acid because it contained $H^+(aq)$ ions: in the new view we are saying the *HCl molecule* is the acid, providing it is in a solvent environment in which it can perform as a proton donor.

Question

6 Hydrogen sulphide is a near relative of water in Periodic Table terms. It is a liquid at temperatures below −61 °C, so we can consider what sort of solution might be obtained if hydrogen chloride had to dissolve in hydrogen sulphide instead of water.

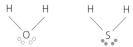

a Which of the two atoms, O in H_2O or S in H_2S, is more negatively polarised?

b So which of the two will be more 'pushy' with its lone pairs?

c Which liquid would be the more willing receiver of the hydrogen ions offered by hydrogen chloride – in other words, in which solvent will hydrogen chloride be able to *behave more acidically*?

d How do you think the two solutions might compare in conductivity? Justify your answer.

12.3 The Brønsted–Lowry definition of a base, and the concept of conjugate pairs

It is in the definition of a base that the Brønsted–Lowry theory differs most sharply from its predecessor. By a nice symmetry, if an acid is defined as a proton *donor* then a base can be seen as a proton *acceptor*. But if we accept this idea, then the water in a solution of hydrogen chloride becomes a base, despite the fact that it obviously has not neutralised the acid.

However, this need to readjust our ideas is a small price to pay for the insight which the Brønsted–Lowry picture of basicity gives. Consider the following ideas which flow from it.

1 The basicity of the solvent becomes a variable

Your answer to question 6 should have suggested that hydrogen chloride is less able to show its acidity in liquid hydrogen sulphide. This highlights the point that the ionic population in an acidic solution depends on *the acidity of the acid and the basicity of the base*. It could even be that not every hydrogen chloride molecule can donate a hydrogen ion to the solvent, leaving the resulting solution as an equilibrium mixture:

$$HCl + H_2S(l) \rightleftharpoons H_3S^+ + Cl^- \qquad (2)$$

So an acid cannot be rubber-stamped once and for all as a 'good proton donor' – the solvent environment has to be specified as well.

2 The idea of conjugate pairs

Seeing an acid in a reversible relationship with the solvent forces us to look at the reverse reaction. We realise that it too is a proton donation, and that therefore by Brønsted–Lowry theory equations such as (2) contain *two* acids and *two* bases.

Question

7 Which is the acid and which the base on the reverse path of equation (2)?

Brønsted and Lowry added to the vocabulary of acid–base chemistry by recognising the symmetry of equations such as (2). If two substances differ by the loss or gain of a hydrogen ion, they are called a **conjugate pair**. Thus in (2), HCl and Cl^- are a conjugate pair, in which HCl is the 'conjugate acid' and Cl^- the 'conjugate base':

$$\underset{\text{acid 1}}{HCl} + \underset{\text{base 2}}{H_2S} \rightleftharpoons \underset{\text{acid 2}}{H_3S^+} + \underset{\text{base 1}}{Cl^-}$$

Question

8 Identify the conjugate acids and bases in equation (3), using the same system as shown above.

$$\underset{}{HCl(aq) + H_2O(l) \rightarrow}$$
$$H_3O^+(aq) + Cl^-(aq) \qquad (3)$$

Notice that the equation does not have to be reversible to be able to be analysed into conjugate pairs.

3 Definitions of 'weak' and 'strong' as applied to acids and bases

In the pre-Brønsted–Lowry theory there were two sorts of acids, divided by the ability to donate all, or hardly any, of their hydrogen ions. In the Brønsted–Lowry version, two new views emerge. One is the idea, already mentioned, that it is not just the responsibility of the acid, but also the solvent which determines how many ions 'break off'. The second idea is that, if you do stay with one base/solvent, there is a sliding scale among acids of ability to give hydrogen ions. All acid–water interactions can be written in the general form:

$$\underset{\text{acid 1}}{HA(aq)} + \underset{\text{base 2}}{H_2O(l)} \rightleftharpoons \underset{\text{acid 2}}{H_3O^+(aq)} + \underset{\text{base 1}}{A^-(aq)}$$
$$(12.1)$$

where HA is a general formula for an acid which gives its hydrogen ion away as a proton to water, and where the A^- left behind represents the conjugate base of the acid in question.

If the equilibrium in (12.1) is overwhelmingly to the right, as it is when A is chlorine, then the acid is described as **strong** (at least with respect to water-as-base/solvent). If the equilibrium is well to the left, as it is with, say, the organic acid ethanoic acid (CH_3CO_2H), then the acid is described as **weak**. For instance, the collision depicted below will only result in successful H^+ transfer in a small fraction of cases.

Among the hundreds of acids in existence, there is a wide spread of 'strengths'.

Question

9 The word 'strength' is in inverted commas above because it has acquired a new shade of meaning. Hitherto you might have said that 'strength' was synonymous with 'concentration', but now 'strength' is being used to mean something fixed and inherent in the nature of a particular acid–solvent combination and is not dependent upon any human-set variable like concentration. So give examples of the following apparently contradictory classes:

a A dilute strong acid.

b A concentrated weak acid.

Dissociation constants

The spread of strengths can be given quantitative form in a league table by using a special sort of equilibrium constant. The equilibrium constant for equation (12.1) would be:

$$K_c = \frac{[H_3O^+][A^-]}{[HA][H_2O]} \qquad (12.2)$$

However, the solvent water is always present in a large excess, so its change in concentration will be proportionately small. (Even in, say, 1 dm^3 of 2 mol dm^{-3} hydrochloric acid, the two moles of water which are required to turn themselves into hydroxonium ions are still unimportant compared with the 56 or so moles which are present in the solution.) If $[H_2O]$ is more or less constant, it makes sense to absorb it into the other constant, like this:

$$K_c[H_2O] = \frac{[H_3O^+][A^-]}{[HA]}$$

Let $\quad K_c[H_2O] = K_a$.

$$K_a = \frac{[H_3O^+][A^-]}{[HA]} \qquad (12.3)$$

It is this value which you see in league tables of acid strength. It is called the **acid dissociation constant** and has units of $(conc.)^2/(conc.)$ or mol dm^{-3}. Table 12.1 (opposite) shows the league table for the acids mentioned within this chapter.

Question

10a What would be the K_a value for a completely strong acid, that is, one which in water has managed to donate *all* its hydrogen ions?

b What would be the K_a value of the interaction between ethoxyethane and water? (See the answer to question 4.)

c How might the K_a value for hydrogen chloride in water differ from the equivalent value if H_2S were the solvent?

The answers to these questions show that a higher value of K_a means an acid which is a good hydrogen ion donor in the chosen solvent (in other words, a 'strong' acid), while a lower value of K_a means that there are very few hydrogen atoms polarised enough to come off as hydrogen ions (that is, a 'weak' acid).

Notice that when an acid has two removable hydrogen ions, as in the example of H_2SO_4, we are really dealing with two separate acids, in this case HSO_4^- as well as the 'parent' acid. Sulphuric acid clearly deserves to be called a strong acid, but note that it is strong only in its deployment of the *first* hydrogen ion. The hydrogensulphate ion is a few places lower in Table 12.1.

Questions

11 This pattern is general for all acids with more than one hydrogen ion – HA^- is never stronger than H_2A, because the loss of the first hydrogen ion alters the polarity of bonds throughout the molecule, in a way that makes the remaining —H bond slightly less polar. Can you use this idea to suggest why, in the specific case of sulphuric acid and hydrogensulphate, the second hydrogen ion is harder to remove?

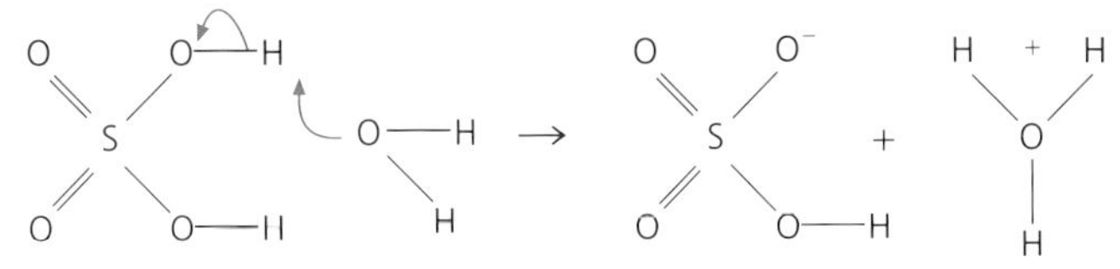

First proton is removed – 100% to products.

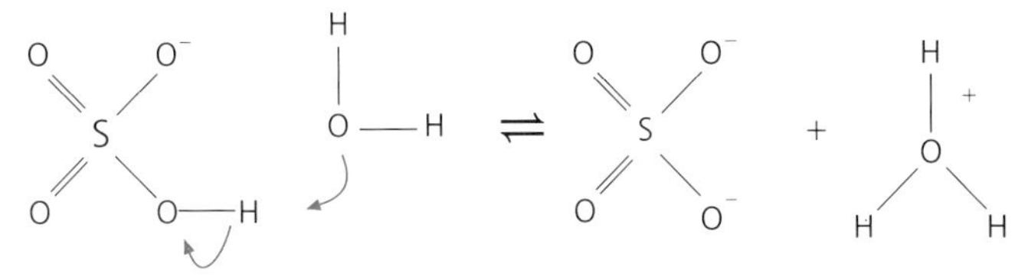

Second proton is removed, but equilibrium lies towards reactants.

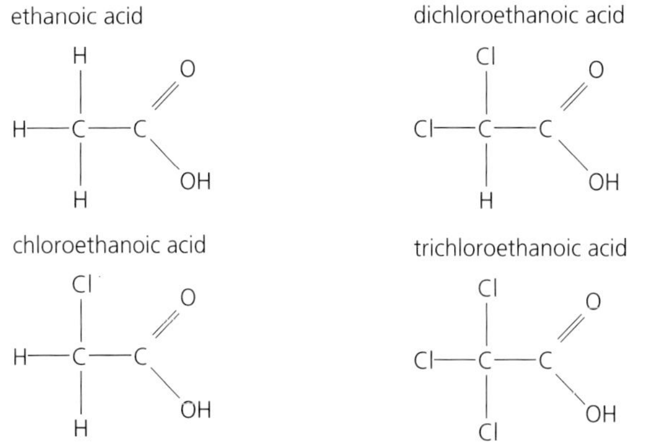

12 Another polarity factor that affects acid strength is when the molecular framework of an acid carries certain substituents. Can you suggest a reason for the differences in strength of the series ethanoic, chloroethanoic, dichloroethanoic and trichloroethanoic acids, as shown in Table 12.1 and on the right?

Table 12.1 Acids in water. In this table $H^+ \equiv H_3O^+$, and the contribution of water has been omitted for brevity. Hence $H_2SO_4 \rightleftharpoons H^+ + HSO_4^-$ is short for $H_2SO_4 + H_2O \rightleftharpoons H_3O^+ + HSO_4^-$ (all aq)

Acid	Equation for dissociation	K_a (mol dm^{-3})	pK_a
hydrochloric	$HCl \rightleftharpoons H^+ + Cl^-$	∞	
sulphuric(VI)	$H_2SO_4 \rightleftharpoons H^+ + HSO_4^-$	∞	
nitric(V)	$HNO_3 \rightleftharpoons H^+ + NO_3^-$	40	−1.6
trichloroethanoic	$CCl_3CO_2H \rightleftharpoons H^+ + CCl_3CO_2^-$	2.3×10^{-1}	0.2
dichloroethanoic	$CHCl_2CO_2H \rightleftharpoons H^+ + CHCl_2CO_2^-$	5×10^{-2}	1.3
sulphuric(IV)	$H_2SO_3 \rightleftharpoons H^+ + HSO_3^-$	1.5×10^{-2}	1.8
hydrogensulphate ion	$HSO_4^- \rightleftharpoons H^+ + SO_4^{2-}$	1×10^{-2}	2.0
phosphoric(V)	$H_3PO_4 \rightleftharpoons H^+ + H_2PO_4^-$	8×10^{-3}	2.1
iron(III) ion	$[Fe(H_2O)_6]^{3+} \rightleftharpoons H^+ + [Fe(H_2O)_5(OH)]^{2+}$	6×10^{-3}	2.2
chloroethanoic	$CH_2ClCO_2H \rightleftharpoons H^+ + CH_2ClCO_2^-$	1.3×10^{-3}	2.9
hydrofluoric	$HF \rightleftharpoons H^+ + F^-$	5.6×10^{-4}	3.3
methanoic	$HCO_2H \rightleftharpoons H^+ + HCO_2^-$	1.6×10^{-4}	3.8
ethanoic	$CH_3CO_2H \rightleftharpoons H^+ + CH_3CO_2^-$	1.7×10^{-5}	4.8
propanoic	$C_2H_5CO_2H \rightleftharpoons H^+ + C_2H_5CO_2^-$	1.3×10^{-5}	4.9
aluminium ion	$[Al(H_2O)_6]^{3+} \rightleftharpoons H^+ + [Al(H_2O)_5(OH)]^{2+}$	1×10^{-5}	5.0
carbonic	$H_2O/CO_2 \rightleftharpoons H^+ + HCO_3^-$	4.5×10^{-7}	6.4
hydrogen sulphide	$H_2S \rightleftharpoons H^+ + HS^-$	8.9×10^{-8}	7.1
dihydrogenphosphate(V) ion	$H_2PO_4^- \rightleftharpoons H^+ + HPO_4^{2-}$	6.2×10^{-8}	7.2
ammonium ion	$NH_4^+ \rightleftharpoons H^+ + NH_3$	5.6×10^{-10}	9.3
phenol	$C_6H_5OH \rightleftharpoons H^+ + C_6H_5O^-$	1.3×10^{-10}	9.9
hydrogencarbonate ion	$HCO_3^- \rightleftharpoons H^+ + CO_3^{2-}$	4.8×10^{-11}	10.3
hydrogenphosphate(V) ion	$HPO_4^{2-} \rightleftharpoons H^+ + PO_4^{3-}$	4.4×10^{-13}	12.4

$K_a = [H^+][A^-]/[HA]$ (See equation (12.3).) $pK_a = -\log_{10} K_a$

ethanoic acid

dichloroethanoic acid

chloroethanoic acid

trichloroethanoic acid

4 Linking the weak/strong idea with the conjugate pair idea

We now have a number of parallel ways of talking about acids. If we say that an acid HA is strong in water, we are also saying:

1 that HA is a successful donor of protons to water
2 that K_a for this dissociation is big
3 that the reverse reaction hardly matters
4 therefore that A^- is poor at retrieving H^+ ions.

The last comment is valuable, in that it leads to a useful generalisation. 'Retrieving H^+ ions' is the same thing as being a base, so we can say that **a strong acid has a weak conjugate base**. In other words, if HA is good at giving protons, A⁻ will be bad at taking protons. The reverse is also true – if it is hard to pull a proton off then the resulting conjugate base is going to be good at getting the proton back.

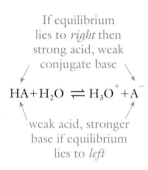

If equilibrium lies to *right* then strong acid, weak conjugate base

$$HA + H_2O \rightleftharpoons H_3O^+ + A^-$$

weak acid, stronger base if equilibrium lies to *left*

Questions

13a In the case of the dissociation of HCl(aq), equation (*3*), identify the two bases, one on either side of the equilibrium sign. Which base is the stronger – that is, which one ends up as the successful 'proton-grabber'?

b In the case of ethanoic acid:

$$CH_3CO_2H(aq) + H_2O(l) \rightleftharpoons$$
$$H_3O^+(aq) + CH_3CO_2^- \quad (4)$$

identify the two conjugate pairs. Which is the stronger of the two bases?

14 Select five bases from the equilibria in Table 12.1 and put them in order of base strength, strongest first.

As you get used to the Brønsted–Lowry method you will see that their way of looking at acids gives you a number of viewpoints. In particular, you can choose to focus either on the acids or on their conjugate bases – from one point of view an equation like (*4*) can be seen as representing a weak acid struggling to dump its protons on the water molecules, and from another it can be seen as a victory for the reverse reaction – a win for the ethanoate ion $CH_3CO_2^-$ over water in a 'battle of the bases'.

12.4   A Brønsted–Lowry view of conventional acid–alkali neutralisation reactions

Radical new ideas such as water being a base can be very disconcerting. Water cannot turn Universal Indicator paper purple; and yet that has hitherto been the hallmark of a base. Equally obvious is the fact that water cannot neutralise an acid to form a salt. So how does the Brønsted–Lowry theory tie in with your earlier knowledge? One comforting fixed point from the past is this: even in the Brønsted–Lowry theory sodium hydroxide is still a base!

We can analyse the reaction of sodium hydroxide with hydrochloric acid from a Brønsted–Lowry viewpoint. The hydrochloric acid, as we have seen, acts as acid to water's base to set up a solution of ions:

$$HCl(aq) + H_2O(l) \rightarrow$$
$$H_3O^+(aq) + Cl^-(aq) \quad (3)$$

Imagine that mixture pouring from a burette into a solution of sodium hydroxide. Prior to this moment, the strongest base that was present in the hydrochloric acid environment was water. As such, the water had seized the available hydrogen ions to make H_3O^+. But now comes a base which is much better at taking on hydrogen ions, namely $OH^-(aq)$:

$$H_3O^+(aq) + OH^-(aq) \rightarrow$$
$$H_2O(l) + H_2O(l) \quad (12.4)$$

Question

15a Add together equations (*3*) and (*12.4*) as if they were mathematical equations and cancel any species that appear on both sides. Does the resulting equation look familiar?

b In equations (*3*) and (*12.4*) considered together there is a total of three different bases. Identify them and put them in order of base strength, strongest first.

Looking back over this analysis of the acid plus alkali case, you could say that

the Brønsted–Lowry theory has greatly enlarged the number of things that can be called bases, but still includes the strong bases of the old theory. In the titration, water was a sort of temporary base (in both senses of the word) for the hydrogen ion, until the more pushy base ($OH^-(aq)$) came along.

(Notice that Brønsted–Lowry theory concentrates on the word 'base' and does not refer to alkalis. We are free to keep the word alkali to refer to water-soluble bases, if we want to.)

Question

16 Table 12.2 shows the enthalpies of neutralisation of a number of acids.

Table 12.2 Molar enthalpies of neutralisation, ΔH, of some acid–base reactions

Acid	Base	$-\Delta H$ (kJ mol⁻¹)
HCl	NaOH	57.1
HCl	KOH	57.2
HNO₃	NaOH	57.3
HNO₃	KOH	57.3
CH₃CO₂H	NaOH	55.2
HCl	NH₃	52.2
HF	NaOH	68.6

The data relate to the reagents in dilute aqueous solutions, and are expressed as kJ per mol of H^+ neutralised. The striking points about the data are the close similarities between a number of the items and the sudden pattern-breaking difference of others.

a Suggest why the top four entries in Table 12.2 are so similar.

b Account for the noticeable reductions in the reactions featuring ethanoic acid and ammonia.

c Suggest, if you can, the reason for the sharp increase associated with HF.

It is no coincidence that all the strong alkalis are hydroxides. There is a process of self-regulation going on which decrees that any base stronger than the hydroxide ion in a water solution must destroy itself and create hydroxide ions. It is impossible, for instance, to have a soluble oxide ion, as the following question should make clear.

Questions

17a Compare the base strengths (H^+ – grabbing abilities) of O^{2-} and OH^-. Which is the stronger base?

b Sodium oxide, Na_2O, appears to dissolve in water, but in fact there is a reaction. Complete the following equation:

$$O^{2-} + H_2O \rightleftharpoons \qquad (12.5)$$

c On which side will the equilibrium be biased?

d Write out the full equation for the reaction of Na_2O with water, including spectator ions.

18 It may already have occurred to you that the bases we have met so far, such as O^{2-}, OH^- and H_2O, could all equally have been presented as **nucleophiles**, to borrow a well-used concept from organic chemistry (p. 228). What is common to the role of the base and the nucleophile which makes the same species good at both?

12.5 A Law of clockwise circles – competition between bases for hydrogen ions

At the beginning of the chapter the two transfer processes, acid–base and redox, were described as a matched pair. In each case it is possible to present the participants in a league table. In the redox case, the league table (the electrochemical series) helps to decide, in situations where there is competition, the issue of which species should end up with the electrons. (In other words it acts as a predictor of reaction direction.) But it cannot do the job on its own. It has to be used in conjunction with a piece of chemical law known as the **Law of anticlockwise circles** (Section 7.8). The time has now come to look at the full extent of the parallel between the fields of acid–base and redox. We already have an acids' league table in place (Table 12.1), so let us move on to the 'circles' law which turns it into a prediction tool.

Consider the positions in the acids' league table of HCl(aq) and HF(aq).

$$HCl + H_2O \rightleftharpoons H_3O^+ + Cl^- \quad K_a = \text{infinity} \qquad (3)$$

$$HF + H_2O \rightleftharpoons H_3O^+ + F^-$$
$$K_a = 5.6 \times 10^{-4}\,\text{mol dm}^{-3} \qquad (5)$$

Clearly HCl is the better H^+ giver, or to put it another way . . .

Question

19 Which is the strongest base on show?

Let us see how these two equations can predict what happens when the two systems are combined. We will treat the chemical equations like mathematical ones and subtract (5) from (3):

$$HCl + H_2O - HF - H_2O \rightleftharpoons$$
$$H_3O^+ + Cl^- - H_3O^+ - F^- \qquad (3) - (5)$$

$$\text{or} \quad HCl + F^- \rightleftharpoons HF + Cl^- \qquad (6)$$

Equation (6) can be seen as a battle of the bases between F^- and Cl^-, and we know from the answer to question 19 that the result will be an emphatic win for F^-. So in the process of subtracting the lower equation from the upper one we have come up with an equation which actually goes the way it is written. Generalising from this specific example, we can see that the direction of the reaction between any acid and the conjugate base of a second acid can be predicted purely on the grounds of where the two acids are in the league table.

The one rule is:

The higher acid will force a proton (hydrogen ion) on to the conjugate base of the lower acid.

To put it another way:

A stronger acid will 'kick out' a weaker acid's conjugate base.

There is even a third way to visualise this phenomenon. We can view the league table in Table 12.1 as a list of *half-equations*. The water is acting as a 'storage base' for the hydrogen ion but nothing has really happened yet, so to that extent the reactions are half-equations. They are all waiting for the other half to come and take the hydrogen ion away. (This is emphasised by the format in the league table, in which the role of water is ignored.) But since all the half-equations in Table 12.1 are presented with the acid on the left, it follows that the only sort of half-equation which will take a hydrogen ion *away* is one which *runs backwards*, from base to acid. As the HCl/HF example showed, the upper entries can *make* the lower ones go backwards. So we can say (compare the redox rule, p. 135):

Only reactions which lie on *clockwise* circles are allowed.

Note: This is the opposite 'rotation convention' to the anti-clockwise one which applies to the electrochemical series (Chapter 16).

You will see in Figure 12.8 that the reactants lie on the NW/SE diagonal, while the products lie on the NE/SW diagonal.

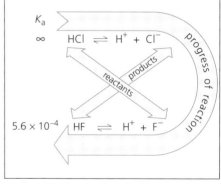

Figure 12.8 Showing how the places in the league table of K_a may be used to predict the direction of proton-transfer reactions using the Law of clockwise circles

Question

20a Give two acids from Table 12.1, other than hydrochloric acid, that would liberate hydrogen fluoride from a solution of sodium fluoride. Write the equation for the reaction between sodium fluoride and one of your choices.

b Give one conjugate base from Table 12.1 to which hydrogen fluoride would be able to *donate* a hydrogen ion (that is, HF as reactant rather than product). Again write the equation.

One of the most famous reactions of acids is their ability to liberate carbon dioxide from carbonates and hydrogencarbonates. We can now see this reaction as an example of the law of clockwise circles. Notice that to get all the way from the carbonate ion, CO_3^{2-}, to CO_2, the acid has to deliver *two* hydrogen ions to two different conjugate bases. So for an acid to be strong enough to make a carbonate fizz, it has not only to be able to make a clockwise circle with CO_3^{2-}, but it must then be able to do it again with the *product* of the first hydrogen ion transfer. So in general, for an acid HA:

$$HA(aq) + CO_3^{2-}(aq) \rightarrow HCO_3^-(aq) + A^-(aq)$$

and then:

$$HA(aq) + HCO_3^-(aq) \rightarrow H_2O(l) + CO_2(g) + A^-(aq)$$

Overall, the reaction is:

$$2HA(aq) + CO_3^{2-} \rightarrow H_2O(l) + CO_2(g) + 2A^-(aq)$$

Question

21a Would the ability to make a carbonate fizz be a reasonable test for a carboxylic acid? Justify your answer.

b Name an acid from Table 12.1 which would be able to give the first hydrogen ion to CO_3^{2-} but not the second one to HCO_3^-.

c Would the ability to make a carbonate fizz be a reasonable test for phenol? Justify your answer.

d Is it true to say of an acid that if it can make a hydrogencarbonate fizz then it will certainly make a carbonate fizz? Justify your answer.

12.6   The uses of sulphuric acid

Sulphuric acid is the premier industrial chemical in the world. The chemical industry is able to make the oxides of sulphur fairly easily, using the Earth's abundance of sulphide ions and elemental sulphur. The only other necessary raw material is air. The resulting sulphur trioxide is readily soluble in water, to generate the final product, sulphuric acid. The sequence of reactions is shown below.

$$\text{sulphide ores} \diagdown$$
$$\xrightarrow[\text{roasting}]{+O_2(g)} SO_2(g) \xrightarrow[\text{Contact process}]{+O_2(g)} SO_3(g) \xrightarrow{+H_2O(l)} H_2SO_4(aq)$$
$$\text{sulphur} \diagup$$

The Contact process is the subject of Section 9.13, p. 183. Figure 12.9 shows how sulphuric acid is used.

The important feature of the sulphuric acid molecule is the way the structure of S=O and S—O groups acts to make the terminal hydrogen atoms heavily positively polarised, and thus very readily released as protons, on to a wide range of bases.

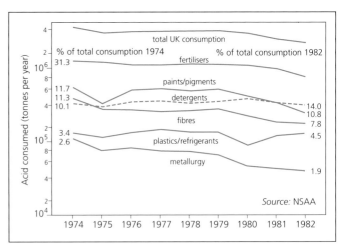

Figure 12.9 The consumption of sulphuric acid in the UK (1974–1982). The y-axis is logarithmic

Sulphuric acid can be thought of as the world's best and cheapest molecular proton-donating machine.

Protons are useful for catalysis, but the bulk uses of sulphuric acid involve the making of one acid from another. Three of the largest uses of sulphuric acid are in the fertiliser, plastics and refrigerants industries.

Some uses of sulphuric acid as a proton donor
The **fertiliser industry** needs to turn the mineral calcium phosphate, $Ca_3(PO_4)_2$, into phosphoric acid, H_3PO_4, as a precursor to phosphate fertilisers (Figure 12.10). Phosphorus is a key mineral for the root growth of plants (Table 12.3).

Table 12.3 Effects of phosphorus in fertilisers on crop yields

Dose of P (kg/hectare)	Yield (tonne/hectare)	
	barley	potatoes
0	2.0	13
14	2.6	29
28	3.0	32
56	3.5	32

Question

22a From Table 12.1, will sulphuric acid give protons to the PO_4^{3-} ion? In particular, can you construct *all three* clockwise circles, for the donation of each of the three protons in the transition from bare PO_4^{3-} to fully protonated phosphoric acid, H_3PO_4?

b Write a chemical equation for the reaction to produce phosphoric(v) acid from calcium phosphate.

c What second reaction would be needed to create ammonium phosphate from phosphoric(v) acid? What is the equation for this reaction?

d You might have noticed, either from part (b) above or from Figure 12.10, that the process for making phosphoric acid has a by-product. Firms usually seek to sell by-products if at all possible, so who do you think might be interested in this one?

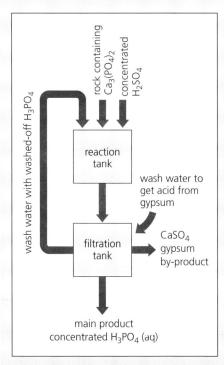

Figure 12.10 The process by which phosphoric acid is made. Notice how wastage is minimised by recycling the washwater. There are both environmental and financial incentives in favour of recycling

Making a fertiliser of correct composition is only part of the problem for the manufacturer. Other matters of importance include two factors that can be easily overlooked.

1 The need to 'pellet' the product into granules of the right size for the client's spreading machinery.

2 The need to make sure the product can be stored for quite long periods.

Question

23 The reason that fertilisers need to be able to be stored safely is because there is a mismatch of needs between maker and user with regard to *time*. As far as the maker is concerned the preference is for a production line running continuously.

a Why might the fertiliser user have a very different time-of-need profile?

b As storage heaps might get quite high, what problem might arise in the contents of sacks at the bottom of the heap (Figure 12.11)?

Figure 12.11 Storage heap of fertiliser sacks

c Why might this problem link up with the pelleting problem mentioned above?

The **refrigerants** and **plastics industries** need hydrofluoric acid, the former as a precursor of the new generation of refrigerants replacing CFCs (see Section 11.5, p. 232) and the latter as a reactant on the path to polyfluoro compounds like poly(tetra fluoroethene), PTFE. This polymer is the surface material of non-stick cooking ware and is the key ingredient in Gore-tex clothing and shoes, which lets $H_2O(g)$ out but not $H_2O(l)$ in (Figure 12.12). As in the case of fertilisers, the raw fluoride-containing material which the sulphuric acid has to protonate is of mineral origin.

Figure 12.12 Gore-tex is a remarkable material in that it allows the passage of water vapour but not water liquid

Question

24a Suggest a possible mineral source of fluoride ions.

b Use Table 12.1 to predict the chances of sulphuric acid forcing a proton on to fluoride ions.

c Write an equation for the reaction you suggest.

d By looking up the relevant physical properties in a data book, suggest how the product HF might be separated from the reaction mixture.

On returning to the central thread of the chapter, we now encounter some rather more mathematical aspects of proton donation.

12.7 The ionic product of water and the pH scale

The water molecule is normally pictured in books as an unbroken covalently bonded unit, with no overall charge, although admittedly with an internal dipole. If this were a completely true picture, water would have a negligible electrical conductivity because of its lack of ions. However, the small but measurable conductivity of even the purest deionised water tells a different story – tells us in fact that there must be some ions present. It turns out that the ions exist by virtue of an occasional event, in which one water molecule 'steals' a proton from another, by 'impaling' it on one of its lone pairs as shown in the curly arrow diagram below.

The equation for the change is:

$$H_2O(l) + H_2O(l) \rightleftharpoons H_3O^+(aq) + OH^-(aq) \quad (12.6)$$

This is called the auto-ionisation reaction. (Note that it is the opposite of (12.4).)

Question

25 Analyse this reaction by Brønsted–Lowry theory into two acids and two conjugate bases.

The equilibrium constant of this change would be given by a four term expression, if it were a normal K_c, but as in the case of the acid dissociation constant K_a, the $[H_2O]$ is considered to be effectively constant. In this case we have wiped *two* terms out of the normal K_c, and we are left with a straight multiplication, or what in the language of mathematics is called a 'product'. The name given to this product is the **ionic product of water**, symbol K_w:

$$K_w = [H_3O^+(aq)][OH^-(aq)]$$
$$= 10^{-14} \text{ mol}^2 \text{ dm}^{-6} \quad (12.7)$$

You will see from (12.7) that the product is a very small number, which shows that there are very few ions in existence at any one time. From what you know about the dynamic nature of equilibria you will realise that the actual ions are constantly being created and destroyed.

Question

26a What can you say about the mathematical relationship between the quantities $[H_3O^+]$ and $[OH^-]$ in a sample of pure water (bearing in mind the way they are made by the same single act of proton transfer)?

b Apply your answer from part (a) into equation (12.7) and calculate the actual values of $[H_3O^+]$ and $[OH^-]$.

c The density of water is 1 g cm^{-3}, or 1000 g dm^{-3}, and its relative molecular mass is 18. Work out how many moles there are in 1 dm^3 of water. Answer (b) has already told you how many moles of H_3O^+ and of OH^- there are in 1 dm^3 of water. Combine the two answers to work out the ratio '[moles of unbroken water molecules] : [moles of broken ones]'.

Like all equilibrium constants, K_w is true for all situations in which its members meet. In other words it has to be obeyed in *all aqueous solutions*, and not just in pure water. But in some solutions the concentrations of either OH^- or H_3O^+ are high, up to 1 mol dm^{-3} or more, as in the bottles of acids and alkalis in your school laboratory. Let us see what that rise in the concentration of one ion does to the concentration of the other ion.

Question

27a Hydrochloric acid as we have seen is a completely strong acid – meaning that every mole of HCl in solution is existing as a mole of H_3O^+ and a mole of Cl$^-$ ions. What therefore is the value of $[H_3O^+]$ in 1 mol dm^{-3} HCl(aq)?

b Inserting that value into equation (12.7), what do you calculate to be the value of $[OH^-]$ in 1 mol dm^{-3} HCl(aq)?

c By similar reasoning applied to the strong base NaOH, what will be the value of $[OH^-]$ in 1 mol dm^{-3} NaOH(aq)?

d So what will be the value of $[H_3O^+]$ in 1 mol dm^{-3} NaOH(aq)?

Your answers to the last two questions may have given you some insight into the vast range of variation of a quantity like $[H_3O^+]$, taking in anything from 10^{-14} to 1 (and even as high as 10 in concentrated hydrochloric acid). Chemists did what scientists often do when confronted with numbers ranging over many powers of ten – they resorted to a **logarithmic scale**. (The decibel scale of loudness is an example of a logarithmic scale.) They defined a quantity called the 'p$[H_3O^+]$' or pH for short, in which the 'p-notation' indicated that a logarithm and a change of sign had operated on $[H_3O^+]$:

$$pH = -\log_{10}[H_3O^+] \quad (12.8)$$

Chemists now had a system that produced more manageable numbers from the 'raw' ones, as the following exercises in conversion will demonstrate. You can do these in your head if you know enough about logarithms, but if you do not, you can carry out a two-button exercise on your calculator – having typed the number into the calculator, press the 'log' (*not* 'ln') button followed by the '+/−' button.

Question

28 Express as a pH the $[H_3O^+]$ in

a pure water (your answer to question 26b should have been 10^{-7} mol dm^{-3})

b 1 mol dm^{-3} HCl(aq) (your answer to question 27a should have been 1 mol dm^{-3})

c 1 mol dm^{-3} NaOH(aq) (your answer to question 27d should have been 10^{-14} mol dm^{-3}).

So you see that the pH scale compresses that range of numbers into the region 0–14 (although those are not the absolute limits). You can also see it works in an upside down sort of way, with low pHs indicating high $[H_3O^+]$ values.

pH – the ultimate style accessory
pH has been a gift to the advertising world (motto – blind them with science). Advertisers use it as a sort of style accessory to give their adverts the cachet of scientific truth. They have found that truth needs packaging just like the product itself.

SOFTENING FOAM BATH

For the respect and natural health of your skin. Even sensitive.

The Dermo-Protection formula.
Neutralia Dermo-Protection provides gentle, effective cleansing that respects the skin's eco-system. It is this natural protective barrier that gives your skin its greatest potential to be healthy and look beautiful.

Pure, pH neutral
All ingredients are selected for their purity. Soap free, Neutralia Softening Foam Bath respects your skin's natural pH balance.

Hypo-allergenic
Dermatologically tested, Neutralia is formulated to minimise the risk of allergic reaction, redness and irritation.

Hard water softener
Neutralia will protect your skin from the drying effects of hard water.

Colourant free
Neutralia is clear and colourant free.

Figure 12.13 pH – the advertisers' friend

Imagine, for instance, an advert which announced that your shampoo was designed to have a hydrogen ion concentration of 3.16×10^{-8} mol dm^{-3} (Figure 12.13). On the other hand what a nice ring there is to 'pH balance 7.5' – just scientific enough without being off-putting.

We also need to get used to conversion in the opposite direction, from pH to $[H_3O^+]$. If, as in (12.8)

$$pH = -\log_{10}[H_3O^+]$$

then we can change that equation to read:

$$[H_3O^+] = 10^{-pH} \qquad (12.9)$$

On the calculator, once the pH has been keyed in you will need the same two buttons as before, but with these two differences:

- The '+/–' button is pressed first.
- The 'inverse' version of the log button is used, the one that reads '10^x'.

Question

29 Convert these pHs to values of $[H_3O^+]$:

a 0 (such as in a quite concentrated solution of hydrochloric acid).

b –0.3 (such as in an even more concentrated solution of hydrochloric acid).

c 14.3 (such as in a concentrated solution of sodium hydroxide).

d 3.5 (such as in a badly acidified lake in Norway – Figure 12.14).

Figure 12.14 Acidified lakes are a major problem in Scandinavia

e 2 (such as in the human stomach).

The p-notation is useful not only in making numbers easier to write and think about, but also by making certain inter-relationships easier to see. For example, the connection between $[H_3O^+]$ and $[OH^-]$ has a much more memorable shape in p-notation.

$$K_w = [H_3O^+][OH^-] = 10^{-14} \qquad (12.7)$$

If we take logs of each side:

$$\log K_w = \log [H_3O^+] + \log [OH^-]$$
$$= -14$$

Now multiply the whole equation through by –1:

$$pK_w = pH + pOH$$
$$= 14 \qquad (12.10)$$

(where 'p' means 'minus log to the base ten of...')

This result makes for quick calculations of the pH of alkalis. For instance, the 'pOH' of 1 mol dm^{-3} NaOH is $-\log_{10} [OH^-]$ or 0. So instantly from (12.10) we can see that the pH must be 14. We will revisit this approach when we do calculations on weak bases.

12.8 The link between pH and K_a for a weak acid

In our calculations of pH so far, we have only considered solutions of hydrochloric acid and sodium hydroxide. This has kept things relatively simple, for this reason – both hydrochloric acid and sodium hydroxide are completely strong (p. 258). So when you read on the bottle the inscription '1M HCl', you can be confident that the whole of that one mole per litre of HCl molecules will exist as ions. But what happens when we deal with *weak* acids and bases? If the acid were ethanoic, say, you could be sure that most of the ethanoic acid molecules were still whole, and so *much less than* one mole of H_3O^+ ions would be floating about free in solution. It is *free* H_3O^+ ions which decide the pH, so the pH would definitely be greater than 0. How much greater depends of course on the *size of the* K_a. Let us see how the mathematical link is made.

Example: Finding the pH of a weak acid

What is the pH of 0.1 mol dm^{-3} ethanoic acid, an acid whose K_a is 1.7×10^{-5} mol dm^{-3}?

First we need to remember what the K_a means. We start with some abbreviations: for $[H_3O^+]$ we can write $[H^+]$, since the hydroxonium ion is really just a proton with a water molecule attached, and for $[CH_3CO_2^-]$ and $[CH_3CO_2H]$, the ethanoate ion and ethanoic acid concentrations, we will use $[eth^-]$ and $[Heth]$.

For ethanoic acid in water:

$$Heth + aq \rightleftharpoons H^+ + eth^- \qquad (7)$$

$$K_a = \frac{[H^+][eth^-]}{[Heth]}$$

$$= 1.7 \times 10^{-5} \text{ mol dm}^{-3} \qquad (8)$$

This equation appears to contain three unknowns, but with some logic and an approximation we can reduce them to one unknown and so solve the equation. The logic is this – by the stoichiometry of (7), which is $-1 : -1 : +1 : +1$, it is apparent that if x moles per dm^3 of Heth are

destroyed on the way to equilibrium, then x moles per dm^3 of H^+ and x moles per dm^3 (*the same*) of eth^- will be created. So the top line is really the same unknown twice. As for the bottom line, if the original on-the-bottle concentration of Heth was 0.1M (0.1 mol dm^{-3}), then the concentration at equilibrium is $(0.1 - x)$. So now,

$$K_a = \frac{x^2}{0.1 - x}$$

$$= 1.7 \times 10^{-5} \text{ mol dm}^{-3} \qquad (12.11)$$

Now for the approximation. When an acid is seriously weak, like ethanoic (as shown by the tiny K_a value), it is justifiable to say that the amount lost to dissociation will be negligible compared with the total on-the-bottle concentration, or 'parent' acid. To put this idea into mathematical symbols:

$$0.1 \gg x$$

So $\quad (0.1 - x) \quad \approx 0.1$

x has now been removed from the bottom line of equation (12.11), so the equation instead of being a full quadratic is a simple square-root exercise:

$$\frac{x^2}{0.1} = 1.7 \times 10^{-5}$$

So $\qquad x^2 = 1.7 \times 10^{-6}$

So $\qquad x = [H^+] = 1.3 \times 10^{-3}$

It only remains to change to p-notation, giving

$$pH = 2.9$$

(Figure 12.15)

Reverse example: Finding the dissociation constant from the pH

Suppose a 0.01 mol dm^{-3} solution of a weak acid had a pH of 4.1. What would be its K_a?

The procedure is an almost exact step-wise reversal of the previous example, except that we can ignore the 'bottom-line approximation'.

First, convert pH into a $[H^+]$ value by the two calculator buttons:

$$[H^+] = 10^{-pH} = 10^{-4.1} = 7.9 \times 10^{-5} \qquad (9)$$

By our previous argument this number can stand also for $[A^-]$, where A stands for whatever the anion is.

So $\qquad K_a = \frac{(7.9 \times 10^{-5})^2}{[HA]} \qquad (10)$

If we use the bottom-line approximation, we are fairly safe in writing the bottom line as 0.01, which was the original 'parent' concentration. However, the right-hand side of (10) is all number, so it is quite easy to use the actual values, that is, the 0.01 moles per dm^3 on the bottle *minus* the 7.9×10^{-5} moles per dm^3 which has dissociated.

So $\qquad K_a = \frac{(7.9 \times 10^{-5})^2}{0.01 - 7.9 \times 10^{-5}}$

$$= 6.4 \times 10^{-7} \text{ mol dm}^{-3}$$

A note on significant figures

If you reproduce the working of the reverse example, you may argue about the decimal place, claiming it to be .3 not .4. The reason for 6.4

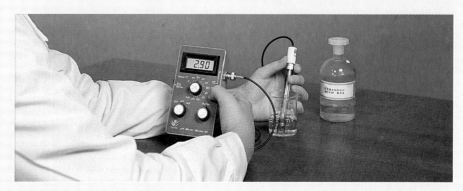

Figure 12.15 pH of ethanoic acid

was that the 7.9 from equation (9) was not rounded off but left in the calculator as 7.94 ... etc. It was then ready to use in the second part of the calculation. The rounding off to two significant figures, which was admittedly always justified by the limitations of the original data, only took place *at the end of a calculation*. You can use your calculator's memory to avoid premature approximation.

(Incidentally, had we used the bottom-line approximation the answer *would* have come out at 6.3.)

In the questions that follow, use the methods outlined above. You will need to consult Table 12.1 (p. 261) for values of dissociation constants.

Questions

30 Predict the pH of the following solutions, all of which have the 'parent' concentration of 0.01 mol dm^{-3}:

a HF (hydrofluoric acid)

b CH_3CO_2H (ethanoic acid)

c NaH_2PO_4 (sodium dihydrogen phosphate, in which the acid is the anion, which still has two possible hydrogen ions left to give)

d $FeCl_3$, in which the acid is one of the water molecules around the $[Fe(H_2O)_6]^{3+}$ ion, so polarised by the charge density of the Fe^{3+} that it is much 'stronger' than ordinary water. It is just another acid in Table 12.1 as far as this problem is concerned.

31 Work out the dissociation constants of the following acids, given that the pHs of their 0.1 mol dm^{-3} solutions are as follows:

a 1.5 (This is a solution of SO_2 in water, sulphuric(IV) acid.)

b 2.4 (This is methanoic acid.)

c 5.1 (This is ammonium chloride, in which the NH_4^+ ion acts as a very weak but discernible acid, thus proving that not everything whose name begins with 'ammoni . . .' is a base.)

Comparing the treatment of equilibria in this chapter with those in Chapter 9

Although the reactions between weak acids and water are equilibria, you may have noticed that we are not using the format of Chapter 9 – the one with the tabular layout and t_0, t_{eqm} as rows (p. 185). This, however, does not mean that acid–base equilibria lie outside the mainstream of equilibrium problems.

In fact it is quite easy to use the earlier format, as this example shows:

$$HA + H_2O \rightleftharpoons H^+ + A^-$$
$$K_a = 6.0 \times 10^{-3}$$

t_0	0.01	0	0
t_{eqm}	0.01 − x	x	x

So
$$K_a \approx \frac{x^2}{0.01}$$

This can be solved for x as in the first example. (This is a re-run of question 30d, the Fe^{3+} question.)

The reason that this format is not used for weak acids is because of the meaning of t_0 and t_{eqm} in this context. Transfers of protons are among the quickest of equilibria to be established, so acid and base solutions reach equilibrium as fast as you can stir them. In a bottle of 0.1 mol dm^{-3} ethanoic acid, say, there was never a moment when the *undissociated* ethanoic acid concentration was equal to 0.1 mol dm^{-3}. Nevertheless, someone had to make up that solution in the bottle, by taking 0.1 mole of pure CH_3CO_2H and dissolving it in 1 dm^3 of water. So the on-the-bottle value represents the concentration which the undissociated acid *would have had* for a hypothetical moment in time. That is the nearest approach to a 't_0' in this branch of equilibrium.

12.9   Calculations with weak bases, and the existence of non-neutral salts

Weak bases require a set of concepts closely parallel to those for weak acids. Once again, in Brønsted–Lowry theory, the solvent is more than a passive bystander. In this case the solvent is the species *from which* a proton is seized, so

the solvent is behaving as an acid. The strength of a particular base will therefore depend upon the acid strength of the solvent. However, as with acids, the vast bulk of our studies will focus on water as the solvent.

A typical weak base in aqueous solution is ammonia, NH_3, using the lone pair that is the trademark of nitrogen. Indeed all nitrogen-based compounds that are non-ionic and single bonded show some basic tendencies in water (for example, hydrazine, NH_2NH_2, and methylamine, CH_3NH_2). However, they cannot be as strong as OH^-, because otherwise they would seize a proton from water and *make* OH^-, destroying themselves in the process.

Question

32 A good example of a really strong 'aggressive' base, of the sort that destroys itself in water, is the nitrogen-only anion, the nitride ion. When magnesium burns in air, it makes magnesium nitride, Mg_3N_2, an ionic compound containing N^{3-} ions. What would be the reaction between the ash from a magnesium burn, and water?

$$Mg_3N_2(s) + 6H_2O(l) \rightarrow ? + ?$$

What would you see if you tested the resulting solution with Universal Indicator?

As for gentler, weak, bases, there is the same problem as with weak acids. That is, you cannot be certain that every molecule of base will seize a proton (or two or three) and make OH^-. Therefore you need information about the balance of the equilibrium mixture – the equilibrium constant. This time the equilibrium constant is called the **base dissociation constant**, K_b.

The generalised chemical equation for a base dissociation shows a base B taking a proton from water:

$$B + H_2O \rightleftharpoons BH^+ + OH^- \quad (12.12)$$

The expression for K_b is therefore:

$$K_b = \frac{[BH^+][OH^-]}{[B]} \quad (12.13)$$

Questions

33 Write equations similar to equations (*12.12*) and (*12.13*) for the following bases:

a NH_3

b PO_4^{3-} (The phosphate ion – show it picking up a single proton.)

c Draw a curly-arrows version of the NH_3 case.

34 Bottles of $NH_3(aq)$ are often still labelled 'ammonium hydroxide'. From your answer to question 33a, would you say that the name was justified? If you are told that the K_b for ammonia is 1.7×10^{-5} mol dm^{-3}, would you still say it was justified?

35 The carbonate ion CO_3^{2-} is a useful base. It is used for curing acidity – by absorbing unwanted surplus protons – in systems as diverse as lakes, swimming pools, farmland, and the crude product you often get after the first stage of a school organic preparation. Its usefulness is in its very weakness – an excess of a strong base like sodium hydroxide would be far too destructive for most environments which need neutralising, and so carbonates and hydrogencarbonates are used instead.

a Which do you suppose is the most ready source of carbonate ions in the Earth's crust?

b Write the equation for CO_3^{2-} picking up one proton, in the style of (*12.12*).

c Write the corresponding K_b expression, in the style of (*12.13*).

d The K_b of CO_3^{2-} is 2×10^{-4} mol dm^{-3}. Using a mirror image of the method of the first example in Section 12.8 (p. 268), work out the 'pOH' of 0.1 mol dm^{-3} $Na_2CO_3(aq)$.

e Now use equation (*12.10*) to deduce the pH.

Non-neutral salts

On several occasions we have referred to, or calculated, the pHs of solutions of what were called **salts** in the pre-Brønsted–Lowry terminology. For example, we have found that sodium carbonate and sodium phosphate are weak bases, while ammonium chloride and iron(III) chloride are weak acids. On the earlier theory these salts were made by neutralising an acid with a base, so *why are they not neutral*? Ammonium chloride (weakly acid) was made from ammonia and hydrochloric acid:

$$NH_3(aq) + HCl(aq) \rightarrow NH_4^+(aq) + Cl^-(aq)$$

The problem arises because this is not an even-handed proton transfer – hydrochloric acid is a great proton-giver but ammonia is not such a great proton-taker. To put it another way, ammonium chloride is the offspring of a union between a *strong* acid and a *weak* base. So while the chloride ion has no interest in having its proton back, the ammonium ion *is* interested in reacting with water:

$$NH_4^+(aq) + H_2O(l) \rightleftharpoons H_3O^+(aq) + NH_3(aq)$$

This reaction, although heavily biased to the left-hand side, is enough to give the small but discernible acidity.

With carbonates (weakly basic) it is the other way round. Here the 'parents' are the strong base sodium hydroxide and the weak acid carbonic acid ($H_2O + CO_2$). In solutions of sodium carbonate, the sodium behaves itself, but the carbonate – whose conjugate acid was reluctant to give away the proton in the first place – indulges in some 'claw-back':

$$CO_3^{2-}(aq) + H_2O(l) \rightleftharpoons HCO_3^-(aq) + OH^-(aq)$$

The equilibrium is heavily biased to the left, but there is enough right-hand side to give weakly alkaline values of pH (as you found in question 35).

To generalise from these two examples, if you want to predict the pH of a salt solution, look at its 'parents'. If one of the parents is weak, the salt will show a small but measurable pH tendency in the direction of the strong parent.

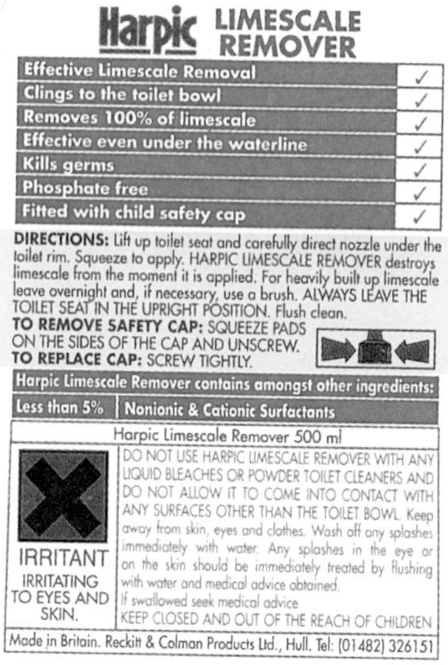

Figure 12.16 Harpic – a solid acid for domestic use

Question

36 When you 'neutralise' ethanoic acid with sodium hydroxide, the salt produced is sodium ethanoate, $CH_3CO_2Na(aq)$.

a Write the chemical equation for this reaction.

b Predict the general trend in pH of solutions of sodium ethanoate. Justify your prediction by use of equations.

c Why do you think the word 'neutralise' is in quotation marks?

Di- and tribasic acids

There is not an acid in existence which gives away *two* protons strongly. Even in sulphuric acid the second proton is reluctant to go, and the ion HSO_4^- appears in the list of weak acids.

$$HSO_4^-(aq) + H_2O(l) \rightleftharpoons H_3O^+(aq) + SO_4^{2-}(aq)$$
$$K_a = 10^{-2} \text{ mol dm}^{-3}$$

Question

37 The HSO_4^- ion finds commercial outlets in the home as a not-too-lethal solid acid. (As the sodium salt $NaHSO_4$ it is marketed as 'Harpic' (Figure 12.16).) What two useful roles might it perform in your lavatory?

An acid with two removable protons is called a '*di*basic' acid. A '*tri*basic' acid,

Harpic LIMESCALE REMOVER

Effective Limescale Removal	✓
Clings to the toilet bowl	✓
Removes 100% of limescale	✓
Effective even under the waterline	✓
Kills germs	✓
Phosphate free	✓
Fitted with child safety cap	✓

DIRECTIONS: Lift up toilet seat and carefully direct nozzle under the toilet rim. Squeeze to apply. HARPIC LIMESCALE REMOVER destroys limescale from the moment it is applied. For heavily built up limescale leave overnight and, if necessary, use a brush. ALWAYS LEAVE THE TOILET SEAT IN THE UPRIGHT POSITION. Flush clean.
TO REMOVE SAFETY CAP: SQUEEZE PADS ON THE SIDES OF THE CAP AND UNSCREW.
TO REPLACE CAP: SCREW TIGHTLY.

Harpic Limescale Remover contains amongst other ingredients:

| Less than 5% | Nonionic & Cationic Surfactants |

Harpic Limescale Remover 500 ml

IRRITANT IRRITATING TO EYES AND SKIN.

DO NOT USE HARPIC LIMESCALE REMOVER WITH ANY LIQUID BLEACHES OR POWDER TOILET CLEANERS AND DO NOT ALLOW IT TO COME INTO CONTACT WITH ANY SURFACES OTHER THAN THE TOILET BOWL. Keep away from skin, eyes and clothes. Wash off any splashes immediately with water. Any splashes in the eye or on the skin should be immediately treated by flushing with water and medical advice obtained. If swallowed seek medical advice. KEEP CLOSED AND OUT OF THE REACH OF CHILDREN

Made in Britain. Reckitt & Colman Products Ltd., Hull. Tel: (01482) 326151

such as phosphoric(v) acid, H_3PO_4, has three removable protons, but the third proton is even more reluctant to go. The third dissociation of phosphoric acid is extremely weak (in water $K_a = 4.4 \times 10^{-13} \, mol \, dm^{-3}$), and only a proper strong base like OH^- can remove the third proton.

$$H_3PO_4 \overset{-H^+}{\rightleftharpoons} H_2PO_4^- \overset{-H^+}{\rightleftharpoons} HPO_4^{2-} \overset{-H^+}{\rightleftharpoons} PO_4^{3-}$$
$$7.9 \times 10^{-3} \quad 6.2 \times 10^{-8} \quad 4.4 \times 10^{-13}$$

(These are values for K_a in $mol \, dm^{-3}$)

Question

38 Why is it not surprising that the multiply-charged anions like carbonate and phosphate are quite passable bases? (Figure 12.17)

12.10 Changes in pH during titrations

The strange and excitable behaviour of the pH value as a titration nears its end-point is one of the more memorable features in standard Advanced-level practicals. The apparatus for doing the practical work is shown in Figure 12.18. For the purposes of this chapter we will treat the electrode purely as a device which dips into a solution and senses its pH. The electrochemical secrets of its operation can remain hidden for now. In the titrations which follow the alkali is in the burette.

Titration of a strong acid against a strong alkali

If we have $0.1 \, mol \, dm^{-3}$ hydrochloric acid, then we know what pH to expect at the start of the titration just by reading

Figure 12.17 pH of washing soda

the bottle. $0.1 \, mol \, dm^{-3}$ 'parent' acid means $0.1 \, mol \, dm^{-3}$ hydrogen ions.

Question

39 What pH would you expect?

Hydrochloric acid is a strong acid and holds nothing in reserve – all its hydrogen ions are in play from the start. So we know that if $\frac{9}{10}$ of all the sodium hydroxide needed for the end-point has been added, $\frac{9}{10}$ of the hydrogen ions will have been removed. To take some hypothetical volumes – suppose $10 \, cm^3$ of $0.1 \, mol \, dm^{-3}$ hydrochloric acid was being neutralised by $0.1 \, mol \, dm^{-3}$ sodium hydroxide. The $\frac{9}{10}$ removed situation would be achieved after $9 \, cm^3$ of alkali had been added.

This attack would have left only $\frac{1}{10}$ of the original moles of acid. Those of you who are used to logarithms will realise that if the concentration has gone from 10^{-1} to 10^{-2}, the pH will have gone from 1 to 2. That is a fair estimate of what really happens, but for a more complete mathematical model, you have to include the fact that the sodium hydroxide has also added extra *volume* to the solution, taking it from $10 \, cm^3$ to $19 \, cm^3$. The actual concentration will be:

$$[H^+] = 0.1 \quad \times \quad \tfrac{1}{10} \quad \times \quad \tfrac{10}{19}$$
reaction dilution
factor factor

$$= 5.3 \times 10^{-3} \, mol \, dm^{-3}$$

So pH = 2.2

So it was hardly worth worrying about the dilution factor, as 2.2 is not much different from the predicted 2. The other fact to notice, which is much more significant for the general shape of the curve, is that we have gone $\frac{9}{10}$ of the way to the end-point and yet *the pH has only changed about one unit*. That is the way logarithms work – to 'damp' the numerical effect of change – and is why they were introduced in the first place.

Figure 12.18 A pH titration in action

Question

40 So the next change of one pH unit will come when the [H⁺] value has dropped by another factor of 10 (from 10^{-2} to 10^{-3}). What volume of NaOH(aq) will have to be added (including the 9 cm³ which is already in) to bring about a pH of about 3?

The story goes on: 9.99 cm³ gives pH 4, 9.999 cm³ gives pH 5, until there are so few hydrogen ions left that the pH is indistinguishable from pH 7 for water. As soon as the sodium hydroxide is in excess, the [OH⁻] starts to climb in a mirror image of the decline of the [H⁺], so that a change of about 8 pH units has happened with the addition of 0.2 cm³ (about four drops of liquid).

The total titration curve is shown in Figure 12.19. Notice that the true end-point is also a point of genuine neutrality, pH 7, as we would expect from our previous discussion about non-neutral salts. Sodium chloride, with two strong parents, will be genuinely neutral.

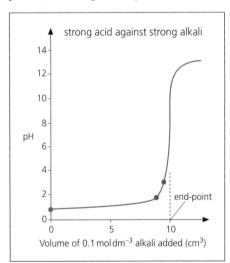

Figure 12.19 The variation of pH as 0.1 mol dm⁻³ sodium hydroxide is added to 10 cm³ of 0.1 mol dm⁻³ hydrochloric acid

Titration of a weak acid against a strong alkali

The first idea you need to be secure about is that 10 cm³ of a 0.1 mol dm⁻³ weak acid will need *exactly the same amount of alkali* for an end-point as 10 cm³ of a 0.1 mol dm⁻³ strong acid. It

is true that there are far fewer free H⁺(aq) ions, and probably it is also true that the incoming OH⁻ ions will mop those up first.

Question

41a If the OH⁻ ions remove the free H⁺ ions from the system like this:

$$HA(aq) \rightleftharpoons H^+(aq) + A^-(aq)$$

which reaction, the forward or the reverse, will be temporarily stopped?

b How will the system react to the removal of H⁺?

c What is the overall equation for the reaction between HA and OH⁻?

d What will have to happen before the OH⁻ ions can build up an excess?

How will this weak acid–strong alkali curve differ from that for the strong acid–strong alkali? We can look at some milestones on the weak–strong curve (Figure 12.20):

• The line will start and travel across at higher values of pH than for strong–strong reactions, because many of the hydrogen ions are kept in reserve, still in the HA molecules.

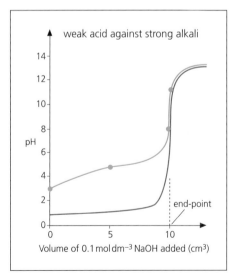

Figure 12.20 The titration of ethanoic acid against sodium hydroxide. Again there is 10 cm³ acid at the start
Key: ▬▬ line from Figure 12.19 (strong acid)

• The true end-point (as shown roughly by the middle of the steep part), will be at a pH *above* 7, as we would expect from a solution of 'NaA(aq)'. (Refer back to question 36 on sodium ethanoate, for instance.)

• There is an interesting point *half way to the end-point* which we can model mathematically:

OH⁻ ions turn HA molecules into A⁻ ions, so half way to the end-point half the HA molecules would have been turned to the conjugate base.

So   [HA] = [A⁻] at this point

But the acid dissociation constant K_a is true for all possible mixtures of H⁺, A⁻ and HA, so that if changes occur in two of them, the third must alter to compensate. By definition:

$$K_a = \frac{[H^+][A^-]}{[HA]}$$

So, in the special case at half way, when [HA] = [A⁻]:

$$K_a = [H^+]$$

Converting to p-notation:

$$pK_a = pH \text{ (still the special case)}$$
$$(12.14)$$

This interesting result means that the half-way point in a weak-acid titration has a pH equal to the pK_a of the acid in question. For ethanoic acid that would be 4.8, and this is the value chosen for Figure 12.20. The strong–strong titration line is shown superimposed for comparison.

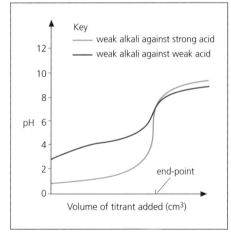

Figure 12.21 Rough graph of the other two types of titration

Questions

42 Why are the upper regions of the curves the same?

43 If you were to titrate a *di*basic acid like sulphuric acid, how do you think the titration curve would look? Bear in mind that once the first H^+ is removed, the first conjugate base (HSO_4^- for instance) begins to act as a new acid.

Titration of a weak base against a strong acid

To cut a long and slightly repetitive story short, the third situation is an inversion of the second, with the acid part of the line following the acid path from case 1, and the weak base part being symmetrical with and opposite to the weak acid part from case 2 (Figure 12.21). As for the fourth combination, the titration of *weak acid and weak base*, you have to superimpose the two 'weak wings' of the existing curves (Figure 12.21 again). There is hardly any steep part at the end-point, and this has implications for the use of indicators in these titrations. The curves of this section provide a guide to the choice of an indicator for any particular titration, as the next section will explain.

12.11   pH indicators

Everyone knows that indicators change colour. That is about the limit of the correct knowledge of many people, because there are many misconceptions and half-truths about indicators. The following statements are the *true* versions in the various areas of uncertainty.

• pH indicators are *not* the only kind – there are indicators which can detect changes in oxidation, complexation and precipitation reactions. The most familiar non-pH indicator is starch, to detect the last vestiges of iodine in a thiosulphate–iodine redox change.

• Universal Indicator is *not* the only pH indicator – in fact it is not even a single entity, as there are multi-coloured systems which cover different pH ranges with different levels of accuracy (Figure 12.22). There are also many indicators which have a single sharp colour change, ideal for use in titrations. The 'universal'

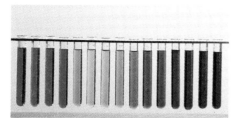

Figure 12.22 Universal Indicator in solution showing the colour range

types are in fact cocktails of the single-change indicators.

• *Not* all indicators are red in acid. This misconception is a tribute to the way that litmus (which *is* red in acid) has penetrated the national consciousness – 'the litmus test' is both a saying in English and a radio quiz game. The indicator phenolphthalein, for example, is very different – in fact it is red in alkaline solutions and colourless in acids (Figure 12.23).

Figure 12.23 Phenolphthalein in acid and alkaline conditions

• Indicators do *not* always show their 'neutral' colour at pH 7.

This last concept is the most surprising. How, you may wonder, can an indicator reflect the progress of a neutralisation if it changes its colour while the solution is still acid or alkaline? The answer comes in two parts:

• As you have seen from 'weak–strong' titrations, the true end-point of a titration – that is, the point at which the exact stoichiometry is satisfied – may not produce an exactly neutral salt, so sometimes you do not *want* an exactly neutral end-point-colour indicator.

• As you have seen from most of the titration curves (Figures 12.19, 12.20, 12.21), there is usually a moment at the true end-point of a titration when the pH changes by several pH numbers in

the course of addition of a few drops of titrant. So as long as the indicator changes its colour at a pH not too far from 7, it may still give the right end-point to within a drop or two.

Nevertheless, there are considerable limits on the choice of an indicator for a titration. We can extend our mathematical model to guide the choice, which will of course depend upon a knowledge of how pH indicators work.

What sort of substances are pH indicators?

• They are *weak acids or their conjugate bases*.

• The two members of the conjugate pair must be different colours.

• They must have colours intense enough so that you can achieve visibility with only a small amount of indicator.

Suppose we represent a generalised pH indicator molecule as HIn in its acid form, with In⁻ as its conjugate base. Let us also suppose, as is the case with phenolphthalein, that HIn is colourless and In⁻ is pink.

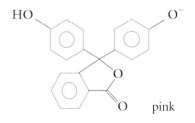

In aqueous solutions the following equilibrium will be set up:

$$HIn(aq) \rightleftharpoons H^+(aq) + In^-(aq) \quad (12.15)$$

colourless   pink

The crucial difference between this and similar-looking equations involving other weak acids is that in this case the $H^+(aq)$ is *not* mainly coming from the acid itself.

After all, the indicator is just a few drops of a solution in a much larger solution. So the position of the equilibrium is dependent on the 'external' or 'background' $[H^+(aq)]$. We can now use Chapter 9 skills to predict the behaviour of the system.

Question

44a If $[H^+(aq)]$ is high, which direction of the reaction will be helped?

b So what colour will the indicator appear in more acidic solutions?

c Use similar reasoning to explain the colour the indicator would have in alkaline solutions.

We do not yet know the exact pH at which our hypothetical indicator will switch between its two colours. HIn might be a strongish acid in which case it will be free with its hydrogen ions, and it will have changed to its alkali colour *before* the titration has reached pH 7 (assuming we began with acid in the flask). On the other hand HIn might hold on hard to its hydrogen ions and not give them up until the pH has gone *above* 7. So we need to know the equilibrium constant for the indicator reaction, called in this special case the 'indicator constant', K_{in}.

Question

45 Before going on to indicator constants, here is a general question. Ideally there should be no more than about 1 molecule of indicator for every 100 molecules of acid in a titration, otherwise it would begin to interfere with the very thing it was helping to measure, namely the amount of alkali needed for the end-point.

a In what way would the indicator 'interfere'?

b How can you link this question with the need for indicators to be intensely coloured?

To return to equilibrium matters, the expression for the indicator constant K_{in} is given by:

$$K_{in} = \frac{[H^+(aq)][In^-(aq)]}{[HIn(aq)]} \quad (12.16)$$

The change-over colour is a special case – it is the moment when there is as much HIn as In^-, so the fraction $[HIn]/[In^-]$ will be 1. Equation (12.26) then collapses to:

$$K_{in} = [H^+(aq)]_{\text{end-point}} \quad (12.17)$$

or using the p-notation:

$$pK_{in} = pH_{\text{end-point}} \quad (12.18)$$

This result is a great help in picking the right indicator for a particular job. It tells us that an indicator will change its colour when the *solution's pH is equal to its own* pK_{in}. The indicator constants of several indicators are shown in Table 12.4.

Table 12.4 Some pH indicators

Indicator	pK_{in}	Colour change range
methyl orange	3.7	3.2–4.4
bromophenol blue	4.0	2.8–4.6
litmus	7.0	6.0–8.0
phenolphthalein	9.3	8.2–10.00

We want to choose an indicator that will 'colour flip' at or very near the stoichiometric end-point. We need to extend the mathematical model a little further, to predict the **range of change**.

Let us make an assumption about the human eye – that if the amount of one of the two coloured forms of the indicator exceeds the other by a factor of 10 or more, then our eyes will see only the majority form, but anything less than 10 might be detected as an in-between-change colour or even an end-point colour. So, again assuming we start with a flask of acid, the indicator will begin its colour change from its acid, HIn, colour when enough alkali has been added to bring the fraction $[In^-]/[HIn]$ up to $\frac{1}{10}$ (Figure 12.24). In this special case:

$$K_{in} = [H^+] \times \frac{1}{10} \quad (12.16, \text{ adapted})$$

Rearranging,

$$[H^+] = K_{in} \times 10$$

and in p-notation (that is, taking logs of each side)

$$pH = pK_{in} - 1 \quad (12.19)$$

For example, litmus would begin to show its alkali colour at pH 6, and be right on its change colour at pH 7. By a similar mathematical argument you can

Figure 12.24 Litmus at pH 2, 4, 6 and 11

show that the other limit of the colour change, when the indicator has gone $\frac{10}{1}$ into its alkaline In^- colour, is at a pH of $pK_{in} + 1$.

Question

46 This model predicts that an indicator will take about 2 pH units to flip between its two colours. What assumptions have we made which might cast doubt on our prediction?

Crude though it is, the ± 1 range prediction holds up well in reality, as Table 12.4 shows. We are now on the brink of our overall goal. To pick the right indicator for a particular titration, we must know the general shape of the titration curve. In particular we must ensure that the 'steep bit' of the titration curve includes the change range of the proposed indicator.

For instance, Figure 12.25 shows the change range of litmus superimposed on

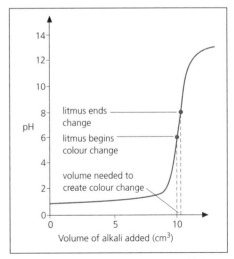

Figure 12.25 Showing how litmus would give a sharp end-point in a strong–strong titration

the titration curve of a strong acid against strong alkali titration. It shows why litmus would indeed be a good choice of indicator, because the change range falls within the addition of a very few drops of alkali.

Question

47 These problems ask you to match the change range of an indicator with the 'steep bit' of a titration curve, in the manner of Figure 12.25.

a Name an indicator which would be suitable for a weak acid–strong alkali titration.

b Name an indicator which would be suitable for a titration whose purpose was to find the amount of ammonia in a solution. Say what the other titrant would be.

c Name three indicators suitable for a hydrochloric acid–sodium hydroxide titration, and say why the choice is less critical than in the first two cases.

d What would you see if you tried to use methyl orange for the titration involving adding sodium hydroxide to a flask of ethanoic acid?

e Why is it difficult to find *any* indicator to work for a titration between a weak acid and a weak alkali?

Extending the scope of the 'indicators' mathematical model

The work we have just done on indicators carries a larger message, which goes beyond a discussion of the pH of colour change. Indicators are, after all, just weak acids and their (different-coloured) conjugate bases, in equilibrium with each other. Apart from that, the characteristic of their situation is that they are in a medium in which they are very much in the *minority* and have to respond passively to changes of pH in that medium. In other words, the presence of the indicator has a negligible *influence on* the pH, but the indicator itself is greatly *influenced by* the pH.

The mathematical model we used on the indicators predicted that a change of 2 pH units in the background medium would be enough to switch the indicator from one state to another. The assumption behind the calculation was that 'to switch' meant to go from a tenfold majority of one species to a tenfold majority of the other. The argument was that tenfold was the sort of factor which would enable one species to swamp the influence of the other, 'colour-wise'. But tenfold is a healthy excess, no matter what terms you are talking in. So if we forget about colour we can propose this general theory.

If there is a situation where *any* protonated and de-protonated conjugate pair are in equilibrium, in a medium with a dominant background pH, then a 2 pH-unit increase in the background pH should be enough to switch our system from 'mostly protonated' (that is, most sites with hydrogen ions attached) to 'mostly de-protonated' (where 'mostly' means 10 : 1).

Let us see this idea in action. Figure 12.26 shows the cleft or **active site** on an

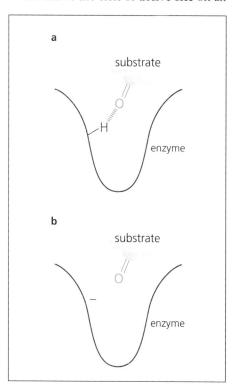

Figure 12.26 Showing why the action of enzymes may be pH-dependent
a Enzyme at low pH. There is a hydrogen bond to bind the substrate to the enzyme's cleft, but as little as 2 pH units higher and that hydrogen could be unavailable.
b Suddenly the cleft is not so inviting

enzyme. The assumption, a perfectly reasonable one in view of the fact that enzymes cling to their substrates by hydrogen bonds, is that the active site includes at least one '$\delta+$' (positively polarised) hydrogen ion. You can see from Figure 12.26 that the substrate is expecting a protonated site, and if it does not find one, it will not 'stick' properly in the cleft. Now, we could visualise the protonated enzyme as being nothing more than a different sort of weak acid, potentially in equilibrium with its conjugate base:

$$\text{cleft-H} \rightleftharpoons \text{cleft}^- + \text{H}^+(\text{aq})$$

This is exactly the same situation as in the indicator case, only without the colour. +2 pH units are enough to switch the enzyme from being 10 : 1 protonated (cleft-H) to 10 : 1 de-protonated (cleft$^-$). And this will, of course, make all the difference between success and failure, as far as the enzyme-catalysed reaction is concerned. We should draw the following lessons from this example.

• The 'indicators' mathematical model is applicable and useful beyond the field of indicators, extending to any proton exchange equilibrium performed against a 'background' pH. It predicts that the proton exchange will be substantially completed in either direction over a range of 2 pH units and, therefore, that any events dependent on the degree of protonation will change their course over that range.

• So, for instance, enzymes are likely to have quite precise needs concerning the pH of their medium, and within about 2 pH units an enzyme will switch from a working to a non-working condition.

Now we need to look again at the part of the mathematical model which tells us *which* 2 pH units.

Question

48a You should recall from the recent indicators work that the 2 pH units cover the range 1 unit either side of a central point, which is obtainable from data on the acid in question. (Equation (*12.18*) captures this idea best.) Assume the hydrogen atom in the enzyme cleft in Figure 12.26 is part of a carboxylic acid group, as in Figure 12.27. Use whatever data

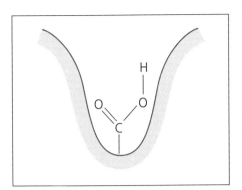

Figure 12.27 Possible molecular grouping in the active site of an enzyme

you think appropriate to estimate the 2 pH range of such a site, and so estimate the *upper* pH limit of the effective working of the enzyme. (Assume that the acidic hydrogen atom has to be *in place* for the enzyme to work properly.)

b Change the assumption so that the enzyme now needs the carboxylate anion (—CO_2^-) in the cleft, rather than the undissociated acid. Under these changed circumstances, what would now be the lower pH limit of effective catalysis?

Swimming pools

The reagent which is most commonly used to kill micro-organisms in swimming-pool water is chloric(I) acid, HOCl, which in most places where commercial chemistry is practised is still called 'hypochlorous acid'. It is what you get if you acidify Domestos – in fact, it is what you get if you ignore the warnings on Domestos and Harpic bottles about the dangers of mixing Harpic with Domestos! The days when chloric(I) acid was added directly to pool water are passing, if not already passed. Nowadays the acid is added indirectly in the form of solid tablets of trichloroisocyanuric acid (which has the trade name 'trichlor') (Figure 12.28). This compound sets up an equilibrium in the water, to release a small amount of free HOCl, but it acts as a reservoir of much more HOCl, if the free molecules are destroyed by

Figure 12.28 Tablets of 'trichlor' for use in swimming pools

action against bacteria. At the same time as it is in equilibrium with its parent trichlor, the HOCl molecule is also in equilibrium with its own conjugate base, as any weak acid would be. The pair of interactive equilibria is shown below.

'trichlor'

isocyanuric acid

$$HOCl(aq) \rightleftharpoons H^+(aq) + ClO^-(aq)$$

Question

49a Use the ideas of Le Chatelier to explain why trichlor acts as a reservoir of available (but not actual) HOCl.

b Can you think of two reasons for adding HOCl indirectly as trichlor, rather than directly as itself?

Before we move on to a question that will link this system up with our indicators' mathematical model, there is one other important fact about trichlor: it so happens that HOCl is a much better 'bug-killer' than ClO^-, being a stronger oxidising agent, with a higher positive $E^\ominus$ value (Chapter 16). The two $E^\ominus$ values are shown in Table 12.5.

Table 12.5 Showing how undissociated HOCl is a more potent oxidising agent than ClO^-. (In case you have never met 'standard electrode potentials', $E^\ominus$, before, a large positive $E^\ominus$ indicates a strong oxidising agent.)

Redox pair	$E^\ominus$(V)
ClO^-, H_2O/Cl^-, $2OH^-$	+0.89
$2HOCl$, $2H^+$/Cl_2, $2H_2O$	+1.63

Question

50 Over which 2 pH units change in the background pH of the pool would the bactericide pass from being mostly in the form of the more-effective HOCl to mostly in the form of the less-effective ClO^-?

The best course of action, as far as the pool operator is concerned, would seem to be to keep the pH at or below the lower end of the range you identified in your answer to question 50.

However, it is not as simple as this. First, the combination of too low a pH and too much HOCl hastens the corrosion of the metal fittings and tile cement of the pool. (In fact, in stainless steel pools, like the one in Holmes Place, Kingston upon Thames, the use of HOCl is abandoned completely in favour of ozone.) Second, excessive amounts of HOCl react with sweat and urine to produce the eye-stinging chemical which we all call 'chlorine' when we are unlucky enough to encounter it in the pool, but which is really NCl_3 (Figure 12.29).

Figure 12.29 Goggles are needed if the pH of a pool gets too low

Questions

51 All things considered, it is best to have the balance between HOCl and ClO⁻ set at about 50:50. Which background pH would achieve this level?

52a When the pool treatment is working properly, any sweat- or urine-based amino compounds are oxidised via chloramines, to nitrogen:

$$NH_3(aq) \xrightarrow{HOCl} NH_2Cl(aq) \xrightarrow{HOCl} N_2(g)$$

or, if the concentration of HOCl is higher:

$$NH_3(aq) \xrightarrow{H} NCl_3(aq. g)$$
eye-stinging
nitrogen trichloride

Write an equation for the oxidation of ammonia to monochloroamine by HOCl.

b Show by another equation that it takes more moles of HOCl to turn ammonia into the eye-stinging trichloroamine (nitrogen trichloride), than it does to make monochloroamine. What, according to the paragraph before question 51, is the best way to keep NCl₃ levels under control?

12.12 Buffer solutions

It should be clear from our work on swimming pools and enzymes that it is sometimes very useful to have a system for *preventing pH changes*. If we can stabilise the pH of a medium at a particular value, we can ensure that the reagent in the system stays as we want it – which in the last section meant the enzyme staying protonated, and the swimming pool's HOCl/ClO⁻ mixture staying at 50:50. A reagent or mixture of reagents which opposes changes in the pH of a medium is called a **buffer**, and its solution is a **buffer solution**. Let us see how a buffer solution might deal with invasions of H⁺ and OH⁻, or in other words how it might prevent those invasions from upsetting the pH too much.

• *Invasion of H⁺*. The buffer solution must take the hydrogen ions out of circulation, so that they cannot affect the pH. It must therefore contain a strongish base. That does not usually mean OH⁻ ions, unless you want a buffer at pH 11 or thereabouts, which very few systems do. Carbonate, CO_3^{2-}, or ethanoate are suitable depending on the required buffer pH.

• *Invasion of OH⁻*. The buffer must have a reserve of hydrogen ions ready to be pulled off to neutralise the incoming OH⁻ ions. We do not want free H⁺ ions (again, unless an extreme pH is required), so a weak acid meets the specification. Ethanoic acid is suitable, since most of its hydrogen ions are held back, but would be mobilised if OH⁻ ions invaded.

'Tuning' a buffer solution

It is one thing to get a buffer solution to buffer and another to get it to buffer at the desired pH. Recourse to a mathematical model is needed. Earlier both ethanoic acid and ethanoate ion were mentioned as performing the twin functions of a buffer solution. Indeed *mixtures* of these two substances (using

sodium ethanoate to provide the ethanoate ion) make good buffer solutions, and we can go further to say that *every weak acid mixed with its conjugate base* (in the form of the sodium salt) *will operate as a buffer* (Figure 12.30).

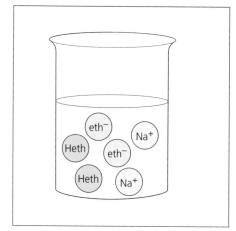

Figure 12.30 A possible buffer solution

Question

53 In the mixture of a weak acid and its conjugate base, which ingredient is responsible for:

a absorbing invading H⁺ ions?

b absorbing invading OH⁻ ions?

At what pH will a particular pair buffer? The answer can, somewhat surprisingly, be lifted straight from our earlier work on titration curves. It may come as a surprise to realise that titrations, or more especially *incomplete* titrations, actually *make* buffer solutions. For what is a half titrated sample of ethanoic acid except a mixture of Heth and eth⁻?

$$2Heth + OH^- \rightarrow Heth + H_2O + eth^-$$

Indeed the shape of the titration curve around the region of half-titration is a *visual image of buffering in action.*

Question

54 By looking at Figure 12.31, explain the meaning of the phrase in italics in the last sentence. In other words, why is the 'horizontality' of that part of the graph a proof that buffering is occurring?

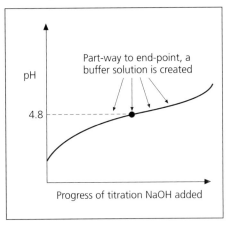

Figure 12.31 Part-titrated weak acids have buffering action

You may remember something else about the half-titration of a weak acid, as captured in equation (*12.14*), namely that the pH of the half-titration point is numerically the same as the pK_a of the acid ($pK_a = pH$). So we know at once that a 1:1 mixture of ethanoic acid and sodium ethanoate will buffer at the pK_a of the conjugate acid member of the mixture, which in this case is about 4.8.

Question

55 Use Table 12.1 to invent mixtures which would buffer at or near the following pHs. In some cases, those where the acid is itself an anion, you will need two sodium salts rather than one. The first example is done for you.

a pH 2 – *Answer*: Table 12.1 shows HSO_4^- has a pK_a of 2, so we want a 1:1 mixture of HSO_4^- and its conjugate base SO_4^{2-}. So the recipe is to mix equimolar amounts of $NaHSO_4$ and Na_2SO_4. (Also, both reagents are quite common chemicals.)

b pH 4

c pH 7

Although so far we have talked only about buffer *mixtures*, there are systems where a single chemical species can act as a pH buffer on its own. A buffer needs to be able to take out both H^+ and OH^-, and if a molecule has bits on it which can do both then a mixture is not needed.

Question

56 Amino acids as a group are an example of this sort of one-substance buffer. The structure of a generalised amino acid is shown below. Where does it get its two-way buffering action from?

$$H_3\overset{+}{N}-\underset{\underset{H}{|}}{\overset{\overset{R}{|}}{C}}-C\underset{O^-}{\overset{O}{\diagup\!\!\!\!\diagdown}}$$

Finally, there is another kind of one-substance two-way buffer, and that is the 'middle anion' of a dibasic acid. For example, the hydrogencarbonate ion, HCO_3^-, is both a proton giver and an acceptor, and performs its buffering function in places like your blood. It is also the chosen buffer for the purpose highlighted in the swimming-pool case – namely, for keeping the pool water within the safe range of pH where it is neither a danger to the bathers (by containing micro-organisms) or a danger to its own fabric (by dissolving its pipes).

Question

57a Show by means of chemical equations how the hydrogencarbonate ion might act as both an acceptor and a donor of protons.

b How do you suppose hydrogencarbonate ions get into blood?

c Pools in the limestone areas of the country often do not need any buffering treatment, since their mains water supply has a natural buffering action. Show how percolation of natural rainwater (rich in dissolved carbon dioxide) through carbonate rocks could produce a self-buffering solution suitable for use in pools.

Summary

• In the Brønsted–Lowry system for defining acidity and basicity, **an acid is a** (potential) **proton donor**, while **a base is a** (potential) **proton acceptor**. (The word 'potential' is meant to signal that they do not actually have to be proton donors or acceptors to merit the names acid and base.)

• Molecules which show acidic tendencies always have **positively polarised covalent bonds to hydrogen atoms**. These bonds cleave heterolytically when the acid releases its H^+ ions.

$$A \overset{\delta-}{\underset{\times}{\circ}}\overset{\delta+}{H}$$

$$HA \rightarrow H^+ + A^-$$

• Molecules which show basic tendencies always have a **lone pair** or **lone pairs** of electrons. These electrons can receive the donated proton with a dative covalent bond.

$$B + H^+ \rightarrow BH^+$$
$$B \overset{\circ}{\underset{\circ}{}} \curvearrowright H^+$$

• In order that potential acids and bases can become actual ones, they must either meet each other or be placed in a solvent medium.

• The equations showing generalised acid and base behaviour in an aqueous medium are as follows. (Conjugate pairs are indicated by 1, 2.)

(i) $\underset{\text{acid 1}}{HA(aq)} + \underset{\text{base 2}}{H_2O(l)} \rightleftharpoons$

$$\underset{\text{acid 2}}{H_3O^+(aq)} + \underset{\text{base 1}}{A^-(aq)}$$

(ii) $\underset{\text{acid 1}}{B(aq)} + \underset{\text{base 2}}{H_2O(l)} \rightleftharpoons$

$$\underset{\text{acid 2}}{BH^+(aq)} + \underset{\text{base 1}}{OH^-(aq)}$$

• A pair of species differing by a single proton is called a **conjugate pair**. The protonated member of the pair is called a **conjugate acid** (symbolically HA or BH^+) and the deprotonated member is called a **conjugate base** (A^- or B respectively).

- A **weak acid** is one in which a high percentage of molecules at any moment in time will have **failed to donate a proton** to a particular medium. For a weak acid, the position of the equilibrium in equation (i) would be well to the *left* and the ionic population would be *low*. (The medium, water, is acting as a 'default base' in the absence of anything stronger.)

- A **weak base** is one in which a high percentage of molecules at any moment in time will have **failed to accept a proton** from a particular acid/medium. For a weak base, the position of the equilibrium in equation (ii) would be well to the *left* and the ionic population would be *low*. (The medium, water, is acting as a 'default acid' in the absence of anything stronger.)

- A **strong acid** is one in which a high percentage of molecules at any moment in time **will have succeeded in donating a proton** to a particular base/medium. For a strong acid, the position of the equilibrium in equation (i) would be well to the *right*, indicating that the acid is *fully ionised*.

- A **strong base** is one in which a high percentage of molecules at any moment in time **will have succeeded in accepting a proton** from a particular acid/medium. For a strong base, the position of the equilibrium in equation (ii) would be well to the *right*, indicating that the base is *fully ionised*.

- Strength or weakness of an acid or base is dependent upon the medium. If an acid or base is referred to as strong or weak with no further qualification, the assumption is that the medium in question is water.

- A strong acid will be a good proton donor, so it follows that its conjugate base must be an ineffective proton acceptor. Hence **strong acids will have weak conjugate bases**, and conversely **weak acids will have strong conjugate bases**.

- The degree of strength of an acid can be quantified by its **acid dissociation constant**, K_a, which is the equi-

librium constant of its reaction with (usually) water. K_a has the standard format:

$$K_a = \frac{[H^+(aq)][A^-(aq)]}{[HA(aq)]}$$

The omission of the reactant water concentration is because it is effectively rendered constant by virtue of its huge molecular majority over the other species. The *larger* the K_a, the *stronger* the acid.

- When an acid has more than one potentially acidic hydrogen atom, they are not all donated at once, nor with equal ease. In fact, once one H^+ ion has cleaved, the next one is less readily donated, and so K_a values for successive donations get smaller.

- The degree of strength of a base can be quantified by its base dissociation constant, K_b, which is the equilibrium constant of its reaction with (usually) water. K_b has the standard format:

$$K_b = \frac{[BH^+(aq)][OH^-(aq)]}{[B(aq)]}$$

The *larger* the K_b, the *stronger* the base.

- When a series of acids is arranged in order of their acid strengths, strongest at the top, a **law of clockwise circles** enables the prediction of the direction of proton donation reactions. In other words, an upper table acid (HX) will donate protons to a lower table base Y^-, to generate HY and X^-.

$$\mathbf{1} \; HX \rightleftharpoons H^+ + X^-$$

$$\mathbf{2} \; HY \rightleftharpoons H^+ + Y^-$$

$$HX + Y^- \rightarrow HY + X^-$$

- **Neutralisation** is the word applied to reactions between an acid and a base, neither of which is merely the solvent medium. The description is not an accurate one unless the acid and base are both equally strong, since otherwise the end product, a **salt**, will show a residual acidity or basicity (according to whichever was the stronger parent).

- Water is one of a number of solvents

which can act as acid and base to itself. An 'auto-ionisation' reaction produces small amounts of $H^+(aq)$ and $OH^-(aq)$ ions as shown below.

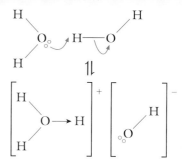

- The auto-ionisation reaction in water has its own special equilibrium constant, called the **ionic product**, K_w, of water:

$$K_w = [H^+(aq)][OH^-(aq)]$$
$$= 10^{-14} \text{ mol}^2 \text{ dm}^{-6} \quad (12.13)$$

where $[H^+(aq)]$ is being used as a shorthand for $[H_3O^+(aq)]$.

- The ionic product's value is a constant for all aqueous solutions, so when either one of $[H^+(aq)]$ or $[OH^-(aq)]$ goes up, the other must come down.

- The acidity of a solution is expressed on the pH scale, which is a logarithmic version of the normal $[H^+(aq)]$ value. Specifically:

$$pH = -\log_{10}[H^+(aq)] \quad (12.14)$$

- Because of the constancy of the ionic product in all aqueous solutions,

$$pH + pOH = 14 \quad (12.16)$$

- The pH of a solution of a strong acid or base can be derived directly from the concentration of the parent species, since every mole of parent will have become a mole of H^+ or OH^- (or two moles in the case of lower-Group-2 hydroxides). So for example the pH of 1.0 mol dm^{-3} HCl is $-\log_{10}(1)$, or 0, and that of 1.0 mol dm^{-3} NaOH is 14.

- The pH of solutions of weak acids and alkalis cannot be directly deduced from the parent concentration, since not all of the parent will be in the ionised state. The pH can be calculated, however, by use of the K_a or K_b value.

• The relationship between the K_a of a weak acid HA and the pH of its solution can be found as follows, where $x = [H^+(aq)]$ (and therefore where $x = [A^-(aq)]$):

$$K_a = \frac{x^2}{[\text{parent HA}]}$$

so $x = \sqrt{([\text{parent HA}] \times K_a)}$

Then $\text{pH} = -\log_{10} x$

This calculation features the 'bottom-line approximation', which says that x is negligible compared with [HA], so $[\text{HA}] \approx [\text{parent HA}]$.

• In the course of an **acid–base titration**, the pH of the solution *changes abruptly* in the region of the end-point. Indeed this feature can be used as an alternative to an indicator to tell when an end-point has been reached. The graph of pH against the volume of burette-titrant added (the titration curve), has different features, depending on whether the participants are strong or weak. For instance, the titration of a strong acid against a strong base gives a titration curve with a long steep vertical section at the end-point (Figure 12.32a).

• A **pH** or **acid–base indicator** is a *weak acid whose conjugate base has a markedly different colour*. A mathematical model can show that an indicator should change from one colour to the other over approximately a 2 pH unit range, either side of its own pK_a.

• Selecting the right indicator for a titration means picking one whose 2 pH unit range *falls within the steep part* of the particular titration curve. A simple indicator selection guide is shown in Table 12.6.

a strong acid/strong base

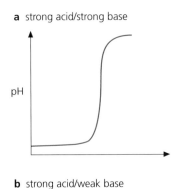

b strong acid/weak base

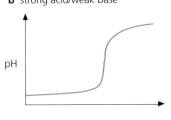

c weak acid/strong base

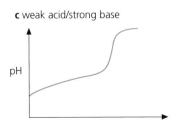

d weak acid/weak base

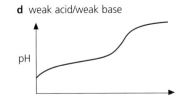

Figure 12.32 Sketches of titration curves. In each case it is assumed that alkali is being added to acid. The *x*-axis is volume of alkali added to a fixed volume of acid

• Many systems in which a proton transfer takes place against a dominant background pH follow the same pattern as indicators. Thus, for example, the effectiveness of an enzyme can decline or increase rapidly over a 2 pH unit range, as it changes between being protonated and de-protonated.

• **Buffers** are chemical systems which impart to a solution a *resistance to changes in pH*. They involve species which are equally good at neutralising incoming H^+ or OH^- ions. Usually this means having a mixture of a weak acid (HA) and its conjugate base (A^-), with each species responsible for neutralising one of the possible 'invaders'. Invasion of H^+ is countered by the proton-accepting tendency of A^-, while invasion of OH^- is countered by release of 'new' protons from HA.

• **Conjugate-pair buffer systems**, when mixed in equimolar proportions, buffer at a pH equal to the pK_a of the acid member of the pair.

• **One-component buffer systems** also exist, using species which have sites on the same molecule for release and for seizure of protons. The half de-protonated anions of dibasic acids act in this capacity. The prime example is HCO_3^-, the hydrogencarbonate ion, which is the half de-protonated ion derived from carbonic acid, H_2CO_3. Invasion of H^+ sends the HCO_3^- back to carbonic acid, while invasion of OH^- sends the HCO_3^- through to CO_3^{2-}. This ion is used both by nature and by human technology, in environments such as blood and swimming pools, to maintain a pH of around 7.

Table 12.6 Choosing an indicator

Type of titration, as in Figure 12.32	Recommended choice of indicator
a	any
b	methyl orange
c	phenolphthalein
d	none (use titration curve itself)

13

Organic chemistry 3: Aldehydes, ketones, carboxylic acids and their derivatives

13.1 Introduction – nomenclature

This chapter deals with three classes of compounds. First, those compounds for whom the carbonyl group (as the C=O bond is often called) is the only functional group in the molecule – the **aldehydes** and **ketones**. Second, a group of molecules which behave in some ways *as if* they were aldehydes or ketones, but have complex patterns of internal rearrangement which disguise their carbonyl character – the **carbohydrates**, molecules of great biological importance. Third, compounds in which the main role of the carbonyl group is to modify the behaviour of a second, adjacent, group – these are the **carboxylic acids**, **acyl halides** and **esters**. Esters include the biochemicals called **fats**, which are of no less significance than carbohydrates, and also the industrial polymers called **polyesters** and **alkyd resins**, which play major roles in textiles and paints.

Examples of the structural formulae of some of these groups, along with their names, are shown in Table 13.1 (overleaf). The following notes are about the nomenclature in Table 13.1.

1 You could think of aldehydes as 'primary carbonyls' (by analogy with the naming of alcohols, for example) and ketones as 'secondary carbonyls'.

Question

1 a Why could there never be a 'tertiary carbonyl'?

b Which is the shortest ketone to require a number in its name?

c Name the unnamed ketone in Table 13.1. (The number goes between the stem-name and the suffix, as in alcohols.)

Table 13.1 Names and formulae of molecules containing the carbonyl bond

Structural formula	Name	Type
$CH_3CH_2C\overset{O}{\underset{H}{<}}$	propanal	aldehyde
$CH_3CH_2CCH_2CH_3$ with $\|$ and O below	pentan-3-one	ketone
$CH_3CH_2CH_2C\overset{O}{\underset{O-H}{<}}$	butanoic acid	carboxylic acid
$CH_3CH_2CH_2C\overset{O}{\underset{Cl}{<}}$	butanoyl chloride	acyl halide
$CH_3C\overset{O}{\underset{OCH_2CH_3}{<}}$	ethyl ethanoate	ester
(skeletal formula)	α-D(+) glucose	monosaccharide carbohydrate
$CH_3CCH_2CH_3$ with $\|$ and O below	ketone (see question 1)	
$CH_3CH_2CH_2CH_2CH_2CH_2CH_2C\overset{O}{\underset{OH}{<}}$	carboxylic acid (see question 2)	
$CH_3C\overset{O}{\underset{Br}{<}}$	acyl halide (see question 3)	

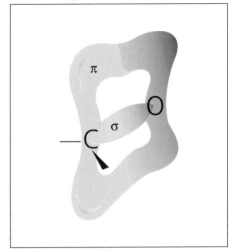

Figure 13.1 Electron density in a carbonyl bond (darker areas represent higher density)

the C=C double bond in alkenes. Perhaps we should, therefore, anticipate a set of electrophilic addition reactions (see p. 209).

2 On the other hand the bond is obviously polarised. In fact, the dipole moments of carbonyl compounds (dipole moment is a measure of the force trying to align a molecule with an electric field, and is therefore a good index of bond polarity) are a good deal larger even than those of alcohols (Table 13.2). Thus far in this book, a molecule's possession of a dipole has led to attack by nucleophiles, resulting in substitution (see p. 228).

Table 13.2 Dipole moments in an alcohol, a ketone and an aldehyde

Substance	Dipole moment (D)
propan-1-ol	1.69
propanone	2.88
ethanal	2.72

Question

4 For what classes of compounds, met earlier in this book, does the last sentence before Table 13.2 hold true?

So the question is – which of the two reaction types (electrophilic addition or nucleophilic substitution) will dominate the chemistry of carbonyl compounds? The answer is *neither*.

2 Notice that although the CO_2H group has the feel of a distinct functional group on the end of a carbon chain, the stem-name 'counts' the carbonyl carbon atom. Thus, $C_6H_{13}CO_2H$ is *hept*anoic acid rather than *hex*anoic acid.

Question

2 Name the unnamed carboxylic acid in Table 13.1.

3 Notice how the names of acyl halides, like the compounds themselves, are derived from the associated acid, by the substitution of '-yl halide' for '-ic acid'.

Question

3 Name the unnamed acyl halide in Table 13.1.

13.2 Compounds containing carbonyl as their only functional group

Figure 13.1 shows the electron-density distribution in a C=O bond. The picture should raise two issues in your mind, drawing on themes from previous chapters.

1 The bond is obviously a double bond, which should invite comparison with

Why are there no electrophilic additions?

It is possible to write a mechanism for the attack by an electrophilic species on a carbonyl bond:

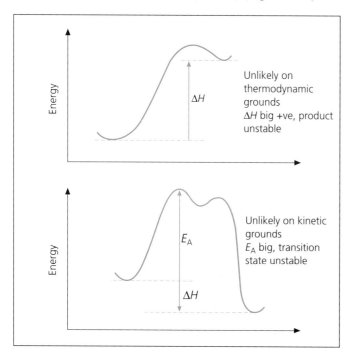

You can draw one of the typical molecules which adds readily to C=C bonds, and put it through the same two-step process at a C=O bond. This does not offend against any common-sense polarity laws – the hydrogen would obviously go to the oxygen end – and yet it just does not happen.

There are two classes of reason to explain why something does not happen: the 'thermodynamics' reason (Chapters 8 and 9) and the 'kinetics' reason (see Section 11.3, with fuller coverage in Chapter 14). In other words, there is either something wrong with the stability of the product, relative to the reactant (thermodynamics), or there is something wrong with the stability of the transition state (kinetics) (Figure 13.2).

![Figure 13.2 Thermodynamic and kinetic barriers in the way of reaction — two energy profile diagrams. Upper: labelled "Unlikely on thermodynamic grounds, ΔH big +ve, product unstable". Lower: labelled "Unlikely on kinetic grounds, E_A big, transition state unstable".]

Figure 13.2 Thermodynamic and kinetic barriers in the way of reaction

Concerning the stability of the *product*, it is quite rare to find any compound in organic chemistry with either two OH groups on the same carbon atom, or one OH group and one halogen atom. It is as if the two electronegative groups destabilise the molecule by fighting with each other for the

electron density around the carbon atom. So it is reasonable to suggest that electrophilic additions to C=O bonds might fail on thermodynamic grounds. Now let's see if the kinetic factors are any more favourable. The next question looks at the stability of the reaction intermediate and draws on some earlier concepts.

Question

5 a Chapter 11 described at some length the stability of tertiary carbocations relative to primary ones (see below). What reasons (to do with inductive effects of alkyl groups) were given then?

$$CH_3CH_2CH_2\overset{+}{C}H_2 \quad \text{primary}$$

tertiary

b Look at the carbocation intermediate being proposed in the mechanism shown before. It is liable to be *un*stable, for reasons which stem from exactly the same line of argument as that used in part (a). Can you explain further?

At the end of this double line of reasoning, we find that electrophilic additions to C=O bonds have neither the thermodynamic nor the kinetic factors in their favour, so it is no wonder they don't happen.

Why are there no nucleophilic substitutions?

The answer is straightforward, and should be obvious from looking at the examples below. These show the *first step* in the nucleophilic attack on an alcohol and on a carbonyl compound, both of which do happen.

Question

6 Why does the attack on the alcohol succeed in dislodging the OH group (leading to substitution) whereas the attack on the carbonyl group does not look likely to lead to substitution?

The answer to question 6 brings home the fact that the double bond has one trick which cannot be taken away – no matter how polar or non-polar – and that is the ability to tolerate the breakage of the π part of the bond while at the same time hanging on to the σ part. The result is going to be addition rather than substitution, and so the reactions of carbonyl compounds give us a new combination of words to add to our broad classes of organic reactions – **nucleophilic addition**.

13.3   Nucleophilic addition to aldehydes and ketones

We have already seen the first stage of the nucleophilic addition reaction (above). The scheme below reveals the whole process, using as an actual example the reaction between carbonyl compounds and cyanide ions, CN^- (from aqueous sodium cyanide). The hydrogen ion in the second step is taken from a water molecule.

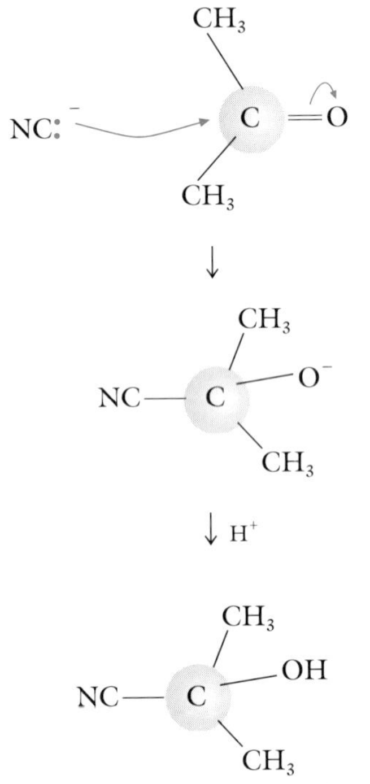

'propanone cyanohydrin'

You can see that the overall effect has been to add H—CN across the C=O double bond, even though the molecule HCN is not itself present to any great extent in aqueous sodium cyanide.

Question

7 The scheme shown above suggests that the sequence of steps is: *first*, a rate-determining attack by CN^- on the positive end of the C=O bond, *then* attack by H^+ (extracted from water) on the negative end. A major piece of evidence for this is that the reaction is much slower in acidic media than it is in neutral or alkaline ones.

a CN^- is a fairly strong base (and therefore a good nucleophile). What does that tell you about the strength of HCN as an acid?

b Which of the two species HCN and CN^- will be the dominant one in strongly acidic media?

c Why therefore should the mechanism shown above be hampered by an acidic medium?

The answer to question 7 gives another example of the way in which a hydrogen ion can 'occupy' a lone pair, and render the possessor of the lone pair impotent as a nucleophile. (You will remember in Section 11.10 that we compared the situation to a fencer with a rubber bung on the end of his/her foil.)

However, acidic media do make an important contribution to the usefulness of this particular reaction, if only as a postscript to the main event. The product of the nucleophilic addition, a **cyanohydrin** (the non-systematic name for a 2-hydroxynitrile), is not very valuable in itself, but it is a useful intermediate. When boiled with aqueous acid, the CN group is hydrolysed and becomes a carboxylic acid group:

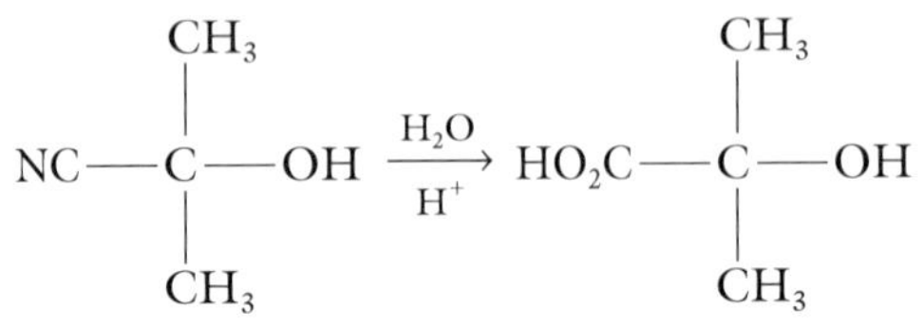

The usefulness of having a carboxylic acid group adjacent to an OH group is highlighted by the next question.

Question

8 2-Methylpropenoic acid (methacrylic acid, below) is a possible precursor of the monomer of the famous polymer Perspex.

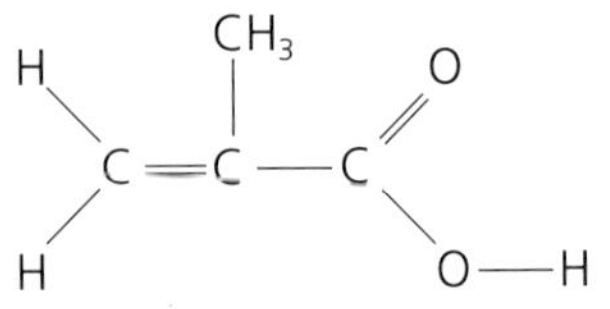

Use the following stages to construct a three-step route from propanone to 2-methylpropenoic acid:

a Make the cyanohydrin of propanone. Show structural formulae, reagents and conditions.

b Boil the product with sulphuric acid, thus hydrolysing the CN group as shown before the question. Show the structural formula of the new product.

c Look back to Chapter 11, p. 245 to find out how to turn the product from (b) into the target molecule shown at the beginning of this question. Show structural formulae for the reaction, and reagents and conditions.

d What fourth step, involving what other *organic* chemical, would actually take us right through to the monomer of Perspex (shown below)?

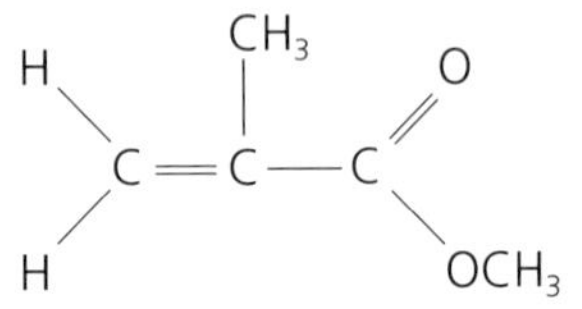

13.4 Reactions of aldehydes and ketones with ammonia derivatives – addition/elimination

Certain molecules allied to ammonia can perform an interesting variation on the theme of nucleophilic addition to carbonyl compounds. One such is the 'double-headed' ammonia derivative, **hydrazine**:

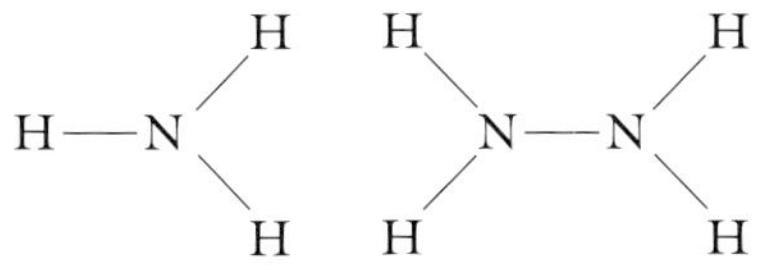

ammonia   hydrazine

Question

9 Hydrazine shares ammonia's basic and nucleophilic properties.

a Why are both molecules basic?

b What would be the structural formula of hydrazine's ammonium chloride equivalent (which is called **hydrazinium chloride**)?

c Hydrazine is usually stored as hydrazinium chloride, because it is more stable in that form. The hydrazine is generated as it is needed. How could that be done readily?

The initial attack by hydrazine on, say, a ketone is shown below.

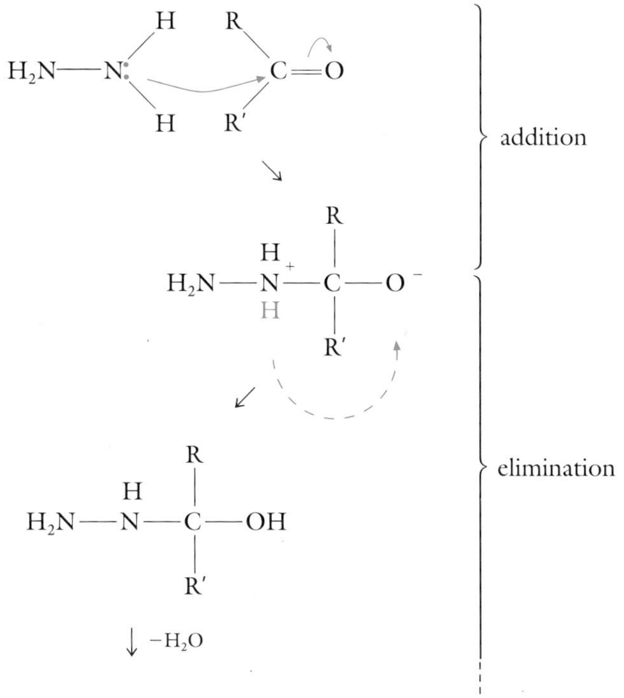

addition

elimination

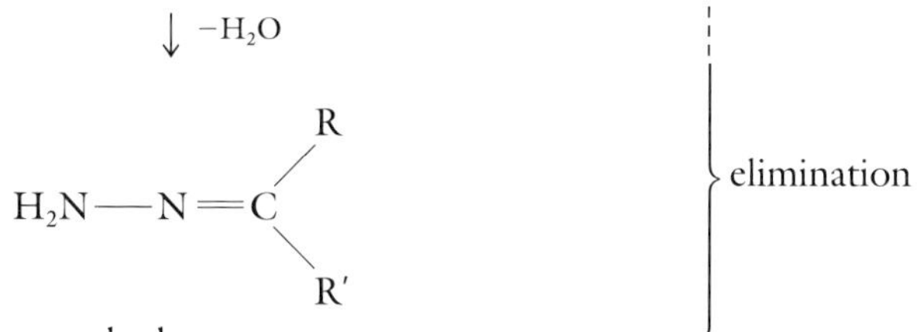

$-H_2O$

a hydrazone

elimination

To begin with, the pattern is the same as that for attack by HCN (opposite). However, this time the addition product turns out to be unstable, and undergoes a second *elimination* step, losing a water molecule and producing a **hydrazone**.

This reaction, and a similar one involving hydroxylamine (NH_2OH), might be seen merely as minor curiosities of the chemical world. (Talking of curiosities, it is strange that ammonia itself does not add to carbonyls.) However, one particular member of this set of addition/elimination reactions is of some importance. This is the one featuring a hydrazine derivative known as **2,4-dinitrophenylhydrazine**, or **Brady's reagent** (below).

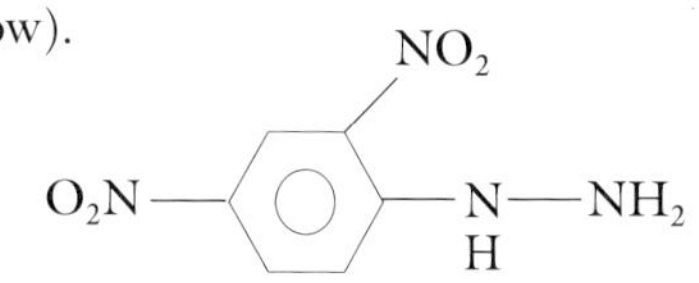

All aldehydes and ketones give orange or yellow precipitates with Brady's reagent, which can be used to identify these two groups of compounds. Better still, the precipitates are easy to recrystallise from alcohol and these crystals give sharp melting points, each melting point uniquely dependent upon the identity of the original aldehyde or ketone. Hence, Brady's reagent can be used both to identify the carbonyl group of compounds in general, and individual members of that group in particular. (These same identifications can be carried out more conveniently by combinations of mass spectrometry, NMR and infrared spectroscopy, but with a massive increase in expense.)

Question

10 You have met Brady's reagent, and the reaction of carbonyls with hydrazine. Can you put the two together, and write the formula for the yellow crystals obtained by adding Brady's reagent to propanal?

13.5 Redox behaviour of aldehydes and ketones

So far we have not met a single example of a reaction which differentiates between aldehydes and ketones. That pattern will change in this section, but not yet. First, there is a reaction which continues the 'nucleophilic addition' theme.

Reduction

There are two main groups of reagents that will reduce carbonyl compounds. One is the 'hydrogen-gas-over-a-transition-metal-catalyst' group, which is rather indiscriminate in that it will reduce C=C double bonds along with C=O ones. (Refer back to Chapter 10.) The other is a group of ionic hydrides with the general formula $M(I)N(III)H_4$, where $M(I)$ and $N(III)$ refer to cations from, respectively, Groups I and III. The reducing action of these hydrides is specific to the C=O bond, leaving C=C double bonds untouched.

Reaction with, say, sodium tetrahydridoborate, $NaBH_4$ ($LiAlH_4$ is the other common member of the group), is equivalent to attack by the nucleophile H^- (hydride ion), and the overall reaction amounts to the addition of H_2 across the C=O bond. The pathway by which this happens can be summarised in the following two-step chemical equation.

$$4CH_3C\begin{matrix}H\\ \\ \| \\ O\end{matrix} + NaBH_4$$

$$\downarrow$$

$$NaB(CH_3CH_2O)_4$$

$$\downarrow H_2O$$

$$4CH_3CH_2OH + NaOH + B(OH)_3$$

Questions

11 Give structural formulae for the products of reactions of:

a propanone with lithium tetrahydridoaluminate, $LiAlH_4$

b butanal with hydrogen gas over a nickel catalyst

c cyclohexanone with sodium tetrahydridoborate, $NaBH_4$.

Can you see any pattern linking the identity of the starting material with the type of alcohol produced?

12 The figure below shows the skeletal formula of the male sex hormone testosterone.

Predict the result of reacting testosterone with:

a hydrogen gas over a nickel catalyst

b sodium tetrahydridoborate, $NaBH_4$.

(Give skeletal formulae as answers.)

Oxidation

Now, at last, we come to something that separates ketones from aldehydes. Ketones are difficult to oxidise, whereas aldehydes are about the most readily oxidised of the common organic compounds. In all cases they are oxidised to **carboxylic acids**. They succumb not only to the typical vigorous agents of oxidation like manganate(VII) and dichromate(VI), but also to much milder oxidising agents such as complexed Cu^{2+} and Ag^+ ions. Once again we will find that the reactions are of both academic and diagnostic value. The two tests that follow are often used after a positive result with Brady's reagent, because once you know you have either an aldehyde *or* a ketone, the next logical step is to find out *which*.

Tollen's (the silver mirror) test

The silver ion Ag^+ oxidises aldehydes in non-acidic media, but if conditions become too alkaline the silver ion precipitates as an insoluble oxide. To keep the silver in solution it is combined with a pair of ammonia ligands, to produce a complex which is in equilibrium with a very low concentration of free silver ions. The resulting mixture is called **Tollen's reagent**, also referred to as ammoniacal silver nitrate. A suspected aldehyde is warmed in a test-tube with Tollen's reagent, and, perhaps because the low free Ag^+ concentration induces slow orderly crystallisation, the silver metal thus liberated deposits itself as a *mirror* on the test-tube:

$$[Ag(NH_3)_2]^+(aq) + CH_3C\begin{matrix}O\\ \| \\ \\ H\end{matrix}(aq) + H_2O(l)$$

aldehyde

$$\downarrow$$

$$Ag(s) + CH_3C\begin{matrix}O\\ \| \\ \\ OH\end{matrix}(aq) + 2NH_4^+(aq)$$

mirror

carboxylic acid

Fehling's test

Fehling's solution is made up immediately prior to use, from two components. These are ordinary $Cu^{2+}(aq)$ ions and tartrate ions (acting as a complexing ligand). It is the resulting Cu(tartrate) complex which brings about the actual oxidation. (The reason for using this obscure version of copper(II) is the same as that in the silver mirror test, namely that the oxidation needs alkaline media to work, but that 'bare' $Cu^{2+}(aq)$ ions would precipitate as copper(II) hydroxide above pH 7.) If an aldehyde is present it will be oxidised by boiling-hot Fehling's solution to a characteristic reddish-orange precipitate of copper(I) oxide. This test came to chemistry from biology, where it was developed as a test for certain sugars, which, as we shall see, have structural similarities to aldehydes and ketones.

To emphasise the point, ketones (unlike aldehydes) will give *negative* results to both Tollen's and Fehling's tests. They can only be oxidised with difficulty, and then only at the expense of the destruction of the carbon framework of the molecule.

Question

13 Can you put into words why it is that aldehydes and ketones differ so sharply in their vulnerability to oxidation? (Your answer should refer to the position of the $C=O$ group in the respective molecules.)

Fehling's test and Tollen's test are part of the repertoire of the analytical chemist, and the next question offers you a realistic context in which they might be used. There is a reaction called **ozonolysis** (meaning 'breaking with ozone') which cleaves an alkene at its $C=C$ double bond, leaving the 'broken ends' of the break as carbonyl groups (see Section 10.14):

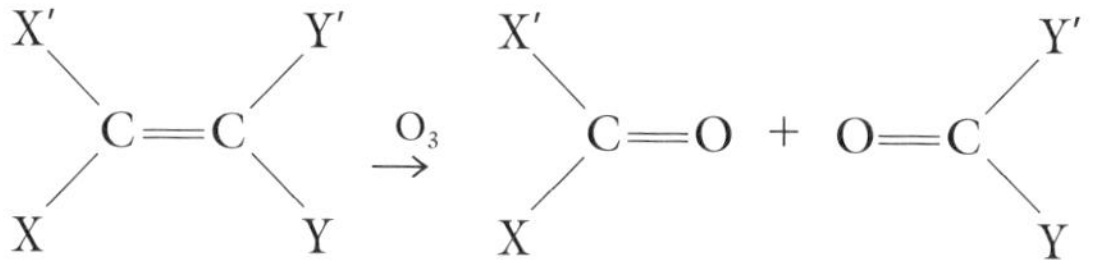

Ozonolysis is often used in conjunction with either a Fehling's or a Tollen's test, with the overall aim of detecting the site of the original $C=C$ bond.

Question

14 A hydrocarbon A has the molecular formula C_4H_8. It is subjected to ozonolysis, whereupon it cleaves into compounds B, C_3H_6O and C, CH_2O. C gives a positive Fehling's test, but B gives a negative one. Both B and C give yellow precipitates with Brady's reagent. Identify the compounds A, B and C.

13.6 Other reactions of aldehydes and ketones

Halogenation

We have already found that carbonyl compounds do not undergo electrophilic addition in the manner of alkenes. In the specific case of halogenation, one factor which would blight such an outcome as shown below is the weakness of the O—X bond.

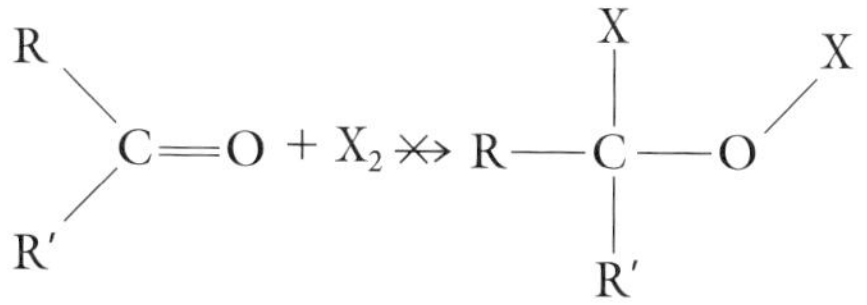

where X = halogen atom

However, carbonyls *do* react with halogens, and one such reaction has a very interesting mechanism which is widely studied at Advanced level. Mechanistic details can be postponed until Chapter 14: for now we will simply focus on the outcome, a typical example of which is shown below.

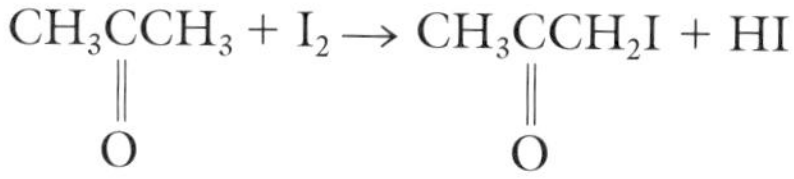

$$CH_3CCH_3 + I_2 \rightarrow CH_3CCH_2I + HI$$
$$\quad\; \| \qquad\qquad\qquad \|$$
$$\quad\; O \qquad\qquad\qquad O$$

Question

15 Oestrone (below) is a female sex hormone.

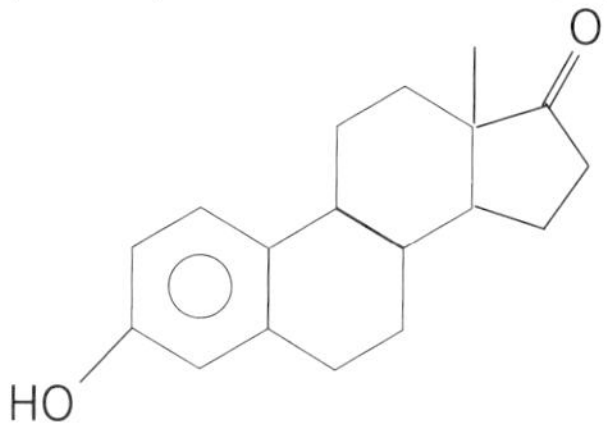

Apply the ideas shown in the simple example above to predict the outcome of the reaction of iodine with oestrone.

The triiodomethane (iodoform) reaction

This is a real 'stand-alone' curiosity of a reaction. The inorganic reagent is a mixture of iodine and dilute potassium hydroxide. You may remember from Section 7.6 that halogens undergo disproportionation in these conditions, so the solution would contain the IO^- ion. This ion reacts with any **methyl ketone**, to produce another of those characteristic yellow precipitates, this time triiodomethane.

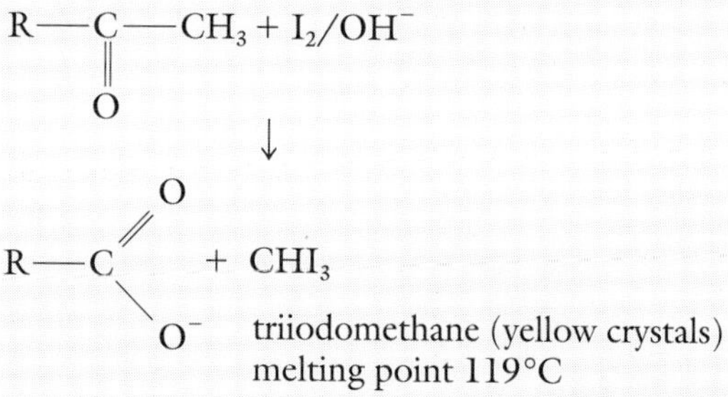

Because the IO^- ion is a strong oxidising agent, it will also generate a positive triiodomethane test in any compound which itself may be *oxidised to* a methyl ketone, as well as with the ketone itself.

Question

16 Give an example of a compound, not itself a methyl ketone, which will nevertheless give a positive triiodomethane test.

13.7   Reactions of aldehydes and ketones – a summary

1   Addition of cyanide – cyanohydrin formation

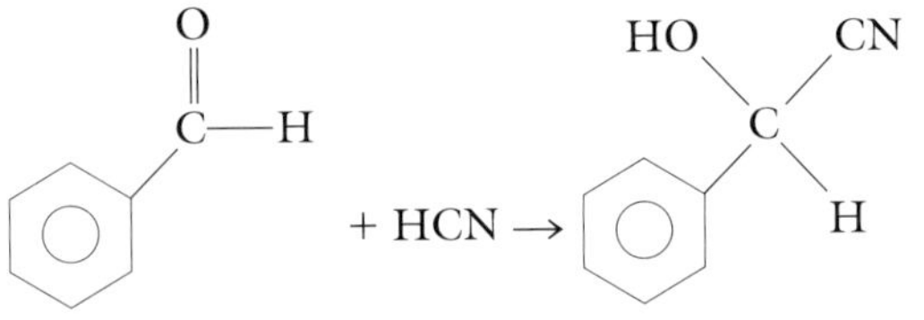

benzaldehyde   a cyanohydrin

2   Addition/elimination of hydrazine and of Brady's reagent

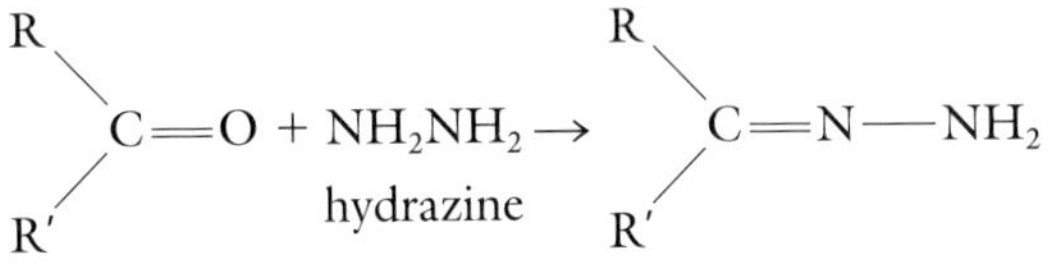

hydrazine

a hydrazone

Brady's reagent

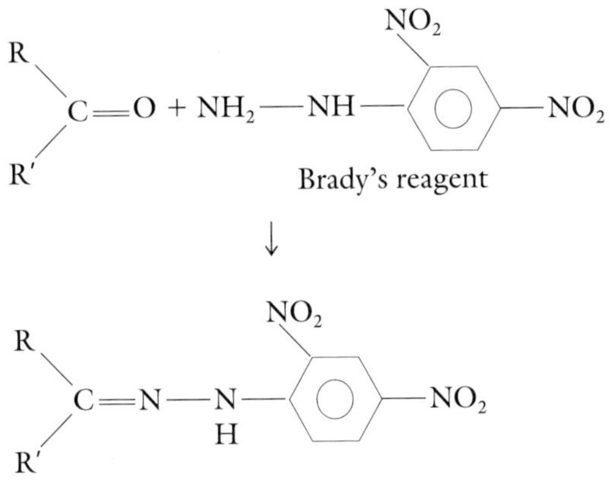

yellow or orange crystalline solid with sharp characteristic melting point

3   Oxidation

a   Aldehydes

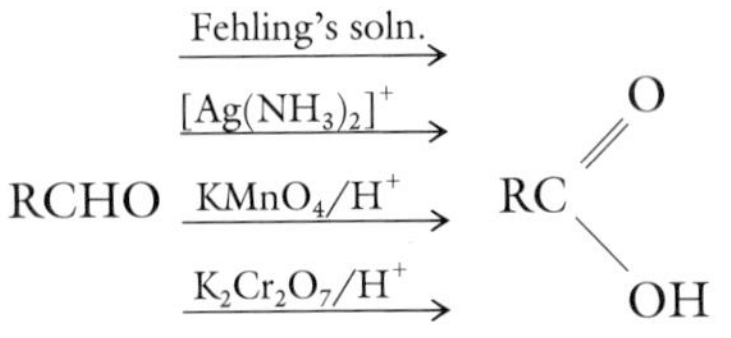

b   Methyl ketones

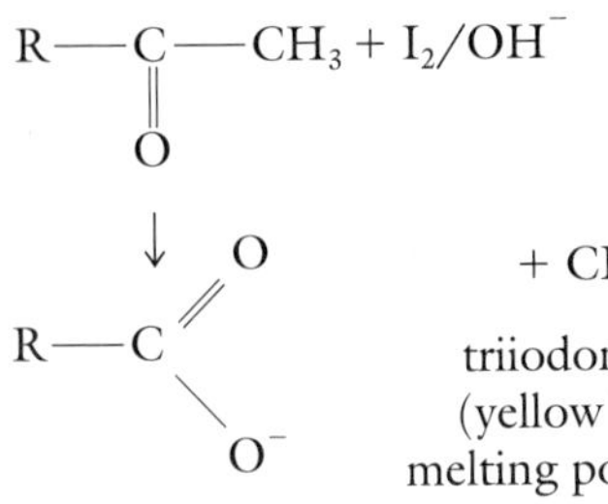

triiodomethane
(yellow crystals)
melting point $119\,^{\circ}C$

4   Reduction

a   Aldehydes

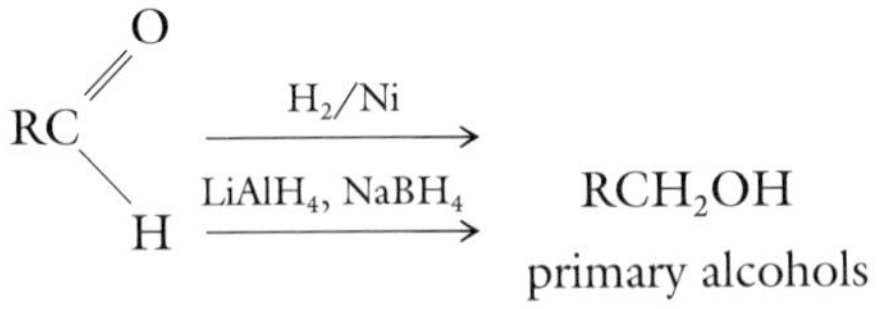

primary alcohols

b   Ketones

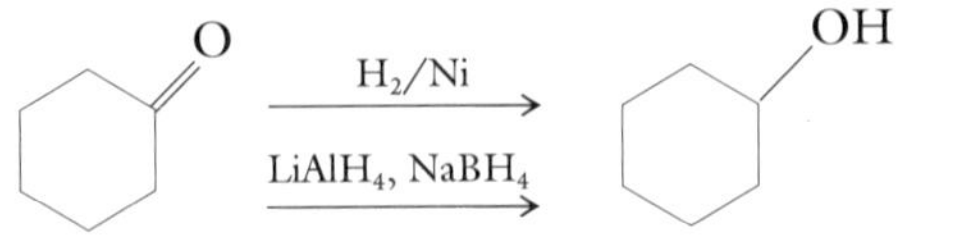

secondary alcohols

13.8   Carbohydrates – structures

Introduction

The Fehling's solution test of the previous section was an idea originally generated in biology. The group of biochemicals which was the focus of this test was the **carbohydrates**, some of which do indeed behave like aldehydes and ketones. But then equally they have structural links to the alcohols, and are best seen as a quite distinct group.

The family of carbohydrates includes sugars, glycogen, starch and cellulose. If we were looking for an example of the importance of carbohydrates, we could remind ourselves of the role they play in the physical reality of this book. Not only is the book *made* of a carbohydrate (**cellulose** in paper), but the energy you are using to turn its pages, and to activate your brain to understand it, comes from carbohydrates (**glucose** in your blood, from digestion of foods containing **starch**). If you are planning a last-ditch revision stint all through the night before your exam, you will certainly need to draw on yet another carbohydrate – **glycogen**, which acts as an energy store for long-term physical activity. It is stored in muscles and the liver.

Carbohydrate structures

The structures of carbohydrates can be discussed at various levels of increasing complexity and fullness. The simplest description sees them as a group of compounds sharing the elements carbon, hydrogen and oxygen and in which the numerical $H:O$ ratio is $2:1$. Hence they have the general formula $C_m(H_2O)_n$. The '-hydrate' part of their name comes from that ratio, indicating that they possess hydrogen and oxygen in the proportions necessary to form water. This fact is emphasised by the behaviour of carbohydrates towards powerful dehydrating agents. From the largest to the smallest, they are all reduced to a steaming mass of carbon by the action of, say, concentrated sulphuric acid.

Any further discussion of the chemical character of the carbohydrates requires a much fuller structural analysis and, indeed, requires us to break the subject up into subsections.

Monosaccharides

The story of carbohydrate structures is a story of building blocks and of the combinations produced by the linking together of those building blocks. However, several of the most significant carbohydrates consist of just a single building block called a **monosaccharide**. One monosaccharide is **glucose**, whose structure and properties will stand as typical of the whole class. The structure of glucose is shown in Figure 13.3, but as we shall see glucose defies attempts to describe it in any one single diagram, to an extent not met before in these organic chemistry chapters. When you try to describe the properties of glucose, there are often aspects that cannot be explained adequately without using *combinations of structures in equilibrium with each other.*

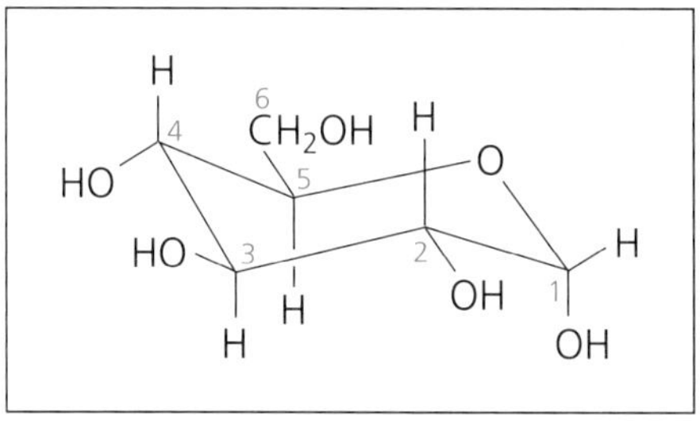

Figure 13.3 The structure of glucose – one of many. (This is the ring-form α-D(+) glucose.)

Properties due to hydrogen bonding

One set of properties of monosaccharides needs no subtle explanation: those that derive from the large number of OH groups within the glucose molecule. These give rise to multiple possibilities for hydrogen bonding, which accounts for the high solubility of monosaccharides in water, their high melting points relative to their molecular size, and the high viscosity of their aqueous solutions (Chapter 6).

Looking at glucose from an evolutionary perspective, it is possible to view the molecule as a tribute to the success of evolution as a *designing* force. The 'design brief' for the molecule is daunting indeed – on the one hand, the target molecule has to be *energy-rich* in order to perform its primary function as a fuel for respiration in nearly all living things. On the other hand, it must be *capable of being transported* in aqueous environments like blood, sap and cell fluids. In the normal way of chemistry, those two

abilities are incompatible; the first is the province of such molecules as hydrocarbons, and the second is more typical of ionic salts. Yet, in glucose, evolution has come up with a marvellous 'Trojan horse' of a molecule, presenting a water-friendly exterior within which is a skeleton similar to a cycloalkane.

Glucose as an aldehyde

It is hard to see how the molecule in Figure 13.3 could behave as an aldehyde – and yet that is what it does, in its reactions with hydrogen cyanide and phenylhydrazine. The explanation for this puzzling behaviour is supplied by the 'auto-equilibrium' idea. It is thought that the glucose molecule spends some of its time in an open-chain, aldehyde, form, and that the interconversion between the open-chain and ring (cyclic) forms is brought about by nucleophilic addition, of a sort typical of carbonyl compounds.

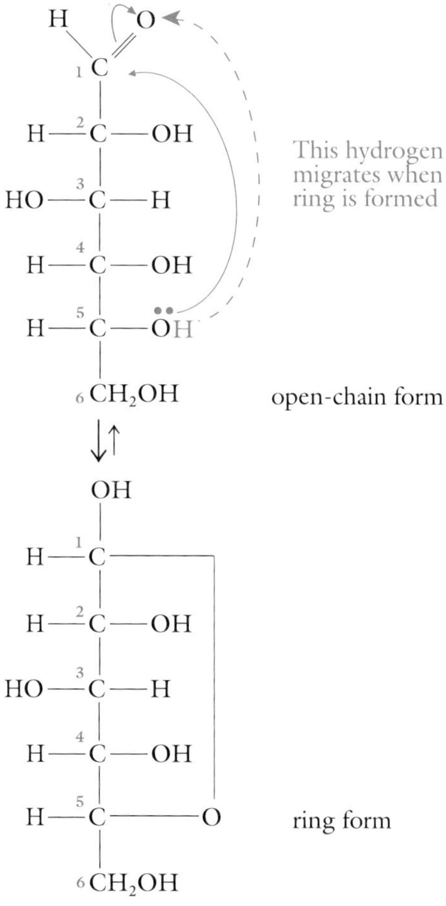

The OH group on the C5 carbon atom of the open-chain form nucleophilically attacks the aldehyde group, thereby pulling the molecule into a ring. In com-

pleting the addition, it offers its hydrogen to the O end of the former C=O bond.

Questions

17 If, as suggested in the last equation, the equilibrium between the ring and the open-chain forms is heavily biased towards the ring form, how is it that aldehyde-style reactions of glucose, such as its reaction with phenylhydrazine, go to completion?

18 The figure below shows the open-chain structure of another commonly occurring monosaccharide, fructose.

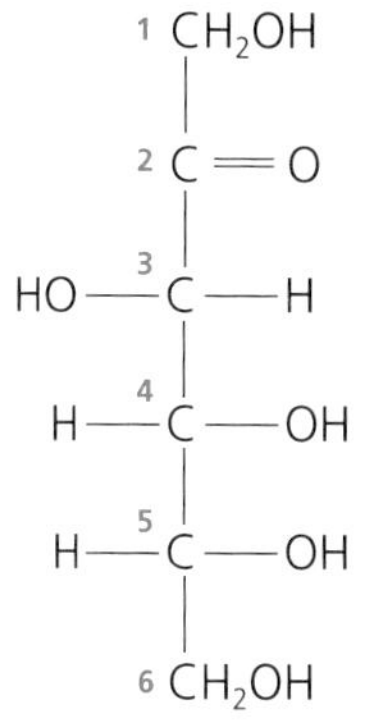

The reaction to produce its ring form is a nucleophilic attack on the C=O group by the OH on the C5 carbon atom. Draw the ring structure, numbering the carbon atoms so as to correlate them with those in the figure above.

19 Which of the following tests would you expect to discriminate between glucose and fructose, and why?

a Fehling's test.

b Tollen's test.

c Brady's reagent test.

Another important characteristic of carbohydrate molecules is their interaction with 'polarised light'. We will need to take a short detour into the science of light itself, if we are to gain some appreciation of this phenomenon.

Chirality

It is extremely difficult to present a convincing mental picture of the nature of light. We have the undeniable evidence of what light *does* (diffract, interfere, etc.), and some rather abstract forms of words such as 'an oscillating magnetic field at right angles to an oscillating electrical one' or 'wave/particle duality'. The teacher is left with slightly lame statements of the kind which begin, 'Light behaves as if . . .'

As a result we have only a rather crude picture of the phenomenon known as the **polarisation** of light. It is as if the wave nature of light were such that its end-on view was like Figure 13.4, in which each ↔ indicates a plane in which an oscillation might take place, and there is really an infinite number of those planes. In other words, the oscillations take place in every conceivable plane intersecting the line of propagation. Certain crystals have the ability to act like a 'letter-box-cum-

filter', producing an output beam in which all oscillations are in *one* plane. In this condition, the light beam is referred to as 'plane polarised' (as in the middle section of Figure 13.5).

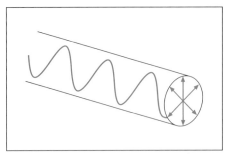

Figure 13.4 A symbolic representation of unpolarised light

If a second filter crystal is now placed in the beam, with its permitted 'letter-box' plane at 90° to that of the first, *no light emerges*. But if certain chemical substances, in solution, are interposed between the first and the second 'letter-boxes', the situation is altered and light gets through. What is more, a new angle can be found for blackout (Figure 13.6). It can only be that the interposed solution has *rotated the plane of polarisation* of the plane-polarised light emerging from the first letter-box. An instrument which is able to measure the angles of rotation produced by various substances is known as a **polarimeter**.

The mysterious causes behind all these effects need not trouble the A-level organic chemistry student, any more than an understanding of gearboxes is necessary to a car driver. What concerns chemists is the information conferred by the discovery that certain molecules rotate the **plane of polarisation of plane-polarised light** (which will be abbreviated to POPOPPL). For a start, the **specific rotation** (that is, the angle produced by a given depth of solution of the test substance at a standard concentration) is a constant unique to that substance (like, say, a refractive index). More importantly still, the ability to rotate the POPOPPL reveals something about the degree of **asymmetry** of the molecule in question.

Specifically, optical rotation ability is possessed by molecules which are asymmetrical to the extent that *a molecule and its mirror image are not the same molecule* – in other words, molecules which can only be formed from their mirror-image molecules by bond breaking and re-making.

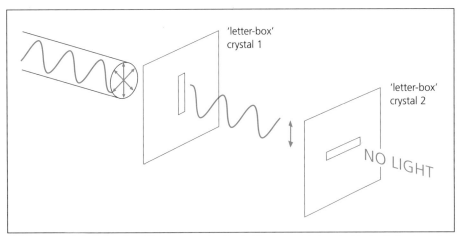

Figure 13.5 Blackout produced by two crystals of polarising material placed with the 'letter-boxes' at 90°

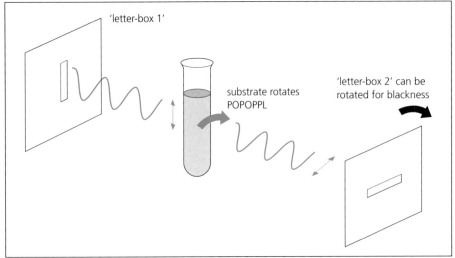

Figure 13.6 Principle of a polarimeter. The rotation of the second 'letter-box' to restore 'blackout' tells you how much the sample has rotated the plane of polarisation of plane-polarised light

Question

20 Draw the mirror images of the following molecules. Which molecule(s) could only be converted into their mirror images by bond-breaking?

ethene

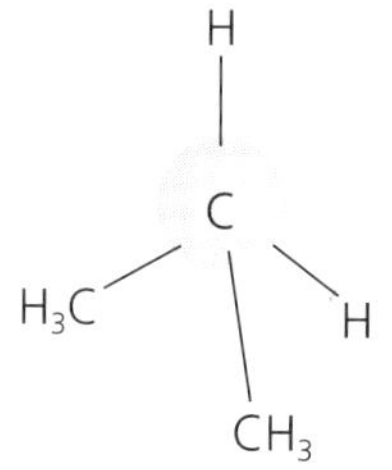

propane

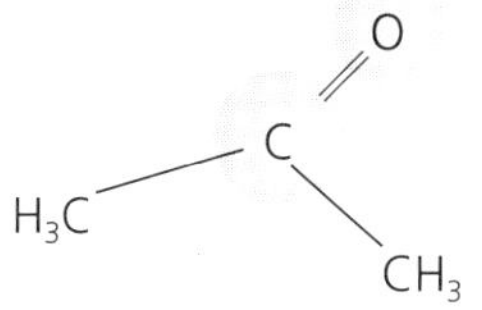

propanone 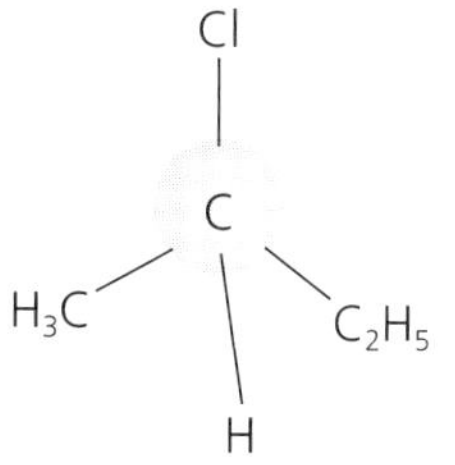   2-chlorobutane

D(+)   L(−)

As your answers to question 20 should have showed, the only carbon compounds which pass the 'non-superimposable mirror-images' test, and which are therefore capable of rotating the POPOPPL, are those which possess *four different groups on a single carbon*. This is clearly a new kind of isomerism, since the molecular formulae of two such mirror-image molecules are bound to be the same. This form of isomerism is a sub-set of **stereoisomerism** (p. 208), in that the differences reside not in the nature of the bonds present, but only in the arrangement of atoms in space. Unfortunately for the student, the field abounds with duplicate terminology: thus pairs of isomers are referred to as **enantiomers** or **optical isomers**, the property possessed by the molecules is called **optical activity** or **chirality**, and the carbon at the heart of the asymmetry is called a **chiral centre** or an **asymmetric carbon** atom.

Returning, after that rather long diversion into the nature of chirality, to the subject of monosaccharides, we can see that both glucose and fructose abound with chiral centres.

Question

21 Draw the open-chain form of glucose, and put an asterisk on any carbon atom which has four different groups attached to it.

Bewilderingly, since every chiral centre produces two isomers, and because there are four chiral centres, there are 2^4 or 16 versions of glucose, all with nearly identical chemistry. Fortunately, only *one* exists in Nature. It is designated D(+) glucose. The D derives from the word 'dextro' for right-handed and carries information of a structural nature which is not in the Advanced syllabus, while the + sign means that this particular isomer rotates the plane of polarisation *clockwise*. The figure in the next column shows D(+) glucose, and its enantiomer L(−) glucose in which all four chiral centres are reversed. This second isomer would rotate the POPOPPL anti-clockwise by an angle equal to the clockwise rotation of D(+) glucose.

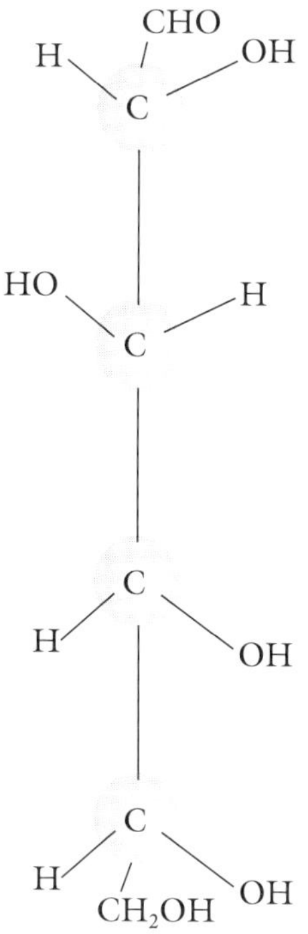

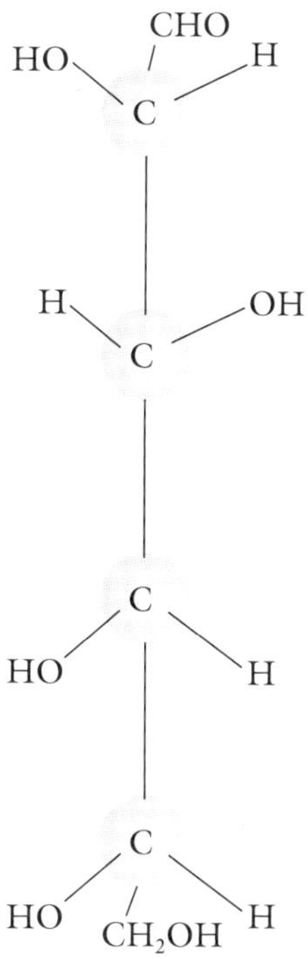

Polysaccharides

Glucose and fructose are the building blocks of carbohydrate chemistry. They are capable of being joined like links in a chain. The chains can be many hundreds of links long (a polysaccharide) or just two (a disaccharide), with most possibilities in between. The joining is brought about by the elimination of water between adjacent monosaccharide units (condensation). This process can be achieved in a number of ways, which may affect the aldehyde/ketone properties of the polymer, its water solubility, its vulnerability to attack by enzymes or any combination of those features. Some examples may make these points clearer.

Sucrose

This is the common sugar of the kitchen, from sugar cane and sugar beet plants. It is a disaccharide and a dimer of D(+) glucose and D(−) fructose (Figure 13.7). The bridge is in the 'α'

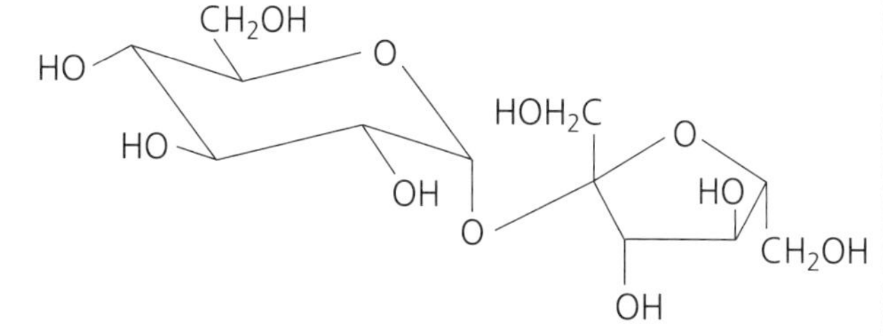

Figure 13.7 (+) Sucrose. (Ring hydrogen atoms are omitted for clarity.)

style, where the Greek letter indicates that the bridging oxygen atom is pointing *down* rather than *out* from the glucose ring (see Figure 13.8). The join is at a point which blocks off the possibility of either molecule reverting to the open-chain format, since the bridging oxygen is the one that would, in a monosaccharide, be carrying the 'migrating' hydrogen.

Question

22 Referring to sucrose's inability to open out in open-chain form, would you expect it to be able to reduce Tollen's or Fehling's reagents? Explain your answer.

Sucrose can be hydrolysed into its monosaccharides by boiling with dilute acid, or under the action of the enzyme **invertase**. This enzyme is possessed by bees, who use it on the sucrose they collect. Honey is mostly **invert sugar**, the name coming from the fact that sucrose has a clockwise rotation of plane-polarised light, whereas the split-up mixture of glucose and fructose is dominated by the larger anti-clockwise rotation of fructose (see below).

$$C_{12}H_{22}O_{11} + H_2O \longrightarrow C_6H_{12}O_6 + C_6H_{12}O_6$$

sucrose	D(+) glucose	D(−) fructose
+66.4	+52.7	−92

(Specific rotations denoted by the numbers)

Starch

Starch is one of the two great energy-storage chemicals of the plant world (fats being the other). Acid hydrolysis, or the enzyme-catalysed equivalent, breaks starch down into its component monosaccharides, and thereby reveals that it is composed entirely of D(+) glucose units. But there is much more to know about polysaccharide structures than the monomer units which make them up. For instance, cellulose is another polysaccharide, also made entirely of D(+) glucose units, yet it is many times more difficult to hydrolyse/digest and is used within the plant for an entirely different purpose, namely as a structural framework.

It has been pointed out that plants and animals use starch and cellulose in analogous ways. Both plants and animals use starch as a biochemical energy source, and we all break it down into glucose in order to do so: and both animals and plants use cellulose for construction. In this second case the comparison has a different flavour – plants build *themselves* out of cellulose, whereas no animal *tissues* are cellulose-based. What we do is employ its physical strength in external structures like nests and houses.

Common sense dictates that if these two materials, starch and cellulose, are chains made of the same monomer units, then any differences between them must reside in the method of linkage.

The differences look surprisingly slight, when seen on the page (Figure 13.8). The contrast resides in nothing more dramatic than the choice of an α (axial, pointing down) or β (equatorial, pointing out) orientation of the bond to the oxygen atom in the bridge linking the rings – all starch links are

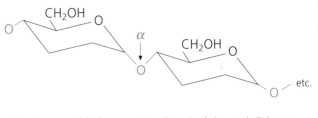

Starch: many D(+) glucose units polymerised via α-style linkages. (many OH and H groups omitted)

Cellulose: many D(+) glucose units polymerised via β-style linkages. (The right-hand ring has been rotated so as to keep the chain straight-ish on the page.)

Figure 13.8 Starch and cellulose. (There is a second form of starch with a branched-chain format.)

α, all cellulose links are β. But what humans see as no more than trivial differences can be crucial to the action of enzymes. To the enzymes of the amylase family the α linkage in starch is a 'soft touch', exactly what they were made for, whereas the β linkage in cellulose is an impossibility. This should of course remind us of the 'lock-and-key' theory of enzyme activity, which asserts that intimacy of contact or fit between the enzyme and its substrate is of paramount importance (Figure 13.9), and which explains why enzymes are so particular about what seem to be small structural differences. A fair analogy would be the way that the two shoes of a pair look very similar to each other, but feel totally wrong when applied to the wrong feet.

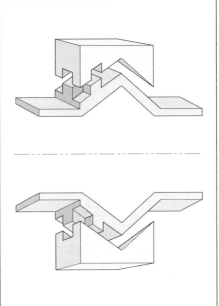

Figure 13.9 Showing how a shape (in colour, playing the enzyme) which fits over one substrate will not fit over the substrate's mirror image – providing that both enzyme and substrate are themselves chiral. Although starch and cellulose are not actually mirror images of each other, a similar principle applies, whereby the success or failure of the enzyme action is dependent on what looks to us like fine detail

A footnote on fibre

As you read this book, your body is processing both these crucial polysaccharides. In the case of cellulose, 'processing' means little more than 'ushering through'. And yet even this non-use is useful, in that dietary fibre (as cellulose is called in a nutritional context) has been shown to be necessary to the healthy function of the colon. The incidence of cancer of the colon, for example, is far greater in societies in which the intake of dietary fibre is low.

Question

23 Biology quiz department.

a The appendix in humans is a useless part of our alimentary canals, relevant to some of our mammalian forebears, but now just a leftover of evolutionary progress. What is it used for in those mammals where it still has a significant job to do?

b Grazing animals have rather more need than we have to make some use of the hard-to-digest cellulose. This is because cellulose makes up not only the wood of stems but also the cell walls of all plant cells. What strategy(ies) is (are) employed by members of the cattle family to break down cellulose?

13.9 Carboxylic acids – the origins of their acidity

Introduction

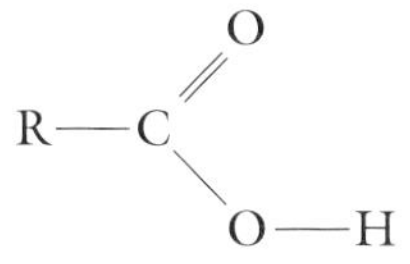

$$R—C\overset{\displaystyle O}{\underset{\displaystyle O—H}{}}$$

The generalised formula of a carboxylic acid is shown above. It is immediately obvious that the group contains *two* functional groups, namely the OH group (as in alcohols) and the C=O group (as in aldehydes and ketones). So the main focus of interest will be on how these two neighbouring functional groups affect each other. In fact, it turns out that the 'alcohol' side of a carboxylic acid's character seems to dominate, and we will spend more time considering how the carboxylic acids behave like modified alcohols than on considering their similarities with aldehydes or ketones.

And yet in some ways (and here is an analogy that looks rather extravagant, but which gives quite a true, if highly fanciful, picture) the carbonyl group acts like the sly child in the playground fight, taking hardly any part in any of the action directly, but having a profound effect behind the scenes on the group which gets 'beaten up', namely the OH group.

The origins of the acidity of carboxylic acids

We have met OH groups which are minutely, just discernibly, acidic, as in the conventional alcohols (even though, as we noted on p. 245, the sodium-metal reaction is really just as much redox as acid–base). Then again there are OH groups in phenols which just about manage to register on the pH scale, and can be neutralised by sodium hydroxide. Finally, we now come to OH groups in the carboxylic acids, which far surpass the phenols in acidity, even to the extent that they can do tricks more closely associated with the inorganic acids.

Question

24 Use your data book to complete Table 13.3. Remember that the table works on the principle that acids higher up in the table can give protons to bases lower down in the table (Law of 'clockwise circles') (p. 263). Armed with these ideas, can you see a familiar test for acids, which ethanoic acid will 'pass' but phenol will 'fail'?

Table 13.3 The relative strengths of three acids

Acid	Equation	pK_a
ethanoic	$CH_3CO_2H \rightleftharpoons CH_3CO_2^- + H^+$	
carbonic	$H_2O + CO_2 \rightleftharpoons H^+ + HCO_3^-$	
phenol	$C_6H_5OH \rightleftharpoons H^+ + C_6H_5O^-$	

Why are the carboxylic acids stronger acids than any other OH-containing molecules in organic chemistry? This is a classic case of behind-the-scenes interference by the C=O group: by tugging electron density away from the carbonyl carbon atom, the *double*-bonded oxygen sets up an altered polarity in the neighbouring C—O single bond. The effect is achieved by a pair of 'knock-ons': the unusually positively polarised carbonyl carbon does not let the *single*-bonded oxygen have its usual dominant share in the C—O bonding electrons, so the oxygen takes more than its usual share of the O—H bonding electrons. Thus, the O—H bond is even more polarised than it is in alcohols, and therefore heterolytic (acidic) breakage is made all the more likely. That is why the carboxylic acids are the strongest acids in the organic field. Below is the visual version of this explanation.

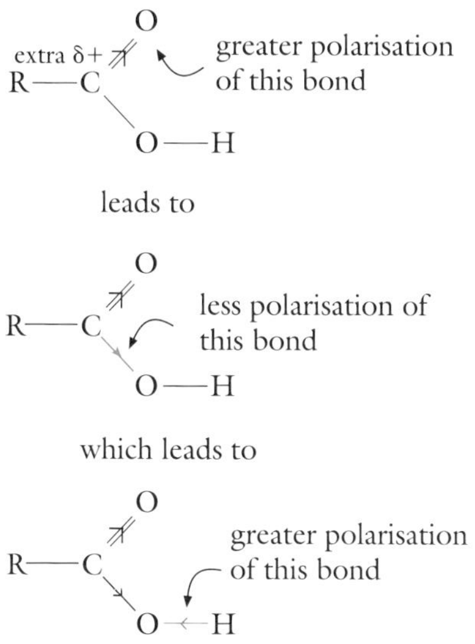

Question

25 In your book of data, find three close relatives of ethanoic acid which are stronger acids still. Identify them and explain their extra acidity, using bond-polarity arguments similar to those used in the previous paragraph.

Part of the typically acidic behaviour of the carboxylic acids is that they react with bases such as sodium hydroxide to form **salts** (for example, the ethanoates). These salts are for the most part perfectly respectable ionic compounds, dissolving in water, undergoing electrolysis, having high melting points, not smelling and generally seeming to belong more to inorganic than to organic chemistry. However, remember that the carboxylic acids are quite weak compared with the main inorganic acids, and that their salts are the salts of strong alkalis with weak acids (see Chapter 12, p. 270). So a solution of, say, sodium ethanoate is slightly alkaline (pH $\approx$ 8), and the ethanoate ion (the quite strong conjugate base of a quite weak conjugate acid) is willing to have the H^+ back again if it is offered it by a strong inorganic acid.

Questions

26 From the discussions above, deduce and write the equations for the reactions of ethanoic acid with sodium hydroxide, and of sodium ethanoate with hydrochloric acid. Also, if you can, write the equation for the equilibrium between the ethanoate ion and water which gives rise to the slight alkalinity of solutions of sodium ethanoate.

27 You already know how sodium metal reacts with water and ethanol. By considering the relative acidities of the three compounds, predict how sodium metal would react with ethanoic acid, in terms of both the actual equation and its relative degree of vigour.

13.10   Carboxylic acids in water

The carboxylic acids are all quite strongly hydrogen-bonded to each other in the pure liquid and the solid state. This accounts for their high melting and boiling temperatures which, comparing molecules of similar molecular mass, are even higher than those of the corresponding alcohols. (Pure ethanoic acid freezes at around 12 °C, so on cold mornings in January it sits in its bottle like a lump of ice. This property is the source of the old name 'glacial acetic acid' for 100% ethanoic acid.) The ability of carboxylic acids to hydrogen bond also explains their high degree of miscibility with water, at least until the carbon chain has more than about four carbon atoms. Beyond this point the story is the same as that of the alcohols, with van der Waals' bonding becoming more significant and with a resulting drop in water solubility.

There is one other quirk in the behaviour of carboxylic acids. For instance, solutions of ethanoic acid in non-polar solvents showed a solute molecular mass of 120, which is exactly double what was expected. Chemists realised that when carboxylic acids could not hydrogen bond to the solvent, they were liable to make hydrogen-bonded dimers with each other (shown below), and this explained the doubling of the molecular mass.

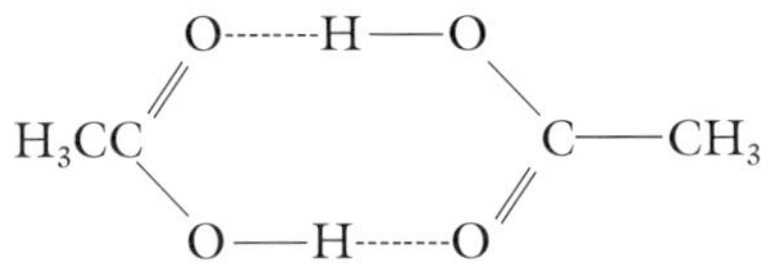

13.11   Carboxylic acids – nucleophilic substitution

From our studies of alcohols (Chapter 11), we know already that the C—O bond acts as an attracter of nucleophiles, and from our studies of aldehydes and ketones we know that the C=O bond does likewise.

However, if you compare the early sections of this chapter with relevant sections of Chapter 11, you will realise that they do not attract the *same* nucleophiles: a C—O bond attracts species such as hydrogen bromide, HBr, phosphorus(v) chloride, PCl_5, and thionyl chloride, $SOCl_2$, while a C=O bond prefers the attentions of the cyanide ion, CN^-, and hydrazine, H_2NNH_2. The question is, which set of nucleophiles is interested in a molecule containing *both* functional groups, or will it turn out that they can *all* do something?

The answer fits the pattern already mentioned (p. 293) – in other words, the C=O bond may play some background role in enhancing the reactivity of the C—O bond, but it is the latter which plays the active part. Accordingly, it is the phosphorus halides and thionyl chloride which react with carboxylic acids, and the result is straight substitution of the OH group by a halogen to give a class of compounds called **acyl halides** (of which more in Section 13.13):

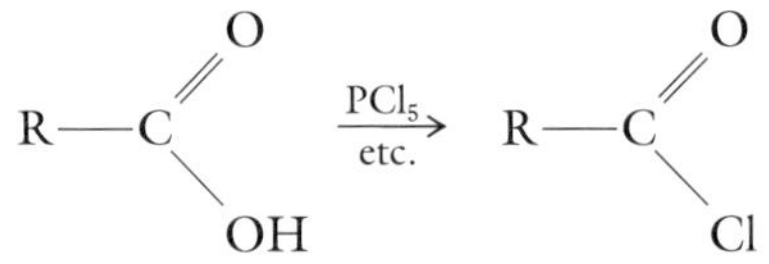

acyl halide

Question

28 Which of the following compounds will *not* react with PCl_5?

C_2H_5OH, C_6H_5OH, $(CH_3)_2C$=O, CH_3CO_2H.

There is one other situation in which a nucleophile attacks a carboxylic acid, and that is during the formation of **esters**. As was pointed out in the section on alcohols (p. 246), it is a lone pair on the oxygen atom of the alcohol which initiates the attack, and the OH group of the acid which leaves:

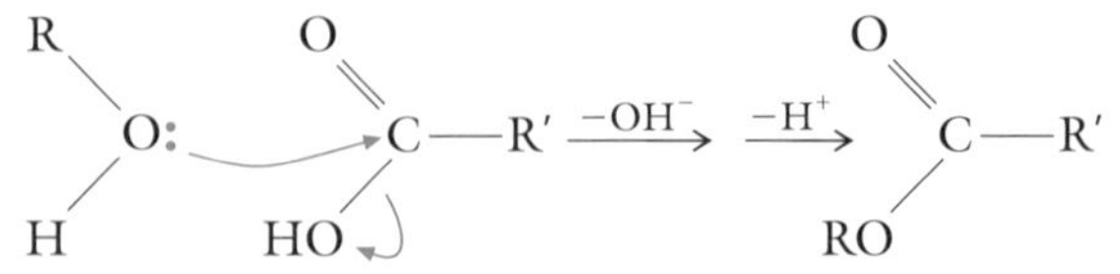

an ester

Question

29 The reaction between benzoic acid and methanol is shown below.

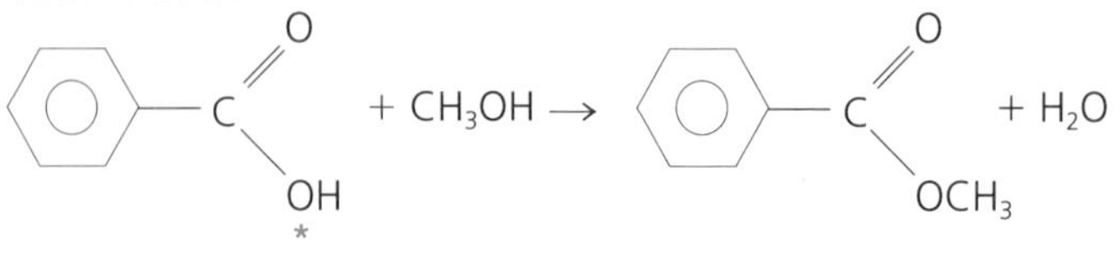

methyl benzoate

a One oxygen atom on the left-hand side of the equation has been asterisked. Show with a second asterisk where that oxygen atom will appear on the right-hand side.

b Which other pair of reactants, in which one or both reagents are different from those shown above, could also be used to produce methyl benzoate?

13.12 Carboxylic acids – reactions of the C=O group

'Reaction' in the singular might have been a better title for this section, since the carbonyl group in carboxylic acids has only a small repertoire. As already mentioned, most of the time it acts as a 'prompter' of the activities of its neighbour, the OH group. However, there is one example of carboxylic acid behaviour which has echoes from the chemistry of aldehydes and ketones, and that is the behaviour towards the *reducing agents* NaBH$_4$ and LiAlH$_4$ (Section 13.5). By means of these reagents it is possible to reduce carboxylic acids back down to primary alcohols, in a reaction which is an exact reversal of the action of prolonged refluxing of primary alcohols with acidified sodium dichromate. Even here, however, the mechanism suggested in the reaction scheme below has an initial attack by H$^-$ ions from the reducing agent kicking out the OH group, followed by a second attack on the aldehyde thus produced.

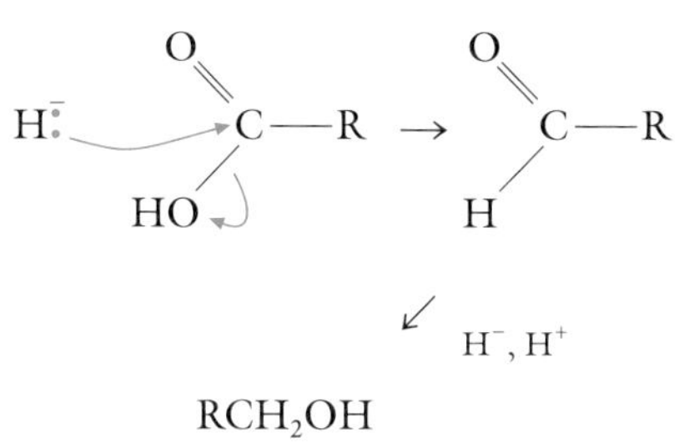

So the OH group in the final product is not the same OH group as in the original acid.

Question

30 How might the idea in the last sentence be tested experimentally?

As a footnote to this section, and as a tribute to the versatility of LiAlH$_4$-type reducing agents, note that these reagents will also reduce esters to alcohols.

Questions

31 Write the equation for this reaction, picking any ester as an example. (Treat the complex hydride LiAlH$_4$ only as a source of H$^-$ ions, as in the reaction shown above.)

32 This question looks back across the entire chemistry of carboxylic acids (and therefore, in this case, forces you to make your own reaction-summary list). State the product likely to be formed, if any, and where relevant the necessary conditions, in the reactions of ethanoic acid with:

a ethanol

b potassium dichromate(VI), K$_2$Cr$_2$O$_7$

c thionyl chloride, SOCl$_2$

d lithium tetrahydridoaluminate(III), LiAlH$_4$

e sodium hydrogencarbonate, NaHCO$_3$

f sodium hydroxide, NaOH.

13.13 Acyl halides

The name **acyl halide** is given to a molecule in which the OH group of a carboxylic acid is replaced by a halogen. In other words, we are talking about the products of the reaction of phosphorus halides or thionyl chloride with carboxylic acids.

The chemistry of acyl halides follows one single theme: one that we have used already to analyse the chemistry of the parent acids. That is: 'The carbonyl group provides the extra polarity to activate the C—X bond but remains untouched by the actual bond changes'. The upshot of this arrangement is to make the X group in the C—X bond of acyl halides one of the most willing and receptive 'leaving groups' in the field of nucleophilic substitution. Thus, a wide range of nucleophiles have been known to react with acyl halides, and in most cases the reactions are fast, complete (as opposed to reversible) and even somewhat violent. Nucleophiles which are too weak to perform substitutions with any other co-reagent can still manage to push out the halogen atom (as X$^-$) from acyl halides. You may remember the role of acyl halides in tempting phenols finally to reveal their alcohol-family connection and succeed in forming esters. In case you have forgotten, the next question relates to this and also asks you to predict outcomes when specific nucleophiles follow the general pattern of reaction outlined above.

Questions

33 Write equations for attacks by the following nucleophiles on ethanoyl chloride: H$_2$O, OH$^-$, ROH, RO$^-$, C$_6$H$_5$O$^-$, NH$_3$, RNH$_2$.

34 Write a three-step reaction scheme to produce ethanamide (CH$_3$CONH$_2$) from ethanol via ethanoyl chloride. If the yield of each step were 50%, how many moles of starting material would be needed to provide 0.1 mole of the amide?

35 Put the following reagents in order of their rate of production of a precipitate when shaken with aqueous silver nitrate, AgNO$_3$ (fastest first): chlorobenzene, 2-chloro-2-methylpropane, butanoyl chloride, 1-chlorobutane. Justify your answer.

36 Give the structural formula of the most probable product of the reaction of the two compounds shown below.

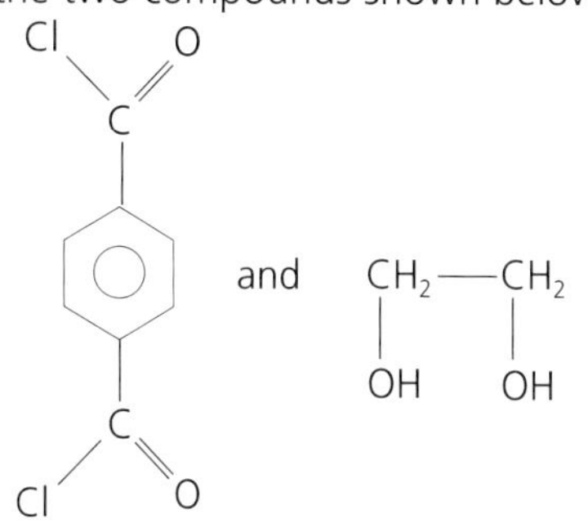

13.14 Esters

In a chapter full of unpleasant smells – the vinegar smell of ethanoic acid, the rancidity of butanoic acid, the violent assault on one's mucous membranes mounted by the acyl halides – the esters come as a wonderful relief. Within their selection of smells they can summon up roses, peardrops, Cox's Orange Pippins.

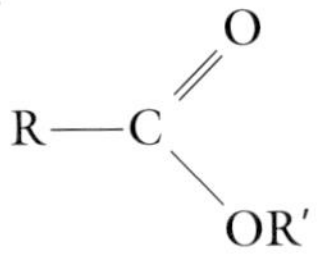

generalised ester
formula

Even the esters of benzoic acid provide the interesting (if not exactly pleasant) smell of a changing room before a sports event – methyl 2-hydroxybenzoate (methyl salicylate) (shown below), for example, is 'oil of wintergreen', an ingredient in liniment rubs such as Deep Heat.

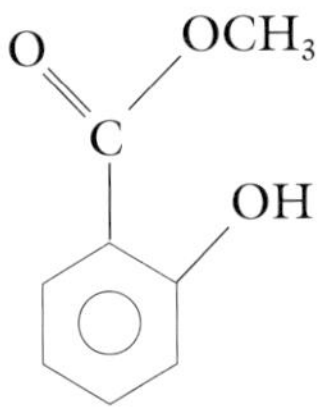

'Deep Heat'
(or oil of wintergreen)

Unfortunately, the only (A-level) reactions that esters actually undergo have the effect of converting them back to their smelly parents. One of these reactions is known as **ester hydrolysis**, which literally means the breakage of an ester by water; it gives the acid and alcohol from which the ester was made. Water alone cannot do the job, however, and an acid catalyst is needed, for exactly parallel reasons to those applying in the ester-*making* process. The H^+ will be picked up by the single-bonded oxygen of the ester (held on a lone pair), and this will make the alcohol fragment a better leaving group (leaving as neutral ROH rather than as RO^-). The reaction is the exact reverse of esterification, which means of course that it is an equilibrium:

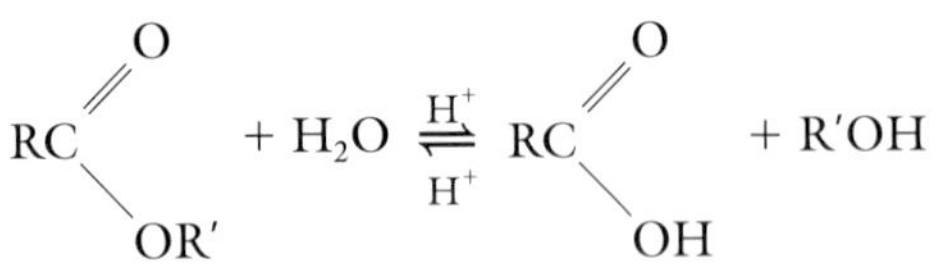

Question

37 Show the acid-catalysed hydrolysis of an ester in terms of structural formulae and curly arrows. Also show clearly where the catalytic H^+ becomes attached and at what stage it is released again.

A better way of carrying out ester hydrolysis, one that does not result in an equilibrium but instead goes all the way, is to carry out the reaction in alkaline conditions. This method sac-

rifices the catalytic power of the H^+ ion, but it more than compensates in two other ways: first, it employs the more powerful nucleophile OH^- instead of water and, second, the acid fragment produced by the break-up of the ester does not appear as itself, but as its anion, in association with, say, Na^+.

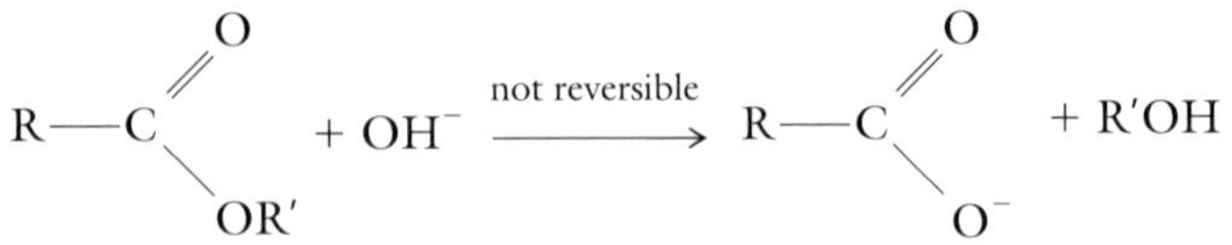

The anion of a carboxylic acid is less prone to attack by alcohol than the parent acid, and so it is not tempted to reverse the hydrolysis. The reason for this stability of the carboxylate anion, compared with the equivalent carboxylic acid, is the anion's *delocalisation*:

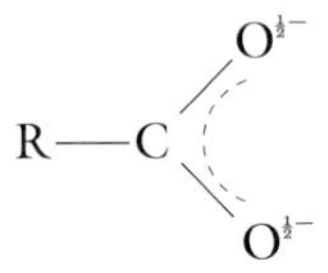

13.15 The ester linkage in Nature – fats

We have already seen, in Chapter 6 on intermolecular bonds, that Nature needs molecules in kit form – molecules that can be attached and detached to order. When assembly and subsequent detachment of the pieces of the kit has to be quick and easy, then weak intermolecular forces are ideal, as we saw in the sections on taste (p. 117), on DNA (p. 114), and on enzyme/substrate links (p. 115). When Nature needs to hold the kit components in rather stronger associations, and yet still be able to snap off the pieces when necessary, the choice is often to use the ester linkage as the joining system. This section is about a class of chemicals which offer a prime example of this principle in action.

Not the sweetest smelling but certainly the most important esters, in terms of everyday life, are the **fats**. They are one of the 'big three' classes of compounds involved in nutrition (the one responsible for long-term stored energy), and people interested in diet pay enormous attention to the amount and type of fats in foods. In addition, fats are the precursors of many important industrial and domestic compounds including soaps, paint-drying oils and plasticisers (compounds which change the degree of flexibility of plastics, see p. 241).

They are a class of esters which *all share the same 'alcohol part'*, and they occur in a wide range of both plants and animals. They are all esters of the trihydric (meaning three OHs) alcohol **glycerol** (propane-1,2,3-triol, shown below).

$$CH_2OH$$
$$|$$
$$CHOH$$
$$|$$
$$CH_2OH$$

The various acid parts show much greater diversity, but are nevertheless limited by some interesting natural 'rules'.

1 The hydrocarbon part of the acid is straight-chain.
2 There are eight or more carbons in the acid part as a whole (that is, including the C in CO_2H).
3 In Nature, only *even* numbers of carbon atoms can occur.
4 Each chain may include up to three double bonds. These are the 'unsaturated fats' mentioned in margarine advertising. *Poly*unsaturated fats are the ones with more than one double bond in their carbon chains.
5 In nearly all cases the orientation of the double bonds is *cis*, as in Figure 13.10.

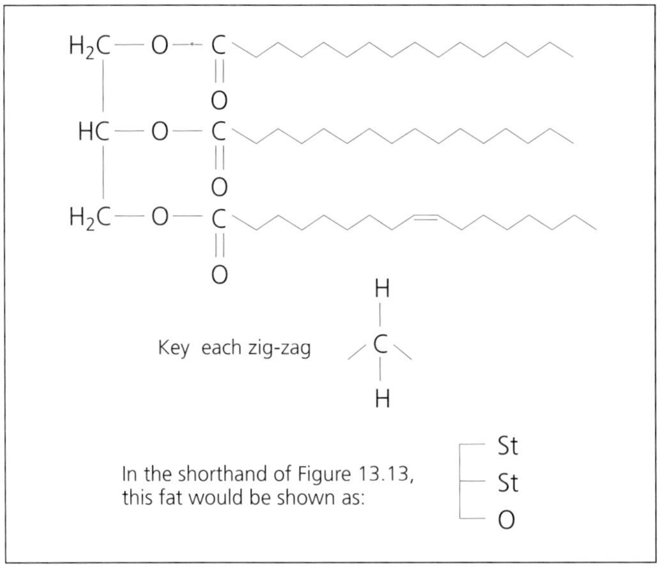

Figure 13.10 A typical fat. This one is the ester formed between glycerol, two C_{18} 'stearic acid' side-chains in the 1 and 2 positions on the glycerol and an unsaturated C_{18} 'oleic acid' side-chain at the 3 position

Nomenclature in 'fats' chemistry

Fats, as a group, are often called **triacylglycerols**. This is not a systematic name, since it does not conform to the 'ethyl ethanoate' method of naming esters, but it does bring out the fact that in all fats there are three acyl side-chains attached to glycerol. However, when a *specific* fat is named, normal rules of nomenclature are resumed. The fat whose computer-generated picture is shown in Figure 13.11 is thus called glyceryl tripalmitate. If the fat esters are cleaved, as happens often in their 'working lives', the cleavage is the normal one into acids and alcohol, and the acids are then referred to as **fatty acids**. While still in the uncleaved state, they're called fatty-acid side-chains or residues.

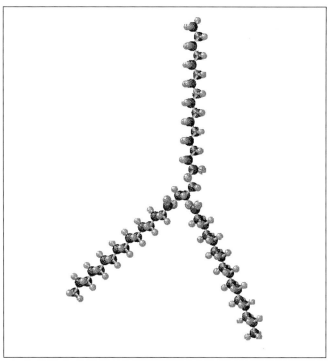

Figure 13.11 Glyceryl tripalmitate

Figure 13.12 shows some of the commonly occurring fatty acids. The abbreviations are used in one of the shorthand ways of representing fats. Figure 13.13a uses this shorthand to show a fat with three palmitic acid residues (as found in the oil from palm trees), while Figure 13.13b shows a different fat in which one of the palmitic side-chains has been replaced by

Trivial name	Systematic name	Abbreviation	Structural formula	Skeletal formula
lauric acid	dodecanoic acid	La	$CH_3(CH_2)_{10}COOH$	COOH
myristic acid	tetradecanoic acid	M	$CH_3(CH_2)_{12}COOH$	COOH
palmitic acid	hexadecanoic acid	P	$CH_3(CH_2)_{14}COOH$	COOH
stearic acid	octadecanoic acid	St	$CH_3(CH_2)_{16}COOH$	COOH
oleic acid	octadec-*cis*-9-enoic acid	O	$CH_3(CH_2)_7CH=CH(CH_2)_7COOH$	COOH
elaidic acid	octadec-*trans*-9-enoic acid	E	$CH_3(CH_2)_7CH=CH(CH_2)_7COOH$	COOH
linoleic acid	octadec-*cis*-9, *cis*-12-dienoic acid	L	$CH_3(CH_2)_4CH=CHCH_2CH=CH(CH_2)_7COOH$	COOH
linolenic acid	octadec-*cis*-9, *cis*-12, *cis*-15-trienoic acid	Le	$CH_3CH_2CH=CHCH_2CH=CHCH_2CH=CH(CH_2)_7COOH$	COOH

Figure 13.12 Structures of some common fatty acids

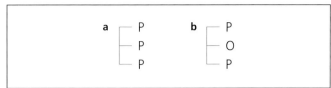

Figure 13.13 A shorthand way of representing fats

an oleic group. A similar symbolism can be used to record less detailed information – the letters S or U are used to indicate simply whether the side-chains are saturated (no double bonds) or unsaturated. For example, the fat in Figure 13.13b is shown in Figure 13.14 using the simpler symbolism.

Figure 13.14 Another shorthand, showing only whether the side-chains are saturated or unsaturated

Finally, in the context of fats chemistry, the word **oil** refers to a fat which is a liquid at room temperature.

What is the biological purpose of fats?

As far as human beings are concerned, there are several answers to that question.

1 As an energy store

Fats have the three main attributes of a good energy-storage chemical – they are rich in chemical energy, easily stored and quite easily mobilised *from* storage. The energy is released in many stages, but the overall reaction is combustion. The figures below show how much more energy is released from a C_{18} fatty-acid side-chain than from three C_6 monosaccharides.

Fatty acid:

$$C_{18}H_{36}O_2 + 26O_2 \rightarrow 18CO_2 + 18H_2O$$

$$\Delta H = -11\,210 \text{ kJ mol}^{-1}$$

Carbohydrate:

$$3C_6H_{12}O_6 + 18O_2 \rightarrow 18CO_2 + 18H_2O$$

$$\Delta H = -8448 \text{ kJ mol}^{-1}$$

Question

38 Why is it meaningful to compare fats with carbohydrates in this context?

As for the storage aspect, our bodies contain specialised **adipose tissue**, up to 90% of whose mass can consist of stored fat droplets. This tissue is situated under the skin and significant amounts are found in the region of the abdomen in men and the hips, thighs and mammary glands in women. When the body needs the energy stored in the fats, hormones trigger the release of enzymes to hydrolyse the fats at the ester linkage, releasing fatty acids into the bloodstream.

Question

39 Write an equation for the overall hydrolysis of a triacylglycerol (you need not be too specific about the exact identity of the fatty-acid side-chains). Bear in mind that triacylglycerols are esters, and that the result of the hydrolysis is to release free fatty acids.

(As a footnote, it has to be said that many people find that the storage of fats is easier to achieve than the release!)

2 As a connective tissue

If you have ever dissected kidneys in biology, you may remember that quite a large amount of fat adheres to them (Figure 13.15). The purpose of this fat is to provide support and protection for the organ within the abdominal cavity.

Figure 13.15 The fat around the kidney acts as a protection and support

Muscles, too, are often well supplied with fatty connective tissue, especially obvious in certain joints of meat, such as shoulder of lamb.

3 As a precursor to other biomolecules

Certain fatty-acid side-chains are the raw materials for biosynthesis within the body. Some are used for the construction of cell membranes, which contain esters called **phospholipids**. In another area of biosynthesis, **linoleic acid** (Figure 13.16) is the precursor of a class of compounds called **prostaglandins**, which are implicated in a number of vital body functions, including matters as diverse as blood-pressure control and allergic response. We cannot synthesise linoleic acid, so it is an essential part of the human diet, a point worth remembering as a counterweight to the popular image of fats as the villains of the dietary scene.

Figure 13.16 Linoleic acid, an essential part of the human diet

Sources of fats for human use

Table 13.4 shows that in the worldwide production of fats, the dominant sources are plants rather than animals; and that the dominant use of fats is in the food industry. Plants and human beings have much in common in the way they use fats. The only differences are that plants have less need of fatty connective tissue, and are capable of synthesising fats from scratch, in keeping with their primary position in the food chain.

Table 13.4 Oils and fats: world production figures in millions of tonnes (1989/90)

Oils and fats	Annual amount produced (millions of tonnes)
Industrial raw materials	
linseed, castor	1.0
tallow (beef fat)	6.6
Edible raw materials	
Vegetable oils	
soya	16.1
palm	10.9
sunflower	7.7
rapeseed	8.0
coconut	3.2
palm kernel	1.4
groundnut	3.8
cottonseed	3.6
olive	1.7
other	2.0
Fish oils	1.4
Animal fats	
butter	6.3
lard	5.3

Question

40 The main purpose for which plants make their own fats has been omitted from the above paragraph. What do you think it is, and why do you think fats are nearly always found concentrated in the *seeds* of a plant (Figure 13.17)?

So fats perform a common function across large parts of the living world. But the basic triacylglycerol blueprint for a fat molecule offers enormous scope for variation – in chain length, in degree of unsaturation, in the relative positioning of the chains in the C1, C2 and C3 sites on the glycerol, and even in the *cis* or

Figure 13.17 An oil palm is an important source of oil

trans orientation of double bonds (Figure 13.12). It should not be a surprise that fats from different sources show some interesting contrasts (Figure 13.18). What is more, it is sometimes possible to get at least an inkling of why certain organisms tailor certain specific fat formulations for themselves.

The next question asks you to recognise and to seek the causes of some of the patterns in Figure 13.18.

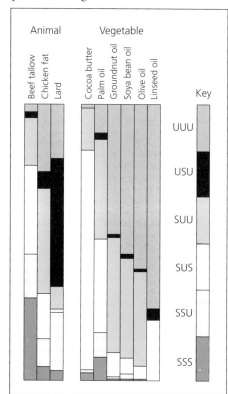
Figure 13.18 Triacylglycerol components of fats and oils, classified according to the position and degree of saturation of their fatty-acid residues. For example, UUU indicates a triacylglycerol in which the three fatty acids are unsaturated

Question

41a What is the general pattern linking the amount of 'triple-unsaturation' (UUU) with whether the source is animal or vegetable?

b Bearing in mind that nearly all the double bonds in these U fats are *cis*, suggest how well the U side-chains will stack together with side-chains from adjacent molecules, compared with the S side-chains (Figure 13.19).

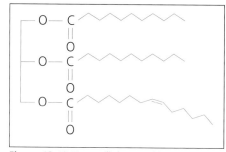
Figure 13.19 How will they stack?

c Suggest what effect this will have on melting temperatures, contrasting, say, a UUU fat with an SSS one.

d Suggest why plants have more need for many U groups in their fats (Figure 13.20), bearing in mind the different degrees of *temperature control* exercised by plants and by animals over their internal environments.

Figure 13.20 Plant-derived fats are usually liquid oils at room temperature, whereas animal-derived fats are usually solid

e Compared with the rest of the animal kingdom, fish have quite a high percentage of U groups in their body fats (Figure 13.21). Again with temperature in mind, offer a reason.

f The fat in Figure 13.22 is very likely to be SSS, and long chain. Can you say why?

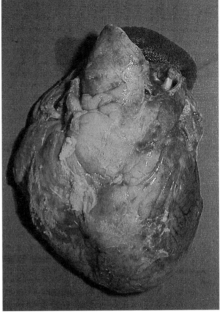

Figure 13.21 Fish oils have a higher proportion of unsaturated side-chains than most animal fats (see question 41e)

Figure 13.22 Animal fat, high in SSS components (see question 41f)

Comparing fats and carbohydrates

The way organisms and their biochemistries have evolved can call to mind the idea of a design and a designer. Sometimes it is illuminating to compare classes of chemical against the 'design brief' which covers their role – as we have already done in the case of sugars. In that case it was:

- an energy-rich molecule

- *but* must be quickly soluble in water

- *and* capable of existing in single-unit and macro-linked versions, for purposes of digestion and storage respectively

- *and* the single-unit and macro-linked versions must be readily interconvertible.

With fats the design brief takes in some of the same requirements, most obviously the one about being energy rich. The difference is one of emphasis, but the shift in emphasis gives rise to a sharply different final product. The emphasis with fats is on *maximising* energy content, and if that involves some sacrifice in the ease of distribution, then so be it. The design brief now says:

- the molecule must have as much hydrocarbon material as possible, compatible with absorption in an aqueous environment (blood)

- with this in mind, arrange the hydrocarbon fragments as three 'plug-in energy cartridges', which can be loaded on to a carrier backbone, and which can be unplugged for mobilisation round the body and re-attached for storage (Figure 13.23)

- when the cartridges become detached, give them a water-loving end to assist assimilation in aqueous environments (Figure 13.23)

- and arrange for some detergent-style molecules to assist in the digestion process

- adjust the shapes and lengths of the hydrocarbon chains to ensure the fats do not solidify within the organism – they will be hard enough to digest as emulsified liquids, let alone as hydrophobic waxy blobs.

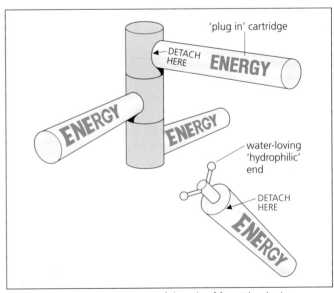

Figure 13.23 Fanciful impression of the role of fat molecules in supplying energy to organisms

The next question asks you to compare the actual features of fats with the design requirements set out in the brief.

Question

42a What is the 'carrier backbone' of a triacylglycerol?

b What are the 'plug-in energy cartridges'?

c Fats are esters – what features of the design brief are fulfilled by virtue of their *being* esters?

d What features of the fat molecule can be adjusted to set its melting temperature at a given value?

e To return to sugars (p. 289), why was the need for water solubility not entirely compatible with maximum energy from combustion? (In other words, what molecular pieces did sugars need to have that contributed nothing to the enthalpy of combustion?) What in contrast is the position in fat molecules with regard to non-contributing molecular pieces?

Here is another question that relates back to a feature of the design brief. It concerns the way fats are digested, or in other words how the molecule is dis-assembled prior to absorption.

Question

43 Figure 13.24 shows the parts of the digestive system which are involved in the digestion of fats. The gall bladder releases a detergent-like solution called bile, while the pancreas releases two chemicals, a pH adjuster (sodium hydrogencarbonate), and an ester-hydrolysis enzyme (lipase). The hydrolysis is not carried to completion, but leaves the former triacylglycerol as a monoacylglycerol, with the surviving side-chain in the 2 position (Figure 13.25).

a Very conveniently, the monoacylglycerol actually acts as a soap (or more correctly a detergent). If you look back to Chapter 6 (p. 111) you will remember that the molecules of soaps and detergents have 'dual personalities', so that one end can dissolve in non-polar and the other in polar solvents. Show how the molecule in Figure 13.25 could act as a detergent.

b If the products of the reaction are detergents, they can actually assist in the digestion of some of their undigested 'comrades'. They do this by causing the undigested fats to break into small droplets called 'micelles', which can disperse in the aqueous medium of the mixture (Figure 13.26, overleaf). What might happen to the rate of reaction as digestion proceeds?

c Suggest why pH adjustment of the overall digestion mixture is a good idea, relating your answer to conditions in the previous section of the alimentary canal and also to the influence of pH on ester hydrolysis.

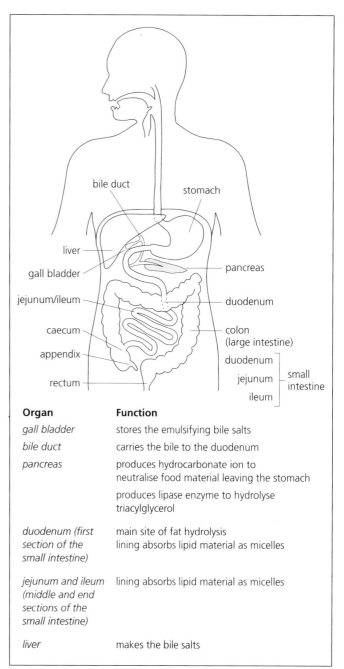

Organ	Function
gall bladder	stores the emulsifying bile salts
bile duct	carries the bile to the duodenum
pancreas	produces hydrocarbonate ion to neutralise food material leaving the stomach
	produces lipase enzyme to hydrolyse triacylglycerol
duodenum (first section of the small intestine)	main site of fat hydrolysis lining absorbs lipid material as micelles
jejunum and ileum (middle and end sections of the small intestine)	lining absorbs lipid material as micelles
liver	makes the bile salts

Figure 13.24 Fat digestion in humans

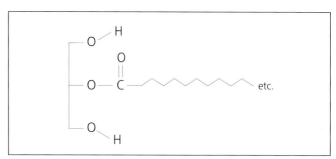

Figure 13.25 A partially hydrolysed fat – a 'monoacylglycerol'

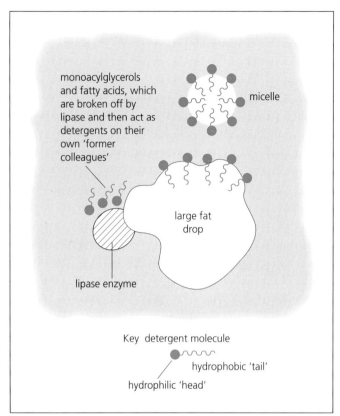

Figure 13.26 Simplified representation of a stage of fat digestion. The detergent molecules are actually the products of the first wave of fat digestion (supplemented by others from bile)

Figure 13.27 Various cooking oils

sites of vulnerability, especially to oxidation by atmospheric oxygen.

The fats industry

It is possible to divide up the fats we consume in modern life into three categories. Some are used more or less *straight from the source*, with extraction and purification being the only processing. *Cooking oils* would come into this category. Others are *lightly altered*, compared with the source material – perhaps by traditional physical churning, as in *butter*, or by more chemical means, as in the *margarine* and *chocolate* industries. Finally, there are products which are made by major alteration to the fat molecule, of which the most notable are *soaps* and *paints*. We can now look at each of these applications in a little more detail.

Cooking oils

Countries in which people eat high levels of *saturated* fats are also by and large those with the highest rates of coronary heart disease. Nowadays, public awareness of this correlation is quite high, and so the demand for *unsaturated* fats has gone up. The biggest-selling brands of cooking oil in British supermarkets are now those based on oils from sunflowers and rapeseed (Figure 13.27). The only problem with unsaturated oils is that the double bonds are

Question

44a There is now an unexpected echo of our work in Section 10.13 on rubber. Look back to the way that sulphur was used to cross-link rubber molecules, by extracting vulnerable hydrogen atoms. Therefore suggest how the fatty-acid side-chains in the oil in, for example, a chip pan, might become cross-linked, using a couple of linoleic acid side-chains as an example.

b Chip pans develop an unpleasant layer of a rather varnish-like gum (Figure 13.28). Suggest how this may have arisen.

c Hot cooking oils are not only at risk from cross-linking reactions but also from ester hydrolysis. The effects of this second reaction will be noticed by the chip-pan user because it causes the oil to smoke much more easily. Explain how the lowering of the smoke point might be caused by the hydrolysis.

Figure 13.28 The sticky gum formed by frying oils is due to polymerisation

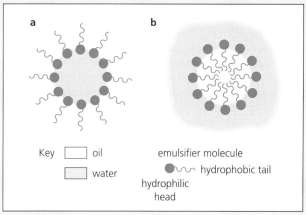

Figure 13.29 Two types of emulsion   a Water in oil   b Oil in water

Butter, margarine and low-fat spreads

Normal butters and margarines are water-in-oil emulsions (Figure 13.29). The tendency for water and oils to separate as two layers is opposed by edible **emulsifying agents**, which in their action could be thought of as edible detergents. Butters are made from the fats present in milk, while margarines use a proportion of oils from vegetable sources. As with cooking oils, the trend in margarines is towards sunflower and other vegetable oils, so as to increase the degree of unsaturation in the fatty-acid side-chains (Figure 13.30).

Question

45a E471 is a popular emulsifier in margarines. For example, it is used in a particular very-low-fat spread. It is actually a monoacylglycerol, called glyceryl monostearate (Figure 13.31). Why can it act as an emulsifier, and where else in this section have we come across monoacylglycerols acting in this role?

b If the margarine manufacturer goes too far in his or her drive towards unsaturation, the margarine has an unacceptably low melting point. Some quite tricky blending and compromising is needed. One of the ploys at the disposal of the margarine maker is to 'fill in' some, but not all, of the C=C bonds with hydrogen atoms. Suggest a chemical reaction which would achieve this aim, and explain why too great a proportion of unsaturated side-chains would give an over-runny product.

The very-low-fat spreads (Figure 13.32) have achieved amazingly low fat contents, around 25% (compared with up to 80% in traditional margarine). They have achieved it by

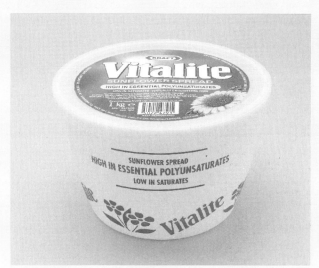

Figure 13.30 A margarine containing unsaturated fats

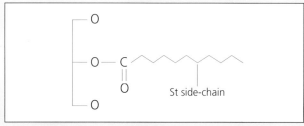

Figure 13.31 A margarine emulsifier: E471, glyceryl monostearate

Figure 13.32
A very-low-fat spread

making the water section of the emulsion into a gel, using gelling chemicals which act like nets and hold large numbers of water molecules in three-dimensional jumbled arrays. These gelling chemicals achieve two objectives – they keep the water content reasonably 'stiff', so that the emulsion does not physically collapse into a sloppy mess and, at the same time, they enable a higher proportion of water (and therefore a lower proportion of fat) to be built in to the emulsion. Next time you experience the rather cold watery 'mouthfeel' (a food industry jargon word) of low-fat spreads, you will know why.

Chocolate

Our discussion of edible fats would be incomplete without a mention of a product which brings so much pleasure (and guilt) to so many people (Figure 13.33). Chocolate is made from cocoa butter, which stands out in Figure 13.18 as unusual – no other listed vegetable oil has such a high percentage of saturated fats (they are the ones, as we have seen, with the higher melting points). That is why choco-

Figure 13.34 The orientation of a fat molecule which leads to the most 'crystalline' solid – as in chocolate (a, top) Model of fat molecule. (b, bottom) Electron density map

Figure 13.33 The delightful (some would say addictive) taste and texture of chocolate is mostly due to the 30% of its mass that is cocoa butter. Vulnerable readers may suddenly feel this is a good time for a break

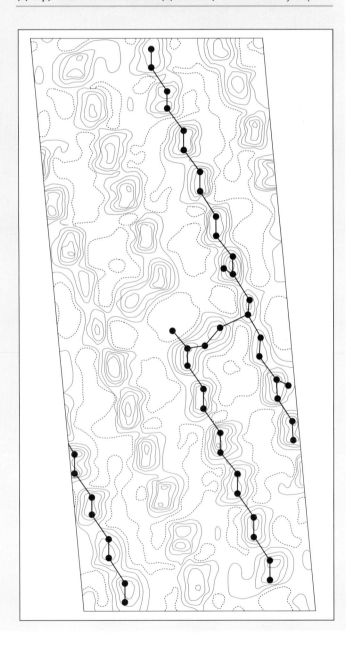

late is so pleasantly crisp at just below room temperature. But surprisingly, the chocolate maker cannot take this crispness for granted. It depends on how the chocolate molecules arrange themselves while cooling from the molten state, in a way which bears some strong comparisons with very un-chocolate-like materials like glass and polymers.

If the molten chocolate, once poured into its mould, cools too fast then the sets of three fatty-acid side-chains on adjacent molecules do not have time to arrange themselves in the neatest way, and end up randomised and jumbled, rather like the amorphous regions in polymers. With slow cooling, the organised structure *is* achieved, and good-quality chocolate is produced. Figure 13.34a shows the 'neat' arrangement of the three side-chains, in which the freedom of rotation in the bonds has been allowed to produce a linear-and-parallel conformation for the chains. The evidence for this model is seen on the electron density map, produced from X-ray crystallography (Figure 13.34b). The whitish bloom on some cheap chocolate is not, as is often thought, a sign of the chocolate going mouldy – it is instead a crazing of the surface layers. This is caused by contraction, as the less stable amorphous chocolate reverts slowly to the more crystalline form.

Question

46a How do you think the cheaper more amorphous types of chocolate would differ in their crispness and melting temperature from properly crystalline chocolate?

b The fact that amorphous chocolate reverts naturally to the more organised form implies that the latter is the more stable. Can you see why, of all the incalculable number of ways of arranging a fat molecule by bond-twisting, it is structures like that in Figure 13.34 which (given time) are preferred by Nature?

c Why does the gradual conversion to the more crystalline form bring about a physical contraction?

Soaps and detergents

'Dual-personality' molecules, that is, molecules in which different regions dissolve in different solvent media, have been cropping up throughout this section on fats. In fact, whenever we have seen a fat molecule cleaved – perhaps, say, into two fatty-acid side-chains and a monoacylglycerol – the products have had the dual-personality feature.

The reasons for this go all the way back to the fundamental design brief for fats – by using the ester linkage to hold the 'energy cartridges' to the 'backbone', Nature has ensured that a broken fat-fragment will end in an OH group at the very least, or, in the case

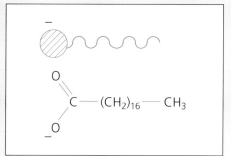

Figure 13.35 The stearate anion, formed from the alkaline hydrolysis of a fat with a stearic acid side-chain. The diagram also shows the link between the real structural formula and the 'head and tail' representation of soap and detergent molecules

of the fatty-acid residues in alkaline media, an O⁻ (Figure 13.35). (Those are the water-soluble ends, of course; the hydrocarbon chains on the fatty acids are the parts which deliver the compatibility with non-polar solvents.)

We have also been prepared by our study of fat chemistry for the *action* of soaps and detergents. The way fats are broken up inside the body for digestion – as when the first batches of product molecules turn the rest of the fat globules into an emulsion of tiny droplets (micelles, look back to Figure 13.26) – is exactly mimicked by the action of artificially created soaps and detergents. The only differences between Figures 13.36 and 13.26 are that the water-loving ends of soaps and detergents are often fully fledged anions and are deployed against external oils and greases rather than internal fats.

Question

47 The micelles in Figure 13.36 are at less risk of coming together and coalescing back into large bodies of water and oil, than are those in Figure 13.26. In other words, Figure 13.36 represents the more stable emulsion. Can you suggest why?

We will now give an account of some of the different types of soaps and detergents.

Soaps. Soaps have been known to humankind for millennia – the process of

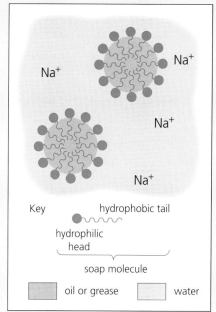

Figure 13.36 Soap in action

Key — hydrophobic tail
hydrophilic head
soap molecule
oil or grease   water

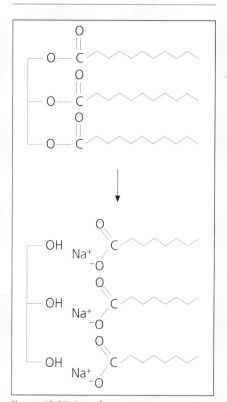

Figure 13.37 From fat to soap

boiling fats with alkalis to hydrolyse the ester linkage is at least as old as Roman times. The hydrolysis is usually taken to completion, so that all three fatty-acid side-chains come off (Figure 13.37). The soap makers' art consists of deciding which fats to use and which fragrances to add. One of the perennial problems with soaps has proved (literally) insoluble. That is the fact that the anions in soap have insoluble calcium salts, which precipitate the soap as scum in hard-water areas (as shown in the equation below).

$$Ca^{2+}(aq) + 2St^- \rightarrow Ca(St)_2$$
$$scum$$

St = the stearate group:

$$CH_3(CH_2)_{16}—C \overset{O}{\underset{O}{<}}$$

This was one of the incentives for finding alternative molecules with detergent action.

Question

48 Soaps are the sodium salts of 'weak' acids. This means that the carboxylic acid anion is not averse to picking up any surplus hydrogen ions. How would this affect the washing action of the soap, if it had to be used in an acidic medium?

Synthetic detergents. Synthetic detergent manufacture uses raw materials from the petrochemical industry, rather than the natural world. Alkylbenzenes with side-chains about ten atoms long are treated with concentrated sulphuric acid, to produce a sulphonic acid group on the ring. The sulphonic acid (which is a strong acid, that is, fully dissociated at the O—H bond into ions) is then neutralised by sodium hydroxide, and the product sold as its sodium salt (shown below).

$$\text{AAAAA}—\bigcirc—\overset{O}{\underset{O}{S}}—O^- Na^+$$

The anion does not precipitate with calcium ions, so the hard-water problem is solved.

Non-ionic detergents. Both the types of cleaning agent discussed so far have carried their dual personality in an anion. A family of **non-ionic** detergents has been developed, largely to comply with the needs of washing machines and dishwashers which require low-lather washing action. They have a long hydrocarbon chain derived from a fat or a crude oil fraction, plus a repeating ethoxy unit, culminating in an alcohol group:

$$\text{AAAAA}—\bigcirc—O\wedge O\wedge O\wedge O\wedge$$

Question

49 What would you predict for the response of non-ionic detergents to hard water?

If you wish to remind yourself of the effects of soaps on the surface tension of water, look back to Chapter 6.

Paints

Paints have three main ingredients – a pigment for the colour, a drying oil to cause the paint to 'set' or cure, and a 'vehicle' in which to deliver the paint, normally a solvent which evaporates away. We mentioned earlier that chip pans had an unfortunate tendency to 'varnish' themselves due to the cooking oil cross-linking (Figure 13.38), and all we need for a drying oil is to find a fat which does the same reaction 'on purpose' (and is good at it). For centuries the prime choice was linseed oil, from the plant flax (flax is the plant with the pretty blue flowers which you see increasingly these days in fields throughout Britain, though not as often as the ubiquitous bright yellow rape. Both plants are raw materials for the fats and oils industry.)

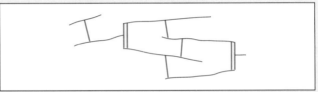

Figure 13.38 Showing how a fat which can cross-link, such as those in linseed oil, could become a 'thermoset' polymer (except that it would be air-set rather than 'thermo-')

Question

50 Linseed oil contains many *linolenic acid* side-chains. Remind yourself from Figure 13.12 of the structure of this acid, and then suggest why it is the best qualified of all the listed acids to undergo the cross-linking reaction when exposed to air.

Linseed-oil paints have a tendency to go brown with age, so paint chemists have sought to find substitutes. They have improvised on Nature's drying oils, but kept close to the underlying theme of ester linkages suspending a series of dangling unsaturated fatty-acid side-chains. The side-chains are still responsible for the actual cross-linking reaction which causes the paint to cure in air.

The synthetic drying oils are a family of compounds known as **alkyd resins**. They have become, since the 1940s, the workhorses of the coatings industry (Figure 13.39). The main new idea in the molecular pattern of the synthetic resins is the use of a dibasic acid (that is, one with *two* acid groups) to extend the 'backbone' of the molecule. The alcohol (still in many cases glycerol) and the fatty acid are still there as in the natural equivalents. Figure 13.40 shows a generalised alkyd resin.

Figure 13.39 The popular range of Dulux gloss paints use alkyd resins as drying oils

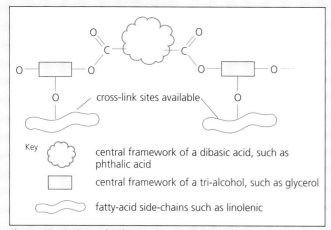

cross-link sites available

Key

central framework of a dibasic acid, such as phthalic acid

central framework of a tri-alcohol, such as glycerol

fatty-acid side-chains such as linolenic

Figure 13.40 Generalised representation of an alkyd resin

Questions

51a One reaction which is used to synthesise the backbone of an alkyd resin is the esterification reaction between the dibasic acid phthalic acid (benzene-1,2-dicarboxylic acid) and glycerol:

phthalic acid
(benzene-1,2-dicarboxylic acid)

glycerol

Show how two or three units of this 'polyester' would link up.

b More commonly in the paint industry the above reaction is done using phthalic anhydride (shown below) instead of phthalic acid.

The product polymer is the same, but there is less frothing. Can you suggest which by-product, present in (a) but absent in this second version of the reaction, might have caused the frothing?

c Another common practice is to replace the more 'natural' glycerol with an alcohol called 'pentaerythritol' (shown below).

$$HOCH_2 - \underset{\underset{CH_2OH}{\overset{CH_2OH}{|}}}{C} - CH_2OH$$

The resulting cured paint film is stronger than those in which the alcohol is glycerol. Can you see, from the contrasts between the two alcohols, why this is?

52 The backbones of alkyd resins are very similar in structure to the polyesters which make useful fibres for incorporating into 'poly-cotton' fabrics. This second sort of polyester, which is marketed in the UK under the name Terylene, is formed by the reaction of 'terephthalic acid' and ethane-1,2-diol:

terephthalic acid

$$HO - CH_2 - CH_2 - OH$$

ethane-1,2-diol

Terylene

a Compared with the alkyd resin backbone, we notice a switch from benzene-1,2-dicarboxylic acid (phthalic acid) to benzene-1,4-dicarboxylic acid (terephthalic acid). Can you suggest why, in a linear polymer designed to be pulled into fibres, the choice is for the latter?

b A very popular mixture for cloth is 65% polyester/35% cotton. This is what you get, for example, in 'poly-cotton' sheets (Figure 13.41). This mixture delivers what is considered to be the best compromise between the need for water/sweat absorption and wear resistance. If sweat absorption is a priority you are still better off with 100% cotton (Figure 13.42). By comparing the two molecules, suggest why this should be so.

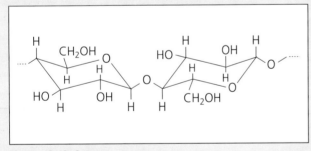

Figure 13.41 Scanning electron micrograph of poly-cotton. The cotton fibres are the rough-looking ones

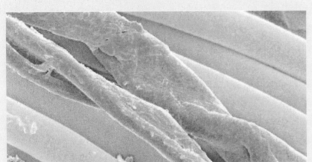

Figure 13.42 Cotton

13.16 Another natural use of the ester linkage – acetylcholine

At the beginning of Section 13.15 we said that Nature used the ester linkage as a strong-but-breakable bonding system. There are plenty of examples from the biochemistry of fats to illustrate that point, but a second example can be found in the workings of the human nervous system. When a message reaches the end of one nerve cell, it has to be transmitted across the gap to the next one. This is achieved by the release of a chemical from nerve cell 1 (the 'sender'), triggered by the arrival of the nerve message itself, which then opens the gates on nerve cell 2 (the 'receiver'). The open gates can now receive the actual messengers, which are cations. One such gate-opening chemical is called **acetylcholine**, and it is an ester. Its structure is shown in Figure 13.43a, and its action symbolised in Figure 13.43b.

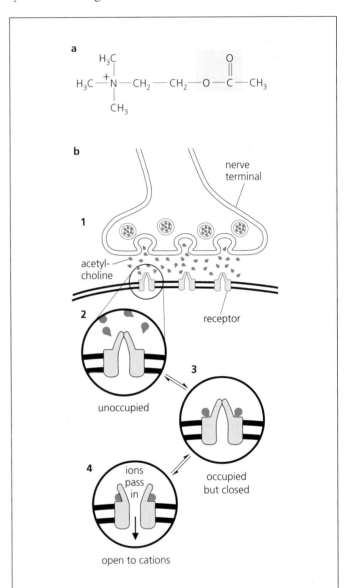

Figure 13.43 a The acetylcholine cation. The ester linkage is shown in the tinted panel. **b** Action of acetylcholine. Molecules of acetylcholine (represented as coloured teardrop shapes) are stored in vesicles in nerve endings. A nervous impulse causes them to be released. The molecules of free acetylcholine diffuse across the space between nerve cells and bind to a receptor in the cell membrane (2 and 3). This causes a change in the shape of the protein molecule (the receptor). It becomes a molecular tunnel through which ions can pass, and this causes the nervous impulse to continue down the nerve

The acetylcholine must be cleared away quickly, to allow the gates to re-close, and thus to leave the path free for new messages. This is achieved by the rapid hydrolysis of the acetylcholine at its ester linkage. We know that ester linkages do not normally hydrolyse quickly, even in strongly acidic or alkaline media, but in the presence of the enzyme **acetylcholinesterase**, the seemingly improbable is achieved (Figure 13.44).

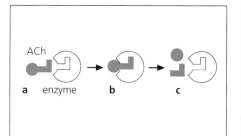

Figure 13.44 Schematic view of the hydrolysis of acetylcholine (ACh) by an enzyme (for example, acetylcholinesterase)

Question

53 Write the equation for the hydrolysis, at the ester linkage, of acetylcholine.

Figure 13.45 Nerve gas. These weapons of war work by inhibiting the enzyme that clears away acetylcholine from the gaps between nerve cells. They are among the most toxic chemicals in existence

Insecticides, nerve gases and disease

The implications for an animal which cannot deploy its acetylcholinesterase are very serious. The unhydrolysed acetylcholine is left stuck in the gates, which are left 'swinging open', and so the nervous system is in a state of uncontrolled overload. The war weapons known as **nerve gases** are designed specifically to interfere with the victims' acetylcholinesterase, resulting of course in death (Figure 13.45). On a slightly less gruesome level we control insect pests by the use of insecticides called **organophosphates**, which inhibit the insects' acetylcholinesterase.

Several voices have been raised to express concern about the use of organophosphates on farms, where they are used to control insect parasites on livestock. The risks of nerve disease for those administering the chemicals on a regular basis are well known, and in theory there are stringent safety precautions which attend the use of concoctions such as sheep dip which include organophosphates. But questions persist over exposure to low levels of organophosphates over long periods of time.

In August 1994 the *Guardian* newspaper published the story of one of the most passionate and committed opponents of the use of organophosphate insecticides. He is an organic farmer called Mark Purdey, who believes that organophosphates, and not infected feeds, are the real cause of the dreaded cow-brain disease BSE (bovine spongiform encephalopathy). (The Government held that BSE had come from sheep's brains and spinal columns, which carried the disease scrapie and which had been incorporated into animal feeds.) In the article Purdey declares that:

> with one possible exception, there hadn't been a case of BSE among home-reared cattle on **fully organic** farms, even though some had used the (supposedly contaminated) feed. (The bold type is mine.)

The article relates what happened when a cow which Purdey had bought in from a non-organic farm contracted BSE:

> Before calling in the ministry, I blood-tested her. The red blood cell acetylcholinesterase was down by about 20% in Damson (the cow), compared with three control cows.

Question

54 Find from the quotations the pieces of evidence which could be used to link BSE as effect to organophosphates as cause.

13.17 NMR spectroscopy in the analysis of fats

There are several wet-chemical techniques for analysing fats, two of which provide the following numerical data.

- The **saponification value** – how much alkali is needed to hydrolyse a given mass of fat.

- The **iodine number** – how much I_2 is taken up by the fat.

Question

55 Indicate which aspect of a fat determines each of the above two parameters.

However, as in most areas of chemistry, wet-chemical tech-niques are being supplanted by modern physical spectroscopic methods. **Nuclear magnetic resonance** is used on fats. Although discussed in the specific context of this chapter, it is one of the most general and powerful analytical tools, often used in collaboration with infrared and mass spectrometry in all branches of chemistry.

Nuclear magnetic resonance (NMR)

Each hydrogen nucleus is a single proton. Each proton behaves as if it is spinning, and spinning charged particles have magnetic fields. To put it at its simplest, each hydrogen nucleus is a tiny magnet. In proton-NMR spectroscopy, sam-ples of hydrogen-containing chemicals are put, in glass tubes, between the poles of a powerful magnet. The response of the 'proton-magnets' within the hydrogen atoms of the test mole-cules is to line up, like millions of obedient plotting com-passes, along the lines of force of the applied external magnetic field. However, the only actual movement is not a movement of the molecules, nor even of the hydrogen atoms within the molecules. Your mental picture should be of the nuclei remaining central within their atoms, but turning to align their magnetic poles with the external field (Figure 13.46).

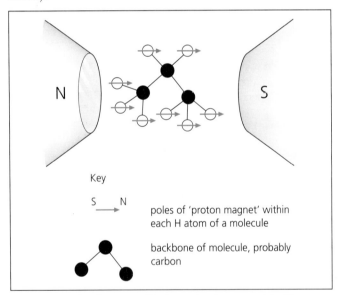

Key

S ——→ N   poles of 'proton magnet' within each H atom of a molecule

backbone of molecule, probably carbon

Figure 13.46 The alignment of hydrogen 'nuclear magnets' with an external magnetic field. The hydrogen atoms are part of the sample molecule

It is at this point that we must introduce a level of explana-tion you might call 'quantum surrealism'. You should by now be used to the idea that very small objects rarely behave exactly like big analogues. Electron energy and bond vibration energy are two quantities which have already required a quan-tum theory to explain them. Here again we find energy restrictions. It appears that proton magnets can line up *with* the external magnetic field, in a stable **low-energy** orienta-tion, or otherwise they can absorb photons and align them-selves *against* the external magnet, in a **high-energy** orientation (Figure 13.47). No other positions are allowed.

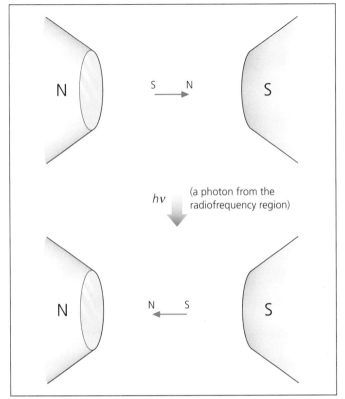

$h\nu$ (a photon from the radiofrequency region)

Figure 13.47 Showing how any proton-magnet within the field can absorb a photon and turn to a high-energy like-poles-facing orientation

The effect was observed as long ago as the 1940s. When very strong magnetic fields of the order of 15 000 gauss were used, samples of hydrogen-containing molecules really did absorb photons, indicating that some of the proton magnets had moved into the high-energy 'against-the-field' orienta-tion. The absorptions occurred at a frequency of 60 MHz, which is in the radiofrequency region – a very much lower energy region than those involved in the other two spectro-scopies. This discovery must have greatly encouraged the physicists who had predicted it, but it turned out that the people who gained the most were practising chemists.

The reason was that not all protons absorbed at exactly the same photon frequency. If the magnetic field was held con-stant and the photon frequencies were varied a little from 60 MHz, it was found that some absorptions of photons occurred a few ppm higher or lower than 60 MHz. Ethanol gave three separate absorptions (Figure 13.48).

The reason for this phenomenon, which became known as **chemical shift**, concerns the chemical environment in which each proton finds itself (Figure 13.49). Specifically, if the hydrogen atom resides near an electronegative atom, such as an oxygen, the electron density around that proton (hydro-gen) will be slightly thinned out (see H1 in Figure 13.49). As a result that proton will experience the influence of the exter-nal magnetic field more keenly, and so the energy difference between being with the external field and against it, will be very slightly greater than would be the case for a proton which was nowhere near an electron-withdrawing 'deshield-ing' atom.

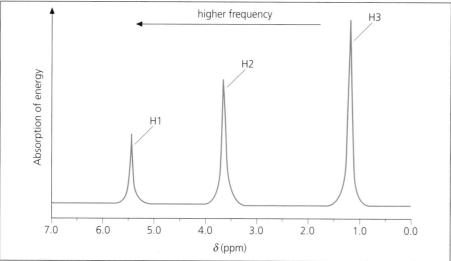

Figure 13.48 Low-resolution spectrum of ethanol. See Figure 13.49 for explanation of H1, H2 and H3

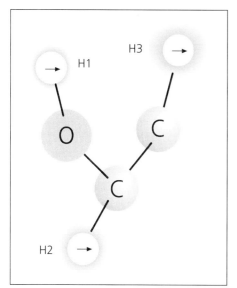

Figure 13.49 The three proton environments in the ethanol molecule (not all hydrogen atoms are shown). The clouds represent the degree of 'cocooning' of each H nucleus. The nearer the oxygen atoms, the less the cocooning. H1 is deshielded relative to H3

In ethanol there are three proton environments:

• Right next to the oxygen atom.

• On a carbon atom 'next-door' to the oxygen atom (the CH_2 group).

• On a carbon atom 'next-door-but-one' to the oxygen atom (the CH_3 group).

The 'deshielding' effect of the oxygen atom wears off the further the hydrogen atoms are away from it. So the three sets of protons listed above give successively lower frequency absorptions, which accounts for the three signals in the NMR spectrum of ethanol.

Ethanol is perhaps a rather blatant case of three different proton environments in one molecule. It was found that NMR spectra would show a slightly different signal position for *every* different proton environment in *any* given molecule. In fact, it was possible to draw up tables of typical chemical shifts which identified certain peaks as being caused by certain groups. These tables had similar usefulness to the tables of infrared absorptions of particular bonds. However, NMR spectra could give two items of extra information beyond the infrared data.

1 Peak splitting

'High resolution', when applied to spectroscopy, means that fine details of peaks with only slightly different positions can be distinguished, or *resolved*. High-resolution NMR revealed a strange phenomenon. Many peaks were found to be split into sub-peaks. For reasons which lie outside the scope of this book, the number of sub-peaks is equal to **the number of hydrogen atoms on next-door carbon atoms, plus one**. For example, the CH_3 peak in ethanol is split into three, (Figure 13.50) because the next-door

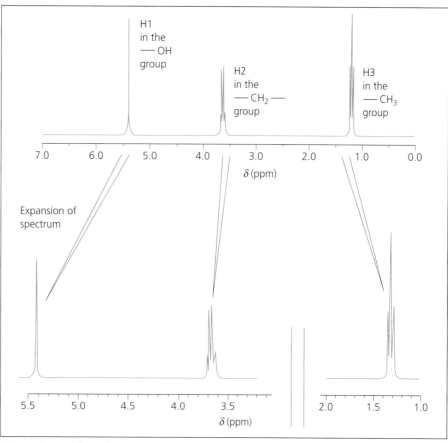

Figure 13.50 High-resolution proton-NMR spectrum of ethanol

carbon atom carries two hydrogen atoms, and three is two plus one.

(If you would like a further explanation of peak splitting, you should consult a specialised text, as in the references at the end of the book. However, just as you can drive a car without being a mechanic, so you can extract the chemical information from an NMR spectrum without knowing how it came about.)

Question

56a Account for the fact that the CH_2 peak in the NMR spectrum of ethanol is split into four sub-peaks.

b The NMR spectrum of propan-1-ol would have four peaks rather than three, and one of them would be split into *six* sub-peaks. Can you explain both these facts, and make a sketch of what the spectrum might look like?

c The reason the peak due to the OH proton is not split is because in pure ethanol the alcoholic protons are constantly being exchanged between molecules. If the ethanol were in a solvent which prevented this exchange, how would the OH proton signal split?

2   Integration

NMR spectrometers are capable of calculating the area under each peak and displaying it as a second line on the graph, called the **integration curve**. These areas are in proportion to the numbers of protons making the peaks, so not only can you tell how many proton environments there are in a molecule, but you can also tell how many equivalent protons reside at each site. An example in which both peak splitting and integration play a part should help you to get these ideas straight.

Figure 13.51 shows the NMR spectrum of an ester, propyl ethanoate. Its features are as follows.

- Four peak clusters in all – four proton environments in the molecule.

- From the integration curve, peaks in ratio 2 : 3 : 2 : 3 showing relative numbers (and absolute numbers in this case) of protons at each site.

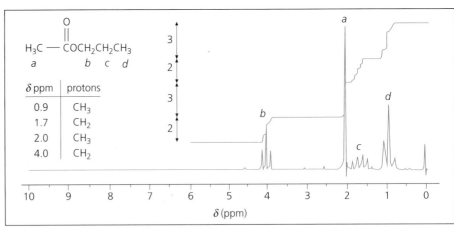

Figure 13.51 NMR spectrum of propyl ethanoate showing the integration curve and how it is used to obtain information on the numbers of protons resonating at each frequency. The ratio of the heights of the steps on the integration curve tells us the ratio of the numbers of protons resonating at each frequency. In this case the ratio of protons (not necessarily the absolute number of protons) $b : a : c : d$ is $2 : 3 : 2 : 3$

- Peak b at 4 ppm, split into a triplet, indicating its closeness to the oxygen atom, and the fact that the adjacent carbon atom carries *two* hydrogen atoms.

- Peak a at 2 ppm, unsplit. The 2 ppm position indicates a greater remoteness from the oxygen than the protons causing peak b, and the lack of splitting shows the absence of any protons at all on adjacent carbon atoms.

- Peak c at 1.7 ppm, split into a sextet. The 1.7 ppm position indicates a proton site even more remote from the oxygen atom than the protons causing peak a, and the splitting pattern indicates *five* nearest-neighbour hydrogen atoms, which must therefore be on both sides.

- Peak d at 0.9 ppm, split into a triplet, indicating maximum remoteness from the oxygen atom and two hydrogen atoms on the adjacent carbon atom.

Question

57 The next step is for you to try the reverse process. Instead of trying to justify the peaks of the spectrum of a known substance, try to do what the professionals do, and identify an unknown substance from its spectrum. The chemical in question has a molecular formula of $C_4H_8O_2$, and has a sweet smell. The spectrum is shown in Figure 13.52.

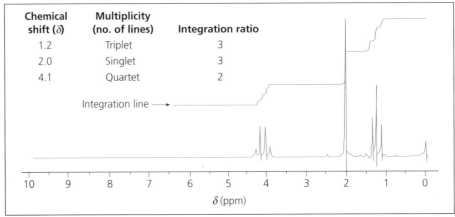

Chemical shift (δ)	Multiplicity (no. of lines)	Integration ratio
1.2	Triplet	3
2.0	Singlet	3
4.1	Quartet	2

Figure 13.52 A colourless mobile liquid at room temperature, with a sweet smell, having the empirical formula $C_4H_8O_2$

NMR and fats

Finally in this section on NMR, we look at its application to analysis of fats. In the following example, the side-chains of a particular fat have been removed by hydrolysis, separated by column chromatography and the methyl esters synthesised. It is in this form that their proton-NMR spectra are recorded. This kind of cleavage treatment gets round the problem of the great complexity of the NMR spectra of complete fats. The spectra of the side-chains are shown in Figure 13.53. To keep the number of peaks manageable, the spectra have been recorded at low resolution (that is, with no peak splitting). Table 13.5 shows the relevant chemical shifts for protons in the environments normally found in fatty-acid side-chains. Question 58 asks for some detective work based on these spectra and Table 13.5.

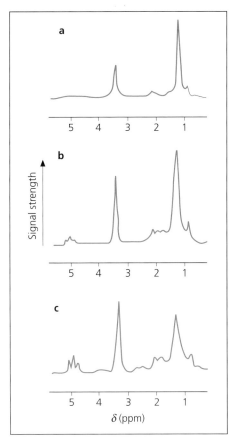

Figure 13.53 NMR spectra of fatty-acid methyl esters

Table 13.5 Typical chemical shifts for hydrogen atoms in different environments (see Figure 13.53)

	Chemical shift (ppm)
CH_3 *hydrogens*	
CH_3—C	0.9
CH_3—O—CO—R	3.7
CH_2 *hydrogens*	
—C—CH_2—C—	1.4
—C—CH_2—C=C	2.3
—C—CH_2—CO—O—R	2.2
CH hydrogens	
—CH=CH—	5.0

Question

58a Why have all three spectra got a big peak around 1.4 ppm?

b Why is that peak a good deal bigger, in each spectrum, than the common peak at 0.9 ppm?

c Which peak would identify an unsaturated side-chain?

d Based on the evidence of this peak, which side-chains are unsaturated?

e Based on the size of this peak, which side-chains might be unsaturated at more than one site?

Summary

• All the compounds in this chapter contain the **carbonyl** group, or C=O bond.

• The carbonyl group is strongly **dipolar**, and the positive, carbon, end of the dipole attracts the attention of **nucleophiles**.

Aldehydes and ketones

• In molecules in which the C=O bond is the only functional group (the aldehydes and the ketones, shown below) the bond tends to undergo **nucleophilic addition reactions**, or nucleophilic addition/elimination reactions.

$$R—C\overset{H}{\underset{O}{=}} \qquad R—C\overset{R'}{\underset{O}{=}}$$

aldehyde   ketone

• **Nucleophilic additions** occur when aldehydes and ketones react with cyanide ions. The effect is to add H and CN across the C=O bond (with the CN going to the carbon atom):

$$\underset{R'}{\overset{R}{>}}C=O \xrightarrow{CN^-, H^+} R'—\underset{CN}{\overset{|}{C}}—OH$$

The result is a class of compounds known as **cyanohydrins**.

• Another reaction which can be seen as a nucleophilic addition is the reaction with *hydride ion donors*. These reactions can equally be classed as **reductions**. Typical reagents for this donation are sodium tetrahydridoborate(III), $NaBH_4$, and lithium tetrahydridoaluminate(III), $LiAlH_4$. The effect is to add a hydrogen atom on each end of the C=O bond, to produce **alcohols**:

$$\underset{R'}{\overset{R}{>}}C=O \xrightarrow[LiAlH_4]{NaBH_4 \text{ or}} R'—\underset{H}{\overset{R}{\underset{|}{\overset{|}{C}}}}—O—H$$

• In these reduction reactions, **aldehydes** are reduced to **primary** alcohols, and **ketones** to **secondary** alcohols:

$$R-\overset{\displaystyle O}{\underset{\displaystyle H}{C}} \rightarrow R-\overset{\displaystyle H}{\underset{\displaystyle H}{\overset{|}{\underset{|}{C}}}}-OH$$

$$R-\overset{\displaystyle O}{\underset{\displaystyle R'}{C}} \rightarrow R'-\overset{\displaystyle R}{\underset{\displaystyle H}{\overset{|}{\underset{|}{C}}}}-OH$$

• Another group of reagents make nucleophilic attacks on the C=O group of aldehydes and ketones, but the addition compound is a reactive intermediate, which goes on to eliminate H_2O. These reactions are often called **addition/elimination reactions**. The reagents in this group are all **derivatives of ammonia** (but not ammonia itself).

• The reagents for these addition/elimination reactions are hydrazine, hydroxylamine and 2,4-dinitrophenylhydrazine, this last one also known as Brady's reagent.

hydrazine H_2N-NH_2

hydroxylamine H_2N-OH

2,4-dinitrophenylhydrazine
(Brady's reagent)

$$H_2N-\underset{\displaystyle H}{\overset{|}{N}}-\overset{}{\bigcirc}-NO_2$$
$$\underset{\displaystyle NO_2}{}$$

• Brady's reagent is used to identify aldehydes and ketones, from the melting temperatures of the crystalline products of the reaction.

• Aldehydes are readily oxidised to carboxylic acids, even under mild conditions. Two oxidising agents that fall into this category are **Fehling's solution** and **Tollen's reagent** (the silver mirror test), both of which contain complexed heavy metal cations which are the actual oxidisers. Both these reagents give highly characteristic and visible products (orange Cu_2O from Fehling's solution and a silver mirror from Tollen's reagent). Since ketones are *not* oxidised under these mild conditions, the two reagents offer a method of discriminating between aldehydes and ketones.

• It is impossible to oxidise ketones without rupturing one of the C—C bonds in the molecule.

• Halogens will react with aldehydes and ketones, resulting in the substitution of a halogen atom for a hydrogen on the carbon atom adjacent to the C=O bond:

$$H_3C-\overset{\displaystyle O}{\overset{\|}{C}}-CH_3 \xrightarrow{X_2} H_3C-\overset{\displaystyle O}{\overset{\|}{C}}-CH_2X + HX$$

• Methyl ketones and 2-hydroxyalkanes have their own identification test (the iodoform test) which involves the production of a yellow precipitate of triiodomethane (iodoform) after treatment with iodine in dilute potassium hydroxide.

Carbohydrates

• Carbohydrates have the general formula $C_m(H_2O)_n$.

• They all undergo dehydration reactions to remove the water content of the molecule and leave carbon.

• Carbohydrates are constructed from unitary building blocks called **monosaccharides**. These monosaccharides may be found as individual molecules, as in glucose, or in two-unit *di*saccharides, as in sucrose, or in polymeric chains, like the *poly*saccharides starch, glycogen and cellulose.

• Monosaccharides exist as **rings**, sometimes five-membered and sometimes six-membered, one member of which is an oxygen atom. They are liberally supplied with hydroxyl (OH) groups.

• Yet some of the behaviour of monosaccharides can only be understood if it is assumed that they spend some time in an **open-chain form**. In this form they have either an aldehyde or a ketone formation, and undergo the corresponding sets of reactions.

• Mono- and disaccharides show the property of **chirality**, which derives from asymmetry within the molecule. Any carbon atom with *four different* tetrahedrally held *substituents* will confer chirality on its molecule. The measurable effect of chirality is the ability to rotate the plane of polarisation of plane-polarised light (POPOPPL).

• Carbohydrates are intimately involved in the energy release and storage systems of all organisms. In plants they are also a key body-building material. Mono- and disaccharide forms of carbohydrates are used when transportation and energy release are required (for example, **glucose** is the carbohydrate which is transported to cells for **respiration**). Polysaccharides are used for **storage** (for example, **starch** in plant bulbs and tubers, and **glycogen** in the liver). When body construction is required, the choice is always a polysaccharide, most notably **cellulose** in plant cell walls.

• The switching between monosaccharide and polysaccharide forms of carbohydrates is achieved biochemically by enzymes. The enzymes can synthesise or digest only one particular chiral form of the various carbohydrate molecules. The inability of humans to digest cellulose is caused by small but crucial differences in the orientation of the bonds linking the monosaccharide units in the polysaccharide.

• Carbohydrates are effective in their biochemical role because they combine the properties of **combustibility** and **compatibility with aqueous media**.

• Monosaccharides and disaccharides are to varying extents sweet-tasting and are called **sugars**.

Carboxylic acids

$$R-C\begin{array}{c}O\\\\O-H\end{array}$$

• The C=O bond in carboxylic acids undergoes very little chemical activity, but it has a major influence on the adjacent OH group.

• The effect is to make the OH group much more active in a number of ways. Most notably, the polarity of the O—H bond is increased, thus stimulating loss of the H atom as H^+. Carboxylic acids are acidic enough to liberate carbon dioxide from carbonates and hydrogencarbonates, and they also give ionic salts when neutralised by alkalis.

• The OH group is also prone to nucleophilic substitution reactions, and nucleophiles which carry out such reactions include the **phosphorus halides** (to give **acyl halides**) and the **alcohols** (to give **esters**).

• The esterification reaction is slow, needs catalysis, and results in equilibrium mixtures of reactants and products.

• Carboxylic acids can be reduced to aldehydes and thence to primary alcohols, by reagents such as sodium tetrahydridoborate(III), $NaBH_4$.

Acyl halides

$$R-C\begin{array}{c}O\\\\X\end{array}$$

• **Acyl halides** are very useful in synthetic chemistry because they are so reactive.

• They will react with any molecule with even the smallest claim to be a nucleophile, including water and phenols. Products include esters (a better route than via carboxylic acids), amides and acids.

Esters

$$R-C\begin{array}{c}O\\\\O-R'\end{array}$$

• Small-molecular esters have a rather monotonous chemistry involving the single reaction of **ester hydrolysis**, but the fact that other molecular units can be linked and un-linked via the ester linkage has given esters tremendous importance in Nature and in industry.

• Esters are hydrolysed back to their *parent acid and alcohol* ((a) below), except when alkaline media have been used, in which case the products are the alcohol and the sodium salt of the acid ((b) at the top of the next column).

$$\text{a}\quad R-C\begin{array}{c}O\\\\O-R'\end{array} + H_2O \rightarrow R-C\begin{array}{c}O\\\\OH\end{array} + R'OH$$

$$\text{b}\quad R-C\begin{array}{c}O\\\\O-R'\end{array} + NaOH \rightarrow R-C\begin{array}{c}O\\\\O^-Na^+\end{array} + R'OH$$

This second option, of an aqueous alkaline medium, is the one normally chosen for laboratory ester hydrolyses.

• The only other significant laboratory reaction of esters is their reduction to alcohols, using lithium tetrahydrido-aluminate(III), $LiAlH_4$, and similar reagents.

• The most important bio-esters are the **fats**. Fats are used by most living organisms as energy-storage compounds.

• The ester linkage suits the role of energy-storage chemical well. The ester linkages in each fat molecule are used to connect three energy-rich hydrocarbon groups to a water-compatible backbone (provided by the tri-alcohol **glycerol**). There is the added advantage that when the linkages are broken, the hydrocarbon groups reveal a hydrophilic end (the —COOH or —CO_2^- group). These ends may not put the molecules in the same league of water-solubility as carbohydrates, but at least they enable the molecules to be emulsified and transported in aqueous media as micelles.

• Digestive enzymes like **lipase** can bring about the ester hydrolysis cleavage reaction under conditions which are much milder than in equivalent laboratory reactions.

• When fats cleave at the ester linkage, the hydrocarbon sections come away as **fatty acids**. Many different fatty acids occur in Nature, and give rise to a wide variety of properties in the corresponding fats. This variety has its origins in differences in chain length and in degree of unsaturation (that is, the degree of inclusion of C=C bonds).

• Variability of fats shows itself in variations of such properties as melting temperature, vulnerability to atmospheric oxidation and tendency to cross-link.

• Fats are an essential part of human diet, but their consumption raises many issues. It is suggested that in Western countries the overall consumption of fats is too high, leading to obesity and on to heart disease. There is also plenty of evidence to suggest that unsaturated fats are less implicated in the incidence of heart disease than are the saturated ones. This latter evidence has been translated into advice to switch from butter (mostly saturated fats, animal origin) to margarine (more unsaturated fats, vegetable origin). This has presented food technologists with a challenge, since unsaturated fats tend to be liquid at room temperature. The challenge has been met with such ploys as partial hydrogenation of C=C bonds using hydrogen gas over a nickel catalyst, and by using gelling agents to pack the margarine with large amounts of water, without sacrificing 'solidity'.

• The sodium salts of the fatty-acid side-chains from fats make ideal **soaps**; that is, compounds designed to encourage the miscibility of oils and water (Figure 13.54). (This is

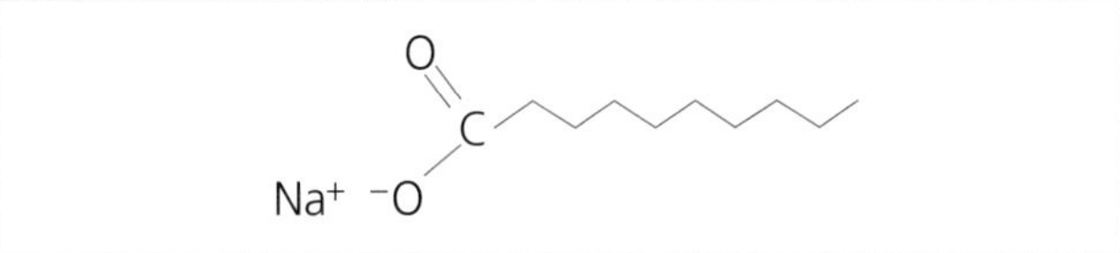

Figure 13.54 Generalised formula for a soap

one of their roles in biochemical environments too, as in the alimentary canal. During digestion, the first fatty acids to be broken off from the fats being digested can then act as 'soaps' towards the rest of the batch.) The hydrocarbon chains of the soap molecules dissolve in oily and greasy deposits on clothes and skin, and break them up into droplets; the ionic ends of the soap molecules enable an aqueous medium to receive the droplets as micelles. Since the exterior of each micelle presents a sphere of negative charges, the micelles are stopped from meeting and re-coalescing. The **emulsion** of micelles in water is thus stabilised.

• **Detergents** are synthetic molecules built to the same design brief as soaps, but which avoid problems of hard-water scum. They are usually the sodium salts of benzene-sulphonic acids, or else are neutral molecules with a repeating ether theme at one end.

• **Alkyd resins** are polyesters with strong resemblances to fats. They, too, have fatty-acid groups attached to 'multi-alcohols' like glycerol, but the chemical industry has improvised on the theme by selecting different backbone-making extenders such as the dibasic acid phthalic acid. Alkyd resins have **unsaturated** fatty-acid side-chains which cross-link when exposed to air. This gives the clue to their function in the chemical industry; it is to provide the **coating/drying** component of **paints and varnishes**.

In this respect they have replaced the natural fat linseed oil.

• **Polyester fibres** are made from more linear formulations of dialcohols and diacids. They find extensive use in the fabrics industry, especially in mixtures with cotton.

• Another biochemical ester, in which the ability of the ester linkage to be cleaved is a crucial feature (as in fats), is the nerve messenger chemical **acetylcholine**:

$$(CH_3)_3 \overset{+}{N}\!-\!\!-(CH_2)_2\!-\!\!-O\!-\!\!-\overset{\displaystyle O}{\overset{\|}{C}}\!-\!\!-CH_3$$

An enzyme called **acetylcholinesterase** is responsible for clearing away 'used' acetylcholine after a message has been sent. Organophosphate nerve gases and insecticides work by inhibiting the action of this enzyme.

• **Nuclear magnetic resonance** (**NMR**) is a very powerful analytical technique which can assist in the determination of the structure of unknown molecules. In proton-NMR, samples of hydrogen-containing molecules are placed in an intense magnetic field. Under these circumstances every hydrogen atom in the molecule is capable of absorbing a photon, and the frequencies of the absorbed photons vary subtly with the chemical environment of the hydrogen atoms within the molecules.

• Thus the NMR absorptions of a molecule appear as a spectrum, with the positions of the peaks (called **chemical shifts**) being characteristic of certain types of chemical environments for hydrogen atoms within the molecule.

• As an example of the power of NMR spectroscopy, we saw its ability to discriminate between singly and multiply unsaturated fats.

14

Chemical kinetics: The study of rates of reaction

- ▷ A simple collision theory model
- ▷ The effect of concentration on reaction rate
- ▷ A mathematical model of a bimolecular collision reaction
- ▷ A mathematical model of a unimolecular process
- ▷ Do real systems fit theoretical models?

- ▷ More diagnostic tests for use on experimental data
- ▷ Initial rates method
- ▷ Practical methods
- ▷ Mechanisms
- ▷ Matching mechanisms to rate laws
- ▷ The effect of temperature on reaction rate

- ▷ A mathematical model linking temperature and reaction rate
- ▷ Testing the model linking temperature and reaction rate
- ▷ Activation enthalpy and reaction profiles
- ▷ Catalysis
- ▷ Styles of catalysis
- ▷ Enzymes and drugs
- ▷ Summary

14.1   A simple collision theory model

Rates of chemical reactions vary enormously (Figure 14.1). Most sixth-form chemistry students have experience of reactions that range in rate from the near-instantaneous to the 'come-back-next-week-and-have-another-look'. An example

Figure 14.1 (left) Explosive reactions take place in a fraction of a second while the rusting of a wrecked ship (right) may take centuries to complete

of the latter is the rusting of iron. An 'instantaneous' reaction might be the reaction between aqueous solutions of alkali and acid.

Although reaction rates are extremely wide ranging, we can make some generalisations to help us study them:

1 All chemical reactions involve either bond making, bond breaking, or most likely both.
2 For the majority of reactions to happen, there must be some sort of **collision** between particles (Figure 14.2).

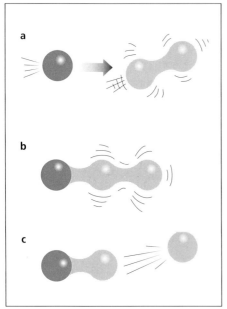

Figure 14.2 A collision picture of a chemical reaction

3 Anything that increases the chance (and therefore the frequency) of collisions will speed up the reaction.
4 Not every collision will be sufficiently violent to rupture the bonds that need breaking. Anything that makes a collision more forceful will speed up the reaction.

Question

1 a Suggest one factor that would increase the chance of collisions happening in a reaction.

b Suggest one factor that would increase the violence of a collision.

In answer to question 1a, you probably named factors such as **concentration** (or **partial pressure** in the case of gaseous reactions), or maybe **surface area** in a heterogeneous system. You may also have cited **temperature**, since faster-moving molecules will collide more often. However, the main influence of temperature is on the violence rather than the frequency of collisions (question 1b).

You may have mentioned **catalysts**. Catalysts create conditions in which the collisions do not have to be so forceful for the particles to react.

Question

2 Suppose someone said 'stirring' in answer to either part of question 1. Would you say the answer was totally wrong, right in some situations or totally right? Explain your answer.

We have focused on two main factors that have a profound effect on the rate of a reaction, namely temperature and concentration. Temperature will be covered in Section 14.11. We begin with the effect of concentration.

14.2 The effect of concentration on reaction rate

Let us begin with a statement that is both important and yet very simple: concentrations of reactants obviously go down during reactions.

Question

3 What does this fact imply about the rate of a reaction as time passes? (You may need to revisit your answer to question 1a.)

Figure 14.3 shows the concentration of a reactant measured against time as the reaction proceeds. The reaction is of the type:

$$A + B \rightarrow C + D$$

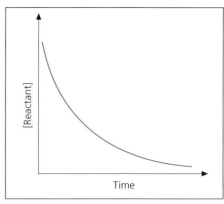

Figure 14.3 Graph showing how a reactant concentration varies during the course of most types of chemical reaction

Questions

4 Does the curved shape of this graph agree with your answer to question 3? How could you get a number from a graph like this which represents the rate of the reaction at any particular point? (*Heavy clue:* Think about tangents.)

5 If the *y*-axis of Figure 14.3 was concentration of A in mol dm^{-3}, and the time was in seconds, what would be the units of the rate of reaction?

6 If you calculated the rate of reaction as in question 4, you would have a number that expressed the rate of disappearance of reactant A. How would that be related to the rate of disappearance of the other reactant B, or indeed the rate of *appearance* of the products C and D? (*Clue:* The answer lies in the stoichiometry of the reaction, which you can see is 1 : 1 : 1 : 1.)

A graph like Figure 14.3 fits in with certain common-sense attributes of reaction systems, for example, reactions slow down because there is progressively less stuff to collide, and gradients of concentration/time graphs should therefore get progressively shallower. But to be more rigorous, we need to propose a mathematical model for the relationship between rate of reaction and concentration, and then to investigate whether the predictions of the model correspond to real experimental data.

14.3 A mathematical model of a bimolecular collision reaction

Figure 14.4a shows equal concentrations of A and B in the reaction:

$$A + B \rightarrow C + D$$

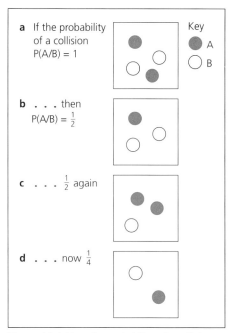

a If the probability of a collision P(A/B) = 1

Key
A
B

b . . . then P(A/B) = $\frac{1}{2}$

c . . . $\frac{1}{2}$ again

d . . . now $\frac{1}{4}$

Figure 14.4 How the probability of a collision between A and B varies as the reactant concentrations change

Let us assume that the reaction is **bimolecular** – that is to say, it proceeds via a simple collision between a molecule of A and a molecule of B. Imagine the effect of halving the number of molecules of A while keeping the number of molecules of B and the volume constant (Figure 14.4b). Clearly the chance of a collision between a molecule of A and one of B would be halved. By the same token, halving the number of molecules of B while keeping the number of molecules of A and the volume constant would have the same effect (Figure 14.4c). So halving the concentration of both while keeping the volume constant should *quarter* the chance of collision, and with it the rate of reaction (Figure 14.4d).

Halving the number of particles while keeping the volume constant is the same as halving the concentration. Our model therefore predicts that the rate should be proportional to each of the concentra-

tions of A and B. We can express this in an equation as shown (to address both mathematicians and non-mathematicians, word equations and formal differential equations will be given together in this chapter):

$$\text{Rate} = k[A][B] \quad \text{or} \quad -\frac{d[A]}{dt} = k[A][B]$$
$$(14.1)$$

Here are a few explanations:

1 The minus sign registers the fact that [A] (along with [B]) is going *down* with time, and therefore d[A] is negative. Otherwise the equation would imply that the rate of *increase* of [A] was proportional to [A][B].
2 k is the constant of proportionality. It is called the **rate constant**. This will be numerically equal to the rate when the two concentrations [A] and [B] are 1, in whatever units are being used. In simple terms, a big rate constant means an inherently quick reaction.
3 If you check back to your answer to question 6 (p. 318), you will realise that
$$\frac{d[A]}{dt} = \frac{d[B]}{dt} = -\frac{d[C]}{dt} = -\frac{d[D]}{dt}.$$

In other words, the same rate could be expressed by referring to any of the participants in the system.

Question

7 Things get a little more complex when the stoichiometry is not 1:1:1:1. Suppose the reaction were:

$$A + B \rightarrow 2C$$

The rate of disappearance of A and B would still be equal, but how would the rate of appearance of C relate to the other two?

4 Let us return to our discussion of equation (*14.1*). The equation is called a **rate equation** or **rate law**. Although this one has been derived by theoretical modelling, we will be meeting plenty that are deduced from experimental data. Whatever their origins, rate laws always have the rate on the left, expressed as a function of various reactant concentrations on the right.

5 Rate laws are described using the term **order**. The word applies to the power of each concentration in a rate law. Thus in the generalised rate law:

$$\text{Rate} = k[A]^a[B]^b \quad (14.2)$$

you would say that the order of reaction with respect to A was a, the order of reaction with respect to B was b, and the overall order of the reaction was $(a + b)$. Turning back to the rate equation (*14.1*), we are predicting that a reaction involving simple bimolecular collision between A and B would be first order with respect to each ingredient separately, and second order overall. (In other words, $a = b = 1$.)

Question

8 We have been referring to reactions that begin:

$$A + B \rightarrow \ldots$$

and concluding that the rate of decline of [A] would be equal to the rate of decline of [B]. Would that conclusion be just as correct if [A] and [B] had been unequal at the start of the reaction, say twice as much A as B? Explain your answer.

As a footnote to this section, note that a single-step bimolecular collision, as modelled here, must always result in a second-order (overall) rate equation. As we shall see later, the reverse is *not* always true – in other words, not all second-order rate equations signify simple bimolecular single-step collisions. This will be covered when we discuss mechanism in Section 14.9.

14.4 A mathematical model of a unimolecular process

Suppose we had a reaction in which a solo reactant simply decomposed, for instance:

$$A \rightarrow B + C$$

We could still imagine the reaction proceeding via collisions, but this time it might be collisions with, say, solvent molecules which acted just as passive 'bumpers', and whose presence therefore remained a constant factor. Common sense leads us to say that if [A] were halved, the chances of collision would be halved too, and therefore the rate would halve. Since there is nothing else on the reactant side that can vary, this produces a rate equation in which the rate is proportional to just a single concentration term, in other words an overall first-order rate equation.

$$\text{Rate} = -\frac{d[A]}{dt} = k[A] \quad (14.3)$$

The decay of a radioactive isotope follows this theoretical model. It hardly qualifies as a chemical system at all, but it does fit perfectly the blueprint of a change that a particle undergoes all on its own. Such a change is called a *unimolecular* process. It would clearly follow equation (*14.3*), on the basis that the fewer the atoms left to decay, the slower will be the rate of that decay.

14.5   Do real systems fit theoretical models?

Now for a crucial stage in the development of any theory: does it fit reality? A more humble question might be: how will we know when it does? We need to think harder about what those equations are telling us. When we write:

$$\text{Rate} = -\frac{d[A]}{dt} = k[A] \quad (14.3)$$

we are really saying that if [A] were to halve, then so would the rate, or what amounts to the same thing as rate, the gradient of the concentration/time plot. So we should be able to plot a real concentration/time graph from an experimental run, and draw two tangents, one at t_0 (when time = 0, i.e. at the start) and one at a later time when [A] had dropped to half its original value (Figure 14.5). The gradients of the tangents would represent the rates at these two moments in the course of the reaction. We could therefore diagnose whether or not the system was following our proposed mathematical scheme.

Question

9 If the real system were behaving in a first-order way, what would you expect to be the relationship between those two gradients?

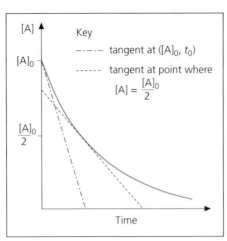

Figure 14.5 Testing equation (*14.3*) by the method of two tangents

If the real system were behaving in a second-order way, the possibilities are a bit more complicated. Let us first consider a simple case in which two reactants begin at equal concentrations of 1 mol dm^{-3}, so:

$$[A]_0 = [B]_0 = 1$$

(where the '0' subscript signifies 'at the start').

Let us furthermore assume that the stoichiometry is simple:

$$A + B \rightarrow C + D$$

Questions

10 What would be the value of [B] at a time when [A] had fallen to 0.5 mol dm^{-3} (i.e. half its original value)?

11 If both reactant concentrations had dropped as in question 10, and if the system were obeying the rate equation:

$$\text{Rate} = k[A][B] \quad \text{or} \quad -\frac{d[A]}{dt} = k[A][B]$$
$$(14.1)$$

by what factor would the rate have fallen at that time, compared with the rate at the start?

12 If we drew the same two tangents as in Figure 14.5, that is, at [A]$_0$ and at [A]$_0$/2, what ratio of gradients would confirm the second-order rate equation?

13 If the experiment had begun with reactant B in a large excess, our two-tangents method would not have worked. Let us see why not. Let us assume we had begun with 1 mol of A and 10 mol of B in 1 dm^3.

a By the time [A] had fallen to 0.5 mol dm^{-3}, what would [B] have fallen to? (*Remember:* They react in the ratio 1:1.)

b The rate at the start was equal to $k \times 1 \times 10$ mol dm^{-3} s^{-1}. What will the rate be when [A] = 0.5 mol dm^{-3}? (Remember k is a constant in the equation and applies to all concentrations.)

c By what factor would the rate have fallen at [A]$_0$/2, compared with the start?

You can see from question 13 that if [B] is in a large excess, the graph of [A] against time would look more like a first-order curve, with the tangent approximately halving. Complications like this make second-order reactions more difficult to diagnose than first-order ones.

One second-order system is easier to deal with. Think about a reaction in which a single reactant decomposes, but not in the way described in Section 14.4. Instead of decomposing on collision with passive bystanders (vessel walls, solvent molecules, etc.), our molecules have to collide with *each other*, whereupon they both decompose. Under these circumstances the rate of the reaction is more sensitively dependent on the decline of [A], because not only are there fewer A molecules to react, but there are also fewer to react *with*. It is as if we had an 'A + B' situation in which the Bs just happen to be As. Perhaps you can see that equation (*14.1*) would read:

$$\text{Rate} = k[A]^2 \quad \text{or} \quad -\frac{d[A]}{dt} = k[A]^2$$
$$(14.4)$$

In a real reaction system, a halving of [A] would lead to a quartering of [A]^2, so the ratio of the two tangents would be 4:1.

By now you should be ready to try out your diagnostic tests on some real systems. You can decide which kind of rate equation they conform to.

Questions

14 The data in this question are taken from the hydrolysis of 1-bromobutane by OH^- ions in a mixed alcohol/water solvent at room temperature. (The reaction is a nucleophilic substitution.) The stoichiometric equation is:

$$C_4H_9Br + OH^- \rightarrow C_4H_9OH + Br^-$$

a What do you suspect would be the rate equation for this reaction?

The experimenter began with both reactants at 0.25 mol dm^{-3}. The data (Table 14.1) refer to the changing OH^- concentration, but remember that the bromobutane concentration will be falling in parallel.

Table 14.1 Data for question 14

Time (hours)	$[OH^-]$ (mol dm^{-3})
0	0.250
10	0.200
20	0.168
30	0.142
40	0.125
60	0.098
80	0.084
100	0.071
120	0.063
140	0.055

b Plot a graph and use it to determine the overall order of the reaction by the 'two tangents' method. Write a rate equation for the reaction.

c For one of your gradient values, find the two concentrations that existed at that moment. Insert them into the rate equation to obtain a value for the rate constant. Check it *is* a constant using the concentrations for the other gradient. What are the units of the rate constant?

d In Section 11.3, p. 228, we said that the mechanisms of nucleophilic substitution fall into one of two categories, known as S_N1 and S_N2. At the time the significance of the numbers 1 and 2 was not fully explained – it was merely said that they expressed the overall order of

the reaction. In the light of your new knowledge, can you say which of the two mechanisms we are dealing with here?

e The fact that 1-bromobutane is an unbranched molecule might have led you to predict the mechanism before diagnosing the data. Thinking back to Chapter 11 again, can you explain the basis for this?

15 The data in this question refer to a reaction in the gas phase, with only one reactant, although there are two molecules of it in the stoichiometric equation. The reaction is the thermal decomposition of dinitrogen pentoxide:

$$2N_2O_5(g) \rightarrow 4NO_2(g) + O_2(g)$$

a What do you suspect, looking at the stoichiometry, will be the rate equation for this reaction?

Table 14.2 Data for question 15

Time (s)	$p(N_2O_5)$ (10^4 Pa)
160	3.45
225	3.13
285	2.83
340	2.64
400	2.41
460	2.16
525	1.97
580	1.80
700	1.52
825	1.28
950	1.05
1060	0.87
1300	0.60
1540	0.43

The data (Table 14.2) are the partial pressures (which you can treat as concentrations) of dinitrogen pentoxide at different times. The temperature was 55 °C.

b Again, plot a graph to find the order of reaction by the 'two tangents' method, and as in question 14, write out the rate equation. Find a value for the rate constant, and its units.

As we leave this section, let me point out an important lesson to be learnt from questions 14 and 15. Your guess as to the rate equation in question 14 may well have been right. If so, your success may have led you to think that rate equations can be predicted with confidence from the stoichiometric equation. However, question 15 demonstrates that this is not the case. The stoichiometry, with two dinitrogen pentoxide molecules on the left, suggested a rate equation of the type:

$$-\frac{d[A]}{dt} = k[A]^2 \qquad (14.4)$$

However, analysis of the data showed the reaction to be first order. So remember:

You cannot necessarily predict rate equations from stoichiometric equations.

14.6 More diagnostic tests for use on experimental data

(It should be said at the outset of this section that doing every single question is very laborious and time consuming. The asterisks (*) indicate paragraphs and questions that may be omitted on at least the first reading.)

So far we have used one rather crude test to determine whether a set of data fits first- or second-order rate laws – namely the method of 'two tangents'. Selective use of such small sections of the total data, along with the error involved in drawing a tangent to a hand-drawn curve, is obviously not very reliable. There has to be a better way – and in fact there are many better ways, described below.

Drawing lots of tangents

This is not much more sophisticated than the 'two-tangents' method, but it does at least use more of the curve. Instead of two set-piece points, tangents are drawn at five or six points (Figure 14.6a). Then a second graph is drawn in which the tangent gradients (i.e. the rates) are plotted against the concentrations at the points where the tangents were drawn (Figure 14.6b). If this second graph gives

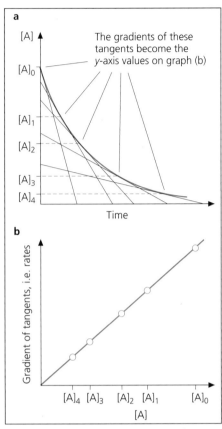

Figure 14.6 a Concentration/time graph.
b Rate/concentration graph

a straight line, this tells you that $d[A]/dt$ is indeed proportional to $[A]$, so the reaction is first order with respect to A.

Questions

16 Do this operation on the data from question 15. (To give you just a flavour of the idea, keeping labour to a minimum, just use the two gradients you already have.)

17 What meaning would you attach to the gradient of your second graph?

18 How could this method be adapted to check for second-order kinetics – i.e. to prove that $d[A]/dt$ was proportional to $[A]^2$ instead of to $[A]$?

*Log(rate) versus log[A]

This method is not usually used at Advanced level, and non-mathematicians may wish to skip it. However, it is very clever and neat.

This method is a variant on the lots-of-tangents method. The drawback with that method is that if you guess the order of the reaction wrongly, you waste a graph. Suppose your graph of $d[A]/dt$ against $[A]$ or $[A]^2$ was *not* linear. You would then have to discard it and plot another graph. (In fact many different orders exist other than first and second order – there are even fractional orders – so you could not assume that failure of one test implied success with the other.) The log method works for any order, provided the rate law can be expressed in terms of one single concentration variable. Suppose the rate law had the form:

$$\frac{d[A]}{dt} = -k[A]^a \qquad (14.5)$$

Taking logs gives:

$$\begin{aligned} \log(\text{rate}) &= \log(-k) + \log[A]^a \\ &= \log(-k) + a\log[A] \end{aligned} \qquad (14.6)$$

(Either 'log' or 'ln' may be used in this method.) So a plot of $\log(\text{rate})$ (rate is obtained from drawing tangents again) against $\log[A]$ (the concentrations at the points where the tangents are drawn) would be of the form $y = c + mx$. The advantage of this method is that this second graph is always linear. Moreover, if $\log[A]$ corresponds to x, then the order a will be given by m, the gradient (and the rate constant k by c, the y intercept).

*Question

19 Try this method on the data from question 14.

*Fitting data to integrated versions of rate equations

Equations such as (14.3) (p. 320) and (14.4) (p. 320) are known to mathematicians as differential equations. They can be integrated to produce equations which tell a different version of the same truth. For example, the differential equation (14.3):

$$\frac{d[A]}{dt} = -k[A] \qquad (14.3)$$

may be integrated:

$$\ln[A] = -kt + \text{a constant} \qquad (14.7)$$

The constant is actually $\ln[A]_0$, where $[A]_0$ indicates the value of $[A]$ at the start of the reaction.

Another version of this equation reads:

$$[A] = [A]_0 e^{-kt} \qquad (14.8)$$

We shall return to this later.

If you are not an Advanced-level mathematician, you can just view 'ln' in equation (14.7) as a button on your calculator. If you record a series of values of $[A]$ at different times, you can plot $\ln[A]$ (y-axis) against time. If this turns out to be a straight-line graph, then you have a positive test for a first-order rate equation. The gradient of the straight line will be k, because equation (14.7) is of the form $y = mx + c$.

*Question

20 Try the $\ln[A]$ versus t plot on the data from question 15. Again, to save work, you need not use every point.

Half-lives

This method is simple and quick, but it only works for first-order reactions. The mathematical justification for the method is given here, but non-mathematicians can skip this and just remember the conclusion.

Let us feed some data into equation (14.8), the first-order integrated rate equation.

$$[A] = [A]_0 e^{-kt} \qquad (14.8)$$

The data come from the time after which the concentration $[A]$ has fallen to half its initial value. We shall call that time $t_{\frac{1}{2}}$ (in words, the **half-life** of the equation (Figure 14.7)). At time $t_{\frac{1}{2}}$, then, the value of $[A]$ is $[A]_0/2$:

$$\frac{[A]_0}{2} = [A]_0 e^{-kt_{\frac{1}{2}}}$$

Cancelling $[A]_0$:

$$\tfrac{1}{2} = e^{-kt_{\frac{1}{2}}}$$

Taking logs of each side:

$$\ln\left(\tfrac{1}{2}\right) = -kt_{\frac{1}{2}}$$

Therefore:

$$t_{\frac{1}{2}} = \frac{\ln 2}{k} \qquad (14.9)$$

Look carefully at the right-hand side of equation (14.9). 'ln 2' is just a number $(0.6931 \ldots$ actually). k is a constant too, for any particular reaction. So the right-

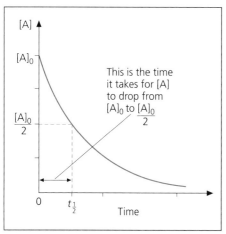

Figure 14.7 Illustrating the concept of half-life

hand side is a constant divided by another constant. The overall message of equation (*14.9*) is therefore that:

The half-life of a first-order reaction is a constant.

You may have said to yourself that if the half-life is the time it takes for [A] to fall from $[A]_0$ to $[A]_0/2$, then of course the half-life is constant. However, the half-life can be measured at *any* period in the reaction during which the concentration halves. So if we start at any value of [A], at any stage in the reaction, and time how long it takes for that value of [A] to halve, that time will be a half-life. Equation (*14.9*) is saying that if you measure the half-life several times at different starting points on the same curve, and get the same half-life from each calculation, then your curve has tested positive for a first-order rate law (Figure 14.8).

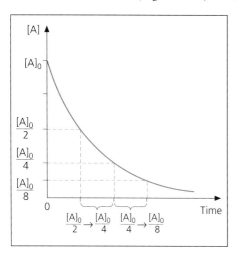

Figure 14.8 (left) Showing how, in the case of first-order reactions, the half-life is a constant. Notice how it takes as long to go from $[A_0]/2$ to $[A_0]/4$ as it does to go from $[A_0]/4$ to $[A_0]/8$. In fact, it would take the same length of time to halve *any* value of [A]

Question

21 Return to your plot of the data in question 15. Find three half-lives on different sections of the curve, and test to confirm a first-order reaction. Find a value for the half-life, and use equation (*14.9*) to derive a value for the rate constant. Compare this with the value you obtained in question 15.

Why investigate rate equations?

Let's just pause to remind ourselves of the purpose of all these calculations. We are trying to identify certain rate laws using the appropriate graphs. Rate laws give an insight into the mechanisms of reactions. For example, a first-order rate law suggests that the reaction involves a single molecule breaking up. A second-order rate law suggests that the reaction involves a simple collision between two molecules. Knowledge about mechanisms enables us to manipulate reactions to our advantage. We shall return to this in Section 14.9.

*Integrated rate equation for a second-order reaction

For simplicity, we shall only consider the case of a reaction that is second order with respect to a single reactant. The reaction happens because one molecule of A hits another molecule of A. The differential rate equation is:

$$-\frac{d[A]}{dt} = k[A]^2 \qquad (14.4)$$

This integrates to:

$$\frac{1}{[A]} - \frac{1}{[A]_0} = kt \qquad (14.10)$$

In this case, therefore, the diagnostic test is to plot $1/[A]$ against t, whereupon a straight line will be proof of second-order kinetics. The gradient will be k.

*Question

22 Carry out this diagnostic test on the data in question 14, and report a rate constant. (Again, do not use all the data.) How do you think the last three methods compare, in both ease of use and reliability, to the 'differential' methods involving tangents?

Remember, from question 14, that the two reagents, OH^- and C_4H_9Br, began at equal concentration. So $[OH^-]$ = $[C_4H_9Br]$ at all times. We have just made the data fit a graph which gives the impression that the rate is proportional to $[OH^-]^2$, but this is really $[OH^-][C_4H_9Br]$ 'in disguise'.

Using data relating to the build-up of product

So far we have used data relating to reactant concentrations going down. Let us suppose that it was more convenient to follow the change of a product concentration over time. (Perhaps the product offered an easier method of concentration measurement, such as an easy titration.) The resulting graph would look like Figure 14.9. Despite the upward trend of Figure 14.9, there are ways of adapting the data so that we can still use the method of half-lives, or any of our other tests.

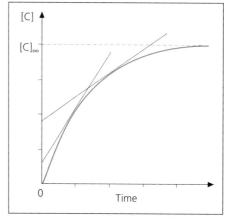

Figure 14.9 Showing what happens to a product concentration during a reaction. The gradients still give a measure of the rate of reaction

Let us suppose the reaction was:

$$A \rightarrow B + C$$

and we were following [C]. Since the rate of appearance of a product is normally equal to the rate of disappearance of a reactant, it would be true to say that $-d[A]/dt = d[B]/dt = d[C]/dt$. So we could adapt the (say) first-order rate law easily enough to read:

$$\frac{d[C]}{dt} = k[A] \qquad (14.11)$$

We can plot a graph of [C] versus t (Figure 14.9) and draw tangents to give a set of values of $d[C]/dt$. Then comes a problem. In order to demonstrate that (14.11) was being followed, we'd have to relate our tangents to the corresponding [A] values. Unfortunately, though, we have only measured [C]. However, the value of [A] at any time is closely related to the simultaneous value of [C], since the remaining [A] plus the newly created [C] were once all together in $[A]_0$ (Figure 14.10).

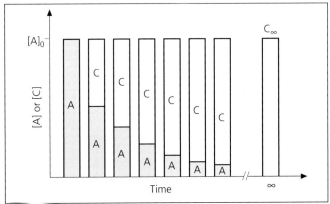

Figure 14.10 Suggesting a way of relating reactant concentrations [A] to product concentrations [C]. This diagram assumes a 1:1 ratio between A destroyed and C created

Question

23a Express these ideas in symbols, and thereby work out a formula to convert a value of [C] to the equivalent value of [A]. (Figure 14.10 may help you.)

b What is the special-case version of this relationship that applies at t_∞ ('t infinity'), i.e. when the reaction has finished and all the molecules of A have gone?

So you see you can still draw your tangents, which still stand for the rates. The values of [C] can be converted to values of [A] by $[A] = [A]_0 - [C]$ (question 23a) or by $[A] = [C]_\infty - [C]$ (which from question 23b you will see amounts to the same thing). You can now plot rates against values of [A], using the method of question 16.

Another approach uses the same conversion formula, but in a slightly different way. Values of [C] are converted to corresponding values of [A] before drawing the first graph. The resulting graph, as you might expect, looks like those we plotted *before* from reactant concentrations. Any of the diagnostic methods can then be used on this graph. Question 24 uses this approach.

Question

24 Benzenediazonium chloride (introduced in an organic context in Chapter 15) decomposes in aqueous solution unless kept below about 5 °C. The chemical equation is:

$$C_6H_5N_2{}^+Cl^-(aq) + H_2O(l) \rightarrow$$
$$C_6H_5OH(aq) + N_2(g) + HCl(aq)$$

The data in Table 14.3 refer to [HCl] at different times. Hydrochloric acid corresponds to C in question 23, while benzenediazonium chloride corresponds to A. So $[C]_\infty$ is [HCl] after all reaction has ceased.

Table 14.3 Data for question 24

Time (min)	[HCl] (10^{-2} mol dm^{-3})
0	0
1	0.83
2	1.45
4	2.90
6	3.88
8	4.60
12	5.56
14	6.05
∞	7.50

a Find reactant concentrations from these figures using the formula

$$[A] = [C]_\infty - [C]$$

(*Note*: Even though there are several products, the ratio ([A] destroyed : [C] created) is still 1:1). Write this new set of values of [A] as a third column in a copy of Table 14.3. Now use one of the standard methods to find the order of reaction with respect to benzenediazonium chloride.

b Report a rate constant.

c Why would these data tell us nothing about the order of reaction with respect to the other reactant? (Remember the other reactant is water, the solvent, which is present in a very large excess.)

Isolation techniques – pseudo-zero-order rate equations

So far we have seen several methods for analysing real data and fitting them to theoretical rate laws. But such methods only work for a relatively restricted set of circumstances, namely:

1 reactions whose overall order is one
2 second-order reactions of the rate type $-\dfrac{d[A]}{dt} = k[A]^2$

3 second-order reactions of the type $-\dfrac{d[A]}{dt} = k[A][B]$ in

which the experimenter has ensured that $[A]_0 = [B]_0$.

We do not have a method for analysing an 'A/B'-type reaction without these special starting conditions, nor do we have methods for overall third- or fourth-order rate equations. To overcome these problems, we can arrange things so that we can isolate the effect on the rate of reaction of one concentration at a time.

Isolation of one reactant is achieved by having all the other reactants in a large excess (greater than $\times 10$).

Question

25 Imagine a reaction between A and B, of 1:1 stoichiometry, which started with $[B]_0 = 10[A]_0$. If $[A]_0$ were 0.1 mol dm^{-3}, what would $[B]$ have been at the start? What would it have dropped to by the time $[A]$ had run out? (Figure 14.11 may help you here.)

Figure 14.11 Showing how two reactant concentrations would vary in a 1:1 reaction if B was in a large excess

$[B]$ remains fairly constant throughout the reaction run. If $[B]$ is approximately constant, then it can be combined with the rate constant k, as in this example from a second-order rate law:

$$\text{Rate} = k[A][B] = k'[A] \quad (14.12)$$
$$(\text{where } k' = k[B])$$

Now we can focus our attention on the variation of the rate with $[A]$ alone, and use any of the diagnostic tests so far developed. (In this case of course, we would have discovered a first-order dependence on $[A]$.) This method eliminates any excess reactants from the effective rate law, irrespective of their real orders. It is as if the reaction had become zero order with respect to these reactants, since they 'disappear' from the rate law in much the same way as a true zero-order reactant would. (Remember, for any x, $x^0 = 1$.) The expression **pseudo-zero order** is used to describe how the rate depends upon the concentration of a reactant in a large excess.

*Question

26 Any reactant that is also the solvent for the reaction is obviously present in a large excess, as was the case with water in question 24. Table 14.4 gives data for a nucleophilic substitution by water onto a branched isomer of 1-bromobutane, called 2-bromo-2-methylpropane. The reaction is carried out in a propanone/water mixture at 20 °C.

Table 14.4 Data for question 26

Time (min)	[2-bromo-2-methylpropane] (10^{-3} mol dm^{-3})
0	10.0
3	9.0
8	7.6
12	6.5
18	5.2
23	4.3
28	3.7
43	2.1
63	1.0

a Select one of the diagnostic tests already mentioned, and establish the order of reaction with respect to the organic substrate. Find a rate constant. Why can you not state the rate equation for the reaction?

b Although it is impossible to derive the overall rate equation from this evidence alone, you could hazard a guess, based on the fact that the substrate molecule is highly branched. With Chapter 11 in mind, suggest whether this mechanism might be S_N1 or S_N2, and thereby suggest an overall rate equation.

14.7 Initial rates method

So far all our work has been directed at reading the message in a whole set of data from a single run of a reaction. The drawbacks in this procedure have already been highlighted, but let's repeat the main one anyway – single-run whole-curve analysis is only really good at giving information about *one ingredient of the reaction mixture at a time*. Therefore the whole-curve-analysis methods usually employ large excesses of concentration, to create a state of 'pseudo constancy' for all but the 'focus' reactant, and therefore if information is needed about other reactants then other runs are going to have to be set up.

But if other runs are going to be needed anyway, there is a dramatically different way of using them, and 'reading' them. The new idea is to forget whole curves altogether. Whole-curve methods use the reaction itself to vary the concentration of a reactant, after which the human experimenter decodes the message in the concentration/time graph. It is as if we had left it to Nature to manipulate our variables for us. In the new method, called the *initial rates* method, all the manipulation of reactant concentrations is human. The experimenter mixes some known concentrations of reactants and then measures the rate of reaction just at that initial moment when the reactants are *at* those concentrations. The reaction mixture is then *thrown away*. The experimenter then mixes a new mixture of reactants, with one of the concentrations doubled or halved or whatever. The new initial rate is measured. Subsequently, run by run, each reactant concentration is varied relative to the first run. Each run is only kept going long enough to measure the initial rate and is then thrown away. By this means the relationship between rate and concentration is found for every reactant in turn. For instance, if an initial rate *quadruples* when a reactant concentration is *doubled* then the rate law is second order with respect to that reactant. Question 27 should test whether you have picked up the gist of this method.

Question

27 The redox reaction between iodate(v) ions and iodide ions proceeds measurably slowly at low concentrations.

$$IO_3^-(aq) + 5I^-(aq) + 6H^+(aq) \rightarrow 3I_2(aq) + 3H_2O(l)$$

Table 14.5 gives some initial rate data for the reaction at 298 K.

a Comparing run 1 and run 2, which initial concentration has varied, and by what factor?

b By what factor did the initial rate change as a result?

c What therefore is the order of reaction with respect to this reactant?

d Compare other pairs of runs to find the effect of changing concentrations of the other reactants.

e Derive a rate equation for the reaction.

f Use any set of corresponding data from the table to calculate the rate constant. Consider carefully its units.

Notice that in question 27 it did not matter which reactant was in excess, or indeed if any reactant was in excess. One of the beauties of the initial rates method is that, by requiring the rate to be known merely at a single moment in time, it eliminates all the tiresome need to create 'pseudo-constant' concentrations.

Question

28 Now try to analyse the initial rate data in Table 14.6. The reaction is in the gas phase, and the 'concentrations' are partial pressures. The reaction is between hydrogen and nitrogen monoxide, at 1100 K.

$$2H_2(g) + 2NO(g) \rightarrow 2H_2O(g) + N_2(g)$$

As in question 27, report a rate equation and a rate constant.

Table 14.6 Data for question 28

Run	$p(NO)$ (Pa)	$p(H_2)$ (Pa)	Initial rate (Pa s^{-1})
1	1580	263	43.4
2	1580	526	88.1
3	1580	789	131
4	263	1580	7.24
5	526	1580	28.9
6	789	1580	65.5

You may be wondering exactly how *initial* rates are measured. After all, finding a rate involves dividing a number representing a change of concentration by a number representing a passage of time. So it follows logically that time must be allowed to pass. It follows in turn that not all the measurements can be taken at that initial moment. However, there is no real paradox – the data are collected

by methods identical to those used in whole-curve runs, but only for as long as it takes to enable an extrapolation back to 'time zero'. Figure 4.12 shows how it needs only an early bit of whole-curve to allow the reading of an initial rate.

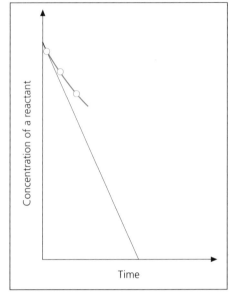

Figure 14.12 Showing how data collection over the early portion of a reaction can be used to give a value for the rate at 'time zero' – the initial rate

Now that we've met the fact that both whole-curve and initial rates methods use the same repertoire of data-gathering techniques, the obvious question is 'What *are* those techniques?'. The next section offers some answers.

14.8 Practical methods

You have so far worked with quite a lot of data, all of it given in terms of concentration or partial pressure units (as well as units of time). As it turns out, the 'concentration' data don't absolutely *need* to be in the classical 'moles per cubic decimetre' format at all.

In fact, raw data are gathered from variables as diverse as optical rotation and volume of solution. In general, anything you can measure that changes during the course of the reaction and that can be related back to a concentration will serve as raw data.

There are many convenient measurements that can be taken, depending on the reaction system.

Table 14.5 Data for question 27

Run	$[H^+]_0$ (10^{-3} mol dm^{-3})	$[I^-]_0$ (10^{-4} mol dm^{-3})	$[IO_3^-]_0$ (10^{-4} mol dm^{-3})	Initial rate (10^{-9} mol dm^{-3} s^{-1})
1	2.0	4.0	0.37	7.1
2	2.0	4.0	0.74	14.2
3	2.0	4.0	1.48	28.4
4	1.0	4.0	1.48	7.1
5	2.0	2.0	1.48	7.1
6	1.0	1.0	1.48	0.44

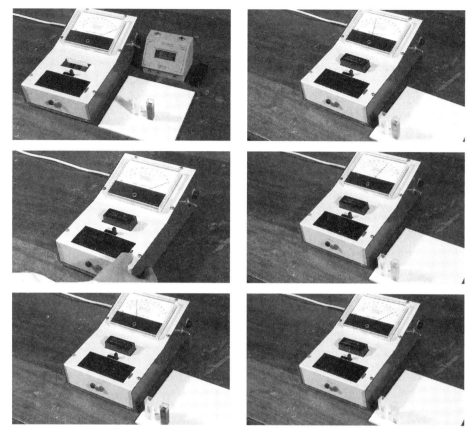

Figure 14.13 How a colorimeter converts depth of colour to an electrical signal (top to bottom, left to right)

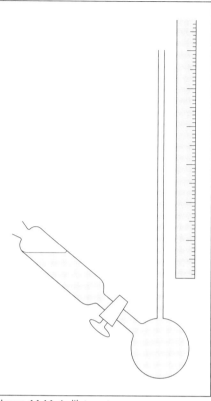

Figure 14.14 A dilatometer

1 If one of the reactants or products is coloured, then the whole reaction can be carried out in a colorimeter (Figure 14.13). (Using appropriate apparatus, 'colour' can mean anything from ultra-violet to infrared emission or absorption.) This method has the advantage that you do not have to disturb the reaction to take measurements.

2 If a reactant or product is chiral (p. 280), then optical rotation can be measured to follow the reaction.

3 If a gas is given off, then its volume can be followed.

4 If gaseous reactions involve changes in the number of molecules, then the total pressure indicates the extent of reaction.

5 Samples can be removed for titration.

6 In the absence of any other measurable change, you can use the fact that nearly all solution reactions are accompanied by small changes in volume. These can be turned into measurable movements by a machine called a **dilatometer** (Figure 14.14).

Methods using variables that are proportional to the concentration of a reactant or product

Colorimeters and spectrophotometers (Figure 14.13) are instruments designed to give readings of **absorbance**. This quantity is directly proportional to the concentration of absorbing species. The two are related by the Beer–Lambert law:

$$A = \varepsilon c l \qquad (14.13)$$

where A is the absorbance, c the concentration, l the length of the path taken by the light through the absorbing medium, and ε a constant of proportionality known as the **extinction coefficient**.

Let us see how data gathered in this way can be fitted to a first-order rate equation. The concentration of the reactant is now represented by c rather than [A] as previously, since A now means the absorbance of the reactant.

$$-\frac{dc}{dt} = kc \qquad (14.3)$$

But, from equation (14.13): $c = \dfrac{A}{\varepsilon l}$

Therefore:

$$-\frac{1}{\varepsilon l}\left(\frac{dA}{dt}\right) = \frac{kA}{\varepsilon l}$$

Cancelling $1/\varepsilon l$:

$$-\frac{dA}{dt} = kA \qquad (14.14)$$

This is in the form of the familiar first-order differential equation (14.3), except with absorbance data in place of the concentrations. This result means that absorbances can be used for diagnostic tests for first-order kinetics in exactly the same way as concentrations themselves, and will give the same rate constant.

Can we use absorbance data to do similar tests for second-order rate equations? Let us see:

$$-\frac{dc}{dt} = kc^2 \qquad (14.4)$$

Substituting:

$$-\frac{1}{\varepsilon l}\frac{dA}{dt} = \frac{kA^2}{\varepsilon^2 l^2}$$

Therefore:

$$-\frac{dA}{dt} = \frac{k}{\varepsilon l}A^2$$

or:

$$-\frac{dA}{dt} = k'A^2, \quad \text{where } k' = \frac{k}{\varepsilon l} \quad (14.15)$$

So if a reaction were second order with respect to the absorbing species, the absorbance data could be used quite satisfactorily for diagnostic tests instead of the concentration, except that any rate constant derived would not be the true one. (It would be k' in equation (14.15).)

Question

29 If you knew k', what extra information would you need to derive the true rate constant k?

So variables that are proportional to reactant concentrations can be used in place of reactant concentrations themselves to find the rate equation. If the variable is proportional to a product concentration, the same is again true. It is a good idea to convert the data so it becomes proportional to [reactant]. Try this example:

*Question

30 The following data refer to the familiar reaction between hydrochloric acid and marble chips (Figure 14.15).

Figure 14.15 Apparatus for collecting carbon dioxide in the reaction between hydrochloric acid and marble chips

In this case the calcium carbonate is in a large excess, and looks almost unchanged after the acid has stopped reacting. The reaction has been followed by measuring the volume of carbon dioxide produced. The results are given in Table 14.7.

Table 14.7 Data for question 30

Time (min)	Volume of carbon dioxide (cm^3)
0	0
1	26
2	49
3	68
4	80
5	91
6	97
7	102
8	106
9	108
10	111
∞	120

Think of the volume of carbon dioxide at time t, V_t, as being proportional to the hydrochloric acid that has reacted. Then see that V_∞ is proportional to the hydrochloric acid

that was there to start with. Now work out a method for combining V_∞ and V_t to get a number that is proportional to the hydrochloric acid present at time t. As before, find the rate equation, order and rate constant.

Using direct titration

In this method, you simply remove small samples of the reaction mixture in a pipette at timed intervals and find out the concentration of a reactant or product by titration. However, there is a practical problem. Unlike, say, absorbance measurements, a titration can take quite a long time to perform.

Question

31 Why is this a problem, and how does this problem limit the range of reactions for which titration is a suitable method?

To overcome this problem, the reaction can be **quenched** as soon as the pipetteful has been removed. Titration can then proceed at your leisure. Quenching means stopping or greatly slowing the rate of reaction, so that concentrations are effectively 'frozen' at the values they had on removal. There are several methods of quenching a reaction:

1 The simplest method is to cool the withdrawn pipette-full quickly, to a temperature at which the reaction rate is substantially slower than the titration rate.
2 Another method is to remove the catalyst if there is one, by dumping the pipette-full into a flask of catalyst remover. For example, a pipetteful from an acid-catalysed reaction could be dumped into excess sodium carbonate solution.
3 A reactant can be removed by reacting it with something that converts it quantitatively to a different, titratable, reagent. For example, samples from a reaction involving an oxidising agent could be dumped into a flask containing excess aqueous iodide ions (and acid if necessary). That would stop the reaction immediately, since one of the reactants would have disappeared. The

oxidising agent would have been replaced by an equivalent amount of iodine. This could be titrated with thiosulphate ions at leisure, knowing that the iodine concentration was proportional to the oxidising agent concentration at the moment of sampling. This method can obviously not be used if both a reactant *and* a product can oxidise iodide ions, or if iodine can carry on oxidising in the reaction mixture as a substitute for the reactant it replaced.

4 Samples can be dumped into a solution containing an excess of a 'reactant killer', and then titration could be used to find how much of the excess survives.

Questions

32 Suggest a quenching method from the above list for:

a propanone + iodine $\xrightarrow{\text{acid catalyst}}$ iodopropanone + hydrogen iodide

b $2MnO_4^-(aq) + 6H^+(aq) + 5(CO_2H)_2(aq) \rightarrow$ $2Mn^{2+}(aq) + 10CO_2(g) + 8H_2O(l)$

c $S_2O_3^{2-}(aq) + C_3H_7Br(aq) \rightarrow C_3H_7S_2O_3^-(aq) + Br^-(aq)$

33 It was suggested that removing samples for titration must slow the reaction down, thereby invalidating the measurement. Admittedly after each removal the system is smaller, but in reality the *rate* is unchanged by the removal. Can you support this view?

*Extension material: Practical methods using variables that are affected by both reactants and products

(This section can be omitted by those not taking S-level or Oxbridge entrance.)

Consider this reaction between gases:

$$A(g) + B(g) \rightarrow C(g)$$

Clearly there are fewer molecules on the right, so overall the pressure will decrease during the course of the reaction (Figure 14.16). However, it will not drop to zero, and therefore the pressure is not directly proportional to the amount of reactants. We need to relate the 'target' variable, p_A, to the two measurable quantities p_0, the pressure at the start, and p_{tot}, the total pressure at any time t (which will vary as the reaction proceeds). We will assume equal initial amounts of A and B, so that p_A is always equal to p_B.

$$p_{tot} = p_A + p_B + p_C = 2p_A + p_C \qquad (14.16)$$

There is a link between p_C and p_A, because the concentration of C is equal to the loss of concentration of A. The 'lost A' is given by $(A_0 - A_{now})$.

Therefore:

$$\text{'lost A'} = \frac{p_0}{2} - p_A$$

(because p_A at the start was $p_0/2$).

Figure 14.16 Apparatus for measuring pressure during a gaseous reaction

Therefore:

$$p_C = \frac{p_0}{2} - p_A$$

Substituting equation (*14.16*):

$$p_{tot} = p_A + \frac{p_0}{2}$$

Therefore:

$$p_A = p_{tot} - \frac{p_0}{2} \qquad (14.17)$$

Question

34 Convert the p_{tot} data in Table 14.8 to p_A data for system

$$A(g) + B(g) \rightarrow C(g).$$

Remember that p_{tot} at the start is p_0. Deduce the order of reaction with respect to A.

Table 14.8 Data for question 34

Time (s)	p_{tot} (atm)	p_A
2.00	0	
1.70	10	
1.50	20	
1.32	30	
1.24	40	
1.17	50	
1.10	60	

The clock technique

Clocks have obviously been involved in all the work so far in this chapter, in that the time t has been measured as the reaction proceeds. However, in a genuine clock technique:

- The clock or stopwatch is only used once per run.
- The time is taken for a certain event to happen.
- This event should happen fairly early in the course of the reaction.
- The rate is then taken to be inversely proportional to the time taken.
- The run is thrown away and a new one, with altered variables, set up.

Figure 14.17 In reactions involving the creation of a precipitate, the 'clock moment' can be the time when the solution becomes completely opaque, after 40 seconds

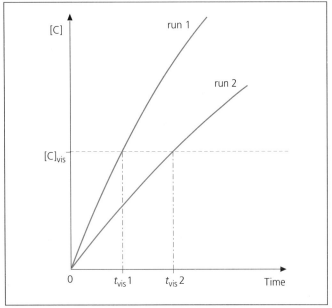

Figure 14.18 Showing how the time taken for an event to happen is inversely proportional to the rate of reaction

There is clearly a link here with the initial rates method. Suppose the event you are waiting for is the development of opacity (turbidity) in the solution (Figure 14.17). In the example in Figure 14.18, this happens at the threshold of visibility, $[C]_{vis}$. The graph shows concentration/time data from the early stages of two runs, up to the point when C becomes visible – that is, when the concentration of C has reached $[C]_{vis}$. This is early enough in the run for the 'curves' to be almost linear, by the same logic that says that if you take a small enough section of the circumference of a circle, it will look like a straight line.

Run 2 started at half the rate of run 1, and so will take twice as long to reach the $[C]_{vis}$ threshold. In symbols:

$$t_{vis}(2) = 2 \times t_{vis}(1)$$

You can see why we said before that the rate is inversely proportional to the time taken. The 'clock' times are in effect an inverse measure of the average rate over the early part of the reaction. As long as the event happens during an early part of the total graph, that rate will be close to the initial rate. (For instance, for a reaction that takes 1 hour to complete, the average rate over the first minute will be approximately the same as the rate at the very start.) The rest of the clock technique is identical to the initial rates method – each 'clocked' run is set up with one different reactant concentration, relative to the baseline of run 1.

Questions

35 One application of clock techniques is to reactions that produce cloudiness (turbidity). The timed event is the moment you can no longer see through the reaction vessel (compare with Figure 14.17 and look forward to Figure 14.29, p. 339). Table 14.9 gives some data for the disproportionation of thiosulphate ion in acidic solution:

$$S_2O_3^{2-}(aq) + 2H^+(aq) \rightarrow SO_2(aq) + S(s) + H_2O(l)$$

Table 14.9 Data for question 35

Run	$[HCl]_0$ (mol dm^{-3})	$[S_2O_3^{2-}]_0$ (mol dm^{-3})	Time taken to go opaque (s)
1	1.0	0.10	63
2	2.0	0.10	32
3	1.0	0.05	130

a Which one of the products is turning the solution opaque?

b From the results given, work out the order of reaction with respect to [HCl] and $[S_2O_3^{2-}]$, and then write out the rate equation.

c The technique described above is a bit rough and ready. Can you think of any refinements of technique or apparatus that would improve accuracy?

36 This question looks back over all the practical methods that have been discussed:

- polarimetry (monitoring the angle of rotation produced by a chiral molecule)
- colorimetry
- titration
- pressure measurement
- clock method (specify the 'clocked' events)

Select from the five techniques listed above those most appropriate to the study of the rate of each of the following reactions. In each case justify your choice.

a $2N_2O_5(g) \rightarrow 4NO_2(g) + O_2(g)$

b $2NO(g) + 2H_2(g) \rightarrow N_2(g) + 2H_2O(g)$

c $BrO_3^-(aq) + 5Br^-(aq) + 6H^+ \rightarrow 3Br_2(aq) + 3H_2O(l)$

d $ClCH_2CH_2Cl(g) \rightarrow CH_2{=}CHCl(g) + HCl(g)$

e ester + water $\xrightarrow{\text{conc HCl catalyst}}$ alcohol + carboxylic acid

f $H_2O_2(aq) + 2I^-(aq) + 2H^+(aq) \rightarrow 2H_2O(l) + I_2(aq)$

g $CH_3CHBrCH_2CH_3 + OH^- \rightarrow$
$CH_3CH{=}CHCH_3 + H_2O + Br^-$ (all in ethanol)

14.9 Mechanisms

We have invested a lot of time and effort into the study of rates. What have we achieved?

We know how to discover a system's rate law. If we know the rate law and rate constant, then we can predict how fast the reaction will be and also which reagent concentrations will have a particular effect on that rate. Both these pieces of information have obvious value to an industrial chemist.

However, the ultimate use to which rate data can be put is the discovery of the detailed way a reaction functions. Such data can tell us how many steps a reaction takes, in what order the steps occur, and which steps are faster and which slower. This information combines to give us the reaction's **mechanism**.

What do mechanisms tell us?

Why bother, you might say, with something called a mechanism when you have an equation that tells you (for example) that:

$$C_4H_9Br + OH^- \rightarrow C_4H_9OH + Br^-$$

and an experimental rate law (worked out in question 14) that tells you that:

$$\text{Rate} = -\frac{d[C_4H_9Br]}{dt} = k[C_4H_9Br][OH^-]$$

Surely the 'mechanism' is the collision of one bromobutane molecule with one hydroxide ion, as written in the equation? The rate law is a logical upshot of this simple fact.

Well, yes, but the point is that not all reactions are quite so transparent in their workings. If we switch from 1-bromobutane to one of its isomers, 2-bromo-2-methylpropane, then this simple picture begins to break up. If you react 2-bromo-2-methylpropane with aqueous hydroxide ions, then although the stoichiometric equation is exactly the same, the rate law is different. The following rate law emerges from the experimental data:

$$\text{Rate} = -\frac{d[C_4H_9Br]}{dt} = k[C_4H_9Br]$$

It has changed from a second-order to a first-order rate law. If the rate does not depend on $[OH^-]$, then it is impossible to view the reaction as happening in a single collision. We are forced to ask a second question:

How do reactants 'go missing' from rate laws?

To answer this question, we're going on a short detour away from mainstream chemistry. Imagine you are on a film set. It is 1966. A crew is filming some crowd scenes for a rather inaccurate bit of prehistory called '1 million years BC' (everyone knows that dinosaurs didn't coexist with humans, especially humans who looked like Raquel Welch, but the director is exercising the right to dramatic licence). There are one hundred extras lining up to be cave men and women. They must go through two stages of make-up (Figure 14.19).

Figure 14.19 Getting to look this rough is a two-stage process

1 One make-up crew blows dust and grime over the actors' clothes and bodies. to give them that 'hard day's hunter/gathering' look. They can spray ten extras per minute.
2 Another make-up crew fits the facial hair and wigs. They can fix up two extras per minute.

Table 14.10 shows how the numbers would vary in the various stages of the queue. The numbers show how many people are in each section of the queue at the end of each minute, for the first 4 minutes. The diagonal lines show how many people get through the 'gates' of the two processes each minute.

Table 14.10 People waiting in make-up queue, grime-spraying first

Time (min)	Extras waiting	Grime-spraying	Extras just grimy waiting	Wig-fitting	Extras grimy and hairy
0	100		0		0
		＼10		＼2	
1	90		8		2
		＼10		＼2	
2	80		16		4
		＼10		＼2	
3	70		24		6
		＼10		＼2	
4	60		32		8
		＼10		＼2	

Question

37a How many minutes will it take to get all 100 extras ready?

b How many minutes will it take for the grime-sprayers to finish their job?

c What is the overall rate of the whole make-up procedure? Give your answer in extras/min^{-1}.

d Can you see a link between the overall rate and the rates of the individual stages?

One day the tea-girl mentions to the cinematographer that it's a bit illogical getting the extras all grimed up and then fitting them with shampooed-and-conditioned wigs and beards. The cinematographer gives a grunt to dismiss the tea-girl, and when she's gone, hurries off to see the director. 'Sid, Sid, I've had this great idea – why don't we spray on the grime *after* we stick on the hair??' 'Don – it's that attention to detail that makes you such a true professional! Let's do it!'

And so the make-up sequence was re-jigged to place the hair artists first. However, the change in order *did not have any effect on the combined rate of the two stages*. Table 14.11 shows how the new arrangement affected the size of the queues. Notice how the diagonal line representing the grime-sprayers shows a 2, although their full capacity is 10.

Table 14.11 People waiting in make-up queue, wig-fitting first

Time (min)	Extras waiting	Wig-fitting	Extras just hairy waiting	Grime-spraying	Extras grimy and hairy
0	100		0		0
		＼2		＼2*	
1	98		0		2
		＼2		＼2*	
2	96		0		4
3	94		0		6
4	92		0		8

*could have handled 10, but only got 2 per minute

Question

38a What do you notice about the numbers of extras waiting for grime-spraying?

b How long would it take to get through the 100 extras? What would be the overall rate in extras min^{-1}? How does this compare with the previous rate?

c If you were Sid, what would you say to Don if he suggested hiring more grime-sprayers to speed up the process?

d If you were Sid, what would you say to the tea-girl if she suggested hiring more wig-fitters to speed up the process?

Thank you for staying with this little fantasy. It actually had a serious purpose as a piece of mathematical modelling. I hope that you will have reached four major conclusions as a result of the last two questions.

1. The overall rate of a process is the rate of the slowest step. The slowest step is therefore the **rate-determining step**.
2. It does not matter where this step occurs in the sequence.
3. Speeding up a step that is not rate determining has no effect on the overall rate.
4. Speeding up the rate-determining step does affect the overall rate.

Now let us apply this model to the reaction between aqueous hydroxide ions and 2-bromo-2-methylpropane. The experimental rate law tells us that $[OH^-]$ has no effect on the overall rate – adding more hydroxide ions is like hiring more grime-sprayers. In other words, hydroxide ions must be involved in a step which, like grime-spraying, is not rate determining. By the same logic, the fact that the overall rate does depend on $[C_4H_9Br]$ means that the substrate molecules, like the wig-fitters, are involved in the slow step. You will notice that in the course of the last two sentences we have tacitly accepted that the reaction takes place in (at least) two steps.

So how do reactants go missing from rate laws? Now we have our answer – because they are involved in one of the fast steps of a multi-step process. In fact, in the reactions we will be studying at Advanced level, we can assume that a missing reactant will be involved in a step *after* the rate-determining step.

14.10 Matching mechanisms to rate laws

Is it a simple system?

Before searching for a mechanism to fit a reaction, it is worth checking whether it is one of those simple systems in which the equation *is* the mechanism. This can be decided by two tests – if the answer to both test questions is yes then you might have a simple system.

1. **Does the stoichiometric equation match the rate equation?**

The answer to this question was yes in the case of the hydrolysis of 1-bromobutane, since both equations could be understood in terms of a straight collision between two particles.

$$A + B \rightarrow \text{products}$$

$$\text{Rate} = -\frac{d[A]}{dt} \left(\text{or} -\frac{d[B]}{dt} \right) = k[A][B]$$

A single-step two-particle collision is called an **elementary step**, in the terminology of chemical kinetics. If one of the reactants has gone missing from the rate equation, you are dealing with a **composite** (multi-step) **mechanism**, because there must be one faster step and one slower step.

2 Does the stoichiometric equation require the meeting of no more than two particles?

The reason for this question is that a collision requiring three or more particles to come together and react is *overwhelmingly unlikely*. Even as simple-looking a reaction as:

$$2H_2(g) + O_2(g) \rightarrow 2H_2O(g)$$

must necessarily be composite, since otherwise it would require two hydrogens and one oxygen to hit simultaneously (Figure 14.20).

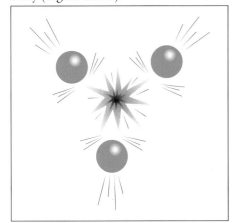

Figure 14.20 An improbable event – a simultaneous collision between *three* reactant molecules

Question

39 Which of the following systems might well be single-step elementary reactions? For those that are not, say which of the above two tests is being failed:

a $3A \rightarrow 2B + C$ Rate $= k[A]^3$

b $2A \rightarrow B$ Rate $= k[A]$

c $A + B \xrightarrow{catalyst} D + E$ Rate $= k[A][C]$

d $2A \rightarrow B$ Rate $= k[A]^2$

Composite mechanisms

If you have established that you are dealing with a composite mechanism, the next stage is to propose a number of steps that together add up to the overall process. Once again there are certain rules:

1 The mechanism must consist of a series of elementary steps. This is no more than a repetition of the fact that

three-way collisions just do not happen. A single molecule decomposing on its own, like a bimolecular collision, counts as an elementary step.

2 The steps must together add up to the overall stoichiometric equation. You can do this test by adding up the equations for the various steps and cancelling anything that appears on both sides. For example, if you were proposing a mechanism that consisted of:

$$A + B \rightarrow C$$
$$C + B \rightarrow D$$

then the overall stoichiometric equation would be:

$$A + 2B \rightarrow D$$

3 The steps, and especially their individual rates, must generate a rate law that agrees with the experimental one. At Advanced level we can say that the reactants missing from the rate law must come into the mechanism after the slow step. The rate law will only include particles that feature in (or sometimes before) the rate-determining step.

Here is a problem that asks you to apply these rules to select an acceptable mechanism from several candidates.

Question

40 There is a reaction between oxides of nitrogen in the gas phase, whose equation is:

$$N_2O_5 + NO \rightarrow 3NO_2$$

The experimental rate law is:

$$\text{Rate} = -\frac{d[N_2O_5]}{dt} = k[N_2O_5]$$

Of the four suggested mechanisms below, three can be rejected on the grounds that they disobey one or more of the above rules, and one is a good candidate for the genuine mechanism. Which is which? Justify your choices.

a $N_2O_5 + NO \rightarrow 3NO_2$

b $N_2O_5 \xrightarrow{slow} NO_2 + NO_3$
$NO_3 + NO \xrightarrow{fast} N_2O_4$

c $N_2O_5 \xrightarrow{slow} NO_2 + NO_3$
$NO_3 + NO \xrightarrow{fast} 2NO_2$

d $N_2O_5 \xrightarrow{slow} NO_3 + NO + O$
$2NO + O \xrightarrow{fast} N_2O_3$
$N_2O_3 + NO_3 \xrightarrow{fast} 3NO_2$

Reactions involving equilibria

Some reactions involve a mixture of equilibrium and one-way steps. For example, consider the following mechanism:

$$A + B \underset{}{\overset{fast}{\rightleftharpoons}} C$$

$$C + D \xrightarrow{slow} product$$

Overall: $A + B + D \rightarrow product$

Because the equilibrium is quick to create C, the first step can remain in equilibrium. C is gradually being used up, but A and B can react to create more C to maintain the equilibrium constant, as would be predicted by using Le Chatelier's principle (p. 179). In this way A and B are gradually used up too, but the equilibrium expression remains valid. Hence we can say:

$$K_c = \frac{[C]}{[A][B]}$$

which rearranges to:

$$[C] = K_c[A][B] \qquad (14.18)$$

But the overall rate is the rate of the slowest step, so:

$$\text{Overall rate} = k[C][D] \qquad (14.19)$$

Substituting from equation (14.18) into equation (14.19), we get:

$$\text{Overall rate} = kK_c[A][B][D] \ (14.20)$$

So we conclude that a reaction with that proposed mechanism would have the third-order rate law shown in equation (14.20).

This is a reversal of our normal way of doing things, in that we have proposed a mechanism for an imaginary rather than a real system. However, it is good practice to think of the more likely possible mechanisms, ready to apply to real systems.

One real system commonly studied at Advanced level which can be interpreted as having an equilibrium step is the acid-catalysed reaction between iodine and propanone. The stoichiometric equation is:

$$CH_3COCH_3 + I_2 \xrightarrow{H^+ \text{ catalyst}}$$
$$CH_2ICOCH_3 + HI$$

The following mechanism has been proposed, and does pass all the tests (the detail of some of the intermediate species has been omitted, and left hidden behind single letters – X and Y):

$$CH_3COCH_3 + H^+ \xrightarrow[]{\text{fast}} X$$

$$X \xrightarrow{\text{slow}} Y + H^+$$

$$Y + I_2 \xrightarrow{\text{fast}} CH_2ICOCH_3 + HI$$

The following question considers this mechanism, and revises some of the practical aspects of previous sections.

Question

41 The aim is to deduce the rate law this mechanism would produce. Follow this suggested path, which runs parallel to the paragraphs above.

a Write the expression for the equilibrium constant of the first step (which is similar to equation (*14.18*)).

b If the overall rate is the rate of the slowest step, write an expression for the overall rate in terms of [X] (which is similar to equation (*14.19*)).

c Combine your two equations to derive a rate law for the overall reaction in terms of concentrations of actual starting ingredients, either catalyst or reactants (similar to equation (*14.20*)).

d Which reactant concentration will be missing from the rate law?

e What is the order of the reaction with respect to the reactant named in (d)?

Now the focus switches to the practical aspects of studying this system.

f An experimenter starts with concentrations as follows:

$[CH_3COCH_3] = 2.0 \text{ mol dm}^{-3}$

$[H^+] = 2.0 \text{ mol dm}^{-3}$

$[I_2] = 0.01 \text{ mol dm}^{-3}$

She follows the progress of [I_2] as the reaction proceeds, and plots a graph of [I_2] against time. Which

reactant is in such a large excess that its concentration would be approximately constant?

g Which ingredient would have an exactly constant concentration over the course of the reaction?

h Which ingredient will be completely consumed?

i What would be the general shape of the graph of [I_2] against time?

j Why could the overall experimental rate law not have been discovered by this one run?

k How would the experimenter have set up other runs to discover the full rate law?

Question 41 established one significant generalisation, reminiscent of our discussions of half-lives and graph shapes:

A zero-order reactant gives a straight-line graph of concentration against time.

Proposing mechanisms to fit the evidence

In this section so far we have either predicted a rate law from a mechanism, or picked a suitable mechanism from several possibilities to best-fit a rate law. Now we shall do what professional chemists do – invent a mechanism from scratch to fit the experimental rate law.

Question

42 Chlorate(I) ions (used in bleach) can undergo a disproportionation reaction in aqueous solution, giving rise to the oxidation states 5 and −1.

Stoichiometric equation:

$$3ClO^-(aq) \rightarrow ClO_3^-(aq) + 2Cl^-(aq)$$

Rate equation:

$$\text{Rate} = -\frac{d[ClO^-]}{dt} = k[ClO^-]^2$$

Suggest a mechanism for this reaction, comprising two elementary steps, which fits the above evidence. Indicate which you think is the rate-determining step. If you can, suggest

any corroborative experiments that might support your proposed mechanism.

Some examples of corroborative evidence

In answering question 42, you probably suggested that the reaction goes via an **intermediate**. This is a species that is created in one step only to be destroyed again in the next. Obtaining evidence that intermediates exist is a strong source of corroborative evidence for a mechanism. Sometimes such evidence may be direct, such as 'seeing' the intermediate by use of spectroscopic methods. On other occasions only indirect evidence is available. The methods for gathering such evidence are varied, and often ingenious. For instance, you may have suggested the following answer to question 42:

Step 1: $2ClO^- \xrightarrow{\text{slow}} ClO_2^- + Cl^-$

Step 2: $ClO_2^- + ClO^- \xrightarrow{\text{fast}} ClO_3^- + Cl^-$

In this case, the intermediate ClO_2^- fortunately exists as a (fairly) stable chemical. So an obvious test is to see whether ClO_2^- really does quickly oxidise ClO^-. If not, that would be the end of the proposed mechanism. (Note that you can often be certain that a mechanism has been *dis*proved, yet rarely if ever feel certain that a mechanism is totally proved. As an example, it was thought for decades that:

$$H_2(g) + I_2(g) \rightarrow 2HI(g)$$
$$\text{Rate} = k[H_2][I_2]$$

took place as an elementary step in a single collision, and yet recently it has been shown to be composite.)

One ingenious experiment used competition to find out more about the intermediate. Alcohols undergo nucleophilic substitution, as was mentioned in Chapter 11. There are two routes, called S_N1 and S_N2, with different rate laws (see Section 11.3).

S_N1:   Rate = $k[ROH]$

S_N2:   Rate = $k[ROH][N]$

where N is a nucleophile like, say, a halide ion. The rate law for the attack by chloride ion on butan-1-ol was known to be second order, so the S_N2 mechanism, with a single-step collision, was suspected. In a clever supporting experiment, butan-1-ol was attacked by two halide ions, Cl^- and Br^-, simultaneously, and the product mixture analysed by gas chromatography. The results are shown in Figure 14.21, and their meaning discussed in the next question.

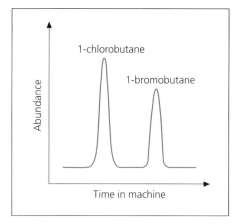

Figure 14.21 The result of gas-chromatographic analysis of the mixture produced by the simultaneous attack by Cl^- and Br^- on butan-1-ol

Question

43a If the mechanism involved the OH^- group dropping off slowly, followed by the carbocation combining quickly with whichever nucleophile it met first (which is the two-step S_N1 mechanism shown in Figure 14.22a), what would have been the ratio of chlorobutane : bromobutane in the product mixture?

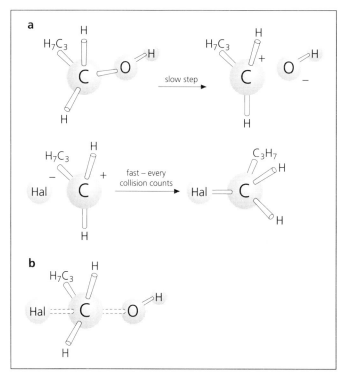

Figure 14.22 a S_N1 mechanism for nucleophilic substitution.   b S_N2 mechanism for nucleophilic substitution. This is the transition state, as the halide ion tries to push the OH group off

b If the mechanism involved a single collision between a complete alcohol molecule and a halide ion (as in the S_N2 case shown in Figure 14.22b), and the degree of nucleophilic 'thrust' of the lone pair of electrons was a factor in the success rate of the collisions, what can you say about the ratio of chlorobutane : bromobutane in the product mixture?

c Which type of mechanism does the corroborative evidence support?

14.11   The effect of temperature on reaction rate

The chapter so far has stayed with a single overall theme – that reaction rates depend, in one way or another, on concentration. Now we switch over to the other major theme – the effect of temperature on rates.

In contemplating molecular collisions, it is important to have a sense of scale. Even in gases, where molecules are far apart, collisions happen often. In fact the collision rate between gas molecules at 400 K and 1 atm is about 10^{32} collisions per litre per second. If these gas molecules were reacting with each other, and if every collision brought about a successful reaction, the reaction would be over in about 10^{-10} seconds.

The fact that the vast majority of chemical reactions are much slower than this tells us that not every collision is effective. The S_N2 mechanism studied in question 43 will provide us with an example of why most collisions fail to effect a change.

Figure 14.23 shows the energy profile of a typical S_N2 mechanism, in the style of Figure 8.13 (p. 147). You will see that the reactants and products have similar potential energies,

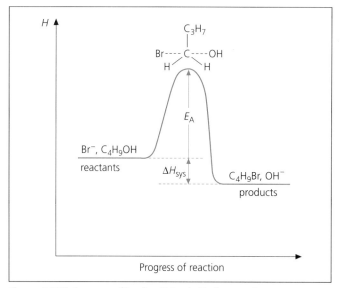

Figure 14.23 Energy profile of an S_N2 substitution reaction

as derived from the strengths of their bonds. However, it is impossible to make the transition from reactants to products without going through the structure labelled A. Structure A has two weak bonds. One is the C—OH bond, which is stretching towards breaking point, its electrons being repelled from around the carbon by the incoming lone pair on the bromide ion. The other weak bond is the half-formed Br—C bond. These two weak bonds give structure A a higher potential energy than either reactants or products. Structure A, the peak on the energy profile, is called the reaction's **transition state**. The quantity E_A, which expresses the enthalpy change (always endothermic) in going from reactants to transition state, is called the **enthalpy of activation**.

Let us take a rather more physical view of the transition state. If structure A has a high potential energy, where has that energy come from? The answer is that it came from a collision. Not every collision will be forceful enough to provide this energy. The incoming halide ion must thrust its electrons sufficiently forcefully to push the C—OH bond to that point of near-breaking seen in structure A. Once past that critical point, the C—O bond will weaken, while the forming Br—C bond takes the new structure back to stability – now, *product* stability.

As an analogy, think what happens when you try to run up a loose sand dune (Figure 14.24). If you take a half-hearted run at the dune, then you might succeed in raising your potential energy a bit, but not enough to surmount the dune. You therefore fall back on the same (reactant) side. If you take a really vigorous run at the dune, you might just be able to get to the top (the maximum of potential energy – the 'transition state'). If you gauged your energy input just right, you would almost stop at the top before accelerating down the other side, back to low energy and to 'products'.

Where does temperature fit into this picture? It has to do with the violence of collisions. The potential energy of the transition state above the baseline of the reactants, the enthalpy of activation E_A, has to come from the kinetic energy of the collision. Collisions do not result in reaction if the molecules do not have a high enough collision velocity.

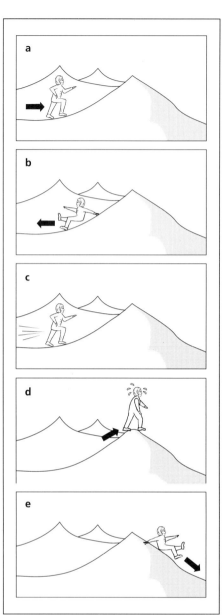

Figure 14.24 The sand-dune analogy. **a** and **b** represent a non-effective collision. **c**, **d** and **e** represent a successful collision. **d** is the transition state

You probably know already that temperature is proportional to molecular kinetic energy. At one temperature the molecules are moving too slowly to provide the enthalpy of activation on collision. At another higher temperature, they are moving fast enough, so collisions result in reaction. However, there are problems with this picture, since it implies that reactions 'turn on' at a particular temperature, whereas in real life they just get gradually faster with increased temperature.

Question

44 Before we discuss the reason for the variation of reaction rate with temperature, we should mention a secondary influence. Imagine the three collisions symbolised in Figure 14.25. The six molecules all have the same kinetic energy x. The activation enthalpy required for the reaction to happen is $2x$, and this energy must be supplied by the collision.

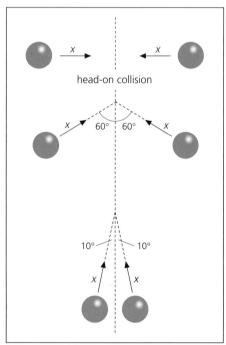

head-on collision

Figure 14.25 Showing how the direction of collision has an influence on whether or not a reaction occurs

a Are all these molecules at the same temperature?

b Which of the three collisions will be successful in bringing about reaction?

The **orientation** of the molecules in a collision is therefore a factor, and might explain why not every collision at any given temperature is successful. However, the main reason is that in a mass of molecules at a given temperature, not all the individual molecules have the same kinetic energy. The overall temperature is really derived from an *average* of kinetic energies over many molecules.

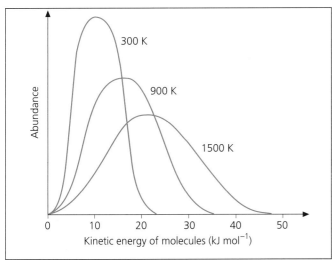

Figure 14.26 Graph showing how kinetic energy is distributed amongst molecules at three different temperatures

Within that average there will be a wide distribution of kinetic energies.

Figure 14.26 shows the distribution of kinetic energies of some molecules at three different temperatures. You can see several things from the graph:

1 At any given temperature, there are a few molecules moving very quickly, a few very slowly, and most are somewhere near the average.
2 At higher temperatures, there are more of the faster molecules than at lower temperatures.
3 At higher temperatures, there is more of a spread in kinetic energies – there is not such a high peak near the average.

The next question involves Figure 14.26. This is a key question, and you should check your answer before going on.

Question

45 Suppose a reaction between the molecules in Figure 14.26 has an activation enthalpy of 50 kJ mol^{-1} (a typical figure for many reactions).

a What minimum energy would a single molecule have to have, in order to hit an identically fast molecule head on and bring about successful reaction?

b Draw this energy value as a vertical line on a copy of the graph.

c Shade the areas under the lines on the graph which represent molecules with this minimum kinetic energy or more.

d Express as an approximate ratio the relative chances of successful collisions at 300 K, 900 K and 1500 K.

The example in question 45 emphasises why temperature is so important to the rates of reactions – because it has a profound effect on the population of the *fastest* molecules, and therefore on the number at or above the minimum kinetic energy on which the reaction depends (as shown by the shaded areas).

14.12 A mathematical model linking temperature and reaction rate

We have arrived at a point at which we can say:

Rate of reaction = total number of collisions per second × fraction of collisions that are successful

Now let us convert that word equation into symbols. We shall assume a simple bimolecular collision mechanism between two reactants X and Y. The total number of collisions per second will be related to the concentration, the molecular speed and even the molecular size (since big molecules are easier targets for collision). We are going to separate out the concentration variables, and group the rest into a quantity called A (or sometimes the 'frequency factor'). So now our equation becomes:

$$-\frac{d[X]}{dt} = A[X][Y] \times \text{fraction of effective collisions}$$

In order to introduce symbols for the 'fraction of effective collisions' term, we must draw on theoretical work known as the Maxwell–Boltzmann distribution law. We already know that the fraction in question is the fraction of collisions having energy greater than or equal to the activation enthalpy E_A. The Maxwell–Boltzmann law gives this fraction as:

$$e^{-E_A/RT} \qquad (14.21)$$

in other words, a function that varies exponentially with temperature, as shown in Figure 14.27. Our equation now becomes:

$$-\frac{d[X]}{dt} = A[X][Y]e^{-E_A/RT} \qquad (14.22)$$

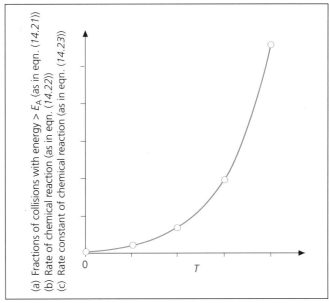

(a) Fractions of collisions with energy > E_A (as in eqn. (14.21))
(b) Rate of chemical reaction (as in eqn. (14.22))
(c) Rate constant of chemical reaction (as in eqn. (14.23))

Figure 14.27 Showing how an exponential mathematical expression works. The general shape of the graph is the same for any of the three possible y-axis variables

When we set up the situation we proposed a simple second-order rate expression, so

$$-\frac{d[X]}{dt} = k[X][Y] \quad (k \text{ is the rate constant}) \quad (14.23)$$

Combining equations (14.22) and (14.23) we arrive at the final mathematical model:

$$k = Ae^{-E_A/RT} \quad (14.24)$$

This describes how a rate constant will vary with temperature. Let us look at this equation.

1 The foundation of this equation is theoretical, based on the collision model. No part of the equation has yet been tested against reality.

2 If we take natural logs of each side of either equation (14.22) or equation (14.24), we have:

$$\ln(\text{rate}) \text{ or } \ln(k) = -\frac{E_A}{R}\frac{1}{T} + \text{constant} \quad (14.25)$$

This is an equation in the form $y = mx + c$, the equation for a straight line. The equivalence between the variables is as follows:

$$y \equiv \ln(\text{rate}) \text{ or } \ln k$$

$$x \equiv \frac{1}{T}$$

$$m \equiv -\frac{E_A}{R}$$

Equation (14.25) therefore predicts that a plot of $\ln k$ or $\ln(\text{rate})$ against $1/T$ would be linear (Figure 14.28).

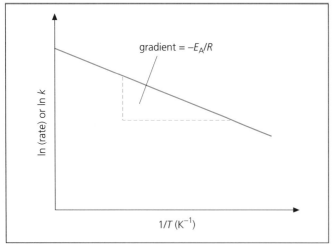

Figure 14.28 If our collision/activation energy model is correct, a plot of ln (rate) or ln k against $1/T$ should be linear, with a highly significant gradient

3 A possible objection is that A, the 'frequency factor', cannot be a constant because it is itself a function of the main variable, temperature. As we have seen, temperature affects the frequency of collisions as well as their violence. However, the advanced form of the model shows that:

$$A \propto T^{\frac{1}{2}}$$

So a temperature increase from icy to boiling (273 K to 373 K) would only increase the collision frequency by:

$$\left(\frac{373}{273}\right)^{\frac{1}{2}} = 1.17$$

In contrast, let us see what would happen to the exponential term in the model for the same temperature rise. Remember, that is the term responsible for describing how the effectiveness of collisions varies with temperature.

Question

46 Assume a reaction has an activation enthalpy of 500 000 J mol^{-1}. Take the gas constant R to have a value of 8.3 J K^{-1} mol^{-1}. Work out the value of the exponential factor e$^{-E_A/RT}$ at:

a 273 K

b 373 K.

c Express these two numbers as a ratio, with the bigger one first. How many times will the rate increase as a result of this factor? Does it swamp the effect of the temperature dependence of the A factor? Are we reasonably justified in saying that the A factor is virtually a constant by comparison?

14.13 Testing the model linking temperature and reaction rate

We shall now take real data for the temperature dependence of a reaction rate, and see if they agree with the model.

Question

47 Table 14.12 shows some students' work on the acid-catalysed disproportionation of thiosulphate ions:

$$S_2O_3^{2-}(aq) + 2H^+(aq) \rightarrow SO_2(aq) + S(s) + H_2O(l)$$

Table 14.12 Data for question 47

Temperature (K)	Time (s)
294	87
303	48
316	24
325	16
333	9

We met this reaction in Section 14.8, question 35, where concentration was varied while the temperature remained constant. We shall now look at varying the temperature, while concentrations are kept constant.

The rate of reaction was measured by a clock method. The measured event was the time for the sulphur precipitate to obscure a black dot viewed through the

Figure 14.29 The clock method for the acid-catalysed disproportionation of thiosulphate ions

reaction vessel (Figure 14.29). Reciprocals of the times can be taken as being proportional to the rates.

a Convert the time data into numbers that are proportional to rate.

b Convert your rate figures into a suitable form to plot on the y-axis. Look back at equation (*14.25*) to check what this is.

c Convert the temperature data into a suitable form to plot on the x-axis.

d Plot the graph. The y-values will all be negative, but make sure they become less negative as the temperature rises.

e Does your graph support equation (*14.25*)? Justify your answer.

f What part does the gradient play in equation (*14.25*)?

g Use the gradient to calculate a value for the activation enthalpy of the reaction.

You will come across many questions like question 46. The rate data may vary, and will not always be 'clock' times. But equation (*14.25*) can be applied to any variable that is proportional to the true rate.

The work of Arrhenius

We have approached the subject of the temperature dependence of reaction rates from the starting point of a theoretical model. But historically the experiments came first. A Swedish chemist called Svante Arrhenius had discovered that many chemical reactions fitted the same experimental pattern – their rate constants were related to temperature by the equation:

$$k \text{ (the rate constant)} = Xe^{-y/RT} \qquad (14.26)$$

For Arrhenius, X and y were merely experimental parameters, which had different values for each reaction he studied. It is only by theory that we have been able to give X and y their present meanings – as a frequency factor and the activation enthalpy respectively. Nevertheless, the theoretical equations (*14.24*) and (*14.25*), based on post-Arrhenius modelling, have in nearly all books been 'married' with their experimental equivalents, so that they are now referred to as two forms of the **Arrhenius equation**. Its success can be gauged by the fact that it applies with fair accuracy to rate/temperature data from nearly all chemical reactions except explosions and enzyme catalysis.

14.14   Activation enthalpy and reaction profiles

In a composite mechanism, each step will probably have stretched-bond high-energy transition states. In a profile with several humps, the question arises as to which hump height is the activation enthalpy, or is it the sum of all of them? Figure 14.30 shows the reaction profile for the S_N1 substitution mechanism.

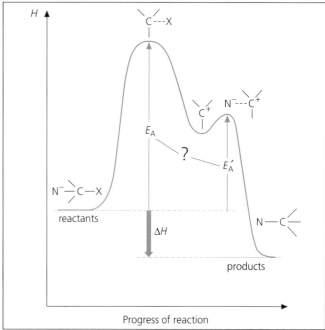

Figure 14.30 Reaction profile for an S_N1 substitution reaction. Which is the true activation enthalpy, E_A, E'_A or $(E_A + E'_A)$? See question 48

The answer comes from thinking once more about rate-determining steps. If the collisions have enough energy for plenty of molecules to get over the big hump, then the small one will present little barrier to the reaction. So the effective activation enthalpy is the distance from the reactants to the highest peak or, in other words, the activation enthalpy of the rate-determining step.

Question

48a In Figure 14.30, which vertical distance represents the activation enthalpy of the *reverse reaction*? (Note that even if the reverse reaction does not happen, it will still have an activation enthalpy, just as every reaction has an enthalpy change in both directions, whether reversible or not.)

b Look at Figure 14.30 and derive a relationship between E_A(forward), E_A(reverse) and ΔH_{sys} (p. 339).

The species that exist in the 'high valleys' of the reaction profiles are called **intermediates**. We have already met these while proposing steps in multi-step mechanisms. The intermediate in the S_N1 mechanism is a carbocation, with an electron-deficient central carbon atom (Figure 14.31). While transition states are extremely short-lived, intermediates can have significant lifetimes. The fact that carbocations like that in Figure 14.31 do last for a while was established by an elegant experiment like that featured in the next question.

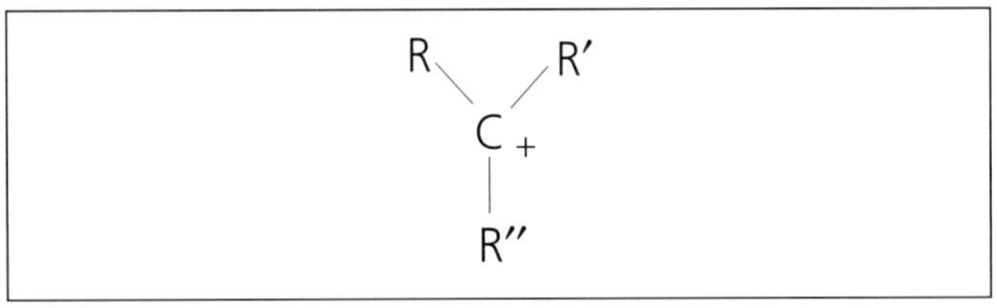

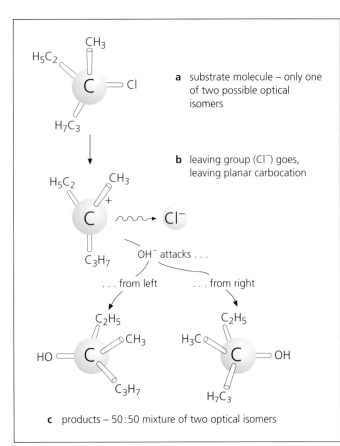

a substrate molecule – only one of two possible optical isomers

b leaving group (Cl^-) goes, leaving planar carbocation

OH^- attacks . . .

. . . from left   . . . from right

c products – 50:50 mixture of two optical isomers

Figure 14.32 Hydrolysis of 3-chloro-3-methylhexane

R, R', C +, R''

Figure 14.31 A carbocation intermediate (see also the 'high valley' in Figure 14.29)

Question

49 The compound 3-chloro-3-methylhexane undergoes hydrolysis with aqueous hydroxide ions, forming the corresponding alcohol (Figure 14.32). The rate law is overall first order (zero order with respect to OH^-) and an S_N1 mechanism is expected. Both reactant and product are chiral, and an experiment is carried out starting with *only one* of the chiral isomers of the chloro-compound (Figure 14.32a). The product alcohol is found to be a 50 : 50 mixture of the *two* chiral isomers (Figure 14.32b). Why does this result give strength to the idea that the intermediate has a significant lifetime, between the end of step (b) and the beginning of step (c)?

14.15   Catalysis

A **catalyst** is a substance that takes part in a reaction system by providing a new pathway with a lower activation enthalpy between the two fixed points of reactant and product (Figure 14.33). Four aspects of catalysis follow from this definition.

1 If a catalyst lowers the activation enthalpy, the reaction will go more quickly.

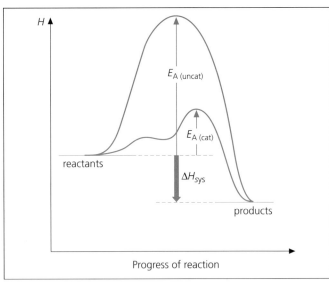

Figure 14.33 A generalised picture of how catalysts work. Note that ΔH_{sys} is unaffected by the catalyst

2 If a catalyst lowers the activation enthalpy, then it follows that different bonds must be being stretched, and therefore it must be offering a different mechanism.

3 If the catalyst operates between the existing thermodynamic points set by the enthalpies of reactant and product, then it follows that the catalyst itself cannot undergo overall change. It is not consumed.

4 If a catalyst lowers the 'hump' on the reaction profile, then it will reduce both the forward and reverse activation enthalpies, and therefore catalyse the reaction in both directions. This becomes a significant factor in equilibria.

Points 3 and 4 above give us a guide to how catalysis, and indeed kinetics generally, interacts with purely thermodynamic branches of chemistry. We can regard thermodynamics and kinetics as having separate domains or areas of responsibility. Those domains are well defined by reference to the reaction profile. Anything relating to the difference between starting and finishing states – e.g. any of the parameters ΔH_{sys}, ΔG_{sys}, ΔS_{sys} and K – are in the realm of thermodynamics. You could say thermodynamics answers the question 'how far?' but not the question 'when?'. Anything that happens between the starting and finishing states, notably the sizes of the humps, is in the realm of kinetics. You could say kinetics answers the question 'when?', but not 'how far?'.

Question

50 Catalysts offer alternative paths with lower activation enthalpies. This question is about what catalysts do to equilibria:

a How will a catalyst affect the composition of the mixture when the system reaches equilibrium – how will it affect K?

b How will a catalyst affect the rate at which the system reaches its equilibrium composition?

The industrial chemist's dilemma

Many of the key reactions of the chemical industry are equilibria. Examples include the routes to the vital raw materials ammonia and sulphuric acid. But if a reaction is exothermic, it presents the operator with a dilemma – whether to operate the reaction at *low* temperatures, under which conditions the equilibrium constant is highest for exothermic reactions and the yield greatest (note that there is nothing about 'when' here), or whether to operate the reaction at *high* temperatures, at which for all reactions equilibrium will be reached faster, but for exothermic reactions the yield is lower.

This dilemma is symbolised graphically in Figure 14.34. The low-temperature yield is inching up slowly to the equilibrium plateau, but the yield in the early part of the reaction is higher using the high-temperature route.

Question

51a Most chemical firms do not wait for their reaction systems to reach equilibrium – why do you think this is?

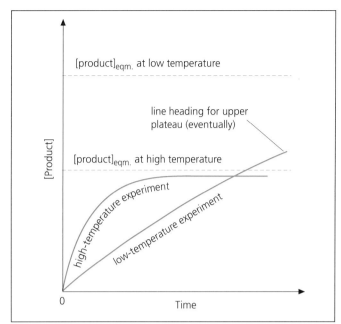

Figure 14.34 The two dotted lines show equilibrium concentrations of product in an exothermic reaction. The higher yield is achieved at low temperature, but the approach to equilibrium is very slow

b Looking at Figure 14.34, which of the two temperatures would you prefer to use if you were responsible for the process? Justify your choice.

c A catalyst can give a high yield, quickly. Explain this by drawing an extra line on a copy of Figure 14.34 to show how the catalyst might affect the low-temperature line.

14.16   Styles of catalysis

Homogeneous catalysis – protonation

The word 'homogeneous' implies that the catalyst is in the same phase as the reactants and products – normally in solution with them.

One of the most simple forms of catalysis falls into this group, and involves the simple addition of a hydrogen ion to a lone pair on one of the reactants, for example:

The effect is to weaken the bond that is to be broken in the reaction, so reducing the activation enthalpy. The reaction rate therefore increases. An example of this is the acid-catalysed nucleophilic substitution reaction that converts alcohols into halogenoalkanes.

Without catalysis (profile (a) on Figure 14.35):

$$ROH + HBr \rightarrow RBr + HOH$$

With catalysis (profile (b) on Figure 14.35):

$$ROH + H^+ \rightleftharpoons ROH_2^+$$
$$ROH_2^+ + HBr \rightarrow RBr + H_2O + H^+$$

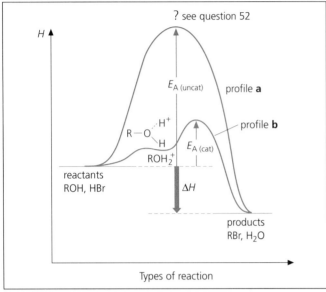

Figure 14.35 Catalysed and uncatalysed routes for the substitution reaction that converts ROH to RBr

Question

52a The bigger hump in profile (b), Figure 14.35, corresponds to the pushing off of the —OH$_2^+$ group. The big hump on profile (a) is the pushing off of the —OH group. Draw transition states for these two situations.

b Why is it easier to reach the transition state in profile (b) than that in profile (a) – in other words, why does it have a smaller activation enthalpy?

c Mechanism (a) is an elementary single step. What would be the rate law for the uncatalysed reaction?

d Mechanism (b) is a fast equilibrium followed by a slow step. How can we tell from Figure 14.35 that the second step is the rate-determining one?

Homogeneous catalysis – the go-between

Suppose I had a large number of unsold copies of this book in my house. If I wanted to deliver them to a bookshop, I could do it myself, but getting the boxes on the roof rack would be a high-energy process. Instead, I call a van firm. They stack the boxes neatly in the van (at a much lower level than the roof rack), deliver them to the shop, and come back to their yard. The 'thermodynamics' of the two routes (with and without the van) are the same (books move to the shop), and the catalyst (the van) is back where it started, so you might say that the second route was 'catalysed' by the use of the van. Many catalysed reactions are of this sort, where a reactant seems to 'employ' another reagent to do its work for it.

One of the oldest examples of industrial catalysis belongs to this category. The oxidation of sulphur dioxide to sulphur trioxide is a step on the path to sulphuric acid. As early as 1746 it was recognised that the reaction went much faster with some nitrogen monoxide, NO, present. The equations are as follows.

Without catalyst: $\quad 2SO_2 + O_2 \rightarrow 2SO_3$

With catalyst:
Step 1 – O$_2$ oxidises NO $\quad 2NO + O_2 \rightarrow 2NO_2$

Step 2 – NO$_2$ oxidises SO$_2$ $\quad 2NO_2 + 2SO_2 \rightarrow 2SO_3 + 2NO$

The two steps produce lower activation enthalpies than the single step. Catalyses like this can be represented by cyclical diagrams as shown in Figure 14.36. The catalyst stays rotating in the middle, and the reactants and products enter and leave the circle.

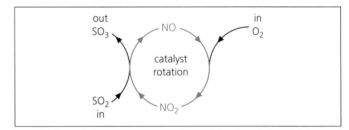

Figure 14.36 Catalytic cycle for the oxidation of sulphur dioxide by oxygen

Questions

53 Returning to the van-hire analogy, let us consider that oxygen plays the part of the boxes of books.

a What plays the part of the empty van?

b What plays the part of the full van?

c What plays the part of the empty shop?

d What plays the part of the full shop?

54 Two molecules of hydrogen peroxide can oxidise and reduce each other, producing water and oxygen – a disproportionation reaction. Iodide, I$^-$, is a go-between catalyst for this reaction. The reactions are as follows.

Without catalyst:

$$2H_2O_2 \rightarrow 2H_2O + O_2$$

With catalyst:
Step 1 $\quad H_2O_2 + I^- \rightarrow H_2O + IO^-$
Step 2 $\quad IO^- + H_2O_2 \rightarrow H_2O + O_2 + I^-$

a (*Revision*) What changes occur to the oxidation number of the oxygen atoms in the overall reaction? (*Remember*: In a disproportionation reaction, oxidation numbers go in two directions.)

b (*More revision*) Between which two values does the iodine oxidation number oscillate?

c Draw a cycle diagram in the style of Figure 14.36 to represent the catalysed mechanism.

d If the rate law for the catalysed mechanism were:

$$\text{Rate} = k[\text{H}_2\text{O}_2]\,[\text{I}^-]$$

suggest which step would have been rate-determining, and what the reaction profile would look like.

Heterogeneous catalysis – transition metals

The word 'heterogeneous' implies that the catalyst is not in the same phase as the reactants. In such cases the catalyst is usually a solid and the reactants are gases or liquids. The catalysis happens at the surface of the catalyst. About 90% of all industrial chemical reactions are catalysed, and of these the majority are heterogeneous, and feature a transition metal.

Transition metals and their compounds readily accept lone pair donations. This means they can often bind molecules onto their surfaces. This results in catalysis because:

1 The binding process can sometimes stretch a bond ready for the transition state.
2 The bound molecules may find themselves lying next to other molecules with which they can react. These other molecules may also be a little stretched. Most important, they may be lying in a perfect orientation for a reaction with their bound neighbour.

Figure 10.21 (p. 210) gave an artist's impression of how the bound reactants might look, using the case of the hydrogenation of an alkene. Figure 14.37 offers a possible reaction profile for such a reaction.

Question

55 The degree of binding of substrates to catalysts is finely balanced – a catalyst that holds too tight, or one that holds too loosely, will not work. Can you suggest the consequences of binding too tightly or too loosely?

Chapter 19, which deals with all aspects of transition metal chemistry, takes another look at catalysis, and includes the Ziegler–Natta catalysts that catalyse the polymerisations of alkenes. These use a mechanism of catalysis featuring co-ordination of the substrate onto the metal (that is, bonding by electron donation into vacant orbitals). However, in this type of catalysis the active site is not the surface of a solid metal – instead, the co-ordination is onto a transition metal compound which is in a homogeneous medium with the substrate. This is similar to the method of catalysis used by

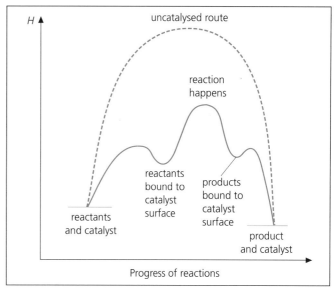

Figure 14.37 Generalised picture of heterogeneous catalysis showing the two stages – binding and reaction

enzymes (see below), in which substrate and catalyst also share the same medium.

Enzyme catalysis

Enzymes are being used as catalysts increasingly in industry, especially in the manufacture of precise chemicals such as drugs and pesticides. Enzymes are the natural catalysts used by living things to carry out their functions, and biochemists have found ways of genetically engineering organisms such as bacteria to produce the required enzymes.

The action of enzymes is covered in detail in Chapter 15. For now let us observe that enzyme action is a variation on transition metal catalysis. The enzymes hold substrate molecules in positions that facilitate reaction. But in contrast to the metals, enzymes hold their substrates in molecular clefts called **active sites**, and they hold them with intermolecular forces, rather than ligand–metal dative bonds.

The shapes of the substrate molecule and the active site of the enzyme must match, which has given rise to a famous descriptive metaphor, the **lock and key**. Figure 14.38 shows this – the enzyme is just the right shape (and has just the right 'adhesive' groups in the right places) to ensure that it binds the substrate in the active site. There are many different enzymes and substrates, but (as with locks and keys) only *that* enzyme and *that* substrate can engage each other properly.

As we shall see in Chapter 15, the enzyme–substrate bonding process often makes long sections of the substrate 'comfortable' within the cleft, but in some key place within the site, a nucleophilic group on the enzyme (say) is waiting to react with the substrate.

The characteristic feature of enzyme catalysis is that the temperature dependence of the rate of reaction does not conform to the Arrhenius equation (Figure 14.39). Enzyme molecules are fragile, and thermal agitation can destroy the delicate 'locks' into which the substrates fit so snugly.

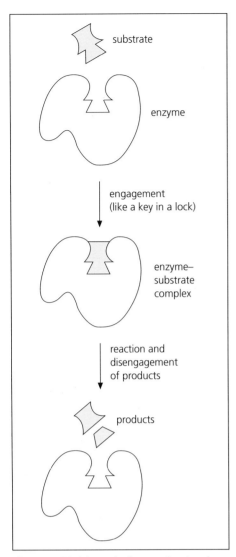

Figure 14.38 Schematic diagram showing the stages of an enzyme–substrate association, illustrating the lock-and-key model

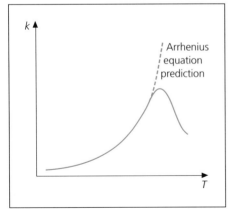

Figure 14.39 Showing how an enzyme-catalysed reaction goes through a maximum of rate constant k at a particular temperature

14.17 Enzymes and drugs

Calculating doses

Without pre-empting Chapter 15, we shall have a look at the kinetics of enzyme reactions. A good context in which to set this study is the action of pharmaceutical drugs on human patients (Figure 14.40). Some of the problems associated with the administration of medical drugs are set out below.

1 If the drug is given orally or by intra-venous injection, it will spread fairly evenly throughout the body, rather than targeting the region at which the treatment is aimed. A doctor can view the patient as the container of a huge aqueous solution, consisting mainly of blood and cell fluid, and can assume that the drug will be dispersed into this solution. The effective 'distribution volume' for an adult woman of average build, for example, is about $25\,dm^3$, and that for a man is about $40\,dm^3$.
2 In consequence, the dose must be larger than if local-tissue targeting were possible, to cope with the dilution effect.
3 All drugs have some degree of toxic side-effects if the dose exceeds a certain threshold level. All the healthy tissue in the body is being exposed to the drug.
4 Drugs themselves are attacked by the body's enzymes, especially in the liver. As a result the concentration of drug molecules dissolved in the body fluids diminishes with time according to a sort of rate law.

Calculating doses

When doctors prescribe drugs, they manipulate two variables – the mass of drug in each tablet, and the frequency of dose. They have standard reference books to guide them in setting these two parameters.

The doctor would look up the **therapeutic range** of the drug, so as not to prescribe too much or too little, and its **half-life** or **decay time** in the body, to decide when the dose should be repeated.

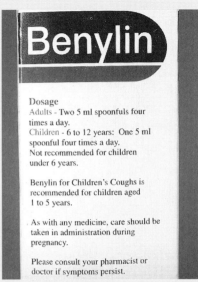

Figure 14.40 The concentration in the body depends on the amount taken, and how often it is taken

The size and age of the patient would be relevant, since a child, for example, has a much smaller distribution volume than an adult. Also, the patient's state of health would be relevant, especially if side-effects were serious.

Now let us look at rate laws. The drug molecules diffuse to the enzyme, 'lock in', get processed (i.e. broken) and then disengage. This is analogous to 'processing' lots of people at the windows of a post office counter (Figure 14.41).

Question

56a If the post office is fairly quiet, one customer can be 'processed' before the next one comes in. Under this condition, what would be the effect on the overall rate, i.e. the number of customers served per unit time, of halving and of doubling the number of people coming in? Assume that there is still spare time at the windows between customers for double the number of people.

b How would you describe the dependence of the overall serving rate on the concentration of customers, using the language of

Figure 14.41 Post office counters

kinetics – zero order, first order or second order?

c The post office is busy and there is a long queue. The counter staff are serving continuously. What would now be the effect on the overall rate of customer service of halving or of doubling the number of customers? What now is the order of overall rate with respect to concentration of customers?

We can apply these ideas to enzyme chemistry. The enzyme active sites are analogous to the staff in the post office, and the substrate molecules to the customers. If the number of available enzyme sites is sufficiently great (there are lots of members of staff), or if the concentration of substrate molecules is sufficiently small (not many customers), then the enzyme active sites will be unoccupied some of the time, and the rate law for the reaction will show a proportional (first-order) dependence on the substrate concentration.

If there are only a few enzyme molecules (staff), or if the substrate concentration is high (lots of customers), or if the enzyme–substrate reaction takes a long time, then the enzyme will be working at saturation. The rate law shows the rate of reaction to be invariable (zero-order)

with respect to substrate concentration. The rate depends only upon the number of enzyme sites (which will itself be a constant, unless more enzyme is added), and the time taken for the reaction to happen (another constant).

Question

57 Pick out the rate-determining step in each of the two contrasting situations outlined above.

Drug companies need to know which sort of rate law will be followed by a particular drug, so that the concentration of the drug in the patient's body fluids can be predicted at various times after administration. Then the correct dose and dose frequency can be calculated for the data books. The next question, in which the real facts have been slightly simplified for the sake of ease of mathematical manipulation, deals with this problem of drugs and rate laws.

Question

58 The drug phenytoin is used to control epilepsy. It can prevent attacks, but has the side-effects of drowsiness, dizziness, slurred speech and rolling of the eyes. The guidance issued with the drug informs the doctor that:

- The **therapeutic range** is 8–20 mg dm^{-3}. In other words, below 8 mg dm^{-3} in tissue fluids it is ineffective, and above 20 mg dm^{-3} the risk of side-effects is significant.

- The **half-life** of the drug, when used in the middle to lower end of its therapeutic range, is a constant, about 12 hours.

a The fact that the half-life is constant tells us which of the two 'post office' situations (and which of the two rate laws) is applying here. Which is it, and how can you tell?

b Assume the patient is a man, with a body-fluid distribution volume of about 40 dm^3. Work out the mass of phenytoin needed to give an initial concentration of 16 mg dm^{-3} in body fluids.

c Assume that the dose you calculated in (b) can be delivered by two tablets. If the half-life of the drug is 12 hours, work out a dosage regime which would keep the concentration of the drug at a safe but effective level in the patient's body fluids over a two-day period. You can administer either one or two tablets at a time.

The breakdown of alcohol in the body

Alcohol (ethanol) is a drug, and like any other drug it is gradually destroyed by the body's enzymes. It differs from phenytoin in that it has a constant rate of decomposition. The average rates of decomposition of ethanol for men and women are 7.3 and 5.3 g h^{-1} respectively, although the rate can vary between individuals by as much as ±50%. 'Dosages' can be estimated from the fact that a pint of normal strength beer or lager contains about 16 g of ethanol.

We looked at the breathalyser in Chapter 11, when the unit being measured was the concentration of ethanol in the breath – the legal limit for driving was 35 μg 100 cm^{-3}. Now we shall consider the ethanol concentration in the blood. The equivalent to 35 μg 100 cm^{-3} in breath is 80 mg 100 cm^{-3} in blood (and in all other aqueous body fluids).

Questions

59a Work out the ethanol concentration (as mg 100 cm^{-3}) in total body fluids (including blood), after the ingestion of one pint of ordinary strength beer, for these drinkers:

  i a woman, distribution volume 25 dm^3

  ii a man, distribution volume 40 dm^3.

b How many pints of beer would bring each of the drinkers in (a) up to or above the legal limit for driving?

c In order to make facts accessible to the general public, the Government uses a 'units' system, in which one unit is equivalent to (among other things) a half-pint of ordinary strength beer (Figure 14.42). The most often quoted limit for men is five units (two and a half pints). Would the man in (a) have been over the limit if he had drunk his five units quickly and then gone straight to his car?

d The 'five units' advice is about right if the man was drinking at a normal rate of consumption. Why does the way drinking is spread over time

have an influence? What would be the risks of the man depending too heavily on the idea that he was safe with five units?

e If an average man can drink five units, how many pints should an averagely built woman regard as 'safe'?

f Now let us focus on the time taken for the ethanol to be decomposed. The rate of decomposition is constant over time. What can you conclude about the rate law for the decomposition of ethanol by our bodies' enzymes? (In other words, what is the order of the reaction, and which of the two 'post office' models does it resemble?)

g The reason for the difference between the rate laws for the decomposition of phenytoin and of ethanol is probably to do with size of dose. Compare, on the same unit scale, the dose levels of the two drugs in the examples used, and then suggest why that factor alone could possibly be the cause of the different kinetic patterns.

h Another piece of generalised advice is that it takes about one hour to rid the body of one unit. Work out from the data above (two paragraphs before the question) how true that is, and check whether it is more true for men or for women.

60 Sometimes people will have a few drinks, and then decide not to drive home that night. The natural assumption is that it will be all right to drive in the morning. We can investigate this assumption. Let us assume that the person in question is a woman, whose distribution volume is 25 dm^3, and whose rate of decomposition of ethanol is 5.3 g h^{-1}. She has seven glasses of wine, at half-hourly intervals, between 9 p.m. and midnight. Each glass of wine contains about 8 g of ethanol (which means it is equivalent to one unit).

Figure 14.42 These drinks all contain one unit of alcohol

Figure 14.43 Both these drinkers are taking in two units, but the alcohol concentration in the woman's blood will be higher, because she has a lower distribution volume than the man

a By how much, in mg 100 cm^{-3}, does each glass of wine increase her blood ethanol concentration?

b By how much, in mg 100 cm^{-3}, does her blood ethanol concentration go down each hour?

c Between 9 p.m. and midnight her blood ethanol level rises on seven occasions, and is reducing steadily for three hours. What will be her blood ethanol concentration, in mg 100 cm^{-3}, by midnight, after her last drink?

d By how much, in mg 100 cm^{-3}, does the value in (c) exceed the legal limit?

e Bearing in mind your answer to (b), how many hours is the minimum time she should wait before driving home?

Summary

Rates of reaction

• The rates of chemical reactions depend on the **frequency** and on the **violence** of collisions between molecules. Any variable that increases either of these factors will increase the rate of reaction.

• Variables that increase the frequency of collisions are **concentration of reactants**, **surface area** (if there is a phase boundary between reactants) and **temperature**, although the dominant influence of temperature is dealt with in the next paragraph.

• The variable that increases the violence of collisions is **temperature**. Temperature has a very great influence on reaction rate, specifically because of this violence factor. The reason is that, although molecules in gas and liquid phases are colliding with great frequency, only those collisions that are violent enough to disturb the chemical bonds in the reactants actually achieve a reaction.

• Leaving the temperature aside, we shall now focus on concentration. It is possible to use simple mathematical models to predict how the frequency of collisions, and therefore the rate of reaction, will depend upon concentrations of reactants. Two cases were looked at in detail:

1 In a reaction that takes place via a simple molecular collision between molecules A and B (a so-called **bimolecular process**), it is possible to predict that the rate will be proportional to the concentrations of both A and B. This gives rise to the differential equation:

$$-\frac{d[A]}{dt} = -\frac{d[B]}{dt} = \text{rate of reaction}$$

$$= k[A][B]$$

(where k is a constant of proportionality known as the **rate constant**). Such an equation is known as a **rate equation**, or **rate law**. This rate law shows a relationship between rate and concentrations which is said to be **first order** with respect to [A] and [B] separately, and **second order** overall.

2 In a **unimolecular process**, in which a molecule A performs a solo decomposition reaction, we predict that the rate will be proportional to the concentration of A. The corresponding differential equation is **first order** with respect to [A] and **first order** overall:

$$-\frac{d[A]}{dt} = \text{rate of reaction} = k[A]$$

Diagnostic tests for rate data

• We can use a set of diagnostic tests on real data from reaction systems to check whether either of the above models is being followed. 'Raw' data from chemical systems are in the form of concentration against time.

• We need to know what the two differential equations would predict for the variation of concentration with time. In the case of the first-order equation, this can either be done by a mathematical process called **integration**, or by a 'common sense inspection'. Integration gives us the equation:

$$\ln[A] = -kt + \text{a constant}$$

so that a first-order reaction can be identified by the fact that the logarithm of [A] gives a linear graph against time. This is a diagnostic test for a first-order rate law which is suitable for Advanced-level mathematicians. For the non-specialist, common sense tells us that a plot of [A] against time should be a concave curve, with the gradient representing the rate of reaction. The gradient should get shallower because as [A] goes down there are fewer A molecules to react, so the reaction slows down. If you were to compare the gradients at any two points on the curve, chosen so that the [A]-value of the second was half the [A]-value of the first, the two gradient values should also be in the ratio 2 : 1. This is described in the text as the 'two-tangents' method (Figure 14.5, p. 320).

• The simplest of the diagnostic tests for detecting a first-order rate law from raw concentration/time data is the method of **half-lives**. Mathematics tells us that a first-order reaction will have **a constant half-life**. This means that the time taken for the concentration of A to halve should be the same at all stages in the course of the reaction (Figure 14.8, p. 323).

• Diagnostic tests for second-order rate laws are not so easy to perform, unless either [A] = [B] throughout the reaction, or the proposed mechanism involves a collision between two molecules of A. In this case the differential equation is:

$$-\frac{d[A]}{dt} = \text{rate} = k[A]^2$$

If either of these simplifying features is expected, then there are diagnostic tests which work. The 'two-tangents' method predicts that gradients at [A] and [A]/2 should be in the ratio 4:1, and the integrated rate law is:

$$\frac{1}{[A]} = kt + \text{a constant}$$

so that a plot of 1/[A] against time should be linear.

Practical methods 1 – whole-run methods

• The actual method used to obtain information about the change of [A] with time will vary. If one of the species in the reaction absorbs light, then the reaction can be done in the cell of a spectrophotometer, and the absorbance can be used instead of concentration. Alternatives to spectrophotometry include removing samples for titration (sometimes 'quenching' them before titration), and recording changes in volumes of gases or of solutions.

• Whatever the method, it is common practice to isolate one reactant at a time. Thus if the reaction features a number of reactants, all but one of them will be held in a large excess. This means that while the one reactant is used up, the others are only marginally reduced, and their concentrations can therefore be considered as constants. (Catalyst species also appear in rate laws, but their concentrations stay constant whether or not they are in excess, because catalysts are not consumed.) That way, only the minority reactant undergoes any significant change in concentration during the reaction, and the resulting data will only show how the rate varies with the concentration of the minority reactant.

• It is possible to follow a reaction by collecting data about the build-up of a product concentration [C]. Such data can be processed by recording the final value, $[C]_\infty$, at the end of the run. Then all the other [C] values are subtracted from this, to give a series of downward-trending data which can be used as values of [A] (Figure 14.44). Diagnostic tests then proceed as before. *Remember*: Although products serve as indicators of reaction rate, only reactants appear in rate laws (unless there is an unusual phenomenon like autocatalysis, in which one of the products catalyses its own creation).

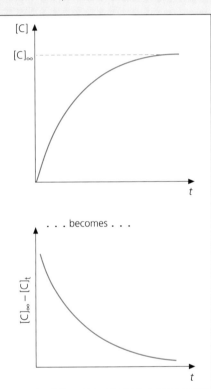

Figure 14.44 Obtaining a graph of reactant concentrations against time from product concentration data

Practical methods 2 – initial rates

• The **initial rates method** can be used with systems containing many reactants. The reactants are mixed and the rate measured (by recording the concentration change in the very early stages of the reaction, using one of the normal methods). This run is then discarded, and a new run is set up with *one* reactant concentration altered, and all the rest unchanged. The new rate is then measured. This process is continued by varying the concentration of each reactant in turn. The initial rates should quickly reveal the rate law relationships for example, if the trebling of [B] causes a ninefold increase in rate, it appears that rate is proportional to $[B]^2$, and that the reaction is second order with respect to [B].

• A version of the initial rates method is the **clock method**. The initial rate is found as the reciprocal of the time taken for some event to happen. The event must be brought about by the reaction, and should happen early in the run (so as to qualify as 'initial'). Typical clock events are the operator's first perception of a coloured product, or the last moment of seeing through an increasingly cloudy solution. In a reaction that was second order with respect to a reactant B, the trebling of [B] would reduce the waiting time for the event by a factor of nine.

Mechanisms

• Sometimes the concentration of a reactant has no effect on a reaction rate. The rate law is then said to be **zero order** with respect to that reactant. The graph of [A] against time in that case would be a downward-trending straight line (Figure 14.45).

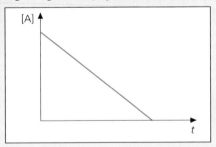

Figure 14.45 Concentration/time graph for a zero-order reactant

• The existence of rate laws that are zero order with respect to a particular reactant raises the question: how can a reactant be used up, and this not have a slowing effect on the overall rate? Two new concepts are necessary to answer this question:

1 Many reactions are not the single-step elementary uni- or bimolecular processes we have considered so far. They may proceed by a number of sequential steps, which we call the **mechanism** of the reaction.

2 In a mechanism, the overall rate of the reaction is the rate of the slowest, **rate-determining** step. Only reactants which participate in the mechanism sequence either in, or before, the rate-determining step, will appear in the rate equation. (This is a slight simplification of a more complicated truth, but it serves the needs of Advanced-level syllabuses.)

So if a reactant is consumed in a fast step, which takes place after the rate-determining step, that reactant will be absent from the rate equation, and hence have a zero-order effect on the rate.

• A proposed mechanism must meet three criteria:

1 Its stages must add up to the overall stoichiometric equation.

2 Its steps must all be elementary – that is, all unimolecular or bimolecular processes. A collision involving more than two particles is highly improbable to the point of impossibility.

3 The arrangement of fast and slow steps must lead to a rate equation that agrees with the experimental one.

• The existence of mechanisms reminds us that the stoichiometric equation is an unsafe guide to the rate equation. Only if the reaction takes place by a single elementary step, as in, say, the S_N2 substitution mechanism, will the reactants A and B of the stoichiometric equation appear simply as $[A] \times [B]$ in the rate equation.

• It may be difficult to prove beyond doubt that a particular proposed mechanism is correct. Researchers often look for **corroborative evidence**. This can include trying to detect short-lived intermediate species, or using specially devised chiral reactants. This second method was valuable in corroborating both the S_N1 and S_N2 mechanisms in organic chemistry.

The influence of temperature on reaction rates

• It is obvious that temperature has a very profound effect on reaction rates. As a rule of thumb, an increase of 10 K will double the reaction rate – this proves to be approximately true for a large number of reactions.

• It can be shown that this degree of sensitivity of reaction rate to temperature changes cannot be accounted for merely by increased collision frequencies.

• An alternative theory recognises that reactions must progress through a stage (or stages) in which the atoms exist in a high-energy arrangement, known as a **transition state**. The enthalpy change from reactants to transition state is called the **enthalpy of activation**, E_A (Figure 14.23, p. 335, shows an example). The theory suggests that this energy can be acquired from the kinetic energy of collisions. Only a fraction of all the millions of collisions per second between the reactants will provide enough energy to match or exceed E_A, and so only a few collisions actually bring about reaction. Because the number of these energetic collisions increases sharply with temperature, so will the rate of reaction (Figure 14.46).

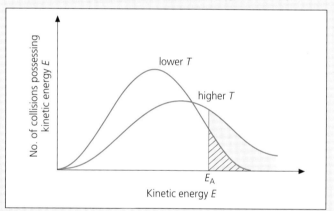

Figure 14.46 Showing how the energetic collisions (shaded areas) increase sharply with temperature

• This theory can be modelled mathematically, and the model predicts a relationship between the rate of reaction and temperature which takes the form:

$$\ln k = -\frac{E_A}{RT} + a \text{ constant}$$

If the experimenter maintains constant initial concentrations throughout a series of runs, then the rate of the reaction can be used instead of the rate constant, k. The equation then becomes:

$$\ln(\text{rate}) = -\frac{E_A}{RT} + \text{a constant}$$

This equation has become known as the **Arrhenius equation**.

• To test the 'transition state model', real reactions can be investigated experimentally by measuring reaction rates at varying temperatures, and plotting a graph of ln (rate) against $1/T$ (Figure 14.28, p. 338). The linearity of this graph would be confirmation that the Arrhenius equation was obeyed by the reaction being studied. A value for the activation energy can be obtained from the gradient of such a graph, since the gradient has the value $-E_A/R$. With the exception of enzyme reactions and explosions, most reactions do indeed follow the equation closely.

• In a multi-step mechanism, the enthalpy changes cause a series of humps and troughs in a diagram called a **reaction profile** (Figure 14.30, p. 339). The effective activation enthalpy of the overall process is the enthalpy difference between the reactants and the *highest* transition state.

• The activation enthalpy of the reverse reaction is the enthalpy difference between the products and the highest transition state. The relationship between $E_A(\text{forward})$ and $E_A(\text{reverse})$ is:

$$E_A(\text{forward}) - E_A(\text{reverse}) = \Delta H_{\text{sys}}$$

Catalysis

• **Catalysts** are reagents which participate in reaction mechanisms to offer paths with lower overall enthalpies of activation. They thus appear in rate laws. Catalysts speed up chemical reactions, but are not consumed.

• This lowering of the enthalpy of activation is achieved by breaking the mechanism down into stages. The stages are general to most forms of catalysis:

1 The catalyst engages with the reactant(s).
2 The main bond breakage or creation occurs.
3 The products disengage from the catalyst.

Although the same bonds are made and broken as on the uncatalysed path, the breaking down of the reaction into stages has the effect of spreading the original activation enthalpy over several 'humps' (Figure 14.37, p. 343).

• In three types of catalysis, the substrate molecule is bound onto a **site** on the catalyst, and the bond changes occur while the substrate is attached. These three classes of catalyst are:

1 **Transitional metal surfaces** – substrate molecules are bound onto the surface, and the binding process may 'pre-stretch' some of the bonds that will break during the reaction. While the substrate is bound on the surface, the bond changes are completed, possibly using vacant orbitals on the metal to assist electron transfers. Finally the product molecules are released. This form of catalysis is **heterogeneous**, with substrates and catalyst in different phases. The substrates are usually gases or liquids, while the catalyst is a solid. A classic example would be the nickel-catalysed hydrogenation of alkenes, but most industrial processes (e.g. ammonia and sulphuric acid production) feature transition metal surface catalysts.

2 **Transition metal compounds** – the substrate is held onto the catalyst by a ligand-to-metal bond (see Chapter 19). The stages and the reaction profile are similar to those with tran-sition metals, the difference being that the system is **homogeneous**. The classic example is in the Ziegler–Natta method for polymerising alkenes.

3 **Enzymes** – these are complex proteins, which have an active site or cleft in which the substrate is bound. The same three-hump reaction profile is seen to apply as in the two catalyst types above. The enzyme is usually in the same homogeneous medium as the substrate, but the binding is done using intermolecular forces. Examples of enzyme catalysis can be drawn from

almost any biochemical pathway, from digestion and respiration to the workings of nerves and the replication of DNA.

• There are two other recognisable types of catalysis, both of them homogeneous:

1 **Protonation** – the attachment of a proton to a substrate molecule requires the substrate to provide a pair of electrons for a dative covalent bond. This disturbs the electron distribution within the molecule, and increases polarity or weakens bonds in a way that facilitates reaction. A classic example is the catalysis of the esterification reaction between a carboxylic acid and an alcohol, in which a proton attaches to either of the oxygen atoms of the acid, and smooths the reaction profile of the nucleophilic attack by the alcohol (Figure 14.47). A related form of catalysis, which also works by disturbing the electron distribution of a substrate molecule, operates in the Friedel–Crafts reaction (see Chapter 10). The aluminium chloride receives a lone pair in a dative bond, and thereby polarises the R—Cl bond in a way that creates an $R^{\delta+}$ polarity, the better to attack the benzene ring.

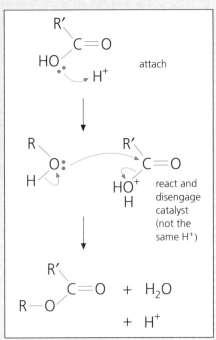

Figure 14.47 Acid catalysis of the reaction between a carboxylic acid and an alcohol

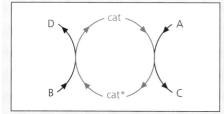

Figure 14.48 Showing the mechanism of go-between catalysis

2 Go-between catalysis – a reagent is found that can oscillate between two states, which we might call 'cat' and 'cat*'. Reactant A is destroyed in the course of converting cat to cat* and product C is created. Meanwhile cat* reacts with the other reactant B, to create the other product D and return itself to the original cat condition (Figure 14.48). The most common application of this type of catalysis is in redox reactions, when the catalyst oscillates between oxidised and reduced forms. Biochemical agents like ATP and ADP perform similar cycles in respiration, and so could be regarded as catalysts on the path to the extraction of energy from glucose.

• Enzyme catalysis in the body can show one of two types of rate law:

1 If the enzyme is **saturated**, with every site working continuously for the entire course of the reaction, then the reaction will have a rate law that is zero order with respect to the substrate concentration. The rate-determining step will be getting the substrate in and out of the cleft. If there is already a 'queue' of substrate molecules, the breakdown will not be speeded up by adding more substrate molecules to the queue. The breakdown of alcohol in the body follows this pattern.

2 If the enzyme's cleft is empty for much of the time, then the rate-determining step will be the diffusion-driven meeting of enzyme and substrate, and the rate law will show a first-order dependence on the substrate concentration. Many pharmaceutical drugs at low dose levels follow this pattern.

15

Organic molecules containing nitrogen

15.1   Introduction

Nitrogen confers quite a distinct 'personality' on the organic molecules in which it occurs. Some features of organic compounds containing nitrogen are reminiscent of the inorganic chemistry of nitrogen. For example, one set of molecules, the **amines**, can be viewed as the 'daughters of ammonia'. Another group (of which TNT is a member) might be dubbed the 'back to dinitrogen' tendency because of their preference for decompositions, often violently exothermic and explosive, in which elemental dinitrogen is expelled. Between these extremes are patterns of behaviour akin to the chemistry of alcohols and esters. Within the organic chemistry of nitrogen, you will find molecules that *make* you up (proteins), molecules that will eventually '*eat*' you up (proteins again, in the form of enzymes, as used by decomposers), *charge* you up (the fight-or-flight hormone adrenaline), *blow* you up (TNT and its relatives), and, even, to quote a recent government advertisement on the perils of heroin, 'screw' you up.

15.2   Trends in the behaviour of the amines

A sample of typical amine structures is displayed in Figure 15.1. As mentioned in the introduction, amines can be viewed as derivatives of ammonia. Where one, two or all three of the hydrogen atoms in ammonia are replaced by alkyl or aryl groups, then the resulting molecules are referred to as primary, secondary or tertiary amines respectively (Figure 15.1).

Let us briefly recap on the chemistry of ammonia, contrasting it with water (which by the same token is the 'parent' of the alcohols). You may recall that the dominant feature of the ammonia molecule is the lone pair of electrons on the nitrogen atom. Nitrogen is by no means the only inorganic molecule with a lone pair, but it is one of the most effective at wielding it.

Molecules with lone pairs of electrons perform two main chemical operations, both of which we have seen already. They may seize electron-deficient species like H^+, thereby creating a dative covalent bond. Alternatively, they may approach an 'electron-complete' site in a molecule (preferably a site which is positively polarised) and forcibly insert their lone pair of electrons, in place of a pair of electrons from an ejected 'leaving group'. These are the behaviours which we have characterised as *basic* and *nucleophilic*, respectively. Ammonia and water show both these patterns of behaviour, but to different extents.

The difference can be appreciated by considering a pair of possible reactions. When a molecule of ammonia uses its lone

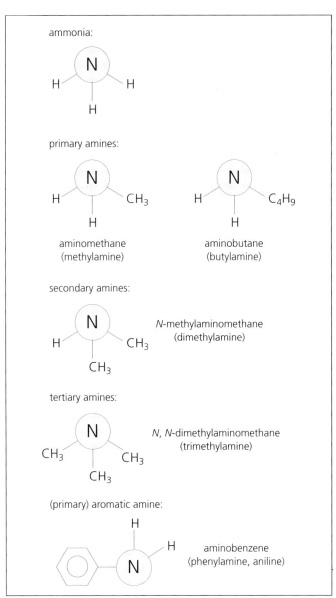

Figure 15.1 A range of amines. The tetrahedral arrangement of bonds at the nitrogen atom has been emphasised

pair to pick up a hydrogen ion from water, an equilibrium is set up:

$$NH_3(aq) + H_2O(l) \rightleftharpoons NH_4^+(aq) + OH^-(aq)$$

or (in displayed form)

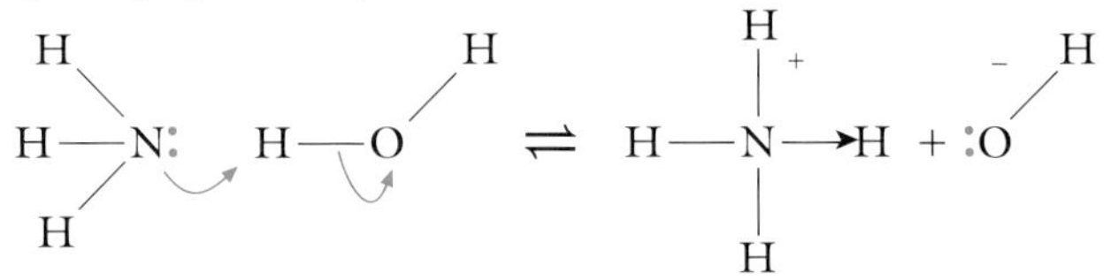

The balance of the equilibrium mixture is about 99% biased to the left-hand side, but the forward reaction goes to a measurable degree, which is more than can be said for the reaction:

$$NH_3(aq) + H_2O(l) \rightleftharpoons NH_2^-(aq) + H_3O^+(aq)$$

or

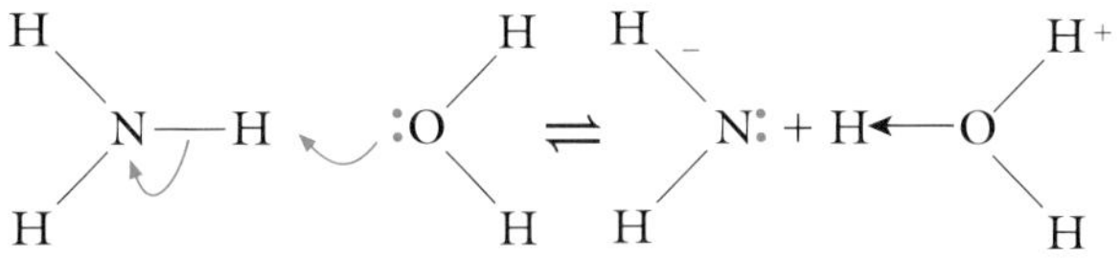

which does not happen at all.

Question

1 What conclusions can you draw about the relative base strengths of ammonia and water, from the fact that the first reaction goes and the second one doesn't?

Returning to the field of organic chemistry, we can imagine the same sort of basicity contest being played out between the organic counterparts of water and ammonia, namely the alcohols and the amines. We would not expect the result of the contest to be unduly altered by the presence in each molecule of an alkyl group or two. So, whereas the alcohols in Chapter 11 were negligible bases and modest nucleophiles, we might expect amines to be significantly basic and nucleophilic (Figure 15.2). And whereas alcohols were just about measurably acidic (at least when reacting with sodium metal), we might perhaps expect amines to show virtually no acidic tendencies. These do indeed turn out to be the major themes of amine chemistry.

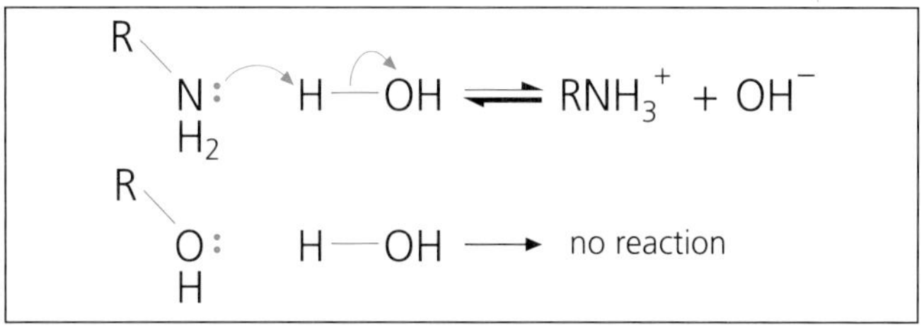

Figure 15.2 Comparison of the relative basicities of amines and alcohols

15.3   The basicity of the amines

Question

2 From the last paragraph, and by comparison with ammonia, construct an equation to show ethylamine behaving as a base in aqueous solution.

So far we have been predicting the properties of amines by direct extension from the properties of ammonia. However, it would be wrong to think that the presence of several alkyl groups has no effect at all on amines. The alkyl groups do in fact affect both the basicity itself, and also the solubility in water needed by any base if it is to remove a hydrogen ion from a water molecule. We shall look at each of these features in turn.

Water solubility

The short-chain amines are fully miscible with water. From about five-carbon chains onwards, the miscibility is partial and diminishing. As the alkyl group chains get longer, the van der Waals' contribution to intermolecular bonding in the amine itself gets more important, and leads to the point where water is ineffective in offering solvent–solute bonding.

Questions

3 Longer-chain amines which are quite insoluble in water may nevertheless be completely soluble in acids. In acids, the amines do not have to 'grab' the hydrogen ions, they have them 'thrust' upon them. Explain this extra solubility by thinking about the number of hydrogen ions present in acids, and the freedom of access to hydrogen ions the amine will have. Use equations to illustrate your answer.

4 Very-long-chain (more than ten-carbon) amines can react with acids to form substances that can act as **cationic detergents**. Repeat the equation from question 3 with a long-chain alkyl group, and suggest why the resulting species would have detergent action. (You may need to look back at Section 13.18 to remind yourself of the equivalent group of anionic soaps that result from the action of alkalis on carboxylic acids.)

5 The molecule adrenaline is shown below.

Decide which of the following statements about adrenaline are likely to be true (you may need to revise the acid–base tendencies of phenols from Section 11.13). Justify your answers.

a Adrenaline would be more soluble in acids than in water.

b Adrenaline would be more soluble in alkalis than in water.

c Adrenaline would be oxidised to an aldehyde.

d Adrenaline would exist as two optically active stereoisomers.

Degree of basicity

This is the second way in which the alkyl group affects an amine. This point concerns the amines' degree of basicity relative to ammonia, and indeed relative to each other. Table 15.1 shows some values of the base dissociation equilibrium constants, K_b, for a range of amines. You met data like these in Chapter 12 – which showed that the bigger the K_b, the stronger the base or, in other words, the greater the tendency to seize hydrogen ions from water.

Table 15.1 Base dissociation constants for amines

Amine	K_b(mol dm^{-3})
Dimethylamine	5.1×10^{-4}
Methylamine	4.4×10^{-4}
Ammonia	1.8×10^{-5}
Phenylamine (aminobenzene)	4.2×10^{-10}

A large K_b means that the molecule is fairly 'pushy' with its lone pair, and the differences in push must be due to effects from the various side-chains. Table 15.1 shows that methyl groups cause the lone pair on the nitrogen atom to be 'pushier' than when only hydrogen atoms are present, while phenyl (C_6H_5) groups produce a less 'pushy' nitrogen lone pair. These two facts are consistent with two pieces of previous work.

1 We have already suggested that methyl groups have a tendency to push electron density away from themselves, relative to hydrogen. This would result in the nitrogen atom being more negatively polarised than normal (Figure 15.3a). The resulting extra electron–electron repulsion would account for the increased base activity of the lone pair, which would be driven to look for bonding alternatives, instead of staying solely on the nitrogen atom. You may remember that we have

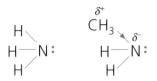

a inductive effect from methyl group increases basicity of aminomethane relative to ammonia . . .

b . . . which calls to mind the stabilising of tertiary carbocations by three such inductive effects

Figure 15.3 Effect of the inductive electron 'push' of the alkyl group

already (p. 230) used this idea of electron push by a methyl group to explain (among other things) the relative stability of the intermediate in S_N1 substitutions (Figure 15.3b).

2 In Chapter 11, while discussing phenol, we noted its lesser nucleophilic activity in ester formation, compared with ordinary alcohols. At the time we attributed this to the lone pair on the oxygen atom being delocalised over the π-electron cloud of the phenyl ring. The same argument accounts for the reduced basicity of phenylamine, except that here the nitrogen atom's lone pair is delocalised round the ring:

Question

6 Phenylamine (also known as aminobenzene or aniline) has two sets of reactions, those characteristic of the amino group and those characteristic of the benzene ring. Reactions of the ring (for example electrophilic attack by bromine) are many times faster than those of benzene itself, for reasons similar to those used to explain why the ring in phenol is more reactive than that in benzene (p. 252). However, in the presence of hydrogen ions, phenylamine loses much of its increased reactivity relative to benzene. Can you explain why this might be?

15.4 Amines as nucleophiles

Reactions with halogenoalkanes

We know already from Chapter 11 that nucleophiles will attack halogenoalkanes. If you react a halogenoalkane with an alcoholic solution of ammonia, the result is a primary amine (Figure 15.4). This reaction is of some value as a path to the preparation of the primary amines, but its use is limited by what happens next. The primary amine is itself capable of imitating ammonia and carrying out its own attack on the halogenoalkane. If you look, for example, at methylamine in Table 15.1 you will see that it is actually slightly better at it than is ammonia.

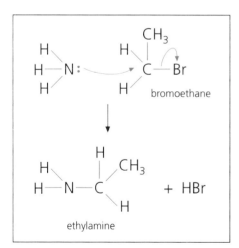

Figure 15.4 Nucleophilic attack by ammonia on bromoethane

Question

7 a Taking up where we left off at the end of Figure 15.4, write an equation for the reaction of ethylamine with bromoethane, and

for any reaction that might happen after that.

b Going back to the original reaction in Figure 15.4 between bromoethane and ammonia, which reagent would you keep in excess if your aim was to prepare only the primary amine?

Reactions with acyl halides

If ammonia and the amines can carry out nucleophilic substitution reactions with halogenoalkanes, then it will come as no surprise that they react even better with acyl halides (Figure 15.5).

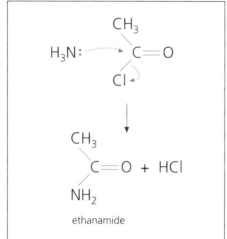

Figure 15.5 Nucleophilic attack by ammonia on ethanoyl chloride, to make an amide

Question

8 What arguments were put forward in Chapter 13 to account for the extra reactivity of acyl halides, relative to halogenoalkanes?

The result of reactions between ammonia and acyl halides is a class of molecules called **amides**, possessing groups typified by ethanamide in Figure 15.5.

Question

9 Once again draw a parallel between ammonia and the amines, to predict what would happen if an acyl halide were attacked by a primary amine. Use

benzoyl chloride and ethylamine as your examples and give your answer in structural formulae, with curly arrows, in the style of Figure 15.5.

Ammonia and amines as ligands in complexes

Before leaving the subject of nucleophilicity in amines, we shall take a preview of Chapter 19, where we shall study **ligands** and **complexes**. There is one further use to which the lone pairs of ammonia and the amines can be put, apart from picking up hydrogen ions and carrying out substitutions. They can provide the dative covalent bonds that hold ligands on to metal ions in complexes.

The bond is made by the lone pair overlapping with, and spending time in, vacant electron orbitals just a little higher in energy than the outermost filled orbitals. The d-block metals have such vacant orbitals in abundance. All the amines resemble ammonia in their ability to form complexes with ions such as $Cu^{2+}(aq)$. One of the more novel ligand molecules is 1,2-diaminoethane, which bears its active lone pairs at just the right spacing to behave like two jaws closing on the central metal ion. Such a ligand is termed a **bidentate ligand**, drawing on the Latin for 'two teeth'. Many metal ions prefer to be surrounded by six lone pairs in an octahedron, so it is not uncommon to find three 1,2-diaminoethane molecules arranged as in Figure 15.6.

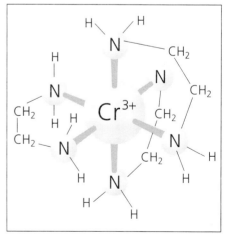

Figure 15.6 A transition metal–amine complex featuring 1,2-diaminoethane (nitrogen atoms emphasised)

Summary of the reactions of the amines

1 As a base, for example:

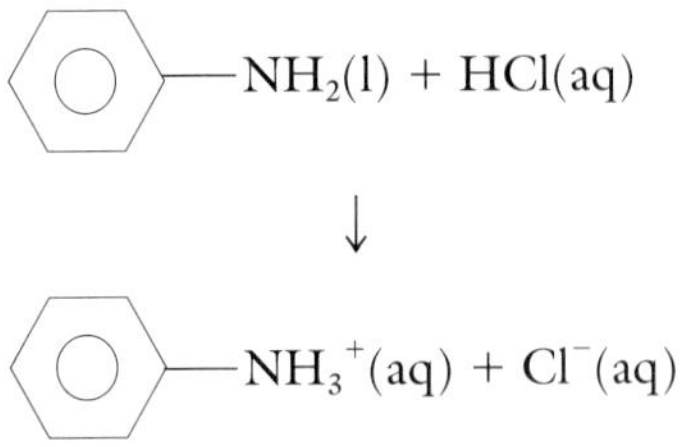

phenylammonium chloride

2 As a nucleophile with alkyl halides, to form secondary and tertiary amines:

$$CH_3CH_2NH_2 + CH_3CH_2I \rightarrow (CH_3CH_2)NH + HI$$
diethylamine

and even:

$$(CH_3CH_2)_2NH + CH_3CH_2I \rightarrow (CH_3CH_2)_3N + HI$$
… if the alkyl halide is in excess.

3 As a nucleophile with acyl halides, to form *N*-substituted amides:

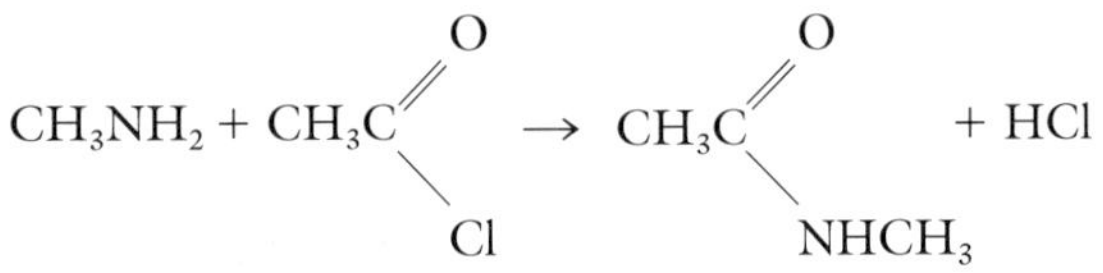

N-methylethanamide

4 With nitric(III) acid (see Section 15.14 on diazonium compounds):

$$CH_3NH_2 + HNO_2 \rightarrow CH_3OH + N_2(g) + H_2O$$
primary   (no diazonium compound)
alkylamine

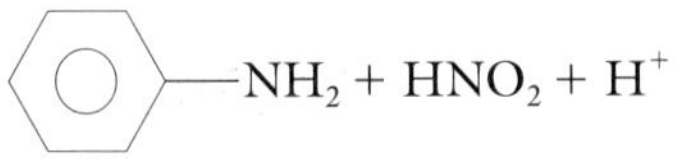

primary aromatic
amine

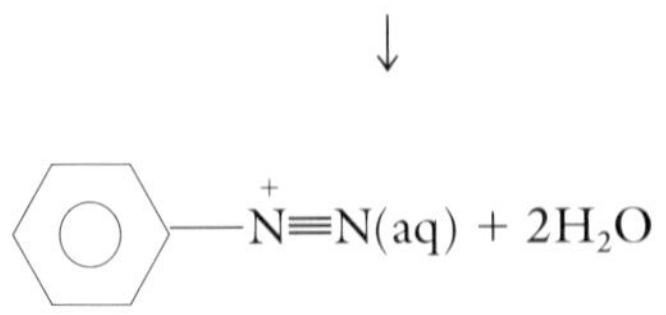

↓

diazonium ion
(a versatile intermediate)

15.5 Amides

The amides, as we have just seen, are the compounds obtained by the reaction of ammonia or amines with acyl halides. Their naming system is derived, like those of many other groups of simple organic chemicals, from the alkane of the same carbon number. Thus the amide in Figure 15.7a is

Figure 15.7 Examples of nomenclature of the amides

called propanamide. Ordinary amides like propanamide are made from ammonia and the acyl halide. If, instead of ammonia, an amine is employed as the nucleophile then, as you saw in question 9, an **N-substituted amide** is the result (Figure 15.7b).

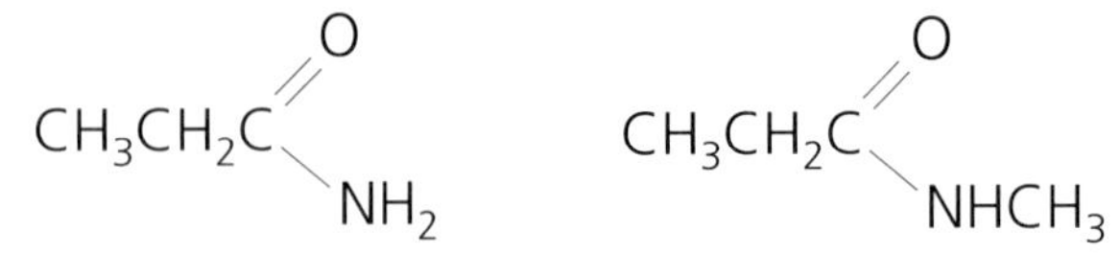

Amides as bases and nucleophiles

Amides are basic in nature, like all ammonia-derived compounds, but not as strongly basic as ammonia or the amines. This is because the electron pull of the carbonyl group reduces the tendency of the lone pair on the adjacent nitrogen atom to be nucleophilic. The main reaction of the amides is hydrolysis, which, like the hydrolysis of esters, returns the amide to its two 'parent' compounds – the carboxylic acid and ammonia or the amine. Also like esters, amides are best hydrolysed in alkaline conditions.

Questions

10 Write an equation for the alkaline hydrolysis of *N*-ethylethanamide. Bear in mind that the alkaline conditions will affect the form of one of the two products.

11 Paracetamol, the well known pain-killing molecule, is in fact an *N*-acyl-substituted amide:

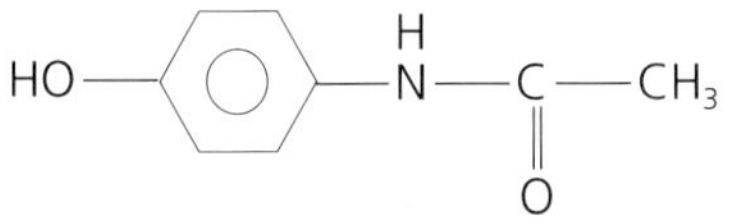

Suggest two molecules that could have been reacted together to make paracetamol.

Polymerisation using the amide linkage – the nylon family of polymers

We have already seen that the ester linkage has been adapted by chemists to create polymers. In Chapter 13 we met the polyester Terylene, and noted its ability to be drawn into fibres. As you might expect, in view of the fact that the properties of amides run parallel to those of esters, there is also an important polyamide, and it too is widely used in fibres.

The synthetic polyamides, which consist of a small family of similar molecules, are called **nylons**. The molecules are constructed by a method analogous to that used in polyester manufacture, namely 'double-headed' parent molecules performing condensation reactions:

. . . with a water molecule eliminated at every join.

Question

12a Which double-headed monomers would be needed for this polymer? Why do you suppose this product is called 'nylon 66'?

b Pick a word from the following pairs to complete the sentence.

The method of making nylon 66 identifies it as a(n) (addition/ condensation) polymer. When heated it would show behaviour typical of a (thermoplastic/thermosetting) polymer, in that it would (melt/char).

c Suggest why, like Terylene, this molecule readily lends itself to being drawn into thin fibres.

d Nylon 66 is somewhat stronger than the equivalent polyester, which suggests that it has superior intermolecular bonding. Can you suggest where in the two molecules, despite their considerable similarities, the source of this difference might lie?

e Another member of the nylon family is nylon 6, made from the polymerisation of caprolactam:

This departs from the pattern established with nylon 66, in that there is only a single monomer, and it is not 'double-headed'. Can you explain how caprolactam operates as a nylon monomer? Suggest what the structure of nylon 6 might be.

The extra strength of nylon means it can withstand higher stresses than polyesters (which explains its use in parachutes). When moulded into blocks, nylon is almost as smooth, hard and slippery as poly(tetrafluoroethene), and is also less expensive. It resists abrasion and needs no lubricating. This property equips nylon for such diverse uses as curtain hooks, gear wheels in small machines and bearings in power tools (Figure 15.8).

Figure 15.8 Nylon products (clockwise from top), gears, fabric, bearings and rope

The downside of nylon

There are two historic failures associated with nylon. As a clothing fibre, nylon gave rise to fabrics which were uncomfortable to wear, due to their lack of sweat absorption. Many survivors of the 1960s now look back with embarrassment on their 'Bri-nylon' shirt period. Nylon shirts were

Figure 15.9 Trying to look chic in a nylon sweater. In a very real sense, 'sweater' was an appropriate word

part of the early love-affair with polymers in the post-war period (Figure 15.9). They were part of a new world which would be transformed by technology. Such an attitude led to some uncritical early use of plastics, before people became aware of their disadvantages, and indeed before real improvements rendered the products more desirable.

Question:

13 Most modern shirts, if not 100% cotton, are made of a mix such as 65%/35% polyester/cotton. Nylon has disappeared completely from shirt fabrics. This is because polyesters like Terylene are better sweat absorbers than nylons such as nylon 66. Can you relate the difference in water absorption to differences in the structure of the two polymers?

The second negative association with nylon is of an altogether more tragic nature. Wallace Hume Carothers (Figure 15.10) was not only the

Figure 15.10 Wallace Carothers puts his invention under stress

inventor of nylon 66 (patented in 1935), but also of neoprene, the first synthetic rubber (1928), and he originated the conceptual distinction between condensation and addition polymers. But these practical and intellectual achievements could not dispel the fits of depression to which he was prone, and in 1937 at the age of 41 he committed suicide. He never saw the launch of his brainchild as a commercial reality, nor the subsequent impact of his work on the world.

15.6 Amino acids

Acidic and basic behaviour

Amino acids are compounds that contain two contrasting functional groups. The general structure of an amino acid is:

$$H_2N-\overset{\displaystyle CO_2H}{\underset{\displaystyle R}{\vphantom{|}C}}-H$$

The presence of one acidic group and one basic group means that **intramolecular** reactions (reactions *within* the molecule) are possible. All the naturally occurring amino acids have the R group and the NH_2 group situated on the carbon next to the carboxylic one, and there are about 20 different R groups involved in protein synthesis in Nature. The properties of these real amino acids suggest that the structure above might not reveal the whole story.

For a start, even when R is just a hydrogen atom, the resulting compound (called glycine) is a solid at room temperature with a fairly high melting point. Furthermore, all the amino acids much prefer water over other solvents, they all exhibit high dipole moments, and they are less acidic than normal carboxylic acids and less basic than normal amines.

The single explanation for all these bits of evidence is that the two ends of the amino acid molecule have already carried out the acid–base proton exchange *with each other*, and so the 'molecule' is really a dipolar ion:

$$^+H_3N-\overset{\displaystyle CO_2^-}{\underset{\displaystyle R}{\vphantom{|}C}}-H$$

This is called a **zwitterion**, from the German for 'hermaphrodite'. So any acidity is due to NH_3^+, rather than to a carboxylic group, and similarly any basicity is due to COO^-, rather than to an amino group.

Having made this point we must accept that we still have a molecule that will behave towards water as both acid and base. The pH of the resulting solution depends upon a 'trial of strength' between the two groups. Glycine, for example, the amino acid mentioned above in which R = H, is slightly more acidic than basic, which means there will be a slight preponderance of ion (b) over ion (c) in the mixture of ions in its aqueous solutions, although zwitterion (a) will vastly outnumber both (Figure 15.11). A more clear-cut acidity or basicity exists in amino acids whose R group contains a second acidic or basic group.

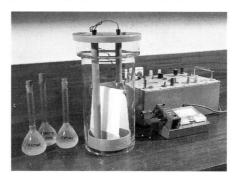

$$H_2O + H_3\overset{+}{N}—CH_2—CO_2^- \quad \textbf{a} \text{ glycine zwitterion}$$

as acid $\nearrow\!\!\!\swarrow$ $\qquad$ $\nwarrow\!\!\!\searrow$ as base

$$H_3O^+ + H_2NCH_2CO_2^- \quad H_3\overset{+}{N}CH_2CO_2H + OH^-$$
$$\textbf{b} \qquad\qquad\qquad \textbf{c}$$

Figure 15.11 Glycine acting as both acid and base towards solvent water

Question

14 With the last sentence in mind, classify the amino acids in Figure 15.12 as either largely neutral, significantly acidic or significantly basic.

a alanine

$$\begin{array}{c} CO_2^- \\ | \\ H_3\overset{+}{N}—C—H \\ | \\ CH_3 \end{array}$$

b

$$\begin{array}{c} CO_2^- \\ | \\ H_3\overset{+}{N}—C—H \\ | \\ CO_2H \end{array}$$

c lysine

$$\begin{array}{c} CO_2^- \\ | \\ H_3\overset{+}{N}—C—H \\ | \\ (CH_2)_4 \\ | \\ NH_2 \end{array}$$

d tyrosine

$$\begin{array}{c} CO_2^- \\ | \\ H_3\overset{+}{N}—C—H \\ | \\ CH_2 \\ | \\ \bigcirc \\ | \\ OH \end{array}$$

e valine

$$\begin{array}{c} CO_2^- \\ | \\ H_3\overset{+}{N}—C—H \\ | \\ CH \\ \diagup\;\diagdown \\ H_3C \quad CH_3 \end{array}$$

Figure 15.12 A selection of amino acids. The name of (b) would have given away the answer to question 14

Separating amino acids

Amino acids are a close-knit family of compounds, and, because they are often encountered as a result of the breakdown of proteins, they are nearly always found in mixtures. Their separation and identification is often an important concern for biochemists studying protein structures. One of the key identification techniques is a method that combines electrolysis with chromatography, and in which the balance between the acidic and basic tendencies of the molecules has an important part to play. The technique is called **electrophoresis**.

Electrophoresis can be carried out in laboratories fairly easily. The amino acid (or more normally a mixture of amino acids) is placed on the 'starting line' on a piece of filter paper as in chromatography, except more usually as a line than as a spot. You will recall that in chromatography the compound

'chooses' between the stationary phase (the paper) and the moving phase (the liquid 'eluent'), to decide its degree of movement (p. 234). In electrophoresis, it is not elution by the solvent that moves the compound across the paper – the solvent has already soaked through the entire paper. Instead, the compound moves under the influence of the forces of **electrolysis** (in other words, electrodes attracting ions of opposite charge). The position of the amino acid after a certain time is found by drying the paper and spraying with a detector substance called ninhydrin, with which almost all amino acids produce a purple coloration (Figure 15.13).

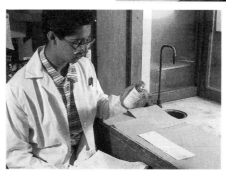

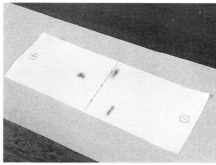

Figure 15.13 An electrophoresis experiment

There are now two molecular factors that affect the movement of the compound across the paper. One is the factor familiar from chromatography of the molecules' 'choice' (based on intermolecular bonding) between solvent and paper, which decides *how far* the 'spot' moves in a certain time. The other is a totally new variable – *whether* to move at all, and if so *in which direction*. Since a solution of an amino acid contains neutral molecules, anions and cations all in equilibrium with each other, there are several possible ways to move.

Question

15 We have already mentioned that glycine has a slight preponderance of $NH_2CH_2CO_2^-$ over $^+NH_3CH_2CO_2H$, and that the majority of molecules are in the zwitterion $^+NH_3CH_2CO_2^-$ form, which would be attracted to neither electrode.

a Why would the zwitterion form not move?

b Despite the majority of zwitterions on the 'start line', and their unwillingness to move, nevertheless, at pH 7, the whole spot does move together, rather than going three ways. Can you suggest a reason why this might be?

You will have realised from question 15 that a full understanding of electrophoresis is quite complicated. This full understanding is less important than an awareness of the usefulness of the results. Every one of the naturally occurring amino acids can be made to travel a different and characteristic distance in chromatographic and electrophoretic experiments, and that is the value of both techniques as identifiers. Figure 15.14 shows how the two techniques can be used in a concerted way.

In electrophoresis, each amino acid provides another identifying 'fingerprint', apart from distance travelled. So far we have assumed the test is conducted in aqueous solution at the pH natural to the particular test substance. However, using pH as a variable, by adding small amounts of hydrochloric acid or sodium hydroxide, it is possible to 'fine tune' each

amino acid so that there are equal numbers of cations and anions. At this pH, no electrophoretic movement occurs. Such a pH is referred to as the **isoelectric point** of the amino acid. The technique may also be applied to polypeptides of all sizes – these are polymers of amino acids that we shall meet in Section 15.7.

Question

16 Reconsider the case of glycine. In which direction would we have to adjust the pH so as to redress the minority of $^+NH_3CH_2CO_2H$ relative to $NH_2CH_2CO_2^-$ (in other words, to create some of the former from the zwitterion 'pool', and send some of the latter back to the 'pool')?

Table 15.2 shows the isoelectric points, in pH units, for a number of amino acids.

Table 15.2 Isoelectric points of selected amino acids. Notice how the amino acids with acidic side-chains (*) require extra proton donation, to suppress their inclination to exist as anions, whereas those with basic side-chains (†) require extra proton removal, to suppress their inclination to exist as cations. All the others have isoelectric points close to 6

Amino acid	R group	Isoelectric point (pH unit)
Alanine (ala)	$—CH_3$	6.1
Arginine (arg)	$—(CH_2)_3—NH—C{\,}^{NH_2}_{NH}$	10.8[†]
Aspartic acid (asp)	$—CH_2—CO_2H$	3.0*
Cystine (cys)	$—CH_2—S—S—CH_2—CH(NH_2)—CO_2H$	5.0
Glutamic acid (glu)	$—CH_2—CH_2—CO_2H$	3.1*
Glycine (gly)	$—H$	6.1
Leucine (leu)	$—CH_2—CH(CH_3)CH_3$	6.0
Phenylalanine (phe)	$—CH_2—C_6H_5$	5.9
Tyrosine (tyr)	$—CH_2—C_6H_4—OH$	5.6*
Valine (val)	$—CH(CH_3)CH_3$	6.0

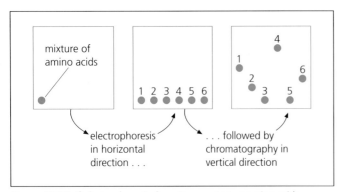

Figure 15.14 If electrophoresis doesn't separate two amino acids, chromatography might. Alternatively, you could use two chromatography stages with different eluents. The original mixture would probably have come from the breakdown of a protein by prolonged acid hydrolysis

Amino acids and chirality

All the naturally occurring amino acids are 2-amino acids, carrying the amino group on the carbon atom next to the carboxylic carbon. Furthermore, apart from glycine, all the amino acids have some sort of substituent group in the R position, so they all have an asymmetric carbon atom. They can therefore rotate the plane of polarisation of plane-polarised light. Just as in the case of the sugars, and in common with many compounds produced under enzyme control in natural systems, only one chiral form of each molecule exists in Nature, in this case always the L-form (Figure 15.15).

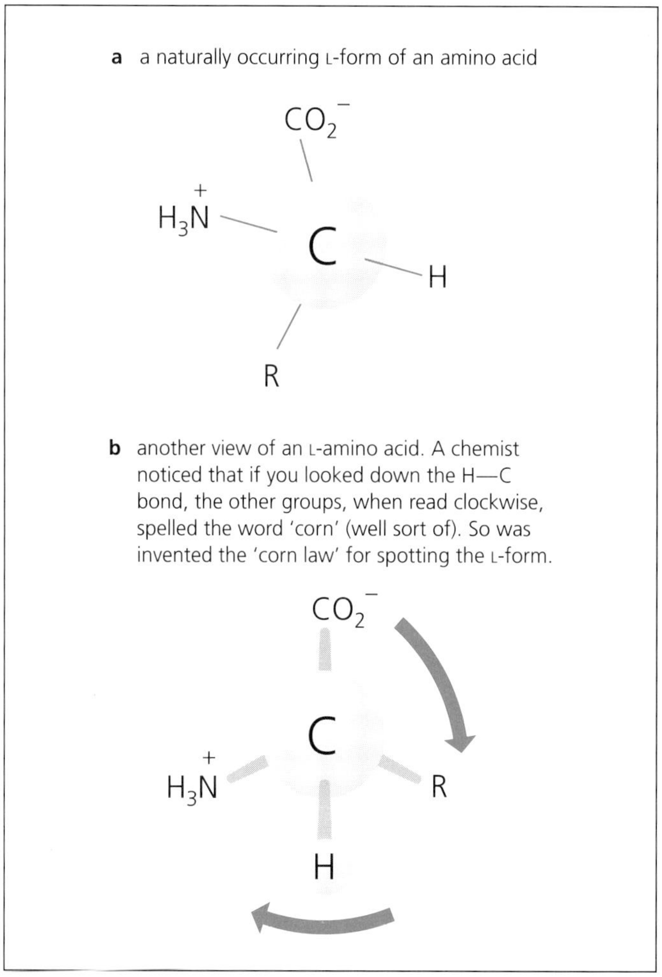

a a naturally occurring L-form of an amino acid

b another view of an L-amino acid. A chemist noticed that if you looked down the H—C bond, the other groups, when read clockwise, spelled the word 'corn' (well sort of). So was invented the 'corn law' for spotting the L-form.

Figure 15.15 Only one stereoisomer of each amino acid exists in Nature

15.7   An introduction to proteins

The possibility of using the amide linkage in polymerisation was realised by Carothers in the 1930s. Several billion years earlier the same bond had been used for the same purpose of polymerisation, in the progress of evolution, and had brought into existence the crucial class of biomolecules called **proteins**. The monomer molecules were amino acids, which do have a sort of 'double head' if not quite in the same way as the dicarboxylic acids and diamines which make up nylon.

The amino group of one amino acid can carry out a nucleophilic attack on the carboxylic acid group of another amino acid, and produce a dimer with an amide linkage in the middle (Figure 15.16). The amide linkage is called, in this context, a **peptide linkage**. This dimer is itself still an amino acid, insofar as it has an NH_3^+ on one end and a CO_2^- on the other, so the process can continue until the growing chain has consumed all the free amino acids in the vicinity. And with 20 natural amino acids to choose from, the number of possible protein polymers is very large.

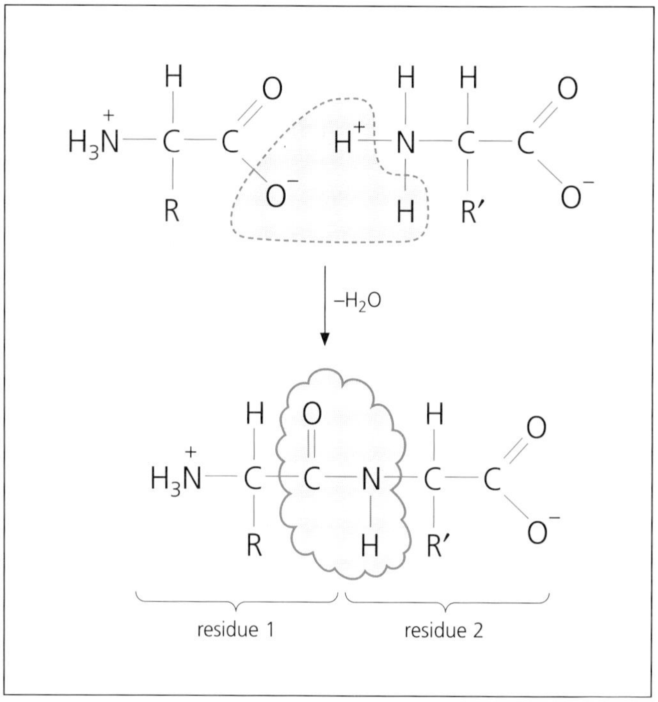

Figure 15.16 Dimerisation of two amino acids to form a dipeptide. The peptide linkage is shown in the 'cloud'

The language of biochemistry

The word 'peptide' grew up in biology, with a different tradition of nomenclature from that of chemistry. There is no chemical difference between the peptide linkage in proteins and the amide linkage in nylon, although of course the monomers between the linkages are very different. The word 'peptide' is firmly ingrained in biochemistry, to the extent that 'short proteins' with two amino acids are called dipeptides, those with three amino acids are called tripeptides and in general all those with one to ten links are called **polypeptides**. Several polypeptides join together to make **proteins**. Another word you will meet in the vocabulary of protein chemistry is **residue**, as in references to amino acid 'residues' or monomers in proteins and polypeptides. The word is used in recognition of the fact that proteins are condensation polymers, and the loss of water in the making of each bond means that they are a little less than the sum of the individual amino acids.

Question

17 The previous paragraph made a comparison between, on the one hand, proteins, which are the natural polymers, and on the other hand, the nylon group, which are synthetic polymers. It was said that the linkage in both classes of polymer was the same but that the two sorts of chains were different. Can you think of two points of contrast between the structures of proteins and nylons?

The importance of proteins

The name 'protein' comes from a Greek word meaning 'first rank', and proteins are indeed of primary importance to us as living things. A large proportion of our bodies is protein. Apart from the structures of human and animal bodies, most of their working systems rely on proteins too – proteins make up enzymes and many hormones. A list of functions carried out by proteins in the bodies of (most) animals is shown below.

1 *Transport* – the oxygen-carrying molecules **myoglobin** and **haemoglobin** are proteins.
2 *Motion* – the contraction of muscles is based on the proteins **actin** and **myosin**.
3 *Structure* – **collagen** is an important fibrous protein in skin and bone. In bone it fulfils a structural role similar to that of the strong graphite fibres in the composite material used for tennis rackets.
4 *Immunity* – **antibodies**, molecules that recognise foreign substances in the body, are proteins.
5 *Sensitivity* – for example, in the receptor protein **rhodopsin**, found in the rod cells of the retina in the eye, which receives photons of light and begins the process of seeing.
6 *Control* – the hormone **insulin** plays a key role in controlling the metabolism of glucose.
7 *Catalysis* – the most widespread, in terms of influence, of all the protein substances in animal physiology, are the **enzymes**, which catalyse a huge range of biochemical reactions, from the 'unzipping' of DNA to the dissolving of carbon dioxide in blood.

The blueprint for life

To extend point 7 a little further, the two-way co-operative relationship between DNA and enzyme proteins is central to the whole shape of life on Earth. Most people are aware from their general knowledge that DNA contains the 'blueprint for life'. When we probe further to ask *how* DNA controls cell activity, and in what language this blueprint is encoded, we discover that it is a code for making proteins.

We will see in Section 15.8 that proteins are made up of strings of amino acids in a particular sequence. DNA is present in chromosomes, thread-like structures in the nucleus of every cell. There are sets of three bases on each chromosome, which are each a signal to call up a particular amino acid on to the 'shop floor' of protein synthesis (Table 15.3). A large proportion of proteins synthesised by a cell are in fact *enzyme* proteins, so if we put these strands of the story together we see that much of what we call the 'blueprint' is a plan not for making the cell directly, but for making the *tools* that make the cell.

Table 15.3 Part of the genetic code. The letters A, T, C, G stand for the bases adenine, thymine, cytosine and guanine. A group of three bases in a particular sequence on the chromosome codes for a particular amino acid. X means that any of the four bases can appear in that position

Code	Amino acid
CGX	Alanine
CCX	Glycine
CTA or CTG	Aspartic acid
TTT or TAC	Lysine
ATA or ATG	Tyrosine
CAX	Valine

The co-operation is two-way because not only does DNA make enzymes, but enzymes help to make new DNA. The very operation of DNA, including its replication and its message-sending ability, is under enzyme control. This begs the chicken-and-egg-type question of which came first, DNA or enzymes. This question is hard to answer, since the solution lies partly in events that took place two billion years ago.

15.8 The primary structure of proteins

Sections 15.8–15.12 deal with the four levels of describing protein structures, and the evidence for, and importance of, each one. They are particularly relevant to those studying biochemistry as a special option. Students who are not can leave out Sections 15.10–15.12 and can think about enzyme action in terms of the straightforward lock-and-key model outlined in Chapter 14.

The first level of analysis, the so-called **primary structure**, is the response to the question: 'What is the amino acid sequence in any given protein?' It is not too difficult to find out which amino acids are *in* the protein – this can be done by complete hydrolysis of the protein under prolonged acidic/reflux conditions (which cleaves all the peptide bonds), followed by paper chromatography of the amino acid fragments.

Two-dimensional paper chromatography

When a protein has been cleaved, there is a mixture of up to 20 components (the number of possible constituent amino acids), in various proportions. Paper chromatography is quite a convenient method to separate them, because paper is a reasonably selective stationary phase. Its cellulose molecules tend to hold back the movement of amino acids with more polar and hydrogen-bonding side-chains, and to allow fairly free movement of those with less polar side-chains. (i.e. R groups).

Some of the 20 components are inevitably going to overlap, even with the best separation, so that some spots on the developed chromatogram may be impossible to resolve (Figure 15.17). In other words, a particular spot could indicate either one, or both, of two amino acids with very similar R_f values.

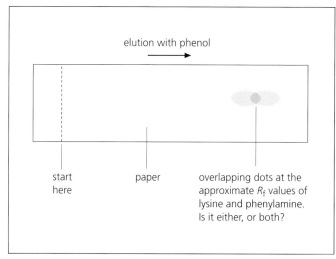

Figure 15.17 Showing how lysine and phenylamine could easily be confused in one-dimensional chromatography

elution with phenol

start here

paper

overlapping dots at the approximate R_f values of lysine and phenylamine. Is it either, or both?

The solution to the problem of unresolved spots is neat and simple. The chromatogram is run again, at *90°* to the first run, and under slightly different conditions. We have met this idea already in Figure 15.14, only that time one separation was done by electrophoresis and the other by chromatography. This time both movements are achieved by chromatography, but using different solvents as the eluent.

How do we choose a solvent for elution? Question 19 addresses this problem. It refers to Figure 15.18, which shows a drawing of a developed two-dimensional chromatogram. This particular one is of the amino acids from the hormone insulin, whose structure determination was one of the milestones of success of mid-twentieth-century biochemistry. (As we shall see, although identifying the amino acid residues in insulin was a major achievement, it merely began the process of understanding the molecule.)

Question

19a Of the two solvents used in Figure 15.18, one is distinctly more polar than the other. Which one is the more polar, and why is it important that there should be a difference in polarity?

b Give two aspects of the structure of insulin which are not conveyed by the results of this experiment.

Question

18 This question is about paper chromatography of amino acids. It is a mixture of revision from this book and from pre-Advanced-level work, and a test of your comprehension of the two previous paragraphs.

a The R_f value, the number that identifies the position of the amino acid 'spot' in the developed chromatogram, is actually a *ratio* of distances. Of which two distances is it the ratio?

b Amino acids, unlike Smartie dyes, cannot indicate their presence by colour. They have to be 'developed' – turned into a coloured derivative – by spraying. Which compound is used in the spray?

c Explain, with reference to intermolecular bonding, why amino acids with more polar side-chains move more slowly across the paper than those with less polar side-chains.

d From the selection of amino acids in Figure 15.12 (p. 359), select one you would expect to have a high R_f value, and one you would expect to have a low one (assuming the stationary phase is paper). Justify your choices.

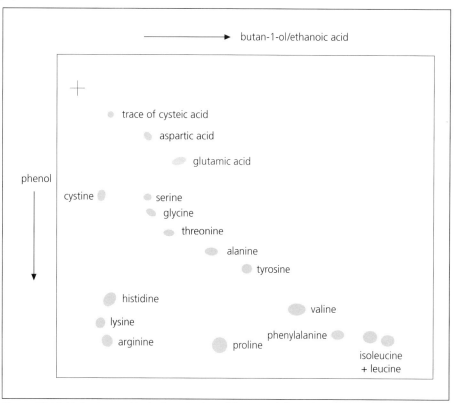

butan-1-ol/ethanoic acid

trace of cysteic acid

aspartic acid

glutamic acid

phenol

cystine

serine

glycine

threonine

alanine

tyrosine

histidine

valine

lysine

arginine

proline

phenylalanine

isoleucine + leucine

Figure 15.18 Notice how, with the introduction of the second dimension, there is a large separation between lysine and phenylamine (compare Figure 15.17)

Figure 15.19 shows the apparatus that is used (in schools at least) for two-dimensional chromatography. The line at the bottom indicates where the amino acid mixture was first dotted on to the paper. The solvent front is just visible at about half-height. The cover on the beaker is necessary to prevent local evaporation of the eluent.

Figure 15.19 Two-dimensional chromatography apparatus. For the second leg, the paper can be turned so that the vertical edge becomes the new bottom

Amino acid sequences in proteins

Having discovered *which* amino acid residues are present in a protein or polypeptide, the next step is finding out in what *order* they occur, information that has been lost in the hydrolysis. Before going further in this area, we need to be clear on the conventional nomenclature and symbolism used to express the formulae of polypeptides.

Question

20 We have not found it necessary to create new systems of symbols for describing other polymers in organic chemistry, like poly(ethene) or poly(phenylethene). What is different about polypeptides?

The system used for describing the structures of polypeptides involves two rules.

1 Each amino acid residue is given a three-letter code which is, where possible, the first three letters of its name (see Table 15.2). Thus glycine is gly while glutamic acid and glutamine are glu and gln respectively.
2 The left-hand end of the structure as written is the bare amino group, while the right-hand end is the bare carboxylic acid group.

Question

21 Use these two rules to draw the structural formulae of these tripeptides:

a gly-ala-val

b val-ala-gly.

Methods for determining amino acid sequences

A huge amount of ingenuity has been invested in this problem, and the result has been the invention of several methods for elucidating residue sequences, and the award of several Nobel prizes for the inventors. The methods fall into two broad groups – those that use a tailor-made organic molecule to interact with the chain, and those that use enzymes whose natural function is protein cleavage. An example of each style follows.

Edman degradation

This is a 'special molecule' method, which uses phenyl isothiocyanate (PTH). This molecule latches on to the *amino end* of the polypeptide, and not only thereby 'labels' it, but crucially renders the final amino acid residue more susceptible to being hydrolysed off from the rest of the chain, in conditions which leave the rest of the polypeptide intact. The PTH-residue can be separated from the rest of the chain by solvent extraction, and the shortened chain can then be fully cleaved into its amino acids. Whichever amino acid is now *missing* (compared to the full cleavage of the original chain) must have been the one on the end. The process is illustrated in Figure 15.20.

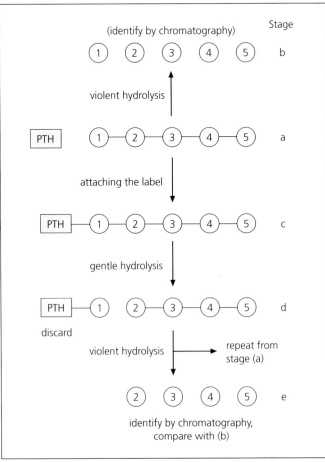

Figure 15.20 Symbolic representation of the stages of the Edman degradation

Question

22 A tripeptide is fully hydrolysed and found to be composed of ala, arg, asp (where the commas indicate lack of information about order). One round of the Edman degradation leaves a dipeptide, which hydrolyses to ala, asp.

a Give two possible structures for the original tripeptide.

b Suggest how the complete structure of the tripeptide might be found.

Selective enzyme breakage of polypeptides

Trypsin is an enzyme with a specific ability to cleave a polypeptide on the *carboxyl side of the lysine and arginine residues*. This is a help, since the smaller fragments are quicker to analyse by Edman's method. For example, the polypeptide:

thr-trp-val-lys-ala-ala-trp-gly-lys

would break into two smaller fragments:

thr-trp-val-lys   and   ala-ala-trp-gly-lys

But the researcher will not know which way round the two fragments fitted together. The answer is supplied by a second enzyme, which breaks the original chain in a different place, so that the new fragment selection includes a fragment that is still complete across the trypsin break site. One enzyme that fits this requirement is **chymotrypsin**, which cleaves on the *carboxyl side of tryptophan* (trp).

Question

23 Give the structures of the three polypeptides that would result from cleavage of the original polypeptide by chymotrypsin. Show how this information, together with the trypsin break pattern from the earlier experiment, establishes the whole original sequence.

The above account of analytical methods is only a sample from a very large specialised field. What we as general chemists need is on the one hand a flavour of the techniques of the biochemist, and on the other hand the realisation that all this work has only told us the **primary structure** of polypeptides. We have reached a stage where all the covalent-bond information of any given polypeptide is revealed, and for many molecules elsewhere in organic chemistry that would constitute a complete description. But much of the action of proteins depends upon the shapes the chains adopt, and it is to these levels of description that we must now turn.

15.9 The secondary structure of proteins

Looking at a polypeptide structure drawn flat on the page, as in Figure 15.21, you might think that, with freedom of rotation about every bond in the backbone, the number of likely

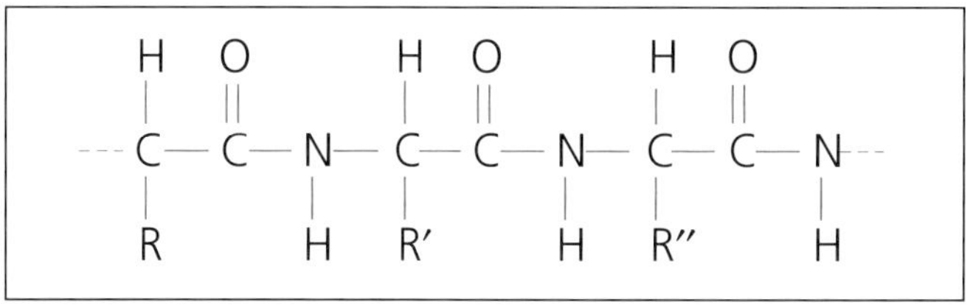

Figure 15.21 Primary structure of a polypeptide, giving no information about shape

conformations in space is almost infinite, in the range between the extremes of tightly clumped and linear. This assumption turns out to be false. First, free rotation is *not* possible about every bond in the structure. Second, the wealth of opportunity for inter- and intramolecular hydrogen bonds imposes a discipline that greatly restricts the choice of spatial layout of the chain.

It is quite usual to find that, under both these influences, protein molecules adopt some orderly pattern, like, for instance, a single-chain spiral, or a sheet made of lined-up chains. These organised formations of protein molecules are called **secondary structures**. Let us look at how the twin factors of restricted rotation and hydrogen bonding exert control over secondary structures. We shall take the lack of free rotation first.

The geometry of the peptide bond

There is strong evidence to suggest that the C—N bond in the peptide linkage has partial double-bond character. Its length is 0.132 nm, and X-ray diffraction studies of the four atoms coming from the peptide bond show them always lying in the same plane, just like the four groups around a C=C bond. Figure 15.22 suggests a resonance structure which, in contributing to the delocalisation, may be the cause of the alkene-like geometry and the lack of free rotation.

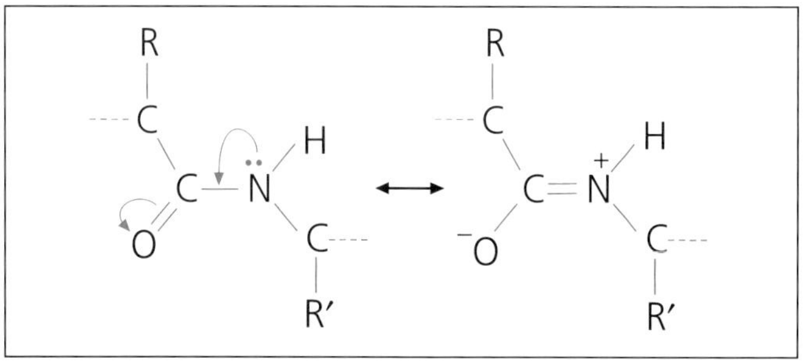

Figure 15.22 Covalent structures which may contribute to the delocalisation in, and the 'stiffness' of, the peptide bond

Question

24 Why does the bond length quoted above lend support to the idea of the C—N bond having partial double character? Quote from other pieces of relevant data to justify your answer.

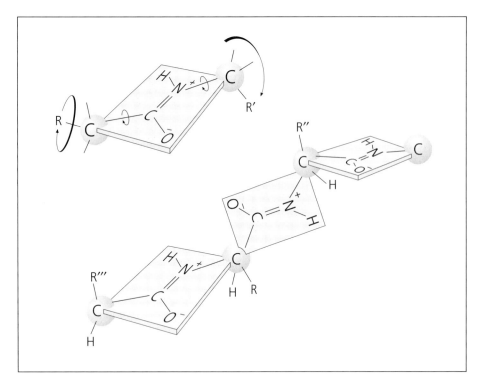

Figure 15.23 a A single peptide linkage, held in an alkene-like plane, but with free rotation at each encircled carbon. **b** The rigidity of each peptide linkage (shown here in its 'alkene-like' form) creates the effect of a chain of plates

Questions

25 What do you notice about the arrangement in space of the R groups in the α helix in Figure 15.25?

26 What kind of instrument would have been used to give the sort of information that would have led to the elucidation of the geometry of the α helix? (Bear in mind that it is a repeating structural pattern, of molecular size, in a solid material.)

It turns out that the α helix is only one of the stabilising strategies that are adopted in real protein structures. There is a second major method for the systematic ordering of chains (the β-pleated sheet, see below), and also there are additional arrangements unique to the circumstances of individual proteins. However, a number of important proteins do feature the α helix as a major theme. Haemoglobin and myoglobin (Figure 15.26) have more than half their residues in the helix format, while keratin, the hair protein, is a coil of two α helices wound round each other like a rope.

The flexibility of the polypeptide chain is affected by this regular periodic 'stiffness of the joints'. As Figure 15.23b shows, the chain behaves like a real metal chain, in which each 'peptide plane' acts like a metal link. Like a real chain, the structure is flexible up to a point, but cannot cope with sharp bends.

The α helix

You would think that, even stiffened at the joints, the polypeptide chain could still take up a very large number of orientations, just like a real metal chain. However, our polypeptide chain has links that are attractive to each other, and so at this point we shall leave the metal chain analogy behind.

If the backbone of the polypeptide molecule is twisted into a regular α (alpha) helix with a repeat distance of 0.54 nm, there is a convenient juxtaposition of hydrogen atoms from the N—H groups with oxygen atoms from the C=O groups of the next spiral up (Figure 15.24). Intramolecular hydrogen bonds can form between these atoms. This extra bond-making gives the helix a lower free energy than a random chain, and thus any random chain might find it energetically favourable to coil spontaneously.

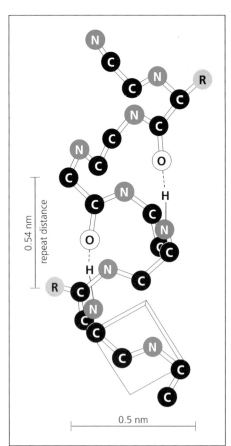

0.54 nm repeat distance

0.5 nm

Figure 15.24 Incomplete picture of the α helix. Occasional R groups are shown, as are some of the O····H bonds that hold the spiral

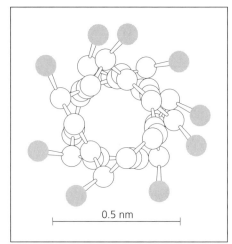

0.5 nm

Figure 15.25 Looking up the 'hole' in the α helix with the R groups shown in grey

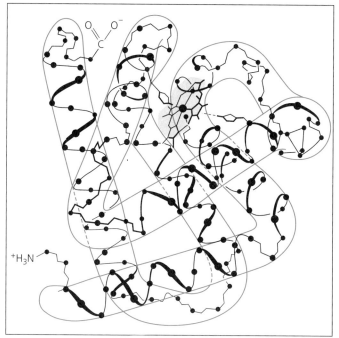

Figure 15.26 Model of myoglobin, showing the high proportion of α helix in the molecule. The grey shaded part high on the right with the fused hexagons is the haem + iron group. Only alternate carbons are shown (*after R E Dickerson*)

Figure 15.27
Flyaway hair

Figure 15.28
Oily hair

Figure 15.29
a A Brylcreem tester telephones headquarters to announce the results of the wind tunnel test

b The disciplined look, circa 1964

Hair conditioners

The chemistry of hair conditioners gives us a double opportunity to apply the ideas and knowledge of this chapter, and at the same time put to rest the uncomfortable suspicion that conditioners owe more to the creative minds of the marketing department than to the laws of chemistry. As this short article will show, they work on perfectly sound chemical principles.

Problems with hair

Immediately after washing, hair is generally 'flyaway' – wispy, windblown and ill-disciplined (Figure 15.27). On the other hand, if it hasn't been washed for a long time, it is readily disciplined, but also lank and oily (Figure 15.28).

For the first half of the twentieth century, *men* solved the discipline problem by exaggerating the naturally greasy state. By applying extra grease or oil, they achieved the 'Brylcreem' look, in which the individual hairs became part of a shiny skullcap with the look of patent leather (Figure 15.29a). *Women* used either permanent wave chemicals (a story in themselves), or systems of clips, plaits and pleats (Figure 15.29b).

The present modern-day hair conditioners give us hair with the extra 'body' and weight to be disciplined, without having to resort to plastering it or trapping it. To understand how conditioners work, we must begin with the chemistry of hair itself.

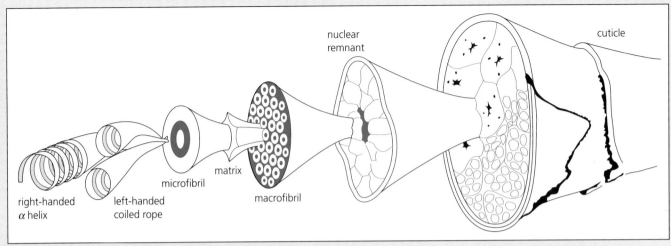

Figure 15.30 The internal structure of hair

The chemistry of hair

The principal hair protein is keratin, which is mainly α helix (Figure 15.30). In the mix of amino acid residues in keratin, there are significant amounts of the basic arginine and the acidic glutamic acid. The latter outnumber the former, so a typical length of hair protein might have a primary structure which could be represented as in Figure 15.31.

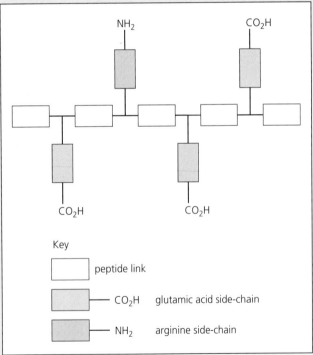

Figure 15.31 Schematic drawing of a portion of keratin chain (primary structure)

Question

27 Soaps and detergents tend to produce neutral or slightly alkaline aqueous media, by virtue of reactions such as:

$$Na^+ O^- \diagdown C \diagup\diagdown\diagup\diagdown\diagup\diagdown + H_2O$$
$$\quad\quad\quad soap$$

$$HO \diagdown C \diagup\diagdown\diagup\diagdown\diagup\diagdown + Na^+ OH^-$$

Small excesses of OH^- ions result. Show what ionisations might occur to the keratin chain in Figure 15.31 in the presence of OH^- ions.

The hair protein will become anionic in alkaline solution, with sodium cations as electrostatic counterbalance, probably located randomly in the tertiary structure of the protein. Conditioners aim to provide cationic species to bind to the hairs by ionic bonds, and these species give extra body with the minimum amount of extra material (unlike hair oils). The cation in question clearly is not just some ordinary metallic one – after all, the washed hair has those already with sodium ions from the soap. It needs to be a species with an almost detergent-like versatility – with long organic side-chains to give body, and a cationic site somewhere on the molecule to bond to the keratin. The answer comes in the shape of a class of amine derivatives, the **quaternary alkylammonium compounds**. Figure 15.32 shows an example of a quaternary ammonium ion which is commonly found as an ingredient in hair conditioners.

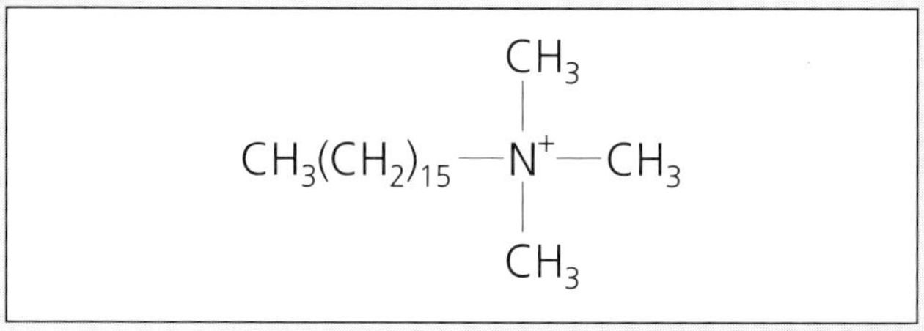

$$CH_3(CH_2)_{15} - \overset{\overset{\displaystyle CH_3}{|}}{\underset{\underset{\displaystyle CH_3}{|}}{N^+}} - CH_3$$

Figure 15.32 Cetyltrimethylammonium ion, used in hair conditioners

Quaternary alkylammonium compounds

If ammonia meets hydrogen ions, the lone pair on the nitrogen atom is employed to create the ammonium ion:

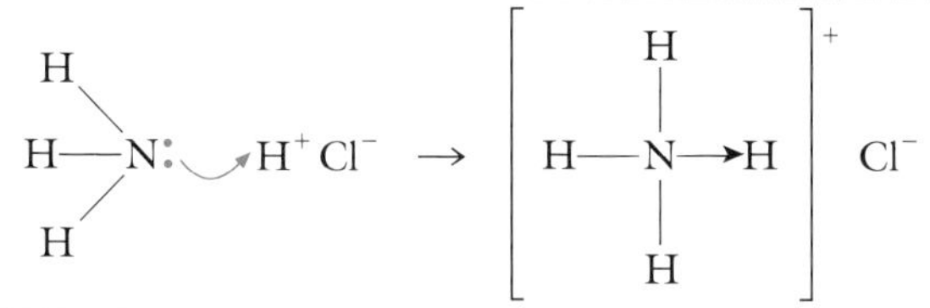

Ammonium compounds are fully ionic, and share some of the characteristics of corresponding Group I compounds. Amines too, in acidic media, can make their own equivalents of the ammonium ion:

$$RNH_2 + HCl \rightarrow RNH_3^+Cl^-$$

These are also white crystalline water-soluble ionic solids, called **alkylammonium salts**.

All the cations mentioned so far suffer from a common weakness: if their medium becomes alkaline, as would be the case in soapy water, they have a tendency to re-surrender their hydrogen ion, and regenerate the original amine:

$$RNH_3^+ + OH^- \rightarrow RNH_2 + H_2O$$

This limits their use as hair treatments, as not only would the neutral amines no longer be able to bond ionically with the hair protein, but, being amines, they would leave your hair smelling like a cross between a fish market and a horse box.

However, observe the reaction between a tertiary amine and, not hydrochloric acid this time, but chloromethane. The lone pair carries out a nucleophilic attack, and the result is a quaternary alkylammonium chloride:

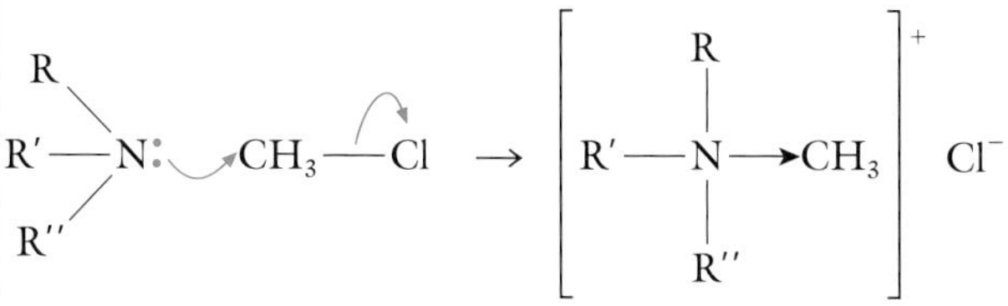

Question

28a Why would this compound not be vulnerable to decomposition in mildly alkaline media?

b In industrial practice, bromoalkanes are preferred to chloroalkanes for the final step of the synthesis of the quaternary alkylammonium ion. Can you suggest why?

c Show a possible final step in the synthesis of the compound shown in Figure 15.32.

d Draw a schematic diagram showing, at about the level of a non-specialist science encyclopaedia, how conditioners add body to hair.

e As well as giving body, conditioners lubricate the movements of individual hairs over each other. Can you suggest what aspect of a molecule like that in Figure 15.32 would deliver this desirable effect?

That is a part of the story of hair conditioners. From the list of ingredients on the bottle, you should be able to detect, sometimes partially disguised by a non-systematic name, the cationic reagent. Try it with the photograph in Figure 15.33.

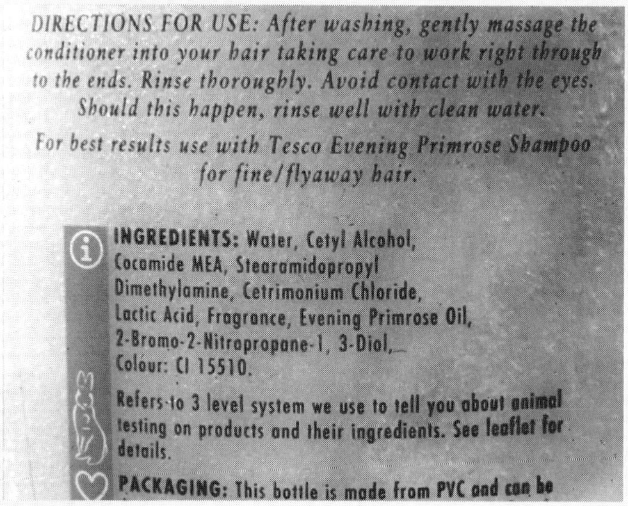

Figure 15.33 Can you find the cationic ingredient?

So next time you see a conditioner advert on television you will know that, surprisingly, *it is all true*. (Well, most of it, anyway.)

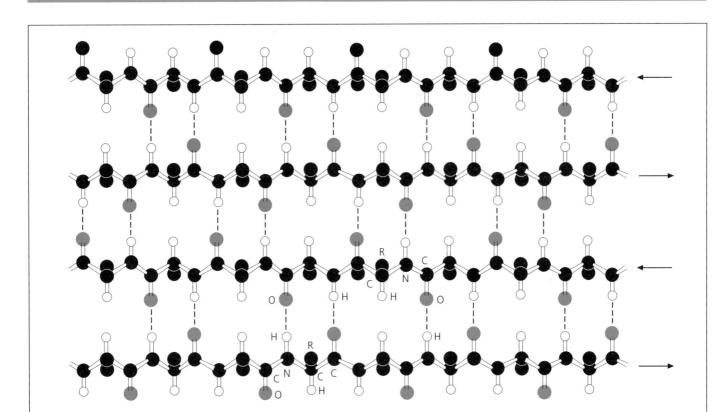

Figure 15.34 The β-pleated sheet. Certain atoms have been picked out, to help with question 29

The β-pleated sheet

The β (beta)-pleated sheet is a very different-looking secondary structure from the α helix, being a flat ribbon of matter rather than a rod (Figure 15.34). However, both structures have in common the use of N—H– – –O=C hydrogen bonds as the 'locating pins' of the formation. The obvious difference is that in the sheet the hydrogen bonds are intermolecular, rather than intramolecular as in the helix.

Despite the sheet metaphor, the structures that occur in Nature in β-pleated sheets are far from square. Commonly there are about five strands side-by-side in the sheet, which may nevertheless be hundreds of residues long (a very long thin oblong). β-pleated sheets, like α helices, often form part of the overall structure of an protein molecule. An example of another fibrous protein, this time containing β-pleated sheet rather than α helix, is silk fibroin.

Question

29 You may have already realised that all the peptide bonds in a polypeptide (if it is laid out straight) run in the same direction (for example, if all the nitrogens are on the right, then all the carbons are on the left). What do you notice about the 'running direction' of adjacent molecules in the β-pleated sheet structure in Figure 15.34?

15.10 The tertiary and quaternary structures of proteins

(If you are not doing a biochemistry special option, you may want to skip Sections 15.10–15.12, and rejoin the narrative at 15.13.) A **tertiary structure** describes how the individual portions of a protein molecule are oriented in space relative to each other. The instrument that plays the major part in determining these orientations is the X-ray diffraction crystallography machine, which produces data that lead to electron density maps (Figure 15.35). The maps might reveal, for instance, a portion of α helix ending in a bend before another portion of helix sets off at another angle. In short, a tertiary structure is an account of the three dimensional shape of a complete polypeptide. It is also the level of structure that must be described for an understanding of the physiological action of most natural proteins, especially enzymes.

Quaternary structure is the term used when the overall structure of a biological molecule comprises several polypeptide chains, and even possibly some non-protein material (as for example in haemoglobin, in which there are four polypeptides, four ligand 'haem' groups and four iron atoms). For instance, the picture of myoglobin in Figure 15.26 would be called a quaternary structure, because although it contains only a single polypeptide, it includes a haem group and an iron atom. The difference between tertiary and quaternary structures rests on a mere technicality – namely whether or not the material involved is a single continuous polypeptide. To the researcher there is no great difference in the type of work, both physical and intellectual, needed to decipher tertiary and quaternary structures.

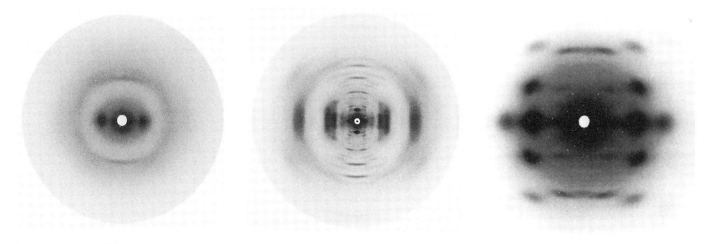

Figure 15.35 X-ray diffraction pictures, like these of (left to right) horn, seagull quill and natural silk, produce data that lead to electron density maps

15.11 The importance of tertiary structures – the action of enzymes

A graphic way to illustrate the importance of tertiary structure in a protein would be to study an enzyme at work, because enzymes *cannot* work unless they have adopted a specific tertiary structure. Lysozyme is an enzyme that has been understood for a long time, and it makes a good case study for the field of enzyme catalysis. It is an enzyme found, among other places, in nasal mucus and tears, and it was first named and studied by Sir Alexander Fleming, the penicillin man. Fleming had a talent not just for discovery but for discovery in a style which harnessed twists of fate and flashes of intuition, and which made for instant folklore. If fungal spores weren't drifting in through open windows (as in the popular story of the penicillin find), then he was letting his nose drip on to agar plates of bacterial cultures to see if the bacteria would be affected.

The nose-drippings did the trick, of course, and the dripped-on bacteria were dissolved away. He called the active ingredient 'enzyme which breaks things', which was rendered into scientific Greek as lysozyme. Its activity comes from its ability to catalyse the hydrolysis of certain polysaccharides (materials similar to starch and cellulose), and the reason for its toxic effect on Fleming's bacteria was that the bacterial cell walls contained just such polysaccharides.

The account that follows, of the enzyme's role in cleaving a polysaccharide molecule, will feature five major aspects, nearly all of which are typical of enzyme catalysis. You should recognise a recurrent theme which links most of the five points, namely the importance of tertiary structure.

1 The overall reaction undergone by the substrate

The reaction lysozyme catalyses is a simple cleavage with water (hydrolysis) of an oxygen bridge between adjacent sugars:

Questions

30 How might researchers have established that the new incoming oxygen atom had gone on to sugar A?

31a By means of a simple calculation of bonds broken and bonds made, estimate the size of the enthalpy of reaction.

b Consider your answer to (a), and the fact that the polysaccharide chain is broken by the reaction. Would the reaction be favoured overall on entropy grounds?

c Account for the fact that the reaction does not happen at all in water at room temperature.

2 The tertiary structure of lysozyme

Figure 15.36 shows the amino acid sequence in lysozyme from hen's egg-white. It shows some important sites at which bridging occurs between nearby cysteine amino acids to form a **disulphide bridge**, cys-cys. Notice that the existence of the bridges forces the diagram to assume a certain number of turns in the chain, but that Figure 15.36

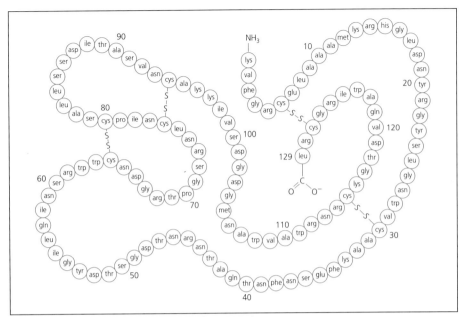

Figure 15.36 The amino acid sequence in the enzyme lysozyme

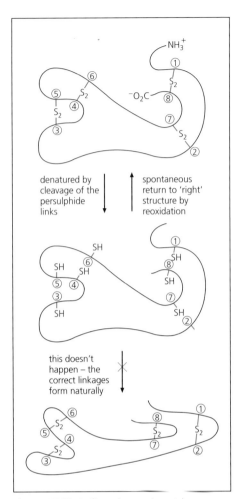

is still little more than a diagram of primary structure. The true tertiary structure does not lend itself very well to paper representations, but Figure 15.37 proves at least that it is a heavily convoluted molecule, with very little clear-cut α helix.

A biochemist called Christian Anfinsen developed a method for breaking and rebuilding the disulphide bridges in proteins such as lysozyme. These experiments shed fascinating light on the forces acting to maintain the tertiary structure. His method, applied to lysozyme, produces a version of the enzyme molecule in which the broken bits of the four disulphide bridges exist as hydrogen sulphides. This molecule would have absolutely no enzyme activity. The second stage of the method is to remove the cleaving reagent, and then allow a gentle reoxidation which rebuilds the bridges.

There are many ways of linking eight things into pairs. Even allowing for some of these combinations being impossible on the grounds of length of chain, it is still remarkable that, when the rebuilding is complete, all the molecules are found to have gone back into the one single, biologically active shape, identical to the original natural lysozyme (Figure 15.38).

This result, and many others like it, has led to the belief that the biologically active tertiary structure of a biochemical molecule is also the most stable version thermodynamically. To add that the sec-

ondary structure holds the key to the tertiary one really only says the same thing in other words. This second version of the idea has been elegantly expressed as '*sequence specifies conformation*'. ('Conformation' means the same as 'tertiary structure' or 'three-dimensional shape'.)

Figure 15.37 Graphic representation of the tertiary structure of lysozyme. The blank space shows where the substrate would sit. Note that glu35 and asp52 have crucial parts to play

Figure 15.38 Anfinsen's experimental technique, applied to lysozyme. When the cys–cys bridges are broken and allowed to reform, they always go back to the same, physiologically active, conformation

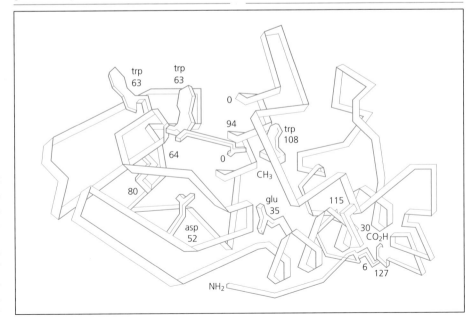

Question

32 Give two factors that might operate to make one conformation of a molecule such as lysozyme more stable than another. Bear in mind that 'stability' here implies stability in its working medium, namely aqueous solution, so it needs to present as 'water-friendly' an exterior as possible.

3   How does lysozyme work on its substrate?

First an enzyme has to bind its target molecule, then catalyse the reaction, and finally release the products. The whole process is over very quickly – a lysozyme molecule processes a substrate molecule every two seconds, and this is at the slow end of the range of enzyme turnover times.

(Periodic warnings must be given against letting your imagination invest the enzyme with the status of a living thing. Despite anthropomorphical language which suggests that enzymes direct their operations with intelligence, all the three processes named above are entirely mechanical pieces of chemistry, requiring no new concepts. Enzyme and substrate meet, for example, by nothing more directed than diffusion.)

Binding

Enzymes always hold their substrates in a **cleft** in their tertiary structure. In the case of lysozyme this cleft is just big enough for six sugar units to fit in at once. Five of the six monosac-

charides are bonded snugly and without stress. However, the other monosaccharide of the six is stretched and bent. These stressed bonds are the bonds that are earmarked for breakage. Because of the five unstressed monosaccharides, the whole process is thermodynamically acceptable.

At the risk (mentioned just now) of your attributing to enzymes a living quality which they do not really have, here is a metaphor for enzyme activity that is a little more dynamic than the lock-and-key picture. Imagine the active site (the cleft) being something like a dentist's chair. In a dentist's chair your four limbs and torso are made very comfortable (the five happy monosaccharides), and even your head is resting securely. Unfortunately the awful side-effect is to orient your mouth (the sixth monosaccharide) at such an angle as to render it vulnerable to invasion by people with drills! Similarly the sixth monosaccharide is waiting, bonds strained in just the right places, for the destructive invasions of crucial parts of the enzyme molecule. Figure 15.39 shows a visual version of the metaphor.

Question

33 It is important to keep in view the strong links between enzyme and 'mainstream' chemistry. To this end, try to draw on familiar themes to suggest how bonding might occur between the groups in Figure 15.40. This is a picture of part of the enzyme–substrate (ES) complex in lysozyme, showing three of the 'comfortable' sugar groups sitting in

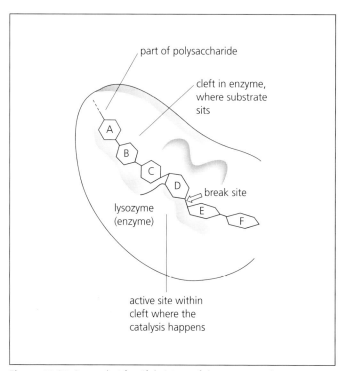

Figure 15.39 Somewhat fanciful picture of the enzyme–substrate complex between lysozyme and a polysaccharide. Sugar residues A, B, C, E and F are 'sitting comfortably', while D, where the break is to be made, is sitting 'on a spike'

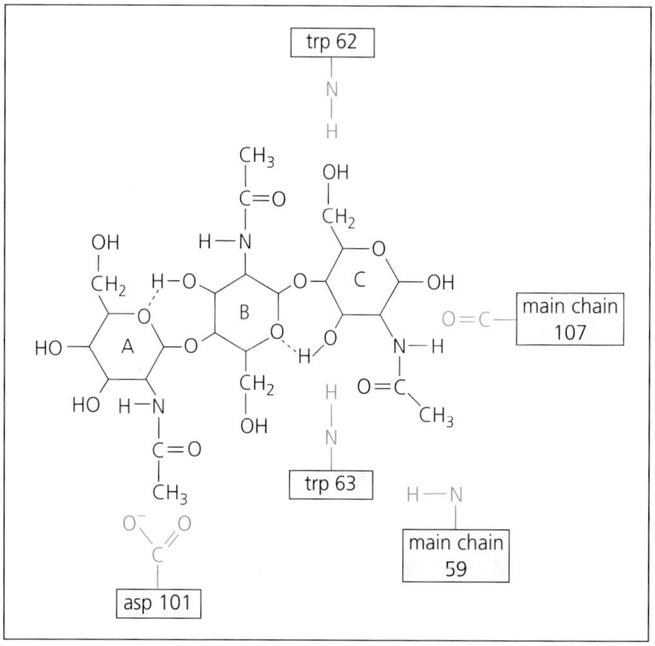

Figure 15.40 Part of the lysozyme–polysaccharide complex. A, B and C are the same monosaccharides as in Figure 15.39

the active site. (The green parts represent parts of the enzyme.) Copy the diagram and draw where you think bonds could be created, and say what kind of bonds they might be.

Catalysing the reaction

The reaction goes via a sort of S_N1 substitution mechanism. In the absence of catalysis, its two steps would look like

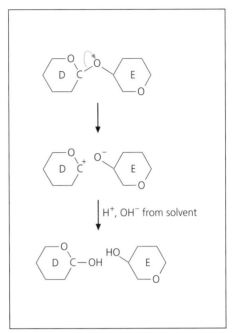

Figure 15.41 What would have to happen in an uncatalysed S_N1-style hydrolysis reaction

Figure 15.41. Acid catalysis would obviously help (Figure 15.42), but the enzyme can improve on ordinary acid catalysis in two ways. First, it can offer electrostatic stabilisation of the intermediate carbocation. Figure 15.43 shows the crucial amino acid residues that are positioned right at the reaction site in the cleft. Notice how glu35 provides the H^+ ion, and then, together with asp52, offers an environment with two negative charges to stabilise the carbocation.

But there is a second factor at work. Monosaccharide D in Figures 15.41, 15.42 and 15.43 is the 'uncomfortable' one. Its discomfort takes a highly specific form – the bridge carbon atom in ring D has been forced into a near-planar conformation before reaction, so it is already strained into the shape it will have to assume in the intermediate carbocation. It has less to lose in forming a carbocation than a conventional tetrahedral carbon. This process has been achieved in a thermodynamically neutral manner by virtue of the five other 'comfortable' monosaccharides. That is to say, despite the bond strain associated with the binding of ring D, the overall process will have ΔH and ΔG close to zero.

The reaction is completed in a third step in which ions from water arrive, complete the hydrolysis, and regenerate the catalytic hydrogen on glu35. The cleaved substrate now diffuses away.

Figure 15.44 shows how the energy profiles of the process would look, assuming respectively uncatalysed, conventional acid-catalysed and enzyme-catalysed mechanisms. Notice how the enzyme route achieves the lowest activation enthalpy.

Questions

34 This question requires you to theorise about the pH dependence of the catalytic action of lysozyme. Figure 15.43a shows that at the beginning of the catalytic cycle the asp52 is in an ionised condition, while glu35 is un-ionised.

a What does this situation tell you about the relative acidities of the glu and asp residues?

b The pH dependence of the rate of reaction of lysozyme hydrolysis is shown in Figure 15.45. Suggest why

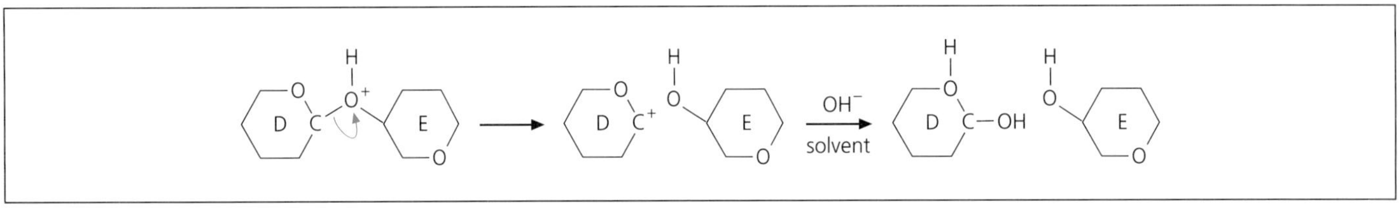

Figure 15.42 Conventional acid-catalysed hydrolysis of a polysaccharide

a glu 35 donates the catalytic proton. The 'D' carbocation forms

b the carbocation intermediate is stabilised by by the presence of *two* CO_2 groups

c OH^- and H^+ from the solvent come in. OH^- completes the hydrolysis, while H^+ regenerates the enzyme

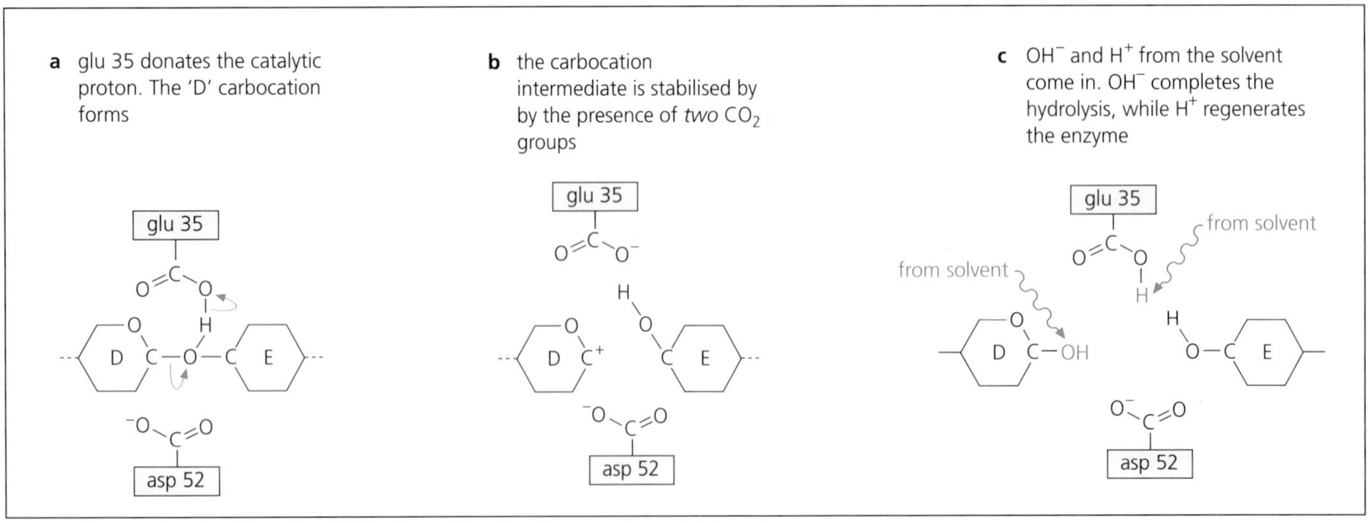

Figure 15.43 The enzyme-catalysed hydrolysis of a polysaccharide

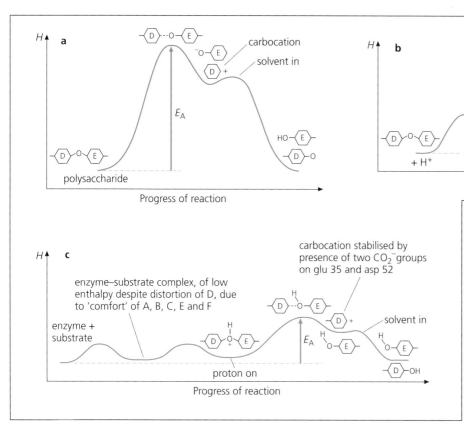

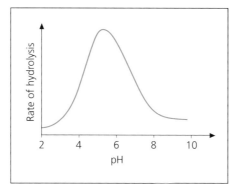

Figure 15.45 pH dependence of the activity of lysozyme

Figure 15.44 a Uncatalysed cleavage – large activation enthalpy. b Conventional H^+ ion catalysis – smaller activation enthalpy. c Enzyme catalysis – smallest activation enthalpy

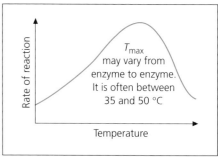

Figure 15.46 Generalised temperature

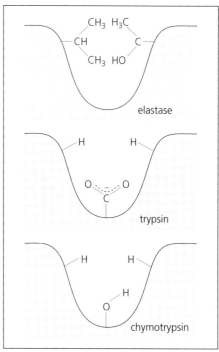

Figure 15.47 See question 37

catalysis is ineffective at pHs below 3 or above 8, and why there is a sharp maximum at about 5.

35 As well as pH dependence, all enzymes show a characteristic temperature dependence, which also shows a maximal pattern (Figure 15.46).

a An overheated enzyme that has lost its ability to catalyse is said to be **denatured**. From the account of lysozyme at work, suggest what denaturing really means on a molecular scale.

b Having explained the decline in activity at high temperatures, account for the slowness at low temperatures, again answering in terms of molecular events.

36 From your study of enzyme activity, suggest why enzymes often only accommodate a single specific substrate, and catalyse a single specific reaction. You could use the lock-and-key theory here, but try to refer to the

chemistry of the enzyme–substrate complex too, as typified by the lysozyme case study.

37 Question 36 was about the phenomenon called **enzyme specificity**. This question offers another angle on this subject.

   Figure 15.47 shows cross-sections through the active site cleft of three well-known enzymes (all simplified). All these enzymes act to cleave (other) proteins. Pick from the list below the type of amino acid side-chain that might be expected to fit comfortably into each cleft:

- small, non-polar
- bulky, with phenolic group
- long, with cationic group.

For each case suggest one particular amino acid that would typify the general description.

4 Inhibition of enzyme activity

Inhibition is the process by which enzymes are deflected from their task by molecules other than their normal substrates. **Competitive inhibition** is achieved when a rival fake substrate occupies the active site of the enzyme and stays there, without being cleaved. Again lysozyme provides us with a classic example.

Questions

38 Recall the lysozyme–polysaccharide enzyme–substrate complex. Five of the six monosaccharide groups were held quite tightly, while the fourth one along was twisted ready for reaction. Account for the fact that the trisaccharide made from three of the same units as lysozyme's usual substrate is a damaging competitive inhibitor, 'poisoning' the enzyme's activity.

39 We can picture the equilibria involved in competitive inhibition as in Figure 15.48. With this model in mind, explain why the poisoning effects of competitive inhibition can sometimes be overcome by massive doses of the proper substrate.

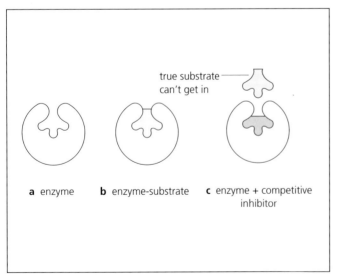

Figure 15.48 Symbolic representation of the action of a competitive inhibitor

Non-competitive inhibition is a little harder to understand, insofar as it challenges one of our conceptions about enzymes. One of our metaphors for the enzyme–substrate relationship has been the lock-and-key model, implying that the cleft has a shape, the substrate has a shape, and one fits the other. However, this is rather too clear-cut a picture. Real clefts and substrates are rather softer and more bendy than locks and keys. We have already touched on this fact in saying that a crucial part of the substrate is bent by the binding process. Now we come to evidence that suggests the cleft itself is flexible. It has been found that inhibition can be caused by molecules which bind at a completely different site from the active one. Heavy metal ions, for example, can use the enzyme as a polydentate ligand. As the enzyme responds to its new relationship with the inhibitor, its tertiary structure is distorted. The distortion transmits itself across the whole molecule and upsets the enzyme–substrate interaction (Figure 15.49).

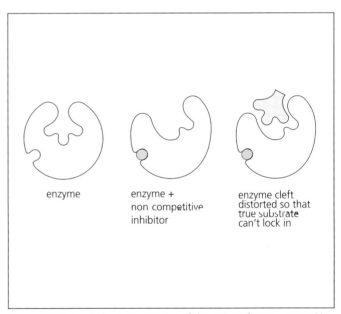

Figure 15.49 Symbolic representation of the action of a non-competitive inhibitor

5 The kinetics of enzyme reactions

This section revises material in Chapter 14 on rates of reactions. Figure 15.50 is a graph of the relationship between the substrate concentration [S], and the reaction rate, which we shall call V, for a typical enzyme-catalysed reaction. Question 40 can be answered by drawing on Chapter 14 work. (The 'post office counters' analogy may need to be revised.)

Question

40a How does the graph in Figure 15.50 compare with the equivalent graph of rate against reactant concentration for an ordinary uncatalysed first-order reaction? (*Remember*: We are not talking about the graph of concentration against time, but the second

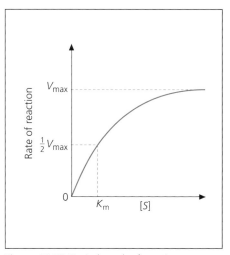

Figure 15.50 Typical graph of reaction rate versus substrate concentration for an enzyme E at constant [E]

graph which was obtained by plotting the gradients against the concentration.)

b Which regions of the graph show first-order kinetics, and which show zero-order kinetics?

c Explain why the rate of this reaction reaches a maximum value V_{max} at high [S].

d Figure 15.50 is drawn assuming a constant value of the enzyme concentration [E]. Indicate how the graph would alter if [E] were increased. Your second line should not diverge from the first until after the point $(K_m, V_{max}/2)$.

The quantity K_m is called the **Michaelis constant**, after one of two people who did pioneering work on enzyme kinetics (Figure 15.51). You were asked to draw the second line in part (d) of question 40 to go through the point $(K_m, V_{max}/2)$ for a particular reason. It seems that the value of [S] at which the rate is half of V_{max} is a constant for most enzyme–substrate associations. Each enzyme–substrate combination has its own unique value of K_m. Two rather different values of K_m are shown in Table 15.4. One refers to the reaction we have just studied in detail, involving lysozyme. The other relates to an enzyme, carbonic anhydrase, which catalyses the equilibrium reaction:

$$CO_2(aq) + H_2O(l) \rightleftharpoons H_2CO_3(aq)$$

Figure 15.51 Leonor Michaelis (1875–1949), who worked on the kinetics of enzyme-catalysed reactions in collaboration with Maud Menten

Table 15.4 Values of K_m, the Michaelis constant

Enzyme	Substrate	K_m (mol dm^{-3})
Lysozyme	A polysaccharide	6×10^{-6}
Carbonic anhydrase	Carbon dioxide	8×10^{-3}

Question

41 Let us think about what a high value of K_m means. A large K_m must mean that the reaction does not reach half-maximum speed until quite a high value of [S] has been reached. In other words, the onset of zero-order kinetics is postponed until the system is very crowded with substrate molecules. What does this indicate about the 'throughput' time of the enzyme in question?

(As a postscript to question 41, it has been estimated that a carbonic anhydrase enzyme molecule can handle 100 000 reactions per second, in either direction.)

15.12 A concluding note on the nature of biochemistry

By the end of Section 15.11, and including sections in Chapter 13 on carbohydrates and fats, we have touched on as much biochemistry as is appropriate to a general Advanced text. I hope this has given you a flavour of this fascinating sub-branch of chemistry. It covers quite simple and vivid 'stories', featuring interlocking shapes, bindings and breakings. Yet the uncovering of these simple stories, and of the structures of the molecules which enact them, has been a tribute to human ability to handle evidence of great complexity. This spirit of tenacity is typified by the career of Dorothy Crowfoot Hodgkin (Figure 15.32), whose discovery of the three-dimensional structure of insulin came 36 years after her first X-ray diffraction work on proteins. In the course of those years she claimed solutions to the structures of cholesterol, penicillin and vitamin B_{12}, and was rewarded by a Nobel prize in 1962.

Figure 15.52 Dorothy Crowfoot Hodgkin

There is no 'magic' new chemistry involved in this work on proteins. The laws of thermodynamics concerning enthalpy and entropy still apply, catalysis still occurs by simple means such as proton transfer, and we still have the same repertoire of covalent, dative, ionic, hydrogen, dipolar and van der Waals' bonds as before. What is staggering is the number of variations on those themes that have evolved in our natural world, and in particular the enormous significance that Nature has invested in the humble-looking amide linkage.

15.13   Nitro-compounds

After our detour into biochemistry, we shall now return to the mainstream organic chemistry of nitrogen. There are two nitrogen-based functional groups still to be considered. As Figure 15.53 shows, the **nitro group** shares some structural features with nitric(v) acid, including the dative covalent bond to oxygen. It also shares the feature whereby both N—O bonds are the same length, despite appearances to the contrary in Figure 15.53.

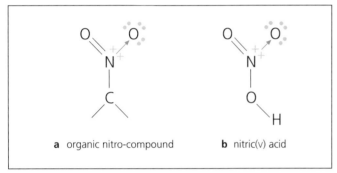

    a organic nitro-compound   **b** nitric(v) acid

Figure 15.53 Comparison of the structure of an organic nitro-compound with that of nitric(v) acid

Question

42 Explain the similarity in the lengths of the two bonds, and reinforce your answer with 'curly arrow' diagrams.

The only organic nitro-compounds studied at Advanced level are those in which the nitro group is attached to a benzene ring. This is partly because the nitroalkanes are very unstable. However, the aromatic nitro-compounds can achieve a degree of delocalisation, involving the π-electrons of both ring and substituent (Figure 15.54), and thus obtain some stabilisation.

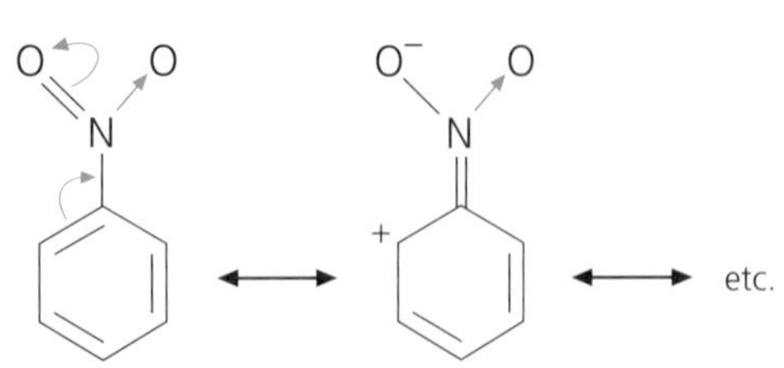

Figure 15.54 Delocalisation in aromatic nitro-compounds. This *de*activates the ring towards electrophilic attack, by withdrawing electrons

Question

43 Suggest how the contribution of the alternative structures in the delocalisation mix might affect the length and strength of the C—N bond in aromatic nitro-compounds, compared with their alkyl equivalents.

Aromatic nitro-compounds are prepared, as mentioned in Section 10.16, via an electrophilic attack on the benzene ring by the nitronium ion, NO_2^+, which is present in mixtures of nitric and sulphuric acids. More active rings, like those of phenol, are readily nitrated by much milder reagent concentrations, and indeed may go on to produce multiply-nitrated products.

The trinitro aromatic compounds are a very interesting family, containing some of the most violent explosives. The structure of TNT (trinitrotoluene, or 1-methyl-2,4,6- trinitrobenzene) is:

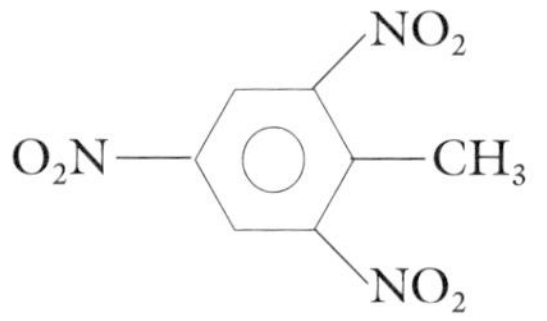

and a possible reaction for its explosion is as follows:

$$2C_7H_5N_3O_6(l) + 2\tfrac{1}{2}O_2(g) \rightarrow 3N_2(g) + 5H_2O(g) + 14CO_2(g)$$

You are referred back to Chapter 8 for a fuller telling of the story.

The chief attribute of the nitro group, apart from its tendency in some circumstances to react with extreme violence, is its ability to be reduced to an amino group. Typically the reducing agent might be a metal/acid mixture (in which hydrogen radicals are the reducing agents), or the versatile hydrogen/platinum or hydrogen/nickel system. This route is the most convenient laboratory means of preparing the aromatic amines:

$$Sn(s) + 2HCl(aq) \rightarrow SnCl_2(aq) + 2H\bullet$$

hydrogen radical

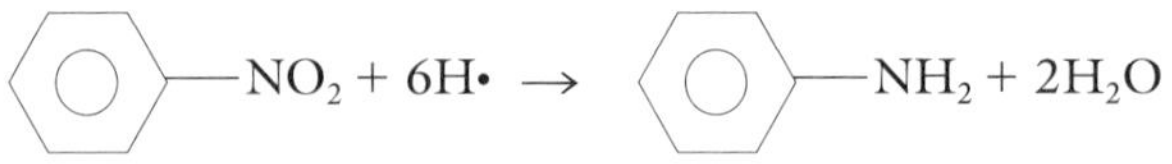

15.14   Diazonium compounds

Once aromatic amines have been made, by the route outlined above, they can be made to perform a reaction that has no real parallel in alkyl chemistry, and which is of considerable importance to synthetic chemists in general and to dyestuff makers in particular. This group of compounds is known as the **diazonium salts**.

These compounds illustrate a general principle that emerged in late nineteenth-century organic chemistry, namely that if you can create a compound which is just stable enough to be isolated and retained, but so reactive that it will react with almost any other reagent you put it with, then you have a versatile intermediate, with a wide range of synthetic uses. So far in this book, the nearest we have come to this type of chemical is the acyl halides, which start reacting as soon as you unscrew the bottle-top. However, the diazonium salts belong to a different league of finely poised instability, since they cannot even be kept in bottles. If you need a diazonium salt you need to make it where and when you intend to use it, and make sure that it stays below about 5 °C.

Preparation of diazonium salts

The standard recipe is to react the aromatic amine with an iced mixture of concentrated hydrochloric acid and sodium nitrate(III) (sodium nitrite). The last two reagents create nitric(III) acid (also known as nitrous acid, another compound too unstable to be bought in bottles), and that reacts with the amine:

$$HCl + NaNO_2 \rightarrow HNO_2 + NaCl$$

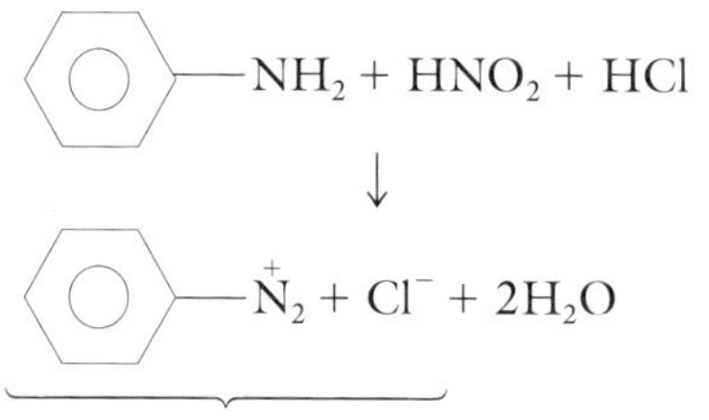

nitric(III)
acid

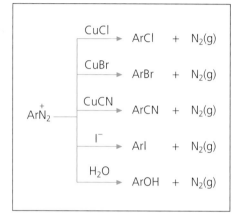

benzenediazonium
chloride

Both stages of the reaction must be carried out at no more than 5 °C. The structure of benzenediazonium chloride is a very curious one, which only makes sense in a dot-and-cross diagram if seen as a dative covalent bond between a nitrogen molecule and an aromatic carbocation:

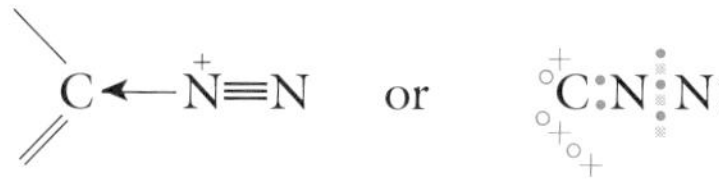

Whatever the actual structure of the diazonium salt, the clear central theme of most of its chemistry is the desire to expel the two nitrogen atoms as an N_2 molecule. This means that the two electrons in the C—N bond are heavily polarised away from the carbon atom. The carbon atom is in such an extreme environment that it develops a tendency which goes against all the norms of aro-

matic chemistry – it accepts nucleophilic attackers. We therefore have the best route to a very large range of organic chemicals, giving high yields, and providing insurance against isomeric mixtures with multiple substitutions. (This is achieved by purifying the original nitrocompound of any di- or trinitro-compounds.)

Figure 15.55 shows a range of the more common uses of a diazonium salt. The route to iodobenzene is almost the only feasible method of synthesising that particular compound.

Figure 15.55 Synthetic uses of the diazonium ion. The use of copper(I) halides is rather curious, and is called the Sandmeyer reaction. The other reactants are conventional nucleophiles

Question

44 Other groups that can replace the diazonium group include —OR, —SH and —SR, the two last-named being the sulphur equivalents of hydroxide and alkoxide. Suggest reagents that could be used as 'carriers' of these groups to the diazonium salt.

Azo coupling reactions

There is one significant reaction in which the nitrogen molecule is not expelled from an aromatic diazonium salt.

Instead, the whole diazonium ion carries out an electrophilic attack on another aromatic ring. The resulting class of compounds is referred to as the **azo-compounds**, and the reaction is known as azo coupling (Figure 15.56). All the azo-compounds are intensely coloured, and the class has become one of the mainstays of the dye industry.

The rings used as the receivers of the attack tend to be amines and phenols, which are, among aromatic compounds, the most reactive towards electrophiles. This is important because if you had to wait too long for the reaction, or to speed it up by heat or increased concentration, the delicately poised diazonium compound might choose to take an OH from the aqueous medium and give you an unwanted phenol. What is more, this new phenol would also receive the diazonium compound along with your intended phenol or amine, and you would get a mixture containing a different azo-compound from the one you wanted.

Questions

45 Dye chemists have found that the pH of the medium is important in coupling reactions. For instance, the coupling of diazonium compounds with phenols is facilitated by a mildly alkaline pH.

a What happens to phenol molecules in alkaline conditions? Draw a structure to support your answer.

b How will this affect the reactivity of the ring towards electrophilic attack? (*Remember*: The greater the electron density in the ring, the more reactive it will be.)

Figure 15.56 A typical azo coupling reaction – really an electrophilic attack by the diazonium ion on phenylamide

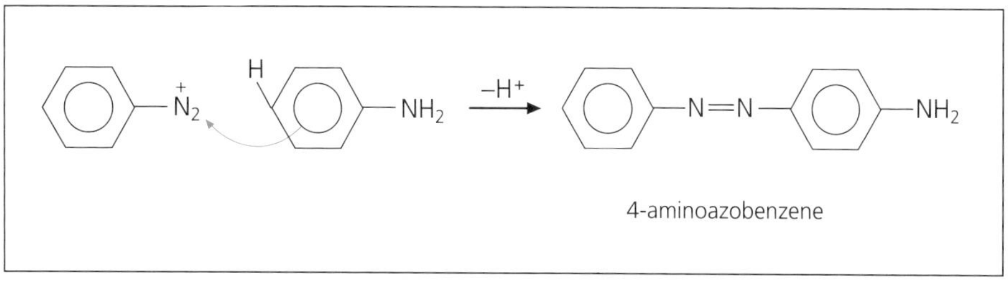

4-aminoazobenzene

c Suggest why only *mildly* alkaline conditions are used. (Think what might otherwise happen to the diazonium ion.)

46 We have just described how to make diazonium salts and how to turn them into dyes. Suggest a way of using this chemistry as the basis of a visual colour-change test which would be able to reveal whether an unknown amine was aromatic (aryl) or aliphatic (alkyl).

Why are molecules coloured?

The simple answer to this question is that a molecule will show a colour if it is capable of absorbing photons of light of any wavelength within the visible spectrum.

Question

47 If a molecule absorbs photons from the red/orange region of the spectrum, what colour will this compound look?

Intensity (strength) of colour is to do with the number of photons absorbed as a proportion of those incident on the sample. This in turn has its roots in the electronic arrangements within the absorbing molecule since, as we found in Chapter 3, absorbed photons interact with electrons and promote them between energy levels. It has been found that intense colour absorption is shown by molecules with an extended delo-

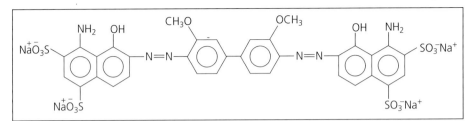

Figure 15.58 One of the more extended delocalised systems. Such a system, giving rise as it does to colour, is called a chromophore

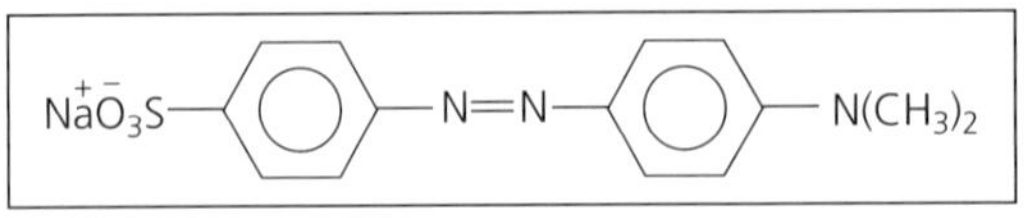

Figure 15.59 An azo acid–base indicator – methyl orange

calised system of π-electrons. The azo-compounds have such a system, since the π-bond between the two nitrogens can act as a 'delocalisation bridge'. Figure 15.57 shows some Kekulé-style structures (p. 220), as a way of demonstrating the delocalisation in (phenylazo)benzene. Any extended delocalised region in a molecule which gives it the property of colour is called a **chromophore**.

Some molecules have extremely long delocalised regions. The dye Chicago Blue 6B (Figure 15.58) is an example. Passing from the bizarre to the familiar, the acid–base indicator methyl orange (Figure 15.59) is an azo-compound. Its use as an indicator shows how quite minor adjustments of electron balance in a molecule (in this case, the loss or gain of an H⁺ ion) can affect the exact wavelength at which the molecule absorbs.

Question

48 (Phenylazo)benzene is orange. It can be reduced to *N,N'*-diphenylhydrazine (hydrazobenzene) by zinc/sodium hydroxide, which is another reducing agent that acts as a source of hydrogen radicals. The equation for the reaction is:

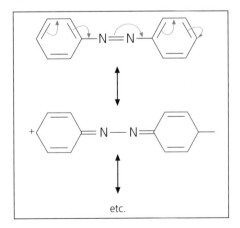

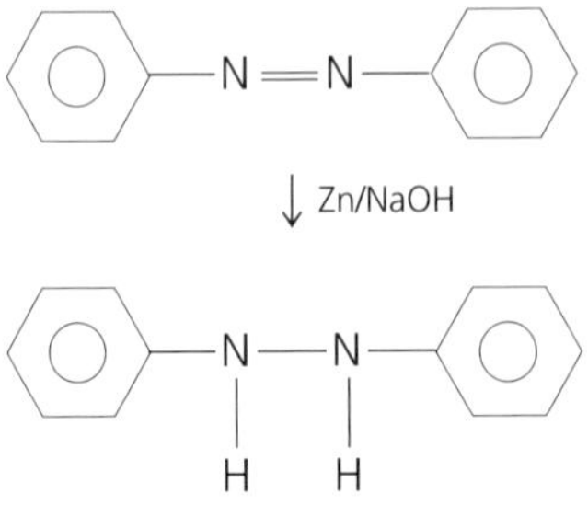

hydrazobenzene

Suggest what you might see as the reaction proceeded, and explain your suggestion.

The dye industry is a fascinating field of applied chemistry. Historically it played an important part in giving momentum to early industrial chemistry in Britain (Figure 15.60) and today it is, like the polymers industry, an excellent example of human ability to 'tailor' molecules. This tailoring is obviously crucial in tuning the colour of the dye to market needs, but no less important is to make the dye molecule compatible with the medium to which it is giving the colour. The panel opposite concerns the problem of making fabric dyes which stay in their fabrics.

Figure 15.60a William Perkin (1838–1907) discovered mauve (often referred to as Perkin's mauve)

Figure 15.57 Delocalisation in (phenylazo)benzene – electrons may move from one end of the molecule to the other

b Medieval dye workers at their labours

Dyeing and wash-fastness of textiles – a practical example of chemical bonds in action

Introduction

Imagine the scene: a load of clothes is being washed in the washing machine using the fast coloureds program. After spin-drying the clothes are taken out and – oh dear, some of the whites are pink. The colour must have run from that new red underwear.

Familiar? Yes, but why does it happen, and what are the dye manufacturers doing about it?

Dye–fibre interactions

The wash-fastness of a dyed textile depends on two things – first, on how well the chemical nature of the dye is matched to that of the textile fibre, in order to achieve a sufficiently strong interaction or bond that the dye molecules are held in the fibre, and second, on the solubility of the dye molecule itself.

There are five broad types of interaction between dye and fibre:

1 physical adsorption
2 solid solution
3 ionic bonds
4 covalent bonds
5 insoluble aggregates within the fibre.

The first four interactions are listed in order of increasing strength. There may be little or no interaction of insoluble aggregates with the fibre, but such a system gives the best wet-fastness because of the insolubility of the dye.

Where it exists, the strength of the interaction between dye and fibre may not be particularly strong and the thermal energy provided by a hot wash may be sufficient to overcome the interaction force for a proportion of the dye molecules, allowing them to escape from the fibre into the wash water. The increase in dye solubility with temperature also tends to reduce wash-fastness. Coloured clothes therefore have to be washed at lower temperatures than whites.

Some of the dye that escapes from the fibre may then transfer to other fibres in the wash. This may not be a problem if those fibres are already highly coloured, e.g. dark brown, blue or black, as the extra dye will probably not be noticed. However, if the fibres are only lightly dyed or are white, then even a very small amount of dye taken up from the wash water will be noticeable.

To understand the different types of interaction one needs to understand the nature of the different types of fibre. These may be classified as follows:

1 protein fibres, e.g. wool, silk
2 synthetic cationic fibres, e.g. nylon
3 synthetic anionic fibres, e.g. acrylics
4 cellulosic fibres, e.g. cotton, linen, rayon
5 synthetic non-ionic fibres, e.g. polyester, cellulose triacetate.

Protein and cationic fibres like wool and nylon contain amino groups. Both types are dyed under mildly acidic conditions with soluble acid dyes containing sulphonate ($-SO_3^-$) groups. The basic amino groups become protonated (positively charged) and form ionic interactions with the negatively charged dyes (Figure 15.61).

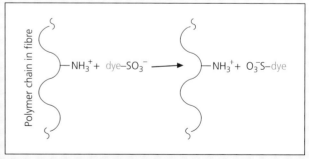

Figure 15.61 How acid dyes bond to wool and nylon

Anionic fibres like acrylics contain carboxylic ($-CO_2^-$) and some sulphonate groups in their structure. They are dyed using soluble **basic dyes** which contain cationic groups, normally quaternary amines. The interaction is again ionic, though this time the charges on dye and fibre are reversed (Figure 15.62).

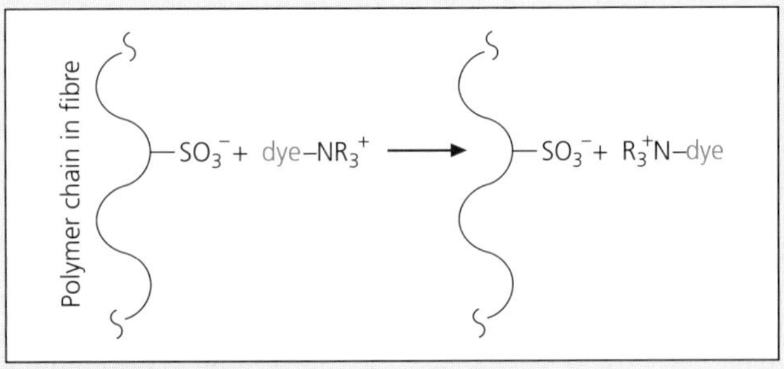

Figure 15.62 How basic dyes bond to acrylics

Polyester and cellulosic fibres such as cotton have a non-ionic structure, and so cannot form ionic interactions with dyes. Cotton fibres are hydrophilic and have a very open structure. Soluble dyes can enter the fibres very easily – the problem is keeping them there! Some dyes for cotton (**direct dyes**) are designed to be long flat molecules which can lie along the cellulose chains and interact via hydrogen bonds and van der Waals' forces. However, these interactions are not very strong and wash-fastness is poor.

A better wash-fastness with cotton is given by **vat dyes**, which are precipitated in pores in the fibres as microscopic aggregates. The dye in a chemically reduced, soluble form enters the fibres, and the system is then treated with an oxidising agent to precipitate the insoluble dye in pores in the fibres. This is how denim is dyed with indigo. But vat dyes are often expensive and are limited in their colour range.

A third option is to use a **reactive dye**. These dyes, invented by ICI in the 1950s, incorporate in their structure a reactive group such as the chlorotriazinyl group:

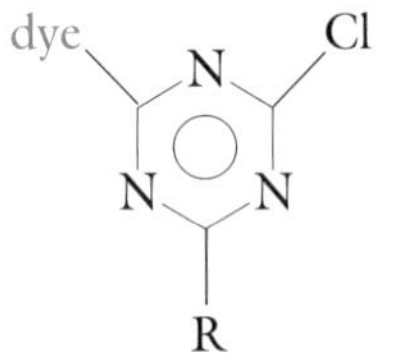

This group can react via the chlorine atom with a hydroxyl group on the cellulose chain to eliminate hydrogen chloride and form a covalent bond between the dye and the fibre (Figure 15.63). Reactive dyes have a very wide colour range and have very good wash-fastness.

Polyester fibres are different from cotton in that they are hydrophobic and the polymer chains are tightly packed. In this case the best dyes are those which form a solid solution in the fibre. To do this, the dyes also have to be hydrophobic, and therefore have a very low solubility in water. They are applied as very fine dispersions in water, hence their name, **disperse dyes.** Despite their low solubility the dyes diffuse into the fibre via aqueous solution. The fibres are dyed at up to 135 °C under pressure, and often in the presence of agents called **carriers** which help to make the dye soluble and swell the fibre, allowing easier diffusion of the dye molecules. Once dyeing is complete, the temperature is reduced, the fibre contracts and the dye is held firmly by van der Waals' forces and often some hydrogen bonding. Unlike the direct dyes in cotton, the very low solubility of the disperse dyes in water means there is little driving force for the dye to diffuse back into the water, and wash-fastness is generally good.

The reason why wash-fastness seems to be a particular problem with red dyes is that the human eye is very sensitive to red and can distinguish a pink stain more easily than one of any other colour. ICI have developed a new class of very bright red dyes for polyester, based on the benzodifuranone ring system shown below, where X and Y are substituents chosen and positioned on the benzene rings to give the desired colour:

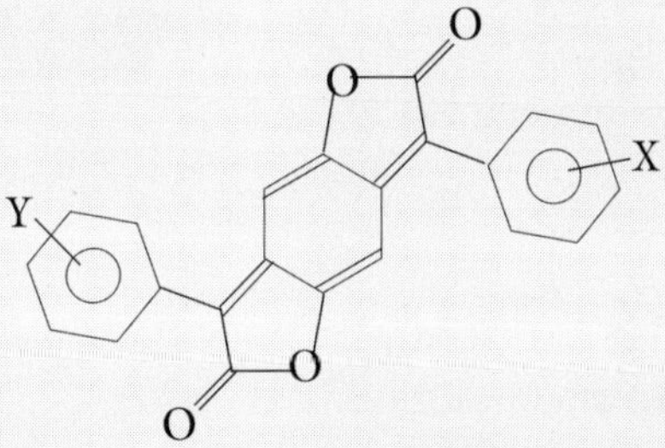

These dyes have extremely good wash-fastness, due partly to their abnormally low solubility in water. The end may soon be in sight for the problem of pink white shirts!

Questions

49 Give short definitions of the following terms:

a solid solution

b hydrophilic

c hydrophobic

d chemically reduced.

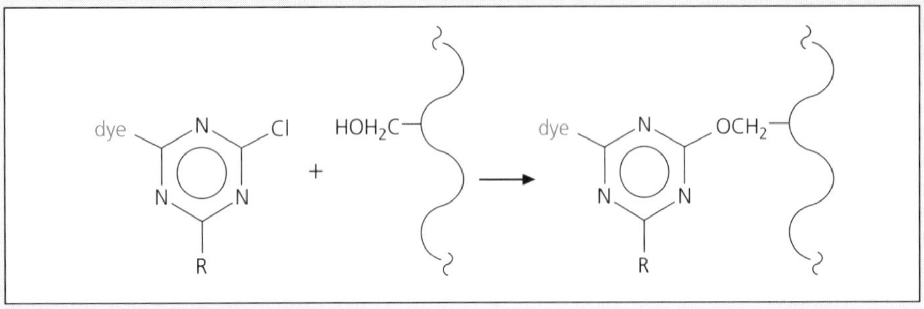

50 The account above contains two lists, the first a list of the ways in which fibre molecules might hang on to dye molecules, and the second a list of fibre types. Make connections between the two methods of classification by completing a copy of Table 15.5, which relates only to cotton and polyester fibres.

Figure 15.63 How reactive dyes bond to cotton

Table 15.5 Classification of dyes

Fibre type	Type of dye	Which interaction?	Fastness (v. good, good, poor)	Other comments
Cotton	(a)			
	(b)			
	(c)			
Polyester				

51 The basic problem with dye fastness can be symbolised as in Figure 15.64. The dye chemist is constantly seeking to keep force 2 as strong as possible relative to force 1, once the dye molecule is in place. However, a degree of compromise is sometimes necessary because if force 1 is too weak, then there is no 'delivery system' for getting the dye molecule to the fabric molecule. From the text:

a Give one example in which force 2 is covalent.

b Give two examples in which force 2 is ionic.

c Give two examples in which force 2 is 'intermolecular'.

d Give one example in which fastness is achieved by having force 1 almost non-existent.

e Give one example where fastness is achieved by changing the dye molecule once it has been delivered.

f For the example in (e), contrast the size of force 1 during delivery and after delivery.

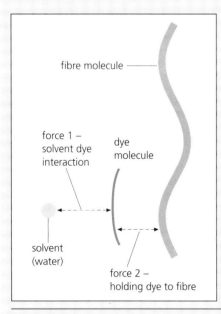

fibre molecule

force 1 – solvent dye interaction

dye molecule

solvent (water)

force 2 – holding dye to fibre

Figure 15.64 Symbolising the 'tug-of-war' between fibre and solvent for the dye

15.15 Organic synthesis

We have now seen a whole landscape of organic chemistry. It is important, though, to be able to draw on any part of that landscape, and perhaps make connections between regions.

In the remaining two sections of this chapter we shall therefore practise two general skills. First we shall look again at constructing synthetic routes between molecules (already considered in Chapter 11). Second, in Section 15.16, we shall turn our attention to analysis, and try to answer questions that begin 'An unknown compound X ...'. We shall establish a 'code book' to help us decipher all the codes associated with results of chemical tests.

You will need to fill in the tables of information in these sections, but you can refer to the answers if necessary.

A repertoire of single synthetic steps

Question

52 Complete a copy of Table 15.6.

Table 15.6 Synthetic steps

Change	Reagents/conditions
—OH to —Cl	PCl_5, HCl(g), $SOCl_2$
—OH to —Br or —I	P(red)/Br_2 or I_2, KBr or KI/sulphuric or phosphoric acid, HBr(g) or HI(g)
—OH to —NH_2	NH_3/ethanol
—CH_2OH to —CO_2H (via aldehyde)	Acidified dichromate (depends on concentration)
—CHOH(sec) to ketone	
—CO_2H to —CH_2OH	
Ester to —CH_2OH	
Acid or alcohol to ester	
Ester to alcohol and acid	
Acid to acyl halide	
Acyl halide to acid	
Acyl halide to ester	
Acyl halide to amide	
Amide to acid and amine	
Alcohol to alkene	
Alkene to alcohol	
Alkene to halogenoalkane	
Halogenoalkane to alkene	(Solvent important)
Halogenoalkane to alcohol	(Solvent important)
Halogenoalkane to amine	
Benzene to nitrobenzene	
Benzene to halobenzene	(Check catalyst)
Benzene to alkylbenzene	(Check catalyst)
Benzene to benzenesulphonic acid	

Synthesis – linking synthetic steps together

These problems have the structure A → ? → B, and so involve working out a feasible intermediate. Begin your search by asking 'What can I make B *from*?', while keeping in the back of your mind the question 'What can I make A *into*?'. When the answers to both questions are the same, it only remains to select the best reagents and conditions.

For example:

ethene → ? → ethylamine

1 Think of precursors to ethylamine – there is only one route to an amine, namely halogenoalkane + ammonia/ethanol.
2 Can we get from ethene to a halogenoalkene? Yes, by reacting it with a hydrogen halide.

Questions

53 Convert ethanoic acid to chloroethane in two steps.

54 Convert ethyl ethanoate to ethene in two steps.

55 Now for two synthesis problems with a purpose. Plan two one-step syntheses to take salicylic acid (2-hydroxybenzoic acid) to aspirin and to oil of wintergreen (see Figure 15.65).

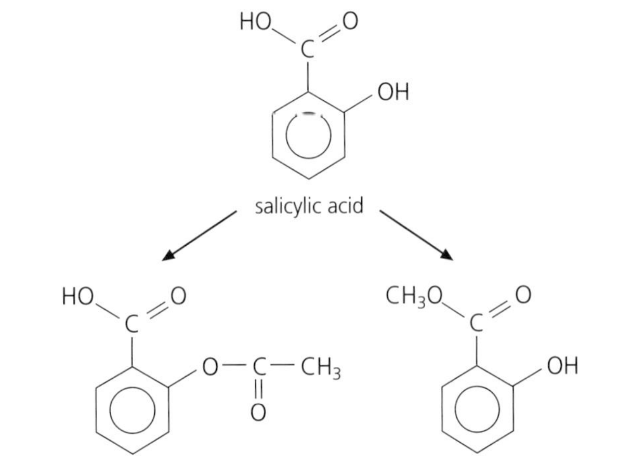

salicylic acid

aspirin – ethanoyl salicylate

oil of wintergreen (a liniment ingredient, methyl salicylate)

Figure 15.65 Two problems in synthesis

Synthesis – practical methods

Crudely speaking, organic synthetic practicals contain some combination of three stages: mixing the ingredients, 'cooking' them, and separating and purifying the product. If the reagents are very reactive, as in an acyl chloride reaction for example, then the middle stage may be omitted.

Mixing

Mixing sounds easy, but sometimes the reagents need to be kept cool and mixed slowly, and at other times you want to make sure one reagent is in excess. As an example, if you want to hold an alcohol oxidation at the aldehyde stage, then the dichromate/concentrated sulphuric acid oxidising mixture has to be mixed and then cooled before the alcohol is dripped in. As an example of the use of reagents in excess, if a primary amine is required it is better to add the halogenoalkane to an excess of ammonia than the other way round. In both these cases, and in many other situations where liquids are added, a separating funnel attached to the second neck of the reaction vessel is usually a good method (Figure 15.66). However, remember that liquids will not run into closed systems.

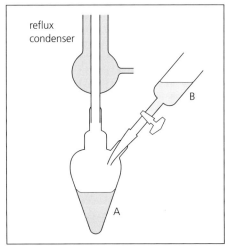

Figure 15.66 Typical set-up for mixing two liquids and refluxing

Refluxing

If a reaction between two liquid reagents is slow, then refluxing is the standard procedure (Figure 15.66), after which the apparatus could be used for distillation if appropriate (see purification below). If the product separates straight away as a solid, then the distillation stages can be left out and it can simply be purified. If (like phosphorus(V) chloride) the reactants decompose in moist air, a guard tube with calcium chloride can be used in the top of the reflux condenser (Figure 15.67).

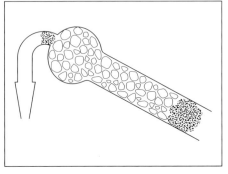

Figure 15.67 A calcium chloride or soda lime guard tube. Calcium chloride preserves a dry atmosphere inside the apparatus. Soda lime absorbs any acidic gases escaping from the condenser

Separation and purification – liquids

If the desired product is one of several non-polar volatile liquids in the reaction vessel you can distil the mixture, and collect the desired boiling-temperature fraction during the distillation. This method only works if the boiling point of the product is distinct from those of the other liquids in the mixture. A fractionating column helps if the boiling points are closer together than about 10 °C.

If your product is, say, the only non-polar liquid, then you can put the whole mixture in a separating funnel and shake it up with 1,1,1-trichloroethane or ethoxyethane, either of which will extract your target chemical.

Your non-polar product may have become a separate layer of its own accord, in which case it can simply be separated off without using a solvent.

Once you have your single non-polar layer or solvent extract, you can use a sequence of washes followed by drying to remove most of the impurities quickly. Sodium hydrogencarbonate solution followed by water, and then drying with anhydrous sodium sulphate, is an appropriate sequence for most purposes. Concentrated hydrochloric acid may be used to remove alcohols from esters.

The last step will be a distillation, first removing any volatile solvent and then collecting the desired fraction within about 2 °C of its predicted boiling point.

Questions

56 Suppose you have just carried out the 'cooking' stage of the preparation of 1-bromobutane from butan-1-ol and a mixture of potassium bromide and sulphuric acid.

a List all the chemicals, including unreacted reagents, by-products and desired products, that might be present in the mixture.

b If you shook the mixture up with ethoxyethane (diethyl ether), which substances would be extracted into this new non-aqueous layer?

c Would this solvent extraction be a sensible stage in the purification process? How would you continue the process?

57 Why are extraction solvents chosen that are volatile (have a low boiling point) and are immiscible with water?

58 What do you think is the purpose of the sodium hydrogencarbonate wash? Why is a stronger base such as sodium hydroxide not used?

59 Why would the wash/dry sequence not work for, say, ethanol?

60 How would you know which layer to throw away at the end of each wash stage?

61 What method would give you an idea of the purity of your final product, other than the closeness of its boiling point or melting point to the expected value?

Purification – solids

If the solid has separated out during the reaction, it can be removed by Buchner filtration. Then it is recrystallised using a suitable solvent. The solvent is chosen such that:

1 the desired product is more soluble in the hot than the cold solvent
2 any impurities stay in solution in the solvent.

If the product is too soluble in the proposed solvent, then the solvent will not be suitable.

The recrystallised product is removed by Buchner filtration, washed with a little clean solvent, and air-dried or oven-dried at a temperature below its melting point. A melting-point test will establish the level of purity. If the yield is too low, more solvent can be evaporated from the filtrate, whereupon more crystals may appear on cooling.

Questions

62 Mixed solvents often prove to be the best for recrystallisation, for example ethanol/water for fairly polar products like amides, and ethanol/cyclohexane for less polar ones. Why do you think mixed solvents allow greater control over the process of recrystallisation?

63 Why is the pure product washed with clean solvent, even though this must inevitably cause slight loss of yield?

64 How do impurities show up in melting point tests?

15.16 Identification of organic chemicals

This, the second of the two sections giving an overview of organic chemistry, could well be called 'reading the codes'. One obvious example of a code is the term 'pleasant-smelling liquid' because to a student with any experience it is synonymous with 'ester'. The codes are not all as easy as that, and the following question requires you to research some of the less obvious ones.

Questions

65 Complete a copy of Table 15.7 (overleaf).

66 Suggest an identity for the compound which behaves as follows in standard school-laboratory tests:

a burns with a smoky flame

b insoluble in cold water, but moderately soluble in boiling water

c liberates carbon dioxide from carbonates

d melts at 121 °C.

Table 15.7 Identification codes

Clue	Inference
Decolorises bromine (two possible answers)	
Decolorises MnO_4^-/H^+	
Oxidised by 'mild' $Cr_2O_7{}^{2-}/H^+$ but product is neither acidic nor gives positive Fehling's test	
Oxidised by Fehling's solution	
Gives a yellow precipitate with Brady's reagent	
Hydrolysed in acidic or alkaline conditions into two organic fragments (two possible answers)	
Light needed to initiate reaction	
Reacts with PCl_5, HCl given off (several possible answers)	
Reacts with H_2/Ni catalyst (three possible answers)	
Acidic enough to liberate CO_2 from carbonates/hydrogencarbonates	
Weakly basic, dissolves in HCl but not in water	
Gives ammonia when heated with NaOH	
Faintly acidic but not enough to give CO_2 with carbonates/ hydrogencarbonates	
Not acidic but reacts with Na(s)	
Gives quick precipitate with $AgNO_3$(aq)	
Gives slower precipitate with $AgNO_3$(aq)	
Burns with a smoky flame	
Dehydrates to alkene	
Reacts with ethanolic KOH to give alkene	
Reacts with acyl halide to give pleasant-smelling liquid	
Gives purple colour with $FeCl_3$	
Gives strong sharp IR band between 1680 and 1750 cm^{-1}	
Gives very broad IR band from 3200 to 3700 cm^{-1}	

67 Repeat for another compound which behaves thus:

a burns with a clean flame

b miscible with water at all times

c reduces acidified dichromate to green Cr^{3+}

d boils at 83 °C.

Summary

- **Amines** are substituted ammonia molecules. Substituents may be alkyl or aryl groups, and there may be one, two or three of them, giving primary, secondary or tertiary amines, respectively (Figure 15.1, p. 353).

- All amines share the basic (H^+-receiving) character of the parent ammonia, which comes from their possession of a lone pair of electrons on the nitrogen atom.

- Because of the 'electron-pushing' tendencies of alkyl groups, alkylamines are stronger bases than ammonia. Because of the electron-withdrawing effect of the benzene ring, aryl amines are weaker bases than ammonia. All amines are **protonated by acids** to give substituted ammonium cations:

$$\left[\begin{array}{c} R \\ H - N \rightarrow H \\ H \end{array} \right]^+$$

- Amines are all **nucleophilic**, using the lone pair of electrons on the nitrogen atom to carry out nucleophilic substitutions, on such substrates as halogenoalkanes and acyl halides (Figure 15.4, p. 355).

- **Amides** are molecules with both the carbonyl group and the amino group sharing the same carbon atom (Figure 15.7, p. 356). As such they can be seen as relatives of carboxylic acids, esters and acyl halides. They are made by the action of amines or ammonia on acyl halides.

- If ammonia itself is used in amide synthesis, the result is a primary amide. If amines are used, then *N*-substituted amides result.

- *N*-substituted amides are similar to esters, insofar as they undergo cleavage to generate the 'parent' acid and amine:

$$R-C{\overset{O}{\underset{\underset{H}{N-R'}}{}}} + H_2O \xrightarrow{OH^-} R-C{\overset{O}{\underset{OH}{}}} + R'NH_2$$

- The amide grouping is the linking unit in many important polymers, both synthetic and natural.

- Molecules in the **nylon** group of polymers consist of hydrocarbon units held together by amide linkages. They are normally made by reacting double-headed amines with double-headed carboxylic acids, in a condensation reaction. The resulting polymers give rise to strong fibres and durable bearings, amongst other items in a wide product range.

- **Proteins** are also polyamides, but made from **amino acid** residues. Twenty amino acids can take part in protein formation, and they are the only ones to concern us. They all feature a carboxylic acid group, an amino group and a side-chain, attached to a single carbon atom.

• Amino acids undergo a form of self-protonation, in which the carboxylic acid group donates its proton to the amino group. The resulting species is known as a **zwitter-ion**.

• Amino acids enter into both acidic and basic equilibria with water (Figure 15.11, p. 359). Each individual amino acid shows a different balance between its acidic and basic tendencies. These tendencies are most marked when the side-chain carries a second acidic or basic group.

• **Electrophoresis** is one way of separating amino acids, and thus of finding out which ones are present in a mixture. It is a technique that combines the principles of electrolysis and paper (or gel) chromatography. The separation, as in chromatography, is based on the difference in willingness of the amino acids to move across the stationary phase. However, in electrophoresis the movement is not prompted by an eluting solvent, but by the influence of an electrical potential difference.

An important variable in electrophoresis is the pH of the medium. At various background pHs amino acids may accept or give protons, or remain unaffected. That means they can exist as cationic, anionic or overall neutral species, which further adds to the variety of their responses. The pH at which an amino acid (or a protein) is electrically neutral, and therefore unable to move in electrophoresis, is called its **isoelectric point**:

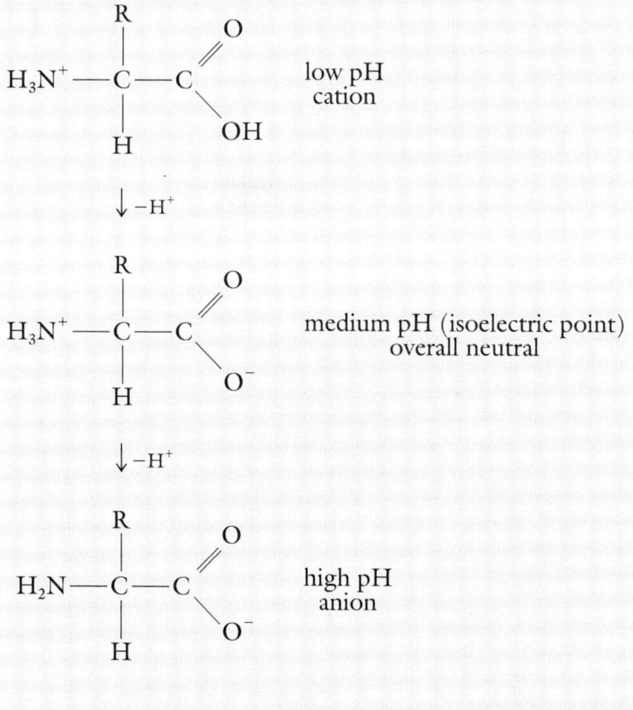

• Proteins are crucially important biochemical molecules, vital to both the anatomy of animals and the way they function.

They are all polyamino acids. With 20 amino acids to choose from, of widely varying character (created by their side-chains), the range of possible protein molecules is almost infinite.

• The character of a protein is set by three structural aspects:

1 its amino acid sequence (its **primary structure**)
2 the way the polymer is arranged into helices, sheets or amorphous chains (the **secondary structure**)
3 the combination of regions of chain, sheet and helix to create an overall three-dimensional protein (the **tertiary structure**).

• These three aspects of structure are not independent of each other. Once the primary structure is set, a protein will always have one unique most stable way of 'curling up' (secondary and tertiary structures). *Sequence specifies conformation.*

• Primary structures are elucidated by techniques such as hydrolysis, followed by two-dimensional chromatography (to find out which amino acid residues were present), and then the chain-cutting techniques of Edman and others (to find out in what order they occur).

• Two-dimensional chromatography consists of two successive acts of chromatographic separation, conducted with different eluents, in directions at 90° to each other. It operates on the principle that amino acids left unseparated by one solvent will be resolved by the second.

• Secondary and tertiary structures are obtained from the results of X-ray crystallographic work, leading to electron density maps.

• Tertiary structures of proteins are crucial to their physiological action. An enzyme, for instance, is inactive if it is prevented from adopting its favoured tertiary structure. Enzymes are complex proteins with special regions known as **active sites**, in which substrates may be temporarily attached for the purposes of reaction. Enzymes lower activation enthalpies and so act as **catalysts**.

• Enzymes usually operate on one substrate only, a fact symbolised in the **lock-and-key** analogy. If a foreign substrate happens to mimic the intended one, the enzyme may be rendered ineffective. This phenomenon is called **competitive inhibition** (because the mimic competes with the real substrate for sites on the enzyme), and is the way some poisons work. Other enzyme poisons engage the enzyme at a site remote from its catalytic cleft (**non-competitive inhibition**), but still cause enough distortion in the molecule to prevent its activity.

• Enzymes are **denatured** by excessive temperature, which distorts the active site by thermal vibration.

• They also have specific needs in terms of pH. This is because protons may be shed or received by the enzyme molecule at different pHs, and the presence or absence of protons on certain sites can have a crucial bearing on the catalytic action (as with lysozyme).

- Enzymes follow rate laws which can pass from first order to zero order with respect to the substrate concentration. The size of the **Michaelis constant** gives an indication of when this switch-over will occur, and also of how quick the enzyme reaction is. The Michaelis constant (symbol K_m) is a constant for a given enzyme–substrate pair.

- **Nitro-compounds** (Figure 15.53, p. 378) include some powerful explosives. They decompose to liberate nitrogen gas. The most stable nitro-compounds are the singly nitrated arenes, which may be easily prepared from benzene and other arenes by reaction of the hydrocarbon with a mixture of concentrated nitric and sulphuric acids. The mechanism involves attack on the ring by the electrophile NO_2^+. Aromatic nitro groups are readily reduced to amino groups, using acidified tin metal as a reducing agent.

- Aromatic amines react with nitrous acid in the cold to give **diazonium cations**. These are unstable compounds, but very convenient intermediates on synthetic pathways.

The presence of the diazonium group renders benzene rings reactive towards a wide range of nucleophiles (which runs counter to the main trend of aromatic character). For example:

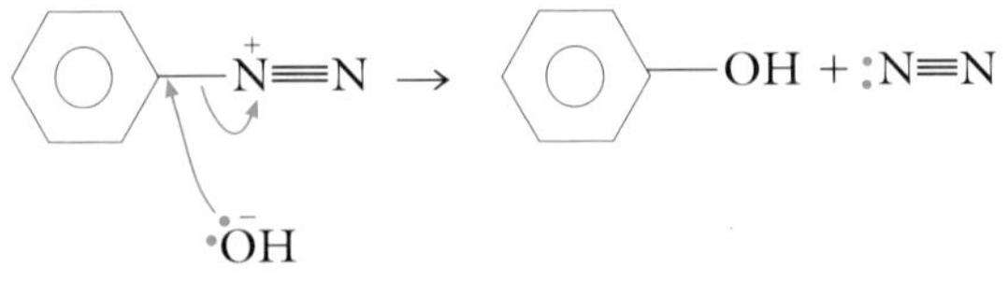

- Diazonium cations also act as electrophiles in their own right, forming **azo-compounds** with other arenes. The azo-compounds are generally intensely coloured, due to extensive networks of delocalised π-bonds, and are staple ingredients of the dye industry (Figures 15.56–58, pp. 379–80).

16

Electrochemistry

16.1 Introduction

Electrochemistry involves the study of the movement of electrons from one chemical species to another. Electron exchange is a fundamental process in chemistry, especially since electrons are the basic unit of chemical bonding. As we saw in Chapter 8, 'electron transfer' is another way of saying 'redox', so all the reactions we have previously classified as redox are material for this chapter too.

The idea of electron transfer can be applied more easily to some reactions than others, even within the definition of redox. It is obvious that:

$$Zn(s) + Cu^{2+}(aq) \rightarrow Zn^{2+}(aq) + Cu(s)$$

is a transfer of electrons from zinc metal to copper ions. The point is emphasised if we divide the single equation into two half-equations, one in which the zinc offers the electrons, and the other in which the copper ions accept them:

$$Zn(s) \rightarrow Zn^{2+}(aq) + 2e^-$$
$$Cu^{2+}(aq) + 2e^- \rightarrow Cu(s)$$

With other redox reactions, the concept of electron transfer is apparently irrelevant, for example in cases when the changing oxidation numbers depend on the polarity of covalent bonds:

$$CH_4 + 2O_2 \rightarrow CO_2 + 2H_2O$$

However, the rules of the redox 'game', as outlined in

Chapter 8, are sufficiently mechanical to ensure that you can write a half-equation for any redox reaction. You just require the right number of water molecules and hydrogen ions, and thus an implied move to an aqueous environment. In this case you would write:

$$CH_4 + 2H_2O \rightarrow CO_2 + 8H^+ + 8e^-$$

and:

$$O_2 + 4H^+ + 4e^- \rightarrow 2H_2O$$

However artificial this exercise looks, we shall see that even these unlikely-looking half-equations do take place under certain circumstances, in devices called fuel cells.

Question

1 To return to the more obvious electron transfers, and to revise Chapter 8, split each of the following redox equations into two half-equations, one showing the giving of electrons and the other the receiving of electrons:

a $Mg(s) + 2H^+(aq) \rightarrow Mg^{2+}(aq) + H_2(g)$

b $2Fe^{3+}(aq) + 2I^-(aq) \rightarrow 2Fe^{2+}(aq) + I_2(aq)$

c $MnO_4^-(aq) + 8H^+(aq) + 5Fe^{2+}(aq) \rightarrow$ $Mn^{2+}(aq) + 4H_2O(l) + 5Fe^{3+}(aq)$

So, if there is so much overlap between the two concepts, what exactly is the *difference* between electrochemistry and redox? The answer is that it is really all a matter of apparatus – electrochemistry is a sub-branch of redox in which normal redox reactions are carried out in an abnormal way. In electrochemistry, the reaction system is linked to an electrical circuit, and this circuit is the only channel by which the electron transfer is allowed to take place. In the reaction between zinc metal and copper ions, for example, the key feature of the layout would be that the zinc and the $Cu^{2+}(aq)$ ions would not be allowed to come into direct chemical contact.

In electrochemistry, as in ordinary redox reactions or indeed any reactions, there are thermodynamic exchanges with the surroundings. The unique feature that is introduced by the apparatus of electrochemistry is that some of those exchanges can be in the form of electrical energy as well as heat, with all the advantages of convenience that are associated with electrical energy.

Chemical and electrical energy can be interconverted in both directions. Some sections of this chapter deal with reactions in which some of the chemical energy *output* from the system is converted to electrical energy, and given to the surroundings in that form (Figure 16.1a). In others, electrical energy *input* to the system is used to drive the system into higher states of chemical energy (Figure 16.1b).

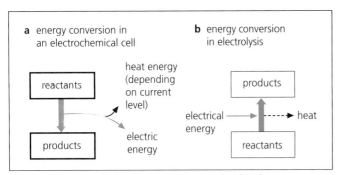

a energy conversion in an electrochemical cell

b energy conversion in electrolysis

Figure 16.1 Types of energy conversion covered in this chapter

Question

2 Both the energy conversions just mentioned were probably present in your GCSE curriculum, and at least one of them is a familiar part of everyday life. This question should prompt you to recognise that fact.

In which direction is the chemical energy/electrical energy interconversion going in the following examples?

a The splitting of dilute sulphuric acid into hydrogen and oxygen.

b A battery.

Now you can see that what we are talking about is **electrolysis** (electrical energy converted *to* chemical energy) and **cells** (electrical energy made *from* chemical energy). The first part of the chapter concerns itself with cells.

16.2 Arranging reaction systems to give electrical energy

For the time being let us return to the Zn/Cu^{2+} system as an example of an electron transfer reaction, and consider how the reactants must be arranged in space to give us electrical energy instead of just heat. The crucial feature must be that the electrons from the zinc must have to travel via an external circuit to reach the copper ions. This in turn demands some sort of compartment structure, of which Figure 16.2 shows the simplest example.

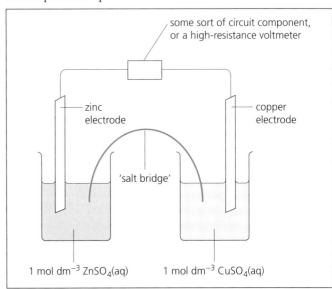

Figure 16.2 Simplest design of an electrochemical cell, in the 'standard' format

The standard cell

Figure 16.2 shows a crude apparatus for converting chemical energy to electrical energy, but it nevertheless represents a **standard cell**. The only requirement for a cell to qualify as 'standard' is that the aqueous ions that are going to take part in the reaction must be at a concentration of 1 mol dm^{-3} and at 298 K and 1 atm.

The standard cell is an academic tool rather than a practical one, and it is normally used to yield a single reading. This is the voltage difference (potential difference) between its terminals when it is doing no work. Given this very limited sphere of use, it does not really matter that a cell of this design is incapable of doing any real job more demanding than, say, lighting an LED (light-emitting diode). We shall come back to the design aspect of cells after the basic principles are in place.

The words 'cell' and 'battery' are often used interchangeably, but strictly speaking Figure 16.2, and any other design that uses a single pair of electrodes, is an electrochemical *cell*. The word *battery* should really be kept for situations where several cells are racked up in banks, working together. A car 'battery', in which each little water-filling hole represents a single cell, is therefore correctly named. On the other hand the cylindrical 'batteries' that supply the energy for torches and radios are really single cells.

How the apparatus works

This description will inevitably be given in stages, but eventually you need to visualise the process occurring as a whole. With this in mind, let us begin with our focus on the zinc electrode.

At the zinc electrode

Zinc atoms become ions and attract the attention of water molecules. The now solvated $Zn^{2+}(aq)$ ions diffuse away into solution, thereby exposing a new 'workface' of zinc atoms on the electrode (Figure 16.3).

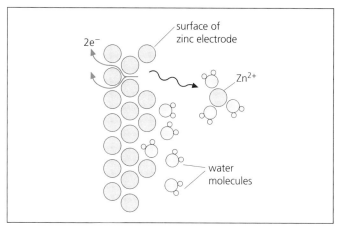

Figure 16.3 How zinc atoms become zinc ions at the electrode surface

Question

3 a Why would the Zn^{2+} ions attract the attention of water molecules?

b Would you expect the cell to function if the electrolyte (that is, the liquid between the electrodes) were hexane instead of water?

In the external circuit

Electrons leave the zinc electrode and make their way round to the Cu/Cu^{2+} side of the cell (Figure 16.4).

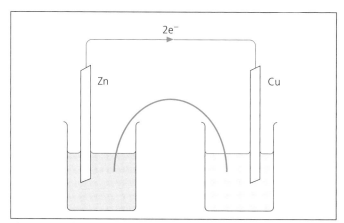

Figure 16.4 Showing the direction of electron flow

At the copper electrode

The electrons arrive from the zinc side of the cell, and use the copper electrode as a solid conductor to bring them in contact with the species that will act as receivers of electrons, namely the $Cu^{2+}(aq)$ ions. The exchange is completed, and the newly formed copper atoms come out of solution and crystallise on to the solid lattice of the electrode material (Figure 16.5).

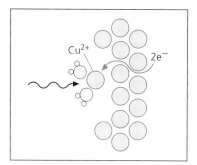

Figure 16.5 A copper ion sheds its solvating water, accepts two electrons and becomes a neutral copper atom on the surface of the electrode

Question

4 From the paragraph above, do you think the cell would have functioned as well if the copper electrode had been replaced by another metal, platinum say?

In the salt bridge

Nearly everyone who studies electrochemistry for the first time is puzzled by the necessity for the salt bridge. A vague explanation like 'it completes the circuit' is true, but not very informative. A fuller picture is provided by saying that it allows for the balancing of charges in the solutions. Perhaps the following question will contain the prompts which bring about understanding of the salt bridge's role.

Question

5 Refer to Figure 16.6.

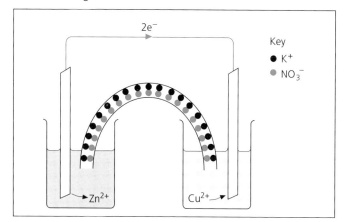

Figure 16.6 Focusing on events in the salt bridge

a If Zn^{2+} ions arrive in the zinc sulphate solution, how will they upset the electrical neutrality of the solution?

b Which sort of ions will need to be attracted off the end of the salt bridge into the solution, to redress the imbalance?

(The choice of potassium nitrate for the salt bridge is governed by the fact that neither ion will get involved chemically with the contents of either cell solution, and so they can be used purely as electrostatic balancers.)

c Meanwhile, back at the copper sulphate solution, what imbalance in the ionic population will have been created by the removal of Cu^{2+} ions?

d How will the salt bridge be able to redress this imbalance?

So a pair of counter-currents run through the salt bridge, anions into the zinc side of the cell, and cations into the copper side. In fact the entire current path, both inside and outside the cell, can only be kept open and flowing by the avoidance of build-ups of one sort of charge or another.

The overall process

Each of the above stages is an artificial division, because the following events happen *simultaneously*:

1 A zinc ion is created and two electrons enter the external circuit.
2 Two different electrons are taken out of the external circuit to make copper atoms.
3 Two potassium ions come off the salt bridge into the copper solution.
4 Two far distant nitrate ions come off the other end of the bridge into the zinc solution.

You can imagine a similar situation involving a line of railway trucks being pushed by an engine at the back – although you might feel justified in talking about A pushing B and then B pushing C, in fact truck Z would be on the move very shortly after the initial move of A. Trucks refuse to be squashed together, and that is why the push gets transmitted rapidly. Similarly, charged particles refuse to be too close to their own kind, and that is what provides the concerted movement of the charge carriers in a cell.

16.3 The driving force for the cell reaction

Competitions and conventions

We have said that the flow of charge carriers is controlled by principles of electrostatic repulsion and attraction, but why did the electrons move in the first place? The answer lies in the chemical character of the four participants – the two metals and the two metal cations. The cell works for exactly the same reason as the reaction works – because the zinc is a better 'loser' of electrons than the copper, and therefore can force copper atoms to keep the electrons it discards.

This concept of a battle between two metals for the right *not* to have possession of electrons is a helpful picture – it introduces to the situation an aspect of competition. However, it was a bit arbitrary to tell the story as a competition between the metals – it can just as correctly be seen as a competition between the two cations, Zn^{2+} and Cu^{2+}.

Question

6 We have just pictured the reaction as being a 'victory' of zinc over copper in the struggle *not* to have the electrons. Rephrase that last sentence, with the ions as the major 'actors', and the struggle being the fight to *have* the electrons.

Both versions of the competition are equivalent and alternative. However, the way we talk about redox systems has been decided to a large extent by rules and conventions. For example, many of our discussions will be based on a list of redox half-equations, which by convention are all written the same way round. Such a list is called the **electrochemical series**, a heavily edited version of which is shown in Table 16.1.

We need to be clear from the start about certain conventions pertaining to the electrochemical series. In all such lists of redox half-equations, the oxidised form is placed on the left, as if it were a reactant. And by another convention, the sequence is always arranged so that zinc's place in the series is above copper's.

With these two conventions in place, the orientation of the series acts as a code

Table 16.1 An extract from the electrochemical series, featuring hydrogen and some common metals

$Na^+(aq) + e^- \rightleftharpoons Na(s)$
$Mg^{2+}(aq) + 2e^- \rightleftharpoons Mg(s)$
$Zn^{2+}(aq) + 2e^- \rightleftharpoons Zn(s)$
$Fe^{2+}(aq) + 2e^- \rightleftharpoons Fe(s)$
$Pb^{2+}(aq) + 2e^- \rightleftharpoons Pb(s)$
$2H^+(aq) + 2e^- \rightleftharpoons H_2(g)$
$Cu^{2+}(aq) + 2e^- \rightleftharpoons Cu(s)$
$Ag^+(aq) + e^- \rightleftharpoons Ag(s)$

that can be read to indicate the direction of the cell reaction. In the case of the copper/zinc cell you can see that the direction of the cell reaction is an **anti-clockwise semi-circle** (Figure 16.7), with strong echoes of the clockwise circles from the acids in Chapter 12, when protons were being exchanged. As then, the two reactants are on one diagonal, and the two products are on the other diagonal.

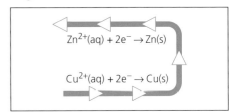

Figure 16.7 The Law of anti-clockwise circles is used to show the direction of a cell reaction

Question

7 a Write down the two half-equations and draw in the diagonals, from what you know to be the reactants and products.

b Which half-equation is being 'driven backwards'?

c Which species is acting as the oxidising agent in the cell reaction? (*Remember*: An oxidant takes electrons.)

d Which is the reducing agent?

e Of the two oxidised forms shown in the two half-equations (Zn^{2+} and Cu^{2+}), which is the stronger oxidant?

f Which is the stronger reducing agent of the two metals?

You can see that it is misleading to think of the reaction being 'initiated' either by the reducing (electron-dumping) power of the zinc, or by the oxidising (electron-grabbing) power of the Cu^{2+} ions. The reaction simply happens because of the combination of the two.

Another point worthy of note is that the two half-cells of Figure 16.2 actually contain the two half-equations of the redox change.

The significance of the equilibrium signs in the half-equations in Table 16.1 is brought home when you find that a half-equation that goes one way in one cell can, in certain other cells, go the other way. For instance, in the electrochemical series of half-equations in Table 16.1, you will find beneath the copper half-equation a half-equation involving silver:

$$Cu^{2+}(aq) + 2e^- \rightleftharpoons Cu(s)$$

$$Ag^+(aq) + e^- \rightleftharpoons Ag(s)$$

Question 8 asks you to apply the law of anticlockwise circles to analyse this pairing, first as a reaction and then as a cell.

Question

8 a Which are the two reactants?

b Which are the two products?

c Which is the dominant oxidising agent? Which is the dominant reducing agent?

d What do you notice about the direction of the copper half-equation, compared with the zinc/copper case?

e What visible events would happen if you mixed aqueous silver ions (as silver nitrate) with copper metal in a test tube? Write an equation for the reaction you suggest.

f Draw the electrochemical cell that could contain and be powered by this reaction, using the basic design of Figure 16.2.

g Indicate on your diagram the polarity of the terminals and the direction of electron flow.

Using the electrochemical series

The electrochemical series is a predictive tool. Admittedly it is constructed from after-the-event experimental data – the reason why Cu^{2+}/Cu is below Zn^{2+}/Zn is because someone saw that copper ions could oxidise zinc. But the important thing is that it has the power to *predict* the results of many more experiments than need ever actually be carried out. For example, we should be able to predict what would happen if we created a cell that paired the silver half-cell with the zinc half-cell, and we could base this prediction solely on the results of the zinc/copper and copper/silver experiments.

To use a sporting analogy, if we see that team A is above team B in a league table, and team B is above team C, then it follows that team A should beat team C. Similarly if zinc can reduce Cu^{2+} to copper, and copper can reduce Ag^+ to silver, then you can be sure that zinc will reduce Ag^+. (I admit the analogy is an imperfect one – if sport were really as predictable as this there would be no such thing as bookmakers.)

Questions

9 Draw an electrochemical cell for the silver/zinc pairing, and indicate direction of electron flow. Give the equation for the chemical change that occurs when electrons flow round the external circuit.

10 In any cell, when will the electrons stop flowing – in other words, what chemical events would bring the electron flow to a halt?

16.4 Quantitative aspects of the electrochemical series

The electrochemical series is more than just a list or a ladder – or at least if it is like a ladder, then there are clearly defined differences between the spacings of the rungs. The reason is that every cell offers one extra piece of information, beyond the fact that a reaction is occurring in one direction. It offers the potential difference between the terminals, which can be measured by a (preferably high-resistance) voltmeter.

Defining terms

We need to establish some electrical definitions, before we return to the main narrative:

Coulomb

A **coulomb** is a unit of charge, invented long before the charge-carrying particles themselves were recognised. It turns out to be equal to the charge on a large number of electrons. For instance, 6.2×10^{18} electrons carries a charge of (minus) one coulomb, as does the same number of, say, Cl^- ions. 6.2×10^{18} protons or Na^+ ions would have a charge of (plus) one coulomb. A *mole* of electrons has a charge of $-96\,500$ C, an amount sometimes referred to as one **faraday**. The charge on a single electron (or proton) is minus (or plus) 1.6×10^{-19} C.

Potential difference

The **potential difference** between two points in a circuit is the energy in joules that would be obtained when one coulomb of charge passes between the points. It is measured in volts, so a volt is a joule per coulomb. The formula linking potential difference, energy and charge is:

$$\text{p.d. (volts)} = \frac{\text{energy (joules)}}{\text{charge (coulombs)}}$$

$$(16.1)$$

Hence if you get a reading of one volt across the terminals of a cell, then you know that one joule of work (or work + heat energy) will be liberated every time you allow one coulomb of charge (that is 6.2×10^{18} electrons) to flow. The popular analogy for potential difference is pressure difference, as in a plumbing circuit. A cell with a high potential difference across its terminals would be like a water pump generating a high pressure at its outlet. Clearly such a pump would give you more work from a given amount of water than a low-pressure pump, and likewise you would get more work or heat from a mole of electrons if that mole was being 'pumped' at a high potential difference (Figure 16.8).

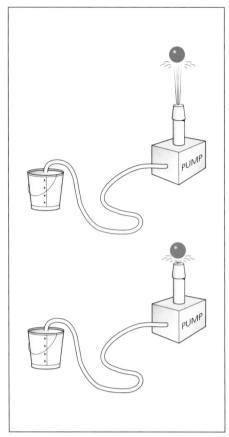

Figure 16.8 A watery analogy for the potential difference of a cell – both pumps have the same amount of water to work with, but the top one does *more work*

Electromotive force (e.m.f.)

The **electromotive force** is the maximum possible potential difference between the terminals of a cell. This maximum is approached as the current driven by the cell approaches zero. So to measure the e.m.f. you must use a very high-resistance voltmeter, while the cell is doing *no work*. Under these circumstances hardly any electrons are actually flowing. When charge *is* passing through the circuit, the imperfections in the design of the cell start to show. They present an internal resistance which causes a wastage of energy as heat in the cell itself, and reduces the energy available between the terminals and the outside circuitry. In most cases discussed in this book, we shall assume that any potential difference has been measured while the cell does no work, so it is the e.m.f. of the cell. Figure 16.9 shows how the potential difference across a cell's terminals drops away from its e.m.f. value as

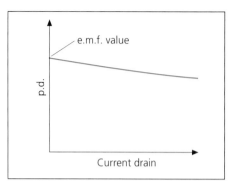

Figure 16.9 The e.m.f. is the p.d. when no current flows

the cell has to start passing electrons. In taking the 'pure' e.m.f. measurement, we are not allowing the cell reaction to happen at all, but we are measuring the 'pressure' of the electrons 'trying' to get round.

Question

11 Figure 16.10 shows the e.m.f.s of two cells containing the same chemicals at the same concentrations, but with a difference in efficiency of design.

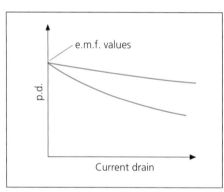

Figure 16.10 See question 11

a Which cell is the more efficient one, and why?

b How can you tell that they contain the same chemicals?

Cell diagrams

When you apply a voltmeter to an unknown cell to measure its e.m.f., you are as likely to get a negative as a positive reading. To recognise this fact there is another convention – you can write a cell either way round on the page, but the

sign of the e.m.f. is the sign of the right-hand electrode as written. This in turn requires a standard way of representing cells on the page, with codes for phase changes and salt bridges. The example in Figure 16.11 should show what I mean, and show how the sign convention works. Confusingly enough the name given to this collection of symbols is **cell diagram**, whereas common sense might have used that name for pictures like Figure 16.2.

$$Mg(s) \,|\, Mg^{2+}(aq) \,\vdots\, Cu^{2+}(aq) \,|\, Cu(s)$$

Figure 16.11 A cell diagram. Electrodes go at either side. Solid lines represent phase changes (in this case solid/liquid). Dotted lines represent salt bridges

Cell e.m.f.s

The significance of cell e.m.f.s can be made clearer by looking at the e.m.f.s of some of the cells we have met already.

$$Zn(s) \,|\, Zn^{2+}(aq) \,\vdots\, Cu^{2+}(aq) \,|\, Cu(s)$$

$$E_{cell}^{\ominus} = 1.1 \text{ V}$$

The $\ominus$ added to the E_{cell} symbol indicates that this is the e.m.f. of the cell in which the conditions are 'standard' – i.e. 1 mol dm^{-3} concentration for electrolyte species, temperature 298 K, pressure of any gases (not relevant here) 1 atm.

This value means that if we were to let one 'coulomb's-worth' of electrons pass round this cell, we would get 1.1 J of heat (or heat and light or heat and work, depending on whether the cell was driving, say, a bulb or a motor) given to the surroundings. But the coulomb is not really a useful unit in chemistry – let us see what happens if we allow a *mole* of cell reaction to happen.

Question

12a If a mole of zinc were to reduce a mole of Cu^{2+} ions, how many moles of electrons would have to pass round the external circuit?

b How many coulombs is that equivalent to? (You may need to look back to the definition of the faraday.)

c Using a rearranged version of formula (16.1), and the value of 1.1 V for the cell e.m.f., calculate how much energy is available from the cell reaction.

Question 12 will have awakened you to the true nature of a cell e.m.f., $E_{cell}^{\ominus}$. It merely required two multiplication sums, and then only by the constants 2 and 96 500 C, to turn it into an energy value. It turned into the energy available per mole of cell reaction, revealing the fact that $E_{cell}^{\ominus}$ is really just another thermodynamic quantity. The cell e.m.f. tells you how many joules you'll get from your coulomb, so it only requires a scale-up of $\times 96\,500$ (known as the Faraday constant) to find how many joules you'd get from a mole of electrons. Whether or not a '2' is needed depends on circumstances particular to each system – in this case it was that two moles of electrons pass per mole of cell reaction. $E_{cell}^{\ominus}$ values are similar to ΔH values, then, but that is a comparison which can be improved upon.

Question

13a Can you always tell whether a reaction is thermodynamically feasible, and in which direction, solely from its ΔH value?

b Can you do this prediction from a ΔG value?

c Can the sign of a cell e.m.f. give you sure predictions of the direction of reaction?

d So which has E_{cell} more in common with, ΔG or ΔH?

As question 13 showed, there must be a direct relationship between ΔG_{sys} and $E_{cell}^{\ominus}$, since they share the same predictive power, and indeed there is. The exact formula is:

$$\Delta G_{sys}^{\ominus} = -zFE_{cell}^{\ominus} \qquad (16.2)$$

where F is the Faraday constant and z is the number of moles of electrons per mole of cell reaction ($z = 2$ in our one example so far.) In fact, you will see that your answer to question 12 *was* $zFE_{cell}^{\ominus}$, so we can identify $\Delta G_{sys}^{\ominus}$ directly with

the *maximum energy available from the cell reaction* (with the minus sign indicating system-loss/surroundings-gain).

Using cell e.m.f.s to predict other cell e.m.f.s

The thermodynamic character of $E_{cell}^{\ominus}$ was emphasised because it puts us in a position to draw a parallel with Chapter 8. In that chapter, if the free energy changes of a couple of reactions were as follows:

$$A \to B \qquad \Delta G = x \qquad (1)$$
$$B \to C \qquad \Delta G = y \qquad (2)$$

then we could be confident in saying that the enthalpy change of:

$$A \to C \qquad (1+2)$$

was:

$$\Delta G = x + y$$

In other words, we could treat chemical equations like mathematical ones, and when we added or subtracted them, their associated ΔHs or ΔGs could also be added or subtracted. So it should be no surprise to find that $E_{cell}^{\ominus}$ values can be treated in the same way. For instance, suppose we know from experiment the $E_{cell}^{\ominus}$ values of the following cells:

$$\text{Zn(s)} \mid \text{Zn}^{2+}\text{(aq)} \vdots \text{Cu}^{2+}\text{(aq)} \mid \text{Cu(s)}$$
$$E_{cell}^{\ominus} = +1.10 \text{ V}$$

Cell reaction:

$$\text{Zn(s)} + \text{Cu}^{2+}\text{(aq)} \to \text{Cu(s)} + \text{Zn}^{2+}\text{(aq)}$$

and:

$$\text{Cu(s)} \mid \text{Cu}^{2+}\text{(aq)} \vdots \text{Ag}^{+}\text{(aq)} \mid \text{Ag(s)}$$
$$E_{cell}^{\ominus} = +0.46 \text{ V}$$

Cell reaction:

$$\text{Cu(s)} + 2\text{Ag}^{+}\text{(aq)} \to 2\text{Ag(s)} + \text{Cu}^{2+}\text{(aq)}$$

Question

14 Predict the e.m.f. of the cell:

$$\text{Zn(s)} \mid \text{Zn}^{2+}\text{(aq)} \vdots \text{Ag}^{+}\text{(aq)} \mid \text{Ag(s)}$$

Cell reaction:

$$\text{Zn(s)} + 2\text{Ag}^{+}\text{(aq)} \to 2\text{Ag(s)} + \text{Zn}^{2+}\text{(aq)}$$

In the light of this result, we can construct a diagram like Figure 16.12. In Figure 16.12 we have entered the $E_{cell}^{\ominus}$ values of the zinc/copper and

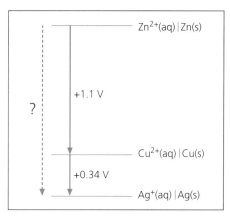

Figure 16.12 Predicting the value of one $E_{cell}^{\ominus}$ from two others (see question 14)

copper/silver cells as scaled distances, and as question 14 showed, we now know that the distance from silver to zinc indicates the $E_{cell}^{\ominus}$ of that third cell. An idea comes immediately to mind – if we were to place *any* half-cell X relative to the copper one, by actually setting up the X/Cu cell and measuring its e.m.f., then we could place the X half-cell on Figure 16.12. Now we can predict the e.m.f. of all possible combinations of other half-cells in our table with X. And along with that, of course, we could predict the direction of the cell reactions, and their ΔG values, and their K values.

16.5 The standard electrode potential $E^{\ominus}$ of a half-cell

The above idea is the basis of the quantitative version of the electrochemical series. There is only one change in the plan outlined above, and that concerns the choice of half-cell against which all the other half-cells are measured. For reasons that could not have included convenience, the world of chemistry decided to use the half-cell:

$$\text{H}^{+}\text{(aq)} + e^{-} \to \tfrac{1}{2}\text{H}_2\text{(g)}$$

as the one against which to measure all the others. The problem with this is the practical set-up of the hydrogen electrode itself.

There is no such thing as a 'rod' of hydrogen, so the convenience of the $\text{Cu}^{2+}\text{(aq)} \mid \text{Cu(s)}$ half-cell is lost. What passes for a rod of hydrogen is hydrogen gas at one atmosphere in contact with

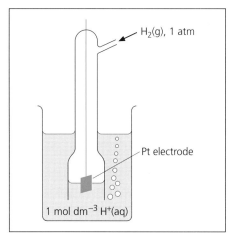

Figure 16.13 A hydrogen electrode

the surface of a transition metal on to which hydrogen gas is adsorbed, namely platinum (Figure 16.13). The cell diagram for the hydrogen half-cell is:

$$2H^+(aq) \mid [H_2(g)]Pt(s)$$

We then define the **standard electrode potential** of a half-cell as the e.m.f. of the cell made by pairing that half-cell and the hydrogen half-cell. The test cell is written on the right in the cell diagram. The symbol for standard electrode potential is $E^{\ominus}$, the $\ominus$ symbol signifying standard conditions.

Figure 16.14 shows an edited electrochemical series with an $E^{\ominus}$ vertical axis. You can see that the $E_{cell}^{\ominus}$ of any cell can be calculated from the distance between the two rungs in the ladder that represent your chosen half-cells. In fact we can symbolise this idea in the formula:

$$E_{cell}^{\ominus} = E_{right}^{\ominus} - E_{left}^{\ominus} \qquad (16.3)$$

where 'left' and 'right' refer to the way the cell diagram has been written on the page.

As a result of formula (16.3), the sign of the cell e.m.f. is always the same as the sign of the right-hand half-cell as written. This idea is easily confirmed by question 15.

Question

15 Write out the cell diagram for the silver/zinc cell both ways round, and apply formula (16.3) to it both times, to find its $E_{cell}^{\ominus}$.

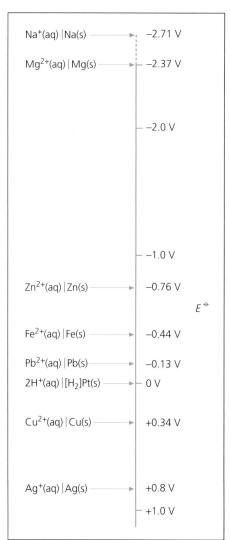

Figure 16.14 The electrochemical series with a quantitative vertical axis

There is often confusion concerning the fact that you never have to double or treble the $E^{\ominus}$ values in formula (16.3), even if one of the half-equations in the pair being matched is a two-electron transfer and the other a one-electron transfer. Yet if $E^{\ominus}$ were a thermodynamic quantity like ΔG, as was suggested earlier, then it would certainly be doubled along with the equation. The answer to the apparent paradox lies in the one big difference between ΔG and $E_{cell}^{\ominus}$. ΔG refers to molar quantities of the equation as written, whereas $E^{\ominus}$ always refers to the energy you would get *per mole of electrons*. So $E^{\ominus}$ values, for both half-cells and whole cells, give an indication of the energy you would get if every cell equation were scaled down to make $z = 1$. Hence there

is no need to do any multiplications to make $E^{\ominus}$ values additive.

Question

16 The electrochemical series is both a qualitative and a quantitative predictive tool. This question uses both facets.

a You may have noticed that only some metals 'fizz' when treated with dilute hydrochloric acid. The electrochemical series shows why. Which metals, according to the law of anticlockwise circles, can liberate hydrogen gas from an acid?

b What is the largest e.m.f. you could get from any pair of standard half-cells in Figure 16.14?

c Why would the cell suggested in (b) not be a practicable proposition?

16.6 Half-cells with a passive electrode

Up to now, all our half-cells have had electrodes that participated in the chemical equation of the cell, either as reactants or products. But this does not have to be the case. All you need for a half-cell is a chemical system capable of receiving or giving electrons, or what amounts to the same thing, undergoing a redox change.

You may at this point be wondering how the electrons are to be conducted in or out of the half-cell other than by a metal electrode. You obviously *do* need a metal (or graphite) electrode, but the metal does not have to be one of the participants in the reaction.

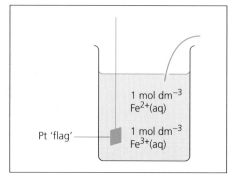

Figure 16.15 The Pt(s) $\mid$ Fe^{2+}(aq), Fe^{3+}(aq) half-cell

For instance, a half-cell can be made up of a mixture of the two aqueous ions Fe^{2+} and Fe^{3+}. This system can accept or donate electrons, according to the reversible half-equation:

$$Fe^{3+}(aq) + e^- \rightleftharpoons Fe^{2+}(aq)$$

The electrical connection can be made by a sheet (or 'flag') of platinum, a metal that is so unreactive it can be guaranteed not to take part in any redox events itself. The arrangement is shown, in the simplest design format, and with reactants at standard concentrations, in Figure 16.15. The use of platinum or graphite in this way means that every conceivable aqueous redox pairing can be made the subject of a half-cell (except where the pair of oxidation states react directly with each other, like say MnO_4^- and Mn^{2+}).

So the electrochemical series can now be revealed to its fullest extent, with half-cells of all kinds arranged by standard electrode potential (Table 16.2). As with the earlier and simpler cases of the metal/metal ion electrodes, the conventions demand that oxidised forms are on the left-hand side of each half-equation, and the most negative half-cells are at the top. 'Standard conditions' now means that every aqueous reagent must be at 1 mol dm^{-3}, so that in for instance the half-cell:

$$\vdots \, [MnO_4^-(aq) + 8H^+(aq)], Mn^{2+}(aq) \mid Pt(s)$$

even the $H^+(aq)$ has to be at 1 mol dm^{-3}.

In this form the electrochemical series acts as a comprehensive predictive tool for the direction of every standard-condition redox reaction under the sun, and if the half-equations are actually arranged into a cell, it gives the cell e.m.f. too (which can be converted to give ΔG and K).

Now we have the complete table, we need to extend some of the conventions that govern the setting down on paper of electrochemical cells, to cope with complications that never arose with simple metal/metal ion electrodes. You have seen how the manganese-based half-cell was set out above, using square brackets where there is more than one chemical species involved in the electron exchange. Notice also the comma, to indicate two oxidation states sharing the same homogeneous aqueous phase.

The final rule is the rule for the complete presentation of the cell diagram. A mnemonic to guide you is EROSORE, meaning that the ingredients of the cell must be set out in the order:

- **E**lectrode
- **R**educed form (if different from electrode)
- **O**xidised form
- **S**alt bridge
- **O**xidised form
- **R**educed form (if different from electrode)
- **E**lectrode.

These conventions, and that of using solid lines for phase boundaries and dotted lines for salt bridges, give you the complete protocol for cell diagrams. A typical example, in which nearly all the possible complications are present, is shown here:

$$Pt(s) \mid Fe^{2+}(aq), Fe^{3+}(aq) \vdots [MnO_4^-(aq) + 8H^+(aq)], Mn^{2+}(aq) \mid Pt(s)$$

All passive-electrode systems still follow formula (*16.3*), and the Law of anti-clockwise circles still holds as a predictor of cell reaction direction, just as in the metal/metal ion cases.

The following questions cover a wide range of uses and applications of the electrochemical series.

Question

17a Draw a picture of the cell that would be made by the pairing of the half-cells:

$$\vdots \, I_2(aq), 2I^-(aq) \mid Pt(s)$$

and:

$$\vdots \, [MnO_4^-(aq) + 8H^+(aq)], Mn^{2+}(aq) \mid Pt(s)$$

b Write out the cell diagram in a way that produces a positive cell e.m.f., and use formula (*16.3*) to calculate it.

c Write out the cell reaction, in the direction that 'goes', and calculate ΔG_{sys} and K_c. (Formulae (*16.2*) and (*9.11*))

You will find more examples of this type of question in the past exam questions at the end of the book. For now, let us look at a problem that begins to consider the practicalities of cells.

Question

18 Imagine there was a way of getting the zinc/copper cell into a handy casing (Figure 16.16). Can we predict how much energy we could pack in to a certain space?

Table 16.2 Conventional version of the electrochemical series

Half-cell	$E^{\ominus}$ (V)
$Na^+(aq) \mid Na(s)$	−2.71
$Mg^{2+}(aq) \mid Mg(s)$	−2.37
$Zn^{2+}(aq) \mid Zn(s)$	−0.76
$Pb^{2+}(aq) \mid Pb(s)$	−0.13
$2H^+(aq) \mid [H_2(g)]Pt(s)$	0.00
$Cu^{2+}(aq) \mid Cu(s)$	+0.34
$[\frac{1}{2}O_2(g) + H_2O(l)], 2OH^-(aq) \mid Pt(s)$	+0.40
$I_2(aq), 2I^-(aq) \mid Pt(s)$	+0.54
$Fe^{3+}(aq), Fe^{2+}(aq) \mid Pt(s)$	+0.70
$[ClO^-(aq) + H_2O(l)], [Cl^-(aq) + 2OH^-(aq)] \mid Pt(s)$	+0.89
$Br_2(aq), 2Br^-(aq) \mid Pt(s)$	+1.09
$Cl_2(aq), 2Cl^-(aq) \mid Pt(s)$	+1.36
$[MnO_4^-(aq) + 8H^+(aq)], Mn^{2+}(aq) \mid Pt(s)$	+1.51
$PbO_2(s) \mid [SO_4^{2-}(aq) + 4H^+(aq)] \mid PbSO_4(s)$	+1.8 (estimated)*
$F_2(g), 2F^-(aq) \mid Pt(s)$	+2.87

*See problems about lead/acid cells later on.

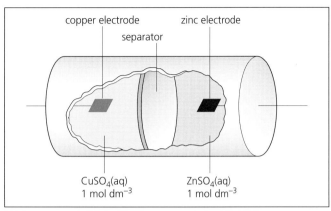

Figure 16.16 Imaginary adaptation of the standard zinc/copper cell, to fit the format of shop batteries

Clearly we shall have to make some assumptions. Let us begin with the size of the cell, using the approximate dimensions of an ordinary SP2.

- volume of cell = 50 cm³
- concentration of $CuSO_4$(aq) and $ZnSO_4$(aq) = 1 mol dm⁻³ (i.e. standard)
- density of solutions ≈ 1 g cm⁻³
- volume of electrodes is negligible compared with volume of solutions.

a How much volume will be occupied by each half-cell?

b How many moles of Cu^{2+} ions will be present in the copper side?

c How much zinc will be needed for reaction, remembering that the stoichiometric ratio of Cu^{2+} : Zn is 1 : 1?

d Estimate the total mass of the two solutions and the two electrodes, stating any assumptions you make.

e Work out the value of $E_{cell}^{\ominus}$, and therefore the $\Delta G_{sys}^{\ominus}$ for the reaction (use formulae (*16.3*) and (*16.2*)), which we shall take as the 'free' or usable energy of the cell per mole.

f Work out the usable energy available from this particular cell (i.e. not per mole, but 'per this cell').

g Express your answer in kilojoules per gram of cell.

h The industry uses the unit 'W h per kg' to measure the energy density of a cell. If a watt-hour is equivalent to 3.6 kJ, find the energy density of your cell by converting answer (g) to this industrial unit.

i How does it compare with the value of 80 W h kg⁻¹ for the common-or-garden SP2 cell?

16.7 Commercial cells

Question 18 suggested that the electrochemical cell industry would not have much use for the kind of cell in Figure 16.16, or indeed for standard cells in general. Their preoccupations are much more with packing a lot of energy into a small space, and cells that run consistently for a long time.

Question

19 Which features of 'batteries' are given highest profile in most of the battery adverts on television?

Our standard zinc/copper cell is riven with practical problems. Apart from having a rather modest energy density, we could also mention the problem of having liquids slopping around inside your radio, and we have not even begun to address the matter of internal resistance, which is to do with ionic conductivity *within* the cell, and with how easily the electrons can get on or off the electrode surface.

Question

20a Can you think of a way of keeping something liquid enough to be an electrolyte, yet not liquid enough to slop around?

b What features of design, of both the electrodes and the solutions, might decrease internal resistance (i.e. increase the ease of passage of charge across the cell)?

c What would happen to the internal resistance of the cell as the Cu^{2+}(aq) ions began to run out?

d In response to part (c), we could keep the other reactant, the zinc, in a stoichiometric minority, so the Cu^{2+} did not run out. However, that introduces its own problems. What are they?

A note on the nomenclature of half-cells

From early in your study of science you have been used to associating the word 'anode' with 'plus' and 'cathode' with 'minus'. You first learn these associations when discussing electrolysis in GCSE science courses, and then go on to confirm their validity by talking about cations and anions. This knowledge will make the next piece of information hard to accept.

In the study of cells, the *negative* terminal of the cell is the *anode* (and the *positive* is the *cathode*). It turns out that the true definition of **anode** is *the electrode from which electrons flow round the external circuit*. In electrolysis, this definition sits happily with the old 'anode = positive' one, since the positive terminal of your labpack is indeed the one into which the electrons flow. But, as Figure 16.17 shows, in an electrochemical cell, the electrons flow from the *negative* pole of the cell.

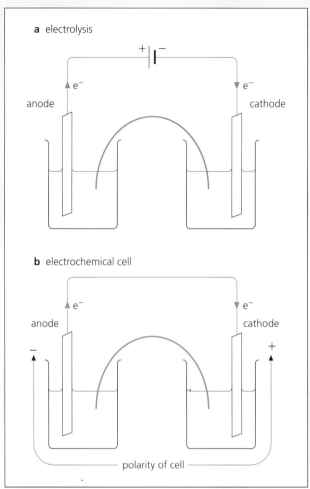

Figure 16.17 Showing how the words anode and cathode relate to *direction of electron flow*, not directly to polarity of supply or cell

Question

21 In the zinc/copper cell just discussed, and in the light of the system for naming the electrodes, which half-equation represents the anode reaction, and which the cathode reaction?

In cell manufacture, the terms 'anode' and 'cathode' are used in this surprising way. For instance, in all the cells in which zinc is used commercially, it is the anode.

Looking back on the problems inherent in the zinc/copper cell, it is clear that a lot of ingenuity has been needed to produce a workable battery. Figure 16.18b shows the end result of about 200 years of development time. We shall look at three examples of presently available commercial cells, and see how the industry has solved the problems.

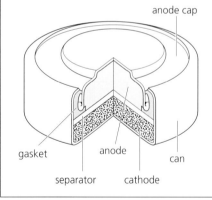

Figure 16.18
a Alessandro Volta (1745–1827), seen here with his 'pile', invented the electric battery. b A modern button-style cell, about six times life size

The zinc/carbon battery

The ordinary zinc/carbon battery is shown in Figure 16.19, and simplified in Figure 16.20. This is the 'silver seal' type of battery. It contains many of the standard ploys of the battery maker:

1 The half-cells are not arranged side by side but one inside the other, concentrically.
2 The entire casing of the cell is an electrode, in this case the zinc one (although outside this there is a plastic film).
3 The salt bridge is replaced by a porous barrier called a separator. Ions can pass through, just as they can with a salt bridge.

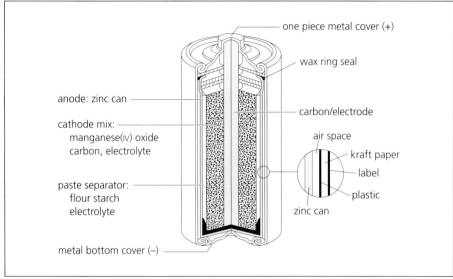

Figure 16.19 Zinc/carbon battery

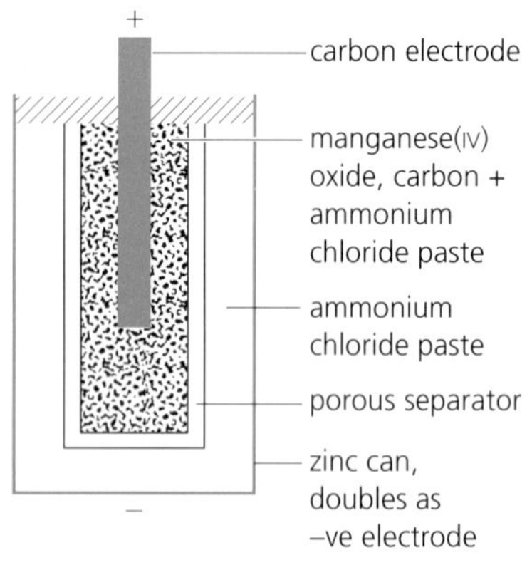

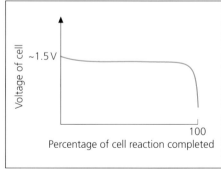

Figure 16.20 Zinc/carbon battery, simplified

4 Instead of $1\,mol\,dm^{-3}$ zinc sulphate around the zinc electrode, the electrolyte is aqueous ammonium chloride on both sides of the cell.

5 The electrolyte does not provide the ions being reduced. That job is done by a paste of insoluble manganese(IV) oxide. Overall the cell reaction is the oxidation of zinc by manganese(IV) oxide.

Each of these features carries some advantage to the operation of the battery, and this will be the focus of the next question.

Question

22a Having the entire casing as an electrode is a bonus for the internal conductivity of the cell. What feature of the electrode does it maximise?

b The concentric half-cells arrangement has advantages over the side-by-side arrangement, not least in the increased surface area of the separator (the substitute for the salt bridge). Show how the side-by-side arrangement would provide a smaller separator in a cylindrical cell, and say what advantage there is in having the larger separator area.

c A separator and a salt bridge have similar functions – that is, to enable the populations of ions to stay

electrically neutral, while at the same time making sure no contact is made between reactants. If the reaction at the anode (zinc electrode) is:

$$Zn(s) \rightarrow Zn^{2+}(aq) + 2e^-$$

and the electrolyte is ammonium chloride, which ion in the electrolyte would have to leave the anode compartment, to prevent an electrical imbalance? (Or alternatively, which ion might have to come in through the separator?)

d Which two species is the separator trying to separate? What are its chances of success?

e What happens in the manganese(IV) oxide (cathode) compartment is a bit obscure, but you could see it like this. Ammonium ions (which are mobile) are the initial receivers of the zinc's donated electrons:

$$NH_4^+(aq) + e^- \rightarrow NH_3(aq) + \tfrac{1}{2}H_2(g)$$

Then the hydrogen gas percolates between the granules of manganese(IV) oxide, and is oxidised back to water, from whence H^+ ions are reclaimed by the ammonia. But to keep things as simple as possible, try to write a single half-equation in which manganese(IV) oxide, MnO_2, is reduced to manganese(III) oxide,

Mn_2O_3. Finally, try to write the *overall* cell reaction (which should contain no electrons, and be the sum of the other stepwise reactions, from both anode and cathode compartments).

f This cathodic (electron-receiving) system, featuring $NH_4^+(aq)$ ions followed by manganese(IV) oxide, confers a big advantage – the receiver population of ions in solution does not 'run down' as it would in an ordinary $Cu^{2+}(aq)$ compartment. Figure 16.21 shows how this benefit is felt in terms of voltage stability – the voltage does not sag until very near the end of the cell's life. Suggest, using the same axes, how a zinc/copper cell might fare.

Figure 16.21 Showing how a zinc/carbon battery maintains its voltage for most of its life

g There is one drawback in having a reactant as a casing, with which anyone who has run their batteries too low or left them in a tape recorder too long will be familiar. What is this drawback, and why do you think it results from using the casing as the electrode?

A note on non-standard conditions

We noted that the ordinary zinc/carbon battery, although it contains the half-cell $Zn^{2+}(aq) \mid Zn(s)$, does not contain the *standard* version of that half-cell. But this turns out to be an advantage. If we consider that the half-cell reaction is an equilibrium, as in:

$$Zn^{2+}(aq) + 2e^- \rightleftharpoons Zn(s) \quad E^\ominus = 0.76\ V$$

then we can see that the effect of having fewer zinc ions would be to hamper the forward reaction as written, and so make the electron-giving (backwards) direction of the equation more unopposed, so the half-cell will become more electron giving.

Question

23a How will this effect alter the E (no longer $E^\ominus$) value of the $Zn^{2+}(aq) \mid Zn(s)$ half-cell?

b Which way would the half-cell move away from its 'standard' position in the electrochemical series?

c Would this change increase or decrease the overall cell e.m.f.?

d If the non-standard electrode potential of the zinc half-cell is now $-0.83\ V$, and the overall cell e.m.f. of an ordinary shop battery is 1.5 V, calculate the E value of the manganese(IV) oxide half-cell.

The zinc/air cell

If the paramount objective in cell design is to maximise energy density, then how about a component that weighs nothing and takes up (almost) no space? If it is not asking too much, it would be even better if the component cost nothing too, and while we are at it let us include the proviso that it should never run out. This apparently supernatural device does actually exist.

At the heart of the design is the idea of having oxygen as the electron receiver. The cathode reaction is therefore:

$$O_2 + 2H_2O + 4e^- \rightleftharpoons 4OH^- \text{ (on surface of catalyst)}$$

The source of the oxygen and the water is simply the air around the battery, which is allowed to percolate in through holes, and meet the electrons on a cathode coated with a finely divided catalyst. The anode half-cell reaction is the

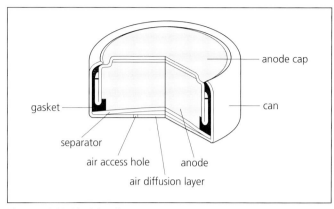

Figure 16.22 Zinc/air cell

familiar one using zinc (shown here going the way it actually goes):

$$Zn + 2e^- \rightarrow Zn^{2+}$$

The cell (shown in Figure 16.22) finds applications in devices such as wristwatches and hearing aids. Let us consider some of the advantages in the design of the zinc/air cell, in the course of the next question:

Question

24a Work out the overall cell reaction of the zinc/air cell. Indicate which species is reduced and which is oxidised.

b You will notice by comparing the zinc/air cell in Figure 16.22 with the more conventional type (shown in 'button' format in Figure 16.18b) that the zinc/air design is able to devote much more of its internal space to the anode compartment. Why is this possible?

c Why would you predict that the zinc/air cell would go on longer than the 'rival' in Figure 16.18b?

d From your knowledge of the sort of catalysts that seem to be effective for catalysing reactions between gases, suggest a material with which to coat the cathode.

e Calculate the $E^\ominus_{cell}$ for the zinc/air cell from data in a data book.

f Why is the e.m.f. of a commercial cell likely to differ a bit from a data book value?

g Can you see any environmental advantages in the zinc/air cell?

h The zinc/air cell has an energy density of 300 W h kg^{-1} compared with 80 W h kg^{-1} for a zinc/carbon/MnO_2 design. What is the major reason for this difference?

i Which features make the zinc/air cell appropriate for watches and hearing aids?

16.8 Rechargeable cells

In our discussion on cells and cell design, we have so far only included the types for which, once the cell reaction has run down, the sole option is disposal. Yet, there is another whole group of cells that are capable of re-use. Indeed, if you had to wake up to a world devoid of one or other of the two major groups of cells (disposables or rechargeables), you would probably give up the disposables. After all you can always plug your radio into the mains, and while this may not be an option for your digital watch, at least you can get out your old 'clockwork' one.

Figure 16.23 Life before car batteries

But do away with rechargeable batteries and you would see streetfuls of people on cold winter mornings all trying to start their cars with old-fashioned starting handles (Figure 16.23). Car engines can run without a battery once they are going, but if you stalled in heavy traffic you would have to get out and use the handle once again. So we cannot complete our study of cells without looking at the familiar lead/acid rechargeable type, and its possible successors.

All cells are rechargeable to an extent, but there are often dire warnings dissuading you from trying to do so on a simple type, since there are dangers of melting metal components and short-circuiting cells. The main feature that confers rechargeability on a cell is that *the oxidised and reduced forms of each half-cell are insoluble in the electrolyte.* They cling to the electrode and are ready for the chemical reversals of the discharging and recharging cycles.

The example of the lead/acid cell will help make this clear. In the fully charged condition, the electrodes of a lead/acid cell are lead and lead(IV) oxide. The anode process during the discharge cycle is:

$$Pb(s) + SO_4^{2-}(aq) \rightarrow PbSO_4(s) + 2e^-$$

The cathode process during the discharge cycle is:

$$PbO_2(s) + SO_4^{2-}(aq) + 4H^+(aq) + 2e^- \rightarrow PbSO_4(s) + 2H_2O$$

So the key idea behind the cell is this: sulphate ions are present, so the Pb^{2+} ions (which are produced in both half-cells)

cannot get into solution, but are trapped on the surface of the electrodes as insoluble lead sulphate. That way they are immediately on hand ready for their part in the recharging cycle. In the recharge cycle, the $PbSO_4(s)$ on one electrode is driven back to oxidation state 0 by the addition of electrons, while the $PbSO_4(s)$ on the other electrode is driven back to oxidation state +4 by the removal of electrons. The reactions are:

$$PbSO_4(s) + 2e^- \rightarrow Pb(s) + SO_4^{2-}(aq)$$
$$PbSO_4(s) + 2H_2O(l) \rightarrow PbO_2(s) + SO_4^{2-}(aq) + 4H^+(aq) + 2e^-$$

. . . and they are of course the exact reversals of 1 and 2.

The electrodes are made in mesh form, and then the gaps in the mesh are filled with a paste containing $PbSO_4(s)$. In other words, the battery, at the moment of first assembly in the factory, has identical electrodes. However, as the two above equations show, the charging-up process generates the differences that create positive and negative plates. The practical reason for smearing a $PbSO_4$ paste into the original mesh, is that whichever chemical is then formed by electrolysis either Pb or PbO_2, it is formed as a spongy mass of large surface area (and thus high efficiency), which derives its physical strength from hanging on to the ribs of the mesh.

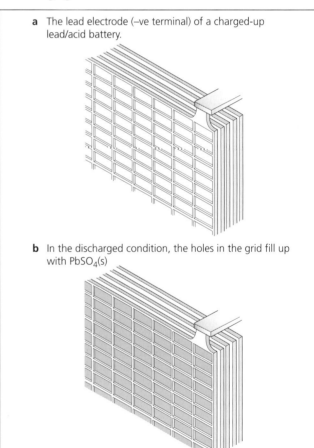

a The lead electrode (–ve terminal) of a charged-up lead/acid battery.

b In the discharged condition, the holes in the grid fill up with $PbSO_4(s)$

Figure 16.24 An electrode in a lead/acid cell: **a** as the bare mesh prior to 'pasting' and **b** after 'pasting' and after the paste has been electrolytically reduced to lead metal. It is now ready to be the negative terminal of the cell

Question

25a Estimate the e.m.f. of a single lead/acid cell, by looking in Table 16.2 (p. 397) for suitable half-equations.

b What oxidation state change occurs in the anode (discharge) reaction?

c What oxidation state change occurs in the cathode reaction?

d Write the overall cell reaction.

e A lead/acid cell gets worn out when the lead sulphate no longer sticks properly to the electrode surface, perhaps falling off altogether. Why would this harm the cell's ability to be recharged?

f The lead/acid cell has been earmarked for replacement for years, and yet it is still the standard car battery. The reason engineers want to replace it is its low energy density (35 Wh kg^{-1}). What factors do you think contribute to this low value?

Figure 16.25 **a** A modern car battery. **b** A large proportion of a milk float is taken up by the battery needed to power it. (**c**, top right) The latest electric car

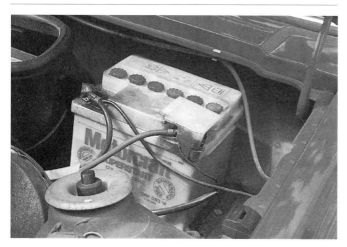

g Energy density is especially important when the battery is the source of traction rather than just lighting the headlights and turning the starter. With lead/acid technology it is hard to design anything faster than a milk float, or that goes for more than a few hours without a recharge (Figure 16.25). The solution would seem to be bigger batteries, but why is this not a real answer?

h Electric cars are a prize design goal because of their ability to reduce urban pollution. And yet you could argue that the pollution still happens, except that it happens in the making of the electricity to power the recharge cycle instead of at the vehicle itself. Can you argue, even so, that electric cars are still more environmentally friendly than our present ones?

Fuel cells

Like a rechargeable cell, a fuel cell is not a disposable cell, but it takes a different approach to the idea of continuity. Instead of undergoing alternate charge/discharge cycles, it receives its reactants in a steady flow whilst actually working. The name 'fuel cell' comes from the fact that the reactants could conceivably take part in burning reactions, as the following example shows.

The 1960s and 1970s saw the prestigious space programmes of the USA and the USSR. The American Gemini and Apollo craft used hydrazine, N_2H_4, itself a rocket fuel, and oxygen, to produce electricity to power the lighting and control functions. As in all cell reactions, the reaction is of a redox kind, and as ever the reactants are not allowed to meet.

In one compartment:

$$N_2H_4(g, \text{ on catalyst}) + 4OH^-(aq, \text{ on catalyst}) \rightarrow N_2(g) + 4H_2O(l) + 4e^-$$

In the other compartment:

$$O_2 (g, \text{ on catalyst}) + 2H_2O(l) + 4e^- \rightarrow 4OH^-(aq)$$

The gases are bubbled into their respective compartments over the surfaces of their catalysts.

Question

26a Which of the above reactions is the anode reaction, and which the cathode reaction?

b Write an equation for the half-cell reaction that might occur if methane were the fuel in a fuel cell, in place of hydrazine, and were being oxidised to carbon dioxide. Be prepared for eight electrons in the half-equation, as the oxidation number of carbon goes from −4 to +4.

c Compare the use of methane in a fuel cell with its use in a gas-fired power station. Can you suggest why people hope that fuel cells might offer a more efficient use of chemical energy?

Figure 16.26 This bus is powered by a fuel cell

Hydrogen-powered vehicles

Vehicles driven by hydrogen/air fuel cells (Figure 16.26) are under development, in which the cell is of the **proton-exchange membrane** (PEM) type. A schematic drawing of the layout is shown in Figure 16.27. The design is notable for its novel electrolyte. The essential task of the electrolyte in a cell is to be a conveyor of ions, either cations passing from anode to cathode, or anions doing the reverse, or both. In the PEM hydrogen fuel cell, there is only a single electrolyte, and no separator. But most remarkably, the electrolyte is a *solid* (membrane), made from a substituted version of Teflon (polytetrafluoroethene, PTFE). The substituent groups are all potential proton-carrying sites, so the electrolyte can act not so much as a conventional fluid ion-carrier, but more like a conveyor belt for the passage of protons across the cell (Figure 16.28). This electrolyte material is the membrane that gives the cell its name.

The fact that protons do have to pass across the cell can be seen from the electrode reactions.

Anode reaction:

$$2H_2(g) \quad \rightarrow \quad 4H^+ \qquad + 4e^-$$

(to cathode region via electrolyte)   (to cathode region via external circuit)

Cathode reaction:

$$O_2(g) \quad + \quad 4H^+ \qquad + 4e^- \quad \rightarrow 2H_2O(l)$$

(from anode region via electrolyte)   (from anode region via external circuit)

A stripped-down fuel cell is shown in Figure 16.29.

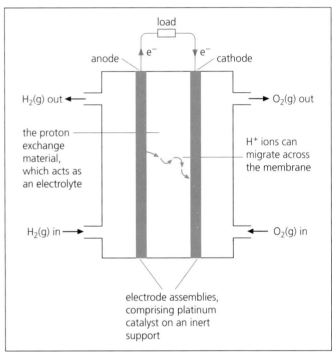

Figure 16.27 Schematic drawing of a proton-exchange membrane fuel cell

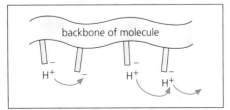

Figure 16.28 Showing how protons can 'hop' from site to site along the polymer chain

Figure 16.29 The layers of a stripped-down fuel cell, with gas compartments, electrodes and membranes visible

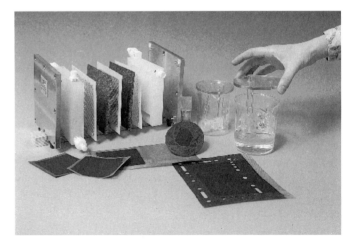

Question

27a Suggest an advantage of having a solid electrolyte in a fuel cell designed for use in vehicles.

b The separator in ordinary cells is needed to keep reactants (like Cu^{2+} ions and zinc) apart. Why is there no need for a separator in a hydrogen/oxygen fuel cell?

c Although the electrolyte is a solid, it needs to be kept just a little moist, or else the protons are unable to pass freely from one site to another. Can you suggest how the presence of a few water molecules would assist proton conductivity?

d Once the cell is running, there should be no problem keeping the electrolyte moist. Can you see why?

e If a gas is used to power a vehicle, there will always be the problem of space. Gases, naturally, take up rather a lot of space. Some designers have proposed that the hydrogen be stored in a solid compound, from which it could be liberated by just adding water. There is a selection of likely candidates for the job of 'solid hydrogen store'. They are capable of a reaction in which the water molecule acts as an acid and the 'solid hydrogen' compound acts as a base, with the result that hydrogen gas is formed. Can you suggest a possible identity for such a compound?

16.9 Microelectrochemical cells – the story of corrosion

The familiar and depressing phenomenon of the corrosion of metals in damp or wet environments is caused by an electrochemical process (Figure 16.30). Corrosion occurs when conditions allow for the setting up of regions on a metal surface

Figure 16.30 Red rust on a car

which can perform the functions of anode and cathode of an electrochemical cell.

This may sound strange since it is by no means obvious why a process like rusting needs to be electrical at all. Why can't the water, oxygen and iron just *meet* and get on with transferring their electrons in a direct way, as with 'normal' redox reactions? After all, one of the first ideas in the chapter was that you only get electrochemistry from redox systems 'frustrated' by their *in*ability to meet. For another thing, there is the idea that bits of *the same metal surface* can act as cathode and anode to each other – surely the minimum requirement for a metal-based cell is that there be *two* metals?

Let us look closely at these points. The one about 'why can't it be direct electron transfer, like "ordinary" redox changes?', can be answered first. So-called 'ordinary' redox changes *are* often electrochemical, as long as you redefine electrochemical on a 'micro' scale. We need to stop thinking about wires and external circuits, and redefine the word electrochemical to describe any process where there is electron flow from an anode to a cathode, even if the flow is between micro-regions of the reactants themselves.

Consider the reaction between copper metal and silver ions. If copper is just dropped into silver nitrate solution, it is logical that the first electron transfer will be of the direct-contact type (Figure 16.31a).

Indeed, a layer of silver atoms might well begin to coat the copper, but if that process were carried to its logical conclusion the reaction would stop itself there and then, since no more copper atoms could get out into the solution as ions

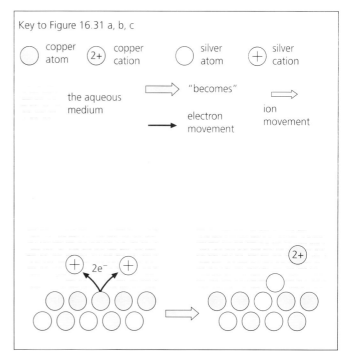

Figure 16.31a A direct electron transfer from Cu(s) to $2Ag^+$(aq). This event might occur in the early stages of the reaction

(Figure 16.31b). Besides, as the copper surface regions shrink in area, it is harder for the silver ions to find sites at which to receive electrons, and so the direct-contact electron-exchange process slows down.

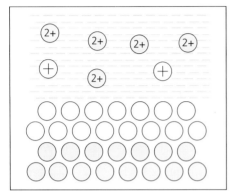

Figure 16.31b Showing how, if the process of Figure 16.31a were to continue, the reaction would stop itself, because Ag⁺ ions couldn't meet Cu atoms, and the latter couldn't escape into solution

Instead another process begins to be favoured. These are metals after all, with excellent electrical conductivity, so it is possible for a current of electrons to flow from the remaining copper surface regions to the silver surface regions, and carry out the reduction of silver ions there. The silver crystals therefore grow on their own kind, outwards from the surface instead of along it. This process ensures that certain regions of copper atoms never get 'silver-coated in', and so maintain access to the solution. These develop into 'pits' as the atoms turn to ions and migrate away (Figure 16.31c). Evidence that this out-growth of silver

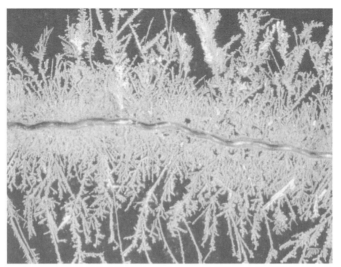

Figure 16.32
Dendritic growth of silver metal crystals, extending from the surface of a piece of copper metal immersed in a solution of silver nitrate

does indeed happen can be got, spectacularly, from looking at this reaction happening under a microscope. Figure 16.32 shows the beautiful tree-like growth of silver metal crystals on a piece of copper dropped in silver nitrate solution.

How has this copper/silver story helped us towards an understanding of iron rusting? The behaviour of the copper/silver system provides an example of a **microelectrochemical process**. There are no wires or voltmeters, but there *is* an electric current, there *is* an ion current in an electrolyte (because the loss of silver ions around the silver 'trees' would draw copper ions into these regions of the solution) and there *are* regions on the metals' surfaces which are doing the characteristic acts of anode (electron donation) and cathode (electron reception). These three features are the characteristic structural units of electrochemical cells, and to emphasise the

point, the parallels between micro- and macroelectrochemical set-ups are shown in Figure 16.33.

The copper/silver case has also given us an example of a logical reason why areas on the same metal surface might take on different roles – the reason being that the reaction would otherwise grind to a halt. So, in the case of rusting, it shouldn't be too big a jump to imagine that there will be micro-regions on a steel surface where the iron atoms are giving up electrons and 'escaping' into solution (anodic regions), and then other micro-regions to which the electron currents flow, where a reduction reaction goes on (cathodes). However, in this case it is not the reduction of another metal ion, but of dissolved oxygen.

Anode reaction:

$$Fe(s) \rightarrow Fe^{2+}(aq) + 2e^-$$

Cathode reaction:

$$O_2(aq) + 2H_2O(l) + 4e^- \rightarrow 4OH^-(aq)$$

Overall reaction:

$$2Fe(s) + O_2(aq) + 2H_2O(l) \rightarrow 2Fe(OH)_2(s)$$

The silver/copper system was presented as a fairly easy-to-understand model for electrolytic corrosion, so as to make the more complex business of rusting more accessible. However, the two systems are by no means a perfect match, and you may be able to spot three differences between them. First, the reduction product in rusting (hydroxide ion) actually forms a compound with the oxidation

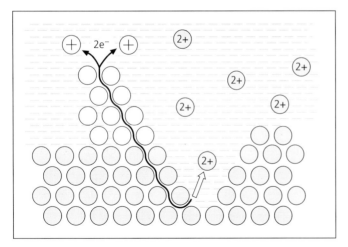

Figure 16.31c
Showing how the reaction can proceed. Copper ions are released at one region and the electrons can travel to a 'spike' (or dendrite) of silver atoms. There they reduce more silver ions from the solution, and the spike grows

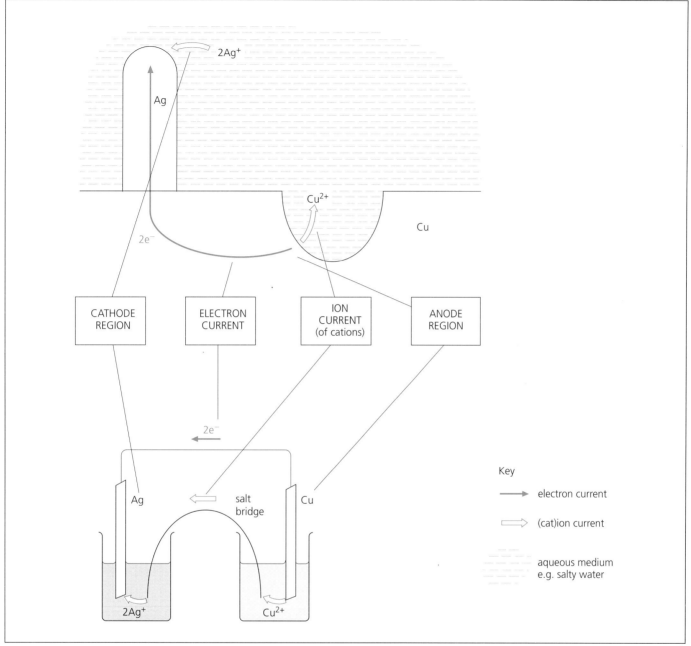

Figure 16.33 Showing that there is equivalence between micro- and macroelectrochemical cells, and where the various bits of the process occur in each system

product (Fe^{2+} ion), and the resulting $Fe(OH)_2$ precipitate accumulates *between* the anode and cathode regions. Second, this initial product undergoes further reactions leading to the actual stuff we call 'rust', which is a complex hydrated oxide of iron in oxidation state +3. A simplified version of this change is:

$$4Fe(OH)_2(s) + O_2(aq) \rightarrow$$
$$2Fe_2O_3(s) + 4H_2O(l)$$

The third difference deserves a heading of its own. It concerns the reasons why any one region takes up the anode or cathode role. In the copper/silver case it was down to random chance (at least in our simple model – although doubtless in reality it involves factors like grain boundaries). But in cases like rusting, where the oxidising agent is dioxygen, the siting of anodes and cathodes can sometimes have a more specific cause.

Differential aeration

This cause is a simple one: if one region of the sample has easier access to oxygen than another, perhaps by being nearer the surface of the aqueous medium, then it will set itself up as a cathode, while other less oxygen-rich regions will have to take the anodic role. That this effect can happen is demonstrated by a contrived **differential aeration cell** on the laboratory bench (Figure 16.34). This is constructed to the blueprint of the standard cells from earlier in the chapter, but *both* electrodes are iron and both electrolytes are brine (salty water). The only difference is that one side is having oxygen blown into it, and so becomes the cathode – the site of oxygen reduction. Figure 16.34 shows the equivalences between the bits of the bench apparatus and the microelectrochemical set-up on a single piece of iron.

Bimetallic corrosion

Right at the start of this section on corrosion it was suggested that two ideas sound strange: the idea of electrochemistry happening on a micro-scale without wires, and the idea that it could all happen on one piece of metal. Hopefully both those paradoxes have been resolved by the above account of events occurring on single pieces of iron and copper. But two-metal cells *do* have a part in our story, both as heroes and as villains, as we shall see.

If two metals are touching each other in an aqueous oxygenated environment then it doesn't even need differential aeration to decide which regions will become anodic and cathodic – the more reactive metal will automatically take the anode (metal-dissolves-as-ion) role. So *it* will be the one to corrode, preferentially, while its electrons will pass as a micro-current to the region occupied by the less reactive metal, where the oxygen reduction (cathodic) reaction will happen. There is thus a double guarantee that the less reactive metal will *not* corrode: first because the more reactive metal is doing all the corroding, and second because even if the less reactive metal were 'tempted' to give up some electrons, it would find it hard to do so in a region already flooded with electrons flowing in from the anode region. Figure 16.35 summarises the ideas in this paragraph.

The last-but-one sentence in the above paragraph reveals how bimetallic corrosion cells have within them the start of an idea for an *anti-corrosion* system. If the metal to be protected is intentionally put into contact with a second metal which can be guaranteed to play the anode and thereby sacrifice itself to corrosion, then

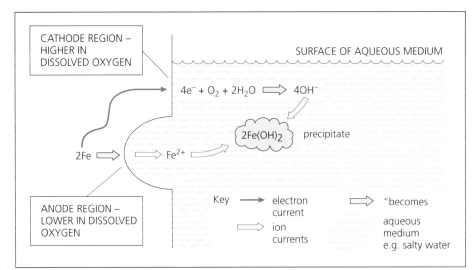

Figure 16.34a A microelectrochemical corrosion cell. The decisions on the location of the cathode and anode regions are settled by a single factor: *access to oxygen*. The *surface* regions of the aqueous medium are naturally richer in dissolved oxygen

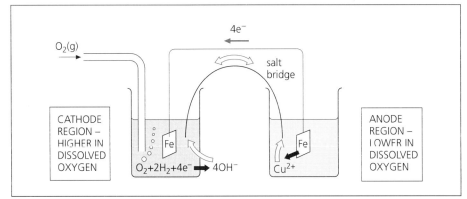

Figure 16.34b An equivalent macroelectrochemical corrosion cell. In this case the differential aeriation factor is contrived by actually blowing oxygen through one half of the cell

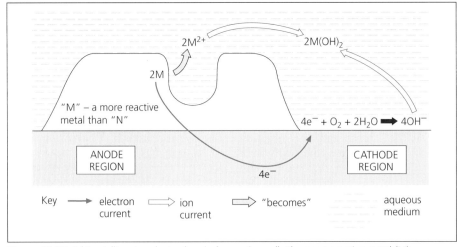

Figure 16.35 A bimetallic microelectrochemical corrosion cell. The more reactive metal (M) automatically becomes the anode, and sends electrons to the less reactive metal (N). In effect, M 'sacrifices' itself and N is thus protected from corrosion

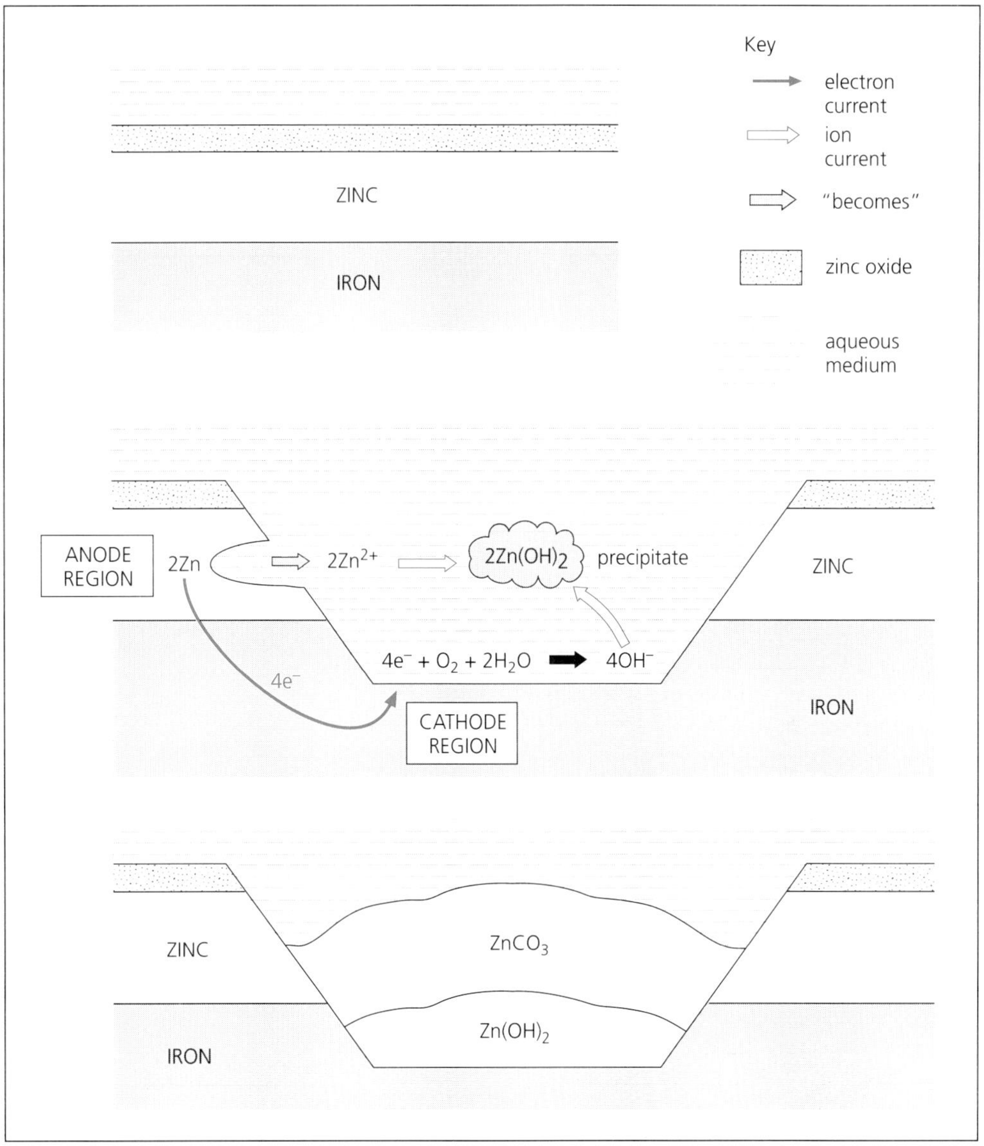

Figure 16.36a
Undamaged galvanised steel.
Corrosion prevented by a layer of zinc oxide

Figure 16.36b
A scratch in the zinc coating exposes the iron, but the zinc acts as a sacrificial anode to protect the iron

Figure 16.36c
The scratch fills up with insoluble zinc hydroxide, which reacts with CO_2 to form $ZnCO_3$. This non-porous layer restores full protection to the system

its sacrifice will ensure the survival of the first metal. The most striking and widespread application of this concept involves coating zinc on to steel, the process known as 'galvanisation'.

Zinc-coated steels can deploy three distinguishable anti-corrosion mechanisms. To start with zinc is one of those metals, like aluminium, which is reactive enough to undergo dry corrosion in air. This really does seem to be a uniform coating process, with no separate regions, and quickly produces a very thin uniform coat of zinc oxide all over the zinc. But again like aluminium, and unlike iron, the oxide

film is very impervious to air and water, so corrosion ceases quickly at the zinc surface. Meanwhile the iron underneath is, of course, completely protected.

The second protection mechanism is activated if the zinc surface is mechanically ruptured, as say by a scratch. This method *is* electrochemical and is exactly the process we discussed above – namely that of the zinc (in the presence of water) acting as a 'sacrificial anode', and pumping electrons towards the exposed steel in the crack. You might expect the zinc coating now slowly to wear away, as it sacrifices itself, but there is even a third

device waiting to come into play. Zinc hydroxide, which forms as the product of zinc corrosion, deposits itself in the crack and reacts with carbon dioxide to form zinc carbonate. This is another of those zinc compounds which has low porosity to air and water, so the crack re-seals itself. The three mechanisms are shown at work in Figure 16.36. The advantage of using zinc-coated steel for car body panels is graphically illustrated in Figure 16.37 (overleaf). Further testimony can be found in the *extent* to which zinc-coated panels are used in modern car manufacture (Figure 16.38).

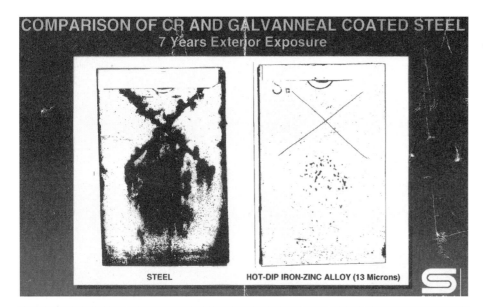

COMPARISON OF CR AND GALVANNEAL COATED STEEL
7 Years Exterior Exposure

STEEL   HOT-DIP IRON-ZINC ALLOY (13 Microns)

Figure 16.37 Painted zinc-coated steel and steel panels after exposure to a corrosive environment

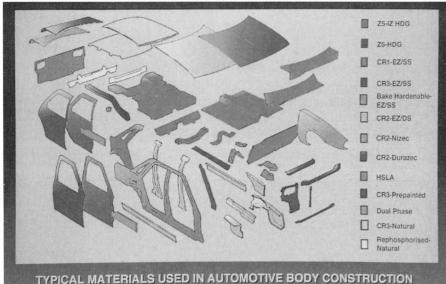

- Z5-IZ HDG
- Z5-HDG
- CR1-EZ/SS
- CR3-EZ/SS
- Bake Hardenable-EZ/SS
- CR2-EZ/DS
- CR2-Nizec
- CR2-Durazec
- HSLA
- CR3-Prepainted
- Dual Phase
- CR3-Natural
- Rephosphorised-Natural

TYPICAL MATERIALS USED IN AUTOMOTIVE BODY CONSTRUCTION

Figure 16.38 (left) Typical coated steel content of a car and (above) Painted coated steel panel for a car

We have now completed the basic story of electrolytic corrosion. The following questions are about various applications of the principles we have established.

Questions

28 One way not yet mentioned of preventing corrosion is to impose a D.C. potential difference between the component being protected and a sacrificial anode. The precious component is always made the cathode of the 'electrolysis' cell, and the arrangement is shown in Figure 16.39 in the specific context of an oil pipeline.

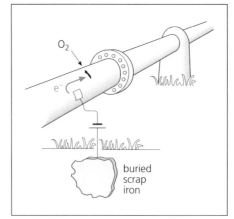

O_2

e^-

buried scrap iron

Figure 16.39 Showing how an imposed voltage can push electrons towards any crack in the paintwork and force the crack to act as a cathode and so not corrode

a Why does making the pipeline negative relative to the 'earthed' lump of iron ensure its protection?

b Pipelines are usually painted too, so the protection current is zero so long as the paint barrier holds. Why would there be no current under these circumstances?

29a The 'tins' used for containing foodstuffs are actually tin-coated steel, called 'tinplate' (Figure 16.40). Can you suggest a reason why tin coatings are preferred to zinc in the environment inside a tin of, say, canned fruit?

Figure 16.40 Various tinplated steel objects

b The market for drinks cans is the scene of stiff competition between the aluminium and the tinplate industries, with the market share being about 50:50 at present. The steel people claim their product has the advantages of being easier to sort from other household rubbish and being easier to recycle. Can you suggest what is the basis for this claim?

30 A famous and very simple laboratory experiment in corrosion chemistry is to leave a blob of salty water standing on a flat steel surface (Figure 16.41), and then predict and observe exactly where corrosion will appear, within the volume of the blob. What would *your* prediction be?

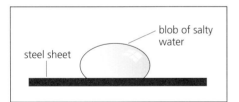

Figure 16.41 Where will corrosion occur? See question 30

31 Modern cars, as we have seen, employ formidable anti-corrosion systems. In addition to using coated steels, manufacturers realised the importance of avoiding mudtraps under wheel arches, and so blanked off the insides of such panels. Older cars of the Cortina/Marina generation suffer badly from rust inside wheel arches and notably up around the headlights (Figure 16.42). Can you apply one of our principles of electrolytic corrosion to suggest why

Figure 16.42 Some of the worst (visible) rust on older cars occurs in places where mud becomes trapped (Scanning electron micrograph of rust on a Cortina)

the precise location of the worst corrosion is where the mud is likely to be thickest (in the deepest nooks and crannies)? Try to take your answer beyond the simple level of saying 'that's where it's wettest'.

32 Car owners know that the conditions which are most corrosive are when the roads are salted. Yet it is not immediately obvious why salt matters. After all it doesn't take part in the stoichiometric equation for rusting. Without trying to achieve a deep understanding of the role of salt can you identify the part of the electrolytic corrosion cycle which must somehow be enhanced by salt?

33 The final shootout in the film *French Connection II* takes place deep in a dry dock in Marseilles. Gene Hackman finds himself forced to dive for cover behind piles of large ingots of zinc, lying around on the floor of the dry dock. Can you suggest what part might be played by ingots of zinc in the refitting of a large ship (Figure 16.43)?

34 Corrosion occurs much faster in acids, of course. The main reason is the

Figure 16.43 Zinc sacrificial anodes on a ship's rudder

existence of a much better alternative cathode reaction to replace the rather slow dioxygen reduction. What do you suppose it is?

35 The Bristol car company was set up just after World War II, as a branch of the aircraft firm based at Filton near Bristol. They made a few cars per year to very high standards of engineering (Figure 16.44). As befits an aircraft firm, they made use of aluminium for body panels, mounting the body on a steel chassis. The one blot on the firm's reputation from those days is that their cars were prone to corrosion. Can you suggest the cause of corrosion and which bit or bits of the car would have suffered?

Figure 16.44 A Bristol 401 car

16.10 Electrolysis – an introduction

Driving an increase in free energy

In the electrochemical cells we have been studying till now in the chapter, a chemical system is allowed to run to a state of lower free energy, and the difference in energy between reactants and products is made available as electrical energy. Electrolysis is the exact opposite of this. It is the conversion of low-energy reactants to products of a higher free energy, driven by an input of energy from an outside electrical source (look back to Figure 16.1b, p. 390).

Question

36 We have already seen one example of electrolysis in the chapter so far. What is it?

Electrolysis is important in chemistry because it can bring about reactions that are uphill in free energy, ΔG_{sys}. For example, with electrolysis you can prepare hydrogen and fluorine from hydrogen fluoride, and aluminium and oxygen from aluminium oxide, in the exact reversal of what Nature would do with the same atoms.

Electrolysis seems therefore to be a way of violating the second law of thermodynamics. That is the law that says that the only thermodynamically feasible processes are those in which ΔS_{uni} is positive and ΔG_{sys} is negative. To that extent, electrolysis is reminiscent of photosynthesis. In photosynthesis it seems that a reaction has occurred in which there has been a clear decrease in entropy (gaseous carbon dioxide and water forming carbohydrate chains), with corresponding increase in free energy ('burnt-out' carbon dioxide and water forming 'fuel-worthy' starches and sugars). This anomaly is solved when we remember that the all-important photons were generated on the Sun by a nuclear reaction whose downhill free energy more than compensates for the little local increase here on Earth.

Question

37 By comparison with the photosynthesis example, why can we say that even a reaction like:

$$2Al_2O_3(s) \rightarrow 4Al(s) + 3O_2(g) \quad \Delta G^{\ominus} = +3164.8 \text{ kJ mol}^{-1}$$

when brought about by electrolysis, is not breaking the second law of thermodynamics? (In other words, what is playing the part of the Sun?)

The mechanism of electrolysis

As in an electrochemical cell, in an electrolysis cell there is an anode and a cathode, each the site of a redox reaction, and there is a liquid containing ions called an electrolyte. The only difference is that instead of the electrons flowing the way Nature intended, this time they are being forced by an external potential difference to go the opposite way. Take as an example the electrolysis of molten sodium iodide. This is a relatively simple case because the absence of water removes complications involving $H^+(aq)$ and $OH^-(aq)$ ions. The ionic population comprises just $Na^+(l)$ and $I^-(l)$ ions (Figure 16.45).

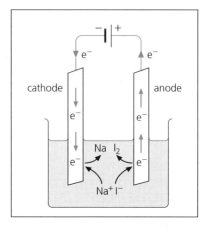

Figure 16.45 Electrolysis of molten sodium iodide between graphite electrodes. The design is very crude and has no means of keeping the products from recombining

At the cathode there is no confusion with names this time, because not only is the cathode the electrode to which electrons flow, it is also linked to the negative terminal of the supply:

$$Na^+(l) + e^- \rightarrow Na(l)$$

At the anode:

$$I^-(l) \rightarrow \tfrac{1}{2}I_2(g) + e^-$$

The overall reaction is:

$$NaI(l) \rightarrow Na(l) + \tfrac{1}{2}I_2(g)$$

The compound has been broken up into its constituent elements. This incidentally is the origin of the word 'electrolysis', which is derived from Greek and roughly translates as 'breakage with electrons'.

Questions

38 Complete this sentence:

For electrolysis to occur, and for the electrolysis current to be sustained, it is necessary that the cathode region contains a species that can approach the electrode and _____ electrons, while the anode region must contain a species that can approach the electrode and _____ electrons.

39a You will already be familiar with the idea of metals plating on to cathodes. Why will that not happen in the case of the electrolysis of molten sodium iodide?

b What would you actually see happening at each electrode, bearing in mind that the cell must be at a temperature of at least 661 °C to ensure the electrolyte is molten?

The energetics of electrolysis

It is easy to estimate the free energy change of the reaction that has been brought about by electrolysis, by looking up the ΔG_f of sodium iodide and reversing its sign. This also gives us an insight into the minimum voltage the electrolysis power source has to deploy:

$$NaI \rightarrow Na + \tfrac{1}{2}I_2 \quad \Delta G_{sys} = +286.1 \text{ kJ mol}^{-1}$$

We know from our work on electrochemical cells that if a cell reaction is *giving out* energy, then the cell voltage is equal to the energy obtained per coulomb-worth of electrons (formula (*16.1*)):

$$\text{Potential difference (volts)} = \frac{\text{energy (joules)}}{\text{charge(coulombs)}}$$

Now we are in the 'looking-glass' world of electrolysis, the formula is the same but the roles (and some of the signs) are reversed. The coulombs are now being pumped *into* the system to raise its energy. So the energy pumped in per coulomb-worth of electrons is equal to the *input* voltage of the supply. As we shall now see, this will only bring about an electrolytic change if the electrons are energetic enough to force themselves where natural chemistry does not want them to go.

The question is, how can we look at a figure for ΔG_{sys} like the one above of $286.1 \text{ kJ mol}^{-1}$, and work out the voltage that will make the reaction go? The following calculation shows you how.

Example

To decompose one mole of sodium iodide we would have to pump one mole of electrons (that is $96\,500\,C$) from the I^- ions back to the Na^+ ions, at an energy cost of $286.1\,kJ$ (the value of ΔG_{sys}).

Therefore one coulomb of electrons would need:

$$286.1\,kJ/96\,500\,C \approx 3\,J\,C^{-1}$$

But 'joules per coulomb' is volts, so we can conclude that it would require a supply of at least $3\,V$ to make the electrolysis happen.

If you attached, say, a 1V supply to the molten sodium iodide, then you could possibly interest the Na^+ ions in drifting towards the negative electrode, but having attracted them you would fail to make them take the vital step of accepting an electron and returning to sodium metal. In general, then, the 'looking-glass' version of formula (*16.2*) (p. 395) applies, where E is now the minimum e.m.f. across the terminals of the electrolytic cell necessary to force the reaction to happen:

$$E = \frac{\Delta G_{sys}}{zF} \quad (16.4)$$

(Notice that compared with formula (*16.2*), the minus sign has disappeared, because we are now talking about an *input* e.m.f. and a *positive* ΔG_{sys}.)

40 We said earlier that $+286.1\,kJ\,mol^{-1}$ was no more than an *approximation* to the true free energy of the change brought about by the electrolysis of sodium iodide. However, $-286.1\,kJ\,mol^{-1}$ is the exact value from a data book of the free energy change of formation of sodium iodide. The source of the discrepancy lies in the fact that electrolysis demands a liquid electrolyte.

a Why can electrolysis not happen with a solid electrolyte?

b Why does the molten condition of the electrolyte render the value $+286.1\,kJ\,mol^{-1}$ an approximation?

16.11 Electrolysis of aqueous solutions

If you offer electrons to a chemical system, the change that occurs will generally be the easiest one – the one with the minimum uphill free energy path. In the case of sodium iodide, there was only one redox change possible, namely the reduction of Na^+ ions and the oxidation of I^- ions. However, there are two ways of preparing an electrolyte for electrolysis, and melting it is only one of them.

The other, of course, is dissolving the electrolyte in water. Although this is often an easier option than melting it, it does introduce a degree of complication, because of the ions present (albeit at a very low concentration) in water itself:

$$H_2O(l) + H_2O(l) \rightleftharpoons H_3O^+(aq) + OH^-(aq) \quad (1)$$

Sometimes these ions offer a reaction of lower free energy than those of the solute, and can therefore frustrate efforts to electrolyse the solute. Take, for example, the effect of shifting the sodium iodide electrolysis from a molten to an aqueous environment. The electrons going to the cathode are now offered two possibilities:

$$Na^+(aq) + e^- \rightarrow Na(s) \quad (2)$$

and:

$$H_3O^+(aq) + e^- \rightarrow H_2O(l) + \tfrac{1}{2}H_2(g) \quad (3)$$

You might object to (*3*) as an option, given that the concentration of free ions in water is so low, but that would be to forget the way equilibria operate.

41 What would be the effect of the electrolytic reduction reaction (*3*) on the equilibrium (*1*), and why would there be no problem in keeping (*3*) going? Give the overall reaction that would result from (*1*) (forward) followed by (*3*).

It is immediately obvious from observing the aqueous electrolysis of sodium iodide that equation (*3*) is indeed the 'winner', and equation (*2*) is the 'loser', as the cathode is seen to evolve hydrogen gas. This should have been no surprise to us given our knowledge of the electrochemical series, as question 42 shows.

Question

42 The standard electrode potential of $H^+(aq) \mid \frac{1}{2}H_2(g)$ is 0.00 V and that of $Na^+(aq) \mid Na(s)$ is -2.71 V.

a From this evidence, which of the two cations is the more willing to accept electrons?

b Which one is the more likely to be reduced in a cathode reaction?

Even if, by some overturning of the natural order of things, sodium metal *had* been made, Nature would have instantly reminded us of its unwillingness to exist while there is anything else around to accept its electron.

Question

43 What would have happened to sodium metal if it had been made at the cathode?

Meanwhile, at the anode, the choice of reaction lies between the oxidation of I^- ions to iodine, and the oxidation of OH^- ions to oxygen:

$$I^-(aq) \rightarrow \tfrac{1}{2}I_2(aq) + e^-$$

and:

$$2OH^-(aq) \rightarrow H_2O(l) + \tfrac{1}{2}O_2(g) + 2e^-$$

Question

44 By considering the positions of these two half-equations (or rather their reverse versions) in the electrochemical series (Table 16.2, p. 397), or else by your general knowledge of the 'electron-grabbing' powers of the elements oxygen and iodine, predict which one of these half-equations is more likely to occur.

So we can see that the introduction of water to the electrolysis of sodium iodide has left the anode process just as it was in the molten state, but has altered the cathode reaction to make hydrogen instead of sodium.

Question

45 Combine your answers for the cathode and anode reactions to show the overall equation for the electrolysis of aqueous sodium iodide.

16.12 An industrial use of aqueous electrolysis – the chlor-alkali industry

Background to the chlor-alkali industry

We saw in Section 16.11 that the electrolysis of aqueous sodium iodide resulted in the overall reaction:

$$2NaI(aq) + 2H_2O(l) \rightarrow 2NaOH(aq) + H_2(g) + I_2(aq)$$

If you substitute sodium chloride for sodium iodide, you have a very valuable industrial chemical process:

$$2NaCl(aq) + 2H_2O(l) \rightarrow 2NaOH(aq) + H_2(g) + Cl_2(g) \quad (4)$$

You have a way of making sodium hydroxide or 'caustic soda' – a key reagent in many processes, notably the manufacture of soaps and paper – using no other starting materials beyond common salt and water (and of course a source of electrical energy). Also useful, and too valuable in their own right to be called mere by-products, are the gases hydrogen and chlorine. Hydrogen is used in the making of ammonia, in the edible fats industry to saturate double bonds, and may, as we have seen, find important future uses in fuel cells, while chlorine is a staple ingredient in the manufacture of PVC, solvents, anaesthetics and water treatment products.

The industry that has grown up around the exploitation of equation (4) is usually called the **chlor-alkali industry**, in recognition of the equality of importance of the two products. Since both chemicals are made industrially almost exclusively by using equation (4), there are close ties between the production of chlorine and that of sodium hydroxide. Figure 16.46 shows how closely the production of the two chemicals have shadowed each other.

Before we go on to consider the finer points of the electrolysis of brine (aqueous salt), let us look at the economics of the process.

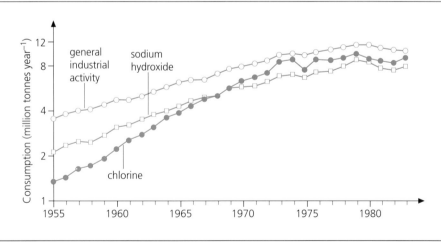

Figure 16.46 Graph showing consumption of sodium hydroxide and chlorine in Western Europe, together with general industrial activity

Question

Figure 16.47 Victorian Widnes

46a The centre of the chlor-alkali industry is in the English county of Cheshire, at Runcorn and Widnes (Figure 16.47). Why do you think it might have grown up there? (*Clue*: Find out about the physical geography of the area.)

b A close look at Figure 16.46 indicates that since the late 1960s the demand for chlorine has outstripped that for sodium hydroxide. Can you suggest what might have caused this reversal? (*Clue*: Think of the industries that use the chemicals.)

c What effect would an increased demand for chlorine have on the world price of sodium hydroxide if, as has been suggested, you cannot make one without the other?

Practical chlor-alkali cells

The central drawback with many electrolytic processes is this – you have just driven a system to a place where, thermodynamically speaking, it did not 'want' to go, so unless you are careful the products of the electrolysis will react together and undo all your good work. In equation (*4*), there are two pitfalls of this kind.

1 The hydrogen and chlorine would recombine explosively if they met. This can be avoided by surrounding the electrodes by hoods through which the gases can be extracted.

2 As the chlorine bubbles off through the solution, it will tend to react with the newly formed sodium hydroxide. This is harder to avoid, and the problem has provoked three interesting technological solutions, described below.

Question

47 (*Revision*) How would the chlorine react with the sodium hydroxide?

The diaphragm cell

In equation (*4*), chlorine is made at the anode, while OH⁻ ions come into being at the cathode with the removal of H⁺ ions from water. (Compare the anode and cathode reactions of the sodium iodide cell.) So the idea arose of restricting the flow of solution between anode and cathode regions, by putting them in separate compartments. The barrier between compartments has to be electrolytically conducting, but able to inhibit free flow. In the diaphragm cell (Figure 16.48), an asbestos diaphragm is used.

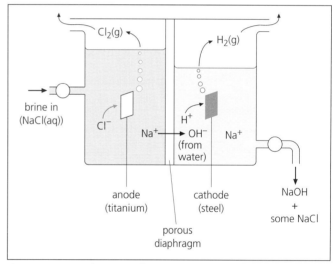

Figure 16.48 Diaphragm cell for the production of sodium hydroxide

A refinement was brought to the cell. This stimulates a bulk flow that takes OH⁻ ions in a direction away from the anode region, away from the chlorine. It is done by a small pressure difference created by having a constant head of liquid in the anode compartment. Maintaining a constant head on one side of a porous barrier requires a steady inflow of electrolyte, but this is not a problem. The whole cell is operated as a continuous-flow process, with fresh brine flowing into the anode compartment, the Cl⁻ ions being 'extracted' (as chlorine gas) there, and the product sodium hydroxide being removed from the cathode region.

Question

48a Not all chloride ions get removed at the anode. Some survive and seep through the diaphragm into the cathode region. The cathode fluid is thus a mixture of

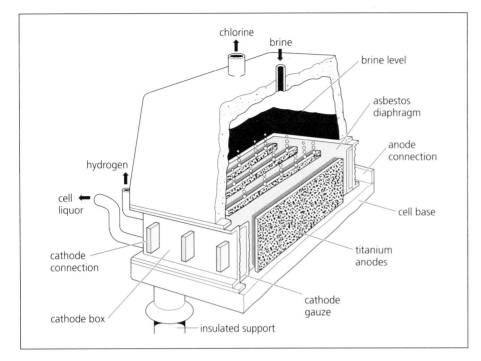

Figure 16.49 Typical diaphragm cell. The wire mesh supports the asbestos diaphragm, and the vertical anodes and cathodes are interleaved

diaphragm is like (to use an agricultural metaphor) a gap in the hedge between two fields, then the membrane is like a gate operated by someone who can tell a sheep from a goat. Question 49 should help you to understand what happens.

Question

49a Which ion would you have to exclude from the cathode compartment, if you wanted pure aqueous sodium hydroxide as a direct output from the cell?

b Which ion exchange (what for what) would the membrane be asked to carry out?

c The purity of the liquid in the cathode compartment is also influenced by what is in there at the start of the electrolysis. Admittedly the raw material of the whole process is brine, but if you begin with brine around the cathode, then the good purification work of the membrane is wasted. What liquid would you suggest having in the cathode compartment at the start?

sodium chloride and sodium hydroxide. Look up their relative solubilities in a data book, and suggest a way of purifying the sodium hydroxide.

b Why does this mixed product reduce the economic viability of the diaphragm process?

c Figure 16.49 shows the rather complicated shape taken by real diaphragm cells. Even if you cannot understand the complete diagram, can you suggest why it is an advantage to have three or four anode and cathode plates interlocking like two combs?

d The removal of Cl⁻ ions in the anode compartment would create an ionic imbalance in that region, with a cation majority. Similarly the removal of H⁺ ions at the cathode threatens to make an anion majority on that side. Neither of these situations would be energetically allowed. The problem is resolved by ions of one particular type moving through the diaphragm faster than the background flow rate. Which ions are they, and how does their movement solve both problems?

The membrane cell

The **membrane cell** features a small change in design relative to the diaphragm cell, but a large increase in effectiveness. The diaphragm is replaced by an **ion-exchange membrane**, apart from which the schematic arrangement is the same as in Figure 16.48. The end result is to remove the need for purification at the end of the process. Pure aqueous sodium hydroxide flows from the cathode compartment, and merely has to be concentrated by evaporation. How does it work? The secret lies in the action of the membrane, which acts more selectively than the diaphragm. If the

The mercury cell

The **mercury cell** (Figure 16.50) relies on a single improbable truth. When the cathode is mercury, the lower-energy

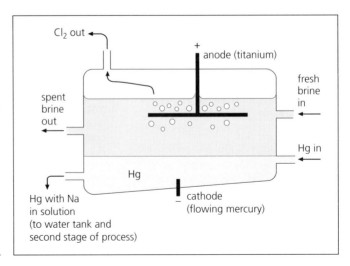

Figure 16.50 Mercury cell for the first stage of sodium hydroxide production. In the second stage, the mercury/sodium mixture simply flows through water, which becomes aqueous sodium hydroxide as a result

option for a cathode reaction ceases to be the reduction of hydrogen, and becomes instead:

$$Na^+(aq) + e^- \rightarrow Na \text{ (in Hg solution)}$$

This is the ultimate way of keeping the chlorine away from the sodium hydroxide, since no sodium hydroxide has been made yet. The mercury (conveniently a liquid) flows out of the cell into a second vessel, where the sodium solute reacts quietly to produce the product:

$$Na(Hg) + H_2O(l) \rightarrow NaOH(aq) + \tfrac{1}{2}H_2(g)$$

Question

50a What would be a sensible thing to do with the mercury at the end of the second stage, and why?

b If the anode reaction is the same as in the other cells, what is the overall equation for the first, electrolytic, part of the process?

c Why would this method produce sodium hydroxide that is free of contamination by salt?

d What extra problems might the mercury cell introduce, which would not be present in the other two designs?

16.13 Electrolysis in the fluorine industry

This application of electrolysis features a chemical change that could not have happened in any other way. We have already met some *unwilling* conversions achievable by electrolysis – chloride ions are not exactly eager to revert to chlorine, for instance. But the chemical oxidation of chloride is not impossible – it can be done by chemical means if necessary. In fact, fluorine is one reagent that will do it. The reaction:

$$F_2(g) + 2Cl^-(aq) \rightarrow Cl_2(g) + 2F^-(aq)$$

is perfectly viable, as confirmed by the electrochemical series (Table 16.2, p. 397) and the Law of anti-clockwise circles. However, there is no chemical on Earth that could provide the other half-equation to drive this one backwards:

$$F_2(g) + 2e^- \rightleftharpoons 2F^-(aq)$$

Question

51 Explain, by reference to fluorine's position in the electrochemical series, why it is impossible to make fluorine chemically.

The great power of electrolysis is that it can make *any* redox change happen. No matter how dauntingly positive the ΔG_{sys} of the proposed reaction, there will always be a voltage that will match it, according to the equation:

$$E = \frac{\Delta G_{sys}}{zF} \qquad (16.4)$$

In the fluorine cell, the electrolyte is a mixture of potassium fluoride and hydrogen fluoride but, not surprisingly, the preferred reaction pathway ignores the potassium ions, and:

$$2HF \rightarrow H_2(g) + F_2(g)$$

is the overall change. The cell in which it is carried out is shown in Figure 16.51.

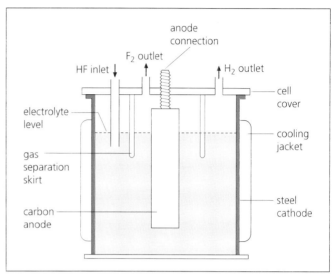

Figure 16.51 Fluorine cell

Question

52a Why is it a good idea to have the steel cathode doubling up as the cell container?

b What factor or factors might have influenced the choice of carbon as the anode?

c Look up the free energy change of formation of hydrogen fluoride, and by suitable sign reversals, and with the help of equation (*16.4*), estimate the size of the minimum cell voltage that would be needed to produce a reaction.

d The actual voltage used to run these cells at ICI is in the region of 8–12 V, considerably more than that calculated in (c). Can you think what advantages this would give?

e Why do you think the cathode reaction was not the reduction of K^+ ions?

f What problems might have occurred had the electrolyte been pure hydrogen fluoride rather than the hydrogen fluoride/potassium fluoride mixture? (*Clue:* Look up the boiling point of hydrogen fluoride.)

g There is another reason for having potassium fluoride in the mixture. The studies of acids in Chapter 12 might have prepared us for the fact that pure hydrogen fluoride is a poor conductor. What is the connection with Chapter 12? Why does the potassium fluoride help?

h Chemical engineers try to avoid wasting heat energy. What could usefully be done with the heat taken out of the cell by the cooling jacket? Why is it necessary to avoid the cell overheating (from its operating temperature of 100 °C)?

i The manufacture of fluorine increased sharply from about 1940, when its major use was in the manufacture of the fluoride of uranium, UF_6. Can you guess what industry was using this compound (and still does)?

16.14 Electrolysis in the extraction and purification of metals

The fact that only copper, silver, gold and platinum are found as elements in the Earth's crust bears testimony to the fact that most metals have a thermodynamic preference for combination with other elements. For metals in the middle of the electrochemical series, it is relatively straightforward to pick a method of reduction from among the chemical possibilities to extract the metal from its ore. Hydrogen and especially carbon are strong enough reducing agents to convert these metal ions to their elemental state, for example:

$$2Fe_2O_3(s) + 3C(s) \rightarrow 3CO_2(g) + 4Fe(l \rightarrow s)$$

and this method is cheaper than the electrolytic alternatives. However, from aluminium upwards in the series, the option of a chemical means of reduction becomes less attractive, and the alternative of electrolysis becomes more favourable.

Question

53 A reaction like:

$$Al_2O_3(s) + 3Mg(s) \rightarrow 3MgO(s) + 2Al(l \rightarrow s)$$

is thermodynamically feasible. Why does it nevertheless make economic sense to prepare aluminium electrolytically?

The extraction of aluminium

Figure 16.52 shows a cell for the preparation of aluminium. Aluminium is the most abundant metal in the Earth's crust, making up about 8% by mass, and there is no shortage of its most usable ore, bauxite. Bauxite is about 60% aluminum(III) oxide, Al_2O_3, with the remaining 40% being iron oxide and silicon oxide. The aluminium oxide ('alumina') must be purified out from the crude ore.

Question

54 What might happen if the alumina were electrolysed without prior purification?

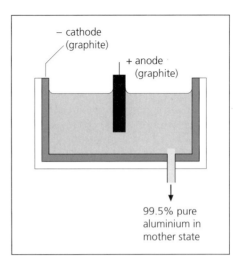

Figure 16.52 Simplified drawing of the cell for preparing aluminium. The electrolyte is an alumina/cryolite (Al_2O_3/Na_3AlF_6) mixture which is molten at about 950 °C. The cell works at a voltage of about 5 V

The actual electrolyte is a molten mixture of alumina and an artificial version of the mineral cryolite, Na_3AlF_6. The addition of cryolite is purely to reduce the melting point of the electrolyte (to around 950 °C).

Question

55a Why does it make economic sense to lower the melting point of the electrolyte?

b Why does the cryolite not interfere with the production of aluminium at the cathode?

c Which of the two anions, O^{2-} or F^-, would be preferentially discharged at the anode? Justify your answer.

d Hence give the overall cell reaction.

e Forty per cent of the world's production of hydrogen fluoride is used in the aluminium industry, purely to make cryolite. Suggest an equation for this process, using alumina and sodium hydroxide as the other reactants.

Aluminium production is generally sited near vigorously flowing rivers. The biggest plant in the world, run by Alcan, is on the Saguenay River in Quebec, Canada, which offers the added attraction of a nearby deep-water port at Port Arthur.

Question

56a Why is it convenient to site aluminium smelters near fast-flowing rivers?

b What two advantages are conferred by the proximity of the deep-water port?

c One of the expenses of running the cell in Figure 16.52 is the need to replace the carbon electrode at regular intervals. Considering what happens at the anode, suggest why the graphite corrodes rapidly.

The purification of copper

There is one anode process we have not yet mentioned in this chapter and that is the dissolving of a metallic anode itself. An anode reaction is one that provides electrons to compensate for those consumed at the cathode. A reaction like:

$$Cu(s) \rightarrow Cu^{2+}(aq) + 2e^- \qquad (1)$$

is therefore perfectly viable as an anode reaction, and may be a better option energetically than the oxidation of an anion from the electrolyte. This type of anode reaction occurs when copper sulphate solution is electrolysed between copper electrodes. The cathode reaction is:

$$Cu^{2+}(aq) + 2e^- \rightarrow Cu(s)$$

The anode reaction is reaction (1) above.

Question

57a What is the overall cell reaction for this process?

b What is the relationship between the loss in mass of the anode and the gain in mass of the cathode?

c How would events in the cell change if the electrodes were made of a substance like graphite, that is not involved in the reactions?

Strangely, such a chemical non-event has a very useful application, not in the direct smelting of copper but in its subsequent purification. The reducing agent used for copper smelting is carbon, as might be expected for a metal low in the electrochemical series. The chief copper ores are sulphides, and the copper sulphides are often found in the same ore bodies (geological seams) as sulphides of silver, gold and platinum. The result is that copper from the smelter is often of less than 90% purity, nowhere near the level of purity required for copper's major use, for electrical wires. So the chemical reduction is followed by an electrolytic purification stage. The idea is to use the impure copper as the anode, and have a slim plate of pure copper as the cathode (Figure 16.53). The electrolyte is a mixed aqueous solution of copper sulphate and sulphuric acid.

Figure 16.53 Copper refining cell in the real industrial apparatus – hundreds of cathodes and anodes are used, in parallel, at once

Question

58a The likely impurities in the anode include the jewellery metals (silver, gold and platinum), all of which are lower than copper in the electrochemical series. Are they more or less likely to undergo the same change as copper at the anode, from metal to cation?

b Are they more likely to go into solution or to drop to the bottom of the cell, as the main copper fabric of the anode dissolves away?

c If more reactive metals go into solution along with the copper, why will they not get discharged on the cathode instead of copper?

d If you were running the copper purification plant, what would you do with the sludge on the floor of the cell?

Anodisation

Another significant process takes place at the anode, which takes advantage of the oxygen released in the course of many aqueous electrolysis experiments. At the moment of liberation, the oxygen appears to be even more reactive than usual. This may be because, although we write the anode reaction as:

$$4OH^-(aq) \rightarrow O_2(g) + 2H_2O(l) + 4e^-$$

it cannot take place in a single step. Logically, only one OH^- ion releases its electron at a time, and the rest of the process must happen in steps. One of these steps may well generate reactive oxygen free radicals. If the anode itself is made of a reactive metal, like aluminium, then quite a thick film of metal oxide can grow on its surface.

As an added bonus, the growing oxide film can incorporate dye molecules, if the latter are included in the electrolyte. So the overall effect of anodisation is to produce an attractive coloured coating on the surface of the anode. The technique finds commercial applications by using everyday objects as the anode. You can see the end product of anodisation in many commercial aluminium products, including window frames (which are often given a white finish), and hi-fi outer cases, for which the fashionable finish is black at present (Figure 16.54).

Figure 16.54 Anodised objects, usually of aluminium, electrolytically coated by its own oxide

Question

59a Not every aqueous electrolyte would be suitable for use in anodisation. The one normally used is sulphuric acid. Suggest an aqueous electrolyte which would not be suitable for anodisation.

b Anodised coatings of aluminium oxide, Al_2O_3, can reach thicknesses of 10^{-5} m. Without attempting any sort of explanation, can you see what is surprising about the fact that electrolysis can proceed with a coating like that in place?

c Anodisation is meant for decoration, but it fulfils a second purpose too. What do you think that might be?

Summary

- Electrochemistry involves the interaction between chemical changes and flows of electrons. It follows therefore that the reactions in this field must be those which involve **electron transfer**, which is to say **redox reactions**.

- The interaction can occur in two directions: chemical reactions can proceed in the natural direction of lower free energy, and thereby generate a flow of electrons (as in **electrochemical cells**), or alternatively a flow of electrons can be used to generate a chemical system of higher free energy (as in **electrolysis**).

- In an electrochemical cell, the reactants that are to exchange electrons are arranged within the apparatus so that they cannot meet. By this means, the electrons cannot be exchanged directly, but instead are delivered via external wires, as an electric current.

- So the apparatus can be seen as housing two separate half-equations, one of which is electron-giving (in which the reactant is oxidised), and the other is electron-taking (in which the reactant is reduced). The electron-giving electrode of the cell is called (surprisingly) the **anode** (despite being the negative terminal), while the electron-receiving electrode is called the **cathode**.

- Any pair of redox half-equations can be employed as the system in a cell. Each half-equation would occur in a half-cell region, connected within the cell by an ion-conducting salt bridge, separator or membrane, and connected externally by an electron-conducting electrical circuit. Half-cells that lack a metallic reactant to serve as an electrode need an inert metal flag to act as a current collector. Platinum is the favoured choice for this role in 'standard cells' (see next column).

- The direction of the cell reaction is predictable from the relative positions of the half-equations, written as half-cells, in the **electrochemical series**. Indeed the electrochemical series is a list compiled by measurements of cell e.m.f.s. Each half-cell in the series is coupled with a hydrogen gas/hydrogen ion half-cell, and the resulting whole-cell e.m.f. is called the **standard electrode potential** $E^{\ominus}$ of the half-cell under test. The word 'standard' implies that all concentrations are $1\,mol\,dm^{-3}$, all gases are at 1 atmosphere, and the temperature is 298 K. If the half-cell gives electrons to $H^+(aq)$, its $E^{\ominus}$ value is negative, while half-cells that take electrons from $H_2(g)$ have positive $E^{\ominus}$ values. The $2H^+(aq)\mid[H_2(g)]Pt(s)$ electrode itself has a zero $E^{\ominus}$ value, by definition.

- The electrochemical series is compiled by listing the half-cells in order of their standard electrode potentials, with the most negative ones at the top. This means that the most electron-giving half-cells, such as $Na^+(aq)\mid Na(s)$, are at the top, while the most electron-taking half-cells, like $F_2(g), 2F^-(aq)\mid Pt(s)$, are at the bottom. Notice that, by convention, the half-cells are always written with their oxidised form on the left.

- The electrochemical series can provide information at two levels. It can *foretell the direction* of any cell (or indeed redox) reaction, because higher half-cells will give electrons to lower ones. This, together with the oxidised-form-on-the-left convention, means that the direction of chemical change follows an **anticlockwise circle** rule, in which the lower half-equation goes from left-hand species to right, while the upper one goes in reverse. This law is illustrated in Figure 16.7 (p. 392).

- The electrochemical series can also *foretell the size and sign of the cell e.m.f.*, via the formula:

$$E^{\ominus}_{cell} = E^{\ominus}_{right} - E^{\ominus}_{left}$$

The designations 'right' and 'left' refer to the arbitrary choice made by the person writing the cell on paper. If you want a positive $E_{cell}^{\ominus}$, then you are required to write the more positive half-cell on the right.

• The cell's $E_{cell}^{\ominus}$ value is related to the free energy of the cell reaction, by the formula:

$$\Delta G_{sys} = -zFE_{cell}^{\ominus}$$

where z is the number of electrons being transferred per mole of cell reaction as written, and F is the faraday, or coulomb-equivalent of a mole of electrons.

• Commercially viable cells are designed to be robust, and to have a high energy density and low internal resistance. Design features that promote these ideals in various individual types of cell include having:

1 the casing as an electrode
2 the electrolytes as pastes (or solids, in some fuel cells)
3 air as a reactant
4 solid reactants whose concentrations do not decline like those in aqueous solution
5 thin, wide separators.

• **Rechargeable cells** make use of insoluble discharge reaction products, which are still therefore in contact with the electrodes ready for the chemical reversal of the charge cycle. The lead/acid cell, although heavy and cumbersome, is still the prime example in use.

• Much hope for the future centres on **fuel cells**. They use reactants that are streamed into the cell continuously while it is in use. Most fuel cells employ the reaction between hydrogen and oxygen, making use of the ready availability of hydrogen gas from the chlor-alkali industry. Each electrode in a fuel cell must include a catalyst, normally platinum, which catalyses the electrode reaction.

• In the process that is the reverse of cell discharge, namely **electrolysis**, electrons are forced into chemical systems to drive them in the redox direction of higher free energy. In the simple electrolysis of a binary ionic compound, the effect is to force the compound to revert to its elements.

• For some elements this is the only commercially viable means of extracting them from their compounds, replacing chemical redox reactions. Indeed, for caesium and fluorine it is the only way.

• **Aqueous electrolysis** introduces the complication of the presence of $H^+(aq)$ and $OH^-(aq)$ ions alongside the solute ions. When more than one ion presents itself for reaction at an electrode, the one with the 'easiest' electrode reaction is discharged. This means that no solute metal whose half-equation is above $2H^+(aq) \mid [H_2(g)]Pt(s)$ in the electrochemical series is liberated at cathodes. Instead hydrogen gas appears. Similarly, oxygen gas will appear at the anode rather than 'difficult' discharges

like fluorine; oxygen also appears when the solute anion is one of the common oxo anions like SO_4^{2-}.

• The release of oxygen at the anode is turned to advantage in the coating of certain reactive metals, notably **aluminium**. The newly liberated oxygen reacts to create a substantial surface oxide layer, which can protect the metal and act as a dye carrier. The process is called **anodisation**, and the object to be anodised is made the anode of the cell.

• In certain situations the easiest anode process can be the **dissolution of the anode** itself. This occurs when copper sulphate is electrolysed between copper electrodes, and forms the basis of the final stage of copper refining. Impure copper is made the anode of a cell, and pure copper serves as the cathode. The latter gradually grows, as the electrolysis has the effect of transferring copper (but not the impurities) across the cell.

• One of the bedrocks of UK industrial chemistry is the **chlor-alkali industry**, which is centred on the electrolysis of brine (containing sodium chloride). The products are sodium hydroxide, chlorine and hydrogen. These three products are raw materials for many branches of the chemical industry, taking in foods, polymers, drugs, solvents, paints, detergents and fuels.

17

The typical elements 1: The s-block

- ▷ **Introduction**
- ▷ **Extraction and uses of the s-block elements**
- ▷ **Reactions and compounds of the s-block elements**

- ▷ **Properties of s-block metal compounds 1: As alkalis**
- ▷ **The industrial preparation of sodium carbonate**

- ▷ **Properties of s-block metal compounds 2: As 'lattice builders'**
- ▷ **Aqueous s-block ions in cell chemistry – the nervous system**
- ▷ **Summary**

17.1 Introduction

Chemistry text books may be structured in a number of ways. They may be 'context led', where a single context like 'the atmosphere' is studied to reveal its chemistry. They may be 'Periodic Table led', where the groups of elements are studied in order and underlying concepts are highlighted as they crop up. Or they may be 'concept led', where the underlying ideas are central, with individual chemicals and situations providing examples.

This book has leant towards the 'concept-led' model, so that by now we should have a working familiarity with the basic ideas such as thermodynamics and bonding, acidity and redox. In the last three chapters, we take a 'Periodic Table' viewpoint.

The block and group structure of the Periodic Table is of course rooted in electron arrangements, via four great principles (discussed in Chapter 3).

1. Outer electron arrangements determine chemical character.
2. Outer electron arrangements have periodic pattern repeats as we scan through the elements.
3. Each column of the table houses elements with the same outer electron arrangement.
4. Each block of the table is named after the type of orbital that is being filled in that block (Figure 17.1).

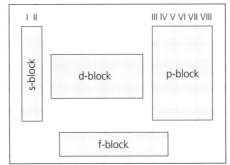

Figure 17.1 The block structure of the Periodic Table. The Roman numbers represent the groups

Within the Periodic Table there are regions in which either:

1. each successive element is noticeably different from its (horizontal) predecessor, while being noticeably similar to its vertical 'relatives', or
2. each successive element is nearly as similar to its horizontal as to its vertical 'relatives'.

Question

1 Point out separate regions in the Periodic Table in which each of the two descriptions (1) and (2) above apply. Explain in terms of electron arrangements why that particular description applies in that region.

Those regions of the Periodic Table that fit description (1) are those that are closely associated with group number. The numbered groups are found in the s-block (Groups I and II) and the p-block (Groups III to VII). Collectively we can think of the elements in those groups as the **typical elements**, since the properties of any member of the group (except sometimes the top one) will be typical of the group, and distinct from other groups. The reason for the sharp differences between groups is that in the s- and p-blocks, the differences between successive elements are differences in electron arrangements in the influential outermost orbitals.

The regions of the Periodic Table that fit description (2) are those in which successive elements have nearly the same outermost electron arrangement. The differences are partially hidden away in the inner orbitals, which do not greatly affect chemical behaviour. This happens in the d- and f-blocks of the Periodic Table.

In Chapters 17, 18 and 19, we shall look at the characteristic chemistry of the s-, p- and d-blocks. In many cases we shall be revisiting concepts covered in previous chapters on bonding, thermodynamics, redox, electrochemistry and acidity and basicity.

17.2 Extraction and uses of the s-block elements

In no other group is there such a close similarity between members as there is in Groups I and II. This comes from a narrowness of options – if you have only got one or two outer electrons overlaying a 'noble gas' structure, there is not much you can do other than lose them.

Question

2 a Complete a copy of Table 17.1 to show the outer electron arrangements of the s-block elements. (The symbols for the noble gases indicate an inner electron arrangement identical to that element's.)

Table 17.1 Outer electron arrangements and ionization enthalpies of s-block elements

Element		Electron arrangement	Ionization enthalpy (kJ mol^{-1})		
			1st	2nd	3rd
Group I	Li	[He] 2s^1	520	>7000	
	Na	[Ne] 3s^1	496	4500	
	K	[Ar]	419	3000	
	Rb				
	Cs				
Group II	Be	[He] 2s^2	900	1750	15 000
	Mg		738	1451	7 500
	Ca		590	1145	
	Sr		550	1064	
	Ba				
	Ra				

b Estimate the ionization enthalpies that are missing from Table 17.1, and justify your estimates.

c Explain why all the s-block elements are metals.

d Consider the sizes of the ionization enthalpies, and what an ionization enthalpy tells you about the reactivity of a metallic element. (The first ionization enthalpy of gold, for comparison, is about 900 kJ mol^{-1}.) Hence explain the generally high reactivity of all the s-block elements towards oxygen, chlorine and water (Figure 17.2).

e Where in each group would you look for the most reactive element, and why?

Figure 17.2 Magnesium burning in oxygen

Extraction

The s-block elements are reactive metals, and are correspondingly difficult to 'un-react', or to extract from those compounds in which they occur naturally. It was not until after 1800 that the first s-block metal was successfully liberated from one of its compounds.

Question

3 Select from Dalton's 1803 list of 'elements' (Figure 17.3) those that are really just compounds of the s-block which no one had yet been able to break down. Suggest which compounds they might have been.

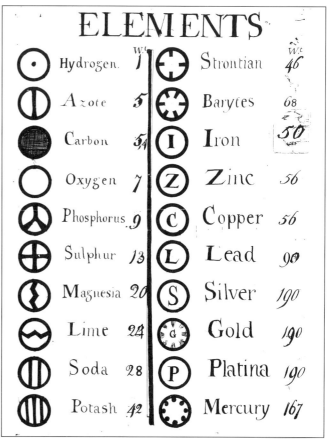

Figure 17.3 An early list of elements by John Dalton, drawn up in 1803 in support of his atomic theory

The nineteenth-century breakthrough that led to the discovery of the s-block elements was the use of electrolysis as a smelting technique, as an alternative to chemical reduction by carbon.

Question

4 This question refers to Figure 17.4. Explain why magnesium, as a representative of the s-block elements, resists conventional carbon-based smelting methods. If this

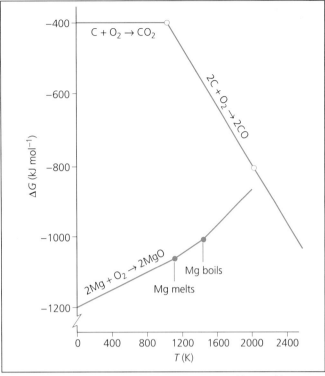

Figure 17.4 Ellingham diagram to assess prospects of being able to smelt MgO using carbon

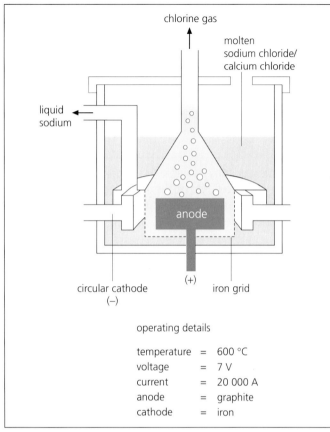

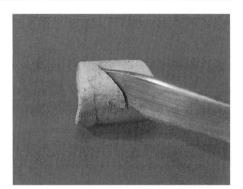

Figure 17.5 The Downs cell for the extraction of sodium from sodium chloride

is true for magnesium, why is it likely to be true for almost the whole block?

The Downs cell for the production of sodium (Figure 17.5) is typical of the industrial plant used in the extraction of the s-block metals. It has several features worthy of note.

1 After initial heating, the molten state is maintained by the effect of the extremely large current (as shown in the data in Figure 17.5).
2 The electrolyte is a mixture of sodium chloride and 40% calcium chloride. This is to reduce the melting point of the electrolyte.
3 A system of hoods funnels the sodium and chlorine away from the cell to separate destinations.

Question

5 This question refers to the Downs cell.

a Why was there an initial heating stage? In other words, why could the electrolyte not be heated from cold by the current alone?

b Sodium iodide has a considerably lower melting point (661 °C) than sodium chloride – why was it not chosen as the electrolyte?

c Assuming that the sodium produced in the Downs process is uncontaminated with calcium, what can you conclude about the electrode potential of $Ca^{2+}(l)|Ca(l)$ compared with that of $Na^+(l)|Na(l)$?

d Why is the system of hoods so important to the success of the operation?

e Write chemical equations for the cathode and anode processes in the manufacture of sodium.

Uses

What can you do with an s-block metal once you have extracted it? Well, metals that are as soft as cheese and that must not get wet (Figure 17.6) are not the most useful of materials. The following lists show some of the uses that do exist.

Figure 17.6 Sodium is a soft metal that reacts with water

Sodium

Sodium is used:

- as a coolant in fast-breeder nuclear reactors, because it is a liquid from about 100 °C to 900 °C, and is free-flowing, light and cheap, with a high thermal conductivity

- as a reducing agent, e.g. in the reduction of titanium(IV) chloride to titanium

- in sodium vapour street lighting.

Question

6 Which property of sodium is being used in street lights?

Potassium

- Potassium is burned in air or oxygen to make potassium superoxide, KO_2. This compound releases oxygen when mixed with water, and so is used as an emergency oxygen supply in submarines (Figure 17.7):

$$2KO_2(s) + H_2O(l) \rightarrow 2KOH(aq) + \tfrac{3}{2}O_2(g)$$

Magnesium

- Magnesium is used as an alloying metal to produce alloys of high tensile strength in relation to weight, for aviation use.

Question

7 Apart from its lightness, why is magnesium the most likely of all the s-block metals to find an out-of-doors application?

Caesium

- Caesium is used as a surface in photocells, as used in photography (Figure 17.8). The surface absorbs photons and emits electrons.

Question

8 Why is caesium uniquely well qualified for this job?

Francium

Francium is rarely mentioned, though students are frequently fired with curiosity about an element that should make even caesium look tame. Francium only exists on Earth as a staging post on certain radioactive decay paths. It is estimated that the amount of francium present at any one time in the whole of the Earth's crust is about one ounce (28 g). No uses have yet been found for francium.

17.3 Reactions and compounds of the s-block elements

Typical reactions

The simplicity of the chemistry of the s-block elements derives from the fact that their electron arrangements leave them no choice. It is either do nothing, or lose everything (all the outer electrons, that is). So (almost) all the s-block element chemistry comprises oxidations of the elements from 0 to +1 or +2 (depending on group number). Typical reactions (Figure 17.9) are:

1 $2Na(s) + 2H_2O(l) \rightarrow 2NaOH(aq) + H_2(aq)$

2 $Mg(s) + 2HCl(aq) \rightarrow MgCl_2(aq) + H_2(g)$

3 $4Li(s) + O_2(g) \rightarrow 2Li_2O(s)$

Figure 17.7 Potassium superoxide is used as an emergency oxygen supply in submarines

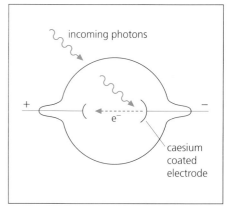

Figure 17.8 Schematic diagram of a photocell. Electron emission is stimulated by photon absorption

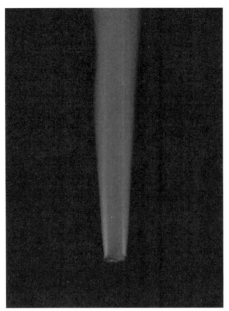

Figure 17.9 a Calcium burning in chlorine   **b** Lithium burning in oxygen

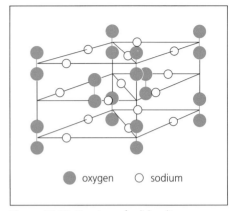

● oxygen   ○ sodium

Figure 17.10 Structure of solid sodium peroxide, Na_2O_2. Note the covalent bond on the O_2^{2-} ions (shown green)

Question

9 a State in each of the above cases which element has been reduced, with the relevant oxidation numbers.

b Is it fair to say that water in (1) and hydrochloric acid in (2) have behaved as acids? Justify your answer.

c If the equivalent of reaction (1) is performed with calcium in place of sodium, the water goes cloudy. Can you explain why? Write the equation for the change.

d Although you can write equations like equation (4) for all the Group I and II metals, you very rarely see the reaction demonstrated with elements except magnesium. Why is this?

Pattern breakers

Not every reaction of the elements of Groups I and II is entirely predictable. In particular, there are several deviations from the norm among the products of combustion. Both sodium and potassium produce unusual oxides when burned in air, and magnesium is said to produce some magnesium nitride. The equations are:

4 $2Na(s) + O_2(g) \rightarrow Na_2O_2(s)$, 'sodium peroxide'

5 $K(s) + O_2(g) \rightarrow KO_2(s)$, 'potassium superoxide'

6 $3Mg(s) + N_2(g) \rightarrow Mg_3N_2(s)$

Question

10a The reason why sodium peroxide is written as Na_2O_2 in equation (6), and not as 2NaO, is because the O_2^{2-}

('peroxide') ion really does exist in this compound (Figure 17.10). What are the oxidation states of the elements in this compound?

b Which compound could be seen as the dibasic conjugate acid of the peroxide ion (that is, with two H^+ ions added)?

c When sodium peroxide is placed in water, it produces not only an alkaline solution (as would a 'normal' Group I oxide of formula M_2O), but also some bubbles. Assume the peroxide ion picks up two protons from water (to give the compound in (b)), and the compound then undergoes a redox disproportionation. From this information, fill in the gaps below to complete the equation for the overall reaction when sodium peroxide meets water.

$$Na_2O_2(s) + H_2O(l) \rightarrow ? + ?$$

d In potassium superoxide, KO_2, the metal is in its normal oxidation state. What therefore must be the apparent oxidation number of oxygen? Go on to show that the O_2^- ion must violate normal rules of bonding, by having an unpaired electron.

Non-existent compounds

A whole book was once written with the title *Non-existent compounds*, and when you get over the surprise of finding that someone devoted a complete publication to chemicals that do not exist, you come to realise that the question of why compounds fail to exist often provides a good test of thermodynamic principles.

Question

11 It is not too hard to decide why the compound $NaCl_2$ does not exist:

a Assuming the chlorine is in its normal oxidation state, what would have to be the charge on the sodium ion in this compound?

b What thermodynamic advantage would the solid lattice of $NaCl_2$ have over that of conventional NaCl? (*Remember*: Lattices are held together by electrostatic forces.)

c And yet which thermodynamic quantity would be so big as to make the creation of this compound unfeasible, despite the advantage mentioned in (b)?

You can check your answer to question 11 against the Born–Haber cycles in Figure 17.11. Notice that the lattice enthalpy is actually bigger (more negative) in the non-existent compound, but that 'IE2[Na]' scuppers the whole plan.

Question

12 The lattice enthalpy of $NaCl_2$ is an estimate (obviously). How do you think it was estimated?

Turning to compounds which fail to exist for a different reason, let us consider MgCl. Again there are 'swings and roundabouts', because although MgCl would have a smaller lattice enthalpy than conventional $MgCl_2$, it would have the advantage of avoiding the reasonably high ($+1451$ kJ mol^{-1}) second ionization enthalpy of magnesium.

Question

13 Construct a Born–Haber cycle using appropriate data to determine $\Delta H_f^{\ominus}$ for MgCl. You may assume that the lattice enthalpy will be approximately -750 kJ mol^{-1}.

The answer to question 13 is quite surprising, in that the formation of MgCl from its elements is exothermic. And while we know that $\Delta G_f^{\ominus}$ is the deciding factor in matters of thermodynamic feasibility, it is still quite a favourable result. It makes the non-existence of a +1 oxidation state in Group II elements a little more open to question than perhaps you expected.

Of course, a +1 state does not exist (outside of freak environments like the mass spectrometer), so perhaps our exothermic result for $\Delta H_f^{\ominus}$[MgCl] is a false clue. And so it turns out. Whether a compound exists depends not on its stability relative to its elements, but on its stability relative to any other possible combination of the elements, as question 14 should reveal.

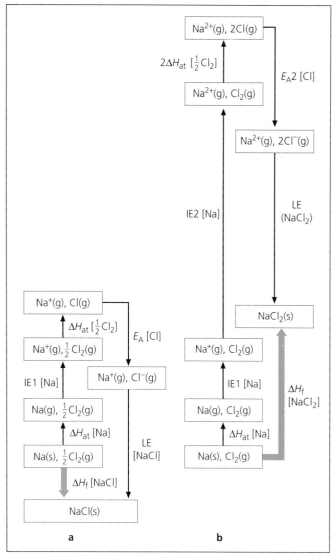

Figure 17.11 a Born–Haber cycle for NaCl(s)   **b** Born–Haber cycle for $NaCl_2$(s). See 'Non-existent compounds', and questions 11 and 12

Question

14 Imagine a disproportionation reaction in which one Mg^+ ion gives its remaining outer electron to another Mg^+ ion.

a Copy and complete the following ionic equation to symbolise the proposed disproportionation:

$$Mg^+ + Mg^+ \rightarrow ? + ?$$

b Now redraft the equation from (a) to include spectator Cl^- ions:

$$2MgCl \rightarrow ? + ?$$

c Construct an energy cycle to find $\Delta H^{\ominus}$ for the reaction in (b). You will need your answer from question 13 for $\Delta H_f^{\ominus}$[MgCl] and some extra data from a data book.

So you see that being exothermic relative to its elements is not enough to save a compound from non-existence. In this case the problem was instability with respect to a disproportionation into magnesium metal and $MgCl_2$. These are the kinds of factors which lead to the great simplicity of s-block element chemistry, and which ensure that 'deviant' oxidation states cannot survive. Even those pattern-breaking oxides of sodium and potassium display their 'oddness' in the anions – the oxidation states of the two metals are as expected.

Question

15 Predict the formulae of the s-block compounds named below. Assume the ionic model is valid to predict stoichiometries, and then be guided by the rule that all ionic compounds display an overall charge of zero.

a sodium sulphate

b calcium nitrate

c sodium carbonate

d sodium hydrogencarbonate

e strontium chloride

f calcium phosphate(v) (the phosphate(v) ion is PO_4^{3-})

g potassium dihydrogenphosphate(v)

I cannot leave this tidy picture of predictable oxidation states without making one small blot on it. All scientific 'laws' are tentative, and even in s-block chemistry you have to be careful when using words like 'never', 'always' and 'without exception'. With this in mind I must mention a compound that does exist against all reasonable expectations of behaviour, with an s-block element being the deviant species.

It turns out that certain cyclic organic compounds offer a 'nest' for an Na^+ ion that is thermodynamically 'cosy'. That is to say that the sodium ion is not only physically the right size to fit perfectly into the nest, but it also does so exothermically. Under these circumstances two sodium atoms enter into a very strange 'deal' – the incentive to create an Na^+ ion to fit into the 'nest' is so great a second sodium atom *accepts* an electron:

$$2Na(s) + \text{'nest'} \rightarrow [Na\text{—}nest]^+ Na^-(s)$$

Crystals of just such a compound were isolated as long ago as 1974 at the University of Michigan (Figure 17.12a). What could be less likely than a sodium *anion*?

Question

16 The 'nest' compound is a sort of triple-barred spherical cage whose formula is given in Figure 17.12b. The Na^+ ion sits closest to the oxygen and nitrogen atoms of the cage. Suggest what sort of bonding might hold the Na^+ ion.

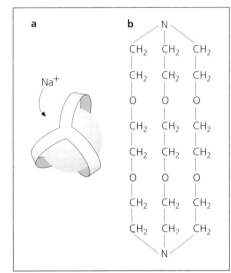

Figure 17.12 a Schematic diagram showing how the Michigan 'cage' molecule fits the Na^+ ion. **b** The structural formula of the 'cage' molecule

Research projects like these are often said to be 'curiosity led', meaning that no direct commercial application is apparent while the work is being carried out. However, apparently useless discoveries often find applications in time (remember that poly(ethene) was once considered a curious lump of waxy stuff). In this particular case, further interest in cages has been stimulated by the finding that they are involved in cell-membrane transport (see Section 17.7).

In the meantime, though, I would advise against going into an exam and putting '−1' as one of the oxidation states of sodium chemistry, at least not unless you give the examiner a few details.

Question

17 Can you think of any other pieces of 'curiosity-led' research which later became the basis of major commercial applications?

17.4 Properties of s-block metal compounds 1: As alkalis

The s-block elements have a wide range of compounds, nearly all of which feature the element in the form of a 1+ or 2+ ion. However, the roles played, especially by the Group I elements, *within* their compounds are, one might say, minimal. In the alkalis, for instance, all the chemical character of the compound resides with the anion. Group III cations have a more influential role in a set of compounds that could be seen as 'lattice builders', of which cement and plaster are everyday examples. Biological systems use these two groups of cations in similar ways – for example, in nerves the Group I ions are simply charge carriers, while in bones the emphasis is on lattice building, this time with Group II compounds. This division of the roles of s-block cations into charge carriers and lattice builders will be used as a framework for the discussion of the s-block compounds. The rest of this section, and the whole of 17.5, is devoted to compounds in which the s-block cations perform the 'minimalist' role of *charge carrier*.

Alkalis in Groups I and II

As we saw in Chapter 12, an alkali is a water-soluble base. This concept translates to mean a soluble hydroxide, or any salt whose anion derives from a weak conjugate acid. However, in normal industrial usage the word 'alkali' usually means hydroxides, hydrogencarbonates or carbonates that dissolve in water. The only s-block bases that are insoluble in water are the Group II carbonates and the hydroxide of magnesium, so all the rest are alkalis (Table 17.2).

We discussed in Chapter 12 what made the OH^- and CO_3^{2-} ions basic. Let us now consider why it is the s-block (and particularly the Group I) versions of these compounds that are alkalis.

Table 17.2 s-block alkalis

Group I	MOH	M_2CO_3	$MHCO_3$
Group II	$N(OH)_2$		$N(HCO_3)_2$

M = any Group I metal.
N = any Group II metal (except magnesium in the case of the hydroxides).

The first reason is solubility in water. Chromium hydroxide (to take an example at random) cannot be an alkali because its solubility is only about $10^{-8}\,mol\,dm^{-3}$. On the other hand, nearly every Group I compound is appreciably water soluble.

The thermodynamic feasibility of solubility depends upon two factors – how easily the lattice breaks, and what sort of reception the free ions get in water. The influence of the first factor can be judged by the size of the lattice enthalpy. The influence of the second factor is embodied in a quantity called the enthalpy of solvation, $\Delta H_{solv}^{\ominus}$, which is the enthalpy change when a mole of gaseous ions becomes a mole of aqueous ions. The balance between these two factors determines the overall enthalpy of solution, $\Delta H_{solution}^{\ominus}$. This balance can be symbolised by an energy cycle, as in Figure 17.13. If the cycle delivers a negative $\Delta H_{solution}^{\ominus}$ then solubility is favoured.

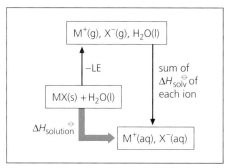

Figure 17.13 Energy cycle to assess the factors affecting the enthalpy of solution $\Delta H_{solution}^{\ominus}$ of an ionic solid (the example shown is exothermic, but this is not necessarily the case)

Question

18a If every Group I compound is soluble in water, then it is likely that all the values of $\Delta H_{solution}^{\ominus}$ tend towards the negative. Which factor in Figure 17.13 must be significantly and consistently small in Group I?

b Group II compounds do not show the same high solubilities. In particular, it is rare to see appreciable solubility when a Group II cation is paired with a 2− anion (sulphate or carbonate, say). In these cases, values of $\Delta H_{solution}^{\ominus}$ would be positive. Explain this, again referring to the thermodynamic cycle in Figure 17.13.

The second reason why the alkalis appear in the s-block is because of the relative 'purity' of ionic bonding in this region. In NaOH(aq), the Na^+ ion will leave the OH^- ion to get on with its job of grabbing protons, but there are other hydroxides in which the cation interacts with the anion to reduce its basicity.

Figure 17.14 shows aluminium hydroxide, $Al(OH)_3$. From the several mental models described in Chapter 5, let us use the charge density argument – the Al^{3+} ion has such a high charge density that it 'reclaims' electron density to

a as an ionic compound (which it is not)

b as a polar covalent compound (which is more like reality)

Figure 17.14 Uncertainty about the best way of describing the bonding in aluminium hydroxide

give the Al—O bond partial covalent character. It now becomes a much more finely balanced decision as to which bond in the string Al—O—H breaks ionically – in fact, the situation depends upon the environment. With plenty of protons around, the compound behaves as a base:

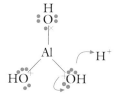

$$Al(OH)_3(s) + 3H^+(aq) \rightarrow Al^{3+}(aq) + 3H_2O(l)$$

but in a proton-grabbing environment, aluminium hydroxide starts to behave like an acid. This stage is followed by loss of another molecule of water, so overall:

$$Al(OH)_3(s) + OH^-(aq) \rightarrow AlO_2^-(aq) + 2H_2O(l)$$

This mode of behaviour, in which a species may act as an acid or as a base depending on its environment, is described by the adjective **amphoteric**.

By comparing the true alkalis with amphoteric hydroxides like that of aluminium, we come to realise the contribution the Group I cations make to the existence of alkalis. Because they do not retain electrons, they never divert the oxygen lone pairs from their central role of proton grabbing.

Question

19 The Group II cations are nearly as effective as the Group I cations in leaving oxide and hydroxide ions free to be bases. Their status as alkalis, however, is less well defined, mainly because of limitations of solubility.

a The decision as to when to call a base an alkali is not clear cut, but as a rule of thumb we could say that for a substance to qualify as an alkali, you should be able to make up a $0.1\,mol\,dm^{-3}$ solution. On this basis, use a data book to say which of the Group II hydroxides (magnesium to barium) would qualify as alkalis.

b A solution of calcium hydroxide is called limewater. Its most renowned action as a base is to react with the faintly acidic gas carbon dioxide, giving a test for the latter. Write an equation for the 'milky limewater' test, and explain why it gives this visual signal.

c It is a little dubious to talk about the 'solubility of oxides', because the word 'solubility' implies that the solute is unchanged by the process of solution. Write an equation to show what really happens when an oxide such as calcium oxide 'dissolves' in water, and explain why the oxide ion cannot survive in aqueous solution.

d Magnesium oxide appears to be nearly insoluble in water, although a little hydroxide formation must occur since the solution turns Universal Indicator slightly blue. On the other hand, magnesium oxide 'dissolves' quite freely in acidic solutions. Explain these observations, with illustrative equations.

17.5 The industrial preparation of sodium carbonate

Sodium hydroxide is the strong alkali of industrial chemistry. We saw in Chapter 16 how it is made by electrolysis, and it is used in industries like papermaking, for the hydrolysis of cellulose (see Section 13.9). The weak alkali of industrial

chemistry is sodium carbonate, which finds enormous use in the making of glass (Figure 17.15):

$$2Na_2CO_3(s) + 7SiO_2(s) + CaO(s) + \tfrac{1}{10}Al_2O_3(s) \rightarrow$$
$$\text{soda glass} + 2CO_2(g)$$

Soda glass has an approximate composition equivalent to $(SiO_2)_7(Na_2O)_2(CaO)(Al_2O_3)_{0.1}$, but in fact is a complex disorderly solid lattice. Over half of all UK production of sodium carbonate is used in the glass industry.

The manufacture of sodium carbonate is a good example of industrial chemistry in action, since it presents problems both environmental and thermodynamic, which have to be overcome by the chemist and chemical engineer.

The Leblanc process

The original method of making sodium carbonate, the infamous **Leblanc process**, must have produced scenes of classic Industrial Revolution environmental degradation and misery (Figure 17.16). The first step involved the high-temperature reaction of salt with concentrated sulphuric acid:

$$2NaCl(s) + H_2SO_4(l) \rightarrow Na_2SO_4(s) + 2HCl(g)$$

The sodium sulphate solid was dug out of the reaction vessel and reacted with limestone and coal to give sodium carbonate:

$$Na_2SO_4(s) + CaCO_3(s) + 2C(s) \rightarrow$$
$$Na_2CO_3(s) + CaS(s) + 2CO_2(g)$$

The sodium carbonate was then dissolved out of the solid product mass, and slag heaps of discarded calcium sulphide quietly decomposed in the rain:

$$CaS(s) + 2H_2O(l) \rightarrow H_2S(g) + Ca(OH)_2(aq)$$

Figure 17.15 (left) A modern sodium carbonate plant. (below) The float glass process – the glass industry is a major consumer of sodium carbonate

Figure 17.16 The Victorian industrial landscape – environmental protection had not yet climbed far up the political agenda

Things must have been bad, because in a society not well known for putting the welfare of its workforce ahead of the demands of trade, Parliament was moved to enact one of the pioneer pieces of legislation for the protection of people's environment – the 'Alkali Act' of 1861.

Question

20a List three environmental problems with the Leblanc process.

b Modern-day chemists try to ensure that chemicals can flow from stage to stage, if possible as part of a continuous process. Why could the Leblanc process not fit this ideal?

c Why would the process therefore have been more labour-intensive?

The modern process of sodium carbonate production is called the **ammonia–soda** or **Solvay process**, and has the overall reaction given by the equation:

$$2NaCl(aq) + CaCO_3(s) \rightarrow CaCl_2(aq) + Na_2CO_3(aq)$$

Question

21 Compare this reaction with the Leblanc one, in terms of availability of starting materials and possible pollutants.

The reaction is environmentally much better than the Leblanc process. It has one drawback, which is unfortunately not a trivial one. And that is that *it is, apparently, impossible.* ($\Delta G_{sys}^{\ominus}$

is about $+70\,kJ\,mol^{-1}$.) The reverse process occurs readily, as anyone who mixes calcium chloride and sodium carbonate solutions will see.

Question

22 Literally what would they immediately see? Why cannot the four ions involved, $Ca^{2+}(aq)$, $Cl^-(aq)$, $Na^+(aq)$ and $CO_3^{2-}(aq)$, just float around in solution independently?

The ammonia–soda process features many clever ideas, but the most ingenious is a thermodynamic concept. It is this – if your reaction does not go, break it up into stages.

You might well feel uneasy about this idea – after all, we know from our thermodynamic studies that if several reactions add up to one overall reaction, then their several $\Delta G^{\ominus}$s will add up to one overall $\Delta G^{\ominus}$, and we have already stated that this quantity is positive for this reaction.

However, the figure of $+70\,kJ\,mol^{-1}$ relates to reactants beginning at 298 K and products ending at 298 K. Other temperatures may deliver different values. This is especially true if some of the individual stages feature a marked increase in entropy, ΔS_{sys}. In that case they can be run at temperatures far above 298 K, because:

$$\Delta G_{sys} = \Delta H_{sys} - T\Delta S_{sys} \qquad (9.7)$$

The high value of T will facilitate a negative ΔG_{sys}. The products of such stages can then be cooled and returned to the main sequence of reactions at more moderate temperatures.

Look out for this ploy in the following sequence. It will be the focus of question 24.

The sequence of reactions in the ammonia–soda process

At the heart of the factory stands a tower, the Solvay tower (Figures 17.17 and 17.18), in which an upward-flowing current of carbon dioxide gas meets a downward-falling cascade of brine, NaCl(aq), saturated with ammonia gas. The reaction in the Solvay tower produces sodium hydrogencarbonate:

$$2NaCl(aq) + 2NH_3(aq) +$$
$$2CO_2(g,aq) + 2H_2O(l) \rightarrow$$
$$2NH_4Cl(aq) + 2NaHCO_3(aq) \quad (1)$$

This is less soluble than ammonium chloride and crystallises out at the bottom of the tower. It is then taken out and roasted to form the target product:

$$2NaHCO_3(s) \rightarrow$$
$$Na_2CO_3(s) + CO_2(g) + H_2O(g) \quad (2)$$
'light ash'

This happens in the light ash calciner (Figure 17.18). That is almost the end of the story, but we began in the middle. What about the origins of some of the reactants in equation (1)? Well, the carbon dioxide comes from the roasting of limestone in a kiln (Figure 17.18):

$$CaCO_3(s) \rightarrow CaO(s) + CO_2(g) \quad (3)$$

and with great economy the calcium oxide is turned into alkali and put with one of the products of equation (1) to regenerate the ammonia:

$$CaO(s) + H_2O(l) + 2NH_4Cl(aq) \rightarrow$$
$$2NH_3(g) + CaCl_2(aq) + 2H_2O(l) \quad (4)$$
This reaction happens in the distiller (Figure 17.18).

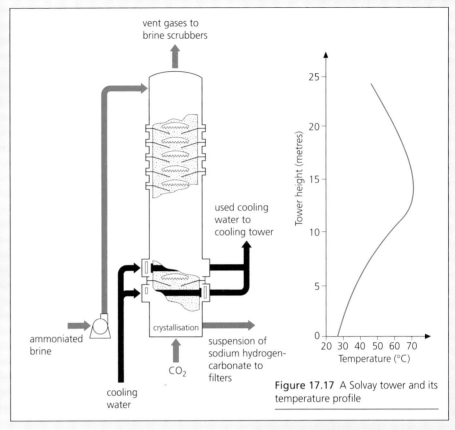

Figure 17.17 A Solvay tower and its temperature profile

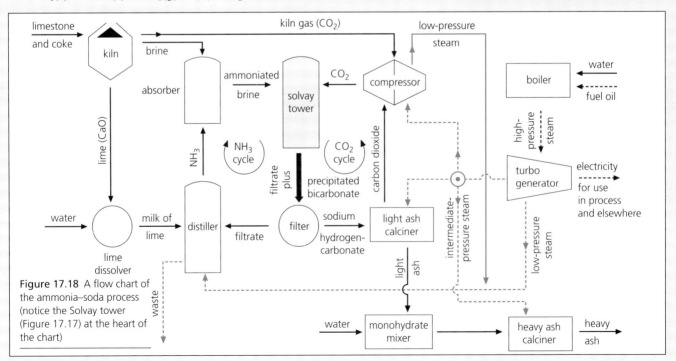

Figure 17.18 A flow chart of the ammonia–soda process (notice the Solvay tower (Figure 17.17) at the heart of the chart)

Questions

23a Put the following list of chemicals involved in the ammonia–soda process into one of four categories:

  i reactant, consumed

  ii main target, product

  iii by-product, for sale or dumping

  iv reagent in constant circulation within the plant, repeatedly destroyed and then regenerated.

Chemicals: salt, ammonia, carbon dioxide, ammonium chloride, sodium hydrogencarbonate, sodium carbonate, calcium carbonate, calcium oxide, calcium chloride.

b What would be the economical thing to do with the carbon dioxide from equation (2)? Does it appear from Figure 17.18 that the industry follows your advice?

c Add up all the equations (1) to (4). What is the overall equation for the process?

d The graph in Figure 17.17 shows that the Solvay tower is cold at the bottom and hot in the middle. What natural events are creating the hot section in the middle, and why is it desirable to cool it down at the bottom?

e Can you think of three major factors to consider when siting a sodium carbonate factory?

f ICI's plants for making sodium carbonate are in Cheshire. Why do you think this is? (Find out about the minerals in or near Cheshire, and the position of a major client, Pilkington Glass.)

g Adding water to the sodium carbonate and re-roasting brings about a conversion in density, rather than any chemical change. You can see this process in Figure 17.18 in the sequence leading from the light ash calciner to the heavy ash calciner. Heavy ash is just another form of sodium carbonate, but unlike light ash it has a particle size similar to, and mixes easily with, sand. Why do you think light ash is converted to heavy ash?

h The one bulk waste product of the ammonia–soda process is calcium chloride, which from the 1950s onwards was put into holes in the ground. Can you think where there might be appropriately sized holes, not too far from the plant?

24 (Needs entropy knowledge from Chapter 9.) This question focuses on the thermodynamics of the ammonia–soda process.

a Select one step from equations (2) to (4) that you would expect to have a large positive ΔS_{sys} (remember that this is the 'chaos factor'), and justify your choice.

b Use a data book to find $\Delta H_{sys}^{\ominus}$ and $\Delta S_{sys}^{\ominus}$ for your chosen stage. (*Hint*: Consult formula (8.2), and for $\Delta S_{sys}^{\ominus}$ use the formula:

$$\Delta S_{sys} = \Sigma S(\text{products}) - \Sigma S(\text{reactants}) \tag{17.1}$$

Then convert ΔS_{sys} from $J\,K^{-1}\,mol^{-1}$ to $kJ\,K^{-1}\,mol^{-1}$ ready for part (c).

c Assume that $\Delta H_{sys}^{\ominus}$ and $\Delta S_{sys}^{\ominus}$ remain fairly constant throughout the temperature range being used, and therefore plot a straight-line graph of ΔG_{sys} against temperature, using:

$$\Delta G_{sys} = \Delta H_{sys}^{\ominus} - T\Delta S_{sys}^{\ominus}$$

(This graph is a sort of Ellingham diagram, except that there is no second line intersecting it.)

d At what temperature does the stage become thermodynamically viable (i.e. have a negative ΔG_{sys})? Would the step have been viable at 298 K?

This rather detailed look into one particular industrial process demonstrates the complex interaction of human skills demanded of the industrial chemist. Developing industrial processes involves fighting for control of thermodynamic feasibility, trying to retain heat energy from exothermic processes to help with endothermic ones, keeping the reagents in flow around the plant with the minimum need for human intervention, and considering costs, prices, markets and buyers, and also environmental concerns to avoid a repeat of the Victorian industrial landscapes.

17.6 Properties of s-block metal compounds 2: As 'lattice builders'

Throughout this chapter there has been a theme of predictability. The cations of Groups I and II tend not to influence their compounds greatly, leaving the anions to dominate their character. This, as we have seen, is the cause of the alkalinity that pervades the compounds of the s-block and also their names – the alkali metals (Group I) and the alkaline earths (Group II).

Another predictable feature of s-block compounds is lattices whose lattice enthalpies are not far removed from those predicted on an electrostatic basis, using a pure ionic bonding model. You will remember that the lattice enthalpy depends on only two factors, both derived from the laws of electrostatic attraction. Those factors are the sizes of the charges on the ions and the distance between the ions.

Question

25 How would you expect lattice enthalpies to vary in the following series? (*Remember*: Lattice enthalpies are negative by definition, so it is less ambiguous to say 'more negative' than 'bigger'.)

a MgO, CaO, SrO, BaO

b NaCl, $MgCl_2$, $AlCl_3$

Solubility in water

It would be convenient if we could correlate lattice enthalpies directly (if inversely) with solubility – in other words, if a more negative lattice enthalpy always went hand-in-hand with low solubility in water. Indeed in the series from question 25a that is just what does happen, with magnesium oxide being the most insoluble. But there are several series in which the *least* negative lattice enthalpies come from the *least* soluble compounds.

Question

26 Look up the solubilities in water of the two series of compounds below. You will not be able to look up the lattice enthalpies, but indicate, on an electrostatic attraction model, which lattices you think will be strongest (LE most negative) and which lattices weakest (LE least negative). Say in which series there is a direct correlation (lattice enthalpy small/solubility low) and in which there is an inverse correlation (lattice enthalpy small/solubility high).

a $Mg(OH)_2$, $Ca(OH)_2$, $Sr(OH)_2$, $Ba(OH)_2$

b $MgSO_4$, $CaSO_4$, $SrSO_4$, $BaSO_4$

A look back at the energy cycle in Figure 17.13 will reveal the factor that we have so far ignored. Solubility depends not only on how strong the lattice is, but also on how strong the interactions are when the ions meet the water. Thus since the anions in these series are a constant factor, what really matters is the interaction between water and the cations.

The relationship between cations and water molecules was described in Section 6.7, and was then called an **ion–dipole interaction**. The strength of this bond lies in the charge on the cation and its size. The most exothermic solvations are between water molecules and ions of high charge and/or small size or, to lump the two factors into a single parameter, high **charge density**.

Questions

27 Sketch how water molecules might cluster round a cation.

28a How does the charge density of the cations vary down Group II? Which Group II ion would have the most exothermic association with water molecules?

b In series (a) of question 26, does it seem that the lattice enthalpy or the solvation enthalpy is having the biggest influence in deciding the overall solubility in water?

c Similarly, which factor seems to be having the biggest influence in series (b)?

d Sulphates give one pattern and hydroxides the other. Can you suggest a major difference between those two anions that might be relevant in this case?

By the end of question 28, we have the makings of a new pattern that is more complex, but that fits all cases. If the anion is small (e.g. hydroxide), then the big (negative) lattice enthalpy (biggest at the magnesium end of the series) works *against* solubility, dominating the high solvation enthalpy (also highest at the magnesium end) working *for* solubility, thus making magnesium hydroxide the least soluble hydroxide. If on the other hand the anion is big (e.g. sulphate), then the high solvation enthalpy of the Mg^{2+} cation seems to be the crucial factor in making magnesium sulphate the most soluble sulphate. Can we offer a model to explain this?

One logical suggestion goes like this: when anions are much larger than cations, the small size of the cation is partially wasted, because even with very low co-ordination numbers (4 say), it is difficult to get four sulphates all tightly clustered around an Mg^{2+} cation. The cation must inevitably rattle around a bit in the cavity in the anion lattice, so a magnesium ion has little advantage over, say, a strontium ion (Figure 17.19). If as a result the lattice enthalpies are not so different down the group, the solvation enthalpies (when the cations are freed to take full advantage of their true size, and the Mg^{2+} ion can make use of its high charge density) become the deciding factor in the overall solubility – so magnesium sulphate is the most soluble sulphate in Group II.

Taking the other case, when the anion is small, the smaller cations like Mg^{2+} are able to co-ordinate more tightly, and can take full benefit from the small cation–anion distance and high charge density. This effect overwhelms the importance of the solvation factor, and magnesium hydroxide is the least soluble Group II hydroxide (Figure 17.20).

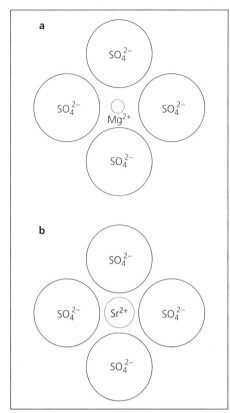

Figure 17.19 When the anion is big and the cation small, its smallness is 'wasted' (a), and a larger cation can achieve a better fit, with no serious loss of lattice enthalpy (b)

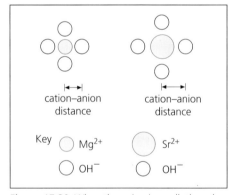

Figure 17.20 When the anion is small, then the smallness of the cation *is* an advantage

Question

29 Predict and justify (and then check) the order of solubility of the carbonates of Group II cations.

Thermal stability

The charge density concept has had a prominent role so far in accounting for the properties of the compounds of Groups I and II. This is to be expected, since the compounds are mostly well described by the ionic/electrostatic model, and charge density is an ionic/electrostatic concept. We can use it further to help our understanding of the way s-block compounds stand up to high temperatures. First let us look at the patterns of thermal behaviour of the s-block metal compounds.

Question

30a (Needs entropy knowledge of Chapter 9.) Copy and complete the data collection in Table 17.3. For the magnesium and barium examples, you will need to use formulae (8.2) and (17.1) to calculate $\Delta H_{sys}^{\ominus}$ and $\Delta S_{sys}^{\ominus}$ for the decompositions.

b Now recall the formula:

$$\Delta G_{sys} = \Delta H_{sys} - T\Delta S_{sys}$$

and use it, together with the data in Table 17.3, to work out the temperature at which each decomposition reaction would become more or less thermodynamically feasible (when ΔG_{sys} becomes zero).

c What assumptions are you making in these calculations about the variability of ΔH_{sys} and ΔS_{sys} with temperature, and about the possibility of changes of state happening to any members of the reaction systems?

d What do your results show about the pattern of thermodynamic stability as we go:

i across a row (from sodium to magnesium)

ii down a group (from magnesium to barium)?

e Are any of the three carbonates likely to be decomposed when heated by a Bunsen burner?

So we can conclude that carbonates are more stable in Group I than in Group II, and more stable at the bottom of a group. These predictions are well borne out by practical results. Why do Groups I and II follow these trends? Can we apply the charge density model?

There appears to be an inverse correlation – the higher the charge density of the cation, the less resistant the anion is to thermal decomposition. So Group I cations with their low charge densities have stable carbonates compared with those of Group II, and in both groups the smallest cation (at the top of the group) seems to have the most destabilising effect on the anion. But why does a high charge density reduce thermal stability?

It is not too difficult to make some plausible suggestions, some of which carry echoes of the solubility data just studied. A high charge density will not greatly affect the lattice enthalpy of the carbonate because, as we have already argued, the closest distance between ions is limited by the large anions. However, the oxide lattice presents a different situation – here the small size of the Mg^{2+} ion, say, is a major asset compared with the bulkier Ba^{2+}, and so the magnesium carbonate/magnesium oxide system has the most to gain by making the transition from carbonate to oxide, in terms of stronger bonding. In summary, the higher the charge density of the cation, the more favourable the products relative to the reactants (Figure 17.21, overleaf).

The above is essentially a thermodynamic argument, since it talks only about the starting and finishing states of the system. There is an alternative picture, which shows that the less stable carbonates will also be those with the most readily reached transition states, and so proposes that they will suffer from kinetic instability (decomposing more quickly) as well as thermodynamic instability (decomposing more completely).

Once again the key is the charge density concept. The high charge density of the Mg^{2+} cation will cause a polarisation of the $C—O^-$ bond, which will predispose it to breaking to give O^{2-}. Hence the enthalpy jump between reactants and the stretched-bond transition state will be less than with a cation of less polarising power. The same effect that gives rise to the thermodynamic stability of the oxide product also explains the relative stability of the partially formed oxide in the transition state. We can therefore imagine a set of enthalpy profiles for the set of decompositions of the Group II carbonates in which higher activation enthalpies were followed systematically by less negative values of $\Delta H_{sys}^{\ominus}$ (Figure 17.22, overleaf).

The pattern of Figure 17.22 is not general for chemistry as a whole. You cannot always predict the height of the activation enthalpy E_A from the size of ΔH_{sys}. There are plenty of cases in which reactions of great thermodynamic feasibility are totally blocked by the size of the activation enthalpy. Without this kinetic stability, this very book would turn to carbon dioxide and water before you could do the next question.

Table 17.3 Thermodynamic data for assessing stability of s-block metal carbonates

	Na$_2$CO$_3$	→	Na$_2$O	+	CO$_2$
$\Delta H_f^{\ominus}$ (kJ mol^{-1})	−1130.7		−414.2		−393.5
$S^{\ominus}$ (J K^{-1} mol^{-1})	+ 135		+ 75.1		+213.6
	$\Delta H_{sys}^{\ominus}$ +323 kJ mol^{-1}		$\Delta G_{sys}^{\ominus,298}$ +277 kJ mol^{-1}		
	$\Delta S_{sys}^{\ominus}$ +153.7 J K^{-1}				
	MgCO$_3$	→	MgO	+	CO$_2$
$\Delta H_f^{\ominus}$ (kJ mol^{-1})					
$S^{\ominus}$ (J K^{-1} mol^{-1})					
	$\Delta H_{sys}^{\ominus}$		$\Delta G_{sys}^{\ominus,298}$		
	$\Delta S_{sys}^{\ominus}$				
	BaCO$_3$	→	BaO	+	CO$_2$
$\Delta H_f^{\ominus}$ (kJ mol^{-1})					
$S^{\ominus}$ (J K^{-1} mol^{-1})					
	$\Delta H_{sys}^{\ominus}$		$\Delta G_{sys}^{\ominus,298}$		
	$\Delta S_{sys}^{\ominus}$				

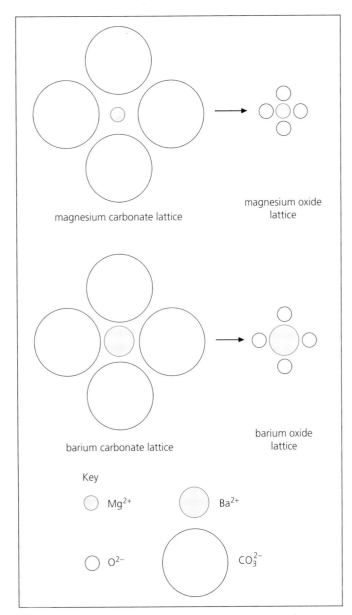

Figure 17.21 The larger size of the Ba^{2+} ion relative to the Mg^{2+} ion has no great effect on the carbonate lattice, but in the oxide lattice the small size of the Mg^{2+} ion is an advantage. Hence the upper reaction is 'product favoured', compared with the lower one

Question

31 The nitrates of Group II are less stable than the carbonates when heated, and quite readily decompose as follows (after any water of crystallisation has been driven off):

$$2Ca(NO_3)_2(s) \rightarrow 2CaO(s) + 4NO_2(g) + O_2(g)$$

The nitrates of Group I are also readily decomposed at Bunsen burner temperatures. With the exception of lithium, they all react thus:

$$2KNO_3(s) \rightarrow 2KNO_2(s) + O_2(g)$$

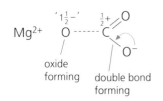

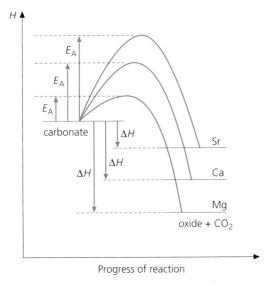

Figure 17.22 The same factor that gives outstanding stability to the product, magnesium oxide, also stabilises the transition state. That factor is the high charge density of the Mg^{2+} ion

a Suggest why potassium nitrate is an ingredient of gunpowder (along with carbon and sulphur).

b Analyse the changes in oxidation number in the equations on the left.

c Predict the relative speeds of decomposition of the nitrates as either of the groups is descended, and give reasons.

d Lithium nitrate decomposes to the oxide, giving off nitrogen dioxide – in other words, more like a member of Group II than Group I. Write an equation for this decomposition, and then use the concept of charge density to explain why lithium is out of line with the rest of its group.

The diagonal relationship

The anomalous behaviour of lithium in question 31d is an example of a larger pattern. The top member of each group is often atypical of the group as a whole, and sometimes more reminiscent of the second member of the next group to the right. Not only do we see the nitrate of lithium giving off nitrogen dioxide, but the compounds of beryllium, at the top of Group II, share the bonding style of Group III. A Be^{2+} ion, if it were to exist, would have such a high charge density (comparable to that of Al^{3+}) that it would influence the electrons of its anion. The ionic model is therefore a poor representation of bonding in compounds like beryllium chloride, which boils at the relatively low temperature of $520\,°C$ and in the vapour phase has linear $BeCl_2$ units, consistent with a covalent molecular style of bonding.

Question

32 It appears the diagonal relationship has to do once again with charge density, because the charge density of the Li^+ ion is similar to that of Mg^{2+}, and the charge density of the Be^{2+} ion is like that of Al^{3+}. But why does the phenomenon mainly affect the character of the *top* member of the group?

Lattices and mechanical construction

Both humans and Nature have exploited the strong lattices of Group II compounds for building structures. Cement, plaster and bone are all compounds containing the cation Ca^{2+}. The final strong lattice is formed slightly differently in each case, with cement and plaster requiring a setting stage triggered by water, while the bone ceramic crystallises straight into its final form, the crystals being grown in special sites on protein fibres.

Plaster

Both plaster of Paris and builders' plaster for skimming walls are the hemihydrate of calcium sulphate, $CaSO_4 \cdot \frac{1}{2}H_2O$. As this formula indicates, there is one mole of water for every two moles of calcium

sulphate in the structure. When wet, the structure is infiltrated by more water molecules which take up co-ordination sites at various points in a strong ionic lattice, and the whole thing recrystallises as $CaSO_4 \cdot 2H_2O$.

Cement

Cement is a complex mixture of the silicate and aluminate of calcium, made from limestone, gypsum and clay (which supplies the aluminium). It has the approximate chemical composition $(2CaSiO_3)(Ca_2Al_2O_5)$, and again the presence of co-ordinated and hydrogen-bonded water is needed to complete a strong three-dimensional ionic lattice.

Bone

The unconscious mimicry of natural structures by human technology is well illustrated in the case of bone. Calling to mind reinforced concrete (as used in load-bearing sections of bridges and buildings), it uses the principle of rods of strong elastic material giving a tensile resilience to an otherwise brittle ceramic. Bone is a mixture of about 70% of a typical Group II 'cement', and 30% of a strong fibrous organic material called collagen. The cement is hydroxyapatite, a mixed phosphate/hydroxide of calcium, formula $Ca_5(PO_4)_3(OH)$, and the collagen is a triple-wound helical protein.

17.7 Aqueous s-block ions in cell chemistry – the nervous system

As we have just seen, humans and Nature make similar use of Group II cations, as strong lattice components. In the case of the Group I cations, the similarity is not so marked. The way that Na^+ and K^+ ions are employed as message carriers in animal nervous systems is highly ingenious (or would be if it were the product of conscious thought), and has no obvious technological equivalent.

Most lay people if asked would say that nerve signals are a form of biological electric current. This is true up to a point, but the current is by no means a conventional one. No electrons or even ions flow the length of a nerve cell. Instead, something like a Mexican wave of potential difference sweeps down the axon of the cell (Figure 17.23). The

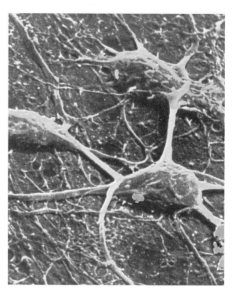

Figure 17.23 Three neurones of the cerebral cortex. The axons are the 'thicker' branches from each cell body (magnification $\times 3924$)

wave is created by movements in and out of the nerve cell of aqueous Na^+ and K^+ ions, in carefully sequenced stages.

When a nerve cell is at rest (i.e. not passing messages), populations of certain ions are maintained at different levels inside and outside the cell, as follows.

1 There is a higher potassium ion concentration inside the axon than out.
2 There is a higher sodium ion concentration outside the axon than in.
3 There are more anions inside the axon than out.

The energy required to maintain this non-equilibrium distribution comes from the body's universal energy carrier, ATP. ATP provides the fuel for a pumping mechanism which achieves the three features above in one complex process. The pump removes three Na^+ ions from the axon for every two K^+ ions it pumps in, so the overall result is not only the aforementioned uneven distributions, but also the creation of a cation shortage, and therefore of an overall negative charge inside the resting axon.

When the nerve cell 'fires', the situation changes rapidly (Figure 17.24). Although the pump goes on working all the time, much more sudden and gross events disrupt the pseudo-equilibrium described above. The first event in the passage of a nerve message or impulse is the opening of 'sodium doors' in the axon membrane at one end of the cell.

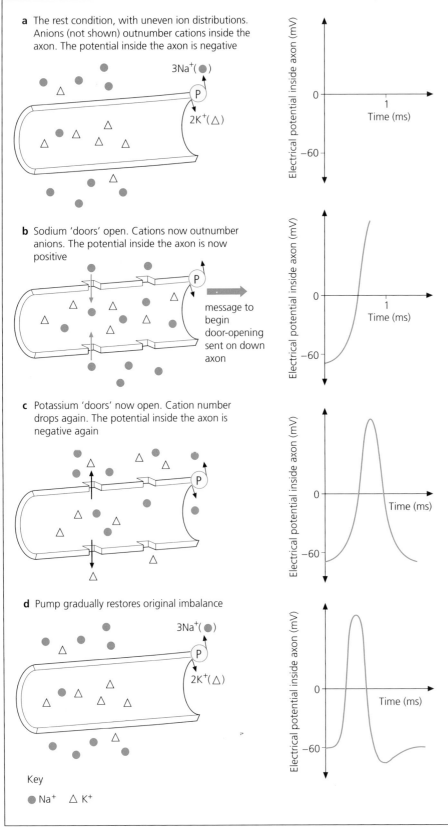

a The rest condition, with uneven ion distributions. Anions (not shown) outnumber cations inside the axon. The potential inside the axon is negative

$3Na^+$ (●)

$2K^+$ (△)

b Sodium 'doors' open. Cations now outnumber anions. The potential inside the axon is now positive

message to begin door-opening sent on down axon

c Potassium 'doors' now open. Cation number drops again. The potential inside the axon is negative again

d Pump gradually restores original imbalance

$3Na^+$ (●)

$2K^+$ (△)

Key

● Na^+ △ K^+

Figure 17.24 Focusing on a particular section of a nerve cell axon, and the changes that happen as the nerve impulse passes

The uneven distribution of Na^+ ions can now be put right by natural diffusion, and the result is an influx of Na^+ ions. The polarity of the interior of the cell is now reversed by this cation migration, and is temporarily positive. This charge condition seems to have two effects.

1 It triggers the nerve cell further along the path of progress of the message to begin *its* sodium-door-opening process.
2 It opens 'potassium doors', which allow the surplus of K^+ ions inside the cell to leak out, returning the interior to a negative polarity, which in turn triggers the shutting of the sodium doors.

As all doors shut again, the pumping mechanism which has been steadily working in the background while the impulse and its attendant upheavals passed by, now needs a few milliseconds to restore all the ions to their correct concentrations, ready for the next bout of action. The 'spike' of positive potential has now swept much further along the cell.

Figure 17.25 gives a highly simplified version of the sweeping influx/Mexican wave idea. The diagram must necessarily exaggerate the proportion of ions that move during any one transmission. In fact, nearly all sodium and potassium cations stay where they are, and the voltages are generated by the movement of a minority. This helps in enabling the system to be pumped back to the 'ready-to-fire' condition quickly.

Much research effort has been directed at the exact nature of the 'doors', with the focus on three questions.

1 How does a door usher an ion through a cell membrane?
2 How does a door select between two such similar chemical species as Na^+ and K^+ ions?
3 How does a door respond to electrical messages in order to open and close?

The first question may not have occurred to you as a problem. GCSE courses leave you with a mental picture of a membrane as a sort of molecular sieve, but in reality it is a rather more intricate structure than that. The cell membrane is a sort of laminated sandwich of sharply variable chemical character.

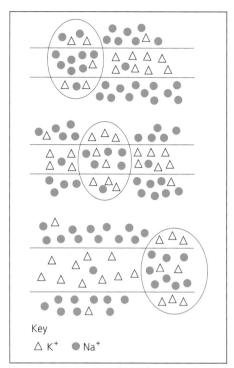

Key
△ K⁺   ● Na⁺

Figure 17.25 The circle highlights the region the sodium influx has reached in each case. This figure does not attempt to show the subtle time lag between the sodium 'doors' and the potassium 'doors' opening

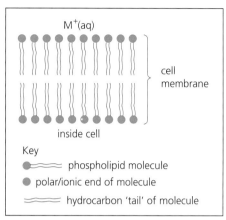

Key
●⟋ phospholipid molecule
● polar/ionic end of molecule
⟋ hydrocarbon 'tail' of molecule

Figure 17.26 A metal cation approaches a cell membrane (seen in much enlarged cross-section). It goes no further than the polar outer coating of the membrane

Like detergent molecules, the phospholipid molecules that make up cell membranes have two distinct ends. One end, which always points outwards into the aqueous medium either inside or outside the cell, is ionic, and hospitable to polar and ionic species. The other end of a phospholipid molecule is a pair of hydrocarbon chains, and all the chains are lined up within the thickness of the membrane. The molecules are arranged in a double sandwich, as in Figure 17.26. Now you can see why the passage of ions across this barrier is not as simple as it would have been with the sieve model. With the sandwich model, a Na⁺ or K⁺ ion would get as far as the outer surface of the cell and then move no further. You might say that the 'door' will have to be more like two doors with a tunnel in between, the tunnel designed to smuggle the ion through an ion-repelling environment. The story of cation transfer and the detail of the 'door' is incompletely understood.

Question

33 a If a 'tunnel' had to be constructed to traverse a cell membrane, it would presumably have to be able to bond into the inner material of the membrane (just as the lining of the Channel Tunnel is bonded to the surrounding chalk rock). What sort of bonds would be needed to secure a tunnel across the middle of the membrane, remembering that membranes are constructed as in Figure 17.26?

b If the tunnel had to allow a sodium ion with, let us assume, a single water molecule solvating it, to pass through, what sort of molecular environment should the inner lining of the tunnel offer?

c Can you suggest a means whereby a tunnel could be designed to allow Na⁺ ions but not K⁺ ions through?

Table 17.4 Preferred Group I oxide formed by reaction with oxygen

Element	Oxide
Lithium	Lithium oxide, Li_2O
Sodium	Sodium peroxide, Na_2O_2
Potassium	Potassium superoxide, KO_2

Summary

- The elements of the s-block of the Periodic Table (Groups I and II) all have their outermost electrons in s sub-shells. They are all soft or fairly soft **metals** of low density.

- All s-block elements have a markedly **electropositive** character, deriving from the relatively weak hold of the nucleus on the outer electrons. So they all show **low ionization enthalpies**, compared with d-block and f-block metals, and therefore are all chemically **reactive**.

- Predictably, given their ease of loss of electrons, in the vast majority of their **compounds** the s-block elements appear as **cationic** species (1+ in the case of Group I, and 2+ in the case of Group II) in combination with non-metals and oxo anions. The lattices of the solid compounds conform closely to predictions based on a simple ionic model of bonding. The few polar-covalent compounds that exist occur mainly in the chemistry of beryllium and magnesium.

- All the s-block metals are **strong reducing agents**, and their cations are correspondingly hard *to* reduce. This means that fairly extreme measures must be taken to extract the metals from their compounds. In all cases the method used commercially is **electrolysis** of a molten ionic compound.

- All the s-block elements react with **water**, with rates of reaction ranging from slow (magnesium) to violently fast (caesium). The product in every case is an **alkaline hydroxide**.

- All the s-block elements react with **oxygen** and air (and some even with nitrogen), the result in each case being an **alkaline** or **basic ionic oxide**. The Group II elements always produce the predictable oxide of formula MO, but an unpredictable set of preferences is shown by the Group I elements. Potassium, for example, produces a superoxide (Table 17.4).

• The **alkalinity** of the s-block hydroxides can be contrasted with the **amphoteric** nature (the ability to act as both acid and base) of the hydroxides of Group III. There is an even sharper contrast with the OH-containing compounds of Groups IV to VII, like sulphuric acid, H_2SO_4, which are totally **acidic** in character. The reasons for the different modes of behaviour can be attributed to the electronegativity differences. The large electronegativity difference between the s-block metal atoms and oxygen results in purely ionic bonds. This leaves the OH group always free to act as the base OH^-. The electronegativity difference between aluminium and oxygen is not so marked. There is a measurable tendency, given certain conditions, for the Al—O bond to be retained, covalently, and instead for the O—H bond to cleave, acidically. In the non-metal OH-compounds, this tendency is extended to the point where O—H cleavage is the only option.

• The most important commercial alkalis are **sodium hydroxide** (whose manufacture was discussed in Chapter 16) and the glass-making alkali **sodium carbonate**. Sodium carbonate is made in the **ammonia–soda** process, which epitomises the ingenuity of industrial chemists, and also the twentieth-century determination to use environmentally benign processes where possible. This process appears to go against the tide of thermodynamic feasibility, which it achieves by splitting the reaction into stages, some of which can be performed under highly non-standard conditions.

• Because the solid compounds of Groups I and II conform closely to the ionic model, they show us how electrostatic principles influence **lattice enthalpies**, and also therefore the associated lattice-breaking processes of solubility and melting. Lattice enthalpy depends on the twin electrostatic factors of **closeness of approach** (in turn dependent on size of ions), and **size of ionic charge**. Differences in structure can cause smaller irregularities in patterns. Notice that the lattice enthalpy is not dependent on how easy the electron removal was in the first place. You should get used to seeing ionization enthalpy and lattice enthalpy as two independent 'players' in the Born–Haber cycle 'game', both influencing the final enthalpy of formation.

• To illustrate these ideas at work, all the Group II/Group VI compounds (like calcium oxide) have much higher melting temperatures than the Group I/Group VII compounds with identical lattice styles (like sodium chloride), because they have double charges and smaller ions, and therefore have much stronger lattices.

• Solubility shows similar trends. Nearly all Group I compounds are soluble in water, whereas many Group II compounds (e.g. carbonates, most sulphates) are insoluble.

• More complex patterns exist within Group I and II compounds of identical anion. If the anion is large (e.g. carbonate), then the upper group member (e.g. magnesium) has the most soluble and most easily decomposed compound, the dominant factor appearing to be the high charge density of the small cation. If the anion is small (e.g. hydroxide), then the upper group member becomes the *least* soluble, the dominant factor being the **stronger lattice**.

• Because these properties of solubility and thermal stability seem to depend upon the charge density (the ratio of charge to size) of cations, the charge density can become the dominant determinant of properties, over-ruling group membership. That is to say, if two cations have similar charge densities, then their compounds may show similarities *despite* the fact that the cations belong to different groups. In particular, compounds of the very smallest member of each group tend to behave like those of the second member of the group to the right. This phenomenon, which is illustrated by the similarities between compounds of lithium and magnesium, beryllium and aluminium, boron and silicon, is known as the **diagonal relationship**.

• The strength of Group II ionic lattices is exploited for building, both in animal skeletons and in human engineering. Bone and concrete, for example, both use ionic lattices containing Ca^{2+} ions.

• In animal nervous systems, the Group I ions Na^+ and K^+ are involved in a complex series of migrations across nerve cell membranes, so that the nerve impulse is like a Mexican wave running down the axon. The equivalent of the 'hands in the air' is the condition where sodium ions have moved into the axon, and it has a positive potential.

18

The typical elements 2: The p-block

18.1 Introduction – general trends in chemical character

We have already seen how the s-block of the Periodic Table contains two of the 'typical' groups of elements, the alkali metals and the alkaline earths. We have also seen that the distinct change in character on moving sideways between typical groups happens because each element differs from its neighbour by one *outer*-orbital electron, in contrast to the situation in the d- and f-blocks.

The rest of the typical groups, from Group III to Group VIII, contain elements in which the orbital being filled up is a p-type orbital, which like the s-type is outer and therefore crucial in deciding character. The electron arrangements of a number of p-block elements are shown in Table 18.1, together with their first ionization enthalpies.

Bonding

Ionization, in the sense of electron removal, does not happen very often in the normal chemistry of the p-block elements. Even so, a lot of the broader characteristics of p-block elements can be inferred from the ionization enthalpies in Table 18.1, as the following question should now reveal. It will serve as revision of Chapters 4 and 5 on bonding.

Question

1 a Why are the ionization enthalpies of most p-block elements higher than their s-block predecessors in the same period (row)?

Table 18.1 The p-block elements, showing outer electron arrangements and first ionization enthalpies

Element	B	C	N	O	F	Ne
A	$2s^2 2p^1$	$2s^2 2p^2$	$2s^2 2p^3$	$2s^2 2p^4$	$2s^2 2p^5$	$2s^2 2p^6$
B	801	1086	1402	1314	1681	2081
Element	Al	Si	P	S	Cl	Ar
A	$3s^2 3p^1$	$3s^2 3p^2$	$3s^2 3p^3$	$3s^2 3p^4$	$3s^2 3p^5$	$3s^2 3p^6$
B	578	789	1012	1000	1251	1521
Element	Ga	Ge	As	Se	Br	Kr
A*	$4s^2 4p^1$	$4s^2 4p^2$	$4s^2 4p^3$	$4s^2 4p^4$	$4s^2 4p^5$	$4s^2 4p^6$
B	579	762	947	941	1140	1351
Element	In	Sn	Sb	Te	I	Xe
A*	$5s^2 5p^1$	$5s^2 5p^2$	$5s^2 5p^3$	$5s^2 5p^4$	$5s^2 5p^5$	$5s^2 5p^6$
B	558	709	834	869	1008	1170
Element	Tl	Pb	Bi	Po	At	Rn
A*	$6s^2 6p^1$	$6s^2 6p^2$	$6s^2 6p^3$	$6s^2 6p^4$	$6s^2 6p^5$	$6s^2 6p^6$
B	589	716	703	812	—	1037

A: electron orbitals occupied in highest principal quantum level, lying outside a noble gas style electron arrangement
A*: electron orbitals occupied in highest principal quantum level, lying outside an electron arrangement which includes full d and/or f sub-shells *plus* a noble gas structure, for example, gallium (Ga) has [Ar] + $3d^{10}$ + $4s^2 4p^1$
B: first ionization enthalpy (kJ mol^{-1}), i.e. the enthalpy change for M(g) → M^+(g) + e^-(g)

b The dividing line between metals and non-metals runs through the p-block. Indicate on a copy of Table 18.1 where it lies.

c Which group is the most likely to show chemistry featuring simple cations of the M^{n+} type? Where within a group are simple cations most likely to be found? Explain each of your predictions. Give an example of a compound in which a p-block element exists as a simple cation (avoid cases where the bonding is 'impure' or ambiguous).

d Which groups are most likely to show simple anions of the X^{n-} type? Give two examples, from different groups within the p-block, of compounds in which a p-block element exists as a simple anion.

e What sort of bond would you expect when elements of the p-block combine? Give two examples.

f What sort of bond would you expect when elements of the p-block combine with elements of the s-block?

g Elements of Groups III and IV make no conventional compounds with s-block metals. Suggest why.

Table 18.2 Redox data on the elements of Periods 2 and 3 of the p-block. (The electronegativity of hydrogen is 2.1, for comparison.)

Element	Stable oxidation numbers	Electronegativity
B	+3	2.0
C	+4 $\overset{\text{all}}{\rightarrow}$ −4	2.5
N	+5 $\overset{\text{all}}{\rightarrow}$ −3	3.0
O	+2*, −1, −2	3.5
F	−1	4.0
Ne	0	—
Al	+3	1.5
Si	+4	1.8
P	+5, +3, −3	2.1
S	+6, +4, +2, +1, −2	2.5
Cl	+7, +5, +1, −1	3.0
Ar	0	—

*Only in compounds with fluorine

d The highest fluorides of the same period of elements have formulae as shown: AlF_3 and SiF_4 (as in (c) above) and PF_5, SF_6, ClF_5. Select one of these compounds that is breaking the rule of eight, and show by a dot-and-cross diagram that it is indeed doing so. Assign oxidation numbers to the elements in these compounds.

e Fluorides were chosen in (d) above because an element is most likely to show its maximal oxidation state when bonded to fluorine. Why do you think this is?

f Why do you think chlorine broke the pattern of the sequence of compounds in (d)? (Iodine *does* form a fluoride IF_7.)

A picture emerges of elements that, apart from the nearly inert members of the noble gas family, mostly take part in covalent or polar covalent bonding. The exceptions are some extreme 'right-wingers' and 'top-of-groupers' (like F) which take part in ionic bonding as the anion, and some extreme 'left-wingers' and 'bottom-of-groupers' (like Tl) which take part in ionic bonding as the cation.

Oxidation numbers and the rule of eight

Those p-block elements with anionic tendencies show negative oxidation numbers in their ionic compounds with metals. Apart from members of Group III and the lower part of Group IV, the only positive oxidation numbers are acquired by polar covalent bonding to other p-block elements of greater electronegativity. Table 18.2 shows the common stable oxidation numbers of two rows of p-block elements, along with their electronegativities. You will notice that the maximal oxidation numbers of many elements correspond to the group number. The next question asks you to consider this and other patterns, and will act as revision for the bonding chapters and Chapter 7 on oxidation number.

Question

2 a Why is it impossible for elements to exhibit a positive oxidation number greater than their group number?

b Why does fluorine, alone among the elements, show no positive oxidation number in any of its compounds?

c Draw dot-and-cross diagrams for a series of fluorides across the period from aluminium to chlorine. Make your compounds obey the rule of eight, where possible. What do you notice about the valencies of the elements in relation to their group numbers? What are the oxidation numbers of the elements in these compounds?

Types of structure in p-block chemistry

The p-block is the home of the small molecule. If a structure is of the small molecular type, it is almost certain to be made solely of p-block elements (or hydrogen). Covalent bonds are inevitable between atoms that share a reluctance to release electrons, and in many cases the results of covalent bonding are small molecular units. However, although all small molecules derive from the p-block, the reverse is *not* always true – not all p-block-only structures are small molecules. Some are giants, in which covalent bonding systems go on 'for ever', as in diamond, graphite and silicon dioxide. Nor is it true that particular groups will always follow a single pattern – the oxides of the top two members of Group IV have opposite structures, with carbon dioxide as the small molecule and silicon dioxide as the giant. The impact of these contrasting structures on physical properties is of course enormous, with carbon dioxide turning to gas at −78 °C, and silicon dioxide remaining solid up to about 1600 °C (depending on lattice type). The reasons behind these and other structural options will be discussed as they arise, in the sections that follow.

18.2 The elements – factors affecting melting and boiling points

If the temperature is cold enough, any element becomes a solid. As such it is merely a collection of identical spheres. Given this common starting point, it is remarkable how such variation in properties is achieved by the elements of the p-block, ranging from neon with a boiling point of $-246\,°C$ to carbon/graphite with a *melting* point near $3700\,°C$. There is also a spectrum of conductivity, from the highly conducting aluminium through the semiconductors silicon and germanium to the insulators of the top right-hand corner of the Periodic Table. (We shall look at this in more detail in Section 18.3.) Where is the source of this great variety?

The answer lies, like many explanations in chemistry, largely with the outer electrons. There are several distinct mechanisms at work, making one element a gas and another a solid, or one a conductor and one an insulator. Taking volatility first, we can isolate two factors that affect melting and boiling points. Both depend on whether bonds between adjacent atoms in the solid element can be used to build a continuous (giant) lattice or not.

1 Valency

If we consider only classical bonds, then it is easy to see that some elements have a much better chance than others of building in two and three dimensions. For instance the noble gases have a natural valency of 0, so they fail to build any structures at all, in any dimension. The only bond available for building lattices is the weak van der Waals' force, which as we have seen is not strong enough to keep the noble gases solid far above $-273\,°C$. This 'dimensional' train of thought continues into the next question.

Question

3 Consider the normal covalent-bond valencies of the elements fluorine, oxygen, nitrogen and carbon. In how many dimensions could they build arrays of atoms using *single* bonds? (Call chains one-dimensional (even if

they zig-zag), layers or plates two-dimensional (even if they are not quite flat), and structures with depth three-dimensional.)

We see from question 3 that carbon's valency of four gives it an obvious advantage in lattice-building. The best nitrogen, for example, could manage would be a lightly undulating layer structure, with covalent intra-layer bonds but weak van der Waals' inter-layer bonds.

2 Choices between single and multiple bonds

This is the second factor affecting the melting and boiling temperatures of the p-block elements. If nitrogen did exist in the layer structure described above, it would probably have melting and boiling temperatures comparable with those of an organic polymer. But of course real nitrogen is not like that, any more than real oxygen exists as chains. They both take another option, which is to fulfil their rule-of-eight valencies via multiple (double or triple) bonds.

Question

4 (*Revision of Chapter 4*) Draw dot-and-cross diagrams of the molecules dioxygen and dinitrogen, showing their multiple bonding.

However, the bonding in the element at the top of each group is not always representative of the group as a whole. Indeed nitrogen and oxygen are the *only* members of their respective groups to occur as multiply bonded diatomics. As a group is descended there is a tendency away from multiple bonds and towards more extended structures with single bonds. For example, Figure 18.1 shows the second member of each of Groups V and VI in one of their allotropic forms. A convincing argument can be advanced to explain this trend, using a familiar and accessible underlying concept – namely electron–electron repulsion.

Figure 18.2 shows the single bonds N—N and P—P, drawn roughly to scale, in the electron-cloud style (with dots and crosses superimposed to show the degree of occupation of each orbital). The orbital shapes show hybrid mixtures of

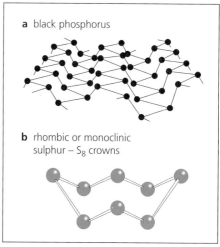

Figure 18.1 Two forms of the elements of the p-block, showing the trend towards extended single-bonded structures in the second and subsequent members of each group

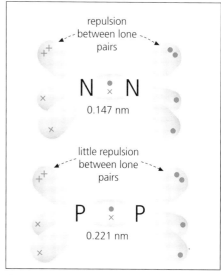

Figure 18.2 Illustrating the reason why the top member of each group finds single bonds a less attractive option (and multiple bonds a more attractive one)

s- and p-type electrons, giving four tetrahedrally arranged orbitals for the outer electrons of each atom (see, for example, Figure 4.45). Overlaps are of the nose-to-nose σ-type. The reason why nitrogen cannot build a lattice on this basis while phosphorus can lies in the fact that there will be considerable repulsion between the orbitals in the case of nitrogen, whereas this will not be significant in phosphorus (these repulsions are symbolised in Figure 18.2).

Question

5 What is different between the two cases in Figure 18.2, causing the repulsion in nitrogen to be a significant destabilising factor?

For nitrogen the solution is simple: instead of allowing those electrons to cause instability, use them as a source of bonding. So the dinitrogen molecule features only one σ-bond, and the other two unpaired electrons on each atom are used to create π-bonds.

Question

6 Table 18.3 shows the bond enthalpies and lengths of some single and multiple bonds in the groups we are considering.

Table 18.3 Lengths and strengths of single and multiple bonds

Group	Bond	Bond enthalpy (kJ mol^{-1})	Length (nm)
V	N—N	159	0.147
	N≡N	933	0.110
	P—P	209	0.221
	P≡P	524	0.190
VI	O—O	142	0.149
	O=O	498	0.121
	S—S	264	0.206
	S=S	428	0.189

a How many times stronger is the N_2 triple bond than the N—N single bond?

b How many times stronger is the P_2 triple bond than the P—P single bond?

c What reason have we just advanced for the relative weakness of the N—N single bond?

d What reason can you suggest for the relative strength of the P—P single bond?

e What reason can you suggest for the much greater success of π-style overlaps (that is, multiple bonds) in N_2 compared with P_2? Remember that π-overlap needs the atoms to approach fairly closely to be effective.

To summarise this section on the choice between single and multiple bonds, we could say it is all a matter of size. The shortness of the N—N bond is responsible for the potential repulsions that act against successful single bonding. But that same shortness is responsible for the success of the π-bonds, which need small interatomic distances in order to be effective. Reversing the story for phosphorus, the length of the P—P bond favours the σ-style and hinders the π-bonding.

Question

7 a Draw an electron cloud picture of dioxygen. Figure 18.3 shows the dot-and-cross equivalent, to remind you how many electrons should be in your picture. Pay attention to the siting of the lone pairs.

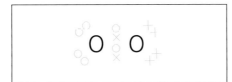

Figure 18.3 Dot-and-cross diagram of O=O, dioxygen

b It may have struck you as odd that two elements of such similar electronegativities as oxygen and nitrogen should be so different in their reactivities as elements. (Nitrogen is nearly as inert as a noble gas, and indeed was once used to fill filament light bulbs to protect the filament from oxygen.) By using data from Table 18.3, and by looking at the electron cloud pictures for the two molecules, and finally by applying our recently coined theories on electron repulsion, suggest why dioxygen is so much more reactive than dinitrogen.

18.3 The elements – factors affecting electrical conductivity

One of the nastier shocks in passing from Advanced level to chemistry in higher education is the realisation that simple models for electrical properties of materials do not quite do the job. So far we have accounted for the conductivity of metals by the simple idea that metals hold their outer electrons loosely, so that to conduct or not to conduct becomes a matter of ionization enthalpy. This level of analysis certainly explains the conductivity of the elements of Group III from aluminium downwards.

Question

8 Can it also explain why boron does *not* show metallic-style conduction?

The problems arise because of a swathe of elements which populate the borderline between the metals and the non-metals, and which exhibit a package of properties collectively called **semiconduction** (Figure 18.4). Characteristics of semiconductors include the following.

- These elements are insulators at low voltage but can become conductors at higher voltages.

- Their conduction is increased by raising the temperature (unlike normal conductors).

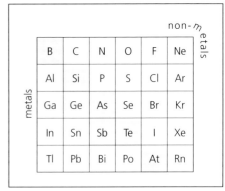

Figure 18.4 Shaded elements are classified as **semi-metals**, which implies that at least one allotropic form will show **semiconduction**. The classification is imperfect, since both carbon and phosphorus have forms that show some conduction

• Conduction may also be improved by absorption of photons (semiconductors are used in light-dependent resistors).

Yet both silicon and germanium, for instance, exist in a diamond-style crystal lattice held together by what appear to be localised covalent bonds (Figure 18.5). Figure 18.6 shows the electron configuration of one atom in the silicon lattice. All orbitals are full and the next orbitals, the 4s and 3ds, are energetically remote. Even if you could tempt an electron out from one of the bonding pairs, there are no viable orbitals on other silicon atoms for it to jump into. With a Group III element the situation is different – there is a spare orbital on each atom because Group III elements cannot complete their rule-of-eight octets. How then do electrons in Group IV elements achieve mobility within their lattice?

The model needed to explain semiconduction, **band theory**, is a branch of molecular orbital theory, and comes from a rigorous wave mechanical view of electron orbital overlap which lies outside Advanced level. Suffice it to say that there *are* other ways in which the atomic orbitals of atoms can overlap, distinct from the straight share-and-bond style we have concentrated on. There is a style of overlap that produces electron density unhelpful to bonding, and the result is an **anti-bonding** molecular orbital. Figure 18.7 shows the two possible orbitals produced by the overlap of two 3sp³ hybrid orbitals on adjacent silicon atoms.

These anti-bonding orbitals offer a new (to us) alternative site for electrons, intermediate in energy between the low-energy stability of the bonding orbitals and the high-energy unattractiveness of the next empty 'normal' orbitals. Providing the surroundings can supply the necessary energy, the bonding electrons can be promoted into these 'half-way house' anti-bonding orbitals. Now mobile electrons are possible, because if a voltage is applied, electrons can be moved from a bonding orbital on one silicon atom to the vacant anti-bonding orbital on the next, while the 'hole' left in the bonding orbital will offer an alternative route to the passage of incoming electrons (Figure 18.8).

The size of the energy gap between bonding and anti-bonding orbitals decides the properties of the semiconductor, or indeed whether a material *is* a semiconductor. For instance, in diamond the gap is too big to allow promotion of electrons, so diamond's electrons are 'stuck' and diamond is an insulator.

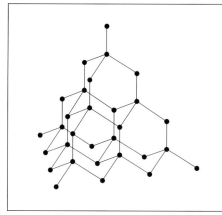

Figure 18.5 The diamond structure

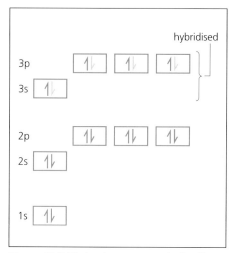

Figure 18.6 Electronic arrangement of a silicon atom in a solid lattice of silicon, in box diagram style. (Green shows 'visiting' electrons.)

Question

9 You may think you have understood little of the above, which would be entirely forgivable given the necessarily superficial depth of the explanation. However, you may be able to answer the following questions, showing the usefulness of a rough-and-ready working model which can always be refined at a later stage in your chemistry career.

a What can you infer about the bonding–antibonding energy gap as Group IV is descended?

b How does our crude model account for the fact that, in semiconductors, conduction increases as the temperature increases?

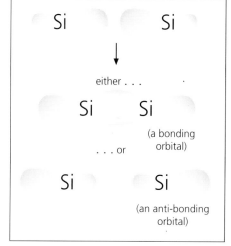

Figure 18.7 Showing that for the overlap of two atomic electron orbitals, there are always *two* possible combinations (this goes back to the wave nature of electron orbitals, and can be equated crudely with in-phase and out-of-phase overlap)

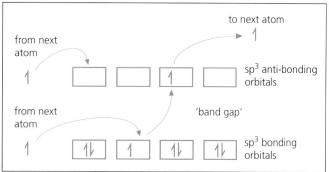

Figure 18.8 Possible movements of electrons (↑) in a semiconductor. The 'band gap' is the energy gap between the bonding and antibonding orbitals

c The bonding–antibonding energy gaps in semiconductors are of the same order of magnitude as the energies of visible-light photons (among others). How might this account for the phenomenon of light-dependent resistance?

d If visible-region photons can interact with electrons in semiconductors, suggest why diamonds are colourless while silicon and germanium are black.

e The fact that silicon is black rather than a particular colour is something our crude theory cannot handle. Can you see where the theory falls down?

Having so far taken a 'broad-brush' approach to the p-block, we now focus on one group at a time.

18.4 The elements – Group IV

The physical data on the Group IV elements are presented in Table 18.4. The points of interest in their chemistries are outlined below.

Table 18.4 Data on the Group IV elements

Element	Electron arrangement	First ionization enthalpy (kJ mol^{-1})	Melting point (°C)	Boiling point (°C)	Crystal type
C	[He]2s^{2}2p^2	1086	3652	4827	Graphite/ diamond
Si	[Ne]3s^{2}3p^2	789	1410	2355	Diamond
Ge	[Ar]3d^{10}4s^{2}4p^2	762	937	2830	Diamond
Sn	[Kr]4d^{10}5s^{2}5p^2	709	232	2270	Diamond
Pb	[Xe] 4f^{14}5d^{10}6s^{2}6p^2	716	328	1740	Face-centred cubic

Carbon

Carbon is of central importance to the biosphere. Its compounds are the building blocks of living things, and also their energy source. In Chapter 17 we saw parallels between natural and human/technological exploitation of the Group II compounds for making hard, insoluble materials. Another example of parallelism applies very clearly in the case of carbon.

The natural world uses carbohydrates of widely variable molecular size as energy-rich compounds. The molecular size usually indicates whether the compound is designed for immediate use or as an energy storage material.

Question

10 Give examples of carbohydrates which organisms can utilise

a for storage

b for immediate use.

The world of human technology has tapped into the natural cycle and borrowed its molecules. We rarely use carbohydrate energy in a technological context, but we are massively dependent for energy on the geological-time-altered versions of natural carbon chemicals, in the forms of coal, oil and gas. Thus we have become major players in the running of our planet's **carbon cycle** (Figure 18.9).

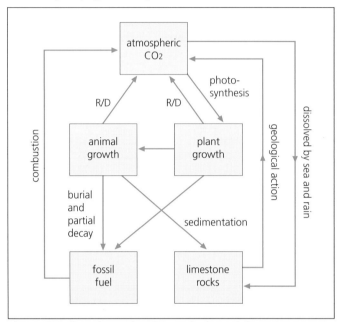

Figure 18.9 The carbon cycle. R/D = respiration and decay. Arrows indicate direction of change, and movement of carbon atoms. Boxes indicate 'resting places' of carbon

The energetics of the carbon cycle can be seen as a field bounded by the extremes of hydrocarbons on the one hand and carbon dioxide and the carbonates on the other. One group contains energy-rich molecules, and the other is the bottom of the 'energy well', exhausted of all free energy.

Question

11 Which group is which?

The passage of carbon compounds through the carbon cycle is a movement up and down this energy spectrum. In fact we can rearrange the carbon cycle into a vertical energy ladder. It is easy enough to see that photosynthesis would be a move up such a ladder, and combustion down, but let us see what happens during one of the more subtle changes (Figure 18.10).

Question

12 Let us model (and simplify) the formation of coal as a reaction by which carbohydrates decay to carbon and water under bacterial influence.

$$C_6H_{12}O_6(s) \rightarrow 6C(graphite) + 6H_2O(l)$$

a Work out the enthalpy change of this reaction.

b The selection of the above reaction owes more to

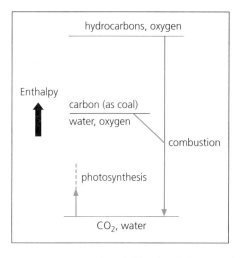

Figure 18.10 Schematic enthalpy diagram for the various forms of carbon in the biosphere. Carbohydrates have been omitted (see question 12)

convenience (available data) than reality. In what ways might it be an oversimplification of the real overall process of plant decay?

c In the light of your answer to (a), place plant carbohydrates on the enthalpy ladder in Figure 18.10.

Charcoal and coke contain carbon in a form that fails to give an X-ray diffraction pattern. Such materials are called **amorphous**, meaning 'shapeless', and it may be inferred that they lack regular repeating units in their solid structures – you could not draw a unit cell.

This lack of order can have two causes – either the material is so finely subdivided that structural patterns are significantly interrupted by grain boundaries, or there is not any long range order even over quite extensive regions of solid. Charcoal belongs to the former group. Members of the latter group, which include compounds of silicon, are collectively known as **glasses** (Figure 18.11).

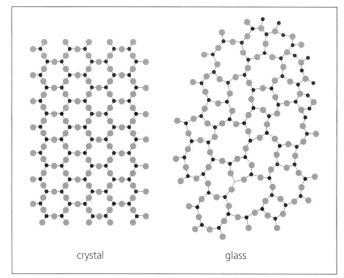

crystal   glass

Figure 18.11 Showing how the glassy state is characterised by a breakdown of orderliness in the structure

You will remember carbon's two genuinely crystalline forms diamond and graphite from GCSE science (Figure 18.12). The contrast between them is impressive – here are two materials that are made from the same atoms, yet have very different properties.

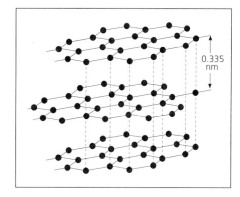

0.335 nm

Figure 18.12 The structure of graphite – compare with the diamond structure in Figure 18.5

Question

13a Make up a table comparing the properties of diamond and graphite.

b The conductivity of graphite is not of the semiconductor type. It is due to an almost benzene-style delocalisation of π-bonding electrons associated with the hexagon units. However, a single crystal of graphite does show one distinctly non-metallic feature, and that is that it does not conduct equally well in all directions. Suggest why.

c Crystals of graphite do not behave in the same way in all directions in response to squashing and shearing forces. This is the cause of graphite's lubrication properties. Explain how its crystal structure leads to these lubricating properties, and why its combination of two properties makes it better than almost any other material for the brushes of electric motors.

Both diamond and graphite occur naturally on Earth, but graphite is actually the more stable thermodynamically (and is the thermodynamic standard state of C(s)). Both can also be made artificially, graphite from coke and diamond from graphite:

$$3C(s) + SiO_2(s) \xrightarrow{2300K} C(s) + Si(s) + 2CO(g)$$
$$\text{coke} \quad \text{sand} \qquad\qquad \text{graphite}$$

$$C(graphite) \rightarrow C(diamond)$$

The conditions for the second reaction are 3000 K and 120 000 atm, and even then it needs a catalyst!

Silicon

Silicon is the second most abundant element on the Earth's crust after oxygen (26% and 46% respectively), and the two in combination, in silicon(IV) oxide (silica, SiO_2) and the silicates, make up the bulk of the minerals in crustal rocks. The element is prepared by the reduction of silica by coke:

$$SiO_2(s) + 2C(s) \rightarrow 2CO(g) + Si(s)$$

Question

14 Calculate the values of $\Delta H_{sys}^{\ominus}$ and $\Delta S_{sys}^{\ominus}$ for this reaction. Assume these values are roughly constant over a large temperature range and calculate the temperature at which this reaction begins to be energetically viable.

Silicon has only one major allotrope, and it has the diamond structure. We can perhaps explain the contrast between silicon and carbon by remembering that to make a graphite-style structure you need π-bonding in each layer.

Question

15 Suggest why a silicon analogue of graphite does not exist.

Silicon is of course used by, indeed is synonymous with, the microelectronics industry. This industry needs silicon with impurity levels less than one part in 10^{12}. Unfortunately silicon atoms find it only too easy to get mixed up with other lattices, especially metals, and indeed alloys of silicon and aluminium have found extensive use in light-alloy car engine blocks.

To get very pure crystals of silicon a technique called **zone refining** is used (Figure 18.13). A 99.9% pure silicon ingot (that is 10^9 times too impure – it has to be 99.9999999999% pure to be acceptable) is placed in a furnace, surrounded by inert gas, and pulled through a region heated by coils. A zone of molten silicon passes through the ingot, collecting impurities as it goes. As the molten part resolidifies, only pure silicon crystals are deposited, with the impurities staying in the melt. After several passes in the same direction, cooling between each, the end is cut off the ingot.

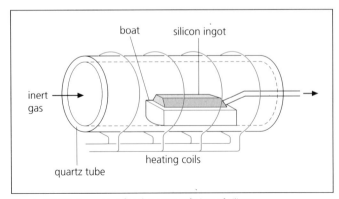

Figure 18.13 Apparatus for the zone refining of silicon

Questions

16 Zone refining works, in part, because the impurities lower the melting point of silicon. So:

a If the impure liquid deposits crystals of pure silicon as it cools, what will that do to the concentration of impurities in the remaining liquid?

b As the liquid zone moves through the ingot, how will the melting point of the liquid be affected?

17 Why is the end cut off the ingot, and why must the passes through the furnace be in the same direction?

Table 18.5 shows the main uses of the Group IV elements and their compounds.

Table 18.5 Uses of Group IV elements and compounds

Element	Used as	Uses
C	Diamond	Abrasives and cutters
	Graphite	Electrodes, pencils, lubricants
Si	Element	Electronic components
	Silica (SiO_2)	Glass
	Silicones	Polymers, e.g. sealants
Ge	Element	Electronic components
Sn	Element	Coating on 'tin cans'
	Alloys	Bronzes (with Cu) and solders (with Pb)
Pb	Element	Batteries, plumbing, roofing
	Alloys	Solders
	$Pb(C_2H_5)_4$	Petrol additive

18.5 The elements – Group V

Nitrogen

The characteristic feature of the chemistry of dinitrogen is its inertness. The N_2 molecule is held together by a very strong triple bond ($945\,kJ\,mol^{-1}$), and the molecule's ionization enthalpy is comparable to that of argon.

Question

18 Draw dot-and-cross and electron-cloud pictures of the N_2 molecule. For the electron-cloud picture, assume that two of the p-electrons on each atom stay in pure p-orbitals, and that both of them enter into π-bonds. The molecule will look a bit like ethene but with an extra 'double sausage'.

Dinitrogen's inertness is a necessary condition for the existence of life on Earth. If dinitrogen were as reactive as dioxygen, then the mixture that makes up our air would have reacted long ago to create seas of nitric acid. As it is, only the extreme conditions in lightning flashes are energetic enough to cause nitrogen to react with oxygen:

$$N_2(g) + 2O_2(g) \rightarrow 2NO_2(g)$$

This is one of two mechanisms in our planet's nitrogen cycle (Figure 18.14) which 'fix' atmospheric dinitrogen in a form available for plants.

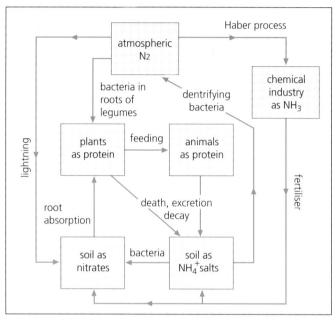

Figure 18.14 The nitrogen cycle

Question

19a Which other natural path fixes atmospheric nitrogen?

b Into which class of compounds do plants build their intake of nitrogen atoms?

c Why have humans tried hard to intervene in the nitrogen cycle to increase the rate of nitrogen fixing?

The Haber process, by which nitrogen is turned into ammonia, is dealt with on pp. 459–63.

Phosphorus

Phosphorus has been known to chemists for over 300 years. It was named for its ability to ignite spontaneously and glow coldly in darkened rooms. The darkened rooms in question were those of the rich élite of the courts of Europe, who had a great appetite for such novelties. The word 'phosphorus' means 'light bearer', and was also the name given by the Greeks to the Morning Star. The romance of the name is slightly at odds with the earliest means of manufacture, which was from large volumes of urine. The phosphate anion from urine was reduced by carbon:

$$4PO_4^{3-}(s) + 10C(s) \rightarrow P_4(s) + 10CO(g) + 6O^{2-}(s)$$

Nowadays the raw material has become the same as that used for the manufacture of phosphate fertilisers, namely the phosphate rock fluorapatite.

Phosphorus exists in a number of allotropic forms, of which white, red and black are the best characterised. Only the white form has a small molecule, existing as a curious equilateral trigonal pyramid (Figure 18.15), and only the white form undergoes spontaneous combustion.

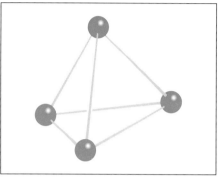

Figure 18.15 The P_4 molecule in white phosphorus

Question

20a Deduce the bond angle in the P_4 molecule, simply by looking at the geometry of the molecule.

b If phosphorus is obeying the rule of eight, it will have four 'bits' (to borrow a piece of jargon invented in Chapter 4) in its outer orbitals, namely three bond pairs and a lone pair. What therefore should be the natural bond angle at trivalent phosphorus atoms?

c Comment on the amount of strain (in the form of electron pair–electron pair repulsion) implied by the difference between your answers to (a) and (b).

d Deduce what this might mean for the activation enthalpy needed to break the P_4 molecule and obtain individual atoms of phosphorus ready to react.

e Hence suggest why P_4 molecules spontaneously combust.

Phosphorus's association with fire has given it a pre-eminent place in the match industry. A combustion reaction whose activation enthalpy was within the scope of a light muscular movement with a person's hand was an obvious boon to humankind. Unfortunately the obvious choice of white phosphorus proved to be disastrous for the workers in match factories, since P_4 initiates a form of tooth decay that ends in extreme cases with rotting of the jaw bone (Figure 18.16). There was also an unhappy tendency for these early matches to ignite at unexpected moments, for example when accidentally rubbed.

The solution has been to switch to a compound that is less volatile, and so does not get into the air in factories. The compound also has a higher activation enthalpy than white phosphorus, yet one that is still attainable from the energy of a strike. The compound in question is phosphorus sulphide,

Figure 18.16 Bryant & May's match factory in Hackney, London, circa 1900

Figure 18.17 a Phosphorus sulphide matches.   b Safety matches

P_4S_3, present in the heads of the style of matches that can be struck on any rough surface (Figure 18.17a). As an extra refinement it has proved possible to separate the elements of this compound so that in safety matches the strike actually creates the compound as well as activating the reaction. Figure 18.17b shows the separation of chemicals on match-box and match head.

Question

21a What is the role of the powdered glass in the energetics of the reaction?

b There is a possible reaction between the chemicals in the head alone. What is it, and why does it not happen when the safety match is struck on an inert rough surface?

c Write the chemical equations for the reactions that happen when the match is struck.

The main uses of nitrogen and phosphorus and their compounds are shown in Table 18.6.

Table 18.6 Uses of Group V elements and compounds

Element/compound	Use
N_2	In Haber process
NH_3	Fertilisers; precursor to nitric acid; refrigerant
HNO_3	Fertilisers; explosives
P	Matches
Phosphoric acids	Fertilisers, detergents

18.6 The elements – Group VI

Oxygen

Oxygen is the most abundant element in the crust of our planet (46% by mass). Apart from its ubiquity, it has a central role to play in the energetics of life on Earth, as it is one of the reactants in normal cellular respiration in all but a few highly unusual living things. So we normally think of dioxygen as reactive, and contrast it with the passivity of dinitrogen.

However, oxygen is not all that reactive, and a good thing too. Several times in the course of this book it has been mentioned that there would *be* no book, and no you or me either, if the oxidations of certain materials like carbohydrates and proteins were not inhibited by quite substantial activation enthalpies.

The point being made here is that the dioxygen molecule itself makes a large contribution to those activation enthalpies. The molecule is quite strongly bonded ($500 \, \text{kJ mol}^{-1}$), and has an ionization enthalpy as high as that of xenon!

Combustion reactions are free-radical reactions, sometimes with branching chains (that is, when a single step can create two radicals from one). Once initiated these reactions can be explosively fast, as they are in gas explosions and in motor car combustion cylinders. Oxygen is well suited to participating in such reactions, for the dioxygen molecule has a property that cannot be explained by our relatively simple bonding theory. Dioxygen has two unpaired electrons. The evidence is incontrovertible, in that liquid oxygen is affected by magnets in just the way that species with unpaired electrons should be.

It is beyond the needs of the Advanced student to follow the story further, except to say that free radicals in the combustion mixture can interact with these unpaired electrons on dioxygen in a way that eases the breaking of the double bond. Once the two oxygen atoms are held together by a single bond, as in the peroxides, then reaction is plain sailing.

Questions

22 The single bonds in N—N, O—O and F—F are all notably weak (Figure 18.18), especially compared with the multiple bonds in nitrogen and oxygen. What theory have we previously advanced to explain this phenomenon?

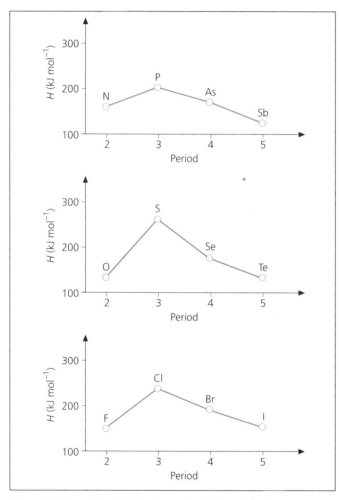

Figure 18.18 Bond enthalpies of X—X bonds (kJ mol^{-1}) plotted against period for elements of Groups V, VI and VII

23 Combustion mixture additives, like the lead compound in leaded petrol, $(C_2H_5)_4Pb$, operate by providing their own free radicals which get involved in the burning mechanism. In the case of $(C_2H_5)_4Pb$ the purpose is to inhibit ignition. Suggest which free radicals might be formed from this compound, and how they might help to obstruct the main free-radical mechanism of burning. (Sensible guesses are all that is required here.)

24 Oxygen has a triatomic allotrope, the infamous ozone, O_3. Suggest how the ozone molecule might be held together (it has a dative covalent bond), and propose a shape for its molecule.

Sulphur

There is a very large amount of sulphur in our planet, 'in' being the operative word. The core of our planet, which is rich in iron, also contains about 15% sulphur, but this is of course inaccessible to us. In the crust the proportion is much lower, around 0.05%, mainly in the form of iron, zinc and lead sulphides and Group II sulphates from evaporated sea water.

There is also an amount of elemental sulphur, some of it the result of volcanic activity and some from bacterial reduction of sulphates. Humans have not yet managed to mimic this bacterial reduction, so our sources of sulphur for industry come from the deposits of either sulphides or elemental sulphur, S_8. These elemental deposits have been exploited by a most unusual and ingenious mining technique, which contrives to bring the sulphur to the surface from depth and purify it at the same time.

The technique is called the Frasch process, after its inventor Herman Frasch, and it relies on the fact that sulphur has a melting point within the range of superheated steam. A set of concentric pipes ensures that the steam (at 165 °C) is delivered down to the sulphur beds, and the molten water/sulphur mixture is pumped up to the surface (Figure 18.19).

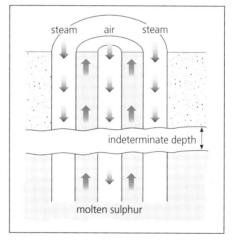

Figure 18.19 Schematic picture of the Frasch process for extracting sulphur from underground strata

Question

25a Explain the need for air in the middle pipe.

b Why is it a good idea to have the steam delivery ring on the outside of the set, with the sulphur delivery ring next in?

c 165 °C is considerably in excess of the melting point of sulphur. Why is it a good idea to work with this apparently generous margin of temperature?

d How would the sulphur be separated from its carrier water, once at the surface?

e If you were designing a Frasch plant, can you think of a heat exchanger stage to make use of the heat contained in the hot water from stage (d)? (Industrial chemists rarely let heat escape into the atmosphere if they can avoid it, for obvious financial reasons.)

f The sulphur emerges from the ground already about 99% pure. How does the process achieve this purification?

g Crude oil is another mineral that does not have to be literally 'mined'. Think of two similarities and two contrasts between the sulphur and oil extraction techniques.

Structures of sulphur

Throughout this chapter we have met examples of the anomalous behaviour of the top elements of groups compared to second and subsequent members, notably in the relative stability of single and double bonds. The same is true in Group VI, so that double-bonded S=S molecules equivalent to dioxygen are only found at extremes of high temperature and low pressure, while chains and rings of S—S single bonds are the norm.

Generations of young chemistry students have been asked by their teachers to experiment with sulphur, to enjoy the trick effects as the element performs its repertoire of natural variations on a single bond theme. The results have been a mixture of fascination, fun, unscheduled burning of sulphur, coughing, classroom evacuation and asthma attacks. The events whose study is supposed to justify all this mayhem are summarised in Figure 18.20.

Question

26a If rhombic and monoclinic sulphur both feature S_8 crowns (as in Figure 18.1b), why do you think they have different crystal shapes?

b Why does sulphur crystallise in the monoclinic form from molten sulphur, but in the rhombic form from solutions in organic solvents at room temperature?

c When newly formed, monoclinic crystals are shiny translucent needles. These go opaque within hours of cooling, although their outer shape remains needle-like. Assuming the sulphur has undergone the change listed in Figure 18.20, explain the opacity.

d Rhombic sulphur is the equilibrium condition of sulphur at room temperature. Why then does plastic sulphur succeed in existing?

e Plastic sulphur can be stretched to 20 times its original length. Obviously chemical bonds cannot stretch this far, so it must be due to a stretching of the secondary structure of the molecule (using the term 'secondary' as it is used to describe the secondary structures of proteins, see Chapter 15). Suggest what sort of secondary structure plastic sulphur chains might have.

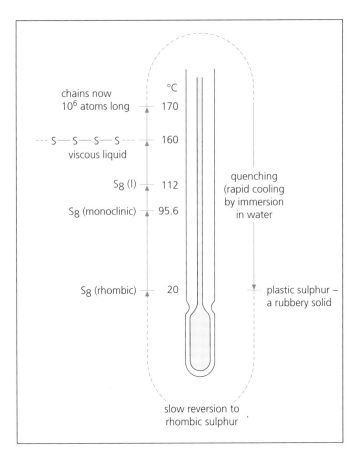

Figure 18.20 The changes undergone by elemental sulphur due to various processes of heating and cooling. The photograph shows (left to right) rhombic, monoclinic and plastic sulphur

Uses of sulphur

We have seen several cases of human mimicry of natural structures, and there is an equivalent in sulphur chemistry. Both human and natural technology use the —S—S— linkage as a kind of pin, to clip chains of structures together (Figure 18.21). In human technology, chains of rubber are modified

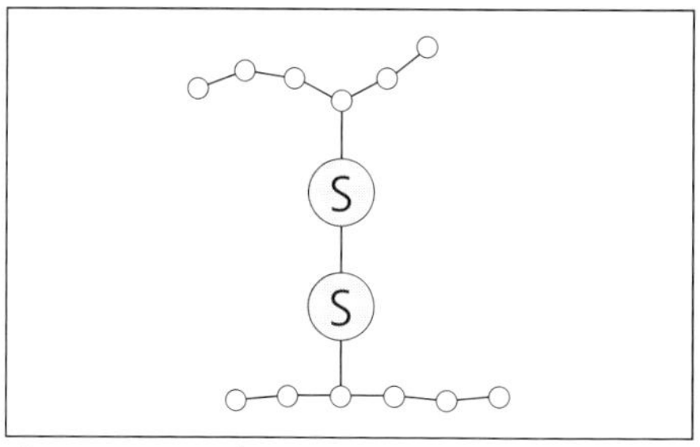

Figure 18.21 Schematic picture of an S—S bridge. This diagram would fit equally well into a discussion of rubber tyres or of enzymes

(**vulcanised**) by sulphur linkage (see Chapter 10), whereas in natural chemistry, the —S—S— bridge in the amino acid cystine holds proteins together. As an example, look back to Chapter 15 to see how sulphur bridges establish the physiologically active conformation of a molecule of the enzyme lysozyme (Figure 15.21).

Table 18.7 shows other uses of Group VI elements.

Table 18.7 Uses of Group VI elements and compounds

Element/compound	Use
Oxygen	For the combustion of fuels, e.g. in gas welding In hospitals, for patients with breathing problems
Sulphur	Precursor of H_2SO_4 Vulcanisation of rubber (by S——S bridging) Gunpowder
SO_2	A bleach for paper
H_2SO_4	Used in practically every branch of industrial chemistry

18.7 The elements – Group VII

The name 'halogen' means 'salt-producer', and signals the tendency of Group VII elements to form ionic salts with metals. In all these compounds the halogen atom is present either as an anion, or at the heavily negatively polarised end of a covalent bond.

The classical rule-of-eight valency of the halogen family is one, but expanded valence shells (breaking the rule of eight) are found in all the members of the group except fluorine.

Question

27 From your knowledge of electron orbitals and their energies, explain

a why halogens have a rule-of-eight valency of one

b why fluorine never breaks the rule of eight

c why fluorine is the most electronegative of all the elements.

Occurrence and extraction of the halogens

The halogens are such a reactive chemical group that they are never found uncombined on Earth.

Question

28 The last sentence seems non-controversial enough, and yet in a way it carries an inconsistency. After all, oxygen is hardly an inert element, and even nitrogen is highly electronegative on the Pauling scale. Yet both nitrogen and oxygen are found in Nature as elements. Can you explain why the halogens, in contrast to nitrogen and oxygen, are always found combined with *other* elements?

The vigorous appetite that all halogens show for gaining electrons means that they are normally found as halide ions. This in turn ensures that, given the solubility of nearly all ionic chlorides, bromides and iodides, their physical resting place is the sea, or deposits left behind by the evaporation of seas. Fluorides are the odd ones out as far as solubility is concerned, and the most common sources of fluorides are rock-forming minerals like fluorite, CaF_2.

Question

29 The occurrence of insoluble fluorides is yet another example of an element at the top of a group being out of step with the rest. As we have seen, these irregularities can usually be traced back to the very small size of the atom or ion. In this case, can you see a link between the small size of the

F^- ion and the relative insolubility of its compounds?

Halide deposits may be abundant, but there is still a formidable barrier in the way of obtaining the halogen elements (especially the top two). The problem is a mirror image of the battle to obtain the elements on the extreme left of the Periodic Table. In the case of the s-block metals, their readiness to lose electrons made it difficult to force electrons back onto their cations. Conversely, in the case of the top halogens we have such strong electron-gainers that it is hard (in the case of fluorine impossible) to find an oxidising agent that will convert Hal^- to Hal_2. But as we saw in Chapter 16, electrochemistry comes to the rescue (as indeed it does in the s-block). The 'oxidising agent' in the production of chlorine and fluorine is a positive electrode.

Question

30a Write a half-equation showing the oxidation of a halide ion at an anode.

b Which electrolytes were used in the production of chlorine and fluorine? (Look back to Chapter 16 if necessary.)

The lower halogens are produced more easily, since their halide ions hold on less tenaciously to their extra electrons. Bromine is made quite simply by oxidising the bromide ions in sea water using chlorine. The reaction is an example of a halogen displacement reaction:

$$Cl_2(g) + 2Br^-(aq) \rightarrow Br_2(aq) + 2Cl^-(aq) \tag{1}$$

The process raises several interesting points, which will feature in the next couple of questions.

Question

31 Halogen displacement reactions (in which a halogen and a halide 'swap roles') are driven by the unequal desire for electrons amongst the halogens.

a From your knowledge of electron-orbital energies, which halogen atom offers the most stable home for an extra electron?

b Therefore which halogen will be the strongest electron remover/oxidising agent?

c Which halide ion will be the *least unwilling* to have the extra electron removed?

d Which halide ion will be the most effective reducing agent? (None of them can be described as a good reducing agent, of course.)

e From the above deliberation, decide which other halogen/halide reactions will happen, in the manner of equation (*1*). (*Clue*: Remember the electrochemical series and the Law of anti-clockwise circles.)

Equation (*1*) may produce bromine quite simply, but the industrial chemist is faced with problems associated with extreme dilution of bromide ions in sea water. The small amount of bromine liberated by equation (*1*) can be blown out of the water by blasts of air, and absorbed into vats of sodium carbonate solution. This stage overcomes the problem of the high dilution: even though the concentration of bromine is low, the amount of bromine absorbed becomes acceptably high if the process is carried on in the same absorption tank for long enough.

The chemistry of the absorption is a characteristic reaction of halogen chemistry. All the halogens have a reaction in aqueous alkalis, which is to undergo redox disproportionation:

$$3Br_2(g) + 6CO_3^{2-}(aq) + 3H_2O(l) \rightarrow$$
$$BrO_3^-(aq) + 5Br^-(aq) + 6HCO_3^-(aq) \qquad (2)$$

Question

32a Show by examining the oxidation numbers of the bromine species in equation (*2*) that the reaction is indeed a disproportionation.

b When the concentration of the bromine-containing species on the products side of equation (*2*) has built up sufficiently, the contents of the absorption solution are subjected to another reaction that has the effect of reversing equation (*2*) and regenerating bromine. Considering that carbonate ions were used to promote the disproportionation, which reagent would you suggest for the job of reversing it?

c It seems fairly likely that bromine made by this process might be contaminated with chlorine. Why is this likely? Suggest at what stage, and how, the chlorine might be removed.

d With or without looking back to p. 132 in Chapter 7, write out the similar disproportionations undergone by chlorine in solutions containing $OH^-(aq)$.

Iodine could in theory be made by a parallel version of the bromine process, but Nature has lent a hand. It is well known that certain seaweeds have the power to absorb iodide ions preferentially from sea water (Figure 18.22), but even more significantly iodide ions are formed further up the food chain in the bodies of sea birds. Massive accumulations of their droppings over long periods of time are thought to have given rise to the deposit of saltpetre in Chile, in which the majority nitrate salts are accompanied by a small but workable amount of sodium iodate(V), $NaIO_3$. Reduction to iodine is carried out with sodium hydrogensulphite.

Figure 18.22 *Laminaria saccharina* – a member of the genus of seaweeds most adept at absorbing I^- ions

Fluoride minerals, fluorine dating and teeth

Fluorine stands apart from the other halogens in its mineralogy as well as in its mainstream chemistry. The major ore is calcium fluoride in the form of the mineral **fluorite** (a subversion of which occurs in Derbyshire and is sold as a decorative item under the name blue john). The fluorite is turned into hydrogen fluoride by action of sulphuric acid, and the hydrogen fluoride is a basic ingredient of the electrolyte for the production of fluorine.

Question

33 Give equations for these two stages.

Another interesting fluorine-containing mineral is apatite, not least because it is also the major mineral ingredient in bones and teeth. The apatites are a series of minerals of formula $Ca_5(PO_4)_3X$, where X can range from 100% hydroxide to 100% fluoride, with all intermediate combinations possible. The version in bone and teeth is high in hydroxide relative to fluoride.

When the hydroxy-version of mineral apatite is subjected to the long-term effects of percolating groundwater rich in fluoride ions, the hydroxide ion is selectively and gradually replaced by fluoride. The mineral slowly turns to fluorapatite:

$$Ca_5(PO_4)_3OH(s) + F^-(aq) \rightarrow Ca_5(PO_4)_3F(s) + OH^-(aq) \quad (3)$$

Question

34a Suggest what might be the thermodynamic driving force that makes this reaction go in this particular direction.

b Levels of chloride ion in groundwater are usually far higher than those of fluoride, and yet the fluoride anion is preferentially chosen to replace the hydroxide ions. Can you suggest a reason for this?

The fact that the same reaction happens in both living and dead bones and teeth has given rise to the technique of fluorine dating, and also to the practice of fluoridation of drinking water. Fluorine dating relies on the fact that the older the bone, the greater will be the fluoride content of its apatite.

Piltdown Man

In the early part of this century, there was growing interest in the study of hominid fossils. This was spurred on by new fossil discoveries, by natural curiosity about who we are, and by the debate between various groups of people both scientific and religious concerning the theory of evolution. The discovery in a Sussex gravel pit of some bone fragments, seemingly from the head of a man/ape creature, caused much excitement, because a misguided interpretation of Darwin's theory led to the belief that Man might be the direct descendant of (modern) apes. The find was called Piltdown Man, and promised to immortalise its finder by virtue of its Latin name, *Eoanthropus dawsonii*, which translates roughly as 'early man found by a person called Dawson'. The newspapers of the time hailed the find as the missing link between apes and people (Figure 18.23).

Question

35 A 260 mg fragment from the jawbone was found to contain 2 mg of fluorine. A 500 mg fragment of the skull contained 15.1 mg of fluorine.

a Explain why these data prove that Piltdown Man was a fake.

b Which sample appears to have been the older?

c Can you suggest why fluorine dating is not very reliable as an independent method of determining an absolute age (as opposed to a relative one) for a fossil? What limitations are there to its use even for relative dating?

Figure 18.23 (left) Reconstruction of the Piltdown discovery. (The dark bits represent the bones found.)   (right) Dawson and a collaborator at work. (bottom) The find caused great interest in the magazines of the day

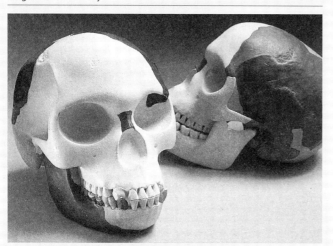

THE ILLUSTRATED LONDON NEWS, DEC. 28, 1912. — 958

THE MOST ANCIENT INHABITANT OF ENGLAND: THE NEWLY-FOUND SUSSEX MAN.

THE JAW OF A KAFFIR.   THE JAW OF A CHIMPANZEE.'   THE JAW OF AN INDIAN.

These photographs show the absence of the chin in the case of a chimpanzee (as in the Heidelberg jaw and that of the Sussex man); its slight development in the African; and the pronounced chin of the Indian.

The partial replacement of hydroxide by fluoride in living teeth can improve resistance to decay. Levels of fluoride of about one part per million in drinking water can achieve the optimum composition of dental apatite.

Question

36a Suggest why fluoridation of water is considered unnecessary in certain areas.

b Fluoride uptake is more effective when the treatment begins in childhood, rather than later in life. Can you suggest why?

c Fluoridation has become an issue with moral and medical debating points. The moral point concerns whether it is right for an outside agency like a water company or a Government department to make health decisions on behalf of the individual citizen. Assemble some arguments that address this problem.

Table 18.8 lists some of the uses of the halogen elements and their compounds.

Table 18.8 Uses of Group VII elements and compounds

Element/compound	Use
Fluorine	Refrigerants, aerosol propellants, polymers (Teflon), separation of ^{235}U and ^{238}U (as UF_6)
Hydrogen fluoride	Etching, making artificial Na_3AlF_6 (for electrolytic preparation of aluminium)
Chlorine	Solvents, PVC, bleaches, disinfectants
Bromine	Precursor to $Pb(C_2H_5)_4$ (petrol additive), $AgBr$ in film and photographic paper
Iodine	AgI in photography

18.8 Hydrogen and the energy for human existence

The Universe, it is believed, was once nearly all hydrogen. Hydrogen is still by far the most abundant element in the Universe (as opposed to the Earth), even though synthesis of heavier elements in the early Universe soon began. This synthesis occurred when gravitational force pulled together great collections of hydrogen gas. The overall masses involved were vast, and the pulling together produced heat by the same principle, albeit greatly magnified, that applies to a hammer hitting an anvil. These gravitationally heated masses of hydrogen were the first stars, and the same ignition process has been giving birth to new stars ever since. Our own Sun is one such hydrogen-powered star.

The initial gravitational heating ignites a thermonuclear reaction between hydrogen atoms, and for the rest of the star's lifetime a sequence of these reactions releases heat and light energy. The synthesis of increasingly heavy elements accompanies the ageing of the star, and at its final demise those elements are spread round the Universe, sometimes to be trapped in orbit around other stars, as planets. That is a very simplified and condensed version of how our planet came to exist. The energy that enables it to support our lives is coming from nuclear reactions on the Sun such as:

$$^1H + {}^1H \rightarrow {}^2H + \beta^+ \qquad (4)$$
$$(\beta^+ \text{ is a positron})$$

$$^1H + {}^2H \rightarrow {}^3He \qquad (5)$$

$$^3He + {}^3He \rightarrow {}^4He + 2{}^1H \qquad (6)$$

Hydrogen, then, is a nuclear energy source for suns. It is an excellent chemical energy source for planets, too, owing to its exothermic reaction with oxygen:

$$2H_2(g) + O_2(g) \rightarrow 2H_2O(l) \qquad (7)$$

The problem is that the vast majority of terrestrial hydrogen has already reacted in this way, and is slopping around in the oceans in the low-energy compound water. Any hydrogen that has not reacted has escaped altogether, since H_2 molecules are too fast and light to be held by the Earth's gravitational field.

Plants, of course, evolved a mechanism of using solar energy to rescue hydrogen atoms from their 'watery grave', and simultaneously to free carbon atoms from their equally energy-exhausted reservoir of carbon dioxide. The products of eons of photosynthesis have come down to us as fossil fuels, and we are gorging on this energy feast, so that the human beings of the late twentieth century enjoy a standard of domination of the environment that is without precedent.

We relentlessly return the fossil-fuel carbon and hydrogen to the 'exhausted' condition by direct reaction with oxygen (by burning), or by a series of more involved indirect routes like the manufacture of polymers. One very significant route, and one that represents the major industrial use of hydrogen gas itself, is this:

$$C_nH_{2m}(g) + nH_2O(g) \xrightarrow{Ni/Al_2O_3} nCO + (n+m)H_2(g) \qquad (8)$$
from oil

$$N_2(g) + 3H_2(g) \rightleftharpoons 2NH_3(g) \qquad (9)$$

Question

37a What is the major use of ammonia, and why is it crucially linked to human population growth?

b By what route would the hydrogen in synthetic ammonia eventually end up in the low-energy compound water?

The human race is just beginning to worry about the fact that we are converting our hydrogen to its energy-exhausted form, water, thousands of times faster than the natural energy-fixing mechanisms of our planet can replace it (not to mention the smaller but hardly trivial worries about global warming and acidic pollution associated with degradation of fossil fuels).

If we as a species are not to undergo a regression to pre-industrial subsistence, we have to find another way, other than photosynthesis, of harnessing solar energy at a speed compara- ble to the rate of consumption. (Alternatively, we may learn to bring the reaction of the solar furnace down to Earth by harnessing nuclear fusion reactions like (4), (5) and (6) above.)

Much hope centres on converting solar energy directly to electricity (as in photovoltaic cells), or on the use of natural Sun-driven engines like tides and wind. This energy will then be processed along pathways in which hydrogen will play a major role. The energy will be used to send hydrogen in water back up (in the energy sense) to hydrogen gas itself.

Question

38a What method might enable solar electricity to be used to turn water to hydrogen?

b Figure 18.24 shows the relative costs of sending various present forms of energy across physical distances. How does the graph support the idea of sending energy as hydrogen gas?

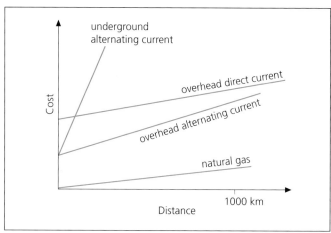

Figure 18.24 Sketch graph showing how the relative costs of various methods of energy transmission vary with distance

c (*Revision*) This vision of the future imagines that hydrogen will be converted to electricity at the consumer end of the network. Fuel cells have been suggested as a method. (You may need to look back to Chapter 16 to revise these.) Hydrogen would become the anode material. (Remember that the anode of a cell is the electrode from which electrons flow, and therefore that anode material is oxidised.) What would be the (readily available) cathode material, and what would be the two electrode reactions? Give an overall equation for the change taking place, including any use of catalysis you feel may be necessary.

d Advocates of a hydrogen economy support their case by citing the cleanness of dihydrogen as a fuel. Point to three ways in which it is less polluting than fuels derived from coal or oil.

e Anyone who has handled hydrogen will be aware of one or two big drawbacks. Give two problems that would have to be faced if it were to be used.

Other uses of hydrogen are shown in Table 18.9.

Now for another big shift of focus. We'll look in turn at the hydrides, oxides and halides of the p-block elements.

Table 18.9 Uses of hydrogen

Making NH_3
Making HCl (for PVC)
Making CH_3OH
Hydrogenating the $C=C$ bonds in vegetable oils to make margarine

18.9 The hydrides of the p-block elements

Table 18.10 shows a range of compounds in which hydrogen is combined with p-block elements. The electronegativity of hydrogen places it to the positive side of most p-block ele- ments. Thus the H—X bond (where X is a p-block element) tends to be of polar covalent character, with the hydrogen showing a positive oxidation number. The degree of polarity varies widely of course, between extremes featuring dipole moments of near zero, to cases in which H—X bonds need only small inducements to break and give H^+ and X^- ions.

Table 18.10 Some properties of hydrides of the p-block elements

Compound	B_2H_6	CH_4	NH_3	H_2O	HF
State at 20 °C	g	g	g	l	l (just)
Bonding	'H-bridge' small molecules	Covalent small molecules			
Polarity of H—X*					

*Where the H end of the bond is δ+

Question

39 Give an example of each of the two extreme styles of behaviour mentioned in the last paragraph.

Hydrides of Group III

The real odd ones out among the p-block hydrides are those of Group III. Here the metals are less electronegative than hydrogen, and the compounds feature hydrogen in a negative oxidation number, even showing some hydride (H^-) ionic character. Group III hydrides all show the predictable empiri- cal formula XH_3.

The empirical formula BH_3 conceals the highly puzzling compound diborane, B_2H_6, whose structure was demon- strated by gas-phase electron diffraction to be as shown in

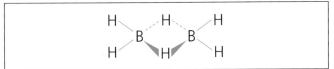

Figure 18.25 The mysterious diborane

Figure 18.25. The use of this compound to humankind has been intellectual rather than practical, because it presents the most formidable challenge to simple theories of bonding.

Question

40 The shape of the diborane molecule is as shown in Figure 18.25. Why is it impossible to describe the bonding in such a compound by normal valency rules? Look up the physical properties of diborane and suggest why no one proposed an ionic lattice description of the bonding.

(It is outside Advanced level to describe the bonds in diborane. Just call them 'multicentre electron-deficient' bonds.)

The other Group III hydrides show more ionic character, but the bonding is of a very impure kind, not surprising when you consider the high charge density (and therefore polarising power) of the cations and the easy polarisability of the anion. (As a reminder, look back to Chapter 5, Figure 5.13.)

Question

41 What *is* surprising is that the hydride ion is so easily polarisable. Look up the size of the H$^-$ ion and compare it with halide ion sizes. Suggest why H$^-$ is so big, and (a related issue) why it is so easily polarised by cations.

A series of complex hydrides of Group III are used in organic chemistry as sources of H$^-$ ions for reductions. In schools you often meet sodium tetrahydridoborate(III), NaBH$_4$. Lithium tetrahydridoaluminate(III), LiAlH$_4$, also finds uses, although it is less stable and less easy to handle.

Question

42a Name a typical organic reduction carried out by sodium tetrahydridoborate(III).

b The BH$_4^-$ ion can be seen as including a dative covalent bond between a BH$_3$ unit and a hydride ion. Draw a dot-and-cross diagram showing this feature.

c These complex hydrides are very unstable in water. The reaction can be seen as a combination of a substitution of OH$^-$ for H$^-$, followed by the ejected H$^-$ ion acting as a base. What would be the conjugate acid of a hydride ion?

d Putting the two stages into one reaction, write an equation for the hydrolysis of lithium tetrahydridoaluminate(III).

Hydrides of Group IV

The hydrides of the upper Group IV members are small molecules with bonds of only slight polarity. Methane and the other alkanes are familiar from organic chemistry, while silanes equivalent to methane and ethane also exist. However, a compound like Si$_2$H$_6$ is considerably less chemically stable than its carbon analogue.

Question

43 Suggest from a study of bond enthalpy terms why the silanes are less stable than the alkanes.

Hydrides of Group V

In Group V hydrides, several characteristics not prevalent in Groups III and IV begin to make their presence felt. Bond polarity, especially at the top of the group, renders the hydrogen atom partially positively charged, giving rise to the first possibility of Brønsted–Lowry acidity. On the other hand, the lone pair characteristic of Group V elements introduces basicity, and in the case of ammonia, NH$_3$, we have physical properties altered by hydrogen bonding.

Acidity and ammonia are not words that one normally associates, because ammonia is a weaker acid than water. So the equilibrium:

$$NH_3(aq) + H_2O(l) \rightleftharpoons H_3O^+(aq) + NH_2^-(aq)$$

is virtually a non-starter. However, the N—H bond is polarised the right way for acid dissociation, and the NH$_2^-$ ion can be coaxed into existence, so long as water is not the solvent medium. For instance, the following reaction takes place in liquid ammonia:

$$2Na(s) + 2NH_3(l) \rightarrow 2NaNH_2 + H_2(g)$$

Question

44a If ammonia is a weaker acid than water, place their two conjugate bases NH$_2^-$ and OH$^-$ in order of base strength.

b Therefore suggest what would happen if you put sodamide, NaNH$_2$, in water.

Ammonia as a base is a familiar part of school laboratory chemistry. It is a moderately weak base (about as weakly basic as ethanoic acid is weakly acidic), and its conjugate acid NH$_4^+$ is a very weak acid. Relevant equilibria are:

$$NH_3(aq) + H_2O(l) \rightleftharpoons NH_4^+(aq) + OH^-(aq)$$
$$K_b = 1.8 \times 10^{-5} \, mol \, dm^{-3}$$

$$NH_4^+(aq) + H_2O(l) \rightleftharpoons H_3O^+(aq) + NH_3(aq)$$
$$K_a = 5.6 \times 10^{-10} \, mol \, dm^{-3}$$

Questions

45 Suggest what colour you would see if you dipped full-range Universal Indicator paper into solutions of:

a ammonia

b ammonium chloride

c ammonium ethanoate.

46 The test for a suspected ammonium compound is to heat it in a boiling tube with a little sodium hydroxide solution. Write a chemical equation focusing on the acid–base aspect of this test reaction, and suggest what would tell you the test was positive.

47 The claim was also made that the physical properties of ammonia would be influenced by hydrogen bonding.

a Suggest three properties that might be affected.

b Use a data book to compare ammonia with phosphine, PH_3, and highlight those differences attributable to hydrogen bonding.

Ammonia production

The compound ammonia is one of the most significant industrial chemicals in the world. Eighty per cent of ammonia production is dedicated to the fertiliser industry, for making ammonium nitrate and ammonium phosphate fertilisers. Figure 18.26 shows the production of ammonia and of fertilisers – the two correlate closely.

Plants, of course, need nitrogen compounds for the synthesis of proteins, and they need them in the form of soluble aqueous salts, normally nitrates. In the natural nitrogen cycle, there is a steady shuttling between proteins, amino acids, ammonia and nitrates, as an accompaniment to the cycles of growth, death and decay. Biological nitrogen is in a sort of equilibrium with atmospheric nitrogen, via agencies such as lightning and bacterial activity. The stability of dinitrogen ensures that the equilibrium is well biased towards the atmospheric gas.

Agricultural scientists soon realised that if we as a race were to support our accelerating population, we needed to intervene in the nitrogen cycle, to transfer more nitrogen into biomass and out of the atmosphere. As was mentioned in the last paragraph, the thermodynamic situation is not helpful. Reactions such as:

$$N_2(g) + 3H_2O(l) \rightarrow 2NH_3(g) + \tfrac{3}{2}O_2(g) \qquad (10)$$

and:

$$N_2(g) + O_2(g) \rightarrow 2NO(g) \qquad (11)$$

(which does in fact happen in lightning strikes) are not, under normal conditions, biased in our favour.

Question

48 Work out the values of $\Delta G^{\ominus}$ for these reactions and comment on their feasibility.

What was needed was a source of energy, to make up for the shortfall in the energetics of reactions like (10) and (11). You will not be surprised that the answer to the problem of ammonia synthesis was looked for in fossil fuels. In most of Britain's ammonia plants (Figure 18.27), the fuel material is methane from the natural gas fields of

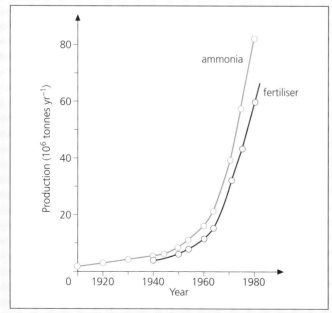

Figure 18.26 World ammonia and fertiliser production, 1910–80

Figure 18.27 'miles and miles of towers and chimneys and stacks and sheds' ICI's ammonia plant at Billingham (see footnote on p. 462):

the North Sea, and that methane is reacted with water to form hydrogen:

$$CH_4(g) + 2H_2O(g) \xrightarrow{\text{Ni, 750 °C}} CO_2(g) + 4H_2(g) \quad (12)$$
$$\Delta H^{\ominus} = +164.9 \text{ kJ mol}^{21}$$

One way of looking at reaction (*12*) is as a transfer of energy from methane to hydrogen. Hydrogen is of course much richer in free energy than water is, and now a new route to ammonia opens up. The direct reaction of the elements to form ammonia is feasible:

$$N_2(g) + 3H_2(g) \xrightarrow{\text{Fe, 400 °C}} 2NH_3(g) \quad (13)$$

synthesis gas
$$\Delta H^{\ominus} = 292 \text{ kJ mol}^{21}$$

The first stage, equation (*12*), is called the **steam reforming reaction**. In between (*12*) and (*13*) some air is introduced, so that a mixture of dinitrogen and dihydrogen of about 1:3 is achieved. There are several intermediate stages dedicated to removing carbon dioxide, and cleaning the synthesis gas of traces of carbon monoxide before it enters the catalyst chamber to undergo reaction (*13*), the **Haber process**.

Question

49a If green plants were to carry out reactions analogous to (*12*) and (*13*) to fix nitrogen (which the legume family do, with the help of bacteria) they would not use methane as an energy source. What would they use?

b Why do you think such pains are taken to remove traces of carbon monoxide before the synthesis gas mixture enters the reaction chamber?

c What is the reason for setting the ratio of moles of hydrogen:nitrogen at 3:1 before reaction?

The above account is an outline of the ammonia synthesis process, but a deeper study offers insights into many aspects of industrial chemistry, including economics, environmental constraints, the manipulation of equilibria and catalysts, and the prospects for a future less dependent on fossil fuels. Let us take these issues in turn.

Manipulating the equilibrium

The reaction:

$$N_2(g) + 3H_2(g) \rightleftharpoons 2NH_3(g)$$

is exothermic to the tune of 292 kJ mol^{21}. Table 18.11 shows various combinations of temperature and pressure, and the resulting yields of ammonia at equilibrium, expressed as percentage of ammonia by volume. This is of course an excellent chance to revise our models of equilibrium systems.

Table 18.11 Percentage of ammonia by volume in the equilibrium mixture of the system $N_2(g) + 3H_2(g) \rightleftharpoons 2NH_3(g)$, under a range of values of temperature and pressure

Pressure (atm)	Temperature (°C)			
	200	300	400	500
10	50.7	14.7	3.9	1.2
100	81.7	52.5	25.2	10.6
200	89.0	66.7	38.8	18.3
300	89.9	71.0	47.0	24.4
400	94.6	79.7	55.4	31.9
600	95.4	84.2	65.2	42.2

Questions

50a Explain why the highest yields are at high pressures.

b Explain why the highest yields are at low temperatures.

c The chosen conditions in ICI's ammonia reactor at Billingham on Teesside are 200 atm, 400 °C. These are not the conditions for maximum yield, as Table 18.11 shows.

  i Explain why temperatures are used which give less than maximum yields.

  ii Explain why pressures are used which give less than maximum yields.

51 Now for a quantitative question on a mathematical model of the system. It is set in the context of the conditions ICI use, namely 200 atm, 400 °C.

a Copy and complete Table 18.12.

Table 18.12 Data relating to the $N_2/H_2/NH_3$ equilibrium system at $p_{total} = 200$ atm, $T = 400$ °C. The operator began with a nitrogen:hydrogen mole ratio of 1:3 (as in the stoichiometric equation)

	$N_2(g)$	$+ 3H_2(g)$	$\rightleftharpoons 2NH_3(g)$
Moles at t_0	1	3	0
Moles at t_{eqm}	$1 - y$	$3 - ?$	?
	Total moles of gas at t_{eqm} = ?		
Mole fractions at t_{eqm}	?	?	?

b If the volume percentage of ammonia at equilibrium is about 40%, what does Avogadro's hypothesis tell us is the mole fraction of ammonia?

c Use your answer to (b) to solve for y.

d Now you can put numbers to the moles of all three reagents at equilibrium, and convert moles to mole fractions to partial pressures. Hence calculate a value for K_p.

e If you have the stamina, repeat the operation for another of the starting pressures in the 400 °C column of Table 18.11, to see how constant the equilibrium 'constant' really is.

f Why would you not *expect* equilibrium constants to be constant across a horizontal row in Table 18.11?

After all that 'number-crunching' you may feel cheated to learn that the industrial process does not wait for equilibrium to be achieved. Once the ammonia content reaches about 15%, the gases go through a refrigeration stage. The ammonia, at these pressures, condenses to a liquid and the unreacted synthesis gases go round for another pass over the catalyst, together with new input nitrogen and hydrogen.

Efficiency

When you add together all the stages in the ammonia synthesis plant, you get this equation:

$$7CH_4(g) + 10H_2O(g) + 8N_2(g) + \underset{\text{from air}}{2O_2(g)} \rightarrow$$
$$7CO_2(g) + 16NH_3(g) \qquad (14)$$

The chemical engineers measure the economy of their process in joules per tonne of ammonia produced, the joules being calculated from the energy that would have been obtained had the methane been burned. (This includes methane which is simply burned to heat reaction systems, as well as that consumed in equation (*14*)). They call this parameter the **fuel energy equivalent** of the process.

Question

52 If $\Delta H_c^{\ominus}$ for methane is -890 kJ mol^{-1}, calculate the fuel energy equivalent in joules per tonne of ammonia, if equation (*14*) were the only consumer of methane in the plant. (You might start by converting one tonne of ammonia into moles.)

The real fuel energy equivalent of an ammonia plant is higher than your answer to question 52, because of the need for heating fuel as well as reactant fuel. One of the ways in which this figure can be kept as low as possible is by the use of **heat exchangers**. These are devices that collect the heat from the exothermic stages of the process (like reaction (*13*) itself) and avoid wasting it all.

Question

53 What use could the heat from an exothermic stage be put to, which would save on the use of fresh fuel?

As well as heat exchange, a crucial role in saving energy is played by making the best choice of catalyst, because although catalysts do not affect the overall thermodynamics of a reaction, they lower the activation enthalpy.

Question

54 Explain as precisely as you can how the use of the best catalyst would contribute towards fuel economy.

Question 54 leads us on to consider the factors guiding the choice of catalysts in this particular process.

Catalysts

The choice of catalysts for industrial chemistry is something of a black art – in the absence of a full understanding of what is going on at a molecular level on the catalyst's surface, trial and error is often the only recourse. However, what is not in doubt is the importance of surface parameters, so much skill is put into the mounting of the right size of catalyst crystals on ceramic supports, and the 'pelletisation' of the final form of the catalyst. Figure 18.28 shows some typical catalyst pellets.

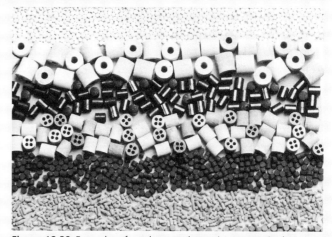

Figure 18.28 Examples of catalysts used at various stages in the manufacture of ammonia. The Fe_3O_4 catalyst for the Haber process itself is in the middle of the photograph

Question

55 One process that shortens the life of a catalyst is **sintering**, whereby tiny granules fuse into bigger ones, normally due to heating. How do you think the mounting of catalyst granules on ceramic supports improves catalyst life?

The nickel catalyst used in the steam reforming stage (*12*) is made by the above method. Recently a new catalyst has been prepared which can do reaction (*12*) using the naphtha fraction of crude oil as a feedstock, instead of natural gas. Strangely enough, given the intensive work done by the chemical industry on catalyst development, reaction (*13*) still uses the same style of iron catalyst that Fritz Haber used – broken lumps of Fe_3O_4 reduced to iron.

Siting

The Billingham plant at the mouth of the Tees (Figure 18.29) was originally sited there because of its proximity to the Durham coalfields. The coalfields are now a shadow of their former selves, but luckily (for ICI) Billingham is not badly situated to cope with the change in energy source.

Figure 18.29 ICI's Billingham complex. Unless you have stood nearby it is impossible to imagine the 'vast unnaturalness of the place'

Question

56a Where would the new raw material for equation (*12*) come from, and why is the situation of the Teesside plant still viable?

b Teesside is good for two other aspects of the ammonia process, one connected to a raw material and one to transport. What are those aspects?

The future of ammonia production

We can divide the future of ammonia production into four categories:

1 more of the same method – but this assumes a steady rate of discovery of new oil and natural gas reserves
2 nuclear (perhaps even fusion) energy used to create the hydrogen
3 coal replacing the oil or gas in reaction (*12*)
4 use of green plants to trap the Sun's energy.

Let us now look at some of these categories within a question.

Question

57a Nuclear energy could not be absorbed into the process in the same direct chemical way as methane is in reaction (*12*). Suggest a way in which nuclear energy could be used to make the dihydrogen for equation (*13*).

b If coal is mainly carbon, write a version of equation (*12*) with coal as the fuel source.

The last idea in the futures list sounds neatest, especially from an environmental protection point of view. You might think the best plan would be to grow lots of legumes and use their root bacteria to make the nitrogen compounds directly. This is indeed the strategy behind growing clover on organic farms and ploughing it straight into the soil. Oddly enough, in this case Nature's way is not the most energy-efficient. Perhaps it is something to do with the fact that the nitrogen-fixing bacteria are separate organisms, and not actually cells of the plant itself. They are after all working to their own evolved agenda. Whatever the reason, the best conversion of solar energy into ammonia is by the path:

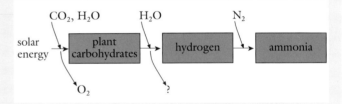

An interesting footnote to the connection between ICI, Teesside and ammonia is that the local football team, who play in the Northern League, are called Billingham Synthonia. This was originally an ICI works team who called themselves 'Billingham Synthetic Ammonia', but it was felt the original name just didn't have the right ring to it. The relationship between the team and the plant is a physically close one:

'ICI's Billingham synthetics plant looms over Central Avenue Stadium like a giant's chemistry set. Unless you have stood near it it is impossible to imagine the vast unnaturalness of the place.

Compared to ICI's Billingham, Hammersmith Broadway is a water meadow, the Birmingham Bull Ring a cottage garden in Kent. It stretches into the distance, miles and miles of towers and chimneys and stacks and sheds. Steel brachia and intestinal aluminium tubing swing from its flanks as if it were some huge metallic beast with a multiple hernia. And at night, on the other side of the Cleveland Hills where I was born, the colours it pumps out dye the sky pink and it throbs like a heart.

'In the social club we sat drinking and eating peanuts.' From *The Far Corner*, by Harry Pearson, publ. Little, Brown & Co. 1994.

and here for comparison is a summary of ICI's Billingham process:

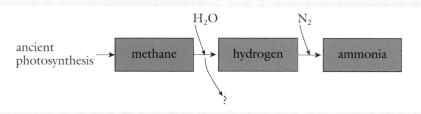

b Identify the gaseous waste product common to both sequences (shown as '?').

This is the end of the ammonia story, told in some detail to give a taste of industrial chemistry, firmly grounded in a real-world, whole-world context. The two great guiding principles are efficiency and profitability on the one hand and environmental protection on the other. However, there is no such thing as technology that is at the same time life-enhancing, and yet has zero effect on our surroundings.

Question

58a Try to write an equation for the conversion of carbohydrates plus water to hydrogen in the first sequence. (Use glucose, $C_6H_{12}O_6$, as a typical carbohydrate.) It will help if you see it as just another variant of the steam reforming process, as carried out on methane in equation (12).

Hydrides of Group VI

Water is another hydride in which the hydrogen is clearly in its positive oxidation state, but water is more acidic and less basic than ammonia – in other words, it is freer with its H^+ ions and less pushy with its lone pairs. This latter point is emphasised by the position of the equilibrium:

$$NH_4^+(aq) + H_2O(l) \rightleftharpoons NH_3(aq) + H_3O^+(aq)$$
$$\text{base 1} \qquad \text{base 2}$$

which is very heavily over to the left (a win for base 2).

Because of the stability of water, oxygen can extract the hydrogen from the hydrides of all other elements except those of the lighter halogens. Furthermore if there is a shortage of oxygen in such a reaction, it is always the other element rather than hydrogen that 'goes short'. This is typified in the familiar incomplete combustion of hydrocarbons:

$$CH_4(g) + O_2(g) \rightarrow C(s) + 2H_2O(g)$$

The reaction between dihydrogen and dioxygen to form water is famously explosive, due to its being a branching-chain free-radical reaction. This is a mechanism in which the propagation steps produce two radicals for every one on the reactant side.

The higher Group VI hydrides may be prepared by the action of acids on the simple 2− anion, for example:

$$FeS(s) + 2H^+(aq) \rightarrow H_2S(g) + Fe^{2+}(aq)$$

Hydrogen sulphide, H_2S, is the poisonous 'bad-eggs' gas, and is spectacularly unlike water in its effect on people. These differences, and many other aspects of the chemistry of water, are accounted for in previous chapters on bonding, intermolecular forces, acidity and electrochemistry.

Hydrides of Group VII

The hydrogen halides are gases (HF boils at 20 °C), but their highly polar bonds give rise to the strong acidity of their aqueous solutions. A name change accompanies the change in medium from gaseous to aqueous. They cease to be called hydrogen halides and become hydrohalic acids. None of the compounds show any basic tendencies, even though they are generously endowed with lone pairs.

The hydrogen halides are a good place to see exhibited the atypical nature of the top-of-group elements and their compounds, since a number of features of hydrogen fluoride are out of step with the other three.

Question

59 The enthalpies of formation of the HX gases are shown in Table 18.13. The value for HF is missing.

Table 18.13 $\Delta H_f^\ominus$ for the hydrogen halide gases (kJ mol^{-1})

HF	HCl	HBr	HI
?	−92	−36	−26

a Since the reaction:

$$H_2 + F_2 \rightarrow 2HF$$

takes place entirely in the gas phase, we can calculate its enthalpy directly from bond energy terms. Which bonds are being broken and made? Calculate the missing value in the table.

b You may have noticed that the ΔH_f of hydrogen fluoride is very much more exothermic than those for the other compounds in Table 18.13. You may also have noticed that some of the data you used in part (a) had unusually large or small values, compared to average bond energies. Indicate which data might explain the anomalous ΔH_f of hydrogen fluoride.

Hydrogen fluoride is also by far the weakest acid of the four in aqueous solution (despite the higher hydration enthalpy of the smaller F^- ion), which we can attribute to the great strength of the H—F bond.

Uniquely strong intermolecular hydrogen bonds give hydrogen fluoride a boiling point (about 20 °C) far in excess of those of the other hydrogen halide gases, and the same phenomenon gives rise to a series of solid compounds with an anion HF_2^-, in which a normal fluoride ion is hydrogen bonded to a molecule of hydrogen fluoride. In fact KHF_2 would inevitably be present in the electrolyte (HF/KF mixture) for the making of fluorine gas.

Question

60 Draw displayed formulae for a loose polymer of hydrogen-bonded hydrogen fluoride molecules, and for the HF_2^- ion.

A last little quirk of hydrogen fluoride is that although it is a weak acid in the Brønsted–Lowry sense of the word, it is very corrosive. It cannot be kept in glass bottles because it dissolves them (polythene has to be used), and this has led to the use of hydrogen fluoride in etching glass so that designs can be picked out by opaque regions eaten into the surface (Figure 18.30). Wax is used to mask the parts of the glass that are to remain clear. Hydrogen fluoride burns on the skin are different too, because while all acid burns are nasty, those from hydrogen fluoride take an unusually long time to heal.

The resistance to oxidation of the four compounds exactly mirrors that of other halide compounds, with hydrogen iodide being easiest to oxidise. In fact, hydrogen iodide can be made to decompose back into its elements by placing a hot nickel or platinum wire in it. This decomposition was one of the earliest equilibrium reactions to be studied in detail:

$$2HI(g) \underset{}{\overset{Ni}{\rightleftharpoons}} H_2(g) + I_2(g)$$

In contrast, it takes a vigorous oxidising agent like acidified manganate(VII) to oxidise hydrogen chloride to chlorine, while hydrogen fluoride is immune to chemical oxidation.

The formation of the hydrogen halides by direct reaction of the elements is, like the water reaction, explosive, and radicals are again involved. A photon of light serves as initiator:

$$Cl_2(g) \overset{h\nu}{\rightarrow} 2Cl\cdot$$

The first propagation step begins:

$$H_2 + Cl\cdot \rightarrow ? + ?$$

Questions

61 As a revision of free radical mechanisms, complete the above sequence with propagation and termination steps.

62 The presence of hydrogen halide gases can be detected by letting the gas interact with ammonia gas, often by just holding a stopper from a bottle of concentrated ammonia solution nearby (Figure 18.31). The appearance of a solid white smoke is the positive outcome. Write a chemical equation for the event taking place.

Figure 18.31 The reaction between ammonia and hydrogen chloride gas

63 The aqueous hydrogen halide solutions differ in their resistance to oxidation. Look back to the electrochemical series and pick out oxidising agents that would discriminate between the hydrogen halides, by oxidising some but not others.

64 The handiest test for discriminating between the hydrohalic acids, and indeed between all ionic solutions containing unknown halide ions, is to add aqueous silver ions, in order to precipitate the insoluble silver halide. What differences between the silver halides allow for discrimination between the halogens?

Figure 18.30 Etched glass, produced using hydrogen fluoride

18.10 The oxides and oxo acids of the p-block elements

Oxygen is more electronegative than all the elements in the Periodic Table except fluorine. The electronegativity difference in the bond to oxygen gets less from Group III to Group VII, so we see a transition in bond type from ionic giant lattice oxides like alumina, Al_2O_3, through polar covalent small molecules like tetraphosphorus decaoxide, P_4O_{10}, to covalent small molecules like difluorine monoxide, F_2O. Nevertheless, except in the latter case oxygen attracts electrons enough to coerce all other elements into their maximum positive oxidation states. Elements that *can* break the rule of eight do break it with oxygen, because bonds to oxygen are sufficiently strong to make the promotion of previously paired electrons into vacant d-orbitals energetically worthwhile.

The next question invites you to give examples of the generalisations in the last paragraph, and revise some bonding theory.

Question

65a Copy and complete Table 18.14, which gives formulae and oxidation numbers of 'normal' and highest oxides.

Table 18.14

Element	Rule-of-eight oxide	Highest oxide	Oxidation number in highest oxide
C	?	—	?
N	?	N_2O_5	?
O	O_2	—	?
F	F_2O	—	?
Si	?	—	?
P	P_4O_6	?	?
S	?*	?	?
Cl	?	Cl_2O_7	?

*Non-existent

b Phosphorus reaches a valency of 5 (and an oxidation state of +5) in some of its compounds with oxygen. Show a box diagram of the energy levels of the electrons in a phosphorus atom, and include the vacant 3d set. Then show how phosphorus rearranges its electrons to accommodate five shares with oxygen electrons.

c Sulphur shows its +6 oxidation state in compounds with oxygen, and with one other element. Can you guess which other element, and why?

d Oxides exist of one of the noble gases. Oxides of general formulae MO_3 and MO_4 have been prepared, with nominal oxidation numbers of +6 and +8. What would you guess is the most likely identity of M? Why?

Since all these oxides feature p-block elements in positive oxidation states, it follows that they are all possible candidates to be oxidising agents. The position in the league table of oxidising power depends on two factors.

1 How willing was the element to give up its electrons? (The more willing the element, the less oxidising will be its compounds.)
2 How many did it give up? (The more it gave up, the more oxidising will be the compound.)

By these criteria, aluminium(III) oxide is the least oxidising, sulphur trioxide is a stronger oxidiser than sulphur dioxide, and Cl_2O_7 and N_2O_5 are very strong oxidising agents. These predictions are more or less borne out in practice, although difluorine monoxide beats them all, with the oxygen atom trying to regain *its* lost electrons.

Reactions of oxides with water – acidity, basicity and amphotericity

This section is in part a revision of Sections 5.6 and 17.4. Figure 18.32 shows a water molecule approaching an M=O bond. In (a) the bond is ionic, and in (b) it is polar covalent. There is no attempt in the figure to show the stoichiometry. What happens next depends on the nature of the bond, and therefore on M's position in the Periodic Table. The water molecule has two possible modes of interaction with the bond, which might be distinguished as a nucleophilic approach and an electrophilic one.

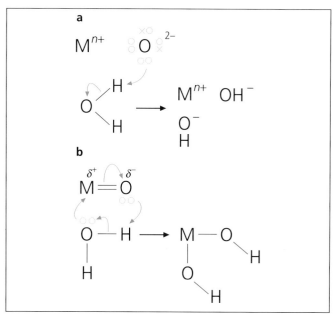

Figure 18.32 Two possible modes of attack of a water molecule on an oxide. **a** A typical s-block oxide and **b** A typical right-upper-p-block oxide (The mode followed depends on the electronegativity of M)

The electrophilic approach is the offer of H^+ ions to an oxide ion (a). We know from Chapter 15 that O^{2-} is a very strong base and nucleophile. The right-hand side of Figure 18.32a shows an ionic hydroxide, obviously still basic since two more hydrogen ions could be accepted. The whole pattern of behaviour is typical of s-block oxides, and something similar also happens in Group III (but see below for complications).

The nucleophilic approach is shown in Figure 18.32b. Here we have a polar covalent M=O bond, but with an M which is still interested in accepting electrons, since it is positively polarised. This then is nucleophilic attack *by* water rather than *on* water (as in (a)). The product still looks superficially like a hydroxide, but if you are used to associating the word hydroxide with alkali, then the impression is misleading. The covalent bond sequence M—O—H on the right-hand side of (b) now has the possibility of ionic fracture at the O—H bond rather than the M—O bond, which means acting as an acid rather than as a source of basic OH^- ions. (In fact, many of the well known inorganic acids are of this 'hydroxide' type.)

Questions

66 The question of whether a 'hydroxide' behaves as a base or as an acid depends upon the character of M. In that context, copy and complete the following sentences.

a The more electronegative M is, the (more/less) likely it is to have an acidic 'hydroxide'.

b The higher the group number of M, the (more/less) likely it is to have an acidic 'hydroxide'.

67a Draw a curly-arrows diagram, in the style of Figure 18.32, to show the reaction of water with:

i calcium oxide, CaO (show one ion pair reacting)

ii sulphur dioxide, SO_2 (show one SO_2 molecule reacting).

b Write normal chemical equations for the two overall changes.

As you saw from question 67, calcium oxide becomes an alkali in water (as limewater), while sulphur dioxide becomes sulphuric(IV) (formerly sulphurous) acid. But this is a chapter on the p-block, so why are we using an s-block oxide to demonstrate alkaline oxide behaviour? As hinted at above, with the Group III oxides, where you might expect to find an oxide behaving as a basic oxide on the calcium model, things are not that simple. It is true that aluminium oxide dissolves in acids, in parallel with calcium oxide:

$$Al_2O_3(s) + 6H^+(aq) \rightarrow 2Al^{3+}(aq) + 3H_2O(l) \quad (15)$$

So it would be reasonable to assume that aluminium oxide would give an alkaline reaction with ordinary water, as in:

$$Al_2O_3 + 3H_2O(l) \rightarrow 2Al^{3+}(aq) + 6OH^-(aq) \quad (16)$$

but two factors stand in the way. One is the high lattice enthalpy of aluminium oxide – it will break up to admit H^+ ions, but not for water. This is not unique to Group III – something similar is observed with magnesium oxide. But the other factor certainly is a trademark of Group III. Aluminium oxide behaves differently depending on the pH of the medium it is in.

The reason is that unless an M=O bond is completely ionic (a state of purity only approached by s-block oxides), there will be some tendency for strong nucleophiles to attack in the style of Figure 18.32b, even though strong electrophiles may still get away with the Figure 18.32a-type attack. As far as aluminium oxide is concerned, water lies between those two extremes. Acids, as equation (15) showed, elicit an s-block-type response, but with alkalis the reaction is:

$$Al_2O_3(s) + 2OH^-(aq) + 3H_2O(l) \rightarrow 2Al(OH)_4^-(s) \quad (17)$$

(This may look a bit different from the sulphur case – the species that would have been the acid, covalent aluminium hydroxide, is insoluble in water, and only dissolves by becoming the complex anion $Al(OH)_4^-$.)

An oxide that behaves like a base towards H^+, but also dissolves in alkalis to give species which look like the oxo anions of a parent acid, is given the adjective **amphoteric**. Group III oxides and hydroxides are classics of the type, and in fact a sort of pH profile can be done of the oxidation state +3 for a Group III element, showing its transformation from the simple cation in acids via the insoluble hydroxide to the complex oxo anion in alkalis (Figure 18.33).

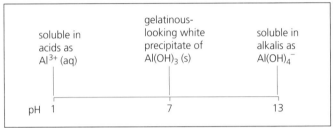

Figure 18.33 What you would see if you subjected aluminium oxide to a range of media of different pH

Questions

68 Compounds containing well-defined Al^{3+} ions, like aluminium sulphate, say, behave as acids in solution, as in the equation below. Say what you would see after *three* OH^- ions had been added.

$$[Al_2(H_2O)_6]^{3+} + OH^- \rightarrow [Al(H_2O)_5OH]^{2+} + H_2O$$

69 The reason why the Al^{3+} ion is able to do this is because it can so polarise its surrounding water molecules that their lone pairs begin to resemble covalent bonds to the aluminium (Figure 18.34). What property of Al^{3+} ions enables them to do this?

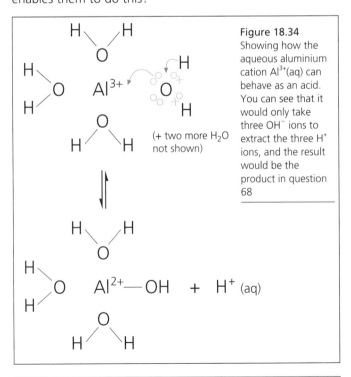

Figure 18.34 Showing how the aqueous aluminium cation $Al^{3+}(aq)$ can behave as an acid. You can see that it would only take three OH^- ions to extract the three H^+ ions, and the result would be the product in question 68

Beyond Group III the amphoteric trend peters out. There are no free ions outside the d-block with a charge greater than +3. From Group IV on, the behaviour of oxides in water is purely to undergo nucleophilic attack, resulting in oxo acids and oxo anions. The behaviour is typified by the oxides of sulphur, as we saw at the end of Chapter 5.

Questions

70 Draw a curly-arrows diagram to show the hydrolysis of N_2O_3 into nitric(III) acid, HNO_2, in the style of Figure 5.38, p. 91.

71 A generalisation concerning the oxo acids is that the one with the highest oxidation number is the strongest. For example, H_2SO_4 is a stronger acid than H_2SO_3, and HNO_3 is stronger than HNO_2. Use theories of bond polarity to suggest why the ease of loss of H^+ correlates with oxidation number.

There are several items of interest in the oxides, oxo acids and oxo anions of the p-block. These special features will be dealt with in group order.

Group IV oxides

Carbon dioxide is faintly acidic in aqueous media, giving a solution called carbonic acid. The equilibrium:

$$CO_2(aq) + H_2O(l) \rightleftharpoons H^+(aq) + HCO_3^-(aq) \qquad (18)$$

lies well over to the left. That is why soda water does not taste very sour. However, the acidity is sufficient to upset the balance of solubility of limestone rocks. Calcium carbonate has very low solubility in pure water, with the equilibrium:

$$CaCO_3(s) \rightleftharpoons Ca^{2+}(aq) + CO_3^{2-}(aq) \qquad (19)$$

being well over to the left. But the right-hand side of equation (19) includes a strong base. So in the presence of naturally acidic rain (in other words, the mixture from equilibrium (18)), equilibrium (19) is moved across to the right. The caves and fissures of limestone country, and hence of the sports of potholing and caving, are dependent on these two equilibria.

Question

72a Most of the carbon dioxide in water is just dissolved rather than reacted. Suggest why the carbon dioxide molecule has appreciable solubility.

b Combine equilibria (18) and (19) in a way that shows the donation of a single proton from (18) to the aqueous base on the right of (19). Hence give an overall equation for the dissolution of calcium carbonate in rainwater.

Your answer to question 72b is also the reason why natural groundwater in limestone areas is 'hard'. Hard water precipitates the long-chain carboxylate anions in traditional soaps.

The calcium salts of these anions are indeed insoluble, so any water with appreciable concentrations of aqueous Ca^{2+} ions will bring about this reaction. The precipitate is the 'scum' that forms the ring round the bath:

$$2C_{15}H_{31}CO_2Na(aq) + Ca^{2+}(aq) \rightarrow$$
$$\text{soap}$$
$$(C_{15}H_{31}CO_2)_2Ca(s) + 2Na^+(aq)$$
$$\text{scum}$$

Question

73 Limestone-induced hardness is called temporary hardness, because it can be removed by boiling. The problem then shifts to inside the boiler or kettle, due to the way calcium hydrogencarbonate decomposes when heated. The equation is in fact the exact reverse of that in question 72b:

$$Ca(HCO_3)_2(aq) \rightarrow CaCO_3(s) + H_2O(l) + CO_2(g)$$

a Can you suggest why boiling might tilt the equilibrium towards carbon dioxide and the carbonate?

b Why is the water used in power station steam generation always softened before use?

c Which would be the best low-energy high-volume way to soften water for this sort of application?

Silicon oxides and silicates lie outside most Advanced syllabuses, but they are worth a mention on the grounds of their social impact and ubiquity. Silicon(IV) oxide or silica, SiO_2, which occurs as the mineral quartz and is a major component of most sand, is of course extremely unlike its carbon analogue carbon dioxide. This is yet another illustration of the way first group members form double-bonded small molecules while second members form single-bonded chains, as discussed when comparing oxygen and sulphur.

The silica structure features indeterminate rafts of tetrahedra (Figure 18.35b). In the liquid phase (which occurs above about $1600\,°C$), such big flexible molecules give rise to very viscous liquids, and viscous liquids do not crystallise easily since it takes too long to achieve the correct stacking orientation. As a result, silica and the silicates solidify as disordered **glasses**, and form the basis of our glass industry.

Question

74 Most naturally occurring silica is not glassy but crystalline – the mineral **quartz** (Figure 18.35a). Why do you think natural silica crystallises on cooling, whereas in an industrial context it is easy to make it into a glass structure?

The ubiquity mentioned above refers to the fact that silicates are by far the biggest group of rock-forming minerals, occurring in association with all the common metals and most of the uncommon ones.

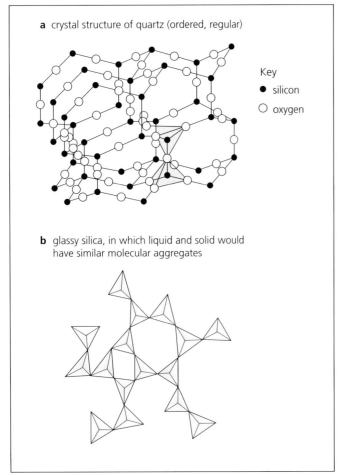

a crystal structure of quartz (ordered, regular)

Key
● silicon
○ oxygen

b glassy silica, in which liquid and solid would have similar molecular aggregates

Figure 18.35 Contrasting forms of silica, SiO_2. (a) is used to define a tetrahedral unit with a silicon atom amid four oxygen atoms. These are the same tetrahedral units as you see sharing oxygen corners in (b)

Group V oxides

Nitrogen oxides are numerous and varied. They are shown in Table 18.15 with their names and points of interest. Several offer a challenge to conventional notions of bonding.

Question

75 Try drawing dot-and-cross diagrams of N_2O and NO. One of these needs only a dative covalent bond, but the other does not fit the idea of shared pairs of electrons at all. This one is actually magnetic when solid for that very reason.

The oxides of phosphorus are a little less diverse and more predictable than those of nitrogen. There are only two, they have straightforward electron arrangements and they fit the general p-block trend in oxides by being unambiguously acidic. However, they are unusual in existing as units of double their empirical formulae (Figure 18.36).

Tetraphosphorus decaoxide, P_4O_{10}, not only undergoes hydrolysis to phosphoric(V) acid, but shows such an appetite

Table 18.15 The range and diversity of the oxides of nitrogen. (*Note:* Skeletal diagrams do not indicate bond order (single/double) or bond length.)

Formula	Skeletal arrangement	Acidic (A) or neutral (N)	Points of interest
N_2O	N—N—O	N	'Laughing gas' – an anaesthetic
NO	N—O	N	Reacts on contact with air to give NO_2
N_2O_3	O═N—N(O)(O) (flat)	A	Reacts with water to give HNO_2
NO_2	O—N(134°)—O	A	The familiar 'brown gas' from heating Group II nitrates
N_2O_4	(O)(O)N—N(O)(O) (flat)	A	Dimer of NO_2, in equilibrium with it
N_2O_5	(O)(O)N—O—N(O)(O)	A	Reacts with water to give HNO_3

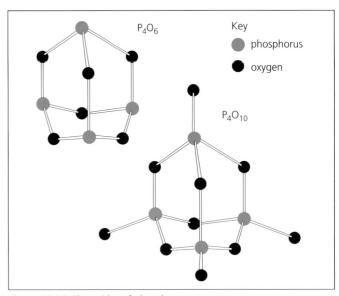

P_4O_6

Key
● phosphorus
● oxygen

P_4O_{10}

Figure 18.36 The oxides of phosphorus

for water that it is one of the most rigorous chemical drying agents. Gases may be passed over P_4O_{10} to remove traces of water vapour. Like concentrated sulphuric acid, it can even remove the elements of water from a molecule, as in:

$$CH_3CH_2C(=O)(NH_2) \xrightarrow[\substack{removes \\ H_2O}]{P_4O_{10}} CH_3CH_2C\equiv N$$

The normal phosphoric(v) acid is shown in Figure 18.37, but it undergoes polymerisation on heating to a long indeterminate chain of repeating HPO_3 units called metaphosphoric acid. (This polymerisation is of the condensation type, since a water molecule has been eliminated at every junction.) The anionic version of a long chain of this kind can wrap itself round certain cations in solution, and prevent them from being precipitated. This has led to the use of these 'polyphosphates', as sodium salts, in water softening. Their trade name is Calgon, and they prevent precipitation of scale or scum (Figure 18.38).

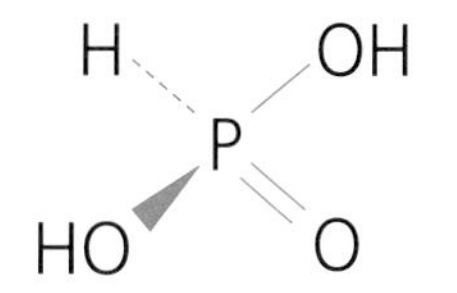

Figure 18.37 Phosphoric(v) acid

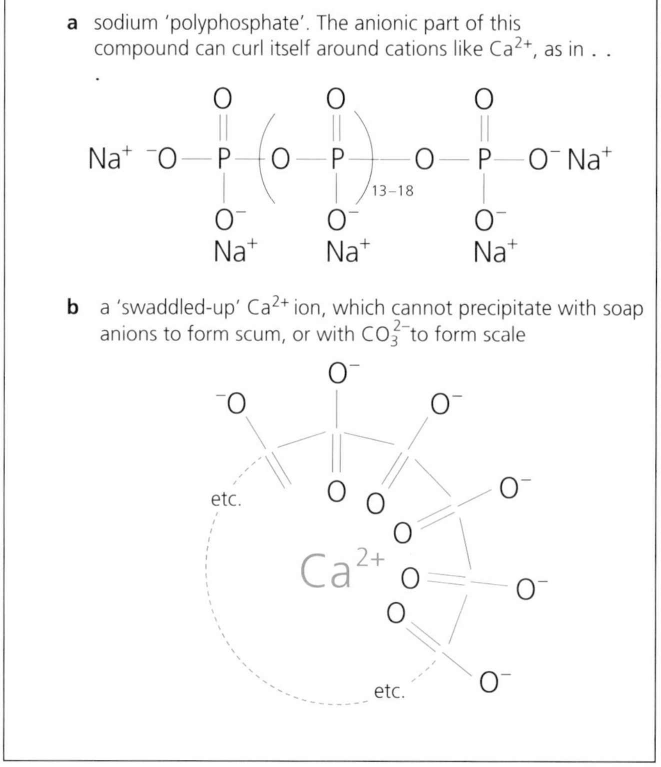

a sodium 'polyphosphate'. The anionic part of this compound can curl itself around cations like Ca^{2+}, as in . . .

b a 'swaddled-up' Ca^{2+} ion, which cannot precipitate with soap anions to form scum, or with CO_3^{2-} to form scale

Figure 18.38 Calgon – a commercial water softener

Questions

76 Assume the Calgon molecule points its P=O ends in towards the Ca^{2+} ion, as in Figure 18.38.

a Why do you think that only molecules in the range 13–18 repeat units are suitable for wrapping round the calcium cation?

b Why do you think a 'swaddled-up' Ca^{2+} ion is unable to form a strong insoluble lattice with, say, CO_3^{2-} ions?

c What kind of interaction do you imagine exists between the ion and its polymer sheath?

77 Versions of the Calgon molecule have seen much use in soap powders, in which they are called 'builders'.

a What job, similar to Calgon's, are they doing in soap powders?

b What unforeseen environmental impact, related to another major use of phosphates, did these builders have when washed into rivers? (Many 'green' washing powders now declare themselves phosphate-free.)

Group VI oxides and oxo acids

Sulphur dioxide is the familiar product of burning sulphur, and in industry is the result of roasting sulphide ores. It is of course the villain of the acid rain story, due to sulphur's original presence in fossil fuels. Its oxidation by dioxygen to sulphur trioxide is simultaneously a blessing and a curse.

On the bad side, it is the step before the production of sulphuric acid in the atmosphere. The reaction is slow at ambient temperatures but is catalysed by nitrogen oxides, which are unfortunately in ready supply from car exhausts.

On the good side it is the step before the making of sulphuric acid in the Contact process (Chapter 9). A catalyst is used, usually vanadium(v) oxide, V_2O_5. Sulphuric acid is the number one industrial chemical, finding uses in all the major branches of the chemical industry. By buying detergents, car batteries, food grown with fertilisers, polymers and paints we in the UK indirectly consume about 20 times as much sulphuric acid as we do beer, per head of population.

Oxides and oxo acids of Group VII

The oxides of the halogens tend to be strongly oxidising, unstable compounds. The corresponding oxo acids and their anions are also strong oxidising agents, as typified by the use of sodium chlorate(I), NaClO, as a domestic bleach. Sodium chlorate(v), $NaClO_3$, is dangerously explosive in mixtures with reducing agents, and rather curiously is actually a better oxidising agent than sodium chlorate(VII), $NaClO_4$. In fact, the latter can be made from the former by a disproportionation reaction.

Question

78 Balance the equation for this disproportionation by the methods introduced in Chapter 7:

$$NaClO_3(s) \rightarrow NaClO_4(s) + NaCl(s)$$

18.11 The halides of the p-block elements

Bond polarity of the M—Hal bond follows the same pattern as in the oxides. Generally the bond is negatively polarised at the halogen end, and thus vulnerable to nucleophilic attack.

The difference compared with oxides is that the halogen is only held by a single bond and thus is totally ejected by the nucleophile.

Question

79 Apply this general pattern to the case of the hydrolysis of silicon(IV) chloride:

$$SiCl_4 + 4H_2O \rightarrow \ ? \ + \ ?$$

Apart from that generalisation, it is worth noting a few curiosities. Nitrogen(III) iodide, NI_3, is a strange and rather nasty impact explosive, P_2Cl_{10} has a peculiar ionic structure, and SF_6 has a very low conductivity which makes it suitable as a gas for surrounding high-voltage apparatus. The halides of the p-block elements have not found enormous application in industrial chemistry.

Summary

General trends in the p-block

• The elements of the p-block generally have **higher ionization enthalpies** than s-block elements. This inclines them towards non-metallic character and anion formation.

• However, as with all the typical groups of the Periodic Table, ionization enthalpies decrease as the group is descended, so a limited degree of metallic character is found in elements towards the bottom and to the left of the block.

• The p-block elements show **medium to high electronegativies**. When they bond with each other their bond type is covalent or polar covalent, with very few exceptions (again, low in Groups III and IV). When they bond with s-block elements, the bond type is predominantly ionic, or at least strongly polar covalent.

• There is a well-established pattern in the p-block, indeed throughout the Periodic Table, that the first element of one of the groups is not all that typical. In the p-block that trend shows up in a tendency for the first member to have strong multiple bonds (if possible), while the second and subsequent members opt for single bonds. This tendency is responsible for some stark contrasts between close relatives, for example between carbon dioxide CO_2 and silicon(IV) oxide SiO_2, and between oxygen O_2 and sulphur S_8.

• The change in character across and down the p-block means that the block houses the dividing line separating the metals from the non-metals. A group of elements that lie *on* this line exhibit semiconductor behaviour, and are sometimes known as the poor metals, for example silicon, antimony and arsenic.

The elements group by group

Group IV

• **Carbon** is one of the four elements most closely associated with life on Earth, and especially with the energetics of life. The food chain is a succession of organisms trying to acquire 'high-energy carbon' (and hydrogen), in forms like carbohydrates and fats. Every organism taps this energy by returning the carbon and hydrogen to their low-energy forms, carbon dioxide and water. This process, and its reversal in photosynthesis, are at the heart of the planet's **carbon cycle**.

• **Fossil fuels** are the result of interruptions in the full rotation of the carbon cycle. They occur when arrested decay of organisms leaves the carbon in high-energy forms like hydrocarbons and elemental carbon. The human race has found that this energy source is so rich that it revolutionises the ability to do work on our environment. As a result, the headlong advance of technology over the last 200 years has used up the larger part of an energy reserve that took 200 million or more years to be formed.

• **Silicon**, in combination with oxygen, is the major rock-forming element on Earth. The variations on the silicate theme are a match for the array of organic chemicals based on carbon.

• Very pure silicon is needed for the microelectronics industry. The high purity is achieved by the technique of zone refining, in which a molten zone passes along a silicon ingot. The impurities are concentrated in the liquid region, which passes to the end of the ingot.

Group V

• **Nitrogen**, along with carbon, hydrogen and oxygen, is a major component of the chemicals essential to life. The nitrogen in the planet is in constant circulation between living and non-living situations, as expressed in the **nitrogen cycle**. Humans have intervened in the nitrogen cycle to bring more nitrogen from the relatively inert form of dinitrogen gas into the form of the compound ammonia, which is a precursor to all important plant fertilisers.

• **Phosphorus** is another element essential to life, and also finds its way into fertilisers. In the inorganic realm it is prized for the low activation enthalpy of its reaction with oxygen, which leads to its use in incendiaries and matches.

Group VI

• **Oxygen** is the most abundant element on Earth, and is found throughout inorganic and organic chemistry, in rocks and in living organisms. It is the second most electronegative element after fluorine, and its main significance and importance lies in its **combustion** reactions with a wide range of other elements and compounds. Combustion of carbon and hydrocarbons lies at the heart of industrial life, via its role in electricity generation. In biology, the rather more complex pathways of respiration achieve the same overall results as combustion, and occur in every living cell.

• **Sulphur** is yet another element essential to life, but in rather smaller amounts than carbon, hydrogen, oxygen, nitrogen and phosphorus. (It occurs in the important amino acids cysteine and cystine.) The elemental form of sulphur occurs on Earth, and is mined by the unusual Frasch steam-pumping process. Elemental sulphur performs one of the most complex and interesting series of molecular rearrangements undergone by any element, passing through two solid allotropes (rhombic and monoclinic), and through liquid forms of sharply variable viscosity, all under the influence of increasing temperature. In addition, when near to its boiling temperature it can be water-quenched to a polymeric solid called plastic sulphur.

• **Sulphur bridges** are used to cross-link polymers. In Nature the links using the amino acid cystine are crucial to maintaining the shapes of proteins, whereas in rubber the links allow for the fine tuning of properties like hardness and elasticity.

Group VII

• **Fluorine** is the most electronegative element of all. As a result fluoride ions will not surrender their electrons to any chemical oxidising agent. Fluorine is therefore made at the anode in electrolysis. It finds uses in making Teflon, the non-stick surface polymer, and refrigerants and propellants, including the new generation of CFC substitutes.

• Fluoride ions are added to drinking water to promote dental health. Fluoride improves the hardness of the tooth material by a lattice substitution reaction, in which hydroxide ions are replaced by fluoride. The same reaction occurs in fossil bone material, and has provided us with the **fluorine dating** method of estimating the relative age of fossils.

• **Chlorine** is the most widely used halogen, being an ingredient in the very versatile plastic PVC. It is also the most widely distributed halogen, being present at quite high concentration in sea water. Chlorine, like fluorine, is very electronegative, and is therefore not easily made by chemical oxidation of chloride ions. Instead it is produced, along with sodium hydroxide, by the electrolysis of brine.

• The other halogens are produced chemically, **bromine** by oxidation of bromide ions in sea water using chlorine. **Iodine** comes to us via a few links of food chain in the coastal ecosystem of Chile. The crucial first step is the accumulation of iodine in seaweeds, and at the other end of the chain the iodine is dumped on sea cliffs in the guano from birds, in the form of iodate ions. Both the heavier halogens find use in the photography industry.

Hydrogen

• **Hydrogen** is the most abundant element in the Universe (as opposed to on Earth), and is the fuel of stars, which derive energy from its fusion reactions. It is also used as a chemical fuel on Earth, and many scientists think that this mode of use will assume increasing importance as a replacement for fossil fuels. Hydrogen-based systems could use solar energy to generate hydrogen by electrolysis, and the gas could then be burnt directly, or transmitted across country and then converted back into electricity in fuel cells. (Hydrogen transmission is cheaper than overhead a.c.) Aside from finding uses as a fuel, hydrogen, available as a by-product of the chlor-alkali industry and by the steam reforming of hydrocarbons, is converted in large amounts to ammonia.

The compounds of the p-block elements

Hydrides

• The **hydrides** of the p-block are all small molecules. The polarity of the bond to hydrogen is nearly always such as to make the hydrogen $\delta+$, and yet the p-block hydrides show a **full range of acid–base behaviour**. This includes strong acids (Group VII, as in hydrochloric acid), very weak acids (Group VI, as in water), very weak bases (Group VI again), weak bases (Group V, as in ammonia), and finally neutral hydrides (Group IV, as in methane).

• Three of the hydrides of the p-block are of monumental importance to humans. But whereas **water** and **methane** are found ready-made in the biosphere and lithosphere, **ammonia** has to be manufactured. The pathway by which this is achieved illustrates many of the principles of industrial chemistry at work.

• At the heart of ammonia production is the **Haber process**, which manipulates an equilibrium system to maximise effective yields. The Haber process is run at **high pressure** (to promote the side of the equation with fewer molecules), and **reasonably low temperature** (to promote the exothermic direction of the equation, while still achieving respectable **reaction rates**). Ammonia is **removed** before equilibrium is achieved (preventing the reverse reaction).

• The hydrogen for ammonia production is made, in the ICI Billingham plant, by the steam reforming of methane. Strenuous attempts are made to keep methane consumption to a minimum, by measures such as **heat exchange**. The use of heat exchangers means that there is less need for extra methane to be burnt merely for the heating of reaction systems. One example of a heat exchange is when the heat from the (exothermic) Haber process is used to pre-heat the reactants of the (endothermic) steam reformation reaction.

• Ammonia production consumes ever greater amounts of energy as the world's population escalates and requires food. Many alternative options for hydrogen generation are being considered, running in parallel with similar researches into the general future of energy utilisation on our planet.

• The properties of water as a solvent, acid, base, hydration agent, temperature buffer, etc., are dealt with in other chapters throughout the book.

Oxides

• The oxides of the p-block vary widely in properties. For instance, they show a variation of approximately 1500 °C in **melting temperatures**, and a range of about 8 pH units of **acidity and basicity**. These variations have their roots in structural differences – the melting temperatures are related to the type of structure (small molecule or giant), while the pH differences relate to variations in bond polarity.

• The variation from **small molecule to giant molecule** is bound up with the differences in the abilities of elements in Periods 2 and 3 to make double bonds. In that respect it resembles the variation in the elements themselves, where for example we met $O{=}O$ but not $S{=}S$. The ability to make double bonds means an inclination towards small molecules, so there are no giant-molecule oxides in Period 2, and instead we find small molecules with double bonds to oxygen, typified by the gases carbon dioxide and nitrogen dioxide. In Period 3, there are both types of molecule, and significantly we see an example of an extended single-bonded lattice, in the very high-melting solid silica, SiO_2.

• The 'ionic' oxides occur mainly in Group III, where the difference in electronegativity between oxygen and the element is greatest (for the p-block). They share, with the giant-molecular oxides, the tendency to have very high melting temperatures. There is a significant disagreement between lattice enthalpy predictions based on pure ionic theory and real lattice enthalpies, so the bonding in Group III oxides is best described as '**impure ionic**'.

• The other main variation in the properties of p-block oxides resides in the field of **acidity**. If the bond between the element and oxygen were purely ionic, then the oxide ion would-deliver its usual package of **basic** properties, familiar from the oxides of Groups I and II. But this only applies, in the p-block, to the bottom elements of Group III, if at all. At the other extreme of polarity, if the bond to oxygen is non-polar then neutral behaviour results, as typified by dioxygen itself, and some of the lower oxides of nitrogen.

Acidic oxides result from a bond between the element and oxygen which is of the polar covalent type, leading to a reaction with water that creates an acidic OH group. The sulphur oxides and the higher oxides of nitrogen are the classic examples here.

Finally, there are those oxides whose oxygen-to-element bond is of intermediate polarity, and which react with water to form 'hydroxide-like' compounds. These compounds can ionize at the element-to-oxygen bond, like a Group I or II hydroxide, to give ordinary basic OH^- ions, or else they can ionize at the oxygen-to-hydrogen bond, to give acids. The choice of behaviour is dependent on the medium in which these oxides find themselves, and these oxides are called **amphoteric**. The classic case is the oxide of aluminium, Al_2O_3.

The acids formed by acidic oxides are the **oxo acids**. They include a number of enormously important industrial chemicals, such as sulphuric acid, which is the world's most convenient proton donor. In one of these proton-donating applications, sulphuric acid makes phosphoric acid, which itself plays a key role in the manufacture of ammonium phosphate fertilisers.

The **anions** of the oxo acids have industrial roles in their own right – the phosphates contribute to water softening, the sulphates to cement and plaster, and the nitrates and chlorates to fertilisers, weedkillers, explosives, pyrotechnics and propellants.

Halides

• The **halides** of the p-block are not as significant a set of compounds as the oxides. The halogenoalkanes are important, but they belong to organic chemistry. Apart from these, the Group III chlorides are catalysts for Friedel–Crafts substitutions of benzene rings, and as such are used in the making of paracetamol and ibuprofen. The phosphorus halides are used for nucleophilic substitutions of hydroxy groups. Nearly all the halides of the p-block undergo hydrolysis in water to liberate the hydrogen halide. The exceptions are the halides of carbon, and all the fluorides.

19

The elements of the d-block

- ▷ Electron arrangements of the d-block elements
- ▷ Electron arrangements and family features
- ▷ Variable oxidation number 1: The 2+ and 3+ cations
- ▷ Variable oxidation number 2: Oxidation states higher than +3
- ▷ Complexes
- ▷ Catalysis by elements and compounds of the d-block
- ▷ Catalysts in automobiles
- ▷ A practical application of d-block chemistry – photography
- ▷ Summary

This chapter, in common with Chapters 17 and 18, provides an opportunity to revise some established concepts (structure and bonding, redox, kinetics, stereochemistry and electrochemistry), while meeting the final set of new ones.

Most notably it introduces the idea that nucleophilic species, referred to in this context as **ligands**, can attach themselves around d-block cations to form a new class of compounds called **complexes** or **co-ordination compounds**.

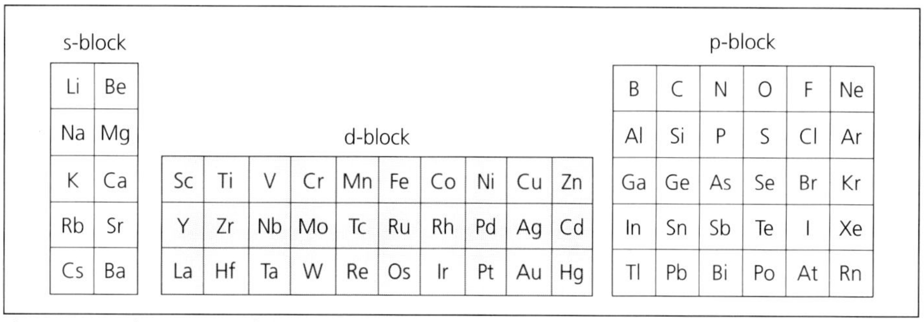

Figure 19.1 The position of the d-block in the Periodic Table

19.1 Electron arrangements of the d-block elements

One of the central themes of chemistry is that chemical behaviour is rooted in electron arrangements, and most significantly in *outer* electron arrangements. As we have seen in Chapters 17 and 18, this principle accounts for the character of the main-group 'typical' elements.

The picture is rather different, however, in the swathe of elements occupying the block that interrupts the main-group sequence, between Groups II and III (Figure 19.1). These are referred to as either the **transition elements** or the **d-block**. The elements in this region of the table, while obviously differing from one another in electron arrangement, nevertheless show quite strong *horizontal* similarity. They are more similar to each other than to most of the elements of the typical groups, leading to the idea that there is some characterising influence coming not so much from the individual electron arrangements as from a feature of electron arrangement common to, and characteristic of, the whole block. The following question should bring home to you the degree of horizontal similarity within the d-block.

Question

1 a Pick any four successive p-block elements, and look up their properties. Fill in the gaps in a copy of Table 19.1.

b Pick any four successive elements from the row of elements scandium to zinc, and fill in another copy of Table 19.1.

c What do you notice?

Table 19.1 Blank table for data from question 1

Element	Metal/non-metal	Melting point (°C)	First ionization enthalpy (kJ mol^{-1})	Density (g cm^{-3})

What is the underlying electronic cause of the strong horizontal similarity in the d-block? You may remember from the end of Chapter 3 that the whole of the d-block suffers from 'deferred outcrop syndrome' – it shows up in the Periodic Table later than expected. One might assume, wrongly, that after the filling of the 3p sub-shell at argon, there would be ten more elements created by the filling of the 3d orbitals. In fact the next element begins a new row, with a new principal quantum number (4), and is a new member of Group I, namely potassium (with the electron arrangement $1s^2 2s^2 2p^6 3s^2 3p^6 4s^1$). It is only after potassium and calcium that the 'forgotten' sub-shell begins to fill up, and the ten new d-block elements appear.

Questions

2 What was the cause, according to Figure 3.44 (p. 46), of 'deferred outcrop syndrome'?

3 The electron arrangements of the elements of the first row of the d-block are shown in Table 19.2, along with some successive ionization enthalpies. Try to pick out some features of electron arrangement common to all or most of the d-block elements, which may be the underlying causes of the horizontal similarity.

Table 19.2 Electron arrangements of the first row of the d-block elements

Element	Electron configuration	First ionization enthalpy (kJ mol^{-1})	Second ionization enthalpy (kJ mol^{-1})	Third ionization enthalpy (kJ mol^{-1})
Sc	[Ar] 3d^{1}4s^2	631	1235	2389
Ti	[Ar] 3d^{2}4s^2	658	1310	2653
V	[Ar] 3d^{3}4s^2	650	1414	2828
Cr	[Ar] 3d^{5}4s^1*	653	1592	2987
Mn	[Ar] 3d^{5}4s^2	717	1509	3249
Fe	[Ar] 3d^{6}4s^2	759	1561	2958
Co	[Ar] 3d^{7}4s^2	758	1646	3232
Ni	[Ar] 3d^{8}4s^2	737	1753	3394
Cu	[Ar] 3d^{10}4s^1*	746	1958	3554
Zn	[Ar] 3d^{10}4s^2	906	1733	3833

[Ar] = $1s^2 2s^2 2p^6 3s^2 3p^6$
*Note the switch of an electron from 4s to 3d, driven by the desirability of full or half-full d sub-shells.

19.2 Electron arrangements and family features

We shall now examine what it is about the electron sequences in Table 19.2 that causes the d-block elements to share each of the following family traits:

- **metallic character**
- **high density**, and (on the whole) **high melting and boiling points**

- **moderate to low reactivity**
- **coloured ions**
- wide **variability of oxidation number**, with the same oxidation numbers cropping up in many elements
- ability to bind a wide range of chemical species around a single central atom or ion – **co-ordination complex formation**
- powerful and broad **catalytic activity**.

Metallic character

Why are all the d-block elements metals? We might speculate what the d-block elements would have been like had the 3d sub-shell been filled before the 4s. Up to argon we have understood the steady general rise in ionization enthalpy throughout the filling of the orbitals of principal quantum level 3, and with it the general decrease in metallic character. This was attributed to the fact that the increase in nuclear pull was dominating the increase in electron–electron repulsion – to put it another way, the shielding effect of the $n = 3$ electrons on each other was not very good. So if the 3d elements had been created immediately after argon, then perhaps they would not have been metals at all.

Question

4 Explain why the d-block elements might, under the fictional circumstances referred to in the last paragraph, have been non-metals.

Of course, all this speculation is swept away by the fact that the outer electrons of the d-block elements are the 4s^2 pair. I must ask you to take it on trust, as I did in the final paragraphs of Chapter 3, that the 'loosest' electrons on any d-block metal atom *are* the 4s^2 pair, despite the apparently contradictory fact that at 'filling-up time' they were the preferred choice. Once we accept this fact then the metallic character, at least of the early d-block elements, is assured.

Question

5 Scandium, for instance, can be seen as quite a close cousin of calcium, having the same outer pair of electrons and just one small difference in an inner sub-shell. Can you suggest therefore why scandium has a first ionization enthalpy 40 kJ mol^{-1} *greater* than calcium's?

Having accepted that scandium will be metallic, by virtue of loosely held outer electrons, we must move on to speculate why all the elements are metals, even after nine more protons have been implanted into the nucleus. The first ionization enthalpies hardly alter across the row (at least not until zinc). The changes in ionization enthalpy across the p-block were considerable. Why is there this difference between the d-block and the p-block?

In the p-block, the electrons involved in ionization were in the sub-shell that was being added to across the row, and we know that electrons in the same sub-shell do not shield each other very well. In contrast, in the d-block the sub-shell being added to is the 3d, while the crucial shielding effect is felt by the outer 4s electrons. To summarise, we could draw the following conclusion from the ionization data:

The shielding effect of the extra 3d electrons on the outer 4s pair in each successive element just about balances out the increased nuclear pull (Figure 19.2).

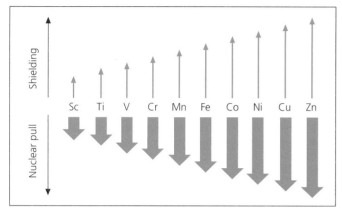

Figure 19.2 Schematic diagram of a model to account for why the ionization enthalpies of the d-block first-row elements are so similar

So we have fairly loosely held outer $4s^2$ electrons all the way across the row, giving rise to ten electrically conducting, shiny, malleable, ductile – **metallic** – elements.

High density and high melting and boiling points

It is unwise to make too many sweeping generalisations about these three factors. The first two depend upon, among other things, the type of crystal lattice adopted, and this shows a wide and apparently random variability (Table 19.3). However, it has been suggested that there is quite a good correlation between melting point and number of unpaired d-electrons (Figure 19.3).

Table 19.3 Crystal structures and melting points of the first row of the d-block elements

Element	Crystal structure	Melting point (°C)
Sc	f.c.c.	1539
Ti	h.c.p.	1660
V	b.c.c.	1890
Cr	b.c.c.	1857
Mn	complex	1244
Fe	b.c.c.	1535
Co	h.c.p.	1495
Ni	f.c.c.	1453
Cu	f.c.c.	1083
Zn	h.c.p.	420

Question

6 Following the examples in Figure 19.3 for scandium and copper, work out the number of unpaired d-electrons in the other elements in the figure. Comment on the degree of correlation between this number and the melting points.

Why might such a correlation between melting point and unpaired d-electrons exist? A reasonable model might suggest that the unpaired d-electrons are free to supplement the metallic bonding being undertaken by the $4s^2$ pair. This would account for the high melting points in certain elements, since extra electrons make stronger bonds.

At a pinch we could attribute high densities to the same root cause, since one might expect a closer clustering from more strongly bonded atoms. However, while this logic might explain why d-block metals in general have higher densities than s-block or p-block metals in general, it does not stand up to closer inspection. For example, vanadium has a very high melting point, consistent with its possession of three unpaired d-electrons, and yet its density is only about $6\,\mathrm{g\,cm^{-3}}$. Zinc on the other hand, with a much lower melting point (consistent with no unpaired d-electrons), has a density of over $7\,\mathrm{g\,cm^{-3}}$. Clearly our models in this field are too crude to support deep and detailed explanations of the observed phenomena.

Moderate to low reactivity

Once again we can use a broad brush approach, which delivers acceptable explanations for gross trends. But once again in order to explain finely detailed variations we need to refine our models. It is of course broadly true that the ionization enthalpies of the d-block elements are higher than those of their more reactive cousins in the s-block or in Group III (see question 5 about calcium and scandium). But if you rely too

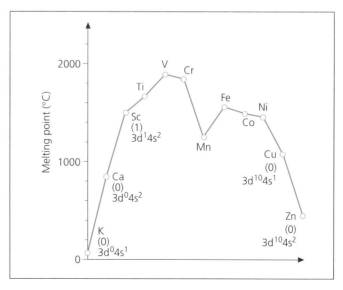

Figure 19.3 Melting points of selected elements (see question 6). The numbers in brackets indicate unpaired d-electrons. To calculate this number for the other elements, feed the available d-electrons into five boxes representing the d-orbitals, in the manner of Figure 3.37 (p. 41)

heavily on the principle that low ionization enthalpy means high reactivity, then the principle starts to creak.

Question

7 A more reliable guide to reactivity is the electrode potential of the half-cell $M^{2+}(aq)|M(s)$. It is more reliable because it relates to the aqueous environment rather than to the gas phase. It predicts accurately the outcome of reactions between metals and acids and between metals and aqueous ions of other metals.

a On the basis of the $E^\ominus$ values of the $M^{2+}(aq)|M(s)$ half-cells, put the metals iron, nickel, copper and zinc in order of reactivity.

b Now find the sum of their first two ionization enthalpies.

c Point to any examples where the sequence of ionization enthalpies does not match the sequence of standard electrode potentials.

Question 7 shows how ionization enthalpy alone is at best a crude guide to reactivity. The standard electrode potential is better because it relates more closely to the actual reaction whose reactivity is being discussed, and to the environment in which it is likely to occur (aqueous). In contrast, ionization enthalpies refer only to the removal of electrons from bare gaseous atoms and ions.

Question

8 The $E^\ominus$ value has built into it a contribution from the strength of the metal lattice, which despite being relevant to reactivity has no input into the ionization enthalpy. By reference to this quantity, whose relative size you can judge from the melting point of zinc compared with those of other d-block metals, suggest why zinc is a good deal more reactive than its ionization enthalpy would imply.

9 You might actually reject most of the above debate, and say that

reactivity is a **kinetic** (rates-of-reaction) concept, so it is not reliable to try to look for causes of reactivity amongst thermodynamic quantities, like ionization enthalpy or standard electrode potential. Which energy quantities would therefore be better guides to reactivity, if only they were listed in data books?

Variable oxidation number

All the first row of d-block elements except scandium and titanium have in their repertoire of chemical species a simple 2+ ion, created by the loss of the $4s^2$ pair of electrons. To that extent they are mimicking the Group II metals whose outer electron arrangement they share. But of course the phenomenon of variable oxidation number, not present in Group II, is due to the existence of the d-electrons. The two exceptions to the M^{2+} pattern, scandium and titanium, start with the 3+ ions, demonstrating that, at this end of the row at least, the nuclear hold over the 3d electrons is not very strong.

Oxidation numbers greater than +2 crop up right across the row, until you reach copper. They are all associated with the loss or part-loss of variable numbers of d-electrons. The very highest oxidation numbers, like +7 in manganese, are not of course bare cations, but polar covalent compounds of the elements with oxygen or fluorine, of which the manganate(VII) ion, MnO_4^-, is a typical example.

Question

10 Suggest why the 3+ cation, which forms so easily for scandium and titanium, is an impossibility for copper and zinc.

The phenomenon of variable oxidation number will be looked at more closely in Sections 19.3 and 19.4.

Co-ordination compounds

The elements of the d-block, and especially their ions, can bind around themselves a wide range of electron-donor species. We have already seen cations

heavily hydrated by water of crystallisation – for instance, there are six water molecules clustered round an Mg^{2+} ion in $Mg(NO_3)_2 \cdot 6H_2O$. However, in d-block chemistry the surrounding electron-donor groups are bound on to the central atom more firmly and irreversibly than the ion–dipole attraction that holds the water of crystallisation in magnesium nitrate.

For example, the binding of cyanide ions, CN^-, on to a central Fe^{2+} or Fe^{3+} cation is so secure as to render these normally lethal agents quite harmless (Figure 19.4). The cyanide ions are called **ligands** (from the same root as the word 'ligature' meaning a tied thong or rope), and the tie between the ligand and the central ion is believed to be a **dative covalent bond** into a vacant orbital of energy not far above the filled ones, such as vacant 3d orbitals and the whole 4p set. Hence we see the connection between the occurrence of these so-called **co-ordination compounds** (or **complexes**) and the presence of partially filled d-orbitals in the central species.

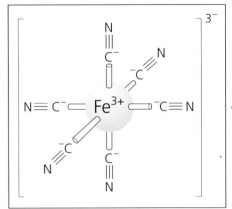

Figure 19.4 The (non-toxic) hexacyanoferrate(III) co-ordination complex

Question

11 If you have to work with sodium cyanide, you should keep it away from aqueous acids.

a What lethal gas would be liberated if these two reagents met?

b What would be a possible antidote to the accidental ingestion of CN^- ions, or the accidental inhalation of the gas from answer (a), if administered quickly enough? (Give your answer as a reagent rather than as an ion.)

A more detailed survey of co-ordination compounds of the d-block will be given in Section 19.5, where we shall also account for the most visible trademark of d-block compounds, namely colour.

Catalytic activity

Catalysis was dealt with as a general concept in Chapter 14, where it was seen to be a means of lowering the activation enthalpy of a reaction, by providing an alternative route. The d-block metals and their compounds can bring about this lowering in three ways, two involving co-ordination of the substrates as temporary ligands, and one involving variable oxidation number. (The word **homogeneous** in the headings below indicates that the catalyst is in the same phase as the substrates, as for example would be the case if substrates and catalyst were all in the same solution. The word **heterogeneous** is applied when the catalyst is in a different phase from the substrates, as for instance is the case for a solid catalyst operating on a gas-phase reaction.)

Homogeneous catalysis by redox intermediate

A look back to Figure 14.48, p. 351, may serve to remind you of this phenomenon. It is only relevant to the catalysis of reactions that are themselves redox. In one version of the process, the oxidising agent oxidises the catalyst, and the oxidised form of the catalyst oxidises the reducing agent (Figure 19.5). Thus the catalyst is repeatedly shuttled between 'cat-ox' and 'cat-red'. (Figure 19.5 is a generalised version of Figure 14.47).

Question

12 Why are d-block compounds well suited to this form of catalysis?

A more detailed analysis of the phenomenon of redox catalysis is to be found in Section 19.6.

Homogeneous co-ordination catalysis

In this style of catalysis, the d-block cation acts as a centre on to which potential substrate molecules can be reversibly bound. While there they meet and react with other substrates under conditions of lowered activation enthalpy (Figure 19.6). It has been found that this method of catalysis has an added advantage – the tight control exercised on the relative positions of the reactant molecules often results in very specific stereochemical outcomes, for instance only one of two possible chiral isomers being synthesised. One has only to think back to the thalidomide tragedy to realise how significant that sort of result can be (Figure 19.7). The matter of stereochemical control is dealt with in more detail in Section 19.6.

Heterogenous co-ordination catalysis

Most large-scale industrial processes, for example the production of nitric acid, ammonia and methanol, are gas-phase reactions catalysed by solid elements or compounds from the d-block (Table 19.4).

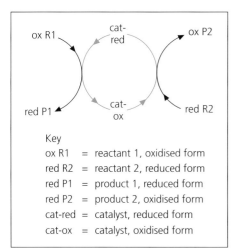

Figure 19.5 Homogeneous catalysis by redox intermediate

Key
ox R1 = reactant 1, oxidised form
red R2 = reactant 2, reduced form
red P1 = product 1, reduced form
red P2 = product 2, oxidised form
cat-red = catalyst, reduced form
cat-ox = catalyst, oxidised form

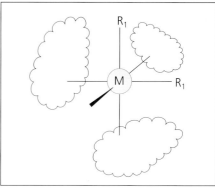

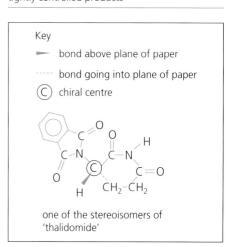

Figure 19.6 Two reactants (R_1 and R_2) whose reaction is catalysed by co-ordination to a d-block metal ion M. The clouds represent steric hindrance by other groups, leading to restricted orientations for R_1 and R_2, and therefore to tightly controlled products

Key
➤ bond above plane of paper
----- bond going into plane of paper
Ⓒ chiral centre

one of the stereoisomers of 'thalidomide'

Figure 19.7 The thalidomide tragedy – human bodies altered by a tiny detail of stereochemistry. The drug was originally prescribed to pregnant women as a sedative to ease feelings of morning sickness. Many victims have fought courageously against their disability

Table 19.4 A sample of the many uses of catalysts in industrial chemistry

Compound manufactured	Step catalysed	Catalyst
Methanol	$CO + 2H_2 \rightleftharpoons CH_3OH$ (and other reactions)	$ZnO/Al_2O_3/CuO$
Sulphuric acid	$2SO_2 + O_2 \rightleftharpoons 2SO_3$	V_2O_5
Nitric acid	$4NH_3 + 5O_2 \rightleftharpoons 4NO + 6H_2O$	Pt or Pt/Rh
Ammonia	$N_2 + 3H_2 \rightleftharpoons 2NH_3$	Fe

Many of the details of how particular catalysts work are either not known, or are secrets. Many variables operate in this field, and sometimes it is modification of the support material, rather than the catalyst itself, that makes the difference. For example, zinc oxide and aluminium oxide along with a copper oxide catalyst in the methanol process:

$$CO(g) + 2H_2(g) \rightleftharpoons CH_3OH(g)$$

were found to keep the catalyst particles from sintering (fusing together under the influence of heat) and maintain the high surface area. The result was a process that could be run at cooler temperatures, and the difference was of crucial commercial significance. It is therefore not surprising that the operation of catalysts is not openly discussed between industrial companies.

What is certain is that the substrate molecules are **adsorbed** on to the surface of the catalyst, using some version of co-ordination to the d-block atoms or ions. So once again, as in the homogeneous case, it is the way the reactant molecules are held in position relative to each other that lowers the activation enthalpy.

Now we have looked at all the ways in which elements and compounds of the d-block show their character. Four of those ways – high melting point, co-ordination, variable oxidation number and catalysis – can be related directly to the existence of d-electrons or d-orbital vacancies. In the other two cases of metallic character and moderate reactivity, the connection is less direct. In the rest of the chapter we shall look more closely at three of these big themes, namely variable oxidation number, complex formation and catalysis. In the course of this review you will perhaps come to see the degree of influence exerted over our everyday lives by the elements and compounds of the d-block.

19.3 Variable oxidation number 1: The 2+ and 3+ cations

Table 19.5 shows data relating to the existence and stability of the 2+ and 3+ ions of the elements of the first-row d-block elements.

There are some inter-related messages to be extracted from Table 19.5, and the next few questions concern these.

Table 19.5 2+ and 3+ cations of the first row of the d-block

	Ion exists?	Aqueous ion exists?	Colour of aqueous ion	$E^{\ominus}$ (V) of $M^{3+}(aq)|M^{2+}(aq)$
Sc^{2+}	✗	✗	—	
Sc^{3+}	✓	✓	Colourless	−2.6*
Ti^{2+}	✓	✗	—	
Ti^{3+}	✓	✓	Purple	−1.2*
V^{2+}	✓	✓	Lilac	
V^{3+}	✓	✓	Green	−0.26
Cr^{2+}	✓	✓	Blue	
Cr^{3+}	✓	✓	Green	−0.41
Mn^{2+}	✓	✓	Pale pink	
Mn^{3+}	✓	✓	Maroon	+1.60
Fe^{2+}	✓	✓	Pale green	
Fe^{3+}	✓	✓	Brown	+0.77
Co^{2+}	✓	✓	Pink	
Co^{3+}	✓	✓	Blue	+1.90
Ni^{2+}	✓	✓	Green	
Ni^{3+}	✗	✗	—	+4.2*
Cu^{2+}	✓	✓	Blue	
Cu^{3+}	✗	✗	—	+4.6*
Zn^{2+}	✓	✓	Colourless	
Zn^{3+}	✗	✗	—	+7.0*

*Estimates, based on ionization enthalpies

Question

13 Where in the row are you most likely to encounter the non-existence of

a the +2 oxidation state

b the +3 oxidation state?

There is a trend running across the row whereby one oxidation state becomes more stable while the other one becomes less stable. This trend should be reflected in the electrode potentials of the $M^{3+}(aq)|M^{2+}(aq)$ half-cells.

Question

14 Refer to Table 19.5.

a Which is the strongest oxidising agent amongst the range of 3+ aqueous ions that do exist?

b Which is the strongest reducing agent amongst the existing 2+ aqueous ions?

c The Ti^{2+} ion is unique among the 2+ cations in that it exists, but never as an aqueous ion (a feature it shares with Cu^+). By looking at the estimate for the $E^{\ominus}$ value for $Ti^{3+}(aq)|Ti^{2+}(aq)$ in Table 19.5, and by considering its place in the electrochemical series relative to the $H^+(aq)|\frac{1}{2}H_2(g)$ half-cell, suggest what reaction Ti^{2+} would undergo in water. Write an equation.

d There are two other redox half-cells in Table 19.5 that lie above the $H^+(aq)|\frac{1}{2}H_2(g)$ half-cell in the electrochemical series, and therefore which it appears could take part in the same anti-clockwise circle as Ti^{2+}. Yet both of them do exist as aqueous ions.

   i Which two ions are these?

   ii In what way does water offer conditions different from those that apply to the *standard* $E^\ominus$ value for the $H^+(aq)|\frac{1}{2}H_2(g)$ half-cell?

   iii In which aqueous medium rather different from pure water would these ions decompose to the 3+ ion? Write one of the relevant equations.

There is a trend that destabilises the 3+ cation while stabilising the 2+ one as you go from scandium to zinc. We can paraphrase this quite neatly in electronic terms – it appears that the first 3d electrons go from being very easily lost (as in Ti^{2+} and V^{2+}) to very firmly held (as in Cu^{2+} and Zn^{2+}).

We can seek confirmation of our ideas by looking at the third ionization enthalpies of the elements. We are switching medium from aqueous to gaseous, but the influence of nuclear pull should be of major significance in both cases. The graph is shown in Figure 19.8. Sure enough we see a steady increase, consistent with the idea that increasing nuclear charge is overriding the shielding of the d-electrons by each other. The situation mimics the change in ionization enthalpy across a short row of the Periodic Table, which we first met as long ago as Chapter 3.

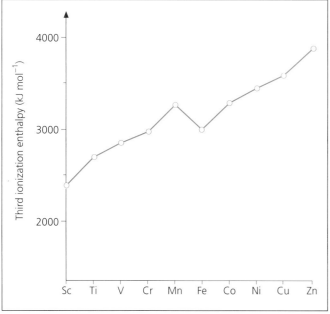

Figure 19.8 Third ionization enthalpies of the first-row d-block elements (i.e. the enthalpy change for the reaction $M^{2+}(g) \rightarrow M^{3+}(g) + e^-$)

Questions

15a The creation of Fe^{3+} and subsequent ions is easier than expected. Which factor is (again) causing this hiccup?

b Which of the following reactions would you expect to work?

   i $Fe^{3+}(aq) + Mn^{2+}(aq) \rightarrow Mn^{3+}(aq) + Fe^{2+}(aq)$

   ii $Mn^{3+}(aq) + Fe^{2+}(aq) \rightarrow Fe^{3+}(aq) + Mn^{2+}(aq)$

16 The following questions require a working knowledge of the law of anti-clockwise circles.

a $V^{2+}(aq)$ ions can be made by the action of a reducing agent on $V^{3+}(aq)$ ions (an exercise worth carrying out to see the lilac colour of $V^{2+}(aq)$). Normally a powdered metal is used as the reducing agent. Suggest a candidate that would be cheap and available.

b $Cr^{2+}(aq)$ ions do exist, but do not last long unless the water has been de-aerated by boiling, and the atmosphere above the solution is an inert gas such as nitrogen, carbon dioxide or argon. Under normal conditions the blue colour of $Cr^{2+}(aq)$ is replaced by a green shade. Suggest what the green species is, and why it occurs when the solution is allowed access to air. Support your answer with reference to the electrochemical series and with an accompanying equation.

17 Aqueous iodine is normally considered a stronger oxidising agent than $Cu^{2+}(aq)$, and yet contrary to expectations Cu^{2+} ions will oxidise $I^-(aq)$ to $I_2(aq)$. To thicken the plot further the mole ratio of [reactant Cu^{2+} : product I_2] is 2 : 1, and when the iodine is re-oxidised by thiosulphate ions it becomes apparent that the other product is a white (not a copper-coloured) precipitate. Suggest what has happened and write a possible equation that fits the evidence. What can you say about the $E^\ominus$ value of the Cu^{2+}-containing half-cell you have proposed?

Preparations and structures of some compounds of oxidation states +2 and +3

The halides provide us with good illustrations of the principles governing the preparation of d-block compounds. It is a general rule that to obtain the lower oxidation state, you react the metal with the HX(aq) acid or with HX(g), whereas the higher halide derives from direct contact between metal and halogen – in other words, you choose a strongly oxidising environment for the +3 halide and a mildly oxidising environment for the +2 halide. Examples of these two modes of preparation are:

$$Fe(s) + 2HCl(aq \text{ or } g) \rightarrow FeCl_2(aq \text{ or } s) + H_2(g)$$

$$2Fe(s) + 3Cl_2(g) \rightarrow 2FeCl_3(s)$$

18 Identify the atom or atoms playing the role of oxidising agent in the above two cases.

None of the solid d-block dihalides shows the standard structure for Group II halides, namely the fluorite (CaF_2) structure. Instead they choose from two others, named after **rutile**, TiO_2, and **cadmium iodide**, CdI_2 (Figure 19.9). Even then, only the most ionic halides (the fluorides) show the rutile structure. The rest choose cadmium iodide, which may reflect the reduced degree of ionic character in these compounds. The following question pursues this point.

Question

19a In both these structures, the metal ion has the same co-ordination number – what is it?

b What is the co-ordination number of the halide ion in both? (The oxide ions in rutile itself stand for halide ions in this context.)

c We should expect these halides to be less purely ionic than those of calcium. Apart from the reduced electropositivity of the d-block elements, what other aspect of their cations inclines them to 'claw back' some of the electron density? (You may need to revise Section 5.2 on impure ionic bonding, and then look up the sizes of the d-block 2+ ions, in comparison with that of the Ca^{2+} ion.)

d An interesting feature of the cadmium iodide structure is that it is arranged in layers, with the metal ions in a halide sandwich (Figure 19.10). Why would it be advantageous for the ions to be less than fully charged in this structure?

e Suggest what kind of forces might be holding adjacent sandwiches together in the cadmium iodide structure (i.e. holding iodide atoms to other iodide atoms in the next sandwich).

f Some compounds with layer structures not too dissimilar to the cadmium iodide structure have found use in car engines as lubricants. Molyslip (Figure 19.11), for example, contains molybdenum sulphide, MoS_2. Suggest why this sort of layer lattice has lubricating properties.

The MCl_3 halides have complex structures that will not be considered here. Iron(III) chloride is quite familiar in several guises from elsewhere in the book. It is a strong enough oxidising agent to be of use in etching the unwanted copper from printed circuit boards (Figure 19.12).

Question

20 Write an equation for the change shown in Figure 19.12.

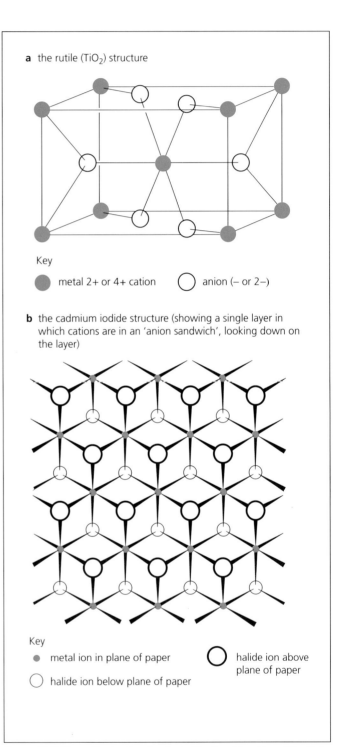

a the rutile (TiO_2) structure

Key

● metal 2+ or 4+ cation   ○ anion (− or 2−)

b the cadmium iodide structure (showing a single layer in which cations are in an 'anion sandwich', looking down on the layer)

Key

● metal ion in plane of paper   ○ halide ion above plane of paper

○ halide ion below plane of paper

Figure 19.9 The rutile structure is adopted by the difluorides of the first-row d-block 2+ ions. Other d-block dihalides show the cadmium iodide structure. Compare diagram (b) with the side view in Figure 19.10

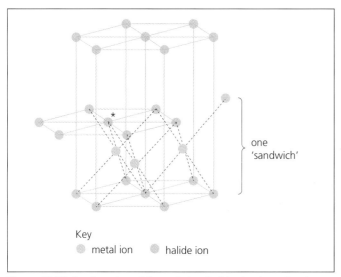

Figure 19.10 Another view of the cadmium iodide structure, showing one sandwich layer and the underside of the next one up. The symbol * shows the viewpoint for Figure 19.9b

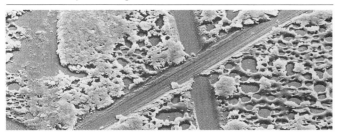

Figure 19.11 The scanning electron micrograph (top) shows a film of MoS_2 on a metal surface. The effect is better lubrication and less engine wear

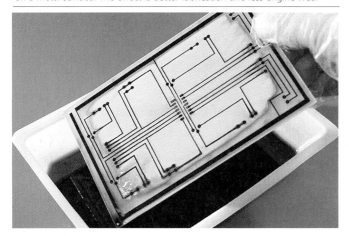

Figure 19.12 Printed circuit boards are etched using aqueous iron(III) chloride. Notice the copper corroding in the corner

Iron(III) chloride is also a potential 'halogen-carrier' catalyst for the electrophilic substitution of benzene by dihalogen molecules:

$$Br\!-\!Br\cdots\!Fe\!\begin{array}{c}\!-Cl\\-Cl\\-Cl\end{array}$$

Question

21 Show how the dihalogen molecule is pre-polarised by iron(III) chloride.

Finally, the aqueous solution of iron(III) chloride is acidic, since one of the co-ordinated water molecules dissociates:

$$[Fe(H_2O)_6]^{3+}(aq) + H_2O(l) \rightleftharpoons$$
$$[Fe(H_2O)_5OH]^{2+}(aq) + H_3O^+(aq)$$

Question

22 This last phenomenon can be put down to the high charge density of the Fe^{3+} ion. Show how that charge density sets up a polarisation that aids acid dissociation in one of the water molecules.

The oxides and hydroxides of the d-block tend also to exist in both the +2 and +3 oxidation states. The higher oxides are made like the higher halides, by direct reaction between elements:

$$4Fe(s) + 3O_2(g) \rightarrow 2Fe_2O_3(s)$$

The lower oxides (oxidation state +2) are more difficult to make by a method parallel to that for the lower halides, namely by reaction with water. This is first because the reactions are very slow, and second because the hydroxides are formed preferentially in water. All the hydroxides, whether of 2+ or 3+ cations, can be made by simple ionic 'partner swap' reactions in solution, where the driving force is the insolubility of the hydroxide:

$$CuSO_4(aq) + 2NaOH(aq) \rightarrow Cu(OH)_2(s) + Na_2SO_4(aq)$$

These hydroxides of either oxidation number offer an alternative route to the corresponding oxide by thermal dehydration:

$$2Cr(OH)_3(s) \rightarrow Cr_2O_3(s) + 3H_2O(g)$$

All the oxides and hydroxides show predominantly basic reactions towards water and acids:

$$Cr_2O_3(s) + 6HCl(aq) \rightarrow 2CrCl_3(aq) + 3H_2O(l)$$

Questions

23 A common compound in the Earth's crust is the magnetic ore magnetite, Fe_3O_4. This formula only makes sense in redox terms if we assume that iron is present in two oxidation states in the same compound. Rewrite the formula of magnetite in a way that shows it as a 'mixed oxide', using the symbols Fe(II), Fe(III) and O.

24 Iron is the fourth most abundant element in the Earth's crust, after oxygen, silicon and aluminium. It appears in the clays, limestones and sandstones of Britain, often staining them the characteristic iron(III) colour and prompting names such as new red sandstone – a formation is exposed on the Devon coast near Sidmouth (Figure 19.13), among other places. Sometimes, especially where there are cracks between bedding planes, the colour has changed to green (Figure 19.14). Suggest what the green might be, and what agency, either living or non-living, might have caused it.

Figure 19.13 New red sandstone cliffs near Sidmouth, Devon

Figure 19.14 New red sandstone near Rye, East Sussex, showing colour change in the cracks

19.4 Variable oxidation number 2: Oxidation states higher than +3

None of the first-row d-block elements has a cation with a charge higher than +3 (except titanium in rutile, TiO_2). However, there are polar-covalent oxides and fluorides with higher oxidation numbers, and oxo acids and oxo anions (even some oxo cations in vanadium's chemistry), all of which are more or less vigorous oxidising agents (again with the exception of rutile). A selection of these is shown in Table 19.6.

Table 19.6 Oxidation numbers higher than +3 of the first row of the d-block. No such compounds are known for the elements cobalt to zinc

Element	Oxidation number			
	+4	+5	+6	+7
V	VO^{2+}	VO_3^-, V_2O_5, VO_2^+, VF_5	—	—
Cr	Obscure	CrF_5	CrO_3, $Cr_2O_7^{2-}$, CrO_4^{2-}	—
Mn	MnO_2	MnO_4^{3-}	MnO_4^{2-}	Mn_2O_7, MnO_4^-
Fe	—	—	FeO_4^{2-}	—

Question

25 What do you notice about the highest available oxidation number as you cross the row?

The oxides of these high oxidation numbers are very different from the basic +2 and +3 oxides. They are covalent compounds that react with water to give acids, for example with chromium(VI):

$$CrO_3(s) + H_2O(l) \rightarrow H_2CrO_4(aq)$$

These are the parent acids of familiar oxidising anions like (in this case) chromate(VI), CrO_4^{2-}, manganate(VII), MnO_4^-, and dichromate(VI), $Cr_2O_7^{2-}$. We have met these ions in our studies of redox reactions, and also in organic chemistry where they are used as oxidising agents. We should be reasonably familiar with their redox half-equations, for instance:

$$MnO_4^-(aq) + 8H^+(aq) + 5e^- \rightarrow Mn^{2+}(aq) + 4H_2O(l) \quad (1)$$

Questions

26 Manganate(VII) ions are such strong oxidising agents that they can oxidise almost anything. Solutions containing these ions are sometimes used to analyse sewage outflows. An excess of manganate(VII) is left in contact with the sewage-contaminated water for a few hours, and then any remaining manganate(VII) is used to oxidise some acidified potassium iodide. The resulting iodine is estimated by a 'thio' titration. Some results are shown in Table 19.7.

Table 19.7 The results of a analysis on sewage outflow. Volume of sewage supplied = 100 cm³; volume of 0.01 mol dm⁻³ potassium manganate(VII) added = 10 cm³

Day	Volume of 0.01 mol dm⁻³ thiosulphate solution, $Na_2S_2O_3$ (cm³)
Monday	40
Tuesday	36
Wednesday	32
Thursday	30
Friday	35

a Recall and write the equation whereby aqueous thiosulphate $S_2O_3^{2-}$ ions 'mop up' aqueous iodine. What indicator would sharpen the end-point?

b Write the equation for the oxidation of acidified aqueous iodide ions by manganate(VII) (that is, combine half-equation (1) above with the I_2/I^- half-equation).

c From answers (a) and (b) above work out the stoichiometric ratio of number of moles of thiosulphate used to original number of moles of manganate(VII) left over.

d On which day was the sewage 'strongest'?

e Turn the raw data in Table 19.7 into numbers directly proportional to the strength of the sewage.

f Why not titrate the sewage directly with manganate(VII), stopping at the first permanent sign of the intense pink colour of manganate(VII), rather than go to the trouble of a back-titration with 'thio'? (There are two possible answers here.)

27 Manganese(VI) is an obscure oxidation state with a rich dark green colour. The reason why it is obscure may become clear when you look at these electrode potential data (MnO_4^{2-} is the manganese(VI) species):

$$MnO_4^-(aq) + e^- \rightarrow MnO_4^{2-}(aq)$$
$$E^\ominus = +0.56 \text{ V}$$

$$MnO_4^{2-}(aq) + 4H^+(aq) + 2e^- \rightarrow MnO_2(s) + 4H_2O(l)$$
$$E^\ominus = +2.26 \text{ V}$$

a Apply the Law of anti-clockwise circles to these data to show what would happen to the manganese(VI) ion and why it would fail to exist.

b These data refer to the standard conditions where all ions present (including H^+) are at $1\,mol\,dm^{-3}$. In strongly alkaline conditions the opposite reaction happens and MnO_2 and MnO_4^- ions can *make* MnO_4^{2-}. Which of the above two half-equations is sensitive to changes in pH, and how would it change its electrode potential value in solutions of very low hydrogen ion concentration? Explain why manganate(VI) ions survive in strong alkalis.

c Alkaline conditions in general tend to make all these strong oxo anion oxidising agents like the manganates and chromates more stable (less oxidising). Looking at their half-equations, can you see why hydroxide ions would affect them like this?

19.5 Complexes

If you pour ammonia solution into sodium sulphate solution, you see one clear liquid mixing with another. The result is (unsurprisingly) a mixed solution of sodium sulphate and ammonia. But if you do the same thing to a solution of copper(II) sulphate, the result is rather more inspiring. First a baby-blue precipitate of copper hydroxide appears. If more ammonia is added, the hydroxide re-dissolves and the resulting solution is a rich shade of indigo/purple (Figure 19.15).

Figure 19.15 The solutions contain **a** $[Cu(H_2O)_6]^{2+}(aq)$, **b** $Cu(OH)_2(s)$ and **c** $[Cu(NH_3)_4(H_2O)_2]^{2+}(aq)$

The reason for the difference lies not just in the fact that copper compounds are coloured, but because a reaction goes on between nucleophiles and d-block cations that does not take place with s-block ions. It appears that nucleophilic species take advantage of the vacant positions in the 3d and 4p orbitals of d-block ions, as sites into which to push their lone pairs to make a type of dative covalent bond. The nucleophilic species are, in this context, known as **ligands**, and the resulting compounds are known as **co-ordination compounds** or **complexes**. The copper–ammonia complex ion is shown in Figure 19.16.

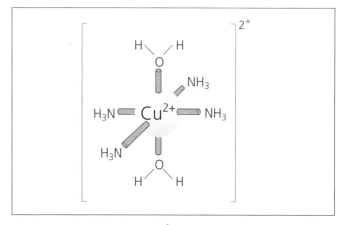

Figure 19.16 The $[Cu(NH_3)_4(H_2O)_2]^{2+}$ complex cation, whose colour is indigo/purple

During the reaction in Figure 19.15, one ligand species has been replaced by another, water by ammonia. We see it as a substitution because what we have previously written as

'Cu^{2+}(aq)' in fact has six water molecules in a ligand-like relationship with the Cu^{2+} ion, so the reaction is really:

$$[Cu(H_2O)_6]^{2+}(aq) + 4NH_3(aq) \rightarrow$$
$$[Cu(NH_3)_4(H_2O)_2]^{2+} + 4H_2O(l) \quad (2)$$

The driving force would appear to be that ammonia is a better ligand than water, which is certainly a reasonable guess, given their relative powers as bases.

Question

28a Why is it logical to expect some correlation between base strength and ligand strength?

b What change in oxidation number has occurred in equation (2)?

The copper(II)/ammonia system is one small part of a wide and varied field. A quick scan across this field gives an indication of its diversity. Complexes can be anions, cations or neutral molecules; they exhibit at least four different geometries (Figure 19.17); they show several types of isomerism; and they encompass a great range of stability. Individual ligands can be neutral molecules or anions, the minimum requirement being a lone pair of electrons. (Even this stipulation is not universal since it appears that dihydrogen can co-ordinate to d-block metals.) Ligands can be singly or multiply tied to

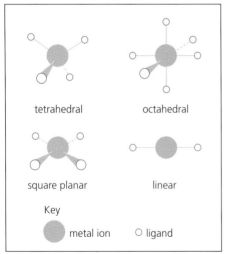

Figure 19.17 Four of the more common geometrical arrangements of ligands around a central ion

tetrahedral   octahedral

square planar   linear

Key

● metal ion   ○ ligand

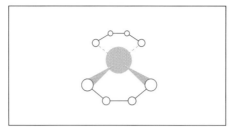

Figure 19.18 A square planar complex featuring two doubly tied **bidentate** ligands

the central cation (Figure 19.18), depending on whether or not they can direct more than one lone pair at the central cation. The whole field of complexes has its own system of names. We shall go on now to look at some examples that illustrate these aspects in more detail.

The complexes of the cobalt 3+ cation

In the early twentieth century the unravelling of the structures of complexes was at the cutting edge of chemical research. The Nobel Prize for chemistry in 1913 went to Alfred Werner for this very task. His pioneering work included a detailed study of the complex ions formed between cobalt 3+ cations and chloride ions and ammonia molecules. Like Mendel's peas and Darwin's Galápagos finches, the original subject matter of the pioneer work still provides one of the best illustrations of the field.

The $Co^{3+}/NH_3/Cl^-$ compounds must certainly have produced turn-of-the-century chemists with the most vexing conundrum. There were five different such compounds, with colours and formulae as shown in Table 19.8. These compounds ignored the established valency rules, and while aqueous

Table 19.8 The range of cobalt(III) chloride/ammonia co-ordination compounds

Molecular formula	Colour
$CoCl_3 \cdot 6NH_3$	Yellow
$CoCl_3 \cdot 5NH_3$	Purple
$CoCl_3 \cdot 4NH_3$	Green
$CoCl_3 \cdot 4NH_3$	Violet
See question 31	Red

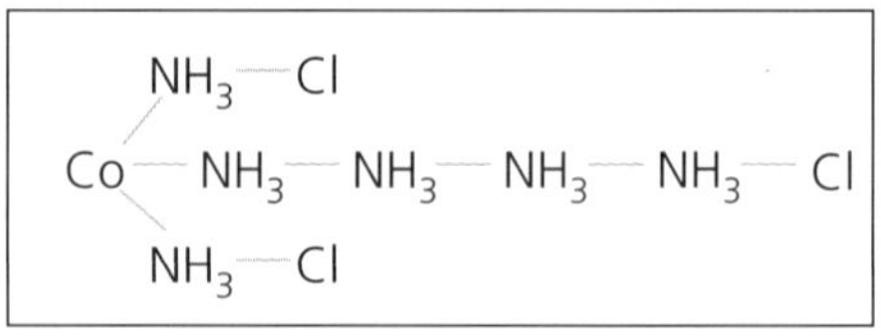

cobalt(III) was known as a very strong oxidising agent, none of these compounds shared that trait to the same extent.

Early workers had offered explanations in the form of chains in different sequences, as in Figure 19.19.

Figure 19.19 An early effort at understanding the structure of $CoCl_3 \cdot 6NH_3$. Various ammonia molecules could be dropped out to account for the other compounds in Table 19.8

Question

29 Raise one objection concerning the structures in Figure 19.19.

Werner's great and prize-winning realisation was that the ammonia molecules could be clustered around the central metal ion. Thus the first structure in Table 19.8 is as shown in Figure 19.20.

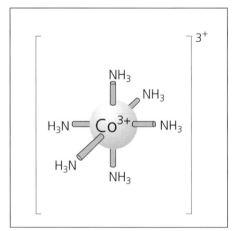

Figure 19.20 The structure of the cation from the yellow complex in Table 19.8. There are three free Cl^- anions

But what about the compounds with five and four ammonia molecules? Could the Co^{3+} ion have six or five or four ligands in its co-ordination group? Werner was of the opinion that the co-ordination number of Co^{3+} was always six, and that the variation was due to chlorides swapping roles and switching from being free ions to ligands. This would have the effect of displacing respectively either one or two ammonia molecules, thus explaining the change in the ammonia 'population'. Thus, for example, the purple complex could be as in Figure 19.21. Once chloride ions have become ligands, they can no longer act like free chloride ions – they could not float off to an anode during electrolysis, for instance. Also, every chloride ligand will bring down by one the charge on the overall complex.

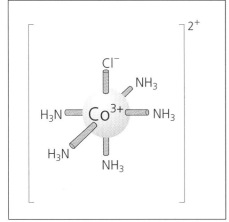

Figure 19.21 The structure of the purple cation in Table 19.8. There are two free Cl^- anions

With these ideas in mind, let us look at some more evidence in the course of the next question. The evidence relates to both the loss of the independent chemical identity of ligand chlorides, and the overall number of free-moving ionic species.

Question

30 a Why is the chloride ion just as reasonable a candidate to be a ligand as ammonia or water?

b In the purple complex (Table 19.8), how many chloride ions would move towards the anode in electrolysis?

c If you treated one mole of both the yellow and the purple complexes with an excess of aqueous silver

nitrate, how many moles of silver chloride precipitate would you get?

d Table 19.9 shows the electrical conductivity of various $0.001 \, mol \, dm^{-3}$ solutions of known electrolytes in water. There appears to be a more important factor than the individual identities of the substances in determining conductivities – what is it?

Table 19.9 Electrical conductivities of some metal chlorides

Compound	Electrical conductivity ($10^4 \, m^2 \, ohm^{-1} \, mol^{-1}$)
NaCl	123
KCl	115 (est.)
$MgCl_2$	260 (est.)
$CaCl_2$	270
$FeCl_3$	400 (est.)

e How many free-moving ionic entities are there respectively in the yellow and purple complexes in Table 19.8?

f The conductivities of the yellow and purple complexes on the same unit system as in Table 19.9 are 430 and 260 respectively. Show how this evidence offers support for the structures in Figures 19.20 and 19.21.

g The two complexes with four ammonias each give a conductivity of about 100, and each gives only one mole of silver chloride precipitate per mole of complex when treated with excess silver nitrate. Draw a structure for the cation part of these complexes that is consistent with both these pieces of evidence.

h There are two complexes of formula $CoCl_3 \cdot 4NH_3$, a green one and a violet one. Look for a form of isomerism that would create a structure different from that in (g) that would still fit the conductivity and silver nitrate data.

Table 19.10 shows how to use square brackets to express the structural differences uncovered in question 30. These formulae are similar to the structural formulae of organic chemistry, carrying more information than molecular formulae such as $CoCl_3 \cdot 5NH_3$. In future we shall use either these structural formulae or full-blown three-dimensional diagrams, and you should be able to interconvert them.

Table 19.10 Structural formulae of the cobalt(III) chloride/ammonia complexes

Colour	Structural formula
Yellow	$[Co(NH_3)_6]Cl_3$
Purple	$[Co(NH_3)_5Cl]Cl_2$
Green	trans-$[Co(NH_3)_4Cl_2]Cl$
Violet	cis-$[Co(NH_3)_4Cl_2]Cl$
Red	$[Co(NH_3)_5H_2O]Cl_3$

Question

31 For instance, try this interconversion. There is another complex featuring cobalt(III), ammonia and chloride, whose structural formula, as shown in Table 19.10, is $[Co(NH_3)_5H_2O]Cl_3$.

a Draw the cation part of this formula as a diagram.

b Say how many moles of silver chloride will be precipitated by excess aqueous silver nitrate, per mole of complex.

c Estimate its conductivity value in the units used previously.

Some complexes of platinum

A brief look at some platinum complexes will revisit some of the cobalt 3+ themes, and at the same time bring in some new variations. In one of its oxidation states platinum resembles Co^{3+} in preferring a co-ordination number of six. Again the result is a set of octahedral complexes. Figure 19.22 (overleaf) shows two of these complexes which exhibit *cis–trans isomerism*.

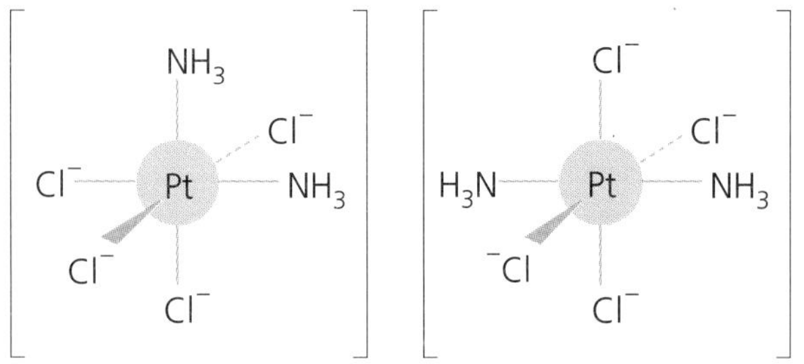

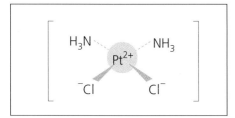

Figure 19.22 Two isomeric octahedral complexes, the *cis*- and *trans*-versions of $[Pt(NH_3)_2Cl_4]$. For the charge on platinum, see question 32

Question

32 Neither of these complexes shows any conductivity at all in solution. Deduce from this fact what would have been the charge on the original bare platinum cation.

The shapes of complex species generally follow our 'bits stay apart' theory from Chapter 4. Thus 6-co-ordinate complexes are octahedral, and 4-co-ordinate ones are generally tetrahedral. An exception is the group of 4-co-ordinate complexes based around another oxidation state of platinum, namely Pt^{2+}. These are **square planar**, and one of them is shown in Figure 19.23.

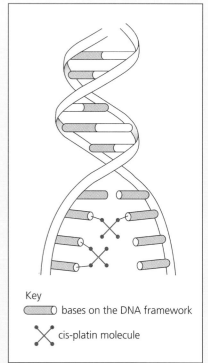

Figure 19.23   A square planar complex of Pt^{2+}

Question

33a Draw a stereoisomer of the structure in Figure 19.23.

b Can you think of any reason why the four ligands might prefer the square planar arrangement, even though they would be further apart if spread tetrahedrally?

c What is the overall charge on the structure in Figure 19.23? What conductivity would you expect from it?

d Which of the two stereoisomers would have a zero overall dipole moment?

e Which of the two isomers might have the greater solubility in solvents such as hexane, and why?

So far we have looked at complexes of copper(II), cobalt(III), platinum(IV) and platinum(II). We have encountered two geometries – octahedral and square planar – and both cationic and neutral complex species. It is also possible to have anionic complexes, which occur whenever the number of anion ligands exceeds the original positive charge on the bare ion. For instance the tetrahedral complex formed between Pt^{2+} and CN^- ions has the formula $[Pt(CN)_4]^{2-}$, and the $[Fe(CN)_6]^{3-}$ ion we saw in Figure 19.4 (p. 476) was another example.

Biomedical uses of platinum complexes

This panel will mainly take the form of a single extended question.

Question

34 The complex in Figure 19.23 is actually a drug. It is marketed under the name cisplatin for the treatment of ovarian and testicular cancers. The mechanism by which the drug works is not fully understood (or is secret). What does seem fairly certain, however, is that the Cl^- ligands are replaced, once the drug is inside the cells of the patient, by a nucleophilic substitution reaction featuring water as the nucleophile.

a Show the molecule that would result from two such substitution events on cisplatin.

b It is then supposed that the newly substituted molecule brings about intra- and inter-strand linkages on DNA molecules (Figure 19.24), thus disrupting the DNA-replication process and thereby halting cell division in the tumour. How would the substituted version of cisplatin attach itself to a single strand of DNA?

Key
⬭ bases on the DNA framework
✕ cis-platin molecule

Figure 19.24 Schematic diagram showing how cisplatin molecules might disrupt DNA replication

c If cisplatin has this effect on cancer cells, then why does it not do the same to all cells, with disastrous results? Fortunately, if there is only a limited dose of cisplatin in the body, it selectively targets those cells that are dividing most rapidly, such as cancer cells. Suggest why the DNA in such cells is more vulnerable than that in ordinary cells.

d Many cancer drugs have severe side effects, and cisplatin is no exception. Patients undergoing cisplatin treatment suffer from, among other things, nausea and a drop in white blood cell count. If you know enough biology, can you suggest why the latter effect occurs?

e Suggest why the dosage level of cisplatin would require even more careful monitoring than that of other medicines.

The search for alternative drugs with less harmful side effects is being urgently pursued. A new patent drug, which is called carboplatin, and which meets some of the specified requirements, is shown in Figure 19.25.

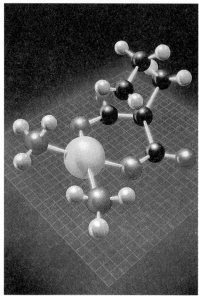

Figure 19.25 Carboplatin – a new anti-cancer drug with fewer side-effects than cisplatin

The species that has replaced the two chloride ligands in carboplatin is a single molecule with two ligand sites. Ligands with this feature represent a distinct sub-group of the overall family of ligands, and it is to this sub-group that we now turn our attention.

Polydentate ligands

Imagine a complex in which two ammonia molecules were adjacent ligands. Then remove one hydrogen from each ammonia and adjust their positions so that the vacancies face each other. Finally build a hydrocarbon chain of suitable length to link the vacancies and thereby bridge the gap between the ligands. Such a structure is shown in Figure 19.26.

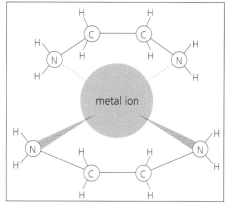

Figure 19.26 Two bidentate ligands producing 4-co-ordination at a metal ion. This particular ligand is 1,2-diaminoethane, often shortened to 'en'

The ligand that emerges from this exercise is 1,2-diaminoethane. This is one of the simplest of a large group of molecules, all of which possess two or more suitably positioned lone pairs – suitable, that is, for simultaneous co-ordination on to the same metal cation. They are called **polydentate ligands**,

with obvious links to teeth and biting – a rather untidy mixed metaphor when you recall the derivation of 'ligand', from ties and tying.

The strict definition of a polydentate ligand is a ligand with two or more binding sites. Even if the sites can only be used to link two different cations, the molecule is still polydentate. A polydentate ligand that can bind all its sites to a single cation is called a **chelating agent**. The majority of cases encountered in Advanced courses will be both polydentate and chelating, and with the exception of question 35, the word 'polydentate' will be assumed to mean 'chelating'.

Question

35 Hydrazine, NH_2NH_2, might well act as a polydentate ligand, but it does not act as a chelating agent. Explain why it would not be able to behave as a chelating agent like the 1,2-diaminoethane in Figure 19.26.

Some commonly met polydentate ligands are shown in Table 19.11. As you can see there are some high numbers of binding sites from the more complex ligands. For example, edta is tailored to act as a one-molecule hexadentate ligand, and encloses its cations in a cage. Polydentate ligands may of course be anions or neutral molecules, and so can affect the overall charge of the complex just as monodentate ligands would.

Table 19.11 Some polydentate ligands

Ligand	Structural formula	Number of binding sites
1,2-Diaminoethane 'en'	H_2N—$(CH_2)_2$—NH_2	2
Bipyridyl 'bipy'	[structure: two pyridine rings, each with N]	2
2-Hydroxybenzoate 'salicylate'	[structure: benzene ring with CO_2^- and OH]	2
Ethylenediamine-tetraacetic acid or its anion 'edta'	$(^-O_2CCH_2)_2N$ —$(CH_2)_2$— $N(CH_2CO_2^-)_2$	6 (octahedrally)

Question

36 Draw out skeletal formulae of the polydentate ligands in Table 19.11. Use asterisks to indicate which atoms are likely to act as the binding sites on each molecule. Make sure the number of sites on each molecule matches the figure in the third column of the table.

Polydentate ligands form more stable complexes than monodentate ligands. For example, the equilibrium constant for the reaction:

$$[Fe(H_2O)_6]^{3+}(aq) + edta^{4-}(aq) \rightleftharpoons$$
$$[Fe(edta)]^-(aq) + 6H_2O(l) \qquad (3)$$

is about $10^{25} \, mol^{-1} \, dm^3$. This stability is at first sight quite unexpected, since the atoms responsible for binding to the metal ion are nitrogens and oxygens, the same ones that monodentate ligands commonly use. Why should, say, the N—metal bond be stronger when there is a bit of hydrocarbon chain in the ligand? Considering bond polarity does not help – the hydrocarbon bits will not have a major impact on the 'pushiness' of the nitrogen lone pairs.

There *is* a model that explains the stability of polydentate ligand complexes quite well. It is based on **entropy**.

Question

37a By considering the number of free-moving species on the right and left of equation (3), comment on the change in 'chaos factor' when the reaction occurs.

b How is this likely to affect ΔS_{sys} and ΔG_{sys}?

c Would this be a satisfactory explanation of the large equilibrium constant for equation (3)?

Polydentate ligand complexes sometimes offer new opportunities for optical isomerism. For example, consider the complex in Figure 19.27. It is shown with its mirror image. If you rotate the mirror image about the vertical axis, one of the 'en' loops will be in the right place, but the other one will not. So these two molecules are non-superimposable mirror images – they are optical isomers.

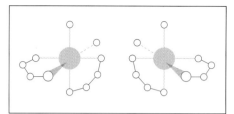

Figure 19.27 Two chiral isomers of the complex $[Cr(NH_3)_2(en)_2]^{3+}$

Question

38 Consider which of the complex ions in Figure 19.28 might exist as optical isomers, and back up your answer with drawings.

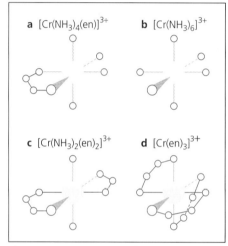

a $[Cr(NH_3)_4(en)]^{3+}$ **b** $[Cr(NH_3)_6]^{3+}$

c $[Cr(NH_3)_2(en)_2]^{3+}$ **d** $[Cr(en)_3]^{3+}$

Figure 19.28 Some 'en' and ammonia complexes of the Cr^{3+} ion

Competition between ligands and ordinary anions for cations

A set of ligands clustered around a cation will of course have a huge effect on the size of the ion, and possibly on its charge as well. Since size is important in determining the strength of ionic bonds in solid lattices, the complexed and uncomplexed equivalents of the same ion pair can have very different lattice enthalpies. This in turn has an impact on solubility. For example, silver chloride is virtually insoluble in water, with a K_{sp} of about $10^{-10} \, mol^2 \, dm^{-6}$, whereas the chloride of $[Ag(NH_3)_2]^+$ is freely soluble. If you add concentrated ammonia solution to a precipitate of silver chloride, it goes into solution as $[Ag(NH_3)_2]Cl(aq)$ (Figure

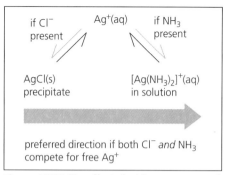

Figure 19.29 The effect of adding concentrated ammonia solution to a silver chloride precipitate

19.29). (Complexes of the Ag^+ ion usually show a co-ordination number of two and are linear.)

In general, it is reasonable to conclude that all complexed salts should be fairly soluble in water. However, it is not always true that the cation will choose the soluble complexed option. For instance, if you take the solution from the procedure outlined in the previous paragraph and add $Br^-(aq)$ ions, then the silver promptly deserts its ligand and re-precipitates as silver bromide (Figure 19.30).

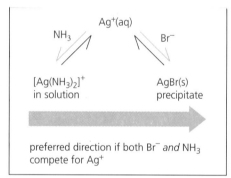

Figure 19.30 The effect of adding bromide ions to the silver/ammonia complex

This behaviour can be rationalised by the following qualitative model. Both the precipitate and the complex are in equilibrium with the free aqueous ion $Ag^+(aq)$. The halide ion and the ammonia molecules compete to take the silver ion to form either $AgX(s)$ or $[Ag(NH_3)_2]^+(aq)$. The winner of the competition may be decided on thermodynamic grounds, with the value of the equilibrium constants guiding us as to the result. Thus chloride gives way to ammonia, while the ammonia complex gives way to bromide. The values of the equilibrium constants are shown in Table 19.12, and the competitions depicted in Figures 19.29 and 19.30. There is no need at Advanced level for a rigorous mathematical treatment of this phenomenon, but you should notice the smaller value of the solubility product for silver bromide relative to that for silver chloride. This can be read as a sign of the greater thermodynamic stability of $AgBr(s)$. The ammonia complex must be intermediate in stability between the silver chloride and silver bromide solids.

Table 19.12 Stability constants K_{stab} and solubility products K_{sp} for some silver species. K_{stab} is the equilibrium constant for:
$[Ag(H_2O)_2]^+ + 2L \rightleftharpoons [AgL_2]^+ + 2H_2O$.
K_{sp} is the equilibrium constant for:
$AgX(s) \rightleftharpoons Ag^+(aq) + X^2(aq)$

Complex	K_{stab} of complex ($mol^{-2}\ dm^6$)
$[Ag(H_2O)_2]^+$	1
$[Ag(NH_3)_2]^+$	10^7

Insoluble salt	K_{sp} of salt ($mol^2\ dm^{-6}$)
AgCl	10^{-10}
AgBr	10^{-13}

Having seen the pendulum swing from precipitate to solution to precipitate, you might be curious to know whether the silver ion can be coaxed back again into solution from the silver bromide lattice. Silver bromide is the light-sensitive chemical at the heart of the photography industry. The entire process of photography depends upon being able to mobilise and remove unwanted silver bromide from photographic film and paper after exposure and development. Section 19.8 looks at the chemistry of photography, and explains how complexation does this 'fixing' job.

Question

39 A parallel case, also concerned with the use of ligands to affect the solubility of cations, involves iron in the soil. Some fruit trees in Florida were showing leaf yellowing, a symptom of a disease called chlorosis which is caused by a deficiency of iron. Another more serious symptom of this disease

is reduced yields. In one experiment to boost yields, the deficiency was treated with one *tonne* of soluble $FeSO_4 \cdot 7H_2O$ per tree. In another, the dose per tree was 50 *grams* of the complex Na[Fe(edta)]. The second method produced better results.

a Why is it important that the iron is in soluble form?

b Iron is quite a common mineral in soils, but in all except the most acidic soils, $Fe^{2+}(aq)$ ions have a tendency to form $Fe(OH)_2(s)$ by loss of two protons from water molecules in the escort group, followed by precipitation of the solid hydroxide. Copy and complete the following equation to show that change:

$[Fe(H_2O)_6]^{2+}(aq) \rightarrow ? + ? + 2H^+(aq)$

(The H^+ ions would be accepted by, for example, limestone particles in the soil.)

c The plant only requires small amounts of iron, and the insolubility of the iron(II) hydroxide is not too excessive. The problem is that iron(II) hydroxide becomes oxidised by air in the soil to iron(III) hydroxide, which is much more insoluble. Use the 'moles up/moles down' redox method to complete and balance this equation for the oxidation:

$Fe(OH)_2(s) + O_2(g) + ? \rightarrow ?$

d Adding edta^{4-} to the soil does not stop the oxidation, but it does help make soil iron available to the plant. By comparison with the ammonia/silver chloride case, suggest why it does this.

e As you can see from the introduction to the question, the method used in Florida was the direct application of Na[Fe(edta)], rather than just adding edta^{4-} to complex with the iron already there. Why does this add an extra degree of certainty to the outcome?

f If a lack of iron causes yellowing of the leaves, can you suggest one

biochemical synthesis within the plant that might depend on iron?

g Gardeners in regions with alkaline soils are advised to buy 'blueing tablets' if they want blue hydrangeas, otherwise the flowers stay pink (Figure 19.31). What would you suspect is the chemical background to this advice? Include in your explanation a guess at what 'blueing tablets' might contain, and at the significance of the reference to alkaline soils.

Figure 19.31 Hydrangeas **a** in acidic soil and **b** in alkaline soil. The use of 'blueing tablets' results in blue flowers even in alkaline soils

The role of iron in haemoglobin

It is not only in the biochemistry of plants that iron plays a key role. Haemoglobin and myoglobin are two iron-containing protein-based molecules that carry oxygen in mammalian blood and muscle tissue. The evolution-driven design process seems to have 'chosen' iron compounds for their effectiveness at complex formation with dioxygen.

The rest of the protein molecule modifies the behaviour of the iron atom in certain ways, one of which is to bias the molecule against binding carbon monoxide. Figure 19.32 shows myoglobin, the oxygen carrier in whales (haemoglobin is even more complex, but has a similar structure). The inset shows how the normally six-co-ordinated iron(II) ion has only five ligands and therefore one vacancy. When the molecule is carrying oxygen, the oxygen molecule binds as shown in Figure 19.33. The bent O=O/Fe orientation is thought to be significant in avoiding carbon monoxide poisoning – the distortion is forced on the oxygen molecule by steric hindrance from other parts of the protein. This bent position is thought to be more tolerable for an oxygen molecule than a carbon monoxide one, and this partially offsets the fact that in a 'fair fight' to co-ordinate iron(II), carbon monoxide is much the better ligand. In other words carbon monoxide would be even more lethal than it is but for this ingenious evolved defence mechanism.

Questions

40 It is possible to offer quite a simple explanation for why carbon monoxide is disfavoured by the bent position. Work out where the lone pairs are on dioxygen and carbon monoxide, and suggest why dioxygen can tolerate the angle.

41 Clearly Nature did not evolve this clever antidote to carbon monoxide poisoning because She foresaw the traffic jam on the M25. Presumably there must be circumstances under which a mammal needs to defend itself against carbon monoxide occurring naturally in its own cells. Can you suggest how?

Colour in d-block compounds

Colour is of course one of the trademarks of the compounds of the d-block elements. The explanation of colour in the d-block is quite involved, and this book will offer a simplified version of the story.

We can begin by saying that colour in d-block compounds is a result of the possession of partially filled d-orbitals. You might have guessed that already if you have spotted which d-block compounds *do not* exhibit colour.

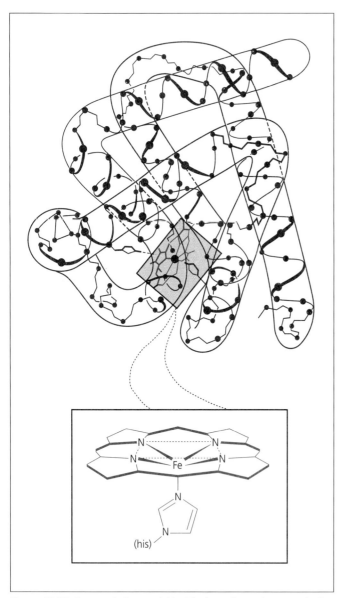

Figure 19.32 Sperm whale myoglobin. The inset shows the atoms in the boxed section – the **haem group**

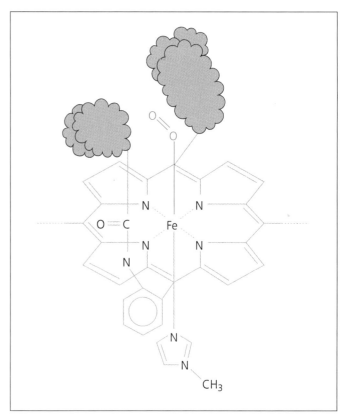

Figure 19.33 Another look at the haem group of myoglobin. The clouds symbolise other parts of the protein which partially hinder access to the sixth co-ordination site on the iron

Question

42 The following compounds are white or colourless:

- titanium(IV) oxide (so white it is used as a white pigment)
- all zinc(II) compounds
- all copper(I) compounds
- all scandium(III) compounds.

What common feature links these species to each other and to the paragraph above?

We saw as long ago as Chapter 3 that colour is associated with changes in the energy levels of electrons. The Cu^{2+}(aq) ion, for example, is blue because there is an electron transition in the ion that absorbs colours *other than* blue. These electron transitions in this and other d-block ions are jumps *between d-orbitals*.

You may think that's odd, because surely the d-orbitals are all at the same energy level. While this is true of gaseous single atoms and ions, it is not true of ions in the midst of the strong influence of a set of ligands. Figure 19.34 shows how one of the d-orbitals would bring the electrons closer to the surrounding ligands of an octahedral complex than another d-orbital. Bringing an electron near to a site which is at the very least nucleophilic and possibly also an anion destabilises it relative to another electron which does not have to make such a close approach. This means that we have an energy level split in the set of five orbitals, and therefore the possibility of electron transitions between the subsets (Figure 19.35).

The main weakness in this model is that it sees the ligands as centres of negative charge surrounding the central ion, which has empty d-orbitals. This concept does not sit too happily with the picture of ligands as donors of dative covalent bonds to those empty orbitals. However, the difference between the two models recedes a little when you consider the polarity of those dative covalent bonds. The amount of electron density that a nitrogen atom, say, will donate to a Cu^{2+} ion is not all that great, so the idea of the ligand as a centre of electron density is not incompatible with the idea of a dative bond from ligand to metal.

Question

43a If ligands are responsible for splitting the energy levels of the d-orbitals, can you explain why different ligands produce different colours or shades? For example, $[Cu(H_2O)_6]^{2+}$ is plain blue, whereas $[Cu(NH_3)_4(H_2O)_2]^{2+}$ is a deeper indigo/purple.

b A common test for water is to drop the liquid on to anhydrous copper(II) sulphate. Anhydrous copper(II) sulphate is colourless (despite having an unfilled d sub-shell), while of course $CuSO_4 \cdot 5H_2O$ is blue. Explain this surprising lack of colour in anhydrous copper(II) sulphate.

Ultraviolet/visible spectroscopy

The colours of copper complexes, and any other species that absorb due to electron transitions, can be studied using a form of spectroscopy. Having covered techniques using visible, infrared and radiofrequency (as in NMR) waves, and mass spectrometry, we find ourselves back in the visible region of the electromagnetic spectrum (with a continuation into the ultraviolet), to look at the absorption of photons by species in solution.

UV/visible spectroscopy shares a feature with atomic absorption spectroscopy from Chapter 3. In both cases, UV/visible photons are used to promote electrons to higher energy levels. But there is a significant difference. The absorption spectrum of, say, sodium vapour is a series of dark lines of **sharply defined frequency** removed from 'white light'. In contrast, the UV/visible spectra of ions in solution are broad bands (Figure 19.36). The reasons are to do with the simplicity of the environment of gaseous

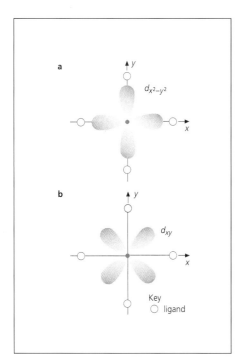

Figure 19.34 Two from the set of five possible 3d orbitals, in a ligand field. Note how an electron in the $d_{x^2-y^2}$ orbital would be brought closer to the surrounding ligands, and therefore be destabilised relative to one in the d_{xy} orbital

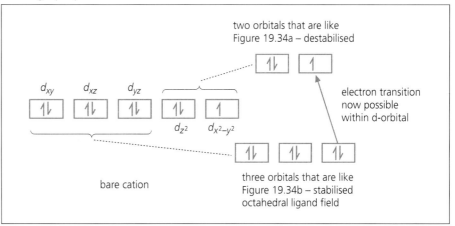

Figure 19.35 Showing, using the example of copper(II), how moving the cation into the ligand field opens the possibility of electron transitions between d-orbitals

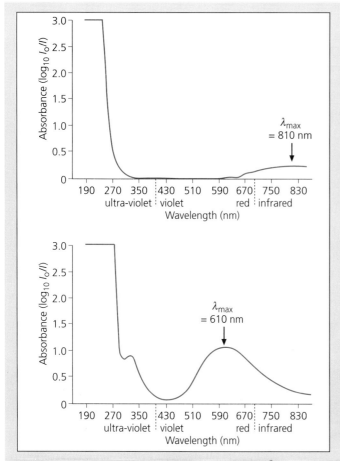

Figure 19.36 a Ultraviolet/visible spectrum of $[Cu(H_2O)_6]^{2+}$
b UV/visible spectrum of $[Cu(NH_3)_4(H_2O)_2]^{2+}$. In contrast to IR spectra, UV/visible absorptions are shown as *upward* peaks

jump, or absorb it, and promote an electron. A single spectral line of well-defined frequency then results.

However, an ion in solution, surrounded by ligands and solvent molecules, possesses energy in two additional ways – by virtue of its **vibrations** as a whole complex, and by virtue of its **rotations**. Let us suppose a photon came along with an energy slightly greater than x, say $x + dy$, where x and dy are defined on Figure 19.37. Instead of rejection, the complex can absorb the photon, using energy x to promote the electron to the higher energy d-orbital, (Figure 19.35), and using the extra energy dy to put the molecule into a more agitated state of vibration.

This explanation would predict a more complicated but still line-based spectrum, with a nest of close lines all clustered around energy x and separated by values like dy, the spacing of the vibrational energy levels. But each complex can have rotational energy as well, so there is a second way of absorbing additional photons, even those that do not quite match energy jumps like $x + dy$. A photon of energy $x + 2dy + 3dz$, for instance, would be accepted on the grounds that energy x can promote the electronic transition, energy $2dy$ can put the electronically excited molecule into its second level of vibrational energy, and finally, the extra $3dz$ can be used to increase rotational energy (see the green arrow on Figure 19.37).

The result of all these energy levels on top of energy levels is that a whole range of photons centred somewhere near the original x can be absorbed. The closeness of the lines is beyond the powers of resolution of most UV/visible absorption spectrophotometers, and the resulting absorption appears as a continuous wide peak.

The spectrochemical series

The spectra for the two copper complexes in Figure 19.36 show this continuous absorption peak. They also confirm the observation that different ligands produce different colours. The ammonia complex absorbs at a slightly different place in the spectrum, taking out more red and leaving a narrower, less white, blue 'window'. A whole

ions, compared to the situation faced by ions in solution. An ion in the gas phase only possesses energy in two ways: kinetic, by virtue of its motion, and electronic, by virtue of its electron arrangement. It can only interact with an incoming photon in two ways – ignore it, because its energy/frequency does not match an allowed electron

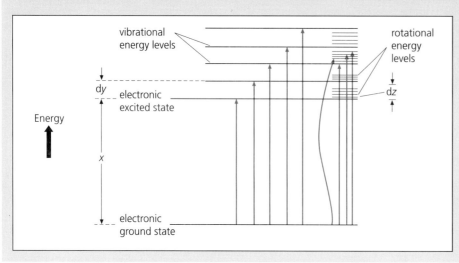

Table 19.13 The spectrochemical series

Ligand	λ_{max}
I^-	Longest wavelength
Br^-	
Cl^-	
H_2O	
NH_3	Shortest wavelength

Figure 19.37 Electronic and vibrational levels of the d-block elements in a complex. The order of magnitude of the energy level separations is: between electronic levels $100\,kJ\,mol^{-1}$; between vibrational levels $1\,kJ\,mol^{-1}$; between rotational levels $0.01\,kJ\,mol^{-1}$

range of ligands on the Cu2+ ion have been studied, and the λ_{max} value is different for each (Table 19.13). Such a list is sometimes referred to as the **spectrochemical series**.

Question

44 The lowest members of the spectrochemical series appear to be also the best bases. The stages of this question are designed to help you to understand the reasons for that correlation.

a If a species is a good base, is it more or less 'pushy' with its lone pair of electrons?

b If a ligand is more 'pushy' with its lone pair of electrons, are the d-electrons on the central metal ion likely to be more or less 'aware' of the presence of the ligand sextet?

c Is it going to become more or less important for a metal electron to get into one of the orbitals further away from the ligand? What will happen to the energy separation between the two subsets of d-orbitals in Figure 19.35, as the ligand becomes a better base?

d As the electron jump gets bigger with the change of ligand, in which direction will the λ_{max} move in the UV/visible spectrum (to longer or shorter wavelengths)?

e Finally, put all the arguments together to predict the general appearance of the visible absorption spectrum of $[CuCl_4(H_2O)_2]^{2-}$(aq).

The x- and y-axes of an ultraviolet/visible spectrum

The x-axis of a UV/visible spectrum is conventionally shown as a series of wavelengths in nanometres (in contrast to the x-axis of an infrared spectrum, which is expressed as wavenumbers in cm^{-1}). The visible region runs from 400 to 700 nm. Conversions between wavelength, frequency and energy are readily carried out by use of the equations $c = \lambda\nu$ and $E = h\nu$.

More trouble is taken over the exact values on the y-axis in UV/visible spectroscopy than in any other form of spectroscopy. In other forms, the most significant fact about a peak is that it is there, as part of the 'fingerprint' of its source molecule. In contrast, the y-axis value of a UV/visible peak is often used as a measure of concentration. The important y-axis parameter, absorbance A, is defined by the equation:

$$A = \log_{10}(I_0/I)$$

where I_0 is the intensity of the light going into the sample, and I is that coming out. The absorbance as defined above is proportional to the concentration of the absorbing species, according to:

$$A = \varepsilon c l$$

where c is the concentration and l is the length of the light path through the cell containing the solution. The quantity ε, the extinction coefficient, is a constant unique to each absorbing species, and expresses the inherent depth of colour of the species, at a particular wavelength.

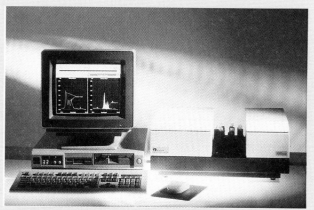

Figure 19.38 The cells used in a UV/visible spectrophotometer often have a path length of 1 cm

ε-values are quoted in data books, and it is assumed that they apply to the wavelength of the highest part of the absorption, at what we have been calling λ_{max}.

Question

45a What would be the absorbance of a solution which absorbed all but a tenth of the light incident upon it?

b Figure 19.36 suggests that the tetraammine copper(II) complex, with its higher line, has a more intense colour than ordinary aqueous copper(II) ions (as indeed it has). Why, given the limited information on those spectra, would it be impossible to reach that conclusion from Figure 19.36 alone?

c Suppose the spectra in Figure 19.36 had been from solutions of unknown concentration. What information would you need to know to determine their concentrations?

A footnote on colour

d-block compounds do not have a monopoly on colour in chemistry: we only have to remember the dye molecules of Chapter 15, where we traced the origin of colour to the existence of extended delocalised π-bonded systems called 'chromophores'. The common link between d-block compounds and dyes is that in both cases, electrons are promoted to higher energy levels by absorbed photons, and in both cases the overlay of vibrational and rotational energy levels ensures broad absorption bands. In the organic chromophores, the electron jumps are of course nothing to do with d-orbital splitting. Instead they are jumps between bonding π-type molecular orbitals and antibonding π-type molecular orbitals, which are almost certainly not in your syllabus.

Nomenclature of complexes

The nomenclature of complexes is quite involved, following a systematic set of rules. Up to now we have used formulae when referring to complexes, though clearly a universally recognised official naming system is needed (even if 'recognised' is not quite the same as 'used' – in the chemical industry there is still lots of mileage left in unofficial chemical names like 'acetone' and 'ferric chloride').

The rules for naming complexes are summarised in Table 19.14.

Additional rules

- Ligand names come before metal names.
- Ligand names are in alphabetical order.
- Overall neutral complexes are named by cation rules (see $[PtCl_2(NH_3)_2]$ in the example below).
- Multipliers are the ordinary di-, tri-, tetra- series for simple ligands, and the bis-, tris-, tetrakis- series for more complicated ligands and for polydentate ligands.

Examples

- $[Co(NH_3)_5Cl]Cl_2$ is pentaamminechlorocobalt(III) chloride.
- $Na_3[Fe(CN)_6]$ is sodium hexacyanoferrate(III).
- $Na_4[Fe(CN)_6]$ is sodium hexacyanoferrate(II).
- *cis*-$[PtCl_2(NH_3)_2]$ is *cis*-diamminedichloroplatinum(II) (which is neutral, and in which the two chlorides are next to each other).
- $[Cr(en)_3]Cl_3$ is tris(1,2-diaminoethane)chromium(III) chloride (where 'en' is the normal abbreviation for the 1,2-diaminoethane ligand).

Question

46 Name the following complexes (you will need to note the number and identity of the ordinary cations or anions to work out the overall charge on the complex ion):

a $Na[FeBr_4]$

b $[Ag(NH_3)_2]Cl$

c $Ca[PtCl_4]$

d *trans*-$[Cr(en)_2Cl_2]Cl$.

Table 19.14 The nomenclature of complexes

Cationic complex ions (i.e. those that end up as positive ions after all the ligand charges are added to the bare metal charge, e.g. $[Co(NH_3)_5Cl]^{2+}$		Anion complex ions (i.e. those that end up as negative ions after all the ligand charges are added to the bare metal charge, e.g. $[CuCl_4]^{2-}$	
Ligand name	Metal name	Ligand name	Metal name
If neutral, keep same name except for: H_2O = aqua- NH_3 = ammine- CO = carbonyl- If anion, '-ide' becomes '-o', e.g. chloro-	Keep same name (but with oxidation number)	'-ide' becomes '-o' '-ate' becomes '-ato' e.g. chloro-, oxalato-, cyano-	Takes on '-ate (oxidation number)' ending (often using archaic word for metal), e.g. ferrate(III), cuprate(II), argentate(I), platinate(IV)

19.6 Catalysis by elements and compounds of the d-block

Redox 'go-between' catalysis

This topic was covered in Chapter 14 on kinetics. You may recall diagrams like Figure 14.48 (p. 351) (and indeed Figure 19.5 in this chapter), in which the catalyst is shown shuttling between two of its own oxidation states while acting as a go-between in the delivery of electrons from reductant to oxidant. Naturally the ions of the d-block can carry out this kind of catalysis, as they have a ready access to variable oxidation states.

In fact it requires more than just the ability to exist in two oxidation states to be a successful catalyst for a redox reaction. There is another qualification which harks back to the electrochemical series and the associated Law of anti-clockwise circles. Both the steps of the catalysed route are themselves normal redox reactions, so like all such reactions they must pass the thermodynamic feasibility test enshrined in the Law of anticlockwise circles.

Question

47 Peroxodisulphate(VI) ions oxidise iodide ions to iodine, according to:

$$S_2O_8^{2-}(aq) + 2I^-(aq) \rightarrow 2SO_4^{2-}(aq) + I_2(aq) \qquad (4)$$

The reaction is perfectly feasible on thermodynamic grounds but is quite slow. It is several times faster in the presence of quite a small amount of Fe^{3+} ions.

a Look up suitable values of standard electrode potential and hence show that the main reaction obeys the Law of anti-clockwise circles.

b How might the Fe^{3+} ion intervene in the reaction with a redox step of its own? Again prove the change is feasible and give a balanced equation for the step.

c Now suggest how the original Fe^{3+} ion can be regenerated in a second step. Again give an equation and prove feasibility.

d Show that the steps you have suggested in (b) and (c) do indeed add up to exactly equation (4) and nothing but equation (4).

e The Fe^{2+} ion also catalyses the reaction. Explain why that was to be expected, and what change in the detail of the catalysed mechanism will occur compared with the Fe^{3+} ion case.

f Generalise from your analysis of the $Fe^{3+}|Fe^{2+}$ system to derive a rule for a feasible catalyst for any redox reaction, basing your rule on positions in the electrochemical series.

g Use your own generalised rule to select possible catalysts for reaction (4) from the following list: $MnO_4^-/H^+(aq)$, $Mn^{2+}(aq)$, $Cu^{2+}(aq)$, $V^{2+}(aq)$.

h 'Feasible' does not necessarily mean effective, and several feasible catalysts do not actually speed up the reaction. Why might this be?

i Show a generalised reaction profile for this style of catalysis, and superimpose the reaction profile for the uncatalysed reaction.

Homogeneous co-ordination catalysis

Sometimes d-block cations can provide catalytic routes by co-ordinating other molecules. One of the most significant pieces of synthetic organic chemistry of the second half of the twentieth century was of this kind. It rejuvenated the polymers industry, created a new field of organometallic catalysis and won for its inventors the 1963 Nobel prize.

Prior to this breakthrough, the discovery of the free-radical polymerisation of ethene had already given a massive boost to the polymer industry after the Second World War, and the bright new plastics were symbolic of the post-war era, when science was seen as improving people's lives after half a decade of deprivation. These early plastics were rather limited in comparison to the array of versatile polymers of the present day. 1950s polythene was a waxy sort of solid that discoloured and went flaky and brittle with time.

The problem was a chemical one. Free-radical polymerisation suffered from several drawbacks – it was a **batch process** conducted at high temperature and pressure. The reagents were mixed, brought up to reacting conditions, brought down again and the product removed from the mixture. During all this there was a tendency for chain branching to occur, when a free radical would steal a hydrogen atom from the middle of a ready-formed chain, instead of waiting for the 'correct' target, namely an unreacted alkene monomer:

$$R\bullet + \ce{C} \longrightarrow RH + \bullet\ce{C}$$

Question

48a Continue the story from the situation shown above, bringing in the next alkene. Hence show how branched-chain hydrocarbons are produced.

b Why would the branched-chain polymer be less hard, less dense and less chemically resistant than a straight-chain equivalent?

It was a discovery by Karl Ziegler that led to the straight-chain **high-density polythene**. The breakthrough was a catalyst in the form of a titanium(IV) complex, in which the titanium atom had a covalently bonded alkyl group along with a vacant co-ordination site. Figure 19.39a shows the catalyst molecule with no substrate in residence. In Figure 19.39b an alkene (propene in the example shown) acts as what can best be described as a π-bonded ligand. In Figure 19.39c the alkyl part of the catalyst itself has transferred to the end of the π-bond, and alkene has gone. Instead there is a new longer alkyl group and a new vacancy on the next-door site (Figure 19.39d).

The cycle continues with a new alkene coming to the new vacant site.

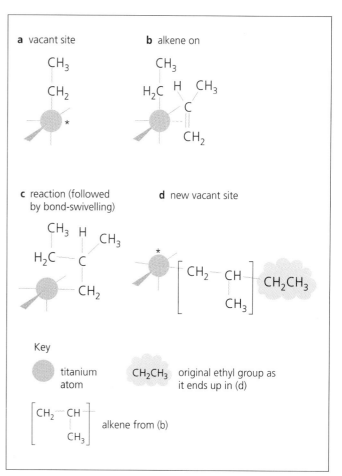

Figure 19.39 A possible model for the mechanism of Ziegler–Natta-style catalysis. *shows the vacant site

Question

49a Draw two more repetitions of the cycle.

b Which polymer is being synthesised in Figure 19.39? (It is not poly(ethene).)

c The original free-radical mechanism allowed the radicals to 'roam free' and react where they chose in the reaction mixture. Explain why this catalyst gives a much stronger guarantee that only straight chains will be synthesised.

It turned out that not only were the chains straight – in addition, the method exercised control over the chiral geometry of the product. In all the polyalkenes except poly(ethene), alternate carbon atoms are chiral.

Question

50 Show by a drawing that alternate carbons in poly(chloroethene) are chiral.

What was remarkable about Ziegler-polymerised polymers was that the orientation of their chiral centres was always the same. Giulio Natta was able to show that Ziegler's titanium catalyst produced polymers in which every chiral centre was the same way round (Figure 19.40). Natta shared Ziegler's Nobel prize in 1963, and the fast-growing group of catalysts that perform polymerisation reactions are called Ziegler–Natta catalysts in their joint honour.

Heterogeneous co-ordination catalysis

Most of the major industrial processes on which the chemical industry depends employ catalysts. While the organometallic catalysts like Ziegler–Natta ones may represent a part of the future, the large-scale bulk productions that have been the basis of industrial life in the twentieth century – sulphuric acid, ammonia, methanol and nitric acid – all feature catalysis of gas-phase reactions on solid surfaces (hence the word heterogeneous). In all the above-named cases the surface in question is that of a d-block metal or one of its compounds.

Why does adsorption on to a surface have a catalytic effect on a reaction? Let us re-open the question for the whole field of catalysis – why does any catalyst, redox, homogeneous or heterogeneous, make a reaction go faster? The answer offered to you so far has been (with the exception of the story of enzyme catalysis) of limited depth – that catalysts lower the activation enthalpy of a reaction. But *how* do catalysts lower activation enthalpies?

From one viewpoint, the whole process is rather mysterious. The same bonds have to be broken in the catalysed reaction as in the uncatalysed reaction, and since bond breaking has to take place on the way to the transition state (albeit offset in part by bond making), you would think the activation enthalpy hump was, on reflection, quite a hard thing to reduce.

a isotactic poly(propene), with every chiral carbon the same way round

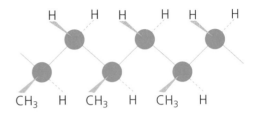

b atactic poly(propene), with methyl groups sticking out both back and front, and with some monomer units embedded backwards

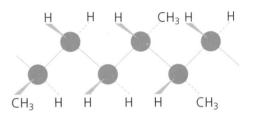

Figure 19.40 Ziegler–Natta polymerisation produces isotactic chains (a) which are capable of better crystalline stacking and van der Waals' bonding than the atactic chains (b). The result is a material that is harder, denser and chemically more stable

In the case of enzyme catalysis, the secret lay in the process of pre-stretching bonds on the substrate that were earmarked for breakage, while making the rest of the molecule 'comfortable' (see the dentist's chair analogy in Section 15.12). Proton catalysis (see, for example, Section 14.16) also involved a relationship between reactant and catalyst which predisposed the target bond to break, this time more by altered polarity than because of any physical 'discomfort'.

The model we can use for surface catalysis stands a little apart from such mechanisms. Here we think not so much in terms of the reactant–catalyst relationship doing a pre-breaking or partial breaking job on the reactant bonds, but instead consider the electron movements that would have to occur in the reaction, and see how the catalyst can ease them. In other words, we see how the catalyst surface might lower the enthalpy of the transition state.

Let us take a well-known example of surface catalysis from organic chemistry. Hydrogen adds to alkenes, and the process is catalysed by a surface of nickel, palladium or platinum. Without the catalyst, the reaction would need to take place via an elementary step, namely the bimolecular collision between the alkene and a hydrogen molecule.

The transition state would be a four-centre species in which both the H—H bond and the π-bond had stretched, and a small amount of C—H bond formation had begun (Figure 19.41).

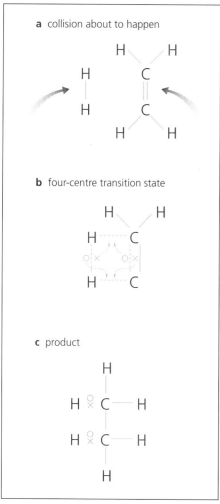

Figure 19.41 Simple bimolecular collision mechanism for the hydrogenation of ethene

a collision about to happen

b four-centre transition state

c product

a possible mode of attachment – hydrogen and ethene molecules co-ordinated on to nickel surface

b electrons from the bonds to be broken can be stabilised by spending time in orbitals of the nickel atom . . .

c . . . and then reform as new C—H bonds

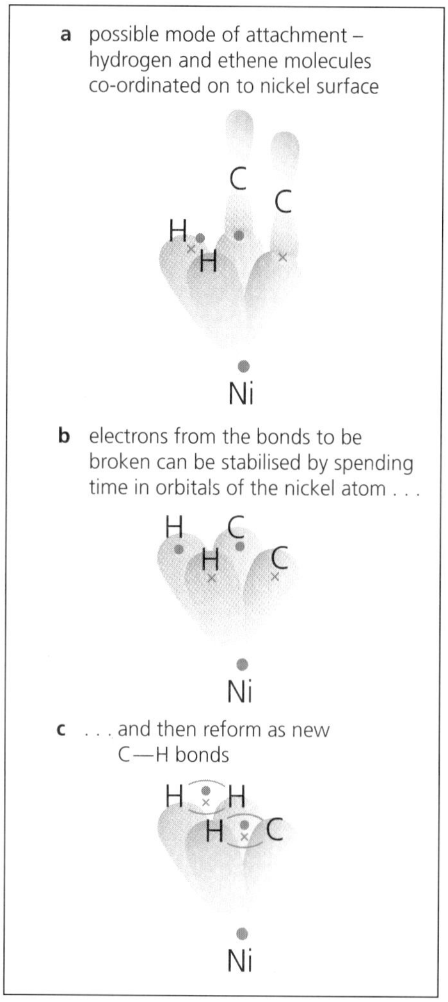

Figure 19.42 Possible model for the role of nickel in the hydrogenation of alkenes

and ethene are adsorbed reversibly on to nickel, palladium and platinum surfaces on their own, even when there is no reaction going on. It is thought that these adsorptions do represent a sort of weak chemical bonding, and therefore some electron sharing, between the gas molecules and the metal surface.

Questions

52 Figures 19.43a and b show the possible reaction profiles of the separate adsorption processes for hydrogen and ethene on to a nickel surface.

a Use these two diagrams, plus your own, to put together the full reaction enthalpy profile of the nickel-catalysed hydrogenation of ethene. Assume the process takes place in a series of steps in this order:

- adsorption of hydrogen
- adsorption of ethene
- reaction between the adsorbed molecules
- desorption of ethane.

In other words, you are to trace the potential energy of the system during the course of these stages and put it on a diagram with several humps.

b Show by an arrow a distance on the diagram that represents the overall activation enthalpy, E_A, of the catalysed reaction.

Question

51 Make a rough quantitative estimate of the size of the activation enthalpy for this process, showing clearly the basis for your estimation.

In the catalysed process, both the reactants are adsorbed on to the surface of the metal. The electrons in the H—H bond and the C=C bond must still execute the same movement in order to create the new C—H bonds, but now we can imagine how their path might be energetically smoothed. The reason is that the movement is not a leap through space, which would require the negotiation of a high point remote from the stabilising influence of the reacting nuclei. Instead it is a very reasonable proposition that the orbitals of the d-block metal atom can act as channels by which the electrons make their voyage, a voyage which is always stabilised by the influence of one nucleus or another. This process is shown imaginatively in Figure 19.42.

Although the above story is no more than reasonable speculation, it is supported by the fact that gases such as hydrogen

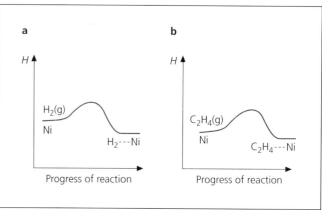

Figure 19.43 Reaction profiles of the separate adsorption steps between nickel and **a** hydrogen gas and **b** ethene gas

c Superimpose a line that might represent the enthalpy profile of the uncatalysed reaction, and show its activation enthalpy.

d What is it about d-block atoms that equips them well for this catalytic role?

53 The above picture of catalysis belongs squarely in the category of 'It may not be true but at least it's plausible'. I hope this book has persuaded you of the rightful place of such speculation in serious science. It is common in cases like this that more than one picture may offer satisfactory explanations of actual events. For instance, it would have been possible to account for surface catalysis by re-using the model that says the catalyst partially breaks the reactant bonds, as we used when looking at enzymes. What clues in Figure 19.43 indicate that a degree of bond breaking has occurred in the reactants as they 'settle down' on the catalyst surface?

A closer look at metal surfaces

The above account of surface catalysis has made the assumption that the metal surface is a flat, uniform landscape of metal atoms. Incoming substrate molecules, on this model, have a near-infinite plane to land on, and it makes little difference where on this plane they land. The truth may be a little more complicated. In fact the metal surface may have more in common with enzyme clefts, because it is thought that there are certain **sites** on the metal surface that are just right for catalysis, and others that are not so good. What makes a site right or wrong may depend upon lattice defects, or upon where the boundaries between crystals occur, or even the angle at which the close-packed planes of atoms in the crystal meet the surface (Figure 19.44).

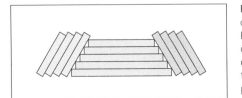

Figure 19.44 Schematic picture of how three crystals might appear at the surface of a catalyst. Each set of layers indicates a close-packed plane of atoms. Note that, by chance, they are oriented differently relative to the surface, so that one crystal offers 'steps' and another a flat plane. Note also the crevices at grain boundaries

Question

54 Suggest why the angle of the close-packed planes of atoms might influence the effectiveness of catalysis.

There is yet another aspect of metal surfaces that affects their catalytic activity, and that is that the surface layer is quite likely, unless measures are taken to prevent it, to be an oxide or hydroxide layer.

A group of scientists are trying to utilise this very aspect of metal surfaces. They have attached calixarene, a complicated organic molecule made of a 'bowl' of benzene rings, to a metal surface. The computer-graphic picture of the reaction is shown in Figure 19.45, with the calixarene performing a nucleophilic substitution to oust the previously resident hydroxide groups. It is hoped that calixarene will be the forerunner of a selection of 'clip-on' molecular attachments to metal surfaces, all designed to enhance some aspect of the metal's performance (not just its catalytic activity).

Question

55 Suggest in general terms how the calixarene method might be adapted to lead to the following:

a metals with in-built self-lubrication

b metals that bond strongly with paints

c metals that resist corrosion.

19.7 Catalysts in automobiles

The last 20 years have seen an escalating effort to reduce the environmental price we pay for the right to own and use petroleum-powered vehicles. The main pollutants which come from our cars and lorries are listed in Table 19.15. They

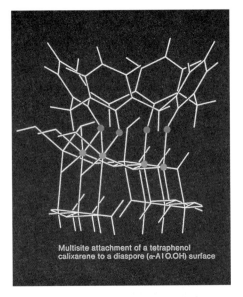

Multisite attachment of a tetraphenol calixarene to a diaspore (α-Al O.OH) surface

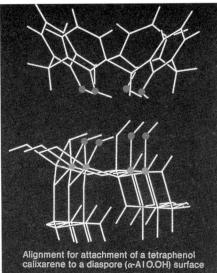

Alignment for attachment of a tetraphenol calixarene to a diaspore (α-Al O.OH) surface

Figure 19.45 Computer-graphic image of an organic molecule (calixarene) being attached to a metal surface

include irritants, poisons and carcinogens.

Unleaded petrol has been made available by the fine-tuning of hydrocarbon mixtures by means such as catalytic reforming (see Section 10.9). The increase in aromatic (i.e. benzene ring-containing) hydrocarbons can be used to produce a fuel with burn kinetics as good as leaded petrol. (However, at the time of writing there is growing concern that by eliminating lead we have begun to make carcinogenic petrol.) There are also efforts under way to reduce the sulphur content of fuels. But the other four

Table 19.15 Pollutants from petrol-powered vehicles

Pollutant	Environmental impact
$CO_2(g)$	Chief greenhouse gas
$CO(g)$	Toxin
Hydrocarbons (HC)	Photochemical smogs, may be carcinogens
Oxides of nitrogen (NO_x)	Toxins, irritants, precursors of acid rain
Lead compounds	Toxins
$SO_2(g)$	Irritant, precursor of acid rain

pollutants are inherent in the very process that goes on in the cylinder during the explosion.

Question

56 Explain how each of the first four pollutants in Table 19.15 comes to be in the exhaust gas stream. (You might find the oxides of nitrogen a bit puzzling, but remember that a spark is like a tiny lightning flash, and then recall the nitrogen cycle.)

Carbon dioxide is an inevitable outcome of any internal combustion engine, but the next three pollutants on the list can all be tackled by preventative counter-measures. The levels of these three pollutants are quite sharply dependent upon the tuning of the air/fuel mixture in an engine. Figure 19.46 shows how they vary. The x-axis parameter λ requires some explanation. The value $\lambda = 1$ refers to an air/fuel mixture which is more or less stoichiometric with respect to the complete combustion of the hydrocarbon. Its real value by weight is approximately $14.7 : 1$. So for instance a λ-value of 1.2 indicates the presence of 20% more air than is needed for complete combustion. When you look at the scale of the units on the y-axes you can see that small but damaging amounts of important pollutants are being produced alongside the stoichometric equation products (carbon dioxide and water).

Question

57 The gases in Figure 19.46 are all there because of what happened or did not happen in the combustion chambers of the engine.

a Suggest why hydrocarbons are high when $\lambda < 1$.

b Suggest why hydrocarbons are lower when $\lambda = 1$.

c Suggest why oxides of nitrogen are low when $\lambda < 1$.

d Suggest why carbon monoxide is high when $\lambda < 1$.

e Suggest why carbon monoxide is low when $\lambda > 1$.

There is a simple idea behind the technology to purge exhaust gases of these three pollutants – to get them to react with each other (and with excess oxygen). Of the four gases in

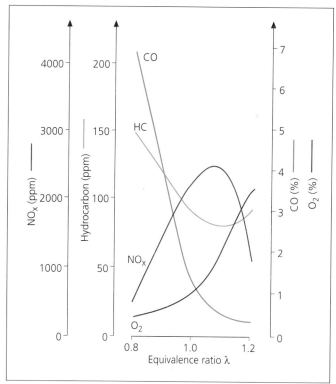

Figure 19.46 Showing how the levels of four pollutants change as a result of variation of the air/fuel ratio in the combustion chamber of a car. See the text for an explanation of λ

Figure 19.46, two are capable of oxidising the other two, so a picture begins to take shape of the possibility of utilising redox reactions between the gases.

Question

58 For the purposes of this question, assume that the oxide of nitrogen is nitrogen dioxide, and take octane, C_8H_{18}, as a typical petrol hydrocarbon.

a Which would be the two oxidising agents, and which the two reducing agents, if we want to produce 'clean' exhaust products?

b Write four equations that show the two oxidising agents oxidising the two reducing agents. Your products should be nitrogen, water and carbon dioxide only.

c It might be said that even if these reactions worked perfectly and to completion, the car would still be a source of pollution. Can you explain why?

The next few paragraphs will explore some of the problems associated with this technology, which keep many of today's chemists busy.

Problem 1 – the reactions do not happen

The reactions in question 58b are not spontaneous at exhaust-pipe temperatures. (If they happened spontaneously there would be no problem, and no Figure 19.46.) But catalysts come to the rescue again. Platinum is an effective catalyst for two of the reactions from question 58b. It catalyses the two reactions in which oxygen is the oxidising agent – it is nowhere near as effective for the other two, in which nitrogen dioxide is the oxidising agent.

A **catalytic converter** contains platinum in a stainless steel can, fitted on the exhaust pipe between the engine and the silencer. Figure 19.47 shows a cutaway catalytic converter, with the catalyst mesh exposed. The mesh consists of three layers:

• an inert support material

• an aluminium(III) oxide covering – this is the ceramic material of spark plugs, and has good thermal stability

• an outer layer of catalyst.

The manufacturers claim to achieve a catalyst surface the size of a football pitch.

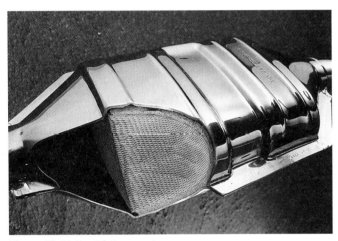

Figure 19.47 A catalytic converter

Question

59 How is it possible to get the area of a football pitch inside a steel can, underneath a family car?

Figure 19.48 shows that over a third of all platinum used by the Western nations is destined for automotive (auto-) catalysts – that is second only to the portion used for jewellery.

Problem 2 – when there is insufficient oxygen

When the air/fuel mixture is richer in fuel than $\lambda = 1$, i.e. when $\lambda < 1$, the oxygen level is quite low. Oxygen is the major oxidising agent in the exhaust stream. The situation is further exacerbated by the fact that just when the oxygen level has gone low, the levels of substances that need to be oxidised have risen (for reasons which you explored in question 57).

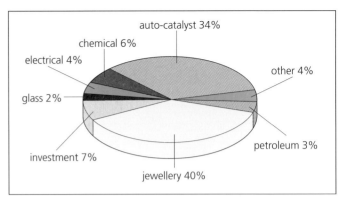

Figure 19.48 Platinum demand in the Western world, 1992. Total demand was 3800 million ounces

Question

60 The problem is not as bad as it looks at first sight from Figure 19.46, since the y-axis scales are different. The oxygen units are per cent, while the hydrocarbons and NO_x units are parts per million.

a If percentages are 'parts per hundred', how many decimal places would the hydrocarbon data need to be shifted, so that they were directly comparable to the oxygen data?

b Hence show that even at $\lambda = 0.8$, there is still enough oxygen in the exhaust gas to oxidise the hydrocarbons. (Assume octane as your typical hydrocarbon in petrol.)

The problem with rich ($\lambda < 1$) mixtures is of course carbon monoxide. The solution is simple – an inlet that allows in more air while the gases are in the exhaust pipe, after the ignition, so the carbon monoxide gets oxidised in the catalyst chamber as it would have been had the value of λ been greater than one.

Question

61 You might think that all these problems associated with rich mixtures could be side-stepped by tuning the carburettor on the 'lean' side of $\lambda = 1$. Indeed that is the direction in which most engine designers are heading, since 'lean-burn' (high air : fuel ratio) engines are more economical in terms of fuel consumption. However, even a lean-burn engine has to start from cold in the morning.

a When a car is started in the morning, the choke is pulled out (either manually or automatically). Find out what the choke does.

b Why is it necessary to have an air inlet in the exhaust stream, even in lean-adjusted engines?

Problem 3 – cold starting

We have just looked at one of the problems associated with cold starting, namely the unavoidable need to use low-air mixtures, with associated increases in carbon monoxide and hydrocarbon emissions. The auto-catalyst industry estimates that an 'unclean' engine emits about 70% of its harmful emissions per journey in the first 80 seconds. The extra air inlet is of course a step in the right direction, but there is another problem – the catalyst is cold too. (Even rates of catalysed reactions show an Arrhenius-style dependence on temperature.) The solutions to the problem cover two areas – catalyst selection and catalyst pre-heating.

As far as selection of catalyst metals is concerned, rhodium seems to be the answer. Alloys of platinum and rhodium start working at about 150 °C, as compared with 240 °C for platinum and 350 °C for copper. Unfortunately the more effective catalysts are also more expensive (Table 19.16). Rhodium is too expensive for us even to wear it as jewellery (Figure 19.49).

Table 19.16 Prices of catalyst metals, 1992 (an ounce is about a teaspoonful)

Metal	Price per ounce ($)
Pt	359.90
Rh	2398

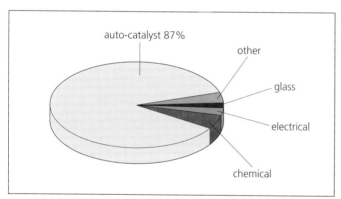

Figure 19.49 Rhodium demand in the Western world, 1992. Total demand was 324 000 ounces

Pre-heating the catalyst to 150 °C before a journey and using a rhodium/platinum catalyst would therefore seem to offer a compromise solution to the cold-start problem.

Question

62 How could the catalyst be pre-heated? Discuss the problems that might arise in the technology of pre-heating. Why do you think scientists in this field have investigated changing the support material of the catalyst mesh to a thin metallic material (Figure 19.50)?

Problem 4 – getting rid of nitrogen oxides under lean-burn conditions

We have already targeted lean-burn engines as a desirable design direction, since they ensure an oxidising environment to get rid of the hydrocarbons and the carbon monoxide, and also result in greater fuel economy. However, there is a problem with the oxides of nitrogen. The reaction that gets rid of them is the one in which they oxidise the other two pollutants and thus reduce themselves to nitrogen. So the oxides of nitrogen and the oxygen compete with each other for molecules to oxidise. In conditions of high excess oxygen, the balance is clearly tilted against the oxides of nitrogen, and the danger is that the hydrocarbons and carbon monoxide will all be oxidised by oxygen, leaving the oxides of nitrogen unchanged (Figure 19.51).

Figure 19.50 A new style of catalyst support using thin metal sheets

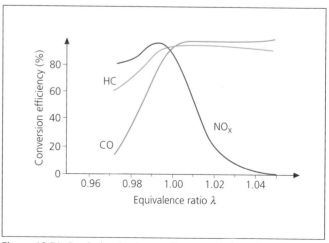

Figure 19.51 Graph showing how well auto-catalysis works in removing noxious emissions, as a function of the λ variable. Note that when oxygen is in even a small excess, the conversion of NO_x drops dramatically. (A high conversion means a low emission.)

63 A platinum-only catalyst is more effective for oxidations by oxygen, while increasing amounts of rhodium favour oxidations by oxides of nitrogen. Suggest how the catalytic material could be sequenced in the exhaust gas flow to maximise the reactions that get rid of oxides of nitrogen before the oxygen uses up all the hydrocarbons and carbon monoxide.

Some workers in the field of catalysis are reporting that Cu^{2+} ions, replacing the natural Na^+ and Al^{3+} ions in **zeolites**, have a future as NO_x reduction catalysts. This would of course be a big breakthrough in cost-saving, compared with using Pt/Rh catalysts. Zeolites are amazing edifices made of silicate (SiO_4^{2-}) tetrahedra intricately linked up until they resemble molecular versions of the bouncy castles you get at fêtes (Figure 19.52). They are best described as **molecular sieves**, and incidentally are also the materials from which cation exchange resins are made. Their ability to act both as supports for very highly dispersed catalyst cations, and also as filters which can select molecules by size, offers the prospect of a range of highly selective catalysts in the future.

Problem 5 – emission control laws

Figure 19.53 shows how the USA plans to restrict emissions from new cars by law. The y-axis parameter of this graph is not the same as the previously used parts per million, so correlation is not easy. However, the general shapes of the graphs are a reflection of what was possible with existing technologies (and were no doubt also shaped by political lobbies from the auto-industry). For example, the first generation of platinum-only catalysts was on stream by 1980.

Question

64 What evidence is there in the graph that between 1983 and 1993 the US Environmental Protection Agency began to recognise the possible contribution of rhodium catalysts?

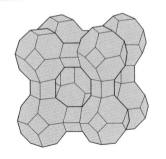

Figure 19.52 A zeolite – a mineral comprising Na^+ and Al^{3+} cations with SiO_4^{2-} tetrahedra together forming a microporous structure. The replacement of Na^+ and Al^{3+} cations by Cu^{2+} ions looks promising for NO_x reduction catalysis

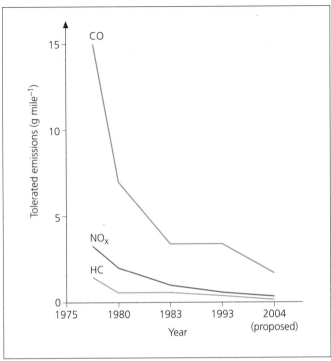

Figure 19.53 Showing our decreasing tolerance of car emissions – US federal emissions limits

19.8 A practical application of d-block chemistry – photography

The following panel explains aspects of silver chemistry, as applied to photography, which take in three of the outstanding characteristics of d-block elements and compounds:

* their redox ability
* their catalytic ability
* their ability to form complexes.

It can serve, therefore, as a fitting end to the chapter.

Chemistry in pictures

You might be surprised to learn that the simple chemical reaction on which modern photography is still based was discovered over 250 years ago. One day Johann Schulze, the professor of anatomy at the University of Altdorf in Germany, went into his laboratory and was astonished to see that something very strange had happened to the silver salts he was experimenting with. He had been working near an open window through which sunlight had been shining, and that part of the silver salt which had been facing the window had darkened, whilst the portion facing away from the window remained white. Schulze was a thorough experimenter and he was able to demonstrate that the darkening was caused by light, rather than other causes such as heat.

Even though Schulze had effectively discovered a crude form of photography, it was not until the 1830s that this laboratory curiosity was turned into a viable commercial product by the work of Louis Daguerre. Daguerre's system employed a copper plate which was coated with metallic silver and exposed to iodine vapour. The precipitated silver iodide is sensitive to light and, after exposing the plate to light in a camera (followed by further chemical processing), an image was obtained on the mirrored surface. Although Daguerre's system had several disadvantages (it was expensive and slow, and copies could not be made of the original photograph), the images obtained were extremely pleasing and the process was a commercial success.

How photographic materials are made

As indicated above, the light-sensitive compounds that are used in photographic materials are halides of silver, in particular silver bromide, silver chloride and silver iodide. These salts are prepared industrially using the same basic chemistry that you might use in the laboratory. First, bars of silver are dissolved in nitric acid to form silver nitrate:

$$Ag(s) + 2HNO_3(aq) \rightarrow AgNO_3(aq) + NO_2(g) + H_2O(l)$$

The silver nitrate is then dissolved in water and mixed with a potassium halide solution (bromide in the equation):

$$AgNO_3(aq) + KBr(aq) \rightarrow AgBr(s) + KNO_3(aq)$$

The silver bromide precipitates out (since it is more or less insoluble in water), but by carrying out the precipitation in the presence of gelatin (the same material from which edible jellies are made) and then allowing it to set, the tiny crystals of silver halide can be held in suspension and washed free of the potassium nitrate which is also formed in the reaction. In the next stage of manufacture the gelatin suspension is melted and coated in a very thin layer on to a base of clear plastic (for film) or paper (for print material). All this must, of course, be done in the dark since the silver halide is sensitive to light.

A piece of photographic film or paper is shown in Figure 19.54.

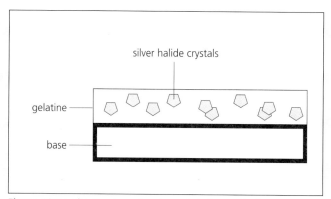

Figure 19.54 The structure of a piece of photographic film or paper

The fundamental photographic reaction

The photochemical reaction that occurs when silver halide is exposed to light is extremely simple:

$$2AgBr(s) \overset{light}{\rightarrow} 2Ag(s) + Br_2(l)$$

Where light falls on silver bromide, metallic silver and bromine are formed by transfer of electrons from the bromide ions to the silver ions. Finely divided metallic silver is grey/black and a dark image is thus formed in proportion to the degree of exposure.

If we were to look down a microscope at a crystal of silver bromide before and after exposure to light, we might see something like Figure 19.55. The silver ions have been photochemically reduced to metallic silver, and the bromide ions have been oxidised to bromine.

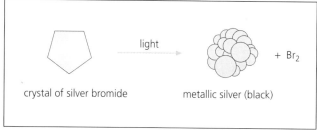

Figure 19.55 The photochemical reaction of silver bromide

Development

The conversion of silver ions to silver metal by shining light on a silver halide is not a particularly efficient process and it takes several minutes of exposure to intense light to obtain a visible image. As you can imagine, this would be very inconvenient in practice since it would require the person or object being photographed to remain still for that time. This is where the chemical process called **development** comes in. If the silver halide is given only a brief exposure to light, only a tiny fraction of the silver ions will be converted to metallic silver. The microscopic speck of silver which is formed is called a **latent image** (since it is so tiny it cannot be seen) and can be as small as four atoms of silver.

Looking down our microscope again at one of the crystals of silver bromide in our film we might imagine the view in Figure 19.56. In the first stage, the crystal of silver

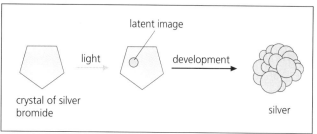

Figure 19.56 The reaction on brief exposure to light

bromide has received a brief exposure to light and a micro-scopic speck of silver (a latent image) has been formed. Although this speck of silver is infinitesimally small, it is sufficient to act as a catalyst in the next stage, when the film is removed from the camera and placed in a solution of a reducing agent called a developer. The developer reacts with (reduces) the silver ions in those crystals of silver bromide that contain latent images many thousands of times more rapidly than in those without. The overall result of development is an amplification of the effect of light by the chemical conversion of the exposed silver bromide to black metallic silver. Hydroquinone is a typical black-and-white developer. This is oxidised by silver halide which has been exposed to light to form quinone and silver metal during development:

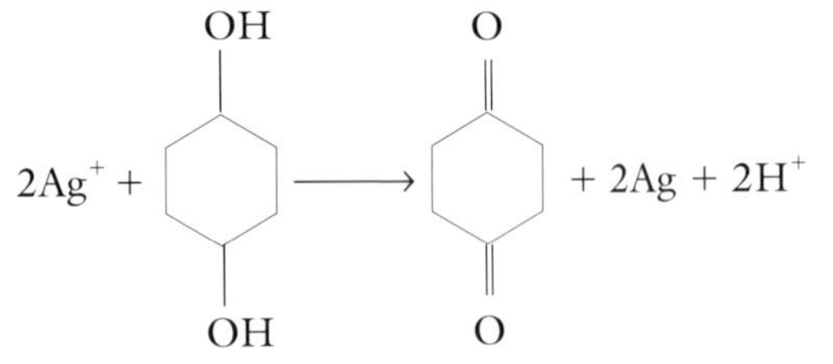

$$2Ag^+ + \text{(hydroquinone)} \longrightarrow \text{(quinone)} + 2Ag + 2H^+$$

hydroquinone   quinone

Colour developers are structurally similar and are, typically, p-phenylenediamines. When these are oxidised by silver halide a reactive intermediate called a quinonediimine is formed:

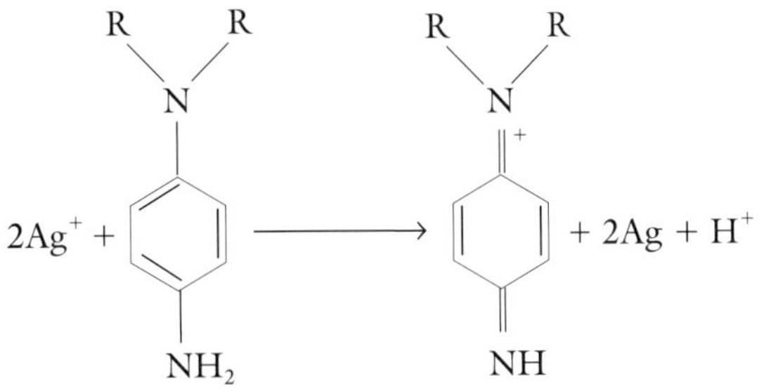

$$2Ag^+ + \text{(p-phenylenediamine)} \longrightarrow \text{(quinonediimine)} + 2Ag + H^+$$

p-phenylenediamine   quinonediimine

In black-and-white photography the image is composed of the developed metallic silver, but in coloured photographs the image is composed of dyes which are formed by an additional process. The reactive oxidised colour developer (quinonediimine) is allowed to react with a compound called a coupler which is incorporated into the photographic layer with the silver halide. A dye results from this reaction:

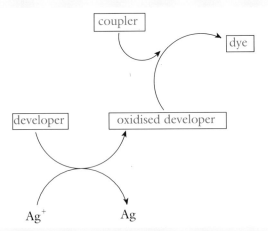

To generate a colour photograph it is necessary to form only three dyes coloured respectively yellow, magenta (bluish red) and cyan (greenish blue). The colour of the dye formed depends on the structure of the coupler. The reaction is illustrated here for the formation of a cyan dye:

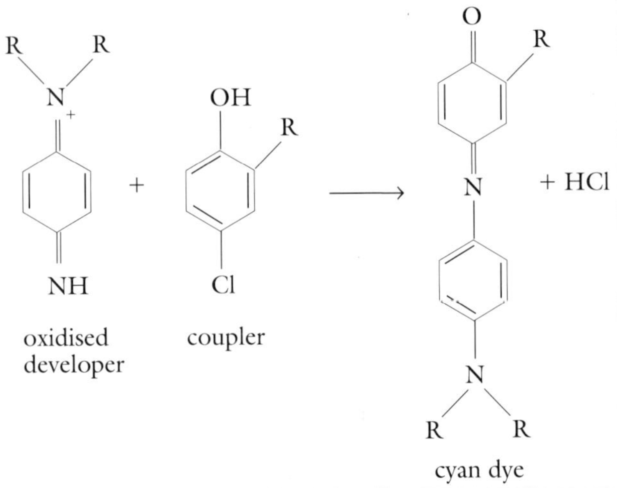

$$\text{(oxidised developer)} + \text{(coupler)} \longrightarrow \text{(cyan dye)} + HCl$$

oxidised   coupler
developer

cyan dye

Fixation

Obviously, those crystals of silver halide that have not been exposed to light do not undergo development and remain substantially unchanged. These unexposed crystals of silver halide must be removed (by a process called **fixation**), otherwise they too would darken when eventually exposed to light and the photographic image would become obliterated. The unexposed crystals of silver bromide are removed by converting them into water-soluble salts by reaction with a solution of sodium thiosulphate:

$$AgBr(s) + 2Na_2S_2O_3(aq) \rightarrow$$
(insoluble)   (fixer)
$$Na_3[Ag(S_2O_3)_2](aq) + NaBr(aq)$$
(soluble)

Bleaching

In black-and-white photography, the image is composed of the metallic silver which is formed during development. In coloured photographs, once dye formation has taken place, the silver must be removed. This is achieved by re-oxidising (**bleaching**) it back to silver ions using a solution of iron(III) ions:

$$Ag(s) + Fe^{3+}(aq) \rightarrow Ag^+(aq) + Fe^{2+}(aq)$$

The silver ions formed are then fixed out in the normal fashion to leave remaining only the yellow, magenta and cyan dyes that make up the colour photograph.

Questions

65 Record the redox changes (if any) which are undergone by silver atoms in the following stages:

a dissolving the bars of silver

b precipitating the silver bromide

c exposing the photographic film to light in a camera

d developing the film (black-and-white)

e fixing the film.

66 Where did catalysis fit into the story of photography? Could you hazard a guess as to which sort of catalysis (heterogeneous, homogeneous redox go-between, etc.) is operating?

67 What is the charge on the complex ion $[Ag(S_2O_3)_2]$?

68 It is known that bromide ions can 'win' in a battle for Ag^+ ions, when the opposition is water or ammonia. For instance, all the following reactions do happen:

$$[Ag(H_2O)_2]^+(aq) + 2NH_3(aq) \rightarrow [Ag(NH_3)_2]^+(aq) + 2H_2O(l)$$

$$[Ag(H_2O)_2]^+(aq) + Cl^-(aq) \rightarrow AgCl(s) + 2H_2O(l)$$

$$[Ag(H_2O)_2]^+(aq) + Br^-(aq) \rightarrow AgBr(s) + 2H_2O(l)$$

$$[Ag(NH_3)_2]^+(aq) + Cl^-(aq) \rightarrow \text{no change}$$

$$[Ag(NH_3)_2]^+(aq) + Br^-(aq) \rightarrow AgBr(s) + 2NH_3(aq)$$

$$AgBr(s) + \text{fixer} \rightarrow \text{as in the text}$$

From the evidence of the above reactions, place the two halide ions chloride and bromide and the three ligands water, ammonia and the thiosulphate ion in order according to their relative power to bind with silver ions.

69 The developed image as described in the text would be a negative image, with the blacks and whites of the original scene reversed. How can we tell this from the text, and how do we end up with a positive picture?

70 As you can visualise from the text, if you looked at a developed and fixed black-and-white photographic negative under a microscope it would consist of dots, each one of which had once been a crystal of silver bromide. You might have gained the impression that there were only two types of outcome for these original crystals – if they had been 'hit' by light they would develop to black, and if they had not been 'hit' by light they would be dissolved away by the fixer. Yet photographs include parts of every shade of grey, as long as you do not develop for too long. Can you suggest how a crystal can go just a bit black?

Summary

- Almost all d-block elements have **electron structures** that feature an outer pair of s-electrons, within which there is **a partially or totally filled d sub-shell**.

- There are minor exceptions to this pattern. Certain d-block elements (e.g. chromium and copper) opt for the arrangements d^5s^1 and $d^{10}s^1$ rather than, respectively, d^4s^2 and d^9s^2. This is thought to reflect the **stability of half-full and full d sub-shells**.

- The d-block elements show **less variation** in properties across successive elements than do the elements in the 'typical' groups. The d-block elements share certain common characteristics, as listed below:

1 They are **metals**.
2 They have **high densities** and **moderate to low reactivities** compared with the s-block metals.
3 They have compounds that show **variable oxidation number** (with the exception of the first and last members of a row).
4 They have a strong tendency to form **complexes** with ions or molecules that donate electron pairs. Such species are called **ligands**.
5 They give rise to **coloured compounds** (again with the exception of the first and last members of a row), in which the colours depend upon the oxidation number and the identity of the ligand.
6 They show a diverse range of **catalytic activity**.

- **Metallic character** derives from the relatively low ionization enthalpies of the outer s^2 pair of electrons. In fact the ionization enthalpies of the elements show surprisingly little variation, considering that there is a difference of nine protons between the nuclei of elements at extreme ends of a row. We must conclude that the increased nuclear attraction, as felt by the s^2 pair, is offset almost exactly by the increased shielding of the inner d sub-shell.

- Nearly all d-block elements have compounds in which they exist as simple cations. With the exception of elements near the beginnings or ends of rows, there exists **both a 2+ and a 3+ cation**. The choice of two charges arises because electron loss does not have to cease after the removal of the outer s^2 pair; it is possible for a 2+ ion to lose one of the inner d-electrons, since they are less firmly held than, say, the inner p-electrons in Ca^{2+}.

- However, the influence of the nucleus over the d-electrons is reflected in the fact that, as the end of a row is approached, the 3+ cation ceases to exist, due to the impossibility of removal of any of the now quite firmly held d-electrons. (Examples of non-existent cations are Cu^{3+} and Zn^{3+}.) Conversely, at the beginning of the row, the third, d, electron is so easily removed that the +2 oxidation state becomes non-existent (for example, there is no such thing as Sc^{2+}).

- In the middle regions of each row, there is a fairly smooth variation between the extremes mentioned above. From left to right the +2 oxidation state gets more stable, while the +3 gets less stable. In other words the 3+ cation passes from being stable on the left (e.g. Cr^{3+}) to being a strong oxidising agent on the right (e.g. Co^{3+}), while the 2+ cation is a strong reducing agent on the left (Cr^{2+}) and a stable species on the right (Co^{2+}).

- To make the **trihalides** of the d-block elements, direct reaction of the element with the halogen is needed. The **dihalides** are obtained by reacting the element with hydrogen halides.

- The structures of the d-block halides imply that the bonding is very impure ionic, or even polar covalent. A widely occurring structure for crystalline d-block dihalides is the **cadmium iodide** CdI_2 layer structure, in which adjacent planes of halogen atoms seem to be attached to each other by van der Waals' forces.

- **Oxidation states higher than +3** occur widely in the d-block, especially in the middles of the rows (where there are enough d-electrons to generate the high numbers, and that are still 'removable'). Electron removal is not total in these compounds; the high oxidation numbers are due to polar-covalent partial removal, as in other molecules with highly oxidised atoms like sulphur trioxide, SO_3, and iodine heptafluoride, IF_7.

- The **high oxidation number compounds** of the d-block are nearly always oxides, oxo acids, oxo anions and fluorides. These oxides have much in common with acidic oxides like those of sulphur. For instance chromium(VI) oxide, CrO_3, mimics sulphur trioxide by giving rise to chromic(VI) acid, H_2CrO_4, in water, and to a set of anionic salts (the chromates) analogous to the sulphates. In contrast chromium(III) oxide, Cr_2O_3, is of basic character, and dissolves in acids to give salts in which the metal is the cation, such as chromium(III) sulphate, $Cr_2(SO_4)_3$.

- The high oxidation states of the d-block elements are associated with compounds used as **oxidising agents**, like the manganate(VII) and dichromate(VI) ions. These agents are particularly potent in acidic media, where the H^+ ions assist by 'mopping up' the oxygen atoms.

Complexes (co-ordination compounds)

- **Complexes** are formed when molecules that donate lone pairs enter into dative covalent bonding with a d-block element or, more normally, its ion. The lone pairs of electrons are received into vacant orbitals in the relevant d, s or p sub-shells. The electron-donating species are called **ligands**, a name which signals the relative permanence of the bonding, as compared to an ion–dipole attraction.

- The **co-ordination number** of most complexes is 4 or 6, and the geometries associated with these numbers are octahedral (for 6 ligands) and either tetrahedral or square planar (for 4 ligands).

- Many possibilities for **isomerism** exist in complex chemistry. Isomerism can derive from three structural variations:

1 which ions are ligands and which are free

2 which ligands are *cis* and which *trans*

3 whether the possibility exists for chirality.

- Isomers can differ quite markedly in properties. They may, for instance, show different colours or different electrical conductivities.

- Certain ligands can offer more than one electron pair to the central cation, from separate sites on the ligand molecule. Such ligands are called **polydentate**. If a polydentate ligand is attached to a single cation (rather than linking two) it is called, in addition, a **chelating agent**.

- Competition can be set up between ligands for the same cations, in which case **substitution reactions** can occur, with one ligand displacing another. Competition can also exist between ligands and conventional anions, so that a ligand can replace a cation from within a precipitate of one of its ordinary binary compounds. This is what happens when silver chloride dissolves in ammonia – the ligand ammonia removes the silver ions from the silver chloride lattice, to make them into $[Ag(NH_3)_2]^+$ ions. (Something similar also occurs during the fixing reaction in photography.)

- The strengths of ligands, as tested in competition, seem to correlate with their strengths as bases. For instance ammonia molecules will normally replace water ligands from around a cation. This correlation is not unexpected, since both basicity and complexation involve lone pair donation. However, polydentate ligands have an added advantage over monodentate ones, even though the base strength of the donating atoms is roughly the same (as for example when 1,2-diaminoethane replaces ammonia). This advantage can be traced to entropy effects, because the replacement of a number of single ligands by one large molecule causes an increase in the number of free molecules, and thus in the entropy of the solution.

- Polydentate ligands are often used for their competitive edge in solubilising cations, either from other ligands or from insoluble ionic compounds. Edta for instance can mobilise iron cations from insoluble hydroxides in the soil, to the great benefit of plants.

- The artificial complex cisplatin, an anti-cancer drug, involves itself with cell division, possibly by shedding its two chloride ligands, replacing them with water, and getting hydrogen bonded to DNA strands. The natural complexes of iron in the blood of mammals can take on and then shed dioxygen ligands, and thus act as a delivery system for taking oxygen to cells all over the organism.

Colour in d-block compounds

- **Colour** in the compounds of d-block elements can be traced to a **split in the energy levels of the d sub-shell**. The model we used to explain this splitting relied on an 'electrostatic' view of the ligand–cation association, rather than a dative covalent one. It proposed that certain d-orbitals within the set of five are more energetically 'comfortable' because they are not so close to, say, a set of octahedrally arranged ligands. The less 'comfortable' d-orbitals have lobes which take their electrons right up to the ligand positions, and these are the high-energy ones, due to electron–electron repulsion.

- The actual colours are caused by the absorption of photons, which are used to promote transitions of electrons from lower energy to higher energy d-orbitals. The exact nature of the absorptions can be studied by means of **ultraviolet/visible spectroscopy**.

- UV/visible spectroscopy reveals that absorptions occur across broad bands of wavelengths, due to the complicating effects of vibrational and rotational energy levels on top of the electronic ones. The significant measurable aspects of a UV/visible spectrum are the **wavelength of maximum absorption** λ_{max}, and the **absorbance**.

- The value of λ_{max} is found to be dependent upon the choice of ligand for a given cation. There seems to be a correlation between the most competitive ligands and the value of λ_{max}. Specifically, the more competitive ligands seem to give rise to bigger splittings in the energy levels of the d-orbitals, and so to shorter wavelengths (higher frequencies) of peak absorption. This order of λ_{max} between ligands is called the **spectrochemical series**.

- The absorbance (y-axis) of a species is proportional to its concentration. The relationship is:

$$A = \varepsilon cl$$

where A is the absorbance, ε is the **extinction coefficient**, c is the concentration of the absorbing species and l is the path length of light through the cell (usually 1 or 10 cm). This equation shows us that a knowledge of the absorbance and the extinction coefficient of a species will enable its concentration to be measured.

Catalysis by d-block compounds

- d-block compounds display three distinct modes of catalysis, which lower the activation enthalpy of the transition state of the catalysed reaction.

- **'Go-between' catalysis** relates only to redox reactions in aqueous solutions, where the d-block catalyst exploits its ability to exist in two oxidation states to act, in effect, as an electron-transfer go-between, as summarised in Figure 19.5 (p. 477). The choice of catalyst is constrained by the positions of the two reactants and the catalyst in the electrochemical series – the standard electrode potential of the catalyst half-cell must lie between those of the two reactants. This does not guarantee catalysis, but it is the necessary condition for thermodynamic feasibility.

- **Homogeneous co-ordination catalysis** – in this method, the reactants are temporarily held as ligands around a central cation, and reaction takes place between ligands. The method has had great application in the polymerisation of alkenes, where its special value is in providing straight chains and consistent stereochemistries. The pioneer workers in the field of alkene polymerisation were Karl Ziegler and Giulio Natta, and their names have become closely linked with this style of catalysis.

- **Heterogeneous catalysis** is the most important form of catalysis in industry. The catalyst is a solid d-block element or compound, and the reactants are adsorbed on to the catalyst surface, from either the liquid or gaseous phase. The mechanisms of these catalyses all follow the same general three-step pattern, whatever the detailed variations:

1 reactants adsorbed on to surface
2 reaction occurs
3 products desorbed.

The activation enthalpy of the reaction is spread over three 'humps', corresponding to the three stages, and is thus lowered relative to the single-stage reaction profile.

- Heterogeneous catalysis is involved in the production of nearly all our bulk chemicals, including ammonia, sulphuric and nitric acids, cracked hydrocarbons, buta-1,3-diene and methanol. It is also the form of catalysis in the automobile catalytic converter.

- **Auto-catalysts** are usually **platinum based**, and are put in the exhaust stream of cars, to bring about reactions between the exhaust gases themselves. The oxides of nitrogen (NO_x) and surplus dioxygen are made to oxidise the carbon monoxide and unburned hydrocarbons.

- Engine designers are aiming to produce **lean-burn engines**, which have a low fuel:air ratio. This regime is economical, and good for suppressing carbon monoxide, carbon dioxide and hydrocarbon emissions, but leaves problems with the oxides of nitrogen. The use of a percentage of **rhodium** along with the platinum provides a more effective catalyst for reactions featuring oxides of nitrogen. There are still problems in the period when the engine runs cold, after start-up.

- **Photography** exemplifies all the aspects of d-block chemistry. The original photochemical reaction, when the light 'hits' the film, is redox. The development reaction is an example of catalysis, and the fixing reaction uses a ligand ($S_2O_3^{2-}$) to draw unused silver ions from within a bromide crystal lattice. Ironically, in view of the links between the d-block and colour, the colour aspect of photography is relatively independent of d-block chemistry.

A-level Exam Questions

The following abbreviations have been used for the examinations boards:

AEB   Associated Examining Board
HK   Hong Kong Examinations Authority
JMB   Joint Matriculation Board
NEAB   Northern Examinations and Assessment Board
NICCEA   Northern Ireland Council for the Curriculum Examinations and Assessment
O & C   Oxford and Cambridge Schools Examination Board
UCLES   University of Cambridge Local Examinations Syndicate
ULEAC   University of London Examinations and Assessment Council

We are very grateful to the above boards for permitting us to reproduce the questions in this section.
M denotes a paper from a modular syllabus and (part) means only a part of the original question is printed here.

*Please note that responses to questions marked by an asterisk are based on the scheme below. For each question, **one** or **more** of the responses given is (are) correct. Decide which of the responses is (are) correct. Then choose:

A   if **1**, **2** and **3** are correct
B   if **1** and **2** only are correct
C   if **2** and **3** only are correct
D   if **1** only is correct
E   if **3** only is correct.

Directions Summarised

A	B	C	D	E
1, 2, 3	1, 2	2, 3	1	3
correct	only	only	only	only

---

1 Industry uses millions of tonnes of sulphuric acid, H_2SO_4, in the U.K. each year. This is used in the manufacture of many important products such as paints, fertilisers, soap, plastics, dyestuffs and fibres. The sulphuric acid may be prepared from sulphur in a 3-stage process.

Stage 1
The sulphur is burnt in oxygen to produce sulphur dioxide, SO_2.

$$S + O_2 \rightarrow SO_2$$

Stage 2
In the presence of a catalyst, the sulphur dioxide reacts with more oxygen to form sulphur trioxide, SO_3.

$$2SO_2 + O_2 \rightarrow 2SO_3$$

Stage 3
The sulphur trioxide is dissolved in concentrated sulphuric acid to form 'oleum', $H_2S_2O_7$, which is then reacted with water.

a i) Write a balanced equation for the formation of sulphuric acid from oleum. (1)
ii) How many tonnes of sulphur are required to produce 70 tonnes of sulphuric acid?
(A_r: H, 1.0; O, 16; S, 32) (2)

b A $50.0\,cm^3$ sample of sulphuric acid was diluted to $1.00\,dm^3$. A sample of the diluted sulphuric acid was analysed by titrating with aqueous sodium hydroxide. The reaction is

$$H_2SO_4 + 2NaOH \rightarrow Na_2SO_4 + 2H_2O$$

In the titration, $25.0\,cm^3$ of $1.00\,mol\,dm^{-3}$ aqueous sodium hydroxide required $20.0\,cm^3$ of sulphuric acid for neutralisation.

i) Calculate how many moles of sodium hydroxide were used in the titration. (1)
ii) Calculate the concentration of the diluted sulphuric acid. (2)
iii) What was the concentration of the original sulphuric acid? (1)

(UCLES, 1994 (M, specimen))

2 A $1.55\,g$ sample of hydrated ethanedioic acid, $H_2C_2O_4 \cdot xH_2O$, was dissolved in water and the solution made up to $250\,cm^3$. When $25.0\,cm^3$ of this diluted solution was titrated with $0.100\,mol\,dm^{-3}$ aqueous sodium hydroxide, $24.6\,cm^3$ of the alkali were required for neutralisation.

The chemical equation for the reaction that occurred is

$$H_2C_2O_4 + 2NaOH \rightarrow Na_2C_2O_4 + 2H_2O$$

a Calculate how many moles of
i) sodium hydroxide were used in the titration, (1)
ii) ethanedioic acid reacted in the titration, (2)
iii) ethanedioic acid were present in the original weighed sample. (1)

b Calculate the relative molecular mass of anhydrous ethanedioic acid. (1)

c Using your answers in (**a**) and (**b**), calculate the value of x in $H_2C_2O_4 \cdot xH_2O$. (2)

(UCLES, 1996/7 (M, specimen))

3 Poly(phenylethene), known as polystyrene, is a widely used polymer. The starting material for its manufacture is benzene which is converted into ethylbenzene and then, by catalytic dehydrogenation, into phenylethene.

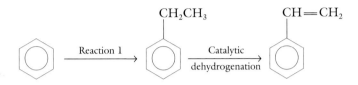

The ion-exchange resin formed from poly(phenylethene) can be used to remove metallic cations from an aqueous solution. Draw a clearly labelled diagram of a suitable apparatus for carrying out this procedure. (3)

(ULEAC, 1990 (Nuffield) (part))

4 A power station produces 55 000 tonnes of gypsum per year: 1 tonne = 10^3 kg.

a How many moles of gypsum are produced per year?
(Relative atomic masses: H = 1, O = 16, S = 32, Ca = 40)

b What volume of sulphur dioxide was absorbed in the production of 55 000 tonnes of gypsum?
(1 mol of sulphur dioxide at this temperature has a volume of 24 dm^3.) (2)

(ULEAC, 1990 (Nuffield) (part))

5 By carrying out appropriate calculations, predict what will happen at each stage of the following experiment and describe what you expect to see.

a 1 g of calcium hydroxide was mixed with 100 cm^3 of water at room temperature.

b 10 cm^3 of 2 mol dm^{-3} sulphuric acid was then added to the mixture.

c 5 drops of a full range universal indicator was then added.

Describe an experimental procedure you could use on your resulting mixture to obtain a value for the solubility of calcium sulphate in water.

(ULEAC, 1991 (Nuffield) (part))

6 A chemical plant releases 2000 mol of hydrochloric acid per week in an effluent stream. Which one of the following is the minimum mass (in kg) of calcium carbonate needed per week in order to neutralise the acid?

A 1

B 2

C 100

D 200

E 1000

(JMB, 1992)

7 a Using hydrogen as an example, explain the processes which give rise to an emission line spectrum. Explain why several series of emission lines exist for atomic hydrogen.(3)

b The energy change ΔE which produces one of the lines in the emission spectrum of hydrogen is 4×10^{-19} J.

Using the equations $\Delta E = hf$; $c = \lambda f$, calculate the wavelength λ of the emission line and state in what region of the spectrum it would occur.

(In these equations, h is the Planck constant, f is the frequency of the emission and c is the speed of light.) (4)

(UCLES, 1996/7 (specimen) (part))

8 A sample of oxygen consisting mainly of the isotope oxygen-16 was enriched with oxygen-18. The composition of the mixture was 75.0% oxygen-16 and 25.0% oxygen-18, by volume.

The oxygen sample reacted with sodium as follows:

$$4Na(s) + O_2(g) \rightarrow 2Na_2O(s)$$

a Complete the table below to show the composition of some of the species involved in the reaction. (A_r: ^{16}O, 16.0; ^{18}O, 18.0; ^{23}Na, 23.0)

species	protons	neutrons	electrons
$^{23}_{11}$Na			
$^{16}_{8}$O			
$^{18}_{8}$O^{2-}			

(4)

b Write down the electronic configuration of:
i) a sodium atom;
ii) an oxide ion. (2)

c State **three** physical properties of sodium oxide. (2)

d Some carbon-12 was burned in another sample of the oxygen mixture. The carbon dioxide produced gave the following mass spectrum.

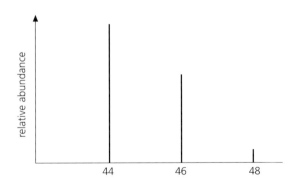

Identify the peaks shown. (2)

(UCLES, 1996/7 (M, specimen) (part))

9 a Describe any one experimental method to determine the first ionization energy of an element.

b Arrange the following elements in order of increasing first ionization energy: F, Ne and Na. Explain your order.

c Describe and explain the variation of the successive ionization energies of potassium.

(8)

(HK, 1992 (part))

10 Five ionization energy values (in $kJ\,mol^{-1}$) are listed below:

834   869   1008   1170   376

These values are most likely to be

A successive ionization energies for the element of atomic number 5

B successive ionization energies for a transition element with 4 electrons in the d sub-level

C the first ionization energies for five elements from Group 1, listed in order of increasing relative atomic mass

D the first ionization energies for elements with atomic numbers 1–5

E the first ionization energies for successive elements in Groups 5, 6, 7, 0 and 1

(ULEAC, 1992 (Nuffield))

11 In which one of the following pairs is the first ionisation energy of element **Y** greater than that of element **X**?

electronic configuration of element **X**	electronic configuration of element **Y**
A $1s^1$	$1s^2$
B $1s^2$	$1s^2 2s^1$
C $1s^2 2s^2$	$1s^2 2s^2 2p^1$
D $1s^2 2s^2 2p^3$	$1s^2 2s^2 2p^4$
E $1s^2 2s^2 2p^6$	$1s^2 2s^2 2p^6 3s^1$

(JMB, 1992)

12 The formula for the amino acid alanine may be displayed as

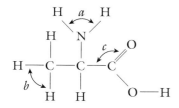

What are the approximate values of the bond angles, *a*, *b* and *c*, in the compound?

	a	*b*	*c*
A	106°	109°	110°
B	106°	109°	120°
C	120°	109°	110°
D	120°	90°	110°
E	120°	90°	120°

(ULEAC, 1990 (Nuffield))

*13 From the dot-and-cross diagram of ethanamide:

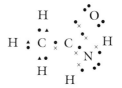

it may be deduced that ethanamide contains

1 single covalent bonds

2 double covalent bonds

3 dative covalent bonds

(ULEAC, 1990 (Nuffield))

*14 The electron density map for the hydrogen molecule ion, H_2^+, is shown below.

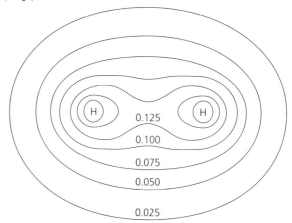

Contours are in electrons per $10^{-30}\,m^3$.

Which of the following statements about the H_2^+ species are correct?

1 The bond between the hydrogen atoms in H_2^+ is an ionic bond.

2 The electron density along a line joining the two centres of the hydrogen nuclei in H_2^+ will always be greater than at corresponding points along a similar line for H_2.

3 The bond dissociation energy of H_2^+ has a positive sign.

(ULEAC, 1991 (Nuffield))

15a This method [of reducing rutile, TiO_2, by heating it with carbon] is unsatisfactory because of the formation of titanium carbide. In this compound the carbon atoms occupy the octahedral sites in the close-packed structure of the metal. Explain the term *octahedral site*, illustrating your answer with a simple diagram. (2)

b The unit cell of rutile, TiO_2, is shown below.

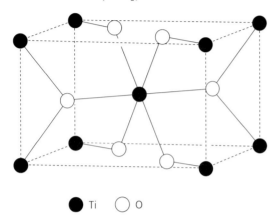

● Ti   ○ O

i)   How many titanium atoms are there in the unit cell?
ii) How many oxygen atoms are there in the unit cell?
iii) Explain whether this is consistent with the formula of rutile. (3)

(ULEAC, 1992 (Nuffield) (part))

*16 The presence of a protein or other polypeptide is shown by the formation of

1 a red precipitate on boiling with Fehling's solution

2 a deep blue colour with iodine solution

3 a purple colour with sodium hydroxide and a very little copper(II) sulphate solution

(ULEAC, 1993 (Nuffield))

17a Complete the table below, in each case giving the number of bonding pairs and non-bonding pairs of electrons around the central atom. (3)

Compound	Formula	Number of bonding pairs of electrons	Number of non-bonding pairs of electrons
Ammonia	NH_3		
Boron trichloride	BCl_3		
Water	H_2O		

b Draw clear diagrams to show the shape of the molecules in each of the compounds in (a). (3)

c Briefly explain why the shapes of the molecules of boron trichloride and ammonia are different. (3)

d Describe, with the aid of a diagram, the nature of the bonding **between** molecules of water. (3)

(AEB, 1994 (part))

18 This question refers to the first 20 elements in the Periodic Table. You should refer to the copy of the Periodic Table in the Data Booklet. All descriptions apply to room temperature and pressure.

From the first 20 elements, those of proton (atomic) numbers 1–20, give the symbol for:

a an element existing as free atoms. (1)

b an element that forms a chloride of formula XCl which dissolves readily in water to form a neutral solution. (1)

c the element with the highest first ionisation energy. (1)

d an element that forms an oxide with a giant molecular structure (1)

e an element that forms two acidic oxides of formulae XO_2 and XO_3. (1)

f an element that forms a liquid chloride which has a tetrahedral molecule and which reacts with water. (1)

g an element that forms two chlorides in which the oxidation states of the element are +3 and +5 respectively. (1)

(UCLES, 1994 (M, specimen))

*19

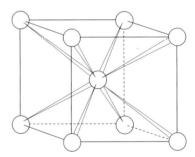

The above diagram of the structure of an element shows:

1 a unit cell containing the equivalent of 3 atoms

2 a unit cell in which each atom has 8 co-ordination

3 a unit cell of a body-centred cubic structure

(ULEAC, 1990 (Nuffield))

*20 On going from Group 1 to Group 3 across a period, the chlorides of the elements become less ionic and more covalent.

Factors which contribute to this include

1 a more negative lattice energy

2 the increase in charge on the positive ion

3 the decrease in size of the positive ion

(ULEAC, 1991 (Nuffield))

*21 The elements of the second short period of the Periodic Table are, in sequence,

Na Mg Al Si P S Cl Ar

In passing across the period from sodium to chlorine

1 the structures of the elements change from giant metallic lattices, through giant structures of covalently bonded atoms, to molecules and atoms

2 the melting points of the elements increase from Na to S

3 the first ionization energies of the elements decrease to a minimum at Si and then increase

(ULEAC, 1993 (Nuffield))

22 Mendeleev is usually regarded as the founder of the modern Periodic Table. His table of 1869 can be represented:

Group / Period	I	II	III	IV	V	VI	VII	VIII
1	H							
2	Li	Be	B	C	N	O	F	
3	Na	Mg	Al	Si	P	S	Cl	
4	K Cu	Ca Zn	* *	Ti *	V As	Cr Se	Mn Br	Fe Co Ni
5	Rb Ag	Sr Cd	Y In	Zr Sn	Nb Sb	Mo Te	* I	Ru Rh Pd

* indicates a blank left by Mendeleev in his table

a i) Which group of elements from a modern Periodic Table is missing from Mendeleev's table? (1)
ii) Suggest a reason for this. (1)

b Mendeleev listed the elements in order of ascending atomic mass, but he reversed the positions of tellurium and iodine. (A_r(Te) = 128; A_r(I) = 127)
i) Explain why he thought it necessary to do this. (1)
ii) What property do we use today to determine the position of an element in the Periodic Table? (1)

c Mendeleev grouped copper and silver with hydrogen and the alkali metals. Describe *one* chemical feature which they all have in common. (1)

d Give the symbol for an element from Mendeleev's period 3 (one in each case) which:
i) forms a chloride XCl_3 (1)
ii) has an outer electron arrangement s^2p^3 in its atoms (1)
iii) forms a basic oxide XO. (1)

e Mendeleev left blanks (denoted by *) in his periodic table. Why did he do this? (1)

f Predict for the element shown by a star (*) in group IV:
i) the structure type of the element (metallic, small molecules, etc.) (1)
ii) a formula for its oxide. (1)

(O & C, 1993 (Salters))

23 Which one of the following ions has the greatest polarising power?

A Cl^-
B S^{2-}
C Li^+
D Mg^{2+}
E Al^{3+}

(JMB, 1992)

24 Clathrates are materials in which molecules are trapped in 'cages' formed within the open crystal lattice structure of such substances as ice and 1,4-dihydroxybenzene. Substances found in clathrates include some noble gases, methane and oxygen.

a i) Complete the structural formula of 1,4-dihydroxybenzene.

ii) Draw a diagram to show how strong intermolecular forces might arise between two adjacent molecules of 1,4-dihydroxybenzene. Mark and give the values of two different bond angles outside the benzene ring.
iii) Explain, in terms of electronic structure, how water molecules combine to form an open crystal lattice structure in ice.
iv) Describe how 1,4-dihydroxybenzene molecules might form a 'cage' to trap a molecule such as oxygen. (8)

b Noble gases, except helium, also form clathrates with 1,4-dihydroxybenzene.
i) Which type of intermolecular force is likely to be involved between the noble gas and the 1,4-dihydroxybenzene molecules in the clathrate?
ii) Suggest why helium does not form such a clathrate. (2)

(ULEAC, 1990 (Nuffield))

25 The boiling point and density of butan-1-ol are 117.9 °C and 0.81 g cm^{-3} respectively, and those of its isomer 2-methylpropan-2-ol are 82.2 °C and 0.79 g cm^{-3} respectively. Account for these differences. (2)

(HK, 1993 (part))

26 The melting point of paraffin wax (a mixture of saturated hydrocarbons which have high relative molar masses) is determined by

A covalent bonds within the hydrocarbon molecules

B covalent bonds between the hydrocarbon molecules

C ionic bonds between the molecules

D hydrogen bonds between the molecules

E van der Waals forces between the molecules

(O & C, 1994)

27a Some elements and compounds have crystal structures composed of discrete molecules held together in a regular array (molecular crystals).

   i) Taking as examples bromine, $Br_2(s)$, and bromomethane, $CH_3Br(s)$ (both crystalline solids at low temperature), describe and explain the various forces which are responsible for holding molecules together in such crystals. (4)

   ii) Give a reason for **one** physical property associated with molecular crystals. (2)

b Some substances have giant lattice crystal structures in which discrete molecules cannot be identified. Examples of such substances are silicon dioxide and sodium chloride. For each of these substances:

   i) sketch or describe their crystalline structures, (4)

   ii) state the type of chemical bonding in the crystals, and (2)

   iii) explain why silicon dioxide adopts one type of bonding and sodium chloride another. (4)

(O & C, 1994)

28a List the main types of intermolecular forces associated with simple molecules. Indicate the relative strength of these forces and state how each type arises. (6)

b The compounds in each of the following pairs have similar relative molecular masses. By consideration of the forces between particles in each compound, predict which member of the pair has the higher boiling point.

   i) butane and 2-methylpropane

   ii) 2-methylpropene and propanone

   iii) propanoic acid and 2-aminoethanoic acid *(glycine)*(12)

c Propanol, propanal and butane have similar relative molecular masses. Explain why propanol is very soluble in water, propanal is less soluble and butane is insoluble. (6)

(NEAB, 1994 (part))

29 Sulphur dioxide is added to wines as an antioxidant as it is oxidized by air in preference to ethanol. It also kills unwanted bacteria. If too little sulphur dioxide is used the wine may become oxidized, or affected by unwanted bacteria, while too much sulphur dioxide may affect the flavour.

It is therefore important to be able to determine the concentration of sulphur dioxide in wine with reasonable accuracy. One method is by titration with a solution of iodine of known concentration, since one mole of sulphur dioxide is oxidized by one mole of iodine molecules.

a Iodine is reduced by the sulphur dioxide.

   i) What is the iodine reduced to by the sulphur dioxide?

   ii) What is the sulphur dioxide oxidized to by the iodine?

   iii) Write a balanced equation, with state symbols, for the reaction of iodine solution with sulphur dioxide. (4)

b 50.0 cm^3 of a white wine was acidified with sulphuric acid, and then titrated against 0.0100 M iodine solution, using a suitable indicator. Exactly 2.50 cm^3 of iodine solution was required.

   i) Suggest a suitable indicator for this titration.

   ii) Calculate the number of moles of sulphur dioxide in the 50.0 cm^3 sample of white wine.

   iii) Hence calculate the concentration of sulphur dioxide in this wine, and the volume of sulphur dioxide dissolved in a 1 dm^3 bottle of this wine at room temperature and atmospheric pressure. (Volume of 1 mole of gas at room temperature and atmospheric pressure is 24.0 dm^3)

   iv) What difficulty might make this method of limited suitability for the analysis of red wines? (5)

(ULEAC, 1989 (Nuffield) (part))

30 This question is about sulphur dioxide. In the laboratory, the adsorption of sulphur dioxide may be demonstrated by passing sulphur dioxide through active charcoal, but this is not a practical method in industry. One way of tackling the acid rain problem, used in West German power stations, is to pass the waste gases containing sulphur dioxide through an aqueous suspension of limestone.

The overall reaction is

$$2SO_2(g) + 2CaCO_3(s) + O_2(g) \rightarrow 2CaSO_4(aq) + 2CO_2(aq)$$

Gypsum is then crystallized out as $CaSO_4 \cdot 2H_2O(s)$. 1.2 million tonnes of gypsum are produced per year in West Germany by this method.

a Describe the changes you would expect to *see* when sulphur dioxide is passed through an aqueous suspension of limestone. (1)

b Sulphur dioxide can be detected by the reduction of $Cr_2O_7^{2-}$ ions to Cr^{3+} ions.

   i) Describe the changes you would expect to see when sulphur dioxide is passed through a solution of $Cr_2O_7^{2-}$ ions.

   ii) What is the oxidation number of chromium in the ion, $Cr_2O_7^{2-}$?

   iii) In the reaction, the oxidation number of sulphur increases from +4 to +6. Suggest the likely product of the oxidation of sulphur dioxide and hence deduce the equation for this reaction. (5)

c **i)** What does the term *adsorption* mean?

   ii) You are provided with a small cylinder containing sulphur dioxide. Draw a labelled diagram to show how you could measure out 100 cm^3 of sulphur dioxide and then find the proportion adsorbed by 10 g of active charcoal.

   iii) How would you know when adsorption was complete? (5)

d The amount of sulphur dioxide adsorbed by the active charcoal can also be determined by a titration method.

The 10 g of active charcoal containing the sulphur dioxide was added to 1000 cm^3 of iodine solution of concentration 0.00500 mol of I_2 per dm^3.

20.0 cm^3 portions of this solution were then titrated with 0.0100 mol dm^{-3} sodium thiosulphate solution: 11.6 cm^3 were required for complete reaction.

The relevant equations are:

$$SO_2(g) + I_2(aq) + 2H_2O(l) \rightarrow 2I^-(aq) + SO_4^{2-}(aq) + 4H^+(aq)$$

$$I_2(aq) + 2S_2O_3^{2-}(aq) \rightarrow 2I^-(aq) + S_4O_6^{2-}(aq)$$

i) Calculate the number of moles of sodium thiosulphate in $11.6\ cm^3$ of its solution.
ii) Calculate the total number of moles of iodine, I_2, which reacted with the sodium thiosulphate.
iii) Deduce the total number of moles of iodine which reacted with the sulphur dioxide.
iv) Hence calculate the number of moles of sulphur dioxide present in 10 g of active charcoal. (4)

(ULEAC, 1990 (Nuffield) (part))

*31 A potassium compound is known with the formula K_2FeO_4.

Which of the following statements about this compound are correct?
1 The iron has an oxidation number of +6.
2 The compound is likely to be a good reducing agent.
3 The compound is likely to be a white solid.

(ULEAC, 1990 (Nuffield))

32 $50.0\ cm^3$ of $0.100\ mol\ dm^{-3}$ potassium peroxodisulphate(VI) ($K_2S_2O_8$) solution was added to excess aqueous potassium iodide:

$$S_2O_8^{2-}(aq) + 2e^- \rightarrow 2SO_4^{2-}(aq)$$

$$2I^-(aq) \rightarrow I_2(aq) + 2e^-$$

What volume of $0.100\ mol\ dm^{-3}$ sodium thiosulphate solution is required to react exactly with the iodine released?

$$I_2(aq) + 2S_2O_3^{2-}(aq) \rightarrow 2I^-(aq) + S_4O_6^{2-}(aq)$$

A $1.5\ cm^3$

D $100.0\ cm^3$

B $25.0\ cm^3$

E $200.0\ cm^3$

C $50.0\ cm^3$

(ULEAC, 1993 (Nuffield))

33 Iodate ions in the presence of acid oxidise sulphite ions to sulphate ions. Which one of the following equations represents this reaction?

A $IO_3^- + 2SO_3^{2-} + 2H^+ \rightarrow I_2 + 2SO_4^{2-} + H_2O$

B $4IO_3^- + SO_3^{2-} + 22H^+ \rightarrow 2I_2 + SO_4^{2-} + 11H_2O$

C $2IO_3^- + SO_3^{2-} + 2H^+ \rightarrow I_2 + SO_4^{2-} + H_2O$

D $2IO_3^- + 5SO_3^{-2} + 2H^+ \rightarrow I_2 + 5SO_4^{2-} + H_2O$

E $2IO_3^- + SO_3^{-2} + 10H^+ \rightarrow I_2 + SO_4^{2-} + 5H_2O$

(NICCEA, 1994)

34a What is meant by the term *standard enthalpy change of combustion*? (2)

b Write a balanced equation for the complete combustion of ethanol, C_2H_6O. (1)

c When 1.00 g of ethanol was burned under a container of water, it was found that 100 g of water was heated from $15\ °C$ to $65\ °C$. The process was known to be only 70% efficient.

Use these data and values from the *Data Booklet* to calculate the enthalpy change of combustion per mole of ethanol. (4)

d Using the value you have calculated in (c) and the following data, calculate the enthalpy change of formation of ethanol from its elements. (3)

enthalpy change of combustion of carbon = $-393.5\ kJ\ mol^{-1}$

enthalpy change of combustion of hydrogen = $-285.8\ kJ\ mol^{-1}$

e Based on the qualitative use of bond energy data given in the *Data Booklet*, suggest briefly, with a reason, whether the combustion of ethanol is an exothermic process. (2)

(UCLES, 1996/7 (specimen))

35 A mixture of 0.50 mol of ethanoic acid and 1.00 mol of ethanol was shaken for a long time to reach equilibrium. The whole mixture was titrated quickly with $1.00\ mol\ dm^{-3}$ sodium hydroxide and $80\ cm^3$ of alkali were required.

a i) Write an equation for the reaction between ethanoic acid and ethanol. (2)
ii) Explain why the reaction mixture was titrated *quickly*. (1)

b By making use of the titration results and the equation in (a) (i), calculate
i) how many moles of ethanoic acid remained at equilibrium,
ii) how many moles of ethanoic acid had reacted,
iii) how many moles of ethanol were left in the equilibrium mixture. (3)

c i) Write the expression for the equilibrium constant, K_c, for the equation you have given in (a) (i).
ii) Calculate a value for K_c for this reaction. (2)

(UCLES, 1996/7 (specimen))

36 Given the following thermochemical data at 298 K:

Compound	$\Delta H^{\ominus}_{combustion}$/kJ mol^{-1}	$\Delta H^{\ominus}_{formation}$/kJ mol^{-1}
cyclopropane (g)	−2091	–
propene (g)	−2058	–
propane (g)	−2220	–
water (l)	–	−285.8

a Calculate the enthalpy change involved in the hydrogenation of cyclopropane to propane.

b Calculate the enthalpy change involved in the conversion of cyclopropane to propene. Comment on the relative stabilities of cyclopropane and propene.   (8)

(HK, 1993 (part))

37 The compound furan, $(CH)_4O$, is unsaturated and has a structure in which the four carbon atoms and the oxygen atom are connected together in a ring. It is a liquid at room temperature with a boiling point of 31 °C. It is used to make solvents and nylon, and is transported in bulk by road tanker.

a Draw a displayed formula for furan, showing all the atoms and bonds.   (2)

b Balance the equations in the Hess cycle below. Use your balanced cycle and the data to calculate the enthalpy change of atomization of furan.

$\Delta H^{\ominus}_{f,298} [(CH)_4O(l)]$ $= -62.4$ kJ mol^{-1}

$\Delta H^{\ominus}_{at,298} [C(graphite)]$ $= +716.7$ kJ mol^{-1}

$\Delta H^{\ominus}_{at,298} [\frac{1}{2}H_2(g)]$ $= +218.0$ kJ mol^{-1}

$\Delta H^{\ominus}_{at,298} [\frac{1}{2}O_2(g)]$ $= +249.2$ kJ mol^{-1}

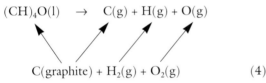

$$(CH)_4O(l) \rightarrow C(g) + H(g) + O(g)$$
$$C(graphite) + H_2(g) + O_2(g)$$ (4)

c Use the bond energies given below to calculate another value for the enthalpy change of atomization of furan.

$E(C\!-\!C) = +347$ kJ mol^{-1}

$E(C\!-\!O) = +358$ kJ mol^{-1}

$E(C\!-\!H) = +413$ kJ mol^{-1}

$E(C\!=\!C) = +612$ kJ mol^{-1} (3)

d By considering the bonding in furan and the definition of bond energy, suggest two reasons for the difference in values calculated in **(b)** and **(c)**.   (2)

e i) What kinds of intermolecular forces will exist between molecules of furan?

ii) Would you expect furan to dissolve in water? Justify your answer.   (3)

f Liquid furan is transported by road tanker. Suggest two hazards involved and explain how the risks could be minimised.   (4)

(ULEAC, 1994 (Nuffield))

38 Which of the following ionic solids has the largest lattice enthalpy?

A NaF

B NaCl

C CaO

D MgO

E KCl

(O & C, 1994)

39 An energy-level diagram is shown below for the dissolving of anhydrous copper(II) sulphate.

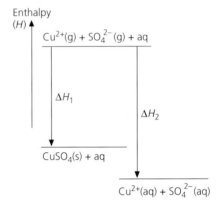

a Choose terms from the following list to describe the enthalpy changes indicated by ΔH_1 and ΔH_2 in the diagram:

enthalpy of activation   hydration enthalpy
bond enthalpy   enthalpy change of
enthalpy change of formation   ionization
lattice enthalpy

ΔH_1 is   (1)
ΔH_2 is   (1)

b Draw an arrow on the diagram above to represent the enthalpy change of solution (ΔH_{soln}) of $CuSO_4(s)$.   (1)

c Would you expect the temperature of the water to rise or fall when anhydrous copper(II) sulphate dissolves? Explain your answer.   (2)

(O & C, 1993 (Salters))

40a Distinguish clearly, by means of examples, between the terms 'bond enthalpy' and 'mean bond enthalpy'.   (4)

You will need the following data for the calculations which follow:
Mean bond enthalpies/kJ mol^{-1}: C=C 612; C—C 348; C—H 412; H—H 436
Enthalpy of atomization of carbon/kJ mol^{-1}: 715

b Using these mean bond enthalpies, calculate the enthalpy change for the dimerization of ethene to give cyclobutane:

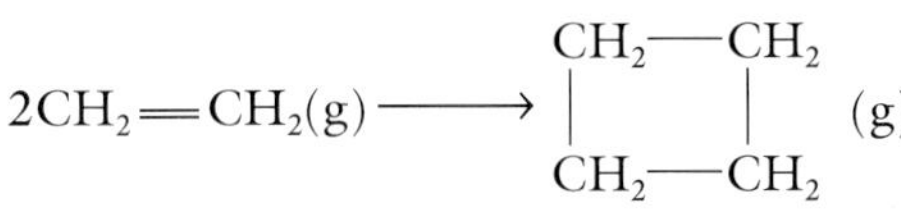

$$2CH_2\!=\!CH_2(g) \longrightarrow \begin{array}{c} CH_2\!-\!CH_2 \\ | \qquad | \\ CH_2\!-\!CH_2 \end{array} (g)$$

c i) Given that the mean enthalpy of the C——C bond in cyclobutane is, in fact, $320 \, kJ \, mol^{-1}$, calculate the enthalpy change for the reaction below.

$$4H_2(g) + 4C(5) \longrightarrow \begin{array}{c} CH_2——CH_2 \\ | \qquad | \\ CH_2——CH_2 \end{array} (g)$$

ii) What does your answer suggest about the above reaction as a way of making cyclobutane? Explain your reasoning.

iii) The C——C bond enthalpy in cyclobutane is $28 \, kJ \, mol^{-1}$ smaller than in butane. Comment on this difference. (10)

(ULEAC, 1992)

41a What is meant by a *dynamic equilibrium*? (3)

b In the Contact process for the preparation of sulphuric acid (H_2SO_4), sulphur dioxide, SO_2, is converted to sulphur trioxide, SO_3, in accordance with:

$$2SO_2(g) + O_2(g) \rightleftharpoons 2SO_3(g) \qquad \Delta H = -94.6 \, kJ \, mol^{-1}$$

i) State *Le Chatelier's Principle* and use the principle to predict the optimum theoretical temperature and pressure to ensure the maximum possible yield of sulphur trioxide, SO_3.

ii) Typical industrial conditions employed are a temperature of around $450 \, °C$ at atmospheric pressure in the presence of vanadium(V) oxide, V_2O_5, as a catalyst. Compare these conditions with the optimum theoretical conditions and justify their use.

iii) At $450 \, °C$, the partial pressures of the gases in the equilibrium mixture are:

p_{SO_2}, 0.090 atm; p_{SO_3}, 4.5 atm; p_{O_2}, 0.083 atm

Write an expression for the equilibrium constant, K_p, and calculate its value under these conditions. (14)

c Gaseous sulphur dioxide is a major pollutant. A possible sequence for the conversion of $SO_2(g)$ into $SO_3(g)$ is:

$$NO(g) + \tfrac{1}{2}O_2(g) \rightarrow NO_2(g)$$

$$NO_2(g) + SO_2(g) \rightarrow NO(g) + SO_3(g)$$

i) State the role played by the $NO(g)$ in this sequence.

ii) Write an equation to show how the $SO_3(g)$ is converted into sulphuric acid and state **two** possible effects of the acid rain produced on the environment. (4)

d The combustion of petrol in cars causes numerous forms of pollution of which $NO(g)$ and $NO_2(g)$ are two. More and more cars are being fitted with catalytic converters which 'convert' these oxides back to their elemental form. The catalytic converter contains a fine-meshed aluminium alloy coated in a platinum–rhodium mixture.

i) Write equations to show the decomposition of the nitrogen oxides into their elements.

ii) Explain why the catalytic converter contains a fine mesh.

iii) Discuss the limitations on the type of petrol that can be used in a car fitted with a catalytic converter. (4)

(UCLES, 1996/7 (M, specimen))

42 Ellingham diagrams, showing the variation of standard free energy change, $\Delta G^{\ominus}$, with temperature, have proved useful in deciding the best conditions for the extraction of metals from their ores. [This is described in Students' Book II, pages 242 to 253.] A simplified Ellingham diagram for the oxides of aluminium, carbon, hydrogen and zinc is shown below.

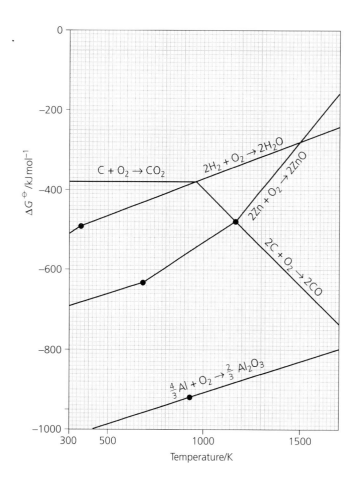

a Discuss the advantages and disadvantages of using aluminium, hydrogen and carbon as reducing agents in the extraction of metals.

b Write the equation for the reaction between zinc oxide and carbon to form zinc and carbon monoxide.

Use the Ellingham diagram above to obtain a value for $\Delta G^{\ominus}$ for this reaction at $1100 \, K$. Would aluminium or hydrogen reduce zinc oxide at this temperature?

c By considering both $\Delta H^{\ominus}$ and $\Delta S^{\ominus}$ explain why $\Delta G^{\ominus}$ varies with temperature for the reactions between

i) zinc and oxygen

ii) carbon and oxygen forming carbon monoxide.

(ULEAC, 1991 (Nuffield))

43 This question is concerned with the element titanium, Ti, and with some of its compounds.

a i) State the electronic configuration of a titanium atom in its ground state.

ii) Is titanium likely to display more than one oxidation number in its compounds? Justify your answer.   (2)

b Titanium occurs naturally as the mineral rutile, TiO_2. One possible method suggested for the extraction of the metal is to reduce the rutile by heating it with carbon:

$$TiO_2(s) + 2C(s) \rightarrow Ti(s) + 2CO(g)$$

i) Calculate ΔH for this reaction given that

$\Delta H^{\ominus}_{f,298} [TiO_2](s) = -940 \text{ kJ mol}^{-1}$ and
$\Delta H^{\ominus}_{f,298} [CO](g) = -110 \text{ kJ mol}^{-1}$

ii) Calculate ΔG at 2200 K for this reaction using your value for ΔH, $\Delta S = +365 \text{ J K}^{-1} \text{ mol}^{-1}$, and the relationship: $\Delta G = \Delta H - T\Delta S$.

iii) Is this reaction feasible at 2200 K? Justify your answer.

iv) Explain the pollution problem which might be caused by this process.   (7)

(ULEAC, 1992 (Nuffield) (part))

*44 In which of the following changes is there likely to be an increase in entropy of the system?

1 $N_2O_4(g) \rightarrow 2NO_2(g)$

2 $C_2H_5OH(l) \rightarrow C_2H_5OH(g)$

3 $Mg(OH)_2(aq) + CO_2(g) \rightarrow MgCO_3(s) + H_2O(l)$

(ULEAC, 1992 (Nuffield))

45 For the oxides of nitrogen tabulate their physical states, standard entropies and standard enthalpy changes of formation at 298 K.

Entropy can be considered as a measure of disorder. Do your data support this view? Explain your answer.

What patterns are there in the values of the standard enthalpy changes of formation of these oxides and what interpretations can you give for these patterns?

Would you expect N_2O_3 to decompose into NO_2 and NO at 298 K? Calculate ΔS_{system}, ΔH, $\Delta S_{surroundings}$ and hence ΔS_{total} for the reaction.

Comment on your value for ΔS_{total}.

(ULEAC, 1993 (Nuffield))

46 Ammonia can be oxidised by air to form nitrogen oxide.

$$4NH_3(g) + 5O_2(g) \rightleftharpoons 4NO(g) + 6H_2O(g)$$
$$\Delta H = -909 \text{ kJ mol}^{-1}$$

This reaction forms the basis of the first stage in the manufacture of nitric acid from ammonia.

a State, and explain in terms of Le Chatelier's Principle, the change in the equilibrium yield of nitrogen oxide caused by:

i) increasing the pressure at constant temperature,   (2)

ii) increasing the temperature at constant pressure.   (2)

b The industrial process is operated at a temperature of about 900 °C. Bearing in mind your answer to **(a)(ii)**, suggest a reason for this temperature.   (1)

c In the industrial process the mixture of gases is passed through gauzes of a platinum–rhodium alloy. This alloy acts as a *heterogeneous* catalyst.

i) Suggest **one** reason for using the catalyst.   (1)

ii) Explain the term *heterogeneous*.   (1)

iii) Give **one** advantage of using a catalyst in gauze form.   (1)

d If nitrogen oxide is mixed with air it can react to form a brown gas. This brown gas is an equilibrium mixture.

$$2NO_2(g) \rightleftharpoons N_2O_4(g)$$

At 77 °C and 700 kPa pressure, an equilibrium mixture contains 48% by volume of N_2O_4.

i) Write the expression for the equilibrium constant, K_p, for this equilibrium. Include units in your expression.   (2)

ii) Calculate the value of K_p at 77 °C.   (2)

(AEB, 1994)

47 When HF is added to water at 298 K, the following equilibrium is established.

$$HF(aq) \rightleftharpoons H^+(aq) + F^-(aq)$$

If the concentrations of HF(aq) and F^-(aq) at equilibrium are $7.7 \times 10^{-3} \text{ mol l}^{-1}$ and $2.3 \times 10^{-3} \text{ mol l}^{-1}$, respectively, which one of the following is the value of the equilibrium constant, in mol l^{-1}, at this temperature?

A 1.45×10^3 **D** 6.87×10^{-4}

B 3.35 **E** 6.87×10^{-7}

C 2.99×10^{-1}

(JMB, 1992)

48a i) Briefly explain how crystallinity arises in plastics.

ii) Explain why crystallinity varies with the size of the polymer side-chain.   (4)

b i) Explain how crystallinity is affected when a plastic is cold-drawn.

ii) How does this treatment affect the properties of nylon-66?   (3)

c Explain why an intermediate range polymer, M_r 10 000 to 20 000, is chosen in the manufacture of nylon-66.   (3)

(UCLES, 1996/7 (specimen))

49 The following diagrams show the structure of four isomers of molecular formula C_4H_8.

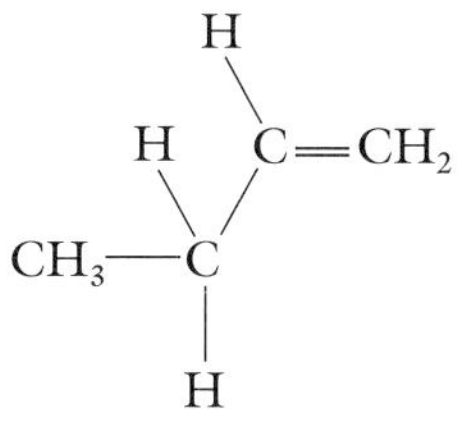

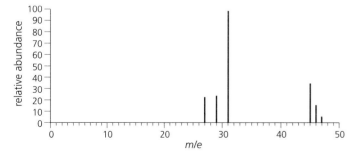

A 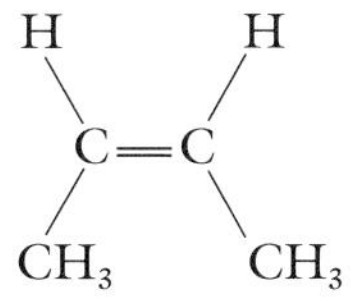   **B**

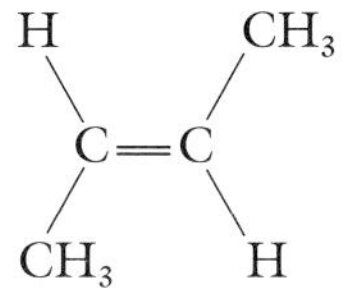

C 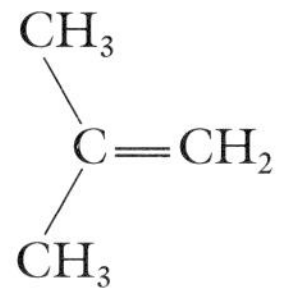   **D**

a i) To which class of compounds do the four isomers belong? (1)

ii) Which two diagrams show compounds which are cis-trans isomers? (2)

iii) Compound **A** reacts with steam to form a pair of optical isomers. Draw the structure of one of these isomers. Draw a circle around the chiral carbon atom. (2)

b i) Which of the above compounds could be formed from 2-methylpropanol by the elimination of water? (1)

ii) State the reagents and conditions by which this reaction could be carried out in the laboratory. (2)

(UCLES, 1996/7 (specimen))

50 The following diagram represents part of the mass spectrum of ethanol, C_2H_5OH.

a Identify on this diagram
i) the M^+ (molecular ion) peak,
ii) the (M + 1) peak. (2)

b Explain why this (M + 1) peak occurs. (1)

c Explain the presence of the other peaks in the diagram. (1)

(UCLES, 1996/7 (M, specimen))

51 A liquid hydrocarbon **X** is found to contain 85.7% carbon by mass, and to have a relative molecular mass of 70. The infra-red spectrum for this compound includes peaks at $3085\ cm^{-1}$ and $1650\ cm^{-1}$.

a Calculate the molecular formula of **X** from the percentage composition. Draw the structural formulae of FIVE possible isomers which might produce these two peaks in the infra-red spectrum. Suggest the structural formula of another isomer, **Y**, which would not have these two peaks in the infra-red spectrum.

b Predict some of the chemical properties of one of your isomers of **X**. For each prediction write an equation or reaction scheme, and indicate the necessary reaction conditions.

Comment on whether all the isomers you have drawn in **(a)** would be expected to have identical physical and chemical properties.

(ULEAC, 1992 (Nuffield))

52a Cyclohexene, C_6H_{10}, reacts with bromine.
i) Write a balanced equation for this reaction using structural formulae.
ii) Name the type and mechanism of this reaction. (3)

b In an investigation into this reaction, cyclohexene (molar mass = 82 g mol^{-1}) and lithium chloride were dissolved in methanol as solvent and then some liquid bromine was added.

When the reaction was complete, the mixture was investigated by chromatography and mass spectrometry. Four different substances were isolated.

Substance	Parent ion peaks m/e
1	82
2	242 (with smaller peaks at 240 and 244)
3	198 (with smaller peaks at 196 and 200)
4	192 and 194

i) Give the molecular formulae of
Substance **2**
Substance **3**
ii) Substance **4** was initially unexpected and later work showed it to be 1-bromo-2-methoxycyclohexane. Draw the displayed formula of substance **4**.
iii) Suggest how substance **4** might have been produced. (5)

(ULEAC, 1994 (Nuffield) (part))

53 The molecular formula C_4H_8 represents four different alkenes. Two are shown below:

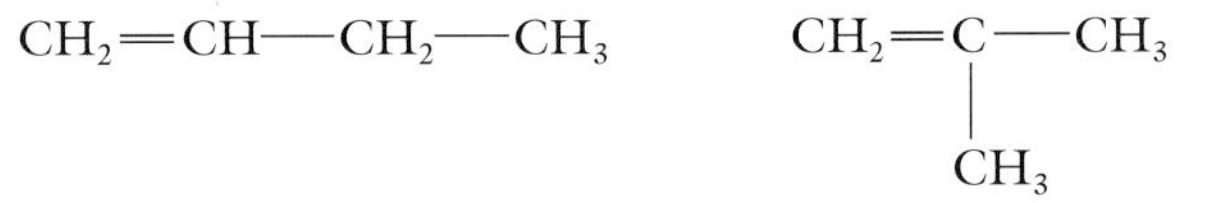

$$CH_2 = CH - CH_2 - CH_3 \qquad CH_2 = \underset{\underset{CH_3}{|}}{C} - CH_3$$

Alkene A Alkene B

a Draw the structures of the two other alkene isomers of C_4H_8, name the type of isomerism shown by these two other isomers and explain why this isomerism is not shown by alkenes A and B.

b Draw the repeating unit of the polymer formed by alkene A above.

c **i)** Name the type of mechanism for the reaction shown by alkenes with concentrated sulphuric acid.
ii) Write a mechanism showing the formation of the major product in the reaction between concentrated sulphuric acid and alkene A.
iii) Explain why this compound rather than one of its isomers is the major product.

(NEAB, 1994)

54 Which one of the following does NOT represent a step in the reaction of chlorine with methane?

A $Cl_2 \longrightarrow Cl\bullet + Cl\bullet$ Initiation

B $CH_4 \longrightarrow CH_3\bullet + H\bullet$ Initiation

C $CH_4 + Cl\bullet \longrightarrow CH_3\bullet + HCl$ Propagation

D $CH_3\bullet + Cl_2 \longrightarrow CH_3Cl + Cl\bullet$ Propagation

E $CH_3\bullet + CH_3\bullet \longrightarrow CH_3CH_3$ Termination

(NICCEA, 1994)

55a What is meant by the term *substitution reaction* in Organic Chemistry?

Give **one** example of **each** of the following types of reaction, by identifying starting materials and products and by describing the conditions used:
i) nucleophilic substitution,
ii) electrophilic substitution. (5)

b When propane reacts with chlorine, two different monochloropropanes are produced.
i) State the type of isomerism shown. (1)
ii) What conditions are used for this reaction? (1)
iii) Of what type of substitution reaction is this an example? (1)
iv) Draw the structural formula of each of the two isomeric products, and suggest (with a reason) in what ratio they might be formed. (4)

(UCLES, 1996/7 (specimen))

56a Mass spectrometry and infra-red spectroscopy are complementary techniques in chemical analysis.

Explain
i) the chemical principles involved in these techniques,
ii) what information may be obtained from the spectra,
iii) the importance of high resolution in mass spectrometry. (15)

b Analysis of an organic compound **Z** containing carbon, hydrogen and oxygen gave the following data:

composition, by mass: C, 66.7%; H, 11.1%;
infra-red spectrum: strong absorption band at 1715 cm^{-1};
mass spectrum: lines of m/e values 72, 57, 43.

Use these data to suggest the identity and molecular structure of **Z**, showing your reasoning. (10)

(UCLES, 1996/7 (specimen))

57 This question is about the primary alcohol pentan-1-ol, $C_5H_{11}OH$, and some of its related compounds.

a Give the structural formula and systematic name of another *primary* alcohol which is an isomer of pentan-1-ol. (2)

b The boiling point of pentan-1-ol is $138\,^\circ C$ and that of pentan-2-ol is $120\,^\circ C$. Suggest an explanation for this difference. (1)

c One method of preparing 1-bromopentane is to put pentan-1-ol, water and sodium bromide into a flask and add concentrated sulphuric acid slowly from a tap funnel. The resulting mixture then needs to be heated for some time in order to obtain a reasonable yield.
i) Draw a labelled diagram of the apparatus you would use for carrying out this process in the laboratory.
ii) The reaction mixture often goes yellow or orange at this stage in the preparation. Name the substance likely to be responsible for this coloration and explain how it forms.
iii) This preparation normally gives a yield of 60%. Calculate the minimum mass of pentan-1-ol you would need to produce 15 g of 1-bromopentane by this method. (Relative atomic masses: H = 1, C = 12, O = 16, Br = 80.) (6)

d Dehydration of pentan-1-ol yields pent-1-ene. Sketch on the diagram below the electron density distribution in the carbon–carbon double bond. (2)

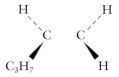

e The two compounds shown below are among those secreted by insects to attract other members of the same species.

$$\underset{I}{CH_3CH_2CH_2CH_2CO_2H} \text{ and } \underset{II}{CH_3CO_2CH_2CH_2CH_2CH_2CH_3}$$

Such compounds are used to control insects, but to do so they need to be made synthetically. Give the reagents and conditions needed to synthesise these two compounds using pentan-1-ol as one of the starting materials. (4)

(ULEAC, 1990 (Nuffield))

58 Oil of peppermint is a plant extract obtained from the leaves and stems of *Mentha piperita*. It is a complex mixture of organic compounds, which is used widely in perfumery, food flavouring, toothpastes and a range of pharmaceutical products. The main component (which contributes about 46% by mass of oil of peppermint) is menthol, which can be separated out as crystals when the oil is cooled. The crystals of menthol are only slightly soluble in water, but they are moderately soluble in warm ethanol.

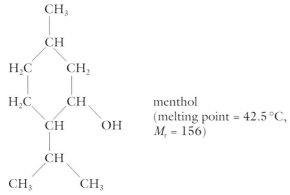

menthol
(melting point = 42.5 °C,
M_r = 156)

a i) Name the functional group present in menthol. (1)
ii) Classify this functional group as primary, secondary or tertiary. (1)
iii) The infra-red (i.r.) spectrum of a sample of Brazilian oil of peppermint is shown in Figure 1.

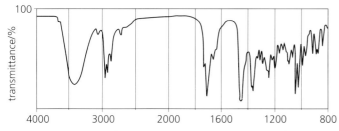

Figure 1 Infra-red spectrum of a sample of Brazilian oil of peppermint

Write down the wavenumber of an absorption in this i.r. spectrum which is characteristic of the functional group present in menthol. [A table of characteristic i.r. absorptions is given at the end of this question.] (1)

b The crystals of menthol can be filtered off from the cold oil of peppermint and purified by recrystallisation. Describe how you would recrystallise a sample of menthol. (4)

c You could check the purity of your sample of recrystallised menthol by the determination of its melting point. Use this example to explain how the temperature at which a substance melts can be interpreted to provide information about its purity. (3)

d Another substance, compound X, can be separated as a colourless liquid from oil of peppermint. Its i.r. spectrum and mass spectrum are shown in Figures 2 and 3. Compound X can be made from menthol in a simple one-step process.

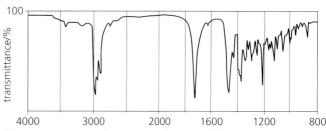

Figure 2   Infra-red spectrum of compound X

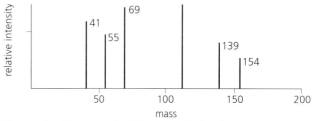

Figure 3   Six most significant peaks in the mass spectrum of compound X

i) Use the information contained in Figures 2 and 3 to suggest a structure for compound X and explain your reasoning.

(O & C, 1993 (Salters) (part))

59 In 1930 the American Thomas Midgeley inhaled a lungful of the CFC with formula CCl_2F_2 and used it to blow out a candle. He did this to demonstrate a new refrigerant which was non-flammable and non-toxic.

a What does CFC stand for? (1)

b State *two* other uses (as well as refrigerants) to which CFCs have been put. (2)

c In the stratosphere, CFCs are broken down to give chlorine atoms. These chlorine atoms react with ozone molecules thus: $Cl^{\bullet} + O_3 \rightarrow ClO^{\bullet} + O_2$ *equation 1.1*

i) Write *equation 1.2* which shows how the $ClO^{\bullet}$ radicals can react further with oxygen atoms to give oxygen molecules and chlorine atoms. (1)
ii) Use *equations 1.1* and *1.2* to give an overall equation. Then explain why a few chlorine atoms can cause the destruction of many ozone molecules. (1)
iii) Give the equation for a *termination* reaction in which chlorine atoms are removed from the stratosphere. (1)
iv) Why is the rate of this termination reaction slow in the stratosphere? (1)

d If the production of CFCs were to be phased out now, it would be a long time before their damaging effects on the ozone layer began to be reduced. Suggest *two* reasons for this. (2)

e Suggest a reason why Midgeley was unaware of the draw-backs associated with CFCs. (1)

f Chemists are now looking for compounds to replace CFCs. Describe *two* properties which these compounds should have, as well as being non-toxic and non-flammable. (2)

g A variety of halogenoalkanes similar to Midgeley's original compound have been synthesised to see whether they might be useful. For example, CF_2I_2 has been synthesised.
  i) Explain why this compound has a higher boiling point than CCl_2F_2. (2)
  ii) Give a reason why CF_2I_2 reacts faster than CCl_2F_2 with sodium hydroxide solution. (1)
  iii) When CF_2I_2 reacts with sodium hydroxide, iodide ions are formed in the alkaline solution. In order to test for the presence of these ions a suitable substance has first to be added to neutralise the alkali.

  If you were doing this test, state:

  the substance you would use to carry out the neutralisation; (1)
  the substance you would add to test for the iodide ions; (1)
  the positive result of the test. (1)

(O & C, 1994 (Salters) (part))

60 Which one of the following is the major product when the compound

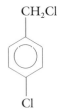

is warmed with aqueous sodium hydroxide?

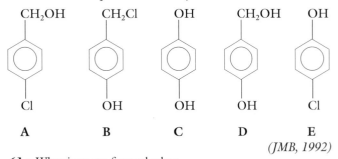

| A | B | C | D | E |

(JMB, 1992)

61a What ions are formed when
  i) *ammonia* is dissolved in HCl(aq),
  ii) *water* has HCl(g) dissolved in it,
  iii) *carbon dioxide* is dissolved in an excess of NaOH(aq),
  iv) *boron trifluoride* reacts with NaF? (4)

b In (a) above, the bond angles in the reactants in *italics* differ from those in the ions they form. By considering, in each case, the changes in the number, and type, of electron pairs around the central atom, explain why. (8)

(UCLES, 1996/7 (specimen))

62 Indicators are substances that change colour with a change in the pH. A weak acid, HIn(aq), can be used to represent an indicator. It is essential that the species HIn(aq) and the anion In⁻(aq) are of different colours.

e.g
$$HIn(aq) \rightleftharpoons H^+(aq) + In^-(aq)$$
blue   yellow

a i) Given the indicator system above, what colour will the indicator be in a strongly acidic solution? (1)

  ii) Explain your answer. (2)

Methyl orange can be used as an acid–base indicator.

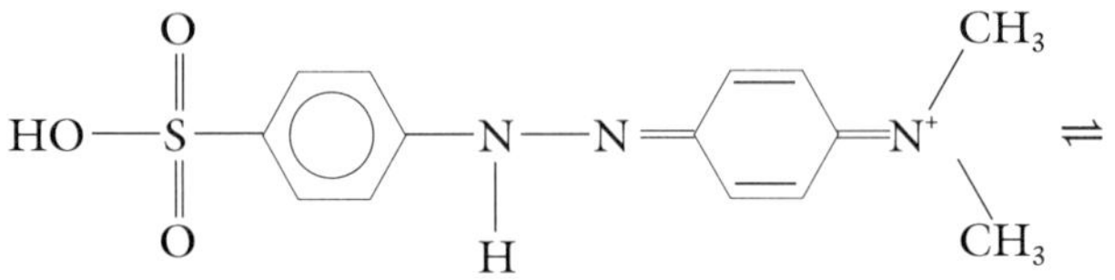

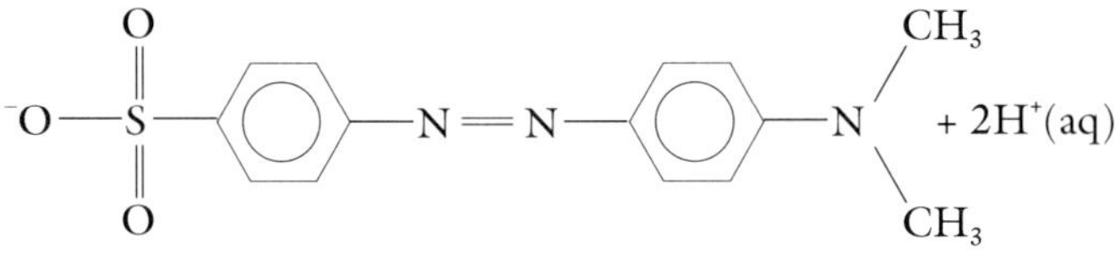

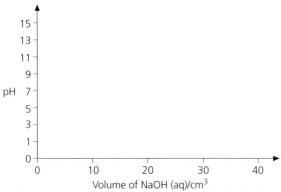

methyl orange (yellow)

b i) Explain why methyl orange turns yellow on the addition of OH⁻(aq). (2)
  ii) Predict the colour, at the end-point of an acid–base titration, using methyl orange as an indicator. (1)
  iii) For a given titration, at the end-point, the [H⁺(aq)] concentration was found to be 4.23×10^{-5} mol dm⁻³. Calculate the pH at the end-point. (2)

c i) Ethanoic acid, $CH_3COOH(aq)$, is a weak acid that can be titrated using aqueous sodium hydroxide, NaOH(aq). Sketch the changes in pH during the titration of 25 cm³ of 0.10 mol dm⁻³ $CH_3COOH(aq)$ using 0.10 mol dm⁻³ NaOH(aq) until the NaOH(aq) is in excess.

  ii) Explain why methyl orange is not a suitable indicator for this titration. (1)
  iii) Suggest an alternative indicator that would be suitable for this titration. (1)

(UCLES, 1994 (M, specimen))

*63 Which of the following dilute aqueous mixtures are examples of buffer solutions?

1 Ammonia and ammonium chloride

2 Ethanoic acid and sodium ethanoate

3 Hydrochloric acid and sodium sulphate

(ULEAC, 1990 (Nuffield))

64 In which of the following reactions is sulphuric acid acting as a base?

A $H_2SO_4 + OH^- \rightarrow HSO_4^- + H_2O$

B $H_2SO_4 + H_2O \rightarrow HSO_4^- + H_3O^+$

C $H_2SO_4 + HNO_3 \rightarrow H_2NO_3^+ + HSO_4^-$

D $H_2SO_4 + HClO_4 \rightarrow H_3SO_4^+ + ClO_4^-$

E $H_2SO_4 + 2H_2O \rightarrow 2H_3O^+ + SO_4^{2-}$

(O & C, 1994)

65 Loch Fleet is a small lake in Scotland which had been suffering from the effects of acid rain. It is fed by several streams draining the surrounding area and it has a single outlet stream. The acidity killed its previously healthy stock of brown trout partly because of the low pH of the water and partly because of the toxic aluminium ions leached out of the soil under acid conditions.

In 1986 and 1987 large quantitites of limestone $(CaCO_3)$ were dropped on the area surrounding the lake. Brown trout were re-introduced and they survived!

a One reason for the acidity of the rain was the presence of sulphur dioxide. This reacts with water thus:

$$SO_2(aq) + 2H_2O(aq) \rightleftharpoons H_3O^+(aq) + HSO_3^-(aq)$$

For this reaction $K_a = \dfrac{[HSO_3^-(aq)] \times [H_3O^+(aq)]}{[SO_2(aq)]}$

$$= 1 \times 10^{-2} \text{ mol dm}^{-3}$$

i) State the relationship between pH and $[H_3O^+(aq)]$. (1)

ii) Calculate the concentration of sulphur dioxide needed to produce a pH of 4 in the rain. (3)

iii) Suggest one large-scale source of the sulphur dioxide pollution. (1)

b Write a balanced equation to show how the limestone removes the acidity from water in the area surrounding the lake. (2)

c Aluminium ions are often held in clay soils, surrounded by oxide ions in an octahedral arrangement.

i) Draw a diagram to illustrate such an octahedron, labelling the aluminium and oxide ions. (2)

ii) Write a balanced equation (with state symbols) to show how the insoluble aluminium oxide Al_2O_3 reacts with acid to give soluble aluminium ions which are leached out of the soil. (3)

d Suggest *one* reason why the insoluble solid calcium carbonate, rather than an alkaline substance, is used to neutralise the acid. (1)

e i) Suggest why the calcium carbonate is placed on the soil around the lake, not in the lake itself. (2)

ii) Suggest an environmental *disadvantage* of placing calcium carbonate on the soil round the lake. (1)

f Another cause of acidity in the supply streams to the lake is found to be the presence of sea-salt in the rain. This causes H_3O^+ ions to be released by cation exchange in the soil. Draw a labelled diagram to illustrate this process.

(O & C, 1993 (Salters))

66 Crude oil fractions often contain unwanted sulphur compounds. The sulphur can be converted to hydrogen sulphide by mixing the oil fractions with hydrogen under pressure and passing over a catalyst.

$$2RSH + H_2 \rightarrow R_2 + 2H_2S \qquad \textit{equation 2.1}$$
$$R\text{-}S\text{-}R + H_2 \rightarrow R_2 + H_2S \qquad \textit{equation 2.2}$$

Hydrogen sulphide gas dissolves in monoethanolamine and this reaction is used to separate hydrogen sulphide from the other gases present:

$$2HOCH_2CH_2NH_2 + H_2S \rightleftharpoons 2HOCH_2CH_2NH_3^+ + S^{2-}$$
$$\textit{equation 2.3}$$

Heating the solution liberates pure hydrogen sulphide and regenerates the amine. The pure hydrogen sulphide is then burned with a limited amount of air to give sulphur dioxide, sulphur vapour and steam. The sulphur is condensed and separated and the remaining mixture of hydrogen sulphide and sulphur dioxide is passed over a bauxite catalyst bed to form more sulphur.

a Draw a dot-cross diagram for H_2S, showing outer-shell electrons only. (2)

b Suggest a source for the hydrogen gas which is used in this process. (1)

c Explain why hydrogen sulphide is acting as an acid in *equation 2.3*. (2)

d Hydrogen sulphide reacts with water thus:

$$H_2S(aq) + H_2O(l) \rightleftharpoons H_3O^+(aq) + HS^-(aq)$$
$$K_a = 8.9 \times 10^{-8} \text{ mol dm}^{-3} \text{ at 298 K}$$

i) Write the expression for K_a in terms of the concentrations of the molecules and ions involved. (2)

ii) Calculate the pH of a 0.10 mol dm^{-3} solution of H_2S at 298 K. (4)

e Write a balanced equation for the reaction between hydrogen sulphide and sulphur dioxide where the products are sulphur and water. Write the oxidation state below each sulphur atom in your equation. (4)

f Explain why the removal of sulphur from crude oil fractions is important for the environment. (2)

g Much of the sulphur recovered from crude oil in this way is used to make sulphuric acid. Calculate the mass of sulphuric acid which could be obtained from 1.0 tonne of crude oil containing 1.0% sulphur by mass, assuming all the sulphur from the crude oil is converted into the acid. (Relative atomic masses: H = 1; O = 16; S = 32) (2)

(O & C, 1994 (Salters))

67a Write an equation to represent the dissociation of water. (1)

b Give an expression for the *ionic product of water*, and show how it is related to the equilibrium constant for the reaction in (**a**). (3)

c Define pH. (1)

d The ionic product of water is $5.6 \times 10^{-14} \text{ mol}^2 \text{ dm}^{-6}$ at 333 K. Calculate the pH of water at this temperature. (3)

e Given that at 298 K the pH of water is 7, state whether the dissociation of water is exothermic or endothermic. Give a reason for your answer. (2)

(AEB, 1994)

68 A well known household cleaner contains an active ingredient **A** which behaves as a monobasic acid. Deduce what you can about **A** from the following information.

A solution contains 10.00 g of **A** per dm^3 and has a pH of 1.50. On treatment with barium nitrate solution, it gives a white precipitate which is insoluble in dilute hydrochloric acid. 20 cm^3 aliquots of this solution of **A** required 16.7 cm^3 of a solution of sodium hydroxide of concentration 0.100 mol dm^{-3} for complete neutralization.

Given that **A** is anhydrous and is very soluble in water, suggest a method for obtaining a dry solid sample of **A** from appropriate starting materials. (12)

(ULEAC, 1991 (part))

69 Adrenalin is a hormone which, when secreted directly into the bloodstream, acts as a stimulant. It has the structure:

a Name **three** functional groups present in the adrenalin molecule. (3)

b By means of an asterisk, identify the chiral centre in the structure of adrenalin drawn above. (1)

c The synthesis of adrenalin includes the following stages:

Give the displayed formulae of **A** and **B**. (2)

(UCLES, 1996/7 (specimen))

70a Polysaccharides are polymers built up from many sugar units.

Name
i) a structural polysaccharide, (1)
ii) a storage polysaccharide. (1)

b Explain, briefly, the difference in structure between the two polysaccharides in (**a**). (You may answer by means of diagrams if you wish.) (2)

c Suggest **one** reason why, prior to hibernation, animals tend to store fat rather than polysaccharide. (2)

d All sugars are related to the triose glyceraldehyde (2,3-dihydroxypropanal, $CH_2(OH)CH(OH)CHO$).

Glyceraldehyde can exist in two optically active forms, D-glyceraldehyde and L-glyceraldehyde.

Draw diagrams to show these two different forms of glyceraldehyde. (You are **not** required to identify which is the D-form and which is the L-form.) (2)

e Other sugars have a similar structure to glyceraldehyde in that a chain of variable length is substituted for the aldehyde group.

The structure of the α-form of D-glucose may be represented as

i) On the diagram above, circle the carbon atom on which the attached groups are interchanged in L-glucose. (1)
ii) Suggest why L-glucose costs over 300 times as much as D-glucose. (2)

(UCLES, 1996/7 (specimen))

71 The following reactions were observed for a compound **G** of formula C_3H_6O.

I) The compound did not react with alkaline aqueous copper(II) ions, even when heated.
II) On adding 2,4-dinitrophenylhydrazine, a yellow-orange precipitate formed.
III) Reaction with hydrogen in the presence of a catalyst produced a colourless liquid **H**. Liquid **H** reacted with sodium to give hydrogen.

a Draw the displayed (full structural) formulae of two compounds of formula C_3H_6O. (2)

b What does the result of reaction **I** show? (1)

c The formation of a yellow-orange precipitate in reaction **II** is a positive test for a particular organic group. Identify this group. (1)

d Using the formula of the compound and the results of reactions **I** and **II**, identify **G**. (1)

e Write balanced equations for
   i) the reduction of **G** to give **H**,
   ii) the reaction of **H** with sodium. (2)

f Draw the displayed (full structural) formula of **H**, and give its systematic name. (2)

g Under suitable conditions, **G** will react with hydrogen cyanide to form a compound of formula C_4H_7ON. What type of reaction is this? (1)

(UCLES, 1966/7 (specimen))

72 Give a **chemical** test to distinguish one compound from the other in each of the following pairs. Your answer should include the reagents required, the observation expected, and the chemical equation(s) for each test.
   i) $CH_3CH_2COCH_2CH_3$ and $CH_3CH_2OCH_2CH_3$
   ii) $CH_3CH_2CH_2CH_2CHO$ and $CH_3CH_2COCH_2CH_3$
   iii) 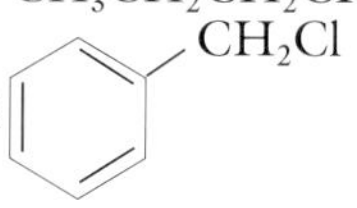 [structure: benzene ring with CH₂Cl] and [structure: benzene ring with Cl]
   iv) 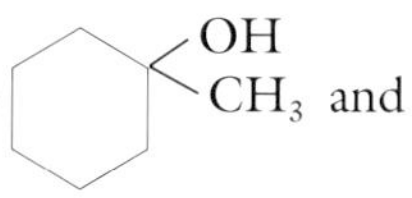 [structure: cyclohexane with OH and CH₃] and [structure: cyclohexane with CH₂OH]

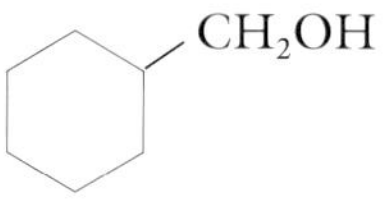

(12)

(HK, 1993 (part))

73 *Musk xylene* is a synthetic perfume that has a pleasant musky smell. It is used widely in soaps and cosmetics. Its structure is very different from that of *muscone*, which is present in the natural musk obtained from the scent gland of the male musk deer, found in the mountains of Central Asia.

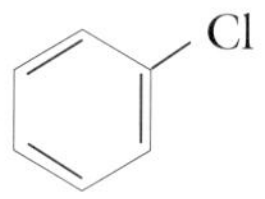

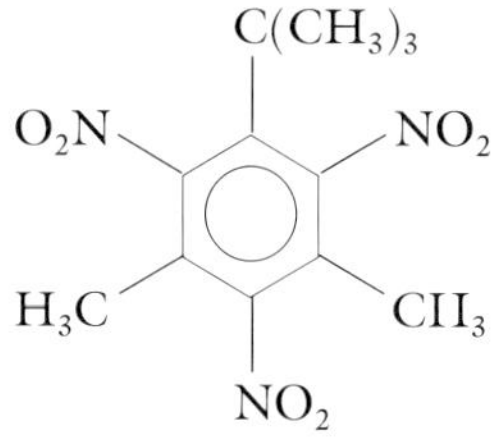

[structure of musk xylene: benzene ring with C(CH₃)₃ at top, O₂N and NO₂ on upper sides, H₃C and CH₃ on lower sides, NO₂ at bottom]   [structure of muscone: CH₃—CH—CH₂—C=O with (CH₂)₁₂ ring]

musk xylene 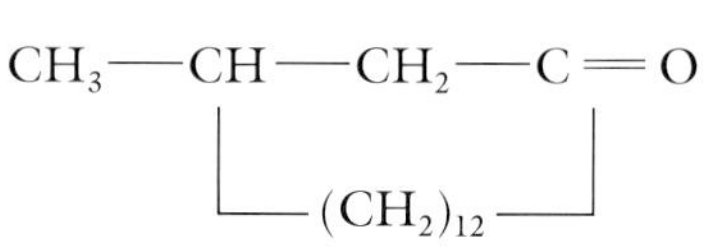   *muscone*

a i) What type of carbonyl compound is *muscone*? (1)
   ii) Which spectroscopic technique could best be used to demonstrate the presence of a $C{=}O$ group in the molecule? (1)

b *Musk xylene* can be prepared from the hydrocarbon, **Compound 1**, which is present in crude oil:

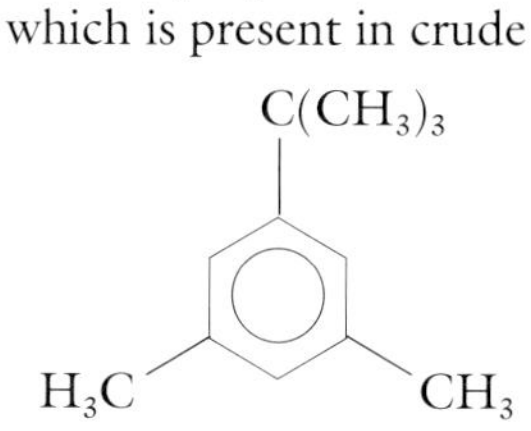

[structure of Compound 1: benzene ring with C(CH₃)₃ at top, H₃C and CH₃ at bottom]

Compound 1

i) Classify **Compound 1** as an alkane, alkene or arene.(1)
ii) What reagents could be used to convert **Compound 1** into *musk xylene*? (3)
iii) The reaction in part **(ii)** is described as *electrophilic substitution*. Referring to this specific reaction, explain what is meant by:

   electrophilic (2)
   substitution (1)

c Use the table of chemical shift values to predict what you would expect the n.m.r. spectrum of *musk xylene* to look like.

Draw in the signals you would expect to see on the spectrum below, indicating their relative heights: (3)

[graph: Absorption (y-axis) vs Chemical shift (x-axis) from 10 to 0, with TMS peak near 0]

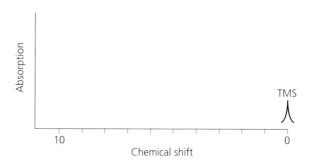

Type of proton	Chemical shift in the region of
[structure: R—C(R)(R)—CH₃]	0.9
$R{-}CH_2{-}R$	1.3
[structure: R—CH(R)—R]	2.0
$R{-}CO{-}CH_3$	2.3
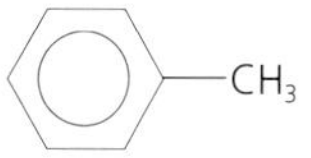 [structure: benzene ring—CH₃]	2.3
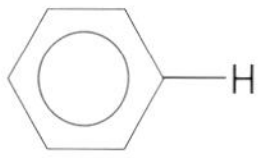 [structure: benzene ring—H]	7.5

d *Muscone* is said to have a *chiral centre* and thus exist as two *stereoisomers*. Draw the 3-dimensional structural formulae of the two isomers around the chiral centre to show how they differ. (3)

(O & C, 1994 (Salters))

74a Draw the structural formula for each of the following compounds **and** state the reagent and reaction conditions that could be used for preparing each from propanoic acid, C_2H_5COOH.
   i) Propanoyl chloride. (3)
   ii) Sodium propanoate. (3)

b i) Give the name **and** structural formula of the organic product of the reaction between propanoyl chloride and sodium propanoate. (2)
   ii) State how the product formed in **(b)(i)** could be converted into propanoic acid. Write an equation for the reaction. (2)

c Draw a structure to represent the organic product of the reaction between:
   i) benzenecarboxylic acid, C_6H_5COOH, and ethanol, C_2H_5OH; (1)
   ii) benzene-1,4-dicarboxylic acid, $HOOCC_6H_4COOH$, and ethane-1,2-diol, $HOCH_2CH_2OH$. (2)

d i) What is the general name given to the type of compound formed in **(c)(ii)**? (1)
   ii) Suggest a use for this type of compound and state a property of the compound which makes it suitable for such a use. (2)

(AEB, 1994)

75 Which one of the following does NOT correctly represent the behaviour of the alcohol?

A $CH_3CH_2OH \xrightarrow{\text{acidified dichromate}} CH_3COOH$

B $CH_3CH_2OH + CH_3COOH \xrightarrow{\text{conc. sulphuric acid}} CH_3COOCH_2CH_3$

C $CH_3CH_2OH + CH_3COCl \rightarrow CH_3CH_2COOCH_3$

D $CH_3CH(OH)CH_3 \xrightarrow{\text{acidified permanganate}} CH_3COCH_3$

E $CH_3CH_2OH + HBr \rightarrow CH_3CH_2Br + H_2O$

(NICCEA, 1994)

76 Which of the following would be produced in the saponification of the fat below:

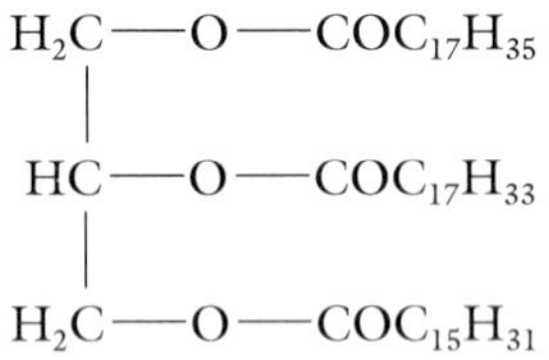

$$H_2C-O-COC_{17}H_{35}$$
$$HC-O-COC_{17}H_{33}$$
$$H_2C-O-COC_{15}H_{31}$$

1 $HOCH_2CH(OH)CH_2OH$
2 $C_{15}H_{31}COONa$
3 $C_{17}H_{33}COOH$

A 1 only   **D** 1 and 2 only

B 2 only   **E** 1 and 3 only

C 3 only

(NICCEA, 1994)

77 At 700 °C, nitrogen monoxide and hydrogen react as follows.

$$2NO(g) + 2H_2(g) \rightarrow N_2(g) + 2H_2O(g)$$

The results of some investigations of the rate of this reaction are shown below.

Experiment number	Initial concentration of nitrogen monoxide /mol dm^{-3}	Initial concentration of hydrogen /mol dm^{-3}	Initial rate of reaction /mol dm^{-3} s^{-1}
1	0.0020	0.012	0.0033
2	0.0040	0.012	0.013
3	0.0060	0.012	0.030
4	0.012	0.0020	0.020
5	0.012	0.0040	0.040
6	0.012	0.0060	0.060

a Explain what is meant by the term *order of reaction*. (2)

b i) Use the above data to determine the order of the reaction with respect to
   1 nitrogen monoxide,
   2 hydrogen. (2)
   ii) Use these answers to write a rate equation for the reaction. This will include the rate constant, k. (1)
   iii) Determine a value for k, stating the units. (2)

c Explain briefly why the initial reaction rate would be expected to increase by increasing each of the following:
   i) the pressure,
   ii) the temperature. (2)
   [Note: a different explanation is required for each of these factors.]

d Suggest, with reasons, whether you would expect the reaction between nitrogen monoxide and hydrogen to be endothermic or exothermic. (2)

e i) Explain why oxides of nitrogen should be eliminated from the exhaust fumes of car engines. (1)
   ii) Suggest one substance likely to be present in exhaust fumes which could convert nitrogen oxide into a harmless product in a car fitted with a catalytic converter. (1)

(UCLES, 1996/7 (specimen))

78 For a gaseous reaction, at a given temperature, the variation of molecular energies can be represented by the Boltzmann distribution:

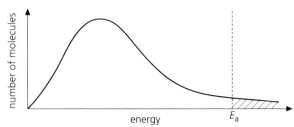

a i) E_a represents the activation energy. Explain what is meant by the term *activation energy*. (2)

ii) In the Boltzmann distribution above, what is represented by the shaded area? (1)

b Explain the role of a catalyst in speeding up a chemical reaction. (2)

c i) Enzymes are often described as biological catalysts and are said to have specific activity. Explain what is meant by *specific activity*. (2)

ii) Immobilised enzymes are used in many industrial processes. In these the enzymes are held on the surface of a gel, and reactants are passed over the gel. Suggest **two** difficulties that immobilisation overcomes compared with a normal enzyme catalysed reaction. (2)

(UCLES, 1996/7 (M, specimen))

79 The rate of reaction between peroxodisulphate ions, $S_2O_8^{2-}$, and iodide ions, I^-, in aqueous solution

$$S_2O_8^{2-}(aq) + 2I^-(aq) \rightarrow 2SO_4^{2-}(aq) + I_2(aq)$$

may be studied by measuring the amount of iodine formed at fixed intervals of time. The data below was obtained from three separate experiments **A**, **B** and **C** carried out at constant temperature.

	Initial conc. of $S_2O_8^{2-}$ (aq)/mol dm^{-3}	Initial conc. of I^-(aq)/mol dm^{-3}	Initial rate /mol dm^{-3} s^{-1}
A	0.01	0.2	4.10×10^{-6}
B	0.02	0.2	8.20×10^{-6}
C	0.02	0.4	1.64×10^{-5}

a i) Show how these data can be used to deduce the order with respect to each reactant and the overall rate equation for the reaction between the $S_2O_8^{2-}$(aq) ions and the I^-(aq) ions. Explain all the terms in the rate equation and calculate a value for the rate constant.

ii) Suggest a suitable method for determining the amount of iodine produced during the reaction and explain why the above kinetic investigation was carried out at constant temperature.

iii) For a mixture of $S_2O_8^{2-}$(aq) ions and I^-(aq) ions, sketch a graph to show how the concentration of iodine, I_2, varies with time. Explain how the initial rate of reaction could be measured. (16)

b Radioactive decay shows first-order reaction kinetics. Archaeologists can determine the age of organic matter by measuring the proportion of radioactive carbon-14 present.

Write the rate equation for the decay of ^{14}C and estimate the age of a piece of wood found to contain $\frac{1}{8}$th as much ^{14}C as living material. [Assume that carbon-14 has a half-life of 5600 years.] (4)

c Catalysts can influence the rate of a chemical reaction. Explain:

i) why transition metals can often function as catalysts;

ii) the difference between a heterogeneous catalyst and a homogeneous catalyst, giving a suitable example of each. (5)

(UCLES, 1996/7 (M, specimen))

*80 Which of the following methods would be suitable for studying the progress of the reaction

$$OH^-(aq) + Cl\!-\!CH_2\!-\!C_2H_5(aq) \rightarrow Cl^-(aq) + HO\!-\!CH_2\!-\!C_2H_5(aq)$$

1 Acid/base titration of samples

2 Colorimetry

3 Polarimetry

(ULEAC, 1990 (Nuffield))

81 In colour photography, the film consists of three layers of 'emulsion', each of which is a suspension containing silver halides in gelatin. Each layer is sensitive to one of the three primary colours in light – red, green and blue.

a The first stage in processing the film is development. Control of development is critical for the correct colours. Any variation in temperature must be allowed for by altering the time allowed for development. The instructions for one make of colour developer include a temperature/time table:

Temperature/°C	Development time
38	3 min
37	3 min 15 sec
36	3 min 40 sec
35	4 min 10 sec
34	4 min 35 sec
33	5 min 15 sec
32	5 min 45 sec

i) Suggest a reason why control of the development is likely to be more accurate at a lower temperature.

ii) At a given temperature:

rate of development $\propto 1/$time required for normal development

Hence $1/$(development time in seconds) may be used as a measure of the reaction rate at that temperature.

In the table below, most of the data in the temperature/time table above has been converted into values of $1/$(temperature/K) and $\ln[1/($development time$)]$. Complete the table by calculating the missing values and writing them into the appropriate spaces:

Temp/°C	$\dfrac{1}{\text{temperature}}$ / K^{-1}	ln[1/development time]
38	3.215×10^{-3}	-5.19
37	3.226×10^{-3}	-5.27
36	3.236×10^{-3}	-5.39
35	3.247×10^{-3}	-5.52
34	3.257×10^{-3}	-5.62
33	3.268×10^{-3}	-5.75
32	—	—

iii) Plot a graph of $\ln[1/(\text{development time})]$ on the y-axis against $1/\text{temperature}$.

iv) The relationship between reaction rate and temperature is given by the equation:

$$\text{rate} \propto A\, e^{-E/RT}$$
$$\text{or} \quad \ln(\text{rate}) = \text{constant} - E/RT$$

where T is the temperature in kelvin, E is the activation energy, and R is the gas constant.

Use your graph from **(iii)** to calculate the activation energy, E, for the development process. ($R = 8.31\,\text{J K}^{-1}\,\text{mol}^{-1}$)   (7)

(ULEAC, 1991 (Nuffield (part)))

82 At temperatures above 1000 K hydrogen reacts with nitrogen monoxide to form nitrogen and steam.

The rate equation for this reaction is: $\text{Rate} = k[\text{NO}(g)]^2[\text{H}_2(g)]$

Which of the following mechanisms is consistent with this rate equation?

A $\text{NO}(g) + 2\text{H}_2(g) \xrightarrow{\text{slow}} \text{NH}_2(g) + \text{H}_2\text{O}(g)$

$\quad \text{NH}_2(g) + \text{NO}(g) \xrightarrow{\text{fast}} \text{N}_2(g) + \text{H}_2\text{O}(g)$

B $2\text{NO}(g) + \text{H}_2(g) \xrightarrow{\text{fast}} \text{N}_2(g) + \text{H}_2\text{O}_2(g)$

$\quad \text{H}_2\text{O}_2(g) + \text{H}_2(g) \xrightarrow{\text{slow}} 2\text{H}_2\text{O}(g)$

C $\text{NO}(g) + \text{H}_2(g) \xrightarrow{\text{slow}} \tfrac{1}{2}\text{N}_2(g) + \text{H}_2\text{O}(g)$

$\quad \tfrac{1}{2}\text{N}_2(g) + \text{NO}(g) + \text{H}_2(g) \xrightarrow{\text{fast}} \text{N}_2(g) + \text{H}_2\text{O}(g)$

D $2\text{NO}(g) + \text{H}_2(g) \xrightarrow{\text{slow}} \text{N}_2\text{O}(g) + \text{H}_2\text{O}_2(g)$

$\quad \text{N}_2\text{O}(g) + \text{H}_2(g) \xrightarrow{\text{fast}} \text{N}_2(g) + \text{H}_2\text{O}(g)$

E $3\text{NO}(g) + \text{H}_2(g) \xrightarrow{\text{slow}} \text{N}_2(g) + \text{H}_2\text{O}(g) + \text{NO}_2(g)$

$\quad 2\text{NO}_2(g) + 4\text{H}_2(g) \xrightarrow{\text{fast}} \text{N}_2(g) + 4\text{H}_2\text{O}(g)$

(ULEAC, 1992 (Nuffield))

83 When halogenoalkanes are heated to high temperatures in the absence of air, they decompose slowly:

where R^1, R^2, R^3, R^4 are alkyl groups or hydrogen atoms.

The kinetics of these reactions can be studied by measuring the pressure changes in sealed tubes containing a pure sample of one of these compounds. The results of such investigations show that most of these thermal decompositions are *first order* reactions.

a Using structural formulae, write down the equation for the thermal decomposition of 1-chloropropane.   (2)

b In an investigation of the thermal decomposition of 1-chloropropane at 720 K, the following results were obtained:

time /minutes	partial pressure of 1-chloropropane/mmHg
0	120
15	92
30	71
45	54
60	42
90	26
120	15

i) The **total** pressure in the sealed tube was 120 mmHg at the start. After 15 minutes the **total** pressure was 148 mmHg. Account for this value for the total pressure after 15 minutes.

ii) Plot a graph of partial pressure of 1-chloropropane against time.

iii) Does your graph confirm that this reaction is first order? Explain your answer, using your graph to do any necessary calculations.

iv) What further investigations would have to be done to determine the activation energy of this reaction?   (9)

c i) State the bonds broken and made in the thermal decomposition of 2-chloropropane.

ii) The activation energies for the thermal decomposition of halogenoalkanes depend on several factors.

Approximate activation energy values at 500 K are:

chloroethane	$237\,\text{kJ mol}^{-1}$
bromoethane	$226\,\text{kJ mol}^{-1}$
2-chloropropane	$211\,\text{kJ mol}^{-1}$
2-bromopropane	$200\,\text{kJ mol}^{-1}$
2-chloro-2-methylpropane	$188\,\text{kJ mol}^{-1}$
2-bromo-2-methylpropane	$170\,\text{kJ mol}^{-1}$

Suggest what factors may be important in determining the magnitude of these activation energies. Justify your answer.

(5)

(ULEAC, 1993 (Nuffield))

84 A solution of hydrogen peroxide decomposes in the presence of a catalyst according to the equation:

$$2\text{H}_2\text{O}_2(aq) \rightarrow 2\text{H}_2\text{O}(l) + \text{O}_2(g)$$

In experiments to determine the rate of this reaction, hydrogen peroxide solutions of different concentrations were used with the same catalyst. The following results were obtained.

Experiment	$[\text{H}_2\text{O}_2(aq)]/\text{mol dm}^{-3}$	Rate of reaction/$\text{mol dm}^{-3}\,\text{s}^{-1}$
1	0.05	0.28×10^{-4}
2	0.15	0.85×10^{-4}
3	0.25	1.43×10^{-4}

a You are provided with $100 \ cm^3$ of a $0.50 \ mol \ dm^{-3} \ H_2O_2$ solution, 10 g of a solid catalyst and any necessary apparatus.
  i) How would you prepare $50 \ cm^3$ of a $0.15 \ mol \ dm^{-3}$ solution of H_2O_2 for use in experiment **2**?
  ii) Draw a labelled diagram of the apparatus you would use to carry out one of these experiments.
  iii) What measurements would you make in your experiments, and how would you use your results to obtain the rate of reaction? (8)

b This decomposition is an example of a *disproportionation reaction*. Explain the meaning of this phrase by identifying the element undergoing disproportionation, and quote its oxidation numbers on both sides of the equation. (3)

c i) Plot the results of the experiments.
  ii) From the graph deduce the order of the reaction.
  iii) Hence write the rate equation for the reaction. (6)

d This reaction is catalysed by solutions of transition metal ions. Explain why transition metal ions make good homogeneous catalysts, and explain how they work. (2)

(ULEAC, 1994 (Nuffield) (part))

85 Sucrose, $C_{12}H_{22}O_{11}$, is a disaccharide which can be readily hydrolysed into glucose and fructose. The structure of sucrose is:

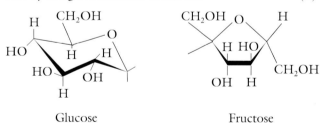

a What do you understand by the terms
  i) *disaccharide*
  ii) *hydrolysed*? (2)

b Complete the structures for molecules of glucose and fructose by using the structure of sucrose. (2)

Glucose   Fructose

c In an experiment the hydrolysis of sucrose was investigated. $5 \ cm^3$ of concentrated hydrochloric acid was added to a solution of 100 g of D(+)-sucrose in $40 \ cm^3$ of distilled water.

The mixture was placed in a polarimeter. The rotation of the plane of plane polarized light was measured at intervals over a 60-minute period.

α_t represents the angle of rotation at time t, α_o is the final value of the angle of rotation.

Time, t /minutes	α_t /degrees	$\alpha_t - \alpha_o$ /degrees
0	+65	+80
10	+19	+34
20	+1	+16
30	−9	+6
40	−13	+2
50	−15	0
60	−15 (α_o)	0

i) Why is a solution of sucrose optically active?
ii) What is the meaning of the (+) and (−) signs in the table of results?
iii) The mixture of glucose and fructose formed in this experiment is called 'invert sugar'. Given that one of the products is D(+)-glucose, what two things can you deduce about the other product, D-fructose?
iv) Plot a graph of $(\alpha_t - \alpha_o)$ against time.
v) Find three values of the half-life, $t_{1/2}$, for this reaction. What can you deduce from them? (10)

(ULEAC, 1992 (Nuffield))

86 A dipeptide, *W*, yields only valine and glycine on hydrolysis.

$$CH_3-CH-CH-CO_2H \qquad H_2NCH_2CO_2H$$
$$\qquad \quad | \qquad |$$
$$\qquad \quad CH_3 \quad NH_2$$
$$\qquad \text{valine} \qquad\qquad\qquad \text{glycine}$$

a How many different structural isomers are possible for *W*? Draw one of these structures. (2)

b Outline a synthesis of valine from 3-methylbutanoic acid. Include all necessary reagents. (2)

c Draw the structure of the predominant form in which glycine exists in aqueous solutions:

  under acidic conditions

  under basic conditions (2)

(HK, 1993 (part))

87 The table below gives data about a number of amino acids which occur in proteins.

name and abbreviation	relative molecular mass	R_f value in Solvent I	R_f value in Solvent II
alanine, ala	89	0.43	0.38
aspartic acid, asp	133	0.13	0.24
glycine, gly	75	0.33	0.26
leucine, leu	131	0.66	0.73
lysine, lys	146	0.62	0.14
phenylalanine, phe	165	0.64	0.68
serine, ser	105	0.30	0.27
valine, val	117	0.58	0.40

A small polypeptide was hydrolysed with concentrated acid and, after neutralisation, the resulting amino acids were separated by two-way chromatography. The chromatogram is shown below.

Solvent front (Solvent I)
B
A
glycine
serine
aspartic acid
Solvent I
Solvent II
Solvent front (Solvent II)

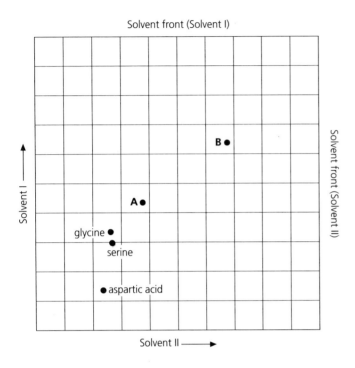

The R_f value is given by the distance travelled by an amino acid spot divided by the distance travelled by the solvent front.

a Determine the R_f values in both solvents of the amino acids labelled **A** and **B** on the chromatogram and, hence, identify them.

amino acid	R_f value in Solvent I	R_f value in Solvent II	identity
A			
B			

(4)

b Mark on the chromatogram where you would find the spot corresponding to lysine. (1)

c By reference to the table of data, explain why two-way chromatography is needed to separate these eight amino acids. (2)

d Explain why the R_f values of an amino acid differ in different solvents. (1)

e When the solvents have moved up the chromatography paper, how is the paper treated to locate the positions of the amino acids? (2)

f Why is it important to avoid fingering the chromatography paper during the experiment? (1)

(UCLES, 1996/7 (M, specimen))

88a Complete the reaction scheme shown below which starts with the compound ethene. For each of **A** to **F** write the structural formula of the principal organic product or intermediate compound.

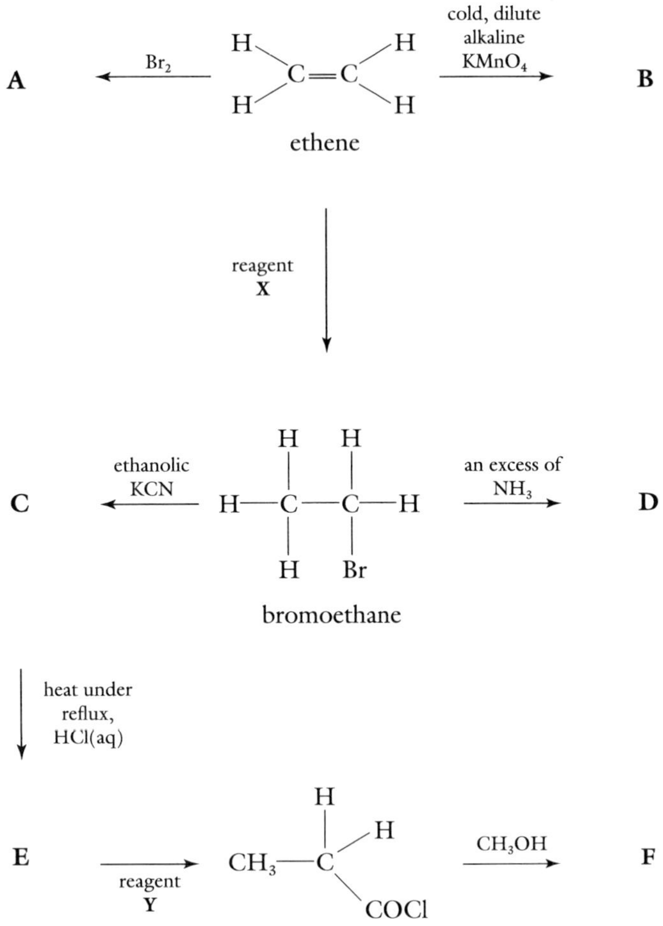

A
Br2
ethene
cold, dilute alkaline KMnO4
B
reagent X
bromoethane
C
ethanolic KCN
an excess of NH3
D
heat under reflux, HCl(aq)
E
reagent Y
COCl
CH3OH
F

b Identify the reagents **X** and **Y**.
X
Y (2)

(UCLES, 1994 (M, specimen))

89a Polymers of biological interest are usually electrically charged and can be effectively studied by electrophoresis.

Describe this technique, including in your answer
i) a description of the apparatus,
ii) an account of the principles involved,
iii) the effect of pH on the results. (14)

b Outline examples of how this technique may be used for
i) detecting and separating organic species,
ii) the analysis of genes and genetic fingerprinting. (11)

(UCLES, 1996/7 (M, specimen))

90 The yellow dye, Disperse Yellow 3, can be prepared using the reaction sequence shown:

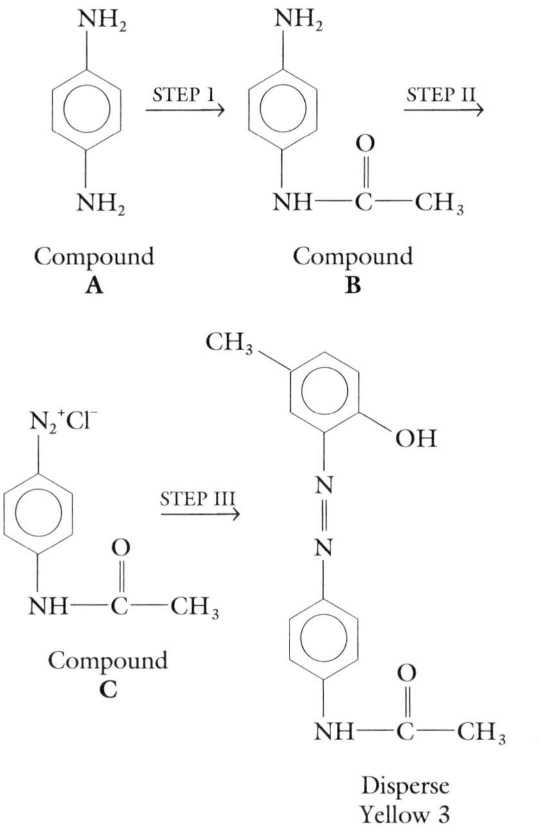

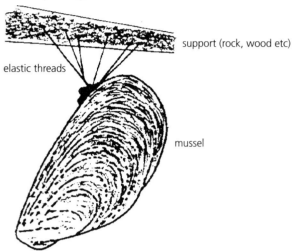

Compound
A

Compound
B

Compound
C

Disperse
Yellow 3

a **i)** Give the name and structural formula of the reagent needed to carry out STEP I.
ii) Give the names of the reagents and the reaction conditions needed to carry out STEP II.
iii) Give the name of the functional group ——N_2^+ in compound C.
iv) Give the structural formula of the reagent used in STEP III.   (6)

b Suggest TWO important properties, apart from being coloured, which are essential for a compound used as a dye.   (2)

(ULEAC, 1990 (Nuffield))

91 For each of the following pairs of compounds, describe some of the similarities and differences in physical and chemical properties between the two compounds, and give explanations where you can.

a Ethanol, C_2H_5OH, and ethylamine, $C_2H_5NH_2$

b Phenol, ⟨◯⟩—OH, and phenylamine, ⟨◯⟩—NH_2

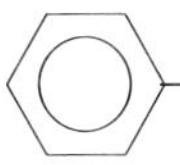

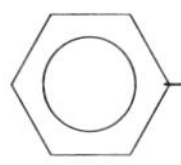

c Propene, CH_3—C=CH_2, and ethanal, CH_3—C=O
         | |
         H   H

(ULEAC, 1990 (Nuffield))

92 Modern synthetic adhesives and coatings have marvellous properties, but they still have limitations. For example, they often do not stick well to damp surfaces. In the search for a solution, chemists have turned to the common mussel (*Mytilus edulis*) for help. Mussels have to tie themselves very firmly onto rocks, and their adhesive has to be applied under water: the mussel cracked the water-resistant coating problem millions of years ago.

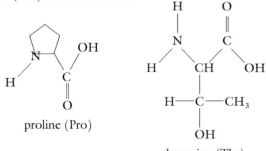

The elastic threads which anchor the mussel to its support are coated all over in a protein which may be the basis for future water-resistant adhesives. The protein has quite a simple composition, being made up from repeating units of just six amino acid residues:

. . . Ala Lys Pro Thr DOPA Lys . . .

There are two unusual aspects to this protein.
1. It appears to have no secondary structure.
2. It contains a large proportion of L-DOPA (L-3,4-dihydroxyphenylalanine), an amino acid more usually associated with the treatment of Parkinson's disease.

a State the term which is used to describe the sequence of amino acid residues in a protein: for example, Ala Lys Pro Thr etc. in this case.   (1)

b Explain the meaning of the term 'secondary structure' as it is applied to proteins.   (2)

c The structures of the amino acids proline (Pro) and threonine (Thr) are shown below.

proline (Pro)

threonine (Thr)

Use these as examples to draw a structure which shows how two amino acids are joined together in a protein.   (2)

d The amino acid composition of a protein can be investigated using the technique of chromatography. Describe in outline how you would analyse a sample of mussel thread protein to show that it was made up from the five amino acids:

Ala, Lys, Pro, Thr and DOPA. (7)

e The structure of phenylalanine is shown below. Draw the structure of DOPA. (2)

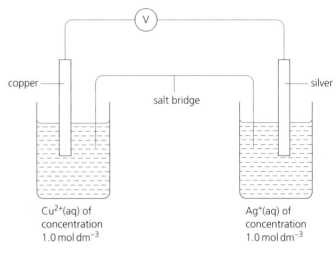

L-phenylalanine   *DOPA*

f The prefix, L, is necessary to distinguish L-phenylalanine from its isomer D-phenylalanine. Describe the feature of the phenylalanine molecule which is responsible for the existence of these two isomers. (2)

g The use of L-DOPA as a treatment for Parkinson's disease relies on the conversion of L-DOPA to the neurotransmitter, dopamine, in the body. A pharmaceutical company was interested in the possibility of producing dopamine itself in *two* steps from the compound, Y, which was available from another of the company's processes.

compound Y

intermediate compound $\xrightarrow{\text{Step D}}$

dopamine

Complete the reaction sequence shown below by giving the reagent(s), conditions and the intermediate compound for the synthetic route which you would suggest to the company for the conversion of Y to dopamine. (6)

Step C   Reagent(s)
         Conditions

Step D   Reagent(s)
         Conditions

Stucture of intermediate compound

(O & C, 1993 (Salters))

93a Draw a labelled diagram of a cell that makes use of the reaction:

$$2H_2SO_4(aq) + PbO_2(s) + Pb(s) \rightarrow 2PbSO_4(s) + 2H_2O(l)$$

Indicate the direction of electron flow on your diagram.

b Write a balanced equation for the reaction at each electrode.

c Using the IUPAC convention, write the cell diagram for the above cell.

(6)

(HK, 1993 (part))

94a Standard electrode potentials are measured against the standard hydrogen electrode. What do you understand by the term *standard*? (4)

b The apparatus in Fig. 4.1 was used to measure the e.m.f. of a cell.

Fig. 4.1

i)   Suggest suitable materials for the salt bridge. (2)
ii) What is the purpose of the salt bridge? (1)
iii) Use the Data Booklet to calculate the standard e.m.f. of the cell. Show your working. (2)
iv) Write an equation for the overall reaction taking place in the cell. (2)
v)   Draw an arrow on Fig. 4.1 to show the direction of electron flow in the external circuit when the cell is used to deliver current. (1)

c Describe a fuel cell and state what advantages it has as an energy source over a cell such as that in Fig. 4.1. (4)

(UCLES, 1996/7 (specimen))

*95 When the two half cells

$$Fe^{2+}(aq)|Fe(s), \qquad E^{\ominus} = -0.44\,V$$
$$Pb^{2+}(aq)|Pb(s), \qquad E^{\ominus} = -0.13\,V$$

are made into the cell

$$Fe(s)|Fe^{2+}(aq) \vdots Pb^{2+}(aq)|Pb(s)$$

1 the e.m.f. of the cell is 0.57 V

2 electrons would move from the iron to the lead in an external circuit

3 lead ions are reduced

(ULEAC, 1991 (Nuffield))

*96

Electrode system	$E^{\ominus}$/volt
$Fe^{2+}(aq)|Fe(s)$	-0.44
$Sn^{2+}(aq)|Sn(s)$	-0.14
$Cu^{2+}(aq)|Cu(s)$	$+0.34$
$Fe^{3+}(aq),Fe^{2+}(aq)|Pt$	$+0.77$
$Cl_2(aq),2Cl^-(aq)|Pt$	$+1.36$

From the standard electrode potentials given above, it can be deduced that

1 iron will displace copper from copper(II) chloride solution

2 tin will reduce iron(III) ions to iron(II) ions

3 the e.m.f. of the cell

$$Pt|Fe^{2+}(aq),Fe^{3+}(aq)\colon\!Cl_2(aq),2Cl^-(aq)|Pt$$

is $+0.59$ volts

(ULEAC, 1992 (Nuffield))

97 Rechargeable batteries in common use are sealed so that their contents will not leak as they are carried about. A student investigated a nickel/cadmium rechargeable battery and found that it had an e.m.f. of 1.3 V. The electrolyte had a pH of 14 and contained potassium ions.

a Suggest what is used as the electrolyte in the rechargeable battery. (1)

b Two standard electrode potentials from the Book of Data are:

$$Cd^{2+}(aq)|Cd(s) \qquad E^{\ominus} = -0.40\,V$$

$$Ni^{2+}(aq)|Ni(s) \qquad E^{\ominus} = -0.25\,V$$

i) Write down a possible cell diagram based on the electrode potential data above. Deduce the standard e.m.f. of the cell you have written.

ii) From your answer to **(i)**, suggest why this cell would be unsuitable for many common uses of batteries. (3)

c The student discovered in a reference book that the cell reaction in the rechargeable nickel/cadmium battery was:

$$2NiO(OH) + Cd + 2H_2O \rightarrow 2Ni(OH)_2 + Cd(OH)_2$$

i) Write down the oxidation number of each metal before and after this reaction. Include positive or negative signs as appropriate.

Nickel: before..........   after..........

Cadmium: before..........   after...........

ii) Use the changes in oxidation numbers to deduce whether the electron flow around the external circuit is in the same direction for this cell reaction as for the cell reaction based on your answer to **(b)(i)**. Explain your answer.

iii) Write down the equation for the reaction that takes place as the cell is recharged. (4)

(ULEAC, 1993 (Nuffield))

98 Some steps in the rusting of iron may be represented as follows:

(*a*) $$O_2(g) + 2H_2O(l) + 4e^- \rightarrow 4OH^-(aq)$$
(*b*) $$Fe(s) \rightarrow Fe^{2+}(aq) + 2e^-$$

In addition to water and oxygen a further reagent is required for rusting to occur.

The steps and the further reagent may be described as follows:

	Step (a)	*Step (b)*	*Reagent*
A	reduction	oxidation	any electrolyte
B	reduction	oxidation	an acid only
C	reduction	oxidation	a base only
D	oxidation	reduction	any electrolyte
E	oxidation	oxidation	an acid or a base

(O & C, 1994)

99a Describe the reactions of the Group II metals, magnesium to barium, with
   i) oxygen,
   ii) water.

   Write equations where appropriate. (5)

b i) Suggest reasons why magnesium gives the nitride, Mg_3N_2, in addition to its oxide when burned in air. (2)
   ii) A 1.00 g sample of the powder obtained from burning magnesium in air was boiled with water. The ammonia that was evolved neutralised 12.0 cm^3 of 0.5 mol dm^{-3} hydrochloric acid.

   Construct balanced equations for the production of magnesium nitride and its reaction with water.

   Calculate the percentage of magnesium nitride in the 1.00 g sample. (5)

(UCLES, 1996/7 (specimen))

100 When the elements of Group 1 are heated in oxygen, the type of oxide formed varies from metal to metal. Lithium oxide, Li_2O, sodium peroxide, Na_2O_2, and potassium superoxide, KO_2, illustrate this variation.

a i) In which one of the above three compounds do *both* elements have their usual oxidation numbers?
   ii) Draw a 'dot-and-cross' electron diagram for the peroxide ion, O_2^{2-}, showing only the outermost electrons.
   iii) What is unusual about the electron arrangement in the superoxide ion, O_2^-? (3)

b Sodium peroxide reacts exothermically with cold water:

$$2Na_2O_2(s) + 2H_2O(l) \rightarrow 4NaOH(aq) + O_2(g)$$

i) Which element is both oxidized and reduced? Give the changes in oxidation number.
ii) Why should sodium peroxide be kept out of contact with material such as paper or organic liquids? (4)

c Sodium peroxide is used to absorb carbon dioxide and release oxygen in a closed environment such as inside a submarine, but lithium peroxide, Li_2O_2, is preferred in space capsules.
   i)   Write an equation for the reaction of sodium peroxide with carbon dioxide.
   ii) Suggest why lithium peroxide is preferred to sodium peroxide in space capsules.
   iii) Suggest three factors which must be taken into account in calculating an appropriate mass of lithium peroxide to be carried on a space voyage.   (5)

(ULEAC, 1990 (Nuffield))

101 This question is about an investigation into the thermal decomposition of the anhydrous nitrates of the Group 2 elements.

The aim of the investigation is to compare the ease with which the nitrates of the different elements decompose.

Some oxides of nitrogen are evolved as brown fumes and can be collected by condensation in the cooled U-tube, while oxygen gas is collected in the syringe.

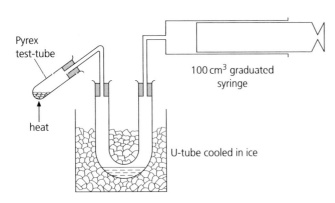

a i) A valid comparison of the ease of thermal decomposition requires that the test-tube in each experiment is heated in an identical manner. How could this be achieved?
   ii) Suggest a valid way in which the ease of decomposition of different nitrates may be compared using this apparatus.   (3)

b Write an equation, including state symbols, for the decomposition of anydrous magnesium nitrate in the heated test-tube. Assume that the oxides of nitrogen formed can be represented by the formula NO_2.   (2)

c The results of an investigation showed that magnesium nitrate decomposes more easily than barium nitrate. Suggest an explanation for this.   (2)

(ULEAC, 1993 (Nuffield) (part))

102 Which of the following statements could be a reason why barium hydroxide has a much higher solubility in water than magnesium hydroxide?

A $Ba(OH)_2$ is more covalent than $Mg(OH)_2$.

B Barium has higher first and second ionisation energies than magnesium.

C The enthalpy of formation of $Ba(OH)_2$ is greater than that of $Mg(OH)_2$.

D The lattice enthalpy of $Ba(OH)_2$ is less than that of $Mg(OH)_2$.

E The enthalpy of hydration of Ba^{2+} is greater than that for Mg^{2+}.

(O & C, 1993)

103 Lithium (atomic number 3) and sodium (atomic number 11) are both in Group 1 of the periodic table. The radius of the Li^+ ion is 0.060 nm and that of the Na^+ ion is 0.095 nm.

Which prediction is **least** likely to be correct?

A The radius of a lithium atom is less than that of a sodium atom.

B The enthalpy change of hydration of Li^+ is more negative than that of Na^+.

C The first ionisation energy of lithium is greater than that of sodium.

D Lithium carbonate is more thermally stable than is sodium carbonate.

E Lithium compounds have more covalent character than the corresponding sodium compounds.

(O & C, 1993)

104 Which one of the following shows the variation in melting points of the elements of the period sodium to chlorine? (The diagrams are not to scale.)

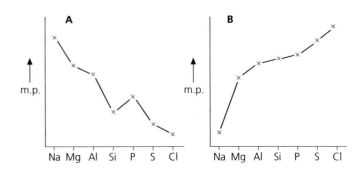

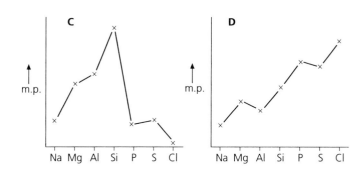

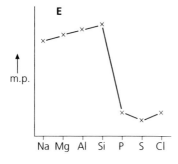

(NEAB, 1994)

105 When the chemical industry wishes to manufacture a desired chemical product, chemists have to solve two basic problems:

1. Is there a chemical reaction which will form that product to an appreciable extent from available starting materials?
2. Is the rate of reaction adequate?

Discuss how these problems have been investigated and overcome by chemists in the case of the synthesis of ammonia.

A further problem is the separation and collection of the desired product. Suggest ways in which ammonia could be separated from excess starting materials: justify your suggestions.

(ULEAC, 1990 (Nuffield))

106 This question concerns hydrazine, which has the molecular structure:

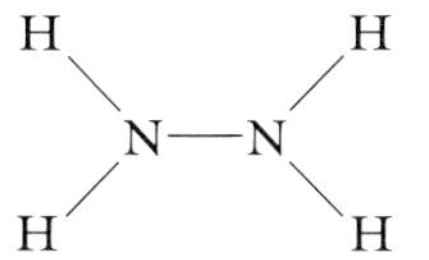

[Some data for hydrazine may be found on pp. 84–85 of the Revised Book of Data; note that hydrazine is a liquid at room temperature.]

a The enthalpy change of formation for gaseous hydrazine at 298 K has a value of +95 kJ mol^{-1}. Using bond energy data (Revised Book of Data p. 51) and an appropriate energy cycle check that this value is consistent with the bond energy data.

b Because of its structure, the properties of hydrazine are likely to be similar to those of ammonia. Describe some of the major physical and chemical properties you would expect hydrazine to have on the basis of this similarity to ammonia.

As a guide, an answer which gives four sound predictions and discusses these fully with supporting data and/or equations is likely to be awarded a high mark. Answers should include both inorganic and organic examples.

You are advised that the description of the use of hydrazine as a rocket fuel (Students' Book 1, pp. 181–188) contains very little information relevant to answering this question.

(ULEAC, 1992 (Nuffield))

107 This question concerns the extraction, properties and uses of aluminium [details of which appear in the Data Book and in Topic 18 in Students' Book II].

a For each stage in the manufacture of aluminium give BALANCED equations. You should include equations starting with powdered raw bauxite up to the manufacture of pure aluminium. The reactions of impurities should also be given. Which equation(s) represent redox reactions?

b Suggest why aluminium is used for electrical conductors, window frames and garden furniture by linking these uses to appropriate data from the Book of Data, and making a comparison with the use of alternative metals.

c Aluminium should be a reactive metal on the basis of its $E^{\ominus}$ value. However, a very thin layer of oxide about 10 nm thick forms on the surface of the aluminium which protects the metal underneath.
  i) Give the equation for the reaction between aluminium and oxygen in the presence of water and calculate the $E^{\ominus}$ value for this system.
  ii) Estimate how many ions make up the thickness of the oxide layer. Justify your estimate.

(ULEAC, 1993 (Nuffield))

108a Write the formula in each case of ONE compound or ion in which nitrogen has the oxidation state +5, +3, −1 and −3.

b Describe the structure and the bonding in nitrogen, N_2, and in white phosphorus, P_4. (6)

(HK, 1992 (part))

109 This question is about Group 7 of the Periodic Table.

a Complete the following table.

Halogen	Physical state at room temperature	Colour
Fluorine Chlorine Bromine Iodine		

(3)

b When potassium chloride is treated with concentrated sulphuric acid, a steamy gas, Y, is evolved.
　i) Write an equation for this reaction.
　ii) When Y is dissolved in water, in which it is readily soluble, a strongly acidic solution, Z, is formed. Describe the reaction that occurs between Y and water in terms of the Brønsted–Lowry theory.
　iii) Silver nitrate solution is added to Z. A white precipitate forms. If silver nitrate in concentrated aqueous ammonia is cautiously added to Z, no precipitate forms. Explain these observations. Why must the second addition be carried out cautiously? (8)

c Aqueous chlorate(I) ions, ClO^-, decompose on warming in a disproportionation reaction. Write the equation for this reaction, and by considering the oxidation state changes involved, explain the term *disproportionation*. (3)

d Chlorine is extensively used in water treatment plants.
　i) Upon which chemical and physical properties of chlorine does its use rely?
　ii) Excess chlorine can be removed by treatment with aqueous sulphur dioxide which is oxidized to sulphate ions. Derive an equation for this reaction. (4)

(ULEAC, 1992)

110a Outline the processes in organic compounds which bring about absorptions of energy which give rise to
　i) ultraviolet spectra,
　ii) infra-red spectra,
　iii) nuclear magnetic resonance spectra. (5)

b Using $[Cr(H_2O)_6]^{3+}$ as an example, explain why transition metal complexes absorb energy in the uv/visible region of the spectrum, and hence appear coloured. (3)

c The uv/visible spectrum shown below is that of a transition metal complex. Predict the colour of the complex, indicating how you arrive at your answer. (2)

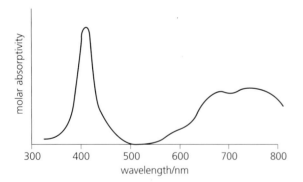

(UCLES, 1996/7 (specimen))

111a With the aid of diagrams, identify and describe the types of isomerism that can exist in octahedral complex ions. (4)

b Palladium(II) salts can form square planar complexes. Successive addition of ammonia and hydrogen chloride to an aqueous palladium(II) salt produces, under different conditions, three compounds with empirical formula $PdN_2H_6Cl_2$. Two of these, **A** and **B**, are non-ionic, with $M_r = 211$. **A** has a dipole moment, whereas **B** has none. The third compound, **C**, is ionic, having $M_r = 422$, and contains palladium in both its cation and anion.

For each of **A**, **B** and **C**, suggest a structure that fits the above data. (6)

(UCLES, 1996/7 (specimen))

112a By using different reagents, aqueous vanadium(V) ions can be reduced to lower oxidation states.

Use the *Data Booklet* to suggest the reagent you could use to obtain aqueous ions of
　i) vanadium(II),
　ii) vanadium(IV).
Write a balanced equation for each reaction.

Explain why the reagent you have chosen to produce vanadium(IV) ions does not further reduce the vanadium even when added in excess. (6)

b A 0.0100 mol sample of an oxochloride of vanadium, $VOCl_x$, required 20.0 cm^3 of 0.100 mol dm^{-3} acidified potassium manganate(VII) for oxidation of the vanadium.

By calculating
　i) how many moles of electrons were removed by the MnO_4^- ions,
　ii) the change in oxidation state of the vanadium, deduce the value of x in the formula $VOCl_x$. (4)

(UCLES, 1996/7 (specimen))

113a The electronic configuration of the Cu^+ ion is $1s^2 2s^2 2p^6 3s^2 3p^6 3d^{10}$.
　i) Write the electronic configuration of the Cu^{2+} ion. (1)
　ii) Usually, copper(II) compounds are coloured: this is not the case for copper(I) compounds. Explain this difference. (1)

b Some reactions of $Ni^{2+}(aq)$ ions are shown below.

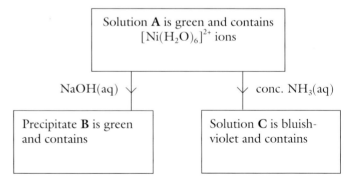

Give the chemical formulae of the nickel-containing species present in **B** and **C**. Write your answers in the boxes. (2)

c In the reaction scheme shown below, solution **D** contains a salt of metal **X**. All the unidentified solutions are aqueous.

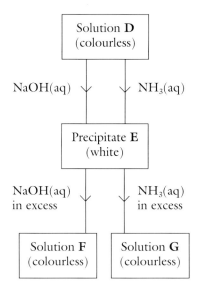

i) Identify **X**. (1)
ii) Write the chemical formula of **E**. (1)
iii) What ion of **X** is present in solution **F** and in solution **G**? (2)

(UCLES, 1994 (M, specimen))

114a Explain, in terms of electronic arrangement, why most transition metal ions are coloured. (2)

b Chromium forms three common oxides: CrO, Cr_2O_3 and CrO_3.

Arrange these oxides in ascending order of basicity. Explain your arrangement. (3)

c There are three compounds of chromium with the same formula $[Cr(en)_2Cl_2]^+Cl^-$, where **en** is ethane-1,2-diamine, $H_2NCH_2CH_2NH_2$.

Draw a three-dimensional structure for one of these three compounds, and give its systematic name. (2)

d Write equations for the reactions described below.

The addition of an aqueous solution of KSCN to a solution containing Fe(III) ions gives a complex with a deep red colour. The deep red colour fades when excess NaF solution is added. On the addition of dilute alkali to the resulting solution, a brown precipitate forms. (3)

(HK, 1993 (part))

115 Car exhaust fumes contain NO_2, NO, unburnt hydrocarbons and carbon monoxide as undesirable pollutants. In an exhaust fitted with a catalytic converter this mixture passes through a 'honeycomb' mesh coated with platinum and rhodium. These metals catalyse, by chemisorption, the reaction of these gases with each other.

a Suggest three likely products of reaction of these gases with each other.

b Why is a 'honeycomb' structure used in the catalytic converter?

c What is the meaning of the term *chemisorption*?

d What stages are involved in reactions that involve chemisorption? (7)

(ULEAC, 1993 (Nuffield) (part))

Answers

Answers

CHAPTER 1   page 1

1   The fact that ice is a crystalline solid tells us that, like all crystals, here is something produced by an organised stacking of regular repeating units rather than a random jumble. We could also add that the creation of water from two distinct ingredients, hydrogen and oxygen, and the reversal of the same process by electrolysis, is supportive of a particle model, if not absolute proof.

2   Einstein's general theory of relativity has established the equivalence of mass and energy, and enshrined it in the famous equation:

$E = mc^2$

which expresses the amount of energy E (in joules) available from the annihilation of mass m (in kg), where c is the speed of light (in m s^{-2}). So we can now no longer say that the amounts of either mass or energy in the universe are separately constant. But in ordinary laboratory chemistry the law of Conservation of Mass is still used. The student comparing masses before and after a chemical reaction in a school laboratory need not worry about how much mass has been destroyed. Any differences in mass are more likely to be found on the floor.

3   Any given amount of a gas (if by 'amount' we mean number of molecules) will have a volume which depends upon two other factors – temperature and pressure. So if we are to talk meaningfully about gas volumes we have to say what the temperature and pressure were when the volume was measured.

4   It is a good approximation to say that the volume of the actual molecules of a gas is negligible compared to the volume they take up by moving around and hitting the walls of their container (just as the volume of the bodies of a football team is negligible compared to the size of the pitch they fill by running around on it – admittedly this is comparing an area with a volume but it makes the point). So the size of the actual molecule is irrelevant.

In a vessel like a syringe (which guarantees a pressure inside of 1 atmosphere) the space taken up depends upon how much force is exerted on the walls per molecular collision. This in turn depends on the kinetic energy ($\frac{1}{2}mv^2$) of the molecules. Kinetic energy of gas molecules depends **only on temperature** – in fact, in a way kinetic energy per molecule *is* temperature – so you can see that a bromine, Br$_2$ and a hydrogen, H$_2$ molecule can only have the same temperature by having the same overall value of $\frac{1}{2}mv^2$, the hydrogen being light-and-fast, and the bromine heavy-and-slow. The effects of their collisions with the container walls are also the same, with the same trade-off between speed and mass.

5   Material may not be destroyed in a chemical change, but volume certainly can be. In this case the matter in the system is more 'clumped together' as water than it is in the form of the two elements, and so the reduction in particle numbers means a reduction in volume too (by a factor of 3:2).

6   The volume ratio of 10:20:10 means there must be a molecular ratio of 1:2:1. The only equation to meet this requirement is **B**.

7   It is clear from the data (especially when presented as a graph, see below) that reaction ceases afer the addition of 25 cm^3 of oxygen. The argument then proceeds thus:

a   50 cm^3 of NO reacts with 25 cm^3 of oxygen, hence the reacting ratio is 2 (NO):1 (O$_2$).

b   So 25 cm^3 of oxygen were left unreacted, to be present in the 75 cm^3 product mixture.

c   The actual product gas must have occupied (75 – 25) or 50 cm^3.

d   The ratio NO:O$_2$:product is 50:25:50 by volume, and therefore 2:1:2 by molecules.

e   This suggests an equation

$2NO(g) + O_2(g) \rightarrow 2NO_x(g)$

which solves to give $x = 2$. The product is **nitrogen dioxide**.

f   The tubing contains unmeasured volumes of O$_2$(g) and NO(g). You could argue that the direction of flow means that the unmeasured tube-full of oxygen on the right remains unaltered at the end, but the tube-full of NO(g) will get involved. So in effect we are adding oxygen to (50 + x) cm^3 of NO, without knowing the value of x. Clearly it would be better if x were negligibly small.

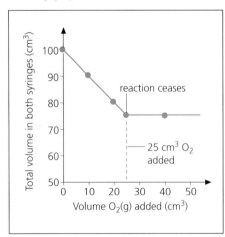

Figure A1.1 Graph for answer 7

8   If 100 cm^3 of air contains 0.3 cm^3 of CO$_2$, the volume ratio is 3:10 000, or 300:1 000 000. It follows that the molecular ratio must be the same, so CO$_2$ is present in air at **300 ppm**.

9a   CH$_4$(g) + 2O$_2$(g) → CO$_2$(g) + 2H$_2$O(g)
So volume ratio is 1:2.

b   2C$_2$H$_2$(g) + 5O$_2$(g) → 4CO$_2$(g) + 2H$_2$O(g)
So volume ratio is 2:5.

c   By downward displacement of water.

10   We are looking for the one equation in which the number of gas molecules on the product side exceeds that on the reactant side, because then the volumes will show the same pattern. The answer is **D**, from 7 molecules to 8 molecules.

11a   C$_{18}$H$_{18}$(g) + $\frac{25}{2}$O$_2$(g) → 8CO$_2$(g) + 9H$_2$O(g)

b   1:12.5.

c   O$_2$ is 20.95 % of air by molecule, which is the same as by volume, so O$_2$:air by volume = 20.95:100 or 1:4.77.

d   So to take enough air to get all the octane burnt we would need a ratio of 1:(12.5 × 4.77) or 1:59.6.

e   Imagine the 1 and the 59.6 were in dm^3. Their masses would be 4.66 g and (1.18 × 59.6) g, a ratio of 1:15.1.

f   Yes, the numbers of molecules do not depend on temperature, so neither will the volume ratios.

g   (i) Mass.
(ii) 1 dm^3 of air contains the same number of molecules as 1 dm^3 of octane, so if the volume occupied by that number expands to 2 dm^3, it will do so for both gases.
(iii) If the masses have stayed the same and the volumes have expanded together, the ratio of densities will have stayed the same.

h   (i) If petrol were a C$_7$ hydrocarbon there would be a fuel:oxygen volume ratio lower in oxygen, which would work out to a fuel:air mass ratio lower in air, so *yes*, that would explain '14.7'. (This assumes that the density of the new petrol is more or less the same as the old.)
(ii) The opposite position cannot be true, so *no*.
(iii) If the petrol eventually vaporises in the course of the burn, it should not matter how it enters. As long as it is all burnt, the mass ratio of part (e) will stand, so *no*.

CHAPTER 2   page 8

1a   Missing figures and graph are shown opposite. The ratio by mass of Mg:O is about 3:2.

b   This ratio does not translate into the relative *numbers* of atoms reacting, since we do not know (in the context of this question at least) their relative atomic masses. One interpretation is that there is the same number of magnesium atoms as oxygen atoms, but that magnesium atoms happen to be heavier than oxygen atoms in the ratio 3 : 2. Another is that there are more magnesium atoms than oxygen atoms.

Waals' force on its neighbours are electrons, and they all have some of those. Dipoles, on the other hand, depend upon the electronegativities of particular atoms.

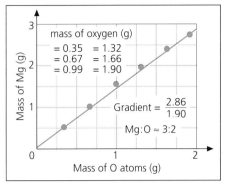

Figure A2.1 Graph for answer 1

2a The mass ratio suggests that helium particles have twice the mass of hydrogen particles.

b Chemists know that elemental hydrogen exists as dihydrogen molecules, H_2, whereas helium is monatomic. This accounts for the factor of two by which hydrogen is heavier than predicted.

3a

Dalton's name	Real identity
Magnesia	MgO
Lime	CaO
Soda	Na_2CO_3
Potash	K_2CO_3
Strontian	$SrCO_3$ probably
Barytes	$BaSO_4$ probably

b Electrolysis.

c He must have assumed that water had the formula HO. Then having measured the reacting masses of oxygen and hydrogen as being in the ratio 8:1, he concluded that oxygen atoms were eight times heavier than hydrogen atoms.

d Had he considered the evidence of reacting volumes, he would perhaps have guessed that *two* atoms of hydrogen combine with *one* atom of oxygen. So eight would be the number of times that one atom of oxygen was heavier than two atoms of hydrogen, giving an atomic mass for oxygen of 16.

e If magnesium atoms are heavier than oxygen atoms in the ratio 3:2, their atomic mass relative to hydrogen would be $\frac{3}{2} \times 16$, or 24.

f The only *fact* we have is the magnesium to oxygen mass-ratio of 24:16, so it could well be the case that the '24' bit was 2×12 in a formula Mg_2O. This would lead us to express the formula of magnesium chloride incorrectly as Mg(12)Cl. So you can see how in this field of early chemistry, one false assumption could lead to a whole string of false conclusions being drawn from perfectly valid experimental data.

4 Their mass ratio would be 118.69:126.9, or about 1:1.07.

5 This mass ratio would be 118.69:(4 × 126.9), that is, 118.69:507.6 or about 1:4.28.

6 Any masses in this ratio would do, so for example 10 g of tin and 42.8 g of iodine.

7 They are the same.

8a There would be 6×10^{23} carbon atoms in 12 g of diamond.

b The mole concept does not necessarily apply only to single continuous molecules. A mole is a mole of anything you choose it to be. So it makes most sense, in the diamond case, to talk about a mole of C(s, diamond) atoms, meaning a mole of carbon atoms in the diamond crystal lattice. And this of course would have a mass of 12 g. As we shall see it is important for the sake of clarity to specify what you are taking a mole **of** – for example, the phrase 'a mole of hydrogen' is ambiguous: is it a mole of H atoms (1 g) or a mole of dihydrogen molecules (2 g)?

9a $3 \times 126.9 = 380.7$ g. In fact this would come in the form of I_2 molecules – you would have no choice in that – but it would *contain* three moles of iodine atoms. We could equally well refer to this amount as 1.5 moles of I_2.

b $0.5 \times (2 \times 126.9) = 126.9$ g
(It would be a sign of sophistication if you did not actually *do* the calculation, having recognised at once that this amount was identical to one mole of I atoms.)

c $0.1 \times (23 + 35.5) = 5.85$ g

d $5 \times (2 \times 12 + 6 \times 1 + 16) = 230$ g

e $2.5 \times (2 \times 23 + 12 + 3 \times 16 + 10 \times 18)$ = 715 g

f $0.2 \times (40 + 32 + 4 \times 16 + \frac{1}{2} \times 18) = 29$ g

g
$0.4 \times (40 + 2 \times (14 + 3 \times 16) + 4 \times 18)$ = 94.4 g

10a 50 g of silver is equivalent to a number of moles given by

Number of moles

$$= \frac{\text{mass in grams}}{\text{relative atomic mass}} \qquad (2.1)$$

$$= \frac{50}{108}$$

= 0.463 mol

b $\frac{1000}{207} = 4.83$ mol of lead

Use (2.2) (p. 000) for parts (c) and (d).

c $\frac{100}{2 \times 63.5 + 12 + 3 \times 16 + 2 \times 17}$

= 0.452 mol of malachite

d $\frac{50}{4 \times 12 + 9 + 80}$

= 0.365 mol of 1-bromobutane

11a One mole of malachite contains two moles of copper atoms, so to get one mole of copper atoms we need half a mole of malachite. This is equivalent to a mass given by:

Mass = number of moles required × RFM
$= 0.5 \times 221$
$= 110.5$ g

b Percentage by mass of copper in malachite
$$= \frac{2 \times 63.5}{221}$$
= 57.5%

12a 10^6 g of chlorine is equivalent to a number of moles of dichlorine molecules given by:

Number of moles $= \dfrac{\text{mass in grams}}{\text{RFM}}$

$$= \frac{10^6}{71}$$

The mole ratio of (Cl_2 destroyed : Br_2 created) is 1:1.

Therefore moles of Br_2 created $= \dfrac{10^6}{71}$.

This is equivalent to a mass of Br_2 given by:

Mass in grams = number of moles × RFM

$$= \frac{10^6}{71} \times 160$$

$= 2.25 \times 10^6$ g

b 50 kg of KIO_3 is equivalent to a number of moles given by:

Number of moles $= \dfrac{\text{mass in grams}}{\text{RFM}}$

$$= \frac{50 \times 10^3}{39 + 127 + 48}$$

= 233.6 mol

The mole ratio of (KIO_3 destroyed : I_2 created) is 2:1.

Therefore moles of I_2 created = 116.8 mol.
This is equivalent to a mass of I_2 given by:

Mass in grams = number of moles × RFM
$= 116.8 \times 254$
$= 29.7 \times 10^3$ g
$= 29.7$ kg

(In the course of this example I left all the figures in the calculator, not rounding off to the appropriate number of significant figures until the end.)

13 (Full formula layout is being abbreviated to prevent repetition.)

Moles of diamine used

$$= \frac{10^3}{2 \times 16 + 6 \times 14} = 8.62 \text{ mol}$$

So moles of chloride = 8.62 mol (we are told the mole ratio is 1:1)

This is equivalent to:

$8.62 \times (8 \times 14 + 2 \times 35.5 + 2 \times 28)$
= 2060 g = 2.06 kg

14 The number of moles of hydrogen chloride created is in a 2:1 ratio with numbers of moles of either of the individual monomers destroyed, that is, 8.62×2 mol.

So mass of hydrogen chloride created
= number of moles × RFM
$= (8.62 \times 2) \times 36.5$
$= 629.3$ g

But total mass of reaction system = mass of combined reactants = (1000 + 2060) g.

Therefore mass of nylon = 3060 − 629.3 = 2431 g = 2.43 kg

The reason we could not go direct to the mass of nylon was because there is no absolute value of n, so no RFM for the polymer.

15 Moles of Fe atoms created

$$= 1000 \times \frac{10^3}{56}.$$

The mole ratio of (Fe_2O_3 destroyed):(Fe created) is 1:2.

The moles of Fe_2O_3 destroyed

$$= 0.5 \times \frac{1000 \times 10^3}{56}$$

= 8928 mol.

This is equivalent to a mass of

$8928 \times (2 \times 56 + 3 \times 16)$
$= 1428 \times 10^3$ g = 1428 kg.

(Some of you will feel the temptation to use *double* the RFM of iron in the first line of the problem, because of the 'two' in front of the iron in the equation. The reason that is wrong is because it would lead to the 'two' being used twice. The correct place to use it is in considering the (moles created):(moles destroyed) ratio. To put that another way, if you *had* used 112 for the RFM of '2Fe' it would be as if you had decided to define the product as '2Fe', in which case the reaction would become a 1:1 reaction between Fe_2O_3 and '2Fe'. This would give you a correct answer via some rather unconventional terminology, but it does emphasise the fact that one way or another the 'two' must only get used once.)

16a The choice depends on the virtual insolubility of barium sulphate, $BaSO_4$, so that every sulphate ion is dragged out of solution to get weighed. The method also relies on Ba^{2+} ions *only* precipitating with SO_4^{2-} ions.
b The excess of Ba^{2+} ions refers to the use of *more than enough* cations to combine with every sulphate ion. Unreacted Ba^{2+} ions just remain in solution.

c 29 mg of barium sulphate is equivalent to

$$\frac{29 \times 10^{-3}}{137.3 + 32 + 64} = 1.24 \times 10^{-4} \text{ mol.}$$

Since the mole ratio of (soluble sulphate):(precipitate) is 1:1, this number also represents the moles of soluble sulphate.

This is equivalent to a mass of sulphate ions given by:

$1.24 \times 10^{-4} \times (32 + 64) = 0.0119$ g
= 11.9 mg per kg of water
(or in percentage terms,

$$\frac{11.9 \times 10^{-3}}{1000} \times 100 = 0.0012\%)$$

17a Number of moles of carbon dioxide is given by:

$$\text{Number of moles} = \frac{\text{mass in grams}}{\text{RFM}} \qquad (2.2)$$

$$= \frac{31.4}{44} = 0.7136$$

Similarly, number of moles of water

$$= \frac{12.9}{18} = 0.716$$

b This is close to being a 1:1 mole ratio.
c So the ratio of (atoms of carbon) : (atoms of hydrogen) = 1:2.

18a 72.8 cm^3 at 37.3 K, 1 atm, is equivalent to a volume at 298 K, 1 atm, given by:

$$V_1 = V_2 \times \frac{T_1}{T_2}$$

(since P does not change)

$$= 72.8 \times \frac{298}{373}$$

$$= 58.2 \text{ cm}^3$$

b This is equivalent to a number of moles given by:

Number of moles

$$= \frac{\text{volume in dm}^3 \text{ at 298 K, 1 atm}}{24.45} \qquad (2.3)$$

$$= \frac{58.2 \times 10^{-3}}{24.45}$$

$$= 2.379 \times 10^{-3}$$

If this fraction of a mole weighs 0.2 g, then a whole mole will have a mass given by

$$\text{RFM} = \frac{\text{mass in grams}}{\text{number of moles}}$$

$$= \frac{0.2}{2.379 \times 10^{-3}}$$

$$= 84 \text{ g mol}^{-1}$$

Notice that a quicker route is to say that a whole mole is bigger than the sample by a factor of 24.450:58.2, so their masses must be in the same ratio. Hence the answer can be obtained in one line:

Mass of one mole $= \dfrac{0.2 \times 24\,450}{58.2} = 84$ g

equivalent to several lines in the previous method.

c If the hydrocarbon has an empirical formula (from question 17) of CH_2, and a RFM of 84, we can say that its molecular formula is $(CH_2)_n$, where

$$n \times 14 = 84$$
and therefore $\qquad n = 6$

The hydrocarbon's molecular formula would be written conventionally as C_6H_{12}, and might be cyclohexane or one of the hexenes.
d $C_6H_{12}(l) + 9O_2(g) \rightarrow 6CO_2(g) + 6H_2O(l)$
e A mass spectrometer could be used to look at the 'molecule ion' peak (Chapter 3, p. 31).

19a They would occupy 24 450 cm^3.
b The ratio gas : liquid is $\frac{24\,450}{18}$. So a mole of water as a gas occupies a space 1360 times greater than a mole of liquid water.

20 See Table A2.1.

21 See Table A2.2.

The last entry was done by realising that by the assumptions in the problem, 1000 cm^3 of the solution would contain 500 cm^3 (that is, 500 × 0.789 grams) of ethanol.

22a Volume $= \dfrac{\text{moles}}{\text{concentration}}$ (2.7, *rearranged*)

$$= \frac{(100/98)}{1}$$

$$= 1.02 \text{ dm}^3 = 1020 \text{ cm}^3$$

Table A2.1 Answer for question 20

Substance	Moles needed given by vol. (dm^3) × conc.	Mass needed given by moles × RFM
NaCl	1 × 1 = 1	1 × 58.5 = 58.5 g
NaOH	0.5 × 1 = 0.5	0.5 × 40 = 20 g
$BaCl_2$	0.1 × 0.25 = 0.025	0.025 × 208 = 5.2 g
$K_2Cr_2O_7$	0.2 × 0.5 = 0.1	0.1 × 294 = 29.4 g
$CuSO_4 \cdot 5H_2O$	0.1 × 2 = 0.2	0.2 × 249.5 = 49.9 g

Table A2.2 Answer for question 21

Substance	Moles used given by $\frac{\text{mass}}{\text{RFM}}$	Concentration (in mol dm^{-3}) given by moles/vol (in dm^3)
KI	1000/166 = 6.02	6.02/10 = 0.602
$FeSO_4$	100/152 = 0.658	0.658/0.25 = 2.63
Ethanol	(500 × 0.789)/46 = 8.58	8.58/1 = 8.58

b $\text{Volume} = \dfrac{\text{moles}}{\text{concentration}}$. (2.7, rearranged)

$$= \dfrac{(100/80)}{8 \times 10^{-4}}$$

$= 1560 \text{ dm}^3$ (to 3 SF)

23a Since every mole of calcium chloride, $CaCl_2$, provides two moles of chloride ions, a solution which is 0.1 mol dm^{-3} with respect to $CaCl_2$ is 0.2 mol dm^{-3} in chloride ions.

b   For a 0.5 mol dm^{-3} solution of sodium ions, you need a solution of Na_2SO_4 which is 0.25 mol dm^{-3} with respect to the parent solute.

24 Take 200 cm^3 of concentrated hydrochloric acid, and make up to 1 dm^3.

25 The total volume of solution which could contain 2.12 moles and have a concentration of just 2.00 mol dm^{-3} is given by:

$\text{Volume (dm}^3) = \dfrac{\text{moles}}{\text{concentration (mol dm}^{-3})}$

(2.7, rearranged)

$$= \dfrac{2.12}{2}$$

$= 1.06 \text{ dm}^3 = 1060 \text{ cm}^3$

So you have to add 60 cm^3 to the original 1000 cm^3 to get the desired round-figure result.

26 Number of moles of sodium hydroxide used is given by:

Number of moles
= concentration (mol dm^{-3}) × volume (dm^3)
= 1 × 0.025
= 0.025

Since the mole ratio is 1:1, the number of moles of ethanoic acid involved in the titration will be the same, except that this time the volume involved is 35.6 cm^3. So its concentration will be given by:

$\text{Concentration (mol dm}^{-3}) = \dfrac{\text{moles}}{\text{volume (dm}^3)}$

(2.7, rearranged)

$$= \dfrac{0.025}{0.0356}$$

$= 0.702 \text{ mol dm}^{-3}$

27a Moles of NaOH used
= concentration (mol dm^{-3}) × volume (dm^3)

(2.7, rearranged)

= 1 × 0.035
= 0.035

b   Since the mole ratio NaOH:H_2SO_4 is 2:1 (from the chemical equation), the number of moles of sulphuric acid present

$= \dfrac{0.035}{2} = 0.0175.$

c   Since the mole ratio H_2SO_4:$CaSO_4$ is 1:1, the number of moles of calcium sulphate in the 25 cm^3 of solution is also 0.0175.

d   So the solubility of calcium sulphate is given by:

$\text{Concentration (mol dm}^{-3}) = \dfrac{\text{moles}}{\text{volume (dm}^3)}$

(2.7, rearranged)

$$= \dfrac{0.0175}{0.025}$$

$= 0.7 \text{ mol dm}^{-3}$

e   The water wash ensures that all the newly released H$^+$ ions have come off the column.

f   We were never interested in the concentration of sulphuric acid in the wash water. The target was the *number of moles*, as an indicator of the number of moles of original Ca^{2+}.

28a There are several objections to this mode of telling an end-point. One is the slow percolation of the acid towards the centre of a lump after the surface layers have surrendered their calcium carbonate content and the fizzing has apparently died down.

b   The 5.13 g lump is not all calcium carbonate, so it is wrong to say its RFM is 100.

c   Moles of sodium hydroxide used
= 1 × 0.0111   (2.7)
= 0.0111

d   Therefore the number of moles of acid in the 10 cm^3 titration batch is 0.0111. The original 100 cm^3 must have contained 10 × 0.0111 = 0.111 moles of hydrogen chloride.

e   Before it came into contact with the limestone the acid had a concentration of 2.00 mol dm^{-3}, which means that 100 cm^3 would have contained moles given by:

Number of moles
= concentration (mol dm^{-3}) × volume (dm^3)   (2.7)
= 2 × 0.1
= 0.2

So the moles of acid destroyed by contact with the lump is given by 0.2 − 0.111 = 0.089.

f   Since the mole ratio of HCl:$CaCO_3$ is 2:1, the number of moles of calcium carbonate in the lump is 0.089/2 = 0.0445.

g   So the mass of calcium carbonate in the lump is given by:

Mass (g) = number of moles × RFM

(2.2 rearranged)

= 0.0445 × 100
= 4.45 g

As a percentage of the lump this is

$\dfrac{4.45 \times 100}{5.13} = 86.7\%.$

29 Moles of sodium hydroxide
= 2 × 0.0137 = 0.0274   (2.7)
(Note the conversion of 13.7 cm^3 to 0.0137 dm^3.)

Moles of the acid

$= \dfrac{2}{6 \times 12 + 10 + 4 \times 16}$ (2.2)

= 0.0137

So the mole ratio of alkali:acid is 2:1, which means that the acid must have *two* acidic hydrogens. Let us imagine you have no organic chemical knowledge at all – the most you could write for an equation would be:

$C_6H_8O_4H_2 + 2NaOH \rightarrow C_6H_8O_4^{2-}Na_2^+ + 2H_2O$

CHAPTER 3   page 24

1a *Experiment A*, opposites attract, and *Experiment B*, evidence from laws of electromagnetic induction. In the latter case the fact of deflection proves the particles are charged, and are therefore behaving like a sort of 'current without a conductor'. The application of Fleming's left-hand rule shows that the classical current direction is from anode to cathode. However, if you assume that the cathode must be the source of the rays, then the actual flow must be of negatively charged particles from cathode to anode.

b   No matter what materials he used for the cathode, anode, glass tube or trace gas, the rays obtained had identical mass:charge ratios. The answer then is *Experiment C*.

c   Currents had been assumed to be a stream of positive particles moving from regions of positive potential to the regions of negative potential. Now it was realised that the mobile entity in most conduction situations was negative electrons moving counter to the traditional current direction. The tradition lives on however, in that we still talk about positive sides of resistors as being at a higher potential, even though the electrons are doing work and losing energy coming the other way. Also conventions like Fleming's rules still use the traditional current direction.

d   The positive rays are the positive ions of the residual gas in the tube. They are created when fast-moving 'free' electrons in the cathode stream collide with gas atoms. The energy of the collisions strips electrons off; for example:

$Ne(g) + e^-(\text{fast}) \rightarrow Ne^+(g) + 2e^-(\text{slow})$

The newly created gaseous cations are now attracted to the cathode, and if it is designed as a grid or a tube, they will become another stream of particles ready to be tested. They will of course give different mass:charge ratios for each gas.

2   Density of helium nucleus = 2 × 10^{15} g cm^{-3}
Density of gold atom = 20 g cm^{-3}
Ratio of densities = 10^{14}:1

(This question shows up whether people know how to type a number like 10^{-23} into their calculators. Students who think you have to type '10 E − 23' are wrong, and have really typed in 10^{-22}. You only have to think how you would type in, say, 6 × 10^{-23} to realise what 10 E −23 really means.)

3   The possibility exists for varying the **accelerating voltage** to create a mass spectrum.

4 An ion X^{2+} would experience double the deflecting force compared to X^+. It would therefore be deflected more. In fact it will be deflected as much as a singly charged ion of **half** the mass. Thus it would be misleading to call the x-axis mass, since particles of different masses could give peaks at the same x-axis position. For example, a 2+ ion of mass 24 would give a peak at 12, the same as a 1+ ion of mass 12. The only safe thing you can say about peaks in that position is that they both have m/z ratios of 12.

5 The bit that has been knocked out is an electron, which as far as low-resolution mass spectrometry is concerned has negligible mass.

6a The peaks at 35 and 37 are the Cl atoms. (In fact they are Cl^+ ions by then, but they derive from Cl atoms.)
b Two versions, one of mass number 35 and one of 37.
c Because the mix of atoms in naturally occurring chlorine is weighed in favour of mass number 35.
d There are **three** possible dichlorine molecules to be made from the two types of atom: $^{35}Cl_2$ (mass number 70), $^{37}Cl_2$ (mass number 74), and $^{35}Cl^{37}Cl$ (the other one).
e This answer is to do with the laws of probability. If the weighted average calculation reveals that ^{35}Cls outnumber 37s by 3 : 1, then:
Chances of finding a ^{35}Cl in position 1 in a Cl_2 molecule
$= \dfrac{3}{4}$
Chances of finding a ^{35}Cl in position 2 in a Cl_2 molecule
$= \dfrac{3}{4}$
So chances of getting a $^{35}Cl_2$ molecule
$= \dfrac{3}{4} \times \dfrac{3}{4} = \dfrac{9}{16}$.
By similar reasoning, chances of a $^{37}Cl_2$
molecule $= \dfrac{1}{16}$.
The chances of a mixed molecule can be found by similar means to the first two lines, except that the chances are doubled because it does not matter which of the two positions each atom is
in. This gives $2 \times (\dfrac{3}{4} \times \dfrac{1}{4}) = \dfrac{6}{16}$.
So finally we can say that the chances of getting dichlorine molecules of mass numbers 70, 72 and 74 are in the ratio 9 : 6 : 1, which shows up in the abundances in Figure 3.18.

7a Because atoms are electrically neutral.
b A change in the number of electrons (and an associated change in proton number) produces a different element with different chemical character.

8 The heights of the peaks give the abundance for each isotope, i.e. the weighting

to give it. The weighted average formula when individual values of a parameter have different weighting factors (or abundances) is:
Average =
$$\dfrac{\sum (\text{weighting factor} \times \text{parameter value})}{\sum (\text{weighting factors})}$$
In this case, we have:
Average =
$$\dfrac{0.3 \times 204 + 2.5 \times 206 + 2.2 \times 207 + 5.0 \times 208}{0.3 + 2.5 + 2.2 + 5.0}$$
= 207.16 or 207 (to a more realistic 3 sf)

9 These samples of lead must have been formed by different radioactive decay paths from different precursors (starting materials) and therefore have different isotopic mixtures.

10a Neutrons will not undergo electrostatic deflections if they come close to nuclei, because they are not charged.
b In a cloud chamber a moving radioactive particle will ionize (knock electrons from) air molecules by impact. α-particles are good at this since they are both relatively massive and attractive to electrons. Neutrons are attractive neither to air molecules nor to their electrons, so tend to pass straight through without causing ionization. Since there are no ions to act as the 'seeds' for droplet formation, no 'vapour trails' are observed.
c There is a high occurrence of water and other hydrogen-containing molecules in bodies.
d The infrastructure would be largely intact, ideal for taking over and running the country after the war.

11 The figure suggests that not all ions are created at exactly the same distance from the accelerating plates, so the 'force × distance' factors for individual ions are similar but not identical.

12a CO = 27.994 91 and N_2 = 28.006 14
b The carbon monoxide molecule might have had slightly more kinetic energy than the nitrogen molecule, and therefore have been slightly harder to deflect than a less energetic nitrogen molecule, the kinetic energy difference in favour of carbon monoxide more than compensating for the mass difference in favour of nitrogen.
c 28 (only 2 sf count).
d Some individual molecules of each kind would fragment in the ionization stage, and then the carbon monoxide would be recognisable from peaks at 12 and 16, as distinct from peaks at 14 in the nitrogen case.
e A collection of particles containing the same number of particles as there are carbon atoms in 12.0000000 g of the isotope ^{12}C.

13

	p^+	e^-	n
^{75}As	33	33	42
$^{79}Br^-$	35	36	44
^{208}Pb	82	82	126
$^{208}Pb^{2+}$	82	80	126

14a It would be a single line of colour.
b White light would be many photons of different wavelengths.

15 He used values for the charge on an electron and proton, and the mass of an electron. These were needed to set up the equations for circular motion. He then worked out the radius at which the force pulling the electron into the nucleus was balanced by the desire of objects to travel in straight lines. Notice that this is still essentially a Newtonian picture.

16 For example, a person (of amazing energy) climbing a very high ladder until they had almost left the Earth's gravitational field. They should find the millionth step easier than the first.

17a Dark (absorption) lines.
b An absorption spectrum.
c Reversing the arrows would represent an emission spectrum.
d The jump from, say, $1 \rightarrow 7$ is only slightly bigger than that from $1 \rightarrow 6$, and hence the wavelengths, frequencies, wavenumbers and therefore positions in the spectrum will be correspondingly close.

18 The convergence frequency represents the frequency of the photon which has enough energy to ionize an electron. So if we apply Planck's equation to this frequency, we will be calculating the energy jump between levels 1 and infinity, or in other words the ionization energy. So:
$$E = h \times \nu \qquad (3.1)$$
$E_{photon} = 6.63 \times 10^{-34} \times 3.27 \times 10^{15}$ J for a one-electron jump from quantum level 1 to 'infinity'
$= 2.168 \times 10^{-18}$ per electron
For a mole of electrons (using the Avogadro constant):
$E = 2.168 \times 10^{-18} \times 6.02 \times 10^{23}$
≈ 1300 kJ mol^{-1}
This compares well with the quoted ionization energy for hydrogen of 1312 kJ mol^{-1}.

19 A series of jumps from level 2 to higher levels would give such a series, and that is what the Balmer series is. The energy intervals would be less and hence the frequency lower (from $E = h \times \nu$).

20 Because the innermost orbital is the one nearest to the nucleus and lowest in energy.

21a 2
b 8
c 18
d The Periodic Table has blocks of width 2, 8, 18 elements.

e On most Periodic Tables it is not clear that principal quantum level 3 is an 18. It looks like another 8, because of the way the 3d orbitals defer their filling-up till after the 4s orbital.

f The 'K shell' is the 1s orbital.

22a Ionization energies would go *up* from helium to neon if nuclear pull were the only factor (that is, as the nucleus gets bigger).

b Ionization energies would go *down* from helium to neon if distance were the only factor.

c Ionization energies would go *down* from H to Ne if electron–electron repulsion were the only factor.

23a See Figure A3.1.

b The orbital is full at helium, whereupon there is a dramatic drop in ionization energy.

c Distance and electron–electron repulsion must have overruled nuclear pull.

24 **A** and **B** must be the cause of the general increase in ionization energy from lithium to neon.

25 See Figure A3.2.

26 Jumps from, say, 2s to 1s would give different lines from jumps from 2p to 1s. As it is, however, jumps from 2s to 1s are the same as 2p to 1s.

27

Element	Electron arrangement
N	$1s^2 2s^2 2p^3$
O	$1s^2 2s^2 2p^4$
F	$1s^2 2s^2 2p^5$
Ne	$1s^2 2s^2 2p^6$
Na	$1s^2 2s^2 2p^6 3s^1$
Mg	$1s^2 2s^2 2p^6 3s^2$
Al	$1s^2 2s^2 2p^6 3s^2 3p^1$
Si	$1s^2 2s^2 2p^6 3s^2 3p^2$
P	$1s^2 2s^2 2p^6 3s^2 3p^3$
S	$1s^2 2s^2 2p^6 3s^2 3p^4$
Cl	$1s^2 2s^2 2p^6 3s^2 3p^5$
Ar	$1s^2 2s^2 2p^6 3s^2 3p^6$

28 See Figure A3.3.

29a $1s^2 2s^2 2p^6 3s^2 3p^6$

b 3p

c There are fewer remaining electrons to provide shielding and/or repulsion (both of which would have helped removal), or you could argue that successive electrons are being pulled away from an entity which is getting increasingly positive, as in Ar^+, Ar^{2+}, etc.

d Up to electron eight, electrons are being removed from principal quantum level 3; electron nine comes from principal quantum level 2.

e That is when there is a move from the 3p to the 3s sub-shell.

f The first three electrons are helped off by 'two-in-a-bed' repulsion.

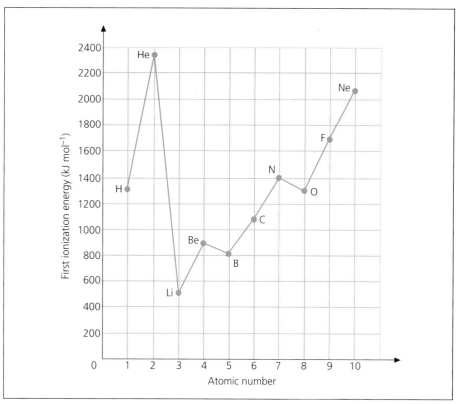

Figure A3.1 (See answer 23a.) A graph of first ionization energy of the first ten elements, plotted against atomic number

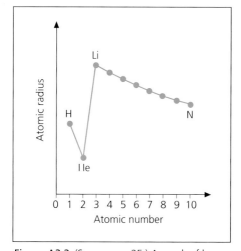

Figure A3.2 (See answer 25.) A graph of how atomic radius varies with atomic number

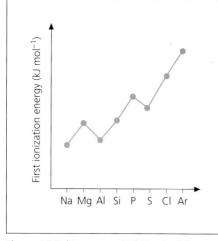

Figure A3.3 (See answer 28.) Graph of first ionization energies for elements from Na to Ar

30

Ge: $1s^2 2s^2 2p^6 3s^2 3p^6 3d^{10} 4s^2 4p^2$

Sr: $1s^2 2s^2 2p^6 3s^2 3p^6 3d^{10} 4s^2 4p^6 5s^2$

Ag: $1s^2 2s^2 2p^6 3s^2 3p^6 3d^{10} 4s^2 4p^6 4d^9 5s^2$
(actually $4d^{10} 5s^1$ but you would not have predicted it).

Sn: $1s^2 2s^2 2p^6 3s^2 3p^6 3d^{10} 4s^2 4p^6 4d^{10} 5s^2 5p^2$

(I have used the convention of putting all sub-shells of a principal quantum number together, so that $3d^{10}$ precedes $4s^2$.)

Mg^{2+} $1s^2 2s^2 2p^6$

Na^+ $1s^2 2s^2 2p^6$

Cl $1s^2 2s^2 2p^6 3s^2 3p^6$

Cu^{2+} $1s^2 2s^2 2p^6 3s^2 3p^6 3d^9$ (note that the $4s^2$ are the ones to go)

31

I	II	III	IV	V	VI	VII	VIII
s^1	s^2	s^2p^1	s^2p^2	s^2p^3	s^2p^4	s^2p^5	s^2p^6

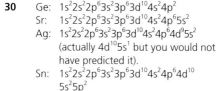

CHAPTER 4 page 49

1a −436 kJ mol⁻¹, that is, the work needed to separate one mole of hydrogen into separate atoms.

b The data book gives 0.037 nm as the covalent radius of a hydrogen atom. This value is found by taking the distance between two covalently bonded hydrogen atoms in a hydrogen molecule and dividing by two. So the value at point X is 0.037 × 2 or 0.074 nm.

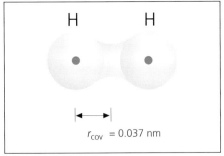

Figure A4.1 Covalent radius of H atom

c As the inter-nuclear distance reduces from point X, the inter-nuclear repulsion factor begins to more than compensate for the attraction of nuclei by electrons. Hence the potential energy of the molecule rises, eventually passing a point (at $y = 0$) where the situation is actually 'worse' than no bond at all.

2a 8.
b 6.
c Two, since each bond allowed the oxygen access to a share in one or more electron.
d One, since hydrogen can accept a share in one more electron.

3 See Figure A4.2.

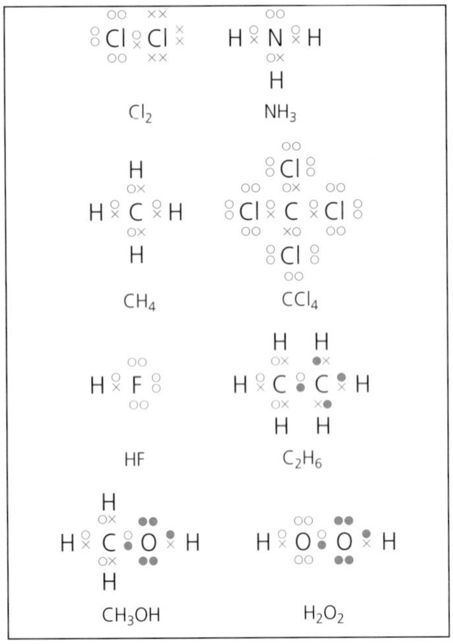

Figure A4.2 Answer to question 3

4a Normal valencies: H = 1, C = 4, N = 3, O = 2, F = 1, Ne = 0.
b Group number + valency number = 8 (or two for hydrogen).

5a Two, just like oxygen.
b SF₆ in which sulphur accepts six covalent bonds.

6 Sulphur can break the rule of eight because the energy cost incurred by unpairing and promoting two electrons to occupy vacancies in the 3d sub-shell is more than recouped by the four extra bonds which become possible. Oxygen cannot do this, because 2d electrons do not exist, and the 3s sub-shell is energetically too far 'uphill' (Figure 4.12, p. 52).

7 Yes, because it cannot do better than use *all* its outer electrons for bonding, and the inner-shell electrons are too stably held to get involved.

8 The maximum valency number and the group number will be the same – see the answer above.

9 In this business of energy pay-backs, the energy of bond formation must compensate for the energy of electron unpairing and promotion. The chances of a successful outcome are therefore best if the new bonds are the stablest available, and this is achieved if the partner in the bonding is either oxygen or fluorine, with their highly desirable vacancies.

10 Carbon promotes one of its 2s pair into its 2p vacancy, at very little energetic cost (2s → 2p), thus providing for two more bonds.

11 See Figure A4.3.

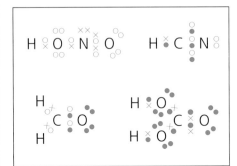

Figure A4.3 For answer 11

12 See Figure A4.4

Figure A4.4 For answer 12

13 See Figure A4.5.

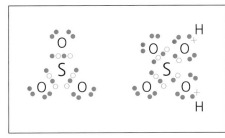

Figure A4.5 For answer 13

14 See Figure A4.6.

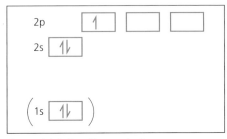

Figure A4.6 Box diagram for boron

b Yes, similar to carbon.
c It can get up to six.
d Less than eight.
e Yes, one completely vacant orbital.
f See Figure A4.7.

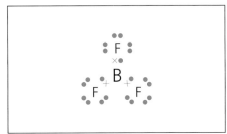

Figure A4.7 Boron trifluoride

15 Any atom from the second row, from a group beyond Group IV, will have at least one lone pair in any of its compounds, unless the atom is already involved in dative covalent bonding. So you could have picked, for example, water, ammonia or hydrogen fluoride as typical examples.

16a See Figure A4.8.

Figure A4.8 Nitrous acid

b   Figure 4.24 would require the nitrogen atom to make *five* covalent bonds, which would be a violation of the rule of eight.

c   See Figure A4.9.

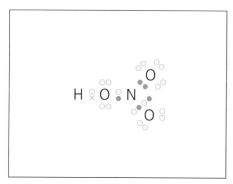

Figure A4.9 Dot-and-cross nitric acid

d   See Figure A4.10.

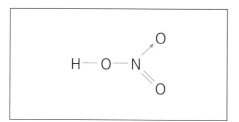

Figure A4.10 Letters-and-sticks nitric acid

e   Instead of two double bonds, there would be a double and a single, and doubles show up on electron density maps by virtue of their *shortness* and *high electron density*, relative to singles. In fact, the two bonds in question turn out rather surprisingly to be the *same* length, which is intermediate between those of double and single bonds. This is due to **delocalisation** (see Section 4.6).

17   Dinitrogen is isoelectronic with carbon monoxide (but with a normal triple bond and no dative).

18   2 = linear, bond angles 180 °
          3 = trigonal planar (a flat triangle), bond angles 120 °
          5 = trigonal bipyramidal (Figure A4.11), bond angles 120 ° and 90 °

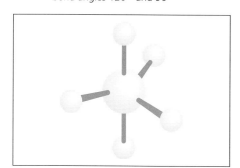

Figure A4.11 Trigonal bipyramid

6 = square bipyramid (octahedron) (Figure A4.12), bond angles 90 °.

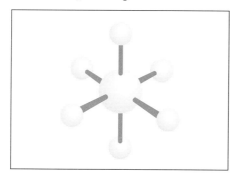

Figure A4.12 Octahedral arrangement of six atoms around a centre

19   Ethane is two fused tetrahedra, as in Figure A4.13.

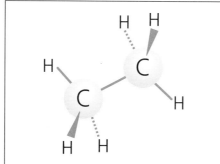

Figure A4.13 Ethane

20 $H_2C{=}O$ Two single bonds, one double, no lone pairs, so *three bits*, so a trigonal plane, bond angles 120 °.

BF_3 Three single bonds, no lone pairs, so *three bits*, so a trigonal plane, bond angles 120 °.

PCl_5 Five single bonds, no lone pairs, so *five bits*, so a trigonal bipyramid, bond angles 120 ° and 90 °.

NH_3 Three single bonds, one lone pair, so *four bits*, so a tetrahedron with one 'invisible' point, which will end up looking like a very shallow trigonal pyramid, bond angles 109.5 ° (Figure A4.14).

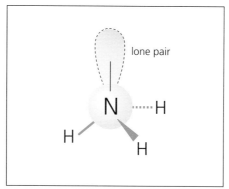

Figure A4.14 Ammonia

SO_2 Two double bonds, one lone pair, so *three bits*, so a trigonal plane with one 'invisible' point, which will end up looking bent, bond angle 120 ° (Figure A4.15).

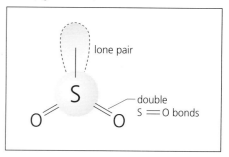

Figure A4.15 Sulphur dioxide

SO_3 Three double bonds, no lone pairs, so *three bits*, so a trigonal plane, bond angles 120 °.

HNO_2 One single bond, one double bond, one lone pair, so *three bits*, so a trigonal plane with one 'invisible' point, which will end up looking bent, bond angle 120 ° (Figure A4.16).

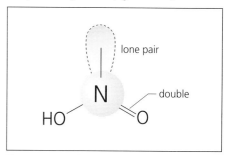

Figure A4.16 Nitrous acid

HNO_3 One single bond, one double bond, one dative bond, no lone pairs, so *three bits*, so a trigonal plane, bond angles 120 ° (but see 'delocalisation' in Section 4.6).

21a BrF_3 Bromine is in Group VII, so it has seven outer electrons. Three must be involved in binding the three fluorine atoms, so the other four must be arranged in two lone pairs. So three bond pairs and two lone pairs make this a *five-bit* molecule, a trigonal bipyramid with two 'invisible' points, leaving a T-shape. (But I don't know why the two lone pairs do not go top and bottom of the bipyramid.) (Figure A4.17.)

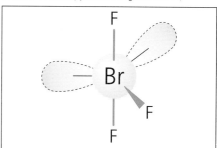

Figure A4.17 BrF_3

b   XeF₄   Xenon is in Group VIII or 'Group 0', and has eight outer electrons. Four are involved in binding the fluorine atoms, so the other four must be arranged in two lone pairs. So four bond pairs and two lone pairs make this a *six-bit* molecule, an octahedron with two 'invisible' points. This time the two lone pairs *are* top and bottom, leaving a square plane (Figure A4.18).

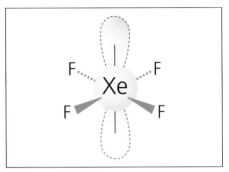

Figure A4.18 XeF₄

22   CH₄   Four bond pairs means a perfectly regular tetrahedron.

NH₃   The one lone pair repels the three bond pairs more than they repel each other, so they close up their angles to 107° (see Figure A4.14).

H₂O   The *two* lone pairs repel each other most forcefully, which in turn causes them to bear down on the two bond pairs. The two bond pairs are forced even closer, to 104.5° (Figure A4.19).

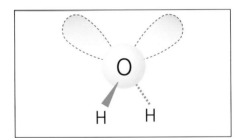

Figure A4.19 Water

23   It seems reasonable to predict that the angles will be similar to ethene, on the basis that again there is a double bond and two singles. So HCH = 118°, and the two HCOs = 121°.

24   The message of the 120° bond angle in SO₂ is that the [lone pair]–[double bond] repulsive effect must be *equal* to the [double bond]–[double bond] repulsive effect.

25   A dot-and-cross diagram is shown in Figure A4.20 from which it can be seen that there is one lone pair, as well as two single bonds and one double, at the central S atom. The shape is therefore that of a *four-bit* molecule, a tetrahedron, but with one 'invisible' point, leaving the same sort of shallow trigonal pyramid as in the NH₃ case. If we allow for small distortions, we might predict that the O—S—O

angle might close up to about 104.5°, since it is being pushed by repulsive forces comparable to those in water. The O=S=O angles might stay about 109.5°, since there are influences working both ways – the extra repulsive effect of the lone pair seeking to close them down, and their own electron density pushing the other way.

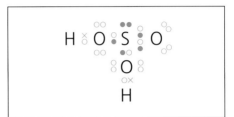

Figure A4.20 Sulphurous acid

26   The fact that the bonds are all the same length is the incompatible feature – the original Kekulé formula would have alternate longer and shorter bonds.

27a   Lengths of 'pure' single and double NO bonds are listed as 0.120 nm and 0.114 nm, so a reasonable guess would be 0.117 nm for the nitrogen–oxygen bonds in HNO₃.
b   It is possible to draw a curly arrow which transfers a lone pair from the OH oxygen into the ON bond (Figure A4.21), but it does not make sense from the energy angle – the main drawback is that it requires the creation of two new charge centres, and charge separation of this sort is not easily achieved. However, if something could be persuaded to make off with the H⁺ from the OH group, then we have a different story – in fact the nitrate (NO₃⁻) ion *is* delocalised over all three NO bonds. This is touched on again in Chapter 5.

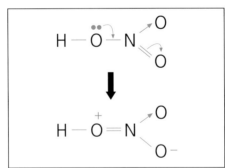

Figure A4.21 Nitric acid

28a   The double bond tolerates no rotation at all, other than a sort of limited 'waggle' when it is energetically excited (see infrared spectroscopy, p. 77). (If you have seen that famous clip of film of the last dying moments of the Tacoma Narrows bridge one windy day in the 1940s, you will be able to picture the sort of waggle I mean – and that case did end in free rotation, unfortunately.)

b   The two-bent-single-bonds model can cope with the isomer evidence, since it predicts zero rotation. Lack of rotation is the root cause of this particular form of isomerism.
c   The two-bent-single-bonds model could explain the shortness of the double bond, since the rather sideways-on overlap brings the carbons closer to each other.

29a   109.5°, since each carbon is in *four-bit* mode.
b   If distortion were to do anything to the HCH angle it would surely *close it up* slightly. In any case the difference between 118° and 109.5° is rather too large to explain away purely by distortion theory. Therefore the 118° angle is incompatible with the two-bent-single-bonds model.

30a   Vacancies.
b   Further from the nucleus and well shielded by full-up inner shells.

31   An ionic bond seems feasible whenever a member of Groups I, II, III (especially a lower-down member) or a d-block element, combines with a member of Groups V, VI or VII (especially a higher-up member). As we shall see, this simple prediction has to be hedged about by many detailed considerations about the degree and 'purity' of ionic bonding.

32   MgCl₂, BaBr₂, Na₂O, Cs₂S, CaO, AlF₃, Rb₃N, Al₂O₃, Mg₃N₂.

33   All compounds with the sodium chloride structure must have a 1:1 stoichiometry.

34   Yes, as long as the anion is 2–, thus maintaining the 1:1 stoichiometry.

35   Each Cl⁻ ion occupies an octahedral site in the Na⁺ lattice, so the two types of ion interlock equivalently.

36   Centres of edges, $12 \times \frac{1}{4}$ = 3
Wholly within cell, 1×1 = 1
Total Na⁺ within cell   = 4.

37   Remember to look up *ionic radii*, not covalent or van der Waals' radii.
a

Table A4.1 Sizes of anions and cations

Compound	Cation radius (nm)	Anion radius (nm)	Cation : anion ratio
NaF	0.102	0.133	0.77
NaCl	0.102	0.180	0.57
NaBr	0.102	0.195	0.52
NaI	0.102	0.215	0.47
CsF	0.170	0.133	1.28
CsCl	0.170	0.180	0.94
CsBr	0.170	0.195	0.87
CsI	0.170	0.215	0.79

b   Apart from the extreme case of caesium fluoride, all the anions are bigger than all their cation partners; so it is harder to get a lot of anions around a cation rather than the other way round.

38a Yes, it is true, they occupy equivalent sites in each other's lattice.

b   8:8

c   Eight Cs^+ ions at corners $= 8 \times \frac{1}{8} = 1$
One Cl^- ion totally within the
cell = 1
So 1:1 stoichiometry is confirmed.

d   All (or almost all) ionic compounds go for the maximum co-ordination number, under the limitation set by the big-anion-little-cation problem. That limitation dictates that they cannot have a co-ordination number so big that the cation is 'rattling around' inside a cavity created by hulking anions touching each other (repulsively). The large size of Cs^+, and the resultingly larger cation:anion radius ratio, means that caesium chloride can have a structure with a higher co-ordination number than can sodium chloride.

e   Caesium chloride has a cation:anion radius ratio in the region of one, so we might expect that other compounds with similar values of the same parameter would adopt the same structure. A prime candidate is caesium bromide (0.87), followed by the other two caesium halides. Of compounds not in the table above, RbF (1.12) seems another reasonable choice.

39a Tetrahedral, in both cases.

b   4:4

c   Four Zn^{2+} ions totally within the cell = 4
Eight S^{2-} ions at corners $= 8 \times \frac{1}{8}$ = 1

Six S^{2-} ions at face-centres $= 6 \times \frac{1}{2}$ = 3
Total S^{2-} ions = 4.
So stoichiometry is confirmed at 1:1.

d   The ionic radii of the Zn^{2+} and S^{2-} ions are 0.075 nm and 0.185 nm respectively. This gives a cation:anion radius ratio of 0.41, which is smaller than any of the Group I halides in the table above. So the switch to an even more modest co-ordination number is consistent with our radius ratio model for predicting structures.

40a A cubic hole.

b   The co-ordination numbers of Ca^{2+} and F^- are eight and four respectively, which is an upside-down version of the stoichiometric ratio.

c Ca^{2+} ions are arranged as were the S^{2-} ions in Figure 4.58, so we can say there are *four* of them in all. But there are *eight* F^- ions totally within the unit cell, so the stoichiometry is confirmed at 1:2.

CHAPTER 5   page 72

1   Both quantities depend on nuclear pull. If two atoms have very unequal ionization energies, it will be because they have very unequal nuclear pulls on the outer electrons. The atom with a high ionization energy (IE) will have a high IE because of strong nuclear pull, so any vacant orbitals on this atom will offer a visiting electron a very 'warm welcome', in the shape of a big (negative) electron affinity. The situation

will be exactly reversed for an atom with a low ionization energy.

2   Caesium and fluorine. This pairing combines the element with the lowest first ionization energy and the element with the biggest (negative) electron affinity. (That ignores the exotic radioactive element francium, of which there is estimated to be about 30 g in the entire Earth's crust at any one time!)

3   See Table A5.1.

Table A5.1 Results for question 3;
ΔN_P = difference in electronegativity
(see Table 5.1, p. 74)

Molecule	ΔN_P of one bond	Approximate percentage ionic character
Cl_2	0	0
CH_4	0.4	5
NH_3	0.9	15
H_2O	1.4	36
$AlCl_3$	1.5	40
$FeBr_3$	1.0	18
NaCl	2.1	66
CsCl	2.3	72
CsF	3.3	>90

Some of these results seem to underestimate the degree of ionic bonding. 66 % seems too low for the ionic character of sodium chloride – if you look ahead to Table 5.2 you will see that the assumption of 'pure' ionic bonding (100 %) allows a very accurate prediction for the lattice energy of sodium chloride.

4   Series 1 (halogenoalkanes) reflects the decreasing electronegativity (and therefore electron-attracting power) of the halogen atoms as you descend Group VII.
Series 2 (chloralkanes) illustrates that the greater *length* of the alkyl group allows greater distance between the poles of the dipole, thus increasing the dipole moment.

5   The overall dipole moment decreases as the number of chlorine atoms increases. This is despite the fact that each C—Cl bond has its own dipole. The reason is of course because of the way the dipoles interact.

6   In the case of a multiple bond, the individual dipoles *are* pointing in the same

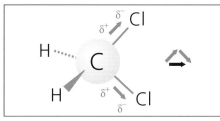

Figure A5.1 a Dipoles in C—Cl bonds in dichloromethane. Each dipole is represented by a vector. **b** Vector addition diagram, showing how the resultant dipole vector is less than the individual bond dipole vectors

direction, and so reinforce each other. And the more bonds the bigger the dipole.

7   It will be a *dative covalent bond* between a lone pair on the oxygen atom of water and the hydrogen ion.

8a   The dipole in the S═O double bond is large, because there is an electronegativity difference of 1.0 between oxygen and sulphur, amplified by the fact that the bond is double. The negative end of the dipole resides on the *oxygen* atom.

b   The sulphur atom is having its electron density drawn away by the S═O double bond, so it will be less able to allow withdrawal by the O of the S—O single bond.

c   So the oxygen atoms will not be getting their customary electron-share from one of their two bonds (the one with a sulphur atom), and so will want more electron-share from bonds to hydrogen atoms, which will be more extremely dipolar than normal O—H bonds.

d   Yes, the increased polarity of the O—H bonds (with the increased positivity of the hydrogen atoms) increases the ease with which the O—H bond can break ionically – that is, acidically.

9 OH^- is a better hydrogen ion receiver than H_2O because the ionic charges will attract the species together, and also because the lone pair on an OH^- will be more pushy, by virtue of greater electron–electron repulsion.

10   *Delocalisation* will render the bonds identical. The molecule-ion is a symmetrical '4-bit' species with no lone pairs, hence the regular tetrahedron. The SO_4^{2-} ion can be thought of as a mixture of these six classical localised structures (Figure A5.2).

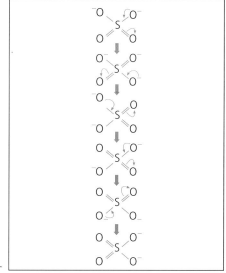

Figure A5.2 Delocalisation makes a regular tetrahedron out of the SO_4^{2-} ion

11   They will both be *trigonal planes*, because they are symmetrical 'three-bit' molecule-ions.

12 The alkalinity comes from the formation of a few OH⁻ ions.

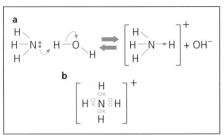

Figure A5.3a The reversible reaction by which ammonium ions are formed from ammonia and water, and **b** dot-and-cross diagram of NH_4^+

13 The water and hydrogen chloride dipoles enable the molecules to align, and then a lone pair from the water's oxygen atom will take off the H^+ on the end of a dative covalent bond. In other words the H—Cl bond, already dipolar, breaks ionically under the influence of water.

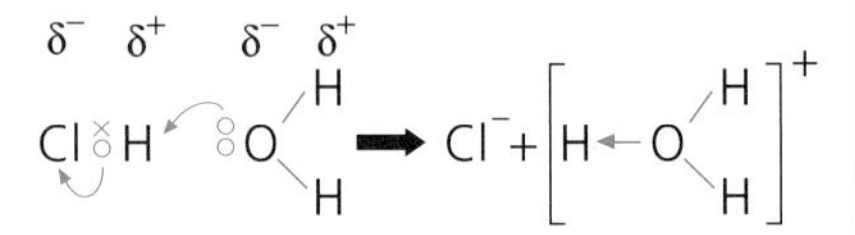

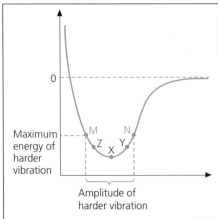

Figure A5.4 How hydrochloric acid forms from hydrogen chloride gas

14a 0.127 nm.
b The bond has extended from X to Y. The bond is in the stretched condition, so is at higher potential energy.
c The force will be seeking to pull the atoms together.
d The atoms will pass *through* the point X, because when they reach that point they will have kinetic energy. This will be reconverted to potential energy as the bond/spring goes into compression at point Z. (This should remind you of the way a pendulum behaves.)
e The bond, pendulum-like, will oscillate continuously between Y and Z, with energy converting between potential and kinetic.

15 Points M and N represent inner and outer limits of a 'harder' vibration. If the bond vibrated

between M and N, the physical movement (*x*-axis) would be greater because M is more compressed than X and N is more extended than Y. Also, M and N are higher in energy than X and Y.

16 The three non-absorbing molecules all lack a *dipole moment*. It seems that a photon will only interact with a vibrating molecule if the electromagnetic field of the photon can resonate with the oscillating electromagnetic field of an oscillating dipole moment.

17 See Table A5.2.

Table A5.2 Oscillating overall dipole set up by bending motion of carbon dioxide

Bond condition 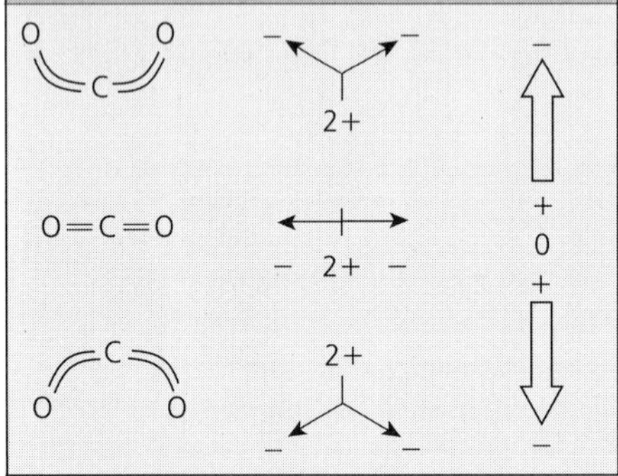	Individual dipoles	Overall dipole
O⟍C⟍O	2+	↑ −
O=C=O	− 2+ −	+ 0 +
O⟋C⟋O	2+	↓ −

18 Because the molecule is bent, there is an overall dipole even in the 'rest' position of the molecule, and therefore also in all the positions of the symmetrical stretch mode.

Table A5.3 Oscillating dipole set up by symmetrical stretch of sulphur dioxide – contrast carbon dioxide

Bond condition 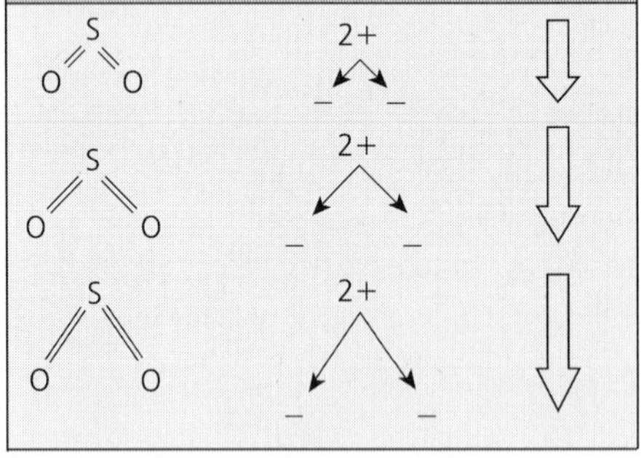	Individual dipoles	Overall dipole
O=S=O	2+	↓
O=S=O	2+	↓
O=S=O	2+	↓

19a The force needed for a bending deformation is less than that for a stretch. So the vibration is of lower frequency ('floppier'), and the absorbed photons are of lower frequency and energy.
b Triple bonds are on the whole stronger ('stiffer') than double bonds, and so absorb at higher frequencies.

20 The peak absorption wavenumbers correlate with bond strength. The HI bond is the 'floppiest' and absorbs at the lowest frequency.

21 The two peaks involving vibrations of the C—H bond will move to lower wavenumber. This is because, now that it is a C—D bond, the heavier mass of the D atom will cause the vibrations to be slower.

22 Absorption/emission spectroscopy (visible/ultraviolet) involves gross movements of electrons, up to the extreme of complete removal. By comparison in infrared spectroscopy the molecular rearrangement involved in a vibration is a relatively minor affair. So the changes in potential energy are much bigger when an electron changes its energy level, and thus require the absorption of more energetic photons.

23 See Figure A5.6.

$$S_8 \quad CH_4 \quad SiCl_4 \quad H_2O \quad AlCl_3 \quad NaCl \quad CsF \longrightarrow$$

most covalent   most ionic

Figure A5.6 Order of increasing dipolarity

24 The best chance of getting a genuine Al^{3+} ion would be in AlF_3, because there is no anion less polarisable than F^-.

25 It could be that the impure, heavily polar bond is ruptured by the pull of the electrodes at either end of the dipole. If it did rupture, it would certainly rupture in such a way as to leave the chlorine atoms with both the electrons from the 'polar covalent' bonds. In other words, in an electrolysing environment, the $MgCl_2$ 'molecule' would rupture into ions.

26a Na → Ar, getting smaller. The increasing nuclear pull overrules the modest amount of electron–electron repulsion between electrons in the same shell.
b $Na^+ \to Al^{3+}$, getting smaller. All three ions are isoelectronic ($1s^2 2s^2 2p^6$), and the nuclear pull is increasing.
c N^{3-}, O^{2-}, F^-, getting smaller. All three ions are isoelectronic ($1s^2 2s^2 2p^6$), and the nuclear pull is increasing.

27 They all have the same electron structure (see (b) and (c) above).

28 Version 1 – the ionization energy needed to remove four successive electrons to make C^{4+} is prohibitively large, with each electron harder to remove than the last. Similarly the electron affinity will get less and less (negative) as the ion has to cope with an increasing negative charge on the way to C^{4-}.
Version 2 – if C^{4+} did exist, it would have such a high charge density that it would polarise any anion which could be put with it. If C^{4-} did exist, it would have such a poor hold over its outer electrons that it would be a polarisation 'push-over'.

Figure A5.5

29a Na_2SO_4, $MgSO_4$, $Al_2(SO_4)_3$.
b $NaNO_3$, $Mg(NO_3)_2$, $Al(NO_3)_3$.
c Na_2CO_3, $MgCO_3$, $Al_2(CO_3)_3$.

30   The outer electrons of metals will be loosely held, and so will be able fairly easily to move between atoms.

31   Malleability and ductility, high boiling temperatures, and high latent heats of vaporisation.

32   The ceramic material's 'tricky moment' involves like charges meeting (and repelling).

33   The 'hole' is lined with like-charge particles.

34a 6
b   3
c   12.

35   Eight atoms at corners $= 8 \times \frac{1}{8} = 1$

Six atoms at centres of faces $= 6 \times \frac{1}{2} = 3$
Total atoms $= 4$

36   See Figure A5.7.

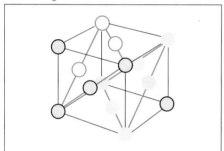

Figure A5.7 Close-packed planes running across FCC unit cell

37a $(1 \times \frac{1}{8}) + (1 \times 1) = 2$ atoms
b   8.

38   Ionization energy goes mainly *up* as you cross a row (although there are the two 'hiccups', p. 45).

39   Electronegativity generally goes *up* across a row.

40   It will emphasise and amplify the difficulty of creating Al^{3+} relative to Na^+.

41   There would be very high electron density in one 'bit', destabilising the bond by repulsion.

42   Top right
43   Groups I and VII

44   One relevant point is that the ionization energy of hydrogen is rather high for a metal. Band theory would tell you about full-up bonding orbitals and the energy gap to the next anti-bonding orbital.

45   Group III, in which *boron* is a non-metal.

46   For example, Group IV, which contains non-metals C and Si and Ge (the latter two semi-conductors), and metals Sn (tin) and Pb (lead).

47   The ionization energies do indeed correlate with being lower in the group and with metallic character. See Table A5.4.

48   Yes, ionization energies of pairs like boron and silicon are indeed quite close. So the diminishing effect on ionization energy of descending a group is compensated for by the increasing effect on ionization energy of moving to the right. It is also interesting that pairs of elements in diagonal relationships have *similar charge densities*. Moving to the right and down puts up both charge and size respectively. For instance Li^+ has a similar charge density to Mg^{2+}. Having seen the role of charge density in setting chemical character, we can see why diagonal pairs might therefore have similar chemical character.

49a Ionic
b   Left
c   Covalent
d   Right

50   See Table A5.5.

51   The O—H bond, in which breakage results in the creation of $H^+(aq)$ or $H_3O^+(aq)$ ions – giving acidic behaviour.

CHAPTER 6   page 94

1   Propanone and propanol have dipoles. They are also the ones with higher boiling points, so there is support for the idea that dipoles are a source of intermolecular forces.

(Dipole moments: propanone 2.95 D, propanal $\approx$ 2.54 D, butane 0 D)

2   The propanone molecules orient to maximise the attraction between them. The lower case delta symbol δ with a plus or minus sign is used to show the position of charge.

$$\underset{CH_3}{\overset{CH_3}{\delta^+ C {=} O\, \delta^-}} \, {-}{-}{-}{-} \, \underset{CH_3}{\overset{CH_3}{\delta^+ C {=} O\, \delta^-}}$$

3   For example, consider the dihalogen molecules, which all have the same dipole moment (0) and the same shape.

Molecule	Boiling temperature (K)
Cl_2	238
Br_2	332
I_2	457

So boiling temperature correlates with number of electrons, suggesting that number of electrons is a factor in van der Waals' forces.

4a The number of electrons is obviously *not* a variable in a trio of isomers. Neither will dipole moment vary much, because the main source of polarity is the C—Br bond which is common to all three molecules.
b

Molecule	Boiling temperature (K)
1-bromobutane	375
2-bromobutane	364
2-bromo-2-methylpropane	346

So shape seems to be a significant variable. (Look back at the molecules in question 45, p. 116.)
c   1-Bromobutane is the most linear molecule, and 2-bromo-2-methylpropane the most globular.

5a Polymers such as poly(ethene) are often linear, and can get physically tangled, so the degree of intermolecular contact is high and therefore ideal for van der Waals' bonding.
b   The better contact in the more ordered high-density poly(ethene) means that van der Waals' forces are higher, and so density, hardness and softening temperature will all be higher.

6   All a particle needs in order to exert a van der Waals' force on its neighbours are electrons, and they all have some of those. Dipoles, on the other hand, depend upon the electronegativities of particular atoms.

Table A5.4 First ionization energies

First ionization energy (kJ mol^{-1})					
Group III		Group IV		Group V	
Al	578	Si	789	P	1012
Ga	579	Ge	762	As	947
In	558	Sn	709	Sb	834

Table A5.5 Predicted properties of chlorides

	Na	Mg	Al	Si	P	S	Cl	Ar
Melting point	high	high	medium	low	low	low	low	–
Does it conduct when liquid?	yes	yes	yes	no	no	no	no	–
Dipole moment	–	–	large	large	medium	medium	zero	–

7a Nuclear hold over electrons decreases as the group is descended, so polarisability and therefore van der Waals' interaction increases.
b Polarisability is at a minimum in HF, where all the electrons are held very firmly in their orbits, even the bonding pair. CH_4 is the most polarisable, since the hold exerted by the carbon nucleus is not so great.
c The sequence of boiling points (lowest first) would be: HF, H_2O, NH_3, CH_4. The real sequence is almost the exact opposite.
d If it were left to van der Waals' forces, each line would show a gradient like the one for inert gases, since each line represents the same decreasing nuclear pull (and therefore increasing polarisability) down a group.

8 A good comparison is the one between butan-1-ol and ethoxyethane (sometimes called diethyl ether or just ether), since they are structural isomers. Both have the molecular formula $C_4H_{10}O$, but the alcohol's ability to form intermolecular hydrogen bonds, and the ether's inability to do so, make for a very marked contrast in volatility:

Substance	Boiling temperature (K)
Butan-1-ol	390
$CH_3CH_2CH_2CH_2OH$	
Ethoxyethane	308
$CH_3CH_2OCH_2CH_3$	

9 Anything with an OH group can hydrogen bond with its own kind, so carboxylic acids qualify along with alcohols.

10 In fact, there are two more dimensions to hydrogen bonding in water, as the ice structure implies. This question just asks you to picture the cross-links between chains.

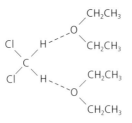

11 In the structure below, all three chlorine atoms work together as electron-withdrawing groups, to create a carbon atom with an unusual degree of electron deficiency. The carbon atom then 'takes it out on' the hydrogen atom, making it electropositive enough to enter into hydrogen bonds.

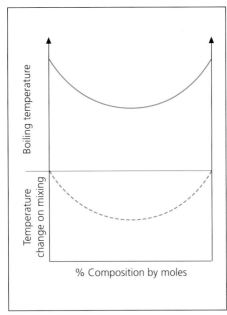

The electronegative 'end' is split over three chlorine atoms, so the net result is a molecule which can meet only one of the criteria for a hydrogen bond, and which, therefore, is incapable of bonding with itself. The δs in the structure shown are not meant to be quantitative. The arrows show the direction of drift of electron density.

12 The hump in the boiling point–composition curve indicates the reduced escaping tendency of the molecules in the mixed condition, caused by strong A–B interactions. This is suggestive of the formation of hydrogen bonds between the molecules of ethoxyethane and trichloromethane.

13 The sulphur equivalent, diethyl sulphide, would prove the point – a collapse or even an inversion of the graph's hump might be expected.

$$CH_3CH_2SCH_2CH_3$$

14 The inversion of the graph shape happens because vapour pressure varies inversely with boiling point. If intermolecular forces are strong, a matched pair of results occurs – first, the escaping tendency is lowered (low vapour pressure) and, second, more thermal agitation is needed to overcome the intermolecular forces (high boiling point).

15a Dipole–dipole interactions predominate in trichloromethane.
b There are dipoles in the individual C—O bonds, but being opposed to each other at about 105°, they partially cancel each other out. The residual dipoles still provide the majority source of intermolecular force.
c Hydrogen bonds.
d The hydrogen bonds being made are stronger than the dipole attractions which are being broken.
e Both maxima indicate the composition of the mixture in which maximum bond making takes place, and the indication is that the intermolecular 'compound' is 1:1.

16 See Figure A6.1.

17a Hydrogen bonds.
b Van der Waals' bonding.
c The only form of possible A–B interaction is a van der Waals' force. This will represent a net weakening of bonding, chiefly because the hydrogen bonds in the alcohol have to be sacrificed in the mixture.

18 The maximum of boiling point would have occurred at a composition of 2 parts trichloromethane : 1 part ethoxyethane (Figure A6.2).

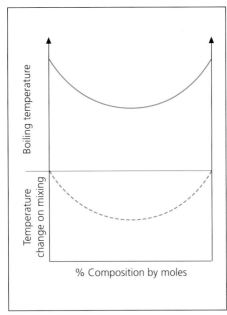

Figure A6.1 The two *y*-axis labels operate above and below the middle line. The reason is that A–B interactions are weaker than A–A and B–B, so the mixture will boil more readily and mix endothermically

19a, b and **c**

a

b

c

d Only (c) would give an off-centre hump. The other two would both give 1:1 stoichiometries for the intermolecular 'compound'.

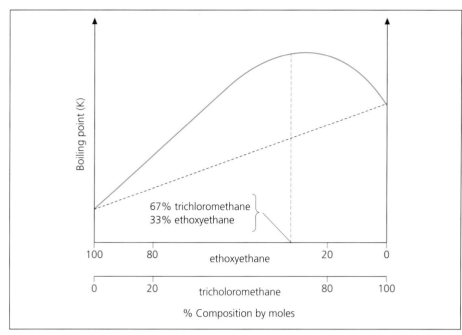

Figure A6.2 Boiling point/composition graph

Table A6.1

Mixture	Intermolecular force A–A	Intermolecular force B–B	Intermolecular force A–B	Boiling point shift	Enthalpy of mixing
a	d–d	d–d	hb	+	exo (−)
b	d–d	d–d	hb	+	exo (−)
c	d–d	vdW	vdW	−	endo (+)
d	d–d	hb	hb	−	endo (+)
e	d–d	d–d	hb	+	exo (−)
f	hb	hb	hb	0	0

d–d = dipole–dipole interaction; hb = hydrogen bonding; vdW = van der Waals' bonding; exo = exothermic; endo = endothermic

20 See Table A6.1.

21a See Figure A6.3.
b $p_A = kx_A$
When $x_A = 1$, $p_{A,\ x=1} = k$.
But p_A when $x_A = 1$ is none other than the vapour pressure of pure component A. (Call it P_A.) So $k = P_A$.
So the equation becomes $p_A = x_A P_A$.

22 As the hydrocarbon chain gets longer, so the contribution of van der Waals' bonding gets greater. So not only does that have to be sacrificed to allow the interposing of water molecules, but also the water's own hydrogen bonds are interrupted. Put beside this, the advantages of hydrogen bonds between water and the alcohol's OH group become quite insignificant.

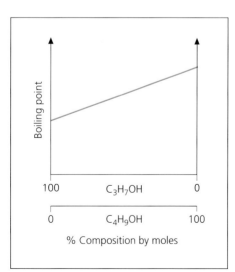

Figure A6.3 Possibilities for hydrogen bonding

23 There are multiple possibilities for hydrogen bonding between water and sucrose molecules. Just a few of the possible hydrogen bonds which glucose might form in aqueous solution are shown below.

24 There are more chances of hydrogen bonding between water and wool, compared to water and acrylic. So in each repeat unit of a protein fibre there are three possible sites for hydrogen bonding (*) with water, whereas in acrylic there is only one (*). (This is an over-simplification in the wool case, since some of the (*) sites will be occupied with intramolecular hydrogen bonds. On the other hand the R groups will offer some chances for hydrogen bonding, depending on the selection of R groups in a given sample.)

25 The strength of the ion–dipole interactions must be similar to the strength of the lattice energy of the crystal.

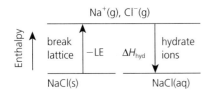

26

Cation	Formula of sulphate
Mg	$MgSO_4 \cdot 7H_2O$
Ca	$CaSO_4 \cdot 2H_2O$
Sr	$SrSO_4$
Ba	$BaSO_4$

There is an obvious correlation between the charge density of the cation (which decreases down the group) and the number of moles of water of crystallisation.

27 The higher cation charges in Group II (relative to 1) provide stronger **crystal lattices** to counterbalance the stronger solvent–cation bonds.

28a Petrol and grease are alike in that their intermolecular forces are van der Waals'.
b Water and sugar both use hydrogen bonds, so they are 'like'.
c Water and grease are very 'unlike' – water uses hydrogen bonds, grease uses van der Waals' forces.
d Water is not exactly 'like' salt, except in so far as they both have electrostatic charge build-up, causing ion–dipole interaction. The difference between them is that in salt the charges are unitary ones (one whole electron's worth), whereas water's charges are only partially developed. So the maxim has rather limited applicability in this case.

29 Viscosity will decrease with increasing temperature, because the thermal agitations will oppose the effects of intermolecular forces.

30 There are multiple possibilities for hydrogen bonding in glycerol.

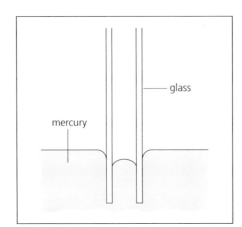

31a Van der Waals' bonding.
b Van der Waals' bonding.
c When it is more tangled with itself it is less able to get involved with the oil molecules.

d As it gets hotter it uncoils and so increases its bonding to the oil, just when most intermolecular forces would be getting less.

32 Spheres have the minimum surface area for a given volume, so this satisfies the first condition above.

33a Force C will tend to coalesce the bubble.
b Even small values of D will be able to rupture the bubble.

34 The hydrogen bonds in water are interrupted all along the hydrocarbon chain region of the soap molecule, thus weakening the intermolecular forces in water (Figure 6.29b).

35a An oily layer might not have too much effect, if the feet went through the layer and rested on the water underneath. Reality might be more complicated, however, with among other things the chance that the oil would affect the chitin of the insect's exoskeleton.
b Detergent lowers surface tension, and so the water surface would no longer be able to support the insect.

36 The spreading behaviour on the clean glass fails to minimise surface area.

37 In a capillary tube, the surface area of the bore determines the force in favour of lifting the water, and the volume (and mass) to be lifted determines the opposition to lifting. As the radius of the bore decreases, the volume goes down faster than the area (the volume being proportional to the square of the radius), and so the balance moves in favour of higher lifting.

38a Mercury, like all liquid metals, consists of separate atoms – except that the 'sea of electrons' is still in place.
b For liquid metals we use the 'sea of electrons' model of metallic bonding described in Chapter 5, p. 85. Not much by way of interatomic bonding is sacrificed in the melting process, so interatomic bonding is still quite formidable.

Figure A6.4 The meniscus of mercury in glass

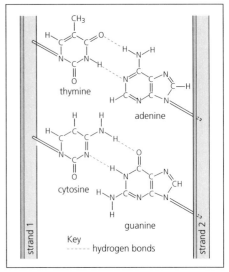

Figure A6.6 Part of the DNA structure

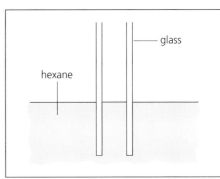

Figure A6.5 The meniscus of hexane in glass (or lack of it)

c Mercury's meniscus, in narrower-bore tubes at least, will be an inverted hemisphere (Figure A6.4) as it uncompromisingly maintains its minimum surface area against the opposition of gravity.

39 Hexane molecules will be so unconcerned with any bonding options, either to each other or to glass, that they will submit to gravity and present a flat surface (Figure A6.5).

40 On a newly cleaned car the layer of wax will prevent any water/paint bonding, and the dominant tendency of the water molecules will become 'own-kind' bonding. This leads to a minimising of the surface area, and hence near semi-spherical droplets.

41 Soapy water has lower surface tension, because the soap molecules have weakened the intermolecular bonding in the water (Figure 6.31). This makes spreading easier, as the tendency to be constrained in a sphere or blob will be reduced. An added incentive to spreading is that the hydrocarbon chains of the soap molecules can actually interact with the hydrocarbons in wax.

42   See Figure A6.6.

43a If enzyme–substrate bonding were too strong, the products would never disengage, and the enzyme would be blocked from further action.

b   If the substrate molecule does not form strong enough bonds with the enzyme, then the enzyme will not be able to distort the substrate and 'nudge' it along the way to full reaction.

44   See Figure A6.7.

45a The dotted lines represent hydrogen bonds.

b $\alpha = 109.5°$ $m = 0.154$ nm
$\beta = 104.5°$ $n = 0.143$ nm
$\gamma = 109.5°$ $p = 0.096$ nm

c   (Scale drawing.)

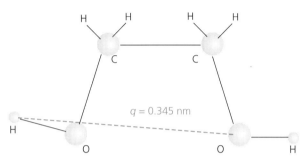

Scale 0.1 nm ≡ 20 mm

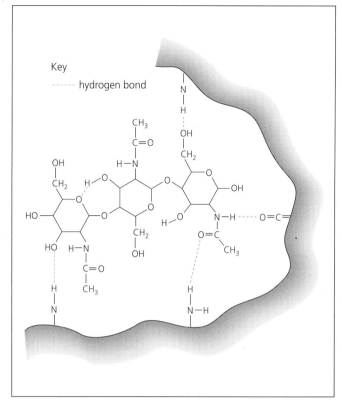

Figure A6.7 Bonds holding enzyme to substrate

d   The model is inadequate because it depicts the crucial bit of the molecule as lying flat on the page. The distance q could be fine-tuned by rotations around the C—C and C—O bonds.

e   The separation of the two OH groups in propane-1,3-diol is too great to allow for a correct q value, even allowing for bond rotations. In propane-1,2,3-triol any two adjacent OH groups can perform like those in ethane-1,2-diol, and confer sweetness on the molecule.

f   C—C bond rotation is impossible, because of the cyclic formation.

g   The group 'C' would have to be non-polar, and the C···Z interaction would be of the van der Waals' type.

h

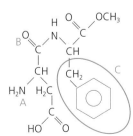

CHAPTER 7   page 121

1 $KClO_3$. Its loss of oxygen is revealed by the presence of KCl on the right-hand side.

2   It is very visibly both an oxidation (of C) and a reduction (of Cu in CuO). However, as we shall see, this oxidation-with-reduction feature is actually present in *all* the reactions in this field. This example is only different in that it shows the feature clearly.

3   Many oxidations and reductions are close to the mainstream of human life. For example:

- Category – oxidation of an element: context – burning of coke, as in domestic fires:

  $C(s) + O_2(g) \rightarrow CO_2(g)$

- Category – oxidation of a compound: context – burning petrol, for example, the internal combustion engine:

  $C_8H_{18}(l) + 12\frac{1}{2}O_2(g) \rightarrow 8CO_2(g) + 9H_2O(l)$

- Category – oxidation using an indirect source of oxygen: context – potassium nitrate in gunpowder, a major historical development in warfare. A possible equation is:

  $4KNO_3(s) + C(s) + S(s) \rightarrow$
  $4KNO_2(s) + CO_2(g) + SO_2(g)$

- Category – reduction *to* an element: context – iron oxide reduced by coke, the foundation of the steel industry:

$2Fe_2O_3(s) + 3C(s) \rightarrow 4Fe(l) + 3CO_2(g)$

4a The oxygen in dioxygen is reduced.

b   The chlorine in dichlorine is reduced.

5   The oxygen in HgO is oxidised.

6   In the formation of sodium chloride from its elements, sodium is oxidised and chlorine reduced.

7

Compound	Oxidised	Reduced
SO_2	S	O
H_2O	H	O
H_2S	H	S
NH_3	H	N
HF	H	F
BF_3	B	F

8   This is because in every individual bond there is (by the rules of this game) a winner and loser of the equivalent of one electron. (What is actually won and lost is a half-share in 2.) So the sum of the changes of oxidation number in each individual bond is itself zero (+1 – 1). Naturally when this process is repeated over the whole molecule the answer is still zero. To put it more shortly, no electron actually gets lost or gained, viewing the molecule as a whole. And this applies to all cases.

9   See Figure A7.1.

10 See Figure A7.2. C has lost its share in all 4 bonds, so is seen to be oxidised to +4 by the 'loss' of its 4 electrons.

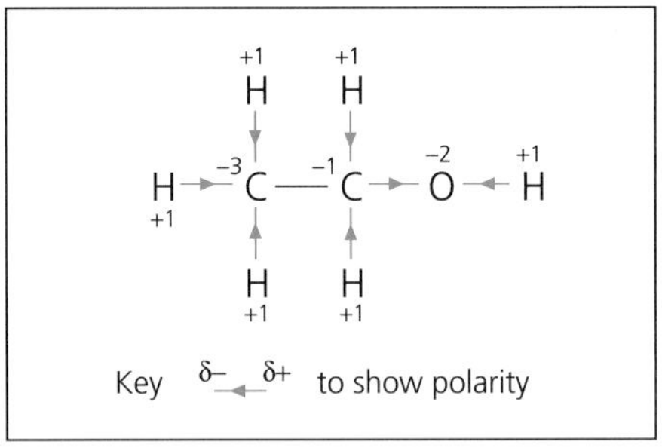

Figure A7.1

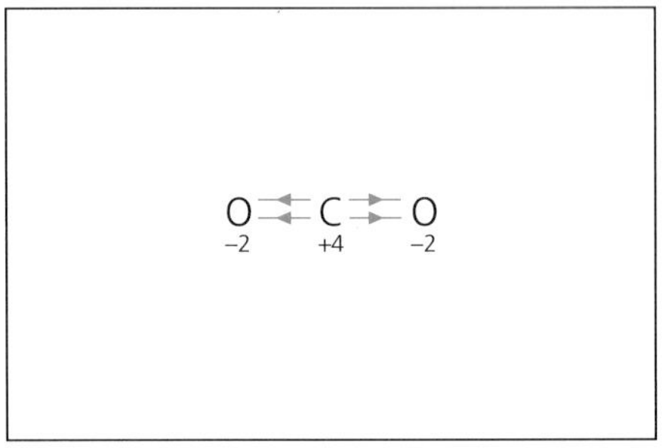

Figure A7.2 Carbon dioxide

11 The hydrogen atom only has one electron, and in most of its bonding partnerships it is the loser – hence +1. The oxygen atom is the winner in all its bonds except those to itself and to fluorine, and always gains two electrons for rule-of-eight reasons – hence −2.

12 C in CH_4 $x + 4(+1) = 0$
                     so $x = -4$

    S in SO_3 $x + 3(-2) = 0$
                     so $x = +6$

    S in Na_2SO_4 $2(+1) + x + 4(-2) = 0$
                     so $x = +6$

    Cr in $Na_2Cr_2O_7$
    $2(+1) + 2x + 7(-2) = 0$ so $x = +6$

    N in HNO_3 $(+1) + x + 3(-2) = 0$
                     so $x = +5$

13 F_2O F at −1, O at +2
    NaH   Na at +1, H at −1
In the latter case $H_2(g)$ is liberated at the *anode*.

14 An oxidation number of +2 is impossible for H, since it only has one electron to lose.

15 See Figure A7.3. The 'reliable' method gives +2, but actually it is a +4 and a 0.

16 C in CO_3^{2-} $x + 3(-2) = -2$ so $x = +4$
    P in PO_4^{3-} $x + 4(-2) = -3$ so $x = +5$
    V in VO_2^+ $x + 2(-2) = +1$ so $x = +5$

17 N in $NH_4^+ = -3$
    O in $[Cu(H_2O)_4]^{2+} = -2$
    N in $N_2O_5 = +5$

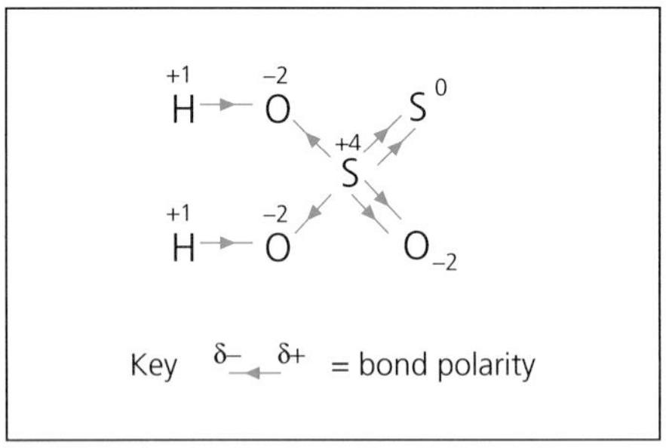

Figure A7.3

18 Sodium sulphate(VI), iron(III) nitrate (V), sodium sulphate(IV).

19 See Figure A7.4.

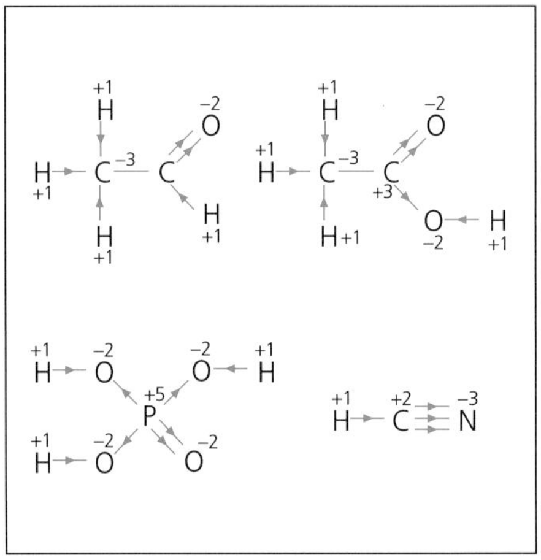

Figure A7.4

20 C +4, P +5, S +6, Cl +7.

21 P in $HPO_3 = +5$
    Mn in $KMnO_4 = +7$ Potassium manganate(VII)
    S in $Na_2S_2O_3 = +2$ (apparently)
    Cl in $HOCl = +1$ Chloric(I) acid

22 Cl in $ClO_4^- = +7$ Chlorate(VII)
    Mn in $MnO_4^- = +7$ Manganate(VII)
    S in $S_2O_5^{2-} = +4$
    S in $SO_3^{2-} = +4$ Sulphate(IV)
    V in $VO^{2+} = +4$
    N in $N_2H_5^+ = -2$
    C in $CN^- = +2$

23a S in $S_2O_8^{2-}$ (apparently) +7
b But maximum oxidation number of sulphur is +6 (since it has only 6 outer electrons).
c See Figure A7.5. You can see that the S atoms *are* in an oxidation state of +6. The source of the anomaly is the 'peroxide-style' bridge, where the two O atoms are at an 'unreliable' oxidation state of −1.

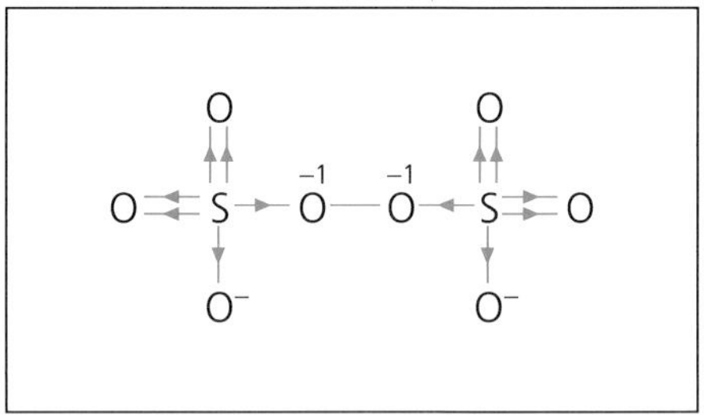

Figure A7.5

24 B in $H_3N{\rightarrow}BF_3$ Although the B atom is the receiver of the bond, the N atom is most definitely the more electronegative of the two atoms, so is seen as 'keeping' the electron pair. So B stays at +3.
    O in H_3O^+ Again O is a donor but being much more electronegative than H, is seen to keep the pair. So it stays at −2.
    Pt in $[PtCl_4]^{2-}$ Again it is a case of 'the donor keeps its gift'. So Pt stays at +2, which could be deducted from 4Cl atoms at −1 each and from the overall charge on the ion.

25 CuO is copper(II) oxide
    Cu_2O is copper(I) oxide
    Fe_2O_3 is iron(III) oxide
    FeO is iron(II) oxide
    NaClO is sodium chlorate(I)
    $NaClO_3$ is sodium chlorate(V)
    $NaClO_4$ is sodium chlorate(VII)
    H_2SO_3 is sulphuric(IV) acid
    H_3PO_4 is phosphoric(V) acid

26 See Figure A7.6.

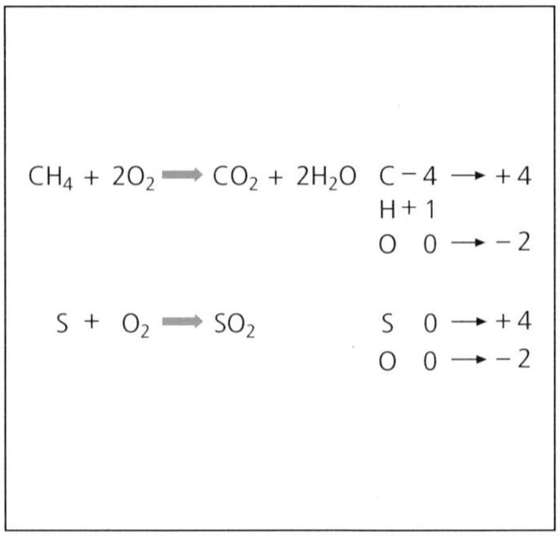

Figure A7.6 Changes in oxidation state

27a Moles of $KMnO_4 = \dfrac{9.92}{1000} \times 0.005$
    $= 4.96 \times 10^{-5}$

So moles of Fe^{2+} = 5 × 4.96 × 10⁻⁵ = 2.48 × 10⁻⁴

That is in a 10 cm³ batch, so in the whole 100 cm³ solution derived from the two tablets there is:

2.48 × 10⁻³ moles of Fe^{2+} and therefore of $FeSO_4$.

b So mass of $FeSO_4$ = 2.48 × 10⁻³
(56 + 32 + 64) = 377 mg

That is in two tablets, so in one tablet there is 188 mg, which is a shade *under* strength.

c The end-point would be the *first* permanent pink colour.

28 (Method of half-equations)

a
$$Cu(s) \rightarrow Cu^{2+}(aq) + 2e^-$$
$$2Fe^{3+}(aq) + 2e^- \rightarrow 2Fe^{2+}(aq)$$
_____
$$Cu(s) + 2Fe^{3+}(aq) \rightarrow Cu^{2+}(aq) + 2Fe^{2+}(aq)$$

b
$$Sn(s) + 2H_2O(l) \rightarrow SnO_2(s) + 4e^- + 4H^+(aq)$$
$$4HNO_3(aq) + 4H^+(aq) + 4e^- \rightarrow 4NO_2(g) + 4H_2O(l)$$
_____
$$Sn(s) + 4HNO_3(aq) \rightarrow SnO_2(s) + 4NO_2(g) + 2H_2O(l)$$

c
$$SO_2(aq) + 2H_2O(l) \rightarrow SO_4^{2-}(aq) + 2e^- + 4H^+(aq)$$
$$Br_2(aq) + 2e^- \rightarrow 2Br^-(aq)$$
_____
$$SO_2(aq) + Br_2(aq) \rightarrow SO_4^{2-}(aq) + 2Br^-(aq) + 2H_2O(l) \quad + 4H^+(aq)$$

29a $KClO_3(s) + 3Mg(s) \rightarrow KCl(s) + 3MgO(s)$
b $2MnO_4^-(aq) + 16H^+(aq) + 10I^-(aq) \rightarrow 2Mn^{2+}(aq) + 5I_2(aq) + 8H_2O(l)$
c $H_2SO_4(aq) + 8HI(aq) \rightarrow H_2S(g) + 4I_2(aq) + 4H_2O(l)$

30a $H_2O_2(aq) + 2I^-(aq) + 2H^+(aq) \rightarrow 2H_2O(l) + I_2(aq)$
b $MnO_2(s) + 2HCl(aq) + 2H^+(aq) \rightarrow Cl_2(g) + 2H_2O(l)$
c $2Fe^{3+}(aq) + 2I^-(aq) \rightarrow 2Fe^{2+}(aq) + I_2(aq)$

31 $5HNO_2(aq) \rightarrow N_2(g) + 3HNO_3(aq) + H_2O(l)$

32a Moles of thiosulphate used
$$= \frac{6.5}{1000} \times 0.001$$
$$= 6.5 \times 10^{-6}$$

b So moles of I_2
$$= \frac{6.5 \times 10^{-6}}{2} = 3.25 \times 10^{-6}$$

c So moles of Cl_2 = 3.25 × 10⁻⁶.
d There are 3.25 × 10⁻⁶ moles of Cl_2 in 250 cm³.
$$Concentration = \frac{moles}{volume\ in\ dm^3}$$
$$= \frac{3.25 \times 10^{-6}}{0.25}$$
$$= 1.3 \times 10^{-5}\ mol\ dm^{-3}$$

e $S_4O_6^{2-}$ has an apparent oxidation number of 2.5! But see Figure A7.7.

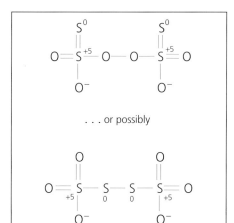

Figure A7.7 Possible structures for the 'tetrathionate' ion, $S_4O_6^{2-}$

33a Silver is giving.
b Copper is giving.
c Copper is the better giver.

34a $Cl_2(aq)$, $MnO_4^-(aq)/H^+(aq)$, and $F_2(aq)$ can oxidise $Br^-(aq)$ to $Br_2(aq)$.
b Cu(s), Zn(s) or Mg(s) would be oxidised by $Fe^{3+}(aq)$.
c $Mg^{2+}(aq)$ can *not* be reduced by Zn(s).

35a $Fe^{3+}(aq) + I_2(aq)$ – *no*, both are oxidised forms.
$Zn^{2+}(aq) + Mg(s)$ – *yes*.
$Br_2(aq) + Cl^-(aq)$ – *no*, it breaks the anti-clockwise rule.
b The first one was impossible irrespective of the electrochemical series (both oxidised forms).

36a F_2 and MnO_4^- are the top oxidising agents.
b Mg and Zn are the top reducing agents.

37a The electrons came from the hydrogen atoms in dihydrogen.
b *Advantages include*: the overall efficiency of energy conversion is greater. Fuel cells produce much less atmospheric pollution than a normal power station, so they can be sited nearer habitations and then the heat by-product can be used for domestic heating. (Most conventional power stations use their spare heat merely to make steam in the big cooling towers.) Depending on how the hydrogen was made, fuel cells might also make less contribution to the greenhouse effect and to acid rain.

Disadvantages include the fact that most hydrogen is currently made by the 'reformation' of hydrocarbons:

$$CH_4(g) + 2H_2O(g) \rightarrow CO_2(g) + 4H_2(g)$$

so there is no real saving in consumption of fossil fuels, or avoidance of the greenhouse effect. Also the platinum catalysts are expensive.

CHAPTER 8   page 139

1 Longbow – Origin: the forces which keep molecules in certain preferred positions (fundamentally, electrostatic). Unstable position: deformed.
Trampoline – Same as longbow.
Magnets – Origin: force of magnetic attraction (fundamentally, derived from unpaired electrons).
Unstable position: apart.

2a Compared to the size of the Earth, a move of a few metres from the surface produces a negligible reduction in gravity. You would have to move kilometres before it became apparent that gravitational force fields get weaker with distance. In other words, if we changed the units on the x-axis to kilometres we would see a graph of the same shape as the 'two attractive magnets' case.
b In cases (1) and (2) the energy would end up in the form of heat, created by the collisions of the pairs of objects.
c See Figure A8.1.

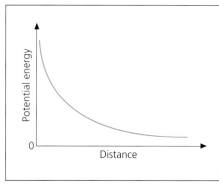

Figure A8.1 Two magnets, like poles facing

3 The enthalpy change, ΔH, would be +436 kJ mol⁻¹, and the surroundings would *lose* this amount of heat. (Note that even positive answers are given signs in this very sign-dominated field of chemistry.)

4 *Note:* Different data books may give different values for bond enthalpies, which gives you an immediate insight into their limitations. This problem will crop up throughout the series of questions. If your answers differ from those below, check whether that is the reason.
a See Figure A8.2.
$$x = +a + b$$
$$= + (+612 + 366.3) + (-347 - 413 - 290)$$
$$= -71.7\ kJ\ mol^{-1}$$
The surroundings receive 71.7 kJ mol⁻¹ as heat.
b See Figure A8.3.
$$x = +a + b$$
$$= + (+358 + 298.3) + (- 464 - 228)$$
$$= -35.7\ kJ\ mol^{-1}$$
The surroundings receive 35.7 kJ mol⁻¹ as heat.

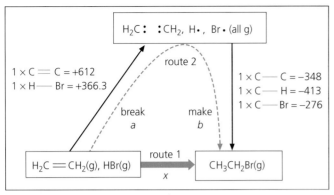

Figure A8.2 All values are in kJ mol⁻¹

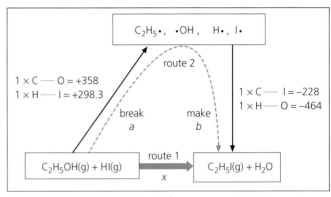

Figure A8.3 All values are in kJ mol⁻¹

c

$$2H_2(g) + O_2(g) \rightarrow 2H_2O(g)$$
$$+2 \times 435.9 \quad +498.3 \quad 4(-464)$$

$$x = + (\{2 \times +435.9\} + 498.3) + (4 \times -464)$$
$$= -486 \text{ kJ mol}^{-1}$$

The surroundings receive 486 kJ mol⁻¹ as heat.

d

$$H_2(g) + Cl_2(g) \rightarrow 2HCl(g)$$
$$+435.9 \quad +243.4 \quad 2(-432)$$

$$x = + (+435.9 + 243.4) + (2 \times -432)$$
$$= -185 \text{ kJ mol}^{-1}$$

The surroundings receive 185 kJ mol⁻¹ as heat.

e

$$C\equiv O(g) + H_2(g) \rightarrow \begin{matrix} H \\ C=O(g) \\ H \end{matrix}$$
$$+1077 \quad +435.9 \quad 2\times(-413) \quad -695$$

$$x = + (+1077 + 435.9) + (\{2 \times -413\} - 695)$$
$$= -8.1 \text{ kJ mol}^{-1}$$

The surroundings receive 8.1 kJ mol⁻¹ as heat.

f

$$CH_4(g) + 2O_2(g) \rightarrow CO_2(g) + 2H_2O(g)$$
$$4(+435) \quad 2(+498.3) \quad 2(-805) \quad 4(-464)$$

$$x = + (\{4 \times +435\} + \{2 \times +498.3\}) + (\{2 \times -805\} + \{4 \times -464\})$$
$$= -729 \text{ kJ mol}^{-1}$$

The surroundings receive 729 kJ mol⁻¹ as heat.

g

$$2C_2H_2(g) + 5O_2(g) \rightarrow 4CO_2(g) + 2H_2O(g)$$
$$2(+838) + 4(+413) \quad 5(+498.3) \quad 8(-805) \quad 4(-464)$$

$$x = + (\{2\times +838\} + \{4 \times +413\} + \{5 \times +498.3\}) + (\{8 \times -805\} + \{4 \times -464\})$$
$$= -2480 \text{ kJ mol}^{-1} \text{ (to 3 sf)}$$

The surroundings receive 2480 kJ mol⁻¹ as heat.

5a *a* is the same reaction as in question 4f.

b *x* is the standard enthalpy of combustion of methane, $\Delta H_c^\ominus$.

c The extra arrow *b* is the enthalpy change ΔH of the reaction:

$$2H_2O(g) \rightarrow 2H_2O(l)$$

The data book gives this as (2×-41) kJ mol⁻¹, and since

$$x = + a + b$$
$$x = + (-729) + (-82)$$
$$= -811 \text{ kJ mol}^{-1}$$

6 −8.1 kJ mol⁻¹ as compared to +1.8 kJ mol⁻¹. The bond enthalpy answer is −8.1 kJ mol⁻¹; whereas the enthalpy of formation method gives +1.8 kJ mol⁻¹.

a The +1.8 kJ mol⁻¹ is more trustworthy.

b Bond enthalpies are *average* values, not specific to the reaction.

c The percentage error looks bad because each answer is a small difference between two big numbers. (The big numbers are the enthalpies of the bonds broken and made, in the one case, and the enthalpies of formation of reactants and products in the other.) So if those big numbers are subject to 1 or 2% variation (which is true of the bond enthalpies in particular), then the final value of −8.1 kJ can vary quite violently, in percentage terms, relative to the 'truer' +1.8 kJ.

d $H_2(g)$ is already the element in its standard state.

7 $2Na(s) + C(s,graphite) + 6\frac{1}{2}O_2(g) + 10H_2(g) \rightarrow Na_2CO_3 10H_2O(s)$

8 $H_2C=CH_2(g) + 3O_2(g) \rightarrow 2CO_2(g) + 2H_2O(l)$

Reasons for inaccuracy:

- Bond enthalpies are only averages.
- Bond enthalpies can only predict the change to $H_2O(g)$.

9a +436 kJ mol⁻¹

b The bond enthalpy of H—H is the ΔH of this reaction:

$$H_2(g) \rightarrow 2H(g)$$

whereas the $\Delta H_{at}[\frac{1}{2} H_2(g)]$ is the ΔH of *this* reaction:

$$\frac{1}{2} H_2(g) \rightarrow H(g)$$

So $2 \times \Delta H_{at}$ = bond enthalpy of H—H. This is confirmed by values in data books of 218 and 436 kJ mol⁻¹.

10a $HCHO(g) \rightarrow 2H(g) + C(g) + O(g)$

b $HCHO(g) \rightarrow H_2(g) + C(s,graphite) + \frac{1}{2}O_2(g)$

$\Delta H = -\Delta H_f = -(-108.7) = +108.7 \text{ kJ mol}^{-1}$

c Because they are always bond breaking only.

11 A hydration.

12 Because lattice energies are always bond making.

13 Because electrons are always being removed.

14 $O^-(g) + e^-(g) \rightarrow O^{2-}(g)$;
$\Delta H_{\text{2nd electron affinity}}^\ominus = +844 \text{ kJ mol}^{-1}$

15 The enthalpies of vaporisation, ΔH_{vap}, of ethanol(l) and $H_2O(l)$.

It could be looked up as the standard enthalpy of combustion, $\Delta H_c^\ominus$, of ethanol(l).

16a See Figure A8.4.

$$x = b - a$$
$$= \{\Delta H_f^\ominus [C_9H_{20}(l)] + \Delta H_f^\ominus [C_2H_4(g)]\} - \{\Delta H_f^\ominus [C_{11}H_{24}(l)]\}$$
$$= \{-274.9 + (+52.2)\} - \{-327.2\}$$
$$= 104.5 \text{ kJ mol}^{-1}$$

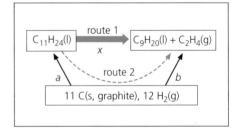

Figure A8.4 A 'cracking' reaction

b First, the industrial version would occur at higher temperatures than 298 K, at which the reagents would be gaseous. Second, the enthalpy values in data books apply at 298 K, and there is a small variation of enthalpy with the temperature at which a reaction is conducted: ΔH is nearly a constant over a range of temperatures, but not quite.

17a See Figure A8.5.

$x = a - b$

$= +716.7 - (+714.8)$

$= +1.9 \text{ kJ mol}^{-1}$

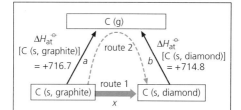

Figure A8.5 The conversion of graphite to diamond

b The interior of the crust offers extreme conditions of temperature and pressure.

18a See Figure A8.6.

$x = b - a = 90.2 + 249.2 - (+33.2)$

$= +306.2 \text{ kJ mol}^{-1}$

See Figure A8.7.

$x = b - a = +142.7 - (+249.2)$

$= -106.5 \text{ kJ mol}^{-1}$

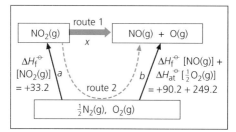

Figure A8.6 All values are in kJ mol^{-1}

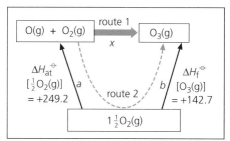

Figure A8.7 All values are in kJ mol^{-1}

b So if the two reactions run sequentially, the overall ΔH is +306.2 – 106.5 or +199.7 kJ mol^{-1}.

c Photons supply an energy input to the system.

19 $\Delta H_{\text{reaction}} = \Delta H_c^{\ominus}(\text{reactants}) - \Delta H_c^{\ominus}(\text{products})$

$\Delta H_{\text{reaction}} = \Delta H_{\text{at}}^{\ominus}(\text{reactants}) - \Delta H_{\text{at}}^{\ominus}(\text{products})$

20 $\Delta H_{\text{reaction}} = \Delta H_f^{\ominus}(\text{products}) - \Delta H_f^{\ominus}(\text{reactants})$

The enthalpies of formation of ethanol(l), HI(g), iodoethane(l) and H$_2$O(l) are: –277.1, +26.5, –15.5 and –285.8 respectively, all in kJ mol^{-1}.

So $\Delta H_{\text{reaction}} = -15.5 - 285.8 - (-277.1 + 26.5) = -50.7 \text{ kJ mol}^{-1}$

21 We can use the combustion version of the formula:

$\Delta H_{\text{reaction}}$
$= \Delta H_c(\text{reactants}) - \Delta H_c(\text{products})$
$= \{\Delta H_c[\text{heptane}]\} - \{\Delta H_c[\text{toluene}] + 4 \times \Delta H_c[\text{H}_2(\text{g})]\}$
$= \{-4816.9\} - \{-3909.8 + 4 \times -285.8\}$
$= +236.1 \text{ kJ mol}^{-1}$

22 The reaction generated by (4) – (5) is:

carboxyhaemoglobin + O$_2$(aq) → oxyhaemoglobin + CO(aq)

Correspondingly,

$\Delta H(4) - \Delta H(5) = -23.1 - (-32.4) = +9.3 \text{ kJ mol}^{-1}$

This endothermic result fits the idea that, once formed, carboxyhaemoglobin is *unwilling* to allow the transport of oxygen.

23 An X-ray diffraction crystallography machine.

24 (See Figure A8.8.)

$\Delta H_{\text{lattice}}[\text{NaCl(s)}]$
$= x = -d - c - b - a + e$
$= -(-348.8) - (121.7) - (496) - (107.3) + (-411.2)$
$= -787.4 \text{ kJ mol}^{-1}$

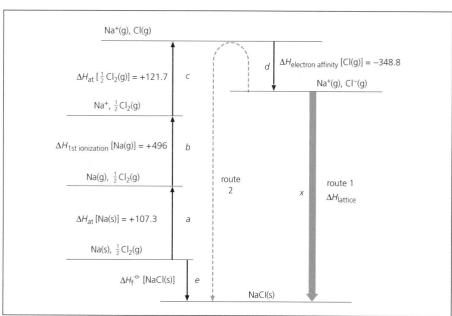

Figure A8.8 Born–Haber cycle for NaCl (s). All values are in kJ mol^{-1}

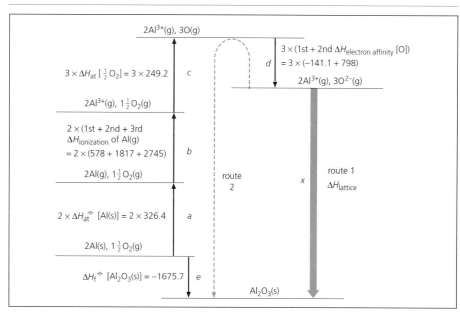

Figure A8.9 Born–Haber cycle for Al$_2$O$_3$ (s). All values are in kJ mol^{-1}

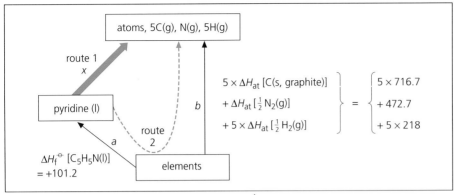

Figure A8.10 (See answer 25a.) All values are in kJ mol^{-1}

(See Figure A8.9.)

$\Delta H_{lattice}[Al_2O_3(s)]$
$\begin{aligned}
&= x = -d - c - b - a + e \\
&= -3(-141.1 + 798) \\
&\quad - 3(249.2) - 2(578 \\
&\quad + 1817 + 2745) \\
&\quad - 2(326.4) + (-1675.7) \\
&= -15\,326.8 \text{ kJ mol}^{-1}
\end{aligned}$

(See Figure A8.10.)

25a $\Delta H_{at}[pyridine]$
$\begin{aligned}
&= x = -a + b \\
&= -101.2 + 5146.2 \\
&= +5045 \text{ kJ mol}^{-1}
\end{aligned}$

b From bond enthalpy data, the enthalpy of atomisation of pyridine is given by:

Break:		
	$2 \times C{=}C$	2×612
	$1 \times C{=}N$	615
	$2 \times C{-}C$	2×347
	$1 \times C{-}N$	286
	$5 \times C{-}H$	$\underline{5 \times 413}$
		$+4884 \text{ kJ mol}^{-1}$

c So the experimental value for real pyridine is 161 kJ mol^{-1} greater (more positive) than the bond enthalpy value based on the molecule on p. 154, a difference that can possibly be attributed to delocalisation-stabilisation. (The only problem here is that the $\Delta H_{at}[pyridine]$ in (b) is that of $C_5H_5N(gas)$, since bond enthalpies do not apply to liquids. The missing factor is $\Delta H_{vap}[C_5H_5N(l)]$, which is about 36 kJ mol^{-1}, which brings the difference between the ΔH_{at} values of real and 'localised' pyridine to 125 kJ mol^{-1}. This is a 2.5% fluctuation on a value of about 5000 kJ mol^{-1}, so there is room for scepticism. A sceptic could argue that the difference might be due to 'wobble' in the bond enthalpy values themselves.)

26a The extra portion of ΔH_c per —CH$_2$— should be:

Break:		
	$1 \times C{-}C$	347
	$2 \times C{-}H$	2×413
	$1\frac{1}{2} \times O{=}O$	$1\frac{1}{2} \times 498.3$
		$+ 1920.45 \text{ kJ mol}^{-1}$
Make:		
	$2 \times C{=}O$	2×-805
	$2 \times O{-}H$	$\underline{2 \times -464}$
		$-2538 \text{ kJ mol}^{-1}$

So the extra enthalpy of combustion for —CH$_2$— is −617.5 kJ mol^{-1}.

b ΔH_c [alkanes] (kJ mol^{-1})

C$_1$ −890		
	}	−670
C$_2$ −1560		
	}	−660
C$_3$ −2220		
	}	−660
C$_4$ −2880		
	}	−630
C$_5$ −3510		

These data lend credence to the bond enthalpies, since they appear to be settling down to a value not so dissimilar to −617.5 kJ mol^{-1}

c ΔH_c [alcohol] (kJ mol^{-1})

C$_1$ −730		
	}	−640
C$_2$ −1370		
	}	−650
C$_3$ −2020		
	}	−660
C$_4$ −2680		
	}	−650
C$_5$ −3330		

It is not so easy to see a pattern in the alcohols, but you could expect the electron-withdrawing OH group to weaken C—H bonds in the shorter chain alcohols. This may account for the extra link in ethanol only being worth −640 kJ mol^{-1}.

27 The heat transfer is helped by the choice of copper (an excellent conductor of heat) for the coil and by the coil's maximum path length through the water.

28a The minus sign registers the fact that heat gained by the surroundings signals an enthalpy *decrease* in the system.

b You find out which reactant had not been in excess, and then divide the enthalpy change by the number of moles of this reactant. (Obviously a factor of 1000 does the joules to kilojoules conversion.)

29a The yellow colour in a flame is made by incandescently hot soot particles. Soot is carbon, which clearly has not been fully oxidised through to carbon dioxide.

b The use of oxygen would reduce the problem, and the gas could be pumped in at the vent in the base of the combustion calorimeter, instead of using suction of air.

30 The ratio of the two results for the apparent heat transfer would be:

$$\frac{\text{water only}}{\text{water + copper}} =$$

$$\frac{600 \times 4.2}{(600 \times 4.2) + (100 \times 0.39)} = 0.985$$

This means that the results are in the ratio 0.985:1. Since 98.5% of this absorption is done by the water, the error introduced by ignoring the copper is only 1.5%. Therefore the matter of how much of the stirrer is immersed is very trivial indeed.

31a The electricity heats exactly the same absorbing medium as the reaction did. So all material absorbers and heat leakages will be the same for both. In other words, if you discover by the electrical heating stage that it takes, say, 2 kJ to raise the temperature of your apparatus with all its different materials and heat leaks by 1 °C, then you can be sure that any other energy source would also have to supply 2 kJ per degree.

b The only thing you can do wrong with electrical compensation is change any single factor between electrical and chemical stages. If the chemical stage raised the temperature of the apparatus by 5 K, then it is good if you make the electrical heater do the same rise, but even better if you make it go up the same 5 K, because you could argue that leaks are more serious when the temperature difference between the apparatus and the room is greater.

c If an endothermic reaction has cooled the heat transfer medium down, then electrical compensation is used to restore it to the original temperature.

32 If there is no temperature differential between the apparatus and the room, there will be no heat loss. The bomb calorimeter rather cleverly contrives to have an artificially heated 'room'.

33 There is a reduction by half a mole of gaseous material in the course of the reaction, so the pressure would decrease.

34 The volcano reaction clearly will do work against the atmosphere, if it is conducted in a syringe or an open vessel. So some of the loss of (true) chemical potential energy (the ΔU factor) will go into work not heat. So the observed ΔH will be less in magnitude than the ΔU (less negative, of course). At constant volume *all* the loss of chemical potential energy will be converted to heat because work is impossible from a constant volume container. In short, the surroundings will get more heat in the constant volume case.

35a In the power data. The very short times of the explosive reactions are what gives them their very high powers, since

$$power = energy/time$$

b Too great a build-up of pressure in the breech of a gun would rupture the metal of the barrel and damage the user.

36a *Methane*: From tables, the enthalpy of combustion $\Delta H_c^{\ominus}$ of methane is -890.3 kJ mol^{-1}. So the enthalpy change per kg will be

$(-890.3)/(16 \times 10^{-3}) = -55\,640$ kJ kg^{-1} (to 4 sf)

Black powder: Using the formula on p. 151, we get:

$\Delta H^{\ominus}_{reaction} = \Delta H_f^{\ominus}(products) - \Delta H_f^{\ominus}(reactants)$

$= \{5\Delta H_f^{\ominus}[CO_2(g)] + 2\Delta H_f^{\ominus}[K_2O(s)]\}$
$- \{4\Delta H_f^{\ominus}[KNO_3(s)]\}$

$= (\{5 \times -393.5)$
$+ (2 \times -361.4)\})$
$- (4 \times -494.6)$

$= -711.9$ kJ mol^{-1}

The 'per mole' here refers in a rather imprecise way to molar quantities of the reaction as written, but the mass meaning is clear enough. Four moles of KNO_3 plus five moles of carbon weigh 464.4 g. So heat per kilogram is given by:

$$\frac{-711.9}{464.4 \times 10^{-3}} = -1533 \text{ kJ kg}^{-1}$$

TNT: $\Delta H_f^{\ominus}$ of TNT will have to be found by using bond enthalpies. This introduces an error due to the fact that the benzene rings will have to be treated as alternate single/double bonds, but we can add on a little correction factor of 125 kJ mol^{-1} for the extra stability, borrowed from the earlier 'pyridine' question (25). The cycle is shown in Figure A8.11.

$x = +a - b$
$= +17\,350 - 20\,032$
$= -2680$ kJ mol^{-1} (to 3 sf)

Again, this is the value for 'molar amounts of the equation as written', which means per *two* moles of TNT. The 'per kilogram' figure is therefore -5800 kJ kg^{-1} (to 2 sf).

We can see looking back over the last three results that explosiveness has little to do with the absolute amount of heat available, in which respect methane is vastly superior to the other two. As we shall see, explosiveness has more to do with volume of gas produced and the speed of the reaction.

b *Methane*: 1 mole of C atoms goes up 8 oxidation numbers
4 moles of O atoms go down 2 oxidation numbers

Black powder: 4 moles of N atoms go down 5 oxidation numbers
5 moles of C atoms go up 4 oxidation numbers

c In *inter*molecular reactions there must be a time delay while reactant molecules meet and collide. No such problem attends *intra*molecular reactions.

37 See Figure A8.12.
$\Delta H_{reaction} = \Delta H_f^{\ominus}(products)$
$- \Delta H_f^{\ominus}(reactants)$

$= + (\{4 \times -1675.7\} + \{3 \times -436.7\})$
$- (3 \times -432.8)$

$= -6714.5$ kJ mol^{-1}

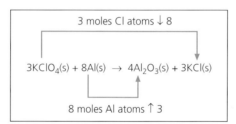

3 moles Cl atoms ↓ 8

$3KClO_4(s) + 8Al(s) \rightarrow 4Al_2O_3(s) + 3KCl(s)$

8 moles Al atoms ↑ 3

Figure A8.12 Reaction involving $KClO_4$ and Al in a photoflash

This is quite heavily exothermic but completely lacking in gaseous products. This inability to create a major blast pressure wave (other than by heating the adjacent air) renders the mixture useless as an explosive.

38a See Figure A8.13.
$x = +a - b$
$= +27\,512 - 34\,770.7$
$= -7260$ kJ mol^{-1} (to 3 sf)

b If it is true that the function pv/nRT assumes values as large as 14, then the gases are in a very non-standard state when first formed. It is therefore as if they possess more of their energy as pressure and less as temperature, compared to the case with an ideal gas. So one would predict that, compared to the formal ΔH, the result will be more work and less heat.

c The greatest care and attention has to be paid to this task. The one-legged stool ensures that, if the hill man 'drops off', he . . . well . . . drops off. (And he was not encouraged to have a pint at lunchtime either.)

d Gravity feeds remove the risks associated with electrical pumping. All electric motors create sparks at their brushes and a spark in an explosives factory could be very dangerous.

e If the worst came to the worst, there would only be a small explosion from one kilogram of nitroglycerine. And water pastes remove the risk of sparks getting at the material, or of it getting overheated in transit through the process.

39a With the slurry truck system, the explosive only exists in the immediate run-up to the detonation. The water gel system again, as in the nitroglycerine case, will be a fire retardant and heat absorber.

b The staggering of the explosions in time avoids the build-up of a single pressure wave, travelling at the speed of sound in the ground, and likely to disrupt foundations in buildings. If the time for energy release were extended from, say, 10 microseconds to 10 milliseconds, the power of the blast (= energy/time) would be reduced one thousandfold.

40a The metal conducts heat from the flame, so that less heat is left to travel down the flame itself to ignite new gas. Hence the gas in the barrel fails to ignite by virtue of being too cold. If the barrel were too wide the centre of the gas flow would no longer be within reach of its cooling effect, and would be able to reach ignition temperature. Then the reaction front would be able to pass downwards into the barrel, and if it attained a speed greater than the speed of gas flow upwards, the gas in the barrel would ignite.

b In the gauze trick, the gauze is acting as heat conductor, just like the barrel itself as described in (a).

c The speed of reaction front movement downwards has been reduced, again by cooling.

d The skirt is meant to deflect the cold convected air away from the base of the flame.

e The confinement within the plastic bottle allows pressure to build up, and ignition can be by the passage of the pressure wave, as in a true explosion, rather than by heat conduction.

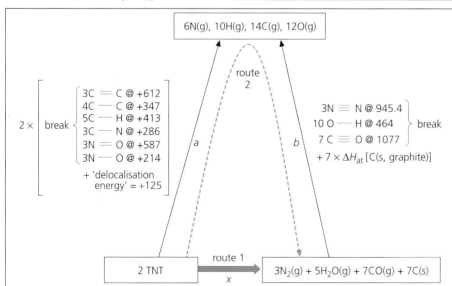

6N(g), 10H(g), 14C(g), 12O(g)

route 2

$2 \times$ break
$\begin{cases} 3C = C @ +612 \\ 4C - C @ +347 \\ 5C - H @ +413 \\ 3C - N @ +286 \\ 3N = O @ +587 \\ 3N - O @ +214 \end{cases}$
+ 'delocalisation energy' = +125

a

b

break
$\begin{cases} 3N \equiv N @ 945.4 \\ 10 O - H @ 464 \\ 7 C \equiv O @ 1077 \end{cases}$
$+ 7 \times \Delta H_{at} [C(s, \text{graphite})]$

route 1

2 TNT x $3N_2(g) + 5H_2O(g) + 7CO(g) + 7C(s)$

Figure A8.11 All values are in kJ mol^{-1}

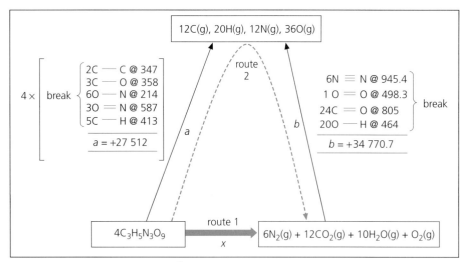

$4 \times$ break
$\begin{cases} 2C - C @ 347 \\ 3C - O @ 358 \\ 6O - N @ 214 \\ 3O - N @ 587 \\ 5C - H @ 413 \end{cases}$

$a = +27\,512$

12C(g), 20H(g), 12N(g), 36O(g)

route 2

a

b

$\begin{cases} 6N \equiv N @ 945.4 \\ 1 O = O @ 498.3 \\ 24C - O @ 805 \\ 20O - H @ 464 \end{cases}$ break

$b = +34\,770.7$

$4C_3H_5N_3O_9$

route 1
x

$6N_2(g) + 12CO_2(g) + 10H_2O(g) + O_2(g)$

Figure A8.13 All values are in kJ mol^{-1}

CHAPTER 9   page 164

1a In Figure 9.1a the bucket will fall, opposed only by the mass of the rope and friction in the pulley, at an acceleration close to *g*. In Figure 9.1b the bucket will also fall but the acceleration will be less, because the accelerating force will be the difference between the weights of bucket and pallet (neglecting friction this time).
b In Figure 9.1a the original potential energy ends in heat (and a little sound), as the bucket hits the ground. In Figure 9.1b there is a smaller release of heat due to the lesser impact, and also an increase in the potential energy of the pallet.

2a The situation in Figure 9.1a would be at the extreme left (all energy ends up as heat), while that of Figure 9.1b would be somewhere in the middle (some heat, some work done on the pallet/surroundings).
b If the pallet were only *slightly* less heavy than the bucket, then the bucket would fall very slowly, landing very gently on the ground and therefore giving it only a tiny portion of kinetic energy as heat. In this arrangement the heaviest possible pallet would have been lifted, which means a large proportion of the original energy of the bucket ending as work on the 'surroundings'. The limiting condition is when the two masses become equal.

3a When you neutralise an acid with an alkali in solution, the solution gets warmer and no work is done. You might have suggested the burning of a fuel on an open fire, but strictly speaking the product gases will do work pushing back the atmosphere, as was mentioned in Section 8.9.
b When a fuel is burned in an internal combustion engine, some of the energy is converted to kinetic energy of the car.

4a No.
b Yes, for instance the solution of most ionic salts in water.
c No – from petrol engines to muscles, there is no such thing as a cold motor.

5 The molecules will spread themselves evenly over the enlarged box.

6 'Downhill' in a chemical context means that bond making is dominant, which in turn implies that there has been more building than destruction. More building of molecules suggests fewer free small fragments, so less 'chaos'. Conversely, high 'chaos' actually requires more small fragments, so bonds will have to be (mainly) broken, and the result will be endothermic, 'uphill' reactions.

7a $2H_2(g) + O_2(g) \rightarrow 2H_2O(l)$
$S^{\ominus}$ per mole   130.6   205   69.6
(of molecules)
So $\Delta S = 2 \times 69.6 - (2 \times 130.6 + 205)$
       $= -326.8$ J K^{-1}

You would expect a negative ΔS due to the chaos loss associated with 3 moles of gases going to liquid.
b $C_3H_8(g) + 5O_2(g) \rightarrow 3CO_2(g) + 4H_2O(g)$
$S^{\ominus}$ 296.9   205   213.6   188.7
per mole
(all J K^{-1})

So $\Delta S = 3 \times 213.6 + 4 \times 188.7 -$
       $(296.9 + 5 \times 205)$
       $= +73.7$ J K^{-1}

You would expect a positive ΔS due to the increase in overall numbers, although admittedly you might not have guessed how very positive it would be.
c $Cu^{2+}(aq) + Zn(s) \rightarrow Zn^{2+}(aq) + Cu(s)$
$S^{\ominus}$ per mole -99.6 $+41.6$ -112.1 $+33.2$
(all J K^{-1})

So $\Delta S = (-112.1 + 33.2) - (-99.6 + 41.6)$
       $= -20.9$ J K^{-1} mol^{-1}

You would expect a small ΔS since the chaos factor of each side is very similar (1 mole of aqueous ions, 1 mole of solid).

8a The chemical system.
b The cold surroundings.
c The hot surroundings.
d The enthalpy change, or in other words the heat given by surroundings to system.

9a $CaCO_3(s) \rightarrow CaO(s) + CO_2(g)$
$\Delta H_f^{\ominus}$ per mole -1206.9 -635.1 -393.5
(all kJ)
$S^{\ominus}$ per mole   92.9   39.7   213.6
(all J K^{-1})

$\Delta H_{sys}^{\ominus} = \Delta H_f^{\ominus}$ (products) $- \Delta H_f^{\ominus}$ (reactants)
       $= (-393.5 + -635.1) - (-1206.9)$
       $= +178.3$ kJ mol^{-1}
$\Delta S_{sys}^{\ominus} = \Sigma S_{products} - \Sigma S_{reactants}$ (9.4)
       $= 39.7 + 213.6 - 92.9 = +160.4$ J K^{-1} mol^{-1}

$\Delta S_{uni} = \Delta S_{sys} - \dfrac{\Delta H_{sys}}{T}$ (9.7a)

$= +160.4 - \left(\dfrac{+178\,300}{298}\right) = -438$ J K^{-1} mol^{-1}

So no, it would not go.
b $2NO(g) + O_2(g) \rightarrow 2NO_2(g)$
$\Delta H_f^{\ominus}$ per mole 90.2   0   33.2 (all kJ)
$S^{\ominus}$ per mole 210.7   102.5   240 (all J K^{-1})
$\Delta H_{sys}^{\ominus} = 2 \times 33.2 - 2 \times 90.2 = -114$ kJ mol^{-1}
$\Delta S_{sys}^{\ominus} = 2 \times 240 - (2 \times 210.7 + 102.5)$
       $= -43.9$ J K^{-1} mol^{-1}
$\Delta S_{uni} = -43.9 - \left(\dfrac{-114\,000}{298}\right) = +339$ J K^{-1} mol^{-1}

So yes, it would go.
c The next problem will be done in the same format, but without the full layout.
         $Fe_2O_3(s) + 2Al(s) \rightarrow Al_2O_3(s) + 2Fe(s)$
$\Delta H_f^{\ominus}$ -824.2 0 -1675.7 0   (all kJ)
per mole
$\Delta S^{\ominus}$ 87.4   28.3   50.9   27.3
per mole   (all J K^{-1})
$\Delta S_{uni} = +2819$ J K^{-1} mol^{-1}

So a big yes, despite the fact that it does not actually go due to 'kinetic control'.

10 The creation of a mole of gas where only solid existed before would ensure a positive ΔS_{sys}.

11a At 500 K, ΔS_{uni}
$= +160.4 - \left(\dfrac{+178\,300}{500}\right) = -196.2$ J K^{-1} mol^{-1}
       (9.7a)
So no, the reaction would not proceed.
b At 1000 K, by similar methods, $\Delta S_{uni} = -17.9$ J K^{-1} mol^{-1}. The reaction nearly goes.
c At 1500 K, $\Delta S_{uni} = +41.5$ J K^{-1} mol^{-1}. Yes, the reaction will proceed.

12 At 1000 K the calcium carbonate decomposition clearly would not go, but data for MgCO$_3$ are:
         $MgCO_3(s) \rightarrow MgO(s) + CO_2(g)$
$\Delta H_f^{\ominus}$ per mole -1095.8 -601.7 -393.5
(all kJ)
$S^{\ominus}$ per mole   65.7   26.9   213.6
(all J)

From which, by the method of question 9, $\Delta S_{uni} = +74.2$ J K^{-1} mol^{-1}.

So the magnesium carbonate reaction goes comfortably within the range of a Bunsen burner, and would at least enable the teacher to show a white powder losing mass.

13a None of the conditions in question 11 were standard.
b $\Delta G_{reaction} = \Delta G_f$ (products) $- \Delta G_f$ (reactants)
Relevant data are:

Substance	ΔG_f (kJ mol^{-1})
CaO (s)	−604
H_2O (l)	−237
$Ca(OH)_2$ (s)	−897

$\therefore \Delta G_{reaction} = (-897) - ((-604) + (-237))$
$= -56$ kJ mol^{-1}

So the reaction is thermodynamically viable (and in this case does actually go spontaneously, because it is kinetically viable too).

14 y is equivalent to ΔG_{sys}.

m (which gives the gradient of the line) is equivalent to $-\Delta S_{sys}$.

x is equivalent to T.

15 Ellingham diagram lines change gradient if any of the members of the reaction system *change state*. This would alter the chaos factor change of the system, and therefore the $-\Delta S_{sys}$ which is playing the part of gradient.

16a Around 780°C, at which point the carbon and iron lines cross.
b At this temperature the carbon line is at about −500 kJ mol^{-1}, while the iron one is at about −400. So the reduction of iron by carbon would have a free energy of
$-500 - (-400) = -100$ kJ mol^{-1}

c They cross at such a high temperature that the chemical reduction of aluminium by coke is not a practicable idea.

17 Using data from question 9, we have:

$\Delta S_{uni} = +160.4 - \left(\dfrac{+178\,300}{T}\right)$ (9.7a)

If we now set ΔS_{uni} at zero, the unknown T works out as 1112 K.

18 $K_c = \dfrac{[SO_3(g)]_{eqm}{}^2}{[SO_2(g)]_{eqm}{}^2 \times [O_2(g)]_{eqm}}$

19a All the ^{18}O would remain in the top layer.
b ^{18}O would be distributed over both layers.
c As many molecules must be leaving the top layer as are returning there.

20 At equilibrium, the rates of forward and reverse reactions must be the same, so:
$k_f[A][B] = k_b[C][D]$
So $\dfrac{[C][D]}{[A][B]} = \dfrac{k_f}{k_b} = K_c$

which agrees with the formula version for the equilibrium constant K_c of the reaction
$A + B \rightleftharpoons C + D$.

21a The volume of the new mixture is 2 dm^3.
b Concentrations in mol dm^{-3} are:

$[A] = 1$, $[B] = [C] = [D] = \frac{1}{2}$

c k_b [C][D] (the reverse rate) is the more affected.
d Temporarily the forward rate is faster, so the concentrations of A and B are reduced and more C and D are made.

22a $\dfrac{[C]_{eqm}[D]_{eqm}}{[A]_{eqm}[B]_{eqm}} = K_c$

It has no units because concentration units cancel out.
b 1
c The temporary false K_c, symbol K_c', is

$\dfrac{(\frac{1}{2})^2}{1 \times \frac{1}{2}} = \frac{1}{2}$

That is, it has got smaller.
d More C and D must be produced at the expense of A and B, to restore the value of 1.
e The changes in the other species' concentrations would be:

B: $-x$ C: $+x$ D: $+x$
f $[A] = 1 - x$ $[B] = \frac{1}{2} - x$ $[C] = \frac{1}{2} + x$
$[D] = \frac{1}{2} + x$
g So $1 = (\frac{1}{2} + x)^2 \backslash (1 - x)(\frac{1}{2} - x)$

which rearranges to a quadratic equation in x, whose sensible root is $x = 0.1$. So the final concentrations are:

$[A] = 0.9$ $[B] = 0.4$ $[C] = 0.6$ $[D] = 0.6$
(all in mol dm^{-3})

23a There must have been 2 moles of A and B.
b So originally there was 50% conversion of B to C.
c Moles of B = concentration × volume
$= 0.4 \times 2 = 0.8$ moles.
d So this time 1.2 moles of the original 2 moles was converted, or 60%.
e Yes and yes.

24a The back reaction will be stopped.
b More C will be created.
c Gradually, despite the equilibrium status of the system, all the A and B will have reacted.
d If supplies of A and B are not replenished, the successive passes through the reaction vessel will be chasing smaller and smaller returns from diminishing amounts of product.

25a $K_c = \dfrac{cV}{ab}$

b $K_c = \dfrac{1 \times 1}{1 \times 1} = 1$

(That is the beauty of making up your own data.)

c $K_c = \dfrac{1 \times \frac{1}{2}}{1 \times 1}$

d More product will be needed, to increase c at the expense of a and b.

26 All the Vs would cancel, so there would be no variable in the 'moles' version of K_c susceptible to changes in pressure.

27 The volume is not affected by pressure. To generalise, the only equilibrium systems which can respond to pressure are those in which the

volume is (1) 'hidden' in the equilibrium constant, and (2) capable of being changed by pressure. So AB/CD equilibria fail the first test, and all aqueous-solution equilibria fail the second, by virtue of the incompressibility of liquids.

28 The reaction is A(g) + B(g) $\rightleftharpoons$ C(g)

So $K_c = \dfrac{[C]}{[A][B]}$, which in units is $\dfrac{(\text{conc.})}{(\text{conc.})^2}$,

or (conc.)$^{-1}$. So the actual units are the reciprocal of normal concentration units.

29 Greater pressure would increase yield, but it also requires higher specification reaction vessels and higher running costs.

30 If you halved the volume, the concentrations would all double, so the forward rate, given by: $k_f[A][B]$
would quadruple, whereas the reverse rate, given by: $k_b[C]$
would merely double. So now the forward reaction would outstrip the reverse, and the equilibrium would give a lurch towards more C.

31 If $\Delta S_{uni} = 0$, then $\Delta G_{sys} = -T\Delta S_{uni} = 0$

Then $RT \ln K_c = 0$, so $\ln K_c = 0$, so $K_c = 1$.

32 At 750 K, $\Delta S_{surr} = -\left(\dfrac{-100\,000}{750}\right)$
$= +133$ J K^{-1} mol^{-1}

So $\Delta S_{uni} = \Delta S_{sys} + \Delta S_{surr}$
$= -200 + 133$
$= -67$ J K^{-1} mol^{-1}

So $\Delta G_{sys} = -T\Delta S_{uni}$
$= -750 \times (-67)$
$= +50\,000$ J mol^{-1}

So $+50\,000 = -RT \ln K_c$
$= -8.3 \times 750 \times \ln K_c$

This solves to give $K_c = 3.25 \times 10^{-4}$.
Remembering that K_c was 1 at 500 K, we see that the *lower* temperature favoured the *higher* yield, a situation which is general for exothermic reactions. However, low temperatures might mean very slow (perhaps infinitely slow) yields, which is not much good either.

33 See Figure A9.1.

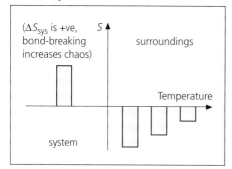

Figure A9.1 Showing that ΔS_{surr} gets less *un*helpful at higher temperatures

This time the ΔS_{surr}

$(= -\dfrac{\Delta H}{T})$ is

inevitably negative, but at least it gets less negative at high temperatures. So yields are bigger when the system is hot.

34 $\quad \Delta G_{sys} = \Delta H_{sys} - T\Delta S_{sys} \qquad (9.9)$
But $\quad \Delta G_{sys} = -RT \ln K_c \qquad (9.11)$
So $\quad -RT \ln K_c = \Delta H_{sys} - T\Delta S_{sys}$

Therefore $\quad \ln K_c = -\dfrac{\Delta H_{sys}}{RT} + \dfrac{\Delta S_{sys}}{R}$

But ΔS_{sys} is approximately constant with varying T, and R is a constant. So:

$\ln K_c = -\dfrac{\Delta H_{sys}}{RT} + \text{constant} \qquad (9.13)$

If we take the second approximation that ΔH_{sys} is also invariable with T, then we can apply (9.13) twice, varying K_c and T, but using the same ΔH_{sys}. (Suffixes indicate the two Ks at the two Ts.)

(1) $\quad \ln K_1 = -\dfrac{\Delta H}{RT_1} + \text{constant}$

(2) $\quad \ln K_2 = -\dfrac{\Delta H}{RT_2} + \text{constant}$

(3) = (1) − (2) $\quad \ln\left(\dfrac{K_1}{K_2}\right) = \dfrac{\Delta H}{R}\left(\dfrac{1}{T_2} - \dfrac{1}{T_1}\right)$

(9.13 adapted)

35a The system will respond to the increase in pressure by decreasing its own pressure, i.e., by going more to the side which takes up less space. The right-hand side shows one mole of gas compared to the left's two, so more moles of C will form at the expense of moles of A and B.
b The system will respond by absorbing heat, that is, by going in its endothermic direction. Hence more moles of A and B will be formed.
c The system will respond by making more moles of C.

36 This follows the pattern mentioned in the answer to question 40, with the equilibrium shifting in its backward, endothermic, direction, and thereby reducing the yield of SO_3.

37a The stoichiometric ratio of SO_2 to O_2 is 2 : 1, so 10 : 1 clearly represents an excess of oxygen compared to 10 : 5.
b The system will respond by reducing the excess of oxygen, which will result in greater yield of SO_3 (and incidentally greater percentage conversion of SO_2). A similar situation was dealt with quantitatively in questions 22 and 23.
c The reaction is exothermic and so heats itself.
d If the reaction system is allowed to get hot, then the equilibrium yield goes down, as described by the thin curved line. So cooling stages will increase the yield.

38a The system will respond to the removal by making (even) more SO_3.

b The system will respond to high pressure by going in the direction of the side with lower volume. Since there are three moles of gas on the left and only two on the right, this side is the product side.
c The reaction between SO_3 and water to form H_2SO_4 is too highly exothermic, and brings about boiling in the liquid.

39a 2 cm³ of 10 mol dm⁻³ HCl(aq) contain 0.002×10 moles of HCl. So the same number of moles of NaOH will be needed to neutralise it. The volume is given by:

$\dfrac{\text{number of moles}}{\text{concentration}} = \dfrac{0.02}{2} = 0.01$ dm³ or 10 cm³

b So (64 − 10), or 54 cm³, of 2 mol dm⁻³ NaOH(aq) was neutralising the new organic acid. If the acid and alkali reacted 1 : 1, the value for moles of acid was 0.054×2, or 0.108 moles.

c

	ester	+ water	⇌ acid	+ alcohol
Moles at t_0	0.150	$\dfrac{15 + 1.5}{18}$	0	0
	↓−0.108	↓−0.108	↓+0.108	↓+0.108
Moles at t_{eqm}	0.042	0.809	0.108	0.108

d $\quad$ 0.042/V $\quad$ 0.809/V $\quad$ 0.108/V $\quad$ 0.108/V
e All the Vs cancel, so K_c is given by:

$\dfrac{0.108^2}{0.042 \times 0.809} = 0.343$ (no units)

f 0.108 moles of ester were converted, from an original 0.150 moles, so percentage conversion is

$\dfrac{0.108}{0.150} \times 100 = 72\%$.

40 A high water : ester ratio would maximise the conversion of ester to alcohol (water is playing a role like that of air in the contact process).

41a

	ester	+ water	⇌ acid	+ alcohol
Moles at t_0	0.15	$\dfrac{30 + 1.5}{18}$	0	0
	↓−x	↓−x	↓+x	↓+x
Moles at t_{eqm}	(0.15 − x)	(1.75 − x)	x	x

b Since K_c is 0.343,

$\dfrac{x^2}{(0.15 - x)(1.75 - x)} = 0.343$

c This can be rearranged into the quadratic format as

$0.657x^2 + 0.6517x - 0.09 = 0$

This gives a sensible root for x as 0.123 moles.
d So moles of ester at t_{eqm} = 0.15 − 0.123 = 0.027. This corresponds to a percentage conversion of 0.123/0.15 = 82%. In question 39, 0.108 moles was converted, equivalent to a percentage conversion of 72%, so it was

worthwhile beginning with a higher water : ester ratio.
e If you went to extremes of high water : ester ratios, you would get products which were very diluted and dispersed over a large volume. This would mean escalating problems of extraction and purification.

42 The titration is quick compared to the rate of response of the system. This will be all the more true since the titration knocks out the catalyst as well as estimating the product.

43 One method would be the absorption of light by one or other of the members of the system. If an ultraviolet/visible spectrophotometer is used, the absorption of the mixture can be plotted against the wavelength of light. If one ingredient has an absorption at a wavelength at which all the others are transparent, and if it is known that certain degrees of absorption correspond to certain concentrations, then the absorption at that wavelength can be converted to a concentration.

44a If x moles of the parent dissolve in 1 dm³, then the concentrations of the daughter ions will be: $[Fe^{2+}(aq)] = x$, $[OH^-(aq)] = 2x$
b $\quad$ So $K_{sp} = 6 \times 10^{-15}$
$\quad\quad\quad = x \times (2x)^2 = 4x^3$
So $x = 1.14 \times 10^{-5}$ mol dm⁻³, which is the solubility of $Fe(OH)_2$.

45a If $[OH^-(aq)] = 10^{-6}$ mol dm⁻³, then, when the iron hydroxide is at the limit of its solubility,

$K_{sp} = 6 \times 10^{-15} = [Fe^{2+}(aq)] \times (10^{-6})^2$

So maximum $[Fe^{2+}(aq)] = 6 \times 10^{-3}$ mol dm⁻³
b Repeating the calculation at the new pH, the allowable value becomes 60 mol dm⁻³. This is unrealistic, because there would not be so much iron available and anyway other anions like sulphate would come into play. However, there is much less limitation on its solubility and so much more available soluble Fe^{2+}, in acid soils.

46a The syringe will expand as B is added, eventually doubling in volume. So A, originally at 1 atmosphere, will now occupy twice the volume and exert half the pressure. Hence $p_A = 0.5$ atm.
b Since both gases 'ignore each other' and since there has been no volume change, p_A is still 1 atmosphere.
c If 300 of every million molecules are carbon dioxide, then the mole fraction of CO_2 is

$x_{CO_2} = \dfrac{300}{1\,000\,000}$

Therefore
$p_{CO_2} = x_{CO_2} \times p_{tot}$
$\quad = \dfrac{300}{1\,000\,000} \times 1 = 3 \times 10^{-4}$ atm

47 In this case, $K_c = \text{mol}^{-2}\,\text{dm}^6$; $K_p = \text{atm}^{-2}$.

48a M_r' is the mass of 24 000 cm³ of the gas mixture, given by $0.29 \times 24\,000/100 = 69.6$ g per mole.
b Some of the molecules are NO_2 (M_r 46) and

some are N_2O_4 (M_r, 92), so since the mixture consists of both types of molecule, it will display an average molecular mass intermediate between 46 and 92.

c   The mole fraction of NO_2 would be $(1 - x)$, since all molecules not N_2O_4 are NO_2.

d   Weighted average formula:
Average value

$$= \frac{\Sigma \text{ (weighting factor} \times \text{quantity)}}{\Sigma \text{ (weighting factors)}}$$

In this context,
Average molecular mass M_r'

$$= \frac{\Sigma \text{ (mole fraction} \times M_r)}{1}$$

Specifically, $69.6 = \frac{(1 - x)46 + x \times 92}{1}$

$46 - 46x + 92x = 69.6$

$46 + 46x = 69.6$

$x = \frac{23.6}{46}$

So x, mole fraction of N_2O_4, is 0.51.
So mole fraction of $NO_2 = 1 - x = 0.49$.

e   Since p = mole fraction $\times P$, where $P = 1$ atm $p_{N_2O_4} = 0.51$ atm and $p_{NO} = 0.49$ atm

f $K_p = \frac{p_{NO}^2}{p_{N_2O_4}} = \frac{0.49^2}{0.51} = 0.46$ atm

g   If pressure were increased, the mole fraction of N_2O_4 in the mixture would increase, since that is on the side of the reaction with fewer gas molecules, and the mole fraction of NO would decrease.

h   The value of K_p would increase, since endothermic equilibria increase the proportion of *products* as T is increased.

i   You never lose mass in a chemical equation, so the apparent molecular mass of the gas only changes if the density (that is, the volume in gas reactions) changes. That cannot happen in the hydrogen/iodine equilibrium, but does happen in the other one, since there are three moles of gas on the left, and only two on the right.

49a If there had been no reaction, then the pressure would have remained at 10 atm. Total reaction would have halved the number of molecules, so the pressure would have been 5 atm.

b	A	+	B	⇌	C
Moles at t_0	1		1		0
	↓$-x$		↓$-x$		↓$+x$
Moles at t_{eqm}	$1 - x$		$1 - x$		x

c   Total moles $= 1 - x + 1 - x + x = 2 - x$

d   So $\frac{10}{7} = \frac{2}{2 - x}$, which gives $x = 0.6$ mole.

e   Mole fractions:

$A = \frac{0.4}{1.4}$ $B = \frac{0.4}{1.4}$ $C = \frac{0.6}{1.4}$

$= 0.286$ $= 0.286$ $= 0.429$

Partial pressures:
$A = 0.286 \times 7$ $B = 0.286 \times 7$
$= 2$ atm $= 2$ atm
$C = 0.429 \times 7$
$= 3$ atm

f $\frac{p_C}{p_A \times p_B} = K_p = \frac{3}{2 \times 2} = 0.75$ atm^{-1}

g   At the higher pressure, the yield of C would be higher.

h   No, because the total pressure would remain the same irrespective of the position of the equilibrium.

CHAPTER 10   page 191

1   Relevant bond enthalpies are:

Bond	Average bond enthalpy (kJ mol^{-1})
C—C	348
C=C	612
O—O	146
O=O	496
N—N	163
N=N	409
N≡N	944
Si—Si	176

The patterns that emerge from these data are detailed below.

C—C bonds are by far the strongest of the single bonds, perhaps because rivals in the same Periodic Table row (nitrogen and oxygen) have lone pairs in places where carbon atoms have single electrons. Lone pairs create more repulsion between atoms than would be found from, say, a C—H bond pair, especially over the short bond lengths involved here, as shown below.

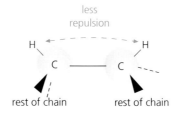
less repulsion

H   H
C ———— C
rest of chain   rest of chain

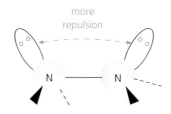
more repulsion

N ———— N

The option of multiple bonding seems to be much more relatively attractive in the case of nitrogen and oxygen than in the carbon case. Again the lone-pair repulsions may be a factor, in that multiple bonding is a good way to keep lone pairs as far apart as possible.

N ≡ N

Finally, note that the Si—Si bond has a lower bond enthalpy than C—C (176 against 348 kJ mol^{-1}).

So the conclusion of bond enthalpies is that relative to carbon chains, all the others are weak, and in the case of oxygen and nitrogen multiple bonding is preferred.

2a   One (CO_2) exists as small discrete molecules with weak intermolecular forces, whereas the other is a giant lattice (sometimes crystalline/orderly, sometimes glassy/random) with strong covalent bonds running throughout the structure.

The root cause is that, by opting for *double* bonds to oxygen atoms the carbon atom has produced a self-contained unit, whereas by opting for *single* bonds to oxygen atoms, the silicon atom has produced an open-ended lattice-building unit.

b   The Si=O bond is weak relative to an alternative single-bonded structure, using Si—O—Si chains.

c   Enthalpies of atomisation will be the sum of the bond enthalpies of each molecule.

Table A10.1

Bond	Number	Average bond enthalpy (kJ mol^{-1})	Total (kJ mol^{-1})
C—C	1	348	348
C—H	6	412	2472
		$\Delta H_{at}^{\ominus}[C_2H_6(g)] = 2820$	
Si—Si	1	176	176
Si—H	6	318	1908
		$\Delta H_{at}^{\ominus}[Si_2H_6(g)] = 2084$	

So ethane is the more strongly bonded molecule, relative to its gaseous atoms.

d   Life on Earth is carbon-based because the carbon-based small-molecular building block CO_2 was available in solution and in the gas phase, from whence it was accessible to the organisms at the start of the food chain. No silicon was available in these formats. Secondly, carbon satisfies the requirement for a 'kit' of life-building molecules made of stable chains, whereas a similar kit made out of silicon chains would fall apart.

3 (d) is the same as (a) – there is a single unbranched chain in each, and the bend is illusory. (c) turns into (b) when you rotate it by 180°. (e) and (g) are the same as (b) and (c), thinly disguised by false bends on their ends. (f) is unique. In fact, there are only three isomers, and their differences can be expressed well in words.

- (d) and (a) – one continuous chain of five carbon atoms, no branches.

- (b), (c), (e), (g) – chain of four carbon atoms, with a single branch at the second-from-the-end carbon atom.

- (f) – chain of three carbon atoms with a double branch in the middle.

4

Six in a line:

$CH_3CH_2CH_2CH_2CH_2CH_3$

Five in a line:

$CH_3CH_2CHCH_2CH_3$ and $CH_3CHCH_2CH_2CH_3$
            | |
          CH_3 CH_3

Four in a line:
                                  CH_3
                                    |
$CH_3CHCHCH_3$ $CH_3CH_2CCH_3$
    | \ |
  CH_3 CH_3 CH_3

5

a $CH_3CHCH_2CH_3$ **b** $CH_3CHCHCH_3$
         | | |
       CH_3 H_3C CH_3

       CH_3 H_3C CH_3
         | | /
c $CH_3CCHCH_2CH_3$ **d** $CH_3CH_2CCHCH_2CH_2CH_3$
         | \ |
       H_3C CH_3 CH_2CH_3

6a Butane
b 3-ethylhexane
c 3,5-dimethylheptane.

7 Carbon dioxide is faintly acidic due to the (equilibrium) reaction:

$CO_2(g) + H_2O(l) \rightleftharpoons H^+(aq) + HCO_3^-(aq)$

So it will dissolve freely in aqueous alkalis. Practically this can be achieved by bubbling the gas through pre-weighed 'absorption tubes' containing KOH(aq). (KOH is slightly better than NaOH for this job because the product Na_2CO_3 is moderately insoluble and might precipitate and clog up the tubes.) The tubes gain in mass by an amount equal to the CO_2 absorbed. The overall reaction is:

$CO_2(g) + 2KOH(aq) \rightarrow K_2CO_3(aq) + H_2O(l)$

(Subtle questions like 'What happened to the HCO_3^- ion?' can wait till Chapter 12.)

Water can be absorbed by a drying agent such as P_2O_5, also in a pre-weighed piece of glassware.

8 100 g of the compound contains 80 g of carbon and 20 g of hydrogen. So the numbers of moles are in the ratio

$$\frac{80}{12} : \frac{20}{1} = 6.667 : 20 = 1 : 3$$

So the empirical formula is CH_3.

9 1.285 g of CO_2 represents 1.285/44 moles, or 0.0292 moles of CO_2. A mole of CO_2 contains a mole of carbon atoms, so the number of moles of carbon atoms in the original sample must also have been 0.0292.

0.631 g of water represents 0.631/18 moles, or 0.035 moles of water, but now we have the situation in which one mole of water contains two moles of hydrogen atoms, so the number of moles of hydrogen atoms in the original sample must have been 0.07.

The moles of carbon atoms and hydrogen atoms are in the ratio

$0.0292 : 0.07 = 1 : 2.4 = 5 : 12$

(That was found by entering 2.4 on the calculator and 'constant-adding' it to itself till it came to a whole number – five lots of 2.4 were needed.)

So the empirical formula is C_5H_{12}.

10 (Using the above method but without the commentary.)

Moles-of-atoms ratio is

$$C : H = \frac{9.25}{44} : 2 \times \frac{4.25}{18}$$

$$= 0.21 : 0.472$$
$$= 1 : 2.24$$
$$= 4 : 9$$

So the empirical formula is C_4H_9. (Notice that the information about 3 g of sample was not used.)

The volume 75 cm³ at 373 K is equivalent to an stp volume of

$$75 \times \frac{273}{373} = 54.9 \text{ cm}^3$$

So if 54.9 cm³ had a mass of 0.28 g, then 22 400 cm³ (the volume of a mole of any gas at stp) would have had a mass of (0.28 × 22 400/54.9) or 114.2 g which is the relative molecular mass of the actual molecule. (This is the molecule, remember, whose empirical formula is C_4H_9.)

We have to reconcile this mass with the empirical formula. We know that some multiple of C_4H_9 must be the target molecule, so it follows that some multiple of $(4 \times 12 + 9 \times 1)$, or 57, must have the target mass. To be formal we can set up the equation:

$n \times 57 \approx 114$, from which $n = 2$

So the molecular formula must be C_8H_{18} (one of the isomers of the petroleum hydrocarbon **octane**).

11 About 1.1% of carbon atoms are carbon-13s. That means that for every carbon atom in the molecule there is a 1.1% chance of its being a C-13. The peak at 99 must have been caused by a single C-13 somewhere in the molecule. If the molecule had been, say, a two-carbon one, then the chance of a single C-13 would have been 2.2%. From the data presented, it seems that the chance of a single C-13 in the molecule (as calculated from the sizes of the peaks at 98 and 99) is 2 : 29, or 2 in 31, or 6.5%. Since this is about six times higher than the chance associated with a single carbon atom, it suggests the molecule has *six* carbon atoms in it.

The chance of the 99 peak being due to the presence of deuterium atoms in the molecule can be discounted. Deuterium accounts for only 0.015% of hydrogen atoms, and so will make only a tiny contribution to the molecules of mass 99.

12 The molecule $(CH_3)_4C$ (2,2-dimethylpropane) could not have been the test molecule, since all its hydrogens are in equivalent positions.

$$CH_3$$
$$|$$
$$CH_3—C—CH_3$$
$$|$$
$$CH_3$$

Hence, it will only give a single monosubstituted product.

13 Ethylbenzene would give a peak of mass 29, caused by $C_2H_5^+$, which would be absent from the spectrum of the other compound.

14

cis-pent-2-ene

trans-pent-2-ene

cyclopentene

15 Homolytic bond fission (into free radicals) would be a favoured option if the bond to be broken had no pronounced polarity – that way there would be no good reason for either member of a bond to lay claim to the electron pair. Molecules like the dihalogens would be likely candidates for this style of breakage, as would molecules containing C—H, C—C or C=C bonds.

16 *Nucleophiles*: H—Cl (Cl end), CN^-, OH^-, NH_2NH_2 – because they all have lone pairs available.
Electrophiles: H—Cl (H end), NO_2^+ – because they both could accept a pair of electrons.

Free radicals: Cl•, C₆H₅COO• – because they are both offering unpaired single electrons.

17   *Nucleophiles* are likely to react with chloroethane and ethanol – because they both contain a polar covalent bond.
Electrophiles are likely to react with propene and the ring of ethylbenzene – because they both have regions of high electron density (their C=C bonds).
Free radicals are likely to react with propene (both C=C bond and methyl group), methane, cyclohexane and the side chain of ethylbenzene – because their bonds are all of the non-polar covalent type.

18   The attack would take place at the single-bonded region on the molecule (the side-chain), to result in bromomethylbenzene, $C_6H_5CH_2Br$. (It is true that free radicals react with π-bonds in alkenes (see Section 10.12), but they do not affect aromatic π-bonded systems, due to the lower reactivity of the latter.)

19

   Initiation

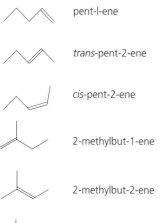

   Cl—Cl $\xrightarrow{h\nu}$ 2Cl•

   Propagation

   Cl• + H₂ ⟶ HCl + H•
   H• + Cl₂ ⟶ HCl + Cl•

   Termination

   H• + H• ⟶ H₂
   Cl• + Cl• ⟶ Cl₂
   Cl• + H• ⟶ HCl

20a   Substitution
b   Substitution
c   Addition

21   In a straight-chain alkane, all the carbon atoms carry *two* hydrogen atoms but the two end carbon atoms carry one extra hydrogen atom each. A general formula based on that reasoning is C_nH_{2n+2}. If there is a branch in the carbon framework, the branch takes the place of a single hydrogen atom, but to compensate for this it also supplies an extra hydrogen atom at its chain-end. So in fact the above formula is correct for all alkanes, irrespective of degree of branching. Cycloalkanes differ from ordinary chains in that they have cancelled out two ends by creating a ring, so the general formula in this case is C_nH_{2n}.

22a   Alkane bonds are likely to break homolytically due to lack of polarity.
b   For the same reason, attacking species will be free radicals.
c   The quite high bond strengths will ensure a comparative lack of reactions and reactivity.

23   Relevant data:
Planck constant = 6.63×10^{-34} J Hz⁻¹

Avogadro constant = 6.02×10^{23} mol⁻¹
Cl—Cl bond strength = 242×10^3 J mol⁻¹
To cleave a single Cl₂ molecule would require

$$\frac{\text{bond strength}}{\text{Avogadro constant}} = \frac{242 \times 10^3}{6.02 \times 10^{23}}$$

$$= 4.02 \times 10^{-19} \text{ J per molecule}$$

The Planck equation allows us to calculate the photon frequency which would deliver this energy:

$$\nu = \frac{E}{h} = \frac{4.02 \times 10^{-19}}{6.63 \times 10^{-34}}$$

$$= 6.1 \times 10^{14} \text{ Hz (equivalent to a wavelength of 492 nm)}$$

The visible region of the spectrum runs from about 400–700 nm, so the photon in question would be from the shorter wavelength (blue) end of the range. Clearly, the idea of initiation by sunlight is feasible.

24   This encounter would create product species identical to the reactants.

25   *First route:*
Break: C—H   +435 kJ mol⁻¹
Make: H—Cl   −432 kJ mol⁻¹
Overall enthalpy change   +3 kJ mol⁻¹

Second route:
Break: C—H   +435 kJ mol⁻¹
Make: C—Cl   −346 kJ mol⁻¹
Overall enthalpy change   +89 kJ mol⁻¹
So route 1 looks the more likely.

26   Each photon absorbed creates two Cl• radicals, and each Cl• radical can set up a chain reaction featuring many thousands of circuits of the cycle on p. 204.

27   They might turn their attention to the product molecules of chloromethane. If a Cl• radical can abstract one of the four hydrogen atoms from methane, then presumably it can also abstract one of the three hydrogen atoms from chloromethane. (In fact it would be an easier task, since the C——H bond energy is a maximum in methane.) The new product would be dichloromethane, by the cycle below, and even tri- and tetrachloromethanes are possible.

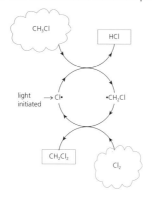

28a   If you wanted to isolate one pure compound from this kind of reaction you would have the job of separating it from the mixture of chloroalkanes with single, double and multiple substitutions.
b   The solvent properties of a mixture of chloroalkanes might be just as effective as those of a single compound.

29   See the scheme below.

Initiation

   Pb(C₂H₅)₄ ⟶ Pb + 4C₂H₅•

Propagation

main   C₂H₅• + Cl₂ ⟶ C₂H₅Cl + Cl•
propagation   ⌈ Cl• + CH₄ ⟶ HCl + CH₃• ⌉
cycle   ⌊ CH₃• + Cl₂ ⟶ CH₃Cl + Cl• ⌋

Termination
   Any two radicals meet

It does not work below 140 °C because there is no means of initiation in the dark at this temperature.

30   Substitution.

31a   Variables in the cracking process are: *dwell time* in the furnace, and *temperature* of the furnace.
b   The hot gases from the furnace are used as a source of heat to create high-pressure steam for the compressors.
c   The separations are all on the basis of *differences in boiling point*.

32   The generalised formula for an alkene is the same as that of a cycloalkane, namely C_nH_{2n}. Both groups have lost two hydrogen atoms compared with the corresponding alkane.

33   Ethene and propene need no number in their names.

34

   pent-l-ene

   trans-pent-2-ene

   cis-pent-2-ene

   2-methylbut-1-ene

   2-methylbut-2-ene

   3-methylbut-1-ene

35   Bond energies are as follows (in kJ mol^{-1}):

C—C   348

C=C   612

which suggests a bond energy of 264 kJ mol^{-1} for the π-bond alone. But you could argue that the C—C σ-bond has been pulled shorter than normal by the π-bond, so that it is not contributing its listed 348 kJ mol^{-1}. Presumably it would be destabilised, so it follows that the π-bond must be slightly stronger than it looks from the simple subtraction sum above.

The relevant bond lengths are (in nm):

C—C   0.154

C=C   0.134

36   Addition.

37   Markovnikov addition products:

a
$$CH_3-\underset{\underset{Br}{|}}{\overset{\overset{CH_3}{|}}{C}}-CH_3$$

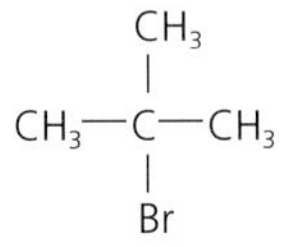

b
$$CH_3-CH_2-\underset{\underset{Br}{|}}{CH}-CH_3$$

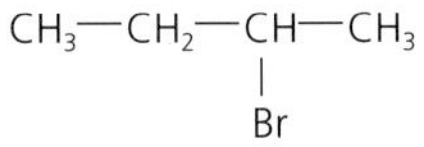

c
$$CH_3-\underset{\underset{CH_3}{|}}{\overset{\overset{Br}{|}}{C}}-CH_2-CH_3$$

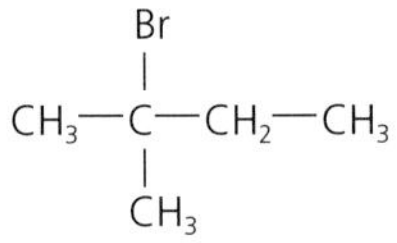

38   As always, termination will occur when two radicals meet. As unreacted alkene molecules begin to run out, there is an increased chance that one chain will meet another and create an extended non-radical chain.

39

a   ---- CH$_2$—CH$_2$—CH$_2$—CH$_2$—CH$_2$ ----

poly(ethene), 'polythene'

b   ---- CH$_2$—$\underset{\underset{CH_3}{|}}{CH}$—CH$_2$—$\underset{\underset{CH_3}{|}}{CH}$—CH$_2$—$\underset{\underset{CH_3}{|}}{CH}$ ----

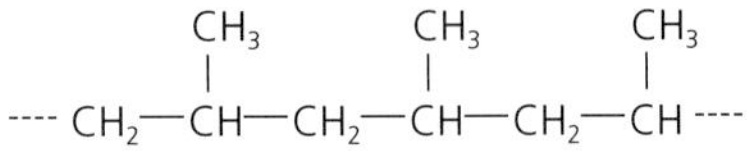

poly(propene), 'polypropylene'

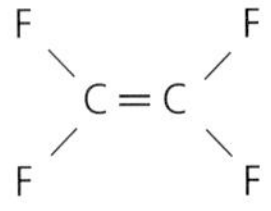

c   ---- CH$_2$—$\underset{\underset{Cl}{|}}{CH}$—CH$_2$—$\underset{\underset{Cl}{|}}{CH}$—CH$_2$—$\underset{\underset{Cl}{|}}{CH}$ ----

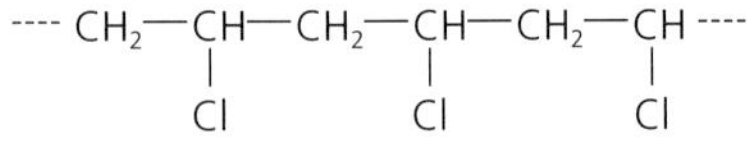

poly(chloroethene), 'PVC'

40

a
$$\underset{H}{\overset{H}{}}C=C\underset{C\equiv N}{\overset{H}{}}$$

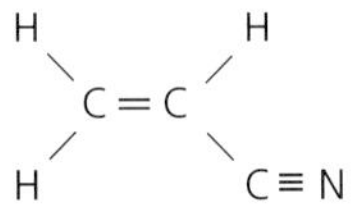

b
$$\underset{F}{\overset{F}{}}C=C\underset{F}{\overset{F}{}}$$

c
$$\underset{H}{\overset{H}{}}C=C\underset{}{\overset{H}{}}$$

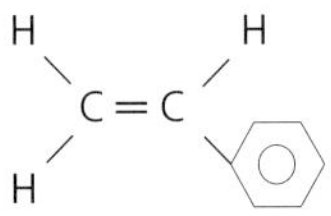

41   Crystallinity would be decreased by random cross-linking, since the somewhat organised van der Waals' structures like those in Figure 10.22 would be less likely. (As an analogy, it would be harder to make a neat sheaf of a bit of goal netting (cross-linked) than of the single fibres which make the net up.)

42   Ziegler–Natta polythene is called *high-density* polypropene, which gives away one part of the answer. Second, a high degree of intermolecular bonding will resist intermolecular movement, which shows up in the macroworld as a resistance to plastic deformation – that is, the material will be *harder* than less crystalline samples. Third, it will have a *higher softening temperature* than low-density polythene. (This third factor might seem odd in the light of the idea that low-density polythene is partially cross-linked, but this partial cross-linking produces bits of 'netting' rather than a whole network. The overriding advantage of the Ziegler–Natta polymer is its stereospecificity, which means the chains sit together in maximum orderliness, with maximum van der Waals' bonding.) Apart from these three clear-cut examples, every other physical and chemical property is affected to some extent by the structural differences.

43   The crystal would behave elastically – the only choice for bonds in ionic crystals is stretch or break. Sliding of crystal planes is not an option. So if they have not undergone a catastrophic fracture the bonds will relax back to their ideal lengths when the force is removed.

44a   Mode (b) deformations would be plastic. The molecules once displaced would feel no restoring force.

b   Mode (c) requires freedom of C—C bond rotation.

c   Mode (b) requires work against intermolecular forces.

d   First the molecules will unravel, then their bond angles will open, and then it must be a close thing whether the bonds themselves will stretch or whether the sliding will set in first. Only the sliding will cause plastic deformation, and the final breakage will also be by failure of intermolecular bonds in mode (b).

45   For elastomeric properties you would want *cross-linking* to be present, so there would be resistance to molecular sliding (you do not want elastics to behave like plastics). But you would also want very little resistance to movements like unravelling and straightening, because these are the movements which make stretching easy. So you want the molecules to find it *hard to take up linear conformations* (that is, have poor van der Waals' forces).

46

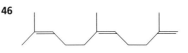

cis- isomer in natural rubber

trans- isomer – 'gutta percha'

From the structures above you can see that both isomers can be laid in conformations which are straight overall, with internal zig-zags. But the *trans* isomer can make zig-zags with a shorter 'wavelength'.

47   Stress crystallisation is caused by molecules lining up under stress and thus making significant new intermolecular bonds. These bonds introduce irreversibility into the stretching process.

48a   Butadiene rubber, unlike the plastic polyalkenes, still has a series of double bonds left in the final polymer, one per original monomer unit.

b   The double bonds, all in the *cis* conformation, ensure a low level of linearity in the chains and thus low van der Waals' forces. This, you will remember from question 45, is one of the two factors making rubbers rubbery. On top of that, these double bonds remaining in the chains make cross-linking possible, either by vulcanisation or by a free-radical mechanism as shown below. That cross-linking provides the restraint which prevents plastic slip – the second requirement for rubberiness.

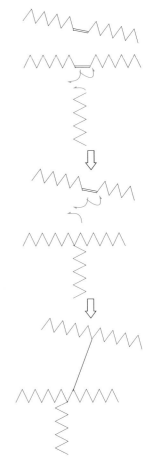

49a As heat in the rubber molecules.

b You could imagine the molecules returning from stretched to unstretched condition, with bond angles relaxing back to 109.5°, and bond rotations bringing about the re-ravelling of the chains. It is feasible that the benzene rings would collide quite often during these changes, and therefore that some of the potential energy in the stretched bonds would be diverted into heating by this process of intermolecular collision.

50 Lorry tyres are of course bigger than car tyres. So with their lower area:volume ratio and greater loading, they are more prone to overheating. The switch from a more shock-absorbing rubber to a more bouncy one is designed to combat the heat problem.

51 Ozonolysis would chop the rubber polymer up into smaller molecular fragments:

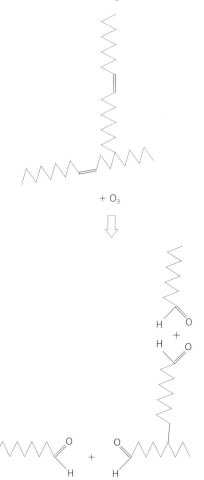

52 The random orientations of the methyl groups minimise the capacity for neat van der Waals 'stacking', thus achieving one of the two main criteria for rubberiness – the other being the cross-linking itself.

53 The molecular masses of the fragments point to an unequal cleavage of the molecule, so we must assume the fragments contain one and three carbon atoms. That limits the possibilities to those shown below.

but-1-ene   methylpropene   cyclobutane

The isomer which would not undergo ozonolysis would be cyclobutane, although given the strain in bond angles forced to stay at 90° when they would rather be at 109.5°, some sort of ozone attack, to spring the ring open, could not be entirely ruled out.

54 We have to reject structure (a), because it has *two* different hydrogen atom environments (on the end, and not on the end).

55 Two disubstituted isomers could be made – one in which both chlorine atoms were on the *same* terminal carbon atom, and one in which they were on different ends.

56 Two adjacent substituents could straddle either a C—C bond or a C=C bond:

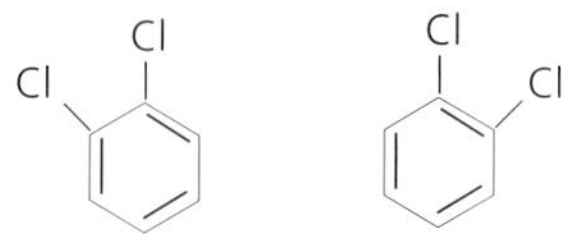

57a

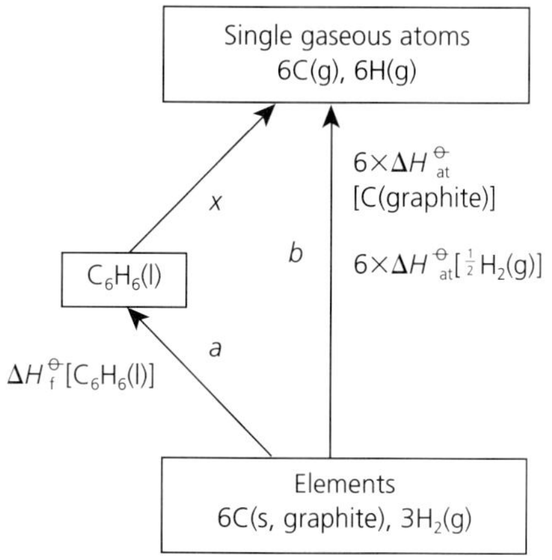

Single gaseous atoms
6C(g), 6H(g)

$x = -a + b$

$= -[-49.0] + [6 \times 715 + 6 \times 218]$

$= 5647$ kJ mol^{-1}

b $6 \times$ C—H $= 6 \times 412 = 2472$

$3 \times$ C—C $= 3 \times 348 = 1044$

$3 \times$ C=C $= 3 \times 612 = 1836$

$\Delta H_{at} = +5352$ kJ mol^{-1}

c So real benzene appears to be more stable than the Kekulé structure by 295 kJ mol^{-1}. (Actually the difference is about 30 kJ mol^{-1} less than this since we are making an unfair comparison between real C_6H_6(liquid) and Kekulé C_6H_6(gas). The use of bond enthalpies is the source of the problem, since they cannot predict atomisation of *liquids*. 30 kJ mol^{-1} is the enthalpy for the change C_6H_6(l) $\rightarrow C_6H_6$(g).)

58 Electrophiles.

59 The $AlCl_3$ acts as an electron-pair receiver, provoking a lone pair on the chlorine atom to proffer a dative covalent bond as shown.

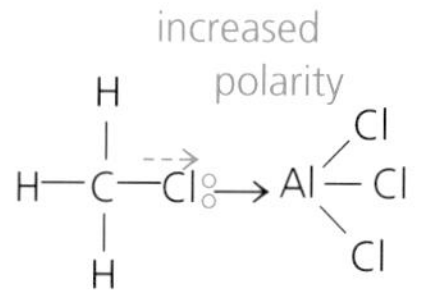

The chlorine atom experiences an electron shortage as a result of this 'gift', and 'takes it out on' the C—Cl bond, from which it attracts more than its usual share of electron density. So the overall effect is to generate a more positive CH_3 group than in normal chloromethane, which then behaves like a methyl carbocation and mounts an electrophilic attack on the ring.

60 The catalyst would still be aluminium chloride, but the desired carbocation would be CH_3C^+=O. Therefore the other reagent must be ethanoyl chloride, CH_3COCl.

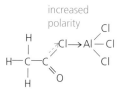

61a They found they had made iodobenzene.

b The I—Cl bond is already polarised, with the iodine end the potential electrophile, so this supported their proposed mechanism of electrophilic addition.

CHAPTER 11   page 227

1a *Electrophile*: a chemical species which is attracted to regions of high electron density. It is electron-deficient itself, and will bond with a pair of electrons from the substrate. Molecules attractive to electrophiles include alkenes and arenes.

Free radical: a chemical species with an unpaired electron. For instance, it may be a bare atom. It will seek to create a new two-electron bond with a substrate, to which the substrate will have contributed the second electron. This can happen in two ways: either an atom will come away from the substrate and join the now-ex-free-radical, or the now-ex-free-radical will join the substrate. Either way, a new free-radical site will have been created, somewhere on the substrate. Free radical attackers have no special preference for polar bonds. Molecules attractive to free radicals include alkanes and alkenes.

b Nucleophiles played no part in hydrocarbon chemistry.

c Electrons on offer from attacking species are as follows: electrophiles 0, free radicals 1, nucleophiles 2.

2a 2-chloro-2-methylpropane

b 2-bromo-2-chloro-1,1,1-trifluoroethane

c 3-ethyl-3-iodopentane

(All show the rules of alphabetic precedence,

and 'keep the numbers as low as possible'. See p. 195.)

3a

$$\overset{\delta+\ \ \delta-}{C\!-\!X}$$

b Cleavage is likely to leave the halogen atom in possession of both the electrons from the broken bond.

c The attacking species would have to offer *two* electrons.

d We call it a nucleophile.

e

N⋮ ⟶ C ⋮ X → N⋮C + ⋮X⁻

f The leaving group has become a halide ion.

4a Water is the best nucleophile in $AgNO_3(aq)$.

b The visible sign that a halide ion has been ejected is the precipitation of silver halide.

c $C_4H_9I(aq) + H_2O(l) \rightarrow C_4H_9OH(aq) + HI(aq)$

$HI(aq) + Ag^+(aq) \rightarrow AgI(s) + H^+(aq)$

5 The fact that 1-iodobutane reacts fastest and 1-chlorobutane slowest suggests that the dominant influence on reaction rate is the weakness of the C—X bond.

6 The OH^- ion is more nucleophilic than H_2O, because it is much more 'pushy' with its lone pairs being an anion, which accounts for the extra rate of reaction. However, in the presence of OH^- ions, the silver ions would precipitate as AgOH or Ag_2O.

7 Both transition states require the partial breakage of the C—I bond, but the one-step mechanism can offset that positive enthalpy change by the partial bond-making of the O—C bond. So you would expect the hump on the one-step reaction profile to be lower than the hump on the two-step one.

8 Both explanations rely on the idea that a carbocation is more stable than it would otherwise be, if the positive charge can be situated at a 'tertiary site' – that is, on a carbon atom which is joined to three other carbon atoms.

9

H H H* H
| | | |
H—C—C—C—Br
| | | |
H H H* H

1-bromobutane

      H*
    H | H
     H C H
    | | |
H—C—C—C—H
    | | |
    H Br H

2-bromo-2-methylpropane

There are *nine* hydrogen atoms (shown *) which can fulfil the role of 'adjacent hydrogen' in 2-bromo-2-methylpropane, whereas there are only two in 1-bromobutane.

10a There are actually four possible organic products to the reaction: they are butan-2-ol, but-1-ene, and the *cis* and *trans* isomers of but-2-ene.

$CH_3\!-\!CH_2\!-\!\underset{\underset{OH}{|}}{CH}\!-\!CH_3$

butan-2-ol

$CH_3CH_2CH\!=\!CH_2$

but-1-ene

$$\underset{H}{\overset{CH_3}{}}C\!=\!C\underset{H}{\overset{CH_3}{}}$$

cis-but-2-ene

$$\underset{H}{\overset{CH_3}{}}C\!=\!C\underset{CH_3}{\overset{H}{}}$$

trans-but-2-ene

b The boiling points of the reactant and possible products are as follows (in K), with certain melting points in brackets:

but-1-ene	267	(134)
trans-but-2-ene	274	(167)
cis-but-2-ene	277	
butan-2-ol	375	
2-bromobutane	364	

So the alkenes will escape as gases (and if necessary could be separated by cooling to liquefy them, and then fractionally distilling them). The alcohol would be very difficult to separate from the water in the solvent, at least by distillation alone, because their boiling points are so close. So a better plan would be to shake the product liquid with ethoxyethane (ether), which would form a second layer and be a preferential solvent for the organic compounds. The ether layer could then be dried over anhydrous $MgSO_4$, and the alcohol fraction collected from the distillate at 375 K.

c The inorganic product is KBr, which would stay dissolved in the aqueous layer (and anyway is non-volatile).

11a The C—Cl bond is much weaker than the C—F bond (338 kJ mol^{-1}, compared with 484), so it breaks much more easily.

b Butane has a boiling temperature of 272 K, which means it is one of those gases which, at room temperature, is only a few degrees above its boiling point, and which is therefore readily liquefied by pressure. This is a prime qualification for an aerosol propellant. A second point in its favour is its chemical stability, and it is also relatively non-toxic, and cheap and available. Against it are its flammability and its potential for 'solvent abuse'.

c It is obviously incapable of creating Cl• free radicals, since it has no Cl atoms in it. As for the chance of other radicals being broken from the molecule by ultra-violet photons, we already know the C—F bond is safe, and the C—H bond is another strong one (at 412 kJ mol^{-1}).

12 The solute mixture is the mixture of dyes that make up the colour of the ink/Smartie. The solvent is the water you dripped on to the paper to spread the dyes and the paper itself is the stationary phase.

13 OH^- is a more reactive nucleophile than water, so reaction 1 has a higher rate of reaction than reaction 2.

14 The OR^- ion is created when sodium metal is reacted with the alcohol ROH. The equation is:

$ROH(l) + Na(s) \rightarrow \frac{1}{2}H_2(g) + NaOR(s)$

NaOR contains the OR^- ion.

15a The Friedel–Crafts reaction is an electrophilic substitution from the point of view of benzene, towards which the R end of the R—X molecule acts as electrophile.

b It is no less true to see it is as a nucleophilic substitution *by* benzene (as nucleophile), from the point of view of the R—X molecule.

c The X is helped to leave by an offer of a dative covalent bond to the electron-deficient $AlCl_3$ catalyst.

16 The nitro group has the effect of pulling electron density *out* of the π-bonding system of the ring, thus making it less hostile to approaching nucleophiles (and at the same time less attractive for conventional electrophilic attackers). The figure below shows one of the 'hybrid' structures which can be drawn to illustrate the withdrawal of electrons from the ring.

17 Taking C—Cl as the example, the average bond length in chloroalkanes is 0.177 nm, while that in chlorobenzene is significantly shorter (and presumably stronger), at 0.169 nm.

18a Ethene would be obtained from the cracking of crude oil, or of other oil-derived fractions and feedstocks.

b Alkenes are hostile to nucleophilic attack, due to the electron density of the π-bond (and anyway HCl would bring about an addition reaction).

19a i Addition.

ii Elimination (although not by a mechanism which we have studied).

b Electrophilic.

20a The by-product is HCl(g).

b The combination of the reactions gives:

$2C_2H_4(g) + Cl_2(g) + \frac{1}{2}O_2(g) \rightarrow 2C_2H_3Cl(g) + H_2O(g)$

c The combined reaction wastes no hydrogen chloride, and vents only water into the atmosphere.

d The second reaction alone uses hydrogen chloride as a raw material, which is more expensive than using chlorine. The reason for this is that hydrogen chloride is made *from* chlorine (which in turn comes from the electrolysis of seawater).

21 The three stages are as follows:

Initiation

$$R—O—O—R \rightarrow 2R—O\bullet$$

(peroxide)

Propagation

a RO• +

b

Termination

Any two radicals meet.

22a Structures like that shown in Figure 11.19 really have two sets of intermolecular forces – there are those in the amorphous regions, which give way at T_g, and those in the crystalline regions; after these give way, complete liquefaction is possible.

b Not every molecule in a polymer is exactly the same length, nor does it sit next to its neighbours in an entirely regular way, so the solid structure gives way gradually, over a range of temperatures.

23a Rubber's molecular 'memory' is helped by its high degree of cross-linking (as in vulcanisation, see Chapter 10). The cross-links ensure that the molecular chains in rubber, despite stretching, do not change their overall orientation with respect to each other.

b A combination of van der Waals' forces (present in all solid and liquid substances), and a degree of dipole–dipole attraction featuring C—Cl bonds on adjacent chains (see below).

c If the plasticiser molecules were too similar to the PVC ones, they might set up intermolecular forces which were as strong as those they interrupted.

d Too 'foreign' a molecule might actually refuse to dissolve in the PVC structure.

e The hydrocarbon bits will ensure it is not *too* similar to its host, while the dipolar bits should maintain solubility. The globular shape will ensure that it achieves a good degree of separation of the chain-like PVC molecules, acting rather like ball-bearings in assisting the chains to move.

f If dioctyl phthalate were used to plasticise poly(ethene), we might find that, in contrast to the PVC case, there was greater intermolecular attraction between the hyrocarbon 'arms' of the plasticiser and the base polymer. The 'arms' might unravel and indulge in full van der Waals' bonding with the polymer; and the resulting material might be stiffer than an ideally plasticised one.

24a Copper wire would go in the stiff/plastic box.

b Glass over very long timespans also acts as stiff/plastic.

c **i** uPVC = stiff/elastic.

ii pPVC when very cold = stiff/elastic.

iii A plastics material near its melting temperature = flexible/plastic.

iv A metal spring = flexible/elastic.

25a False – below their T_g they are stiff/elastic.

b True – all plastics materials would be stiff/elastic if cold enough, and if they were thermoplastics they would be flexible/plastic near their melting point.

c True – all plastics are long-chain molecules (and even possibly cross-linked), but some polymers are naturally occurring (for example, proteins), and so cannot be called plastics.

d False – glasses, especially near their softening point, show plastic behaviour, despite not being plastics materials.

e True – all synthetic elastomers are polymers, but as we have seen many stiff materials (for example, metals, ceramics) also behave in a way which can be called elastic.

26 This answer is to some extent a matter of opinion, but you might certainly mention the 'drop in the ocean' argument which says that banning PVC packaging would make no significant difference to oil consumption. On the other side of the argument, bright shiny PVC packaging could hardly be described as essential, so governments could act to ban its use in much the same way as they have with CFCs.

27 *Primary*:   pentan-1-ol,
          3-methylbutan-1-ol
    Secondary: butan-2-ol, cyclohexanol
    Tertiary: 2-methylbutan-2-ol

28 With the proton in place, the whole leaving group is temporarily positively charged, which increases the polarity in the bond which is to break, and facilitates the leaving of what becomes a neutral water molecule:

29a A tertiary carbocation is stabilised by three electron 'pushes' from its three substituent alkyl groups.

b The benzene ring can stabilise the positive charge by delocalising it, using hybrid structures as below.

30 The mechanism, being two-step and leaving-group led, more closely resembles the S_N1 mechanism.

31 The tertiary alcohol will give rise to a reaction intermediate which is a tertiary carbocation, which we have identified several times before as a more stable species than its primary equivalent. So the high-energy bits of the reaction profile for 2-methylpropan-2-ol will not be as high, so the reaction rate will be greater. See Figure A11.1.

32a The electron 'push' of the ethyl group would partially 'satisfy' the inclination of the oxygen atom in ethanol to attract electrons to itself. So it would not apply such a strong polarising 'pull' on the hydrogen atom, which would therefore not be so inclined to cleave to form a hydrogen ion.

b On the argument above, tertiary alcohols should have the least acidic hydrogen atoms, and therefore react slowest with sodium. The order would be:

butan-1-ol > butan-2-ol > 2-methylpropan-2-ol.

33a It would be better, because it would benefit from the electron 'push' of the R (alkyl) group.

b RO^- is the stronger of the two bases on show (the other being OH^-), so it should win the tug-of-war for the hydrogen ion, which means the reaction will go *forwards*.

c We can answer this one by 'parallelism'. NaOH reacts with bromoethane to give C_2H_5OH, and so $NaOC_2H_5$ gives $C_2H_5OC_2H_5$, called ethoxyethane or diethyl ether.

34

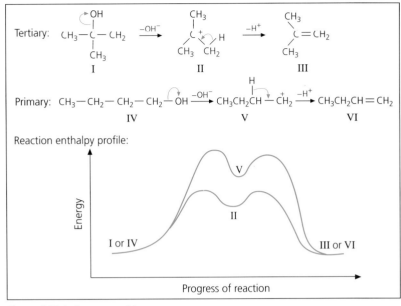

Figure A11.1 See answer 31

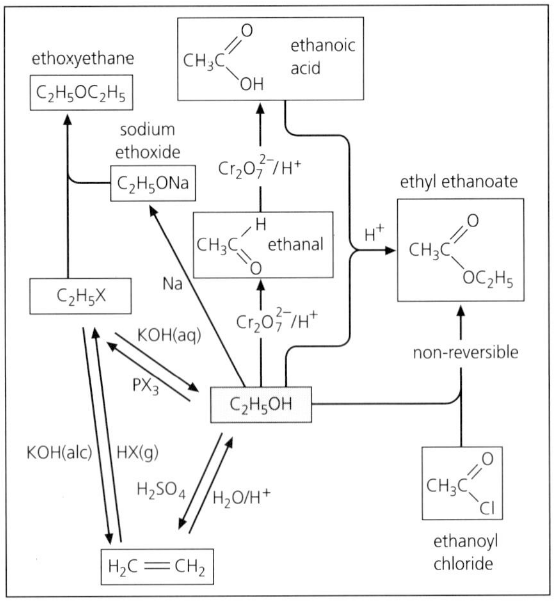

Figure A11.2

35 The oxygen lone pair impacts at the C end of the C=O bond, in which course of action it is encouraged by the well-developed dipole in this bond. The dipole has created a partially positive C atom, which attracts the lone pair.

36a The normal place for an acid to cleave would be in the O—H bond.
b In the mechanism shown, it is the C—O bond in the acid that cleaves.
c In the alcohol, the O—H bond cleaves.
d The labelled oxygen atom would end up in the ester:

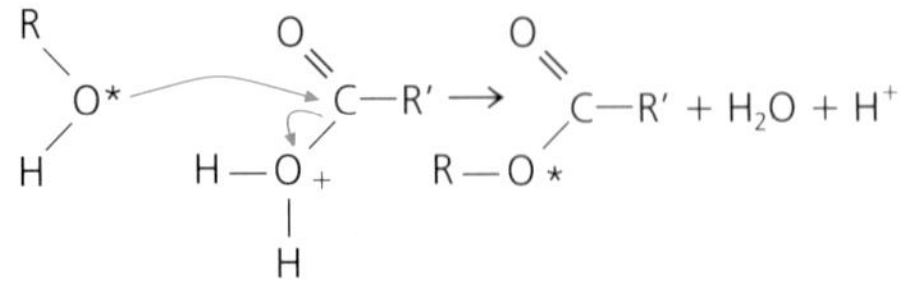

37a The leaving group is the Cl⁻ ion, and it has established its credentials as a good leaver by being the anion of hydrochloric acid. Hydrochloric acid is completely dissociated in aqueous solution, which suggests Cl⁻ has rather 'unsticky' (un-nucleophilic) lone pairs.
b

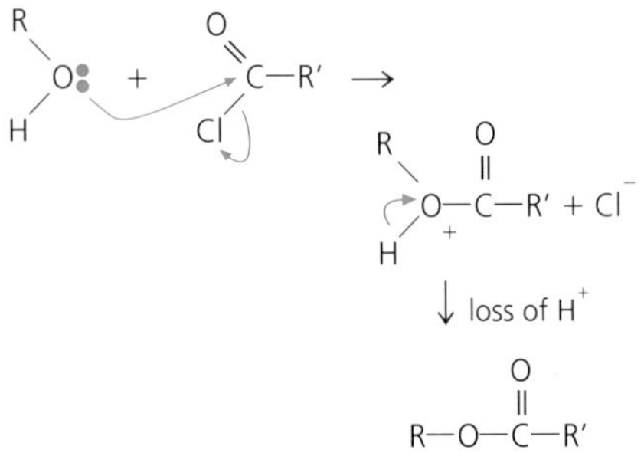

38 Once the oxidation has reached the ketone stage, it is impossible to put more oxygen atoms on to the relevant carbon atom without rupturing the backbone of the molecule.

39a +6
b $Cr_2O_7^{2-}(aq) + 14H^+(aq) + 6e^- \rightarrow 2Cr^{3+}(aq) + 7H_2O(l)$
c −1
d +3
e $CH_3CH_2OH(aq) + H_2O(aq) \rightarrow CH_3CO_2H(aq) + 4e^- + 4H^+(aq)$
f $3CH_3CH_2OH(aq) + 2Cr_2O_7^{2-}(aq) + 16H^+(aq) \rightarrow 3CH_3CO_2H(aq) + 4Cr^{3+}(aq) + 11H_2O(l)$

40 By the rules of redox balancing, if the carbon atom was only going up by 2 oxidation numbers instead of 4, then you would need half the number of chromium atoms going down 3.

41 In sunny countries, solar energy has been collected by banks of photovoltaic cells, by straight head absorption into solar panels and by the use of larger parabolic reflectors to produce high-pressure steam at the focal point.

42a We can use the formula (p. 172):
$\Delta G = \Delta G_f(\text{products}) - \Delta G_f(\text{reactants})$
$= -750 - (6 \times (-395) + 6 \times (-229))$
$= +2994 \text{ kJ mol}^{-1}$
b The reaction can still happen, because the energy to push the system 'uphill' is coming from the Sun. If we included the changes on the Sun which gave rise to the photons which the plant absorbed, then we would find that the entropy of the universe *had* gone up.

43 See Figure A11.2.

44a Using the 'look at the product first' tactic, we can list the things that could be *made into* 2-bromobutane:

• Butan-2-ol, reaction with HBr or PBr₃.
• But-1-ene, reaction with HBr.
• But-2-ene, reaction with HBr.

Of these, only the second can be made in a single step from the starting material, butan-1-ol. So the synthesis proceeds as follows:

Step 1: react butan-1-ol with concentrated sulphuric acid, to make but-1-ene:

$CH_3CH_2CH_2CH_2OH \rightarrow CH_3CH_2CH=CH_2 + H_2O$

Step 2: bubble through HBr(aq) to add HBr across the double bond. Nature (as recognised in Markovnikov's rule) will ensure the product carries the bromine atom in the 2-position. (Some authors suggest using a non-aqueous solvent, such as ethanoic acid, to avoid the side reaction in which water adds across the double bond.)

$CH_3CH_2CH=CH_2 + HBr \rightarrow CH_3CH_2CHBrCH_3$

b Precursors of 1,2-dibromocyclohexane:

• Cyclohexene, reaction with Br₂.
• Cyclohexane-1,2-diol, reaction with HBr.

Of these, only the first can be made easily from cyclohexanol, so the synthesis proceeds as follows:

Step 1: dehydrate the original alcohol with concentrated sulphuric acid, as in problem (a).
Step 2: add Br₂(l) to the alkene, with stirring.
c Precursors of ethyl ethanoate:

• Ethanol/ethanoic acid, with catalyst.
• Ethanol/ethanoyl chloride.

Ethanoic acid can be made in one step from ethanol, whereas ethanoyl chloride would take two. It is a tough choice, because the trouble of the extra step is offset by the better yield of the acyl halide route. Still, if we stay with the one that is easier to think about, the synthesis proceeds as follows:

Step 1: divide ethanol into two portions, and oxidise one by reaction with acidified potassium dichromate(VI).

Step 2: mix the newly formed ethanoic acid with the held-back batch of ethanol, in the presence of sulphuric acid or hydrochloric acid as catalyst. After a few days to allow for the attainment of equilibrium, distil the ester from the equilibrium mixture.

45a The reference path would include the pure solvent, so that any absorptions by the solvent would occur equally in both paths, and therefore not be recognised by the sensing system. So no solvent peaks would appear in the spectrum.
b Glass cells are not used because glass has its own broad absorption in the infrared region. The problem with salt windows is that they have to be kept scrupulously dry (and they are brittle).

46a In a liquid alcohol there is a constantly shifting pattern of hydrogen bonding between OH groups on adjacent molecules. O—H bonds whose oxygen or hydrogen members (or both) are indulging in hydrogen bonding vibrate with different frequencies from those which are temporarily uninvolved in hydrogen bonding. So photons are absorbed over a range of frequencies, giving a broad peak.
b Breath contains water vapour, which also gives a peak from O—H stretches. On the other hand ethanol is probably the only source of C—H bonds.

47a The size of the peak in an infrared spectrum depends upon how many absorbing molecules are in the path of the beam. Since gases are more diffuse than liquids, they are housed in longer cells, to bring molecule numbers up to comparable levels. The intoximeter is designed only to measure a single peak, so needs only to admit photons at that frequency to its detector.
b Propanone is a second source of C—H bonds, so the tester would not be sure how much of the absorption had come from genuine ethanol.

48a Blood and urine samples are not as transient as breath. They enable a snapshot in time to be preserved long after the client has sobered up.
b The accuracy of the intoximeter breath test may not be perfect, but it is good enough to guarantee that a reading of anything over 50 micrograms per 100 cm^3 of breath couldn't possibly be a misread 40.

49a The willingness to release H$^+$ ions will be increased.
b The benzene ring in phenol would become *more* attractive to electrophiles.

50

$$C_2H_5OH + Na \longrightarrow C_2H_5O^- Na^+ + \tfrac{1}{2}H_2$$

$$C_2H_5OH + NaOH \longrightarrow \text{no change}$$

51 *Order of acidity*: phenol > water > ethanol
Order of basicity:
ethoxide > hydroxide > phenoxide

52 Esterifications.

53

54a Nucleophiles.
b Nucleophilic substitutions.

55 The 2, 4 and 6 positions on the ring are the most attractive.

CHAPTER 12   page 256

1 The idea must have grown from the fact that certain elements such as sulphur and phosphorus had to be reacted with oxygen to 'bring out' their acidity.

2 Organic chemistry is full of non-acidic hydrogen compounds. Methane and glucose both fit the bill, despite their mutual differences. (See also Chapter ?.)

3a The appearance of dihydrogen gas at the *cathode* points to the presence of H$^+$ ions in acids.
b Sugars all fit this description, as do the shorter-chain alcohols. Their interaction with water is by means of hydrogen bonds, rather than ion–dipole attraction.

4 The sequence, in order of increasing acidity, is:

ethane (no H atoms polarised) < ethoxyethane (O atom in molecule, but no H atoms directly attached) < ethanol (H atom directly linked to O atom) < sulphuric(IV) acid (H atoms directly linked to O atoms, with additional polarising influence from an S=O bond) < sulphuric(VI) acid (H atoms directly linked to O atoms, with additional polarising influence from two S=O bonds).

Key   * – just detectably acidic H atom
       ** – weakly acidic H atom
      *** – strongly acidic H atom

5 The bond is of the dative covalent type.

6a O in H$_2$O is the more negatively polarised, as can be seen from electronegativity data.
b So the O atom in water will be the more 'pushy' with its lone pairs of electrons.
c Water is the more 'willing' receiver of HCl protons, so HCl will behave more acidically in water.
d HCl in H$_2$S as solvent could well be less fully ionised, and less conducting as a result. (A fair comparison would be hard, however, since water is a solid at –61 °C.)

7 H$_3$S$^+$ is the acid and Cl$^-$ the base.

8 $HCl(aq) + H_2O(l) \longrightarrow H_3O^+(aq) + Cl^-(aq)$
   acid 1   base 2   acid 2   base 1

9a 0.001 mol dm^{-3} HCl(aq) would be a dilute strong acid.
b 2.0 mol dm^{-3} CH$_3$CO$_2$H(aq) would be a concentrated weak acid.

10a With no surviving HCl molecules, the bottom line of the K_a expression would be 0, so K_a would be infinite.
b With no ionisation happening at all, the K_a would be zero.
c The K_a would be smaller (and therefore finite) in the less ionising solvent H$_2$S(l).

11 The change in ease of removal of the second proton can be put down to 'knock-on' changes in polarity right across the molecule, brought about by the removal of the first proton. The newly created O$^-$ site becomes less electron-withdrawing, so the central S atom become less 'δ+'; so the O atom on the *other* OH group can get more of its electron density from the S atom, and so takes less from the H atom. The less polarised H atom is therefore less acidic (see below).

1 This O atom is now more electron-rich

2 so this S atom is less δ+

3 so this O atom can obtain more electron density from the S atom

4 so the O atom does not need to draw as much electron density from the H. The less polar O—H bond is therefore less acidic

12 As the methyl 'head' of the molecule becomes more heavily substituted with electron-withdrawing Cl atoms, the electron distribution in the molecule changes to increase the positive polarity of the hydrogen, and therefore its acidity.

13a The two bases are water and $Cl^-(aq)$ and, from the fact that the equation goes to completion, we can conclude that water is the stronger base.

b

$$CH_3CO_2H(aq) + H_2O(l) \rightleftharpoons H_3O^+(aq) + CH_3CO_2^-(aq)$$
$$\text{acid 1} \qquad \text{base 2} \qquad \text{acid 2} \qquad \text{base 1}$$

From the fact that the position of equilibrium lies well to the left, we can conclude that $CH_3CO_2^-(aq)$ is the stronger base.

14 The idea is to look for the conjugate bases of the weakest acids, so your sequence, whatever it is, should feature right-hand side species, 'bottom first'. For example, this sequence answers the question:

PO_4^{3-} (strongest base) $> NH_3 > HCO_3^- > F^- > SO_4^{2-}$

15a $HCl(aq) + OH^-(aq) \rightarrow H_2O(l) + Cl^-(aq)$

This is almost the familiar
$HCl + NaOH \rightarrow NaCl + H_2O$, except for the omission of the 'spectator' Na^+ ions.

b The three bases, in order of strength, are:
$OH^-(aq) > H_2O(l) > Cl^-(aq)$

16a In these four combinations, the ions are all free agents in solution at the time of mixing, so the only real event taking place is the same in every case:

$$H_3O^+(aq) + OH^-(aq) \rightarrow 2H_2O(l) \qquad (1)$$

with the chloride, nitrate, sodium, or potassium ions going through the whole process as 'spectators'. So the enthalpy change in each case is that of reaction 1.

b When the acid is weak, the acidic hydrogen atoms will mostly be still attached to the parent acid. So the hydroxide ion will have to 'extract' them, in a reaction which involves *more bond breaking* than reaction 1 (and which is thus less exothermic):

$$CH_3CO_2H(aq) + OH^-(aq) \rightarrow CH_3CO_2^-(aq) + H_2O(l) \qquad (2)$$

When the base is weak, the reaction is again less exothermic than reaction 1, but this time we can put it down to reduced strength of bonds *made*. The reason is that NH_3 is a less good 'proton-grabber' than OH^-. In this case, the reaction to be compared to reaction 1 is:

$$H_3O^+(aq) + NH_3(aq) \rightarrow NH_4^+(aq) + H_2O(l) \qquad (3)$$

c HF is a weak acid, and the H—F bond is strong, so the result is surprising. The relevant equation is:

$$HF(aq) + OH^-(aq) \rightarrow F^-(aq) + H_2O(l) \qquad (4)$$

If we search this equation for a source of unexpected 'exothermicity', the prime candidate is the *hydration of F^-* (i.e. the ion–dipole bonds

between F^- ions and water molecules). This is credible, because F^- is the *smallest* anion that can exist in aqueous solution, and small ions are strongly hydrated. Neither can we discount the possibility of some sort of hydrogen-bonded structure, associated with fluorine's unique ability in this field. After all, [F—H–F]$^-$ exists, so why not [F—H–O—H]$^-$?

17a The oxide ion will clearly, by virtue of its electron-repulsion problems, be the more ready to accept a proton, thus dropping its four lone pairs to three and cancelling one negative charge.

b $O^{2-}(s) + H_2O(l) \rightleftharpoons 2OH^-(aq)$

c This 'equilibrium' will actually lie completely on the right. This is an example of the principle that any stronger base than OH^-, in water, will destroy itself and create OH^-.

d $Na_2O(s) + H_2O(l) \rightarrow 2NaOH(aq)$

18 As has been noted on several previous occasions, being a nucleophile and being a base require exactly the same ability, namely a tendency to donate a lone pair of electrons.

19 F^-. It 'wins' the proton-accepting contest against water, while Cl^- 'lost' against the same 'opponent'. (We can tell it 'wins' by the small size of the K_a.)

20a Any acid above the HF line will do, so HNO_3 and HSO_4^- would be possible correct answers.

$NaF(aq) + HNO_3(aq) \rightarrow HF(aq) + NaNO_3(aq)$

b Any base *below* the HF line will do. A significant observation is that HF would make a hydrogencarbonate fizz:

$HF(aq) + HCO_3^-(aq) \rightarrow H_2O(l) + CO_2(g) + F^-(aq)$

21a Yes. All the carboxylic acids in Table 12.1 are more than 20 times more acidic (as judged by K_a values) than HCO_3^-, so it seems reasonable to assume that most, if not all, of the group should be comfortably above HCO_3^- in the table. Hence they should be able to make clockwise circles with HCO_3^- and pass the 'fizz test'.

b For this answer we need an acid that lies between the two carbonate species. So, for example, NH_4^+ ions can protonate CO_3^{2-} ions but not HCO_3^- ions.

c No. Phenols are not capable of the second donation.

d Yes. It is harder to donate the proton to hydrogencarbonate ions, so if an acid can do that, it would certainly be able to protonate carbonate ions.

22a Yes. Sulphuric acid lies six places above $H_2PO_4^-$ in Table 12.1, so it *can* donate the final proton.

b $3H_2SO_4(aq) + Ca_3(PO_4)_2(s) \rightarrow 2H_3PO_4(aq) + 3CaSO_4(s)$

c You would need to react phosphoric acid with ammonia:

$H_3PO_4(aq) + 3NH_3(aq) \rightarrow (NH_4)_3PO_4(aq)$

d The by-product is calcium sulphate, which is the main chemical of the plaster industry serving

the building trade. It is also an ingredient in cement.

23a Farmers need fertilisers only in the growing seasons.

b The fertiliser under compression might 'weld' itself into lumps and so not be spreadable by machinery.

c Well-pelleted fertiliser would not collapse into a powder and 'weld' into a single mass.

24a The main mineral source is fluorite, CaF_2.

b Sulphuric acid is comfortably above F^- in Table 12.1, so the reaction should proceed without trouble.

c $H_2SO_4(aq) + CaF_2(s) \rightarrow 2HF(aq) + CaSO_4(s)$

d HF has a boiling temperature of 20 °C (about room temperature), so it is certainly the most volatile of the components of the system and can be distilled off by quite gentle warming.

25 $H_2O(l) + H_2O(l) \rightleftharpoons H_3O^+(aq) + OH^-(aq)$
$\quad\ \text{acid 1} \qquad \text{base 2} \qquad \text{acid 2} \qquad \text{base 1}$

In other words, one water molecule acts as acid to the other's base. (See equation (12.8).)

26a They should be equal, since they are created together, 1:1.

b Let $[H_3O^+] = [OH^-] = x$
$\qquad\quad$ Then $x^2 = 10^{-14}$
$\qquad\qquad$ So $x = 10^{-7}$ mol dm^{-3}

Therefore the concentrations of H_3O^+ and of OH^- are 10^{-7} mol dm^{-3}.

c In 1 dm^3 of water there are 1000/18 moles, or 55.6 moles. So the ratio of 'unbroken to broken' water molecules at any one time is $56 : 10^{-7}$ or about 560 million to 1.

27a 1 mole of HCl parent would have become 1 mole of each ion, so $[H_3O^+] = 1.0$ mol dm^{-3}.

b It follows that $[OH^-] = 10^{-14}$ mol dm^{-3}.

c Following the pattern of (a), $[OH^-] = 1.0$ mol dm^{-3}.

d Following the previous pattern, $[H_3O^+] = 10^{-14}$ mol dm^{-3}.

28a $pH = -\log_{10}(10^{-7}) = 7$

b $pH = -\log_{10}(1) = 0$

c $pH = -\log_{10}(10^{-14}) = 14$

29a $[H_3O^+] = 10^{-0} = 1$ mol dm^{-3}

b $[H_3O^+] = 10^{-(-0.3)} = 2.0$ mol dm^{-3}

c $[H_3O^+] = 10^{-14.3} = 5 \times 10^{-15}$ mol dm^{-3}

d $[H_3O^+] = 10^{-3.5} = 3.2 \times 10^{-4}$ mol dm^{-3}

e $[H_3O^+] = 10^{-2} = 0.01$ mol dm^{-3}

30a Let the value of $[H^+] = [F^-] = x$. Use the bottom-line approximation.

$$K_a = \frac{x^2}{0.01 - x} \approx \frac{x^2}{0.01} = 5.6 \times 10^{-4}$$

So $\quad x^2 = 5.6 \times 10^{-6}$
$\qquad\ x = 2.37 \times 10^{-3}$
$\quad pH = -\log_{10} x$
$\qquad\quad = 2.6$

b Let the value of $[H^+] = [CH_3CO_2^-] = x$. Use the bottom-line approximation.

$$K_a = \frac{x^2}{0.01 - x} \approx \frac{x^2}{0.01} = 1.7 \times 10^{-5}$$

So $x^2 = 1.7 \times 10^{-7}$

$x = 4.12 \times 10^{-4}$

pH $= -\log_{10} x$

$= 3.4$

c   Let the value of $[H^+] = [HPO_4^{2-}] = x$. Use the bottom line approximation. (Assume that the final proton release from HPO_4^{2-} is negligible compared to this one.)

$$K_a = \frac{x^2}{0.01 - x} \approx \frac{x^2}{0.01} = 6.2 \times 10^{-8}$$

So $x^2 = 6.2 \times 10^{-10}$

$x = 2.49 \times 10^{-5}$

pH $= 4.6$

d   Let the value of $[H^+] = [Fe(H_2O)_5(OH)]^{2+} = x$. Use the bottom-line approximation.

$$K_a = \frac{x^2}{0.01 - x} \approx \frac{x^2}{0.01} = 6.0 \times 10^{-3}$$

So $x^2 = 6.0 \times 10^{-5}$

$x = 7.75 \times 10^{-3}$

pH $= 2.1$

31a Let $[H^+] = [HSO_3^-] = x$.

$x = 10^{-pH} = 10^{-1.5} = 0.0316$

$$K_a = \frac{x^2}{0.1 - x} = \frac{0.0316^2}{0.1 - 0.0316}$$

$= 1.5 \times 10^{-2}$ mol dm^{-3}

b   Let $[H^+] = [HCO_2^-] = x$

$x = 10^{-pH} = 10^{-2.4} = 0.00398$

$$K_a = \frac{x^2}{0.1 - x} = \frac{0.00398^2}{0.1 - 0.00398}$$

$= 1.65 \times 10^{-4}$ mol dm^{-3}

Note: the methanoate ion HCO_2^- is easily confused with the hydrogencarbonate ion HCO_3^-. Their structures are quite distinct however:

methanoate ('formate')   hydrogencarbonate

So the second H atom in methanoate is *not* acidic, while that in hydrogencarbonate *is*.

c   Let $[H^+] = [NH_3] = x$.

$x = 10^{-pH} = 10^{-5.1} = 7.9 \times 10^{-6}$

$$K_a = \frac{x^2}{0.1 - x} = \frac{(7.9 \times 10^{-6})^2}{0.1 - 7.9 \times 10^{-6}}$$

$= 6.3 \times 10^{-10}$ mol dm^{-3}

32   The nitride ion will seize protons from water and make hydroxide ions:

$Mg_3N_2(s) + 6H_2O(l) \rightarrow 2NH_3(aq) + 3Mg(OH)_2(s)$

You would see a purple coloration with Universal Indicator.

33a $NH_3(aq) + H_2O(l) \rightleftharpoons NH_4^+(aq) + OH^-(aq)$

$$K_b = \frac{[NH_4^+(aq)][OH^-(aq)]}{[NH_3(aq)]}$$

b $PO_4^{3-}(aq) + H_2O(l) \rightleftharpoons HPO_4^{2-}(aq) + OH^-(aq)$

$$K_b = \frac{[HPO_4^{2-}(aq)][OH^-(aq)]}{[PO_4^{3-}(aq)]}$$

c

34   From question 33a the name looks reasonable, since it matches the product side of the equation. However, when you consider that the K_b is only 1.7×10^{-5} mol dm^{-3}, the name looks misleading because there are very few actual ammonium and hydroxide ions. A better name would be 'an aqueous solution of ammonia'.

35a Limestone rocks.

b $CO_3^{2-}(aq) + H_2O(l) \rightleftharpoons HCO_3^-(aq) + OH^-(aq)$

c $K_b = \dfrac{[HCO_3^-(aq)][OH^-(aq)]}{[CO_3^{2-}(aq)]}$

d   Let the value of $[OH^-] = [HCO_3^-] = x$. Use a bottom-line approximation.

$$K_b = \frac{x^2}{0.01 - x} \approx \frac{x^2}{0.01} = 2 \times 10^{-3}$$

$x^2 = 2 \times 10^{-5}$

$x = 4.5 \times 10^{-3}$

So pOH $= -\log_{10} x = 2.3$

e   So since pH + pOH = 14 (equation (*12.16*)), it follows that:

pH = 14 − pOH = 11.7

36a $CH_3CO_2H(aq) + NaOH(aq) \rightarrow$
$CH_3CO_2Na(aq) + H_2O(l)$
sodium ethanoate

b   Sodium ethanoate is the salt of a weak acid and a strong base, so there should be a slight alkalinity in its solution. Since the ethanoate ion is the conjugate base of a weak acid, it is quite a respectable base in its own right. So the following equilibrium would exist (heavily biased to the left) between ethanoate ions and water:

$CH_3CO_2^-(aq) + H_2O(l) \rightleftharpoons CH_3CO_2H(aq) + OH^-(aq)$

c   The stoichiometric end-point is a mixture of sodium ethanoate and water, which as you can see would not be literally 'neutral' due to traces of the equilibrium above, the pH would be above 7.

37   Harpic's acidity would enable it to carry out the dual role of limescale ($CaCO_3$)-remover and bactericide.

38   Since the second proton removal of a dibasic acid is nearly always very weak, it follows

that the conjugate base – that is, the *di*-anion – must be quite strong.

39   1.

40   For the $[H^+]$ to drop by another factor of 10, we would have to perform $\frac{9}{10}$ of the remaining titration. Since 1 cm^3 of the titration mixture remains to react, this is 0.9 extra cm^3. So a pH of about 3 would occur after 9.9 cm^3 had been added.

41a The reverse reaction would be stopped by the absence of H^+ ions.

b   The forward reaction would proceed uninterrupted, so more HA molecules would decompose to replace the removed H^+ ions. When these, too, are removed, the decomposition of HA would proceed to completion.

c   So, overall,

$HA + OH^-(aq) \rightarrow H_2O(l) + A^-(aq)$

d   No build-up of OH^- ions is possible until all HA molecules have gone.

42   The upper regions of both curves show the pH due to excess sodium hydroxide.

43   The titration curve would look as in Figure A12.1. The first steep bit shows the end-point of the reaction to 'neutralise' the first set of hydrogen ions. At that stage, the solution contains HSO_4^- ions, and in effect a new titration begins, to remove the second hydrogen ion. This second titration produces a second typical S-shaped curve attached to the end of the first.

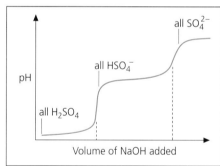

Figure A12.1 See answer 43

44a If the background $[H^+]$ is high, the reverse reaction will be helped (as is predictable by an application of Le Chatelier's principle) and the indicator will be mostly in its HIn form:

HIn (colourless)

b   The indicator will be colourless.

c   In alkaline solutions, the H atoms will be stripped from the indicator molecules, and In$^-$

will dominate the mixture. It will therefore be pink:

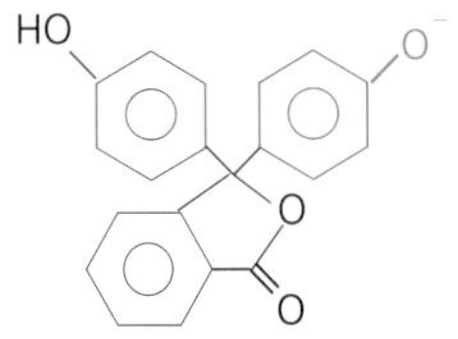

In⁻ (pink)

45a If there were many indicator molecules, then a significant amount of alkali would be needed simply to 'titrate' the indicator acid.
b If the indicator is still to be visible at very low concentrations (because there are very few molecules of it), at least *one* of its forms needs to be intensely coloured.

46 The model depends upon assumptions about people's eyes. It also assumes that each indicator colour is equally intense. In the case of phenolphthalein, where one of the species is colourless, this second assumption does not hold.

47a Phenolphthalein would be ideal, since the 'steep bit' is skewed upwards from pH 7.
b Ammonia would be titrated against sodium hydroxide. This is the opposite of the situation in part (a). The end-point would be slightly acidic, so we need a low-change indicator such as methyl orange.
c Any of the indicators in Table 12.4 could be used, since the 'steep bit' of a strong acid–strong base titration covers so many pH units.
d Methyl orange changes in the pH 4 region, which is the pH for most of the run-up to the end-point of an ethanoic acid–NaOH titration. You would therefore see a long gradual colour change, which would be no use for judging the end-point.
e 'Weak versus weak' titrations do not have a 'steep bit', so any indicator would give a gradual colour change, some more gradual than others.

48a Most carboxylic acids have values of pK_a in the region of 5, so we could assume that the loss of the proton would occur in the pH range from 4 to 6. If the proton needed to be in place, then the enzyme would be seriously hampered at pHs above about 5.5.
b This situation features the same range, but this time the enzyme malfunctions as the pH gets *lower*. Around pH 4.5 there would be too much protonation.

49a As the free HOCl is consumed by the pool, more trichlor molecules decompose to top up the HOCl concentration.
b Tablets are very easy for the staff to handle, compared with unpleasant and corrosive liquids. The tablet system absolves the staff from having to make accurate liquid measurements.

50 HOCl has a pK_a of around 7.4, so the change range would be from 6.4 to 8.4 approximately.

51 The 50:50 mixture would be achieved at a pH equal to the acid's pK_a, namely 7.4.

52a $NH_3(aq) + HOCl(aq) \rightarrow NH_2Cl(aq) + H_2O(l)$
b $NH_3(aq) + 3HOCl(aq) \rightarrow NCl_3(aq) + 3H_2O(l)$
and the best way to avoid this is to keep the overall amount of available HOCl within limits.

53a The A⁻ species (the conjugate base) is responsible for absorbing H⁺ ions.
b The HA species (the conjugate acid) is responsible for absorbing OH⁻ ions.

54 The 'horizontality' of the graph symbolises the flatness of the pH response, even though NaOH is being added. Flatness of pH response is what buffering is all about.

55b Methanoic acid has a pK_a of nearly 4, so a mixture of methanoic acid (HCO_2H) and sodium methanoate (HCO_2Na) would buffer at about 4.
c Dihydrogenphosphate ion has a pK_a of nearly 7, so a mixture of sodium dihydrogenphosphate (NaH_2PO_4) and disodium hydrogenphosphate (Na_2HPO_4) would buffer at about 7.

56 The NH_3^+ group will deal with invasions of OH⁻, and the CO_2^- group will deal with H⁺ ions.

57a As an acid (proton donor) coping with an OH⁻ invasion:
$HCO_3^-(aq) + OH^-(aq) \rightleftharpoons H_2O(l) + CO_3^{2-}(aq)$
As a base (proton acceptor) coping with an H⁺ invasion:
$HCO_3^-(aq) + H^+(aq) \rightleftharpoons H_2O(l) + CO_2(aq)$
b Blood contains dissolved carbon dioxide, which would be a source of the HCO_3^- ion.
c The reaction which puts HCO_3^- ions into water is one between carbonated rainwater and calcium carbonate:

$H_2O(l) + CO_2(aq) + \underset{\substack{\text{in limestone} \\ \text{rocks}}}{CaCO_3(s)} \rightleftharpoons Ca(HCO_3)_2(aq)$

It is the soluble calcium hydrogencarbonate which provides the natural buffering system.

CHAPTER 13   page 281

1a The word 'tertiary' means that three substituent alkyl groups are attached to the carbon atom under consideration. Since the carbonyl group alone takes up *two* bonds, there are only two bonds left to carry alkyl groups, so there cannot be a *tertiary* carbonyl.
b Since ketones must have the =O group attached to an inner carbon, the end carbons become irrelevant. So there is only one propanone, and even in butanone the only possible position is 'next to the end'. Therefore you need *five* carbon atoms before there is any choice. Then, it is possible to have the isomers pentan-2-one and pentan-3-one.
c Butanone (there is no need for a number).

2 Octanoic acid.

3 Ethanoyl bromide.

4 The statement is true of halogenoalkanes (p. 228) and of alcohols (p. 243).

5a Tertiary carbocations are stabilised because there are three alkyl groups all pushing electron density towards the central positive charge, helping to reduce and disperse it.
b Far from having electron-pushing groups, this carbocation has an electron-withdrawing group (OH) attached to the central carbon, which intensifies charge build-up, and destabilises the ion.

6 Nucleophilic attack can dislodge *one* pair of electrons on the 'leaving group' (as in alcohols), but if the 'leaving group' is attached by *two* pairs of electrons, nucleophilic attack is unsuccessful.

7a HCN is a fairly weak acid (p. 260).
b HCN will predominate in strongly acidic media, because CN⁻ ions will be protonated (p. 284).
c In acidic media the lone pair of electrons which could carry out the attack on the carbonyl group will not be a lone pair; instead they will be playing the dative covalent role to a hydrogen ion.

8

a $HCN + (CH_3)_2C=O \xrightarrow[\text{(aq)}]{NaCN} NC-C(CH_3)_2-OH$

b (with $H_2SO_4(aq)$) → $HO-C(=O)-C(CH_3)_2-OH$

($Al_2O_3, -H_2O$)

c $HO-C(=O)-C(CH_3)=CH_2$

(CH_3OH, H^+ (esterification))

d $CH_3O-C(=O)-C(CH_3)=CH_2$

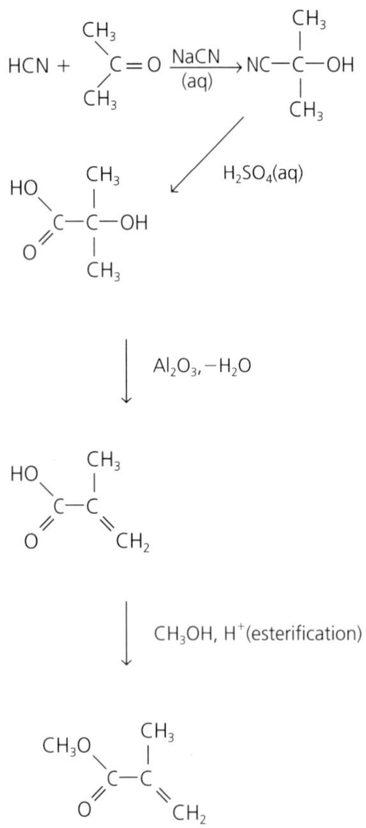

9a Their basicity comes from their possession of a lone pair of electrons, which is the essential requirement for being a base.
b $H_2N-\overset{+}{N}H_3Cl^-$

c Hydrazinium chloride could be readily reconverted to its base hydrazine, by a method parallel to that used in generating ammonia from ammonium salts, namely addition of sodium hydroxide:

$$H_2N\!-\!\overset{+}{N}H_3 + OH^- \longrightarrow H_2N\!-\!NH_2 + H_2O$$

10

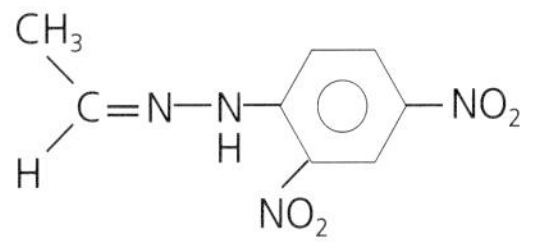

11

a $CH_3\overset{O}{\overset{\|}{C}}CH_3 \xrightarrow{LiAlH_4} CH_3\overset{OH}{\overset{|}{C}}HCH_3$
        propan-2-ol

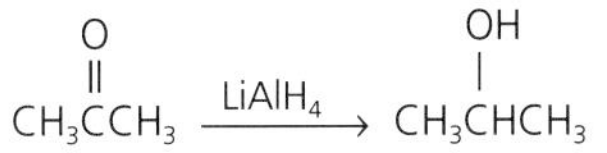

b $CH_3CH_2CH_2\overset{H}{\underset{O}{C}} \xrightarrow[Ni]{H_2} CH_3CH_2CH_2CH_2OH$
        butan-l-ol

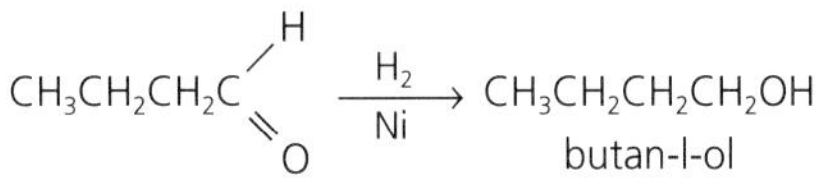

c (cyclohexanone) $=O \xrightarrow{NaBH_4}$ (cyclohexanol)—OH
        cyclohexanol

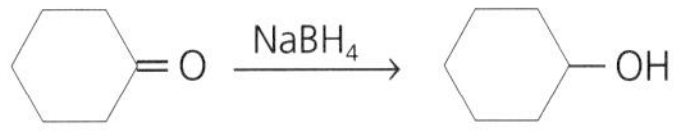

The general pattern is that aldehydes are reduced to primary alcohols, while ketones are reduced to secondary alcohols.

12 The hydrogen gas/nickel catalyst method will reduce both the carbonyl group *and* the alkene group, whereas the $NaBH_4$ will only reduce the carbonyl group (as shown below).

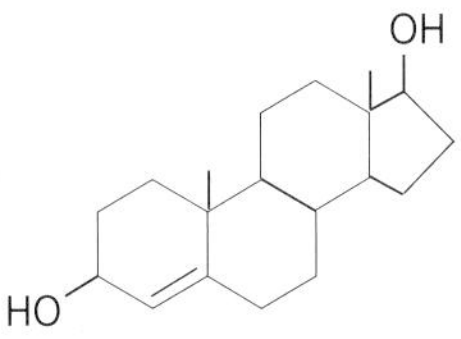

13 Aldehydes owe their vulnerability to oxidation to the fact that they have a terminal C—H bond next to the C=O bond. This bond is readily oxidised up to C—O—H, whereas to oxidise a ketone the whole carbon framework would have to be broken. (It is difficult to offer a better explanation than that without knowing the details of the mechanism.)

14 The evidence for each statement is in brackets. Both B and C are carbonyl compounds (positive Brady's test), but B is a ketone (negative Fehling's test), whereas C is an aldehyde (positive Fehling's test). There is only one aldehyde of formula CH_2O, methanal, and there is only one ketone of formula C_3H_6O, propanone. So A, B and C are as shown next.

A (methylpropene)   **B** (propanone)
    methylpropene   propanone

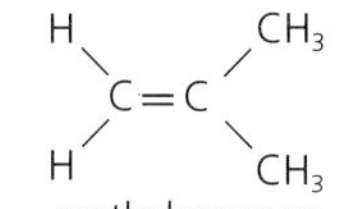

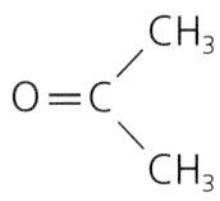

C (methanal)
    methanal

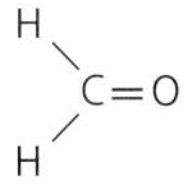

15 An iodine atom will replace a hydrogen atom on a carbon adjacent to the carbonyl group. But since one of the two carbon atoms adjacent to the carbonyl group does not carry *any* hydrogens, the product must be as shown below.

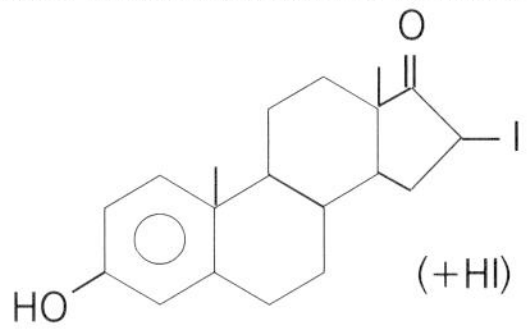

(+HI)

16 Propan-2-ol can be oxidised to a methyl ketone (propanone), so it would give a positive triiodomethane (iodoform) test.

(By the way, the name iodoform looks very strange, but it derives from the non-systematic name of the much more well-known chlorine equivalent, **chloroform**, $CHCl_3$ (or trichloromethane).)

17 All equilibria are in a state of dynamic balance between the forward and back reactions, as discussed in Chapter 9. If the reaction with, say, phenylhydrazine removes the open-chain aldehyde form, then the back reaction will cease. However, the forward reaction will keep on delivering the open-chain form until the stock of ring-form glucose has run out. This is an example of the application of Le Chatelier's principle. The 'interference' has been the removal of the aldehyde form, and the compensatory response of the system has been to try to top up the aldehyde form.

18

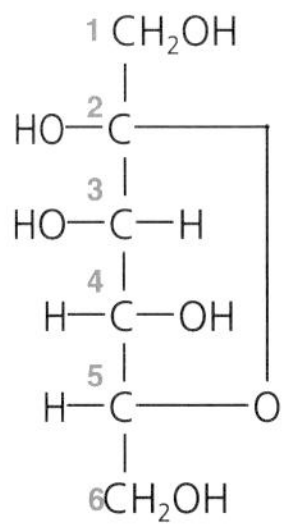

19 You would expect the first two tests (Fehling's and Tollen's) to discriminate between glucose and fructose, since glucose is an aldehyde-sugar (sometimes called an **aldose**), whereas fructose is a ketone-sugar (or **ketose**). Brady's test should be positive for both. Oddly enough, the logical answers to this question turn out to be wrong since, for reasons outside the Advanced syllabus, α-hydroxyketones (as in fructose's open-chain form) *are* oxidised by the two complexed heavy metals.

20
                                    'mirror'

ethene

propane

propanone

2-chlorobutane

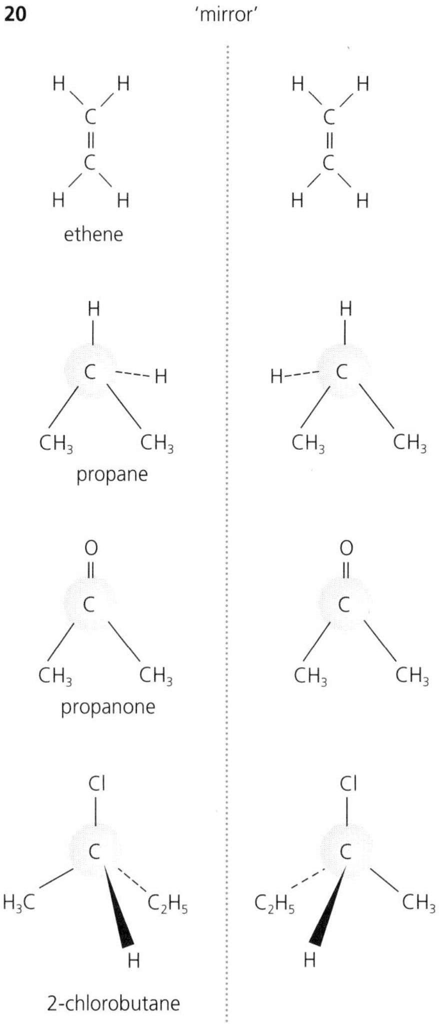

21

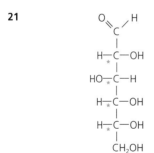

22 You would expect sucrose to give negative Fehling's and Tollen's tests since it cannot manipulate itself into a carbonyl-containing arrangement.

23a It is involved in the digestion of cellulose. Animals that do digest cellulose have big appendices. Rabbits are a case in point.
b Cattle have more than one stomach, so they give the cellulose a double dose of digestive enzyme action. They also bring partially-digested material (called 'cud') back into the mouth from stomach 1 for a second round of chewing and mixing with yet more enzymes.

24 See Table A13.1. Ethanoic acid is a stronger acid than carbonic acid, H_2CO_3. So it has the ability to force protons on to the HCO_3^- or CO_3^{2-} ions, and regenerate carbonic acid. This then decomposes to CO_2 and water, so the visual signal of the acidity of ethanoic acid is that it can make carbonates and bicarbonates fizz. Phenol cannot do this.

Table A13.1 A test for the relative strengths of acids

Acid	Equation	pK_a
ethanoic	$CH_3CO_2H \rightleftharpoons CH_3CO_2^- + H^+$	4.76
carbonic	$H_2O + CO_2 \rightleftharpoons H^+ + HCO_3^-$	6.38 (pK_1)
		10.32 (pK_2)
phenol	$C_6H_5OH \rightleftharpoons H^+ + C_6H_5O^-$	10.0

25 The successively more heavily substituted mono-, di- and trichloroethanoic acids are successively stronger acids. The reasons can be found in the electron-withdrawing effects of the chlorine atoms on the end carbon atom. This heavily positively polarised carbon atom then biases the electron density in the C—C bond, and by a series of knock-on effects the oxygen atom in the O—H bond ends up demanding more electron density from the hydrogen atom. Hence, the cleavage of a hydrogen ion is assisted.

26

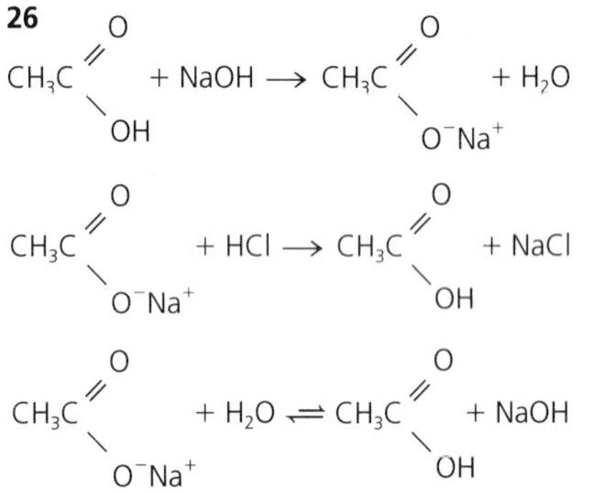

27 The equation is shown below.

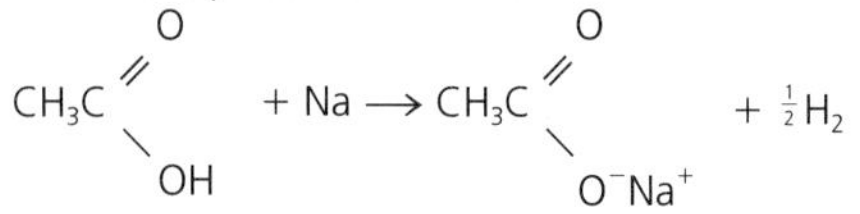

Its reaction should be more vigorous even than water's, on the basis that ethanoic acid is a freer hydrogen ion donor.

28 Phenol and propanone will not react with PCl_5. Phenol is hostile to attacks by nucleophiles, while propanone cannot offer the phosphorus halide the target for which it seems to be a specialised reagent, namely —OH.

29a

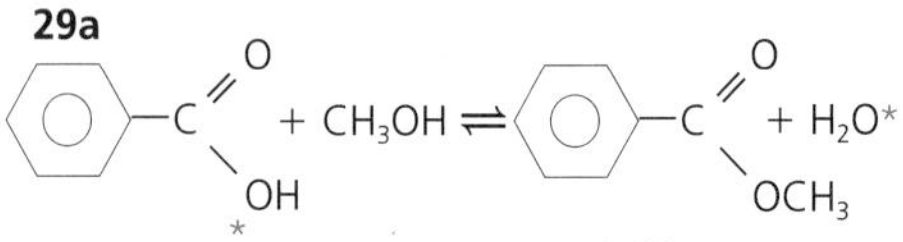

methyl benzoate

b The same product could be formed if benzoyl chloride were used instead of benzoic acid.

30 The matter could be decided by using an acid with some 'labelled' ^{18}O atoms in the OH group and by investigating where they end up.

31

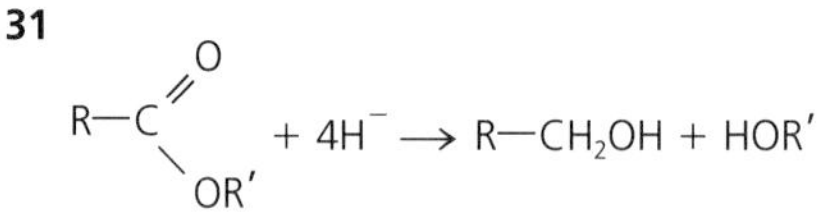

$$R-C{\overset{O}{\underset{OR'}{}}} + 4H^- \rightarrow R-CH_2OH + HOR'$$

32a

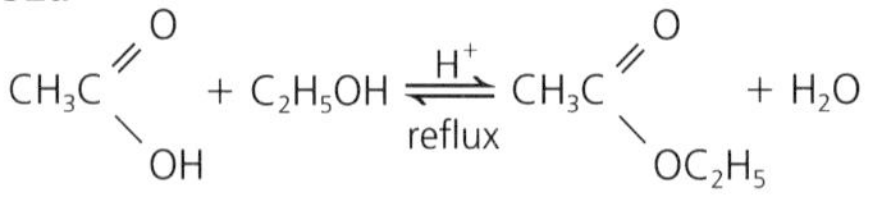

ethyl ethanoate
(an ester)

b No reaction.

c

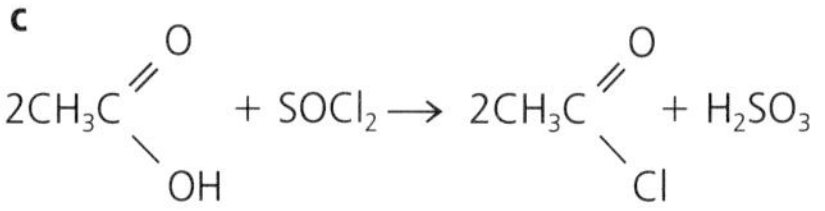

ethanoyl chloride
(an acyl halide)

d

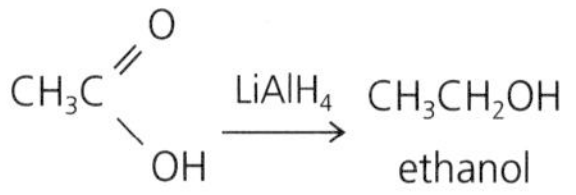

ethanol

e

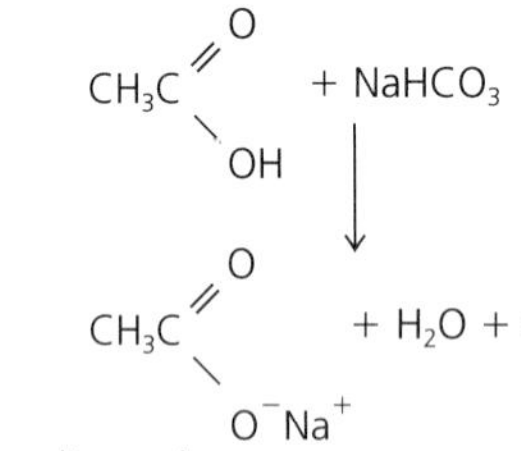

sodium ethanoate

f

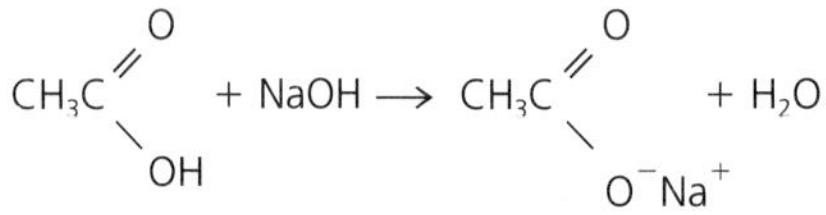

33

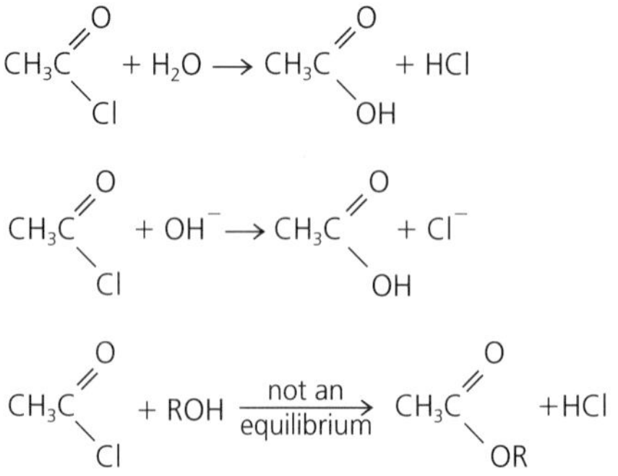

an ester

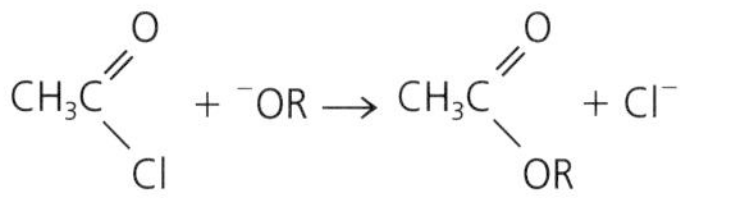

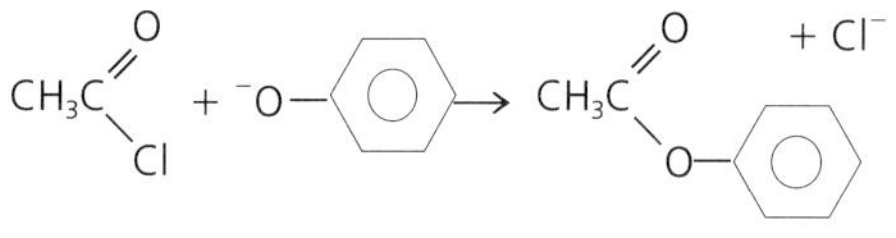

phenyl ethanoate
(an ester)

(This is the only method for esters of phenol)

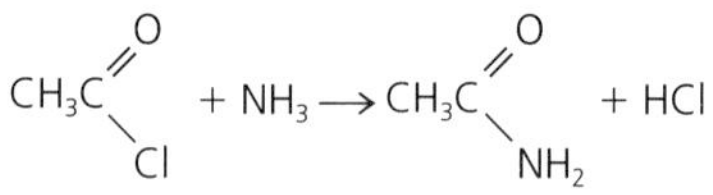

an amide

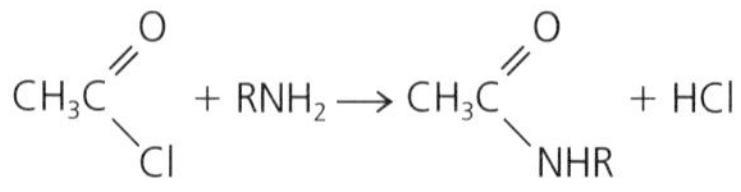

an N-substituted amide

34

Step 1

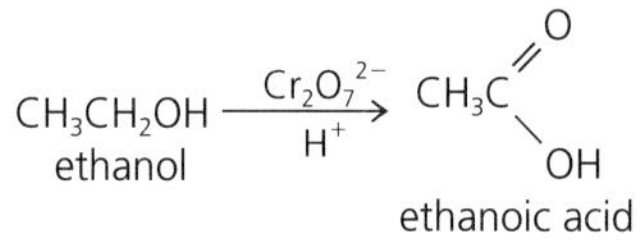

ethanol → ethanoic acid

Step 2

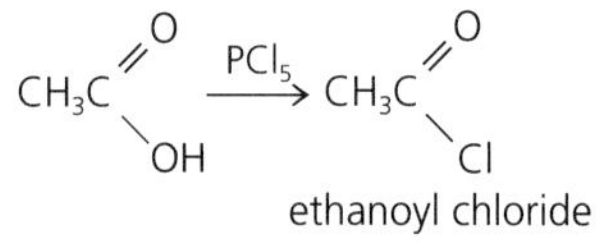

ethanoyl chloride

Step 3

ethanamide

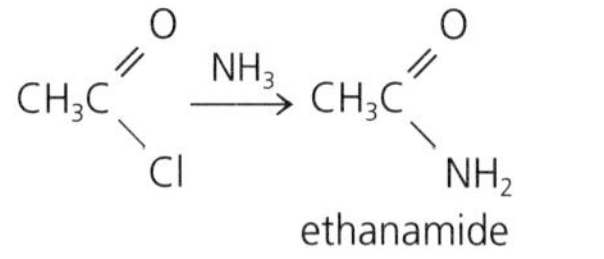

If the yield was one half of the starting material at each stage, you would need 0.8 mole of ethanol to make 0.1 mole of the amide.

35 Butanoyl chloride > 2-chloro-2-methylpropane > 1-chlorobutane > chlorobenzene. The acyl halide's halogen atom is the most reactive. After that the branched halogenoalkane will react faster than the straight-chain one (and by a different mechanism), as explained in Chapter 11. Finally the halogenoarene will resist any nucleophilic approach, at least under these relatively mild conditions.

36

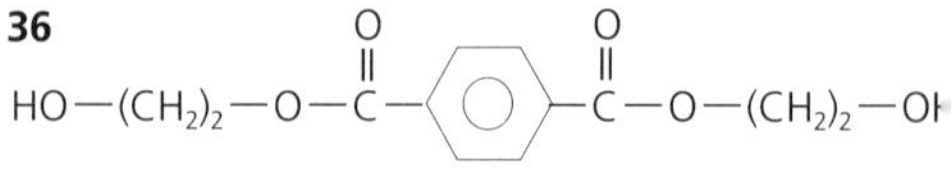

Further reaction is possible to form a **polyester** polymer.

37

catalyst 'on'

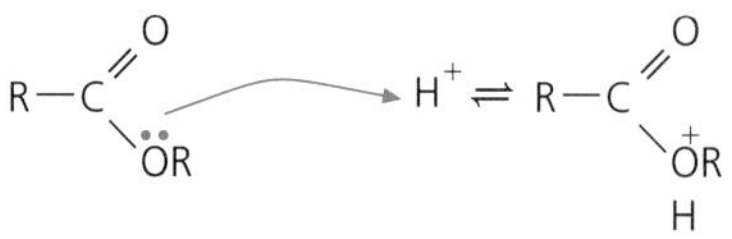

hydrolysis

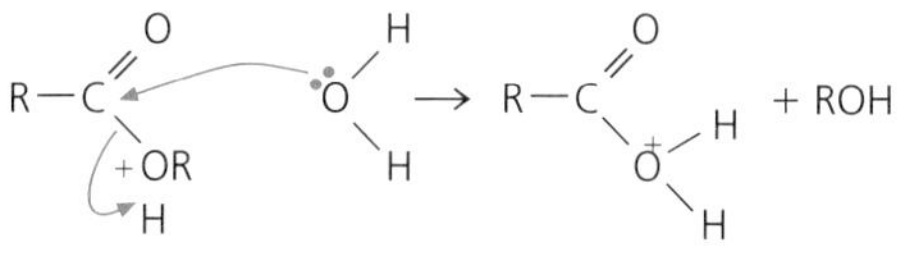

R—C(=O)(+OR)(H) ... ·Ö—H(H) → R—C(=O)(+Ö—H(H)) + ROH

catalyst 'off'

R—C(=O)(+Ö—H(H)) → R—C(=O)(OH) + H⁺

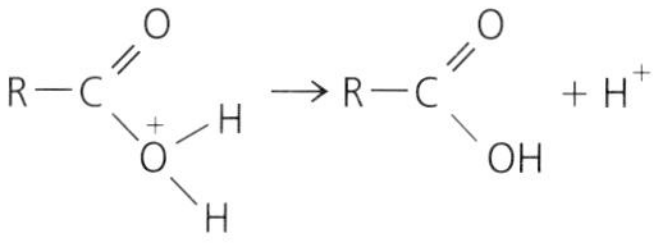

38 They are both energy-supplying chemicals.

39

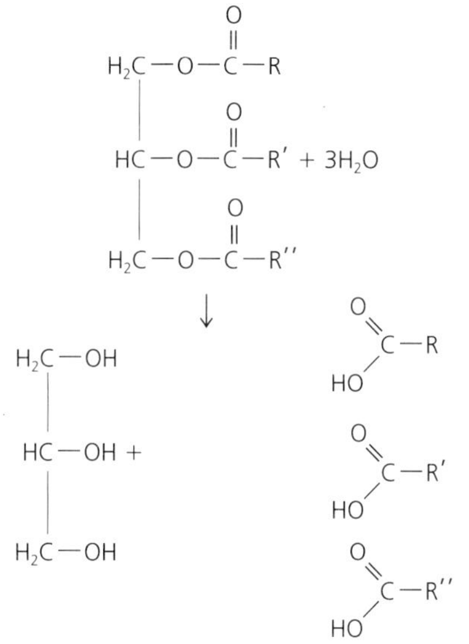

$$H_2C—O—C(=O)—R$$
$$HC—O—C(=O)—R' + 3H_2O$$
$$H_2C—O—C(=O)—R''$$

↓

$$H_2C—OH$$
$$HC—OH +$$
$$H_2C—OH$$

$$C(=O)—R, HO$$
$$C(=O)—R', HO$$
$$C(=O)—R'', HO$$

40 Plants, like animals, use fats as a source of stored energy. A germinating seed has a particular need for stored energy, given its rapid growth rate and its initial inability to photosynthesise its own starch.

41a Fat mixtures from animal sources generally have a lower percentage of triple-unsaturated fats than those from vegetable sources.
b The presence of double bonds (U) should *hinder* the neat stacking of side-chains, because it prevents a completely linear arrangement of the chains.
c Intermolecular bonding will be impeded in unsaturated fats because they are not stacked neatly together. You would therefore expect a UUU fat to have a much lower melting temperature than an SSS fat.
d Plants are sometimes subjected to very low external temperatures, and they have very little ability to maintain their own internal temperatures. So if their fat supplies are to remain liquid and mobile, the fats need to have low melting temperatures – which demands a high degree of U fats.
e This is a fat with a purely structural function, so it needs to keep solid, and to do so

despite an internal temperature of about 37 °C. This demands a fat with the maximum amount of intermolecular bonding, which would require the choice of a long linear chain.
f Fish in temperate and sub-Arctic waters might well have to endure lower core-body temperatures than mammals. An increase in unsaturated side-chains lowers freezing points of their fat molecules, and maintains them in a liquid state.

42a The backbone is the glycerol.
b The 'plug-in cartridges' are the fatty acid side-chains.
c The ester linkages ensure that the 'cartridges' are detachable.
d The length and degree of unsaturation of the various side-chains will determine its melting temperature.
e The oxygen atoms of sugars are not really functional as fuel atoms – they will end up attached to carbon or hydrogen atoms, but then that is how they started out. The proportion of oxygen atoms in fats is much lower (compare the formulae on p. 298).

43a See Figure A13.1.

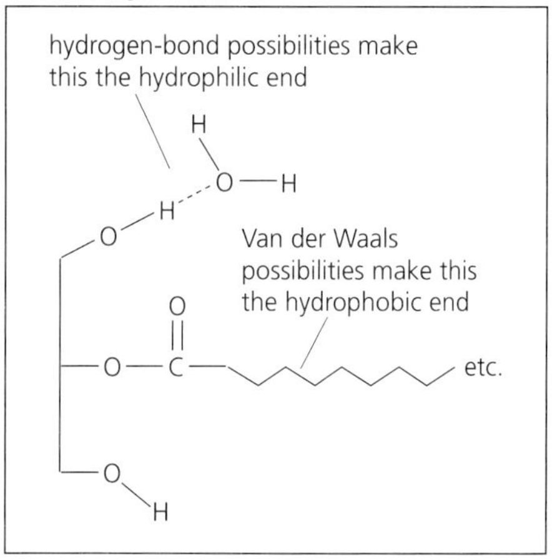

hydrogen-bond possibilities make this the hydrophilic end

Van der Waals possibilities make this the hydrophobic end ... etc.

Figure A13.1

b The rate of reaction should speed up, because the creation by product 'detergents' of more micelles will increase the surface area of the remaining fat.
c The stomach (where the foodstream has come from) is an extremely acid medium, and the ester hydrolysis reaction works best in alkaline media.

44a The message from studying vulcanisation is that the cross-links seem to seek out and replace hydrogen atoms on carbons *adjacent* to double bonds. So cross-linking might occur as in Figure A13.2. No one seems really sure about the actual structures of these air-cured polymers.
b The gum on chip pans is simply polymerised cross-linked fat molecules.
c The smoke from chip pans comes from the boiling off of fragments, which occurs at lower temperatures than the boiling temperature of the unhydrolysed oil.

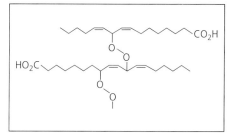

Figure A13.2

45a It has hydrophilic and hydrophobic sections, so it mimics the action of the natural detergents in digestion.
b This could be done by reacting the margarine with controlled amounts of hydrogen gas over a nickel catalyst. You sometimes see margarines labelled 'hydrogenated' when this reaction has been used. The over-runniness is caused by the *cis*-orientation of the double bonds in unsaturated fats. As we have seen, *cis* C=C bonds give rise to 'kinked' chains, which stack poorly and thus have weaker van der Waals' forces.

46a The more amorphous chocolates would have lower melting temperatures and be less crisp.
b The most organised structure is also the stablest because it allows for the maximum degree of intermolecular bonding.
c The 'neatest' arrangement will also be the one which takes up the least space, since intermolecular space is at a minimum.

47 The micelles will be inhibited from re-coalescing by the repulsions of their negatively charged surfaces.

48 In acidic media, the soap anions would be protonated:

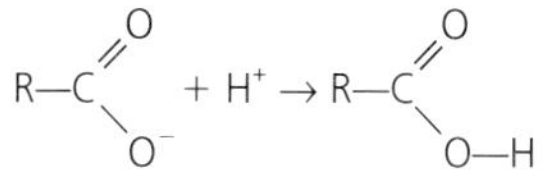

$$R—C(=O)(O^-) + H^+ → R—C(=O)(O—H)$$

which reduces the hydrophilic character of that end of the molecule.

49 You would predict no reaction, since the calcium cations have no anion with which to precipitate.

50 Linoleic acid is the most unsaturated of all the listed side-chains.

51a

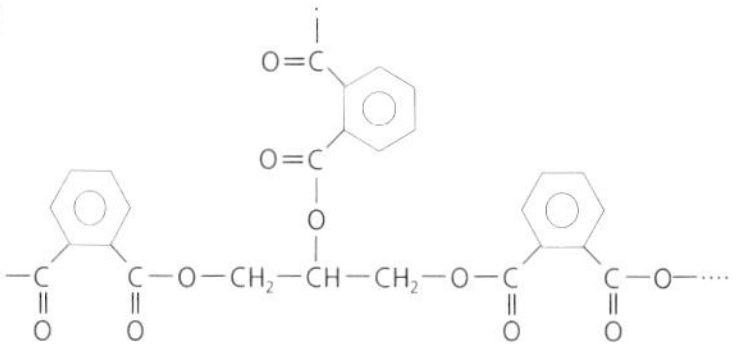

O=C ... O=C ... (benzene rings)
····C—C—O—CH₂—CH—CH₂—O—C—C—O·····

b In the ordinary acid-plus-alcohol reaction, there is a molecule of *water* liberated for every ester linkage made. It is this water which causes the frothing. Phthalic anhydride undergoes esterification with no liberation of water.

c   Pentaerythritol can make four linkages in tetrahedral directions, as opposed to glycerol's three. So the alkyd resin made from pentaerythritol will be more fully cross-linked.

52a The use of terephthalic acid makes for a more linear polymer molecule, which is more suitable for a linear fibre.
b   Cotton (which is a polysaccharide) has more oxygen atoms, and what is more they are in OH groups, so the possibilities for hydrogen bonding are much greater with cotton.

53

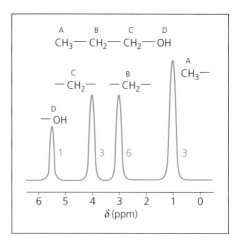

54 Organophosphates attack and disable the enzyme acetylcholinesterase, and this was a chemical which was at a low level in the infected cow. What is more, the absence of BSE in 'organic' cows *despite* their exposure to the allegedly infected feeds, seems to cast doubt on the rival theory that BSE comes from scrapie in sheep. Damson, remember, was the only cow on Mark Purdey's farm which had been exposed to organophosphates, whereas all the cows had had the suspect feeds. It has to be said that the numbers of animals involved are rather small to form the basis of hard-and-fast conclusions.

55 The longer the chain parts relative to the link parts, the less alkali per gram the fat will need. So the saponification value is a measure of the average chain length. The iodine value measures the number of double bonds, and therefore the degree of unsaturation.

56a It has three neighbours on the CH_3 group and by the sub-peak rule the number of sub-peaks is $3+1 = 4$.
b   Every new carbon, at a different remoteness from the oxygen atom, introduces a new proton environment and therefore a new signal in the NMR spectrum. So including the OH hydrogen atom there will now be *four* proton signals. The 'sextet' of sub-peaks would come from the hydrogens on the middle carbon atom, which have three neighbours on one side (the CH_3 group) and two on the other (the other CH_2 group), a total of 5 (and $5 + 1 = 6$). The spectrum might look like Figure A13.3.
c   If the ethanol OH proton signal were split, it would be split by the neighbouring CH_2 group, so it would be a triplet.

57 The sweet smell indicates that the substance is probably an ester. The integration curve shows that it has three protons at one site and three at another, with two at a third site. The splitting pattern of a triplet (at 1.0–1.5) and

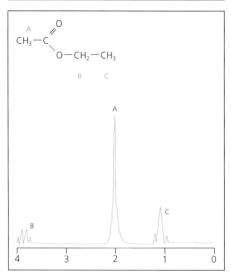

Figure A13.3 Low-resolution proton-NMR of propan-1-ol. The numbers in green indicate the number of sub-peaks which would appear at high resolution

Figure A13.4 See answer 57

a quartet (at 3.9–4.3) is strongly suggestive of a CH_3CH_2— group. This all suggests ethyl ethanoate. See Figure A13.4.

58a All the spectra have a big peak at 1.4 ppm because all the molecules have lots of CH_2 groups.
b   The peak at 0.9 ppm is due to the protons in the terminal methyl, and all the molecules have much more chain-middle (1.4 ppm) than chain-end (0.9 ppm).
c   The peak at 5 ppm identifies an unsaturated side-chain.
d   So spectra (b) and (c) seem to show the unsaturated side-chains.
e   The 5.0 ppm peak in spectrum (c) looks bigger, suggesting that it may be the more unsaturated of the two side-chains.

CHAPTER 14   page 317

1a   Frequency of collision can be increased by an increase in any of the following: temperature, concentration (or partial pressure for gaseous

reactions), surface area of heterogeneous phases and even the viscosity of the solvent.
b   Increased temperature will increase the violence of collisions. (It turns out that despite the presence of temperature in the previous list, this is its main contribution to deciding reaction rates.)

2   Stirring helps to homogenise a badly mixed reactant system. However, once the reactants have free access to each other in a homogeneous mix, stirring makes no difference, because collisions are caused by random molecular motions. Stirring has little or no effect on the violence of molecular collisions.

3   As the concentrations of reactants fall, the chances of their meeting to react, or simply of being there to react, will correspondingly decline. So reactions should slow down as they proceed. (But beware zero-order rate laws, and autocatalysis, of which more later.)

4   It does agree with question 3. The steepness of the curve at any point (as found from the gradient of the tangent) indicates the rate of reaction at that point. The reaction is slowing down, so the gradients get progressively shallower, giving the curved shape.
   The rate of reaction at any point in time could be measured by calculating the numerical value of the gradient of the tangent at that point (as y/x).

5   The units of rate would be the same as the units of the gradient, which is (y-axis units)/ (x-axis units), or $mol\ dm^{-3}\ s^{-1}$.

6   The rate of decline of [A] will be equal to the rate of decline of [B]. It will be equal to the rate of *increase* of [C] and [D]. In differential algebra:

$$\frac{d[A]}{dt} = \frac{d[B]}{dt} = \frac{-d[C]}{dt} = \frac{-d[D]}{dt}$$

7   The rate of appearance of C will now be double the rate of disappearance of A and B, because there are two moles of C being created for every mole of A and B destroyed.

$$\frac{d[A]}{dt} = \frac{d[B]}{dt} = \frac{1}{2}\frac{d[C]}{dt}$$

8   Yes. Even if there were 10 times as much A as B, the number of moles of each that disappeared in unit time would still be the same.

9   You would expect the gradient at $[A]_0$ to be double the gradient at $[A]_0/2$.

10 Because moles of A and B disappear together, when $[A] = 0.5\ mol\ dm^{-3}$, $[B] = 0.5\ mol\ dm^{-3}$.

11 Since both [A] and [B] will have halved, the reaction rate will have dropped to one-quarter of its initial value.

12 This time the gradients of the tangents at $[A]_0$ and $[A]_0/2$ would reflect this $4:1$ decline in the rate.

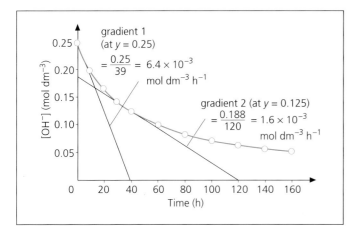

Figure A14.1

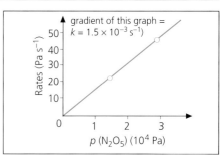

Figure A14.2

13a [B] would have fallen by the same amount as [A], so [B] would be 9.5 mol dm^{-3}. But proportionately the effect is that [B] remains almost constant.

b The rate now would be $k \times 0.5 \times 9.5$, or 4.75k.

c The rate would have fallen from $k \times 1 \times 10$, or 10k, to 4.75k, a decline of 2.1 : 1. The reaction has behaved more like a first-order reaction than a second-order one. As we shall see, the ploy of keeping one reactant in an excess (in this case B) can be used to focus attention on the order with respect to the other reactant. What we have in effect exposed is the first-order dependence on [A].

14a It looks like a reaction of the 'A + B' type, so it seems likely that the rate would depend on [A] × [B]. The equation would be:

Rate = $k[C_4H_9Br][OH^-]$

b See Figure A14.1.
Gradient 1 = 6.4×10^{-3} mol dm^{-3} h^{-1}
Gradient 2 = 1.6×10^{-3} mol dm^{-3} h^{-1}

So the two gradients are in the ratio 4 : 1, which conforms to the second-order model. The rate equation would therefore be:

Rate = $\dfrac{d[C_4H_9Br]}{dt} = \dfrac{-d[OH^-]}{dt} = k[C_4H_9Br][OH^-]$

c At the beginning, [A] = [B] = 0.25 mol dm^{-3}, and gradient 1 = 6.4×10^{-3} mol dm^{-3} h^{-1}. So, inserting these values into the rate equation, we get:

$6.4 \times 10^{-3} = k \times 0.25^2$

So:

$k = 0.10$ mol^{-1} dm^3 h^{-1} (to 2 sf)

From the gradient 2 data:

$1.6 \times 10^{-3} = k \times 0.125^2$

So:

$k = 0.10$ mol^{-1} dm^3 h^{-1} (to 2 sf)

so k is confirmed as a constant.

d The mechanism is of the S_N2 type, where the '2' indicates that the crucial (in fact the only) step is a collision between *two* reactants.

e The alternative, S_N1, mechanism would have required the system to go through the transition state $CH_3CH_2CH_2CH_2^+$, and we have seen in Chapter 11 that straight-chain primary carbocations are less stable than branched tertiary ones like:

$$\underset{H_3C}{\overset{CH_3}{\underset{}{\underset{}{\overset{|}{C^+}}}}}\quad CH_3$$

So this particular substrate would be expected to choose the S_N2 route, which avoids the unfavoured transition state.

15a A first guess might be that rate = $k[N_2O_5]^2$, since you might suppose that the reaction proceeded by a collision between two reactant molecules.

b See Figure A14.2.
Gradient 1 = 46 Pa s^{-1}
Gradient 2 = 23 Pa s^{-1}

So the ratio of the gradients is 2 : 1, suggesting a first- not second-order dependence on [N_2O_5]. The rate equation seems therefore to be rate = $-d[N_2O_5]/dt = k[N_2O_5]$. To get the rate constant we could use the data from gradient 1:

$46 = k \times 3.00 \times 10^4$

So:

$k = 1.5 \times 10^{-3}$ s^{-1}

This 'time to the minus one' unit is typical of first-order rate constants.

16 As an illustration, Table A14.1 and Figure A14.3 use just the two gradients already calculated. However, it is important to bear in mind that four or five gradients would give more reliable results.

17 The gradient of this graph is rate/[N_2O_5], which is just k. The value of k from question 16 comes out as (not surprisingly) more or less the same as that calculated in question 15, i.e. 1.5×10^{-3} s^{-1}.

18 If you suspected that the rate had a second-order dependence on the concentration, you

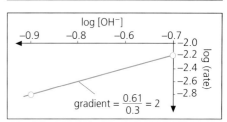

Figure A14.3

Table A14.1

Rate of reaction from gradients 1 and 2) (obtained (Pa s^{-1})	$p(N_2O_5)$ (10^4 Pa)
23	1.5
46	3.0

would plot the second graph as gradients (y-axis) against [A]^2 (x-axis).

19 Again, Table A14.2 and Figure A14.4 use only the two gradients already calculated.
The gradient can be seen as 0.61/0.3, or 2. As

Table A14.2

Rate (gradient) (mol dm^{-3} h^{-1})	[OH$^-$] (mol dm^{-3})	log (rate)	log [OH$^-$]
6.4×10^{-3}	0.25	−2.19	−0.60
1.6×10^{-3}	0.125	−2.80	−0.90

Figure A14.4 See answer 19

Table A14.3

Time (s)	ln ($p(N_2O_5)$)	$p(N_2O_5)$ (10^4 Pa)
160	10.45	3.45
285	10.25	2.83
400	10.10	2.41
525	9.89	1.97
700	9.63	1.52
950	9.26	1.05
1300	8.70	0.60
1780	7.97	0.29

equation *(14.6)* showed, this value confirms a second-order reaction.

20 Table A14.3 and Figure A14.5 use alternate points from the data.

The gradient of the graph is equal to $-k = -1.51 \times 10^{-3}$ s^{-1}. Note that because the graph is linear and does not depend upon drawing tangents, the answer has been given to 3 significant figures, rather than 2 as in the answer to question 15.

21 I have taken half-lives at three places:

• Time for $p(N_2O_5)$ to drop from 4.2 to 2.1 = 460 s
• Time for $p(N_2O_5)$ to drop from 2.1 to 1.05 = 940 − 480 = 460 s
• Time for $p(N_2O_5)$ to drop from 3.0 to 1.5 = 690 −250 = 440 s

Allowing for the uncertainty of the original plot, we have a reasonably constant half-life of around 460 s, and confirmation of a first-order rate law. This half-life would give a value for k as follows:

$$k = \frac{\ln 2}{t_{1/2}} \approx \frac{0.69}{460 \text{ s}} = 1.5 \times 10^{-3} \text{ s}^{-1}$$

which agrees with the value derived from a single piece of data in question 15.

22 Table A14.4 and Figure A14.6 sample the data at 40-hour intervals.

The value of k is given by the gradient, and is 0.1 mol^{-1} dm^3 h^{-1}, in agreement with the value from question 14.

The last three methods are better than the first two, because they do not involve drawing tangents, and they make use of all the data at once to give gradients of best straight lines. The half-life method is quick and convenient, but is of course limited to first-order curves.

23a $[A]_0 = [A] + [C]$

So to convert [C] to [A],

$[A] = [A]_0 - [C]$

b At t_∞, when [A] = 0, the equation breaks down to give:

$0 = [A]_0 - [C]_\infty$

So:

$[C]_\infty = [A]_0$

The concentration of C at the end of the run is equal to the concentration of A at the start.

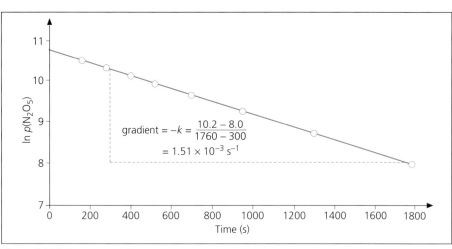

Figure A14.5 See answer 20

24a See Table A14.5. The values of [A] represent concentrations of benzenediazonium chloride calculated using the formula [A] = 7.50 − [HCl]. The graph is plotted in Figure A14.7, and gives a reasonably constant half-life. We would conclude then that the rate is first order with respect to benzenediazonium chloride.

b The rate constant can be found from:

$$k = \frac{\ln 2}{t_{1/2}} = \frac{0.69}{6 \text{ min}} = 0.115 \text{ min}^{-1}$$

Table A14.4 See answer 22

Time (hours	[OH$^-$] (mol dm^{-3})	$\frac{1}{[OH^-]}$ (mol^{-1} dm^3)
0	0.250	4.0
40	0.125	8.0
80	0.084	11.9
120	0.063	15.9
160	0.050	20.0

Table A14.5 See answer 24a

Time (min)	[A] (10^{-2} mol dm^{-3})
0	7.50
1	6.67
2	6.05
4	4.60
6	3.62
8	2.90
12	1.94
14	1.45

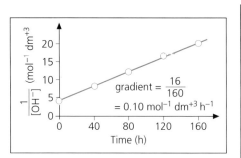

Figure A14.6 See answer 22

Figure A14.7 See answer 24a

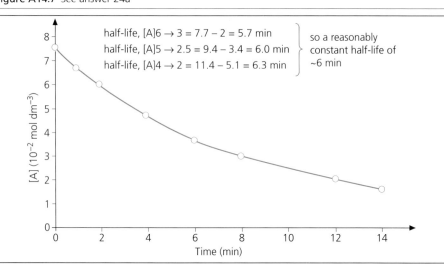

c Since water is present in such a large excess, $[H_2O]$ does not vary appreciably. It is therefore impossible to determine the order of reaction with respect to water, unless we use another solvent and add only the required amount of water. However, such a change of medium might change the mechanism.

25 [B] would have been 1.0 mol dm^{-3} at the start, and would drop to 0.9 mol dm^{-3}.

26a You can see immediately that these data show a constant half-life. The times for decay of the substrate concentration from 10 to 5.2 (18 min), from 7.6 to 3.7 (20 min) and from 9.0 to 4.3 (20 min) all suggest a constant half-life of around 19 minutes. The reaction must be first order with respect to the substrate concentration.

Rate constant $= \dfrac{\ln 2}{t_{\frac{1}{2}}}$ (from equation (14.9))

$= 0.037$ min^{-1}

The reason why you cannot state the rate equation is that one of the reactants (water) is in a huge excess, and so stays effectively constant. It is very difficult to find out how the rate would change if $[H_2O]$ varied.

b The highly branched substrate is likely to undergo substitution via the S_N1 mechanism (Chapter 11). The overall rate equation would be, using S to represent the bromoalkane:

$\dfrac{-d[S]}{dt} = \text{rate} = k[S]$

27a $[IO_3^-]$ has doubled.
b The initial rate has doubled.
c The reaction is first order with respect to iodate(V) ions.
d Comparing runs 3 and 4, we see the $[H^+]$ halve, and the rate drop by a factor of 4, so the reaction is second order with respect to hydrogen ions. Comparing runs 3 and 5, we see the same thing happen when $[I^-]$ is halved, so that must be a second-order relationship too.
e The overall rate equation is thus:
Rate $= k[H^+]^2[I^-]^2[IO_3^-]$
f In general, $k = \text{rate}/([H^+]^2[I^-]^2[IO_3^-])$. Taking the first set of results:

$k =$

$\dfrac{7.1 \times 10^{-9}}{[2.0 \times 10^{-3}]^2 \times [4.0 \times 10^{-4}]^2 \times [0.37 \times 10^{-4}]}$

$= 3 \times 10^8$ mol^{-4} dm^{12} s^{-1}

28 Considering runs 1 and 2, doubling $p(H_2)$ doubles the rate, so the reaction is first order with respect to hydrogen. Considering runs 4 and 5, doubling $p(NO)$ quadruples the rate, so the reaction is second order with respect to nitrogen monoxide. The overall rate equation is:
Rate $= k[NO]^2[H_2]$
Using the first set of data,
$k = \dfrac{43.4}{1580^2 \times 263}$
$= 6.6 \times 10^{-8}$ Pa^{-2} s^{-1}

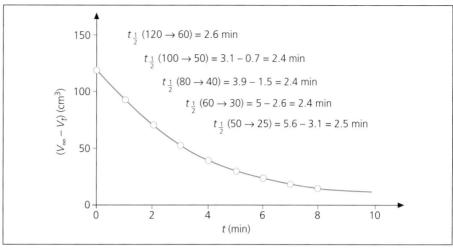

Figure A14.8 See answer 30

Table A14.6 See answer 30

Time (min)	$V_\infty - V_t$ (cm^3)
0	120
1	94
2	71
3	52
4	40
5	29
6	23
7	18
8	14
9	12
10	9

29 You would need to know the extinction coefficient ε for the absorbing species at the wavelength used, and the path length l of the cell.

30 The expression $V_\infty - V_t$ was used to generate a set of data that are proportional to the declining concentration of hydrochloric acid. Table A14.6 and Figure A14.8 show the results. The half-life on this graph is reasonably constant at about 2.5 minutes. So the rate equation is:

$\dfrac{-d[HCl]}{dt} = \text{rate} = k[HCl]$

The reaction is first order with respect to hydrochloric acid. The value of k is given by $k = \ln(2)/t_{\frac{1}{2}} = 0.69/2.5 = 0.28$ min^{-1}.

31 It is a problem because the reaction is going on all the time the reaction mixture is in the pipette, and also during the titration. This leaves an uncertainty as to what to record as the time t. Titration only works for very slow reactions, for which the degree of reaction going on within a typical titration time is negligible.

32a Quench by neutralising the catalyst. A typical quencher might be a carbonate or hydrogencarbonate.
b Dump into excess aqueous iodide ions. The manganate(VII) ions will liberate iodine, which can then be titrated with thiosulphate. Fortunately iodine does not interfere with the other reagents, and the reaction stops at once.
c Here the nucleophile *is* the thiosulphate ion. Quenching with excess iodine would carry the risk that the iodide thus formed would carry on the substitution. Sudden cooling is probably the best method.

33 Removal of samples does not affect the *concentrations* of the reactants, so the rate, which is dependent on concentrations, remains unaltered.

34 The initial total pressure $p_0 = 2$ atm, so if we use the formula $p_A = p_{tot} - p_0/2$, we just subtract 1 atm from each p_{tot} value to get p_A (Table A14.7).

Table A14.7

Time (s)	p_A (atm)
0	1.00
10	0.70
20	0.50
30	0.32
40	0.24
50	0.17
60	0.10

You can see that it takes the same time for the pressure to drop from 1.00 to 0.50 atm (20 seconds) as it does to drop from 0.5 to 0.24 atm. The constant half-life tells us that the reaction rate has a first-order dependence on p_A or $[A]$.

35a Sulphur, as a finely divided precipitate, causes the turbidity.

b Comparing runs 1 and 2, when the [HCl] is doubled, the time halves, so the rate doubles. The dependence of the rate on [HCl] is therefore first order. Comparing runs 1 and 3, when the $[S_2O_3^{2-}]$ is halved, the time doubles, so the rate has halved, so this too is a first-order relationship. The rate equation is therefore:

$$\frac{-d[S_2O_3^{2-}]}{dt} = \text{rate} = k[\text{HCl}][S_2O_3^{2-}]$$

c You could use a colorimeter instead of judging opacity by eye. (Precipitates simply absorb all wavelengths.) Some specific value of absorbance could be used as the 'event'.

36a Since the number of product gas molecules exceeds the number of reactant gas molecules, this reaction can be monitored by pressure change. Colorimetry would also work, because nitrogen dioxide is coloured (brown).
b There are no coloured species here, so pressure change looks the best option, with four reactant molecules becoming three product molecules.
c Bromine is coloured, so a colorimetric method could be used. Alternatively, portions of the reaction mixture could be quenched with an excess of iodide ions, and the liberated iodine titrated with 'thio'.
d One molecule of reactant gas is forming two molecules of product, so a pressure change method would be used.
e This is a slow reaction, so quenching is not necessary. Portions could be removed for titration with alkali, to find the newly formed carboxylic acid (having allowed for the catalyst acid). In fact this is an identical procedure to that used on this system to determine its equilibrium constant in Chapter 8, except that here the titration is being repeated many times, on a series of removed samples.
f Iodine is coloured, so a colorimeter would be used.
g The reactant is chiral, but the product is not, so the reaction could be monitored by the decline in optical rotation, as measured by a polarimeter, assuming there was only one enantiomer present in the reactant.

37a 50 minutes.
b The grime-sprayers will be finished after 10 minutes, so they will be idle for $\frac{4}{5}$ of the time.
c The overall rate is 2 extras min^{-1}.
d This is the same as the rate of the wig-fitting stage.

38a There are none. Because the slow process is now first, it cannot supply enough extras to take up the full capacity of the grime-sprayers. They are idle for $\frac{4}{5}$ of the time (as they were in Table 14.10, except that there their idle time came in one continuous period at the end).
b Again the whole process would take 50 minutes, at an overall rate of 2 extras min^{-1}, the same as in Table 14.10.

c This would not help, as they are idle for 80% of the time already.
d This would speed the process up.

39a Composite – the rate law does fit the stoichiometry, but there is a three-way collision, which is impossible.
b Composite – if the reaction were an elementary single step the rate law would be second order with respect to [A].
c Composite – there must be a second, fast, step to explain the non-appearance of [B] in the rate law.
d Possibly elementary. It could well proceed by the single-step collision of two molecules of A, which would give the observed second-order rate law.

40a This single-step mechanism would give the rate law:

Rate = $k[N_2O_5][NO]$

and so must be wrong.
b This mechanism generates the wrong products, so must be wrong. It produces N_2O_4 and NO_2, whereas the true products are $3NO_2$.
c This is right. The slow step featuring the unimolecular destruction of N_2O_5 gives rise to the correct rate law, and the two equations add up to the correct overall stoichiometric equation. One molecule of NO_2 is produced in the first step and two molecules in the second step, giving a total of $3NO_2$.
d The rate law would be right, since it shares the same rate-determining step as (c), and the equations all add up to the right overall equation. However, step 2 of the mechanism requires an impossible three-way collision.

41a $K = [X]/[CH_3COCH_3][H^+]$ where all concentrations are 'eqm'.
b Overall rate = rate of step 2 = $k[X]$.
c Rearranging the expression in (a), we get $[X] = K[CH_3COCH_3][H^+]$. Substituting this expression for [X] in the answer to (b) gives:

Overall rate = $kK[CH_3COCH_3][H^+]$
$= k'[CH_3COCH_3][H^+]$

where $k' = kK$. We see that the reaction is second order overall, and first order with respect to propanone and hydrogen ions individually.
d $[I_2]$ is missing from the rate law, because it is not involved in the mechanism until after the rate-determining step.
e The rate law is zero order with respect to iodine.
f Propanone is in such a large excess that it will effectively be at constant concentration throughout the run.
g $[H^+]$ will be constant anyway, because H^+ is a catalyst.
h Iodine will be completely consumed.
i The graph of $[I_2]$ against time would be a downward-sloping straight line.
j This run would only reveal the $[I_2]$ dependence of the rate, since nothing else was varied.

k Further runs would have to be done varying $[H^+]$ and $[CH_3COCH_3]$, and using the initial rates method. (There is a useful side-effect of the reaction being zero order with respect to iodine. This is that the initial rate *is* the rate, so the whole reaction goes at the initial rate. Practically, it means you can use the total disappearance of the iodine as the clock event.)

42 Three ClO$^-$ ions are used in two steps. So one step must consume two ClO$^-$ ions, and the other must consume one (since three-way collisions are not allowed). The rate law is second order, so the slow step must be the bimolecular one. We are ready to start proposing mechanisms:

$2\text{ClO}^- \xrightarrow{\text{slow}} \text{Cl}^- + \text{ClO}_2^-$

$\text{ClO}_2^- + \text{ClO}^- \xrightarrow{\text{fast}} \text{ClO}_3^- + \text{Cl}^-$

This adds up to the correct overall stoichiometry.

For corroborative evidence, the intermediate ClO$_2^-$ can be isolated as an almost-stable species. You could look at its UV spectrum, and see if the same absorption appears fleetingly in the course of the reaction. It would also be worth checking that ClO$_2^-$ is capable of oxidising Cl$^-$ ions.

43a The chances of meeting either nucleophile would be equal, and the ratio of products would be 1 : 1.
b If the 'quality' of the nucleophile mattered, then the two would have unequal success in carrying out the substitution, leading to unequal amounts of products.
c The gas chromatography result, with its unequal peaks, points to the S_N2 mechanism.

44a Yes, because they all have the same kinetic energy.
b Only the head-on collision would result in reaction, because it is the only one that converts the whole $2x$ to collision energy, rather than leaving some of it as kinetic energy.

45a 25 kJ.
b See Figure A14.9.
c See Figure A14.9.
d Judged by eye, the shaded areas appear to be in the ratio (almost 0) : 1 : 4.

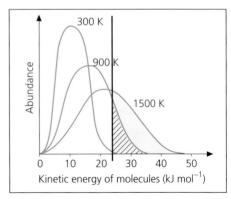

Figure A14.9

46a At 273 K, the exponential factor, that is the fraction of molecules having energy greater than E_A, is:

$e^{-50\,000/(273 \times 8.3)} = 2.6 \times 10^{-10}$

b   At 373 K, a similar calculation gives 9.7×10^{-8}.

c   The ratio of these two factors is about 373 : 1.

While the frequency of collisions (the A factor) has gone up by a factor of 1.17, the fraction of successful collisions has gone up by 373 times. So there is justification for the claim that the overwhelming influence of temperature is its effect on the violence of collisions, not their frequency.

47a In Table A14.8, the reciprocals of time ($1/t$) are presented as rates.

Table A14.8

Temperature, T (K)	Time, t (s)	$\frac{1}{t}$ ($\propto$ rate) $(10^{-2}\,s^{-1})$	$\ln\left(\frac{1}{t}\right)$ (ln (rate))	$\frac{1}{T}$ $(10^{-3}\,K^{-1})$
294	87	1.1	−4.5	3.4
303	48	2.1	−3.9	3.3
316	24	4.2	−3.2	3.2
325	16	6.3	−2.8	3.1
333	9	11.1	−2.2	3.0

b   The next column shows the ln (rate) values, calculated by $\ln (1/t)$.

c   The next column shows values of $1/T$.

d   See Figure A14.10.

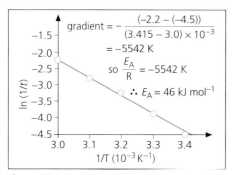

$\text{gradient} = -\dfrac{(-2.2 - (-4.5))}{(3.415 - 3.0) \times 10^{-3}}$

$= -5542\ K$

so $\dfrac{E_A}{R} = -5542\ K$

$\therefore E_A = 46\ kJ\ mol^{-1}$

Figure A14.10

e   The linearity of the graph does support equation (*14.25*).

f   The gradient is $-E_A/R$.

g   See Figure A14.10. $E_A = 46\ kJ\ mol^{-1}$ (to 2 sf).

48a The activation enthalpy of the reverse reaction is the vertical distance from the products to the transition state, which is:

$\Delta H_{sys} + E_A$

b $\Delta H_{sys} = E_A(\text{forward}) - E_A(\text{reverse})$

49   The creation of both chiral isomers of the alcohol suggests that the flat planar carbocation existed for long enough for the leaving group to have lost all influence, so that there was an equal chance of the incoming group attacking from either side.

50a Catalysts have no influence on the thermodynamic parameters in a reaction profile. The enthalpies and free energies of reactants and products stay the same. So K, and with it the composition of the equilibrium mixture, will be unchanged.

b   Both $E_A(\text{forward})$ and $E_A(\text{reverse})$ will be reduced, so the equilibrium mixture will be attained more quickly in the presence of the catalyst.

51a With a constant flow process, the product can be removed on each circuit. Any unreacted reactants go back for another pass through the apparatus. Whether or not the system has reached equilibrium is academic.

b   The higher-temperature option looks better, because a respectable yield is achieved in a reasonable time.

c   See Figure A14.11.

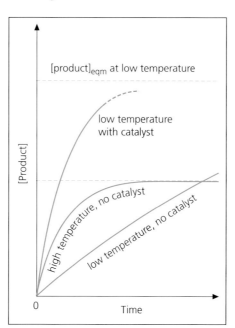

Figure A14.11 Showing how a catalyst can allow access to the higher yields available in exothermic systems at lower temperatures – *and* do it in a reasonable timespan

52a

profile (a)

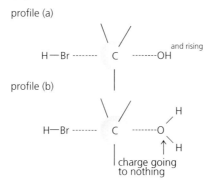

profile (b)

b   The bond to be stretched to breaking point in profile (b) will be weaker than that in profile (a). This is because the oxygen atom in (b) will 'pull harder' on the electrons, because it has acquired an H^+. From another point of view, more (undesirable) charges will be built up in getting to the (a) transition state than the (b).

c   Rate $= k[ROH][HBr]$.

d   The second step is obviously rate determining, because it has the higher activation enthalpy.

53a The nitrogen monoxide, NO, plays the part of the empty van.

b   The nitrogen dioxide, NO_2, plays the part of the full van.

c   The sulphur dioxide, SO_2, plays the part of the empty shop.

d   The sulphur trioxide, SO_3, plays the part of the full shop.

54a All four oxygen atoms in the two moles of hydrogen peroxide are in oxidation state −1. Two oxygen atoms go from −1 to 0 (in O_2), while the other two go from −1 to −2 (in H_2O).

b   The oxidation number of the iodine atoms oscillates between −1 (in I^-) and +1 (in IO^-).

c

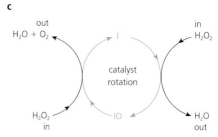

d   Step 1 would be rate determining, since both reactants in step 1 appear in the rate law. The reaction profile would have been as in Figure A14.12.

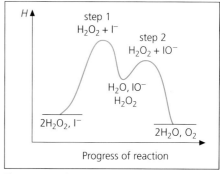

Figure A14.12

55   If a catalyst binds its substrate too tightly, then there may be trouble freeing the active site of products, in order to bind the next reactant molecule. In reaction-profile terms (see Figure 14.37) this would mean an oversized third 'hump'. If the catalyst binds too loosely, there will not be sufficient bond disturbance in the substrate, and the activation enthalpy of the main, middle step (see Figure 14.37) will be too high.

56a Halving or doubling the number of customers coming in would halve or double the rate of serving respectively.
b This is a first-order relationship between rate and customer concentration.
c If all the staff are working at saturation, then the number of customers in the queue will have no effect. The overall rate will depend only on how fast the staff can process each customer. This is a zero-order relationship between rate and customer concentration.

57 In the first-order situation, the rate is determined by the frequency of collisions between substrate and enzyme. In the zero-order case, the rate is determined by how fast the enzyme can carry out its operation on the substrate.

58a A constant half-life implies a first-order rate law for the decomposition of the drug, because constant half-lives are unique to first-order reactions.
b Concentration = $\dfrac{\text{mass (mg)}}{\text{volume (dm}^3)}$ (mg dm^{-3})

So mass = concentration × volume
= 16 × 40 = 640 mg.

c Let the patient take two 320 mg tablets to start with. This will give the required initial concentration of 16 mg dm^{-3} in the body. If the half-life of the drug is 12 hours, then after this time the concentration will have dropped to 8 mg dm^{-3}, which is at the bottom end of the therapeutic range. If another single tablet is now taken, this will add another 8 mg dm^{-3} to the drug concentration, and bring it back up to the starting condition. This pattern can be repeated, with a single tablet being administered every 12 hours. The graph of concentration against time would be as shown in Figure A14.13.

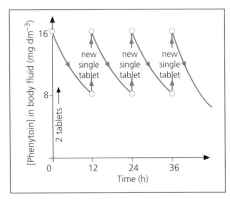

Figure A14.13

59a A pint of beer contains 16 g of ethanol.
i The ethanol concentration in the woman's body after one pint would be
(16 × 10^3)/(25 × 10) = 64 mg 100 cm^{-3}.
ii The equivalent figure for the man is
(16 × 10^3/(40 × 10) = 40 mg 100 cm^{-3}.
b For the woman, another pint of beer, drunk immediately, would take her blood ethanol

concentration to 128 mg 100 cm^{-3}, well over the limit. The man would be exactly on the limit if he drank two pints straight off.
c Five units would give our man 2.5 times the concentration of ethanol he had in (a), equivalent to a blood ethanol concentration of 100 mg 100 cm^{-3}. If he had ingested all 2½ pints in one go, he would be 20 mg 100 cm^{-3} over the limit.
d If the five units had been drunk over a longer period of time, then the first portions of ethanol would already have begun to be decomposed by the body's enzymes. This would probably have reduced the ethanol concentration to below the crucial 80 mg 100 cm^{-3} level. However, this depends upon the man having a 'normal' build and a 'normal' rate of decomposition of ethanol. A small individual, or one with a slow rate of ethanol decomposition, could well be over the limit after five units.
e The blood ethanol levels in men and women are in the ratio 25: 40, given the average body distribution volume figures, so five units for a man would be equivalent to 5 × 25/40, or about three units for a woman.
f The constant rate means that the decomposition of ethanol is a zero-order process, so it would seem that the enzymes in the liver that decompose it are working at saturation all the time, like a busy post office.
g If you compare ethanol and phenytoin on the same mg 100 cm^{-3} scale, the phenytoin therapeutic range is from 1.6 to 0.8 mg 100 cm^{-3}, compared with tens or even hundreds of mg 100 cm^{-3} for ethanol. This factor alone could be the reason why the enzymes that break down ethanol are fully saturated, and therefore show zero-order kinetics.
h Women get rid of ethanol at 5.3 g h^{-1}. Since a unit is 8 g, it would take a woman 1.5 hours to rid her body of one unit. The man's liver processes ethanol at 7.3 g h^{-1}, so it takes 1.1 hours to deal with one unit. The advice is aimed at men.

60a Each glass of wine increases her blood ethanol concentration by (8 × 10^3)/(25 × 10) = 32 mg 100 cm^{-3}.
b The overall mass of ethanol goes down by 5.3 g each hour, so the concentration goes down by (5.3 × 10^3)/(25 × 10) = 21.2 mg 100 cm^{-3} h^{-1}.
c The midnight blood ethanol level will be (7 × 32 − 3 × 21.2) = 160.4 mg 100 cm^{-3}.
d This is 80 mg/100 cm^3 above the legal limit.
e If she breaks down 21.2 mg 100 cm^{-3} per hour, she should wait at least 80/21.2 = 3.8 hours.

CHAPTER 15   page 352

1 We conclude that ammonia is a more effective donor of a lone pair than water.

2 Aminoethane would remove hydrogen ions from water as follows:

$CH_3CH_2NH_2$(aq) + H_2O(l) $\rightleftharpoons$
$CH_3CH_2NH_3^+$(aq) + HO^-(aq)

3 There are of course many more free hydrogen ions in acids than in water. So the —NH_2 group on any amine, irrespective of chain length, would accept a hydrogen ion and become an alkylammonium group, —NH_3^+. (The longer-chain amines result in ions similar to those in detergent molecules, with at least a hydrophilic end, and this facilitates their solubility in water, Figure A15.1.)

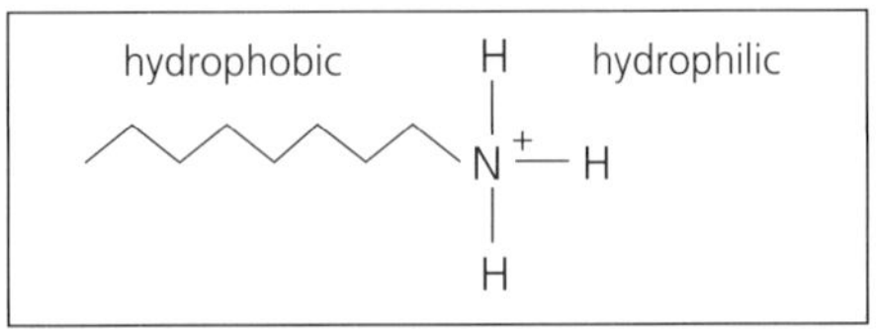

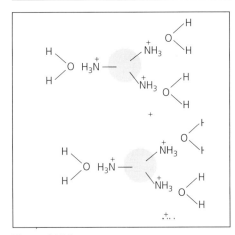

Figure A15.1 Detergent character shown by long-chain alkylammonium ion

Figure A15.2

4 Figure A15.2 shows a cationic detergent acting to emulsify some oil droplets. As we shall see, it is sometimes useful to have a cationic detergent rather than an anionic one, as for instance when washing hair (see the panel on hair conditioners in Section 15.9).

5a True – acids would protonate the NH group:

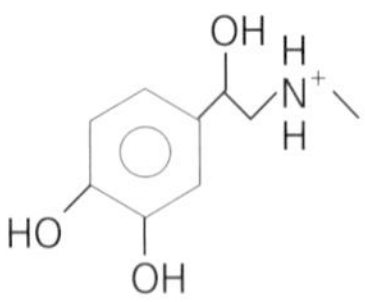

b True – alkalis could abstract a hydrogen ion from the phenolic OH groups, and again the ionic site would boost solubility:

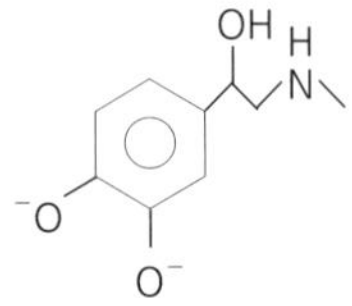

c False – the OH group is in a secondary position, so it would oxidise to a ketone:

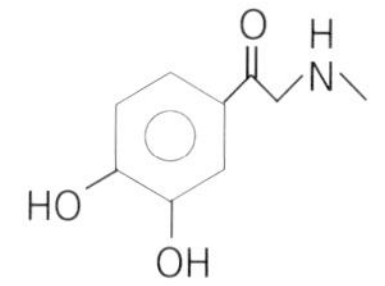

d   True – the carbon atom with the asterisk has four different tetrahedrally arranged groups, and so is chiral:

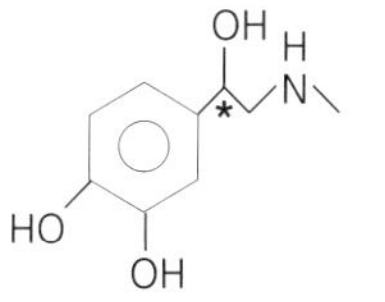

6   The extra activity of the ring in phenylamine depends entirely on the lone pair of electrons on the nitrogen atom being partially delocalised into the ring's π-electron system, making the ring more susceptible to electrophilic attack. If that pair of electrons becomes occupied holding onto a hydrogen ion, the pair of electrons is no longer delocalised into the ring, and the reactivity of the ring towards electrophiles declines accordingly.

7a   The ethylamine can act as a nucleophile in its own right, as follows:

$CH_3CH_2NH_2 + CH_3CH_2Br \rightarrow (CH_3CH_2)_2NH + HBr$

The synthesis is now at the stage of a secondary amine, diethylamine, and that too can attack more bromoethane. The product mix will acquire two more ingredients, triethylamine, $(CH_3CH_2)_3N$, and the **quaternary alkylammonium salt**, $(CH_3CH_2)_4N^+Br^-$. More will be said about these salts in Section 15.9, where they appear as hair cosmetics.

b   To maximise the production of the primary amine, you would make it difficult for the original ammonia molecules to find more than one ethyl group each. In other words, you would start with the ammonia in excess.

8   The reactivity of acyl chlorides is due to the presence of the C=O group. This causes extra positive polarity in the carbonyl carbon atom, which becomes an attractive site for nucleophiles.

9   The product would be a molecule with an amino group adjacent to the carbonyl group. The generic name for these compounds is amides.

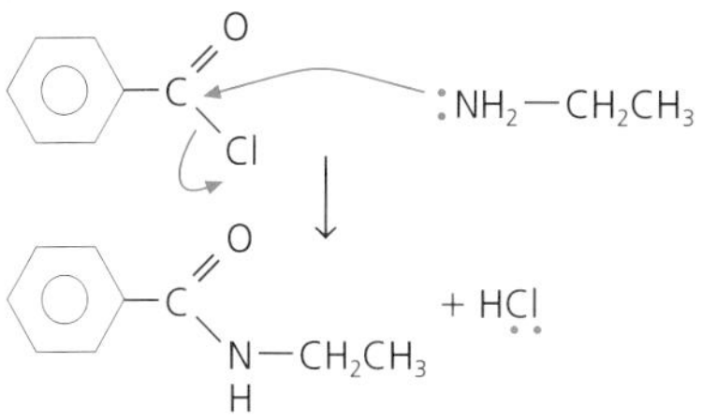

10

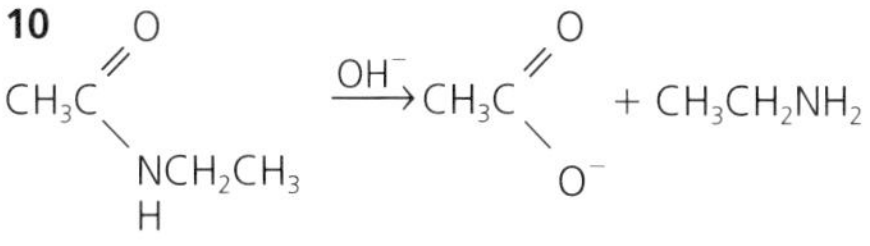

The alkaline conditions mean that the carboxylic acid appears in the form of its anion.

11   Paracetamol could be made by the reaction of 4-aminophenol with ethanoyl chloride:

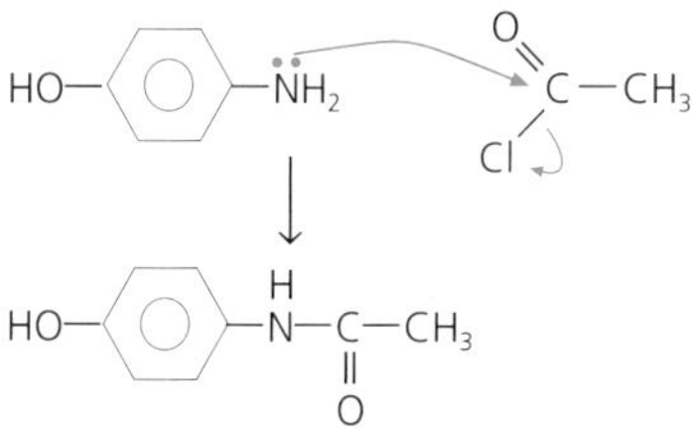

paracetamol

12a   The name comes from the fact that there are six carbon atoms from N to N, and then six carbon atoms from C=O to C=O (including the carbonyl carbons).

b   Condensation (since water is eliminated at each join), thermoplastic (because there is no cross-linking, other than by intermolecular forces, between chains), and melt (because the chains would become free to move relative to each other).

c   The shape of the molecule is linear.

d   Nylon 66 has an extra hydrogen atom on each nitrogen. There is no equivalent in the polyester, so there is extra inter-chain hydrogen bonding in the nylon.

e   The ring 'springs open', revealing an amine and a carbonyl group, which then perform the polymerisation. The reaction is actually initiated by HO^- ions:

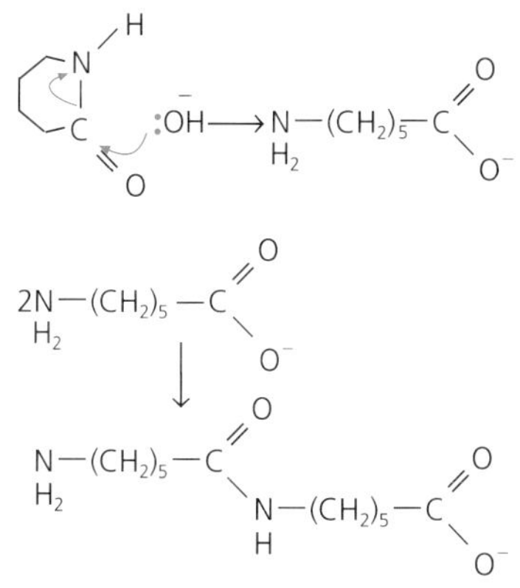

and so on

All the amide linkages in nylon 6 point the same way, in contrast to nylon 66.

13   The hydrocarbon sections of the nylon chain represent a higher proportion of the total length than is the case in Terylene, so proportionately less of the molecule is available for water-absorption by hydrogen bonding.

14   Those with a neutral R group (side-chain) will be overall largely neutral – alanine and valine.

Those with an extra acidic group on the side-chain will be overall acidic – (b), which is called

---

aspartic acid, and tyrosine (by virtue of the slight acidity of the phenolic OH group).

Those with an extra basic group on the side-chain will be overall basic – lysine.

15a   The zwitterion is not charged, so would not be attracted to either electrode.

b   The movement of the majority $H_2NCH_2CO_2^-$ ion will in effect remove a product molecule from an equilibrium system. So more zwitterions will have to shed hydrogen ions to compensate (Figure A15.3). This is an electrically prompted illustration of Le Chatelier's principle at work.

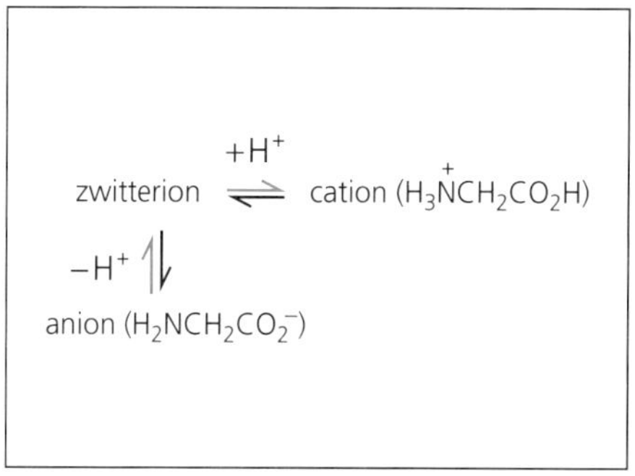

$$zwitterion \underset{-H^+}{\overset{+H^+}{\rightleftharpoons}} \text{cation } (H_3\overset{+}{N}CH_2CO_2H)$$

$$-H^+ \updownarrow +H^+$$

$$\text{anion } (H_2NCH_2CO_2^-)$$

Figure A15.3 The equilibrium is dragged in the green direction by removal of anions

16   If we want more of the protonated species and less of the deprotonated one, then we need to add more protons. The pH would therefore have to come down.

17   To generalise about the contrast between nylons and proteins, the former is mainly main chain, whereas the latter is minimum main chain with lots of side-chains. To be precise, there is only a single carbon atom in the chain between the amide linkages in a protein, but either six or four carbon atoms in nylon 66. Also, the amide groups in proteins all face the same way, whereas in nylon 66 they face alternately:

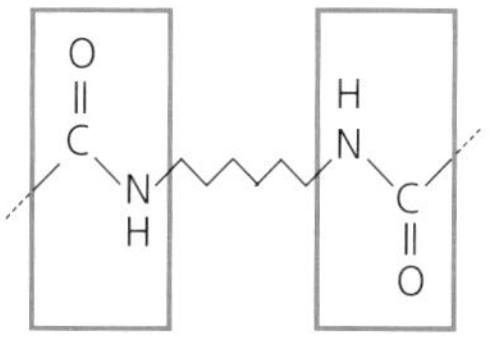

nylon

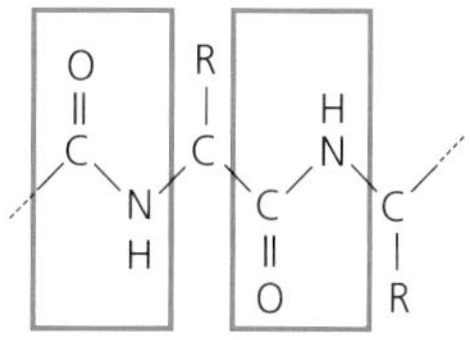

protein

18a   The R_f value is the ratio of the distance moved by the spot (on top in the fraction) to the total distance moved by the solvent front (underneath). Its maximum value is therefore 1.

b   The developer is ninhydrin, which produces a purple coloration with most amino acids.

c The cellulose molecules in paper contain lots of OH groups, which are potential sites for either hydrogen bonding or dipole attractions. An amino acid with, say, an OH group in its own side-chain will be able to make intermolecular bonds at these sites, and so will be less mobile across the paper. Conversely an amino acid with a hydrocarbon side-chain will not be so delayed.

d We would expect valine, with its hydrocarbon R group, to move freely, while aspartic acid, with both OH and C=O, should be firmly held on the stationary phase.

19a The butan-1-ol–ethanoic acid mixture has a higher proportion of polar or hydrogen-bonding groups, compared with phenol. If both solvents had similar polarities, the spots that stayed together on the first run would also stay together on the second.

b From the results of the two-dimensional chromatography, we still do not know how many of each residue are present in the protein, nor in what order they appear.

20 Polypeptides have such great potential variety, because of their use of 20 different possible side-chains, in any order.

21a

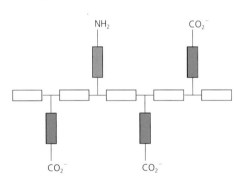

(Note: The amino acids are not shown in zwitterion form.)

22a The removal of arg after one round of degradation shows that arg must have been on the amino end. So the original tripeptide might have been arg-ala-asp or arg-asp-ala.

b The total structure could be obtained by doing another round of Edman degradation on the dipeptide.

23 Chymotrypsin would cleave the polypeptide into thr-trp, val-lys-ala-ala-trp and gly-lys. The largest fragment could be matched up to the two fragments from the other experiment, because it contains both sides of the original break site:

big fragment from experiment 2

val—lys┼ala—ala—trp

thr—trp—val—lys┼ala—ala—trp—gly—lys

break site, experiment 1

24 The C—N bond length is less, at 0.132 nm, than the average value for a C—N single bond, which is 0.147 nm. A short bond is an indication of multiple-bond character.

25 The R groups are all on the outside.

26 An X-ray diffraction machine would have been used. Proteins are not exactly crystalline, as a rule, but they do have repeating structural patterns, and can be studied using X-ray diffraction.

27 Hair becomes anionic in alkaline solution:

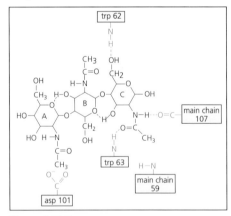

28a It would be safe from attack, because there is no hydrogen ion to be removed by the base.

b As we saw in Chapter 11, the C—Br bond is weaker than the C—Cl one, so bromine makes a better leaving group.

c

CH_3—$(CH_2)_{15}$

CH_3—$N:$ CH_3—Br

CH_3

↓

CH_3—$(CH_2)_{15}$ CH_3

N $:Br^-$

H_3C CH_3

d See Figure A15.4.

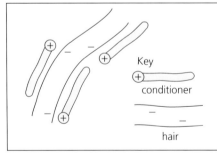

Figure A15.4 Showing how conditioners add body to hair

e The effect of applying conditioner is to coat the hair with hydrocarbon side-chains. These side-chains would not have especially strong intermolecular bonds to other chains on adjacent hairs, so the hairs would run freely over each other.

29 If one chain runs in the order —N—C(R)—C(O)— from left to right, then the chains either side of it run —(O)C—(R)C—N—.

30 They could have used water enriched with labelled ^{18}O atoms, and then detected where the label ended up, by mass spectrometry.

31a C—O and O—H break; C—O and O—H make, so the enthalpy change would be approximately zero.

b The chaos factor change ΔS_{sys} would be positive, since chains are being fragmented, so, with a zero enthalpy change, the overall ΔS_{uni} should be positive too.

c When thermodynamically viable reactions do not happen, the reason is that the activation enthalpy must be too high.

32 Reason 1 – a more stable conformation in an aqueous medium would be one that contrived to present more of the hydrophilic R groups on the outside of the tertiary structure, thus helping intermolecular bonding with water.

Reason 2 – stability might also derive from having compatible R groups meeting in the middle of the structure, again gaining the benefit of extra bonding. This bonding would be *intra*molecular, between R groups on the same molecule.

33 There would be hydrogen bonds as shown in Figure A15.5.

Figure A15.5

34a The message to be drawn is that asp must be more acidic than glu.

b Below pH 3, carboxylate sites on both asp and glu would be protonated. Above pH 8, they would both be deprotonated (anionic). At pH 5 there is the combination that works catalytically – namely the proton on the glu but off the asp.

35a A hot lysozyme molecule would not be able to maintain the proper shape of its cleft. The relatively weak hydrogen bonds, which hold together the secondary and tertiary aspects of the structure, would give way to thermal agitations, and the precise shape of the molecule would be lost.

b   Just as in conventional reactions, enzymes and substrates must meet, and they do so by diffusion. This process will clearly be slower at lower temperatures, when molecular movements are slower.

36   With the great variety of protein structures, active sites exist in great variety. So there is a whole range of enzymes with different active sites, differing both in shape and in the positioning of catalytic groups. Shifting our focus to the substrate, it too has a unique pattern of certain groups in certain positions. The enzyme–substrate complex can only be created if the pattern of group-positions on the substrate matches those on the enzyme. These perfect matches have come about through evolution, and the result is that for every substrate there is just one perfect enzyme.

37a   Elastase – the cleft is crowded, and has hydrocarbon groups for van der Waals' bonding

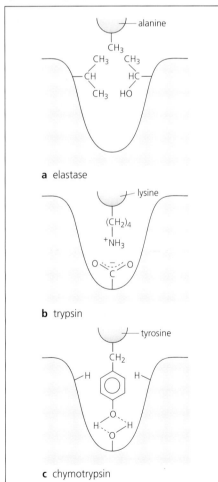

a elastase

b trypsin

c chymotrypsin

Figure A15.6

with the substrate. The amino acid in question could be alanine (Figure A15.6a)

b   Chymotrypsin – the cleft is fairly empty, and has an OH group at the bottom of the cleft for hydrogen bonding to the phenol group on the substrate. The amino acid in question could be tyrosine (Figure A15.6b)

c   Trypsin – the cleft is fairly empty, and has an anionic group at the bottom of the cleft for ionic bonding with the cationic group on the substrate. The amino acid in question could be a protonated lysine (Figure A15.6c)

38   If the trisaccharide were just the A-B-C portion of the substrate shown in Figure 15.39, it would form a substrate with all the favourable bonding aspects, but without the unfavourable reaction associated with the D. (It would have all the 'comfort' and none of the 'strain'.) So it would stay in the active site and render it useless.

39   If the enzyme–substrate and enzyme–inhibitor reactions are both equilibria, then there will always be some unbound enzyme sites. With a large number of substrate molecules, unbound sites are more likely to bind substrate than inhibitor molecules. Figure A15.7 shows that this is an example of Le Chatelier's principle at work, with excess substrate being the perturbing factor in the system.

Figure A15.7 Extra substrate S shifts the equilibrium to the right

40a   The corresponding graph for a first-order reaction would be a straight line through the origin.

b   The initial sloping part is the first-order region, and the plateau is the zero-order region.

c   At high [S] the enzymes will be **saturated** – all their active sites will be bound, and the rate will be unaffected by any more or less substrate. (We are back in the busy post office, Chapter 14.)

d   More enzymes would lift the plateau of V_{max}. (This would be like the post office opening more windows.)

41   If it takes a high concentration of substrate molecules before the graph starts to level off, this suggests that the enzyme is difficult to saturate, which in turn points to a very fast throughput time.

42   The bonding electrons are delocalised:

43   A degree of double bonding in the C—N bond should shorten it.

44   The 'carriers' could be, respectively, an alcohol (ROH) or sodium alkoxide (NaOR), hydrogen sulphide (H_2S) or sodium hydrogensulphide (NaHS), and a thioalcohol (RSH) or sodium alkylsulphide (NaSR).

45a   In alkaline conditions, phenols become phenoxide ions:

b   This will put extra electron density into the ring, and thus activate it towards reaction with the diazonium ion.

c   Too high a concentration of OH^- ions would result in the decomposition of the diazonium ion to give phenol and liberate nitrogen.

46   An unknown amine could be tested with a cold solution of acidified sodium nitrite. Then a reactive ring like phenol could be added. If the amine was aromatic, the azo dye would form, as a visual confirmation. The aliphatic amine would have decomposed to an alcohol, and no dye would form.

47   The compound will be seen to be the colours that are not absorbed, i.e. blue/green.

48   The reaction system would go colourless, since the N=N bridge would be reduced to a single bonded NH—NH group, and the two delocalised systems at either end of the molecule would be isolated from each other.

49a   Solid solution – a dispersion of one chemical species (atom, molecule or ion) within another, in the solid state. No specific style of bonding is implied by the use of the term, except that it would not apply to molecules which were ionically or covalently bonded into the structure.

b   Hydrophilic – able to bond intermolecularly, by means of hydrogen bonds or ion–dipole attractions, with water molecules.

c   Hydrophobic – the reverse of hydrophilic, with the extra implication of ability to bond intermolecularly, by means of van der Waals' forces, with non-polar molecules.

d   Chemically reduced – having undergone a reduction in oxidation number.

50   See Table A15.1.

Table A15.1

Fibre type	Type of dye	Which interaction?	Fastness (v. good, good, poor)	Other comments
Cotton	(a) Direct	Physical adsorption	Poor	
	(b) Vat	Insoluble aggregates	Good	Poor colour range
	(c) Reactive	Covalent bonds	V. good	Good colour range
Polyester	Disperse	Solid solution	Good	

51a Covalent – reactive dyes/cotton.

b Ionic – acid dyes/nylon or wool; basic dyes/acrylics.

c Intermolecular – direct dyes/cotton; disperse dyes/polyester.

d Force 1 almost non-existent – disperse dyes/water.

e Changed after delivery – vat dyes.

f With vat dyes, force 1 is high when they are in their reduced, hydrophilic form, and weak when they are in their oxidised, insoluble form.

52 See Table A15.2.

Table A15.2

Change	Reagents/conditions
—OH to —Cl	PCl_5, HCl(g), $SOCl_2$
—OH to —Br or —I	P(red)/Br_2 or I_2, KBr or KI/sulphuric or phosphoric acid, HBr(g) or HI(g)
—OH to —NH_2	NH_3/ethanol
—CH_2OH to —CO_2H (via aldehyde)	Acidified dichromate (depends on concentration)
—CHOH(sec) to ketone	$Cr_2O_7^{2-}$/H^+, distil off product as soon as formed
—CO_2H to —CH_2OH	$NaBH_4$ or $LiAlH_4$, in ether
Ester to —CH_2OH	$NaBH_4$ or $LiAlH_4$, in ether
Acid or alcohol to ester	Acid + alcohol, reflux with acid
Ester to alcohol and acid	Reflux with NaOH(aq)
Acid to acyl halide	PCl_5 or $SOCl_2$
Acyl halide to acid	Just add water (carefully)
Acyl halide to ester	Add alcohol, reacts in cold
Acyl halide to amide	Add NH_3 or amine, reacts in cold
Amide to acid and amine	Reflux with NaOH(aq), as with ester
Alcohol to alkene	Pass over concentrated H_2SO_4 or Al_2O_3
Alkene to alcohol	Boil with dilute H^+(aq)
Alkene to halogenoalkane	Hydrogen halide, gas or in alcoholic solvent
Halogenoalkane to alkene	KOH in alcoholic solvent
Halogenoalkane to alcohol	KOH in aqueous/alcoholic mixed solvent
Halogenoalkane to amine	NH_3 in alcohol
Benzene to nitrobenzene	Mixed concentrated sulphuric/nitric acids
Benzene to halogenobenzene	Dihalogen in presence of $FeCl_3$
Benzene to alkylbenzene	Halogenoalkane in presence of $AlCl_3$
Benzene to benzenesulphonic acid	Concentrated sulphuric acid, reflux

53 Step 1: ethanoic acid → ethanol, using $NaBH_4$ as in Table A15.2.

Step 2: ethanol → chloroethane, using, say, HCl(g).

54 Step 1: ethyl ethanoate → ethanol, using $NaBH_4$ as in Table A15.2.

Step 2: ethanol → ethene, using Al_2O_3, as in Table A15.2.

55

56a Bromobutane, butan-1-ol, potassium bromide, sulphuric acid, hydrogen bromide, bromine, sulphur dioxide, water.

b Mainly bromobutane, butan-1-ol and bromine.

c The ether extraction has eliminated most of the inorganic reagents. The alcohol could then be extracted by a wash with concentrated hydrochloric acid (in which it is soluble), and the bromine would disproportionate with a sodium hydrogencarbonate wash (which you would do anyway to remove hydrogen chloride and hydrogen bromide residues). Then dry using sodium sulphate and distil.

57 The volatility of an extraction solvent makes it easy to remove, and its immiscibility with water enables it to take its target solute away from the aqueous medium of the reaction mixture.

58 Sodium hydrogencarbonate is used to remove acids. It is used in preference to sodium hydroxide because the latter, a vigorous nucleophilic reagent, might bring about new reactions.

59 Ethanol could not be solvent-extracted by ether, because it is too soluble in water.

60 You can find out the density of your target layer and compare it with that of water. An ether layer will always be less dense than an aqueous layer, for example, so the lower layer would be discarded.

61 Chromatography provides a good test of purity. A pure product will give a one-peak chromatogram.

62 The product being purified should not be too soluble in the recrystallisation solvent, otherwise it will be difficult to crystallise it out. The solubility also has to be highly temperature dependent, so you want a solvent in which the solubility when hot will be boosted by the $-T\Delta S$ term, but which perhaps has an endothermic solution reaction, giving low cold solubility.

63 The final wash is to remove any solution that may be clinging to the crystals, containing dissolved impurities.

64 The presence of impurities lowers melting points, relative to that of the pure compound.

65 See Table A15.3.

66a Suggests arene.

b Probably has regions of both high and low polarity within the molecule.

c Carboxylic acid.

d Benzoic acid:

67a Not an arene, probably quite a small molecule.

Table A15.3

Clue	Inference
Decolorises bromine (two possible answers)	Alkene/phenol
Decolorises MnO_4^-/H^+	Alkene/alcohol/aldehyde
Oxidised by 'mild' $Cr_2O_7^{2-}$/H^+ but product is neither acidic nor gives positive Fehling's test	Secondary alcohol
Oxidised by Fehling's solution	Aldehyde
Gives a yellow precipitate with Brady's reagent	Carbonyl compound
Hydrolysed in acidic or alkaline conditions into two organic fragments (two possible answers)	Ester/amide
Light needed to initiate reaction	Free radical mechanism
Reacts with PCl_5, HCl given off (several possible answers)	Anything containing an OH group
Reacts with H_2/Ni catalyst (three possible answers)	Alkene/aldehyde/ketone
Acidic enough to liberate CO_2 from carbonates/hydrogencarbonates	Carboxylic acid
Weakly basic, dissolves in HCl but not in water	Longer-chain amine/amide
Gives ammonia when heated with NaOH	Primary amide
Faintly acidic but not enough to give CO_2 with carbonates/hydrogencarbonates	Phenol
Not acidic but reacts with Na(s)	Alcohol
Gives quick precipitate with $AgNO_3$(aq)	Acyl halide
Gives slower precipitate with $AgNO_3$(aq)	Halogenoalkane
Burns with a smoky flame	Probably arene
Dehydrates to alkene	Alcohol
Reacts with ethanolic KOH to give alkene	Halogenoalkane
Reacts with acyl halide to give pleasant-smelling liquid	Alcohol
Gives purple colour with $FeCl_3$	Phenol
Gives strong sharp IR band between 1680 and 1750 cm^{-1}	Aldehyde, ketone, carboxylic acid, acyl halide or ester (i.e. compound containing C=O bond)
Gives very broad IR band from 3200 to 3700 cm^{-1}	Alcohol or carboxylic acid (i.e. compound containing O—H bond)

b Suggests one of the short-chain alcohols, amines or carbonyl compounds.

c Primary or secondary alcohol, or aldehyde.

d Propan-2-ol:

CHAPTER 16   page 389

1a Electron-giving: $Mg(s) → Mg^{2+}(aq) + 2e^-$
Electron-taking: $2H^+(aq) + 2e^- → H_2(g)$

b Electron-giving: $2I^-(aq) → I_2(aq) + 2e^-$
Electron-taking: $2Fe^{3+}(aq) + 2e^- → 2Fe^{2+}(aq)$

c Electron-giving: $5Fe^{2+}(aq) → 5Fe^{3+}(aq) + 5e^-$
Electron-taking: $MnO_4^-(aq) + 8H^+(aq) + 5e^- → Mn^{2+}(aq) + 4H_2O(l)$

2a Electrical energy is being converted to chemical energy.

b Chemical energy is being converted to electrical energy.

3a Water molecules would be attracted by Zn^{2+}, a cation of quite considerable charge density, in an ion–dipole interaction.

b No – hexane could not solvate the ions, due to its lack of a dipole. The electrolyte in a cell must allow ionic conduction, and with no dissolved ions this would be impossible.

4   The cell should be able to work with a platinum electrode, if copper crystals can begin to grow on the platinum surface.

5a   The arrival of the Zn^{2+} ions would cause a surplus of cations.
b   Anions (NO_3^-) would have to come in from the salt bridge, perhaps in concert with the departure of some Zn^{2+} ions onto the bridge.
c   In the other half-cell, there will be a surplus of anions.
d   Cations (K^+) will come off the bridge, perhaps with SO_4^{2-} ions going onto it.

6   The reaction can be seen equally as a 'victory' for Cu^{2+} over Zn^{2+} in the contest to *gain* electrons.

7a

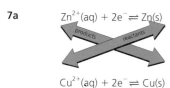

$$Zn^{2+}(aq) + 2e^- \rightleftharpoons Zn(s)$$

$$Cu^{2+}(aq) + 2e^- \rightleftharpoons Cu(s)$$

b   The zinc half-equation is being driven backwards.
c $Cu^{2+}(aq)$ is the oxidising agent.
d $Zn(s)$ is the reducing agent.
e $Cu^{2+}(aq)$ is the stronger oxidant.
f $Zn(s)$ is the stronger reducing agent.

8a   The two reactants are $Ag^+(aq)$ and $Cu(s)$.
b   The two products are $Cu^{2+}(aq)$ and $Ag(s)$.
c   The dominant oxidant is $Ag^+(aq)$ and the dominant reductant is $Cu(s)$.
d   The copper half-equation is going in the other direction – in the zinc case the copper was reduced to the metal from the cation.
e   Silver metal would appear on the surface of the copper (under a microscope, it looks like a snow-covered coniferous forest, see Figure 16.32).

$$2Ag^+(aq) + Cu(s) \rightarrow Cu^{2+}(aq) + 2Ag(s)$$

f,g   See Figure A16.1.

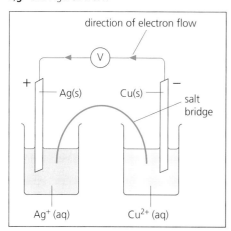

Figure A16.1 The reagents and polarities of a copper/silver cell

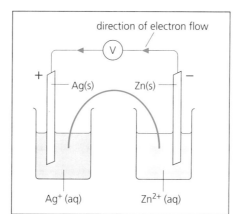

Figure A16.2 Note that the direction of electron flow is opposite to that of conventional current

9   See Figure A16.2. The equation for the cell reaction is:

$$2Ag^+(aq) + Zn(s) \rightarrow Zn^{2+}(aq) + 2Ag(s)$$

10   Electrons will stop flowing, just as chemical reactions will stop happening, when one of the reactants is exhausted.

11a   The cell that gave rise to the upper graph has the better design. The shallower gradient of this graph indicates that less potential difference is being lost within the cell, and that therefore the cell has a lower internal resistance.
b   The e.m.f. depends only on the chemicals and their concentrations, and not on design parameters, so the identical e.m.f.s indicate (without perhaps being a total proof) that the chemicals are identical.

12a   When one mole of zinc reduces one mole of Cu^{2+} ions, there is a movement of two moles of electrons.
b   The coulomb equivalent is $2 \times 96\ 500$, or $193\ 000$ C.
c   Energy   = potential difference × charge
   (J)   (V)   (C)
      $= 1.1 \times 2 \times 96\ 500$
      $= 212.3$ kJ
         per mole of cell reaction

13a   Enthalpy changes are unreliable predictors of reaction direction. Endothermic reactions *do* happen.
b   Free energy changes are reliable. The sign and size of the ΔG_{sys} value relates directly to the value of the equilibrium constant K. Putting it crudely, reactions with negative ΔG_{sys} values are 'goers'.
c   The sign of the cell e.m.f. has the same reliable predictive power.
d   So E_{cell} seems more closely allied to ΔG_{sys}.

14 $E_{cell} = 1.1 + 0.46 = +1.56$ V

15 $Zn(s)|Zn^{2+}(aq) \vdots Ag^+(aq)|Ag(s)$
$E_{cell} = E_r - E_l = +0.80 - (-0.76) = +1.56$ V
$Ag(s)|Ag^+(aq) \vdots Zn^{2+}(aq)|Zn(s)$
$E_{cell} = E_r - E_l = +(-0.76) - (0.80) = -1.56$ V

16a   By the Law of anti-clockwise circles, only metals that lie above the hydrogen half-cell can liberate hydrogen gas from acids. The list includes iron, zinc and magnesium.
b   The largest e.m.f. would be obtainable (in theory) from:
$$Na(s)|Na^+(aq) \vdots Ag^+(aq)|Ag(s)$$
$$E_{cell} = E_r - E_l = +0.80 - (-2.71) = +3.51\ V$$
c   The sodium metal would react with water.

17a   See Figure A16.3.

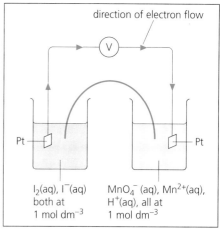

Figure A16.3

b $Pt|2I^-(aq),I_2(aq) \vdots [MnO_4^-(aq) + 8H^+(aq)], Mn^{2+}(aq)|Pt$
$E_{cell} = E_r - E_l = +1.51 - (+0.54) = +0.97\ V$
c $MnO_4^-(aq) + 8H^+(aq) + 5I^-(aq) \rightarrow Mn^{2+}(aq) + \frac{5}{2}I_2(aq) + 4H_2O(l)$
$\Delta G_{sys} = -zFE_{cell} = -5 \times 96\ 500 \times 0.97$
$\qquad\quad = -468\ kJ\ mol^{-1}$
$\Delta G_{sys} = -RT\ln K_c$
so:
$$\ln K_c = -\left(\frac{-468\ 000}{8.3 \times 298}\right)$$
so:
$$K_c = 1.5 \times 10^{82}$$
So it would be safe to say that this reaction goes to completion.

18a   Each half-cell would occupy $25\ cm^3$.
b   Moles of Cu^{2+} = concentration × volume in $dm^3 = 1 \times 0.025 = 0.025$ moles.
c   Mass of zinc = RAM × moles = $65 \times 0.025 = 1.625$ g.
d   If we assume densities of $1\ g\ cm^{-3}$ for the solutions, and if we assume that the copper electrode has a similar mass to the zinc one, the total mass of the cell will be in the region of $25 + 25 + 1.6 + 1.6 + $ mass of case = 53.2 g plus the mass of the case.
e $E_{cell} = E_r - E_l = +0.34 - (-0.76) = +1.10\ V$
So:
$\Delta G_{sys} = -zFE_{cell} = -2 \times 96\ 500 \times 1.1$
$\qquad\quad = -212.3\ kJ\ mol^{-1}$

f The energy available from the cell will be 212.3 × 0.025 = 5.31 kJ.

g If the cell has a mass of ≈ 53 g, the per-gram value is 5.31/53 ≈ 0.1 kJ g⁻¹.

h A joule is a watt-second, so 0.1 kJ = 100 watt-seconds, or 100/3600 watt-hours. If that is the per-gram value, then the per-kilogram value is 1000 × 100/3600 = 28 Wh kg⁻¹.

i This does not compare very well with 80 Wh kg⁻¹ for an 'SP2'.

19 Most 'battery' advertisements stress the lifetime of the product.

20a The liquid could be in the form of a gel or paste.

b Internal resistance would be reduced by high surface areas of electrodes and separator, plus concentrated solutions.

c As the concentration of Cu^{2+}(aq) dropped, the half-cell reaction would slow down, and the internal resistance of the cell would go up.

d If the copper ions did not run out, the zinc metal would, so there would be a loss of electrode surface area and the result would still be an increase in internal resistance.

21 The oxidation of zinc is the anode reaction, and the reduction of Cu^{2+} ions is the cathode reaction.

22a Having the case as the electrode maximises electrode surface area.

b See Figure A16.4. A high surface area of separator ensures good ionic conductivity, again minimising internal resistance.

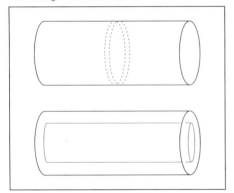

Figure A16.4 Showing how the arrangement of concentric half-cells increases the surface area of the separator, relative to a plain diaphragm separator

c Cations (NH_4^+ or Zn^{2+}) would have to leave, while Cl^- ions would enter.

d The separator is separating zinc from manganese(IV) oxide, and since they are both insoluble solids it should be an easy job.

e The half-equation featuring manganese(IV) oxide is:

$2MnO_2(s) + 2H^+(aq) + 2e^- \rightarrow Mn_2O_3(s) + H_2O(l)$

and the overall equation for the cell's reaction is:

$2MnO_2(s) + 2H^+(aq) + Zn(s) \rightarrow Mn_2O_3(s) + H_2O(l) + Zn^{2+}(aq)$

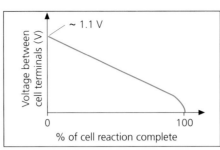

Figure A16.5 Showing how the voltage of the zinc/copper cell would decline in the course of its working life

f See Figure A16.5.

g The unpleasant effect is the leakage of electrolyte from the battery, because the zinc casing has developed pin-sized holes. The holes are the result of the anode reaction, which is the dissolution of zinc.

23a If the half-cell is more electron giving, its E value will become more negative.

b It would move up the electrochemical series.

c The overall cell e.m.f. would increase, since the two half-cells would have moved further apart in the table.

d Rearranging $E_{cell} = E_r - E_l$, and calling the manganese half-cell 'right', we get

$E_r = E_{cell} + E_l = 1.5 + (-0.83) = +0.67$ V

24a The overall cell reaction is:

$2Zn(s) + O_2(g) + 2H_2O(g) \rightarrow 2Zn(OH)_2(s)$

Oxygen is reduced and zinc is oxidised.

b It can give more space to the anode reactant because the cathode reactant (air) needs no compartment.

c This extra zinc means longer life (since it will be loss of zinc that stops the reaction, rather than loss of air).

d Platinum would probably be a good catalyst.

e If we call the air electrode 'right', then

$E_{cell} = E_r - E_l = +0.40 - (-0.76) = 1.16$ V

f These E data were from the standard electrochemical series, so the non-standard conditions in the real cell will cause a difference in E_{cell} from the calculated value.

g Air is certainly preferable to some of the alternative cathode materials in use like mercury or silver, which are both scarce and toxic.

h The high energy density comes from the 'weightless' cathode material – 'weightless' within the cell casing, at least.

i This high energy density means that the cell can pack plenty of energy into minimum mass for devices that have to be very light, like watches and hearing aids.

25a The relevant $E^{\ominus}$ values are on the second from bottom and fourth rows of Table 16.2, p. 397.

$E_{cell} = E_r - E_l = 1.8 - (-0.13) = 1.93$ V

b Lead goes from 0 (as lead metal) to +2 (as $PbSO_4$).

c Lead goes from +4 (as PbO_2) to +2 (as $PbSO_4$).

d $Pb(s) + PbO_2(s) + 2H_2SO_4(aq) \rightarrow 2PbSO_4(s) + 2H_2O(l)$

e If the lead sulphate had lost contact with the electrodes, it would not be in position to give and take electrons, in the reversals of the discharge half-equations.

f The high mass of lead contributes towards the low energy density.

g The more energetic battery would need its extra energy to propel its own weight.

h With battery cars, the burning of the fossil fuels does not happen in busy streets, and neither is lead produced in a volatile form.

26a The oxidation of hydrazine is the anode reaction.

b $CH_4(g) + 2H_2O(l) \rightarrow CO_2(g) + 8H^+(aq) + 8e^-$

or if the medium were alkaline,

$CH_4(g) + 8OH^-(aq) \rightarrow CO_2(g) + 6H_2O(l) + 8e^-$

c In the fuel cell, the energy conversion is directly from chemical to electricity, whereas in the power station we have the intermediate stages of heat and kinetic energy. Fewer stages should mean greater efficiency.

27a A solid electrolyte cannot spill or leak.

b The reactants are the two gases, which are kept remote from each other without the need for a special separator.

c The protons could make H_3O^+(aq) ions with water molecules, which could act as carriers across the gas from site to site (Figure A16.6).

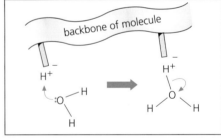

Figure A16.6 Showing how a water molecule can help to carry protons from site to site

d The cell makes its own water.

e Any Group I or II hydride, in which the H^- ion is present, would do. The acid–base reaction is:

$H^-(s) + H_2O(l) \rightarrow H_2(g) + OH^-(aq)$

28a Making the pipeline negative involves pumping electrons into it. This will have the same effect as when electrons arrive from a sacrificial anode – namely that it will be harder for the iron of the pipeline to give away electrons when there is a potential surplus of them.

b As long as the paint barrier holds, there will be no possibility of dioxygen getting access to electrons, so no current.

29a The aqueous environment inside tins might be quite acidic (in fruit for example). Tin is much less reactive towards acids than zinc is. The only drawback with tin is that if the layer breaks, then the exposed iron would be forced, by its position in the electrochemical series, to play the anode. So a scratched piece of tinplate would rust much faster than a galvanised plate.

b   Steel cans can be sorted from general household waste by magnets, while aluminium ones cannot. That does still leave the problem of the tops of tinplate drinks cans, which are still made of aluminium, despite the steel industry's efforts to develop an easy-open tinplate top.

30   The edges of the blob will be preferentially supplied with oxygen, so the *middle* will play the anode and corrode.

31   The deepest nooks and crannies will be oxygen-poor, and so, like the middle of the blob in question 30, will play the anode and corrode.

32   It must be something to do with the ion current in the electrolyte. The presence of salt ions must somehow increase the conductivity for the participating ions, $Fe^{2+}(aq)$ and $OH^-(aq)$.

33   The ingots of zinc were destined to be new sacrificial anodes on the hull of the ship. During the refit the corroded remains of the old ones would be unbolted, and replaced by the new ones. The same system is used on the legs of oil rigs (Figure A16.7).

Figure A16.7

34   The cathode reaction in acids is a much faster reaction – the reduction of $H^+(aq)$ ions:

$H^+(aq) + e^- \rightarrow \frac{1}{2}H_2(g)$

... always assuming the metal in question is above hydrogen in the electrochemical series, of course. Electrochemical corrosion reactions have to live by the same tests of thermodynamic feasibility, enshrined in the Law of anti-clockwise circles, as any others.

35   If the aluminium body panels came close enough to the steel chassis for pathways to be opened for ion and electron currents to pass between them, then we would predict that the aluminium, as the more reactive member of a bimetallic corrosion cell, would be the one to suffer.

36   The recharge cycle of a lead/acid cell is electrolysis.

37   The system includes the charging source,

which will suffer a loss in free energy greater than the local rise in the 'sub-system' represented by the aluminium oxide.

38   The words are 'receive' and 'give'.

39a   At the temperature necessary to melt sodium iodide, sodium metal will be a liquid.

b   At the cathode you would see a little yellow flame – the sodium would burn away as soon as it was made. At the anode you would see the purple colour of iodine vapour.

40a   An electrolyte must be an ion conductor. Apart from unusual cases like the proton-exchange membranes, this demands the liquid state.

b   The electrolysis will occur under non-standard conditions. For one thing, none of the participants will be in the same physical state as they would be at 298 K.

41   According to Le Chatelier's principle, the removal of $H_3O^+(aq)$ ions would drag the equilibrium in (1) to the right, to make more $H_3O^+(aq)$. The overall cathode reaction is:

$2H_2O(l) + 2e^- \rightarrow H_2(g) + 2OH^-(aq)$

42a $H^+(aq)$ is the more willing to accept electrons.

b $H^+(aq)$ is the more likely to be reduced.

43   Sodium metal would have reacted with water to produce hydrogen gas.

44   Oxygen atoms are better 'grabbers' of electrons than iodine atoms, so OH^- ions would be less willing to 'surrender' electrons. Hence the conversion of iodide to iodine looks the easier prospect.

45   The overall equation for the electrolysis of aqueous sodium iodide is:

$2NaI(aq) + 2H_2O(l) \rightarrow H_2(g) + I_2(aq) + 2NaOH(aq)$

46a   Cheshire has large underground salt deposits.

b   The polymers industry (using chlorine in PVC) has grown faster than the alkali-using industries like soapmaking.

c   If there were an unwanted surplus of sodium hydroxide, its price would go down.

47   Chlorine disproportionates in hot strong sodium hydroxide as follows:

$3Cl_2(g) + 6OH^-(aq) \rightarrow$
$5Cl^-(aq) + ClO_3^-(aq) + 3H_2O(l)$

48a   Sodium chloride is considerably less soluble than sodium hydroxide, so a cooled concentrated solution of the two solutes should deposit sodium chloride crystals.

b   The cost of a purification stage must be added to the overall cost of the process.

c   Both the large surface areas and the small inter-electrode distances of the two sets of plates reduce the internal resistance of the cell, thus saving on electricity costs.

d   Sodium ions would cure the ionic

imbalance. They would leave the anode region, curing the cation majority, and by the same action, make up for the cation shortage in the cathode compartment.

49a   Chloride ions would have to be excluded.

b   You would need an ion-exchange membrane that absorbed Cl^- ions and gave OH^- ions in exchange.

c   You would need to start with sodium hydroxide in the cathode compartment – in other words, you use a little sodium hydroxide to kick-start the process, in order to make a lot of it.

50a   Recycle the mercury, since it is much too expensive and toxic to discard.

b   The overall reaction for stage 1 is:

$2NaCl(aq) \rightarrow 2Na(Hg) + Cl_2(g)$

It is the same (except for the mercury) as you would get from the simple electrolysis of molten sodium chloride.

c   The sodium hydroxide would be pure because the sodium is physically removed from the chlorine at the end of stage 1, in solution in mercury.

d   Any leakage of mercury would be an environmental (and financial) blow.

51   Fluorine is the most electron-withdrawing element of all. No other chemical species would be able to remove electrons from F^- ions.

52a   As we have seen with electrochemical cells, a container/electrode offers maximum surface area.

b   Carbon (as graphite) combines economy with relative inertness.

c ΔG_f for hydrogen fluoride is -273.2 kJ mol^{-1}. So $+273.2$ kJ mol^{-1} would have to be supplied to the system to reverse the reaction. This means that one mole ($z = 1$) of electrons must be transferred from F^- to H^+, at an energy of 273 200 joules per 96 500 coulombs. Using formula (16.4):

$E = \dfrac{\Delta G}{zF} = \dfrac{273\,200}{96\,500} = 2.83$ V

d   The excess voltage would mean a faster rate of reaction.

e   The reduction of H^+ ions is by far the easier option.

f   Hydrogen fluoride boils at 20 °C, so the heat generated by the current would have boiled the electrolyte.

g   Hydrogen fluoride is a weak acid, so the ionic population would be low, and the internal resistance high. Potassium fluoride helps to increase the ionic conductivity of the electrolyte.

h   The 'waste' heat could be used to pre-heat incoming electrolyte. Excessive temperatures would cause boiling of the electrolyte and either high pressures or leaks of hydrogen fluoride gas.

i UF_6 is used to create a gaseous uranium compound. The diffusion rates of $^{235}UF_6(g)$ and $^{238}UF_6(g)$ through porous membranes are slightly different, so the fluoride gives the nuclear industry a means of separating the crucial fissile isotope, ^{235}U, from non-fissile ^{238}U.

53 Magnesium is far too expensive in its own right to be used for making aluminium.

54 Iron would appear as a cathode product, contaminating the aluminum.

55a The heating bills are lower with a more fusible electrolyte.

b The introduced cation is sodium, which will not be discharged preferentially to aluminium.

c Nothing exceeds F^- in its 'grip' on electrons, so oxide ions would be discharged.

d The overall cell reaction is:

$$2Al_2O_3 \rightarrow 4Al + 3O_2$$

and the cryolite is not consumed.

e $12HF + Al_2O_3 + 6NaOH \rightarrow 2Na_3AlF_6 + 9H_2O$

56a The electrolytic smelting of aluminium consumes vast amounts of electrical energy, and the close river provides a cheap high-energy hydroelectric source.

b The bauxite can be delivered and the product distributed conveniently.

c The graphite corrodes by a combustion reaction with the oxygen.

57a There *is* no overall reaction. The only change is the transfer of copper from anode to cathode.

b The loss in mass of the anode ought to equal the gain in mass of the cathode.

c A graphite anode would not dissolve, so oxygen would be discharged instead. The overall change would now be:

$$2CuSO_4(aq) + 2H_2O(l) \rightarrow$$
$$O_2(g) + 2Cu(s) + 2H_2SO_4(aq)$$

58a The jewellery metals are all less ready to lose electrons than copper, so copper is the most likely to dissolve from the anode.

b The silver and gold should be present in the sludge at the bottom of the cell.

c The metal that gets discharged at the cathode is the most readily reduced, and if copper is the least reactive metal in solution, then copper is the preferred candidate.

d It would make sense to try to extract silver and gold from it.

59a Sodium chloride would not be suitable, because chlorine instead of oxygen would be released at the anode.

b 10^{-5} m contains a large number of ions of Al_2O_3. It seems surprising that this coating is able to support further growth, because further growth requires both that electrons from the oxide ions reach the electrode, and that aluminium atoms from the electrode join the film as ions. The film must therefore have some conductivity for both ions and electrons.

c The oxide layer offers protection to the aluminium, both against mechanical shock like scratching, and against further corrosion.

CHAPTER 17　　　　　page 422

1 Horizonal differences are most apparent in, and between, the groups of the s- and p-blocks. The contrast between carbon and nitrogen exemplifies this. The d-block features strong horizontal similarities as well as vertical ones. Iron, for example, has a similar chemical character to cobalt.

The underlying reason for this difference is that d-block elements in the same period nearly all have the same outermost electron configuration – in Period 4 it is $4s^2$ (with slight deviations at chromium and copper, which show $4s^1$) – whereas s- and p-block elements in the same row all have different outer electron configurations.

Table A17.1

	Element	Electron arrangement	Ionization enthalpy (kJ mol^{-1})		
			1st	2nd	3rd
Group I	Li	[He] $2s^1$	520	7000	N/A
	Na	[Ne] $3s^1$	496	4500	N/A
	K	[Ar] $4s^1$	419	3000	N/A
	Rb	[Kr] $5s^1$	400	2600	N/A
	Cs	[Xe] $6s^1$	380	2400	N/A
Group II	Be	[He] $2s^2$	900	1750	15 000
	Mg	[Ne] $3s^2$	738	1451	8 000
	Ca	[Ar] $4s^2$	590	1145	5 000
	Sr	[Kr] $5s^2$	550	1064	4 000
	Ba	[Xe] $6s^2$	510	950	3 700
	Ra	[Rn] $7s^2$	490	900	3 500

Numbers shown in green are estimates

2a See Table A17.1.

b See Table A17.1. The estimates are made on the basis that ionization enthalpy gets less as a group is descended.

c They are metals because they all have readily lost outer electrons.

d The metals are reactive because their reactions with oxygen, chlorine and water all feature electron loss, which is favourable.

e The most reactive element would be found where the lowest ionization enthalpy is, at the bottom of the group.

3 Magnesia and lime were the oxides of magnesium and calcium respectively, while soda and potash were the carbonates of sodium and potassium. Strontian and barytes were probably the mineral forms of the sulphates of strontium and barium.

4 In order to complete the anti-clockwise circle that would correspond to carbon reducing calcium oxide, the temperature would have to be above the point where the two lines cross, well in excess of 2000 K. Apart from the expense, there would be the added snag of the calcium being formed in the gaseous state.

5a No current could pass in the cold, because the solid electrolyte could not conduct.

b Sodium chloride is much cheaper than sodium iodide, owing to its abundance in sea water.

c We would assume that sodium ions are more easily discharged than calcium ions, or in

other words, the sodium half-cell has a less negative electrode potential, under these conditions.

d Without the hoods, the products could meet and recombine.

e At the cathode: $Na^+(l) + e^- \rightarrow Na(l)$

At the anode: $Cl^-(l) \rightarrow \frac{1}{2}Cl_2(g) + e^-$

6 In street lights we are using the property of sodium atoms to emit photons in the yellow region of the visible spectrum, due to electrically stimulated electron transitions.

7 Being near the top of Group II, magnesium is the least reactive of the commonly found s-block elements.

8 No other element is so ready to release electrons, so the minimum photon energy necessary to displace an electron is less than for any other element. This means that caesium is sensitive to a wide range of photons, from across the entire visible region of the electromagnetic spectrum.

9a The reduced elements are as follows:
(1) hydrogen ($+1 \rightarrow 0$)
(2) oxygen ($0 \rightarrow -2$)
(3) chlorine ($0 \rightarrow -1$)
(4) hydrogen ($+1 \rightarrow 0$)
(5) oxygen ($0 \rightarrow -2$)

b They have behaved as acids, because their conjugate bases are among the products. However, the whole story comprises proton donation followed by reduction.

c The cloudiness is a precipitate of the partially soluble calcium hydroxide. The reaction is:

$$Ca(s) + 2H_2O(l) \rightarrow Ca(OH)_2(s) + H_2(g)$$

d The other elements react vigorously enough with water, so you can imagine how violent they would be with aqueous acid.

10a The sodium atoms are at +1, and each of the peroxide oxygen atoms is at -1.

b The conjugate acid of the peroxide ion is hydrogen peroxide, H_2O_2.

c Clearly the reaction can be seen as going via the creation of hydrogen peroxide, and its subsequent disproportionation into water and oxygen.

$$Na_2O_2(s) + 2H_2O(l) \rightarrow 2NaOH(aq) + H_2O_2(aq)$$
$$H_2O_2(aq) \rightarrow H_2O(l) + \frac{1}{2}O_2(g)$$

So overall:

$$Na_2O_2(s) + H_2O(l) \rightarrow 2NaOH(aq) + \frac{1}{2}O_2(g)$$

d The O_2^- ion must carry a single negative charge as shown, and be like a peroxide (O_2^{2-}) ion except for a missing electron. The peroxide ion obeys the rule of eight, so the superoxide ion must have an unpaired electron.

11a The charge on the sodium ion would have to be 2+.

b The electrostatic force between the ions would be higher, due to the larger charge.

c   The second ionization enthalpy of sodium would be prohibitively large, due to its being concerned with the removal of an inner shell electron.

12   The estimate could have been made by a comparison with calcium chloride, $CaCl_2$, whose lattice enthalpy might be expected to be similar. The loss of a single electron from the $2p^6$ subshell should not affect the size of the ion too much, and the charge would be the same.

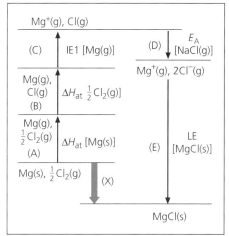

Figure A17.1 Born–Haber cycle for MgCl(s)

13   See Figure A17.1. Relevant data are as follows:

Enthalpy of atomisation of Mg(s) =
$$+148 \text{ kJ mol}^{-1} \quad (A)$$

Enthalpy of atomisation of $\frac{1}{2} Cl_2(g)$ =
$$+122 \text{ kJ mol}^{-1} \quad (B)$$

First ionization enthalpy of Mg(g) =
$$+738 \text{ kJ mol}^{-1} \quad (C)$$

Electron affinity of Cl(g) = -349 kJ mol^{-1} (D)
Lattice enthalpy of MgCl(s) = -750 kJ mol^{-1} (E)
Enthalpy of formation of MgCl(s) = X kJ mol^{-1}
By Hess's Law:

$X = A + B + C + D + E$
$= 148 + 122 + 738 - 349 - 750$
$= -91 \text{ kJ mol}^{-1}$

so MgCl would be formed exothermically from its elements.

14a $Mg^+ + Mg^+ \rightarrow Mg^{2+} + Mg$
b $2MgCl(s) \rightarrow MgCl_2(s) + Mg(s)$
c $\Delta H_{reaction} = \Delta H_f(\text{products}) - \Delta H_f(\text{reactants})$
$$= -641.3 - 2(-91)$$
$$= -459.3 \text{ kJ mol}^{-1}$$

So the disproportionation looks, on this limited evidence, more feasible than the formation of MgCl.

15a Na_2SO_4
b $Ca(NO_3)_2$
c Na_2CO_3
d $NaHCO_3$

e $SrCl_2$
f $Ca_3(PO_4)_2$
g KH_2PO_4

16   The bonding must be of the electrostatic · ion–dipole type, just as in solvated Na^+(aq) ions.

17   For example, the phenomenon of nuclear magnetic resonance was originally seen as a discovery in physics, and the all-important phenomenon of chemical shift came as an unexpected spin-off. At the time of writing there is no obvious application of the 'buckyball' molecular family of spherical carbon cages (Figure A17.2), but people are searching hard.

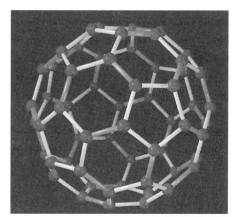

Figure A17.2 The buckminsterfullerenes are allotropes of carbon in which the carbon atoms form a polyhedral framework. There is a whole family of such molecules, dubbed 'buckyballs'

18a The lattice enthalpies of Group I compounds are relatively small, which tilts the $\Delta H_{solution}^{\ominus}$ values towards negativity.
b   Group II compounds have very large (negative) lattice enthalpies, so the left-hand arrow in Figure 17.13 will be long. It is true that the smaller, more highly charged ions will also have larger solvation enthalpies (the right-hand arrow in Figure 17.13). The message to be drawn from the insolubility of Group II compounds is that the increase in lattice enthalpies must be the dominant influence.

19a The solubilities of the Group II hydroxides at 298 K are given in Table A17.2. Only the barium compound could be called an alkali.

Table A17.2

Group II hydroxide	Solubility (mol dm⁻³)
$Mg(OH)_2$	2×10^{-4}
$Ca(OH)_2$	1.53×10^{-2}
$Sr(OH)_2$	3.4×10^{-2}
$Ba(OH)_2$	1.5×10^{-1}

b   The reaction is:
$Ca(OH)_2(aq) + CO_2(aq) \rightarrow CaCO_3(s) + H_2O(l)$
The milkiness is the precipitate of calcium carbonate.
c $CaO(s) + H_2O(l) \rightarrow Ca(OH)_2(aq)$

This is a reaction rather than a dissolution. Oxides cannot be literally 'soluble', since any base stronger than the OH^- ion (which the O^{2-} ion is) will destroy itself by abstracting a proton from water.
d   The oxide is freely soluble in acids, for one of two reasons. Either aqueous protons can react directly with the surface of the lump of oxide:

$MgO(s) + 2H^+(aq) \rightarrow Mg^{2+}(aq) + H_2O(l)$

or else the acid can react with the few hydroxide ions and pull the equilibrium:

$MgO(s) + H_2O(l) \rightleftharpoons Mg(OH)_2(aq)$

to the right.

20a The first stage of the Leblanc process released hydrogen chloride gas. The second stage produced a useless slag which had to be dumped in piles. Finally the piles gave off noxious hydrogen sulphide gas. The local soils would also become heavily alkaline due to leaching of calcium hydroxide from the decomposing slag heaps.
b   The sodium hydroxide had to be dug out at the end of the first stage, because it was a fused lump. Had it been a powder it might have been possible to propel it from stage to stage in a flow.
c   The digging out is clearly labour intensive, compared with the automated process of flow.

21   The Solvay process scores in three respects: it uses harmless limestone rather than sulphuric acid; the reaction takes place in solution so there is no need for digging out; the waste product is a soluble salt, calcium chloride, rather than the toxic calcium sulphide slag.

22   They would see a precipitate of calcium carbonate. This is caused by the low solubility product of calcium carbonate.

23a i Rock salt, NaCl, and limestone, $CaCO_3$.
ii Sodium carbonate, Na_2CO_3.
iii Calcium chloride, $CaCl_2$.
iv Ammonia, NH_3, carbon dioxide, CO_2, calcium oxide, CaO, sodium hydrogencarbonate, $NaHCO_3$, ammonium chloride, NH_4Cl, and water, H_2O.
b   It would make sense to transfer carbon dioxide from the light ash calciner to the Solvay tower, as indeed the plant does.
c   When the four equations are added up as written, they make:

$2NaCl(aq) + CaCO_3(s) \rightarrow CaCl_2(aq) + Na_2CO_3(s)$

d   The hot middle is due to the exothermic reaction (1). The cold at the bottom reduces the solubility of the sodium hydrogencarbonate so that it crystallises out and can be filtered off.
e   The site for a sodium carbonate plant would have to take into account: overall ease of transport; availability of salt and limestone; and location of major users.
f   Cheshire is the site of the UK's major salt deposits, and limestone deposits are nearby in Derbyshire and Yorkshire. The glass industry is sited in St Helens.

g Sodium carbonate needs to be mixed with sand to make glass, so the similar densities and particle sizes will facilitate that mixing.

h The calcium chloride can be put in the holes from which the salt has come.

24a Any of steps (2), (3) or (4) would fit the bill, since they all produce a gas from non-gaseous reactants. We shall analyse stage (2).

b Relevant data are given in Table A17.3.

Table A17.3

	ΔH_f (kJ mol^{-1})	S (J K^{-1} mol^{-1})	
NaHCO$_3$(s)	−950.8	101.7	reactant
Na$_2$CO$_3$(s)	−113.7	135.0	} products
CO$_2$(g)	−393.5	213.6	
H$_2$O(g)	−241.8	188.7	

Using formula (8.2),

$\Delta H_{sys} = \Delta H_f$(products) − ΔH_f(reactants)

$= -1130.7 + (-393.5) + (-241.8) - 2(-950.8)$

$= +135.6$ kJ mol^{-1}

$\Delta S_{sys} = \sum S$(products) − $\sum S$(reactants)

$= 135.0 + 213.6 + 188.7 - 2 \times 101.7$

$= +333.9$ J K^{-1} mol^{-1}

$= 0.334$ kJ K^{-1} mol^{-1}

c See Figure A17.3. If the graph is a straight line, then only two points are needed. When $T = 0$, $\Delta G_{sys} = \Delta H_{sys} = 135.6$ kJ mol^{-1}. For the second point, let $T = 500$ K. Then $\Delta G_{sys} = \Delta H_{sys} - T\Delta S_{sys} = 135.6 - 500 \times 0.334 = -31.4$ kJ mol^{-1}.

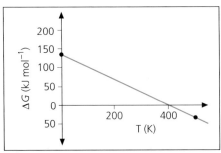

Figure A17.3 Free energy of stages of the Solvay process, as a function of temperature

d The temperature at which ΔG_{sys} becomes zero is $(\Delta H_{sys}/\Delta S_{sys}) = 135.6/0.334 = 406$ K. So we can say that the reaction becomes viable at temperatures from about 500 K upwards (when ΔG_{sys} would have an appreciably negative value). The reaction would not be viable at 298 K.

25a You would expect the lattice enthalpies to become less negative (less exothermic) as the cations get bigger.

b You would expect the lattice enthalpies to become more negative (more exothermic) as the cations become more highly charged – although this prediction is not quite so safe, since the actual lattices are not the same.

26a See data in Table A17.4. In the series of hydroxides, there is an inverse correlation – barium hydroxide has the highest solubility and

Table A17.4

		Lattice enthalpy (kJ mol^{-1})	Solubility (mol dm^{-3})
a	Mg(OH)$_2$	More negative (lattice strong)	2×10^{-4}
	Ca(OH)$_2$		1.5×10^{-2}
	Sr(OH)$_2$	Less negative (lattice weak)	3.4×10^{-2}
	Ba(OH)$_2$		1.5×10^{-1}
b	MgSO$_4$	More negative (lattice strong)	1.8×10^{-1}
	CaSO$_4$		4.7×10^{-3}
	SrSO$_4$	Less negative (lattice weak)	7.0×10^{-5}
	BaSO$_4$		9.4×10^{-7}

the least negative lattice enthalpy.

b The sulphates show a direct correlation – barium sulphate has the lowest solubility and the least negative lattice enthalpy.

27 See Figure A17.4.

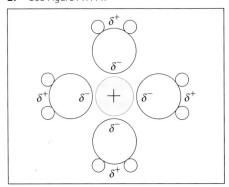

Figure A17.4 Water molecules solvating a cation, showing ion–dipole interactions

28a The charge density decreases down Group II. The magnesium ion, Mg^{2+}, has the most exothermic association with water molecules.

b The lattice enthalpy must have the bigger influence. The strongest lattice is indeed the least soluble.

c The solvation enthalpy must control the solubility here. The weakest lattice (barium sulphate) is the least soluble, so it must be that the stability of the Mg^{2+}—water bonds overcome the stability of the Mg(OH)$_2$(s).lattice.

d Sulphate is a much bigger anion than hydroxide.

29 Carbonates belong firmly in the 'big anions' camp, so we might expect their solubilities to decrease down the group. See data in Table A17.5.

Table A17.5

Carbonate	Predicted trend	Solubility (mol dm^{-3})
MgCO$_3$	Highest	1.5×10^{-4}
CaCO$_3$		1.3×10^{-5}
SrCO$_3$		7.4×10^{-6}
BaCO$_3$	Lowest	9.0×10^{-6}

30a See Table A17.6.

b See Table A17.7.

c We are assuming that ΔH_{sys} and ΔS_{sys} are constants over temperature, and that no change of state occurs.

d **i** Stability goes down across a row.

Table A17.6

	Na$_2$CO$_3$	→	Na$_2$O	+	CO$_2$
$\Delta H_f^{\ominus}$ (kJ mol^{-1})	−1130.7		−414.2		−393.5
$S^{\ominus}$ (J K^{-1} mol^{-1})	+ 135		+ 75.1		+213.6
	$\Delta H_{sys}^{\ominus}$ +323 kJ mol^{-1}		$\Delta G_{sys}^{\ominus,298}$ +277 kJ mol^{-1}		
	$\Delta S_{sys}^{\ominus}$ +153.7 J K^{-1}				

	MgCO$_3$	→	MgO	+	CO$_2$
$\Delta H_f^{\ominus}$ (kJ mol^{-1})	−1095.8		−601.7		−393.5
$S^{\ominus}$ (J K^{-1} mol^{-1})	+ 65.7		+ 26.9		+213.6
	$\Delta H_{sys}^{\ominus}$ +100.6 kJ mol^{-1}		$\Delta G_{sys}^{\ominus,298}$ +48.5 kJ mol^{-1}		
	$\Delta S_{sys}^{\ominus}$ +174.8 J K^{-1}				

	BaCO$_3$	→	BaO	+	CO$_2$
$\Delta H_f^{\ominus}$ (kJ mol^{-1})	−1216.3		−553.5		−393.5
$S^{\ominus}$ (J K^{-1} mol^{-1})	+ 112.1		+ 70.4		+213.6
	$\Delta H_{sys}^{\ominus}$ +269.3 kJ mol^{-1}		$\Delta G_{sys}^{\ominus,298}$ +218 kJ mol^{-1}		
	$\Delta S_{sys}^{\ominus}$ +171.9 J K^{-1}				

Table A17.7 Decomposition temperatures (temperature at which $\Delta G = 0$) for reactions in Table A17.6

	T (K)
Na$_2$CO$_3$	2101
MgCO$_3$	576
BaCO$_3$	1567

ii Stability goes up down a group.

e Magnesium carbonate is the only one that could be decomposed by heating with a Bunsen burner.

31a Potassium nitrate is included in gunpowder to provide a supply of oxygen.

b Nitrogen ($+5 \rightarrow +3$), oxygen ($-2 \rightarrow 0$).

c The quickest decompositions will occur at the tops of the groups, owing to the higher charge densities of those cations.

d $4\text{LiNO}_3(s) \rightarrow 2\text{Li}_2\text{O}(s) + 4\text{NO}_2(g) + \text{O}_2(g)$
This is a classic case of the diagonal relationship. The Li$^+$ ion is so small that it has an appreciable charge density, comparable to that of an Mg^{2+} ion. So lithium and magnesium nitrates behave similarly, and lithium conforms to the Group II pattern.

32 The top cation of the group has the electron configuration 1s^2, so it is exceptionally tiny. The increase in size from the first to the second member of a group is, proportionately, by far the biggest increase, and that is why the top member is the most likely to act unrepresentatively.

33a The middle of the membrane is made of hydrocarbon material, so the tunnel needs to have an outer lining that can interact by van der Waals' forces.

b The tunnel needs an inner lining of polar groups, so that the sodium ion can pass along it by a series of ion–dipole interactions.

c A tunnel could be made to allow selective passage of Na$^+$ ions by making it too small for the larger K$^+$ ions.

CHAPTER 18   page 441

1a The p-block elements lie to the right-hand side of their periods, where the nuclear pull is tending towards a maximum while the size and shielding factors stay fairly constant.

b   The dividing line runs across each row as shown in Table A18.1.

Table A18.1

Row	Between elements
1	Be and B
2	Al and Si
3	Ga and Ge
4	Sb and Te
5	Po and At

c   The simple cations will coincide with the lowest ionization enthalpies, which occur to the bottom and left of the p-block. An example of a fairly 'pure' ionic compound would be aluminium fluoride, AlF_3.

d   The simple anions will coincide with those elements with high ionization enthalpies, combined with electron vacancies. This means elements to the right and top of the p-block (excluding the noble gases). Examples of compounds containing simple anions are sodium chloride, NaCl and calcium oxide, CaO.

e   Bonds between p-block elements of roughly similar electron-attracting character will inevitably be covalent or polar covalent. Examples include methane, CH_4 and sulphur dioxide, SO_2.

f   p-block–s-block compounds will feature ionic bonds.

g   Elements from the left of the p-block are not sufficiently electron-attracting to receive electrons from s-block elements, and the s-block elements themselves are unsuited to covalent bonding.

2a   An oxidation number greater than the group number would require the (impossible) loss of inner shell electrons.

b   Fluorine is the ultimate electron-withdrawing element. No other element can attract electrons more strongly than fluorine in a covalent bond, so fluorine atoms can never be positive.

c

(AlF₃ cannot obey the rule of eight, unless it dimerises as Al_2F_6, shown ball-and-stick)

(only outer electrons are shown)

When the compounds obey the rule of eight, the valencies are equal to eight minus the group number. The oxidation numbers are +(valency).

d ClF_5, for example, has oxidation numbers of +5 for the chlorine atom and −1 for each fluorine atom. Notice that the chlorine atom has a share in 12 electrons (including one lone pair).

e   The attainment of high oxidation numbers is achieved by the partial removal of large numbers of electrons from an atom. This is only worthwhile, energetically, if the new orbital being offered to the electrons is very desirable. As we have noted before, there are no sites more desirable than the vacancy on a fluorine atom.

f   As was implied in (e) above, an element being oxidised by combination with fluorine is not 'over-keen' to yield lots of electrons. There comes a point, reached at chlorine, when the oxidised element is sufficiently electronegative to resist even the pull of the fluorine atom, especially when it comes to the last one or two removals.

3   See Figure A18.1.

4

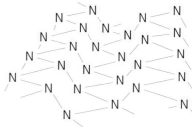

5   The difference is the shorter N—N bond length, which brings the electrons, especially the lone pairs, too close for comfort.

6a   The N_2 triple bond is nearly 6 times stronger than the single bond.

b   The P_2 triple bond is about $2\frac{1}{2}$ times stronger than the single bond.

c   The weakness of the N—N single bond is attributable to repulsion between electrons on the two atoms.

d   The relative strength of the P—P single bond is attributable to a lack of repulsion between electrons on the two atoms.

e   The shortness of the N—N bond enables good overlap between adjacent orbitals, to create a strong π-bond.

7a   See Figure A18.2.

b   One possible reason is that the N_2 molecule keeps its lone pairs as remote as possible from each other, in contrast to the O_2 molecule, which may be partially destabilised by lone pair repulsions. There is probably a major contribution to reactivity from two unpaired electrons that the O_2 molecule has. These electrons are there for reasons unaccounted for by our simple bonding theories, and they are not shown in Figure A18.2. But many of oxygen's

Fluorine can build a single-bonded diatomic molecule, in one dimension (but very short).

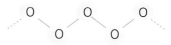

Oxygen can (in theory) build a one-dimensional chain of single-bonded atoms.

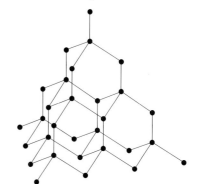

Nitrogen can (in theory) build a two-dimensional sheet of single-bonded atoms.

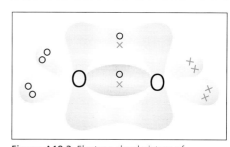

Carbon can build a three-dimensional network of single-bonded atoms.

Figure A18.1

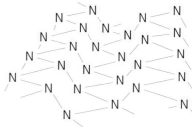

Figure A18.2 Electron-cloud picture of dioxygen, with a dot-and-cross overlay

violent combustion reactions are of the free radical type, and it seems certain that these electrons are implicated in some way.

8   Yes – if there is a crude correlation between low ionization enthalpy and conductivity, then boron's significantly high ionization enthalpy value, for Group III, could explain its non-conductivity.

9a The energy gap must decrease, judging by the change from non-conducting diamond via semiconducting silicon and germanium to the metallic conductors tin and lead.
b The temperature dependence could be due to the absorption of infrared photons, causing the promotion of more electrons into the conducting orbitals.
c Visible photons could promote electrons, in the same way as infrared photons.
d Diamond fails to absorb visible light, so the energy gap between the bonding orbitals and the antibonding ones must be significantly larger than those in silicon and germanium.
e Our simple band theory seems to imply that there is a single energy jump between non-conducting and conducting orbitals, giving rise to a single photon frequency for absorption (as in atomic absorption spectra of sodium, etc.). But the fact that semiconductors can absorb light all through the visible region and into parts of the infrared region gives the lie to this idea.

10a Storage – large molecules like starch.
b Immediate use – small molecules like glucose.

11 The hydrocarbons are the energy-rich pinnacle, and carbon dioxide and carbonates are the energy-exhausted trough.

12a $\Delta H = \Delta H_f(\text{products}) - \Delta H_f(\text{reactants})$
$= 6 \times -285.8 - (-1273.3)$
$= -441.5 \text{ kJ mol}^{-1}$

b Most plant carbohydrates are in polymer form (starch and cellulose) rather than glucose, and there are proteins and fats to decay as well. Furthermore, decay would not be to graphite – coal, after all, is a complex mixture of amorphous carbon and hydrocarbons (along with lots of other substances like ammonia and sulphur compounds).
c Plant carbohydrates lie above carbon in the enthalpy ladder (Figure A18.3).

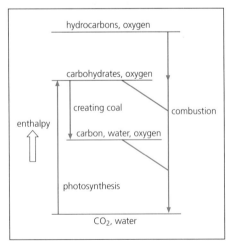

Figure A18.3 Enthalpy ladder for carbon compounds in the biosphere

13a See Table A18.2.

Table A18.2

Property	Diamond	Graphite
Hardness	Very hard	Very soft
Electrical conductor?	No	Yes
Cleaves into flakes?	No	Yes
Opaque?	No	Yes

b Graphite conducts better in the plane of its plates (using the delocalised π-bonding system) than across them.
c The plates slide over each other with ease, as the inter-plate bonds are relatively weak. This property combined with its electrical conductivity makes graphite an ideal material for motor brushes – its softness causes it to mould itself to the contours of the spinning commutator, and its lubricating properties provide minimum friction.

14 $\Delta H = \Delta H_f(\text{products}) - \Delta H_f(\text{reactants})$
$= 2 \times (-110.5) - (-910.9)$
$= +689.9 \text{ kJ mol}^{-1}$

$\Delta S = \Sigma S(\text{products}) - \Sigma S(\text{reactants})$
$= (2 \times 197.6 + 18.8) - (41.8 + 2 \times 5.7)$
$= +360.8 \text{ J K}^{-1} \text{ mol}^{-1}$

The reaction will begin to be viable at the temperature that gives a ΔG value of approximately zero – in other words, when:

$\Delta H = T\Delta S$

or: $T = \dfrac{\Delta H}{\Delta S} = \dfrac{689\,900}{360.8}$

$= 1912 \text{ K (or 1639 °C)}$

15 It has been suggested that the graphite structure includes π-bonds within planes. We have already noted that π-bonds are far weaker in lower elements than in the top member.

16a The loss of pure silicon to the solid state will serve to increase the concentration of impurities in the remaining liquid.
b As the level of impurities builds up, the melting point of the liquid phase will get lower.

17 All the impurities finish up at the end of the ingot. If the passes were made in opposite directions, the impurities would be shuttled back and forth.

18 See Figure A18.4.

19a The second natural method of nitrogen fixation is the conversion of the gas into proteins by bacteria in the root nodules of plants in the legume family (which includes peas, beans, clover and pulses).
b Nitrogen is incorporated into (plant) proteins.
c Humans have tried to influence nitrogen fixation in an urgent attempt to feed themselves. The nitrogen is used to increase soil fertility, in the form of compounds like ammonium phosphate. The alarming rate of population growth demands that we get more and more food from every hectare under cultivation.

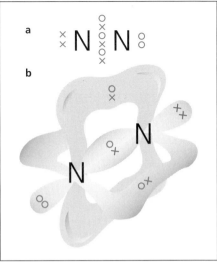

Figure A18.4 a Dot-and-cross diagram of dinitrogen. **b** Electron-cloud picture of dinitrogen, with a dot-and-cross overlay. π-bonds are shown in green

20a The equality of bond lengths demands that the triangles all be equilateral. So the bond angles must be 60°.
b The best point of comparison is with ammonia. The bond angles in ammonia are about 107°.
c The enforced 60° bond angles must incorporate a lot of destabilising strain, in the form of electron–electron repulsion.
d This strain would reduce the activation enthalpy of reactions involving the P_4 molecule. In other words, less severe collisions will be needed to jolt the molecule into action.
e Even at room temperature, the collisions of P_4 molecules with oxygen molecules must presumably be capable of providing the activation enthalpy of the combustion reaction.

21a The glass increases the frictional drag of the match's motion across the box, enabling greater energy to be converted to heat at the match head.
b The potassium chlorate could oxidise the sulphur:

$2KClO_3(s) + 3S(s) \rightarrow 3SO_2(g) + 2KCl(s)$

The fact that it does not happen shows that the activation enthalpy of the reaction must lie outside the range of energies that can be generated by a strike.
c There is probably quite a mixture of reactions, including the direct oxidation of both the elements by the chlorate. But the main reaction sequence is presumably:

$P_4(s) + 3S(s) \rightarrow P_4S_3(s)$

$3P_4S_3(s) + 16KClO_3(s) \rightarrow$
$3P_4O_{10}(s) + 9SO_2(g) + 16KCl(s)$

22 We have attributed the weakness of the single bonds in the first long period to the shortness of the bonds, and the resulting repulsions between lone pairs of electrons on adjacent atoms.

23 The bonds between carbon and lead will be of low polarity. (Although lead is a metal and has a set of ionic compounds containing Pb^{2+} ions, its four-valent covalent compounds are fairly typical of Group IV chemistry as a whole. On the Pauling scale its electronegativity relative to carbon (1.8 to 2.5) would predict a percentage ionic character in the bond of 12%.) So we might expect the molecule to cleave into free radicals:

$$(C_2H_5)_4Pb \rightarrow (C_2H_5)_3Pb + C_2H_5\bullet$$

It seems reasonable to assume that they might get caught up in the combustion mechanism like this:

$$C_2H_5\bullet + O_2 \rightarrow C_2H_5{-}O{-}O\bullet$$

24 We can borrow an idea from nitric acid. In that molecule we saw how a single oxygen atom can be held by a dative covalent bond. The ozone molecule repeats the trick, except with oxygen as the donor (Figure A18.5). The oxygen in the middle is in 'three-bit' formation, so we should expect a bond angle not far from 120°, closed up a little by the repulsion between the lone pair and the double bond. The actual angle of 117° fits the prediction very well.

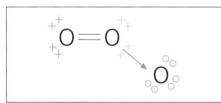

Figure A18.5 The ozone molecule, drawn in a mixture of formats. Bonds are shown as letters and sticks, while lone pairs are shown as dots and crosses

25a The air supplies the force that pushes the molten sulphur to the surface.
b The steam delivery ring can act as a heat-loss barrier around the inner channels. Thus it can help the sulphur in the inner ring to stay molten on its journey to the surface.
c The excess temperature difference allows for some cooling on the steam's two-way journey.
d The sulphur would separate as a solid when the water/sulphur mixture cooled back to ambient temperature.
e A useful heat exchange would be to allow the hot water to make a contribution to pre-heating the incoming water, or even the pumping blast of air.
f The likely impurities would include metal sulphides and sulphates. Ionic sulphides are among the most insoluble compounds known (excepting the Group I sulphides), so they would be left undissolved by the hot water. Even if impurities did dissolve in the water (for instance, the common mineral gypsum, $CaSO_4$), it is

probable that, once the mixture had reached the surface, the sulphur would solidify (at 112 °C) before any of the impurity solutes crystallised.
g Similarities: both minerals can be pumped to the surface as liquids. Neither mineral requires people to descend to the level of the deposits.
Contrasts: crude oil is under naturally generated pressure, so there is no need to push it to the surface. Crude oil is a complex mixture of compounds, whereas the product of the Frasch process is a single element in a high state of purity.

26a The different crystal shapes must be the result of different stacking patterns of the S_8 crowns.
b If solid sulphur crystallises from the melt, it is doing so at 112 °C, a temperature at which the stable form of sulphur is monoclinic. If the crystallisation is from a solvent, then the process is likely to be taking place at a temperature less than 95.6 °C, which means that the rhombic form will be more stable.
c The monoclinic crystals keep their outer shape, but recrystallisation takes place inside the needles. New crystal boundaries appear inside the needles, which cause the opacity.
d Plastic sulphur exists at room temperature because it is formed by fast cooling. This rapid crystallisation gives plastic sulphur no time to rearrange to rhombic sulphur. Furthermore, as with any other reaction, the change from plastic to rhombic sulphur has an activation enthalpy, and at room temperature the thermal agitations of the plastic sulphur molecules are insufficient to supply the necessary energy.
e Plastic sulphur molecules are coiled, and the stretching is due to the uncoiling process (as in rubbers).

27a The valency of one derives from the fact that halogen atoms have a single outer-orbital vacancy, and so can form a single covalent bond.
b Fluorine obeys the rule of eight for the same reason that all the members of that period obey the rule of eight – namely that there are no energetically available empty orbitals to receive promoted electrons, so no possibility of opening up the atom for new bonds. In addition, fluorine's electrons are the most securely held in the period, and would be unlikely to be promoted, even if there were such a thing as a 2d set of orbitals.
c As the needle-like crystals cool, they pass through the transition temperature below which rhombic is the stable form. This is the signal for the S_8 molecules to rearrange themselves, so that a single original monoclinic crystal becomes 'host' to many new rhombic crystals. The effect of many new crystals growing inside the old is like that of a broken car windscreen, which retains the shape of the original object, but is crazed and opaque.

28 There are two possible answers to this question. One centres on the inability of the halogens to form multiple bonds. The inertness of dinitrogen derives from the great strength of the triple N≡N bond. Double-bonded dioxygen

is, admittedly, quite a reactive element, but here a different argument can be applied. There is so much oxygen on the planet that every element that could be oxidised has been oxidised, and the oxygen in the air is, if you like, the surplus.

29 The small size of the F^- ion will have an effect on lattice enthalpies (which depend, you may recall, on the two electrostatic parameters of charge size and charge separation). For instance, the lattice enthalpy of calcium fluoride is 16% more negative than that of the chloride. This strength of the fluoride lattices will oppose their solubility in water. (You might offer the counter-argument that the smallness of the F^- ion would confer a correspondingly favourable hydration enthalpy when it did dissolve in water, but the insolubility of fluorides is a fact, so we must conclude that the lattice enthalpy dominates the hydration enthalpy.)

30a $F^-(aq) \rightarrow \frac{1}{2}F_2(aq) + e^-$
b Chlorine was produced from brine, $NaCl(aq)$. Fluorine was produced from a mixture of potassium fluoride and hydrogen fluoride.

31a Fluorine's vacancy is the most desirable.
b Fluorine again, for the above reason.
c The iodide ion will be the least unwilling to release an electron, since the iodine atom has the weakest hold over the extra electron.
d Reduction is the donation of electrons, so it follows from answer (c) that the best reducing agent is the iodide ion.
e Other feasible halogen–halide reactions are:
$$F_2(g) + 2Cl^-(aq) \rightarrow Cl_2(aq) + 2F^-(aq)$$
$$F_2(g) + 2Br^-(aq) \rightarrow Br_2(aq) + 2F^-(aq)$$
$$F_2(g) + 2I^-(aq) \rightarrow I_2(aq) + 2F^-(aq)$$
$$Cl_2(g) + 2I^-(aq) \rightarrow I_2(aq) + 2Cl^-(aq)$$
$$Br_2(g) + 2I^-(aq) \rightarrow I_2(aq) + 2Br^-(aq)$$

32a Oxidation numbers: reactant, $Br_2 = 0$; products, $BrO_3^- = +5$, $Br^- = -1$. So atoms in one oxidation state have become atoms in two others – the reaction is a disproportionation.
b The carbonate ion is being used as a base, so if we remove it by adding acid, we should drag the equilibrium back to the left-hand side.
c Unreacted chlorine could be left over from reaction (1). A neat way of removing it would be to have a small supply of concentrated $Br^-(aq)$, which would take the chlorine impurity through a repeat of equation (1).
d In cold dilute alkali:
$$Cl_2(aq) + 2OH^-(aq) \rightarrow$$
$$ClO^-(aq) + Cl^-(aq) + H_2O(l)$$
In hot concentrated alkali:
$$3Cl_2(aq) + 6OH^-(aq) \rightarrow$$
$$ClO_3^-(aq) + 5Cl^-(aq) + 3H_2O(l)$$

33 First stage:
$$CaF_2(s) + H_2SO_4(l) \rightarrow 2HF(g) + CaSO_4(s)$$
Second stage:
$$2HF(\text{with } KF, l) \rightarrow H_2(g) + F_2(g)$$

34a We can revisit the argument we used in answer to question 29 for explaining the low solubility of fluorides: the driving force for the reaction must be the greater lattice enthalpy of the fluoride version of apatite.
b   Referring yet again to the link between ion size and lattice enthalpy, we must assume that the bigger chloride ions do not provide a more stable lattice compared with hydroxide ions, and therefore that the chloride substitution reaction is not thermodynamically favoured.

35a If the two samples had genuinely been part of the same skeleton, then the fluorine uptake of all parts of the skeleton should have been very similar. The wide difference in this case clearly points to a fake.
b   The skull appears to have been the older, on account of its having taken up more fluorine.
c   Fluorine uptake by bone is dependent on the level of fluoride in the groundwater. Thus bones of the same age from sites with widely differing fluoride concentrations would give contrasting results. Safe comparisons of age are only possible on samples from the same locality.

36a Fluoridation of water would be unnecessary where natural fluoride concentration is high.
b   In children, the crystalline material is being built up, so the rate of fluoride uptake would be much higher than when, in later life, the amount of tooth material is fairly constant.
c   From the Government's point of view, the most persuasive argument must be that large amounts of money would be saved by reducing the need for dental treatment, and that the money could then be used on other therapies. They could also argue that science was only mimicking what Nature had achieved in certain high fluoride localities (also the localities with the best dental health). The opposing arguments are centred on the moral point about the individual's right to choose, in the light of available knowledge, what to do with his or her own body. You can perhaps see some parallels between this debate and the one about the freedom to smoke tobacco.

37a Ammonia's major use is in the manufacture of fertilisers, needed to support the enormous growth in human numbers.
b   The route taken by the hydrogen atoms in ammonia would be from plant proteins (maybe to animal proteins) via death or excretion plus decay to ammonium ions in the soil, finally to be released as water when the ammonium ions are oxidised to nitrates:

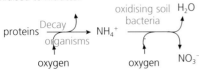

38a Solar electricity could be used to electrolyse water, liberating dihydrogen.
b   Hydrogen transmission would use technology similar to natural gas transmission, which is the cheapest method in Figure 18.24.

c   The cathode material would be dioxygen in the air, and the overall reaction would be:
$2H_2(g,\text{ Pt catalyst}) + O_2(g,\text{ Pt catalyst}) \rightarrow 2H_2O(l)$
d   Hydrogen is cleaner by virtue of:
• no contribution to the greenhouse effect
• no release of sulphur oxides
• no release of carbon monoxide, oxides of nitrogen, unburnt hydrocarbons.
e   The two main problems with hydrogen are its low density, necessitating spacious storage, and its explosive reaction with air, as demonstrated by the Hindenburg disaster.

39   The H—C bond in methane is almost non-polar, while the H—F bond in hydrogen fluoride is strongly polarised.

40   No dot-and-cross diagram could cope with a 2-valent hydrogen atom. However, the boiling point of $-92\,°C$ rules out the idea that the hydrogen atoms might be H^- ions.

41   The H^- ion is, incredibly, bigger than a Br^- ion (0.208 nm as opposed to 0.195 nm). The size must be due to the poor nuclear pull exerted by the single-proton nucleus over two electrons. This poor nuclear attraction would also mean that hydride ion electrons would be easily polarised by adjacent cations.

42a Sodium tetrahydridoborate(III) is used for the reduction of acids, aldehydes, ketones and esters to alcohols.
b

$$\left[\begin{array}{ccccc} & & H & & \\ & & \overset{\circ}{\underset{\times}{}} & & \\ H & \overset{\circ}{\underset{\times}{}} & B & \overset{\circ}{\underset{\times}{}} & H \\ & & \overset{\circ}{\underset{\times}{}} & & \\ & & H & & \end{array}\right]^{-}$$

c   The conjugate acid of H^- is H_2.
d $LiAlH_4(s) + 4H_2O(l) \rightarrow$
$LiOH(aq) + Al(OH)_3(s) + 4H_2(g)$

43   The Si—Si bond (226 kJ mol^{-1}) compares unfavourably with C—C (347 kJ mol^{-1}), and the same is true of the bonds to hydrogen atoms:
Si—H = 318 kJ mol^{-1} while
C—H = 413 kJ mol^{-1}.

44a The base strengths of the conjugate bases must be opposite to the acid strengths, so NH_2^- is a stronger base than OH^-.
b   The amide ion in water would react to give ammonia and a hydroxide ion:
$NH_2^-(aq) + H_2O(l) \rightarrow OH^-(aq) + NH_3(aq)$

45a Purple, pH of about 10 caused by the weak base NH_3.
b   Orange, pH of about 5 caused by the weak acid NH_4^+.
c   Green, neutrality caused by the combination of the weak acid NH_4^+ and the equally weak base $CH_3CO_2^-$.

46 $NH_4^+(aq) + OH^-(aq) \rightarrow NH_3(aq) + H_2O(l)$
You would be able to smell the ammonia.

47a Hydrogen bonding in ammonia would affect its melting and boiling temperatures, and its enthalpies of fusion and vaporisation.
b   We see that ammonia melts and boils at $-78\,°C$ and $-33\,°C$ respectively, while the equivalent values for phosphine are $-133\,°C$ and $-88\,°C$.

48   For equation (*10*),
$\Delta G = \Delta G_f(\text{products}) - \Delta G_f(\text{reactants})$
$= 2 \times (-16.5) - 3 \times (-237.2)$
$= +678.6\text{ kJ mol}^{-1}$

For equation (*11*),
$\Delta G = \Delta G_f(\text{products})$ (since reactants are elements)
$= 2 \times 86.6 = +173.2\text{ kJ mol}^{-1}$

Neither reaction is thermodynamically feasible at 298 K.

49a They would use light energy to synthesise carbohydrates from water and carbon dioxide, which in turn could be the direct source of energy for the synthesis of ammonia and proteins.
b   Carbon monoxide is a catalyst poison, 'stealing' the catalyst sites meant for (the isoelectronic) N_2 molecules.
c   The ratio of 3 : 1 matches the stoichiometric ratio for the reaction.

50a We can base our answer on Le Chatelier's principle. The right-hand side of the equilibrium occupies less space (fewer molecules), so an increase in pressure will prompt the system to shift in that direction.
b   As we saw in Chapter 9, an increase in temperature causes an equilibrium to go in its endothermic direction. In this case the desired outcome is the exothermic direction, so lower temperatures give better yields.
c   **i** It is no good having a reaction giving high yields but taking weeks to reach equilibrium, so the reaction has to be run at a temperature that delivers at least some yield in a reasonable time.
ii High pressures have to be paid for in energy costs and robust equipment, so here too there is a limit to the extent to which it is worthwhile chasing maximum equilibrium yields.

51a See Table A18.3.

Table A18.3

	$N_2(g)$	+	$3H_2(g)$	$\rightleftharpoons$	$2NH_3(g)$
Moles at t_0	1		3		0
Moles at t_{eqm}	$1-y$		$3-3y$		$2y$
	Total moles of gas at $t_{eqm} = 4-2y$				
Mole fractions at t_{eqm}	$\dfrac{1-y}{4-2y}$		$\dfrac{3-3y}{4-2y}$		$\dfrac{2y}{4-2y}$

b   According to Avogadro's hypothesis, if 40% of the space is taken up by ammonia, then 40% of the molecules are ammonia, so the mole fraction of ammonia = 0.4.

c So $2y/(4 − 2y) = 0.4$, which gives $y = 0.57$.
d See Table A18.4 for numerical data for mole fractions and partial pressures.

$$K_p = \frac{p^2_{NH_3}}{p_{N_2}p^3_{H_2}} = \frac{80^2}{30 \times 90^3} = 2.9 \times 10^{-4}\ atm^{-2}$$

Table A18.4

Mole fractions at t_{eqm}	0.15	0.45	0.4
Partial pressures at t_{eqm} (atm)	30	90	80

e Let us call the 100 atm result 25% ammonia. This means that $2y/(4 − 2y) = 0.25$, so $y = 0.4$. The mole fractions and partial pressures are shown in Table A18.5, and the value of K_p is given by:

$$K_p = \frac{25^2}{18.95 \times 56.25^3}$$
$$= 1.9 \times 10^{-4}\ atm^{-2}$$

So we see that equilibrium 'constants' are subject to quite a degree of variation in real life. This is because of factors like the non-ideal behaviour of gases at very high pressures, but a fuller account is outside A level.

Table A18.5

Mole fractions at t_{eqm}	0.1875	0.5625	0.25
Partial pressures at t_{eqm} (atm)	18.75	56.25	25

f If we go across a row in Table 18.11, then the temperature is changing, and equilibrium constants are not expected to be constant under those circumstances.

52 One tonne (10^3 kg) of ammonia is equivalent to $10^6/17$ mol, or 58 824 mol. So if 16 mol of ammonia are produced from 7 mol of methane, then 58 824 mol of ammonia will require the destruction of $58\ 824 \times 7/16$ mol of methane, or 25 736 mol. The total fuel energy equivalent is $25\ 736 \times 890 = 22.9 \times 10^6$ kJ tonne^{-1}. That is just the methane consumed in the reaction, and excludes energy used to heat reactors.

53 The heat from the exothermic ammonia synthesis stage (reaction (*13*)) could be transferred to heat the reactants of the endothermic steam reforming stage (reaction (*12*)).

54 The best possible catalyst would cause the lowest activation enthalpy, and so require the lowest temperatures for a given rate of reaction. There would be the added bonus that lower temperatures would produce larger values of K_p and so bigger yields.

55 The inert support can hold the grains apart, and inhibit sintering, thus maintaining a greater surface area for a longer time.

56a The new energy source would be methane from the North Sea (which Billingham is near to).
b The River Tees can supply the water needed in reaction (*12*), while the deep water allows easy export of finished fertiliser.

57a Nuclear-generated electricity could be used to electrolyse water (or brine) to make hydrogen.
b $C(s) + 2H_2O(g) \rightarrow CO_2(g) + 2H_2(g)$

58a $C_6H_{12}O_6(s) + 6H_2O(g) \rightarrow 6CO_2(g) + 12H_2(g)$
b Both sequences produce carbon dioxide.

59a For hydrogen fluoride, bonds broken are H—H and F—F, and bonds made are 2H—F. So the missing value, per mole of hydrogen fluoride, is:

$(435.9 + 158 − 2 \times 568)/2 = −271$ kJ mol^{-1}

b The reasons for the extremely exothermic ΔH_f of hydrogen fluoride seem to reside in the pronounced weakness of the F—F bond and the extreme strength of the H—F one.

60 Hydrogen fluoride:

H—F---H—F---H—F

HF_2^- ion:

[F---H—F]$^-$

61 Propagation:

$H_2 + Cl\bullet \rightarrow HCl + H\bullet$

$H\bullet + Cl_2 \rightarrow HCl + Cl\bullet$

Termination:

$H\bullet + H\bullet \rightarrow H_2$

$Cl\bullet + Cl\bullet \rightarrow Cl_2$

$H\bullet + Cl\bullet \rightarrow HCl$

62 The smoke is the solid ammonium halide, finely divided. A typical reaction would be:

$NH_3(g) + HBr(g) \rightarrow NH_4Br(s,\ smoke)$

63 Nothing would oxidise hydrogen fluoride to fluorine. Acidified manganate(VII) would oxidise hydrogen chloride, bromide and iodide. Acidified dichromate(VI) would oxidise hydrogen bromide and iodide only. Aqueous iron(III) would oxidise hydrogen iodide only.

64 The colours allow discrimination (silver chloride white, silver bromide pale yellow, silver iodide yellow). The chloride is soluble in concentrated ammonia solution, with the bromide showing slight solubility and the iodide none.

65a See Table A18.6.

Table A18.6

Element	Rule-of-eight oxide	Highest oxide	Oxidation number in highest oxide
C	CO_2	—	+4
N	N_2O_3	N_2O_5	+5
O	O_2	—	0
F	F_2O	—	−1
Si	SiO_2	—	+4
P	P_4O_6	P_4O_{10}	+5
S	*SO	SO_3	+6
Cl	Cl_2O	Cl_2O_7	+7

*Non-existent

b See Figure A18.6.
c Fluorine – fluorine will always incline an element to show its maximum oxidation state, because it offers such 'tempting' vacancies for incoming electrons.

d The lowest members of the noble gases are most likely to have stable compounds, since they have the least hold over their outer electrons.

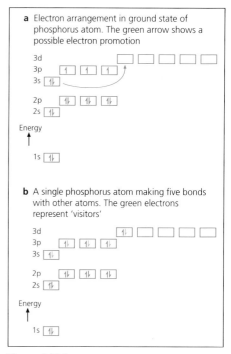

a Electron arrangement in ground state of phosphorus atom. The green arrow shows a possible electron promotion

b A single phosphorus atom making five bonds with other atoms. The green electrons represent 'visitors'

Figure A18.6

Radon is radioactive and not much experimented with (beyond the necessity of detecting it in people's homes), but the oxides of xenon have been prepared.

66a More.

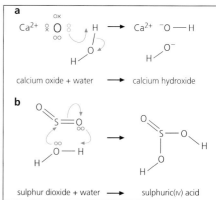

calcium oxide + water $\longrightarrow$ calcium hydroxide

sulphur dioxide + water $\longrightarrow$ sulphuric(IV) acid

Figure A18.7

b More.

67a See Figure A18.7.
b $CaO(s) + H_2O(l) \rightarrow Ca(OH)_2(aq)$

$SO_2(g) + H_2O(l) \rightarrow H_2SO_3(aq)$

68 You would see the gelatinous white precipitate of aluminum hydroxide.

69 Al^{3+} ions polarise water molecules by virtue of their high charge density.

70

Two molecules of nitric(III) acid are produced by the hydrolysis of a molecule of N_2O_3.

71 The more oxygen atoms there are pulling electron density away from the central atom, the less the central atom will be able to give electron density to any one oxygen atom. So the oxygen atoms that are part of OH groups proceed to take more electron density from the hydrogen atoms, rendering them more acidic.

72a It can hydrogen bond with water, using its oxygen atoms.
b $CaCO_3(s) + CO_2(aq) + H_2O(l) \rightarrow$
$Ca(HCO_3)_2(aq)$

so one component of the 'hardness' in water is dissolved calcium hydrogencarbonate.

73a Boiling might have the effect of driving off the carbon dioxide as it forms, thus tilting the equilibrium to the right. The right-hand side appears to be of higher entropy, which would mean that higher temperatures would favour it.
b If the tubes in the boiler became clogged with limescale, then the fuel would be wasted heating limestone, and blockages and pressure build-ups might occur.
c Running the water through an ion-exchange resin avoids the need for heating, and this is what actually happens in power stations. Calcium cations and various anions are replaced with H^+ and OH^- ions, which form water. Occasionally the resins are regenerated by flushing through with solutions of H^+ and OH^-.

74 Naturally occurring silica was able to crystallise very slowly, by slow cooling within veins in rocks.

75

'laughing gas'

76a This number of repeat units must give a ring of the correct radius to fit neatly around the Ca^{2+} ion.
b The swaddled-up Ca^{2+} ion is too big to make a strong lattice.

c The interaction between ion and polymer would be of the ion–dipole type.

77a These soap powders have their own built-in water-softening facility.
b Phosphates can cause excess fertility in waterways, leading to overproduction of plants, excessive amounts of decaying vegetable material, overpopulation of bacteria and the eventual loss of dissolved oxygen from the water. This is the process known as eutrophication.

78 One mole of chlorine atoms going down six oxidation numbers is balanced by three moles of chlorine atoms going up two oxidation numbers. Hence:

$4NaClO_3(s) \rightarrow 3NaClO_4(s) + NaCl(s)$

79 $SiCl_4(l) + 4H_2O(l) \rightarrow Si(OH)_4(s) + 4HCl(g)$

CHAPTER 19 page 473

1 One possible answer is given.
a See Table A19.1.

Table A19.1

Element	Metal/ non-metal	Melting point (°C)	First ionization enthalpy (kJ mol⁻¹)	Density (g cm⁻³)
Al	Metal	660	578	2.7
Si	Non-metal	1410	789	2.3
P	Non-metal	44*	1012	2.3
S	Non-metal	119†	1000	2.0

*Red phosphorus
†Monoclinic sulphur

b See Table A19.2.

Table A19.2

Element	Metal/ non-metal	Melting point (°C)	First ionization enthalpy (kJ mol⁻¹)	Density (g cm⁻³)
Sc	Metal	1541	631	2.99
Ti	Metal	1660	658	4.50
V	Metal	1890	650	5.96
Cr	Metal	1857	653	7.20

c With the possible exception of densities, it is clear that there is much less variation in the properties of the d-block elements.

2 The 4s orbital is of lower energy than the 3d orbitals (at least, when both are empty).

3 Most of the elements have a partially filled 3d sub-shell and a $4s^2$ pair. (Zinc is an exception to both patterns.)

4 The nuclear pull would be increasing, but the outer electrons would still be in the same principal quantum level. So the d-block would have been an extension of the same row of the Periodic Table, and we would expect the ionization enthalpies to go on increasing.

5 There are two factors at work – the increased nuclear pull due to one extra proton, and the increased shielding due to one extra 3d electron. From the value given for the ionization enthalpy of scandium, we conclude that the first factor must dominate the second.

6 Table A19.3 shows the occupancy of the 3d sub-shell.

Table A19.3

Element	3d occupancy	Unpaired electrons	Melting point (°C)
Ti		2	1660
V		3	1890
Cr		5	1857
Mn		5	1244
Fe		4	1535
Co		3	1495
Ni		2	1999

The correlation between number of unpaired electrons and melting temperature is unconvincing. Figure 19.3 does peak roughly in the middle, but there is a heavy skew to the left, giving vanadium a very high melting point considering its three unpaired electrons. The most glaring irregularity occurs with manganese, whose melting point is unexpectedly low.

7a See Table A19.4. The predicted order of reactivity (which is the real order as far as, say, reaction with acids is concerned) is (most reactive first) Zn > Fe > Ni > Cu.

Table A19.4

Half-cell	$E^\ominus (V)$
$Fe^{2+}(aq)|Fe(s)$	−0.44
$Ni^{2+}(aq)|Ni(s)$	−0.25
$Cu^{2+}(aq)|Cu(s)$	+0.34
$Zn^{2+}(aq)|Zn(s)$	−0.76

b See Table A19.5. So the predicted order of reactivity is (most reactive first) Fe > Ni > Zn > Cu.

Table A19.5

Metal	First + second ionization enthalpies (kJ mol⁻¹)
Fe	2320
Ni	2490
Cu	2704
Zn	2639

c Zinc is much more reactive than would be predicted from ionization enthalpy data alone.

8 Zinc has an unusually low melting temperature (420 °C) for a d-block metal. This weakness of the zinc lattice would assist the forward progress of the half-equation $Zn(s) \rightarrow Zn^{2+}(aq) + 2e^-$, so would contribute to the reactivity of real lumps of zinc in aqueous media. In contrast, the ionization enthalpy is based purely on the gaseous reaction $Zn(g) \rightarrow Zn^{2+}(g) + 2e^-$.

9 If by reactivity we mean rate of reaction, then the most relevant quantity would be enthalpy of activation. However, these data tend not to appear in data books.

10 By the end of the row, the influence of increasing nuclear charge has clearly dominated the modest degree of mutual shielding which the 3d electrons exert on each other. So the 3d electrons of copper and zinc cannot be removed.

11a Hydrogen cyanide gas is liberated by the reaction:

$NaCN(s) + HCl(aq) \rightarrow HCN(g) + NaCl(aq)$

(This reaction is used for legal executions in some states of the USA.)

b Iron(III) sulphate might be a good choice. The CN^- ions would cluster round the Fe^{3+} ions to form a stable complex:

$12NaCN(aq) + Fe_2(SO_4)_3(aq) \rightarrow$
$2Na_3[Fe(CN)_6](aq) + 3Na_2SO_4(aq)$

12 d-block compounds are qualified for this role because of their ability to exist at more than one oxidation number.

13a The +2 oxidation state is non-existent at the start of the row.

b The +3 oxidation state is non-existent at the end of the row.

14a The strongest oxidising agent is $Co^{3+}(aq)$, as shown by the high positive $E^\ominus$ value.

b The strongest reducing agent is $Cr^{2+}(aq)$, as shown by the high negative $E^\ominus$ value.

c $Ti^{2+}(aq)$, if it did exist, would reduce hydrogen atoms in water to dihydrogen:

$Ti^{2+}(aq) + H_2O(l) \rightarrow Ti^{3+}(aq) + \frac{1}{2}H_2(g) + OH^-(aq)$

d **i** $V^{2+}(aq)$ and $Cr^{2+}(aq)$ would appear to be unable to exist in the presence of $H^+(aq)$.

ii Water is far from being a standard solution of $H^+(aq)$ – it contains 10^{-7} mol dm^{-3} rather than 1.0 mol dm^{-3}.

iii In 1.0 mol dm^{-3} acid the law of anticlockwise circles predicts that both ions should be oxidised to the 3+ version, for example:

$V^{2+}(aq) + H^+(aq) \rightarrow V^{3+}(aq) + \frac{1}{2}H_2(g)$

(Whether this happens or not would depend upon kinetic factors – it *is* thermodynamically feasible.)

15a The electron being removed from Fe^{2+} and all subsequent members of the row is coming from a doubly occupied d-orbital, so an electron–electron repulsion factor assists removal. From Sc^{2+} to Mn^{2+} the d-orbitals are singly occupied.

b The hiccup means that Fe^{3+} is actually easier to create than, and more stable than, Mn^{3+}. The latter is therefore the stronger oxidising agent. Accepting the limitations of applying ionization enthalpy data to aqueous media, the expected direction of reaction is:

$Mn^{3+}(aq) + Fe^{2+}(aq) \rightarrow Fe^{3+}(aq) + Mn^{2+}(aq)$

16a We need a metal whose $M^{2+}(aq)|M(s)$ electrode potential lies above that of $V^{3+}(aq)|V^{2+}(aq)$, which is -0.26 V. A likely candidate would be $Zn^{2+}(aq)|Zn$, at -0.76 V.

b The green colour is caused by $Cr^{3+}(aq)$, and the oxidising agent is oxygen. The relevant electrode potentials are:

$Cr^{3+}(aq)|Cr^{2+}(aq)$ -0.41 V
$[O_2(g) + 2H_2O(l)]|4OH^-(aq)$ $+0.40$ V

The reaction is:

$4Cr^{2+}(aq) + O_2(g) + 2H_2O(l) \rightarrow$
$4Cr^{3+}(aq) + 4OH^-(aq)$

17 We have 2 moles of iodine atoms going up one oxidation number (finishing as 1 mole of $I_2(aq)$), so that must be balanced by 2 moles of copper atoms coming down one oxidation number (starting as $Cu^{2+}(aq)$). So the copper atoms must end up in the +1 oxidation state. $Cu^+(aq)$ does not exist, but insoluble copper(I) salts do, so the white product must be copper(I) iodide, CuI(s). The reaction is:

$2Cu^{2+}(aq) + 4I^-(aq) \rightarrow 2CuI(s) + I_2(aq)$

The fact that this reaction does go means that the drive to form copper(I) iodide has made $Cu^{2+}(aq)$ into a more formidable oxidising agent than normal. The half-cell $[Cu^{2+}(aq)+I^-(aq)]|CuI(s)$ must have a more positive $E^\ominus$ value than $I_2(aq)|2I^-(aq)$, at +0.54 V.

18 The oxidising agents are, in the first instance, $H^+(aq)$ or the hydrogen atom in HCl(g), and in the second, $Cl_2(g)$.

19a The co-ordination number (the number of nearest anion neighbours) of the metals in cadmium iodide and rutile is 6.

b The halide ions show a co-ordination number of 3.

c The d-block cations are smaller than the Ca^{2+} ion, due to the greater nuclear pull. For example, Ca^{2+} has an ionic radius of 0.1 nm, compared with Mn^{2+} at 0.067 nm. This gives them a greater charge density, which, as we learned in Chapter 5, is a cause of impure ionic bonding.

d If the halide atoms were 'pure' ions, there would be repulsion between adjacent layers of 'sandwiches'.

e They may be van der Waals' forces. This view is supported by the inter-layer iodine–iodine distance in cadmium iodide itself, which is close to double the van der Waals' radius of an iodine atom.

f The weak inter-layer bonds would enable the lattice to behave like graphite, with layers sliding over each other.

20 $2FeCl_3(aq) + Cu(s) \rightarrow 2FeCl_2(aq) + CuCl_2(aq)$

21 The dihalogen molecule is polarised by the iron(III) chloride:

$$X\!-\!X\!\rightarrow\!\overset{\delta^+}{Fe}\begin{matrix} \overset{\delta^-}{Cl} \\ \!-Cl \\ Cl \end{matrix}$$

22 The charge density of the Fe^{3+} ion attracts a lone pair from an oxygen atom on one of the hydrating water molecules, and acidic cleavage results:

$$Fe^{3+}\ :\!O\begin{matrix}H \\ \\ H\end{matrix} \longrightarrow [Fe(OH)]^{2+}(aq) + H^+(aq)$$

23 Magnetite can be seen as $Fe(II)O.Fe(III)_2O_3$.

24 Green is the colour of iron(II) compounds, so we can assume that the iron has been reduced. This is caused by the action of micro-organisms in percolating water. Anaerobic respiration produces reducing gases like hydrogen sulphide, which change the Fe^{3+} ions to Fe^{2+}.

25 Atoms as far as manganese seem able to use every 3d and 4s electron for bonding, at least to oxygen and fluorine. Beyond manganese, due to increasing nuclear pull on the d sub-shell, the maximum oxidation number decreases. The maximum oxidation numbers are given in Table A19.6.

Table A19.6 Maximum oxidation numbers for the first row of the d-block elements

Element	Maximum oxidation number (+ve)
Sc	3
Ti	4
V	5
Cr	6
Mn	7
Fe	6
Co	3
Ni	2
Cu	2
Zn	2

26a $2S_2O_3^{2-}(aq) + I_2(aq) \rightarrow S_4O_6^{2-}(aq) + 2I^-(aq)$
Starch would sharpen the end-point.

b $2MnO_4^-(aq) + 16H^+(aq) + 10I^-(aq) \rightarrow$
$2Mn^{2+}(aq) + 8H_2O(l) + 5I_2(aq)$

c $MnO_4^- : S_2O_3^{2-} = 1 : 5$

d The greater the level of sewage, the less manganate(VII) is left over, and also the less iodine, and the less 'thio'. So Thursday was the day when the sewage was 'strongest'.

e If there had been no sewage at all, the volume of 0.01 mol dm^{-3} $S_2O_3^{2-}(aq)$ needed to titrate 10 cm^3 of 0.01 mol dm^{-3} MnO_4^- would have been 50 cm^3, because if the solutions are of the same concentration, they will react in a volume ratio equal to their mole ratio, in this case 1 : 5. So if we subtract the titration value from 50 cm^3, we will have a number proportional to the loss of MnO_4^-. Monday would score 10, by this method, and Thursday would score 20.

f Solutions of sewage will not be transparent and colourless, so it would be difficult to see the end-point. Also, an observed end-point might gradually disappear, as the manganate(VII) ions sought out more oxidisable material located in the middle of small particles.

27a Remembering that the law of anticlockwise circles presents the reactants on the bottom left/top right diagonal of two correctly arranged half-equations, we see that the reactants are $MnO_4^{2-} + 2MnO_4^{2-}$. In other words, what we have here is a disproportionation:

$3MnO_4^{2-}(aq) + 4H^+(aq) \rightarrow$
$MnO_2(s) + 2H_2O(l) + 2MnO_4^-(aq)$

b The second half-equation has H⁺ as a reactant, so will be pH sensitive. As the alkali concentration rises, the equation as written will be inclined backwards (by Le Chatelier's principle), so its electrode potential will become less positive. Eventually there must be a level of alkali concentration so high that the electrode potential of the second half-equation becomes *less* positive than +0.56 V, and then the disproportionation will go into reverse.

c All their half-equations have H⁺(aq) on the left, so they are all encouraged to go forwards in acidic media.

28a Both bases and ligands operate as lone-pair donors, so you would expect good bases to be good ligands.

b None – complexing hardly ever affects oxidation states.

29 Five-valent covalent bonding to nitrogen atoms does not happen – it violates the rule of eight, to which nitrogen conforms in all its other compounds.

30a To be a ligand, a species needs a lone pair. The chloride ion has the necessary qualification.

b Only two of the chloride ions are free to behave as anions. The ligand chlorides have to go wherever the complex goes, and since the complex has an overall 2+ charge, its destination will be the cathode.

c The number of moles of silver chloride precipitated would be the same as the original number of moles of free chloride – three for the yellow complex, and two for the purple one.

d The conductivities seem to depend simply on how many free ions there are, rather than which ions are free.

e In the yellow complex there are four free ions in total, whereas in the purple complex there are three.

f The yellow complex has a conductivity not far from that of iron(III) chloride, which also has four free ions. The purple complex has a conductivity that matches that of magnesium chloride, which has three free ions. So if conductivity is a guide to number of free ions, the proposed structures are validated.

g $[Co(NH_3)_4Cl_2]^+$ – it has two free ions, with a conductivity similar to that of sodium chloride, and a single free chloride ion to account for the single mole of silver chloride.

h These two complexes exhibit *cis–trans* isomerism (Figure A19.1).

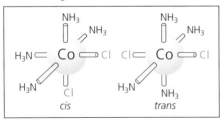

Figure A19.1 *Cis-* and *trans*-isomers of the $[Co(NH_3)_4Cl_2]^+$ ion

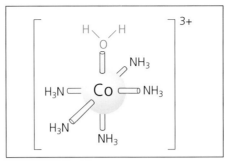

Figure A19.2

31a See Figure A19.2.

b Three moles of silver chloride would be formed.

c There are four free ions, so you would expect a conductivity close to 400 in the units of Table 19.9 (10^4 m² ohm⁻¹ mol⁻¹).

32 No conductivity implies no ions. So the complex must be neutral, despite its inclusion of four chloride ligands. So we deduce that the original platinum species must have been Pt^{4+}. Such a highly charged cation from a metal of such low reactivity is unlikely, so we must assume a good deal of covalent character in the Pt—Cl bonds. However, the oxidation number of platinum is still +4.

33a This is another case of *cis–trans* isomerism (Figure A19.3).

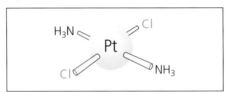

Figure A19.3 The *trans*-isomer of $[Pt(NH_3)_2Cl_2]$

b There may be interference from lone pairs on the platinum atom, as in the square planar xenon tetrafluoride from Chapter 4.

c It has no overall charge, so it should have a negligible conductivity.

d The *trans*-isomer should have a zero dipole moment, since the individual dipoles in the two Pt—Cl bonds are symmetrically opposed.

e The zero dipole moment of the *trans*-isomer should favour its solubility in non-polar solvents like hexane, relative to the *cis*-form.

34a See Figure A19.4.

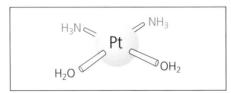

Figure A19.4 The cisplatin molecule after water substitution

b Hydrogen bonds would form between the water ligands and oxygen or nitrogen atoms on the bases of DNA.

c If cells are in rapid division, the DNA will be spending much time in the vulnerable single-strand 'unzipped' condition, when the cisplatin can get at its bases more easily.

d The bone marrow is another site of rapid cell division, due to the high-turnover production of white blood cells. These cells would be affected just like the cancer ones.

e If cisplatin were given in overdose, it would turn its attention to healthy cells, and prevent their DNA functioning normally.

35 Hydrazine is not big enough to situate its two lone pairs so they can offer two co-ordination positions at 90° or 109.5° (Figure A19.5).

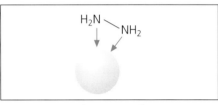

Figure A19.5 Hydrazine's two lone pairs acting as a bidentate chelating ligand

36 See Figure A19.6.

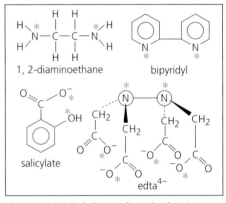

Figure A19.6 Polydentate ligands, showing binding sites (*)

37a One big ligand loses its 'freedom', and six small ones gain theirs. The 'chaos factor' has clearly increased.

b ΔS_{sys} is positive, which will incline ΔG_{sys} to the negative.

c It is difficult to predict the size of ΔS_{sys}, but the liberation of the water molecules would certainly favour a big positive K_c. Using $\Delta G_{sys} = -RT\ln K_c$, an equilibrium constant of 10^{25} leads to a ΔG_{sys} of $-142\,380$ J mol⁻¹. Assuming that most of this is due to the $-T\Delta S$ factor, we have a ΔS_{sys} of about 480 J K⁻¹ mol⁻¹. A data book tells us that the creation of six moles of $H_2O(l)$ creates 420 J K⁻¹ mol⁻¹, so we appear at least to be in the right 'ball park'.

38 Only (d) is non-superimposible upon its own mirror image. Figure A19.7 shows the two chiral isomers.

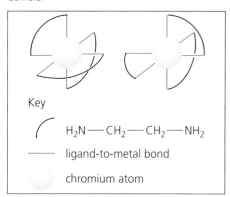

Key

H_2N—CH_2—CH_2—NH_2

ligand-to-metal bond

chromium atom

Figure A19.7 The two chiral isomers of $[Cr(en)_3]^{3+}$

39a Only soluble iron would be able to diffuse into root cells.
b $[Fe(H_2O)_6]^{2+}(aq) \rightarrow$
$Fe(OH)_2(s) + 4H_2O(l) + 2H^+(aq)$
c $4Fe(OH)_2(s) + O_2(g) + 2H_2O(l) \rightarrow 4Fe(OH)_3(s)$
d The edta^{4-} must solubilise the iron, so that it cannot precipitate with hydroxide ions:
$Fe(OH)_3(s) + edta^{4-}(aq) \rightarrow$
$[Fe(edta)]^-(aq) + 3OH^-(aq)$
e By adding Na[Fe(edta)], you are making certain that the edta does not get sidetracked in reactions with other soil cations, and that the iron is available in soluble form.
f The synthesis of chlorophyll seems likely, on this evidence, to depend on iron.
g The biosynthesis of the blue pigment in hydrangeas requires iron. Soluble iron cations are available in acidic soils, but in alkaline soils the iron would be tied up in solid iron(III) hydroxide. The tablets add edta to the soil, which makes the iron soluble and therefore available to the plant.

40 The lone pairs of dioxygen stick out at an angle that suits the vacancy in Figure 19.33. Those on carbon monoxide stick out straight (Figure A19.8). So the carbon monoxide must accept reduced orbital overlap, or approach too closely to the 'guard' parts of the protein.

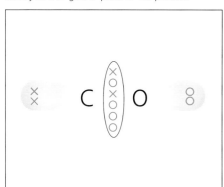

Figure A19.8 The orientation of lone pairs on dioxygen and carbon monoxide molecules

41 Carbon monoxide might occur naturally from the incomplete respiration of carbohydrates, rather as it derives in traffic from the incomplete combustion of hydrocarbon fuels.

42 The colourless compounds in d-block chemistry generally have either completely empty d sub-shells, as in titanium(IV) oxide and scandium(III) compounds, or completely full ones, as in zinc(II) and copper(I) compounds.

43a A more electrically negative ligand would have more effect on the cation, so the energy difference would be greater between those d-orbitals that carry electrons close to the ligands and those that do not. Greater energy splitting within the d sub-shell would mean more energetic photons to promote the transitions, hence a different colour (Figure A19.9).

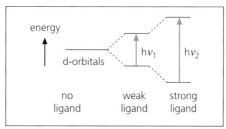

Figure A19.9 Showing how the energy difference between the d-orbitals depends on the nature of the ligand. Since $v_1 > v_2$, the absorptions will be at different places in the spectrum

b When the Cu^{2+} ion is anhydrous, all the d-orbitals, in the absence of ligands, are of the same energy. The water molecules surround the cation octahedrally, so energy splitting occurs, and with it colour.

44a Good bases are more 'pushy' with their lone pairs.
b They will be more aware.
c It becomes more important – the energy difference increases.
d As the jump gets bigger, the frequency of absorbed photons will be higher, so the wavelength will be shorter.
e Cl^- is a 'weak' ligand, so it will cause a small d sub-shell energy gap. The absorbed photons will be of longer wavelength, so λ_{max} will move to the right. In fact the 'window' in the spectrum of the $[CuCl_4]^{2-}$ complex moves so as to make the complex yellow (Figure A19.10).

45a $A = \log_{10} (10/1) = 1$
b The ammonia complex could simply have been at higher concentration.
c You would need to know the path length l and the ε-value at a given wavelength, probably λ_{max}.

46a Sodium tetrabromoferrate(III).
b Diamminesilver chloride.
c Calcium tetrachloroplatinate(II).

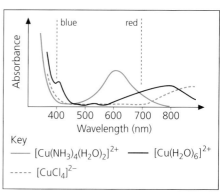

Key
—— $[Cu(NH_3)_4(H_2O)_2]^{2+}$ —— $[Cu(H_2O)_6]^{2+}$
---- $[CuCl_4]^{2-}$

Figure A19.10 Showing how the identity of the ligand affects absorption

d trans-dichloro-bis(1,2-diaminoethane)-chromium(III) chloride.

47a

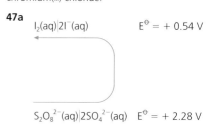

$I_2(aq)|2I^-(aq)$ $E^\ominus = + 0.54$ V

$S_2O_8^{2-}(aq)|2SO_4^{2-}(aq)$ $E^\ominus = + 2.28$ V

The reaction is thermodynamically feasible.
b The Fe^{3+} ion could oxidise the I^- ion, which would also be feasible, given the $Fe^{3+}(aq)|Fe^{2+}(aq)$ electrode potential of +0.77 V (which drives $I_2(aq)|2I^-(aq)$ at +0.54 V backwards):
$2Fe^{3+}(aq) + 2I^-(aq) \rightarrow 2Fe^{2+}(aq) + I_2(aq)$
c The Fe^{3+} ion could be regenerated from the Fe^{2+} ion by an oxidation employing the other reactant:
$S_2O_8^{2-}(aq) + 2Fe^{2+}(aq) \rightarrow 2SO_4^{2-}(aq) + 2Fe^{3+}(aq)$
and +2.28 V easily 'beats' +0.77 V.
d Here are the two steps added up:
$2Fe^{3+}(aq) + 2I^-(aq) \rightarrow 2Fe^{2+}(aq) + I_2(aq)$
$S_2O_8^{2-}(aq) + 2Fe^{2+}(aq) \rightarrow 2SO_4^{2-}(aq) + 2Fe^{3+}(aq)$
$\overline{S_2O_8^{2-}(aq) + 2I^-(aq) \rightarrow 2SO_4^{2-}(aq) + I_2(aq)}$
which is the overall reaction.
e If the Fe^{2+} ion were to be the catalyst, the two steps of the above mechanism need to operate in reverse order – in other words, the $S_2O_8^{2-}(aq)$ ion starts things off by oxidising the Fe^{2+}, and the Fe^{3+} oxidises the $I^-(aq)$.
f In general, a feasible catalyst would be one whose standard electrode potential lay between those of the two reaction-system pairs. That way the catalyst can oxidise the reducing member of the reaction system, and be oxidised back to its start condition by the oxidising member of the reaction system, all within the rules of the Law of anti-clockwise circles (Figure A19.11).
g The $E^\ominus$ values for the proposed catalysts are:
$[MnO_4^-(aq) +8H^+(aq)]|[Mn^{2+}(aq)+4H_2O(l)]$
+1.51 V
$Cu^{2+}(aq)|Cu(s)$ +0.34 V
$V^{3+}(aq)|V^{2+}(aq)$ −0.26 V

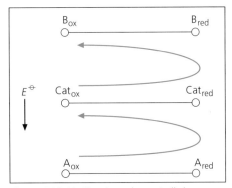

Figure A19.11 Showing schematically how a catalyst can catalyse the reaction: $A_{ox} + B_{red} \rightarrow A_{red} + B_{ox}$ providing its $E^{\ominus}$ value lies between those of the two main reactants

Only the first one lies between the two reactant values of +2.28 V and +0.54 V, so the two viable catalysts would be acidified MnO_4^-(aq) and Mn^{2+}(aq).

h As always, if a thermodynamically feasible reaction fails to occur, the reason must be that the enthalpy of activation is too high relative to the ambient temperature.

i See Figure A19.12.

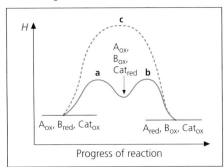

Figure A19.12 Generalised reaction profile for 'go-between' catalysis
a $cat_{ox} + B_{red} \rightarrow cat_{red} + B_{ox}$
b $A_{ox} + cat_{red} \rightarrow A_{red} + cat_{ox}$
c $A_{ox} + B_{red} \rightarrow A_{red} + B_{ox}$

48a See Figure A19.13.
b Branching would reduce the intimacy of contact and therefore the van der Waals' forces between molecules relative to the equivalent straight chains.

49a See Figure A19.14.
b Poly(propene) is being synthesised.
c With this method, the new monomer unit can only join the chain at the end by the titanium atom, so there is no chance of branching.

50 See Figure A19.15.

51 The total enthalpy change of bond breaking is 432 kJ for the mole of H—H bonds, plus about 300 kJ for the π-bond between the carbon atoms. Assuming that bond-making has begun, a rough estimate might be 200 kJ mol^{-1}.

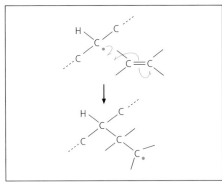

Figure A19.13 Showing how branching can occur in alkene polymerisation if a hydrogen atom is abstracted in the middle of a chain

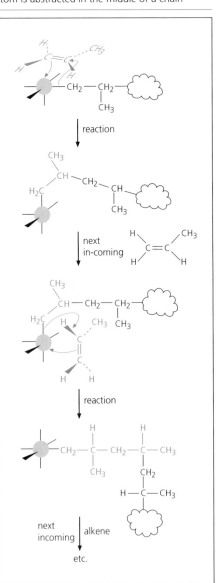

Figure A19.14 Two steps in the Ziegler–Natta polymerisation sequence. Each fresh propene monomer is green. The cloud is the original ethyl group

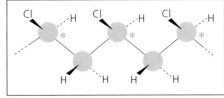

Figure A19.15 Every * carbon atom is a chiral centre, because it has four different groups attached to it (except the one in the exact centre of the chain)

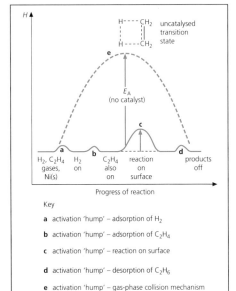

Key
a activation 'hump' – adsorption of H_2
b activation 'hump' – adsorption of C_2H_4
c activation 'hump' – reaction on surface
d activation 'hump' – desorption of C_2H_6
e activation 'hump' – gas-phase collision mechanism

Figure A19.16 Reaction profile of the hydrogenation of ethene on nickel

52a, b and **c** See Figure A19.16.
d d-block atoms can offer vacant orbitals of quite low energy for the use of the reactant electrons.

53 The activation enthalpies in Figure 19.43 suggest that some bond stretching has had to happen prior to the reactant molecules settling down on the catalyst surface.

54 It could be the case that the 'stepped' bits of crystal face do not allow the reactant molecules to get next to each other properly (Figure A19.17).

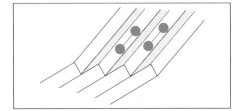

Figure A19.17 Showing how, if the metal crystal presented a 'stepped' surface to the reactants, then they might find themselves trapped on adjacent steps and prevented from reacting

55a Attach molecules that resemble oils to the metal.

b   You may recall from Chapter 13 that paints contain 'drying oils' which cross-link to dry the paint to a solid film (e.g. alkyd resins and linseed oil). So molecules which can cross-link into the drying oil could be attached to the metal.

c   Attach hydrophobic molecules to the metal, like the unwettable poly(tetrafluoroethene).

Table A19.7

Pollutant	Source
$CO_2(g)$	Combustion of hydrocarbons
$CO(g)$	Incomplete combustion of hydrocarbons
Hydrocarbons	Unburned fuel
NO_x	Reaction between nitrogen and oxygen in air, caused by the spark

56   See Table A19.7.

57a When $\lambda < 1$, the low level of oxygen leaves some unburned fuel.

b   At $\lambda = 1$, most of the fuel is being burned.

c   The oxygen reacts preferentially with the fuel, so at low λ, the dinitrogen stays unoxidised.

d   When $\lambda < 1$, the low level of oxygen favours incomplete combustion.

e   When $\lambda > 1$, the high level of oxygen favours more complete combustion.

58a Dioxygen and nitrogen dioxide are the oxidising agents, and octane and carbon monoxide are the reducing agents.

b $25O_2(g) + 2C_8H_{18}(g) \rightarrow 16CO_2(g) + 18H_2O(g)$

$25NO_2(g) + 2C_8H_{18}(g) \rightarrow 16CO_2(g) + 18H_2O(g) + 12\frac{1}{2}N_2(g)$

$O_2(g) + 2CO(g) \rightarrow 2CO_2(g)$

$NO_2(g) + 2CO(g) \rightarrow 2CO_2(g) + \frac{1}{2}N_2(g)$

c   Nothing (except using hydrogen as a fuel) will stop the enormous carbon dioxide emissions, and carbon dioxide is a greenhouse gas.

59   The convoluted layers of catalyst must be very thin. For instance, in Figure A19.18, as d approaches zero, the surface area approaches infinity.

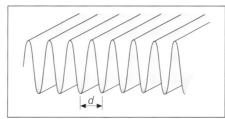

Figure A19.18 Relationship between surface area and degree of convolution – important in solid catalyst supports

60a The hydrocarbon data would need to be reduced by 10 000 (or four decimal places) to put them on the same scale as the oxygen and carbon monoxide axes.

b   At $\lambda = 0.8$, the oxygen level is about 0.5%, or 5000 ppm. At the same value of λ, the level of hydrocarbon stands at 150 ppm. So even allowing for a 25:2 stoichiometric ratio (see question 58b), that is still a big excess of dioxygen.

61a A choke blocks the air inflow.

b   When the choke is operating, even on lean-burn engines, the value of λ is temporarily less than one, so all the attendant problems of rich mixtures are still present.

62   The car's battery is one solution, but it would be a big drain on the battery. Overnight mains charging or mains heating via an external socket on the car might be possible. In any case, a catalyst of low thermal capacity would be an advantage, hence the trials on thin metal catalyst supports.

63   It might be helpful to place the rhodium first, so the NO_x reactions can proceed without the oxygen taking all the oxidisable material.

64   There appears to have been a sharp increase in control of NO_x emissions, as the power of rhodium became appreciated.

65a $0 \rightarrow +1$.

b $+1 \rightarrow +1$, no change.

c $+1 \rightarrow 0$.

d $+1 \rightarrow 0$.

e   Ligands do not change oxidation number, so no change.

66   Silver metal catalyses the reduction of silver bromide to more silver metal, so the catalysis is probably of the heterogeneous/surface type.

67   The total charge on the two thiosulphate ligands is $4-$, so if the silver is in its $+1$ oxidation state, the overall charge on the complex must be $3-$, $[Ag(S_2O_3)_2]^{3-}$.

68   The order of binding power, greatest first, is:

$S_2O_3{}^{2-} > Br^- > NH_3 > Cl^- > H_2O$

69   The parts of the original scene that were light are now rendered as black silver metal. So we need to use the negative as the new object, to achieve a second black–white reversal. This is achieved at the enlarging stage.

70   A crystal that had not received many photons would have less silver metal on it, so there would be less catalyst at the developing stage, so the development of a black crystal would not proceed so far in a given time, resulting in grey.

Index

Acknowledgements

Chapter 1

p. 1 *Fig. 1.1 top* A. Van der Vaeren/The Image Bank, *centre* GSO Images/The Image Bank, *bottom* Robert Kristoflik/The Image Bank; **p. 2** *Fig. 1.2* Peter Gould; **p. 3** *Fig. 1.3* Mary Evans Picture library; **p. 4** *Fig. 1.6* Science Photo Library, *Fig. 1.7* The Mansell Collection, *Fig. 1.8* Peter Gould.

Chapter 2

p. 9 *Fig. 2.1 left* Ann Ronan at Image Select, *right* The Science Museum/Science & Society Picture Library, *Fig. 2.2* The Science Museum/Science & Society Picture Library; **p. 11** *Fig. 2.3* Laurence Hughes/The Image Bank; **p. 13** *Fig. 2.4 top* Martin Bond/Science Photo Library, *bottom* Dr. Jeremy Burgess/Science Photo Library; **p. 15** *Fig. 2.5 top* Gary Gladstone/The Image Bank, *centre* John Burbridge/Science Photo Library, *bottom* Peter Menzel/Science Photo Library, *Fig. 2.6* John Townson; **p. 16** *Fig. 2.7* P & G Bowater/The Image Bank; **p. 19** *Fig. 2.9a and b* Peter Gould; **p. 21** *Fig. 2.10* Doran & Dyble Photograph/Ecowater.

Chapter 3

p. 24 *Fig. 3.1* Royal College of Physicians; **p. 25** *Fig. 3.2* Ann Ronan at Image Select, *Fig. 3.4* Mary Evans Picture Library; **p. 27** *Fig. 3.7* Ann Ronan at Image Select, *Fig. 3.8a* Prof. Peter Fowler/Science Photo Library, *Fig. 3.8b* Mary Evans Picture Library; **p. 29** *Fig. 3.10a* Ann Ronan at Image Select; **p. 35** *Fig. 3.23 top* Peter Gould, *bottom* Damien Lovegrove/Science Photo Library; **p. 37** *Fig. 3.27* Mary Evans Picture Library; **p. 40** *Fig. 3.34* The Science Museum/Science & Society Picture Library, *Fig. 3.35* Ann Ronan at Image Select.

Chapter 5

p. 73 *Fig. 5.2* Thomas Hollyman/Science Photo Library; **p. 83** *Fig. 5.23* Dr. Harold Rose/Science Photo Library; **p. 84** *Fig. 5.25 top left* Alex Bartel/Science Photo Library, *bottom left* Martípié/The Image Bank, *top right* Michael Short/Robert Harding Picture Library, *centre right* Jeff Smith/The Image Bank, *bottom right* Andrew Syred/Science Photo Library; **p. 86** *Fig. 5.29 top* Andrew Lambert, *bottom* Andrew Syred/Science Photo Library; **p. 89** *Fig. 5.35 top left* Mike McNamee/Science Photo Library, *top right* Ferranti Electronics SA/A Sternberg/Science Photo Library, *bottom left* Richard Megna/Science Photo Library.

Chapter 6

p. 95 *Fig. 6.4a* John Cleare/Mountain Camera; **p. 96** *Fig. 6.4b* Robert Harding Picture Library; **p. 97** *Fig. 6.8* Ann Ronan at Image Select; **p. 108** *Fig. 6.25b* Dr. Harold Rose/Science Photo Library; **p. 109** *Fig. 6.26b* Courtesy of Castrol (U.K.); **p. 111** *Fig. 6.29a* David Parker/Science Photo Library; **p. 112** *Fig. 6.32* Claude Nuridsany & Maria Perennou/Science Photo Library, *Fig. 6.33 top and bottom* Peter Gould; **p. 113** *Figs. 6.35, 6.36, 6.37 left and right, 6.38 left and right* Peter Gould; **p. 114** *Fig. 6.39a* BBC Libraries & Archives; **p. 115** *Fig. 6.41a* Physics Department, Imperial College, London/Science Photo Library; **p. 116** *Fig. 6.42a* A. Barrington Brown/Science Photo Library, *Fig. 6.42b* The Science Museum/Science & Society Picture Library.

Chapter 7

p. 121 *Fig. 7.1* Simon Fraser/Science Photo Library; **p. 122** *Fig. 7.2* Peter Menzel/Science Photo Library, *Fig. 7.3* Science Photo Library, *Fig. 7.4* E.T. Archive; **p. 131** *Fig. 7.13* John Townson; **p. 133** *Fig. 7.14* Lovibond/The Tintometer Ltd.; **p. 135** *Fig. 7.16* Solvay Interox Ltd.

Chapter 8

p. 140 *Fig. 8.1 top* Bill Sanderson/Science Photo Library, *bottom* Nikki Rain/Science Photo Library, *Fig. 8.2* Range/Bettmann, *Fig. 8.3* Colorsport; **p. 154** *Fig. 8.18* Peter Gould; **p. 158** *Fig. 8.23* E.T. Archive; **p. 160** *Fig. 8.26* Illustration by David Lidell, Courtesy of John E. Dolan; **p. 161** *Fig 8.27* David Glass/AP Photo *Fig. 8.28* Location Photographics/Exchem Explosives Ltd.

Chapter 9

p. 165 *Fig. 9.3* Ken Cooper/The Image Bank; **p. 169** *Fig. 9.8* The Mansell Collection; **p. 171** *Fig. 9.10 top, centre top and bottom* Blue Circle Industries PLC., *centre bottom* Kay Chernush/Image Bank; **p. 172** *Fig. 9.12* Science Photo Library; **p. 173** *Fig. 9.13* Science Photo Library; **p. 182** *Fig. 9.26* ICI; **p. 184** *Fig. 9.28* ICI; **p. 186** *Fig. 9.31* Levington Horticulture.

Chapter 10

p. 192 *Fig. 10.2* Science Photo Library; **p. 195** *Fig. 10.7* Ed Barber, 'Chemistry in Context' (p. 441), by G. Hill & J. Holman, Thomas Nelson Publishers; **p. 196** *Fig. 10.8 left and right* VG Organic; **p. 197** *Fig. 10.10 left and right* Andrew Lambert; **p. 205** *Fig. 10.15 left* Ann Ronan at Image Select, *right* Simon Fraser/Science Photo Library, *Fig. 10.16a* Allsport; **p. 206** *Fig. 10.16b* Andrew Lambert; **p. 207** *Fig. 10.18* ICI, *Fig. 10.19 left* Tadao Kimura/Image Bank, *right* Rex Features; **p. 212** *Fig. 10.23* Science Photo Library; **p. 213** *Fig. 10.24 top and bottom* Last Resort Picture Library; **p. 214** *Fig. 10.25* Science Photo Library; **p. 215** *Fig. 10.27* The Mansell Collection; **p. 216** *Fig. 10.29 top* Regent Hospital Products; *centre* Robert Harding Picture Library, *bottom* Arnold Crane/Tony Stone Images; **p. 217** *Fig. 10.30 left and right* Last Resort Picture Library, *Fig. 10.31* The Malaysian Rubber Producers Association; **p. 218** *Fig. 10.32* Andrew Lambert, *Fig. 10.33* William Salaz/The Image Bank; **p. 220** *Fig. 10.34* Ann Ronan at Image Select; **p. 221** *Fig. 10.36* Clive Freeman/Science Photo Library.

Chapter 11

p. 230 *Fig. 11.4* Andrew Lambert; **p. 232** *Fig. 11.6* Andrew Lambert; **p. 233** *Fig. 11.9 bottom left* Vencel Resil, *bottom right* David Gardner/Shell; **p. 235** *Fig. 11.13* Dr. Ann Smith/Science Photo Library; **p. 236** *Fig. 11.15* Geoff Lane/Science Photo Library; **p. 238** *Fig. 11.16 all* Zooid Pictures; **p. 239** *Fig. 11.17* Hydro Chemicals Ltd., *Fig. 11.23 top and bottom* Last Resort Picture Library; **p. 247** *Fig. 11.26 top* Tony Stone Images, *bottom* Rex Features; **p. 249** *Fig. 11.31* Geoff Lane/CSIRO/Science Photo Library; **p. 250** *Fig. 11.34* Automobile Association.

Chapter 12

p. 256 *Fig. 12.1* Ann Ronan at Image Select, *Fig. 12.2 top* Jean-Loup Charmet/Science Photo Library, *bottom* The Mansell

Collection; **p. 257** *Fig. 12.3* Andrew Lambert, *Fig. 12.4 top and bottom* Last Resort Picture Library; **p. 258** *Fig. 12.7 top* Royal Danish Ministry of Foreign Affairs, *bottom* The Science Museum/Science & Society Picture Library; **p. 265** *Fig. 12.11* ICI, *Fig. 12.12* W.L. Gore & Associates; **p. 267** *Fig. 12.13* L'Oreal/Foote, Cone & Belding, *Fig. 12.14* Martin Bond/Science Photo Library; **p. 268** *Fig. 12.15* Andrew Lambert; **p. 270** *Fig. 12.16* Reckitt & Colman Products; **p. 271** *Figs. 12.17 and 12.18* Andrew Lambert; **p. 273** *Figs. 12.22 and 12.23* Andrew Lambert; **p. 274** *Fig. 12.24* Andrew Lambert **p. 276** *Fig. 12.28* Robert R. Bartletts & Sons Ltd., Yate, Bristol; **p. 277** *Fig. 12.29* William Salaz/Image Bank.

Chapter 13

p. 297 *Fig. 13.11* Unilever Research; **p. 298** *Fig. 13.15* Geoscience Features Picture Library; **p. 299** *Fig. 13.17* Palm Oil Research Institute of Malaysia, *Fig. 13.20* Andrew Lambert; **p. 300** *Fig. 13.21 left* Fredrik Ehrenstrom/Oxford Fig. Scientific Films, *right* Andrew Lambert, *Fig. 13.22* Biophoto Associates; **p. 302** *Fig. 13.27* Andrew Lambert; **p. 303** *Fig. 13.28* Andrew Lambert, *Fig. 13.30* Jacobs Krafft Suchard, *Fig. 13.32* Van den Burgh; **p. 304** *Fig. 13.33* Cadbury Ltd., *Fig. 13.34* Andrew Lambert; **p. 307** *Fig. 13.39* J. Walter Thompson; **p. 308** *Fig. 13.41* Science Photo Library; **p. 309** *Fig. 13.45* David Higgs/Tony Stone Images.

Chapter 14

p. 317 *Fig. 14.1 left* Tony Waltham/Robert Harding Picture Library, *right* Sinclair Stammers/Science Photo Library; **p. 327** *Fig. 14.13* Last Resort Picture Library; **p. 328** *Fig. 14.15* Andrew Lambert; **p. 329** *Fig. 14.16* Last Resort Picture Library; **p. 330** *Fig. 14.17* Last Resort Picture Library; **p. 331** *Fig. 14.19* Warner Pathé/Ronald Grant Archive; **p. 339** *Fig. 14.29* Andrew Lambert; **p. 344** *Fig. 14.40* Andrew Lambert; **p. 345** *Fig. 14.41* Hulton Deutsch Collection; **p. 346** *Fig. 14.42* Zooid Pictures, *Fig. 14.43 top centre* Andrew Lambert, *bottom centre* Rex Features, *bottom* Zooid Pictures.

Chapter 15

p. 357 *Fig. 15.8 top left* Colonel Mario/Robert Harding Picture Library, *top right* Andrew Lambert, *bottom left* DSM Engineering Plastic Products Ltd., *bottom right* Robert Harding Picture Library; **p. 358** *Fig. 15.9* Retrograph Archive, *Fig. 15.10* Science Photo Library; **p. 359** *Fig. 15.13* Last Resort Picture Library; **p. 364** *Fig. 15.19* Andrew Lambert; **p. 367** *Fig. 15.27* Andrew Lambert, *Fig. 15.28* Camera Press, *Fig. 15.29a* Retrograph Archive, *Fig. 15.29b* Rizzoli/Camera Press; **p. 369** *Fig. 15.33* Andrew Lambert; **p. 371** *Fig. 15.35* The University of Leeds; **p. 377** *Fig. 15.51* Courtesy of Rockefeller Archive Center, *Fig. 15.52* Ralph Crane/Camera Press; **p. 380** *Fig. 15.60a* Science Photo Library; **p. 381** *Fig. 15.60b* Hulton Deutsch Collection.

Chapter 16

p. 399 *Fig. 16.18* Range/Bettmann; **p. 402** *Fig. 16.23* Rex Features; **p. 403** *Fig. 16.25a* Andrew Lambert, *Fig. 16.25b* Rex Features, *Fig. 16.25c* Peugeot Talbot; **p. 404** *Fig. 16.26* Ballard Power Systems, *Fig. 16.29* Johnson Matthey Technology Centre; **p. 405** *Fig. 16.30* John Mead/Science Photo Library; **p. 406** *Fig. 16.32* Andrew Syred/Science Photo Library; **p. 410** *Fig. 16.37* British Steel, *Fig. 16.38 left* British Steel, *right* James Holmes/Rover/Science Photo Library; **p. 411** *Fig. 16.40* British Steel, *Fig. 16.42* Dr. Jeremy Burgess/Science Photo Library, *Fig. 16.43* Ajax News & Feature Service, *Fig. 16.44* Bristol Cars Ltd.; **p. 415** *Fig. 16.47* Widnes Library; **p. 419** *Fig. 16.53* IMI Refiners Ltd.; **p. 420** *Fig. 16.54 top and centre* Zooid Pictures, *bottom* Robert Harding Picture Library.

Chapter 17

p. 423 *Fig. 17.2* Andrew Lambert, *Fig. 17.3* Ann Ronan at Image Select; **p. 424** *Fig. 17.6 top and bottom* Andrew Lambert; **p. 425** *Fig. 17.7* © Crown Copyright/Ministry of Defence; **p. 426** *Fig. 17.9a and b* Andrew Lambert; **p. 430** *Fig. 17.15 left* Carl Frank/Science Photo Library, *right* Pilington plc; **p. 431** *Fig. 17.16* Ann Ronan at Image Select; **p. 437** *Fig. 17.23* Secchi-Lecaque/Roussel-Uclaf/CNRI/Science Photo Library.

Chapter 18

p. 450 *Fig. 18.16* Bryant & May/Hackney Archives Dept., *Fig. 18.17 left and right* Last Resort Picture Library; **p. 452** *Fig. 18.20a* Geoscience Features Picture Library, *Fig. 18.20b and c* Andrew Lambert; **p. 454** *Fig. 18.22* Warren Williams/Planet Earth Pictures; **p. 455** *Fig. 18.23a and b* John Reader/Science Photo Library, *Fig. 18.23c* Illustrated London News, 28 Dec. 1912 p. 958; **p. 459** *Fig. 18.27* ICI/BPL Photographic Sources; **p. 461** *Fig. 18.28* ICI Katalco/BPL Photographic Sources; **p. 462** *Fig. 18.29* ICI/BPL Photographic Sources; **p. 464** *Fig. 18.30* Suzanne & Nick Geary/Tony Stone Images, *Fig. 18.31* Andrew Lambert.

Chapter 19

p. 477 *Fig. 19.7* Rex Features; **p. 481** *Fig. 19.11 left* Dr. Jeremy Burgess/Science Photo Library, *right* Molyslip Atlantic Ltd., *Fig. 19.12* Andrew Lambert; **p. 482** *Fig. 19.13* Harold Taylor Asipp/Oxford Scientific Films, *Fig. 19.14* J. Feltwell/Garden & Wildlife Matters Photo Library; **p. 483** *Fig. 19.15a, b and c* Last Resort Picture Library; **p. 487** *Fig. 19.25* Chris Stammers/Johnson Matthey Technology Centre; **p. 489** *Fig. 19.31a and b* J. Feltwell/Garden & Wildlife Matters Photo Library; **p. 493** *Fig. 19.38* Pharmacia Biotech (Biochron) Ltd.; **p. 498** *Fig. 19.45 top and bottom* Zeneca Specialties; **p. 500** *Fig. 19.47* Johnson Matthey Technology Centre; **p. 501** *Fig. 19.50* Emitec GmbH.

Answers

p. 593 *Fig. A16.7* Keith Scholey/Planet Earth Pictures.
p. 595 *Fig. A17.2* Science Photo Library.

The publishers are grateful to the following for permitting reproduction of material within this book:

p. 28 Macmillan New York for the essay by Ernest Rutherford in *Background to Science*, J Needham and W Pagel, Eds. New York: Macmillan, 1938.
p. 220 Gerald Duckworth & Co Ltd for the extract from August Kekute's diary translated by Alexander Finlay in *100 Years of Chemistry*. London: Gerald Duckworth & Co Ltd, 1937.
p. 234 The *Independent*.
p. 309 The *Guardian* ©.
p. 432 ICI Educational Publishers for the reproduction of Figs. 17.17 and 17.18, which were based on figures in *Steam* issue 7.
p. 462 Little, Brown for the extract from *The Far Corner*, Harry Pearson. London: Little, Brown, 1994.

THE PERIODIC TABLE

Key

Atomic number
Symbol
Name
Relative atomic mass

Group

Period	I	II													III	IV	V	VI	VII	VIII
1	1 H Hydrogen 1																			2 He Helium 4
2	3 Li Lithium 7	4 Be Beryllium 9													5 B Boron 11	6 C Carbon 12	7 N Nitrogen 14	8 O Oxygen 16	9 F Fluorine 19	10 Ne Neon 20
3	11 Na Sodium 23	12 Mg Magnesium 24													13 Al Aluminium 27	14 Si Silicon 28	15 P Phosphorus 31	16 S Sulphur 32	17 Cl Chlorine 35.5	18 Ar Argon 40
4	19 K Potassium 39	20 Ca Calcium 40	21 Sc Scandium 45	22 Ti Titanium 48	23 V Vanadium 51	24 Cr Chromium 52	25 Mn Manganese 55	26 Fe Iron 56	27 Co Cobalt 59	28 Ni Nickel 59	29 Cu Copper 63.5	30 Zn Zinc 65.4			31 Ga Gallium 70	32 Ge Germanium 73	33 As Arsenic 75	34 Se Selenium 79	35 Br Bromine 80	36 Kr Krypton 84
5	37 Rb Rubidium 85	38 Sr Strontium 88	39 Y Yttrium 89	40 Zr Zirconium 91	41 Nb Niobium 93	42 Mo Molybdenum 96	43 Tc Technetium (99)	44 Ru Ruthenium 101	45 Rh Rhodium 103	46 Pd Palladium 106	47 Ag Silver 108	48 Cd Cadmium 112			49 In Indium 115	50 Sn Tin 119	51 Sb Antimony 122	52 Te Tellurium 128	53 I Iodine 127	54 Xe Xenon 131
6	55 Cs Caesium 133	56 Ba Barium 137	57 La ▲ Lanthanum 139	72 Hf Hafnium 178	73 Ta Tantalum 181	74 W Tungsten 184	75 Re Rhenium 186	76 Os Osmium 190	77 Ir Iridium 192	78 Pt Platinum 195	79 Au Gold 197	80 Hg Mercury 201			81 Tl Thallium 204	82 Pb Lead 207	83 Bi Bismuth 209	84 Po Polonium (210)	85 At Astatine (210)	86 Rn Radon (222)
7	87 Fr Francium (223)	88 Ra Radium (226)	89 Ac ▲▲ Actinium (227)	104 Unq Unnil-quadium (261)	105 Unp Unnil-pentium (262)	106 Unh Unnil-hexium (263)														

▲ Lanthanoid elements

58 Ce Cerium 140	59 Pr Praseo-dymium 141	60 Nd Neodymium 144	61 Pm Promethium (147)	62 Sm Samarium 150	63 Eu Europium 152	64 Gd Gadolinium 157	65 Tb Terbium 159	66 Dy Dysprosium 163	67 Ho Holmium 165	68 Er Erbium 167	69 Tm Thulium 169	70 Yb Ytterbium 173	71 Lu Lutetium 175

▲▲ Actinoid elements

90 Th Thorium 232	91 Pa Protactinium (231)	92 U Uranium 238	93 Np Neptunium (237)	94 Pu Plutonium (242)	95 Am Americium (243)	96 Cm Curium (247)	97 Bk Berkelium (245)	98 Cf Californium (251)	99 Es Einsteinium (254)	100 Fm Fermium (253)	101 Md Mendelevium (256)	102 No Nobelium (254)	103 Lr Lawrencium (257)